World States

Barents Sea
SIBERIA
RUSSIA
URAL MTS.
See Europe inset map below
KAZAKHSTAN
MONGOLIA
GOBI (DESERT)
UZBEKISTAN
TIAN SHAN
KYRGYZSTAN
GEORGIA
ARMENIA
AZER.
TURKEY
TURKMENISTAN
TAJIKISTAN
NORTH KOREA
SOUTH KOREA
JAPAN
SYRIA
LEBANON
ISRAEL
JORDAN
IRAQ
IRAN
AFGHANISTAN
CHINA
HIMALAYAS
BHUTAN
NEPAL
PAKISTAN
KUWAIT
BAHRAIN
QATAR
Arabian Peninsula
SAUDI ARABIA
UNITED ARAB EMIRATES
OMAN
YEMEN
Thar Desert
INDIA
Deccan Plateau
BANGLADESH
BURMA (MYANMAR)
East China Sea
PACIFIC OCEAN
TROPIC OF CANCER
TAIWAN
South China Sea
LAOS
THAILAND
VIETNAM
CAMBODIA
PHILIPPINES
Arabian Sea
Bay of Bengal
TUNISIA
LIBYA
SAHARA
EGYPT
SAHEL
NIGER
CHAD
SUDAN
ERITREA
DJIBOUTI
Ethiopian Highlands
ETHIOPIA
SOMALIA
NIGERIA
TOGO
BENIN
CENTRAL AFRICAN REPUBLIC
SOUTH SUDAN
CAMEROON
GUINEA
GABON
REPUBLIC OF THE CONGO
UGANDA
RWANDA
BURUNDI
KENYA
DEMOCRATIC REPUBLIC OF THE CONGO
TANZANIA
SEYCHELLES
COMOROS
ANGOLA
MALAWI
ZAMBIA
MADAGASCAR
NAMIBIA
ZIMBABWE
BOTSWANA
MAURITIUS
Kalahari Desert
MOZAMBIQUE
SWAZILAND
SOUTH AFRICA
LESOTHO
MALDIVES
SRI LANKA
BRUNEI
MALAYSIA
PALAU
MICRONESIA
EQUATOR
SINGAPORE
INDONESIA
PAPUA NEW GUINEA
TIMOR LESTE
SOLOMON ISLANDS
INDIAN OCEAN
TROPIC OF CAPRICORN
AUSTRALIA
Great Victoria Desert
ANTARCTICA
ICELAND
Arctic Circle
SWEDEN
FINLAND
NORWAY
RUSSIA
ESTONIA
LATVIA
LITH.
RUSS.
DEN.
UNITED KINGDOM
IRELAND
NETH.
BELG.
GERMANY
POLAND
BELARUS
LUX.
CZECHIA
SLVK.
UKRAINE
LIECH.
AUS.
HUNGARY
MOLDOVA
SWITZ.
SLOVE.
ATLANTIC OCEAN
FRANCE
ITALY
CRO.
BOS. & HERZ.
SERB.
ROMANIA
Black Sea
KOS.
BULGARIA
MONT.
MAC.
ALB.
PORTUGAL
SPAIN
ANDORRA
GREECE
TURKEY
EUROPE
MOROCCO
ALGERIA
TUNISIA
MALTA
CYPRUS

Strengthens Students' Connection to Geography through Active, Discovery-Based Learning

RUBENSTEIN THE Cultural Landscape TWELFTH EDITION

THE Cultural Landscape

An Introduction to Human Geography

TWELFTH EDITION

James M. Rubenstein

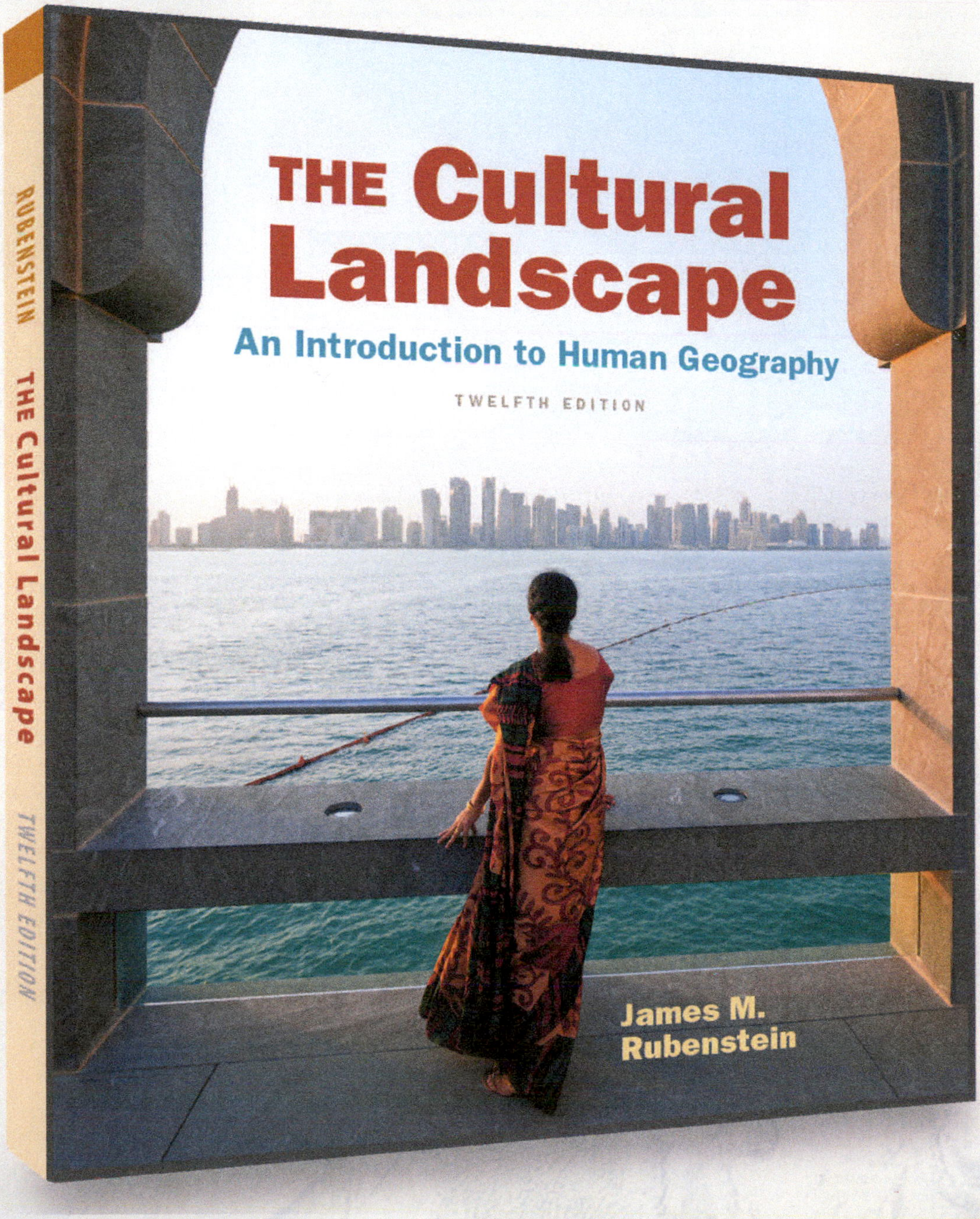

PEARSON

Active, Discovery-Based Learning

DOING GEOGRAPHY California Agriculture and Water

California's extended extreme drought is stressing agriculture, which uses 80 percent of the state's distributed water. Homeowners and businesses in California have been required to make substantial cuts in their water usage. California farmers produce one-third of U.S. vegetables and two-thirds of fruits and nuts. It takes a lot of water to grow these fruits and vegetables. So if you are living in any of the 50 U.S. states, you are consuming California water indirectly through consuming produce. In fact, the average American consumes around 40 gallons of California water per day. Table 9-2 has examples of the amounts of California water that go into growing some fruits and vegetables.

TABLE 9–2 Amount of Water Needed to Grow Selected Fruits and Vegetables in California

Fruits and nuts	Gallons	Your produce consumption	Your water consumption
1 apple, peach, pear, or plum, ¼ melon	7.0		
5 strawberries	3.0		
1 almond	1.0		
1 walnut	5.0		
3 grapes	1.0		
1 lemon, orange, grapefruit, or clementine	20		
1 avocado	40		
Vegetables			
1 broccoli or cauliflower floret	0.5		
Lettuce, cabbage, spinach [salad portion]	1.0		
1 carrot or celery stalk	0.5		
1 slice tomato, onion, or potato	0.5		

What's Your Food and Agriculture Geography?

Your California Water Consumption

How much California water did you consume today in your fruits and vegetables?

1. Determine from Table 9-2 the quantities of the listed fruits and vegetables that you have consumed today (or another day specified for your class).
2. What was your total consumption of California water from eating produce?
3. How does your total consumption compare to the national average of 40 gallons?
4. What factors might account for having consumption that is higher or lower than the national average?

NEW! Doing Geography and the accompanying **What's Your Geography** features discuss the geographic tools, techniques, and skills used to address real-world problems, and then ask students to put themselves in the role of geographers by applying these skills and techniques to their real-world experiences and environments, helping students connect the relevance of human geography to their everyday lives.

DEBATE IT! Immigration reform: Tougher controls or legal status?

Debate over authorized immigration centers on border security and on appropriateness of a path to legal status for unauthorized immigrants in the United States.

TIGHTEN SECURITY AND DO NOT OFFER A PATH TO LEGAL STATUS

- **THE WRONG MESSAGE.** People breaking the law by crossing the U.S. border without proper documentation sends the wrong message to people who obey the law.
- **ENCOURAGE OTHERS.** Rewarding people for illegal behavior will encourage others to enter without documents.
- **POOR SECURITY.** The border is not sufficiently secure, especially in small towns and rural areas.

◀ **FIGURE 3-42 MINIMAL SECURITY AT THE BORDER** Crossing from Palomas, Mexico, to Columbus, New Mexico.

OFFER A PATH TO LEGAL STATUS; SECURITY IS ALREADY TIGHT ENOUGH

- **IMPRACTICAL.** It would be a practical impossibility for law enforcement officials to actually find the 11 million unauthorized immigrants.
- **ECONOMIC IMPACT.** Pulling unauthorized immigrants out of their jobs would cripple the U.S. economy.
- **AGENTS.** The numbers of border agents and deportations of unauthorized immigrants have doubled since 2000.
- **LAW-ABIDING.** Unauthorized immigrants are productive and otherwise law-abiding members of U.S. society.

◀ **FIGURE 3-43 BORDER AGENTS** Rio Grande near Laredo, Texas.

NEW! Debate It presents two sides of a complex topic using a two-column pro vs. con format to help engage students in active debate and decision-making. **Debate It** can be used for homework, group work, and discussions.

A Focus on Sustainability

SUSTAINABILITY & OUR ENVIRONMENT Rising Oceans and the Future of Nauru

The sustainability of the world's smallest island state, Nauru, as well other island microstates, is in danger due to rising ocean levels. Sea levels rose around 17 centimeters (6.7 inches) during the twentieth century. Scientists working for the United Nations forecast another rise of between 18 and 59 centimeters (between 7 and 23 inches). The rising oceans will submerge a large percentage of the tiny island. Another Pacific Ocean microstate, Kirbati, a collection of approximately 32 small islands, has already witnessed the disappearance of two of its islands under rising oceans.

Nauru, Kiribati, and other Pacific island microstates are atolls—that is, islands made of coral reefs (Figure 8-16). A coral is a small sedentary marine animal that has a horny or calcareous skeleton. Corals form colonies, and the skeletons build up to form coral reefs. Coral is very fragile. Humans are attracted to coral for its beauty and the diversity of species it supports, but handling coral can kill it. The threat of climate change to the sustainability coral is especially severe: Coral stays alive in only a narrow range of ocean temperatures, between 23°C and 25°C (between 73°F and 77°F), so global warming threatens the ecology of the portions of the islands that remain above sea level.

▲ FIGURE 8-16 **NAURU: WORLD'S SMALLEST ISLAND MICROSTATE** Rising sea level because of climate change threatens the future of the island, whose area is only 21 square kilometers (8 square miles).

SUSTAINABILITY & OUR ENVIRONMENT Remanufacturing

Remanufacturing contributes to a more sustainable environment. The principal challenge is to increase its economic sustainability.

- **Paper.** Most types of paper can be recycled. Newspapers have been recycled profitably for decades, and recycling of other paper, especially computer paper, is growing. Rapid increases in virgin paper pulp prices have stimulated construction of more plants capable of using waste paper. The key to recycling is collecting large quantities of clean, well-sorted, uncontaminated, dry paper.
- **Plastic.** The plastic industry has developed a system of numbers marked inside triangles. Symbols 2 (milk jugs), 4 (shopping bags), and 5 (such as yogurt containers) are considered to be safest for recycling. The plastics in symbols 3 (such as food wrap), 6 (Styrofoam), and 7 (such as iPad cases) may contain carcinogens. Symbol 1 (soda and water bottles) can allow bacteria to accumulate.
- **Aluminum.** The principal source of recycled aluminum is beverage containers. Aluminum cans began to replace glass bottles for beer during the 1950s and for soft drinks during the 1960s. Aluminum scrap is readily accepted for recycling, although other metals are rarely accepted.

▲ FIGURE 11-85 **REMANUFACTURING** Junked cars await shredding so that the steel can be reused.

- **Glass.** Glass can be used repeatedly with no loss in quality and is 100 percent recyclable. The process of creating new glass from old is extremely efficient, producing virtually no waste or unwanted by-products. Though unbroken clear glass is valuable, mixed-color glass is nearly worthless, and broken glass is hard to sort.

NEW! Sustainability & Our Environment relates the principal topics of human geography to overarching issues of economic, social, and environmental sustainability for our planet.

A Refined Learning Path

KEY ISSUE 2

Why Is Each Point on Earth Unique?

- Place: A Unique Location
- Region: A Unique Area
- Culture Regions

LEARNING OUTCOME 1.2.1

Identify the distinctive features of a place, including toponym, site, and situation.

UPDATED! ***Key Issues*** highlight the four critical "big questions" around which each chapter is organized.

UPDATED! ***Learning Outcomes*** emphasize the skills and knowledge students should gain from each section.

PAUSE & REFLECT 1.2.1

What is the origin of the toponym of your hometown?

UPDATED! ***Pause & Reflect*** questions allow students to check and apply their understanding as they read each section.

CHECK-IN KEY ISSUE 2

Why Is Each Point on Earth Unique?

- ✔ Location is identified through name, site, and situation.
- ✔ Regions can be formal, functional, or vernacular.
- ✔ Culture encompasses what people care about and what people take care of.

UPDATED! ***Check Ins*** conclude each section, summarizing the main points of each Key Issue.

UPDATED! ***Key Issues*** Issues are summarized at the end the chapter, followed by **NEW!** ***Thinking Geographically*** questions and **NEW!** ***Explore*** activities using Google Earth.

KEY ISSUE 2

Why is each point on earth unique?

Geographers identify unique places and regions distinguished by distinctive combinations of cultural as well as economic and environmental features. Location is the position something occupies on Earth. A region is an area characterized by a unique combination of features. The distribution of features helps explain why every place and every region is unique.

THINKING GEOGRAPHICALLY

3. What are elements of the site and situation of your hometown?
4. Can you name another place to which your hometown has strong connections?
5. What is an example of a feature that connects your town to another?

◄**FIGURE 1-59 SITE AND SITUATION OF BOSTON** The site is Boston Harbor and several rivers. Logan Airport is an example of the connections found in Boston to other places.

Visualizing Earth's People & Places

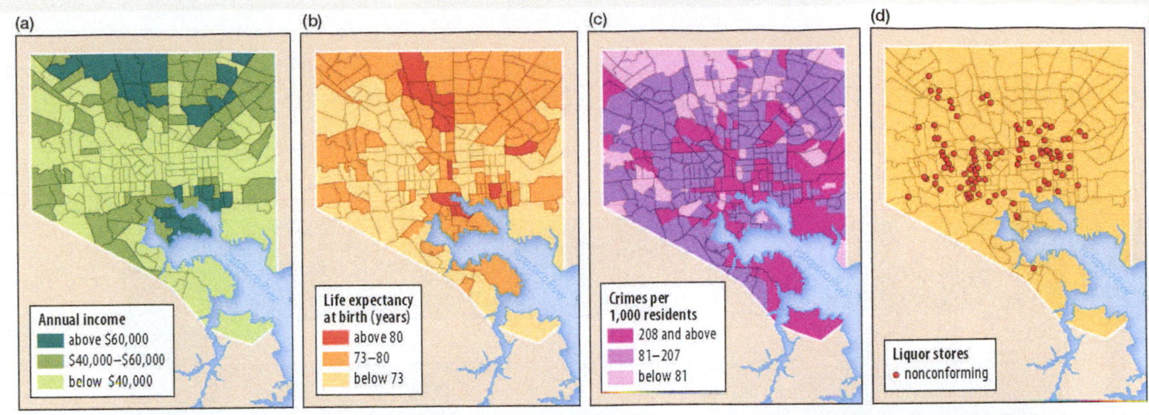

▲ **FIGURE 1-24 SPATIAL ASSOCIATION IN BALTIMORE** **(a)** Income, **(b)** life expectancy at birth, **(c)** crime, **(d)** nonconforming liquor stores.

NEW & REVISED! Cartography. All maps have been thoroughly updated with current data and contemporary cartographic styles, for optimal spatial visualization and analysis.

Regime type, 2014
Full democracy
Democracy
Open anocracy
Closed anocracy
Autocracy
Failed/occupied
Not included

ARCTIC OCEAN
PACIFIC OCEAN
ATLANTIC OCEAN
INDIAN OCEAN
TROPIC OF CANCER
EQUATOR
TROPIC OF CAPRICORN

▶ **FIGURE 8-39 REGIME TYPE, 2014** Most states are either democratic, autocratic, or anocratic. In a few "failed" states, such as Libya and South Sudan, government institutions have broken down because of civil war, extreme poverty, or natural disasters—or some combination of the three.

▼ **FIGURE 3-46 IMMIGRANTS IN EUROPE** Africans trying to reach Italy are rescued by the Italian navy after their boat sunk trying to cross the Mediterranean Sea.

▲ **FIGURE 10-36 WORLD TRADE ORGANIZATION PROTEST** Protestors outside the Department of Agriculture in the Philippines demonstrate during a speech delivered by the director of the World Trade Organization in 2015.

UPDATED! The latest science, statistics, and associated imagery. Data sources include the *2015 Population Reference Bureau World Population Data* and the *2015 United Nations Human Development Report*. Recent world political events are covered, including the rise of the Islamic State, Russia's takeover of Crimea, and the Syrian refugee crisis.

Continuous Learning
Before, During, and After Class

BEFORE CLASS

Mobile Media and Reading Assignments Ensure Students Come to Class Prepared.

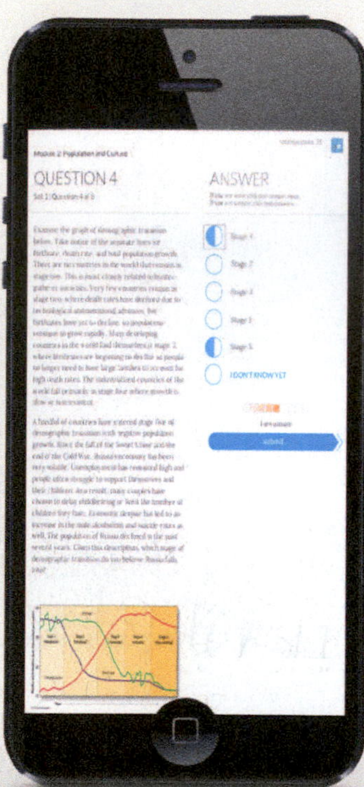

NEW! Dynamic Study Modules personalize each student's learning experience. Created to allow students to acquire knowledge on their own and be better prepared for class discussions and assessments, this mobile app is available for iOS and Android devices.

Pearson eText in MasteringGeography gives students access to the text whenever and wherever they can access the internet. eText features include:

- Now available on smartphones and tablets.
- Seamlessly integrated videos and other rich media.
- Fully accessible (screen-reader ready).
- Configurable reading settings, including resizable type and night reading mode.
- Instructor and student note-taking, highlighting, bookmarking, and search.

Pre-Lecture Reading Quizzes are easy to customize & assign

NEW! Reading Questions ensure that students complete the assigned reading before class and stay on track with reading assignments. Reading Questions are 100% mobile ready and can be completed by students on mobile devices.

with MasteringGeography

DURING CLASS

Learning Catalytics and Engaging Media

What has Professors and Students excited? Learning Cataltyics, a 'bring your own device' student engagement, assessment, and classroom intelligence system, allows students to use their smartphone, tablet, or laptop to respond to questions in class. With Learning Cataltyics, you can:

- Assess students in real-time using open ended question formats to uncover student misconceptions and adjust lecture accordingly.
- Automatically create groups for peer instruction based on student response patterns, to optimize discussion productivity.

"My students are so busy and engaged answering Learning Catalytics questions during lecture that they don't have time for Facebook."

Declan De Paor, *Old Dominion University*

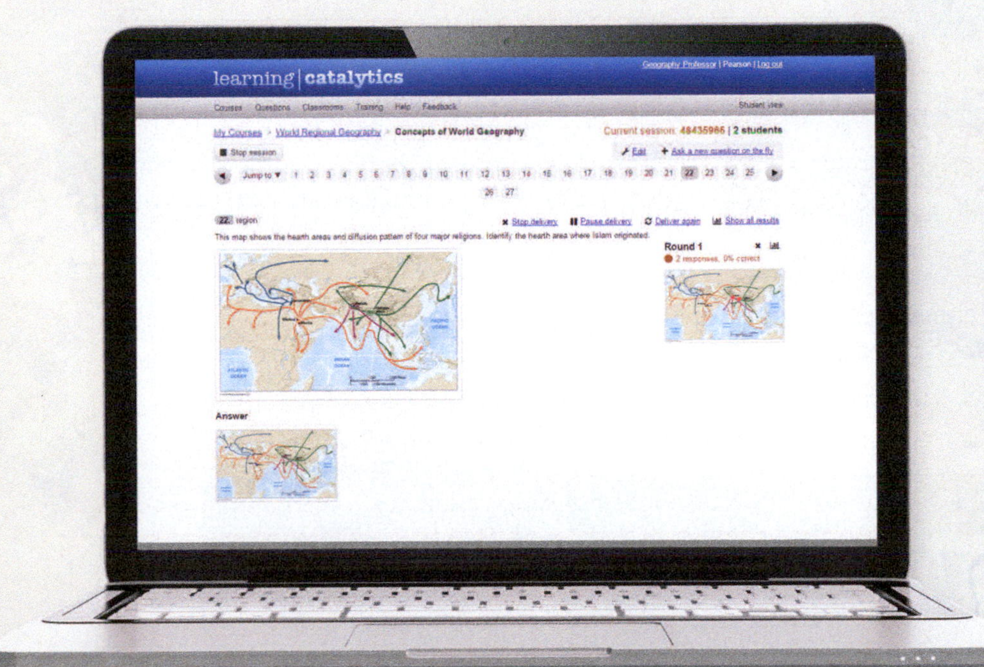

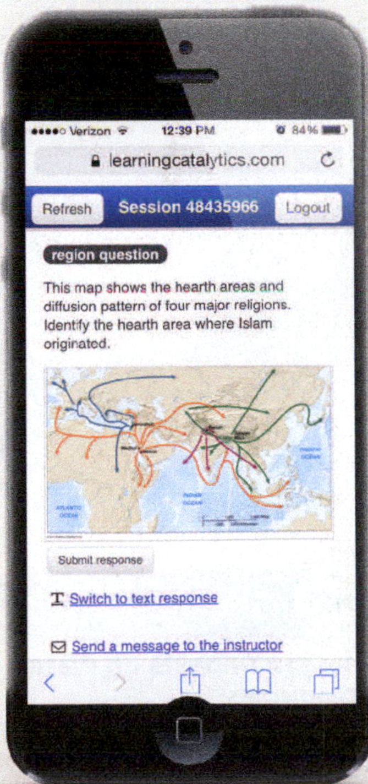

Enrich Lecture with Dynamic Media

Teachers can incorporate dynamic media into lecture, such as Videos, MapMaster Interactive Maps, and Geoscience Animations.

Mastering Geography™

MasteringGeography delivers engaging, dynamic learning opportunities—focusing on course objectives and responsive to each student's progress—that are proven to help students absorb human geography course material and understand challenging geography processes and concepts.

AFTER CLASS

Easy to Assign, Customizable, Media-Rich, and Automatically Graded Assignments

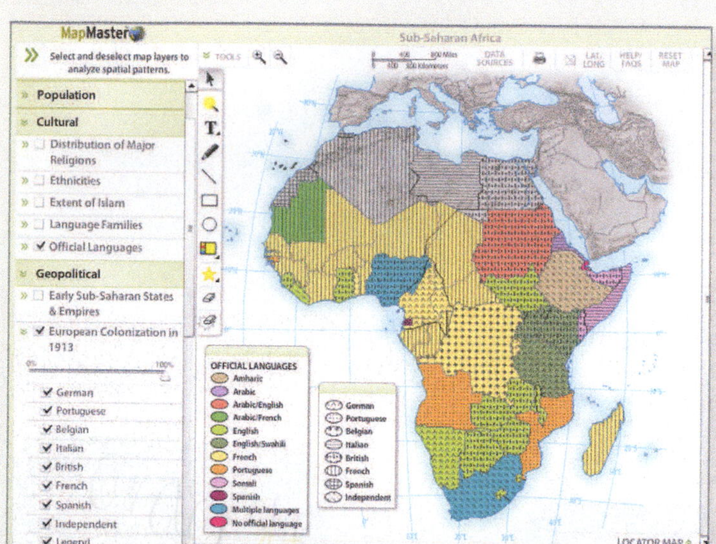

UPDATED! MapMaster Interactive Map Activities are inspired by GIS, allowing students to layer various thematic maps to analyze spatial patterns and data at regional and global scales. This tool includes zoom and annotation functionality, with hundreds of map layers leveraging recent data from sources such as NOAA, NASA, USGS, United Nations, and the CIA.

NEW! Geography Videos from such sources as the BBC and *The Financial Times* are now included in addition to the videos from Television for the Environment's Life and Earth Report series in **MasteringGeography**. Approximately 200 video clips for over 25 hours of video are available to students and teachers and **MasteringGeography**.

 **GeoVideo** | Log in to the **MasteringGeography** Study Area to view this video.

Human Impacts on Water Resources

Humans use water for many purposes, including manufacturing, agriculture, and recreation, as well as direct consumption. Access to fresh clean water is not possible for many people in the world. The poor condition of infrastructure restricts access to fresh clean water for some people. Other people live in arid locations.

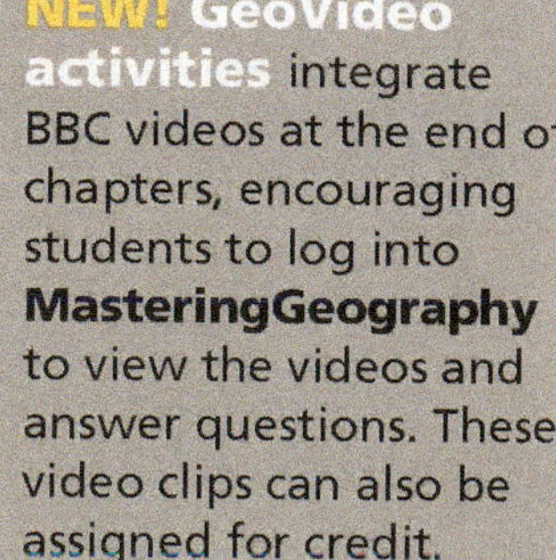

1. What are the principal uses of water resources other than direct consumption by people and animals?
2. Given that the world's total supply of water is constant, how might we increase the world's supply of water suitable as a resource for use by people?
3. What steps, if any, are being taken in your school or community to conserve water?

NEW! GeoVideo activities integrate BBC videos at the end of chapters, encouraging students to log into **MasteringGeography** to view the videos and answer questions. These video clips can also be assigned for credit.

www.MasteringGeography.com

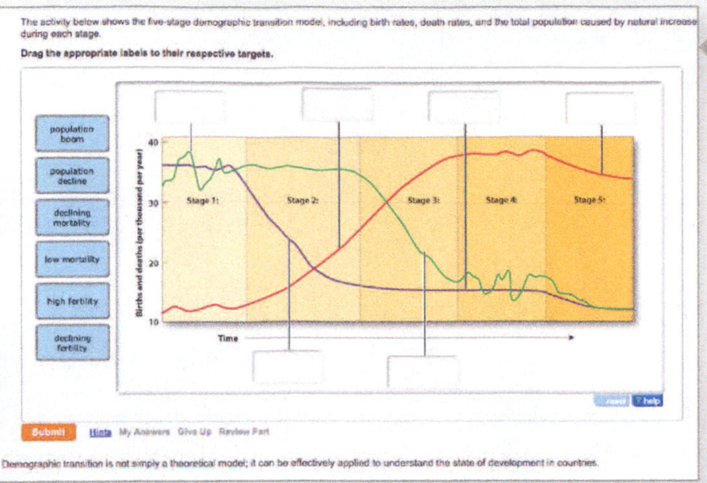

NEW! GeoTutors. Highly visual coaching items with hints and specific wrong answer feedback help students master the toughest topics in geography.

UPDATED! Encounter (Google Earth) activities provide rich, interactive explorations of human geography concepts, allowing students to visualize spatial data and tour distant places on the virtual globe.

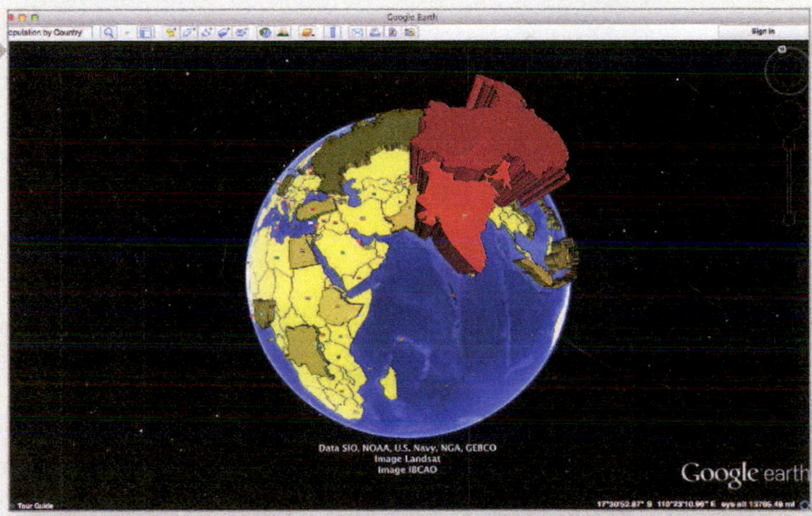

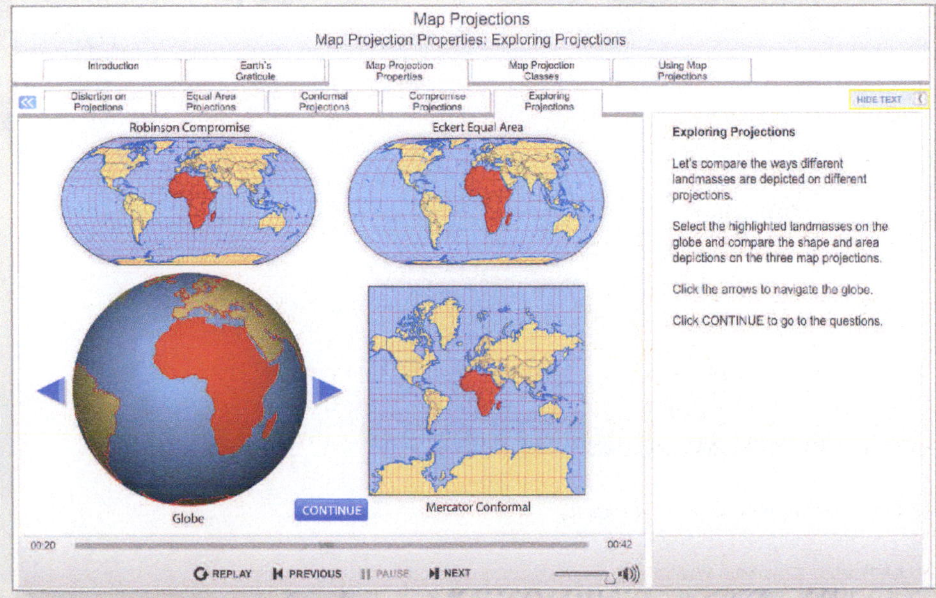

Map Projections interactive tutorial media helps reinforce and remediate students on the basic yet challenging introductory map projection concepts.

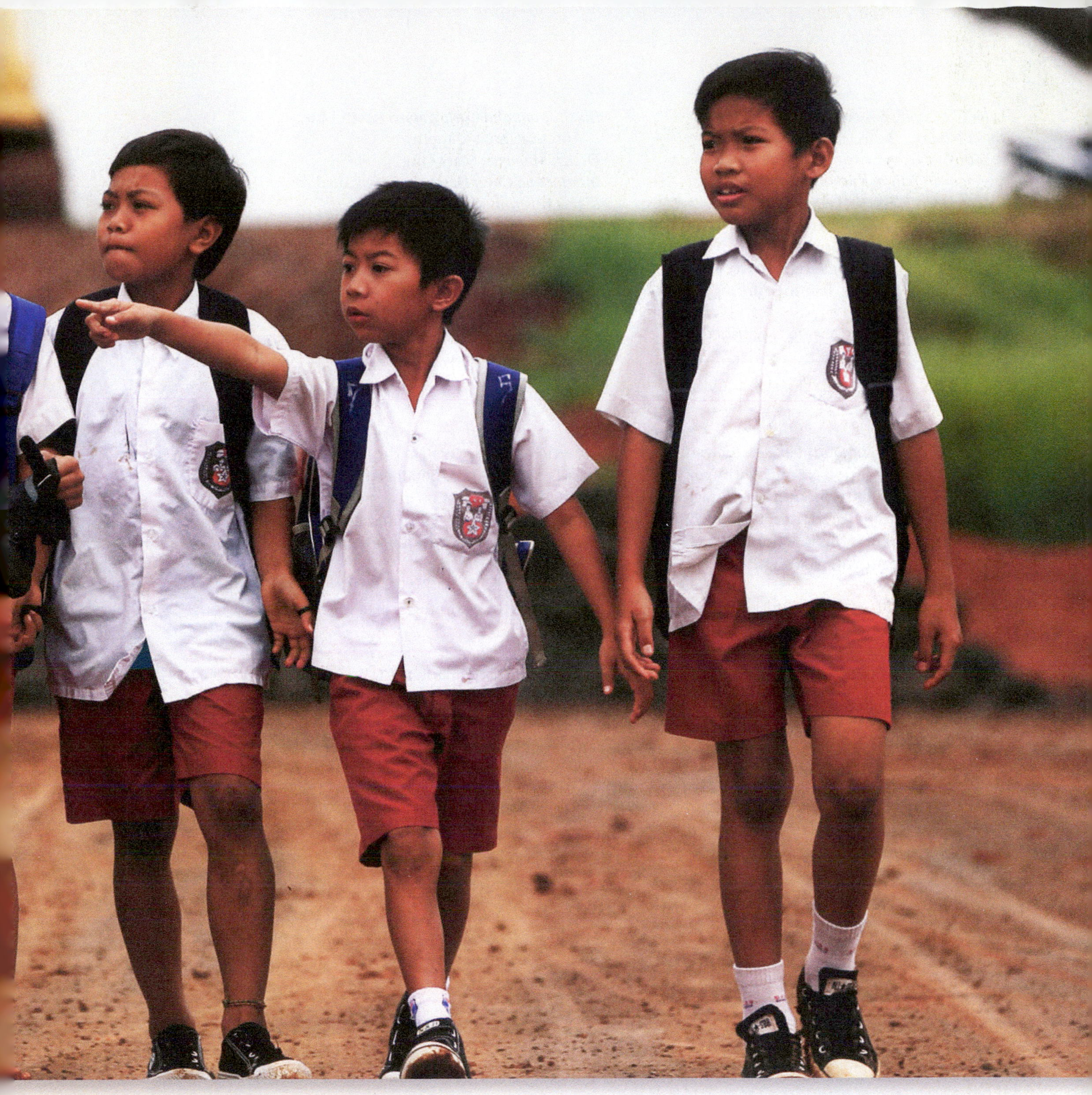

THE CULTURAL LANDSCAPE

AN INTRODUCTION TO HUMAN GEOGRAPHY | TWELFTH EDITION

James M. Rubenstein MIAMI UNIVERSITY, OXFORD, OHIO

PEARSON

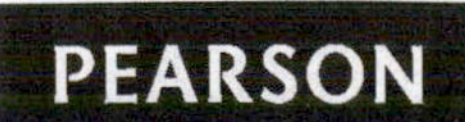

Senior Geography Editor: Christian Botting
Project Manager: Sean Hale
Program Manager: Anton Yakovlev
Development Editor: Karen Gulliver
Media Producer: Ziki Dekel
Director of Development: Jennifer Hart
Program Management Team Lead: Kristen Flathman
Project Management Team Lead: David Zielonka
Production Management: Jeanine Furino, Cenveo® Publisher Services
Copyeditor, Compositor: Cenveo® Publisher Services
Illustrations and Cartography: Kevin Lear, International Mapping
Design Manager: Mark Ong
Interior and Cover Designer: Elise Lansdon
Rights & Permissions Project Manager: Rachel Youdelman
Photo Researcher: Eric Schrader
Manufacturing Buyer: Maura Zaldivar-Garcia
Executive Product Marketing Manager: Neena Bali
Senior Field Marketing Manager: Mary Salzman
Marketing Assistant: Ami Sampat
Cover Photo Credit: Adam Eastland/AGE Fotostock

Library of Congress Cataloging-in-Publication Data
Names: Rubenstein, James M.
Title: The cultural landscape : an introduction to human geography / James M. Rubenstein.
Description: Twelfth edition. | Upper Saddle River, NJ : Pearson, 2015.
Identifiers: LCCN 2015041630| ISBN 9780134206141 | ISBN 0134206142
Subjects: LCSH: Human geography. | Human geography—Textbooks. | Human geography—Study and teaching.
Classification: LCC GF41 .R82 2015 | DDC 304.2—dc23
LC record available at http://lccn.loc.gov/2015041630

7

www.pearsonhighered.com

Student edition:
ISBN 10: 0-134-20623-1
ISBN 13: 978-0-134-20623-3

Instructor's Review Copy:
ISBN 10: 0-134-28625-1
ISBN 13: 978-0-13428625-9

BRIEF CONTENTS

CONTENTS

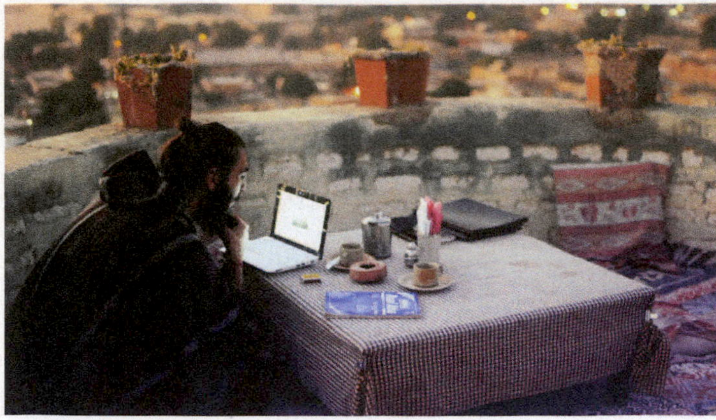

4 Folk and Popular Culture 110

5 Languages 144

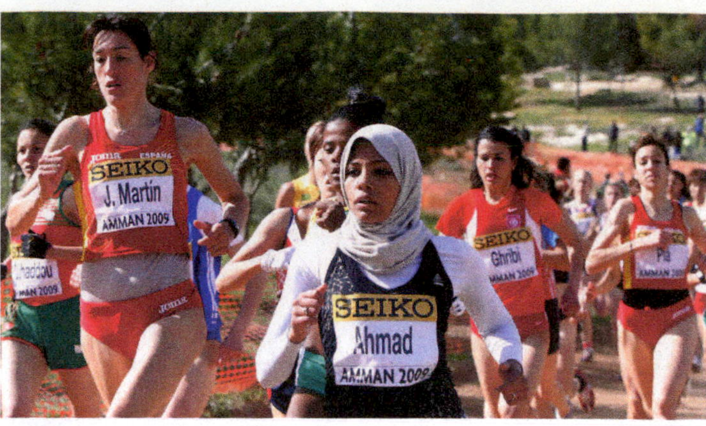
SEIKO
J. Martín
AMMAN 2009
SEIKO
Ahmad
AMMAN 2009

8 Political Geography 264

9 Food and Agriculture 306

GO

12 Services and Settlements 430

13 Urban Patterns 458

PREFACE

Geography is the study of where things are located on Earth's surface and the relationships between people and those locations. The word *geography*, invented by the ancient Greek scholar Eratosthenes, is based on two Greek words. Geo means "Earth," and graph means "to write." According to the National Geography Standards, geographers ask two simple questions: "Where is it?" and "Why is it there?" In other words, where are people and activities located across Earth's surface? Why are they located in particular places? *The Cultural Landscape* seeks to answer these questions as they relate to our contemporary world. The book provides an accessible, in-depth, and up-to-date introduction to human geography for majors and non-majors alike.

New to the 12th Edition

This edition brings substantial changes in both organization and content and, updated data and statistics.

NEW ORGANIZATION

A long-time strength of this book has been its clear, easy-to-use organization and outline. Electronic versions of the books now coexist with traditional paper format, formatted to facilitate reading on tablets and computers without compromising the pedagogic strengths of traditional paper formats. Valuable organizational features established in previous editions have been retained and considerably strengthened for this electronic age through the addition of several new features. The new elements can be grouped into two types:

- New informational features included in each chapter include the following:
  - **Doing Geography** is a new feature that discusses various geographic tools, techniques, and skills used to address real-world problems related to each chapter's concepts.
  - **What's Your Geography,** a feature that accompanies *Doing Geography,* asks students to put themselves in the role of geographers by applying these skills and techniques to their real-world experiences and environments, thereby connecting the global to the local, helping students connect the relevance of human geography to their everyday lives.
  - **Debate It** is a new feature that presents two sides of a complex human geography topic to help engage students in active debate and decision making. Readers may find that they agree with one side of the debate, or they may find merits in both perspectives.
  - **Sustainability & Our Environment** is a new feature that relates the principal topics of human geography to overarching issues of economic, social, and environmental sustainability for our planet.
  - **Interactive image** is a caption that accompanies an image in each chapter. The caption encourages students to interpret the geographic meaning and significance of the image.
  - **GeoVideo** is a new activity at the end of the chapter that integrates a BBC video with the subject of the chapter. Students are encouraged to log into MasteringGeography to view videos that present dynamic applications of chapter topics. Teachers can assign videos for credit.
  - **Explore** is a new end-of-chapter activity that uses Google Earth to investigate a chapter concept or application.
  - **Thinking Geographically** end-of-chapter questions now include images designed to illuminate chapter concepts or suggest directions for reflection.
- New outlining and arrangement of chapters include the following:
  - Each chapter continues to follow an outline based on four Key Issues, as in previous editions. Following each Key Issue title are several bulleted statements that outline the main topics discussed in that key issue.
  - Every two-page spread now begins with either one of the four Key Issues or one of the main bulleted statements.
  - Each two-page spread is now self-contained. As a result, maps and photos appear next to where they are discussed in the text. No more going through a chapter to find a figure that has been referenced on one page but actually appears on another page.
  - Two features initiated in the previous edition are now extended to all two-page spreads. One is a **Learning Outcome** that appears at the beginning of the spread and summarizes the principal purpose of the spread. And each spread now contains a **Pause and Reflect** feature that is presented as a question. Each question is designed to stimulate further reflection or discussion on the material being presented in the spread.
  - At the end of each section, **Key Issue Check-Ins** confirm for students the main issues and themes they should understand before continuing on in the chapter.

NEW CONTENT

Human geography is a dynamic subject. Topics that were central to the discipline a generation ago have faded in importance, while new ones take their place. Each chapter naturally provides updates of the most recently available data. Below are examples of entirely new material included in each chapter.

Chapter 1 (This Is Geography) has a new title. New topics include electronic mapping, geotagging, Volunteered Geographic Information (VGI), and geographic approaches to cultural identity such as gender and sexual orientation. Geography's five most basic concepts are introduced through the example of Luxembourg. The discussion of sustainability includes new information on the drought in the U.S. West.

Chapter 2 (Population and Health) includes an expanded discussion of health issues in a new Key Issue 3. As the rate of population growth declines from its peak during the second half of the twentieth century, population geography is increasingly concerned with the health of humans, not just their fertility and mortality. A new Key Issue 4 addresses future scenarios for world population and health.

Chapter 3 (Migration) includes recent controversies concerning U.S. borders and the surge of migration into Europe from Africa and Asia. The *What's Your Geography* feature helps students consider their own family's migration stories.

Chapter 4 (Folk and Popular Culture) includes new material about the diffusion of social media, as well as the distribution of various types of limits on Internet freedom. The chapter also has a new *Debate It* feature concerning clothing worn by observant Muslim women.

Chapter 5 (Languages) uses the leading authority Ethnologue's latest 5-point classification of languages as institutional, developing, vigorous, in trouble, and dying. The final Key Issue expands a discussion of new, revived, and growing languages. A new *Sustainability & Our Environment* feature focuses on gender differences in languages.

Chapter 6 (Religions) has been substantially reorganized and rewritten, and includes input from some of the nation's leading authorities on the geography of religions. Religion is especially important to many students. A chapter on the geography of religions can foster understanding of the diversity of religions in the world.

Chapter 7 (Ethnicities) includes new material on ethnic diversity in countries other than the United States, including Brazil. Also included is a new section on urban ethnic enclaves.

Chapter 8 (Political Geography) addresses current conflicts such as islands disputed between China and Japan, Russia's annexation of the Crimea from Ukraine, and the rise of terrorist organizations such as the Islamic State and Boko Haram. The chapter also includes a new *Debate It* feature on gerrymandering.

Chapter 9 (Food and Agriculture) now precedes the chapter on development, in accordance with the order suggested by the Advanced Placement Human Geography course syllabus. Key Issue 4 includes expanded discussions of genetically modified foods and food safety.

Chapter 10 (Development) reflects recent changes in United Nations development indexes. The chapter includes an expanded discussion of inequalities in development both among and within countries. In addition to development challenges faced by developing countries, the inequality discussion also considers Europe's current difficulties in attempting to promote economic growth through austerity.

Chapter 11 (Industry and Energy) has a new title that reflects inclusion of material on energy that had been in the previous edition's Development chapter. New material is included on U.S. transportation networks. Readers are asked to identify the national origin of their t-shirts.

Chapter 12 (Services and Settlements) includes a discussion of the new sharing economy, such as Uber and Airbnb. New *Doing Geography* and *What's Your Geography* features include an interactive study of food deserts.

Chapter 13 (Urban Patterns) includes an expanded discussion of the structure of nonwestern cities today, as well as in the past. A new case study illustrates the CBD of Louisville, Kentucky. Also expanded is a discussion of the relationship between transportation and urban patterns.

MasteringGeography has also evolved since the last edition, now featuring a broad library of BBC video clips, a new next generation of GIS-inspired MapMaster interactive maps, Dynamic Study Modules for Human Geography, a responsive-designed eText 2.0 version of the book, and more.

Geography Matters

The main purpose of this book is to introduce students to the study of human geography as a social science by emphasizing the relevance of geographic concepts to human problems. It is intended for use in college-level introductory human or cultural geography courses, as well as the equivalent advanced placement course in high school. At present, human geography is the fastest-growing course in the AP curriculum.

GEOGRAPHY IN OUR ELECTRONIC AGE

Many speculated that geography would be irrelevant in the twenty-first century. Geography's future was thought to be grim because the diffusion of electronic communications, such as the Internet and social media, would make it easier for human activities to be conducted remotely. If any piece of information could be accessed from any place in the world (at least where electronic devices work), why live, shop, work, or establish a business in a crowded city or a harsh climate?

In reality, geography has become more, not less, important in people's lives and the conduct of business. Here are several ways that location matters more now than in the past, because of—not despite—the diffusion of electronic devices:

1. Smartphones and other electronic devices match specific demand to supply in a particular locality. For example:
   - Restaurant apps match hungry people to empty seats in a locality's restaurants.
   - Real estate apps help people find housing for sale or for rent in a locality.
   - Social apps let people know where their friends in a particular locality are hanging out that night.
   - Transportation apps match vehicles with available seats to people trying to get to specific locations.

   These sorts of apps generate data on people's preferences in space, which in turn helps even more location-based business get started and grow. Instead of looking for restaurants in the Yellow Pages, we find places to eat that are mapped on our device and in our locations. No wonder that geography apps, in the form of maps (including navigation) and travel (including transportation), rank as two of the five most frequently used services on smartphones.

2. Electronic devices are essential to the smooth movement of people and goods. For example:
   - Turn-by-turn information can prevent you from getting lost or steer you back if you do get lost.
   - Traffic jams on overcrowded roads can be avoided or minimized.
   - Vehicles in the future will be driverless, so you can spend driving time working, learning stuff, or social networking.
   - Instead of turning on a radio to hear traffic information, we look at the red and green traffic flow patterns on an electronic map.
   - Instead of waiting for a TV weather report, we look at storm patterns on our device's map.

   Images from Google Earth and others that you see throughout this book will become more detailed and accurate as technologies advance. Mapping is expanding into indoor spaces and into three dimensions.
3. The people who make all of these new location-based apps are themselves highly clustered in a handful of places in the world, such as Silicon Valley.
   - Ideas—both brilliant and far-fetched—are still easier to communicate face-to-face than across long distances.
   - Living and working in places like Silicon Valley, despite high expenses and choking traffic jams, put people next to other like-minded innovators in the electronic-based geography of the twenty-first century.
4. Electronic devices also impact the changing geography of cultural diversity.
   - What if you searched for an available restaurant table in a foreign language? Would you find the same places?
   - What if you conducted an Internet search in a foreign country? Would you find the same information?

A central theme in this book explores the tension between two important themes—globalization and cultural diversity. In many respects, we are living in a more unified world economically, culturally, and environmentally. The actions of a particular corporation or country affect people around the world. For example, geographers examine the prospects for an energy crisis by relating the distributions of energy production and consumption. Geographers find that the users of energy are located in places with different social, economic, and political institutions than are the producers of energy. The United States and Japan consume far more energy than they produce, whereas Russia and Saudi Arabia produce far more energy than they consume.

This book argues that after a period when globalization of the economy and culture has been a paramount concern in geographic analysis, local diversity now demands equal time. People are taking deliberate steps to retain distinctive cultural identities. They are preserving little-used languages, fighting fiercely to protect their religions, and carving out distinctive economic roles. Local diversity even extends to addressing issues, such as climate change, that at first glance are considered global. For example, the "greenest" cars for motorists to drive in Oregon are different than the "greenest" cars for Ohio.

Outline of Main Topics

The book discusses the following main topics:

- What basic concepts do geographers use? Chapter 1 provides an introduction to ways that geographers think about the world. Geographers employ several concepts to describe the distribution of people and activities across Earth, to explain reasons underlying the observed distribution, and to understand the significance of the arrangements.
- Where are people located in the world? Chapters 2 and 3 examine the distribution and growth of the world's population, as well as the movement of people from one place to another. Why do some places on Earth contain large numbers of people or attract newcomers while other places are sparsely inhabited?
- How are different cultural groups distributed? Chapters 4 through 8 analyze the distribution of different cultural traits and beliefs and the problems that result from those spatial patterns. Important cultural traits discussed in Chapter 4 include food, clothing, shelter, and leisure activities. Chapters 5 through 7 examine three main elements of cultural identity: language, religion, and ethnicity. Chapter 8 looks at political problems that arise from cultural diversity. Geographers look for similarities and differences in the cultural features at different places, the reasons for their distribution, and the importance of these differences for world peace.
- How do people earn a living in different parts of the world? Human survival depends on acquiring an adequate food supply. One of the most significant distinctions in the world is whether people produce their food directly from the land or buy it with money earned by performing other types of work. Chapters 9 through 12 look at the three main ways of earning a living: agriculture, manufacturing, and services. Chapter 13 discusses cities, the centers for economic as well as cultural activities.

Divisions within Geography

Because geography is a broad subject, some specialization is inevitable. At the same time, one of geography's strengths is its diversity of approaches. Rather than being forced to adhere rigorously to established disciplinary laws, geographers can combine a variety of methods and approaches. This tradition stimulates innovative thinking, although students who are looking for a series of ironclad laws to memorize may be disappointed.

HUMAN AND PHYSICAL GEOGRAPHY

Geography is both a physical science and a social science. When geography concentrates on the distribution of physical features, such as climate, soil, and vegetation, it is a physical science. When it studies cultural features, such as language, industries, and cities, geography is a social science. This division is reflected in some colleges, where physical geography courses may carry natural science credit while human and cultural geography courses carry social science credit.

While this book is mostly concerned with geography from a social science perspective, one of the distinctive features of geography is its use of natural science concepts to help understand human behavior. The distinction between physical and human geography reflects differences in emphasis, not an absolute separation. The integration of physical and human geography is especially important when studying sustainability issues.

TOPICAL AND REGIONAL APPROACHES

Geographers face a choice between a topical approach and a regional approach. The topical approach, which is used in this book, starts by identifying a set of important cultural issues to be studied, such as population growth, political disputes, and economic restructuring. Geographers using the topical approach examine the location of different aspects of the topic, the reasons for the observed pattern, and the significance of the distribution.

The alternative approach is regional. Regional geographers select a portion of Earth and study the environment, people, and activities within that selected area. The regional geography approach is used in courses on Europe, Africa, Asia, and other areas of the world. Although this book is organized by topics, geography students should be aware of the location of places in the world. A separate index section lists the book's maps by location. One indispensable aid in the study of regions is an atlas, which can also be used to find unfamiliar places that pop up in the news.

DESCRIPTIVE AND SYSTEMATIC METHODS

Whether using a topical or a regional approach, geographers can select either a descriptive or a systematic method. Again, the distinction is one of emphasis, not an absolute separation. The descriptive method emphasizes the collection of a variety of details about a particular location. This method has been used primarily by regional geographers to illustrate the uniqueness of a particular location on Earth's surface. The systematic method emphasizes the identification of several basic theories or techniques developed by geographers to explain the distribution of activities.

This book uses both the descriptive and systematic methods because total dependence on either approach is unsatisfactory. An entirely descriptive book would contain a large collection of individual examples not organized into a unified structure. A completely systematic approach suffers because some of the theories and techniques are so abstract that they lack meaning for the student. Geographers who depend only on the systematic approach may have difficulty explaining important contemporary issues.

Suggestions for Use

This book can be used in an introductory human or cultural geography course that extends over one semester, one quarter, or two quarters. An instructor in a one-semester course could devote one week to each of the chapters, leaving time for examinations. In a one-quarter course, the instructor might need to omit some of the book's material. A course with more of a cultural orientation could use Chapters 1 through 8. If the course has more of an economic orientation, then the appropriate chapters would be 1 through 3 and 9 through 13. A two-quarter course could be organized around the culturally oriented Chapters 1 through 8 during the first quarter and the more economically oriented

Chapters 9 through 13 during the second quarter. Topics of particular interest to the instructor or students could be discussed for more than one week.

Acknowledgments

For a book that has been through many editions to maintain its leadership position, stale and outdated material and methods must be cleared out to make way for the fresh and contemporary. It is all too easy for an author in the twenty-first century to rely on practices that brought success in the twentieth century. Strong proactive leadership is required from the publisher to push an already strong book to loftier aspirations. This leadership is especially critical during a period when the teaching and learning environment is changing much more rapidly than even in the late twentieth century. A major reason for the long-term success of this book has been the quality of leadership in geography at Pearson Education.

Christian Botting, Senior Editor for Geography, Meteorology, and Geospatial Technologies at Pearson Education, has now led the team through six of my book projects. Because Pearson Education is the dominant publisher of college geography textbooks, the person in charge of geography wields considerable influence in shaping the nation's geography curriculum. Christian expertly balances the challenges of leading the market and listening to customers, of pushing ahead with innovations and sticking with what works.

Anton Yakovlev, Program Manager at Pearson Education, has now been involved with me on five book projects. Anton not only keeps impeccable control of what has to be done, he has been more proactive than previous project managers in initiating many great ideas.

Sean Hale, Project Manager at Pearson Education, ably handles day-to-day movement of materials and ideas. This project has a nontraditional flow of work among the principal actors, and I am grateful to Sean for keeping everything moving in a timely and an accurate manner.

I have had the great fortune to work with only three editors for most of my long association with Pearson and its predecessors. Paul F. Corey, who is now president of Science, Business and Technology at Pearson, guided development of the third, fourth, and fifth editions of this book. Dan Kaveney guided development of the sixth, seventh, eighth, and ninth editions. I will always value the sound judgment, outstanding vision, and friendship of Paul and Dan, and now Christian.

In this age of outsourcing, Pearson works with many independent companies to create books. This edition has been the beneficiary of a top-notch team:

Karen Gulliver, the development editor, has had lots of great ideas. Because the book has been a success for so long, it is a challenging job to make a great product even stronger.

Jeanine Furino, at Cenveo Publisher Services, smoothly managed the flow of copyediting and other production tasks for this project. This is an especially important task because of the unusual flow of work, especially the unique construction of each two-page spread.

Kevin Lear, Senior Project Manager at International Mapping, and his team, produced outstanding maps for this book. Back in the 1980s, when he was getting started as a professional cartographer, Kevin was the first cartographer to figure out how to produce computer-generated full-color maps for the second edition of this book. That was the first time that either GIS or full color had been used in a geography text.

I am grateful for the great work done on a variety of ancillaries.

I would also like to extend a special thanks to all of my colleagues who have, over the years, offered a good deal of feedback and constructive criticism. Colleagues who served as reviewers as we prepared the 12th edition are: Victoria Alapo (Metropolitan Community College), Christiana Asante-Ashong (Grambling State University), Becky Bruce (Southwestern Oklahoma State Univ), Tom Chapman (Old Dominion University), Xueming Chen (Virginia Commonwealth University), Marcia England (Miami University), Steven Graves (California State University Northridge), Chris Hall (Davis School District, Utah), Institute for Curriculum Services, Gordon Newby (Emory University), William Pitts (Baylor University), Benjamin Ravid (Brandeis University), James Saku (Frostburg State University), Debra Sharkey (Cosumnes River College), Jill Stackhouse (Bemidji State University), John Voll (Georgetown University), Margath Walker (University of Louisville), and Pam Wolfe (Yeshiva of Greater Washington).

DIGITAL & PRINT RESOURCES

For Students and Teachers:

This edition provides a complete human geography program for students and teachers.

Masteringgeography™ with Pearson eText for *The Cultural Landscape*

The Mastering platform is the most widely used and effective online homework, tutorial, and assessment system for the sciences. It delivers self-paced coaching activities that provide individualized coaching, focus on course objectives, and are responsive to each student's progress. The Mastering system helps teachers maximize class time with customizable, easy-to-assign, and automatically graded assessments that motivate students to learn outside of class and arrive prepared for lecture. MasteringGeography offers:

- Assignable activities that include GIS-inspired MapMaster™ interactive maps, *Encounter Human Geography* Google Earth™ Explorations, GeoVideos, GeoTutors, Thinking Spatially & Data Analysis activities, end-of-chapter questions, reading quizzes, Test Bank questions, map labeling activities, and more.
- Student study area with GIS-inspired MapMaster interactive maps, Geoscience Animations, web links, geography videos, glossary flash cards, "In the News" RSS feeds, reference maps, an optional Pearson eText and more. **www.masteringgeography.com**

Instructor Resource DVD (0134259424) The *Instructor Resource DVD* provides high-quality electronic versions of photos and illustrations from the book in JPEG, pdf, and PowerPoint formats, as well as customizable PowerPoint lecture presentations, Classroom Response System questions in PowerPoint, and the *Instructor Resource Manual* and *Test Bank* in MS. Word and TestGen formats. For easy reference and identification, all resources are organized by chapter.

For Teachers

Instructor Resource Manual (Download Only) (0134259416) Updated for the twelfth edition, the *Instructor Resource Manual*, is intended as a resource for both new and experienced instructors. It includes lecture outlines, additional source materials, teaching tips, advice about how to integrate online media, and various other ideas for the classroom. **http://www.pearsonhighered.com/irc.**

TestGen® Computerized Test Bank (Download Only) (0134259408) TestGen is a computerized test generator that lets instructors view and edit *Test Bank* questions, transfer questions to tests, and print the test in a variety of customized formats. This *Test Bank* includes over 1,000 multiple choice and short answer/ essay questions. Questions are correlated to the revised U.S. National Geography Standards and Bloom's Taxonomy to help instructors better map the assessments against both broad and specific teaching and learning objectives. The questions are also tagged to chapter specific learning outcomes. The Test Bank is available in Microsoft Word, and is importable into Blackboard. **http://www.pearsonhighered.com/irc**

For Students

Teaching College Geography: A Practical Guide for Graduate Students and Early Career Faculty (0136054471) This two-part resource provides a starting point for becoming an effective geography teacher from the very first day of class. Divided in two parts, Part One addresses "nuts-and-bolts" teaching issues. Part Two explores being an effective teacher in the field, supporting critical thinking with GIS and mapping technologies, engaging learners in large geography classes and promoting awareness of international perspectives and geographic issues.

Aspiring Academics: A Resource Book for Graduate Students and Early Career Faculty (0136048919) Drawing on several years of research, this set of essays is designed to help graduate students and early career faculty start their careers in geography and related social and environmental sciences. *Aspiring Academics* stresses the interdependence of teaching, research, and service—and the importance of achieving a healthy balance of professional and personal life—while doing faculty work. Each chapter provides accessible, forward-looking advice on topics that often cause the most stress in the first years of a college or university appointment.

Practicing Geography: Careers for Enhancing Society and the Environment (0321811151) This book examines career opportunities for geographers and geospatial professionals in business, government, nonprofit, and educational sectors. A diverse group of academic and industry professionals share insights on career planning, networking, transitioning between employment sectors, and balancing work and home life. The book illustrates the value of geographic expertise and technologies through engaging profiles and case studies of geographers at work.

Goode's World Atlas, 23rd Edition (0133864642) *Goode's World Atlas* has been the world's premiere educational atlas since 1923, and for good reason. It features over 250 pages of maps, from definitive physical and political maps to important thematic maps that illustrate the spatial aspects of many important topics. The 23rd edition includes digitally produced reference maps, as well as new thematic maps on demography, global climate change, sea level rise, CO_2 emissions, polar ice fluctuations, deforestation, extreme weather events, infectious diseases, water resources, and energy production.

Television for the Environment Earth Report Geography Videos on DVD (0321662989) This three-DVD set is designed to help students visualize how human decisions and behavior have affected the environment and how individuals are taking steps toward recovery. With topics ranging from the poor land management promoting the devastation of river systems in Central America to the struggles for electricity in China and Africa, these 13 videos from Television for the Environment's global *Earth Report* series recognize the efforts of individuals around the world to unite and protect the planet.

Encounter Human Geography Workbook & Website by Jess C. Porter (0321682203) For classes that do not use MasteringGeography, *Encounter Human Geography* provides rich, interactive explorations of human geography concepts through Google Earth. Students explore the globe through themes such as population, sexuality and gender, political geography, ethnicity, urban geography, migration, human health, and language. All chapter explorations are available in print format as well as online quizzes, accommodating different classroom needs. All worksheets are accompanied with corresponding Google Earth KMZ media files, available for download for those who do not use MasteringGeography, from **http://www.mygeoscienceplace.com.**

Dire Predictions: Understanding Climate Change, 2nd edition, by Michael Mann and Lee R. Kump (0133909778) Periodic reports from the Intergovernmental Panel on Climate Change (IPCC) evaluate the risk of climate change brought on by humans. But the sheer volume of scientific data remains inscrutable to the general public, particularly to those who may still question the validity of climate change. In just over 200 pages, this practical text presents and expands upon the essential findings of the IPCC's 5th Assessment Report in a visually stunning and undeniably powerful way to the lay reader. Scientific findings that provide validity to the implications of climate change are presented in clear-cut graphic elements, striking images, and understandable analogies.

The **Second Edition** covers the latest climate change data and scientific consensus from the IPCC Fifth Assessment Report and integrates links to online media. The text is also available in various eText formats, including an upgrade option from MasteringGeography courses.

ABOUT THE AUTHOR

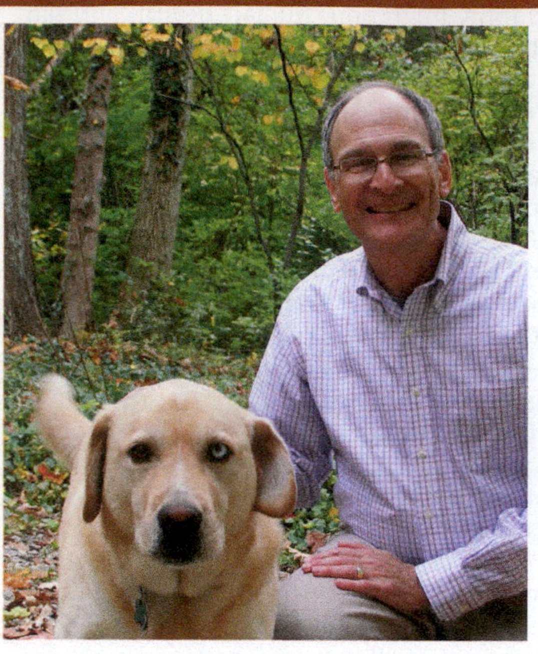

Dr. James M. Rubenstein received his B.A. from the University of Chicago in 1970, M.Sc. from the London School of Economics and Political Science in 1971, and Ph.D. from Johns Hopkins University in 1975. He is Professor of Geography at Miami University, where he teaches urban and human geography. Dr. Rubenstein also conducts research in the automotive industry and has published three books on the subject—*The Changing U.S. Auto Industry: A Geographical Analysis* (Routledge); *Making and Selling Cars: Innovation and Change in the U.S. Auto Industry* (The Johns Hopkins University Press); and *Who Really Made Your Car? Restructuring and Geographic Change in the Auto Industry* (W.E. Upjohn Institute, with Thomas Klier). Dr. Rubenstein is also the author of *Contemporary Human Geography*, as well as Introduction to *Contemporary Geography*. He also writes a weekly column about local food for the *Oxford Press*. Winston, a lab-husky mix with one brown eye and one blue eye, takes Dr. Rubenstein for long walks in the woods every day. Thanks to Ursula Roma for the photo.

DEDICATION

This book is dedicated to Bernadette Unger, Dr. Rubenstein's wife, who has been by his side through many books, as well as to the memory of his father, Bernard W. Rubenstein. Dr. Rubenstein also gratefully thanks the rest of his family for their love and support.

ABOUT OUR SUSTAINABILITY INITIATIVES

Pearson recognizes the environmental challenges facing this planet, as well as acknowledges our responsibility in making a difference. This book is carefully crafted to minimize environmental impact. The binding, cover, and paper come from facilities that minimize waste, energy consumption, and the use of harmful chemicals. Pearson closes the loop by recycling every out-of-date text returned to our warehouse.

Along with developing and exploring digital solutions to our market's needs, Pearson has a strong commitment to achieving carbon-neutrality. As of 2009, Pearson became the first carbon- and climate-neutral publishing company, having reduced our absolute carbon footprint by 22% since then. Pearson has protected over 1,000 hectares of land in Columbia, Costa Rica, the United States, the UK and Canada.

In 2015, Pearson formally adopted *The Global Goals for Sustainable Development,* sponsoring an event at the United Nations General Assembly and other ongoing initiatives. Pearson sources 100% of the electricity we use from green power and invests in renewable energy resources in multiple cities where we have operations, helping make them more sustainable and limiting our environmental impact for local communities.

PEARSON

The future holds great promise for reducing our impact on Earth's environment, and Pearson is proud to be leading the way. We strive to publish the best books with the most up-to-date and accurate content, and to do so in ways that minimize our impact on Earth. To learn more about our initiatives, please visit www.pearson.com/social-impact/sustainability/environment.html.

THE CULTURAL LANDSCAPE

ames M. Rubenstein

THE CULTURAL LANDSCAPE

1

This Is Geography

What do you expect from this geography course? You may think that geography involves memorizing lists of countries and capitals. Perhaps you associate geography with photographic essays of exotic places in popular magazines. Contemporary geography is the scientific study of where people and activities are found across Earth's surface and the reasons they are found there.

Luxembourg City, including St. John Church, built in 1606.

LOCATIONS IN THIS CHAPTER

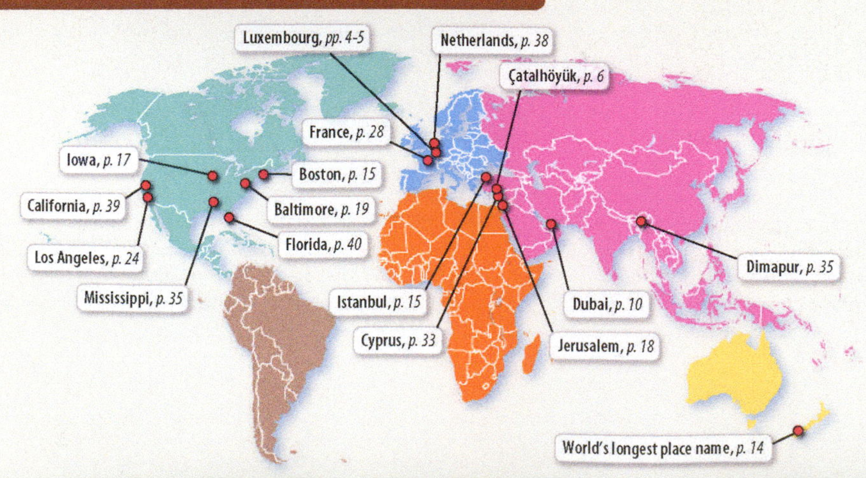

KEY ISSUES

1

Why Is Geography a Science?

Prehistoric humans were the first people to make maps. Contemporary tools enable cartographers—and anyone else who has access to electronic devices—to make precise maps.

2

Why Is Each Point on Earth Unique?

Geographers understand why each ***place*** on Earth is in some ways unique. Each area or ***region*** on Earth also possesses a unique combination of features.

3

Why Are Different Places Similar?

Many features are organized in a regular manner across ***space***. Some regularities are global in ***scale***, whereas others have distinctive local character.

4

Why Are Some Actions Not Sustainable?

Distinctive to geography is the importance given to **connections** between human activities and the physical environment. Some human activities are sustainable, but others are not.

KEY ISSUE 1

Why Is Geography a Science?

- Introducing Geography
- Cartography: The Science of Mapmaking
- Contemporary Geographic Tools
- Interpreting Maps
- The Geographic Grid

LEARNING OUTCOME 1.1.1

Summarize differences between geography and history.

Thinking geographically is one of the oldest human activities. Perhaps the first geographer was a prehistoric human who crossed a river or climbed a hill, observed what was on the other side, returned home to tell about it, and scratched the route in the dirt. The second geographer may have been a friend or relative who followed the dirt drawing to reach the other side.

The word *geography*, invented by the ancient Greek scholar Eratosthenes, is based on two Greek words. *Geo* means "Earth," and *graphy* means "to write." Geography is the study of where things are found on Earth's surface and the reasons for the locations. Human geographers ask two simple questions: Where are people and activities found on Earth? Why are they found there?

In his framework of all scientific knowledge, the German philosopher Immanuel Kant (1724–1804) compared geography and history:

Geographers . . .	Historians . . .
identify the location of important places and explain why human activities are located beside one another.	identify the dates of important events and explain why human activities follow one another chronologically.
ask *where* and why.	ask *when* and why.
organize material spatially.	organize material chronologically.
recognize that an action at one point on Earth can result from actions at another point, which can consequently affect conditions elsewhere.	recognize that an action at one point in time can result from past actions that can in turn affect future ones.

History and geography differ in one especially important manner: A geographer can drive or fly to another place to study Earth's surface, whereas a historian cannot travel to another time to study other eras firsthand. This ability to reach other places lends excitement to the discipline of geography—and geographic training raises the understanding of other spaces to a level above that of casual sightseeing.

Introducing Geography

To introduce human geography, we will concentrate on two main features of human behavior: culture and economy. The first half of the book explains why the most important cultural features, such as major languages, religions, and ethnicities, are arranged as they are across Earth. The second half of the book looks at the locations of the most important economic activities, including agriculture, manufacturing, and services.

This chapter introduces basic concepts that geographers employ to address their "where" and "why" questions. To explain where things are, one of geography's most important tools is a map. Ancient and medieval geographers created maps to describe what they knew about Earth. Today, accurate maps are generated from electronic data.

Geographers employ several basic concepts to explain why every place on Earth is in some ways unique and in other ways related to other locations. Many of these concepts are commonly used English words, but they are given particular meaning by geographers.

To explain why every place is unique, geographers have two basic concepts:

- A **place** is a specific point on Earth, distinguished by a particular characteristic. Every place occupies a unique location, or position, on Earth's surface.
- A **region** is an area of Earth defined by one or more distinctive characteristics. Geographers divide the world into a number of regions, such as North America and Latin America.

To explain why different places are interrelated, geographers have three basic concepts:

- **Scale** is the relationship between the portion of Earth being studied and Earth as a whole. Geographers study a variety of scales, from local to global. Many processes

▼ **FIGURE 1-1 PLACE** The place of the City of Luxembourg is atop a hill overlooking the Alzette River.

▲ **FIGURE 1-2 REGION** Luxembourg is part of the region of Europe.

that affect humanity's occupation of Earth are global in scale, such as climate change and depletion of energy supplies. At the same time, local-scale processes—such as preservation of distinctive cultural and economic activities—are increasingly important.

- **Space** refers to the physical gap or interval between two objects. Geographers observe that many objects are distributed across space in a regular manner, for discernible reasons.
- **Connection** refers to relationships among people and objects across the barrier of space. Geographers are concerned with the various means by which connections occur.

Luxembourg can be used to illustrate the five concepts. The City of Luxembourg is a place located on a hillside perched above the Alzette River (Figure 1-1). The City of Luxembourg is the capital of the country of Luxembourg, located in the world region of Europe (Figure 1-2). Luxembourg plays a major role at a global scale, as one of the principal headquarters of the European Union, which unites 28 countries (Figure 1-3a). At the same time, Luxembourg, like other places, has a distinctive local scale; one example is the availability of distinctive local products not available elsewhere (Figure 1-3b). The space occupied by Luxembourg has distinctive features; for example, most people live in the south of the country, whereas the north is sparsely inhabited (Figure 1-4). Connections between Luxembourg and other places are provided by road, rail, and river (Figure 1-5).

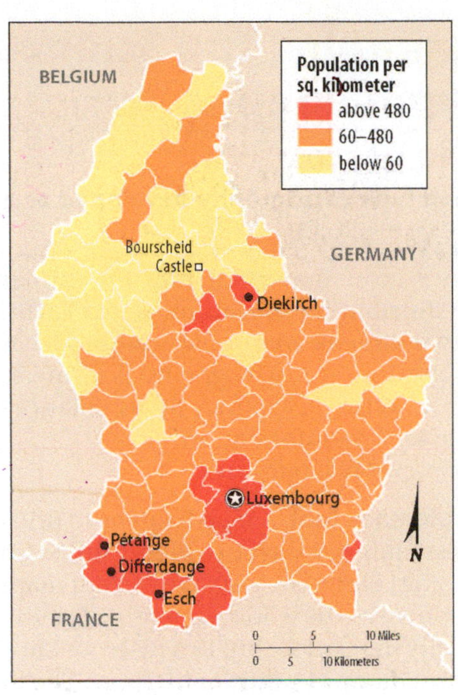

▲ **FIGURE 1-4 SPACE** The space occupied by Luxembourgers is primarily houses built close together in cities in the southern half of the country.

PAUSE & REFLECT 1.1.1

What are the principal connections from your hometown to other places?

▼ **FIGURE 1-3 SCALE (a)** Regional scale: high-rise buildings in the background house offices of the European Union; **(b)** Local scale: vendor at farmers' market sells food products made in Luxembourg.

(a)

(b)

▼ **FIGURE 1-5 CONNECTION** Luxembourg is connected to other places in Europe by train. European Union offices are in the background.

Cartography: The Science of Mapmaking

LEARNING OUTCOME 1.1.2

Understand how cartography developed as a science.

Geography's most important tool for thinking spatially about the distribution of features across Earth is a map. A **map** is a two-dimensional or flat-scale model of Earth's surface, or a portion of it. Geography is immediately distinguished from other disciplines by its reliance on maps to display and analyze information.

A map serves two purposes:

- **As a reference tool.** A map helps us to find the shortest route between two places and to avoid getting lost along the way. We consult maps to learn where in the world something is located, especially in relationship to a place we know, such as a town, body of water, or highway. The maps in an atlas or a smart phone app are especially useful for this purpose.
- **As a communications tool.** A map is often the best means for depicting the distribution of human activities or physical features, as well as for thinking about reasons underlying a distribution.

A map is a scale model of the real world, made small enough to work with on a desk or computer. It can be a hasty here's-how-to-get-to-the-party sketch, an elaborate work of art, or a precise computer-generated product. For centuries, geographers have worked to perfect the science of mapmaking, called **cartography.** Contemporary cartographers are assisted by computers and satellite imagery.

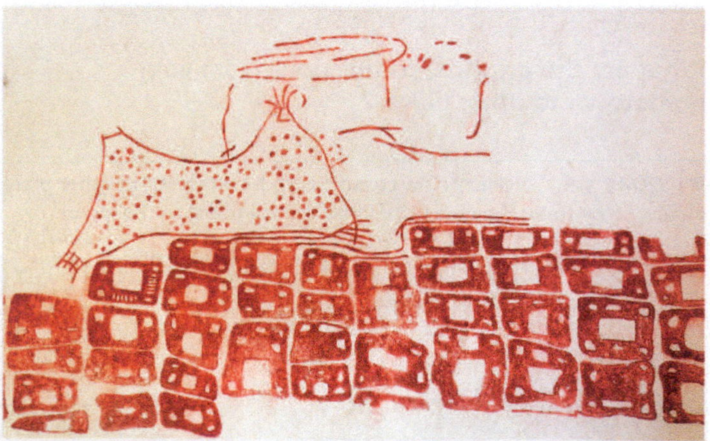

▲ **FIGURE 1-6 EARLIEST SURVIVING MAP** This map, dating from 6200 B.C., depicts the town of Çatalhöyük, in present-day Turkey, and the eruption of the Hasan Dağ (Mount Hasan) twin-peaks volcano, which is actually located around 140 km northeast of the town. Archaeological evidence indicates that the volcano did erupt around the time that the map was made. The map is now in the Konya Archaeological Museum.

GEOGRAPHY IN THE ANCIENT WORLD

The science of geography has prehistoric roots. The earliest surviving fully authenticated map, depicting the town of Çatalhöyük, located in present-day Turkey, dates from approximately 6200 B.C. (Figure 1-6). Archaeologists found the map on the wall of a house that was excavated during the 1960s. Major contributors to geographic thought in the ancient eastern Mediterranean included:

- Thales of Miletus (ca. 624–ca. 546 B.C.), who applied principles of geometry to measuring land area.
- Anaximander (610–ca. 546 B.C.), a student of Thales, who made a world map based on information from sailors and argued that the world was shaped like a cylinder.
- Pythagoras (ca. 570–ca. 495 B.C.), who may have been the first to propose a spherical world and argued that the sphere was the most perfect form.
- Hecateus (ca. 550–ca. 476 B.C.), who may have produced the first geography book, called *Ges Periodos* ("Travels Around the Earth").
- Aristotle (384–322 B.C.), who was the first to demonstrate that Earth was spherical on the basis of evidence.
- Eratosthenes (ca. 276–ca. 195 B.C.), the inventor of the word *geography*, who accepted that Earth was round (as few others did in his day), calculated its circumference within 0.5 percent accuracy, accurately divided Earth into five climatic regions, and described the known world in one of the first geography books.
- Strabo (ca. 63 B.C.–ca. A.D. 24), who described the known world in a 17-volume work titled *Geography*.
- Ptolemy (ca. A.D. 100–ca. 170), who wrote the eight-volume *Guide to Geography*, codified basic principles of mapmaking, and prepared numerous maps that were not improved upon for more than 1,000 years (Figure 1-7).

China was another center of early geographic thought. Ancient Chinese geographic contributions included:

- "Yu Gong" ("Tribute of Yu"), a chapter in a book called *Shu Jing* ("Classic of History"), which was the earliest surviving Chinese geographical writing, by an unknown author from the fifth century B.C., described the economic resources of the country's different provinces.
- Pei Xiu, the "father of Chinese cartography," who produced an elaborate map of the country in A.D. 267.

GEOGRAPHY'S REVIVAL

After Ptolemy, little progress in mapmaking or geographic thought was made in Europe for several hundred years. Maps became less mathematical and more fanciful, showing

▲ **FIGURE 1-7 WORLD MAP BY PTOLEMY, CA. A.D. 150** The map shows the known world at the height of the Roman Empire, surrounding the Mediterranean Sea and Indian Ocean.

▲ **FIGURE 1-9 WORLD MAP BY ORTELIUS, 1571** This was one of the first maps to show the considerable extent of the Western Hemisphere, as well as the Antarctic landmass.

Earth as a flat disk surrounded by fierce animals and monsters. Geographic inquiry continued, though, outside Europe. Contributors outside of Europe included:

- Muhammad al-Idrisi (1100–ca. 1165), a Muslim geographer who prepared a world map and geography text in 1154, building on Ptolemy's long-neglected work (Figure 1-8).
- Abu Abdullah Muhammad Ibn-Battuta (1304–ca. 1368), a Moroccan scholar, who wrote *Rihla* ("Travels") based on three decades of journeys covering more than 120,000 kilometers (75,000 miles) through the Muslim world of northern Africa, southern Europe, and much of Asia.

Making maps as reference tools revived during the Age of Exploration and Discovery. Columbus, Magellan, and other explorers who sailed across the oceans in search of trade routes and resources in the fifteenth and sixteenth centuries required accurate maps to reach desired destinations without wrecking their ships. In turn, cartographers used information collected by the explorers to create more accurate maps. Influential European cartographers included:

- Martin Waldseemuller (ca. 1470–ca. 1521), a German cartographer who was credited with producing the first map to use the label "America"; he wrote on the map (translated from Latin) "from Amerigo the discoverer . . . as if it were the land of Americus, thus America."
- Abraham Ortelius (1527–1598), a Flemish cartographer, who created the first modern atlas and was the first to hypothesize that the continents were once joined together before drifting apart (Figure 1-9).
- Bernhardus Varenius (1622–1650), who produced *Geographia Generalis*, which stood for more than a century as the standard treatise on systematic geography.

PAUSE & REFLECT 1.1.2

What is one main difference between the world maps of Ptolemy (Figure 1-7) and of Ortelius (Figure 1-9)?

► **FIGURE 1-8 WORLD MAP BY AL-IDRISI, 1154** Al-Idrisi built on Ptolemy's map, which had been neglected for nearly a millennium.

Contemporary Geographic Tools

LEARNING OUTCOME 1.1.3

Identify geography's contemporary analytic mapping tools.

Maps are not just paper documents in textbooks. They have become an essential tool for contemporary delivery of online services through smart phones, tablets, and computers.

PINPOINTING LOCATIONS: GPS

Our smart phones, tablets, and computers are equipped with **Global Positioning System (GPS),** which is a system that determines the precise position of something on Earth. The GPS in use in the United States includes three elements:

- Satellites placed in predetermined orbits by the U.S. military (24 in operation and 3 in reserve).
- Tracking stations to monitor and control the satellites.
- A receiver that can locate at least 4 satellites, figure out the distance to each, and use this information to pinpoint its own location.

GPS is most commonly used for navigation. Pilots of aircraft and ships stay on course with GPS. On land, GPS detects a vehicle's current position, the motorist programs the desired destination into a GPS device, and the device provides instructions on how to reach the destination. GPS can also be used to find the precise location of a vehicle, enabling a motorist to summon help in an emergency or a customer to monitor the progress of a delivery truck or position of a bus or train.

Thanks to GPS, our electronic devices provide us with a wealth of information about the specific place on Earth we currently occupy. Cell phones equipped with GPS allow individuals to share their whereabouts with others. Geographers find GPS to be particularly useful in coding the precise location of objects collected in fieldwork.

The locations of all the information we gather and photos we take with our electronic devices are recorded through

DOING GEOGRAPHY Data Collection & Mental Mapping

Most of the maps and other information fed into handheld electronic devices is provided by three companies. Google supplies Android devices, TomTom (formerly Tele Atlas) supplies Apple devices, and Nokia (formerly Navteq, now owned by Microsoft) supplies Microsoft products. These companies get their information from what they call "ground truthing." Hundreds of field researchers drive around, building the database. One person drives, while the other feeds information into a notebook computer (Figure 1-10). Hundreds of attributes are recorded, such as crosswalks, turn restrictions, and name changes. Thus, electronic navigation systems ultimately depend on human observation.

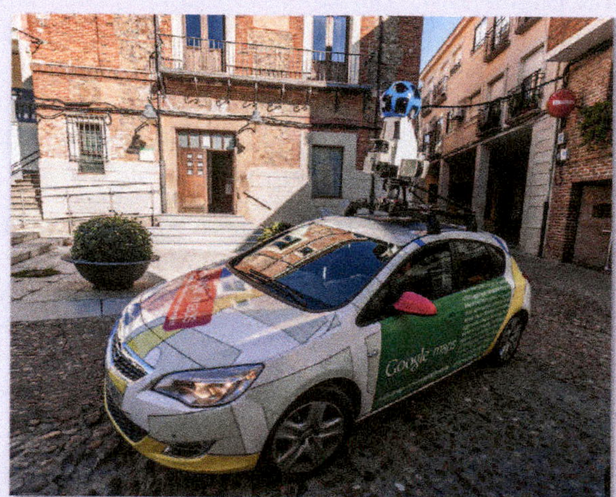

▲ **FIGURE 1-10 GOOGLE STREET MAPPING** Jaraíz de la Vera, Spain.

What's Your Geography?

A **mental map** is a personal representation of a portion of Earth's surface. A mental map depicts what an individual knows about a place, and it contains personal impressions of what is in the place and where the place is located.

1. Draw a mental map depicting your route between two familiar places, such as between home and school or dorm room and geography class. Show the paths (roads or walkways) and important landmarks along the route, such as buildings or shops.
2. Compare your mental map to those made by others in your class. How detailed is your depiction of paths and landmarks compared to those of others? At school, for example, a senior is likely to have a more detailed map than a newcomer.
3. Compare your mental map to a map of the same area from Google Maps. How accurate is your map? Did you forget something important or put something in the wrong place?
4. At OpenStreetMap, see if your route has been mapped. If so, are important landmarks included? If your route has not been mapped, or if important landmarks are not included, you are free to place them on the map by following OpenStreetMap instructions.

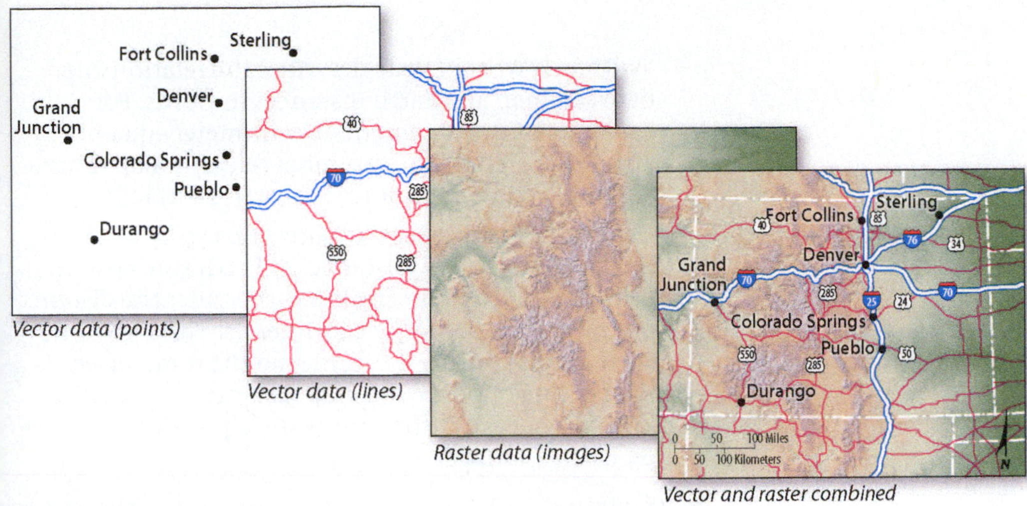

▲ **FIGURE 1-11 GIS** Geographic information systems involve two types of data: vector and raster. Vector data consists of points (for example, for cities) and lines (for example, for highways). Raster data consists of images such as landforms.

geotagging, which is identification and storage of a piece of information by its precise latitude and longitude coordinates. Geotagging has led to concerns about privacy.

ANALYZING DATA: GI-SCIENCE

Geographic Information Science (GIScience) is analysis of data about Earth acquired through satellite and other electronic information technologies. A **geographic information system (GIS)** captures, stores, queries, and displays the geographic data. GIS produces maps (including those in this book) that are more accurate and attractive than those drawn by hand. A map is created by retrieving a number of stored objects and combining them to form an image. Each type of information is stored in a layer (Figure 1-11). For example, separate layers could be created for boundaries of countries, bodies of water, roads, and names of places. A simple map might display only a single layer by itself, but most maps combine several layers, and GIS permits construction of much more complex maps than can be drawn by hand.

The acquisition of data about Earth's surface from a satellite orbiting Earth or from other long-distance methods is **remote sensing.** Remote-sensing satellites scan Earth's surface and transmit images in digital form to a receiving station on Earth's surface. At any moment, a satellite sensor records the image of a tiny area called a picture element, or pixel. Scanners detect the radiation being reflected from that tiny area. A map created by remote sensing is essentially a grid containing many rows of pixels. The smallest feature on Earth's surface that can be detected by a sensor depends on the resolution of the scanner. Geographers use remote sensing to map the changing distribution of a wide variety of features, such as agriculture, drought, and sprawl.

GIScience helps geographers create more accurate and complex maps and measure changes over time in the characteristics of places. Layers of information acquired through remote sensing and produced through GIS can be described and analyzed. GIScience enables geographers to calculate whether relationships between objects on a map are significant or merely coincidental. For example, a map showing where life expectancy is low (such as in Figure 1-24) can be combined with layers showing the location of people with various incomes and the location of crimes.

COLLECTING AND SHARING DATA: VGI

Smart phones, tablets, and computers enable individuals to make maps and share them with others. **Volunteered geographic information (VGI)** is the creation and dissemination of geographic data contributed voluntarily and for free by individuals. VGI is part of the broader trends of **citizen science**, which is scientific research by amateur scientists, and **participatory GIS (PGIS)**, which is community-based mapping. Citizen science and PGIS collect and disseminate local knowledge and information through electronic devices. For example, OpenStreetMap (OSM) is VGI intended to develop a free base map of the world. Individuals can contribute to OSM at OpenStreetMap.org (see Doing Geography and What's Your Geography? feature).

A **mashup** is a map that overlays data from one source on top of a map provided by a mapping service, such as Google Maps or Google Earth. The term *mashup* refers to the practice of overlaying data from one source on top of one of the mapping services; the term comes from the hip-hop practice of mixing two or more songs.

A mashup map can show the locations of nearby pizza restaurants, the locations of commercial airplanes currently in flight, or traffic conditions on highways. Individuals can create mashups on their personal computers because mapping services provide access to the application programming interface (API), which is the language that links a database such as an address list with software such as mapping software. An API for mapping software is available at such sites as developers.google.com/maps.

PAUSE & REFLECT 1.1.3

State a question you have about the area where you live. Describe a mashup that you could create using GIS that would answer your question.

Interpreting Maps

LEARNING OUTCOME 1.1.4

Understand the role of map scale and projection in reading maps.

To make any map, a cartographer must make two decisions:

- How much of Earth's surface to depict on the map (map scale).
- How to transfer a spherical Earth to a flat map (projection).

For more details about cartography, see Appendix A.

MAP SCALE

The first decision a cartographer faces is how much of Earth's surface to depict on the map. Is it necessary to show the entire globe, or just one continent, or a country, or a city? To make a scale model of the entire world, many details must be omitted because there simply is not enough space. Conversely, if a map shows only a small portion of Earth's surface, such as a street map of a city, it can provide a wealth of detail about a particular place.

The level of detail and the amount of area covered on a map depend on its map scale. When specifically applied to a map, **map scale** refers to the relationship of a feature's size on a map to its actual size on Earth. Map scale is presented in three ways (Figure 1-12):

- **Ratio.** A ratio or fraction shows the numerical ratio between distances on the map and Earth's surface. A scale of 1:1,000,000 means that 1 unit (for example, inch, centimeter, foot, finger length) on the map represents 1 million of the same unit on the ground. The 1 on the left side of the ratio always refers to a unit of distance on the map, and the number on the right always refers to the same unit of distance on Earth's surface.
- **Written.** A written scale describes the relationship between map and Earth distances in words. For example, in the statement "1 centimeter equals 10 kilometers," the first number refers to map distance and the second to distance on Earth's surface.
- **Graphic.** A graphic scale usually consists of a bar line marked to show distance on Earth's surface. To use a bar line, first determine with a ruler the distance on the map in inches or centimeters. Then hold the ruler against the bar line and read the number on the bar line opposite the map distance on the ruler. The number on the bar line is the equivalent distance on Earth's surface.

Maps often display scale in more than one of these three ways.

The appropriate scale for a map depends on the information being portrayed. A map of a neighborhood, such as Figure 1-12c, may have a scale of 1:10,000, whereas a map of a city (Figure 1-12a) may have a scale of 1:1,000,000. One inch represents about 1/6 mile on Figure 1-12c and 16 miles on Figure 1-12a.

At the scale of a small portion of Earth's surface, such as a downtown area, a map provides a wealth of details about the place. At the scale of the entire globe, a map must omit many details because of lack of space, but it can effectively communicate processes and trends that affect everyone.

PROJECTION

Earth is very nearly a sphere and is therefore accurately represented with a globe. However, a globe is an extremely limited tool with which to communicate information about Earth's surface. A small globe does not have enough space to display detailed information, whereas a large globe is too bulky and cumbersome to use. And a globe is difficult to write on, photocopy, display on a computer screen, or carry in the glove box of a car. Consequently, most maps—including those in this book—are flat. Three-dimensional maps can be made but are expensive and difficult to reproduce.

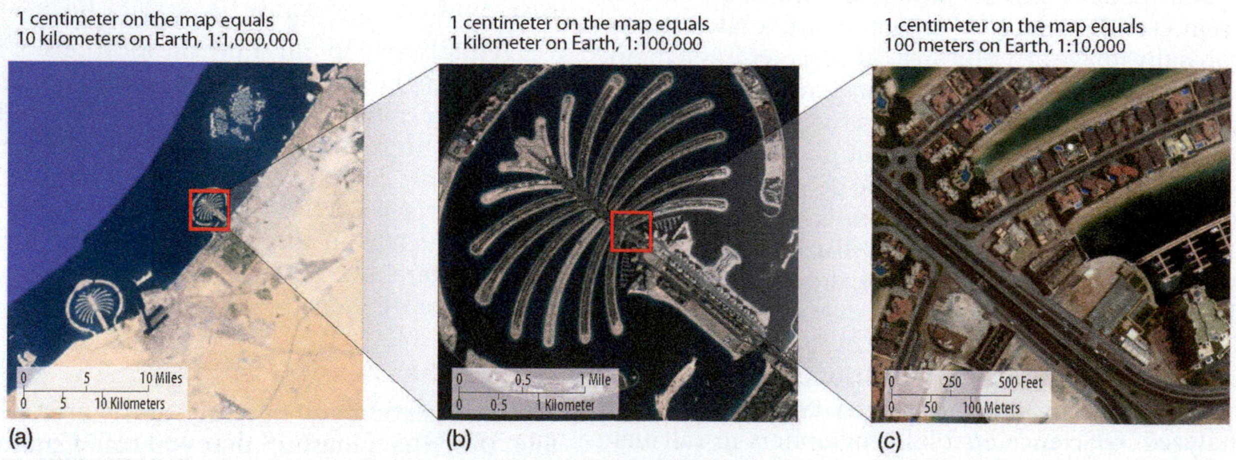

▲ **FIGURE 1-12 MAP SCALE** The three images show the city of Dubai, in the United Arab Emirates, Emirates at three scales.

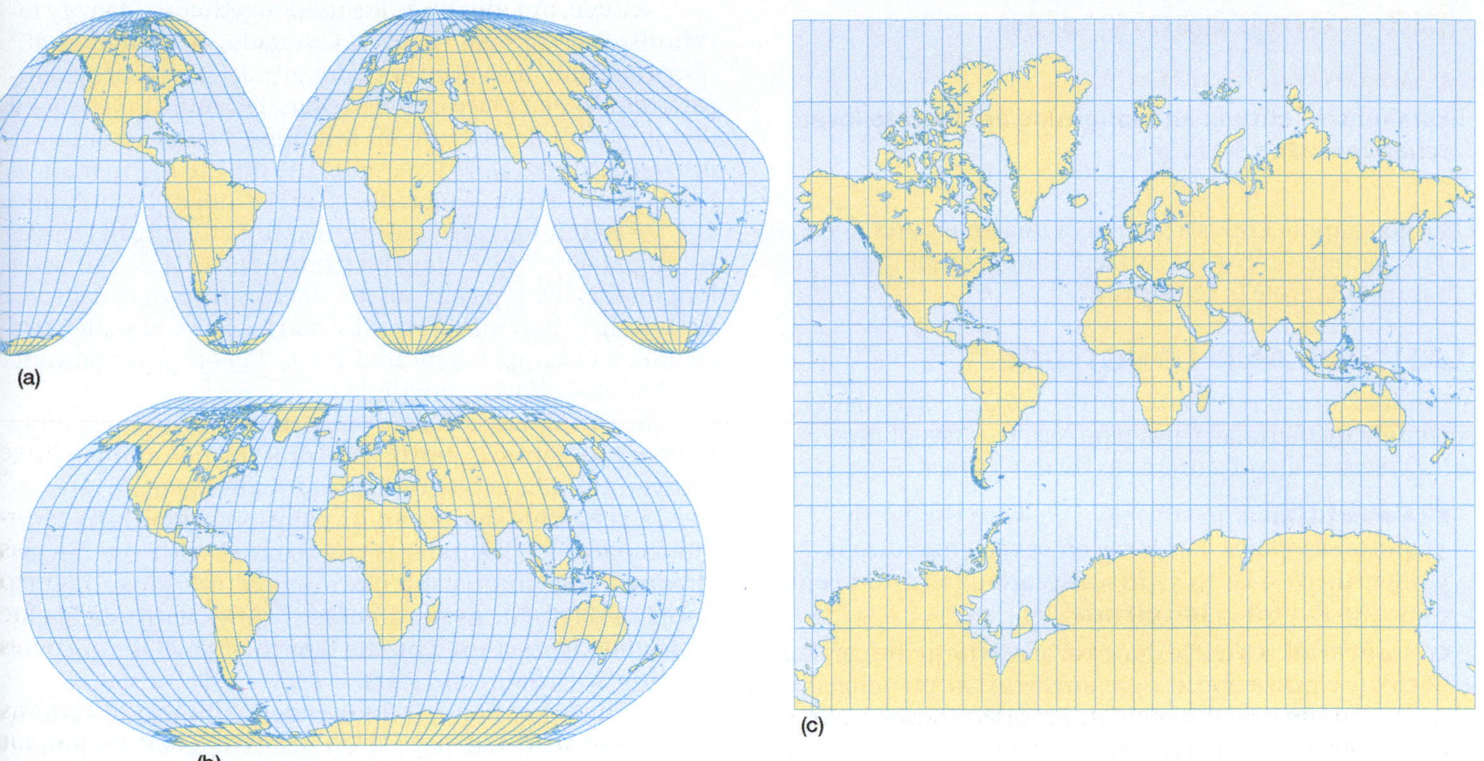

▲ **FIGURE 1-13 PROJECTIONS** (a) Goode Homolosine, an equal area projection; (b) Robinson, an uninterrupted projection; (c) Mercator, featuring accurate shapes and directions.

Earth's spherical shape poses a challenge for cartographers because drawing Earth on a flat piece of paper unavoidably produces some distortion. Cartographers have invented hundreds of clever methods of producing flat maps, but none has produced perfect results. The scientific method of transferring locations on Earth's surface to a flat map is called **projection** (Figure 1-13).

The problem of distortion is especially severe for maps depicting the entire world. Four types of distortion can result:

1. The *shape* of an area can be distorted, so that it appears more elongated or squat than it is in reality.
2. The *distance* between two points may become increased or decreased.
3. The *relative size* of different areas may be altered, so that one area may appear larger than another on a map while it is in reality smaller.
4. The *direction* from one place to another can be distorted.

Most of the world maps in this book, such as Figure 1-13a, are equal area projections. The primary benefit of this type of projection is that the relative sizes of the landmasses on the map are the same as in reality. The projection minimizes distortion in the shapes of most landmasses. Areas toward the North and South poles—such as Greenland and Australia—become more distorted, but they are sparsely inhabited, so distorting their shapes usually is not important.

To largely preserve the size and shape of landmasses, however, the projection in Figure 1-13a forces other distortions:

- The Eastern and Western hemispheres are separated into two pieces, a characteristic known as interruption.
- The meridians (the vertical lines), which in reality converge at the North and South poles, do not converge at all on the map. Also, they do not form right angles with the parallels (the horizontal lines).

The Robinson projection (Figure 1-13b) is useful for displaying information across the oceans. Its major disadvantage is that by allocating space to the oceans, the land areas are much smaller than on interrupted maps of the same size.

The Mercator projection (Figure 1-13c) has several advantages: Shape is distorted very little, direction is consistent, and the map is rectangular. Its greatest disadvantage is that relative size is grossly distorted toward the poles, making high-latitude places look much larger than they actually are.

PAUSE & REFLECT 1.1.4

Compare the sizes of Greenland and South America on the three maps in Figure 1-13. Which of the two landmasses is actually larger? How do you know?

The Geographic Grid

LEARNING OUTCOME 1.1.5

Explain how latitude and longitude are used to locate points on Earth's surface.

The geographic grid is a system of imaginary arcs drawn in a grid pattern on Earth's surface. The geographic grid plays an important role in telling time.

LATITUDE AND LONGITUDE

The location of any place on Earth's surface can be described precisely by meridians and parallels, two sets of imaginary arcs drawn in a grid pattern on Earth's surface (Figure 1-14):

- A **meridian** is an arc drawn between the North and South poles. The location of each meridian is identified on Earth's surface according to a numbering system known as **longitude.**
- A **parallel** is a circle drawn around the globe parallel to the equator and at right angles to the meridians. The numbering system to indicate the location of a parallel is called **latitude.**

The meridian that passes through the Royal Observatory at Greenwich, England, is 0° longitude, also called the **prime meridian.** The meridian on the opposite side of the globe from the prime meridian is 180° longitude. All other meridians have numbers between 0° and 180° east or west, depending on whether they are east or west of the prime meridian. For example, Belo Horizonte, Brazil, is located at 44° west longitude and Baghdad, Iraq, at 44° east longitude.

The equator is 0° latitude, the North Pole 90° north latitude, and the South Pole 90° south latitude. Nicosia, Cyprus, is located at 35° north latitude and Buenos Aires, Argentina, at 35° south latitude.

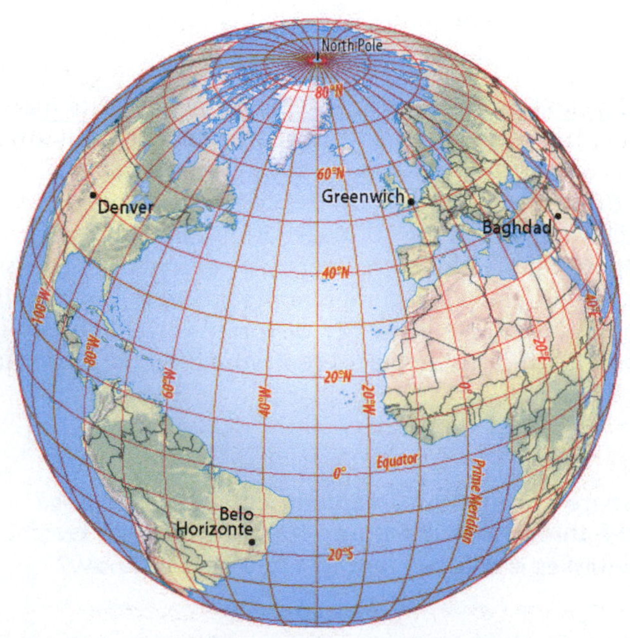

▲ **FIGURE 1-14 GEOGRAPHIC GRID**

Latitude and longitude are used together to identify locations. For example, Denver, Colorado, is located at 40° north latitude and 105° west longitude. The mathematical location of a place can be designated more precisely by dividing each degree into 60 minutes (') and each minute into 60 seconds ("). For example, the official mathematical location of Denver, Colorado, is 39°44' north latitude and 104°59' west longitude. The state capitol building in Denver is located at 39°42'2" north latitude and 104°59'04" west longitude. GPS typically divide degrees into decimal fractions rather than minutes and seconds. The Colorado state capitol, for example, is located at 39.714444° north latitude and 84.984444° west longitude.

Measuring latitude and longitude is a good example of how geography is partly a natural science and partly a study of human behavior. Latitudes are scientifically derived via Earth's shape and its rotation around the Sun. The equator (0° latitude) is the parallel with the largest circumference and is the place where every day has 12 hours of daylight. Even in ancient times, latitude could be accurately measured by the length of daylight and the position of the Sun and stars.

On the other hand, 0° longitude is a human creation. Any meridian could have been selected as 0° longitude because all meridians have the same length and all run between the poles. The 0° longitude runs through Greenwich because England was the world's most powerful country when longitude was first accurately measured and the international agreement was made.

TELLING TIME

Longitude is the basis for calculating time. Earth as a sphere is divided into 360° of longitude (the degrees from 0° to 180° west longitude plus the degrees from 0° to 180° east longitude).

As Earth rotates daily, these 360 imaginary lines of longitude pass beneath the cascading sunshine. If we let every fifteenth degree of longitude represent one time zone, and divide the 360° by 15°, we get 24 time zones, or one for each hour of the day. By international agreement, **Greenwich Mean Time (GMT)**, or Universal Time (UT), which is the time at the prime meridian (0° longitude), is the master reference time for all points on Earth.

Each 15° band of longitude is assigned to a standard time zone (Figure 1-15). The eastern United States, which is near 75° west longitude, is therefore 5 hours earlier than GMT (the 75° difference between the prime meridian and 75° west longitude, divided by 15° per hour, equals 5 hours). Thus when the time in New York City in the winter is 1:32 p.m. (or 13:32 hours, using a 24-hour clock), it is 6:32 p.m. (or 18:32 hours) GMT. During the summer, many places in the world, including most of North America, move the clocks ahead one hour; so in the summer when it is 6:32 p.m. GMT, the time in New York City is 2:32 p.m.

When you cross the **International Date Line,** which, for the most part, follows 180° longitude, you move the clock back 24 hours, or one entire day, if you are heading eastward, toward America. You turn the clock ahead

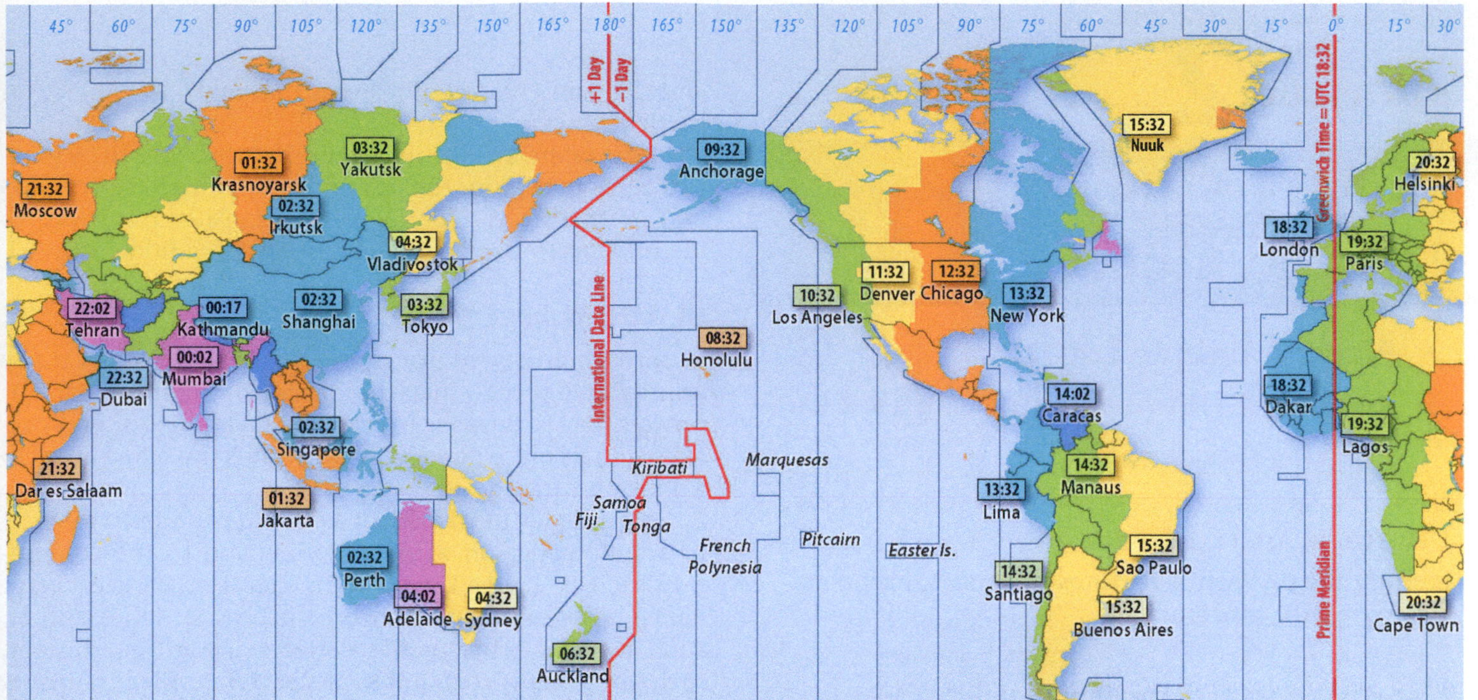

▲ **FIGURE 1-15 TIME ZONES** The United States and Canada share four standard time zones: Eastern (5 hours earlier than GMT, near 75° west), Central (6 hours earlier than GMT, near 90° west), Mountain (7 hours earlier than GMT, near 105° west), and Pacific (8 hours earlier than GMT, near 120° west). The United States has two additional standard time zones: Alaska (9 hours earlier than GMT, near 135° west) and Hawaii-Aleutian (10 hours earlier than GMT, near 150° west). Canada has two additional standard time zones: Atlantic (4 hours earlier than GMT, near 60° west) and Newfoundland (3½ hours earlier than GMT). The residents of Newfoundland assert that their island, which lies between 53° and 59° west longitude, would face dark winter afternoons if it were in the Atlantic Time Zone and dark winter mornings if it were 3 hours earlier than GMT.

24 hours if you are heading westward, toward Asia. To see the need for the International Date Line, try counting the hours around the world from the time zone in which you live. As you go from west to east, you add 1 hour for each time zone. When you return to your starting point, you will reach the absurd conclusion that it is 24 hours later in your locality than it really is. Therefore—if it is 6:32 a.m. Monday in Auckland, when you get to Honolulu, it will be 8:32 a.m. Sunday because the International Date Line lies between Auckland and Honolulu.

The International Date Line for the most part follows 180° longitude. However, several islands in the Pacific Ocean belonging to the countries of Kiribati and Samoa, as well as to New Zealand's Tokelau territory, moved the International Date Line several thousand kilometers to the east. Samoa and Tokelau moved it in 2011 so that they could be on the same day as Australia and New Zealand, their major trading partners. Kiribati moved it in 1997 so that it would be the first country to see each day's sunrise. Kiribati hoped that this feature would attract tourists to celebrate the start of the new millennium on January 1, 2000 (or January 1, 2001, when sticklers pointed out the new millennium really began). But it did not.

Inability to measure longitude was the greatest obstacle to exploration and discovery for many centuries. Ships ran aground or were lost at sea because no one on board could pinpoint longitude. In 1714, the British Parliament enacted the Longitude Act, which offered a prize equivalent to several million in today's dollars to the person who could first measure longitude accurately.

Most eighteenth-century scientists were convinced that longitude could be determined only by the position of the stars. English clockmaker John Harrison won the prize by using the connection between longitude and time. He invented the first portable clock that could keep accurate time on a ship—because it did not have a pendulum. When the Sun was directly overhead of the ship—noon local time—Harrison's portable clock set to Greenwich time could say it was 2 p.m. in Greenwich, for example, so the ship would be at 30° west longitude because each hour of difference was equivalent to traveling 15° longitude.

PAUSE & REFLECT 1.1.5

Where in the world, other than Newfoundland, is standard time on the half-hour rather than the hour? Why might that country prefer not to be on the hour?

CHECK-IN KEY ISSUE 1

Why Is Geography a Science?

- ✔ **Geography has ancient and medieval roots.**
- ✔ **Maps are tools of reference and increasingly tools of communication.**
- ✔ **Reading a map requires recognizing its scale and projection.**
- ✔ **Contemporary mapping utilizes electronic technologies, such as GPS and GIS.**

KEY ISSUE 2

Why Is Each Point on Earth Unique?

- Place: A Unique Location
- Region: A Unique Area
- Culture Regions

LEARNING OUTCOME 1.2.1

Identify the distinctive features of a place, including toponym, site, and situation.

Place was defined at the beginning of the chapter as a specific point on Earth distinguished by a particular characteristic. Every place occupies a unique location, or position, on Earth's surface. Although each place on Earth is in some respects unique, in other respects it is similar to other places. The interplay between the uniqueness of each place and the similarities among places lies at the heart of geographic inquiry into why things are found where they are.

Place: A Unique Location

Humans possess a strong sense of place—that is, a feeling for the features that contribute to the distinctiveness of a particular spot on Earth—perhaps a hometown, vacation destination, or part of a country. Describing the features of a place is an essential building block for geographers to explain similarities, differences, and changes across Earth. Geographers think about where particular places are located and the combination of features that make each place on Earth distinct.

Geographers describe a feature's place on Earth by identifying its **location,** the position that something occupies on Earth's surface. In doing so, they consider three ways to identify location: place name, site, and situation.

PLACE NAMES

Because all inhabited places on Earth's surface—and many uninhabited places—have been named, the most straightforward way to describe a particular location is often by referring to its place name. A **toponym** is the name given to a place on Earth.

A place may be named for a person, perhaps its founder or a famous person with no connection to the community, such as George Washington. Some settlers select place names associated with religion, such as St. Louis and St. Paul, whereas other names derive from ancient history, such as Athens, Attica, and Rome, or from earlier occupants of the place.

A place name may also indicate the origin of its settlers. Place names commonly have British origins in North America and Australia, Portuguese origins in Brazil, Spanish origins elsewhere in Latin America, and Dutch origins in South Africa. Some place names derive from features of the physical environment. Trees, valleys, bodies of water, and other natural features appear in the place names of most languages (Figure 1-16).

The Board of Geographical Names, operated by the U.S. Geological Survey, was established in the late nineteenth century to be the final arbiter of names on U.S. maps. In recent years the board has been especially concerned with removing offensive place names, such as those with racial or ethnic connotations.

PAUSE & REFLECT 1.2.1

What is the origin of the toponym of your hometown?

▲ **FIGURE 1-16 WORLD'S LONGEST PLACE NAME** This place in New Zealand is recognized as the world's longest one-word place name. It translates from the Maori language as "The summit where Tamatea, the man with the big knees, the climber of mountains, the land-swallower who travelled about, played his nose flute to his loved one."

▲ **FIGURE 1-17 CHANGING SITE OF BOSTON** The site of Boston has been altered by filling in much of Boston Harbor. Colonial Boston was a peninsula connected to the mainland by a very narrow neck. During the nineteenth century, a dozen major projects filled in most of the bays, coves, and marshes. A major twentieth-century landfill project created Logan Airport. Several landfill projects continue into the twenty-first century.

SITE

The second way that geographers describe the location of a place is by **site,** which is the physical character of a place. Important site characteristics include climate, water sources, topography, soil, vegetation, latitude, and elevation. The combination of physical features gives each place a distinctive character.

Site factors have always been essential in selecting locations for settlements, although people have disagreed on the attributes of a good site, depending on cultural values. Some have preferred a hilltop site for easy defense from attack. Others have located settlements near convenient river-crossing points to facilitate communication with people in other places.

Humans have the ability to modify the characteristics of a site. Central Boston is more than twice as large today as it was during colonial times (Figure 1-17). Colonial Boston was a peninsula connected to the mainland by a very narrow neck. During the nineteenth century, a dozen major projects filled in most of the bays, coves, and marshes. A major twentieth-century landfill project created Logan Airport. Several landfill projects continue into the twenty-first century. The central areas of New York and Tokyo have also been expanded through centuries of landfilling in nearby bodies of water, substantially changing these sites.

SITUATION

Situation is the location of a place relative to other places. Situation is a valuable way to indicate location, for two reasons:

- **Finding an unfamiliar place.** Situation helps us find an unfamiliar place by comparing its location with a familiar one. We give directions to people by referring to the situation of a place: "It's down past the courthouse, on Locust Street, after the third traffic light, beside the yellow-brick bank." We identify important buildings, streets, and other landmarks to direct people to the desired location.
- **Understanding the importance of a place.** Situation helps us understand the importance of a location. Many locations are important because they are accessible to other places. For example, because of its situation, Istanbul is a center for the trading and distribution of goods between Europe and Asia (Figure 1-18). Istanbul is situated along the Bosphorus strait, which connects the Mediterranean and Black seas. The Bosphorus is especially important route for ships traveling to and from Russia.

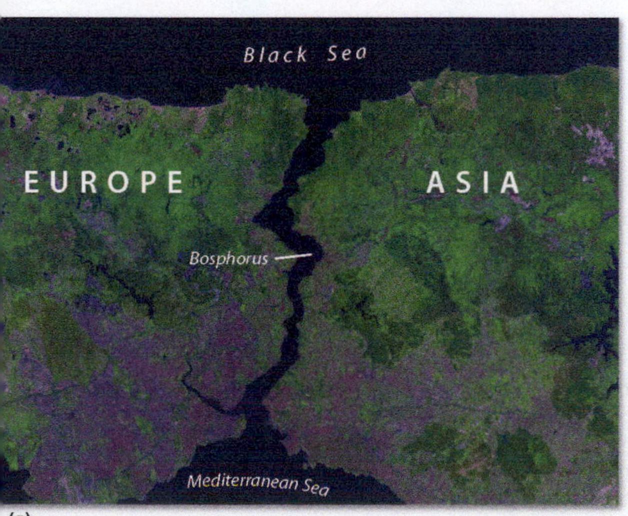

(a)

(b)

▲ **FIGURE 1-18 SITUATION OF ISTANBUL** Istanbul is situated along the Bosphorus, a waterway between the Black and Mediterranean seas. urope lies to the west and Asia to the east. **(a)** Satellite image of Istanbul; **(b)** Istanbul looking east from Asia toward Europe.

Region: A Unique Area

LEARNING OUTCOME 1.2.2

Identify the three types of regions.

The "sense of place" that humans possess may apply to a larger area of Earth rather than to a specific point. *Region* was defined at the beginning of the chapter as an area of Earth defined by one or more distinctive characteristics. Geographers identify three types of regions: formal, functional, and vernacular.

A particular place can be included in more than one region, depending on how the region is defined. The designation "region" can be applied to any area larger than a point and smaller than the entire planet. Geographers most often apply the concept at one of two scales:

- Several neighboring countries that share important features, such as those in Latin America.
- Many localities within a country, such as those in southern California.

A region derives its unified character through the **cultural landscape**—a combination of cultural features such as language and religion, economic features such as agriculture and industry, and physical features such as climate and vegetation. The southern California region can be distinguished from the northern California region, for example.

The contemporary cultural landscape approach in geography—sometimes called the regional studies approach—was initiated in France by Paul Vidal de la Blache (1845–1918) and Jean Brunhes (1869–1930). It was later adopted by several American geographers, including Carl Sauer (1889–1975) and Robert Platt (1880–1950). Sauer defined cultural landscape as an area fashioned from nature by a cultural group. "Culture is the agent, the natural area the medium, the cultural landscape is the result."

People, activities, and environment display similarities and regularities within a region and differ in some way from those of other regions. A region gains uniqueness from possessing not a single human or environmental characteristic but a combination of them. Not content to merely identify these characteristics, geographers seek relationships among them. Geographers recognize that in the real world, characteristics are integrated.

FORMAL REGION

A **formal region,** also called a **uniform region,** is an area within which everyone shares in common one or more distinctive characteristics. The shared feature could be a cultural value such as a common language, an economic activity such as production of a particular crop, or an environmental property such as climate. In a formal region, the selected characteristic is present throughout.

Some formal regions are easy to identify, such as countries or local government units. Montana is an example of a formal region, characterized with equal intensity throughout the state by a government that passes laws, collects taxes, and issues license plates. The formal region of Montana has clearly drawn and legally recognized boundaries, and everyone living within them shares the status of being subject to a common set of laws.

In other kinds of formal regions, a characteristic may be predominant rather than universal. For example, we can distinguish formal regions within the United States characterized by a predominant voting for Republican candidates, although Republicans do not get 100 percent of the votes in these regions—nor in fact do they always win (Figure 1-19).

A cautionary step in identifying formal regions is the need to recognize the diversity of cultural, economic, and environmental factors, even while making a generalization. Problems may arise because a minority of people in a region speak a language, practice a religion, or possess resources different from those of the majority. People in a region may play distinctive roles in the economy and hold different positions in society based on their gender or ethnicity.

FUNCTIONAL REGION

A **functional region,** also called a **nodal region,** is an area organized around a node or focal point. The characteristic chosen to define a functional region dominates at a central focus or node and diminishes in importance outward. The region is tied to the central point by transportation or communications systems or by economic or functional associations.

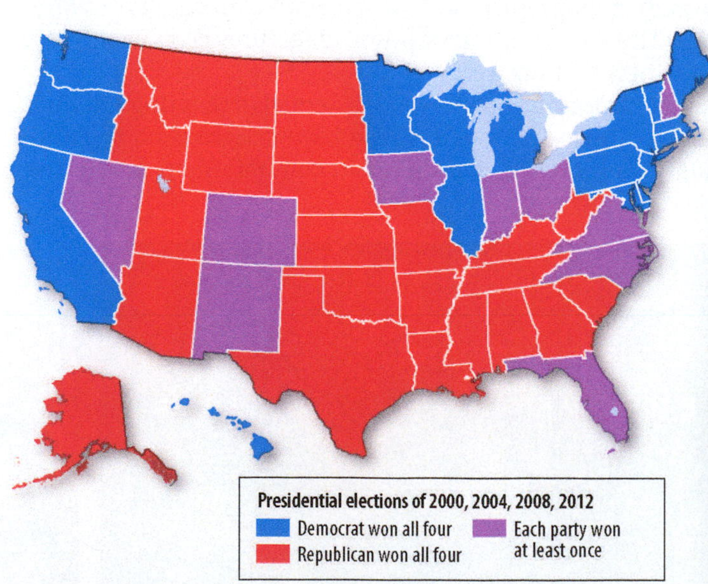

▲ **FIGURE 1-19 FORMAL REGION** The U.S. presidential election map is divided into formal regions based on party preference. The South and Plains states typically give a majority to the Republican candidate and the Northeast and West Coast states typically give a majority to the Democratic candidate. Although Republican George W. Bush won in 2000 and 2004 and Democrat Barack Obama won in 2008 and 2012, 40 of the 50 states (as well as the District of Columbia) gave a majority to the same party in all four elections.

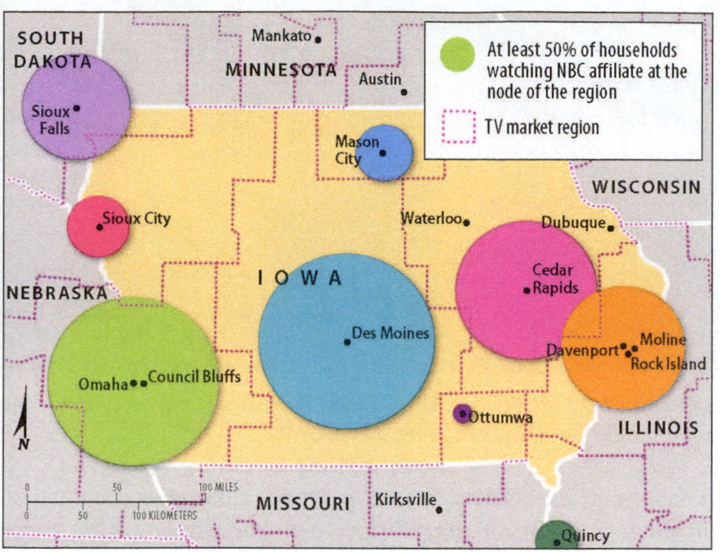

▲ **FIGURE 1-20 FUNCTIONAL REGION** Designated market areas (DMAs) within Iowa are examples of functional regions. In several of the functional regions, the node—the TV station—is in an adjacent state. Functional regions frequently overlap the formal regions delineated by state or national boundaries. The state of Iowa is an example of a formal region.

Geographers often use functional regions to display information about economic areas. A region's node may be a shop or service, with the boundaries of the region marking the limits of the trading area of the activity. People and activities may be attracted to the node, and information may flow from the node to the surrounding area.

An example of a functional region is the reception area of a TV station. A TV station's signal is strongest at the center of its service area (Figure 1-20). At some distance from the center, more people are watching a station originating in another city. That place is the boundary between the nodal regions of the two TV market areas. Similarly, a department store attracts fewer customers from the edge of a trading area, and beyond that edge, customers will most likely choose to shop elsewhere.

New technology is breaking down traditional functional regions. TV stations are broadcast to distant places by cable, satellite, or Internet. Through the Internet, customers can shop at distant stores.

VERNACULAR REGION

A **vernacular region,** or **perceptual region,** is an area that people believe exists as part of their cultural identity. Such regions emerge from people's informal sense of place rather than from scientific models developed through geographic thought.

As an example of a vernacular region, Americans frequently refer to the South as a place with environmental and cultural features perceived to be quite distinct from those of the rest of the United States (Figure 1-21). Many of these features can be measured. Environmentally, the South is a region where the last winter frost occurs in March and rainfall is more plentiful in winter than in summer. Cultural features include relatively high adherence to the Baptist religion and preference for Republican presidential candidates, relatively low rates of high school graduation, and joining the Confederacy during the Civil War.

PAUSE & REFLECT 1.2.2

What environmental and cultural features might help to define the region of the U.S. Midwest?

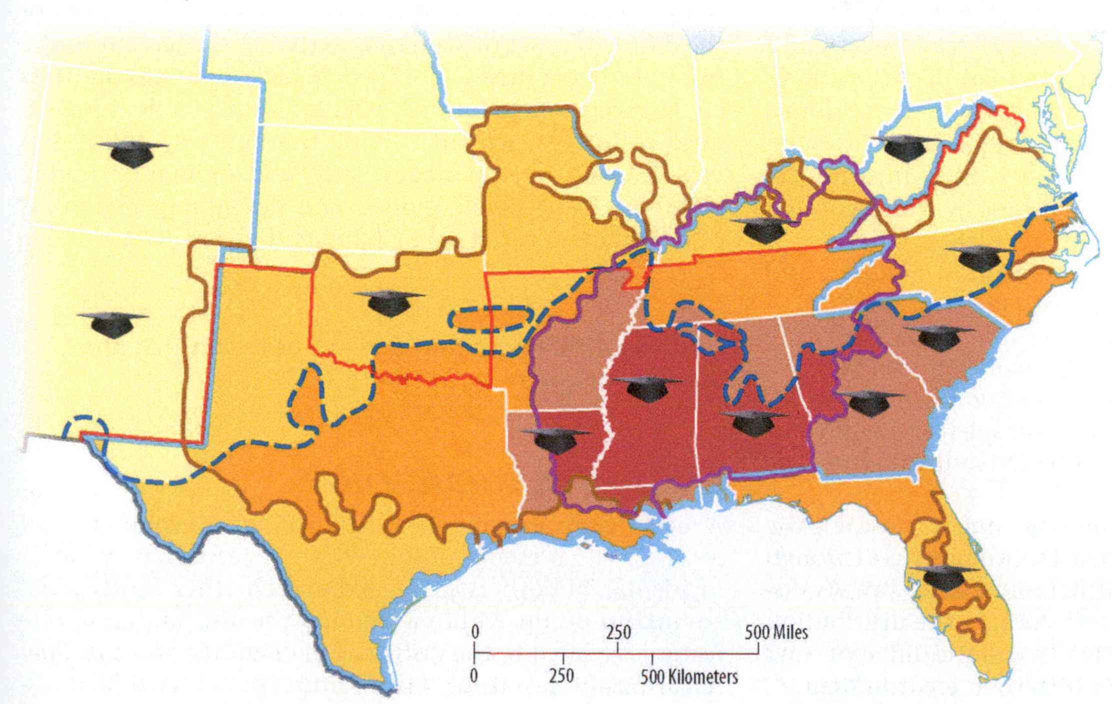

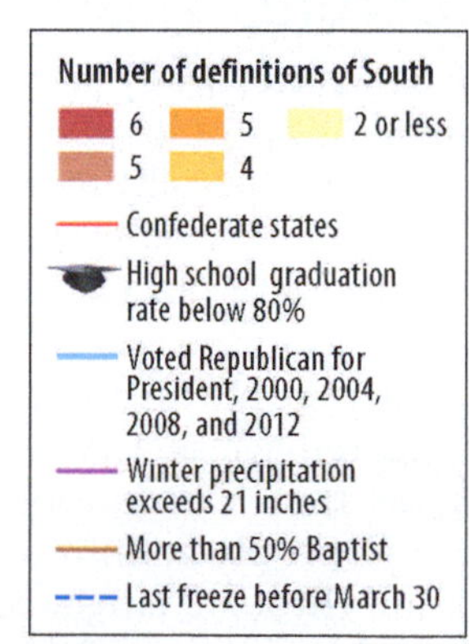

◀ **FIGURE 1-21 VERNACULAR REGION** The South is popularly distinguished as a distinct vernacular region within the United States, according to a number of factors.

Culture Regions

LEARNING OUTCOME 1.2.3

Describe two geographic definitions of culture.

In thinking about why each region on Earth is distinctive, geographers refer to **culture,** which is the body of customary beliefs, material traits, and social forms that together constitute the distinct tradition of a group of people. Geographers distinguish groups of people according to important cultural characteristics, describe where particular cultural groups are distributed, and offer reasons to explain the observed distribution.

In everyday language, we think of *culture* as a collection of novels, paintings, symphonies, and other works produced by talented individuals. A person with a taste for these intellectual outputs is said to be "cultured." Intellectually challenging culture is often distinguished from popular culture, such as TV. *Culture* also refers to small living organisms, such as those found under a microscope or in yogurt. *Agriculture* is a term for the growing of living material at a much larger scale than in a test tube.

The origin of the word culture is the Latin *cultus*, which means "to care for." Culture is a complex concept because "to care for" something has two very different meanings:

- To care about—to adore or worship something, as in the modern word cult.
- To take care of—to nurse or look after something, as in the modern word cultivate.

Geography looks at both of these facets of the concept of culture to see why each region in the world is unique.

CULTURE: WHAT PEOPLE CARE ABOUT

Geographers study why the customary ideas, beliefs, and values of a people produce a distinctive culture in a particular place. Especially important cultural values derive from a group's language, religion, and ethnicity. These three cultural traits are both an excellent way of identifying the location of a culture and the principal means by which cultural values become distributed around the world.

Language is a system of signs, sounds, gestures, and marks that have meanings understood within a cultural group. People communicate the cultural values they care about through language, and the words themselves tell something about where different cultural groups are located (Figure 1-22). The distribution of speakers of different languages and reasons for the distinctive distribution are discussed in Chapter 5.

Religion is an important cultural value because it is the principal system of attitudes, beliefs, and practices through which people worship in a formal, organized way. As discussed in Chapter 6, geographers look at the distribution of religious groups around the world and the different ways that the various groups interact with their environment.

Ethnicity encompasses a group's language, religion, and other cultural values, as well as its physical traits. A group

▲ **FIGURE 1-22 CULTURAL DIVERSITY** Languages, religions, and cultures mix in a Jerusalem street market.

possesses these cultural and physical characteristics as a product of its common traditions and heredity. As addressed in Chapter 7, geographers find that problems of conflict and inequality tend to occur in places where more than one ethnic group inhabits and seeks to organize the same territory.

CULTURE: WHAT PEOPLE TAKE CARE OF

The second element of culture of interest to geographers is production of material wealth—the food, clothing, and shelter that humans need in order to survive and thrive. All people consume food, wear clothing, build shelter (Figure 1-23), and create art, but different cultural groups obtain their wealth in different ways.

Geographers divide the world into regions of developed countries and regions of developing countries. Various shared characteristics—such as per capita income, level of education, and life expectancy—distinguish developed regions and developing regions. These differences are reviewed in Chapter 9.

Possession of wealth and material goods is higher in developed countries than in developing countries because of the different types of economic activities carried out in the two types of countries. Most people in developing countries are engaged in agriculture, whereas most people in developed countries earn their living through performing services in exchange for wages. This fundamental economic difference between developed and developing regions is discussed in more detail in Chapters 10 through 13.

PAUSE & REFLECT 1.2.3

Describe differences that you see between U.S. and South Africa suburbs in Figure 1-23.

SPATIAL ASSOCIATION

A region gains meaning through its unique combination of features. The presence of some of these features may be coincidental, but others are related to each other. **Spatial association** occurs within a region if the distribution of one feature is related to the distribution of another feature. Spatial association is strong if two features have very similar distributions, and spatial association is weak if two features have very different distributions.

(a)

(b)

▲ **FIGURE 1-23 DEVELOPED AND DEVELOPING COUNTRIES**
(a) U.S. suburb; (b) Khayelitsha, the largest and fastest-growing township in South Africa.

Figure 1-24 displays the distribution of four features in Baltimore City:

- **Income.** The highest-income neighborhoods are in downtown Baltimore and on the north side, and the lowest-income neighborhoods are in west and east Baltimore (Figure 1-24a).
- **Life expectancy at birth.** As explained in Chapter 2, this measures the number of years that a baby born this year is expected to live. In downtown Baltimore and in north-side neighborhoods, babies are expected to live at least 77 years, whereas in western and eastern Baltimore neighborhoods, life expectancy is less than 70 years. The distribution of life expectancy in Baltimore displays a strong spatial association with income. The areas with the highest income (downtown and north side) are also the areas with the highest life expectancy (Figure 1-24b).
- **Crime.** The spatial association between crime and income is neither very strong nor very weak (Figure 1-24c). The crime rate is relatively low in the high-income northern neighborhoods, but it is not relatively low in the high-income downtown neighborhoods.
- **Liquor stores.** Baltimore, like many other communities, tries to disperse liquor stores evenly across the city on the basis of the number of residents in each neighborhood (Figure 1-24d). However, a liquor store in existence when the current law went into effect is allowed to continue to operate. It is called a nonconforming liquor store because it does not conform to current law but was legal when originally opened. The presence of a large number of nonconforming liquor stores indicates that a neighborhood has more liquor stores than is considered appropriate. Figure 1-24d shows the distribution of nonconforming liquor stores. Comparing this map with Figure 1-24a shows that the distribution of nonconforming liquor stores is closely spatially associated with lower-income neighborhoods.

CHECK-IN KEY ISSUE 2

Why Is Each Point on Earth Unique?

- ✓ **Location is identified through name, site, and situation.**
- ✓ **Regions can be formal, functional, or vernacular.**
- ✓ **Culture encompasses what people care about and what people take care of.**

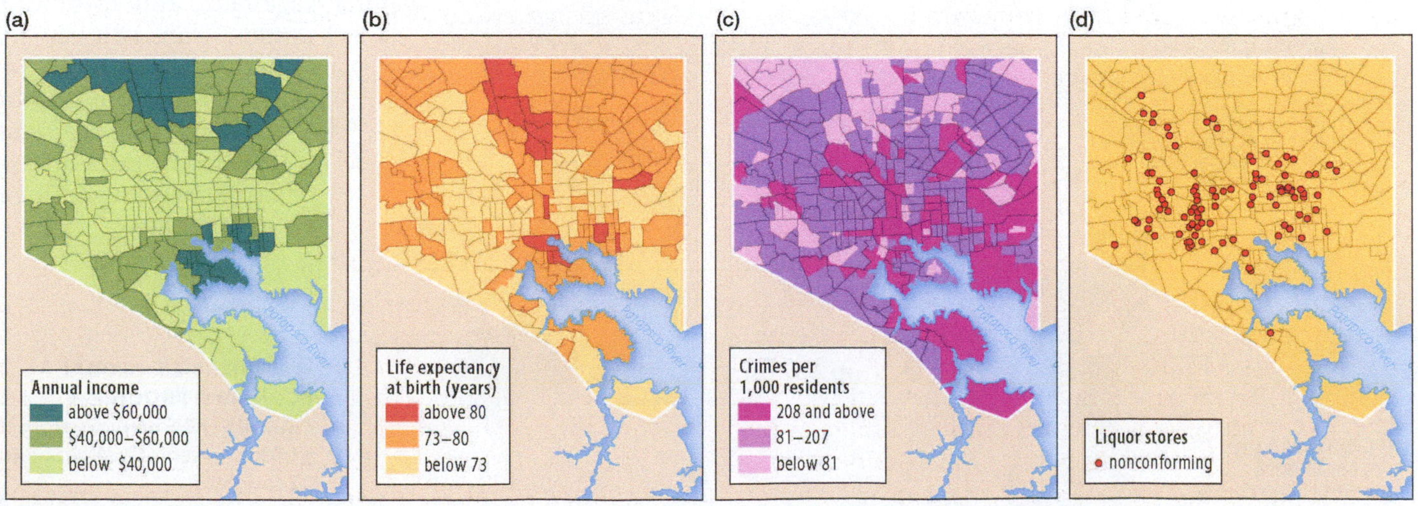

▲ **FIGURE 1-24 SPATIAL ASSOCIATION IN BALTIMORE** (a) Income, (b) life expectancy at birth, (c) crime, (d) nonconforming liquor stores.

KEY ISSUE 3

Why Are Different Places Similar?

- Scale: Global and Local
- Space: Distribution of Features
- Space: Cultural Identity
- Space: Inequality
- Connections: Diffusion
- Connections: Spatial Interaction

LEARNING OUTCOME 1.3.1

Understand global- and local-scale changes in economy and culture.

Geographers recognize that each place or region on Earth is in some ways unique, but they also recognize that human activities are rarely confined to one location. Three basic concepts—scale, space, and connections—help geographers explain why similarities among places and regions result from regularities rather than coincidences.

Scale: Global and Local

Scale was defined at the beginning of the chapter as the relationship between the portion of Earth being studied and Earth as a whole. Geographers are especially concerned with contrasts between the local scale and the global scale.

Scale is an increasingly important concept in geography because of **globalization,** which is a force or process that involves the entire world and results in making something worldwide in scope. Globalization means that the scale of the world is shrinking—not literally in size, of course, but in the ability of a person, an object, or an idea to interact with a person, an object, or an idea in another place.

At the same time, geographers recognize the increasing importance of the local scale. In the face of globalization, groups of people are preserving and reviving distinctive cultural characteristics and implementing distinctive economic practices.

ECONOMIC GLOBALIZATION AND LOCAL DIVERSITY

A few people living in very remote regions of the world may be able to provide all of their daily necessities. But most economic activities undertaken in one region are influenced by interaction with decision makers located elsewhere. The choice of crop is influenced by demand and prices set in markets elsewhere. A factory is located to facilitate bringing in raw materials and shipping out products to the markets.

Globalization of the economy has been led primarily by transnational corporations, sometimes called multinational corporations. A **transnational corporation** conducts research, operates factories, and sells products in many countries, not just where its headquarters and principal shareholders are located. Examples include Procter & Gamble (Figure 1-25) and McDonald's (Figure 1-26).

Every place in the world is part of the global economy, but globalization has led to more specialization at the local level. Each place plays a distinctive role, based on its local assets, as assessed by transnational corporations. A locality may be especially suitable for a transnational corporation to conduct research, to develop new engineering systems, to extract raw materials, to produce parts, to store finished products, to sell them, or to manage operations. In a global economy, transnational corporations remain competitive by correctly identifying the optimal location for each of these activities. Factories are closed in some locations and opened in others.

Changes in production have led to a spatial division of labor, in which a region's workers specialize in particular tasks. Transnationals decide where to produce things in response to characteristics of the local labor force, such as level of skills, prevailing wage rates, and attitudes toward unions. Transnationals may close factories in locations with high wage rates and strong labor unions. Particular production tasks are concentrated in specific geographic areas.

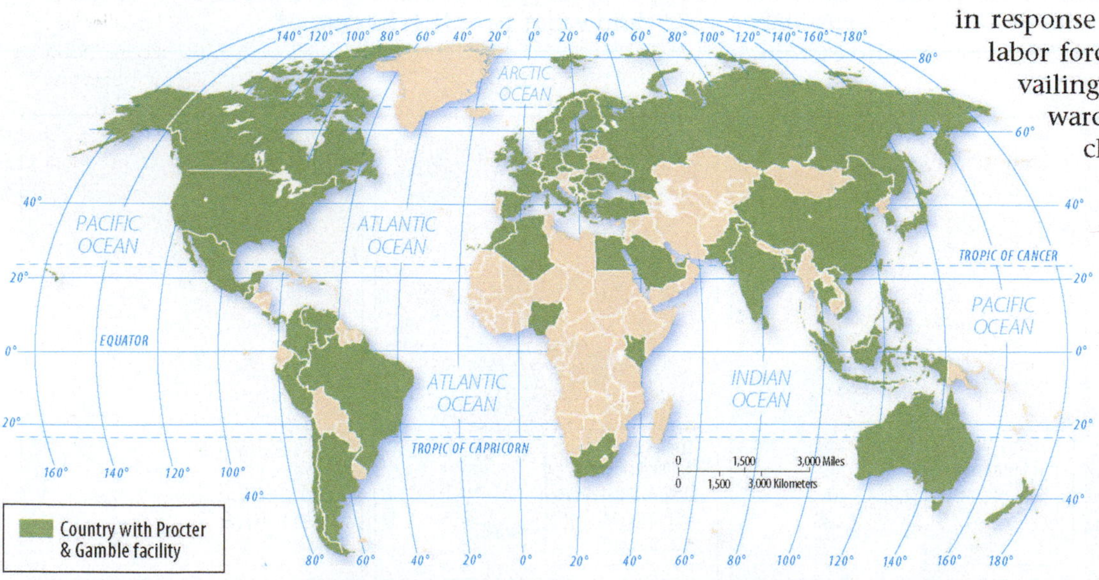

◀ **FIGURE 1-25 GLOBAL ECONOMY: PROCTER & GAMBLE** P&G employs around 61,000 workers at 134 factories in 42 countries.

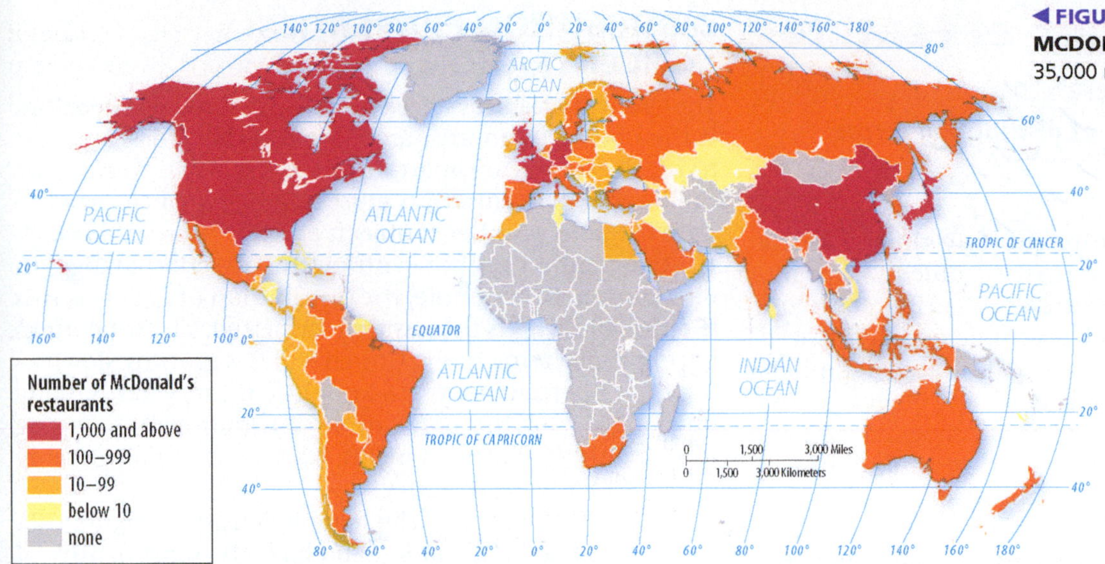

◀ **FIGURE 1-26 GLOBAL ECONOMY: MCDONALD'S** McDonald's has around 35,000 restaurants in 120 countries.

hamburgers, and communicating using cell phones and computers.

Yet despite globalization, cultural differences among places not only persist but actually flourish in many places. Global standardization of products does not mean that everyone wants the same cultural products. The communications revolution that promotes globalization of culture also permits preservation of cultural diversity. TV, for example, was once limited to a handful of channels displaying one set of cultural values. With the distribution of programming through cable, satellite, and Internet, people now can choose from hundreds of programs in many languages.

With the globalization of communications, people in two distant places can watch the same TV program. At the same time, with the fragmentation of the broadcasting market, two people in the same house can watch different programs. Groups of people on every continent may aspire to wear jeans, but they might live with someone who prefers skirts. In a global culture, companies can target groups of consumers with similar tastes in different parts of the world.

CULTURAL GLOBALIZATION AND LOCAL DIVERSITY

Geographers observe that increasingly uniform cultural preferences produce uniform "global" landscapes of material artifacts and cultural values. Fast-food restaurants, service stations, and retail chains deliberately create a visual appearance that varies among locations as little as possible (Figure 1-27). That way, customers know what to expect, regardless of where in the world they happen to be.

Underlying the uniform cultural landscape is globalization of cultural beliefs and forms, especially religion and language. Africans, in particular, have moved away from traditional religions and have adopted Christianity or Islam, religions shared with hundreds of millions of people throughout the world. Globalization requires a form of common communication, and the English language is increasingly playing that role.

As more people become aware of elements of global culture and aspire to possess them, local cultural beliefs, forms, and traits are threatened with extinction. The survival of a local culture's distinctive beliefs, forms, and traits may be threatened by interaction with such social customs as wearing jeans and Nike shoes, consuming Coca-Cola and McDonald's

PAUSE & REFLECT 1.3.1

Give examples of changes in economy and culture occurring at global and local scales.

(a)

(b)

◀ **FIGURE 1-27 GLOBAL CULTURE** (a) Youths in China wear Western jeans and hoodies. (b) Youths in Shanghai, China, patronize a McDonald's.
1. What might be the attraction of jeans and McDonald's for young people in China? **2.** If you visit another country, would you wish to wear jeans? Would you wish to go to McDonald's? Why or why not? **3.** How do you know from the image on the right that the store is a McDonald's?

Space: Distribution of Features

LEARNING OUTCOME 1.3.2

Identify the three properties of distribution across space.

Space was defined at the beginning of the chapter as the physical gap or interval between two objects. Geographers observe that many objects are distributed across space in a regular manner, for discernible reasons. Spatial thinking is the most fundamental skill that geographers possess to understand the arrangement of objects across Earth. Geographers think about the arrangement of people and activities found in space and try to understand why those people and activities are distributed across space as they are.

Look around the space you currently occupy—perhaps a classroom or a bedroom. Tables and chairs are arranged regularly, perhaps in a row in a classroom or against a wall at home. The room is located in a building that occupies an organized space—along a street or a side of a quadrangle. Similarly, the community containing the campus or house is part of a system of communities arranged across the country and around the world.

Geographers explain how features such as buildings and communities are arranged across Earth. On Earth as a whole, or within an area of Earth, features may be numerous or scarce, close together or far apart. The arrangement of a feature in space is known as its **distribution.** Geographers identify three main properties of distribution across Earth—density, concentration, and pattern.

DISTRIBUTION PROPERTIES: DENSITY

Density is the frequency with which something occurs in space. The feature being measured could be people, houses, cars, trees, or anything else. The area could be measured in square kilometers, square miles, hectares, acres, or any other unit of area.

Remember that a large number of a feature does not necessarily lead to a high density. Density involves two measures—the number of a feature and the land area. China is the country with the largest number of people—approximately 1.4 billion—but it does not have the world's highest density. The Netherlands, for example, has only 17 million people, but its density of around 500 persons per square kilometer is much higher than China's 140 persons per square kilometer. The reason is that the land area of China is 9.3 million square kilometers, compared to only 42,000 square kilometers for the Netherlands.

High population density is also unrelated to poverty. The Netherlands is one of the world's wealthiest countries and Mali one of the world's poorest. Yet the Netherlands' density of around 500 persons per square kilometer is much larger than Mali's density of 13 persons per square kilometer (see Chapter 2 for more about density).

DISTRIBUTION PROPERTIES: CONCENTRATION

The extent of a feature's spread over space is its **concentration.** If the objects in an area are close together, they are clustered; if relatively far apart, they are dispersed. To compare the level of concentration most clearly, two areas need to have the same number of objects and the same size area.

Geographers use concentration to describe changes in distribution. For example, the distribution of people across the United States is increasingly dispersed. The total number of people living in the United States is growing slowly—less than 1 percent per year—and the land area is essentially unchanged. But the population distribution is changing from relatively clustered in the Northeast to more evenly dispersed across the country.

Concentration is not the same as density. As Figure 1-28 shows, two neighborhoods could have the same density of housing but different concentrations. In a dispersed neighborhood, each house has a large private yard, whereas in a clustered neighborhood, the houses are close together and the open space is shared as a community park.

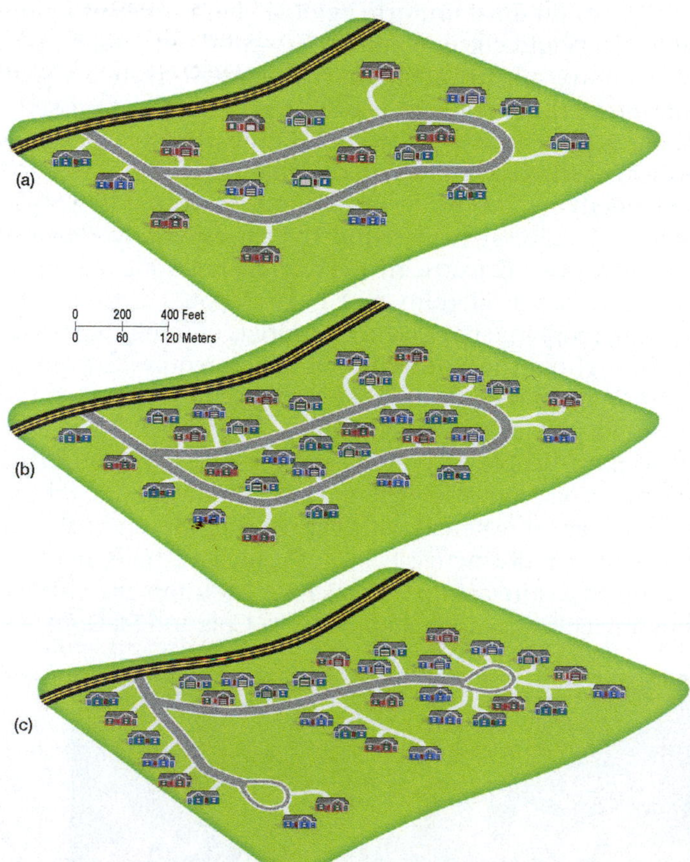

▲ **FIGURE 1-28 DENSITY AND CONCENTRATION OF HOUSES** Neighborhood **(a)** has a lower density than neighborhood (b)—18 houses compared to 32 houses on the same 82-acre piece of land—but both have dispersed concentrations. Neighborhoods **(b)** and **(c)** have the same density (32 houses on 82 acres), but the distribution of houses is more clustered in plan (c). Neighborhood (c) has shared open space, whereas plan (b) provides a larger, private yard surrounding each house.

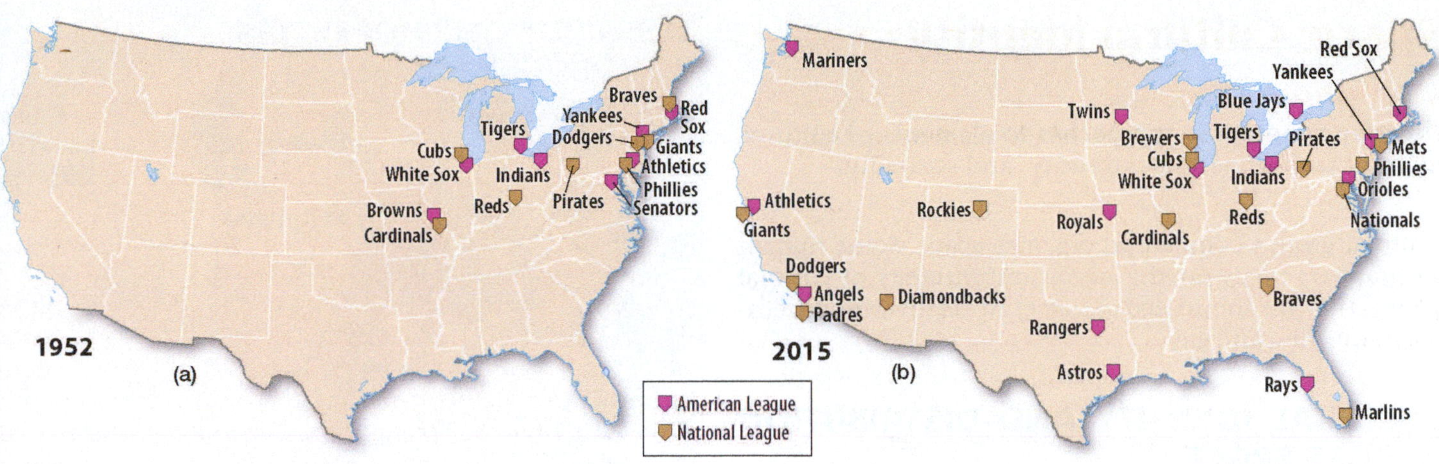

▲ **FIGURE 1-29 CHANGING DENSITY AND CONCENTRATION OF BASEBALL TEAMS** These 6 teams moved to other cities during the 1950s and 1960s: Braves—Boston to Milwaukee in 1953, then to Atlanta in 1966 • Browns/Orioles—St. Louis (Browns) to Baltimore (Orioles) in 1954 • Athletics—Philadelphia to Kansas City in 1955, then to Oakland in 1968 • Dodgers—Brooklyn to Los Angeles in 1958 • Giants—New York to San Francisco in 1958 • Senators/Twins—Washington (Senators) to Minneapolis (Minnesota Twins) in 1961 These 14 teams were added between the 1960s and 1990s: • Angels—Los Angeles in 1961, then to Anaheim (California) in 1965 • Rangers—Washington (Senators) in 1961, then to Arlington (Texas) in 1971 • Mets—New York in 1962 • Astros—Houston (originally Colt .45s) in 1962 • Royals—Kansas City in 1969 • Padres—San Diego in 1969 • Nationals—Montreal (Expos) in 1969, then to Washington (Nationals) in 2005 • Brewers—Seattle (Pilots) in 1969, then to Milwaukee (Brewers) in 1970 • Blue Jays—Toronto in 1977 • Mariners—Seattle in 1977 • Marlins—Miami (originally Florida) in 1993 • Rockies—Denver (Colorado) in 1993 • Rays—Tampa Bay (originally Devil Rays) in 1998 • Diamondbacks—Phoenix (Arizona) in 1998 As a result of these relocations and additions, the density of teams increased, and the distribution became more dispersed.

The distribution of Major League Baseball teams illustrates the difference between density and concentration (Figure 1-29). After remaining unchanged during the first half of the twentieth century, the distribution of Major League Baseball teams changed during the second half of the twentieth century. The Major Leagues expanded from 16 to 30 teams in North America between 1960 and 1998, thus increasing the density. At the same time, 6 of the 16 original teams moved to other locations. In 1952, all the teams were clustered in the northeastern United States, but the moves dispersed several teams to the West Coast and Southeast. These moves, as well as the spaces occupied by the expansion teams, resulted in a more dispersed distribution.

DISTRIBUTION PROPERTIES: PATTERN

The third property of distribution is **pattern,** which is the geometric arrangement of objects in space. Some features are organized in a geometric pattern, whereas others are distributed irregularly. Geographers observe that many objects form a linear distribution, such as the arrangement of houses along a street or stations along a subway line.

Objects are frequently arranged in a square or rectangular pattern. Many American cities contain a regular pattern of streets that intersect at right angles at uniform intervals to form square or rectangular blocks; this is known as a grid pattern. The system of townships, ranges, and sections established by the Land Ordinance of 1785 is another example of a square or grid pattern (Figure 1-30).

PAUSE & REFLECT 1.3.2

How would you describe the density, concentration, and pattern of chairs in your classroom?

◀ **FIGURE 1-30 PATTERN: TOWNSHIP AND RANGE** The U.S. Land Ordinance of 1785 divided much of the United States into a checkerboard pattern, which is still visible in agricultural areas.

Space: Cultural Identity

LEARNING OUTCOME 1.3.3

Describe geographic approaches to elements of cultural identity such as gender, ethnicity, and sexuality.

Cultural groups compete to organize space. Some human geographers focus on the needs and interests of cultural groups that are dominated in space, especially women, ethnic minorities, and gays.

CULTURAL IDENTITY AND DISTRIBUTION ACROSS SPACE

Patterns in space vary according to gender, ethnicity, and sexuality. Geographers study these cultural traits because they are important in explaining why people sort themselves out in space and move across the landscape in distinctive ways.

DISTRIBUTION BY ETHNICITY. The distribution of ethnicities in the United States varies considerably at all scales, as discussed in detail in Chapter 7. African Americans are clustered in the Southeast, whereas Hispanics are clustered in the Southwest. Ethnicities are also highly clustered within cities, as discussed in detail in both Chapters 7 and 13. In Los Angeles, which contains large percentages of African Americans, Hispanics, and Asian Americans, the major ethnic groups are clustered in different areas (Figure 1-31). African Americans are located in south-central Los Angeles and Hispanics in the east. Asian Americans are located to the south and west, contiguous to the African American and Hispanic areas.

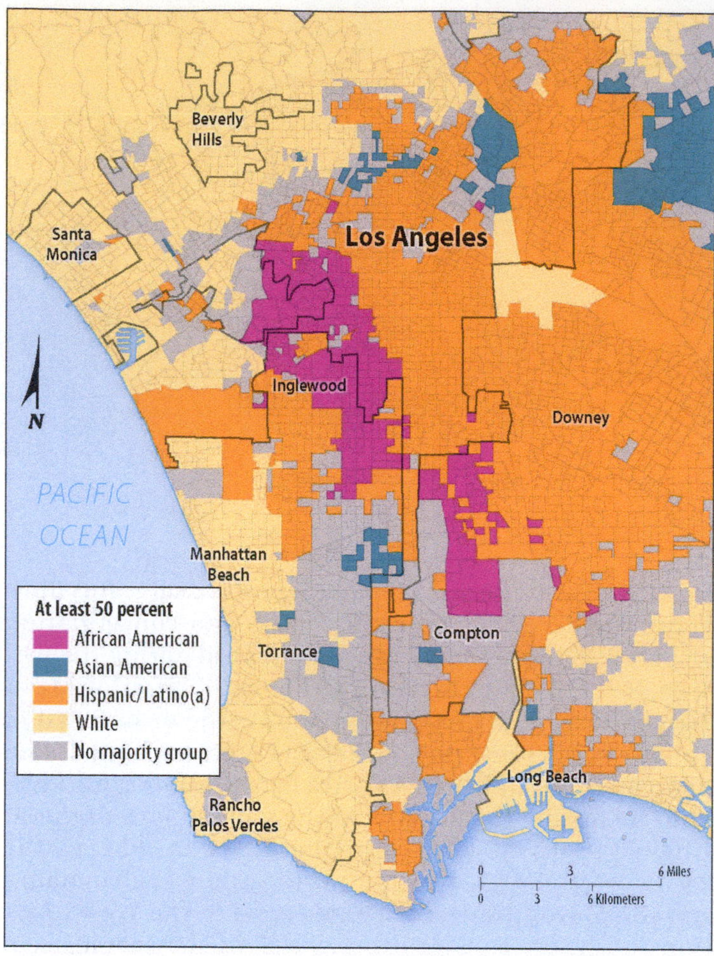

▲ **FIGURE 1-31 DISTRIBUTION OF ETHNICITIES IN LOS ANGELES** African Americans are clustered to the south of downtown Los Angeles and Hispanics to the east. Asian American neighborhoods are contiguous to the African American and Hispanic areas.

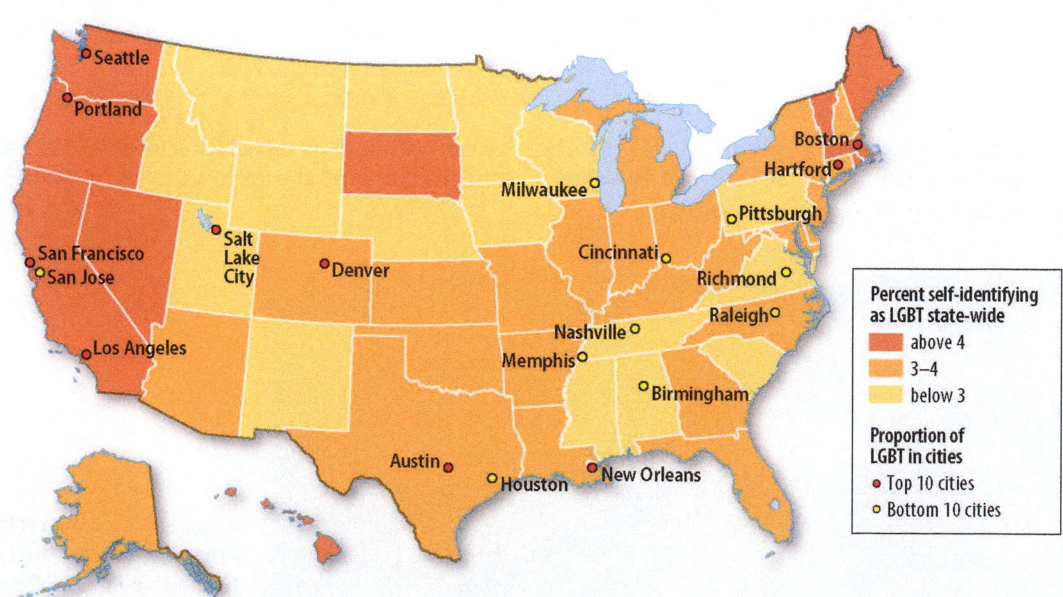

▲ **FIGURE 1-32 LGBT POPULATION PERCENTAGE** Figures are the percentages of people who self-identified as LGBT in a 2015 Gallup survey.

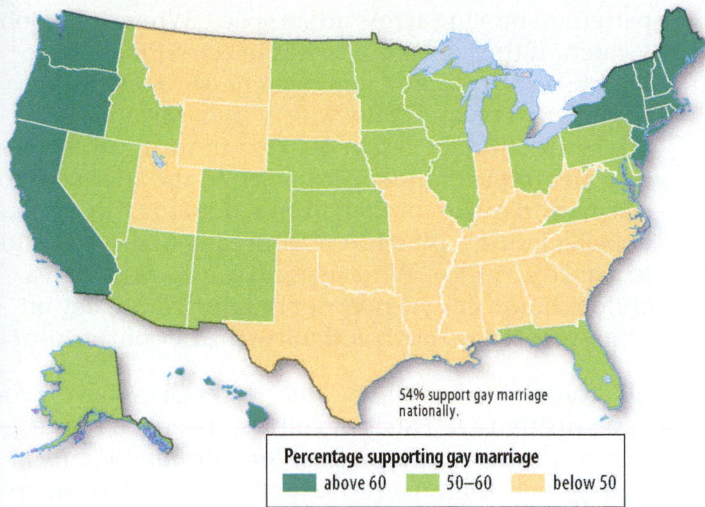

▲ **FIGURE 1-33 SUPPORT FOR GAY MARRIAGE PRIOR TO NATIONWIDE LEGALIZATION** The percentage of people who supported gay marriage prior to the U.S. Supreme Court legalizing it in 2015 varied widely among the 50 states.

DISTRIBUTION BY SEXUAL ORIENTATION. Openly lesbian, gay, bisexual, and transgender (LGBT) people may be attracted to some locations to reinforce spatial interaction with other LGBT people. Among the 50 largest U.S. cities, the percentage of LGBT people varies from 6.2 percent in San Francisco to 2.6 percent in Birmingham, Alabama. San Francisco, which has a reputation for being an especially hospitable city for LGBT people, not surprisingly has the highest percentage of openly LGBT people among the 50 largest U.S. cities, although the percentage is not significantly higher than in other cities (Figure 1-32).

At the scale of the 50 U.S. states, the percentage of people who self-identify as LGBT does not vary widely. According to a 2015 Gallup poll, the percentage varies among states from a high of 5.1 percent in Hawaii to a low of 1.7 percent in North Dakota. (The District of Columbia has 10.0 percent.) However, the percentage of people who supported same-sex marriage in 2015, a few months before the U.S. Supreme Court legalized it, varied much more widely among states, from a high of 75 percent in New Hampshire to a low of 32 percent in Mississippi (Figure 1-33).

PAUSE & REFLECT 1.3.3

Using your own campus or school as the example, describe how movement across space varies during the day for students and faculty.

DISTRIBUTION BY GENDER. The distribution of women does not vary across space like ethnicity and sexual orientation, although some communities such as those with military bases may have a lot more men than women. As discussed in Chapter 2, in some countries, notably the two most populous, China and India, the number of female babies is suspiciously much lower than the number of male babies.

With gender, geographers focus on the distribution of inequality. As discussed in detail in Chapter 9, gender inequality is reflected in numerous factors. The United Nations has not found a single country in the world where the average income earned by women exceeds that earned by men (Figure 1-34). Worldwide, the average income of women is around 50 percent that of men. The income gap between women and men is approximately the same in both developed and developing countries. At best, women in some countries have achieved near-equality with men. The countries with near-equality are primarily in poor countries where both male and female incomes are low, and in some European countries. The gap is especially large in Southwest Asia and North Africa, where few women are permitted to work.

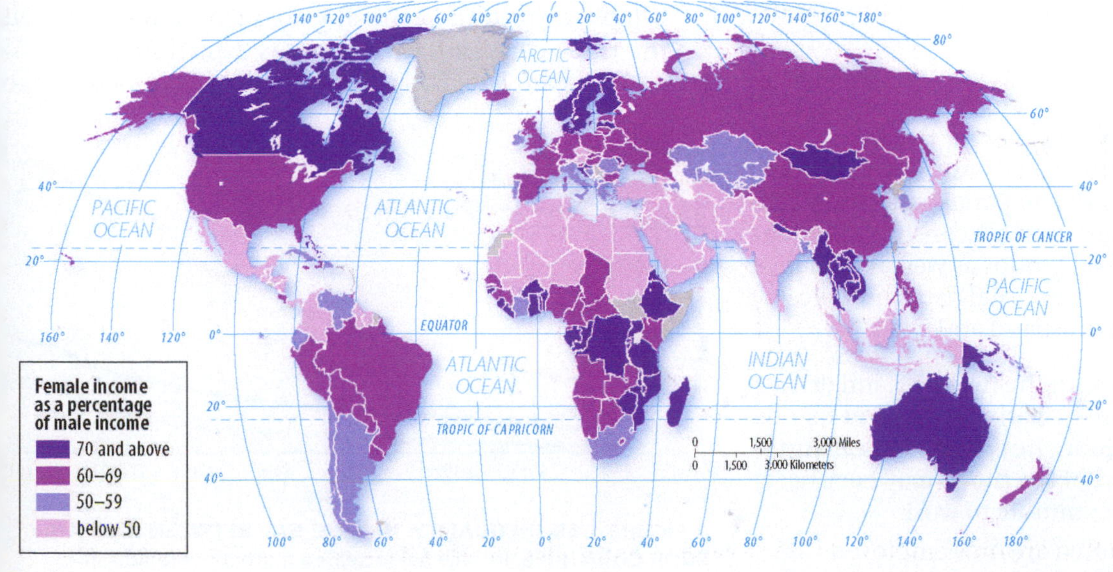

◀ **FIGURE 1-34 FEMALE INCOME AS A PERCENTAGE OF MALE INCOME** According to the United Nations, the average income earned by women is less than that of men in every country of the world.

Space: Inequality

LEARNING OUTCOME 1.3.4

Summarize geographic thought, with application to the geography of inequality.

Cultural characteristics, especially gender, ethnicity, and sexuality, influence the distribution and movement of people across space. The experiences of women differ from those of men, blacks from whites, gays from straights, and boys from girls.

CULTURAL IDENTITY AND CONTEMPORARY GEOGRAPHIC THOUGHT

Geographers take a range of approaches to cultural identity and space, including those of poststructuralist, humanisitic, and behavioral geography.

Poststructuralist geography examines how the powerful in a society dominate, or seek to control, less powerful groups, how the dominated groups occupy space, and confrontations that result from the domination. Poststructuralist geographers understand space as the product of ideologies or value systems of ruling elites. For example, some researchers have studied how, although it is illegal to discriminate against people of color, local governments have pursued policies that impose hazardous, polluting industries on minority neighborhoods.

Humanistic geography is a branch of human geography that emphasizes the different ways that individuals form ideas about place and give those places symbolic meanings. For example, openly homosexual men and lesbian women may be attracted to places such as Christopher Street in New York City because they perceive them as gay-friendly spaces where they can interact socially with other gays. Christopher Street may be seen as offering an accepting location for gay men and lesbians through inclusive policies and business practices. But the street also has symbolic meaning for gays: In 1969, the Stonewall Inn was the site of protests that began the gay liberation movement. Tolerance for gays has increased, although legal discrimination persists in some regions.

Behavioral geography emphasizes the importance of understanding the psychological basis for individual human actions in space. Distinctive spatial patterns by gender, ethnicity, and sexual orientation are constructed by the attitudes and actions of cultural groups as well as the larger society. For example, consider the spatial patterns typical of a household that consists of a husband and wife:

- **Husband.** He gets in his car in the morning and drives from home to work, where he parks the car and spends the day. In the late afternoon, he collects the car and drives home. The location of the home may have been selected to ease his daily commute to work.
- **Wife.** Most American women are now employed at work outside the home, resulting in a complex pattern of moving across urban space. Where is her job located? If the family house was selected for access to her husband's place of employment, she may need to travel across town. Yet the wife is often the one who drives the children to school in the morning, walks the dog, and drives to the supermarket. In the afternoon, she may drive the children from school to Little League or ballet lessons. Who leaves work early to drive a child to a doctor's office? Who takes a day off work when a child is home sick yet may not be entitled to a day off from work to give birth and nurse the newborn child?

All academic disciplines and workplaces have proclaimed sensitivity to issues of cultural diversity. For geographers, concern and deep respect for cultural diversity are not merely politically correct expediencies but lie at the heart of geography's understanding of space. Geographers have deep respect for the dignity of all cultural groups.

UNEQUAL ACCESS

Electronic communications have played an especially important role in removing barriers to interaction between people who are physically far from each other. Physical barriers, such as oceans and deserts, can still retard interaction among people. In the modern world, barriers to interaction are more likely to derive from unequal access to electronics.

Instantaneous expansion diffusion, made possible by electronic communications, was once viewed as the "death" of geography because the ease of communications between distant places removed barriers to interaction. In reality, because of unequal access, geography matters even more than before.

People have unequal access to interaction in part because the quality of electronic service varies among places. Internet access depends on availability of electricity to power the computer and a service provider. Seconds count. Broadband service requires proximity to a digital subscriber line (DSL), a cable line, or other services. Most importantly, a person must be able to afford to pay for the communications equipment and service.

Global culture and economy are increasingly centered on the three core, or hearth, regions: North America, Europe, and Japan. These three regions have a large percentage

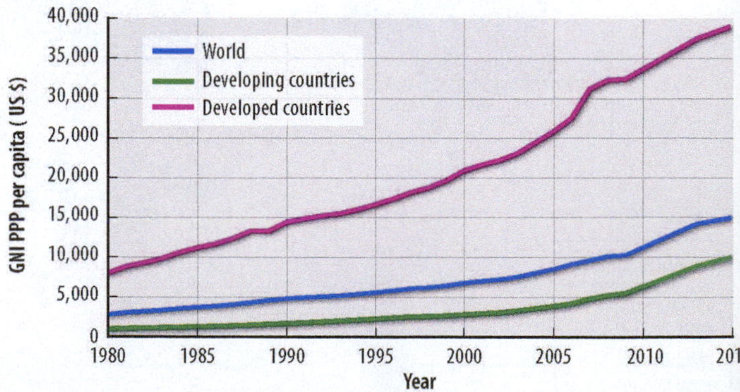

▲ **FIGURE 1-35 INEQUALITY: INCOME GAP BETWEEN RICH AND POOR COUNTRIES** Income has increased much more rapidly in developed countries than in developing ones.

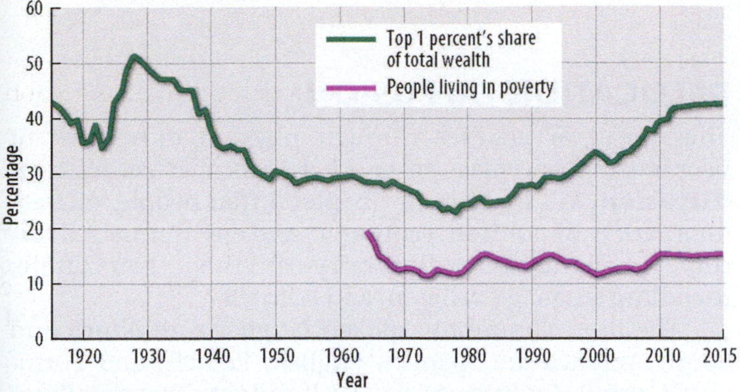

▲ **FIGURE 1-36 INEQUALITY: U.S. WEALTH AND POVERTY** The share of U.S. wealth held by the wealthiest 1 percent declined from the 1930s through the 1970s and began rising in the 1980s. The percentage of people living in poverty declined during the 1960s but has remained relatively unchanged since then.

of the world's advanced technology, capital to invest in new activities, and wealth to purchase goods and services. From "command centers" in the three major world cities New York, London, and Tokyo, key decision makers employ modern telecommunications to send orders to factories, shops, and research centers around the world—an example of hierarchical diffusion. Meanwhile, "nonessential" employees of the companies can be relocated to lower-cost offices outside the major financial centers. For example, Fila maintains headquarters in Italy but has moved 90 percent of its production of sportswear to Asian countries. Mitsubishi's corporate offices are in Japan, but its electronics products are made in other Asian countries.

Countries in Africa, Asia, and Latin America contain three-fourths of the world's population and nearly all of its population growth. However, these countries find themselves on a periphery, or outer edge, with respect to the wealthier core regions of North America, Europe, and Japan. Global investment arrives from the core through hierarchical diffusion of decisions made by transnational corporations.

People in peripheral regions, who once toiled in isolated farm fields to produce food for their families, now produce crops for sale in core regions or have given up farm life altogether and migrated to cities in search of jobs in factories and offices. As a result, the global economy has produced greater disparities than in the past between the levels of wealth and well-being enjoyed by people in the core and in the periphery. The increasing gap in economic conditions between regions in the core and periphery that results from the globalization of the economy is known as **uneven development** (Figure 1-35).

In a global culture and economy, every area of the world plays some role intertwined with the roles played by other regions. Workers and cultural groups that in the past were largely unaffected by events elsewhere in the world now share a single economic and cultural world with other workers and cultural groups. The fate of an autoworker in Detroit is tied to investment decisions made in Mexico City, Seoul, Stuttgart, and Tokyo.

Unequal access and economic inequality have also increased within countries. In the United States, the share of the national income held by the wealthiest 1 percent increased from 24 percent in 1979 to 42 percent in 2014, according to the Congressional Budget Office (Figure 1-36). The share of Americans living in poverty increased during that same period, from 12 percent to 15 percent.

Within the United States, prospects for children of low-income families to escape poverty depend on the community where they spend their childhood. Children who grew up in low-income households in the Midwest and Plains states earn several thousand dollars more per year at age 26 than do children who grew up in low-income households in the Southeast and Southwest (Figure 1-37).

PAUSE & REFLECT 1.3.4

How does the region where you live compare in the percentage of children in poverty?

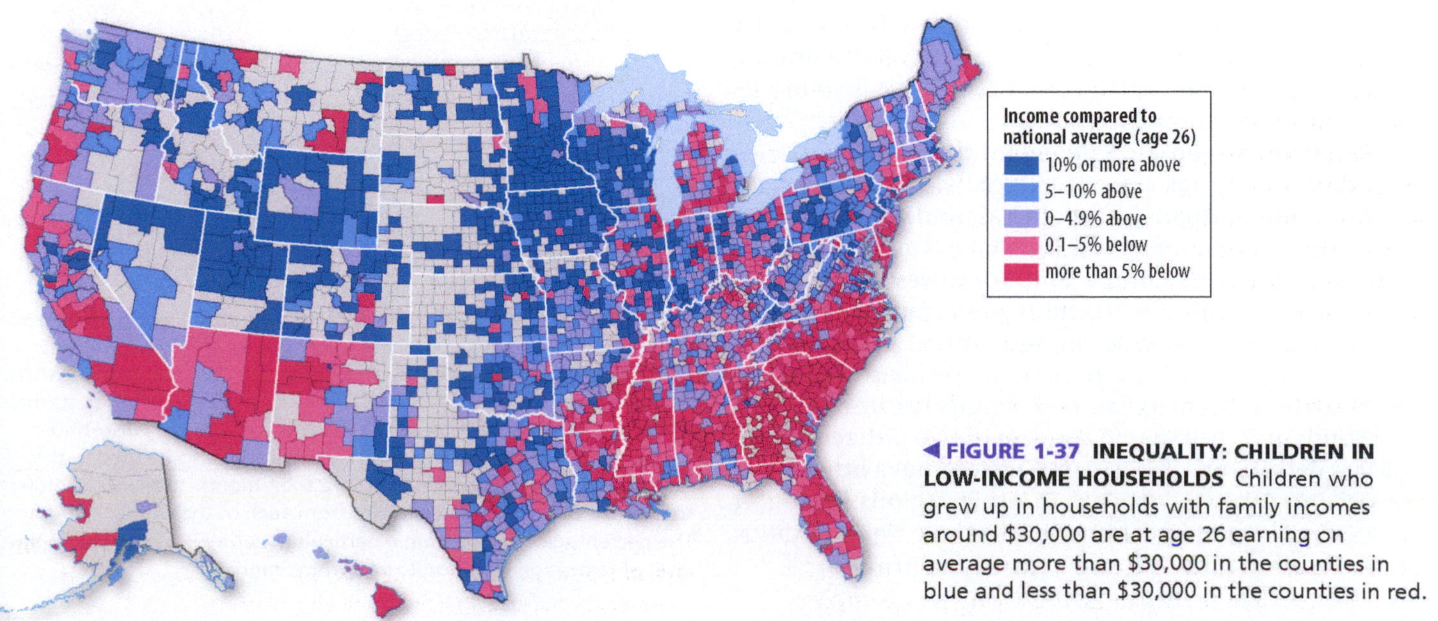

◀ **FIGURE 1-37 INEQUALITY: CHILDREN IN LOW-INCOME HOUSEHOLDS** Children who grew up in households with family incomes around $30,000 are at age 26 earning on average more than $30,000 in the counties in blue and less than $30,000 in the counties in red.

Connections: Diffusion

LEARNING OUTCOME 1.3.5

Describe the various ways that features can spread through diffusion.

Connection was defined at the beginning of the chapter as relationships among people and objects that cross the barrier of space. Geographers are concerned with the various means by which connections occur. More rapid connections have reduced the distance across space between places—not literally in miles, of course, but in time.

Connections between cultural groups can have several results:

- **Assimilation** is the process by which a group's cultural features are altered to resemble those of another group. The cultural features of one group may come to dominate the culture of the assimilated group.
- **Acculturation** is the process of changes in culture that result from the meeting of two groups. Changes may be experienced by both of the interacting cultural groups, but the two groups retain two distinct culture features.
- **Syncretism** is the combination of elements of two groups into a new cultural feature. The two cultural groups come together to form a new culture.

DIFFUSION

Diffusion is the process by which a feature spreads across space from one place to another over time. Something originates at a hearth and diffuses from there to other places. A **hearth** is a place from which an innovation originates. Geographers document the locations of nodes and the processes by which diffusion carries things elsewhere over time.

How does a hearth emerge? A cultural group must be willing to try something new and must be able to allocate resources to nurturing the innovation. To develop a hearth, a group of people must also have the technical ability to achieve the desired idea and the economic structures, such as financial institutions, to implement the innovation.

As discussed in subsequent chapters, geographers can trace the dominant cultural, political, and economic features of the contemporary United States and Canada primarily to hearths in Europe and Southwest Asia. Other regions of the world also contain important hearths. In some cases an idea, such as an agricultural practice, may originate independently in more than one hearth. In other cases, hearths may emerge in two regions because two cultural groups modify a shared concept in two different ways.

For a person, an object, or an idea to have interaction with persons, objects, or ideas in other regions, diffusion must occur. Geographers observe two basic types of diffusion—relocation diffusion and expansion diffusion.

RELOCATION DIFFUSION

The spread of an idea through physical movement of people from one place to another is termed **relocation diffusion.** We shall see in Chapter 3 that people migrate for a variety of political, economic, and environmental reasons. When they move, they carry with them their culture, including language, religion, and ethnicity.

The most commonly spoken languages in North and South America are Spanish, English, French, and Portuguese, primarily because several hundred years ago Europeans who spoke those languages comprised the largest number of migrants. Thus these languages spread through relocation diffusion. We will examine the diffusion of languages, religions, and ethnicity in Chapters 5 through 7.

Introduction of a common currency, the euro, in 12 European countries in 2002 gave scientists an unusual opportunity to measure relocation diffusion from hearths (Figure 1-38). Although a single set of paper money was issued, each of the 12 countries minted its own coins in

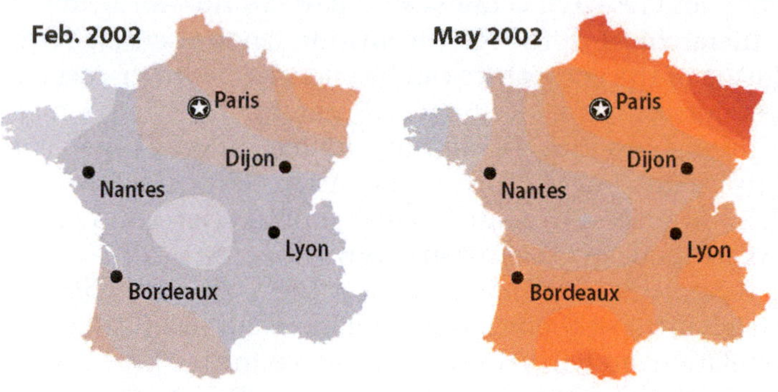

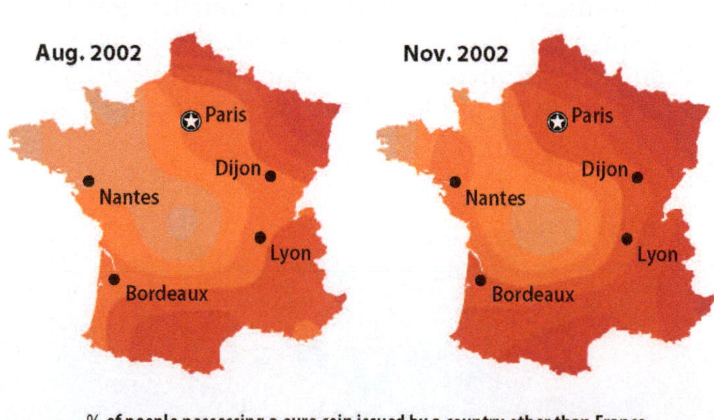

▲ **FIGURE 1-38 RELOCATION DIFFUSION** Introduction of a common currency, the euro, in 12 European countries in 2002 gave scientists an unusual opportunity to measure relocation diffusion from hearths. Each of the 12 countries minted its own coins in proportion to its share of the region's economy. A country's coins were initially distributed only inside its borders, although the coins could also be used in the other 11 countries. Scientists in France took month-to-month samples to monitor the proportion of coins from each of the other 11 countries. The percentage of coins from a particular country is a measure of the level of relocation diffusion to and from France.

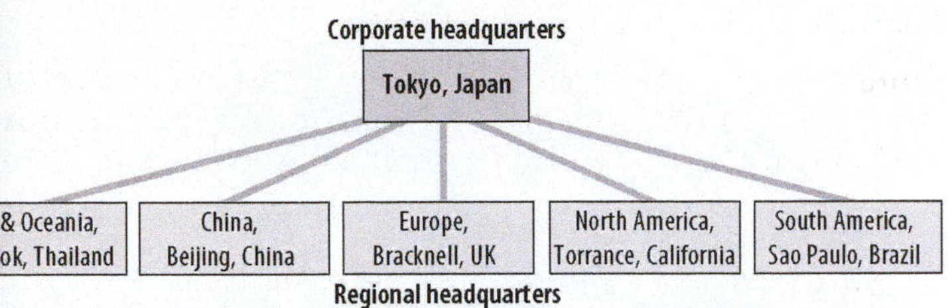

▲ **FIGURE 1-39 HIERARCHICAL DIFFUSION** Honda's worldwide operations are controlled from headquarters in Tokyo, Japan. Officials in regional headquarters oversee operations in several regions and report to corporate headquarters.

proportion to its share of the region's economy. A country's coins were initially distributed only inside its borders, although the coins could also be used in the other 11 countries. Scientists in France took month-to-month samples to monitor the proportion of coins from each of the other 11 countries. The percentage of coins from a particular country is a measure of the level of relocation diffusion into France.

EXPANSION DIFFUSION

The spread of a feature from one place to another in an additive process is **expansion diffusion.** This expansion may result from one of three processes:

- **Hierarchical diffusion** is the spread of an idea from persons or nodes of authority or power to other persons or places (Figure 1-39). Hierarchical diffusion may result from the spread of ideas from political leaders, socially elite people, or other important persons to others in the community. Innovations may also originate in a particular node or core region of power, such as a large urban center, and diffuse later to isolated rural areas on the periphery. Hip-hop or rap music is an example of an innovation that originated in urban areas, though it diffused from low-income African Americans rather than from socially elite people.
- **Contagious diffusion** is the rapid, widespread diffusion of a characteristic throughout the population. As the term implies, this form of diffusion is analogous to the spread of a contagious disease, such as influenza. Contagious diffusion spreads like a wave among fans in a stadium, without regard for hierarchy and without requiring permanent relocation of people. New music or an idea goes viral because web surfers throughout the world have access to the same material simultaneously (Figure 1-40).
- **Stimulus diffusion** is the spread of an underlying principle even though a characteristic itself apparently fails to diffuse. For example, innovative features of Apple's iPhone and iPad have been adopted by competitors.

Expansion diffusion occurs much more rapidly in the contemporary world than it did in the past. Hierarchical diffusion is encouraged by modern methods of communication, such as computers, texting, blogging, Twitter, and e-mail. Contagious diffusion is encouraged by use of the Internet, especially the World Wide Web. Stimulus diffusion is encouraged by all of the new technologies.

PAUSE & REFLECT 1.3.5

U.S. coins are minted in either Denver or Philadelphia. Take a look at your coins. Do you have more from the mint closer to you?

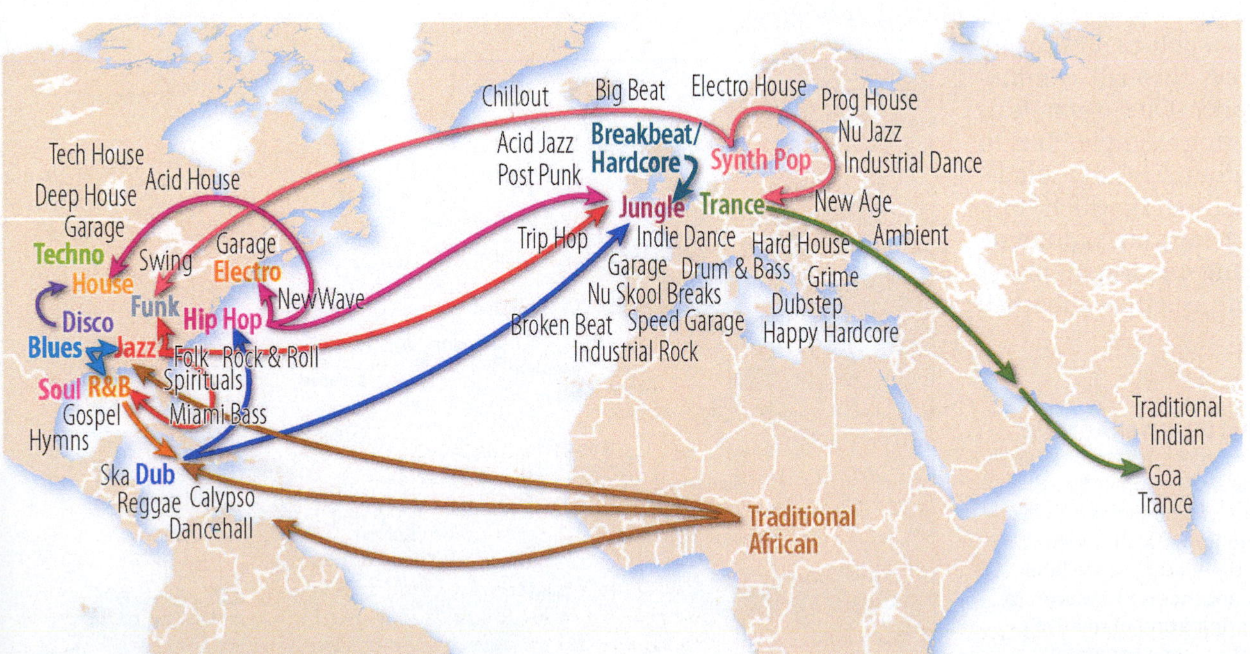

▲ **FIGURE 1-40 CONTAGIOUS DIFFUSION: WESTERN DANCE MUSIC** Popular dance music originated in the Western Hemisphere and diffused to Europe and Asia beginning in the 1980s.

Connections: Spatial Interaction

LEARNING OUTCOME 1.3.6

Explain how places are connected through networks, though inequality can hinder connections.

Interaction takes place through a **network,** which is a chain of communication that connects places. A major airline, for example, typically has a network known as "hub-and-spoke" (Figure 1-41). With a hub-and-spoke network, an airline flies planes from a large number of places into one hub airport within a short period of time and then a short time later sends the planes to another set of places. In principle, travelers originating in relatively small towns can reach a wide variety of destinations by changing planes at the hub airport.

The farther away someone is from another, the less likely the two are to interact. Contact diminishes with increasing distance and eventually disappears. This trailing-off phenomenon is called **distance decay.** In the contemporary world, distance decay is much less severe because connection between places takes much less time. Geographers apply the term **space–time compression** to describe the reduction in the time it takes for something to reach another place (Figure 1-42).

In the past, most connections among cultural groups required the physical movement of settlers, explorers, and plunderers from one location to another. As recently as A.D. 1800, people traveled in the same ways and at about the same speeds as in 1800 B.C.—they were carried by an animal, took a sailboat, or walked.

To be connected with another place in the modern world, we do not need to travel at all. Ideas that originate in a hearth are now able to diffuse rapidly to other areas through communications networks. One example is the TV network (for example, BBC in the United Kingdom, CBC in Canada, NBC in the United States), which comprises a chain of stations simultaneously broadcasting to distant places the same program, such as a football game. Through a communications network, diffusion from one place to another is instantaneous in time, even if the physical distance between places—as measured in kilometers or miles—is large.

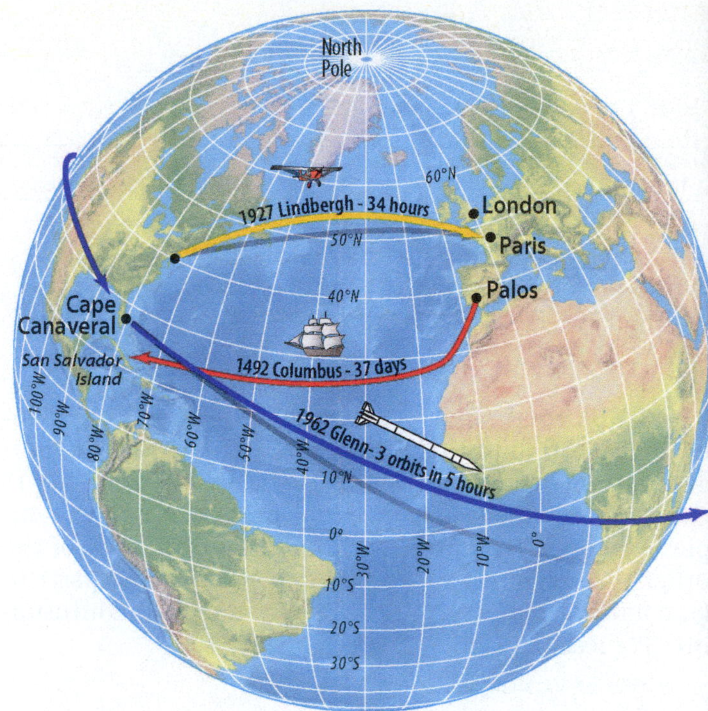

▲ **FIGURE 1-42 SPACE-TIME COMPRESSION** Transportation improvements have shrunk the world. In 1492, Christopher Columbus took nearly 900 hours (37 days) to sail across the Atlantic Ocean. In 1927, Charles Lindbergh was the first to fly nonstop across the Atlantic, taking 33.5 hours. In 1962, John Glenn, the first American to orbit in space, crossed above the Atlantic in about ½ hour and circled the globe three times in 5 hours.

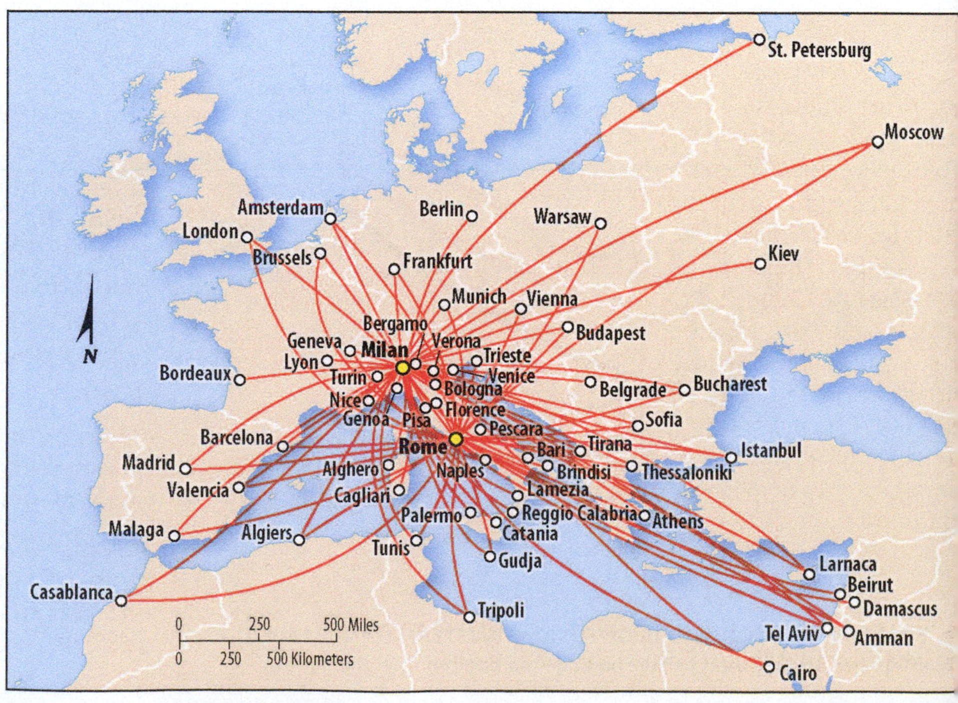

▶ **FIGURE 1-41 SPATIAL INTERACTION: AIRLINE HUB-AND-SPOKE NETWORK** With a hub-and-spoke network, Alitalia flies planes from a large number of places into hub airports at Milan and Rome within a short period of time and then a short time later sends the planes to another set of places. In principle, travelers originating in relatively small towns can reach a wide variety of destinations by changing planes at a hub airport.

DEBATE IT! GPS location service: On or off?

Most of our cell phones have a GPS tracking device that can pinpoint our precise location. By default, most phones have this geotagging feature turned on. Should you leave location service on or turn it off (Figures 1-43 and 1-44)?

LEAVE IT ON

- Emergency services will be able to find you.
- You can access maps, get driving directions, and check traffic.
- You can learn when the next bus or train will arrive.
- You can find nearby restaurants, gas stations, and dog parks.

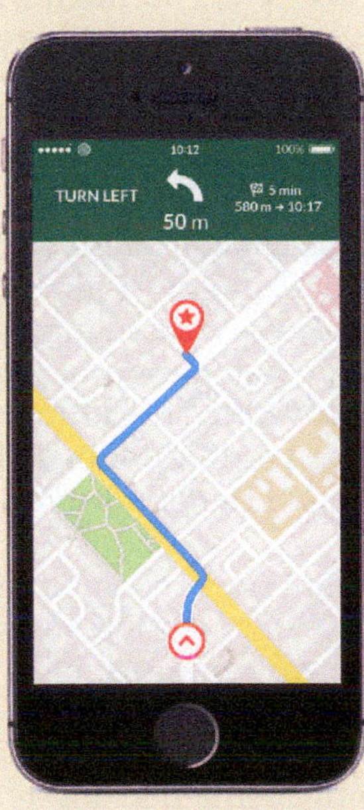

◀ **FIGURE 1-43 GPS TRACKING WITH LOCATION SERVICES ON** Your precise location is known.

TURN IT OFF

- Information that you wish to keep private is shared with others.
- Your movements, preferences, and friends will be known to others.
- Tracking information can be used in legal proceedings.
- Unwanted advertisements and messages will be sent to your device.

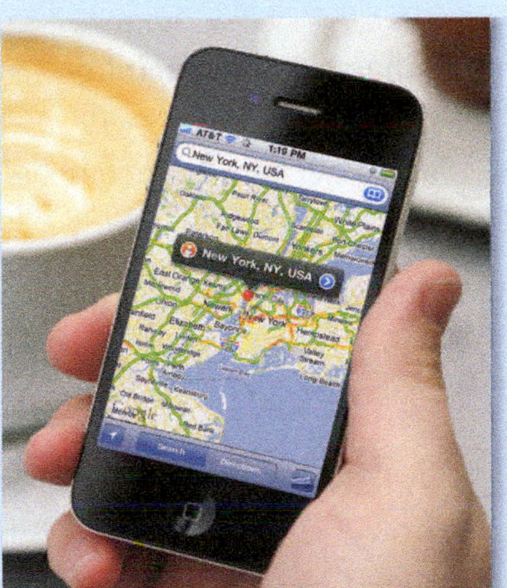

▲ **FIGURE 1-44 LOCATION SERVICES OFF** By default, most phones have the location service on, but it can be turned off.

Computers, tablets, and smart phones make it possible or individuals to set up their own connections through ndividually constructed networks such as Facebook and Twitter. At the click of a button, we can transmit images and nessages from one part of the world to our own personalzed network around the world.

Modern networks make it possible for us to know nore about what is happening elsewhere in the world, and pace–time compression makes it possible for us to know t sooner. Distant places seem less remote and more accesible to us. With better connections between places, we are exposed to a constant barrage of cultural traits and economic initiatives from people in other regions, and perhaps we may adopt some of these cultural and economic elements.

PAUSE & REFLECT 1.3.6

s your nearest airport a hub? If not, to what hub do nost flights go from your nearest airport?

CHECK-IN KEY ISSUE 3

Why Are Different Places Similar?

- ✔ **Geographers examine at all scales, though they are increasingly concerned with the global scale.**
- ✔ **Distribution has three properties—density, concentration, and pattern—and different cultural groups display different distributions in space.**
- ✔ **Places are connected through networks, and phenomena spread through relocation and expansion diffusion.**
- ✔ **In spite of space–time compression, peripheral regions in the global economy often have unequal access to the goods and services available in core regions.**

KEY ISSUE 4

Why Are Some Actions Not Sustainable?

- Geography, Sustainability, and Resources
- Sustainability and Earth's Physical Systems
- Geography, Sustainability, and Ecology
- Sustainable Environmental Change

LEARNING OUTCOME 1.4.1

Describe the three pillars of sustainability.

Geography is distinctive because it encompasses both social science (human geography) and natural science (physical geography). This book focuses on human geography but doesn't forget that humans are interrelated with Earth's atmosphere, land, water, and vegetation, as well as with its other living creatures.

Geography, Sustainability, and Resources

A **resource** is a substance in the environment that is useful to people, economically and technologically feasible to access, and socially acceptable to use. **Sustainability** is the use of Earth's resources in ways that ensure their availability in the future.

From the perspective of human geography, nature offers a large menu of resources available for people to use. A substance is merely part of nature until a society has a use for it. Food, water, minerals, soil, plants, and animals are examples of resources.

Earth's resources are divided between those that are renewable and those that are not:

- A **renewable resource** is produced in nature more rapidly than it is consumed by humans.
- A **nonrenewable resource** is produced in nature more slowly than it is consumed by humans.

Geographers observe that the sustainability of resources is being damaged by human actions:

- Humans deplete nonrenewable resources, such as petroleum, natural gas, and coal.
- Humans destroy otherwise renewable resources through pollution of air, water, and soil.

Geographers also document efforts to improve the sustainability of resources. Recycling paper and plastic, developing new industrial processes, and protecting farmland from urban sprawl are examples of practices that contribute to a more sustainable future.

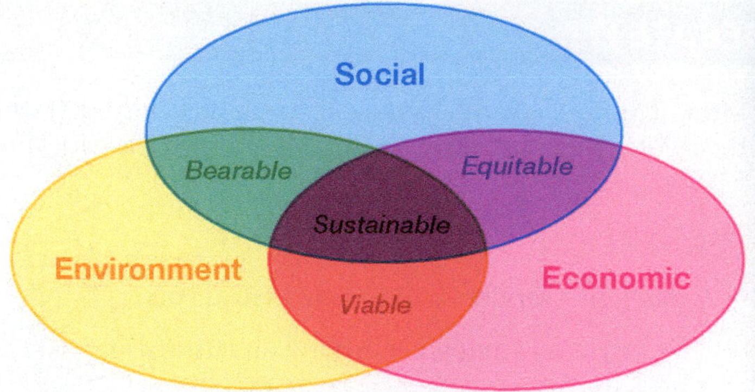

▲ **FIGURE 1-45 THREE PILLARS OF SUSTAINABILITY** The U.N.'s Brundtland Report considers sustainability to be a combination of environmental protection, economic development, and social equity.

THREE PILLARS OF SUSTAINABILITY

According to the United Nations, sustainability rests on three pillars: environment, society, and economy (Figure 1-45). The 1987 U.N. report *Our Common Future* is a landmark work in recognizing sustainability as a combination of natural and human elements. It is frequently called the Brundtland Report, named for the chair of the World Commission on Environment and Development, Gro Harlem Brundtland, former prime minister of Norway.

Sustainability requires curtailing the use of nonrenewable resources and limiting the use of renewable resources to the level at which the environment can continue to supply them indefinitely. To be sustainable, the amount of timber cut down in a forest, for example, or the number of fish removed from a body of water must remain at a level that does not reduce future supplies.

The Brundtland Report argues that sustainability can be achieved only by bringing together environmental protection, economic growth, and social equity. The report is optimistic about the possibility of promoting environmental protection at the same time as economic growth and social equity.

THE ENVIRONMENT PILLAR. The sustainable use and management of Earth's natural resources to meet human needs such as food, medicine, and recreation is **conservation.** Renewable resources such as trees and wildlife are conserved if they are consumed at a less rapid rate than they can be replaced. Nonrenewable resources such as petroleum and coal are conserved if we use less today in order to maintain more for future generations (Figure 1-46a).

Conservation differs from **preservation,** which is the maintenance of resources in their present condition, with as little human impact as possible. Preservation takes the view that the value of nature does not derive from human needs and interests but from the fact that every plant and animal living on Earth has a right to exist and should be preserved, regardless of the cost. Preservation does not regard nature as a resource for human use. In contrast, conservation is

(a)

(b)

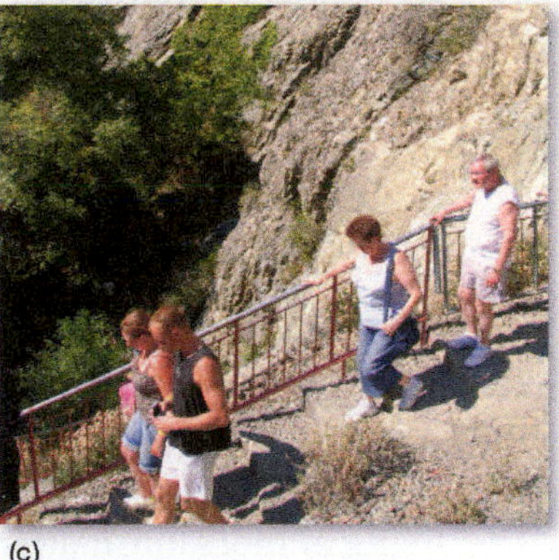
(c)

▲ **FIGURE 1-46 THREE PILLARS OF SUSTAINABILITY: CYPRUS** The Troodos Mountains in Cyprus illustrate the three pillars of sustainability. **(a)** The environment pillar. The area is known for its outstanding rock formations. Much of the area is protected as national forests and U.N. World Heritage sites. **(b)** The social equity pillar. Local residents watch tourists pass by. **(c)** The economy pillar: Tourism is a major economic activity. Some of the money generated by relatively wealthy tourists helps make life more comfortable for residents living in a rugged environment.

compatible with development but only if natural resources are utilized in a careful rather than wasteful manner.

THE SOCIETY PILLAR. Humans need shelter, food, and clothing to survive, so they make use of resources to meet these needs. Homes can be built of grass, wood, mud, stone, or brick. Food can be consumed by harvesting grains, fruit, and vegetables or by eating the flesh of fish, cattle, and pigs. Clothing can be made from harvesting cotton, removing skins from animals, or turning petroleum into polyester.

Consumer choices can support sustainability when people embrace it as a value. For example, a consumer might prefer clothing made of natural or recycled materials to clothing made directly from petroleum products. They can also choose products that benefit people living in a particular place (Figure 1-46b). Society's values are the basis for choosing which resources to use.

THE ECONOMY PILLAR. Natural resources acquire a monetary value through exchange in a marketplace (Figure 1-46c). In a market economy, supply and demand are the principal factors that determine affordability. The greater the supply, the lower the price; the greater the demand, the higher the price. Consumers will pay more for a commodity if they strongly desire it than if they have only a moderate desire. However, geographers observe that some goods do not reflect their actual environmental costs. For example, motorists sitting in a traffic jam do not have to pay a fee for the relatively high level of pollution their vehicles are emitting into the atmosphere.

The price of a resource depends on a society's technological ability to obtain it and to adapt it to that society's purposes. Earth has many substances that we do not use today because we lack the means to extract them or the knowledge of how to use them. Things that might become resources in the near future are potential resources.

SUSTAINABILITY'S CRITICS

Some environmentally oriented critics have argued that it is too late to discuss sustainability. The World Wildlife Fund (WWF), for example, claims that the world surpassed its sustainable level around 1980. The WWF Living Planet Report reaches its pessimistic conclusion by comparing the amount of land that humans are currently using with the amount of "biologically productive" land on Earth. "Biologically productive land" is defined as the amount of land required to produce the resources currently consumed and handle the wastes currently generated by the world's 7 billion people at current levels of technology.

The WWF calculates that humans are currently using about 13 billion hectares of Earth's land area, including 3 billion hectares for cropland, 2 billion for forest, 7 billion for energy, and 1 billion for fishing, grazing, and built-up areas. However, according to the WWF, Earth has only 11.4 billion hectares of biologically productive land, so humans are already using all of the productive land, and none is left for future growth.

Others criticize sustainability from the opposite perspective: Human activities have not exceeded Earth's capacity, they argue, because resource availability has no maximum, and Earth's resources have no absolute limit because the definition of resources changes drastically and unpredictably over time. Environmental improvements, they say, can be achieved through careful assessment of the outer limits of Earth's capacity.

Critics and defenders of sustainable development agree that one important recommendation of the U.N. report has not been implemented—increased international cooperation to reduce the gap between more developed countries and developing countries. Only if resources are distributed in a more equitable manner can we reduce the gap between poorer countries and richer ones.

PAUSE & REFLECT 1.4.1

What do you think might be the reaction of the men in Figure 1-46b to tourists such as those in Figure 1-46c?

Sustainability and Earth's Physical Systems

LEARNING OUTCOME 1.4.2

Describe Earth's three abiotic physical systems.

Geographers classify natural resources as part of four interrelated systems. These four physical systems are classified as either biotic or abiotic. A **biotic** system is composed of living organisms. An **abiotic** system is composed of nonliving or inorganic matter. Three of Earth's four systems are abiotic:

- The **atmosphere**: A thin layer of gases surrounding Earth.
- The **hydrosphere**: All of the water on and near Earth's surface.
- The **lithosphere**: Earth's crust and a portion of upper mantle directly below the crust.

One of the four systems is biotic:

- The **biosphere**: All living organisms on Earth, including plants and animals, as well as microorganisms.

The names of the four spheres are derived from the Greek words for "air" (*atmo*), "water" (*hydro*), "stone" (*litho*), and "life" (*bio*).

ATMOSPHERE. A thin layer of gases surrounds Earth at an altitude up to 480 kilometers (300 miles). Pure dry air in the lower atmosphere contains approximately 78 percent nitrogen, 21 percent oxygen, 0.9 percent argon, 0.036 percent carbon dioxide, and 0.064 percent other gases (measured by volume). As atmospheric gases are held to Earth by gravity, pressure is created. Variations in air pressure from one location to another are responsible for producing such weather features as wind blowing, storms brewing, and rain falling.

The long-term average weather condition at a particular location is **climate.** Geographers frequently classify climates according to a system developed by German climatologist Vladimir Köppen. The modified Köppen system divides the world into five main climate regions that are identified by the letters A through E, as well as by names:

- A: Humid low-latitude climates.
- B: Dry climates.
- C: Warm mid-latitude climates.
- D: Cold mid-latitude climates.
- E: Polar climates.

The modified Köppen system divides the five main climate regions into several subtypes (Figure 1-47). For all but the B climate, the basis for the subdivision is the amount of precipitation and the season in which it falls. For the B climate, subdivision is made on the basis of temperature and precipitation. In addition, "H" denotes highland areas in which climates are cold due to elevation, or local climate variations that are too great to show on the map.

Humans have a limited tolerance for extreme temperature and precipitation levels and thus avoid living in places that are too hot, too cold, too wet, or too dry. Compare the map of global climate to the distribution of population in Figure 2-4. Relatively few people live in the Dry (B) and Polar (E) climate regions, or in the Highlands.

HYDROSPHERE. Water exists in liquid form in oceans, lakes, and rivers, as well as groundwater in soil and rock. It can also exist as water vapor in the atmosphere and as ice in glaciers. Over 97 percent of the world's water is in the oceans. The oceans supply the atmosphere with water vapor, which returns to Earth's surface as precipitation, the most important source of fresh water. Consumption of water is essential for the survival of plants and animals, and a large quantity and variety of plants and animals live in it. Because water gains and loses heat relatively slowly, it also moderates seasonal temperature extremes over much of Earth's surface.

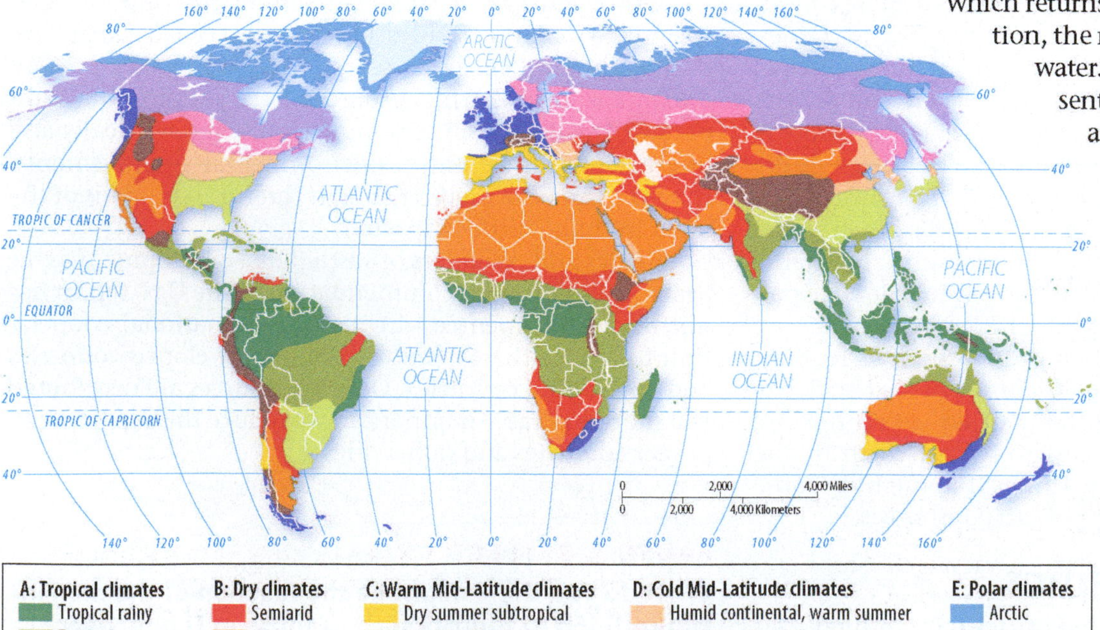

FIGURE 1-47 CLIMATE REGIONS Geographers frequently classify global climates according to a system developed by Vladimir Köppen. The modified Köppen system divides the world into five main climate regions, represented by the letters A, B, C, D, and E, plus H for highlands.

▲ **FIGURE 1-48 MONSOON IN INDIA** People wade through flood water in Dimapur, India, after heavy monsoon rain.

The climate of a particular location influences human activities, especially production of the food needed to survive. People in parts of the A climate region, especially southwestern India, Bangladesh, and the Myanmar (Burma) coast, anxiously await the annual monsoon rain, which is essential for successful agriculture and provides nearly 90 percent of India's water supply (Figure 1-48). For most of the year, the region receives dry, somewhat cool air from the northeast. In June, the wind direction suddenly shifts, bringing moist, warm, southwesterly air, known as the monsoon, from the Indian Ocean. The monsoon rain lasts until September. In years when the monsoon rain is delayed or fails to arrive—in recent decades, at least one-fourth of the time—agricultural output falls, and famine threatens in the countries of South Asia, where nearly 20 percent of the world's people live. The monsoon rain is so important in India that the words for "year," "rain," and "rainy season" are identical in many local languages.

LITHOSPHERE. Earth is composed of concentric spheres. The core is a dense, metallic sphere about 3,500 kilometers (2,200 miles) in radius. Surrounding the core is a mantle about 2,900 kilometers (1,800 miles) thick. The crust is a thin, brittle outer shell 8 to 40 kilometers (5 to 25 miles) thick. The lithosphere encompasses the crust, a portion of the mantle extending down to about 70 kilometers (45 miles). Powerful forces deep within Earth bend and break the crust to form mountain chains and shape the crust to form continents and ocean basins.

Earth's surface features, or landforms, vary from relatively flat to mountainous. Geographers find that the study of Earth's landforms—a science known as geomorphology—helps explain the distribution of people and the choice of economic activities at different locations. People prefer living on flatter land, which generally is better suited for agriculture. Great concentrations of people and activities in hilly areas may require extensive effort to modify the landscape.

Topographic maps, published for the United States by the U.S. Geological Survey (USGS), show details of physical features, such as bodies of water, forests, mountains, valleys, and wetlands. They also show cultural features, such as buildings, roads, parks, farms, and dams. "Topos" are used by engineers, hikers, hunters, people seeking home sites, and anyone who really wants to see the lay of the land (Figure 1-49). The brown lines on the map are contour lines that show the elevation of any location. Lines are further apart in flatter areas and closer together in hilly areas.

PAUSE & REFLECT 1.4.2

Why would maps of Earth's hydrosphere, lithosphere, and biosphere be important in the quest for sustainability?

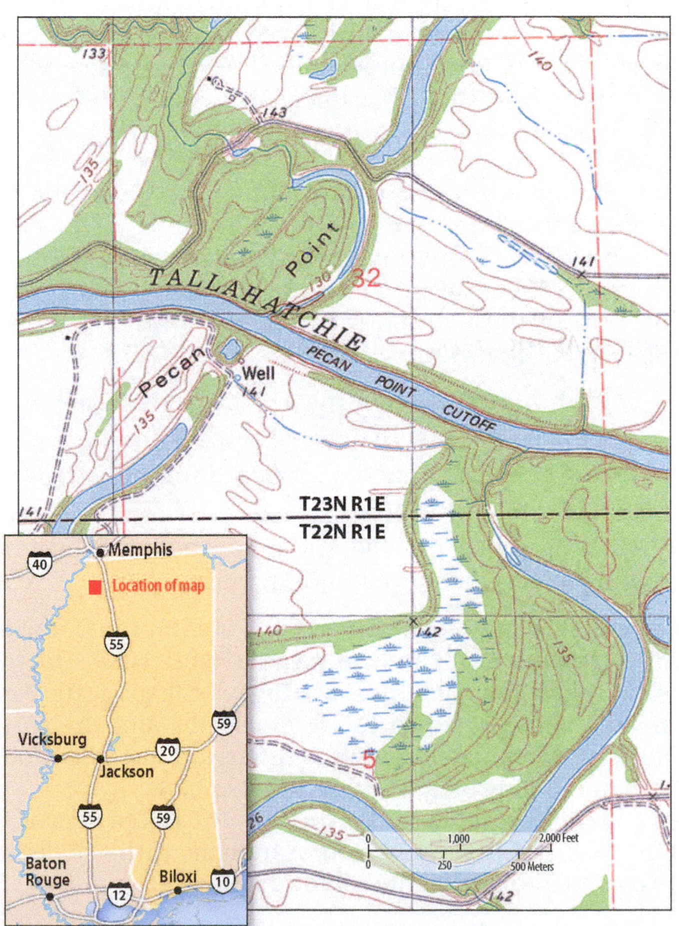

▲ **FIGURE 1-49 TOPOGRAPHIC MAP** A portion of a topographic map published by the U.S. Geological Survey shows physical features in northwestern Mississippi. The brown lines are contour lines that show the elevation of any location. The portion of the topo map shown here is part of sections 29 and 32 of township T23N R1E and section 5 of township T22N R1E.

Geography, Sustainability, and Ecology

LEARNING OUTCOME 1.4.3

Explain how the biosphere interacts with abiotic systems.

Modern technology has altered the historic relationship between people and the environment. People are now the most important agents of change on Earth, and they can modify the environment to a greater extent than in the past. Geographers are concerned that people sometimes use modern technology to modify the environment insensitively. Human actions can deplete scarce environmental resources, destroy irreplaceable resources, and use resources inefficiently.

ECOLOGY AND THE BIOSPHERE

The fourth natural resource system, the biosphere, encompasses all of Earth's living organisms. Because living organisms cannot exist except through interaction with the surrounding physical environment, the biosphere also includes portions of the three abiotic systems near Earth's surface. Living organisms in the biosphere interact with each of the three abiotic systems. For example, a piece of soil may comprise mineral material from the lithosphere, moisture from the hydrosphere, pockets of air from the atmosphere, and plant and insect matter from the biosphere.

Most of the living organisms interact within the top 3 meters (10 feet) of the lithosphere, the top 200 meters (650 feet) of the hydrosphere, and the lowest 30 meters (100 feet) of the atmosphere:

- The lithosphere is where most plants and animals live and where they obtain food and shelter.
- The hydrosphere provides water to drink and physical support for aquatic life.
- The atmosphere provides the air for animals to breathe and protects them from the Sun's rays.

A group of living organisms and the abiotic spheres with which they interact is an **ecosystem** (Figure 1-50). The scientific study of ecosystems is **ecology.** Ecologists study interrelationships between living organisms and the three abiotic environments, as well as interrelationships among the various living organisms in the biosphere.

Human geographers are especially interested in ecosystems involving the interaction of humans with the rest of the biosphere and the three abiotic spheres:

- If the atmosphere contains pollutants, or its oxygen level is reduced, humans have trouble breathing.
- Without water in the hydrosphere, humans waste away and die.
- A stable lithosphere provides humans with materials for buildings and fuel for energy.
- The rest of the biosphere provides humans with food.

Human actions are sustainable if they preserve and conserve elements of the four spheres and unsustainable if they cause destruction. For example, human actions contribute to the destruction of soil, the material that forms on Earth's surface at the thin interface between the air and the rocks. Two sustainability issues arise from the destruction of soil:

- **Erosion.** Erosion occurs when soil washes away in the rain or blows away in the wind. Farmers contribute to erosion by making inappropriate choices. To reduce erosion, farmers can avoid steep slopes, plow less, and plant crops whose roots help bind the soil.
- **Depletion of nutrients.** Soil contains the nutrients necessary for successful growth of plants, including those that are useful to humans. Nutrients are depleted when plants withdraw more nutrients than natural processes can replace. Each type of plant withdraws certain nutrients from the soil and restores others. To minimize depletion, farmers can plant different crops from one year to the next so that the land remains productive over the long term.

CULTURAL ECOLOGY: INTEGRATING CULTURE AND ECOLOGY

Human geographers are especially interested in the fact that different cultural groups modify the four spheres in distinctive ways. The geographic study of human–environment relationships is known as **cultural ecology.** The roots of cultural ecology reach back more than 200 years, to an era when early scientists traveled the globe, observing how people lived in different environments.

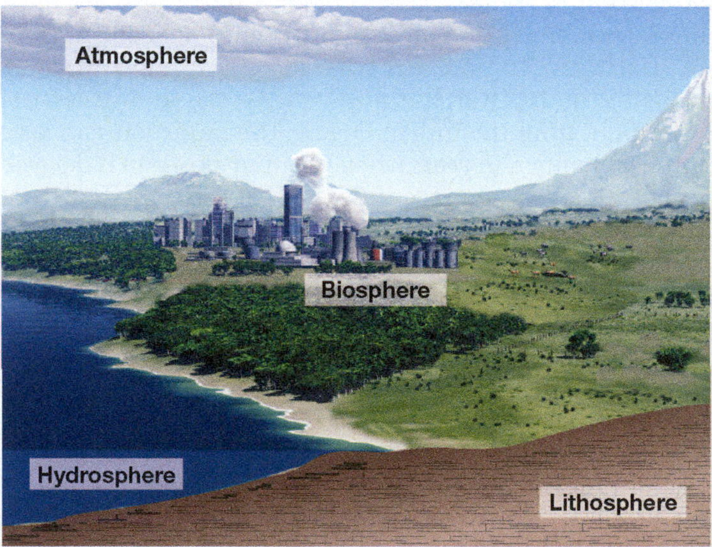

▲ **FIGURE 1-50 ECOSYSTEMS** Geographers are especially interested in the ecosystem of a city because approximately half of Earth's humans live in urban areas. The lithosphere provides the ground and the materials to erect homes and businesses. The hydrosphere provides the water for urban dwellers to consume. The atmosphere is where urban dwellers emit pollutants. Some plants and other animals of the biosphere thrive along with humans in the cities, whereas others struggle.

ENVIRONMENTAL DETERMINISM. Pioneering nineteenth-century German geographers Alexander von Humboldt (1769–1859) and Carl Ritter (1779–1859) believed that the physical environment caused social development, an approach called **environmental determinism.** According to Humboldt and Ritter, human geographers should apply laws from the natural sciences to understanding relationships between the physical environment and human actions. They argued that the scientific study of social and natural processes is fundamentally the same. Natural scientists have made more progress in formulating general laws than have social scientists, so an important goal of human geographers is to discover general laws. Humboldt and Ritter urged human geographers to adopt the methods of scientific inquiry used by natural scientists.

Other influential geographers adopted environmental determinism in the late nineteenth and early twentieth centuries. Friedrich Ratzel (1844–1904) and his American student Ellen Churchill Semple (1863–1932) claimed that geography was the study of the influences of the natural environment on people.

Another early American geographer, Ellsworth Huntington (1876–1947), argued that climate was a major determinant of civilization. For instance, according to Huntington, the temperate climate of maritime northwestern Europe produced greater human efficiency, as measured by better health conditions, lower death rates, and higher standards of living.

POSSIBILISM. To explain relationships between human activities and the physical environment, modern geographers reject environmental determinism in favor of possibilism. According to **possibilism,** the physical environment may limit some human actions, but people have the ability to adjust to their environment. People can choose a course of action from many alternatives in the physical environment.

(a) (b)

▲ **FIGURE 1-51 POSSIBILISM: ALTERNATIVE BEHAVIORS** **(a)** Some humans prefer to mow their lawn. **(b)** Others prefer to let wildflowers grow.

For example, the climate of any location influences human activities, especially food production. From one generation to the next, people learn that different crops thrive in different climates—rice requires plentiful water, whereas wheat survives on limited moisture and actually grows poorly in very wet environments. On the other hand, wheat is more likely than rice to be grown successfully in colder climates. Thus, under possibilism, people can choose the crops they grow and yet be compatible with their environment.

Some human impacts on the environment are based on deep-seated cultural values. Why do we plant our front yard with grass, water it to make it grow, mow it to keep it from growing tall, and impose fines on those who fail to mow often enough? Why not let dandelions or wildflowers grow instead (Figure 1-51)? Why does one group of people consume the fruit from deciduous trees and chop down the conifers for building materials, whereas another group chops down the deciduous trees for furniture while preserving the conifers as religious symbols? Are some of these actions more sustainable than others?

A people's level of wealth can also influence its attitude toward modifying the environment. A farmer who possesses a tractor may regard a hilly piece of land as an obstacle to avoid, but a poor farmer with a hoe may regard hilly land as the only opportunity to produce food for survival through hand cultivation.

POSSIBILISM AND SUSTAINABILITY. Human geographers use the cultural ecology, or human–environment, approach to understand whether particular patterns and processes are sustainable. For example, world population growth is a problem if the number of people exceeds the capacity of the physical environment to produce food. However, people can adjust to the capacity of the physical environment by controlling their numbers, adopting new technology, consuming different foods, migrating to new locations, and taking other actions.

The physical environment is not always the most significant factor in human decisions. People can fashion a landscape by superimposing new forms on the physical environment. For example, the critical factor in selecting a site for a cotton textile factory is not proximity to a place where cotton is grown. A more important factor in selecting a suitable location is access to a supply of low-cost labor. Economic systems, political structures, living arrangements, religious practices, and human activities can produce distinctive landscapes that do not stem primarily from distinctive physical features. The geographer's job is to sort out the associations among various social characteristics, each of which is uniquely distributed across Earth's surface.

PAUSE & REFLECT 1.4.3

How might a bird interact with each of the four spheres?

Sustainable Environmental Change

LEARNING OUTCOME 1.4.4

Compare ecosystems in the Netherlands and California.

The Netherlands and California are ecosystems that are heavily dependent on human modification of the hydrosphere. The two regions face opposite water issues: the Netherlands has more water than it wants and California not enough. At this time, the ecological modifications in the Netherlands appear sustainable, whereas those in California appear unsustainable.

SUSTAINABLE ECOSYSTEM: THE NETHERLANDS

The Dutch have modified their environment with two distinctive types of construction projects—polders and dikes. A **polder** is a piece of land that is created by draining water from an area. The Netherlands has 6,500 square kilometers (2,600 square miles) of polders, comprising 16 percent of the country's land area (Figure 1-52).

Polders, first created in the thirteenth century, were constructed primarily by private developers in the sixteenth and seventeenth centuries and by the government during the past 200 years (Figure 1-53). The Dutch government has reserved most of the polders for agriculture to reduce the country's dependence on imported food. Some of the polders are used for housing, and one contains Schiphol, one of Europe's busiest airports.

The second distinctive modification of the landscape in the Netherlands is the construction of massive dikes to prevent the North Sea, an arm of the Atlantic Ocean, from flooding much of the country. The Dutch have built dikes in two major locations—the Zuider Zee project in the north and the Delta Plan project in the southwest.

The Zuider Zee, an arm of the North Sea, once threatened the heart of the Netherlands with flooding. A dike completed in 1932 caused the Zuider Zee to be converted from a salt-water sea to a fresh-water lake called Lake IJssel. Some of the lake has been drained to create several polders.

A second ambitious project in the Netherlands is the Delta Plan. Several rivers that flow through the Netherlands to the North Sea split into many branches and form a low-lying delta that is vulnerable to flooding. After a devastating flood in January 1953 killed nearly 2,000 people, the Delta Plan called for the construction of several dams to close off most of the waterways.

▶ **FIGURE 1-52 DUTCH POLDER** Kalverpolder in northern Holland has been created by pumping the water from the site into the canal, originally by means of the windmills.

▲ **FIGURE 1-53 SUSTAINABLE ECOSYSTEM: THE NETHERLANDS** The Dutch have considerably altered the site of the Netherlands through creation of polders and dikes.

Once these two massive projects were finished, attitudes toward modifying the environment changed in the Netherlands. The Dutch scrapped plans to build additional polders in the IJsselmeer in order to preserve the lake's value for recreation. The Dutch are deliberately breaking some of the dikes to flood fields. A plan adopted in 1990 called for returning 263,000 hectares (650,000 acres) of farms to wetlands or forests. Widespread use of insecticides and fertilizers on Dutch farms has contributed to contaminated drinking water, acid rain, and other environmental problems.

Global warming could threaten the Netherlands by raising the level of the sea around the country by between 20 and 58 centimeters (8 and 23 inches) within the next 100 years. Rather than build new dikes and polders, the

Dutch have become world leaders in reducing the causes of global warming by acting to reduce industrial pollution and increase solar and wind power use, among other actions.

UNSUSTAINABLE ECOSYSTEM: CALIFORNIA

California and neighboring states in the U.S. Southwest have grown rapidly and prospered despite limited supplies of water. An extended drought in recent years has called into question the region's ability to sustain its residents' current lifestyles.

In normal times, California gets around 70 percent of its water from surface water sources, such as melting snow from mountains. Aqueducts and pipes carry water from the Colorado River to cities and farmland hundreds of kilometers away. A 1922 agreement determines how much of the Colorado River is allocated to California, Arizona, and Nevada. The other 30 percent of California's water comes from groundwater sources, which are underground. In recent years, the level of precipitation has been extremely low, so less water has been available from surface sources. As a result, groundwater consumption has been supplying more than 60 percent of California's water demand. Groundwater is being removed more rapidly than it is being replenished.

California's residents and businesses have been required to reduce water usage by 25 percent. Each of the state's 400 local water supply agencies has determined how to meet the reduction in its service area. Half of the residential demand for water is for lawns and gardens. Homeowners and municipalities are replacing grass lawns and annual flowers with native landscapes with rocks and desert plants.

However, residents and businesses use only 20 percent of California's water. The other 80 percent goes to agriculture. The biggest challenge posed to the sustainability of California's ecosystems by the drought is for agriculture. Much of the land used for agriculture in California does not get enough rainfall even in normal times to grow crops. The counties with the highest per capita use of water are the major agricultural counties (Figure 1-54); the fields are irrigated with water brought in from elsewhere (Figure 1-55).

PAUSE & REFLECT 1.4.4

Which is better positioned to face future threats to sustainability, the Netherlands or California? Why?

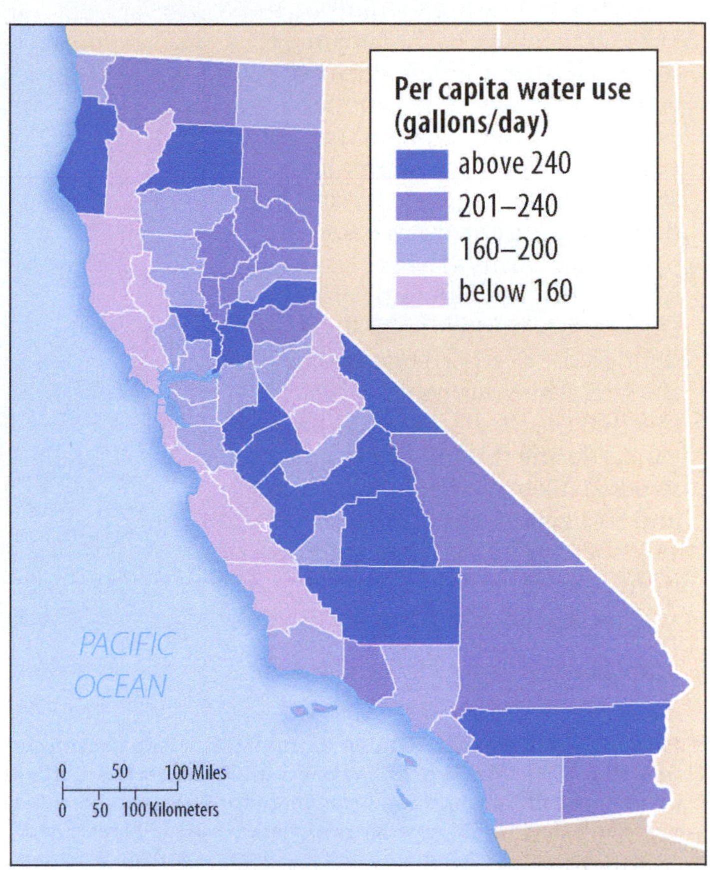

▲ **FIGURE 1-54 CALIFORNIA PER CAPITA WATER USE BY COUNTY** The lowest per capita water use is in the major cities along the Pacific Coast. The highest demand is in the agricultural counties of the Central Valley.

◀ **FIGURE 1-55 UNSUSTAINBLE ECOSYSTEM: CALIFORNIA** Onion field in the Central Valley.

SUSTAINABILITY & OUR ENVIRONMENT The Everglades

The Everglades was once a very wide and shallow fresh-water river 80 kilometers (50 miles) wide and 15 centimeters (6 inches) deep, slowly flowing south from Lake Okeechobee to the Gulf of Mexico (Figure 1-56). A sensitive ecosystem of plants and animals once thrived in this distinctive landscape, and a portion became a national park, but much of it has been destroyed by human actions.

The U.S. Army Corps of Engineers built a levee around Lake Okeechobee during the 1930s, drained the northern one-third of the Everglades during the 1940s, diverted the Kissimmee River into canals during the 1950s, and constructed dikes and levees near Miami and Fort Lauderdale during the 1960s (Figure 1-57). These modifications opened up hundreds of

▲ **FIGURE 1-56 SOUTH FLORIDA ECOSYSTEM**

thousands of hectares of land for growing sugarcane and protected farmland as well as the land occupied by the growing South Florida population from flooding.

But the modification also had unintended consequences for South Florida's ecosystem. Polluted water, mainly from cattle grazing along the banks of the canals, flowed into Lake Okeechobee, which is the source of fresh water for half of Florida's population. Fish in the lake began to die from the high levels of mercury, phosphorous, and other contaminants. The polluted water then continued to flow south into the national park, threatening native vegetation such as sawgrass and endangering rare birds and other animals. Recent plans are attempting to restore a healthy ecosystem to inland South Florida.

▼ **FIGURE 1-57 KISSIMMEE RIVER** To control flooding in central Florida, the U.S. Army Corps of Engineers straightened the course of the Kissimmee River, which had meandered for 160 kilometers. The water was rechanneled into a canal 90 meters wide (300 feet) and 9 meters deep (30 feet), running in a straight line for 84 kilometers (52 miles).

California accounts for 12 percent of all U.S. agriculture, 21 percent of all milk supplies, and 99 percent of such crops as almonds, artichokes, grapes, olives, peaches, rice, and walnuts. The system of industrial agriculture centered on California provides an abundant supply of low-priced food to the United States and other countries. As discussed in Chapter 10, the U.S. industrial agriculture system depends on large quantities of chemicals and energy, as well as water; these practices may not be sustainable in the future.

CHECK-IN KEY ISSUE 4

Why Are Some Actions Not Sustainable?

- ✔ **Sustainability combines environment, economy, and society.**
- ✔ **Earth's resources encompass three abiotic systems and one biotic system.**
- ✔ **Ecology is the study of living organisms and the abiotic spheres with which they interact.**
- ✔ **Ecosystems that may or may not be sustainable.**

Summary & Review

KEY ISSUE 1

Why is geography a science?

Geography is most fundamentally a spatial science. Geographers use maps to display the location of objects and to extract information about places. Early geographers drew maps of Earth's surface based on exploration and observation. Contemporary GIScience, including remote sensing, GPS, VGI, geotagging, and GIS, assist geographers in understanding reasons for observed regularities across Earth.

THINKING GEOGRAPHICALLY

▲ **FIGURE 1-58 CZECH TELEPHONE**

1. Using geographic tools such as maps and GIS is not simply a mechanical exercise. Nor are decisions confined to scale, projection, and layers. For example, should the European country be labeled Czech Republic or Czechia? Czech authorities and citizens do not agree on the proper translation of the country's Czech name Česky into English. 2. What criteria should geographers use to label maps?

KEY ISSUE 2

Why is each point on earth unique?

Geographers identify unique places and regions distinguished by distinctive combinations of cultural as well as economic and environmental features. Location is the position something occupies on Earth. A region is an area characterized by a unique combination of features. The distribution of features helps explain why every place and every region is unique.

THINKING GEOGRAPHICALLY

3. What are elements of the site and situation of your hometown? 4. Can you name another place to which your hometown has strong connections? 5. What is an example of a feature that connects your town to another?

◄ **FIGURE 1-59 SITE AND SITUATION OF BOSTON** The site is Boston Harbor and several rivers. Logan Airport is an example of the connections found in Boston to other places.

KEY ISSUE 3

Why are different places similar?

Geographers work at all scales, from local to global. The global scale is increasingly important because few places in the contemporary world are totally isolated. Because places are connected to each other, they display similarities. Geographers study the interactions of groups of people and human activities across space, and they identify processes by which people and ideas diffuse from one location to another over time.

THINKING GEOGRAPHICALLY

▲ **FIGURE 1-60 HARRY POTTER'S TRANSPORTATION**

6. If you could live anyplace on Earth, where would it be? Why? 7. How might your choice be altered if you had access to a transportation device (such as available to Harry Potter) that enabled you to travel instantaneously to any place on Earth?

KEY ISSUE 4

Why are some actions not sustainable?

Sustainability is the use of Earth's resources in ways that ensure their availability in the future. Sustainability is based on three interrelated pillars: environmental, economic, and social action. An ecosystem comprises a group of living organisms in the biosphere and their interaction with the atmosphere, lithosphere, and biosphere. The sustainability of some of Earth's resources is being damaged by human actions.

THINKING GEOGRAPHICALLY

▲ **FIGURE 1-61 SUSTAINABLE TRANSPORTATION: BIKING AT THE UNIVERSITY**

8. What activities in your community appear to promote sustainability? Which ones do not?

KEY TERMS

Abiotic *(p. 34)* Composed of nonliving or inorganic matter.

Acculturation *(p. 28)* The process of changes in culture that result from the meeting of two groups, each of which retains distinct culture features.

Assimilation *(p. 28)* The process by which a group's cultural features are altered to resemble those of another more dominant group.

Atmosphere *(p. 34)* The thin layer of gases surrounding Earth.

Behavioral geography *(p. 26)* The study of the psychological basis for individual human actions in space.

Biosphere *(p. 34)* All living organisms on Earth, including plants and animals, as well as microorganisms.

Biotic *(p. 34)* Composed of living organisms.

Cartography *(p. 6)* The science of making maps.

Citizen Science *(p. 9)* Scientific research by amateur scientists.

Climate *(p. 34)* The long-term average weather condition at aparticular location.

Concentration *(p. 22)* The spread of something over a given area.

Connection *(p. 5)* The relationships among people and objects across the barrier of space.

Conservation *(p. 32)* The sustainable management of a natural resource.

Contagious diffusion *(p. 29)* The rapid, widespread diffusion of a feature or trend throughout a population.

Cultural ecology *(p. 36)* A geographic approach that emphasizes human–environment relationships.

Cultural landscape *(p. 16)* An approach to geography that emphasizes the relationships among social and physical phenomena in a particular study area.

Culture *(p. 18)* The body of customary beliefs, social forms, and material traits that together constitute a group's distinct tradition.

Density *(p. 22)* The frequency with which something exists within a given unit of area.

Diffusion *(p. 28)* The process of spread of a feature or trend from one place to another over time.

Distance decay *(p. 30)* The diminished importance and eventual disappearance of a phenomenon with increasing distance from its origin.

Distribution *(p. 22)* The arrangement of something across Earth's surface.

Ecology *(p. 36)* The scientific study of ecosystems.

Ecosystem *(p. 36)* A group of living organisms and the abiotic spheres with which they interact.

Environmental determinism *(p. 37)* A nineteenth- and early twentieth-century approach to the study of geography which argued that the general laws sought by human geographers could be found in the physical sciences. Geography was therefore the study of how the physical environment caused human activities.

Expansion diffusion *(p. 29)* The spread of a feature or trend among people from one area to another in an additive process.

Formal region (or uniform region) *(p. 16)* An area in which everyone shares in common one or more distinctive characteristics.

Functional region (or nodal region) *(p. 16)* An area organized around a node or focal point.

Geographic information science (GIScience) *(p. 9)* The development and analysis of data about Earth acquired through satellite and other electronic information technologies.

Geographic information system (GIS) *(p. 9)* A computer system that stores, organizes, analyzes, and displays geographic data.

Geotagging *(p. 9)* Identification and storage of a piece of information by its precise latitude and longitude coordinates.

Global Positioning System (GPS) *(p. 8)* A system that determines the precise position of something on Earth through a series of satellites, tracking stations, and receivers.

Globalization *(p. 20)* Actions or processes that involve the entire world and result in making something worldwide in scope.

Greenwich Mean Time (GMT) *(p. 12)* The time in the zone encompassing the prime meridian, or 0° longitude.

Hearth *(p. 28)* The region from which innovative ideas originate.

Hierarchical diffusion *(p. 29)* The spread of a feature or trend from one key person or node of authority or power to other persons or places.

Humanistic geography *(p. 26)* The study of different ways that individuals form ideas about place and give those places symbolic meanings.

Hydrosphere *(p. 34)* All of the water on and near Earth's surface.

International Date Line *(p. 13)* An arc that for the most part follows 180° longitude, although it deviates in several places to avoid dividing land areas. When the International Date Line is crossed heading east (toward America), the clock moves back 24 hours, or one entire day. When it is crossed heading west (toward Asia), the calendar moves ahead one day.

Latitude *(p. 12)* The numbering system used to indicate the location of parallels drawn on a globe and measuring distance north and south of the equator (0°).

Lithosphere *(p. 34)* Earth's crust and a portion of upper mantle directly below the crust.

Location *(p. 14)* The position of anything on Earth's surface.

Longitude *(p. 12)* The numbering system used to

Log in to the **MasteringGeography** Study Area to view this video.

How GPS Works

The spaced-based Global Positioning Systems (GPS) can almost instantaneously determine geographic location and time information.

1. Describe the infrastructure that supports GPS devices.
2. According to the video, how accurate is the current GPS system? What are the limits of GPS technology?
3. Based on what you learned in the video, why do you think the U.S. Navy has decided to reinstate celestial navigation? Explain.

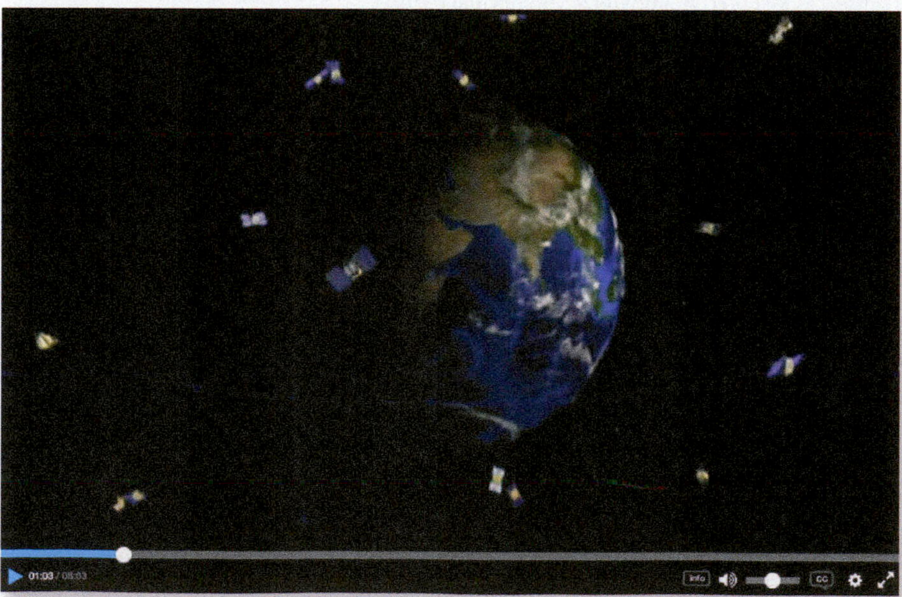

indicate the location of meridians drawn on a globe and measuring distance east and west of the prime meridian (0°).

Map *(p. 6)* A two-dimensional, or flat, representation of Earth's surface or a portion of it.

Map scale *(p. 10)* The relationship between the size of an object on a map and the size of the actual feature on Earth's surface.

Mashup *(p. 9)* A map that overlays data from one source on top of a map provided by a mapping service.

Mental map *(p. 8)* A representation of a portion of Earth's surface based on what an individual knows about a place that contains personal impressions of what is in the place and where the place is located.

Meridian *(p. 12)* An arc drawn on a map between the North and South poles.

Network *(p. 30)* A chain of communication that connects places.

Nonrenewable resource *(p. 32)* Something produced in nature more slowly than it is consumed by humans.

Parallel *(p. 12)* A circle drawn around the globe parallel to the equator and at right angles to the meridians.

Participatory GIS (PGIS) *(p. 9)* Community-based mapping, representing local knowledge and information.

Pattern *(p. 23)* The geometric or regular arrangement of something in a particular area.

Place *(p. 4)* A specific point on Earth, distinguished by a particular characteristic.

Polder *(p. 38)* Land that the Dutch have created by draining water from an area.

Possibilism *(p. 37)* The theory that the physical environment may set limits on human actions, but people have the ability to adjust to the physical environment and choose a course of action from many alternatives.

Poststructuralist geography *(p. 26)* The study of space as the product of ideologies or value systems of ruling elites.

Preservation *(p. 32)* The maintenance of resources in their present condition, with as little human impact as possible.

Prime meridian *(p. 12)* The meridian, designated as 0° longitude, that passes through the Royal Observatory at Greenwich, England.

Projection *(p. 11)* A system used to transfer locations from Earth's surface to a flat map.

Region *(p. 4)* An area distinguished by a unique combination of trends or features.

Relocation diffusion *(p. 28)* The spread of a feature or trend through bodily movement of people from one place to another.

Remote sensing *(p. 9)* The acquisition of data about Earth's surface from a satellite orbiting the planet or from other long-distance methods.

Renewable resource *(p. 32)* Something produced in nature more rapidly than it is consumed by humans.

Resource *(p. 32)* A substance in the environment that is useful to people, is economically and technologically feasible to access, and is socially acceptable to use.

Scale *(p. 4)* Generally, the relationship between the portion of Earth being studied and Earth as a whole.

Site *(p. 15)* The physical character of a place.

Situation *(p. 15)* The location of a place relative to another place.

Space *(p. 5)* The physical gap or interval between two objects.

Space–time compression *(p. 30)* The reduction in the time it takes to diffuse something to a distant place as a result of improved communications and transportation systems.

Spatial association *(p. 18)* The relationship between the distribution of one feature and the distribution of another feature.

Stimulus diffusion *(p. 29)* The spread of an underlying principle even though a specific characteristic is rejected.

Sustainability *(p. 32)* The use of Earth's renewable and nonrenewable natural resources in ways that do not constrain resource use in the future.

Syncretism *(p. 28)* The combining of elements of two groups into a new cultural feature.

Toponym *(p. 14)* The name given to a portion of Earth's surface.

Transnational corporation *(p. 20)* A company that conducts research, operates factories, and sells products in many countries, not just where its headquarters or shareholders are located.

Uneven development *(p. 27)* The increasing gap in economic conditions between core and peripheral regions as a result of the globalization of the economy.

Vernacular region (or perceptual region) *(p. 17)* An area that people believe exists as part of their cultural identity.

Volunteered geographic information (VGI) *(p. 9)* Creation and dissemination of geographic data contributed voluntarily and for free by individuals.

EXPLORE

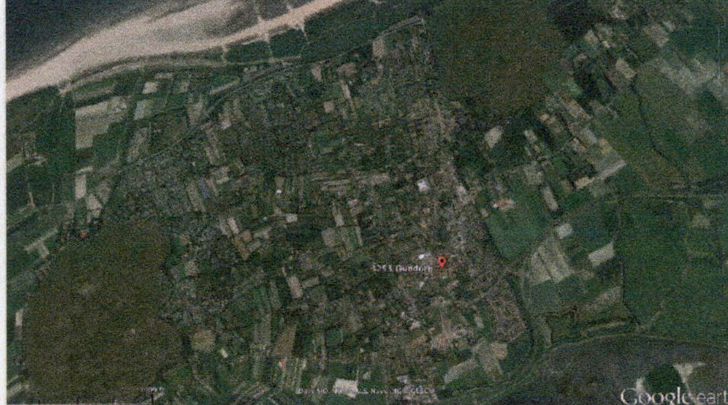

The Netherlands Coast

Use Google Earth to explore the Netherlands coast.
Fly to *Ouddorp, Netherlands.*

1. Using the ruler, what is the distance (in kilometers or miles) from the pin marking Ouddorp to the coast?
2. How many dams do you see in the image?
3. Drag to enter street view on the dam to the north of the image. What occupies the top of the dam?
4. Which side of the dam is the open sea, and which side is the protected inland waterway? Which side of the dam contains the marinas filled with boats? Why would the boats be on that side?
5. What is the elevation of Ouddorp? What is the elevation of Visschershoek to the west of Ouddorp right on the coast? What is the elevation of Melissant to the east of Ouddorp further inland? Based on these observations, is the elevation further inland higher or lower than on the coast? Why is that the case?

2

Population and Health

More people are alive at this time than at any other point in Earth's history, and most of the growth is concentrated in poor countries. Can Earth sustain more than 7 billion people now, let alone the added billions in the future? Geographers have unique perspectives on the ability of people to live on Earth. Indonesia, shown in this image, is the world's fourth most populous country, so it's future population growth will have a major impact on the future population of the world as a whole.

Students walk home from school near Jakarta, Indonesia.

LOCATIONS IN THIS CHAPTER

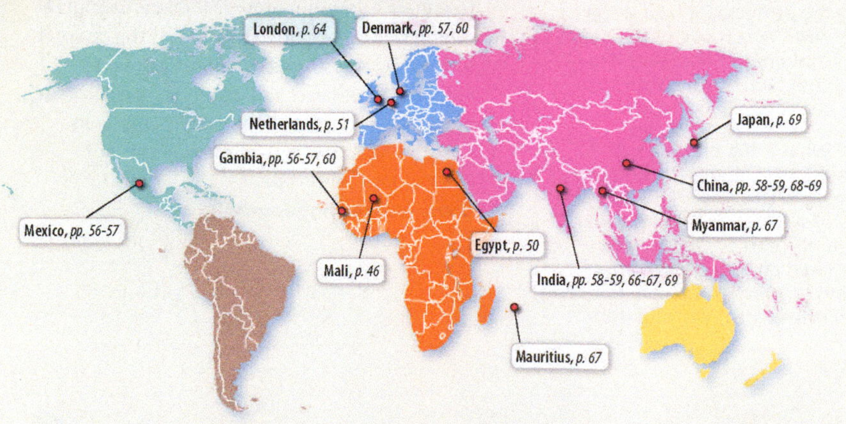

KEY ISSUES

1 Where Are the World's People Distributed?

Humans are not distributed uniformly across Earth. Rather, they are highly clustered in particular *places*, whereas other places are sparsely inhabited.

2 Why Is World Population Increasing?

Population is growing at different rates in different *regions*. Every place is at some stage in a process known as the demographic transition.

3 Why Do Some Places Face Health Challenges?

Patterns of health and provision of medical care vary across *space*. Health and medical care have improved in the world as a whole, but at the local *scale*, some people are at increased risk.

4 Why Might Population Increase in the Future?

Geographers see strong *connections* between the size and growth of population and the adequacy of resources. Geographers find that overpopulation may be a threat to resources in some regions of the world but not in others.

KEY ISSUE 1

Where Are the World's People Distributed?

- **Introducing Population and Health**
- **Population Concentrations**
- **Population Density**

LEARNING OUTCOME 2.1.1

Understand the distribution of the world's peoples.

As explained in Chapter 1, geographers ask "where" and "why" questions. As we begin our study of the major topics in human geography, note the wording of the four key issues that organize the material in this chapter. The first issue asks a "where" question, and the other three ask "why" questions. These four issues rely on the five basic concepts presented in Chapter 1.

▲ **FIGURE 2-1 OVERPOPULATION IN MALI** The Sahel region of Africa, including much of the country of Mali, is threatened by overpopulation. The number of people living here is not very high, but the capacity of the environment to support life is extremely low. These people are chopping down one of the few remaining trees in this region. 1. **For what purpose might these people be chopping down the tree?** 2. **Why might people in Mali not use the same tools as in North America to chop down trees?** 3. **In what ways might chopping down the tree reduce this area's capacity to support life?**

Introducing Population and Health

Geographers study population problems by first describing where people are found across Earth. The location of Earth's 7 billion people forms a regular distribution. The chapter then turns to explaining why population is growing at different rates in different places. With the rate of world population growth slowing in the twenty-first century, geographers are increasingly concerned with differences in the health of people in different places and with the medical care available to them.

The study of population geography is especially important for three reasons:

- More people are alive at this time than at any other point in Earth's long history.
- Virtually all global population growth is concentrated in developing countries.
- The world's population increased at a faster rate during the second half of the twentieth century than ever before in history; the rate has slowed in the twenty-first century but is still high by historical standards.

Geographers are interested in the relationship between population and Earth's resources. **Overpopulation** occurs when the number of people exceeds the capacity of the environment to support life at a decent standard of living. From the perspective of globalization, some demographers argue that the world is already overburdened with too many people, or it will be in the near future. At the scale of local diversity, geographers find that overpopulation is a threat in some regions of the world but not in others (Figure 2-1). The capacity of Earth as a whole to support human life may be high, but some regions have a favorable balance between people and available resources, whereas others do not. Further, the regions with the most people are not necessarily the same as the regions with an unfavorable balance between population and resources.

The single most important data source for population geography is the **census**. In the United States, a census of population and a census of housing take place once a decade, in years ending in zero, including 2010. Despite its importance, the census is controversial in many countries, for two reasons:

- **Nonparticipation.** Homeless people, ethnic minorities, and citizens of other countries who do not have proper immigration documents may be less likely to complete census forms. These individuals may fear that the Census Bureau could turn over the forms to another government agency, such as the FBI or the Department of Homeland Security in the United States.
- **Sampling.** Statistical sampling techniques can be utilized to get a more accurate count, as well as to identify detailed characteristics of people, housing, and businesses. The U.S. Supreme Court has ruled that sampling may not be used to redraw Congressional district boundaries. Politicians sympathetic to the needs of the homeless and immigrants have been especially vocal in support of sampling, whereas those from small towns and rural areas, where the census count is more accurate, are more inclined to oppose it.

DISTRIBUTION OF THE WORLD'S PEOPLES

Human beings are not distributed uniformly across Earth. We can understand how population is distributed by examining two basic properties: concentration and density. Geographers identify regions of Earth's surface where population is clustered and regions where it is sparse. Several density measures help geographers explain the relationship between the number of people and available resources.

The concentration of the world's population can be displayed by dividing Earth's land area into seven portions, each containing around 1 billion people (Figure 2-2). The combined population of North America, Latin America, the South Pacific, and Greenland (an area of 53 million square kilometers [20 million square miles]) is about the same as the population of eastern China (5 million square kilometers) or of southern India (3 million square kilometers).

Concentration can also be displayed on a cartogram, which depicts the sizes of countries according to population rather than land area, as is the case with most maps (Figure 2-3). When compared to a more typical equal-area map, such as in Figure 2-2, the population cartogram displays major population clusters as much larger. As you look at maps of population growth and other topics in this and subsequent chapters, pay special attention to Asia and Europe because global patterns are heavily influenced by conditions in these regions, where two-thirds of the world's people live.

PAUSE & REFLECT 2.1.1

Compare Figures 2-2 and 2-3. Which depicts the shape of countries more accurately? Why?

FIGURE 2-2 WORLD POPULATION PORTIONS Each of the seven portions indicated by color in this figure contains approximately 1 billion inhabitants.

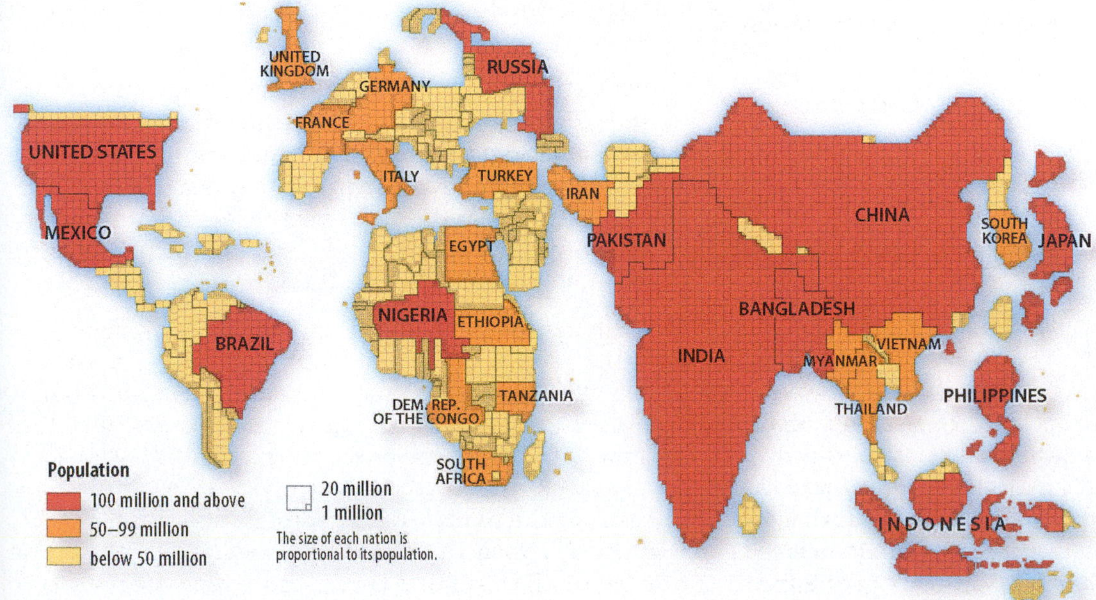

FIGURE 2-3 WORLD POPULATION CARTOGRAM In a cartogram, countries are displayed by size of population rather than land area. Canada is the world's second-largest country in land area, so it appears quite large on most maps. But it ranks only 37th in population so appears small on this map. Conversely, Bangladesh ranks only 92nd in land area but is 8th in population so appears large on the cartogram. Countries with populations over 50 million are labeled.

Population Concentrations

LEARNING OUTCOME 2.1.2

Understand why some regions have clustered populations and other regions are sparsely inhabited.

Two-thirds of the world's inhabitants are clustered in four regions (Figure 2-4). The four population clusters occupy generally low-lying areas, with fertile soil and temperate climate. Most live near the ocean or near a river with easy access to an ocean rather than in the interior of major landmasses.

FOUR CLUSTERS

The four major population clusters—East Asia, South Asia, Europe, and Southeast Asia—display differences in the pattern of occupancy of the land.

EAST ASIA. Nearly one-fourth of the world's people live in East Asia. The region, bordering the Pacific Ocean, includes eastern China, the islands of Japan, the Korean peninsula, and the island of Taiwan. The People's Republic of China is the world's most populous country and the fourth-largest country in land area. The Chinese population is clustered near the Pacific Coast and in several fertile river valleys that extend inland, though much of China's interior is sparsely inhabited mountains and deserts. Nearly one-half of the people live in rural areas where they work as farmers. In sharp contrast to China, 93 percent of Japanese and 80 percent of South Koreans are clustered in urban areas and work at industrial or service jobs.

SOUTH ASIA. Nearly one-fourth of the world's people live in South Asia, which includes India, Pakistan, Bangladesh, and the island of Sri Lanka. The largest concentration of people within South Asia lives along a 1,500-kilometer (900-mile) corridor from Lahore, Pakistan, through India and Bangladesh to the Bay of Bengal. Much of this area's population is concentrated along the plains of the Indus and Ganges rivers. Population is also heavily concentrated near India's two long coastlines—the Arabian Sea to the west and the Bay of Bengal to the east. Most people in South Asia are farmers living in rural areas.

EUROPE. Europe includes four dozen countries, ranging from Monaco, with 1 square kilometer (0.7 square miles) and a population of 33,000, to Russia, the world's largest country in land area when its Asian part is included. In contrast to the three Asian concentrations, three-fourths of Europe's inhabitants live in cities, and fewer than 5 percent are farmers. The highest population concentrations in Europe are near the major rivers and coalfields of Germany and Belgium, as well as historic capital cities such as London and Paris.

SOUTHEAST ASIA. Around 600 million people live in Southeast Asia, mostly on a series of islands that lie between the Indian and Pacific oceans. Indonesia, which consists of 13,677 islands, is the world's fourth-most-populous country. The largest population concentration is on the island of Java, which is inhabited by more than 100 million people. Several islands that belong to the Philippines contain high population concentrations, and population is also clustered along several river valleys and deltas at the southeastern tip

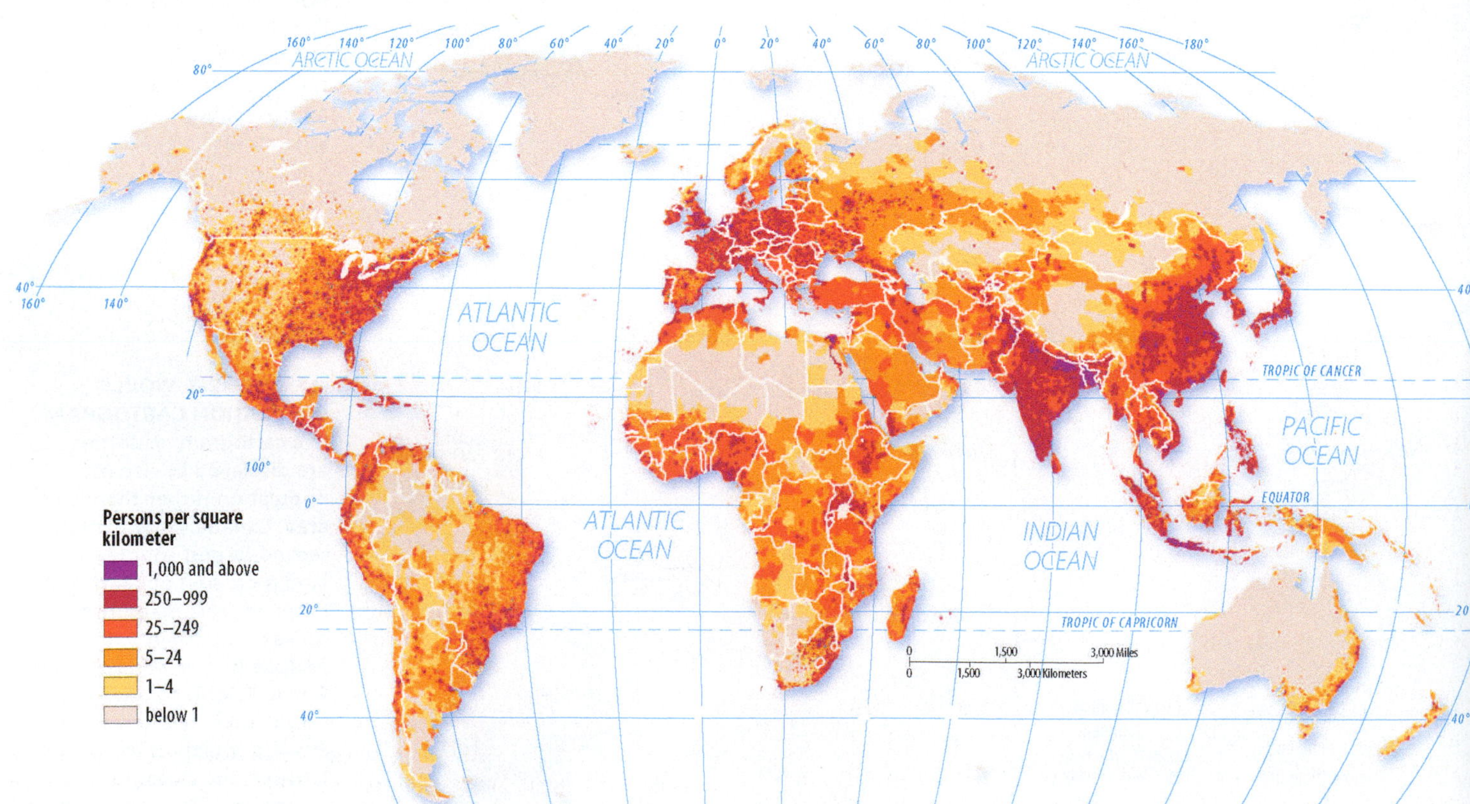

▲ **FIGURE 2-4 WORLD POPULATION DISTRIBUTION** People are not distributed uniformly across Earth's surface.

of the Asian mainland, known as Indochina. The Southeast Asia concentration is characterized by a high percentage of people working as farmers in rural areas.

OTHER CLUSTERS. Africa has several population clusters. The two largest—both with around 300 million people—are along the west coast between Senegal and Nigeria and along the east coast between Eritrea and South Africa. As in the three Asian concentrations, most Africans work in agriculture. The largest population concentration in the Western Hemisphere is in the northeastern United States and southeastern Canada. This cluster, containing around 100 million people, extends along the Atlantic Coast from Boston to Newport News, Virginia, and westward along the Great Lakes to Chicago.

SPARSELY POPULATED REGIONS

Human beings avoid clustering in certain physical environments. Relatively few people live in regions that are too dry, too wet, too cold, or too mountainous for activities such as agriculture. The areas of Earth that humans consider too harsh for occupancy have diminished over time, whereas the portion of Earth's surface occupied by permanent human settlement—called the **ecumene**—has increased (Figure 2-5).

PAUSE & REFLECT 2.1.2

Why are some land areas not part of the ecumene?

DRY LANDS. Areas too dry for farming cover approximately 20 percent of Earth's land surface. Deserts generally lack sufficient water to grow crops that could feed a large population, although some people survive there by raising animals, such as camels, that are adapted to the climate. Dry lands contain natural resources useful to people—notably, much of the world's petroleum reserves.

WET LANDS. Lands that receive very high levels of precipitation, located primarily near the equator, are often inhospitable for human occupation. The combination of rain and heat rapidly depletes nutrients from the soil and thus hinders agriculture.

COLD LANDS. Much of the land near the North and South poles is perpetually covered with ice or the ground is permanently frozen (permafrost). The polar regions are unsuitable for planting crops, few animals can survive the extreme cold, and few humans live there.

HIGH LANDS. The highest mountains in the world are steep, snow covered, and sparsely settled. However, some high-altitude plateaus and mountain regions are more densely populated, especially at low latitudes (near the equator) where agriculture is possible at high elevations.

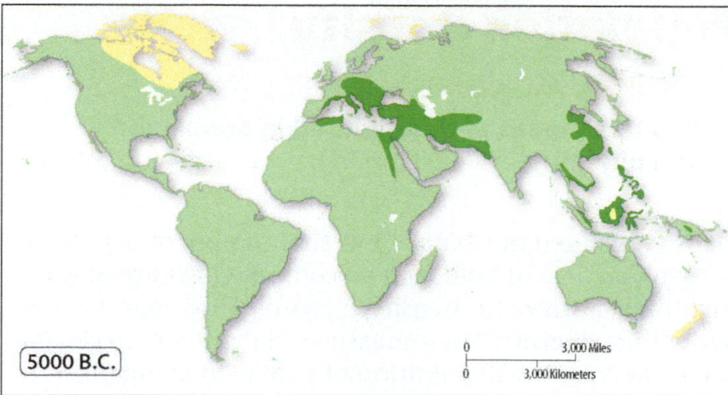

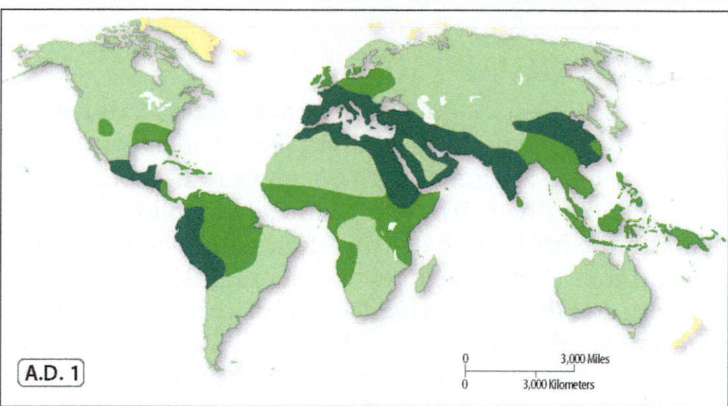

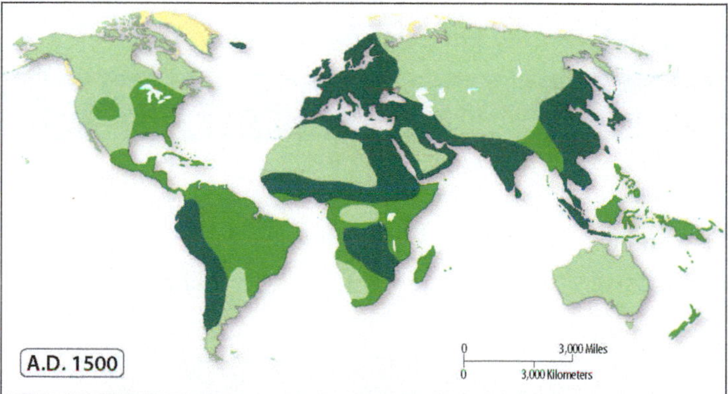

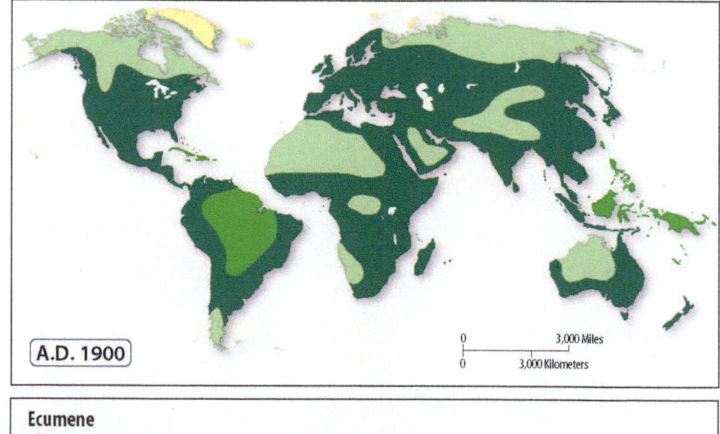

Ecumene
Intensive settlement
Small-scale agriculture
Hunting and gathering
Uninhabited (mainly ice)

▲ **FIGURE 2-5 ECUMENE** Seven thousand years ago, humans occupied only a small percentage of Earth's land area, primarily in Southwest Asia, Eastern Europe, and East Asia. Even 500 years ago, much of North America and Asia lay outside the ecumene. Still, approximately three-fourths of the world's population live on only 5 percent of Earth's surface. The balance of Earth's surface consists of oceans (about 71 percent) and less intensively inhabited land.

Population Density

LEARNING OUTCOME 2.1.3

Define three types of density used in population geography.

Density, defined in Chapter 1 as the number of people occupying an area of land, can be computed in several ways, including arithmetic density, physiological density, and agricultural density. These measures of density help geographers describe the distribution of people in comparison to available resources.

ARITHMETIC DENSITY

Geographers most frequently use **arithmetic density**, which is the total number of objects in an area. In population geography, arithmetic density refers to the total number of people divided by total land area (Figure 2-6). To compute the arithmetic density, divide the population by the land area. Table 2-1 shows several examples.

Arithmetic density enables geographers to compare the number of people trying to live on a given piece of land in different regions of the world. Thus, arithmetic density answers the "where" question. However, to explain why people are not uniformly distributed across Earth's surface, other density measures are more useful.

TABLE 2–1 Density Measures for Four Countries

Country	Arithmetic Density	Physiological Density	Agricultural Density	Percentage Farmers	Percentage Arable Land
Canada	4	83	1	2	5
United States	35	199	3	2	18
Netherlands	498	1,610	26	3	31
Egypt	87	3,011	273	29	3

PHYSIOLOGICAL DENSITY

Looking at the number of people per area of a certain type of land in a region provides a more meaningful population measure than arithmetic density. Land suited for agriculture is called arable land. In a region, the number of people supported by a unit area of arable land is called the **physiological density** (Figure 2-7).

Comparing physiological and arithmetic densities helps geographers understand the capacity of the land to yield enough food for the needs of the people. In Egypt, for example, the large difference between the physiological density and arithmetic density indicates that most of the country's land is unsuitable for intensive agriculture. In fact, all but 5 percent of Egyptians live in the Nile River valley and delta because it is the only area in the country that receives enough moisture (by irrigation from the river) to allow intensive cultivation of crops.

AGRICULTURAL DENSITY

Two countries can have similar physiological densities but produce significantly different amounts of food because of different economic conditions. **Agricultural density** is the ratio of the number of farmers to the amount of arable land (Figure 2-8). Table 2-1 shows several examples.

Measuring agricultural density helps account for economic differences. Developed countries have lower agricultural densities because technology and finance allow a few people to farm extensive land areas and feed many people.

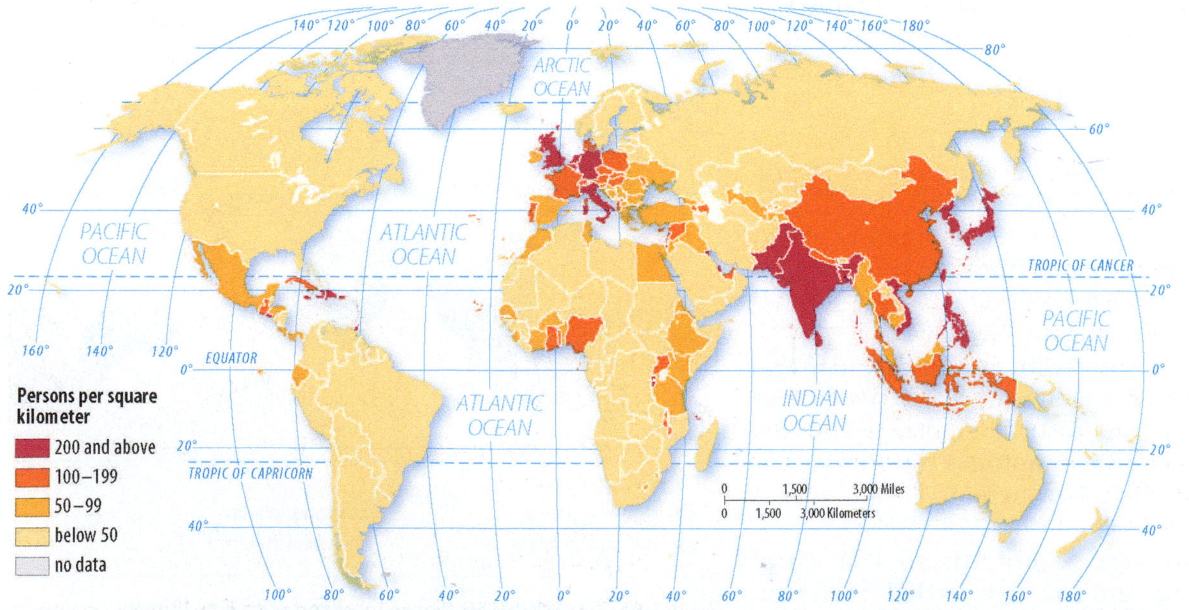

▲ **FIGURE 2-6 ARITHMETIC DENSITY** Geographers rely on the arithmetic density to compare conditions in different countries because the two pieces of information—total population and total land area—are easy to obtain. The highest arithmetic densities are found in Asia, Europe, and Central America. The lowest are in North and South America and the South Pacific.

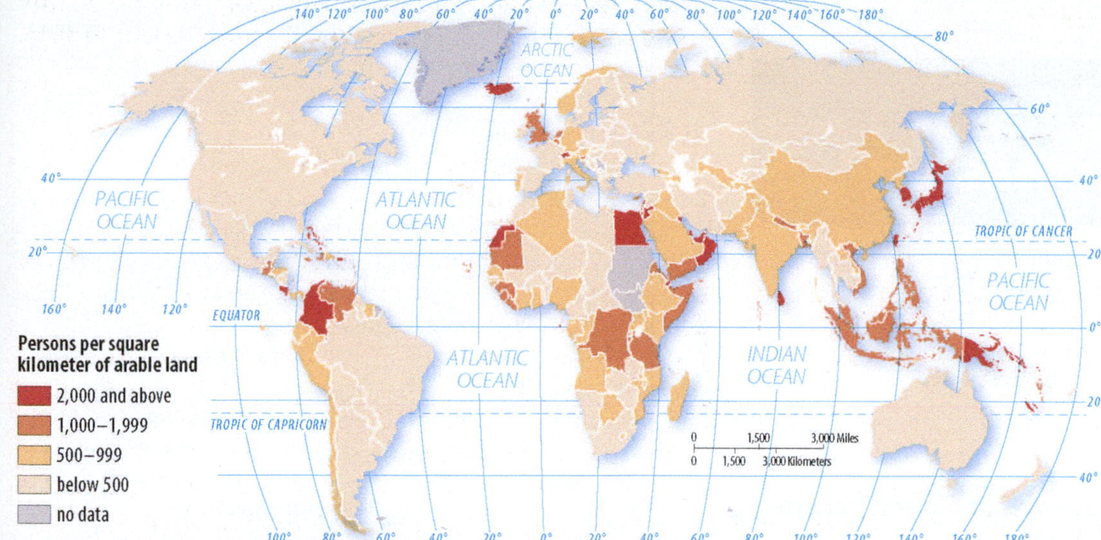

▲ **FIGURE 2-7 PHYSIOLOGICAL DENSITY** Physiological density provides insights into the relationship between the size of a population and the availability of resources in a region. The relatively large physiological densities of Egypt and the Netherlands demonstrate that crops grown on a hectare of land in these two countries must feed far more people than in the United States or Canada, which have much lower physiological densities. The highest physiological densities are found in Asia, sub-Saharan Africa, and South America. The lowest are in North America, Europe, and the South Pacific.

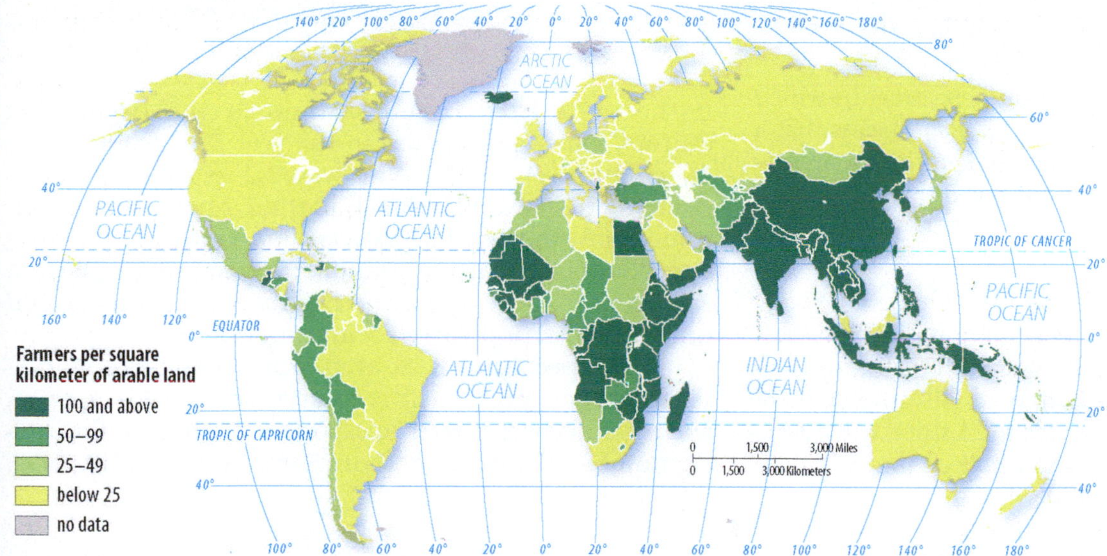

▲ **FIGURE 2-8 AGRICULTURAL DENSITY** The highest agricultural densities are found in Asia and sub-Saharan Africa. The lowest are in North America, Europe, and the South Pacific.

To understand relationships between population and resources in a country, geographers examine a country's physiological and agricultural densities together. For example, the physiological densities of both Egypt and the Netherlands are high, but the Dutch have a much lower agricultural density than the Egyptians. Geographers conclude that both the Dutch and Egyptians put heavy pressure on the land to produce food, but the more efficient Dutch agricultural system requires fewer farmers than does the Egyptian system.

PAUSE & REFLECT 2.1.3

Which density measure differs most between Egypt and Ethiopia? What might account for this difference?

Where Is the World's Population Distributed?

- ✔ The world's population is highly clustered in four regions.
- ✔ The physical environment discourages population concentrations in some regions.
- ✔ Arithmetic, physiological, and agricultural densities describe the distribution of people.

KEY ISSUE 2

Why Is World Population Increasing?

- Natural Increase
- Births and Deaths
- The Demographic Transition

LEARNING OUTCOME 2.2.1

Understand historical and recent rates of natural increase.

For most of human history, Earth's population was unchanged at perhaps a half-million. In contrast, about 75 million people are now being added to the population of the world annually.

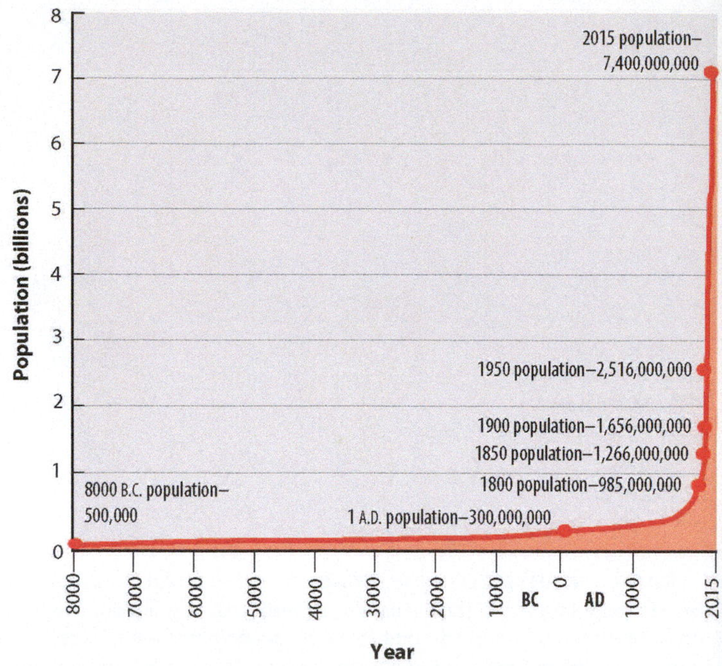

▲ **FIGURE 2-9 WORLD POPULATION THROUGH HISTORY** Through most of human history, population growth was virtually nil. Population increased rapidly beginning in the eighteenth century.

Natural Increase

The **natural increase rate (NIR)** is the percentage by which a population grows in a year. The term *natural* means that a country's growth rate excludes migration. During the twenty-first century, the world NIR has been 1.2, meaning that the population of the world has been growing each year by 1.2 percent.

POPULATION GROWTH IN HISTORY

The NIR was essentially zero for most of humanity's several-hundred-thousand-year occupancy of Earth (Figure 2-9). The world NIR is lower today than its all-time peak of 2.2 percent in 1963, and it has declined since the 1990s. However, the NIR during the second half of the twentieth century was high by historical standards.

The number of people added each year has dropped from a historic peak of 88 million in 1989 to the current 75 million. The decline of this number is less sharp than the decline of the NIR because the population base is much larger now than in the past. World population increased from 3 to 4 billion in 14 years, from 4 to 5 billion in 13 years, from 5 to 6 billion in 12 years, and from 6 to 7 billion in 12 years. As the base continues to grow in the twenty-first century, a change of only one-tenth of 1 percent can produce very large swings in population growth (Figure 2-10).

The natural increase rate affects the **doubling time**, which is the number of years needed to double a population, assuming a constant rate of natural increase. At the current rate of 1.2 percent per year, world population would double in about 54 years. If the same NIR continued through the twenty-first century, global population in the year 2100 would reach 24 billion. When the NIR was 2.2 percent in 1963, doubling time was 35 years. Had the 2.2 percent rate continued into the twenty-first century, Earth's population would currently exceed 10 billion instead of 7 billion. A 2.2 percent NIR through the twenty-first century would produce a total population of more than 50 billion in 2100.

Life expectancy is the average number of years an individual can be expected to live, given current social, economic, and medical conditions. Life expectancy at birth is the average number of years a newborn infant can expect to live.

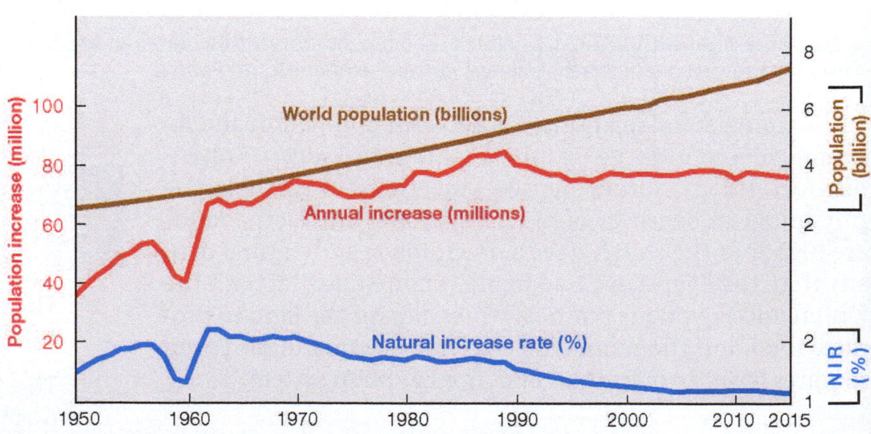

▲ **FIGURE 2-10 WORLD POPULATION GROWTH, 1950–2015** The NIR declined from its historic peak in the 1960s, but the number of people added each year has not declined very much because with world population increasing from 2.5 billion to more than 7 billion people during the period, the percentage has been applied to an ever larger base.

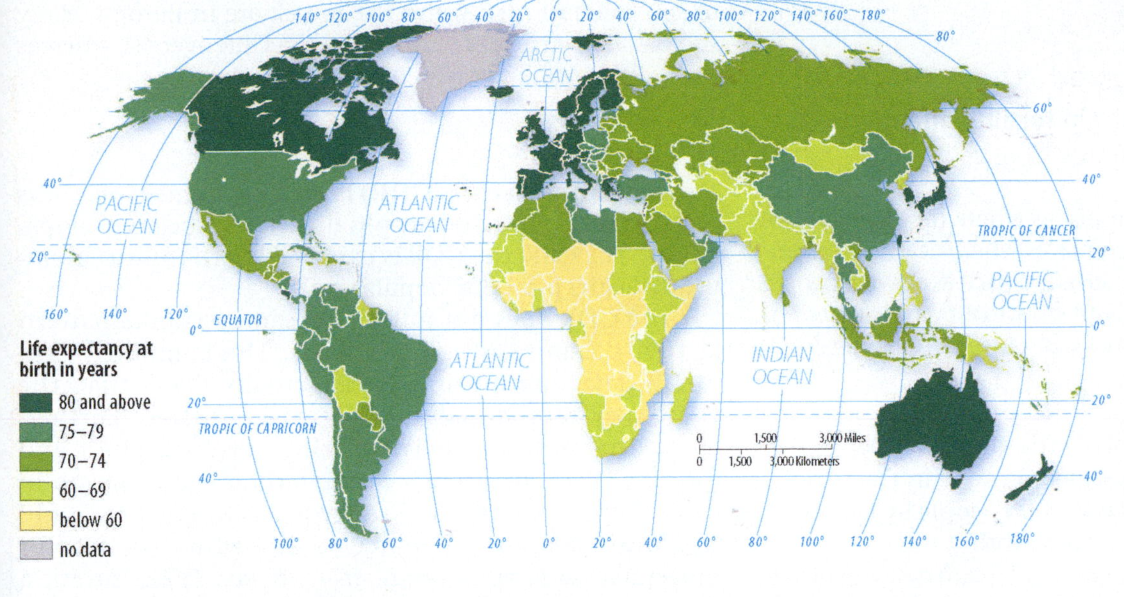

◀ **FIGURE 2-11 LIFE EXPECTANCY AT BIRTH** The highest life expectancies are in Europe and the lowest are in sub-Saharan Africa.

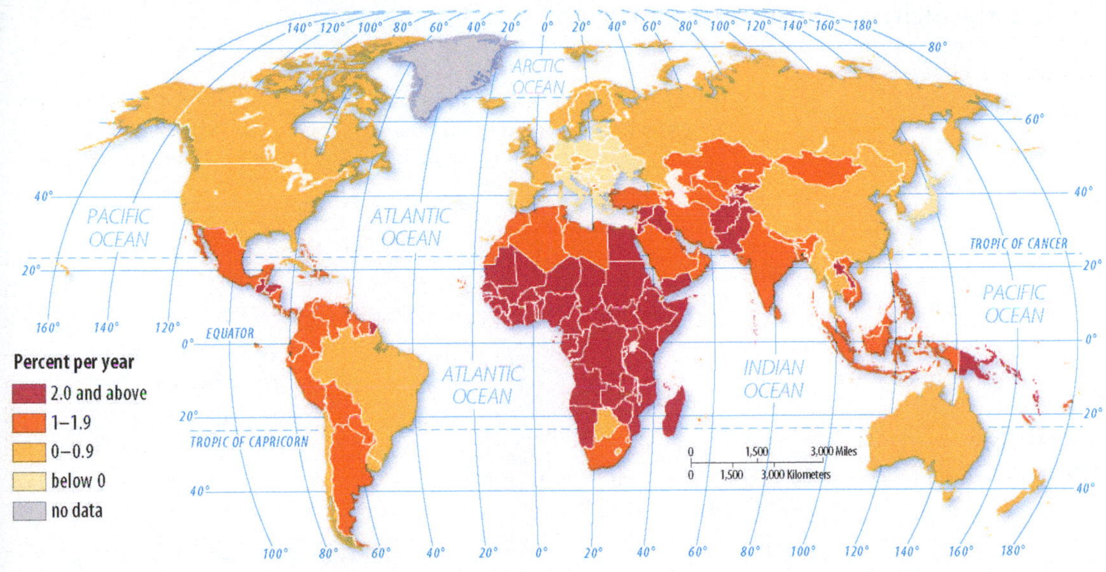

◀ **FIGURE 2-12 NATURAL INCREASE RATE** The world average is currently 1.2 percent. The countries with the highest NIRs are concentrated in Africa and Southwest Asia.

Babies born today can expect to live to around 80 in the developed countries of Europe but only to around 57 in the developing countries of sub-Saharan Africa (Figure 2-11).

REGIONAL VARIATIONS IN NIR

More than 95 percent of the natural increase is clustered in developing countries (Figure 2-12). The NIR exceeds 2.0 percent in most countries of sub-Saharan Africa, whereas it is negative in Europe, meaning that in the absence of immigrants, population actually is declining.

PAUSE & REFLECT 2.2.1

Which region other than sub-Saharan Africa appears to have the highest natural increase rate?

Since 1980, 67 percent of the world's population growth has been in Asia, 20 percent in Africa, 9 percent in Latin America, and 4 percent in North America (Figure 2-13). Europe (including Russia) has had declining population since 1980. Regional differences in NIRs mean that most of the world's additional people live in the countries that are least able to maintain them. To explain these variations in growth rates, geographers point to regional differences in fertility and mortality rates.

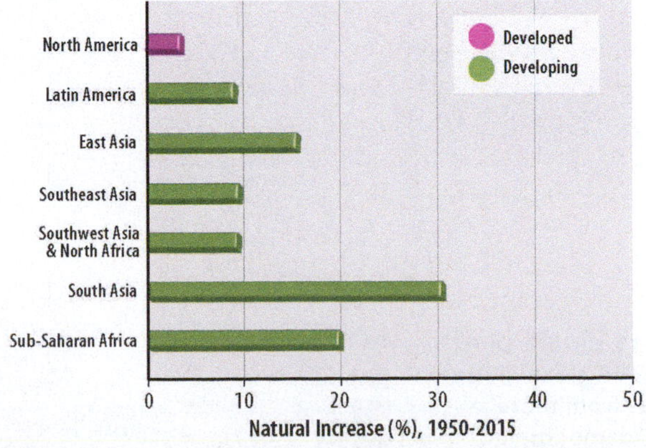

▲ **FIGURE 2-13 REGIONAL DISTRIBUTION OF NATURAL INCREASE, 1950–2015** Europe (including Russia) does not appear on the chart because population has declined in the region.

Births and Deaths

LEARNING OUTCOME 2.2.2

Recognize regional variations in fertility and mortality.

Population increases rapidly in places where more people are born than die, and it declines in places where deaths outnumber births. The population of a place also increases when people move in and decreases when people move out. This element of population change —migration—is discussed in Chapter 3.

FERTILITY

The **crude birth rate (CBR)** is the total number of live births in a year for every 1,000 people alive in the society. A CBR of 20 means that for every 1,000 people in a country, 20 babies are born over a one-year period.

The world map of CBR (Figure 2-14) mirrors the distribution of NIR. As is the case with NIRs, the highest CBRs are in sub-Saharan Africa, and the lowest are in Europe. Many sub-Saharan African countries have a CBR over 40, whereas many European countries have a CBR below 10.

MORTALITY

The **crude death rate (CDR)** is the total number of deaths in a year for every 1,000 people alive in the society. Comparable to the CBR, the CDR is expressed as the annual number of deaths per 1,000 population.

The CDR does not follow the same regional pattern as the NIR and CBR (Figure 2-15). The combined CDR for all developing countries is actually lower than the combined rate for all developed countries. Furthermore, the variation between the world's highest and lowest CDRs is much less extreme than the variation in CBRs. The highest CDR in the world is 21 per 1,000, and the lowest is 1—a difference of 20—whereas CBRs for individual countries range from 6 per 1,000 to 50, a spread of 44.

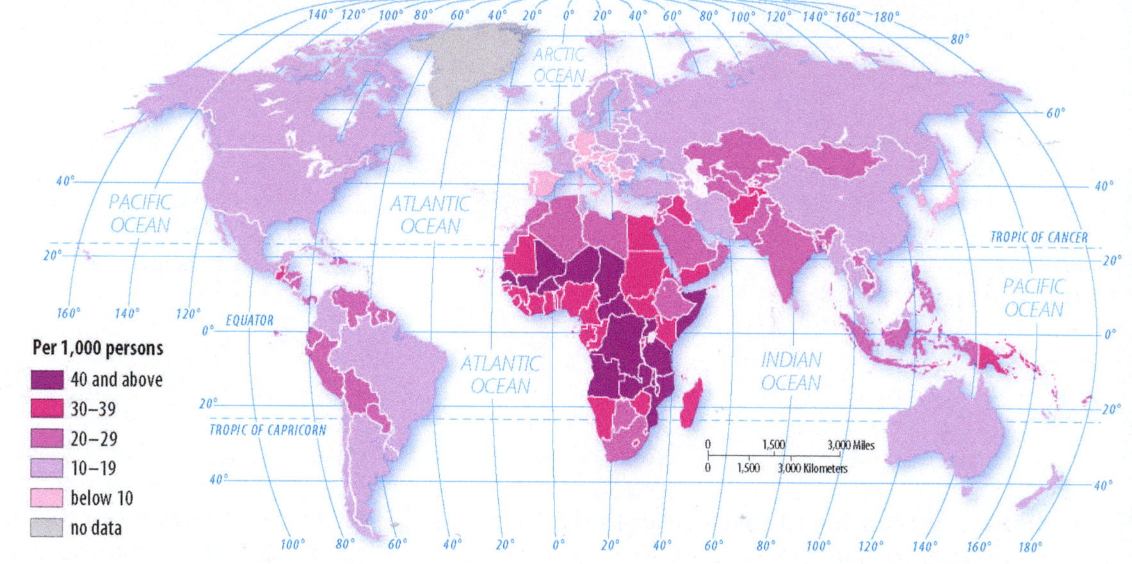

▶ **FIGURE 2-14 CRUDE BIRTH RATE (CBR)** The global distribution of CBRs parallels that of NIRs. The countries with the highest CBRs are concentrated in Africa and Southwest Asia.

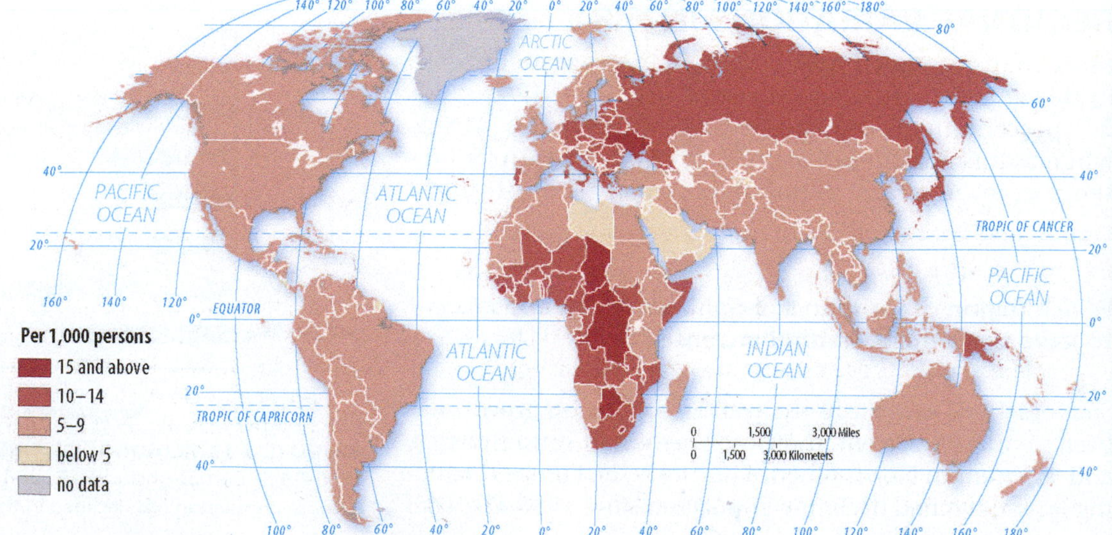

▶ **FIGURE 2-15 CRUDE DEATH RATE (CDR)** The global pattern of CDRs varies from those for the other demographic variables already mapped in this chapter. The demographic transition helps to explain the distinctive distribution of CDRs.

PAUSE & REFLECT 2.2.2

What region of the world appears to have the lowest CDR?

Why does Denmark, one of the world's wealthiest countries, have around the same CDR as The Gambia, one of the poorest? Why does the United States, with its extensive system of hospitals and physicians, have a higher CDR than Mexico and nearly every other country in Latin America? The answer is that the populations of different countries are at various stages in an important process known as the demographic transition, discussed in the next section.

DOING GEOGRAPHY The Future Population of the World

The scientific study of population characteristics is **demography**. Demographers look statistically at how people are distributed spatially by age, gender, occupation, fertility, health, and so on. Geographers are part of the team at organizations that do demographic studies, such as the Population Reference Bureau, the U.S. Bureau of the Census, and the United Nations Population Division.

Demographers can tell us with some precision the current population of the world and of particular places, as well as population in the recent past. Forecasting the future population is much more challenging for demographers, but we need these estimates in order to plan for the future.

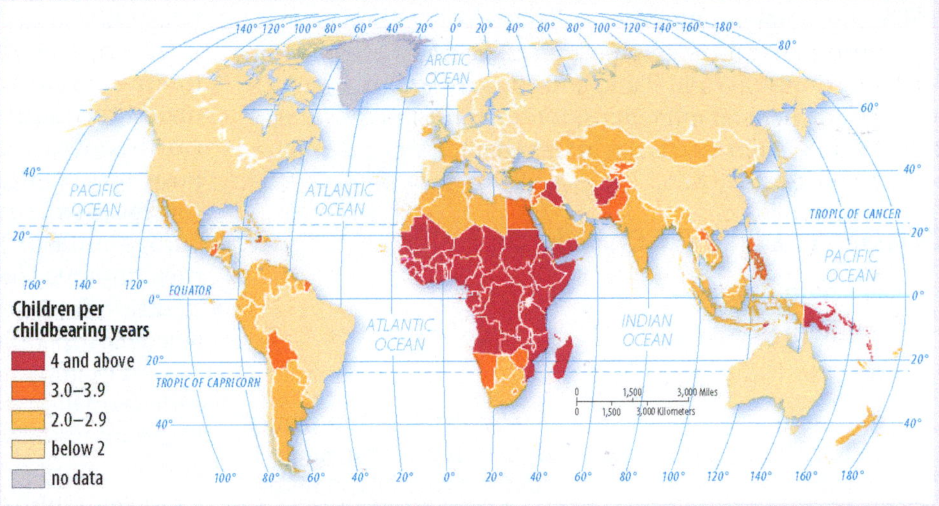

▲ **FIGURE 2-16 TOTAL FERTILITY RATE (TFR)** As with NIRs and CBRs, the countries with the highest TFRs are concentrated in Africa and Southwest Asia.

Geographers use the **total fertility rate (TFR)** to measure the number of births in a society (Figure 2-16). The TFR is the average number of children a woman will have throughout her childbearing years (roughly ages 15 through 49). To compute the TFR, demographers assume that a woman reaching a particular age in the future will be just as likely to have a child as are women of that age today. Thus, the CBR provides a picture of a society as a whole in a given year, whereas the TFR attempts to predict the future behavior of individual women in a world of rapid cultural change.

The TFR for the world as a whole is 2.5, and, again, the figures vary between developed and developing countries. The TFR exceeds 5.0 in sub-Saharan Africa, compared to 2 or less in nearly all European countries.

What's Your Population Geography?

The size of the world's future population ultimately depends on decisions made by people of childbearing age concerning the number of children they will have.

1. What is the TFR of your mother? How does that figure compare with those for your grandmothers?
2. How many children do you expect to have? Is that number more, less, or the same as your mother? How confident are you in your answer? What factors make you feel confident or unsure of your answer?
3. Compare your answers with others in your class or with your friends. Calculate the mean response for your group by totaling all the individual answers and dividing by the number of respondents. Compute the TFR separately for the men and the women. Do the figures differ? If so, what might account for the difference?
4. The TFR in the United States is currently 1.9. How does the TFR of your classmates or friends compare with the overall U.S. figure?
5. Demographers explain that a TFR of 2.1 is needed to maintain the same size of population. A TFR of 1.9, as in the United States, means that the number of births is not sufficient to maintain the current size of the U.S. population. Yet the U.S. population is actually increasing by 0.4 percent per year. What accounts for the increasing population, if not the number of births?

The Demographic Transition

LEARNING OUTCOME 2.2.3

Describe the stages of the demographic transition.

All countries have experienced some changes in NIR, CBR, and CDR, but at different times and at different rates. The demographic transition helps us understand these differences. The **demographic transition** is a process of change in a society's population from high crude birth and death rates and low rate of natural increase to a condition of low crude birth and death rates, low rate of natural increase, and higher total population.

Authoritative demographic sources, such as the Population Reference Bureau, World Bank, and United Nations, currently consider the demographic transition to consist of four stages, (Figure 2-17). As discussed later in this chapter, some demographers expect a stage 5.

STAGE 1: LOW GROWTH

- Very high CBR
- Very high CDR
- Very low NIR

Most of human history was spent in stage 1 of the demographic transition, but today no country remains in stage 1. Every nation has moved on to at least stage 2 of the demographic transition, and, with that transition, has experienced profound changes in population. For most of this period, people depended on hunting and gathering for food (see Chapter 10). When food was easily obtained, a region's population increased; the population declined when people were unable to locate enough animals or vegetation nearby.

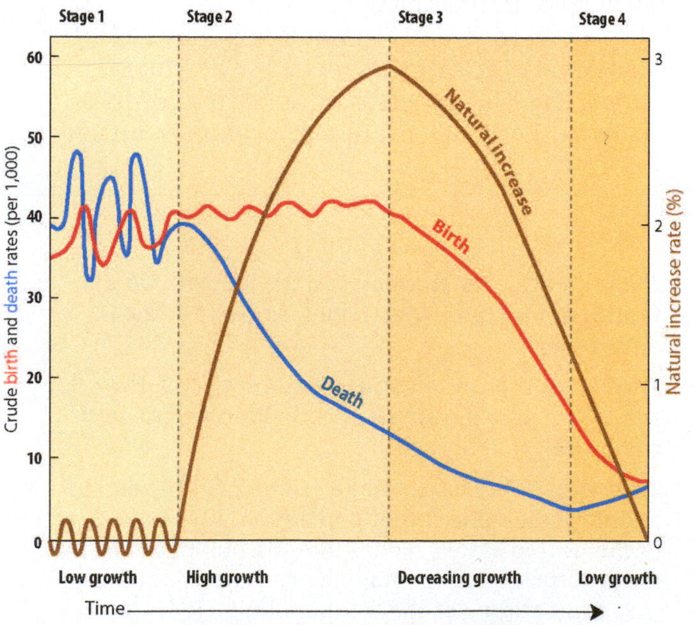

▲ **FIGURE 2-17 DEMOGRAPHIC TRANSITION MODEL** The model consists of four stages. A possible fifth stage is discussed in the section on Population Futures.

STAGE 2: HIGH GROWTH

- High CBR
- Rapidly declining CDR
- Very high NIR

Rapidly declining death rates and very high birth rates produce very high natural increase. Europe and North America entered stage 2 of the demographic transition after 1750, as a result of the **Industrial Revolution**, which involved major improvements in manufacturing goods and delivering them to market (see Chapter 11). The result of this transformation was an unprecedented level of wealth, some of which was used to make communities healthier places to live.

Stage 2 of the demographic transition did not diffuse to Africa, Asia, and Latin America until around 1950, and it made that transition for a different reason than in Europe and North America 200 years earlier. The late-twentieth-century push of developing countries into stage 2 was caused by the **medical revolution**. Medical technology invented in Europe and North America has diffused to developing countries. Improved medical practices have eliminated many of the traditional causes of death in developing countries and enabled more people to experience longer and healthier lives.

An example of a country in stage 2 is The Gambia, one of Africa's smallest and poorest countries. As a colony of the United Kingdom until 1965, The Gambia was in stage 1 of the demographic transition. The World Health Organization launched a program during the 1970s to immunize children in a number of countries, including The Gambia. This sent The Gambia into stage 2 because the CDR declined rapidly, whereas the CBR remained high, thus raising the NIR (Figure 2-18a).

STAGE 3: MODERATE GROWTH

- Rapidly declining CBR
- Moderately declining CDR
- Moderate NIR

A country moves from stage 2 to stage 3 of the demographic transition when the CBR begins to drop sharply. The CDR continues to fall in stage 3 but at a much slower rate than in stage 2. The population continues to grow because the CBR is still greater than the CDR. But the rate of natural increase is more modest in countries in stage 3 than in those in stage 2 because the gap between the CBR and the CDR narrows.

A society enters stage 3 when people have fewer children. The decision to have fewer children is partly a delayed reaction to a decline in mortality. Economic changes in stage 3 societies also induce people to have fewer offspring. People in stage 3 societies are more likely to live in cities than in the countryside and to work in offices, shops, or factories rather than on farms. Farmers often consider a large family to be an asset because children can do some of the chores. Urban homes are relatively small and may not have space to accommodate large families.

Most countries in Europe and North America (including the United States) moved from stage 2 to stage 3 of the demographic transition during the first half of the twentieth century. The movement took place during the second half of the twentieth century in many countries of Asia and Latin America, including Mexico.

Colonial Mexico was in stage 1 of the demographic transition. Mexico entered stage 2 of the demographic transition during the twentieth century, through a combination of a lower CDR and a higher CBR (Figure 2-18b). The government of Mexico believed that higher birth rates would be good for the country's economic growth. A dramatic decline in the CBR and transition into stage 3 came after 1974, when a constitutional amendment guaranteed families the legal right to decide on the number and spacing of children, and the National Population Council was established to promote family planning through education.

STAGE 4: LOW GROWTH

- **Very low CBR**
- **Low or slightly increasing CDR**
- **Zero or negative NIR**

A country reaches stage 4 of the demographic transition when the CBR declines to the point where it equals the CDR and the NIR approaches zero. This condition is called **zero population growth (ZPG)**, a term often applied to stage 4 countries.

ZPG may occur when the CBR is still slightly higher than the CDR because some females die before reaching childbearing years, and the number of females in their childbearing years can vary. To account for these discrepancies, demographers more precisely define ZPG as the TFR that results in a lack of change in the total population over a long term. A TFR of approximately 2.1 produces ZPG.

Social customs again explain the movement to stage 4. Increasingly, women in stage 4 societies enter the labor force rather than remain at home as full-time homemakers. People who have access to a wider variety of birth-control methods are more likely to use some of them.

Denmark, like most other European countries, has reached stage 4 of the demographic transition (Figure 2-18c). Denmark's CDR has actually increased somewhat in recent years because of the increasing percentage of elderly people. The CDR is unlikely to decline unless another medical revolution, such as a cure for cancer, keeps elderly people alive much longer.

PAUSE & REFLECT 2.2.3

Name a country in Latin America that appears to be in stage 2, according to Figures 2-12, 2-14, and 2-15.

A country that has reached stage 4 of the demographic transition has in some ways completed a cycle—from little or no natural increase in stage 1 to little or no natural increase in stage 4. Two crucial demographic differences underlie this process:

- **Total population.** The total population of the country is much higher in stage 4 than in stage 1.
- **CBR and CDR.** At the beginning of the demographic transition, the CBRs and CDRs are high (35 to 40 per 1,000), whereas at the end of the process, the rates are very low (approximately 10 per 1,000).

The four-stage demographic transition is characterized by two big breaks with the past. The first break—the sudden drop in the death rate that comes from technological innovation—has been accomplished everywhere. The second break—the sudden drop in the birth rate that comes from changing social customs—has yet to be achieved in many countries.

CHECK-IN KEY ISSUE 2

Why is World Population Increasing?

- ✔ **The NIR measures population growth as the difference between births and deaths.**
- ✔ **The crude birth rate and crude death rate are the principal measures of population change in a society as a whole.**
- ✔ **The demographic transition has four stages, characterized by varying rates of births, deaths, and natural increase.**

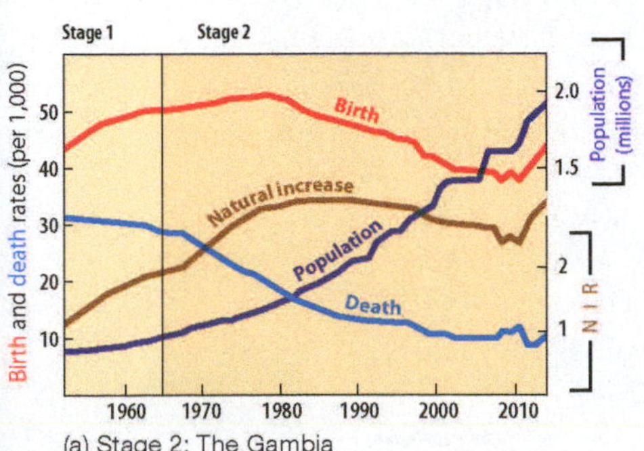

(a) Stage 2: The Gambia

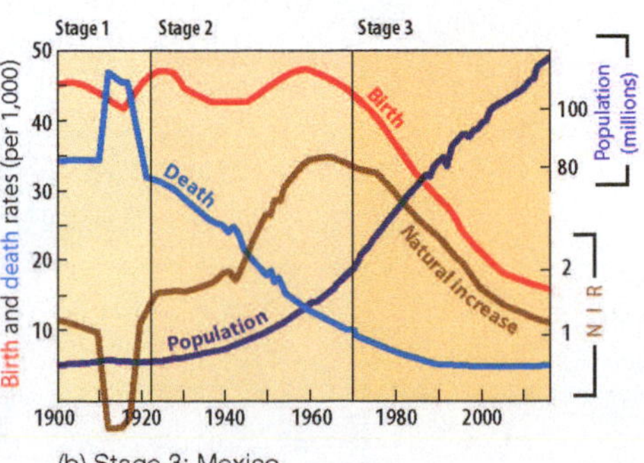

(b) Stage 3: Mexico

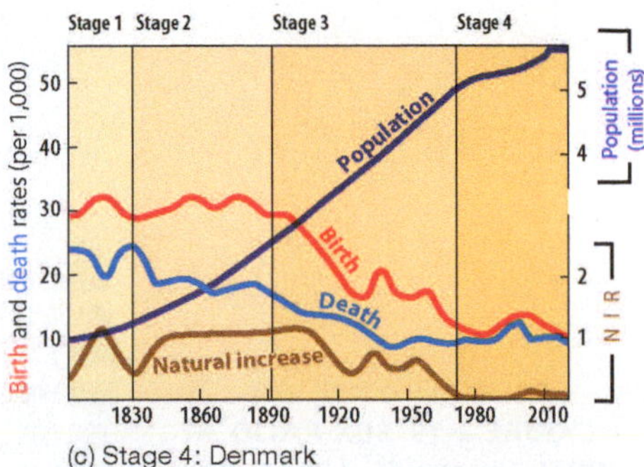

(c) Stage 4: Denmark

▲ **FIGURE 2-18 DEMOGRAPHIC TRANSITION FOR THREE COUNTRIES** **(a)** Stage 2: The Gambia, **(b)** Stage 3: Mexico, **(c)** Stage 4: Denmark.

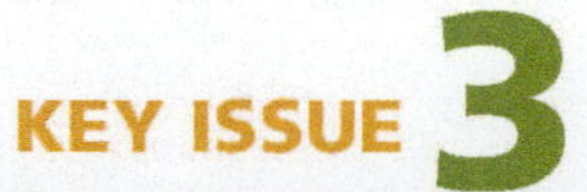

KEY ISSUE 3

Why Do Some Places Face Health Challenges?

- ▸ **Health and Gender**
- ▸ **Health and Aging**
- ▸ **Medical Services**
- ▸ **The Epidemiologic Transition**

As world NIR slows and more countries move into stage 3 or 4 of the demographic transition, geographers increasingly turn their attention to the health of the record number of people who are alive. Countries in different stages of the demographic transition possess different resources to meet the needs of women, children, elderly people, and those with medical needs.

Health and Gender

LEARNING OUTCOME 2.3.1

Understand reasons for varying sex ratios and for reduced birth rates.

Females face especially challenging health risks that profoundly affect the size and composition of the population of individual countries and the world as a whole. These health risks derive from the biological fact of being born female. The risk for females is especially acute at childbirth. Both the mother and the baby girl are at risk.

BABY GIRLS AT RISK

Around 700,000 female babies are "missing" every year in China and India, as a result of gender-based selection. The United Nations Population Fund estimates that overall 117 million females have gone "missing" throughout Asia over the past several decades. The females are "missing" either because the fetus was aborted before birth, the female baby was killed in infancy, or the newborn female was whisked away to somewhere remote and not reported to census and health officials.

We know about the large number of "missing" females because of the **sex ratio**, which is the number of males per 100 females in the population. The standard biological level for humans at birth is around 105 male babies for every 100 female babies. Scientists are not sure why a few more males than females are born.

The standard biological ratio at birth of 105:100 (105 males to 100 females) is characteristic of the developed regions of North America and Europe, as well as in the developing regions of Latin America and sub-Saharan Africa (Figure 2-19). However, in the world's two most populous countries—China and India—the sex ratio at birth is 112:100 (112 males to 100 females). The percentage of newborn females in the world's two most populous countries is much too low to be random (Figure 2-20).

The extremely low percentage of female babies in China and India results from cultural preference on the part of parents to have sons rather than daughters. Government policies in these two countries to promote smaller families have encouraged parents to bring into life male babies rather than female babies. Access to ultrasound machines has enabled parents to know the sex of the fetus, and access to medical procedures has enabled them to abort unwanted female fetuses.

The governments of China and India have attempted to prohibit the use of technology and medical procedures. But the bans are ineffective because people instead seek out unregulated providers of these services. The United Nations concludes that a more effective approach is to address what it calls the "root cause" of sex selection—gender inequality. Many Chinese and Indians conclude that if they are going to have a small number of children, they want them to be boys.

Aside from the ethical question, the widespread practice of sex selection in China and India is creating a practical

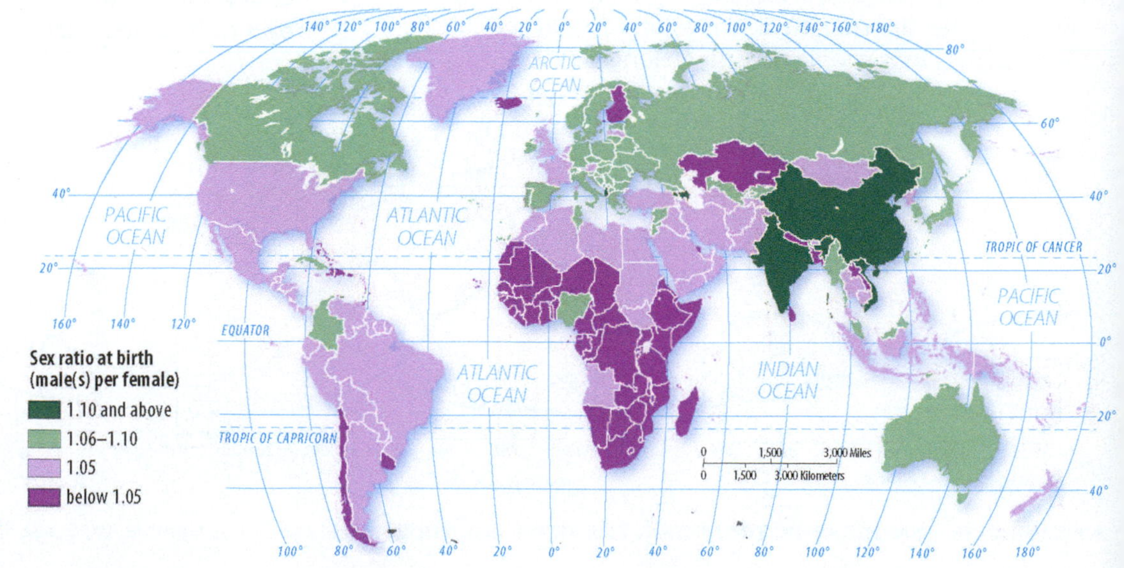

▶ **FIGURE 2-19 SEX RATIO AT BIRTH** The standard biological ratio is 105:100—that is, 105 male babies for 100 female babies.

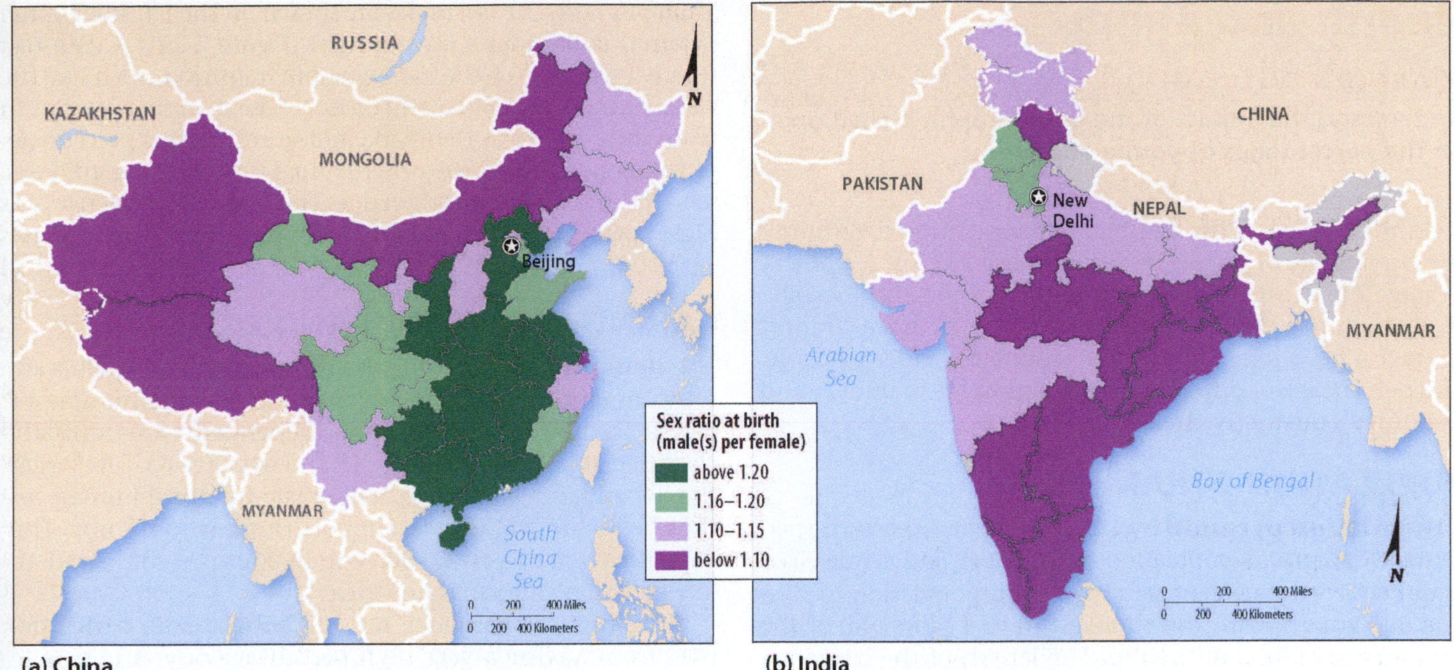

(a) China (b) India

▲ **FIGURE 2-20 SEX RATIO AT BIRTH, CHINA AND INDIA** The sex ratio, which is 112:100 in the two countries as a whole, is considerably higher in some regions, especially rural areas.

problem for society. As the babies grow to adulthood, these countries are left with an enormous surplus of men who are unable to find women to marry.

PAUSE & REFLECT 2.3.1

What other countries, in addition to China and India, appear to have "missing" females?

MOTHERS AT RISK

The **maternal mortality rate** is the annual number of female deaths per 100,000 live births from any cause related to or aggravated by pregnancy or its management (excluding accidental or incidental causes). The rate exceeds 100 deaths per 100,000 mothers (in other words, 1 percent) in much of Africa and Asia, compared to fewer than 10 (0.1 percent) in most European countries (Figure 2-21).

According to the United Nations, the most common cause of maternal death in poor countries is obstetrical hemorrhage, followed by hypertensive disorders of pregnancy. Developed countries have medical facilities, advanced technologies, and trained personnel to limit the incidence of life-threatening conditions during childbirth.

The maternal mortality rate in the United States (28) is higher than in other developed countries. The higher rate is attributable to difficulties faced by low-income people in the United States in gaining access to health care. In addition, the United States is one of the few countries that do not mandate paid leave for mothers of newborn infants.

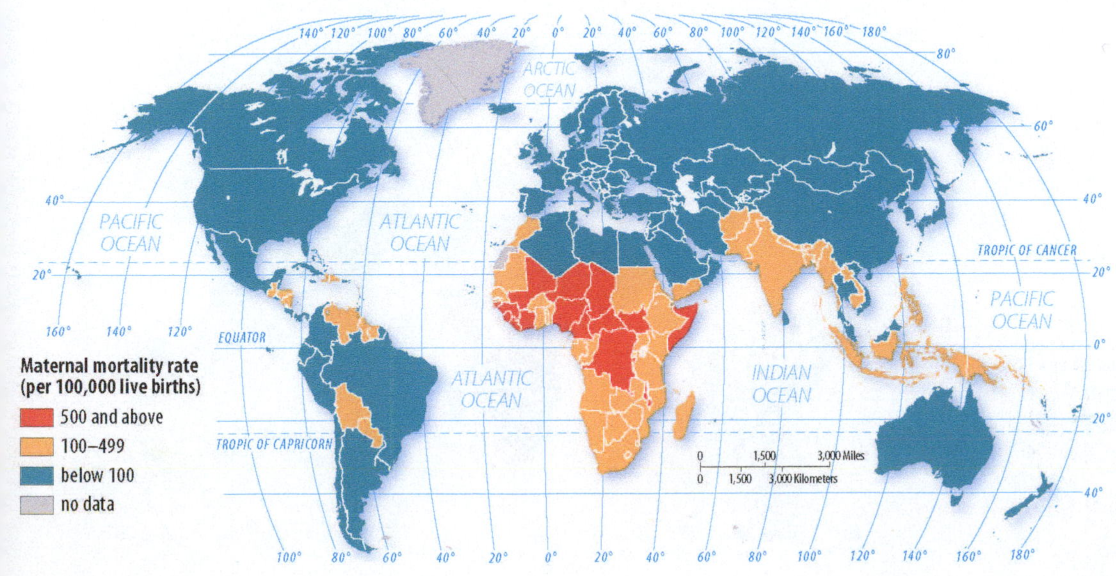

◀ **FIGURE 2-21 MATERNAL MORTALITY RATE** The rate exceeds 100 deaths per 100,000 mothers (in other words, 1 percent) in much of Africa and Asia, compared to fewer than 10 (0.1 percent) in most European countries.

Health and Aging

LEARNING OUTCOME 2.3.2

Understand the impact of the demographic transition on the percentages of young and old.

A country's stage of the demographic transition determines the percentage of people in different age groups. A country in stage 2 of the demographic transition typically has a relatively high percentage of young people, whereas a country in stage 4 has a relatively high percentage of elderly people. The percentage of different age groups helps to understand a country's distinctive health challenges.

POPULATION PYRAMIDS

A **population pyramid** is a bar graph that displays the percentage of a place's population for each age and gender. The graph shows the percentage of the total population in five-year age groups, with the youngest (0 to 4 years old) at the base and the oldest at the top. The length of the bar represents the percentage of the total population contained in that group. By convention, males are shown on the left side of the pyramid and females on the right (Figure 2-22). A country that is in stage 2 of the demographic transition, such as The Gambia, has a pyramid with a broader base than that of a country in stage 4, such as Denmark, indicating a higher percentage of young people. On the other hand, Denmark's pyramid extends farther to the right than does The Gambia's, evidence of a higher percentage of women in Denmark than in The Gambia.

CARING FOR YOUNG AND OLD

The **dependency ratio** is the number of people who are too young or too old to work, compared to the number of people in their productive years. Dependents are normally classified as people under age 15 and over age 65. The larger the dependency ratio, the greater the financial burden on those who are working to support those who do not. The dependency ratio is 47 percent in Europe, compared to 85 percent in sub-Saharan Africa.

The high dependency ratio in sub-Saharan Africa derives from having a very high percentage of young people (Figure 2-23). Young dependents outnumber elderly ones

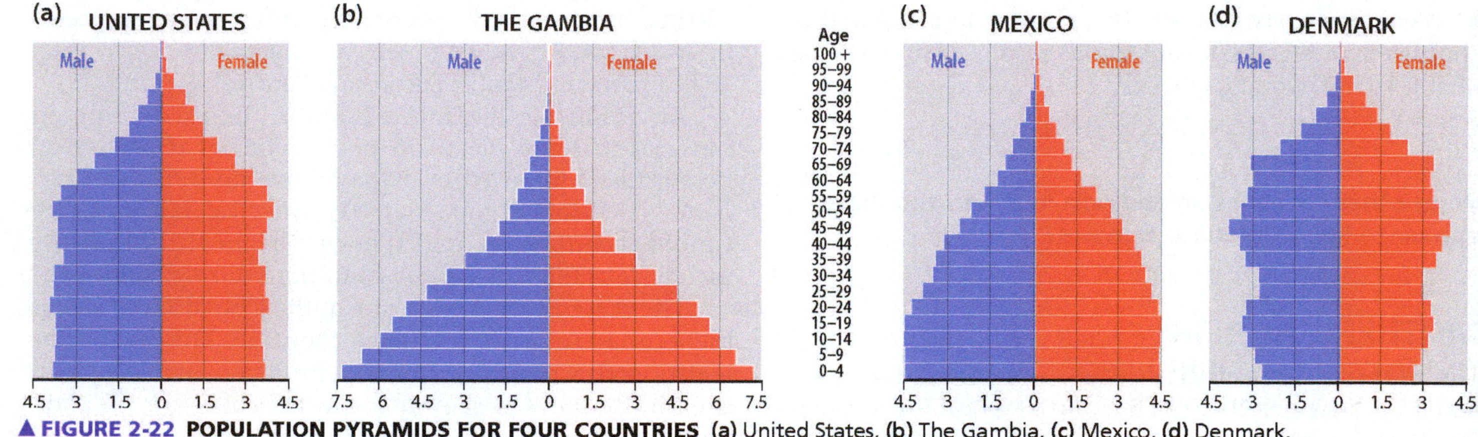

▲ **FIGURE 2-22 POPULATION PYRAMIDS FOR FOUR COUNTRIES** (a) United States, (b) The Gambia, (c) Mexico, (d) Denmark.

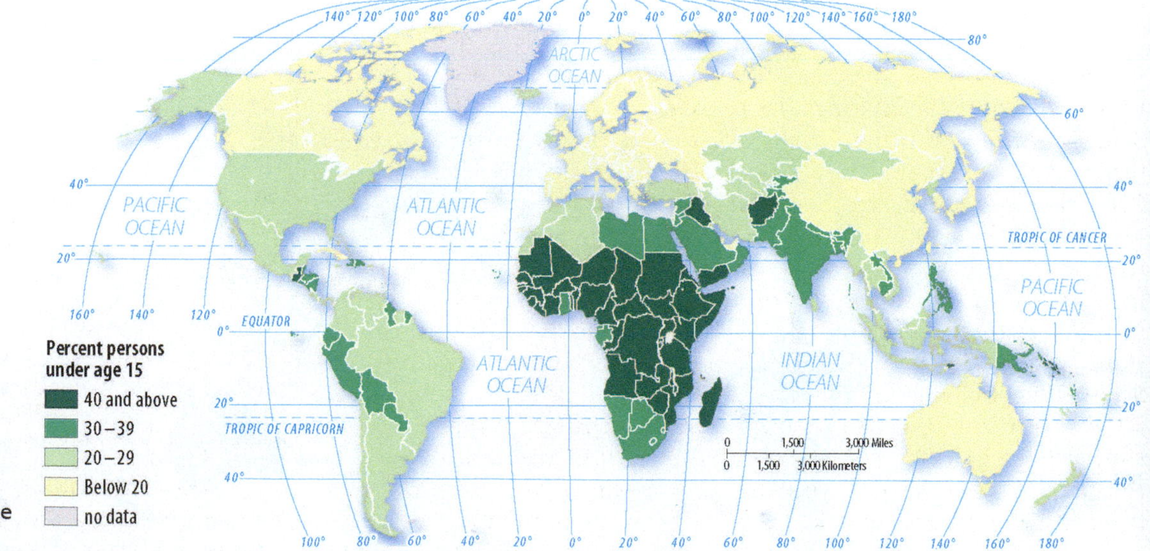

► **FIGURE 2-23 POPULATION UNDER AGE 15** Sub-Saharan Africa has the highest percentage of persons under age 15.

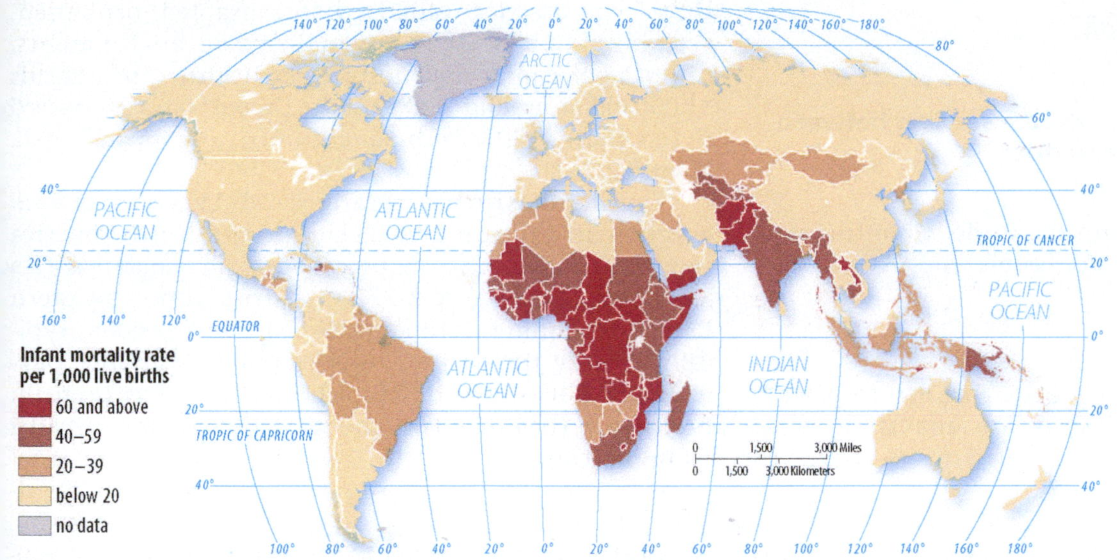

◀ **FIGURE 2-24 INFANT MORTALITY RATE (IMR)** The highest IMRs are in sub-Saharan Africa, and the lowest are in Europe.

by more than 14:1 in sub-Saharan Africa, whereas the numbers of people under 15 and over 65 are roughly equal in Europe. The large percentage of children in sub-Saharan Africa strains the ability of these relatively poor countries to provide needed services such as schools, hospitals, and day-care centers. When children reach the age of leaving school, jobs must be found for them, but the government must continue to allocate scarce resources to meet the needs of the still growing number of young people. On the other hand, the "graying" of the population places a burden on developed countries to meet their needs for income and medical care after they retire from jobs.

The **infant mortality rate (IMR)** is the annual number of deaths of infants under 1 year of age, compared with total live births (Figure 2-24). As is the case with the CBR and CDR, the IMR is usually expressed as the number of deaths among infants per 1,000 births rather than as a percentage (per 100). In general, the IMR reflects a country's health-care system.

The global distribution of IMRs follows the pattern that by now has become familiar. Lower IMRs are found in countries with well-trained doctors and nurses, modern hospitals, and large supplies of medicine. The IMR is 4 in European countries in stage 4, compared with 64 in sub-Saharan Africa. In other words, before reaching their first birthday, 1 in 15 babies die in sub-Saharan Africa and 1 in 250 babies die in Europe.

Even if they survive infancy, children remain at risk in developing countries. For example, 17 percent of children in developing countries are not immunized against measles, compared to 7 percent in developed countries. More than one-fourth of children lack measles immunization in South Asia and sub-Saharan Africa.

As countries pass through the demographic transition, they face increasing percentages of older people, who must receive adequate levels of income and medical care after they retire from their jobs. The "graying" of the population places a burden on governments in developed countries to meet these needs. The **elderly support ratio** is the number of working-age people (ages 15 to 64) divided by the number of persons 65 and older (Figure 2-25).

PAUSE & REFLECT 2.3.2

If the elderly support ratio is declining, does that mean the percentage of elderly people is increasing or decreasing?

The world's elderly support ratio is currently around 9, meaning that there are 9 people of working age for every elderly person. In 2050, the ratio is expected to decline to around 4, meaning that there will be only 4 people of working age available to support elderly people who have retired from work. Thus, as the ratio gets smaller, fewer workers are available to contribute to pensions, health care, and other support that older people need.

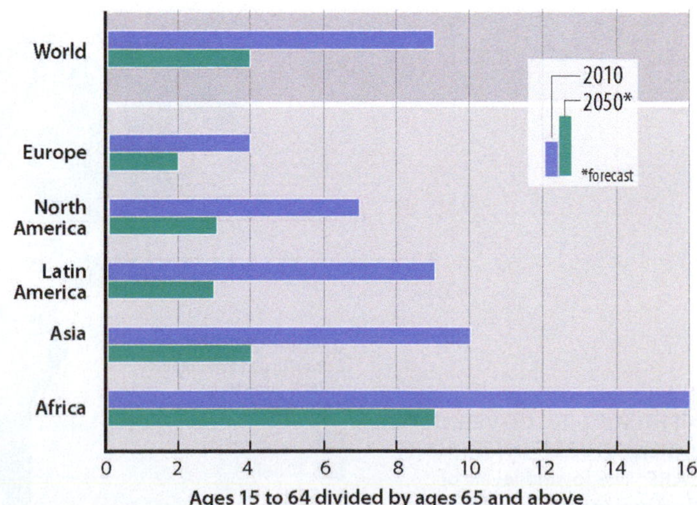

▲ **FIGURE 2-25 ELDERLY SUPPORT RATIO** A smaller number means that fewer workers are available to support elderly people.

Medical Services

LEARNING OUTCOME 2.3.3

Understand variations in health-care services between developed and developing countries.

Health conditions vary around the world. Countries possess different resources to care for people who are sick.

HEALTH CARE

Developed countries use part of their wealth to protect people who, for various reasons, are unable to work. In these countries, some public assistance is offered to those who are sick, elderly, poor, disabled, orphaned, veterans of wars, widows, unemployed, or single parents. Annual per capita expenditure on health care exceeds $1,000 in Europe and $5,000 in the United States, compared to less than $100 in sub-Saharan Africa and South Asia (Figure 2-26).

Expenditures on health care exceeds 15 percent of total government expenditures in Europe and North America compared to less than 5 percent in sub-Saharan Africa and South Asia (Figure 2-27). Countries in northwestern Europe, including Denmark, Norway, and Sweden, typically provide the highest level of public-assistance payments. So not only do developed countries spend more on health care, they spend a higher percentage of their wealth on health care.

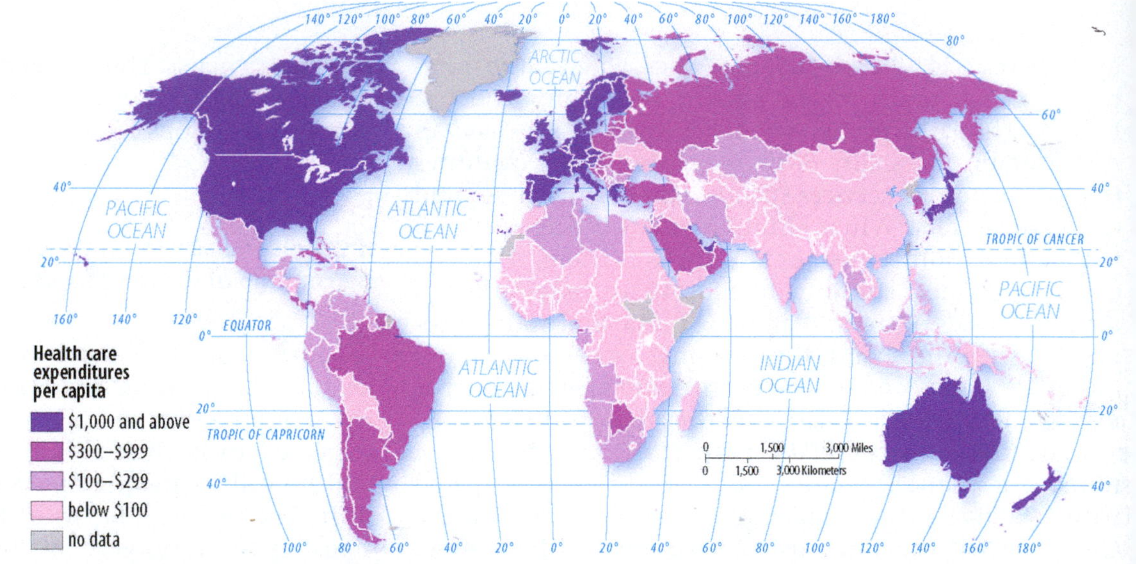

▶ FIGURE 2-26 HEALTH-CARE EXPENDITURES The lowest levels of per capita health-care expenditures are in sub-Saharan Africa and South Asia.

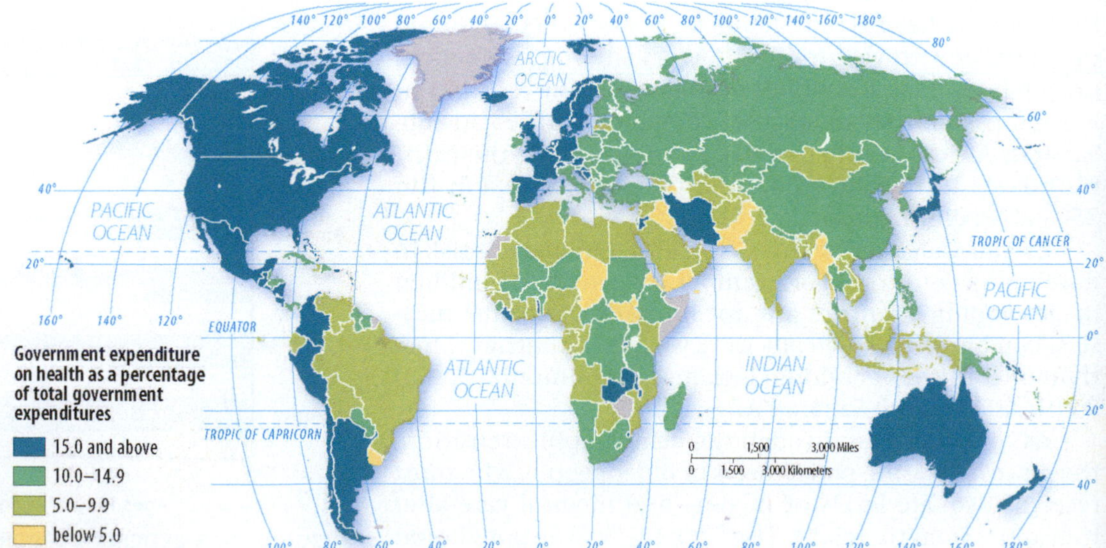

▶ FIGURE 2-27 GOVERNMENT EXPENDITURES ON HEALTH CARE The lowest levels of government support for health care are in sub-Saharan Africa and South Asia.

MEDICAL FACILITIES

The high expenditure on health care in developed countries is reflected in medical facilities. Most countries in Europe have more than 50 hospital beds per 10,000 people, compared to fewer than 20 in sub-Saharan Africa and South and Southwest Asia (Figure 2-28). Europe has more than 30 physicians per 10,000 population, compared to fewer than 5 in sub-Saharan Africa (Figure 2-29).

In most developed countries, health care is a public service that is available at little or no cost. Government programs pay more than 70 percent of health-care costs in most European countries, and private individuals pay less than 30 percent. In developing countries, private individuals must pay more than half of the cost of health care. An exception to this pattern is the United States, a developed country where private individuals are required to pay an average of 55 percent of health care, more closely resembling the pattern in developing countries.

PAUSE & REFLECT 2.3.3

Why might levels of hospital beds and physicians in developed countries of Europe be higher than in North America?

Developed countries are hard-pressed to maintain their current levels of public assistance. In the past, rapid economic growth permitted these states to finance generous programs with little difficulty. But in recent years economic growth has slowed, while the percentage of people needing public assistance has increased. Governments have faced a choice between reducing benefits and increasing taxes to pay for them. In some of the poorest countries, threats to health and sustainability are not so much financial as environmental.

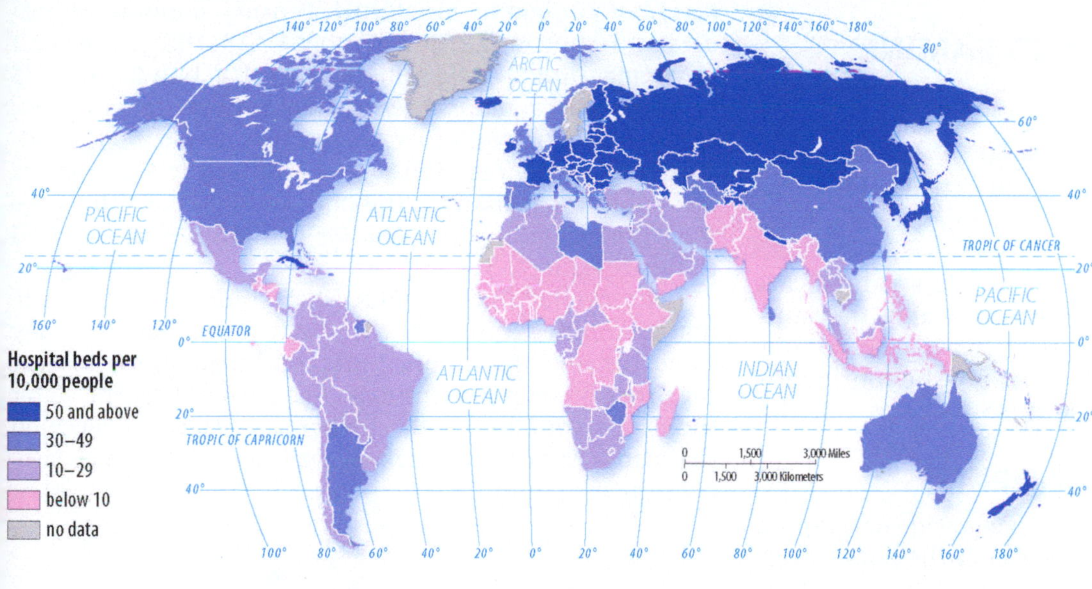

FIGURE 2-28 HOSPITAL BEDS PER 10,000 PEOPLE The lowest numbers of available beds are in sub-Saharan Africa and South Asia.

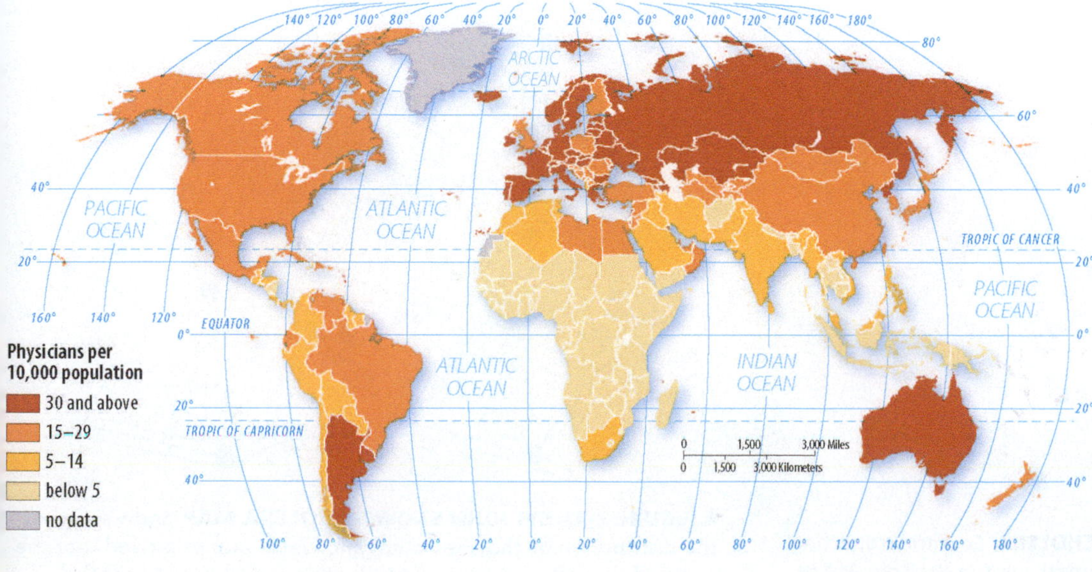

FIGURE 2-29 PHYSICIANS PER 10,000 PEOPLE The fewest available doctors are in sub-Saharan Africa.

The Epidemiologic Transition

LEARNING OUTCOME 2.3.4

Summarize the four stages of the epidemiologic transition.

Epidemiology is the branch of medical science concerned with the incidence, distribution, and control of diseases that are prevalent among a population at a particular time and are produced by some special causes not generally present in the affected place.

The **epidemiologic transition** focuses on distinctive health threats in each stage of the demographic transition. Epidemiologists rely heavily on geographic concepts such as scale and connection because measures to control and prevent an epidemic derive from understanding its distinctive distribution and method of diffusion. The concept was originally formulated by epidemiologist Abdel Omran in 1971.

STAGE 1: PESTILENCE AND FAMINE

In stage 1 of the epidemiologic transition, infectious and parasitic diseases were principal causes of human deaths, along with accidents and attacks by animals and other humans. Thomas Malthus called these causes of deaths "natural checks" on the growth of the human population in stage 1 of the demographic transition.

History's most violent stage 1 epidemic was the Black Plague (bubonic plague), which was probably transmitted to humans by fleas from migrating infected rats. About 25 million Europeans—at least one-half of the continent's population—died between 1347 and 1350.

STAGE 2: RECEDING PANDEMICS

A **pandemic** is disease that occurs over a wide geographic area and affects a very high proportion of the population. Stage 2 of the epidemiologic transition is the stage of receding pandemics because improved sanitation, nutrition, and medicine during the Industrial Revolution reduced the spread of infectious diseases.

Death rates did not decline immediately and universally during the early years of the Industrial Revolution. Poor people crowded into rapidly growing industrial cities had especially high death rates. Cholera, which was uncommon in rural areas, became an especially virulent epidemic in urban areas during the Industrial Revolution. Construction of water and sewer systems had eradicated cholera in Europe and North America by the late nineteenth century.

Cholera persists in several developing regions in stage 2 of the demographic transition, especially sub-Saharan Africa and South and Southeast Asia, where many people lack access to clean drinking water (Figure 2-30). Cholera has also been found on Hispaniola, the island shared by Haiti and the Dominican Republic, especially in the wake of an earthquake in 2010 that killed 200,000 and displaced 1 million.

A century before the invention of computers, a mapping technique that anticipated geographic information systems (GIS) helped to explain and battle the stage 2 pandemic of cholera in nineteenth-century London. To understand cholera, Dr. John Snow (1813–1858), a British physician (not a geographer) created a handmade GIS in 1854. On a map of London's Soho neighborhood, Snow overlaid two other maps, one showing the addresses of cholera victims and the other the locations of water pumps—which for the poor residents of Soho were the principal source of water for drinking, cleaning, and cooking (Figure 2-31).

The overlay maps showed that cholera victims were not distributed uniformly through Soho. Rather, they were clustered around one pump on Broad Street. Tests at the Broad Street pump subsequently proved that the water there was contaminated. Further investigation revealed that contaminated sewage was getting into the water supply near the pump.

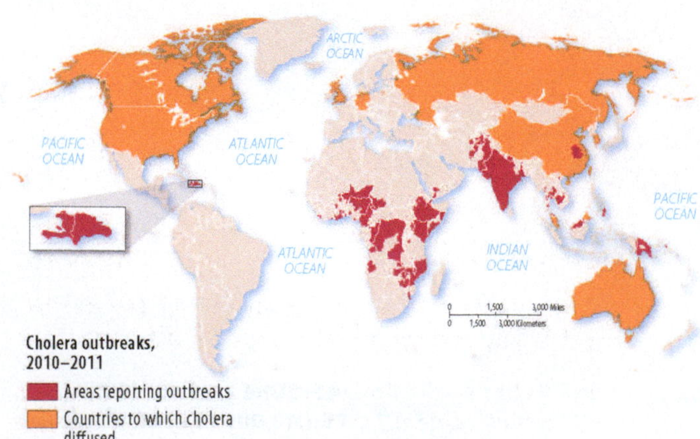

▲ **FIGURE 2-30 STAGE 2 DISEASE: CHOLERA** Countries reporting cholera in recent years are found primarily in sub-Saharan Africa and South Asia.

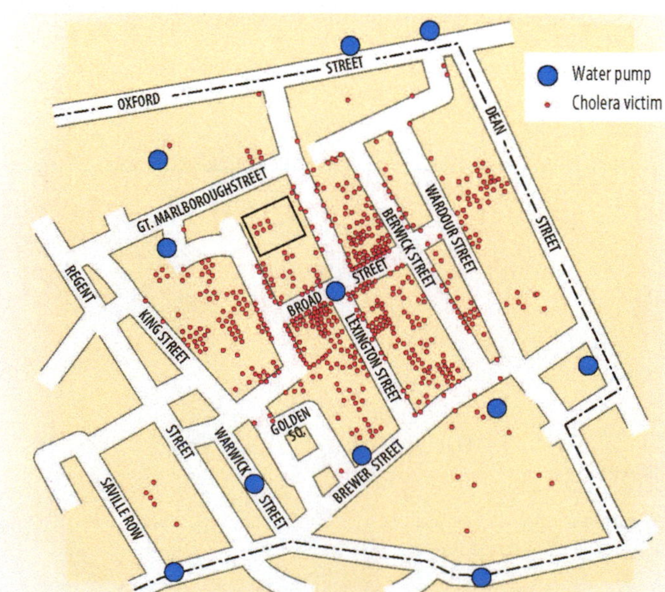

▲ **FIGURE 2-31 SIR JOHN SNOW'S CHOLERA MAP** Snow's map of the distribution of cholera victims and water pumps proved that the cause of the infection was contamination of the pump near the corner of Broad and Lexington streets.

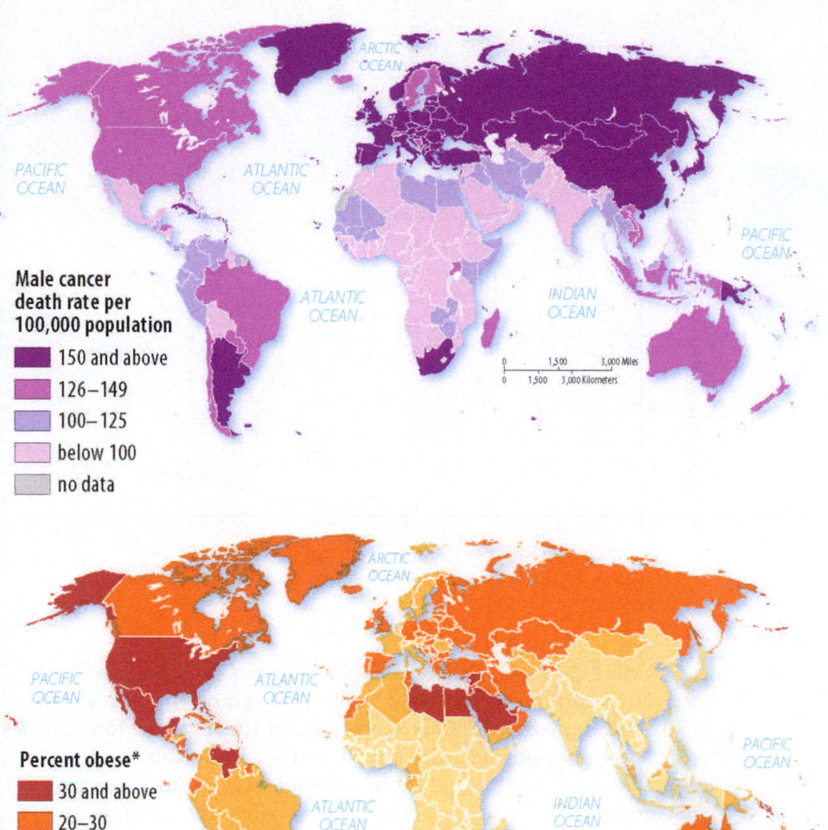

◀ **FIGURE 2-32 STAGE 3 DISEASE: MALE CANCER** Cancer is an example of a cause of death for men that is higher in developed countries than in developing ones.

◀ **FIGURE 2-33 STAGE 4 DISEASE: OBESITY** Obesity is a health problem in the United States and in Southwest Asia & North Africa.

Before Dr. Snow's geographic analysis, many believed that epidemic victims were being punished for sinful behavior and that most victims were poor because poverty was considered a sin. Now we understand that cholera affects the poor because they are more likely to have to use contaminated water.

STAGE 3: DEGENERATIVE DISEASES

Stage 3 of the epidemiologic transition is characterized by a decrease in deaths from infectious diseases and an increase in chronic disorders associated with aging. The two especially important chronic disorders in stage 3 are cardiovascular diseases, such as heart attacks, and various forms of cancer. The global pattern of cancer is the opposite of that for stage 2 diseases; sub-Saharan Africa and South Asia have the lowest incidence of cancer, primarily because of the relatively low life expectancy in those regions (Figure 2-32).

STAGE 4: DELAYED DEGENERATIVE DISEASES

Omran's epidemiologic transition was extended by S. Jay Olshansky and Brian Ault to stage 4, the stage of delayed degenerative diseases. The major degenerative causes of death—cardiovascular diseases and cancers—linger, but the life expectancy of older people is extended through medical advances. Through medicine, cancers spread more slowly or are removed altogether. Operations such as bypasses repair deficiencies in the cardiovascular system. Also improving health are behavior changes such as better diet, reduced use of tobacco and alcohol, and exercise. On the other hand, consumption of non-nutritious food and sedentary behavior have resulted in an increase in obesity in stage 4 countries (Figure 2-33).

PAUSE & REFLECT 2.3.4

How prevalent are the stage 4 causes of death in your family?

CHECK-IN KEY ISSUE 3

Why Do Some Places Face Health Challenges?

- ✔ **Birth rates have declined in nearly all countries through a variety of family-planning approaches.**
- ✔ **The percentage of younger and older people in a country impacts its provision of health care.**
- ✔ **The provision of health care varies sharply between developed and developing countries.**
- ✔ **The epidemiologic transition has four stages of distinctive diseases.**

KEY ISSUE 4

Why Might Population Increase in the Future?

- **Population and Resources**
- **Population Futures**
- **Epidemiologic Futures**
- **Family Futures**

Geographic concepts offer insights into future population and health trends. In view of the current size of Earth's population and the NIR, will there soon be too many of us? Or will Earth's population decline in the future?

Population and Resources

LEARNING OUTCOME 2.4.1

Summarize arguments supporting and opposing Malthus's theory of the connection between population and resources.

English economist Thomas Malthus (1766–1834) was one of the first to argue that the world's rate of population increase was far outrunning the development of food supplies. In *An Essay on the Principle of Population*, published in 1798, Malthus claimed that the population was growing much more rapidly than Earth's food supply because population increased geometrically, whereas food supply increased arithmetically (Figure 2-34). According to Malthus, these growth rates would produce the following relationships between people and food in the future:

Today:	1 person, 1 unit of food
25 years from now:	2 persons, 2 units of food
50 years from now:	4 persons, 3 units of food
75 years from now:	8 persons, 4 units of food
100 years from now:	16 persons, 5 units of food

Malthus stated these conclusions several decades after England had become the first country to enter stage 2 of the demographic transition, in association with the Industrial Revolution. He concluded that population growth would press against available resources in every country unless "moral restraint" produced lower CBRs or unless disease, famine, war, or other disasters produced higher CDRs.

PAUSE & REFLECT 2.4.1

Calculate the units of population and food that Malthus predicted would exist in 200 years.

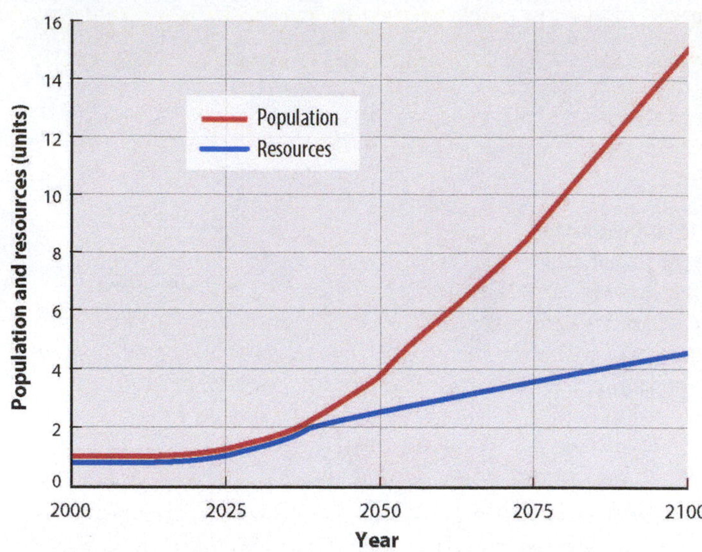

▲ **FIGURE 2-34 MALTHUS'S THEORY** Malthus expected population to grow more rapidly than food supply. The graph shows that if in 2000, the population of a place were 1 unit (such as 1 billion people) and the amount of resources were 1 unit (such as 1 billion tons of grain), then according to Malthus's theory, in 2100 the place would have around 15 billion people but only 5 billion tons of grain).

CONTEMPORARY NEO-MALTHUSIANS AND CRITICS

Malthus's views remain influential today. Contemporary geographers and other analysts are taking another look at Malthus's theory because of Earth's unprecedented rate of natural increase during the twentieth century. Neo-Malthusians argue that characteristics of recent population growth make Malthus's thesis even more frightening than when it was first written more than 200 years ago. Malthus's theory has been severely criticized from a variety of perspectives. Criticism has been leveled at both the population growth and resource depletion sides of Malthus's equation (see Debate It feature and Figures 2-35 and 2-36).

Evidence from the past half-century lends support to both Neo-Malthusians and their critics. Malthus was fairly close to the mark on resources but much too pessimistic on population growth.

Overall food production has increased during the past half-century somewhat more rapidly than Malthus predicted. In India, for example, rice production has followed Malthus's expectations fairly closely, but wheat production has increased twice as fast as Malthus expected (Figure 2-37). Better growing techniques, higher-yielding seeds, and cultivation of more land have contributed to the increase in the food supply (see Chapter 9). However, neo-Malthusians point out that production of both wheat and rice has slowed in India since 2000, as shown in Figure 2-37. Without new breakthroughs in food production, India might not be able to keep food supply ahead of population growth.

On the population side of the equation, recent evidence indicates that Malthus has been less accurate. His model expected population to quadruple during a half-century, but

DEBATE IT! Can Earth's resources support our growing population?

Some geographers argue that Malthus was right, and that population growth is depleting key resources. Other geographers do not accept this view.

MALTHUSIANS SAY NO

- **STAGE 2 AND DEVELOPMENT.** In Malthus's time only a few relatively wealthy countries had entered stage 2 of the demographic transition, characterized by rapid population increase. Malthus failed to anticipate that relatively poor countries would have the most rapid population growth because of transfer of medical technology (but not wealth) from developed countries. As a result, the gap between population growth and resources is wider in some countries than even Malthus anticipated.
- **RESOURCE DEPLETION.** World population growth is outstripping a wide variety of resources, not just food production. Neo-Malthusians paint a frightening picture of a world in which billions of people are engaged in a desperate search for food, water, and energy.

▲ **FIGURE 2-35 POPULATION OUTSTRIPS RESOURCES** Obtaining drinking water from a lake, Myanmar.

CRITICS OF NEO-MALTHUSIANISM SAY YES

- **EXPANDING RESOURCES.** The world's overall supply of resources is actually expanding rather than fixed, as neo-Malthusians believe. While the supply of some resources is decreasing, other resources are available to replace them.
- **ECONOMIC GROWTH.** Population growth stimulates economic growth, and therefore production of more food. More consumers generate more demand for goods, which results in more jobs. More people means more brains to invent good ideas for improving life.
- **INEQUALITY.** Poverty and hunger result from an unjust society and economic inequality, not population growth. The world possesses sufficient resources to eliminate hunger and poverty, if only these resources are shared equally.

▲ **FIGURE 2-36 EXPANDING RESOURCES** Market, Mahebourg, Mauritius.

even in India—a country known for relatively rapid growth—population has increased more slowly than that, and it has also increased more slowly than the country's food supply.

On the other hand, the track toward overpopulation may already be irreversible in Africa. Rapid population growth has led to the inability of the land to sustain life in parts of the region. As the land declines in quality, more effort is needed to yield the same amount of crops. This extends the working day of women, who have the primary responsibility for growing food for their families. Women then regard having another child as a means of securing additional help in growing food.

Sub-Saharan Africa was not classified in Key Issue 1 as one of the world's population concentrations. Geographers caution that the size, density, or clustering of population in a region is not an indication of overpopulation. Instead, overpopulation is a relationship between population and a region's level of resources. The capacity of the land to sustain life derives partly from characteristics of the natural environment and partly from human actions to modify the environment through agriculture, industry, and exploitation of raw materials.

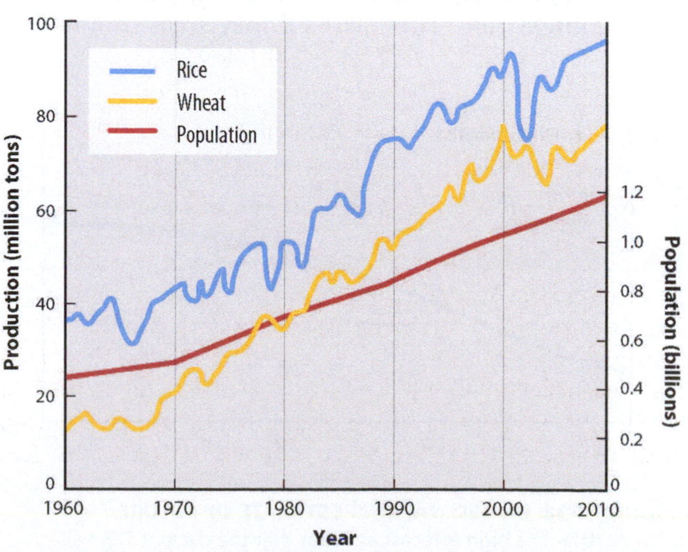

▲ **FIGURE 2-37 MALTHUS'S THEORY APPLIED TO INDIA** Production of wheat and rice has increased more rapidly than has population—the opposite of Malthus's theory in Figure 2-34.

Population Futures

LEARNING OUTCOME 2.4.2

Understand the future population of the world's most populous countries and elements of a possible stage 5 of the demographic transition.

How many people will inhabit Earth in the future? Future population depends primarily on fertility. How many babies will women in the future bear in their lifetimes? Despite the importance of this question, the answer is unknowable.

The United Nations estimates that world population in 2100 could increase to 15.8 billion or decline to 6.2 billion. If the UN's high variant is followed, world population would more than double by 2100. If the low variant is followed, world population would actually decline (Figure 2-38).

DEMOGRAPHIC TRANSITION POSSIBLE STAGE 5: DECLINE

- **Very low CBR**
- **Increasing CDR**
- **Declining NIR**

Demographers predict a possible stage 5 of the demographic transition for some developed countries. Stage 5 would be characterized by a very low CBR, an increasing CDR, and therefore a negative NIR (Figure 2-39). After several decades of very low birth rates, a stage 5 country would have relatively few young women aging into childbearing years. As those in the smaller pool of women each chooses to have fewer children, birth rates would continue to fall even more than in stage 4.

Several European countries, notably Russia and other former Communist countries, already have negative NIRs. Russia's high CDR and low CBR are a legacy of a half-century of Communist rule. The low CBR may stem from a long tradition of strong family-planning programs and a deep-seated pessimism about having children in an uncertain world. The high CDR may be a legacy of inadequate pollution controls and inaccurate reporting by the Communists.

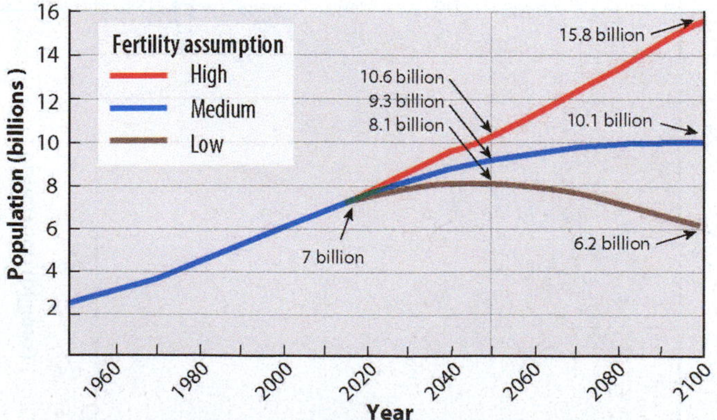

▲ **FIGURE 2-38 UNITED NATIONS ESTIMATE OF FUTURE POPULATION** The high forecast assumes that the current TFR will continue in the future, the medium forecast anticipates a modest decline in the TFR, and the low forecast assumes a sharp decline in the TFR.

CHINA AND INDIA

The world's two most populous countries, China and India, will heavily influence future prospects for global overpopulation. These two countries—together encompassing more than one-third of the world's population—have adopted different family-planning programs. As a result of less effective policies, India adds 12 million more people each year than does China. Current projections show that India could surpass China as the world's most populous country around 2030.

CHINA'S POPULATION POLICIES. China has made substantial progress in reducing its NIR. The core of the Chinese government's family-planning program has been the One Child Policy, adopted in 1980. Under the One Child Policy, a couple needed a permit to have a child. Couples received financial subsidies, a long maternity leave, better housing, and (in rural areas) more land if they agreed to have just one child. To further discourage births, people receive free contraceptives, abortions, and sterilizations. Rules were enforced by a government agency.

Largely as a result of the One Child Policy, China's CBR declined from 18 in 1980 to 12 in 2015, and consequently the NIR declined from 1.2 to 0.5. Since 2000, China has actually had a lower CBR than the United States. The number of people added to China's population each year has dropped by one-half, from 14 million to 7 million, during the past quarter-century.

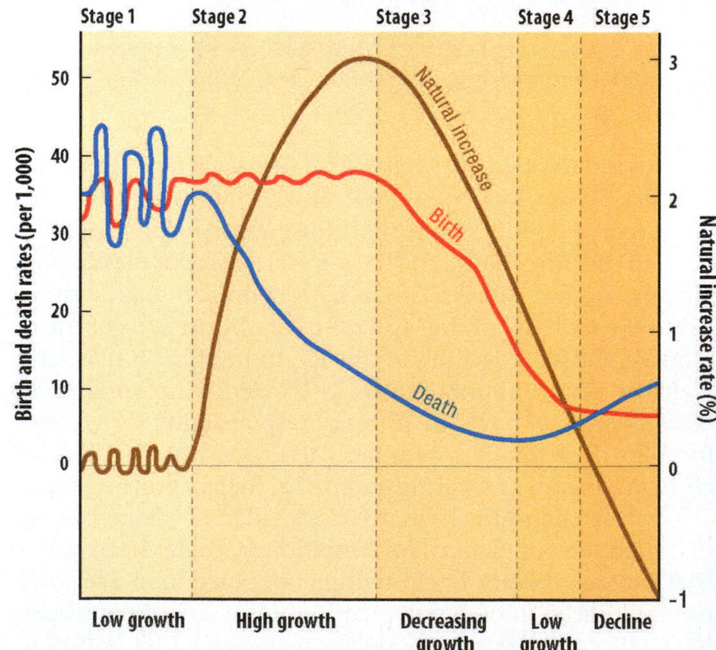

▲ **FIGURE 2-39 FIVE-STAGE DEMOGRAPHIC TRANSITION** Stage 5 of the demographic transition would be characterized by a negative NIR because the CDR would be greater than the CBR.

SUSTAINABILITY & OUR ENVIRONMENT Japan's Population Decline

If the demographic transition is to include a stage 5, Japan will be one of the world's first countries to reach it. Japan's population hit a historic high of 127 million in 2010 and is now starting to decline. The United Nations forecasts Japan's population to fall to 84 million in 2100. With the population decline will come an increasing percentage of elderly people (Figure 2-40).

For planet Earth, a more modest natural increase rate in the coming years reduces the pressure on expanding resources to feed more people. For Japan as a specific country, stage 5 brings economic challenges of sustaining a high standard of living with an aging population and a declining number of people of working age.

Japan faces an unsustainable shortage of workers. Instead of increasing immigration, Japan is addressing its labor force shortage primarily by encouraging more Japanese people to work. Rather than combine work with child rearing, Japanese women are expected to make a stark choice: either marry and raise children or remain single and work. According to Japan's most recent census, the majority of women have chosen to work.

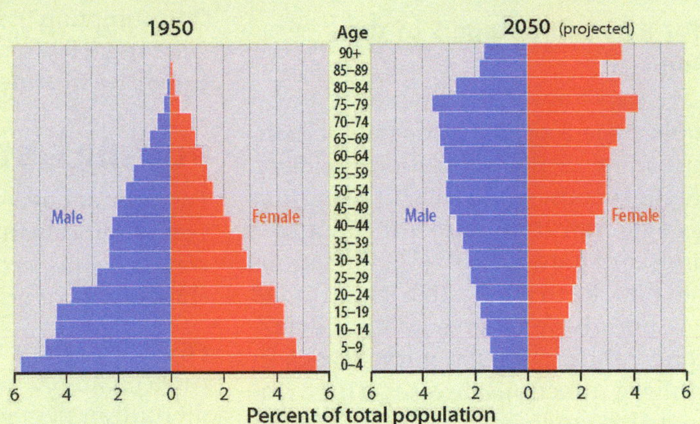

▲ **FIGURE 2-40 JAPAN'S CHANGING POPULATION PYRAMIDS** Japan's population pyramid has shifted from a broad base in 1950 to a rectangular shape today. In the future, the bottom of the pyramid is expected to contract and the top to expand.

With the United Nations now forecasting China to lose population by 2100, the government abandoned the One Child Policy in 2015. But China's CBR is unlikely to increase much because after three decades of intensive educational programs, as well as coercion, most Chinese have accepted the benefits of family planning (Figure 2-41).

INDIA'S POPULATION POLICIES. India was the first country to embark on a national family-planning program, in 1952. The government established clinics, provided information about alternative methods of birth control, distributed free or low-cost birth control devices, and legalized abortions.

Most controversially, during the 1970s India set up camps to perform sterilizations—surgical procedures in which people were made incapable of reproduction. Widespread opposition to the sterilization program grew in the country because people feared that they would be forcibly sterilized, and it increased distrust of other family-planning measures as well.

In the past several decades, government-sponsored family-planning programs in India have emphasized education, including advertisements on national radio and TV networks and information distributed through local health centers. Still, the dominant form of birth control continues to be sterilization of women, though in many cases after the women have already borne several children. Family-planning measures lowered the CBR from 34 in 1980 to 21 in 2015, but India's population increased by 16 million in both 1980 and 2015.

PAUSE & REFLECT 2.4.2

For every 10 baby boys, India has only 9 baby girls. How might population policies be contributing to India having so many more baby boys than girls?

▲ **FIGURE 2-41 CHINA'S ONE CHILD POLICY** Poster in Jiangxi advocates one child.

Epidemiologic Futures

LEARNING OUTCOME 2.4.3

Understand reasons for a possible stage 5 of the epidemiologic transition.

Recall that in the possible stage 5 of the demographic transition, CDR rises because more of the population is elderly. Some medical analysts argue that the world is also moving into stage 5 of the epidemiologic transition, brought about by a reemergence of infectious and parasitic diseases. Infectious diseases once thought to have been eradicated or controlled have returned, and new ones have emerged. Other epidemiologists dismiss recent trends as a temporary setback in a long process of controlling infectious diseases. Three reasons help explain the possible emergence of a stage 5 of the epidemiologic transition: evolution, poverty, and increased connections.

POSSIBLE STAGE 5 CAUSE: EVOLUTION

Infectious disease microbes have continuously evolved and changed in response to environmental pressures by developing resistance to drugs and insecticides. Antibiotics and genetic engineering contribute to the emergence of new strains of viruses and bacteria. Malaria was nearly eradicated in the mid-twentieth century by spraying DDT in areas infested with the mosquito that carried the parasite. However, malaria caused an estimated 118,648 deaths worldwide in 2014, including 30,918 in the Democratic Republic of the Congo. A major reason is the evolution of DDT-resistant mosquitoes.

POSSIBLE STAGE 5 CAUSE: POVERTY

Infectious diseases are more prevalent in poor areas than other places because unsanitary conditions may persist, and most people can't afford the drugs needed for treatment. Tuberculosis (TB) is an example of an infectious disease that has been largely controlled in developed countries but remains a major cause of death in developing countries (Figure 2-42). An airborne disease that damages the lungs, TB (often called "consumption") spreads principally through coughing and sneezing. TB is more prevalent in poor areas because the long, expensive treatment poses a significant economic burden.

POSSIBLE STAGE 5 CAUSE: CONNECTIONS

Pandemics have spread in recent decades through the process of relocation diffusion, discussed in Chapter 1. As they travel, people carry diseases with them and are exposed to the diseases of others.

AIDS. The most lethal pandemic in recent years has been AIDS (acquired immunodeficiency syndrome). Worldwide, 39 million people died of AIDS from the beginning of the epidemic through 2014, and 37 million were living with HIV (human immunodeficiency virus, the cause of AIDS). The impact of AIDS has been felt most strongly in sub-Saharan Africa, home to 26 million of the world's 37 million HIV-positive people (Figure 2-43).

AIDS diffused from sub-Saharan Africa through relocation diffusion, both by Africans and by visitors to Africa returning to their home countries. AIDS entered the United States during the early 1980s through New York, California, and Florida. Not by coincidence, the three leading U.S. airports for international arrivals are in these three states (Figure 2-44). Though AIDS diffused to every state during the 1980s, these three states, plus Texas (a major port of entry by motor vehicle), accounted for half of the country's new AIDS cases in the peak year of 1993.

The number of new AIDS cases has dropped sharply because of the rapid diffusion of preventive methods and medicines such as AZT. The rapid spread of these innovations is an example of expansion diffusion rather than relocation diffusion.

EBOLA. Several dozen "new" pandemics have emerged over the past several decades and have spread through the process

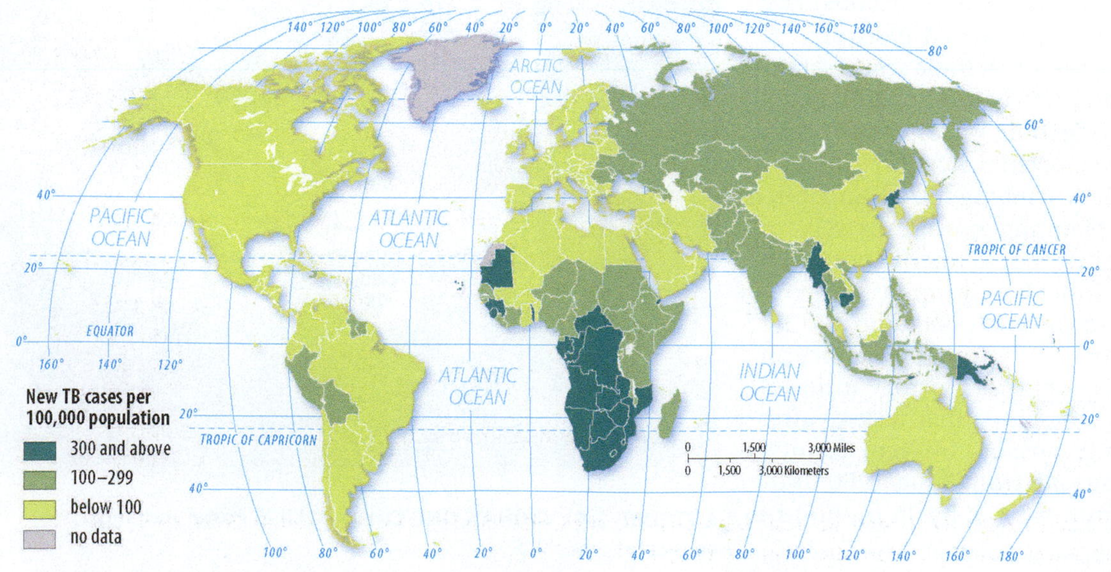

▲ **FIGURE 2-42 TUBERCULOSIS (TB) DEATHS** Deaths from TB are found primarily in poorer countries that are unable to pay for the expensive treatment.

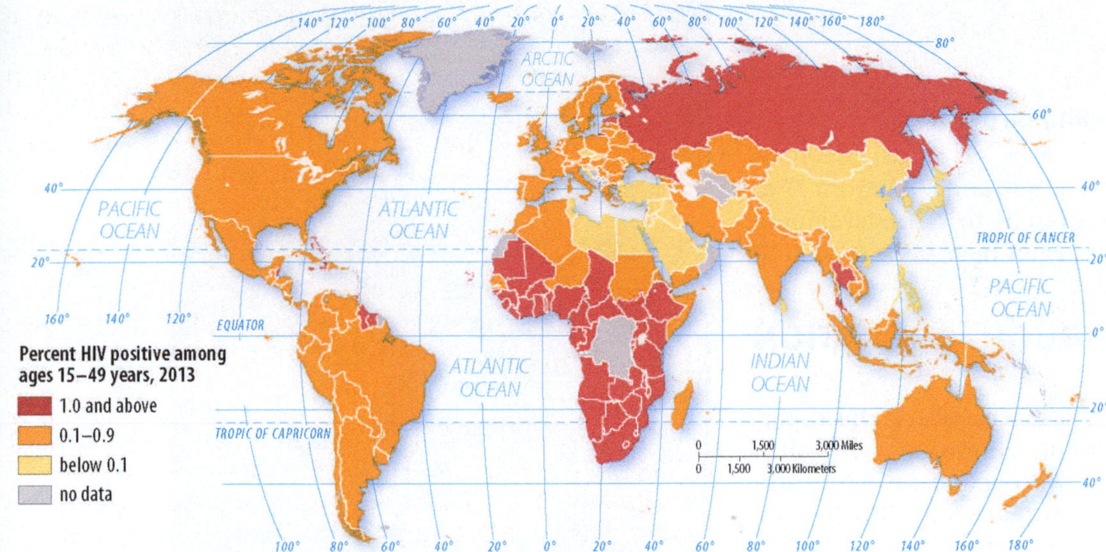

▲ **FIGURE 2-43 AIDS** The highest rates of infection are in sub-Saharan Africa and Russia.

of relocation diffusion. Among them is Ebola. The disease is named for the Ebola River in the Democratic Republic of Congo, where the first known victim was identified in 1976. Mabalo Lokela, the victim, returned to his home village of Yambuku, where he was the school headmaster, after a visit to the Ebola River 150 kilometers from the village. He was originally diagnosed with and treated for malaria.

The first known victim in West Africa was a 2-year-old boy in the village of Meliandou, Guinea, who died in December 2013. Ebola is thought to have been carried from the Congo to Guinea by fruit bats. The virus quickly spread in early 2014 to isolated villages in Guinea and the neighboring countries Sierra Leone and Liberia, which are within walking distance of Meliandou. These are some of the poorest places within some of the poorest countries of the world. They lacked the medical care necessary to prevent the spread of the virus and to save those who contracted it.

Ebola was transported from West Africa by health-care workers who traveled to other places while unknowingly infected. However, these places possessed health-care systems that were able to treat patients.

PAUSE & REFLECT 2.4.3

Which region within the United States has the lowest number of AIDS cases? What geographic factors might explain this low level?

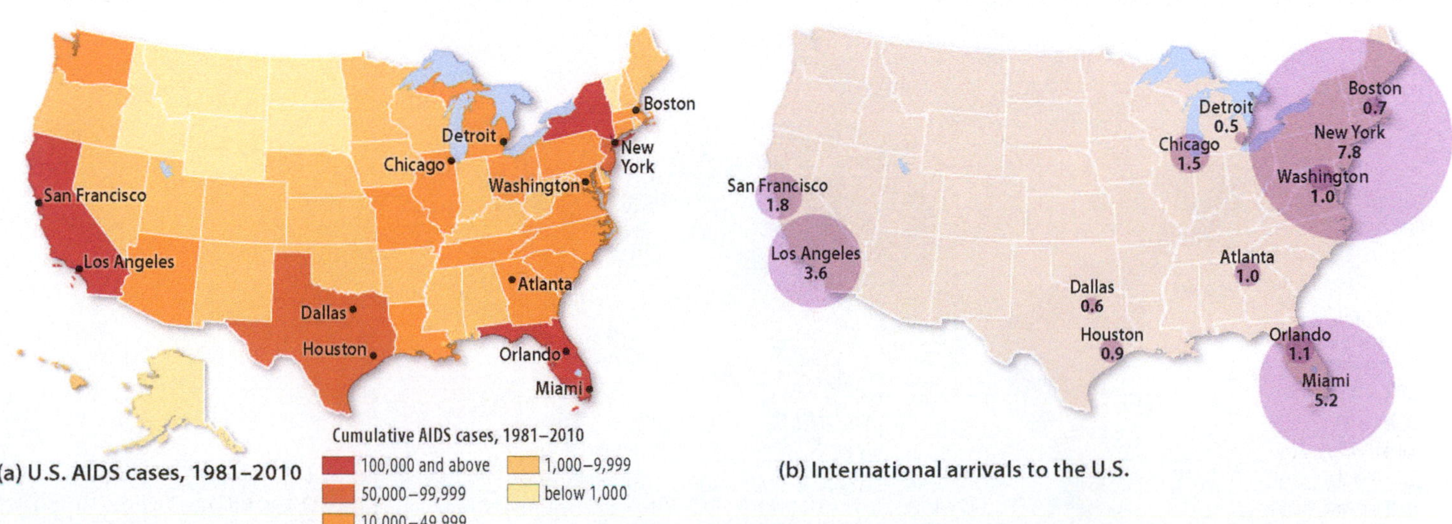

▲ **FIGURE 2-44 U.S. AIDS AND INTERNATIONAL ARRIVALS** Because AIDS arrived in the United States primarily through air travelers, the distribution of AIDS closely matches the distribution of international air passengers.

Family Futures

LEARNING OUTCOME 2.4.4

Understand reasons for declining birth rates.

The CBR declined rapidly between 1990 and 2015, from 27 to 20 in the world as a whole and from 31 to 22 in developing countries (Figure 2-45). Two strategies have been successful in reducing birth rates.

LOWERING CBR THROUGH EDUCATION AND HEALTH CARE

One approach to lowering birth rates emphasizes the importance of improving local economic conditions. A wealthier community has more money to spend on education and health-care programs that promote lower birth rates. According to this approach:

- With more women able to attend school and to remain in school longer, they would be more likely to learn employment skills and gain more economic control over their lives.
- With better education, women would better understand their reproductive rights, make more informed reproductive choices, and select more effective methods of contraception (Figure 2-46).
- With improved health-care programs, IMRs would decline through such programs as improved prenatal care, counseling about sexually transmitted diseases, and child immunization.
- With the survival of more infants ensured, women would be more likely to choose to make more effective use of contraceptives to limit the number of children.

▲ **FIGURE 2-46 INDIA FAMILY PLANNING** Family-planning center, Kolkata, India.

LOWERING CBR THROUGH CONTRACEPTION

The other approach to lowering birth rates emphasizes the importance of rapidly diffusing modern contraceptive methods (Figure 2-47). Economic development may promote lower birth rates in the long run, but the world cannot wait around for that alternative to take effect. Putting resources

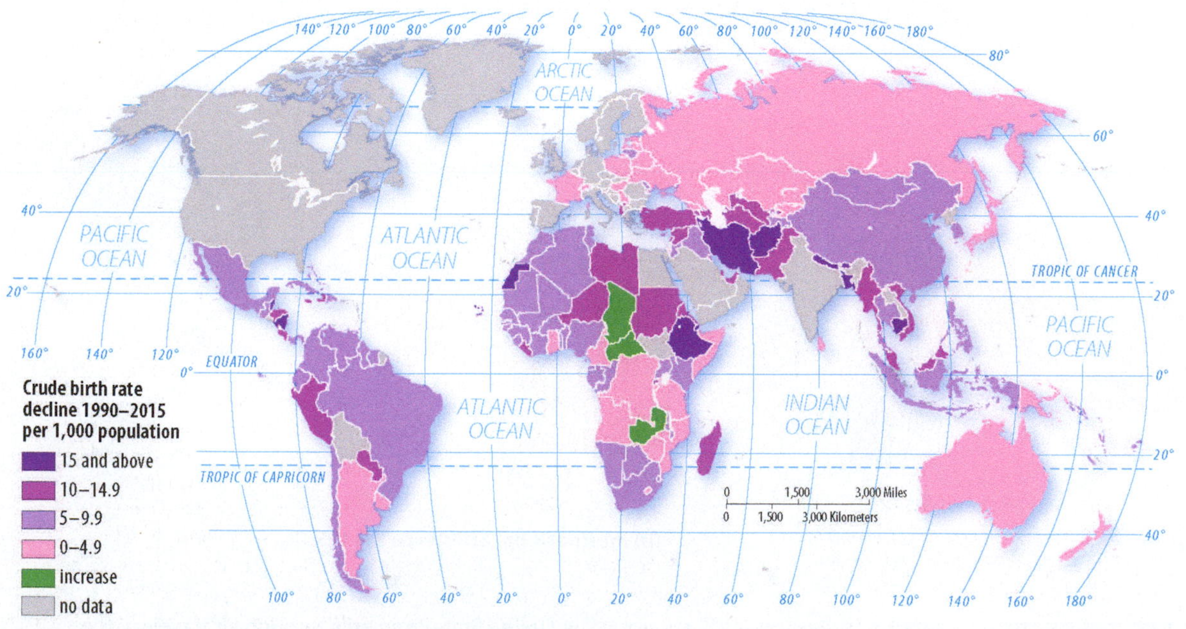

▲ **FIGURE 2-45 CBR CHANGE 1990–2015** The CBR has declined in all but a handful of countries. Declines have been most rapid in Latin America and South and Southwest Asia.

into family-planning programs can reduce birth rates much more rapidly. In developing countries, demand for contraceptive devices is greater than the available supply. Therefore, the most effective way to increase their use is to distribute more of them cheaply and quickly. According to this approach, contraceptives are the best method for lowering the birth rate.

Bangladesh is an example of a country that has had little improvement in the wealth and literacy of its people, but 62 percent of married women in the country used contraceptives in 2014 compared to 6 percent in 1980. Similar growth in the use of contraceptives has occurred in other developing countries, including Colombia, Morocco, and Thailand. Rapid growth in the acceptance of family planning is evidence that in the modern world, ideas can diffuse rapidly, even to places where people have limited access to education and modern communications. The percentage of women using contraceptives is especially low in sub-Saharan Africa, so the alternative of distributing contraceptives could have an especially strong impact there. Around 30 percent of married women in sub-Saharan Africa employ contraceptives, compared to 73 percent in Latin America and 66 percent in Asia (Figure 2-48).

PAUSE & REFLECT 2.4.4

Why might birth rates have declined in most of the world, but not in North America?

CHECK-IN KEY ISSUE 4

Why Might Population Increase in the Future?

- ✔ **Neo-Malthusians argue that population is outstripping resources, but critics do not agree.**
- ✔ **The demographic transition may display a possible stage 5 of population decline.**
- ✔ **A resurgence of infectious diseases may signal a possible stage 5 of the epidemiologic transition.**
- ✔ **Birth rates have declined in most countries through use of two strategies.**

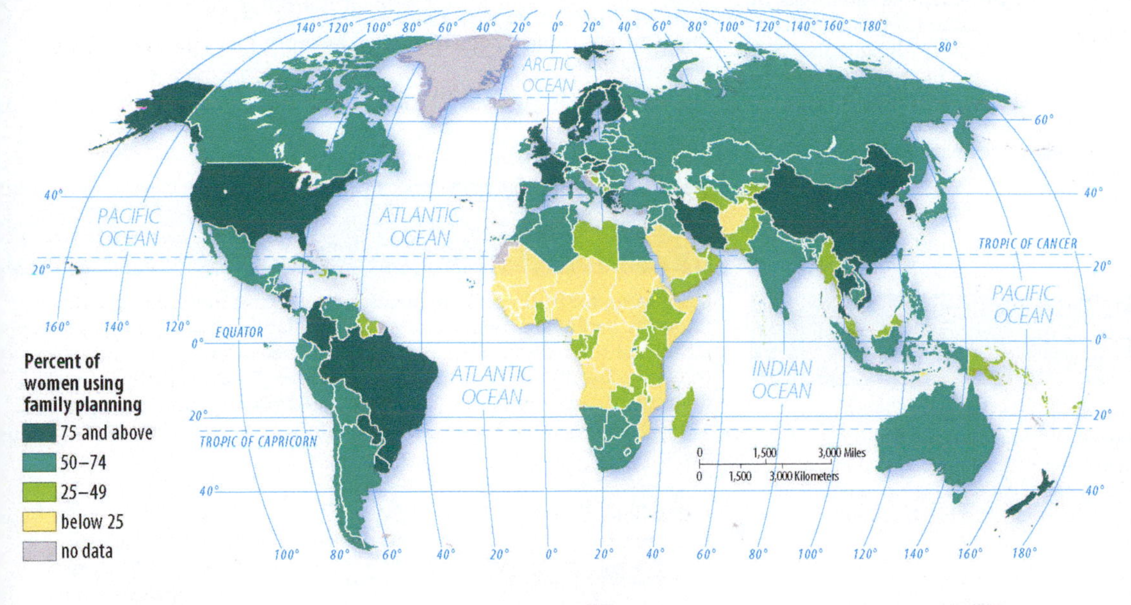

◀ FIGURE 2-47 WOMEN USING FAMILY PLANNING More than two-thirds of couples in developed countries use a family-planning method. Family-planning varies widely in developing countries. China reports the world's highest rate of family planning; the lowest rates are in sub-Saharan Africa.

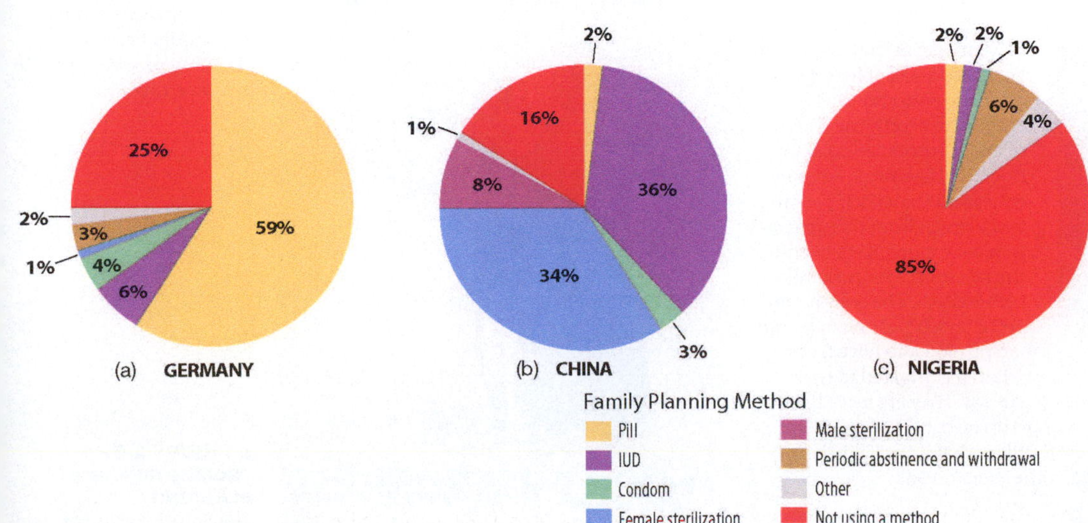

◀ FIGURE 2-48 FAMILY-PLANNING METHODS (a) The principal family-planning methods in developed countries like Germany are condoms and birth-control pills. **(b)** The principal methods in China are intrauterine devices (IUDs) and female sterilization. **(c)** People in sub-Saharan African countries such as Nigeria make minimal use of family planning.

Summary & Review

KEY ISSUE 1

Where are the world's people distributed?

Two-thirds of the world's inhabitants are clustered in four regions. Human beings tend to avoid parts of Earth's surface that they consider to be too wet, too dry, too cold, or too mountainous. Several measures of density are used to describe where people live in the world, and the relationship between people and natural resources.

THINKING GEOGRAPHICALLY

▲ **FIGURE 2-49** **CENSUS** A woman in the United Kingdom completes that country's 2011 census form.

1. The current method of counting a country's population by requiring every household to complete a census form once every 10 years has been severely criticized as inaccurate. The undercounting produces a geographic bias because people who are missed are more likely to live in inner cities, remote rural areas, or communities that attract a relatively high number of recent immigrants. Given the availability of reliable statistical tests, should the current method of trying to count 100 percent of the population be replaced by a survey of a carefully drawn sample of the population, as is done with political polling and consumer preferences? Why or why not?

KEY ISSUE 2

Why is world population increasing?

The demographic transition helps to explain why regions have varying rates of population growth. Virtually all the world's natural increase is concentrated in the developing countries of Africa, Asia, and Latin America. The difference in natural increase between developed countries and developing countries results from differences in birth rates rather than in death rates.

THINKING GEOGRAPHICALLY

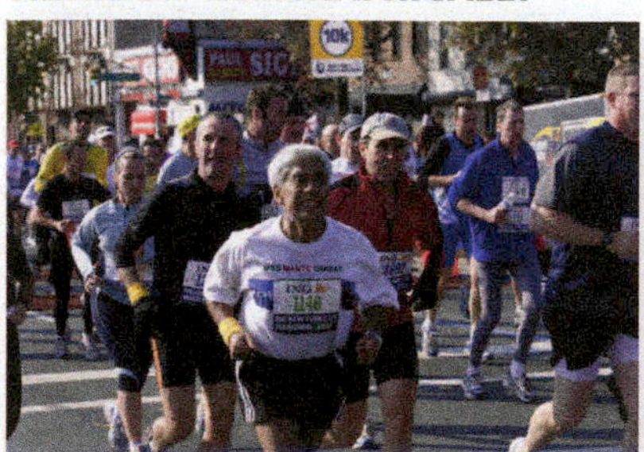

▲ **FIGURE 2-50** **AGING BOOMERS** Older runners participate in New York City Marathon.

2. Members of the baby-boom generation—people born between 1946 and 1964—constitute nearly one-third of the U.S. population. Baby boomers have received more education than their parents, and women from this generation were more likely to enter the labor force than women before them. The baby boomers have delayed marriage and parenthood and have fewer children compared to their parents. They are more likely to divorce, to bear children while unmarried, and to cohabit. As they grow older, what impact will baby boomers have on the American population in the years ahead?

KEY ISSUE 3

Why do some places face health challenges?

Countries have distinctive patterns of gender and age, depending on the stage of the demographic transition, and they display different health conditions and medical services. Health care varies widely around the world because developing countries generally lack resources to provide the same level of health care as developed countries.

THINKING GEOGRAPHICALLY

▲ **FIGURE 2-51** **U.S. HEALTH CARE CHALLENGES** Incidence of obesity is increasing in the United States.

3. Health care indicators for the United States do not always match those of other developed countries. What reasons might explain these differences?

KEY ISSUE 4

Why might population increase in the future?

Malthus argued in 1798 that population would grow more rapidly than resources. Recent experience shows that population has not grown as rapidly as Malthus forecast. Birth rates have declined in some places primarily through education and health care and in other places primarily through diffusion of contraception. Japan and some European countries may be in a possible stage 5, characterized by a decline in population, because CDR exceeds CBR.

THINKING GEOGRAPHICALLY

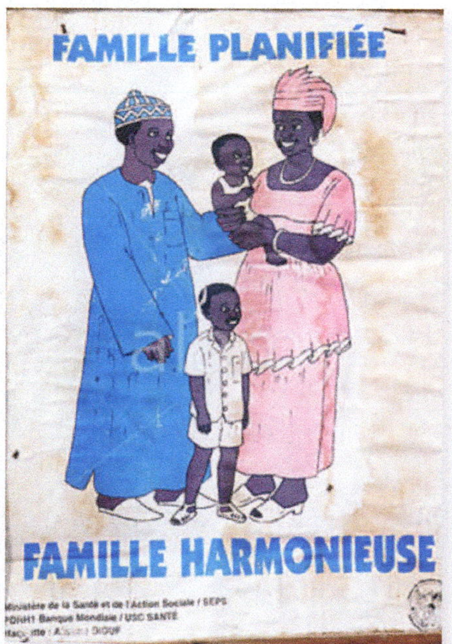

▲ **FIGURE 2-52** **PROMOTING FAMILY PLANNING** The government of Senegal issued this poster to promote family planning.

4. Given the evidence of declining birth rates around the world, what role should family-planning programs play in developing countries?

KEY TERMS

Agricultural density *(p. 50)* The ratio of the number of farmers to the total amount of arable land (land suitable for agriculture).
Arithmetic density *(p. 50)* The total number of people divided by the total land area.
Census *(p. 46)* A complete enumeration of a population.
Crude birth rate (CBR) *(p. 54)* The total number of live births in a year for every 1,000 people alive in the society.
Crude death rate (CDR) *(p. 54)* The total number of deaths in a year for every 1,000 people alive in the society.
Demographic transition *(p. 56)* The process of change in a society's population from a condition of high crude birth and death rates and low rate of natural increase to a condition of low crude birth and death rates, low rate of natural increase, and higher total population.
Demography *(p. 55)* The scientific study of population characteristics.
Dependency ratio *(p. 60)* The number of people under age 15 and over age 64 compared to the number of people active in the labor force.
Doubling time *(p. 52)* The number of years needed to double a population, assuming a constant rate of natural increase.
Ecumene *(p. 49)* The portion of Earth's surface occupied by permanent human settlement.
Elderly support ratio *(p. 61)* The number of working-age people (ages 15 to 64) divided by the number of persons 65 and older.
Epidemiologic transition *(p. 64)* The process of change in the distinctive causes of death in each stage of the demographic transition.
Epidemiology *(p. 64)* The branch of medical science concerned with the incidence, distribution, and control of diseases that are prevalent among a population at a special time and are produced by some special causes not generally present in the affected locality.
Industrial Revolution *(p. 56)* A series of improvements in industrial technology that transformed the process of manufacturing goods.
Infant mortality rate (IMR) *(p. 61)* The total number of deaths in a year among infants under 1 year of age for every 1,000 live births in a society.
Life expectancy *(p. 52)* The average number of years an individual can be expected to live, given current social, economic, and medical conditions. Life expectancy at birth is the average number of years a newborn infant can expect to live.
Maternal mortality rate *(p. 59)* The annual number of female deaths per 100,000 live births from any cause related to or aggravated by pregnancy or its management (excluding accidental or incidental causes).
Medical revolution *(p. 56)* Medical technology invented in Europe and North America that has diffused to the poorer countries in Latin America, Asia, and Africa. Improved medical practices have eliminated many of the traditional causes of death in poorer countries and enabled more people to live longer and healthier lives.
Natural increase rate (NIR) *(p. 52)* The percentage growth of a population in a year, computed as the crude birth rate minus the crude death rate.
Overpopulation *(p. 46)* A situation in which the number of people in an area exceeds the capacity of the environment to support life at a decent standard of living.
Pandemic *(p. 64)* Disease that occurs over a wide geographic area and affects a very high proportion of the population.
Physiological density *(p. 50)* The number of people per unit area of arable land, which is land suitable for agriculture.
Population pyramid *(p. 60)* A bar graph that represents the distribution of population by age and sex.
Sex ratio *(p. 58)* The number of males per 100 females in the population.
Total fertility rate (TFR) *(p. 55)* The average number of children a woman will have throughout her childbearing years.
Zero population growth (ZPG) *(p. 57)* A decline of the total fertility rate to the point where the natural increase rate equals zero.

Log in to the **MasteringGeography** Study Area to view this video.

White Horse Village Urbanization

New cities being constructed in rural China will bring better jobs and living conditions to millions of farmers and their families but also end an ancient way of life.

1. In general, would the village farmers prefer to work in a new factory or remain on their land? Explain.
2. Do village residents have a choice about whether their land becomes a new industrial city? Explain.
3. In the view of China's economic planners, why is urbanization of the countryside essential?

EXPLORE

Mahāmīd, Egypt

Use Google Earth to explore Mahāmīd, a town of 45,000 near the banks of the Nile River.
Fly to: *Mahāmīd, Luxor, Egypt*. Zoom in.

1. What color is most of the land immediately in and around the town? Does this indicate that the land is used for agriculture, or is it desert? Zoom out until you see the entire band of green surrounded by tan.
2. How wide is the green strip? What does the tan color represent? What feature is in the middle of the green strip?

MasteringGeography™

Looking for additional review and test prep materials?

Visit the Study Area in **MasteringGeography™** to enhance your geographic literacy, satial reasoning skills, and understanding of this chapter's content by accessing a variety of resources, including MapMaster interactive maps, videos, *In the News* RSS feeds, flashcards, web links, self-study quizzes, and an eText version of *The Cultural Landscape*.
www.masteringgeography.com

3

Migration

Migration is the permanent move to a new location. Geographers study migration in part because it is increasingly important in explaining changes in population in various places and regions.

Migration is also important because when people migrate, they take with them to their new home cultural values and economic practices. At the same time, they become connected with the cultural and economic patterns of their new place of residence.

A family forced by war to migrate from Syria wait in Budapest, Hungary, for a train to Germany.

LOCATIONS IN THIS CHAPTER

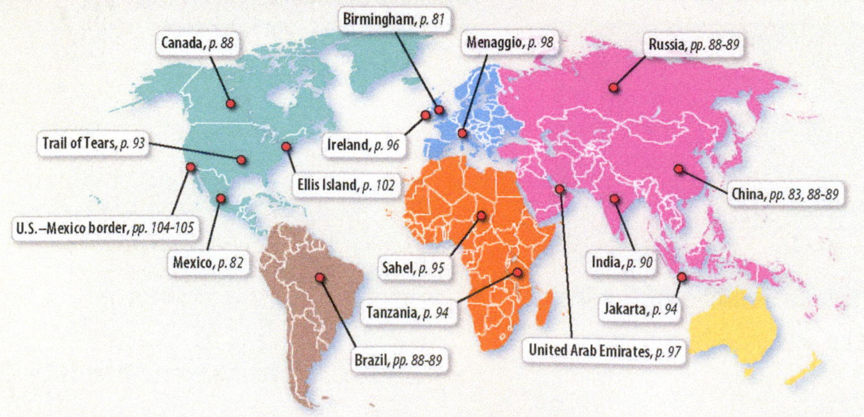

KEY ISSUES

1

Where Are the World's Migrants Distributed?

Looking at where migrants are distributed across the ***space*** of Earth, we see that some of Earth's ***regions*** are gaining migrants and some are losing them. For more than two centuries, the United States has been the most important destination for migrants.

2

Where Do People Migrate Within a Country?

Within a country, some ***places*** are gaining migrants and some are losing them. This key issue describes migration at two ***scales***: between places in two regions of a country and between places within a particular region.

3

Why Do People Migrate?

The characteristics of migrants and the reasons they migrate are especially interesting to geographers. Why people migrate profoundly influences the places and regions that are the sources and the destinations of migrants.

4

Why Do Migrants Face Challenges?

What are the key issues that arise as a result of migration? In a world of improved ***connections***, it is easier than in the past to be transported from one place to another. Yet the ability of people to migrate is more limited than in the past by legal obstacles and the hostility of people at the place of destination.

KEY ISSUE 1

Where Are the World's Migrants Distributed?

- Introducing Migration
- International Net Migration
- International and Internal Migration
- Changing U.S. Immigration

LEARNING OUTCOME 3.1.1

Understand the difference between immigration, emigration, and net migration.

Migration is a permanent move to a new location. Geographers document the migration of people across Earth and reasons for the migration (Figure 3-1). Migration is a specific type of relocation diffusion, which was defined in Chapter 1 as the spread of a characteristic through the bodily movement of people from one place to another.

Introducing Migration

Refer back to Figure 2-5 (ecumene) for a moment. Humans have spread across Earth during the past 7,000 years. This diffusion of human settlement from a small portion of Earth's land area to most of it resulted from migration. To accomplish the spread across Earth, humans have permanently changed their place of residence—where they sleep, store their possessions, and receive legal documents.

Migration is a form of **mobility**, which is a more general term covering all types of movements from one place to another. People display mobility in a variety of ways, such as by journeying every weekday from their homes to places of work or education and once a week to shops, places of worship, or recreation areas. These types of short-term, repetitive, or cyclical movements that recur on a regular basis, such as daily, monthly, or annually, are called **circulation**. College students display another form of mobility—seasonal mobility—by moving to a dormitory each fall and returning home the following spring.

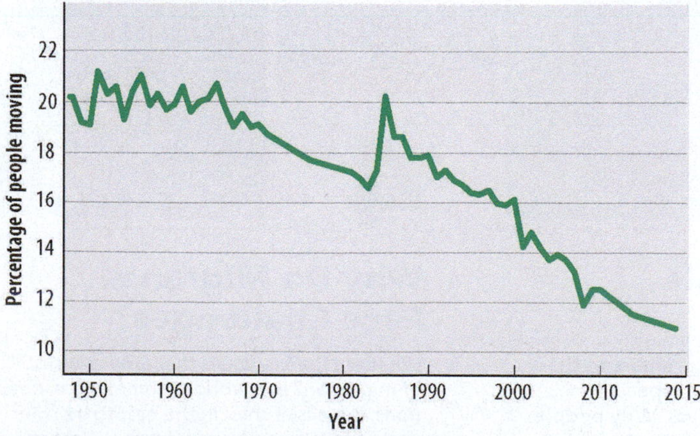

▲ **FIGURE 3-1 MIGRATION OF AMERICANS** Around 11 percent of Americans migrate in a year. The percentage has declined from around 20 percent during the 1980s.

The flow of migration always involves two-way connections. Given two locations, A and B, some people migrate from A to B, while at the same time others migrate from B to A. **Emigration** is migration *from* a location; **immigration** is migration *to* a location.

The difference between the number of immigrants and the number of emigrants is the **net migration** (Figure 3-1). If the number of immigrants exceeds the number of emigrants, the net migration is positive, and the region has net in-migration. If the number of emigrants exceeds the number of immigrants, the net migration is negative, and the region has net out-migration.

Figure 3-2 is a cartogram showing emigration by country, and Figure 3-3 is a cartogram showing immigration by country. A country, such as Mexico, that appears larger in Figure 3-2 than in Figure 3-3 has net out-migration, whereas a country that appears smaller in Figure 3-2 than in Figure 3-3, such as the United States, has net in-migration.

PAUSE & REFLECT 3.1.1

Do the developed countries of Europe and North America appear in Figures 3-2 and 3-3 to have net in-migration or net out-migration?

Geographers are especially interested in why people migrate, even though migration occurs much less frequently than other forms of mobility, because it produces profound changes for individuals and entire cultures. A permanent move to a new location disrupts traditional cultural ties and economic patterns in one region. At the same time, when people migrate, they take with them to their new home their language, religion, ethnicity, and other cultural traits, as well as their methods of farming and other economic practices.

The changing scale generated by modern transportation systems, especially motor vehicles and airplanes, makes relocation diffusion more feasible than in the past, when people had to rely on walking, animal power, or slow ships. However, thanks to modern communications systems, relocation diffusion is no longer essential for transmittal of ideas from one place to another. Culture and economy can diffuse rapidly around the world through forms of expansion diffusion.

Why would people make a perilous journey across thousands of kilometers of ocean? Why did the pioneers

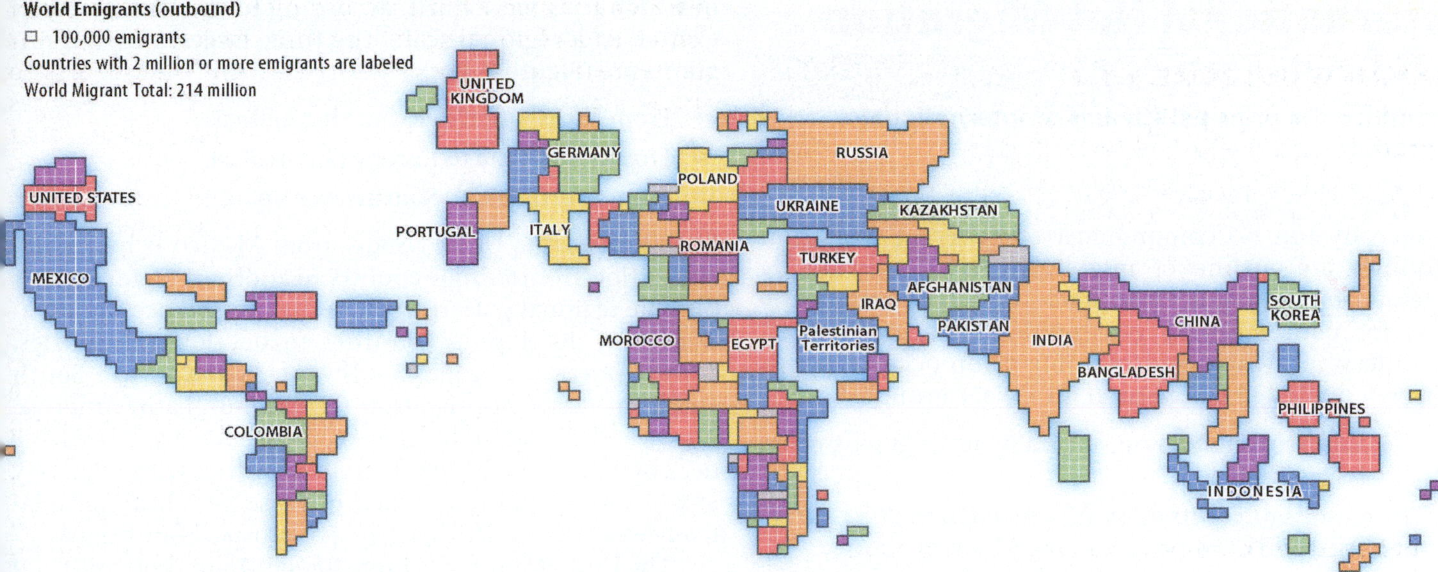

▲ **FIGURE 3-2 CARTOGRAM OF EMIGRANTS** In this cartogram, the size of the country is based on the number of emigrants in 2010.

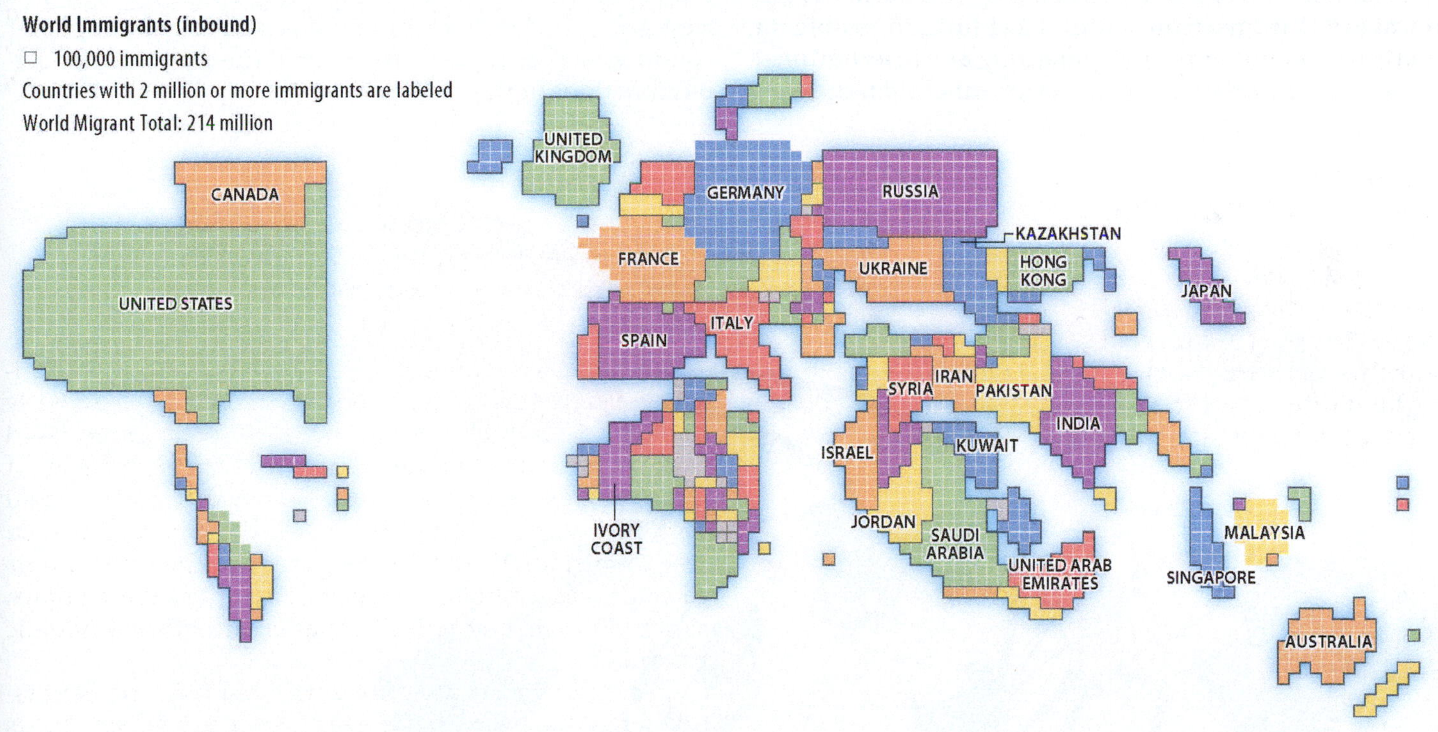

▲ **FIGURE 3-3 CARTOGRAM OF IMMIGRANTS** In this cartogram, the size of the country is based on the number of immigrants in 2010.

cross the Great Plains, the Rocky Mountains, or the Mojave Desert to reach the American West? Why do people continue to migrate by the millions today? The hazards that many migrants have faced are a measure of the strong lure of new locations and the desperate conditions in their former homelands. Most people migrate in search of three objectives: economic opportunity, cultural freedom, and environmental comfort. This chapter will study the reasons people migrate.

International Net Migration

LEARNING OUTCOME 3.1.2

Recognize the principal streams of international migration.

Geography has no comprehensive theory of migration although an outline of migration "laws" written by nineteenth-century geographer E. G. Ravenstein is the basis for contemporary geographic migration studies. To understand where and why migration occurs, Ravenstein's "laws" can be organized into three groups:

- The distance that migrants typically move (discussed in Key Issues 1 and 2).
- The reasons migrants move (discussed in the first part of Key Issue 3).
- The characteristics of migrants (discussed in the second part of Key Issue 3).

INTERNATIONAL MIGRATION FLOWS

A permanent move from one country to another is **international migration.** Around 214 million people, or 3 percent of the world's population, are international migrants and currently live in countries other than the ones in which they were born, according to the Pew Research Center. At a regional scale, the three largest flows of migrants are (Figure 3-4):

- From Latin America to North America.
- From South Asia to Europe (Figure 3-5).
- From South Asia to Southwest Asia.

Migration to the United States from Mexico is by far the largest flow from a single country to another single country.

The regional pattern reflects the importance of migration from developing countries to developed countries. North America, Europe, Southwest Asia, and the South Pacific have net in-migration (Figure 3-6). Latin America, Africa, and all regions in Asia except for Southwest Asia have net out-migration. Migrants from countries with relatively low incomes and high natural increase rates head for relatively wealthy countries, where job prospects are brighter.

The United States has more foreign-born residents than any other country, approximately 42 million as of 2015, and growing annually by around 1 million. Russia is a distant second, with 11 million immigrants. Australia and Canada, which are much less populous than the United States, have higher rates of net in-migration. The highest in-migration rates of all are in petroleum-exporting countries of Southwest Asia, which attract immigrants primarily from poorer countries in Asia to perform many of the dirty and dangerous functions in the oil fields.

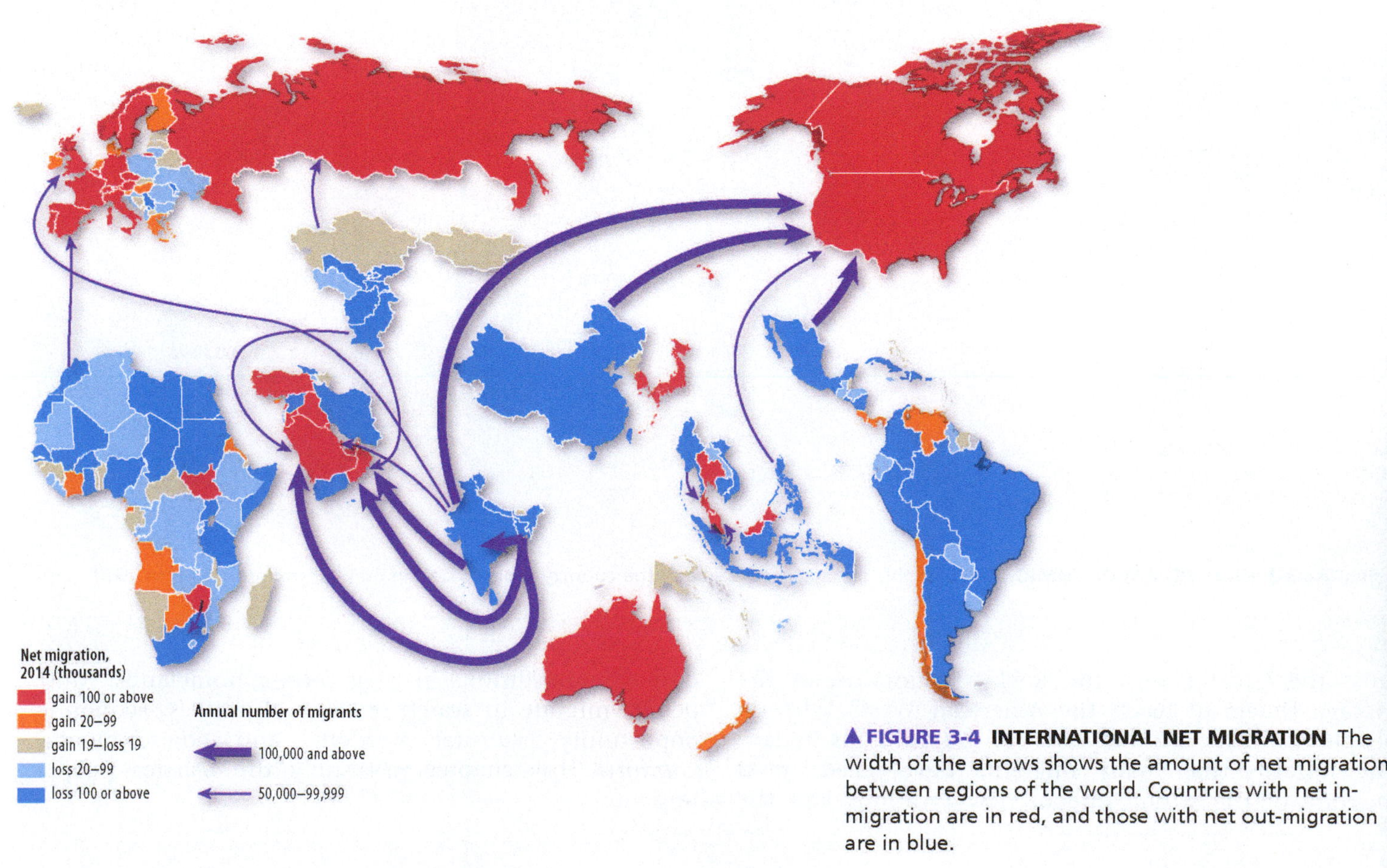

▲ **FIGURE 3-4 INTERNATIONAL NET MIGRATION** The width of the arrows shows the amount of net migration between regions of the world. Countries with net in-migration are in red, and those with net out-migration are in blue.

◀ **FIGURE 3-5 IMMIGRATION, UNITED KINGDOM** An immigrant from Latin America walks past an anti-immigration political poster, Birmingham, United Kingdom.

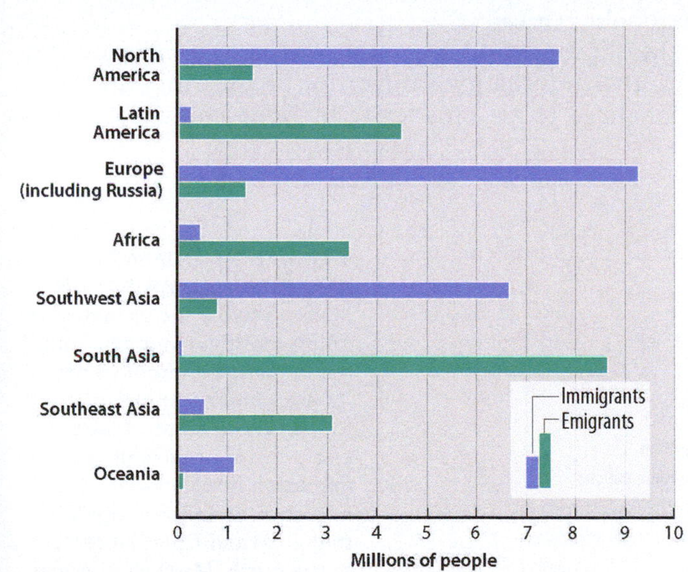

▲ **FIGURE 3-6 IMMIGRANTS AND EMIGRANTS BY WORLD REGION** Europe, North America, and Southwest Asia have substantially more immigration than emigration.

MIGRATION TRANSITION

Geographer Wilbur Zelinsky identified a **migration transition**, which consists of changes in a society comparable to those in the demographic transition (Table 3-1). The migration transition is a change in the migration pattern in a society that results from the social and economic changes that also produce the demographic transition. According to the migration transition, international migration is primarily a phenomenon of countries in stage 2 of the demographic transition, whereas internal migration is more important in stages 3 and 4.

PAUSE & REFLECT 3.1.2

If the demographic transition has a stage 5 in the future, what might be key features of a stage 5 of the migration transition?

TABLE 3–1 Comparing the Demographic Transition and Migration Transition

Stage	Demographic Transition	Migration Transition
1	Low NIR, high CBR, high CDR	High daily or seasonal mobility in search of food
2	High NIR, high CBR, rapidly falling CDR	High international emigration and interregional migration from rural to urban areas
3	Declining NIR, rapidly declining CBR, declining CDR	High international immigration and intraregional migration from cities to suburbs
4	Low NIR, low CBR, low CDR	Same as stage 3

International and Internal Migration

LEARNING OUTCOME 3.1.3

Understand the difference between internal and international migration.

Geographer E.G. Ravenstein developed a set of laws that govern human migration. According to Ravenstein:

- Most migrants relocate a short distance and remain within the same country.
- Long-distance migrants to other countries head for major centers of economic activity.

DISTANCE OF MIGRATION

Migration can be either international or internal (Figure 3-7).

INTERNATIONAL MIGRATION. A permanent move from one country to another is international migration. In Mexico, for example, international migration consists primarily of immigration from Central America and emigration to the United States (Figure 3-8). International migration is further divided into two types:

- **Voluntary migration** means that the migrant has chosen to move, usually for economic reasons, though sometimes for environmental reasons.
- **Forced migration** means that the migrant has been compelled to move by cultural or environmental factors.

The distinction between forced and voluntary migration is not clear-cut. Migrants for economic reasons may feel forced by pressure inside themselves to migrate, such as to search for food or jobs, but they have not been explicitly compelled to migrate by the violent actions of other people.

INTERNAL MIGRATION. A permanent move within the same country is **internal migration**. Consistent with the distance–decay principle presented in Chapter 1, the farther away a place is located, the less likely that people will migrate to it. Thus, internal migrants are much more numerous than international migrants.

Internal migration can be divided into two types:

- **Interregional migration** is movement from one region of a country to another. Historically, the main type of interregional migration has been from rural to urban areas in search of jobs.
- **Intraregional migration** is movement within one region. The main type of intraregional migration has been within urban areas, from older cities to newer suburbs.

In Mexico, for example, the principal interregional migration is from southern states to northern ones, and the principal intraregional migration is within the Mexico City metropolitan area.

Most people find migration within a country less traumatic than international migration because they find familiar language, foods, broadcasts, literature, music, and other

FIGURE 3-7 INTERNATIONAL AND INTERNAL MIGRATION Mexico's two principal patterns of international migration are net in-migration from Central America and net out-migration to the United States. Mexico's two principal interregional migration flows are net migration from the south to the north and from the center to the north. Mexico's principal intraregional migration flow is from Mexico City to outer states in the Mexico City metropolitan area.

◀ **FIGURE 3-8 INTERNATIONAL MIGRANTS** Migrants without legal papers travel through Mexico from Central America to the United States because they don't have enough money to pay for their travels. **1.** Why do you think that these people want to reach the United States? **2.** Why are they traveling by holding on to the side of the freight car? **3.** What do you think would happen to them if they tried to cross the border in a car or truck?

social customs after they move. Moves within a country also generally involve much shorter distances than those in international migration. However, internal migration can involve long-distance moves in large countries, such as in the United States, China, and Russia (Figure 3-9).

PAUSE & REFLECT 3.1.3

When you or your family last moved, was it voluntary international, forced international, interregional internal, or intraregional internal?

◀ **FIGURE 3-9 INTERNAL MIGRATION** The Chinese Lunar New Year is the most important holiday for immigrants in the major cities to visit their families still living in other regions of China.

Changing U.S. Immigration

LEARNING OUTCOME 3.1.4

Describe the different sources of immigrants during the three main eras of U.S. immigration.

The United States plays a special role in the study of international migration, because it is inhabited overwhelmingly by direct descendants of immigrants. About 80 million people migrated to the United States between 1820 and 2015, including 42 million who were alive in 2015.

The United States has had three main eras of immigration:

- Colonial settlement in the seventeenth and eighteenth centuries.
- Mass European immigration in the late nineteenth and early twentieth centuries.
- Asian and Latin American immigration in the late twentieth and early twenty-first centuries.

U.S. IMMIGRATION AT INDEPENDENCE

The U.S. population in 1790, the first census after independence, was 3.9 million, including 950,000 who had immigrated to one of the colonies now part of the United States. Immigration to the American colonies and the newly independent United States came from two principal places:

- **Europe.** According to the 1790 census, 62 percent of immigrants came from Europe, and of those, 45 to 50 percent came from the lands comprising the modern-day United Kingdom and Republic of Ireland. Colonies had been established by British immigrants along the Atlantic Coast, beginning with Jamestown, Virginia, in 1607, and Plymouth, Massachusetts, in 1620.
- **Sub-Saharan Africa.** Most African Americans are descended from Africans forced to migrate to the Western Hemisphere as slaves (see Chapter 7). At the time of independence, 360,000 people living in the United States—38 percent of immigrants—had been shipped as slaves from Africa to the colonies, primarily by the British. The importation of Africans as slaves was made illegal in 1808, though another 250,000 Africans were brought to the United States during the next half-century.

Most of the Africans were forced to migrate to the United States as slaves, whereas most Europeans were voluntary migrants. However, harsh economic conditions and religious persecution in Europe blurred the distinction between forced and voluntary migration for many Europeans.

U.S. IMMIGRATION: MID-NINETEENTH TO EARLY TWENTIETH CENTURIES

Between 1820 and 1920, approximately 32 million people immigrated to the United States. Nearly 90 percent of them emigrated from Europe (Figure 3-10). For European migrants, the United States offered a great opportunity for economic success. Early migrants extolled the virtues of the country to friends and relatives back in Europe, which encouraged still others to come.

Migration from Europe to the United States peaked at several points during the nineteenth and early twentieth centuries:

- **1840s and 1850s: Ireland and Germany.** Annual immigration jumped from 20,000 to more than 200,000. Three-fourths of all U.S. immigrants during those two decades came from Ireland and Germany. Desperate economic push factors compelled the Irish and Germans to cross the Atlantic. Germans also emigrated to escape political unrest.

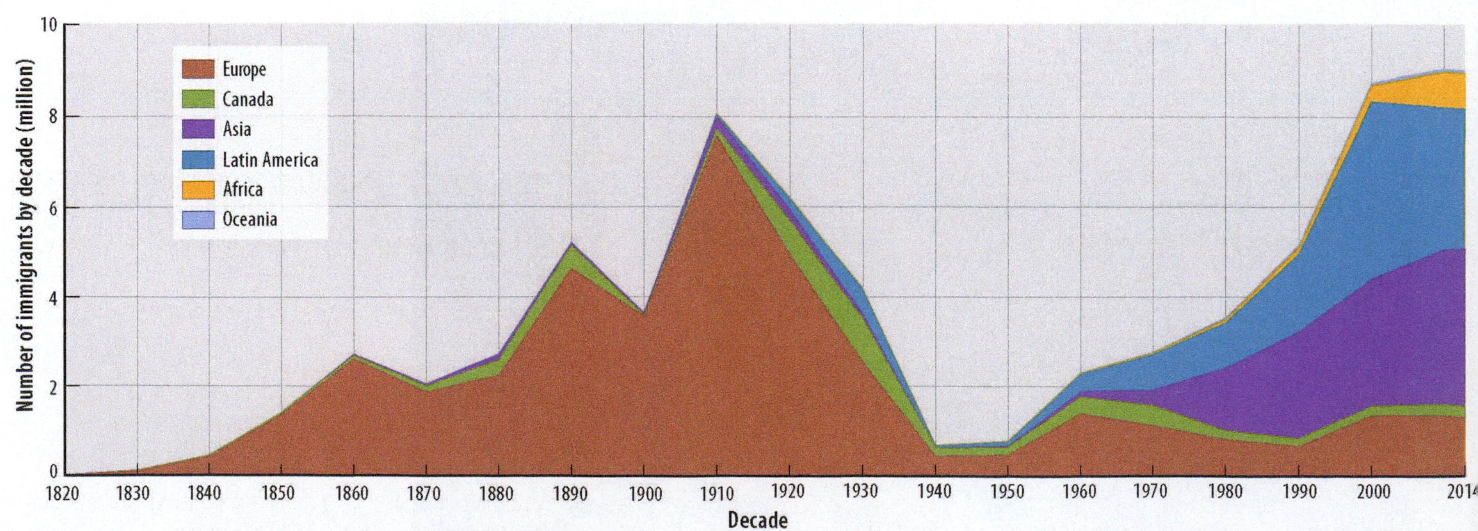

▲ **FIGURE 3-10 TWO CENTURIES OF IMMIGRATION TO THE UNITED STATES** Europeans comprised more than 90 percent of immigrants to the United States during the nineteenth century. Since the 1980s, Latin America and Asia have been the dominant sources of immigrants.

- **1870s: Ireland and Germany.** Emigration from Ireland and Germany resumed following a temporary decline during the U.S. Civil War (1861–1865).
- **1880s: Scandinavia.** Immigration increased to 500,000 per year. Increasing numbers of Scandinavians, especially Swedes and Norwegians, joined Germans and Irish in migrating to the United States. The Industrial Revolution had diffused to Scandinavia, triggering a rapid population increase.
- **1905–1914: Southern and Eastern Europe.** Annual immigration to the United States reached 1 million. Two-thirds of all immigrants during this period came from Southern and Eastern Europe, especially Italy, Russia, and Austria-Hungary. The shift in the primary source of immigrants coincided with the diffusion of the Industrial Revolution to Southern and Eastern Europe, along with rapid population growth.

Among European countries, Germany has sent the largest number of immigrants to the United States, 7.2 million. Other major European sources include Italy, 5.4 million; the United Kingdom, 5.3 million; Ireland, 4.8 million; and Russia and the former Soviet Union, 4.1 million. About one-fourth of Americans trace their ancestry to German immigrants, and one-eighth each to Irish and English immigrants.

Frequent boundary changes in Europe make precise national counts impossible. For example, most Poles migrated to the United States at a time when Poland did not exist as an independent country. Until the end of World War I in 1918, Austria-Hungary encompassed portions of present-day Austria, Bosnia & Herzegovina, Croatia, Czechia, Hungary, Italy, Poland, Romania, Slovakia, Slovenia, and Ukraine, and many immigrants are recorded as coming from Austria-Hungary rather than the present-day countries.

U.S. IMMIGRATION: LATE TWENTIETH TO EARLY TWENTY-FIRST CENTURIES

Immigration to the United States dropped sharply in the 1930s and 1940s, during the Great Depression and World War II. The number steadily increased beginning in the 1950s and then surged to historically high levels during the first decade of the twenty-first century.

More than three-fourths of the recent U.S. immigrants have emigrated from two regions:

- **Latin America.** Around 13 million Latin Americans have migrated to the United States in the past half-century, compared to only 2 million in the two preceding centuries. Nearly one-half million Latin Americans migrate annually to the United States, more than twice as many as during the entire nineteenth century.
- **Asia.** Around 7 million Asians have migrated to the United States in the past half-century, compared to only 1 million in the two preceding centuries. The leading sources of U.S. immigrants from Asia are China (including Hong Kong), the Philippines, India, and Vietnam.

Officially, Mexico passed Germany in 2006 as the country that has sent to the United States the most immigrants ever. Because of the large number of undocumented immigrants, Mexico probably became the leading source during the 1980s. In the early 1990s, an unusually large number came from Mexico and other Latin American countries as a result of the 1986 Immigration Reform and Control Act, which issued visas to several hundred thousand people who had entered the United States in previous years without legal documents.

Although the source of immigrants to the United States has changed from predominantly Europe to Asia and Latin America, the reason for immigration remains the same: Rapid population growth has limited prospects for economic advancement at home. Europeans left when their countries entered stage 2 of the demographic transition in the nineteenth century, and Latin Americans and Asians began to leave in large numbers in recent years after their countries entered stage 2. With poor conditions at home, immigrants were lured by economic opportunity and social advancement in the United States.

The motives for immigrating to the country may be similar, but the United States has changed over time. The United States is no longer a sparsely settled, economically booming country. In 1912, New Mexico and Arizona were admitted as the forty-seventh and forty-eighth states. Thus, for the first time in its history, all the contiguous territory of the country was a "united" state (other than the District of Columbia). This symbolic closing of the frontier coincided with the end of the peak period of emigration from Europe.

PAUSE & REFLECT 3.1.4

In which stage of the demographic transition were most countries when they sent the most immigrants to the United States?

CHECK-IN KEY ISSUE 1

Where Are the World's Migrants Distributed?

- ✔ **Migration is the permanent move to a new location.**
- ✔ **Migration can be international (voluntary or forced) or internal (interregional or intraregional).**
- ✔ **The number and place of origin of immigrants to the United States have varied over time.**

KEY ISSUE 2

Where Do People Migrate Within a Country?

- Interregional Migration in the United States
- Interregional Migration in Other Large Countries
- Intraregional Migration

LEARNING OUTCOME 3.2.1

Describe the principal patterns of interregional migration in the United States.

Internal migration for most people is less disruptive than international migration. Two main types of internal migration are interregional (between regions of a country) and intraregional (within a region).

In the past, people migrated from one region of a country to another in search of better farmland. Lack of farmland pushed many people from the more densely settled regions of the country and lured them to the frontier, where land was abundant. Today, the principal type of interregional migration is from rural areas to urban areas. Most jobs, especially in services, are clustered in urban areas (see Chapter 12).

Recent immigrants are not distributed uniformly through the United States. More than one-half immigrate to California, Florida, New York, or Texas (Figure 3-11).

Interregional Migration in the United States

An especially prominent example of large-scale internal migration is the opening of the American West. At the time of independence, the United States consisted of long-established settlements concentrated on the Atlantic Coast and a scattering of newer settlements in the territories west of the Appalachian Mountains. Through mass interregional migration, the interior of the continent was settled and developed.

CHANGING CENTER OF POPULATION

The U.S. Census Bureau computes the country's population center at the time of each census. The population center is the average location of everyone in the country, the "center of population gravity." If the United States were a flat plane placed on top of a pin, and each individual weighed the same, the population center would be the point where the population distribution causes the flat plane to balance on the pin.

The changing location of the population center graphically demonstrates the march of the American people across the North American continent over the past 200 years (Figure 3-12). The center has consistently shifted westward, although the rate of movement has varied in different eras.

1790: HUGGING THE COAST. Virtually all colonial-era settlements were near the Atlantic Coast. Few colonists ventured far from coastal locations because they depended on shipping links with Europe to receive products and to export raw materials. The Appalachian Mountains also blocked western development because of their steep slopes, thick forests, and few gaps that allowed easy passage. The indigenous residents, commonly called "Indians," still occupied large areas and sometimes resisted the expansion of settlement.

1800–1840: CROSSING THE APPALACHIANS. Transportation improvements, especially the building of canals, helped to open the interior. Most important was the Erie Canal, which enabled people to travel inexpensively by boat between New York City and the Great Lakes. In 1840, the United States had 5,352 kilometers (3,326 miles) of canals. Encouraged by the opportunity to obtain a large amount of land at a low price, people moved into forested river valleys between the Appalachians and the Mississippi River. They cut down the trees and used the wood to build homes, barns, and fences.

1850–1890: RUSHING TO THE GOLD. The population center shifted westward more rapidly during this period. Rather than continuing to expand agriculture into the next available westward land, mid-nineteenth-century pioneers kept going all the way to California. The principal pull to California was the Gold Rush, which began in the late 1840s. Pioneers during this period also passed over the Great Plains because of the physical environment. The region's dry climate, lack of trees, and tough grassland sod

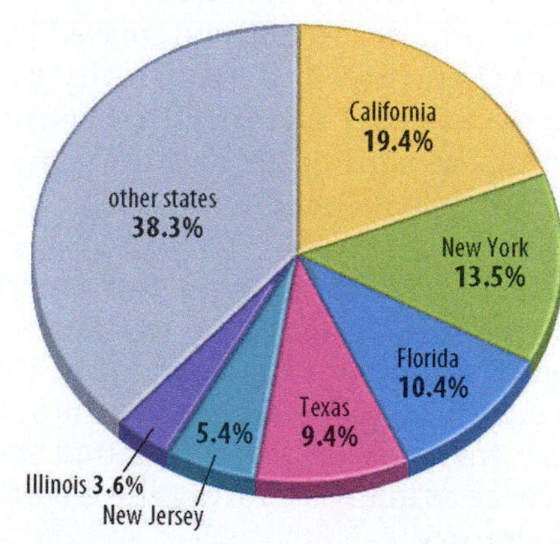

▲ **FIGURE 3-11 DESTINATION OF IMMIGRANTS BY U.S. STATE** California, New York, and Florida are the leading destinations for immigrants.

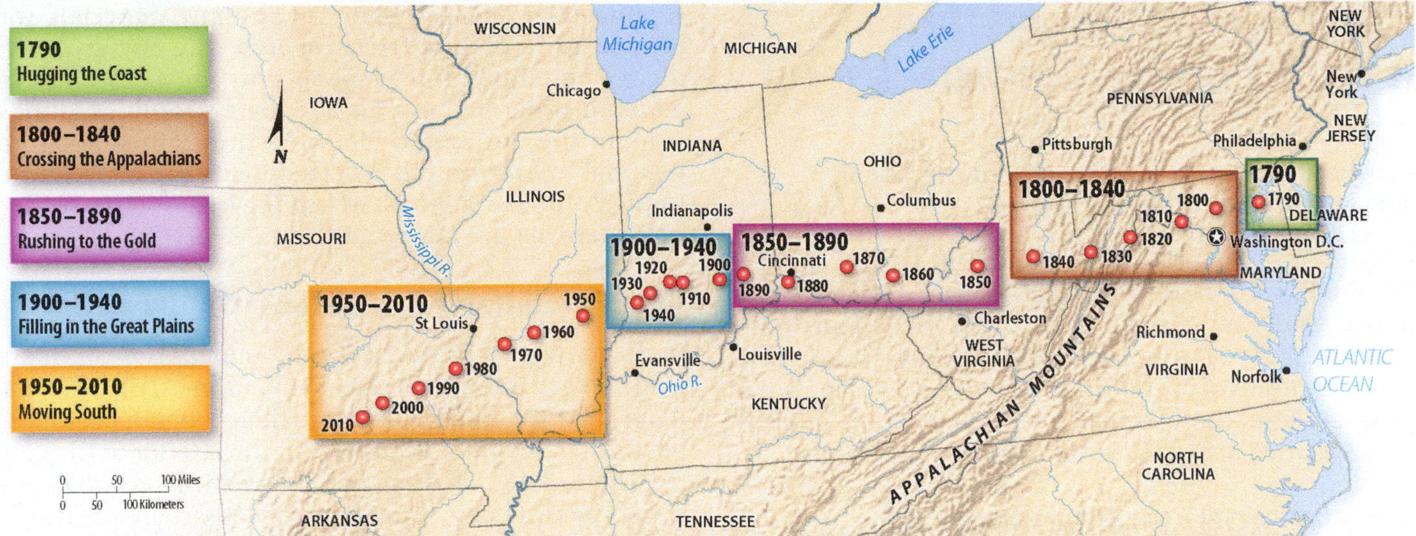

▲ **FIGURE 3-12 CHANGING CENTER OF U.S. POPULATION** The population center is the average location of everyone in the country, the "center of population gravity." If the United States were a flat plane placed on top of a pin, and each individual weighed the same, the population center would be the point where the population distribution causes the flat plane to balance on the head of a pin.

convinced early explorers such as Zebulon Pike that the region was unfit for farming, and maps at the time labeled the Great Plains as the Great American Desert.

1900–1940: FILLING IN THE GREAT PLAINS. The westward movement of the U.S. population center slowed during this period because emigration from Europe to the East Coast offset most of the emigration from the East Coast to the U.S. West. Also, immigrants began to fill in the Great Plains that earlier generations had bypassed. Advances in agricultural technology enabled people to cultivate the area. Farmers used barbed wire to reduce dependence on wood fencing, the steel plow to cut the thick sod, and windmills and well-drilling equipment to pump more water. The expansion of the railroads encouraged settlement of the Great Plains. The federal government gave large land grants to the railroad companies, which financed construction of their lines by selling portions to farmers. The extensive rail network then permitted settlers to transport their products to the large concentrations of customers in East Coast cities.

1950–2010: MOVING SOUTH. The population center resumed a more vigorous westward migration. It also moved southward, as Americans migrated to the South for job opportunities and warmer climate. The rapid growth of population and employment in the South has aggravated interregional antagonism. Some people in the Northeast and Midwest believe that southern states have stolen industries from them. In reality, some industries have relocated from the Northeast and Midwest, but most of the South's industrial growth comes from newly established companies. Interregional migration has slowed considerably in the United States into the twenty-first century (Figure 3-13). Regional differences in employment prospects have become less dramatic. The severe recession of 2008–2009 discouraged people from migrating because of limited job prospects in all regions.

PAUSE & REFLECT 3.2.1

How might climate change affect patterns of interregional migration in the United States?

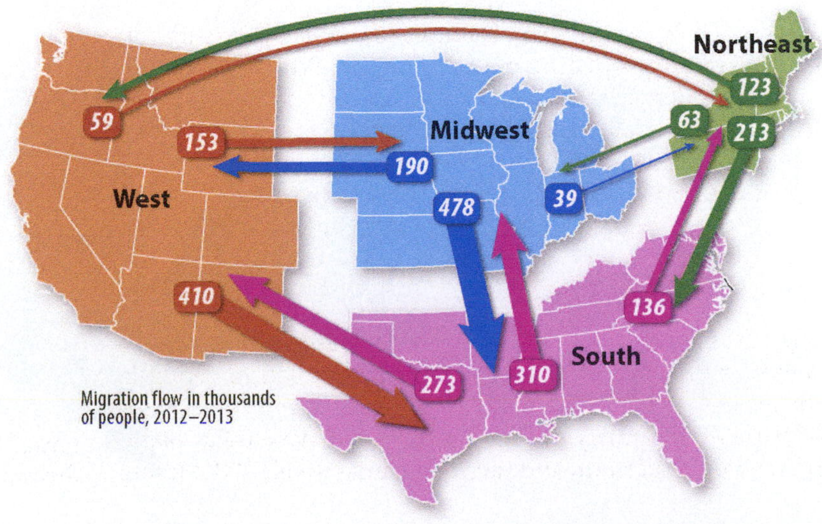

▶ **FIGURE 3-13 INTERREGIONAL MIGRATION IN THE UNITED STATES** Figures show the flow of migrants in thousands of people in 2012 and 2013.

Interregional Migration in Other Large Countries

LEARNING OUTCOME 3.2.2

Describe the principal patterns of interregional migration in several large countries.

The world's three largest countries in land area are Russia, Canada, and China; the United States ranks fourth and Brazil fifth. Long-distance interregional migration has been an important means of opening new regions for development in Russia and Canada in the past and in Brazil more recently. China has tried to discourage interregional migration.

INTERREGIONAL MIGRATION IN CANADA

Canada, like the United States, has had interregional migration primarily from east to west for nearly two centuries. Since 2001, the two westernmost provinces—Alberta and British Columbia—have had nearly all of Canada's net in-migration, whereas Ontario has had the largest net out-migration (Figure 3-14). The three largest interprovincial flows in Canada are from Ontario to Alberta, from Ontario to British Columbia, and from Alberta to British Columbia.

INTERREGIONAL MIGRATION IN RUSSIA

The population of Russia is highly clustered in the western, or European, portion of the country. Interregional migration has been an important tool to promote development in the sparsely inhabited Asian portion of the country. During the Soviet Union period, Communist policy encouraged factory construction near raw materials rather than near major population concentrations (see Chapter 11). Because many of the raw materials were located in remote portions of Asia, the Soviet government sometimes forced people to move to these regions in order to have an adequate supply of labor to work in the mines and factories. In recent years, interregional migration has reversed, with net in-migration to the European regions, where the largest cities and job opportunities are clustered (Figure 3-15).

INTERREGIONAL MIGRATION IN CHINA

In developing countries, the predominant flow of interregional migration is from rural to urban areas, where jobs are more likely to be available. More than 150 million Chinese have emigrated from rural areas in the interior of the country (Figure 3-16). They are headed for the large urban areas along the east coast, where jobs are especially plentiful in factories. The government once severely limited the ability of Chinese people to make interregional moves, but restrictions have been lifted in recent years.

INTERREGIONAL MIGRATION IN BRAZIL

As in China, most Brazilians live in a string of large cities near the East Coast. Brazil's tropical interior is sparsely inhabited. To increase the attractiveness of the interior, the government moved its capital in 1960 from Rio to a newly built city called Brasília, situated 1,000 kilometers (600 miles) from the Atlantic Coast. Development of Brazil's interior has altered historic migration patterns. The coastal areas now have net out-migration, whereas the interior areas have net in-migration (Figure 3-17).

PAUSE & REFLECT 3.2.2

Does Russia's interregional migration pattern more closely represent that of the United States and Canada or that of Brazil and China?

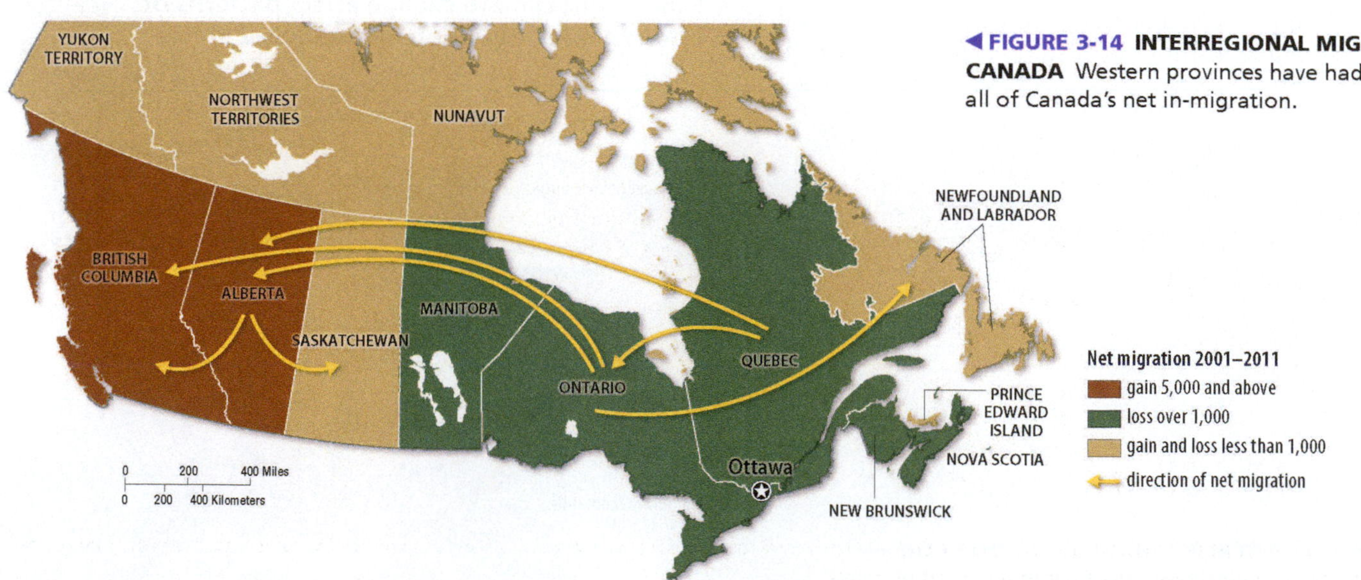

◀ FIGURE 3-14 INTERREGIONAL MIGRATION: CANADA Western provinces have had nearly all of Canada's net in-migration.

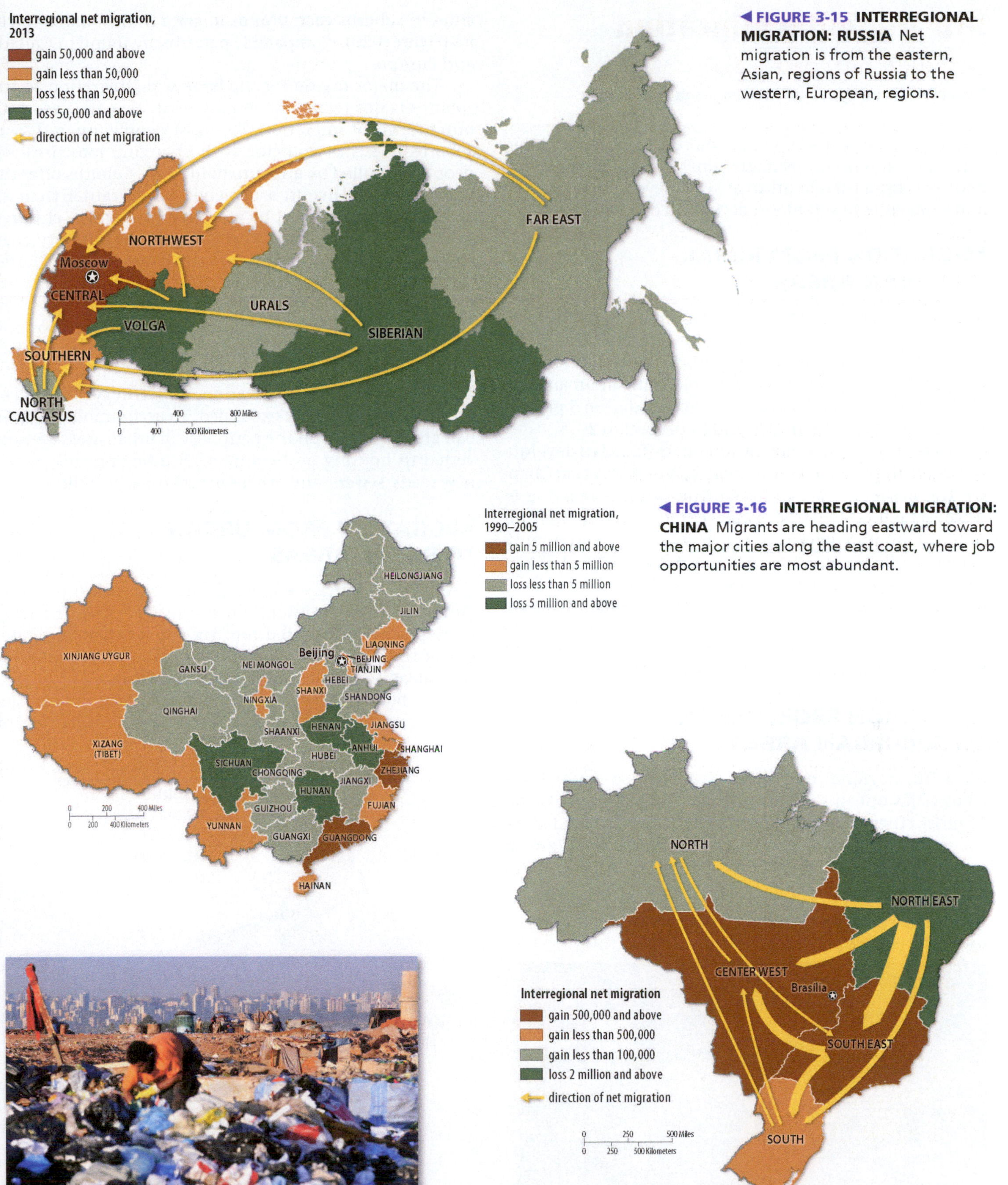

◀ **FIGURE 3-15 INTERREGIONAL MIGRATION: RUSSIA** Net migration is from the eastern, Asian, regions of Russia to the western, European, regions.

◀ **FIGURE 3-16 INTERREGIONAL MIGRATION: CHINA** Migrants are heading eastward toward the major cities along the east coast, where job opportunities are most abundant.

▲ **FIGURE 3-17 INTERREGIONAL MIGRATION: BRAZIL (a)** Net migration in Brazil is from coastal regions to interior ones. **(b)** A recent immigrant living on the outskirts of Brasília picks through a garbage dump.

Intraregional Migration

LEARNING OUTCOME 3.2.3

Describe three types of intraregional migration.

Intraregional migration is much more common than interregional or international migration. Most intraregional migration is from rural to urban areas in developing countries and from cities to suburbs in developed countries.

MIGRATION FROM RURAL TO URBAN AREAS

Migration from rural (or nonmetropolitan) areas to urban (or metropolitan) areas began in the 1800s in Europe and North America as part of the Industrial Revolution (see Chapter 11). The percentage of people living in urban areas in the United States, for example, increased from 5 percent in 1800 to 50 percent in 1920 and 81 percent in 2015.

In recent years, urbanization has diffused to developing countries of Asia, Latin America, and Africa (see Chapter 12). Between 1950 and 2015, the percentage living in urban areas increased from 40 percent to 80 percent in Latin America, from 15 percent to 47 percent in Asia, and from 10 percent to 38 percent in sub-Saharan Africa (Figure 3-18). As with interregional migrants, most people who move from rural to urban areas seek economic advancement. They are pushed from rural areas by declining opportunities in agriculture and are pulled to the cities by the prospect of work in factories or in service industries (Figure 3-19).

MIGRATION FROM URBAN TO SUBURBAN AREAS

Most intraregional migration in developed countries is from cities out to surrounding suburbs. The population of most cities in developed countries has declined since the mid-twentieth century, while suburbs have grown rapidly. Nearly twice as many Americans migrate from cities to suburbs each year as migrate from suburbs to cities (Figure 3-20). Comparable patterns are found in Canada and Europe.

The major reason for the large-scale migration to the suburbs is not related to employment, as is the case with other forms of migration. For most people, migration to suburbs does not coincide with changing jobs. Instead, people are pulled by a suburban lifestyle. Suburbs offer the opportunity to live in a detached house rather than an apartment, surrounded by a private yard where children can play safely. A garage or driveway on the property guarantees space to park cars at no extra charge. In the United States, suburban schools tend to be more modern, better equipped, and safer than those in cities. Cars and trains enable people to live in suburbs yet have access to jobs, shops, and recreational facilities throughout the urban area (see Chapter 13).

As a result of suburbanization, the territory occupied by urban areas has rapidly expanded. To accommodate suburban growth, farms on the periphery of urban areas are converted to housing and commercial developments, where new roads, sewers, and other services must be built.

MIGRATION FROM URBAN TO RURAL AREAS

Developed countries witnessed a new migration trend beginning in the late twentieth century. For the first time, more people immigrated into rural areas than emigrated out of them in some years. Net migration from urban to rural areas is called **counterurbanization**.

The boundary where suburbs end and the countryside begins cannot be precisely defined. Counterurbanization

▼ **FIGURE 3-18 RURAL TO URBAN MIGRATION: INDIA** Poor-quality housing for immigrants in Delhi.

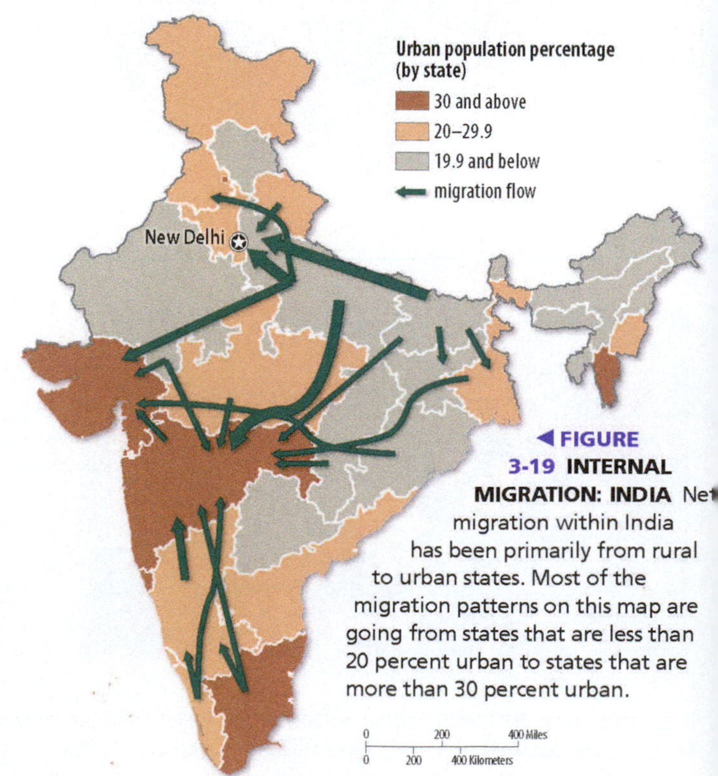

◀ **FIGURE 3-19 INTERNAL MIGRATION: INDIA** Net migration within India has been primarily from rural to urban states. Most of the migration patterns on this map are going from states that are less than 20 percent urban to states that are more than 30 percent urban.

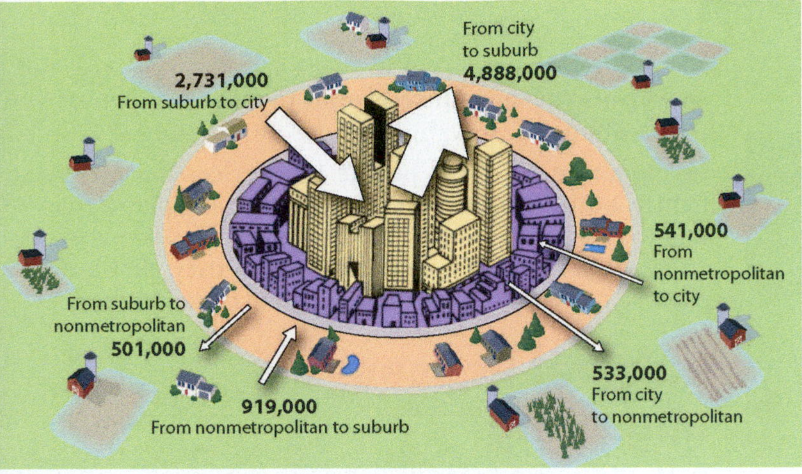

▲ **FIGURE 3-20 INTRAREGIONAL MIGRATION: UNITED STATES** Intraregional migration is primarily from cities to suburbs. Figures are total U.S. intraregional migrants in 2013.

results in part from very rapid expansion of suburbs. But some counterurbanization represents genuine migration from cities and suburbs to small towns and rural communities.

As with suburbanization, people move from urban to rural areas for lifestyle reasons. Some are lured to rural areas by the prospect of swapping the frantic pace of urban life for the opportunity to live on a farm, where they can own horses or grow vegetables. Others move to farms but do not earn their living from agriculture; instead, they work in nearby offices, small-town shops, or other services. In the United States, evidence of counterurbanization can be seen in the Rocky Mountain states. Rural counties in states such as Colorado, Idaho, Utah, and Wyoming have experienced net in-migration (Figure 3-21).

With modern communications and transportation systems, no location in a developed country is truly isolated, either economically or socially. We can buy most products online and have them delivered within a few days. We can follow the fortunes of our favorite teams anywhere in the country, thanks to cable, satellite dishes, and webcasts.

Intraregional migration in the United States has slowed considerably since the 1980s (refer to Figure 3-1). Most intraregional migration in the United States continues to be between cities and suburbs. Since 2010, the number of Americans moving from cities to suburbs has decreased, whereas the number moving from suburbs to cities has increased. Counterurbanization is not a phenomenon every year. Cities have become more attractive especially to younger people (see Chapter 13).

PAUSE & REFLECT 3.2.3

Why might rural to urban migration be most intense in countries in stage 2 of the demographic transition?

CHECK-IN KEY ISSUE 2

Where Do People Migrate Within a Country?

- ✔ **Large-scale interregional migration has resulted in the movement of the U.S. center of population to the West and South.**
- ✔ **Other large countries have experienced substantial interregional migration.**
- ✔ **Intraregional migration has been primarily from rural to urban areas in developing countries and from cities to suburbs in developed countries.**

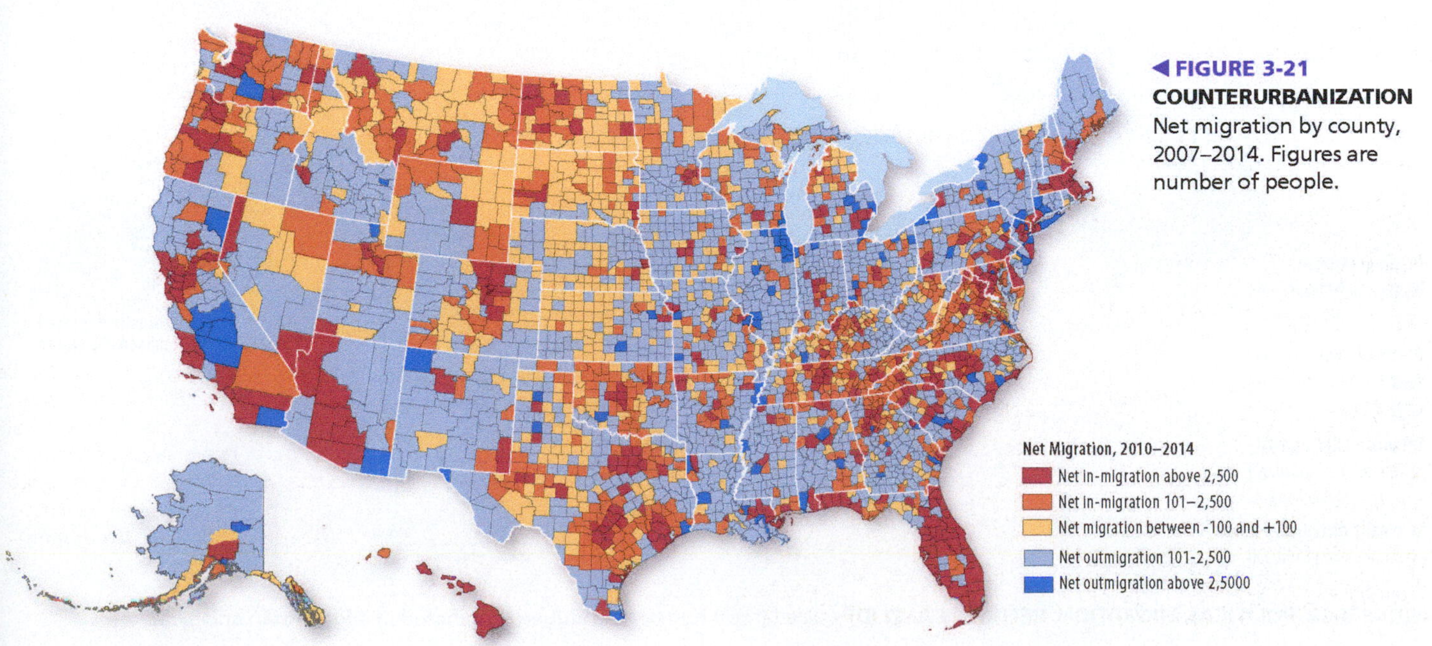

◀ **FIGURE 3-21 COUNTERURBANIZATION** Net migration by county, 2007–2014. Figures are number of people.

KEY ISSUE 3

Why Do People Migrate?

- ▸ Cultural Reasons for Migrating
- ▸ Environmental Reasons for Migrating
- ▸ Migrating to Find Work
- ▸ Gender and Age of Migrants

LEARNING OUTCOME 3.3.1

Explain cultural and environmental reasons for migration.

Ravenstein's laws help geographers explain the reasons people migrate:

- Most people migrate for economic reasons (see the next section).
- Cultural and environmental reasons (discussed in this section) also induce migration, although not as frequently as economic reasons.

One of these reasons usually emerges as most important, although elements of more than one reason may be detectable. Ranking the relative importance of these reasons may be difficult and even controversial.

People migrate because of push factors and pull factors:

- A **push factor** induces people to move out of their present location.
- A **pull factor** induces people to move into a new location.

As migration for most people is a major step not taken lightly, both push and pull factors typically play a role. To migrate, people view their current place of residence so negatively that they feel pushed away, and they view another place so positively that they feel pulled toward it.

Cultural Reasons For Migrating

Cultural migration can occur for a number of personal reasons, such as family status and schools. At the international scale, cultural migration frequently occurs because of political conflict. The United Nations High Commission for Refugees (UNHCR) recognizes three groups of people who are forced to migrate for political reasons:

- A **refugee** has been forced to migrate to another country to avoid the effects of armed conflict, situations of generalized violence, violations of human rights, or other disasters and cannot return for fear of persecution because of race, religion, nationality, membership in a social group, or political opinion.
- An **internally displaced person (IDP)** has been forced to migrate for similar political reasons as a refugee but has not migrated across an international border.
- An **asylum seeker** is someone who has migrated to another country in the hope of being recognized as a refugee.

The UN counted 19.5 million refugees, 38.2 million IDPs, and 1.8 million asylum seekers in 2014 (Figure 3-22).

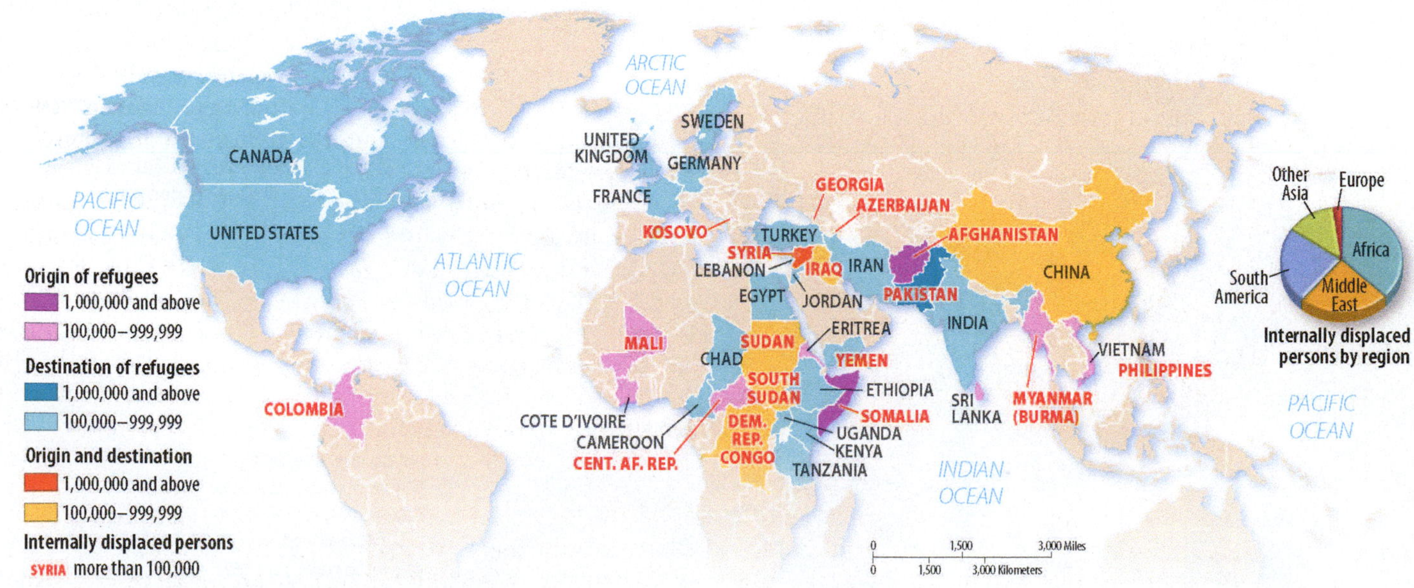

▲ **FIGURE 3-22 POLITICAL MIGRATION: REFUGEES AND IDPs** The largest number of refugees originated in Afghanistan and Syria.

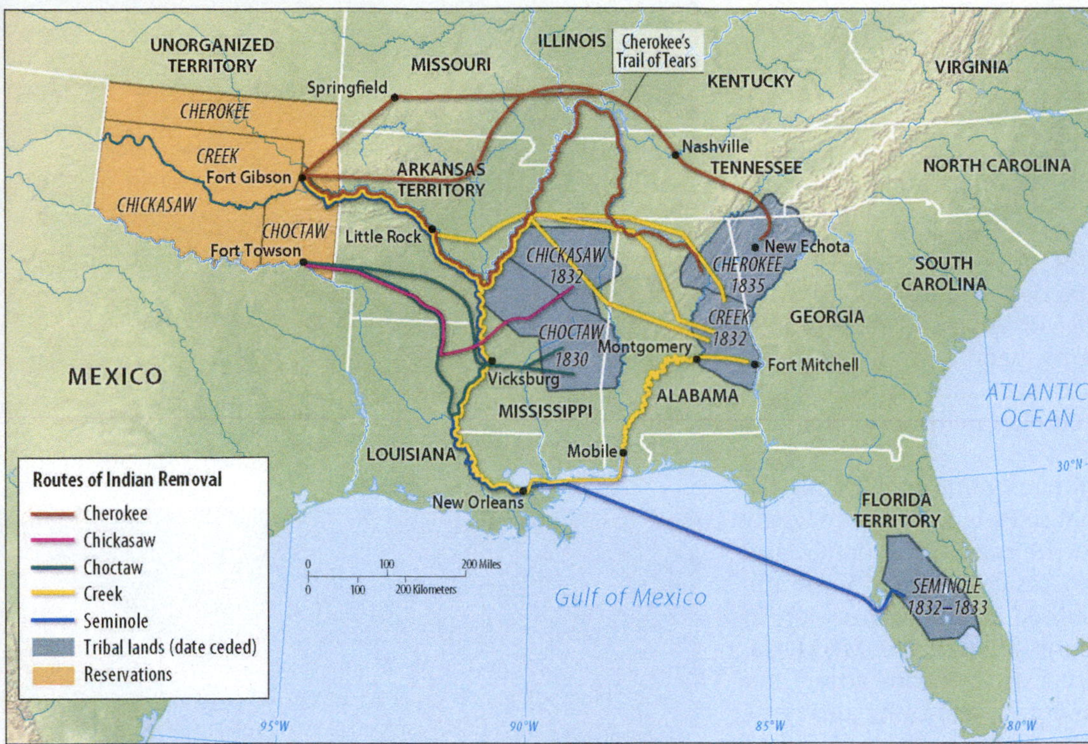

(a)

(b)

◀ **FIGURE 3-23 TRAIL OF TEARS (a)** These are the routes that the Cherokee, Chickasaw, Choctaw, Creek, and Seminole tribes took when they were forced to migrate westward in the early nineteenth century. **(b)** Trail of Tears, Pea Ridge National Military Park, Garfield, Arkansas.

The largest numbers of refugees in 2014 were forced to migrate from Afghanistan and from Syria because of continuing civil wars in those two countries. Neighboring countries received the most refugees—Pakistan and Iran from Afghanistan and Lebanon and Turkey from Syria.

TRAIL OF TEARS

Like many other people in North America, Native Americans also migrated west in the nineteenth century. But their migration was forced rather than voluntary. This inequality was written in law, when the Indian Removal Act of 1830 authorized the U.S. Army to remove five Indian tribes from their land in the southeastern United States and move them to Indian Territory (now the state of Oklahoma). The Choctaw were forced to emigrate from Mississippi in 1831, the Seminole from Florida in 1832, the Creek from Alabama in 1834, the Chickasaw from Mississippi in 1837, and the Cherokee from Georgia in 1838 (Figure 3-23).

The five removals opened up 100,000 square kilometers (25 million acres) of land for whites to settle and relocated the tribes to places that were too dry to sustain their traditional ways of obtaining food. Approximately 46,000 Native Americans were estimated to have been uprooted, and many of them died in the long trek to the west. The route became known as the Trail of Tears; parts of it are preserved as a National Historic Trail.

PAUSE & REFLECT 3.3.1

What similarities and differences can be seen between the interregional migration patterns of Native Americans and of migrants of European ancestry, as shown in Figure 3-12?

Environmental Reasons For Migrating

LEARNING OUTCOME 3.3.2

Explain environmental reasons for migration.

People sometimes migrate for environmental reasons, pulled toward physically attractive regions and pushed from hazardous ones. In this age of improved communications and transportation systems, people can live in environmentally attractive areas that are relatively remote and still not feel too isolated from employment, shopping, and entertainment opportunities.

Attractive environments for migrants include mountains, seasides, and warm climates. Proximity to the Rocky Mountains lures Americans to the state of Colorado, and the Alps pull French people to eastern France. Some migrants are shocked to find polluted air and congestion in such areas. The southern coast of England, the Mediterranean coast of France, and the coasts of Florida attract migrants, especially retirees, who enjoy swimming and lying on the beach. Of all elderly people who migrate from one U.S. state to another, one-third select Florida as their destination. Regions with warm winters, such as southern Spain and the southwestern United States, attract migrants from harsher climates.

Migrants are also pushed from their homes by adverse physical conditions. Water—either too much or too little—poses the most common environmental threat. Many people are forced to move by water-related disasters because they live in a vulnerable area, such as a floodplain (Figure 3-24).A **floodplain** is an area subject to flooding during a specific number of years, based on historical trends. People living in the "100-year floodplain," for example, can expect flooding on average once every century (Figure 3-25). Many people are unaware that they live in a floodplain, and even people who do know often choose to live there anyway.

A lack of water pushes others from their land. Hundreds of thousands have been forced to move from drylands in Africa because of drought conditions. Deterioration of land to a desertlike condition typically due to human actions is called **desertification**, or more precisely semiarid land degradation (Figure 3-26). The capacity of portions of Africa to sustain human life—never very high—has declined recently because of population growth and several years of unusually low rainfall (Figure 3-27). Consequently, many of these nomads have been forced to move into cities and rural camps, where they survive on food donated by the government and international relief organizations.

▲ **FIGURE 3-24 FLOODING, JAKARTA, INDONESIA**

▲ **FIGURE 3-25 100-YEAR FLOODPLAIN** Flooding along the Mississippi River in 2011 covers farmland.

An environmental or political feature that hinders migration is an **intervening obstacle**. The principal obstacle traditionally faced by migrants to other countries was environmental: the long, arduous, and expensive passage over land or sea. Transportation improvements that have promoted globalization, such as motor vehicles and airplanes, have diminished the importance of environmental features as intervening obstacles.

PAUSE & REFLECT 3.3.2

Why might people choose to build houses in floodplains?

◀ **FIGURE 3-26 DRYLANDS, TANZANIA** People are attempting to obtain drinking water from a dry river bed.

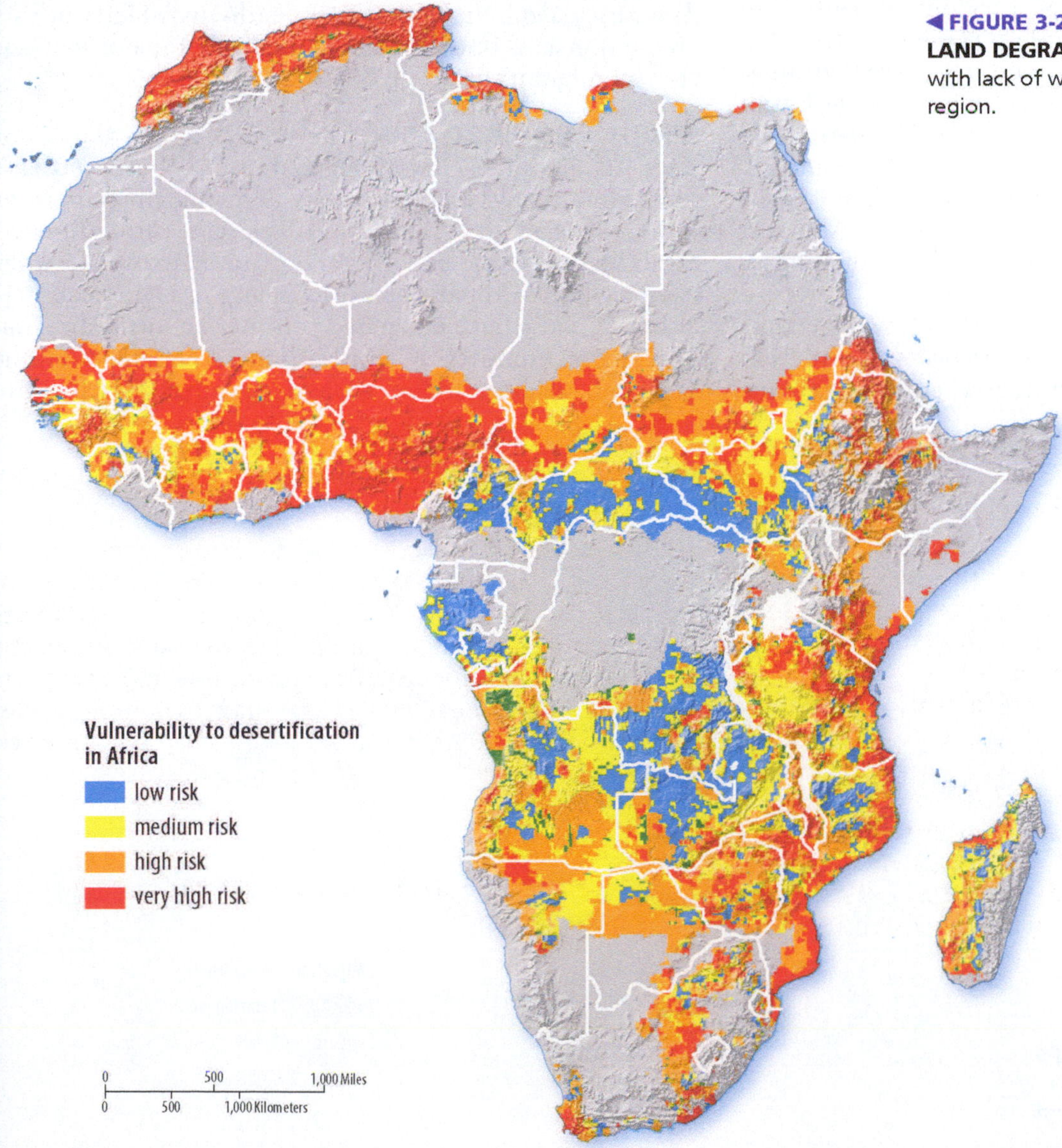

◀ **FIGURE 3-27 DESERTIFICATION (SEMIARID LAND DEGRADATION)** The most severe problems with lack of water in Africa are in the Sahel region.

Vulnerability to desertification in Africa
- low risk
- medium risk
- high risk
- very high risk

0 500 1,000 Miles
0 500 1,000 Kilometers

Migrating to Find Work

LEARNING OUTCOME 3.3.3

Understand economic reasons for international migration.

Most people migrate for economic reasons. People often emigrate from places that have few job opportunities and immigrate to places where jobs seem to be available. Because of economic restructuring, job prospects often vary from one country to another and within regions of the same country.

ECONOMIC REASONS FOR MIGRATING

The United States and Canada have been especially prominent destinations for economic migrants. Many European immigrants to North America in the nineteenth century truly expected to find streets paved with gold. While not literally so gilded, the United States and Canada did offer Europeans prospects for economic advancement. This same perception of economic plenty now lures people to the United States and Canada from Latin America and Asia.

The relative attractiveness of a region can shift with economic change. Ireland was a place of net out-migration through most of the nineteenth and twentieth centuries. Dire economic conditions produced net out-migration in excess of 200,000 a year during the 1850s. The pattern reversed during the 1990s, as economic prosperity made Ireland a destination for immigrants, especially from Eastern Europe. However, the collapse of Ireland's economy as part of the severe global recession starting in 2008 brought a return to net out-migration (Figure 3-28).

It is sometimes difficult to distinguish between migrants seeking economic opportunities and refugees fleeing from government persecution. The distinction between economic migrants and refugees is important because the United States, Canada, and European countries treat the two groups differently. Economic migrants are generally not admitted unless they possess special skills or have a close relative already in the new country, and even then they must compete with similar applicants from other countries. However, refugees receive special priority in admission to other countries.

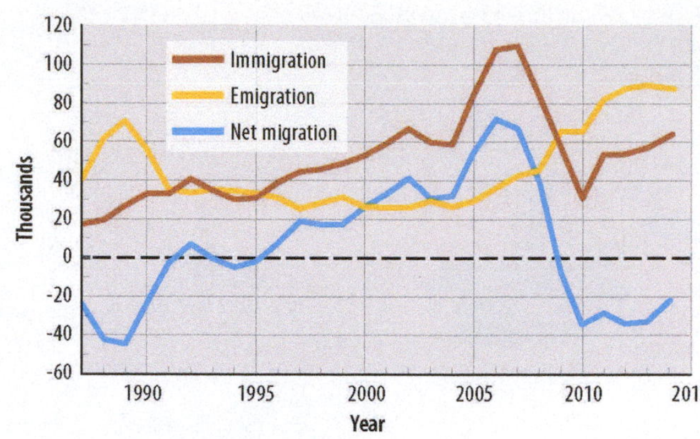

▲ **FIGURE 3-28 NET MIGRATION IN IRELAND** With few job prospects, Ireland historically had net out-migration until the 1990s. The severe recession of the early twenty-first century brought net out-migration back to Ireland.

ASIA'S MIGRANT WORKERS

People unable to migrate permanently to a new country for employment opportunities may be allowed to migrate temporarily. Prominent forms of temporary work are found in Asia (discussed here) and in Europe (discussed later in this chapter). Asia is both a major source and a major destination for migrants in search of work (Figure 3-29).

SOUTH AND EAST ASIA. The world's largest sources of migrants in search of work emigrate from South and East Asia. More than 2 million people annually emigrate from India, Bangladesh, China, and Pakistan. An estimated 50 million Chinese and 25 million Indians live in other countries. The United States is a leading receiving country, although most have emigrated to other countries in Asia. The largest numbers of ethnic Chinese emigrants are in Thailand, Malaysia, Indonesia, and Myanmar, as well as the

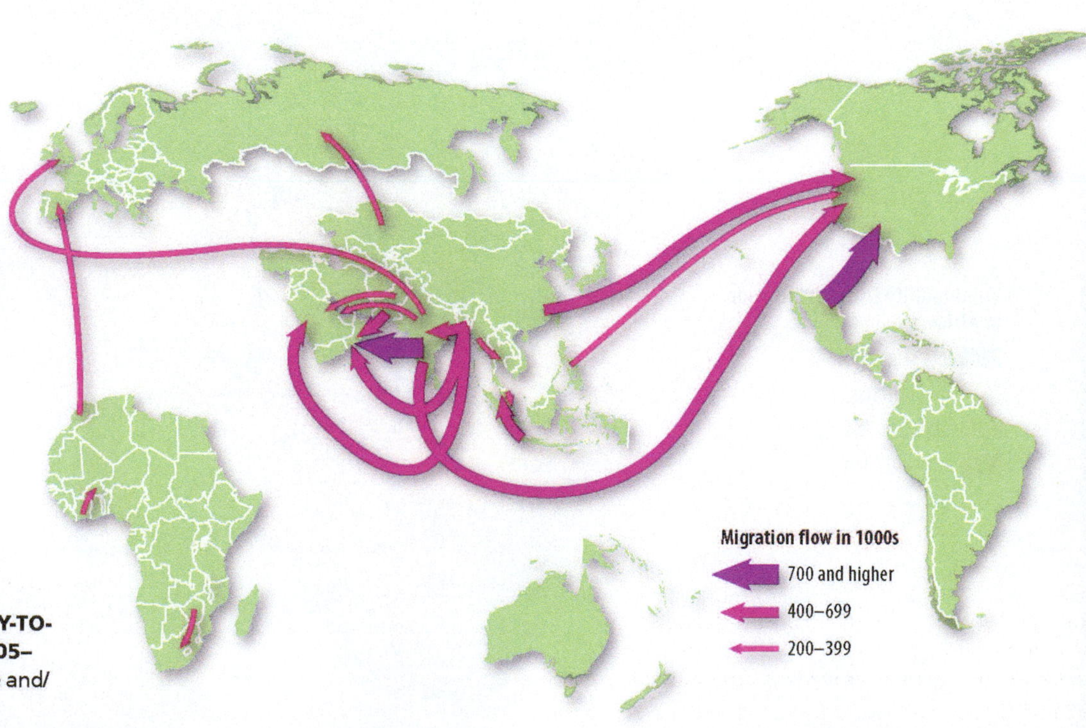

▶ **FIGURE 3-29 LARGEST COUNTRY-TO-COUNTRY MIGRATION FLOWS, 2005–2010** Most migration flows originate and/or end in Asia.

▲ **FIGURE 3-30 IMMIGRANT TO UNITED ARAB EMIRATES FROM BANGLADESH**

United States. Chinese people comprise one-half of the population of Singapore, one-fourth of Malaysia, and one-sixth of Thailand. The largest numbers of Indian emigrants are in Nepal, Myanmar, and Malaysia, as well as the United States.

SOUTHWEST ASIA. The wealthy oil-producing countries of Southwest Asia have been major destinations for people from the South Asian countries of India, Bangladesh, and Pakistan, as well as the Philippines, Thailand, and other countries in Southeast Asia. In addition, citizens of poorer countries in Southwest Asia have emigrated to the region's wealthier countries. Immigrants comprise 84 percent of the population of the United Arab Emirates (UAE), 74 percent of Qatar, 60 percent of Kuwait, and 55 percent of Bahrain. Saudi Arabia and the UAE have the largest numbers of immigrants in the region (Figure 3-30).

Working conditions for immigrants have been considered poor in some of these countries. The Philippine government determined in 2011 that only two countries in Southwest Asia—Israel and Oman—were "safe" for their Filipino migrants, and the others lacked adequate protection for workers' rights. For their part, oil-producing countries fear that the increasing numbers of immigrants will spark political unrest and abandonment of traditional Islamic customs.

REMITTANCES

Migrants who find work in another country frequently send a portion of the wages they have earned to relatives back home. The transfer of money by workers to people in the country from which they emigrated is a **remittance**.

The total amount of remittances worldwide was $550 billion in 2013. The figure has been increasing by nearly 10 percent annually. Remittances are an increasingly important source of wealth for people in developing countries, especially following cutbacks in official assistance from foreign governments and international aid agencies.

People in India received the most remittances in 2013 ($71 billion), followed by people in China ($60 billion). Nearly one-half of the GDP of Tajikistan and one-third of Kyrgyzstan comprised remittances, primarily from emigrants living in Russia (Figure 3-31).

The cost of transferring money is high in many places. Banks and firms such as Western Union that specialize in money transfers charge high fees for the service, an average of 9 percent worldwide. To transfer $200 from the United States, it costs an average of $6 to Mexico and $12 to Haiti; it costs around $20 to transfer $200 between many African countries.

PAUSE & REFLECT 3.3.3

Before becoming the leading company for transferring remittances from the United States, what was the principal business of Western Union?

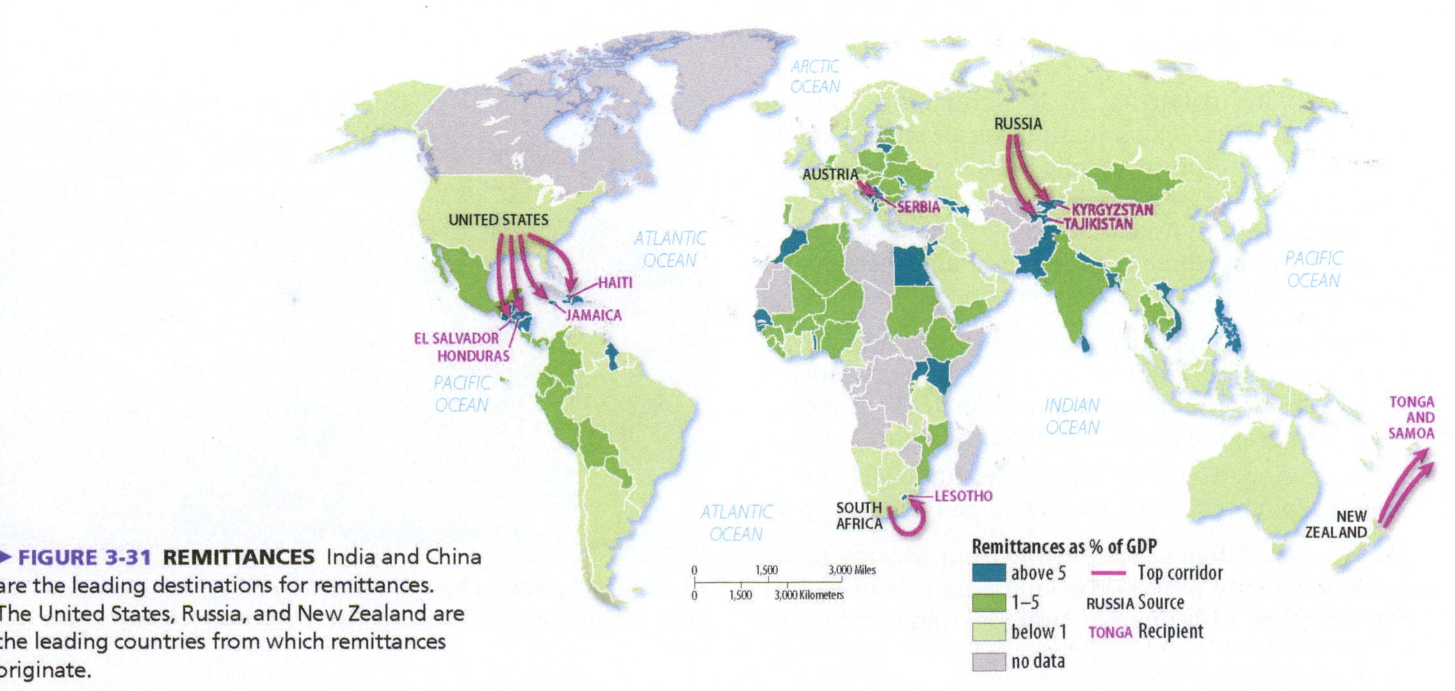

▶ **FIGURE 3-31 REMITTANCES** India and China are the leading destinations for remittances. The United States, Russia, and New Zealand are the leading countries from which remittances originate.

Gender and Age of Migrants

LEARNING OUTCOME 3.3.4

Describe the demographic characteristics of international migrants.

Ravenstein noted distinctive gender and family-status patterns in his migration theories:

- Most long-distance migrants were male.
- Most long-distance migrants were adult individuals rather than families with children.

Adult males may have constituted the majority in the past, but that pattern has changed. In reality, women and children have constituted a high percentage of migrants for a long time.

GENDER OF MIGRANTS

Ravenstein theorized that males were more likely than females to migrate long distances to other countries because searching for work was the main reason for international migration, and males were much more likely than females to be employed. This held true for U.S. immigrants during the nineteenth and much of the twentieth centuries, when about 55 percent were male. But female immigrants to the United States began to outnumber male immigrants around 1970, and they now comprise 55 percent of the total. Female immigrants also outnumber males in other developed countries (Figure 3-32).

The gender mix of Mexicans who come to the United States without authorized immigration documents—currently the largest group of U.S. immigrants—has changed sharply. In the 1980s, males constituted 85 percent of the Mexican migrants arriving in the United States without proper documents, according to U.S. census and immigration service estimates. But since the 1990s, women have accounted for about half of the unauthorized immigrants from Mexico.

In developing countries, male immigrants still outnumber female ones. The situation varies widely among regions, however. Approximately one-half of emigrants from East and Southeast Asia are women, compared to only 44 percent from South Asia.

Two factors contribute to the larger share of females migrating to developed countries than theorized by Ravenstein:

- Because most people migrate to developed countries for job opportunities, the high percentage of women in the labor force in these countries logically attracts a high percentage of female immigrants (Figure 3-33).
- Some developed countries have made it possible for wives to join husbands who have already immigrated.

The increased female migration from Mexico to the United States partly reflects the changing role of women in Mexican society. In the past, rural Mexican women were obliged to marry at a young age and to remain in the village to care for children. Now some Mexican women are migrating to the United States to join husbands or brothers already in the United States, but most are seeking jobs. At the same time, women feel increased pressure to get jobs in the United States because of poor economic conditions in Mexico.

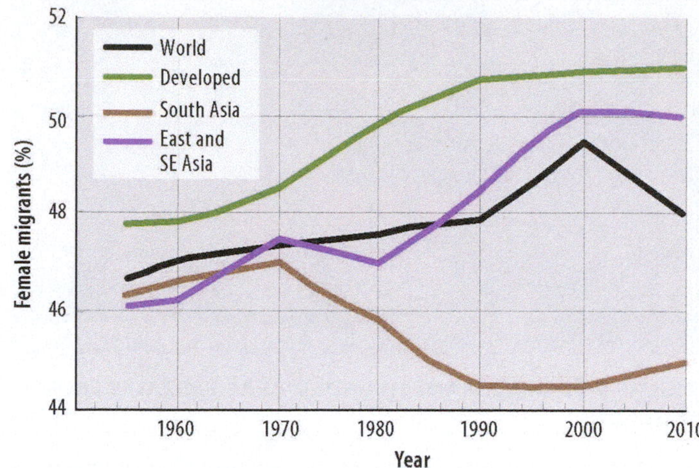

▲ **FIGURE 3-32 FEMALE IMMIGRANTS** Slightly more than one-half of immigrants are female in the developed regions of Europe and North America. The gender ratio varies more widely in developing regions.

AGE OF MIGRANTS

Ravenstein theorized that most long-distance migrants were young adults seeking work rather than children or elderly people. Recent migration patterns in the United States match the theory in some respects but not in others (Figure 3-34):

- A relatively high share of U.S. immigrants are young adults, as Ravenstein expected. People between the ages of 20 and 39 comprise 49 percent of recent

▲ **FIGURE 3-33 FEMALE IMMIGRANTS ITALY** An immigrant from Africa sells to tourists in Menaggio, along Lake Como, Italy.

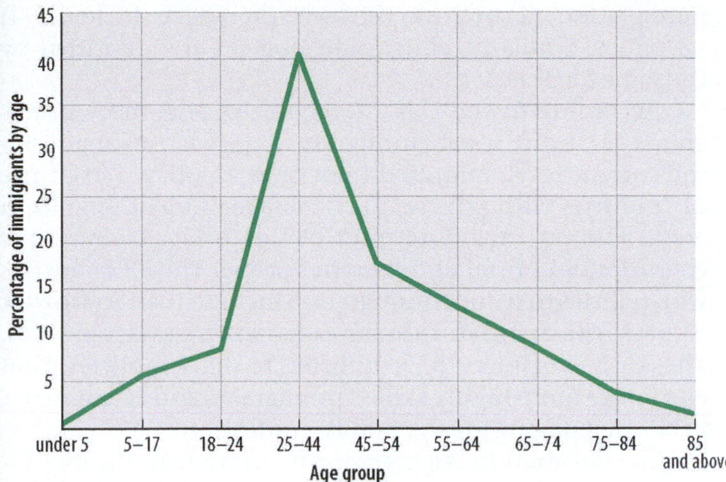

▲ **FIGURE 3-34 AGE OF IMMIGRANTS** Adults between ages 20 and 39 represent a disproportionately large share of immigrants to the United States.

immigrants, compared to only 27 percent of the entire U.S. population.

- Immigrants are, as expected, less likely to be elderly people. Only 5 percent of recent U.S. immigrants are over age 65, compared to 14 percent of the entire U.S. population. However, in developing countries, immigrants are more likely to be elderly—only 6 percent of the total population but 8 percent of immigrants.
- Children under age 20 comprise 21 percent of immigrants, only slightly lower than the 26 percent share in the total U.S. population. In developing countries, immigrants are much less likely to be children; people under age 20 comprise 35 percent of the total population but only 23 percent of the migrants.

The number of unaccompanied minors trying to cross into the United States without proper documentation has increased sharply in recent years. Nearly 90 percent have been males between 12 and 17 (Figure 3-35). As with other migration flows, the large increase in teenage boys trying to reach the United States stems from a mix of push and pull factors. Most are pushed out of Honduras and El Salvador because of increased gang violence there and are pulled to the United States because of rumors that they won't be deported if caught.

PAUSE & REFLECT 3.3.4

Why might elderly people be more likely than average to migrate in developing countries but less likely than average to do so in developed countries?

CHECK-IN KEY ISSUE 3

Why Do People Migrate?

- ✓ **People migrate for a combination of political, environmental, and economic push and pull factors.**
- ✓ **Most people migrate in search of work.**
- ✓ **Most migrants are young adults.**

▼ **FIGURE 3-35 AGE OF IMMIGRANTS: LATIN AMERICA** Teenagers aged 14 to 17 from Honduras, El Salvador, and Guatemala are detained in Juarez, Mexico, after attempting to cross the border into the United States.

KEY ISSUE 4

Why Do Migrants Face Challenges?

- ▸ Government Immigration Policies
- ▸ U.S. Quota Laws
- ▸ U.S.–Mexico Border Issues
- ▸ Europe's Immigration Crisis

LEARNING OUTCOME 3.4.1

Describe government policies that affect immigration.

Transportation improvements that have promoted globalization, such as motor vehicles and airplanes, have diminished the importance of environmental features as intervening obstacles. Today, the major obstacles faced by most immigrants are political. A migrant needs a passport to legally emigrate from a country and a visa to legally immigrate to a new country.

Government Immigration Policies

Most countries have adopted selective immigration policies that admit some types of immigrants but not others. The United States is no exception. The two reasons that most visas are granted are for specific employment placement and family reunification.

The U.N. classifies countries according to four types of immigration policies: (1) maintain the current level of immigration, (2) increase the level, (3) reduce the level, (4) no policy. Similarly, emigration policies are identified by the same four classes.

According to the U.N., 21 countries seek more immigrants, 32 want fewer immigrants, 116 wish to maintain the current level, and 25 do not have a policy. Ten of the 21 countries with policies to encourage more immigration are in Europe, including most of the former Communist countries of Central and Eastern Europe. The 32 countries with policies to reduce immigration include 10 in Southwest Asia & North Africa and 8 in sub-Saharan Africa (Figure 3-36). The U.N. considers 67 countries to have policies that encourage more highly skilled immigrants and 14 countries to be encouraging greater family unification.

The distribution of emigration policies is considerably different. The U.N. found policies to increase emigration in 18 countries, to decrease emigration in 46 countries, to maintain the current level in 43 countries, and 88 with no policy. The 18 countries wishing to increase emigration include 5 each in South Asia, Southeast Asia, and the South Pacific. Sub-Saharan Africa had the most countries seeking to lower emigration.

UNAUTHORIZED IMMIGRATION

The number of people allowed to immigrate into the United States is at a historically high level, but the number who wish to come is even higher. Many who cannot legally enter the United States immigrate illegally. Those who do so are entering without proper documents and thus are called **unauthorized immigrants**.

Controversy extends to what to call the group of immigrants:

- **Unauthorized immigrant** is the term preferred by academic observers, including the authoritative Pew Hispanic Center, as a neutral term.
- **Undocumented immigrant** is the term preferred by some of the groups that advocate for more rights for these individuals.
- **Illegal alien** is the term preferred by some of the groups that favor tougher restrictions and enforcement of immigration laws.

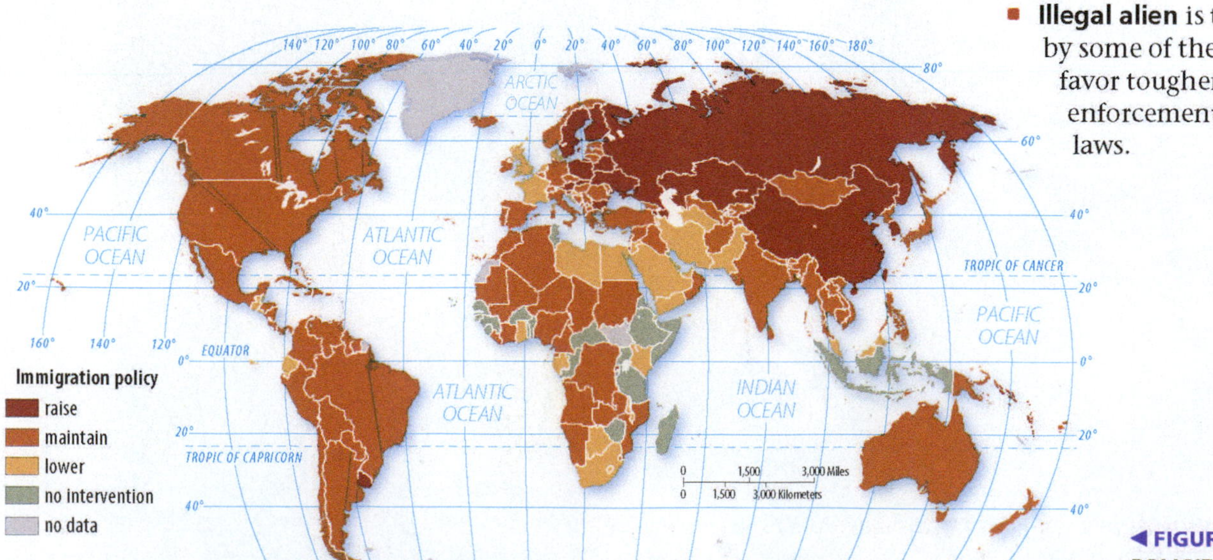

◀ **FIGURE 3-36 IMMIGRATION POLICIES**

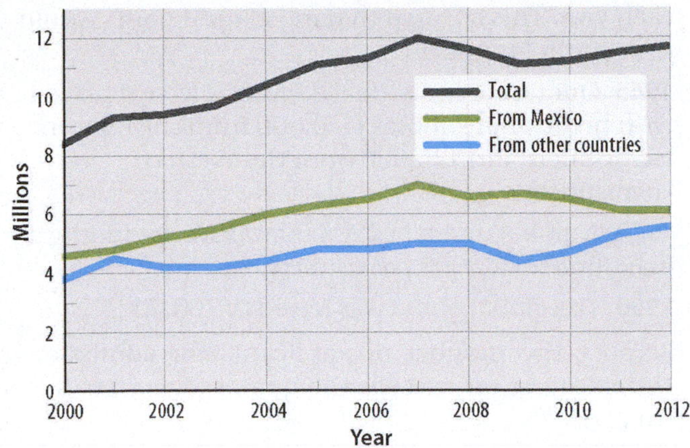

▲ **FIGURE 3-37 UNAUTHORIZED IMMIGRANTS IN THE UNITED STATES** Most are from Mexico.

The Pew Hispanic Center estimated that there were 11.3 million unauthorized immigrants living in the United States in 2014. The number increased rapidly during the first years of the twenty-first century (Figure 3-37). After hitting a peak of 12.2 million in 2007, the number declined because of reduced job opportunities in the United States during the severe recession that started in 2008. In other words, the number of unauthorized immigrants entering the United States is now less than the number leaving.

Other information about unauthorized immigrants, according to Pew Hispanic Center:

- **Distribution.** California and Texas have the largest numbers of unauthorized immigrants. Nevada has the largest percentage (Figure 3-38).
- **Source country.** More than one-half of unauthorized immigrants emigrate from Mexico. The remainder are about evenly divided between other Latin American countries and other regions of the world.
- **Children.** The 11.3 million unauthorized immigrants included 1 million children. In addition, while living in the United States, unauthorized immigrants have given birth to approximately 4.5 million babies, who are legal citizens of the United States.
- **Years in the United States.** The duration of residency in the United States has been increasing for unauthorized immigrants. In a 2013 Pew survey, 61 percent of unauthorized adult immigrants had resided in the United States for 10 years or more, 23 percent for 5 to 9 years, and 16 percent for less than 5 years. A similar survey in 2003 showed a different distribution: 38 percent had been in the United States for less than 5 years, compared to 37 percent for more than 10 years.
- **Labor force.** Approximately 8 million unauthorized immigrants are employed in the United States, accounting for around 5 percent of the total U.S. civilian labor force. Unauthorized immigrants were much more likely than the average American to be employed in construction and hospitality (food service and lodging) jobs and less likely to be in white-collar jobs such as education, health care, and finance.

PAUSE & REFLECT 3.4.1

Why are most unauthorized immigrants in the Southwest?

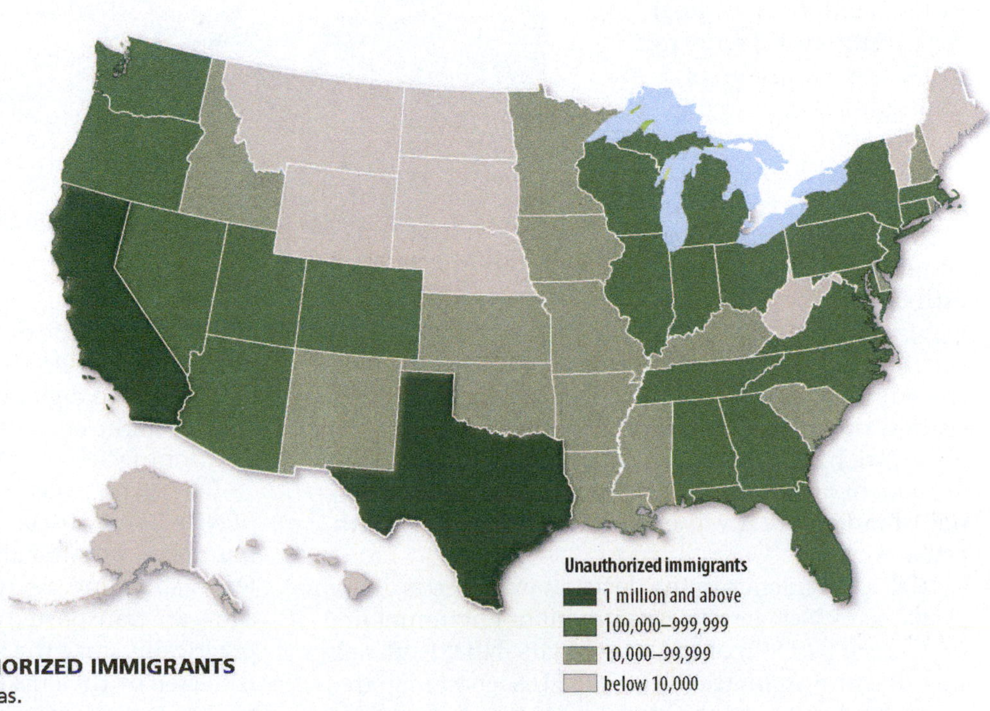

▶ **FIGURE 3-38 DISTRIBUTION OF UNAUTHORIZED IMMIGRANTS** The largest numbers are in California and Texas.

U.S. Quota Laws

LEARNING OUTCOME 3.4.2

Understand features of U.S. Quota laws.

The era of unrestricted immigration to the United States ended when Congress passed the Quota Act in 1921 and the National Origins Act in 1924. These laws established **quotas**, or maximum limits on the number of people who could immigrate to the United States during a one-year period. Key modifications in the U.S. quotas have included:

- **1924.** For each country that had native-born persons already living in the United States, 2 percent of their number (based on the 1910 census) could immigrate each year. This ensured that most immigrants would come from Europe.
- **1965.** Quotas for individual countries were replaced with hemisphere quotas (170,000 from the Eastern Hemisphere and 120,000 from the Western Hemisphere).
- **1978.** A global quota of 290,000 was set, including a maximum of 20,000 per country.
- **1990.** The global quota was raised to 700,000.

Because the number of applicants for admission to the United States far exceeds the quotas, Congress has set preferences:

- **Family reunification.** Approximately three-fourths of immigrants are admitted to reunify families, primarily spouses or unmarried children of people already living

DOING GEOGRAPHY Claiming Ellis Island

Twelve million immigrants to the United States between 1892 and 1954 were processed at Ellis Island (Figure 3-39). Once accepted into the United States, immigrants were transported by ferry from Ellis Island 1.6 kilometers (1 mile) across New York Harbor to New York City.

Though no longer used to process immigrants, Ellis Island became part of the Statue of Liberty National Monument in 1965, and the buildings were restored and reopened in 1990 as a museum of immigration.

▲ FIGURE 3-39 **ELLIS ISLAND**

The State of New Jersey, located only 400 meters (1,300 feet) from Ellis Island, long argued that Ellis Island was actually part of New Jersey rather than New York, as was generally believed. After decades of dispute, New Jersey took its case to the U.S. Supreme Court. In 1998, the Supreme Court ruled 6–3 that New Jersey had jurisdiction over 8.7 hectares (22.8 acres) of Ellis Island. New York had jurisdiction over only 1.9 hectares (4.7 acres), encompassing the low waterline of the original island before it was expanded in the nineteenth century.

Critical evidence in the decision was a series of maps prepared by New Jersey Department of Environmental Protection (NJDEP) officials using GIS. NJDEP officials scanned into an image file an 1857 U.S. coast map that was considered to be the most reliable map from that era. The image file of the old map was brought into ArcView, and then the low waterline shown on the 1857 map was digitized using a series of dots. The perimeter of the current island was mapped using Global Positioning System (GPS) surveying.

The victory was partly a matter of pride on the part of New Jersey officials to stand up to their more glamorous neighbor. After all, Ellis Island is closer to the New Jersey shoreline, yet tourists—like immigrants a century ago—are transported by ferry to New York City. More practically, since the favorable ruling, the sales tax collected by the Ellis Island museum gift shop now goes to New Jersey instead of New York.

in the United States. The typical wait for a spouse to gain entry is currently about five years.

- **Skilled workers.** Exceptionally talented professionals receive most of the remainder of the quota.
- **Diversity.** A few immigrants are admitted by lottery under a diversity category for people from countries that historically sent few people to the United States.

The quota does not apply to refugees, who are admitted if they are judged genuine refugees. Also admitted without limit are spouses, children, and parents of U.S. citizens. The number of immigrants can vary sharply from year to year, primarily because numbers in these two groups are unpredictable.

Other countries charge that by giving preference to skilled workers, immigration policies in the United States, as well as other developed countries, contributes to a **brain drain**, which is a large-scale emigration by talented people. Scientists, researchers, doctors, and other professionals migrate to countries where they can make better use of their abilities.

Asians have made especially good use of the priorities set by the U.S. quota laws. Many well-educated Asians enter the United States under the preference for skilled workers. Once admitted, they can bring in relatives under the family reunification provisions of the quota. Eventually, these immigrants can bring in a wider range of other relatives from Asia, through a process of **chain migration**, which is the migration of people to a specific location because relatives or members of the same nationality previously migrated there.

PAUSE & REFLECT 3.4.2

How are changes in the quota laws reflected in changing U.S. immigration patterns, as shown in Figure 3-10?

What's Your Migration Geography?

- When did you or your ancestors immigrate to the United States? From which region of the world or country? If they came from Europe between 1892 and 1954, they most likely came through Ellis Island. But if they came at a different time or from a different region of the world, they probably entered the United States somewhere else.
- Some sort of international or interregional migration is a nearly universal experience for families living in North America. At the same time, each of our families has a unique immigration experience. What's your migration geography?
- The uniqueness of each of our immigration experiences has many dimensions:
  - **Timing.** Are you the immigrant, or was it a parent, a grandparent, a great-grandparent, or beyond?
  - **Origin.** From what country or countries did you or your ancestors emigrate?
  - **Diversity.** Do you have similar or different immigration experiences within your family? If different, at what generation did the experiences diverge?
  - **Frequency.** Once in the United States, has your family experienced interregional moves? Or have they stayed in the same U.S. region where they first arrived?
- Families often display immigration records on a family tree. Geographers instead can display records on a family map.

1. On a sheet of paper or in an electronic spreadsheet, record an interregional or international migration taken by each of your family members. If the individual moved many times, record the one that seems most significant to your family, such as the one that brought two of your ancestors together.
2. On a blank base map of the world or only the portion of it that you need for your family geography, draw an arrow that connects where (if anywhere) you immigrated from to reach your current home. Using a second color, draw arrows to show the route of each of your two parents, use a third color to draw arrows for your four grandparents, and a fourth color to draw arrows for all of your eight great-grandparents for whom you have information.

Family Member	From	To
Your mother		
Your mother's mother		
Your mother's mother's mother		
Your mother's mother's father		
Your mother's father		
Your mother's father's mother		
Your mother's father's father		
Your father		
Your father's mother		
Your father's mother's mother		
Your father's mother's father		
Your father's father		
Your father's father's mother		
Your father's father's father		

U.S.–Mexico Border Issues

LEARNING OUTCOME 3.4.3

Understand the diversity of conditions along the U.S.–Mexico border.

The U.S.-Mexico border is 3,141 kilometers (1,951 miles) long (Figure 3-41). Rural areas and small towns are guarded by only a handful of agents. Crossing the border on foot legally is possible in several places. Elsewhere, the border runs mostly through sparsely inhabited regions.

The United States has constructed a barrier covering approximately one-fourth of the border. Several large urban areas are situated on the border, including San Diego, California, and Tijuana, Mexico, at the western end, and Brownsville, Texas, and Matamoros, Mexico, at the eastern end. Driving across the border in the urban areas can be fraught with heavy traffic and delays.

A joint U.S.-Mexican International Boundary and Water Commission is responsible for keeping official maps, on the basis of a series of nineteenth-century treaties. The commission is also responsible for marking the border by maintaining 276 6-foot-tall iron monuments erected in the late nineteenth century, as well as 440 15-inch-tall markers added in the 1970s. Actually locating the border is difficult in some remote areas.

MIGRATION POLICY DISPUTES

Americans are divided concerning whether unauthorized migration helps or hurts the country (see Debate It feature). This ambivalence extends to specific elements of immigration law:

- **Border patrols.** Americans would like more effective border patrols so that fewer unauthorized immigrants can get into the country, but they don't want to see money spent to build more fences along the border. The U.S. Department of Homeland Security has stepped up enforcement, including deportation of a record 438,421 unauthorized immigrants in 2013.
- **Workplace.** Most Americans recognize that unauthorized immigrants take jobs that no one else wants, so they support some type of work-related program to make them legal, and they oppose raids on workplaces in attempts to round up unauthorized immigrants. Most Americans support a path to U.S. citizenship for unauthorized immigrants.
- **Civil rights.** Americans favor letting law enforcement officials stop and verify the legal status of anyone they suspect of being an unauthorized immigrant. On the other hand, they fear that enforcement efforts that identify and deport unauthorized immigrants could violate the civil rights of U.S. citizens.
- **Local initiatives.** Polls show that most Americans believe that enforcement of unauthorized immigration is a federal government responsibility and do not support the use of local law enforcement officials to find unauthorized immigrants. On the other hand, residents of some states along the Mexican border favor stronger enforcement of authorized immigration.

The strongest state initiative has been Arizona's 2010 law that obligated local law enforcement officials, when practicable, to determine a person's immigration status. Under the Arizona law, foreigners are required to carry at all times documents proving they are in the country legally and to produce those documents upon request of a local law enforcement official. In 2012, the U.S. Supreme Court struck down several provisions of the law. Although it does not share a border

SUSTAINABILITY & OUR ENVIRONMENT The Immigration View from Mexico

From the United States, the view to the south may seem straightforward. Millions of Mexicans are trying to cross the border by any means, legal or otherwise, in search of a more sustainable economic and cultural life.

The view from Mexico is more complex. Along its northern border with the United States, Mexico is the source for unauthorized emigrants. At the same time, along its southern border with Guatemala, Mexico is the destination for unauthorized immigrants. When talking with its neighbor to the north, Mexicans urge understanding and sympathy for the plight of the immigrants. When talking with its neighbor to the south, Mexicans urge stronger security along the border.

▲ **FIGURE 3-40 BORDER BETWEEN MEXICO AND GUATEMALA** People and goods are being transported across the Suchiate River from Guatemala to Mexico.

Oceans and wide rivers frequently serve as intervening obstacles to hinder migration. But along the Mexico–Guatemala border, the Suchiate River is not an environmentally sustainable obstacle, because it sometimes only ankle deep (Figure 3-40). Immigrants from other Latin American countries, especially El Salvador and Honduras, travel through Guatemala without need of a passport in order to cross into Mexico. Although a passport is needed to cross the border from Guatemala into Mexico, the Mexican government estimates that 2 million people a year do so illegally. Some migrate illegally from Guatemala to Mexico for higher-paying jobs in tropical fruit plantations. For most, the ultimate destination is the United States.

◀ **FIGURE 3-41 U.S.–MEXICO BORDER** **(a)** The border is 3,141 kilometers (1,951 miles) long. **(b)** Some border crossings (such as Tijuana) are congested. **(c)** Others can be crossed easily by foot.

with Mexico, Alabama enacted a similar measure in 2011. The Alabama law also prohibited or restricted unauthorized immigrants from attending public schools and colleges. On the other hand, Texas, which has the longest border with Mexico, has not enacted harsh anti-immigrant laws, and more than 100 localities across the country have passed resolutions supporting more rights for unauthorized immigrants—a movement known as "Sanctuary City."

PAUSE & REFLECT 3.4.3

Why at border crossings is traffic entering the United States backed up further than traffic entering Mexico?

DEBATE IT! Immigration reform: Tougher controls or legal status?

Debate over authorized immigration centers on border security and on appropriateness of a path to legal status for unauthorized immigrants in the United States.

TIGHTEN SECURITY AND DO NOT OFFER A PATH TO LEGAL STATUS

- **THE WRONG MESSAGE.** People breaking the law by crossing the U.S. border without proper documentation sends the wrong message to people who obey the law.
- **ENCOURAGE OTHERS.** Rewarding people for illegal behavior will encourage others to enter without documents.
- **POOR SECURITY.** The border is not sufficiently secure, especially in small towns and rural areas.

◀ **FIGURE 3-42 MINIMAL SECURITY AT THE BORDER** Crossing from Palomas, Mexico, to Columbus, New Mexico.

OFFER A PATH TO LEGAL STATUS; SECURITY IS ALREADY TIGHT ENOUGH

- **IMPRACTICAL.** It would be a practical impossibility for law enforcement officials to actually find the 11 million unauthorized immigrants.
- **ECONOMIC IMPACT.** Pulling unauthorized immigrants out of their jobs would cripple the U.S. economy.
- **AGENTS.** The numbers of border agents and deportations of unauthorized immigrants have doubled since 2000.
- **LAW-ABIDING.** Unauthorized immigrants are productive and otherwise law-abiding members of U.S. society.

◀ **FIGURE 3-43 BORDER AGENTS** Rio Grande near Laredo, Texas.

Europe's Immigration Crisis

LEARNING OUTCOME 3.4.4

Understand attitudes toward immigrants in Europe.

Of the world's 16 countries with the highest per capita income, 14 are in Northern and Western Europe. As a result, the region attracts immigrants from poorer regions located to the south and east. These immigrants serve a useful role in Europe, taking low-status and low-skill jobs that local residents won't accept. In cities such as Berlin, Brussels, Paris, and Zurich, immigrants provide essential services, such as driving buses, collecting garbage, repairing streets, and washing dishes.

MIGRATION PATTERNS IN EUROPE

European countries together have around 40 million foreign-born residents. This total includes around 20 million who have migrated from one European country to another and 20 million who have emigrated from a country outside of Europe.

Within Europe, the flow of migrants is primarily from east to west (Figure 3-44). Nearly all portions of former Communist countries in Eastern Europe have net out-migration, whereas nearly all of Western Europe has net in-migration. The largest flow is from Romania to Italy. Other large migration channels are from Poland to Germany and to the United Kingdom, from Italy to Germany, from Romania to Spain, and from Portugal to France.

Agreements among European countries, especially the 1985 Schengen Treaty, give a citizen of one European country the right to hold a job, live permanently, and own property elsewhere. The removal of migration restrictions for Europeans has set off large-scale migration flows within the region. The principal flows are from the poorer countries of Eastern Europe to the richer ones, where job opportunities have been greater, at least before the severe recession that started in 2008.

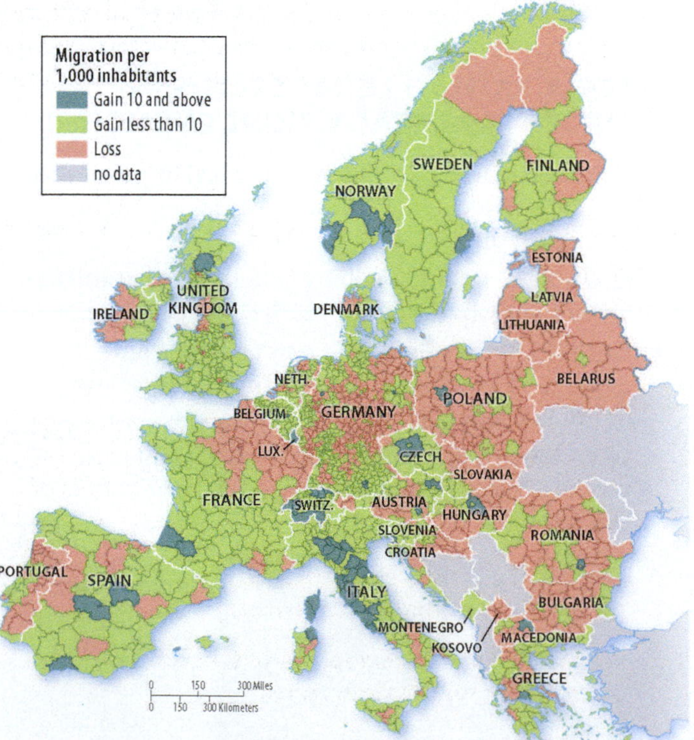

▲ **FIGURE 3-44 NET MIGRATION IN EUROPE** Former Communist countries of Eastern Europe have net out-migration. Most portions of Western Europe have net in-migration.

Prior to 2014, the leading sources of migrants into Europe from elsewhere in the world were countries in close proximity to Europe, such as Turkey and Morocco. The flow of immigrants into Europe escalated rapidly in 2014 and 2015, especially refugees escaping war and persecution in Syria. Large numbers of refugees have also attempted to reach Europe from Afghanistan and North Africa (Figure 3-45). European countries have struggled to figure out how to accommodate the flood of refugees. In an unsuccessful effort to limit the number arriving, several European countries have erected fences, imposed border checks, and shut down train lines (Figure 3-46).

The principal routes into Europe have been by trekking over land from Turkey into Greece or Bulgaria or by sailing across the Mediterranean Sea into Greece or Italy. From Greece, most of the refugees have attempted to reach Germany and other Northern Europe countries by way of Serbia and Hungary. More than 1,000 have died by drowning in the Mediterranean in boats that are not seaworthy, or by suffocating in sealed trucks.

GUEST WORKERS

Germany and other wealthy European countries operated **guest worker** programs, in which immigrants from poorer countries were allowed to immigrate temporarily to obtain jobs. The guest worker programs, operated mainly during the 1960s and 1970s, were expected to be examples of **circular migration**, which is the temporary movement of a migrant worker between home and host countries to seek employment. Guest workers were expected to return to their countries of origin once their work was done.

However, rather than circular migrants, many immigrants who arrived originally under the guest worker program have remained permanently in Europe. They, along with their children and grandchildren, have become citizens of the host country. The term *guest worker* is no longer used in Europe, and the government programs no longer exist.

Although relatively low paid by European standards, immigrants earn far more than they would at home. By letting their people work elsewhere, poorer countries reduce their own unemployment problems. Immigrants also help their native countries by sending remittances back home to their families.

ATTITUDES TOWARD IMMIGRANTS IN EUROPE

Immigrants comprise around 8 percent of Europe's population, including 4 percent who migrated from one European country to another and 4 percent who emigrated to Europe

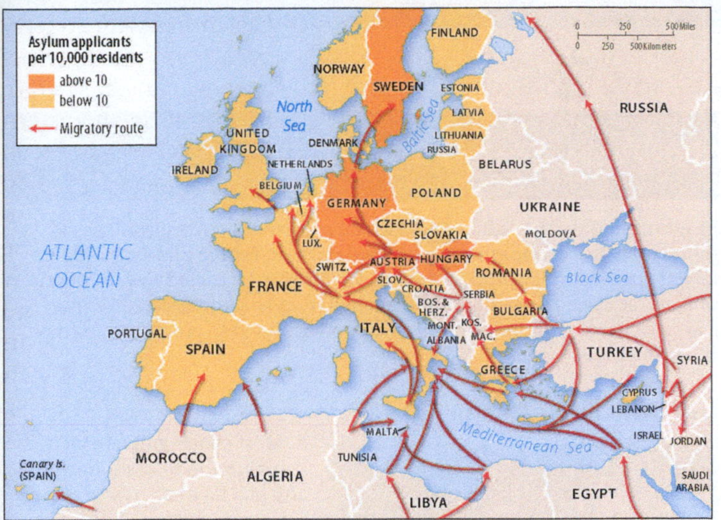

▲ **FIGURE 3-45 REFUGEE FLOWS IN EUROPE** The largest flows of immigrants are from Turkey to Germany and from Romania to Italy.

from elsewhere in the world. In comparison, the foreign-born population is higher in North America, at 13 percent in the United States and 21 percent in Canada.

Despite the relatively modest percentage of immigrants, hostility to immigrants has become a central plank in the platform of political parties in many European countries. These parties blame immigrants for crime, unemployment, and high welfare costs. Above all, the anti-immigration parties fear that long-standing cultural traditions of the host country are threatened by immigrants who adhere to different religions, speak different languages, and prefer different food and other cultural habits. From the standpoint of these parties, immigrants represent a threat to the centuries-old cultural traditions of the host country.

Underlying the hostility toward immigrants in Europe is demographic change. Most European countries are now in stage 4 of the demographic transition (very low or negative NIR). Population growth in Europe is fueled by immigration from other regions, a trend that many Europeans dislike.

The inhospitable climate for immigrants in Europe is especially ironic because Europe was the source of most of the world's emigrants, especially during the nineteenth century. Application of new technologies spawned by the Industrial Revolution—in areas such as public health, medicine, and food—produced a rapid decline in the CDR and pushed much of Europe into stage 2 of the demographic transition (high NIR). As the population increased, many Europeans found limited opportunities for economic advancement.

Migration to the United States, Canada, Australia, and other regions of the world served as a safety valve, draining off some of that increase. The emigration of 65 million Europeans has profoundly changed world culture. As do all other migrants, Europeans brought their cultural heritage to their new homes. Because of migration, Indo-European languages are now spoken by half of the world's people (discussed in Chapter 5), and Europe's most prevalent religion, Christianity, has the world's largest number of adherents (see Chapter 6). European art, music, literature, philosophy, and ethics have also diffused throughout the world.

Regions that were sparsely inhabited prior to European immigration, such as North America and Australia, have become closely integrated into Europe's cultural traditions. Distinctive European political structures and economic systems have also diffused to these regions. Europeans also planted the seeds of conflict by migrating to regions with large indigenous populations, especially in Africa and Asia. They frequently imposed political domination on existing populations and injected their cultural values with little regard for local traditions. Economies in Africa and Asia became based on raising crops and extracting resources for export to Europe rather than on growing crops for local consumption and using resources to build local industry. Many of today's conflicts in former European colonies result from past practices by European immigrants, such as drawing arbitrary boundary lines and discriminating among different local ethnic groups.

PAUSE & REFLECT 3.4.4

How do attitudes toward immigrants differ between Europe and North America?

▼ **FIGURE 3-46 IMMIGRANTS IN EUROPE** Africans trying to reach Italy are rescued by the Italian navy after their boat sunk trying to cross the Mediterranean Sea.

CHECK-IN KEY ISSUE 4

Why Do Migrants Face Challenges?

- ✔ **Immigration is tightly controlled by most countries.**
- ✔ **The United States has more than 11 million unauthorized immigrants, mostly from Mexico.**
- ✔ **Americans and Europeans are divided on attitudes toward immigrants.**

Summary & Review

KEY ISSUE 1

Where are the world's migrants distributed?

Emigration is migration from a location, immigration is migration to a location, and net migration is the difference between the two. Migration can be international (between countries, either voluntary or forced) or internal (within a country, either interregional or intraregional). For most of its history, the United States has been a leading destination for immigrants.

THINKING GEOGRAPHICALLY

1. Compare the cartograms of emigration (Figure 3-2) and immigration (Figure 3-3) with the cartogram of world population (Figure 2-3). Which of the five most populous countries (China, India, United States, Indonesia, and Brazil) appear to have especially high levels of emigration and immigration, and which appear to have especially low levels?

2. What might explain these relatively high or low rates?

▲ **FIGURE 3-47 U.S. IMMIGRANTS** Immigrants to the United States apply for citizenship.

KEY ISSUE 2

Where do people migrate within a country?

Two main types of internal migration are interregional (between regions of a country) and intraregional (within a region). Large countries, including the United States and Canada, have had important patterns of interregional migration. Three intraregional migration patterns are from rural to urban areas (especially in developing countries) and from urban to suburban areas and to rural areas (especially in developed countries).

THINKING GEOGRAPHICALLY

3. Forced migration is considered here as a subset of international migration. What current and historical examples of forced internal migration have been cited in this chapter?

▲ **FIGURE 3-48 FORCED INTERNAL MIGRATION** The Passage, a bridge in downtown Chattanooga, commemorates the Trail of Tears.

KEY ISSUE 3

Why do people migrate?

People undertake migration because of a combination of cultural, environmental, and economic push and pull factors. People may be forced to migrate because of political conflicts. People may be pulled toward physically attractive environments and pushed from hazardous ones. People leave places with limited job prospects and are lured to places where they can find work.

THINKING GEOGRAPHICALLY

4. Most people migrate for a combination of economic push and pull factors. As you consider your personal future, do you expect push factors or pull factors to be more important in your location decisions? Why?

▲ **FIGURE 3-49 COLLEGE GRADUATES** What will they be doing, and where will they be living, after graduation?

KEY ISSUE 4

Why do migrants face challenges?

To immigrate to most countries, people need permission from the government. Opposition to the current level of immigration is high in some countries. The United States has quotas on the number of immigrants and sets preferences for those who are allowed to immigrate. Other countries also have restrictions on immigration. The United States has more than 11 million unauthorized immigrants, who are in the country without proper documents.

THINKING GEOGRAPHICALLY

5. The U.S. border with Mexico has a fence in most places, whereas the U.S. border with Canada does not. What might account for this difference?

▲ **FIGURE 3-50 U.S.–CANADA BORDER** The International Peace Garden is on the border between North Dakota and Manitoba.

KEY TERMS

Asylum seeker *(p. 92)* Someone who has migrated to another country in the hope of being recognized as a refugee.
Brain drain *(p. 103)* Large-scale emigration by talented people.
Chain migration *(p. 103)* Migration of people to a specific location because relatives or members of the same nationality previously migrated there.
Circular migration *(p. 106)* The temporary movement of a migrant worker between home and host countries to seek employment.
Circulation *(p. 78)* Short-term, repetitive, or cyclical movements that recur on a regular basis.
Counterurbanization *(p. 90)* Net migration from urban to rural areas in more developed countries.
Desertification *(p. 94)* Degradation of land, especially in semiarid areas, primarily because of human actions such as excessive crop planting, animal grazing, and tree cutting. Also known as semiarid land degradation.
Emigration *(p. 78)* Migration from a location.
Floodplain *(p. 94)* An area subject to flooding during a given number of years, according to historical trends.
Forced migration *(p. 82)* Permanent movement, compelled by cultural or environmental factors.
Guest worker *(p. 106)* A term once used for a worker who migrated to the developed countries of Northern and Western Europe, usually from Southern and Eastern Europe or from North Africa, in search of a higher-paying job.
Immigration *(p. 78)* Migration to a new location.
Internal migration *(p. 82)* Permanent movement within a particular country.
Internally displaced person *(IDP)* *(p. 92)* Someone who has been forced to migrate for similar political reasons as a refugee but has not migrated across an international border.
International migration *(p. 82)* Permanent movement from one country to another.
Interregional migration *(p. 82)* Permanent movement from one region of a country to another.
Intervening obstacle *(p. 94)* An environmental or cultural feature of the landscape that hinders migration.
Intraregional migration *(p. 82)* Permanent movement within one region of a country.
Migration *(p. 78)* A form of relocation diffusion that involves a permanent move to a new location.
Migration transition *(p. 81)* A change in the migration pattern in a society that results from industrialization, population growth, and other social and economic changes that also produces the demographic transition.
Mobility *(p. 78)* All types of movements between locations.
Net migration *(p. 78)* The difference between the level of immigration and the level of emigration.
Pull factor *(p. 92)* A factor that induces people to move to a new location.
Push factor *(p. 92)* A factor that induces people to move out of their present location.
Quota *(p. 102)* In reference to migration, a law that places a maximum limit on the number of people who can immigrate to a country each year.
Refugee *(p. 92)* Someone who is forced to migrate from his or her home country and cannot return for fear of persecution because of his or her race, religion, nationality, membership in a social group, or political opinion.
Remittance *(p. 97)* Transfer of money by workers to people in the country from which they emigrated.
Unauthorized immigrant *(p. 100)* A person who enters a country without proper documents to do so.
Voluntary migration *(p. 82)* Permanent movement undertaken by choice.

EXPLORE

U.S.–Mexico Border

Use Google Earth to explore the U.S.–Mexico border at Laredo.

Fly to *Nuevo Laredo, Mexico*.

Select *Borders and Labels*.

Zoom in to 5,000 feet.

1. Follow the international border through the built-up area of Nuevo Laredo and Laredo. How many border crossings do you see?
2. What is the means of transport at each of the crossings?

Zoom in to 2,000 feet.

3. The backup of cars is longer trying to enter which country?
4. Based on the Google Earth image, if you had to ship goods across the border (such as car parts), which means of transport appears to be the quickest and easiest to use?

MG GeoVideo

Log in to the **MasteringGeography** Study Area to view this video.

Kenya: The Turkana Way of Life

Many Turkana people in northern Kenya have migrated to the shores of Lake Turkana from the grasslands where they herded cattle and goats.

1. What environmental factor has driven the Turkana to migrate?
2. How do some Turkana gain a livelihood in their new home by the lake?
3. Where else have Turkana migrated to escape the drought?
4. What are the advantages and drawbacks of this choice?

MasteringGeography™

Looking for additional review and test prep materials?

Visit the Study Area in **MasteringGeography™** to enhance your geographic literacy, spatial reasoning skills, and understanding of this chapter's content by accessing a variety of resources, including MapMaster interactive maps, videos, *In the News* RSS feeds, flashcards, web links, self-study quizzes, and an eText version of *The Cultural Landscape*.
www.masteringgeography.com

4

Folk and Popular Culture

What did you do today? What did you wear? After studying or finishing work, what leisure activities did you do? Did you watch TV or play sports? Geographers describe similarities and differences in how people meet their daily needs and make use of their leisure time.

A consideration of culture follows logically from the discussion of migration in Chapter 3. Two locations have similar cultural beliefs, objects, and institutions because people bring along their culture when they migrate. Differences emerge when two groups have limited interaction.

Surfing the web at a cafe in Jaisalmer, India.

LOCATIONS IN THIS CHAPTER

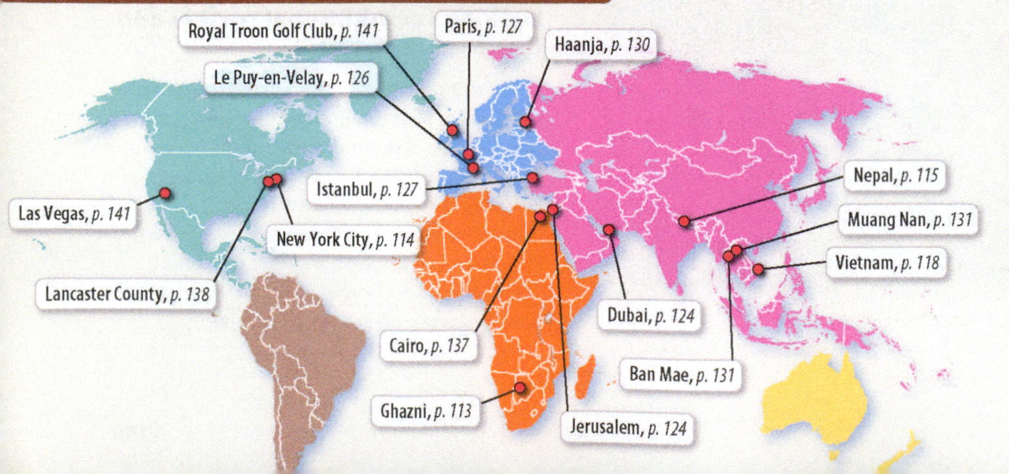

KEY ISSUES

1 Where Are Folk and Popular Leisure Activities Distributed?

Leisure and recreation elements of folk culture and popular culture are distributed across Earth's ***space***. Compared to folk culture, popular culture is more likely to originate at a specific time and place and to diffuse over a wider ***region***.

2

Where Are Folk and Popular Material Culture Distributed?

Material folk culture and popular culture include food, shelter, and clothing. Folk culture is more likely to vary between ***places***, whereas popular culture is more likely to vary between points in time.

3

Why Is Access to Folk and Popular Culture Unequal?

Folk culture and popular culture have different distributions. Popular culture has ***connections*** among people, especially through electronic communications.

4

Why Do Folk and Popular Culture Face Sustainability Challenges?

The global ***scale*** of popular culture raises sustainability concerns. The diffusion of popular culture threatens the maintenance of local diversity in folk customs.

KEY ISSUE 1

Where Are Folk and Popular Leisure Activities Distributed?

- **Introducing Folk and Popular Culture**
- **Origin, Diffusion, and Distribution of Folk and Popular Culture**
- **Geographic Differences Between Folk and Popular Culture**
- **Origin and Diffusion of Folk and Popular Music**
- **Origin and Diffusion of Folk and Popular Sports**

LEARNING OUTCOME 4.1.1

Introduce concepts of folk and popular culture.

Culture was defined in Chapter 1 as the body of material traits, customary beliefs, and social forms that together constitute the distinct tradition of a group of people. Geographers are interested in all three components of the definition of culture:

- The first part of this definition—the visible elements that a group possesses and leaves behind for the future (its material traits)—is discussed in this chapter.
- Two important components of a group's beliefs and values—language and religion—are discussed in Chapters 5 and 6.
- The social forms that maintain values and protect the artifacts (ethnicity and political institutions) are discussed in Chapters 7 and 8.

A consideration of culture follows logically from the discussion of migration in Chapter 3. Two locations have similar cultural beliefs, objects, and institutions because people bring along their culture when they migrate. Differences emerge when two groups have limited interaction.

Geographers search for where various elements of culture are found in the world and for reasons why the observed distributions occur. How does culture influence behavior? To answer this question, habit must be distinguished from custom:

- A **habit** is a repetitive act that a particular individual performs, such as wearing jeans to class every day.

▲ **FIGURE 4-1 HABIT AND CUSTOM** After their election in 2015, Greek Finance Minister Yanis Varoufakis (left) and Prime Minister Alexis Tsipras (right) had a habit of wearing informal clothes. 1. Why do world leaders usually dress in suits? 2. Why might leaders choose to dress informally? 3. How might other world leaders react?

- A **custom** is a repetitive act of a group, performed to the extent that it becomes characteristic of the group, such as many students typically wearing jeans to class.

Unlike custom, habit does not imply that the act has been adopted by most of the society's population. A custom is therefore a habit that has been widely adopted by a group of people (Figure 4-1).

PAUSE & REFLECT 4.1.1

Can you think of an entertainer, a politician, or another public figure who displays a distinctive habit in choice of clothing?

A collection of social customs produces a group's material culture; for example, jeans typically represent American informality and a badge of youth. In this chapter, custom may be used to denote a specific element of material culture, such as wearing jeans, whereas culture refers to a group's entire collection of customs.

Introducing Folk and Popular Culture

Geographers divide culture into two types:

- **Folk culture** is traditionally practiced primarily by small, homogeneous groups living in isolated rural areas (Figure 4-2).
- **Popular culture** is found in large, heterogeneous societies that share certain habits despite differences in other personal characteristics (Figure 4-3).

Each cultural element has a distinctive origin, diffusion, and distribution. Geographers observe that folk culture and

▲ **FIGURE 4-2 FOLK CULTURE** The Gadaba people of eastern India wear jewelry that reflects local folk culture, including large, heavy rings that are not removed until death.

▲ **FIGURE 4-3 POPULAR CULTURE** Jewelry store on 47th Street in New York City. This street is known as Diamond Row because of the clustering of diamond stores.

popular culture typically differ in their processes of origin, diffusion, and distribution.

Landscapes dominated by folk culture change relatively little over time. In contrast, popular culture is based on rapid simultaneous global connections through communications systems, transportation networks, and other modern technology. Rapid diffusion facilitates frequent changes in popular culture. Thus, folk culture is more likely to vary from place to place at a given time, whereas popular culture is more likely to vary from time to time at a given place.

At a global scale, popular culture is becoming more dominant—at least for people with the income to have access to it—threatening the survival of unique folk culture. The disappearance of local folk culture reduces local diversity in the world and the intellectual stimulation that arises from differences in backgrounds.

The dominance of popular culture can also threaten the quality of the environment. Culture and the physical environment are interrelated. Each cultural group takes particular elements from the environment into its culture and in turn constructs landscapes (what geographers call "built environments") that modify nature in distinctive ways. Some of these landscapes are sustainable, and some are not. Folk culture derived from local natural elements may be more sustainable in the protection and enhancement of the environment. Popular culture is less likely to reflect concern for the sustainability of physical conditions and is more likely to modify the environment in accordance with global values.

Two elements of culture are emphasized in this chapter:

- **Daily necessities, including food, clothing, and shelter.** All people must consume food, wear clothing, and find shelter, but different cultural groups do so in distinctive ways (Figure 4-4).
- **Leisure activities, such as arts and recreation.** Each cultural group has its own definition of meaningful art and stimulating recreation. For example, people in the United States and Pakistan do not allocate their leisure time in the same way.

◀ **FIGURE 4-4 MATERIAL CULTURE** Roots are being cooked near Ghazni, Botswana.

Origin, Diffusion, and Distribution of Folk and Popular Culture

LEARNING OUTCOME 4.1.2

Compare processes of origin, diffusion, and distribution of folk and popular culture.

Each cultural element has a unique spatial distribution, but in general, distribution is more extensive for popular culture than for folk culture. Two basic factors help explain the different spatial distribution of popular and folk cultures: the process of origin and the pattern of diffusion.

ORIGIN

Culture originates at a hearth, a center of innovation:

- Folk culture often has anonymous hearths, originating from anonymous sources, at unknown dates, through unidentified originators. It may also have multiple hearths, originating independently in isolated locations.
- Popular culture is typically traceable to a specific person or corporation in a particular place. It is most often a product of developed countries, especially in North America and Europe.

For example, hip hop is considered to have originated on August 11, 1973, at 1520 Sedgwick Avenue, in New York City's Bronx Borough, during a block party with DJ Kool Herc (Figure 4-5). Kool Herc, whose birth name was Clive Campbell, had been born in Jamaica and moved to the Bronx with his family in 1967. Geographers understand that the hearth of hip hop was significant because the music reflected conditions prevailing in the neighborhood. These included violent street gangs, arson, and other crime, as well as extensive demolition of housing and forced relocation of people to build the Cross Bronx Expressway.

Popular music and other elements of popular culture, such as food and clothing, arise from a combination of advances in industrial technology and increased leisure time. Industrial technology permits the uniform reproduction of objects in large quantities (iPods, T-shirts, pizzas). Many of these objects help people enjoy leisure time, which has increased as a result of the widespread change in the labor force from predominantly agricultural work to predominantly service and manufacturing jobs.

DIFFUSION

Folk and popular cultures go through different processes of diffusion:

- Folk culture is transmitted from one location to another relatively slowly and on a small scale, primarily through relocation diffusion (migration).
- Popular culture typically spreads through a process of hierarchical diffusion, diffusing rapidly and extensively from hearths or nodes of innovation with the help of modern communications.

For example, in the late twentieth century, Western dance music diffused rapidly from the United States to Europe, especially Detroit's techno music and Chicago's house music. Techno music was heavily influenced by soul, gospel, and ultimately African folk music. House music was heavily influenced by hip hop that emerged in New York and other urban areas, which in turn diffused from funk, jazz, and again ultimately African folk music.

Hip hop music diffused from the Bronx to nearby Philadelphia during the 1970s and to other U.S. cities during the 1980s. The music was introduced into Western Europe and Japan and diffused back to Caribbean countries, a principal source of inspiration. In more recent decades, hip hop reached Latin America, Asia, and Africa, where local cultural styles influenced the music from the original Bronx hearth. Meanwhile, as sometimes occurs with popular culture, as the style diffuses around the world, it can become less important in its hearth. In fact, sales of hip hop music have declined sharply in the United States since 2005.

▼ **FIGURE 4-5 ORIGIN OF POPULAR CULTURE: HIP HOP** Hip hop is considered to have been founded in 1973 **(a)** by Kool Herc at **(b)** 1520 Sedgwick Avenue, the Bronx, New York.

(a)

(b)

DISTRIBUTION

Popular culture is distributed widely across many countries, with little regard for physical factors. The distribution is influenced by the ability of people to access the material elements of popular culture. The principal obstacle to access is lack of income to purchase the material.

A combination of local physical and cultural factors influences the distinctive distributions of folk culture. For example, in a study of artistic customs in the Himalaya Mountains, geographers P. Karan and Cotton Mather revealed that distinctive views of the physical environment emerge among neighboring cultural groups that are isolated. The study area, a narrow corridor of 2,500 kilometers (1,500 miles) in the Himalaya Mountains of Bhutan, Nepal, northern India, and southern Tibet (China) contains four religious groups: Tibetan Buddhists in the north, Hindus in the south, Muslims in the west, and Southeast Asian folk religionists in the east (Figure 4-6). Despite their spatial proximity, limited interaction among these groups produces distinctive folk customs.

Through their choices of subjects of paintings, each group reveals how its culture mirrors the religions and individual views of the group's environment:

- **Buddhists.** In the northern region, Buddhists paint idealized divine figures, such as monks and saints. Some of these figures are depicted as bizarre or terrifying, perhaps reflecting the inhospitable environment.
- **Hindus.** In the southern region, Hindus create scenes from everyday life and familiar local scenes. Their paintings sometimes portray a deity in a domestic scene and frequently represent the region's violent and extreme climatic conditions.
- **Muslims.** To the west, folk art is inspired by the region's beautiful plants and flowers. In contrast with the paintings from the Buddhist and Hindu regions, these paintings do not depict harsh climatic conditions.
- **Folk religionists.** People from Myanmar (Burma) and elsewhere in Southeast Asia, who have migrated to the eastern region of the study area, paint symbols and designs that derive from their religion rather than from the local environment.

The distribution of artistic subjects in the Himalayas shows how folk customs are influenced by cultural institutions such as religion and by environmental processes such as climate, landforms, and vegetation. These groups display similar uniqueness in their dance, music, architecture, and crafts.

PAUSE & REFLECT 4.1.2

What geographic factors account for the diversity of cultures in the Himalayas?

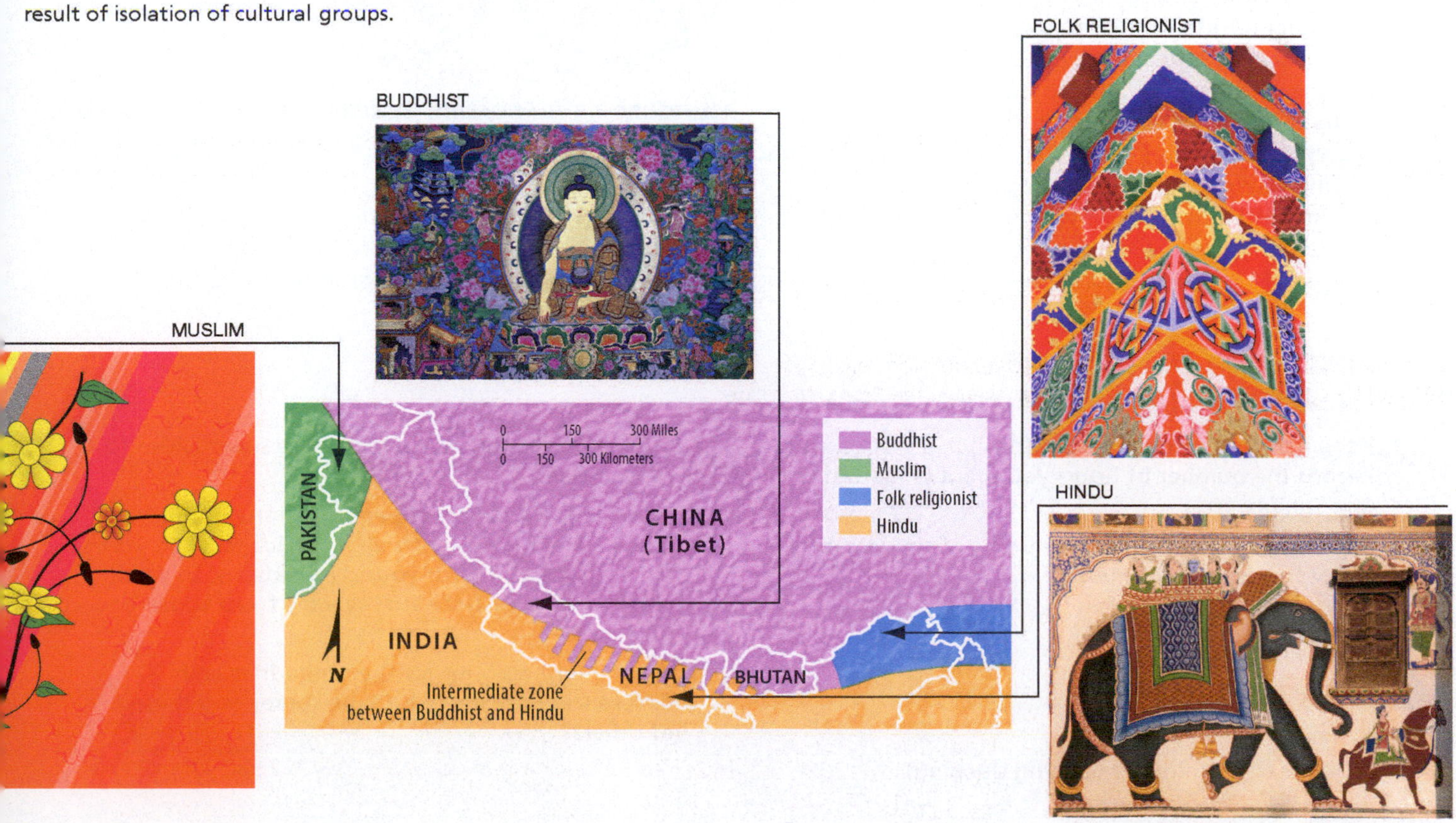

▼ FIGURE 4-6 DISTRIBUTION OF FOLK CULTURE: ART Distinct folk painting traditions are distributed within the Himalayas, a result of isolation of cultural groups.

Geographic Differences Between Folk and Popular Culture

LEARNING OUTCOME 4.1.3

Compare patterns of regions and connections between folk and popular culture.

Geographers observe that cultural features display distinctive regional distributions. Cultural regions are often vernacular, that is perceived by people to exist as part of their cultural identity. In other cases, a cultural region may be formal or functional. Regions of supporters of sports teams may be considered functional, because the percentage of supporters of the team is typically higher near where the team plays (Figure 4-7).

The region covered by a folk culture is typically much smaller than that covered by a popular culture. The reason the distributions are different is connection or lack of it. A group's distinctive culture derives from experiencing local social and physical conditions in a place that is isolated from other groups. Even groups living in close proximity to one another may adopt different folk cultures because of limited connections.

For example, according to Richard Florida, Charlotta Mellander, and Kevin Stolarick, folk musicians once clustered in particular communities according to shared interest in specific styles, such as Tin Pan Alley in New York, Dixieland jazz in New Orleans, country in Nashville, and Motown in Detroit. Now with the globalization of popular music, musicians are less tied to the culture of particular places. As with other elements of popular culture, popular musicians have more connections with performers of similar styles, regardless of where in the world they happen to live, than they do with performers of different styles who happen to live in the same community.

Figure 4-9 illustrates this point by depicting popular musicians on a map like that of the London Underground (subway).

DOING GEOGRAPHY Time Use Surveys

Time use surveys are conducted in a number of countries to determine how people spend their day. The results can be analyzed to identify similarities and differences between people in various countries, as well as among various groups within a country. Obvious differences in the use of time depend on whether an individual is in school or employed.

Cultural geographers are especially interested in variations in the use of leisure time. Distinctive patterns can be observed, depending on age, gender, ethnicity, and place of residence. In the United States, for example, young people are less likely than Americans of all other ages to watch TV and read but more likely to play computer games (Figure 4-8). Young people in Pakistan are less likely than young people in the United States to watch TV or play computer games and are more likely to read, socialize with friends, and listen to the radio.

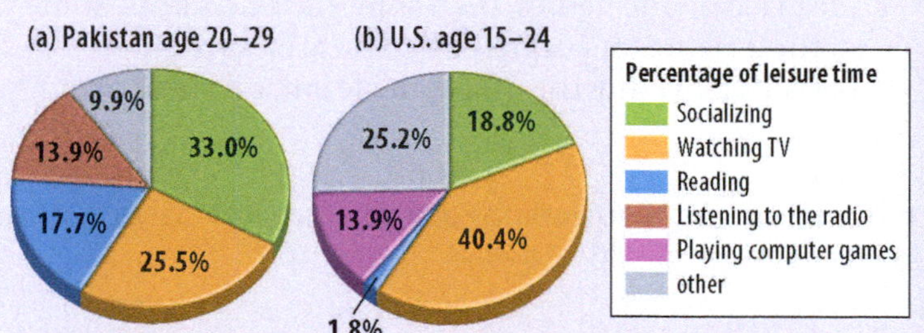

▲ **FIGURE 4-8 ALLOCATING LEISURE TIME** (a) Young Pakistanis and (b) Americans spend time doing different activities.

What's Your Leisure Activity?

1. Keep a diary of your activities through a weekend. Record the number of hours you spent on Saturday and Sunday on:
   - Socializing with friends (including communicating by telephone or social media).
   - Watching TV shows.
   - Reading.
   - Listening to the radio.
   - Playing computer games.
   - Doing other things (excluding sleeping).
2. Convert the hours on each of these six activities to percentages.
3. Compare the percentages you spent on each of these six activities with all Americans, with young Americans, and with young Pakistanis. In what ways does your time survey vary from those of others?
4. What might account for the similarities and differences between your time survey and those of others?

Subway "lines" represent styles of popular music, and "interchanges" represent individuals who cross over between two styles. For example, Kanye West is placed at the interchange between hip hop and soul and Jimi Hendrix at the interchange between rock and blues and country.

PAUSE & REFLECT 4.1.3

What type of music do you like? On what "line" in Figure 4-9 does it fit?

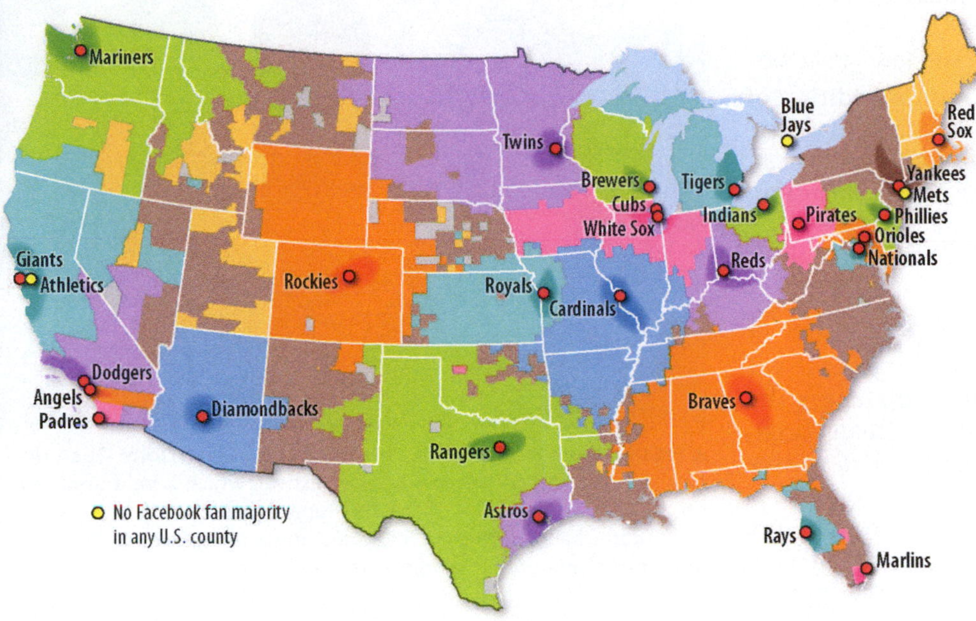

◄ **FIGURE 4-7 REGIONS OF BASEBALL FANS** The area of support for a baseball team, based on the number of Facebook fans, is an example of a functional region. The Yankees and Red Sox have support in regions of the country outside the Northeast. The Athletics, Blue Jays, and Mets do not have a fan majority in any U.S. county. The darker color at the center of some regions, such as the Braves and Rockies, shows the extent of a supermajority supporting the team.

▲ **FIGURE 4-9 POPULAR MUSIC CONNECTIONS** This map showing relationships among musical styles is designed to look like the map of the London Underground (subway) system.

Origin and Diffusion of Folk and Popular Music

LEARNING OUTCOME 4.1.4

Compare differences in geographic dimensions of folk and popular music.

Every culture in human history has had some tradition of music, argues music researcher Daniel Levitan. As music is a part of both folk and popular culture traditions, it can be used to illustrate the differences in the origin, diffusion, and distribution of folk and popular culture.

FOLK MUSIC

According to a Chinese legend, music was invented in 2697 B.C., when the Emperor Huang Ti sent Ling Lun to cut bamboo poles that would produce a sound matching the call of the phoenix bird. In reality, folk songs usually originate anonymously and are transmitted orally. A song may be modified from one generation to the next as conditions change, but the content is most often derived from events in daily life that are familiar to the majority of the people. As people migrate, folk music travels with them as part of the diffusion of folk culture.

Folk songs may tell a story or convey information about life-cycle events, such as birth, death, and marriage, or environmental features, such as agriculture and climate. For example, in Vietnam, where most people are subsistence farmers, information about agricultural technology was traditionally conveyed through folk songs. The following folk song provides advice about the difference between seeds planted in summer and seeds planted in winter:

> Ma chiêm ba tháng không già
>
> Ma mùa tháng ruỗi ắt la'không non.[1]

This song can be translated as follows:

> While seedlings for the summer crop are not old when they are three months of age, Seedlings for the winter crop are certainly not young when they are one-and-a-half months old.

The song hardly sounds lyrical to a Western ear. But when English-language folk songs appear in cold print, similar themes emerge, even if the specific information conveyed about the environment differs.

Festivals throughout Vietnam feature music in locally meaningful environmental settings, such as hillsides or on water. Singers in traditional clothes sing about elements of daily life in the local village, such as trees, flowers, and water sources (Figure 4-10).

[1]From John Blactking and Joann W. Kealiinohomoku, eds., *The Performing Arts: Music and Dance* (The Hague: Mouton, 1979), 144. Reprinted by permission of the publisher.

▲ **FIGURE 4-10 FOLK MUSIC: VIETNAM** Singers perform Quan Ho folk songs as part of the annual Lim Festival, which is held annually on the 13th to 15th days of the first lunar month. Quan Ho folk music dates back more than 500 years and is recognized by UNESCO as part of humanity's intangible heritage.

POPULAR MUSIC

In contrast to folk music, popular music is written by specific individuals for the purpose of being sold to or performed in front of a large number of people. It frequently displays a high degree of technical skill through manipulation of sophisticated electronic equipment.

For example, popular music as we know it today originated around 1900. At that time, the main popular musical entertainment in North America and Europe was the variety show, called the music hall in the United Kingdom and vaudeville in the United States. To provide songs for music halls and vaudeville, a music industry was developed in a district of New York that became known as Tin Pan Alley. The diffusion of American popular music worldwide began in earnest during the 1940s, when the Armed Forces Radio Network broadcast music to American soldiers and to citizens of countries where American forces were stationed or fighting during World War II.

Popular musicians increasingly cluster in communities where other creative artists reside, regardless of the particular style. Nashville has the highest concentration of popular musicians, especially those performing country and gospel (Figure 4-11). New York, Los Angeles, and San Francisco, which are much larger metropolitan areas than Nashville, have relatively high total numbers of musicians, as well as high concentrations. Musicians cluster in these places so they can be near sources of employment and cultural activities that appeal to a wide variety of artists, not just performers of a specific type of music. Popular musicians are also attracted to these places for better access to agencies that book live performances, an increasingly important component of the popular music industry.

Regional variations can be observed in popular music preferences. For example, the favorite artist in each state during 2014 was identified by Music Machinery on the basis of streaming data from The Echo Nest. The most-played artists in 2014 were Jay Z east of the Mississippi River, Drake in the southwest, and Macklemore & Ryan Lewis in the northwest (Figure 4-12).

PAUSE & REFLECT 4.1.4

Do you like your state's favorite artist?

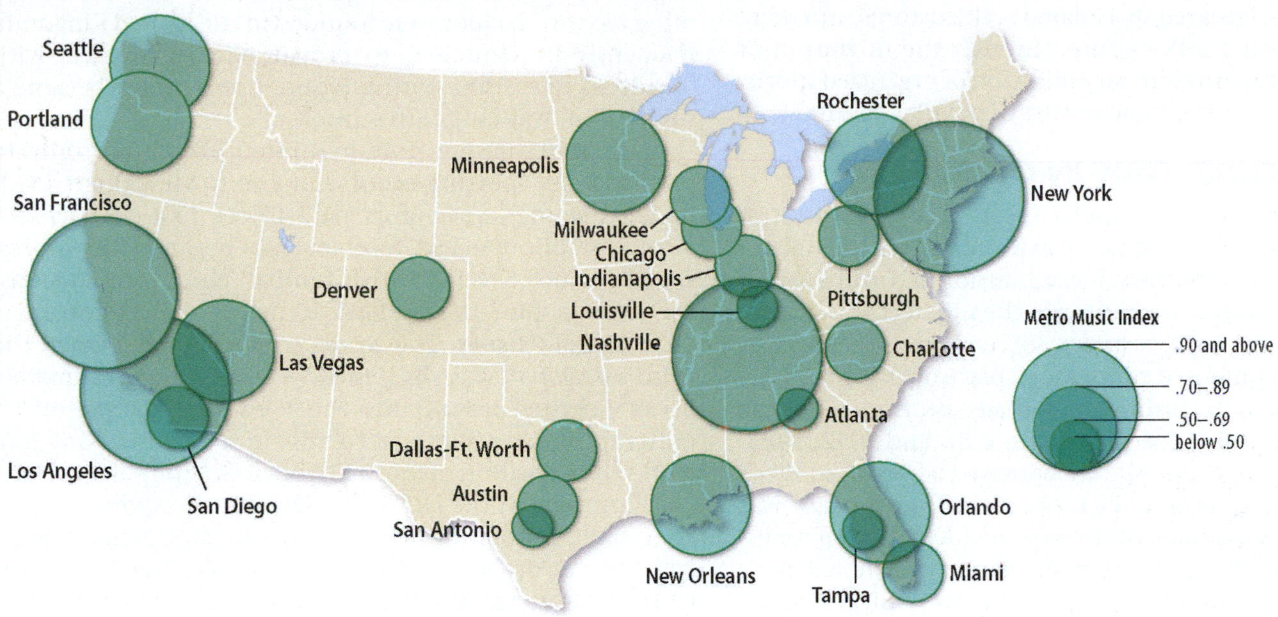

▲ **FIGURE 4-11 POPULAR MUSIC: U.S. CLUSTERS** Nashville has the highest concentration of popular musicians and recording studios.

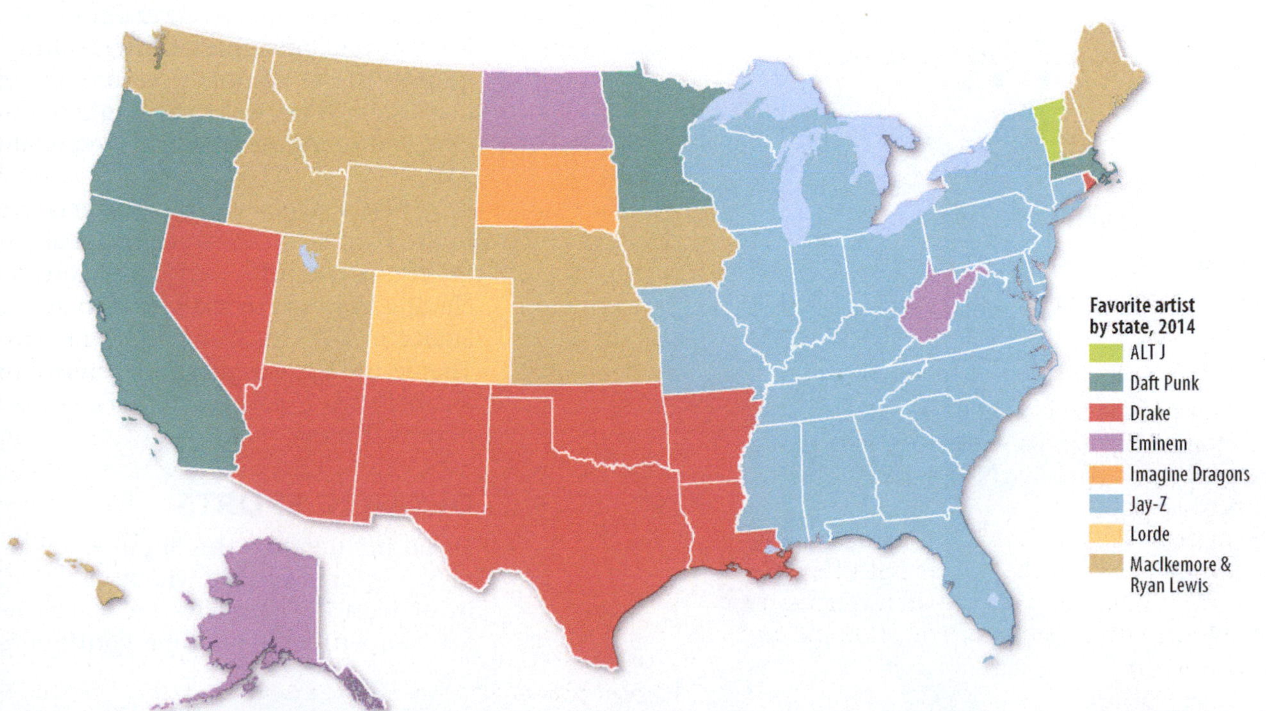

▲ **FIGURE 4-12 FAVORITE ARTIST BY U.S. STATE, 2014** The most-streamed artist varies by region of the country.

Origin and Diffusion of Folk and Popular Sports

LEARNING OUTCOME 4.1.5

Describe the transformation of sports from folk to popular culture.

Many sports originated as isolated folk customs and were diffused like other folk culture, through the migration of individuals. The contemporary diffusion of organized sports, however, displays the characteristics of popular culture.

FOLK CULTURE: ORIGIN OF SOCCER

Soccer, the world's most popular sport—known in most of the world as football—originated as a folk custom in England during the eleventh century. It was transformed into a part of global popular culture beginning in the nineteenth century.

As with other folk customs, soccer's origin is obscure. The earliest documented contest took place in England in the eleventh century. According to football historians, after the Danish invasion of England between 1018 and 1042, workers excavating a building site encountered a Danish soldier's head, which they began to kick. "Kick the Dane's head" was imitated by boys, one of whom got the idea of using an inflated cow bladder. Early football games resembled mob scenes. A large number of people from two villages would gather to kick the ball. The winning side was the one that kicked the ball into the center of the rival village.

POPULAR CULTURE: DIFFUSION OF SOCCER

The transformation of football from an English folk custom to global popular culture began in the 1800s. Football and other recreation clubs were founded in the United Kingdom, frequently by churches, to provide factory workers with organized recreation during leisure hours. Sport became a subject that was taught in school.

Increasing leisure time permitted people not only to participate in sporting events but also to view them. With higher incomes, spectators paid to see first-class events. To meet public demand, football clubs began to hire professional players. Several British football clubs formed an association in 1863 to standardize the rules and to organize professional leagues. The word *soccer* originated when the word *association* was shortened to *assoc*, which ultimately became twisted around into the word *soccer*. Organization of the sport into a formal structure in the United Kingdom marks the transition of football from folk to popular culture.

Beginning in the late 1800s, the British exported association football around the world, first to continental Europe and then to other countries. For example, Dutch students returning from studies in the United Kingdom were the first to play football in continental Europe in the late 1870s. In Bilbao, Spain, miners adopted the sport in 1893, after seeing it played by English engineers working there. British citizens further diffused the game throughout the worldwide British Empire. In the twentieth century, soccer, like other sports, was further diffused by new communication systems, especially TV.

The global popularity of soccer is seen in the World Cup, in which national soccer teams compete every four years, including in Russia in 2018 for men and in France in 2019 for women (Figure 4-13). Thanks to TV, each men's final breaks the record for the most spectators of any event in world history.

OLYMPIC SPORTS

To be included in the Summer Olympics, a sport must be widely practiced in at least 75 countries (50 countries for women) and on four continents.

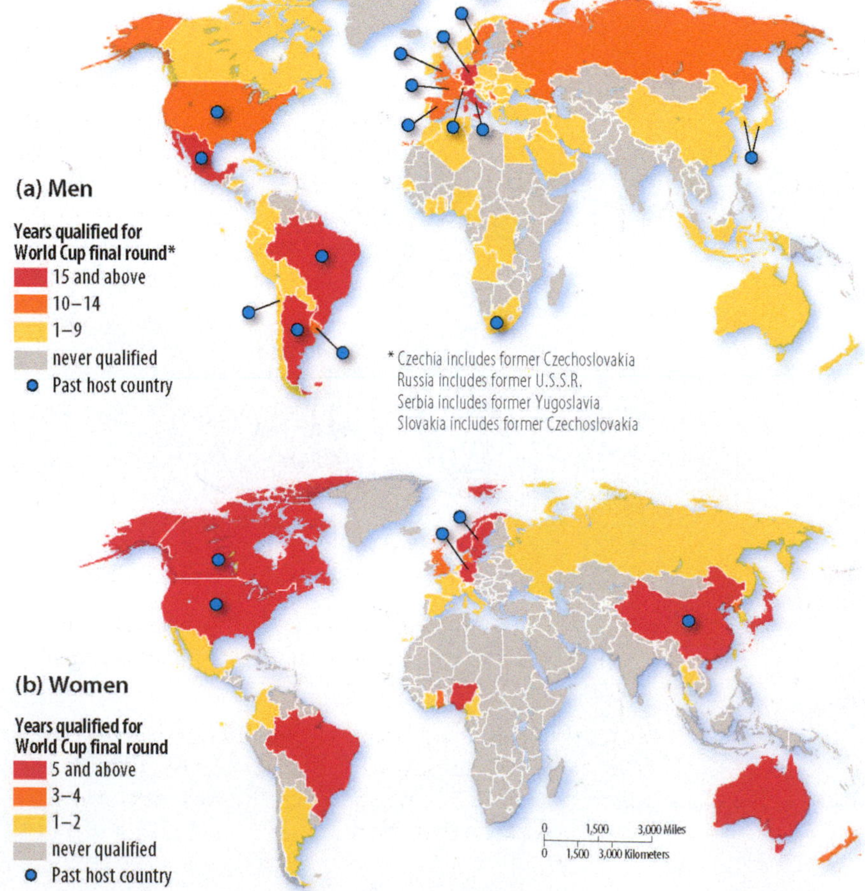

◀ **FIGURE 4-13 POPULAR SPORTS: WORLD CUP SOCCER** Participating countries and past hosts for **(a)** men and **(b)** women.

▲ **FIGURE 4-14 GLOBALIZATION OF SPORTS: OLYMPICS** Practicing for 100-meter hurdles.

The 2016 Summer Olympics included competition in 28 sports: archery, aquatics, athletics, badminton, basketball, boxing, canoeing/kayaking, cycling, equestrian, fencing, field hockey, football (soccer), golf, gymnastics, handball, judo, modern pentathlon, rowing, rugby, sailing, shooting, table tennis, taekwondo, tennis, triathlon, volleyball, weightlifting, and wrestling (Figure 4-14). The two leading team sports in the United States—American football and baseball—are not included.

PAUSE & REFLECT 4.1.5

How many of the Olympic sports are played at your school?

SURVIVING FOLK SPORTS

Most other sports have diffused less than soccer. Cultural groups still have their own preferred sports, which are often unintelligible to people elsewhere. Consider the following:

- Cricket is popular primarily in the United Kingdom and former British colonies, especially in South Asia, the South Pacific, and Caribbean islands.
- Ice hockey prevails, logically, in colder climates, especially in Canada, the northern United States, northern Europe, and Russia.
- Wushu, martial arts that combine forms such as kicking and jumping with combat such as striking and wrestling, is China's most popular sport.
- Baseball, once confined to North America, became popular in Japan in the late nineteenth century after it was introduced by American Japanese returning from studies in the United States, as well as Americans working in Japan.
- Australia rules football is a sport distinct from soccer and the football played in North America. Distinctive forms of football developed in Australia, as well as the United States and Canada, as a result of lack of interaction among sporting nations during the nineteenth century.

▲ **FIGURE 4-15 SURVIVING FOLK SPORT: LACROSSE** In Maryland, lacrosse is played by women in many high schools.

- Lacrosse was traditionally played by the Iroquois, who called it guhchigwaha, which means "bump hips." European colonists in Canada picked up the game from the Iroquois and diffused it to a handful of U.S. communities, especially in Maryland, upstate New York, and Long Island (Figure 4-15).

Despite the diversity in distribution of sports across Earth's surface and the anonymous origin of some games, organized spectator sports today are part of popular culture. The common element in professional sports is the willingness of people throughout the world to pay for the privilege of viewing, in person or on TV, events played by professional athletes.

At the same time, sports can be a strong force for cultural and regional identity. For example, Major League Baseball teams have strong regional identities. Lacrosse has fostered cultural identity among the Iroquois Confederation of Six Nations (Cayuga, Mohawk, Oneida, Onondaga, Seneca, and Tuscarora) because they have been invited by the International Lacrosse Federation to participate in the Lacrosse World Championships, along with teams from sovereign states, such as Australia, Canada, and the United States.

CHECK-IN KEY ISSUE 1

Where Are Folk and Popular Leisure Activities Distributed?

- ✔ **Folk culture and popular culture have distinctive patterns of origin, diffusion, and distribution.**
- ✔ **Folk leisure activities typically have anonymous origins, diffuse through relocation diffusion, and have limited distribution.**
- ✔ **Popular music and sports typically originate with identifiable individuals or corporations, diffuse rapidly through hierarchical diffusion, and have widespread distribution.**

KEY ISSUE 2

Where Are Folk and Popular Material Culture Distributed?

- ▸ Elements of Material Culture
- ▸ Folk and Popular Clothing
- ▸ Folk Food Customs
- ▸ Popular Food Preferences
- ▸ Folk and Popular Housing

LEARNING OUTCOME 4.2.1

Introduce environmental and cultural features of material culture.

Material culture includes the three most important necessities of life: clothing, food, and shelter. As is the case with leisure, material elements of folk culture typically have unknown or multiple origins among groups living in relative isolation, and they diffuse slowly to other locations through the process of relocation diffusion.

Popular clothing, food, and shelter vary more in time than in place. They originate through the invention of a particular person or corporation, and they diffuse rapidly across Earth to locations with a variety of physical conditions. Access depends on an individual having a sufficiently high level of income to acquire the material possessions associated with popular culture.

Some regional differences in food, clothing, and shelter persist in popular culture, but differences are much less than in the past. Go to any recently built neighborhood on the outskirts of an American city from Portland, Maine, to Portland, Oregon: The houses look the same, the people wear jeans, and the same chains deliver pizza.

Elements of Material Culture

Folk culture is more likely to be influenced by environmental conditions, but popular culture is not immune to these influences. Geographers also observe that folk and popular culture can come into conflict with each other.

WINE GEOGRAPHY

The spatial distribution of wine production shows the influence of both environmental and cultural elements. Grapes suitable for making decent wine grow better in some places than in others. At the same time, wine is made today primarily in locations that have a tradition of excellence in making it and people who like to drink it and can afford to purchase it.

WINE PRODUCTION: ENVIRONMENTAL FACTORS. The distinctive character of a wine derives from a vineyard's terroir—the unique combination of soil, climate, and other physical characteristics at the place where the grapes are grown:

- **Climate.** Vineyards are best cultivated in temperate climates of moderately cold, rainy winters and fairly long, hot summers. Hot, sunny weather is necessary in the summer for the fruit to mature properly, whereas winter is the preferred season for rain because plant diseases that cause the fruit to rot are more active in hot, humid weather.
- **Topography.** Vineyards are planted on hillsides, if possible, to maximize exposure to sunlight and to facilitate drainage. A site near a lake or river is also desirable because water can temper extremes of temperature.
- **Soil.** Grapes can be grown in a variety of soils, but the best wine tends to be produced from grapes grown in soil that is coarse and well drained—a soil that is not necessarily fertile for other crops.

WINE PRODUCTION: CULTURAL FACTORS. Although grapes can be grown in a wide variety of locations, the production of wine is based principally on cultural values, both historical and contemporary. The distribution of wine production shows that the diffusion of popular customs depends less on the distinctive environment of a location than on the presence of beliefs, institutions, and material traits conducive to accepting those customs (Figure 4-16).

The social custom of wine production in much of France and Italy extends back at least to the Roman Empire. Wine consumption declined after the fall of Rome, and many vineyards were destroyed. Monasteries preserved the wine-making tradition in medieval Europe for both sustenance and ritual. Wine consumption has become extremely popular again in Europe in recent centuries, as well as in the Western Hemisphere, which was colonized by Europeans. Vineyards are now typically owned by private individuals and corporations rather than religious organizations.

Wine production is discouraged in regions of the world dominated by religions other than Christianity. Hindus and Muslims in particular avoid alcoholic beverages. Thus wine production is limited in South Asia and Southwest Asia & North Africa (other than Israel) primarily because of cultural values, especially religion.

CONFLICTING FOLK AND POPULAR CULTURAL VALUES

Conflicts can arise between folk and popular culture. For example, wearing folk clothing in countries dominated by

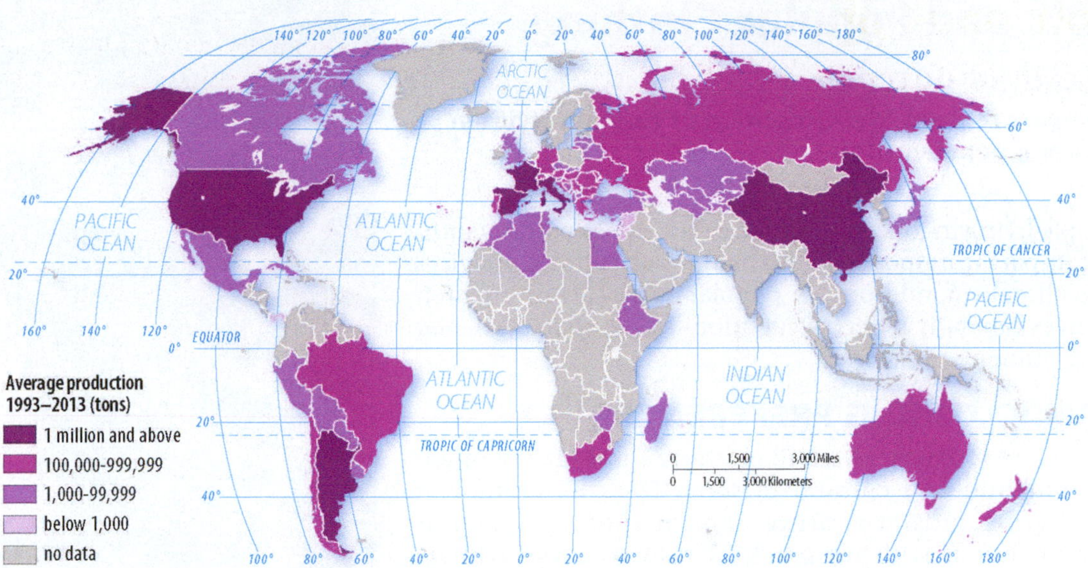

▶ **FIGURE 4-16 WINE PRODUCTION** The distribution of wine production is influenced in part by the physical environment and in part by social customs. Most grapes used for wine are grown near the Mediterranean Sea or in areas of similar climate. Income, preferences, and other social customs also influence the distribution of wine consumption, as seen in the lower production levels of predominantly Muslim countries south of the Mediterranean.

popular culture can be controversial, and conversely so can wearing popular clothing in countries dominated by folk-style clothing. Men must decide whether to wear Western-style suits, especially if they occupy positions of leadership in business or government.

Especially difficult has been the coexistence of the loose-fitting combination body covering, head covering, and veil traditionally worn by women in Southwest Asia & North Africa with casual Western-style popular women's clothing, such as open-necked blouses, tight-fitting slacks, and revealing skirts. Garments that cover the face are typically worn by women who adhere to traditional folk customs in Southwest Asia & North Africa. The practice of covering the head is called hijab (Figure 4-17). The niqab is a veil that covers the bottom half of the face. The burqa covers the entire face and body, leaving a mesh screen to see through. European countries, including France and Belgium, prohibit women from wearing them in public.

PAUSE & REFLECT 4.2.1

Can you think of other restrictions on clothing styles in developed countries, perhaps in schools?

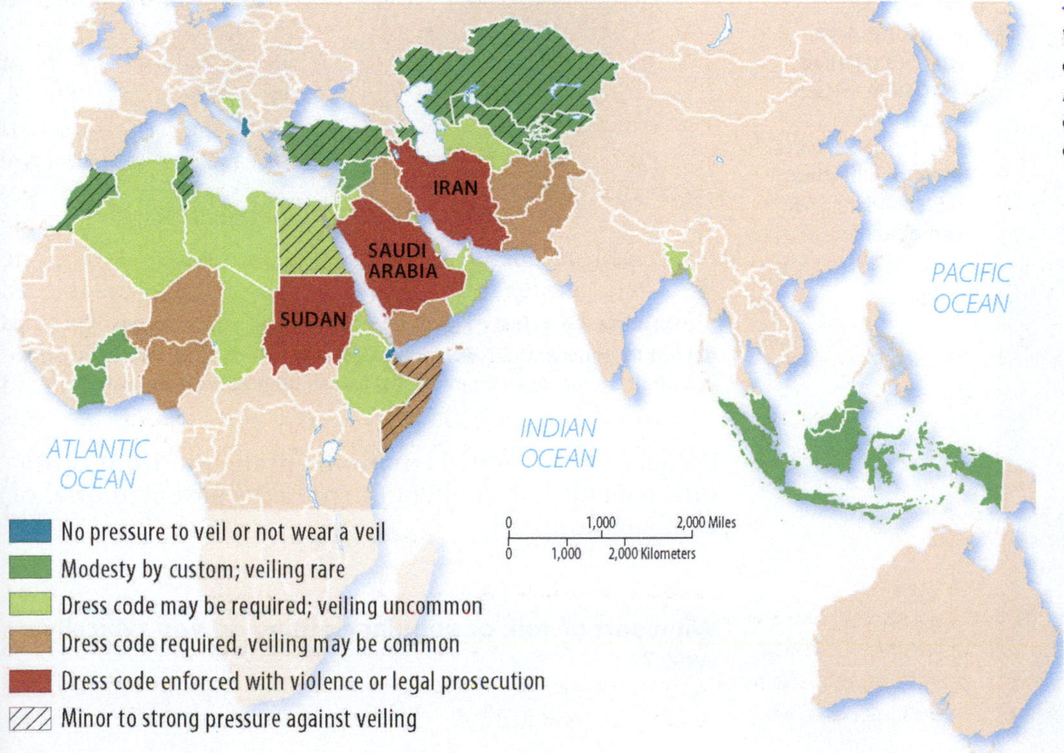

◀ **FIGURE 4-17 FOLK CLOTHING: WOMEN'S DRESS CODES** In some countries of Southwest Asia & North Africa, women are required to wear clothing that partially or completely covers the face.

Folk and Popular Clothing

LEARNING OUTCOME 4.2.2

Compare reasons for distribution of clothing styles in folk and popular culture.

People living in folk cultures have traditionally worn clothing in part in response to distinctive agricultural practices and climatic conditions. In popular culture, clothing preferences generally reflect occupations rather than particular environments.

FOLK CLOTHING PREFERENCES

People wear distinctive folk clothing for a variety of environmental and cultural reasons. The folk custom in the Netherlands of wearing wooden shoes may appear quaint, but it still has practical uses in a wet climate (Figure 4-18). In arctic climates, fur-lined boots protect against the cold, and snowshoes permit walking on soft, deep snow without sinking in. People living in warm and humid climates may not need any footwear if heavy rainfall and time spent in water discourage such use. Cultural factors, such as religious beliefs, can also influence clothing preferences (Figure 4-19).

Increased travel and the diffusion of media have exposed North Americans and Europeans to other forms of dress, just as people in other parts of the world have come into contact with Western dress. The poncho from South America, the dashiki of the Yoruba people of Nigeria, and the Aleut parka have been adopted by people elsewhere in the world. The continued use of folk costumes in some parts of the globe may persist not because of distinctive environmental conditions or traditional cultural values but to preserve past memories or to attract tourists.

◀ **FIGURE 4-18** **FOLK CLOTHING: WOODEN SHOES** A man wearing wooden shoes bikes on a flooded street in Stellendam, the Netherlands.

(a)

(b)

▲ **FIGURE 4-19** **FOLK CLOTHING: RELIGIOUS TRADITIONS** Many devout Muslims and Jews wear modest black clothes. **(a)** Muslim woman in Dubai, United Arab Emirates. **(b)** Jewish teenagers in Jerusalem.

PAUSE & REFLECT 4.2.2

What sort of folk or popular clothing do you typically wear?

RAPID DIFFUSION OF POPULAR CLOTHING STYLES

Individual clothing habits reveal how popular culture can be distributed across the landscape with little regard for distinctive physical features. Instead, popular clothing habits reflect:

- **Occupation.** A lawyer or business executive, for example, tends to wear a dark suit, light shirt or blouse, and necktie or scarf, whereas a factory worker wears jeans and a work shirt. A lawyer in New York is more likely to dress like a lawyer in California than like a factory worker in New York.
- **Income.** Women's clothes, in particular, change in fashion from one year to the next. The color, shape, and design of dresses change to imitate pieces created by clothing designers. For social purposes, people with sufficient income may update their wardrobe frequently with the latest fashions.

Improved communications have permitted the rapid diffusion of clothing styles from one region of Earth to another. Original designs for women's dresses, created in Paris, Milan, London, or New York, are reproduced in large quantities at factories in Asia and sold for relatively low prices in North American and European chain stores. Speed is essential in manufacturing copies of designer dresses because fashion tastes change quickly.

In the past, years could elapse from the time an original dress was displayed to the time that inexpensive reproductions were available in the stores. Now the time lag is only a few weeks because of the diffusion of electronic communications. Buyers from the major retail chains can view fashions electronically and place orders. Sketches, patterns, and specifications can be sent instantly from European fashion centers to American corporate headquarters and then on to Asian factories.

DEBATE IT! Should Europe accept face covers for women?

Some European countries, including France and Belgium, prohibit women from wearing garments such as the burqa and niqab that cover the face. These are typically worn by devout Muslim women who have migrated to Europe from Southwest Asia & North Africa.

PROHIBIT BURQA AND NIQAB IN PUBLIC

- The coverings obliterate personal identity and treat women like second-class citizens.
- The ban protects gender equality and the dignity of women.
- Complete covering of the face poses a security risk by preventing identification of an individual.

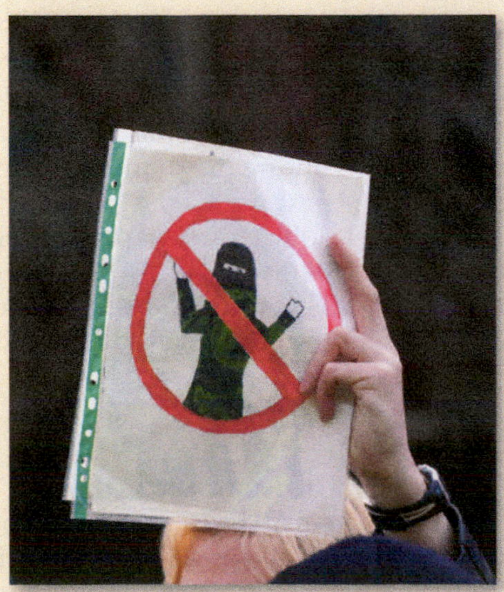

▲ **FIGURE 4-20 POSTER SUPPORTING THE BURQA AND NIQAB BAN**

PERMIT BURQA AND NIQAB IN PUBLIC

- Governments have no business determining clothing preferences.
- The ban shows lack of understanding and intolerance of Muslim cultural traditions.
- The ban infringes on a woman's religious, free speech, and privacy rights.

▲ **FIGURE 4-21 HIJAB SUPPORTERS DEMONSTRATE IN LONDON**

Folk Food Customs

LEARNING OUTCOME 4.2.3

Understand reasons for folk food preferences and taboos.

According to the nineteenth-century cultural geographer Vidal de la Blache, "Among the connections that tie [people] to a certain environment, one of the most tenacious is food supply; clothing and weapons are more subject to modification than the dietary regime, which experience has shown to be best suited to human needs in a given climate."

Food preferences are inevitably affected by the availability of products, but people do not simply eat what is available in their particular environment. Food preferences are strongly influenced by cultural traditions. What is eaten establishes one's social, religious, and ethnic memberships. The surest way to identify a family's ethnic origins is to look in its kitchen.

FOOD AND THE ENVIRONMENT

Folk food habits are embedded especially strongly in the environment. Humans eat mostly plants and animals—living things that spring from the soil and water of a region. Inhabitants of a region must consider the soil, climate, terrain, vegetation, and other characteristics of the environment in deciding to produce particular foods.

The contribution of a location's distinctive physical features to the way food tastes is known by the French term **terroir**. The word comes from the same root as *terre* (the French word for "land" or "earth"), but terroir does not translate precisely into English; it has a similar meaning to the English expressions "grounded" and "sense of place." Terroir is the sum of the effects on a particular food item of soil, climate, and other features of the local environment. For example, a special type of lentil is grown only around the village of Le Puy-en-Velay, France (Figure 4-22). The lentil has a distinctive flavor because of the area's volcanic soil and dry growing season.

▼ FIGURE 4-22 FOLK FOOD CUSTOMS: TERROIR The village of Le Puy-en-Velay, France, is the home of a type of lentil that was the first vegetable to be registered and protected by the French government and the European Union.

People adapt their food preferences to conditions in the environment. For example, in Asia, rice is grown in milder, moister regions, whereas wheat thrives in colder, drier regions. In Europe, traditional preferences for quick-frying foods in Italy resulted in part from fuel shortages. In Northern Europe, an abundant wood supply encouraged the slow stewing and roasting of foods over fires, which also provided home heat in the colder climate.

Soybeans, an excellent source of protein, are widely grown in Asia. In the raw state they are toxic and indigestible. Lengthy cooking renders them edible, but fuel is scarce in Asia. Asians have adapted to this environmental challenge by deriving from soybeans foods that do not require extensive cooking. These include bean sprouts (germinated seeds), soy sauce (fermented soybeans), and bean curd (steamed soybeans).

Bostans, which are small gardens inside Istanbul, Turkey, have been supplying the city with fresh produce for hundreds of years (Figure 4-23). According to geographer Paul Kaldjian, Istanbul has around 1,000 *bostans*, run primarily by immigrants from Cide, a rural village in Turkey's Kastamonu province. *Bostan* farmers are able to maximize yields from their small plots of land (typically 1 hectare) through what Kaldjian calls clever and efficient manipulation of space, season, and resources. In a *bostan*, 15 to 20 different types of vegetables are planted at different times of the year, and the choice is varied from year to year, in order to reduce the risk of damage from poor weather. Most of the work is done by older men, who prepare beds for planting, sow, irrigate, and operate motorized equipment, according to Kaldjian. Women weed, and both men and women harvest.

Certain foods are eaten in folk cultures because their natural properties are perceived to enhance qualities considered desirable by the society. Here are some examples:

- The Abipone people in Paraguay eat bulls, jaguars, and stags to make them strong, brave, and swift. The Abipone believe that consuming hens or tortoises will make them cowardly.
- The Ainu people in Japan avoid eating otters because they are believed to be forgetful animals, and consuming them could cause loss of memory.
- The Mbum Kpau women in Chad do not eat chicken or goat before becoming pregnant. Abstaining from consumption of these animals is thought to help escape pain in childbirth and to prevent birth of a child with abnormalities. During pregnancy, the Mbum Kpau avoid meat from antelopes with twisted horns, which could cause them to bear offspring with deformities.

FOOD TABOOS

According to many folk customs, everything in nature carries a signature, or distinctive characteristic, based on its appearance and natural properties. Consequently, people may desire or avoid certain foods in response to perceived beneficial or harmful natural traits.

People refuse to eat particular plants or animals for a variety of reasons. Such a restriction on behavior imposed

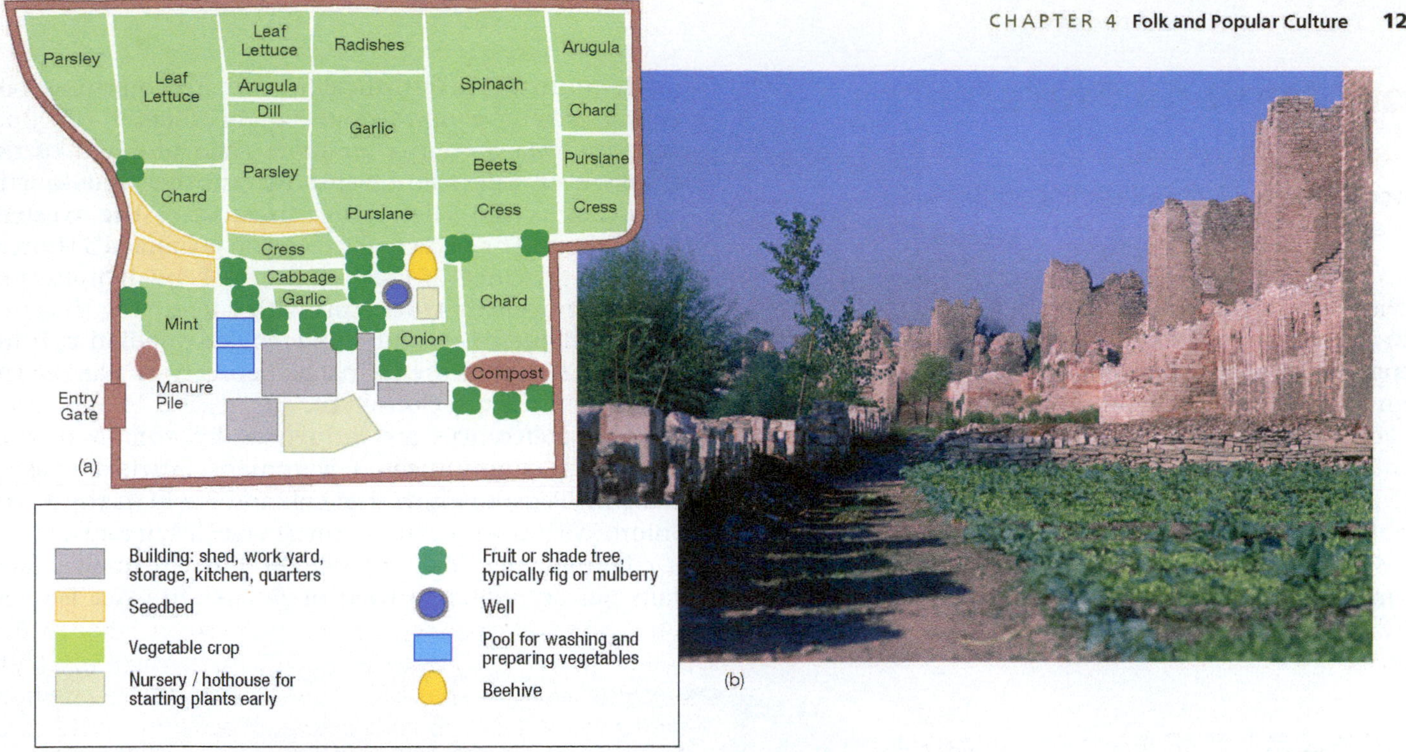

▲ **FIGURE 4-23 ISTANBUL VEGETABLE GARDEN (a)** Geographer Paul Kaldijian sketched a typical *bostan*, a traditional vegetable garden, in the center of Istanbul, Turkey. *Bostans* provide residents of the large city of Istanbul with a source of fresh vegetables. **(b)** Outside the 500-year-old Yedikule Fortress, in Istanbul, a *bostan* dates back to 400 B.C. It is threatened with removal and replacement with a landscaped park.

by religious law or social custom is a **taboo**. Other customs or practices, such as sexual behavior, carry prohibitions, but taboos are especially strong in the area of food. Some folk cultures may establish food taboos because of concern for the natural environment. These taboos may help protect endangered animals or conserve scarce natural resources. To preserve scarce animal species, only a few high-ranking people in some tropical regions are permitted to hunt, and the majority cultivate crops.

Relatively well-known taboos against consumption of certain foods can be found in the Jewish Bible. The ancient Hebrews were prohibited from eating a wide variety of foods, including animals that do not chew their cud or that have cloven feet, and fish lacking fins or scales (Figure 4-24). These biblical taboos were developed through oral tradition and by rabbis into the kosher laws observed today by some Jews.

Muslims embrace a taboo against pork because pigs are unsuited for the dry lands of the Arabian Peninsula. Pigs would compete with humans for food and water, without offering compensating benefits, such as being able to pull a plow, carry loads, or provide milk and wool. Widespread raising of pigs would be an ecological disaster in Islam's hearth.

Hindu taboos against consuming cattle are also partly due to environmental factors. Cows are the source of oxen (castrated male bovine), the traditional choice for pulling plows as well as carts. A large supply of oxen must be maintained in India because every field has to be plowed at approximately the same time—when the monsoon rains arrive. Religious sanctions have kept India's cattle population large as a form of insurance against the loss of oxen and increasing population.

But the taboo against consumption of meat among many people, including Muslims, Hindus, and Jews, cannot be explained primarily by environmental factors. Social values must influence the choice of diet because people in similar climates and with similar levels of income consume different foods. The biblical food taboos helped the Jewish people maintain their identity and communal affiliation. That Christians ignore the biblical food injunctions is consistent with Christianity as a religion that seeks to attract many adherents from many places (see Chapter 6).

▲ **FIGURE 4-24 KOSHER PIZZA RESTAURANT, PARIS**

PAUSE & REFLECT 4.2.3

What foods do you avoid? Do you avoid foods because of taboos or for other reasons?

Popular Food Preferences

LEARNING OUTCOME 4.2.4

Describe regional variations in popular food preferences.

In the popular culture of the twenty-first century, food preferences seem far removed from folk traditions. Popular food preferences are influenced more by cultural values than by environmental features. Still, some regional variations can be observed between and within countries, and environmental influences remain important in selected items.

REGIONAL DIFFERENCES: GLOBAL SCALE

Why do Coca-Cola and Pepsi have different sales patterns (Figure 4-25)? The two beverages are similar, and many people are unable to taste the difference. Yet consumers prefer Coke in some countries and Pepsi in others.

Coca-Cola accounts for more than one-half of the world's cola sales, and Pepsi for another one-fourth. Coca-Cola is the sales leader in most of the Western Hemisphere. The principal exception is Canada's French-speaking province of Québec, where Pepsi is preferred. Pepsi won over the Québécois with advertising that tied Pepsi to elements of uniquely French Canadian culture. The major indoor arena in Québec City is named the Colisée Pepsi (Pepsi Coliseum).

Cola preferences are influenced by politics in Russia. Under communism, government officials made a deal with Pepsi to allow that cola to be sold in the Soviet Union. With the breakup of the Soviet Union and the end of communism, Coke entered the Russian market. Russians quickly switched their preference to Coke because

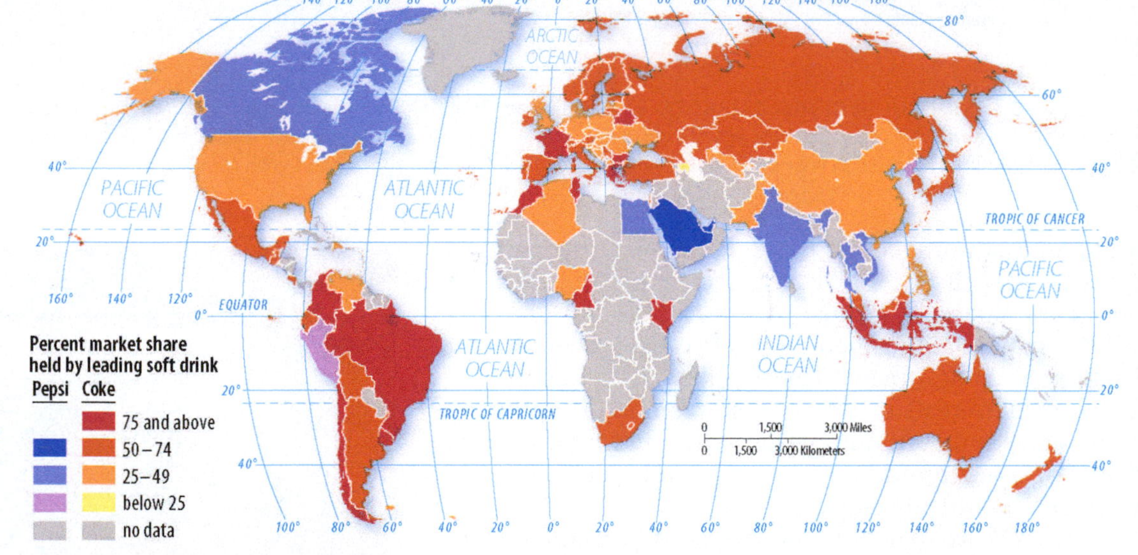

▶ **FIGURE 4-25 POPULAR FOOD PREFERENCES: COKE VERSUS PEPSI** Coca-Cola leads sales in the United States, Latin America, Europe, and Russia. Pepsi leads in Canada and South and Southwest Asia.

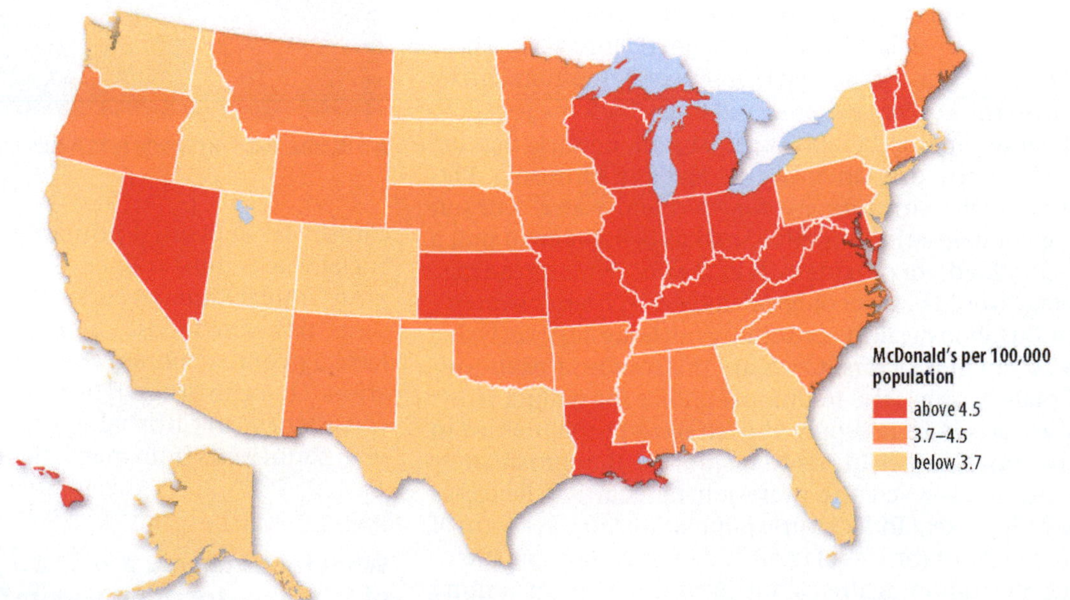

▶ **FIGURE 4-26 REGIONAL VARIATION: CONCENTRATION OF MCDONALD'S** The highest concentration of McDonald's restaurants is in the Midwest.

Pepsi was associated with the discredited Communist government.

At one time, Arab countries in Southwest Asia & North Africa boycotted products that were sold in predominantly Jewish Israel. Because Pepsi was not sold in Israel until 1992, Coke was the only choice in Israel, whereas in most of Israel's neighbors Pepsi was preferred.

REGIONAL DIFFERENCES: U.S. SNACK AND FAST FOOD

Geographers observe regional variations in food preferences within a developed country like the United States. Some of these variations can be attributed to cultural or environmental factors, whereas others do not have a clear explanation.

Here are some examples of the influence of cultural factors on regional variations in food preferences in the United States:

- Utah has a low rate of consumption of all types of alcohol because of a concentration there of members of the Church of Jesus Christ of Latter-day Saints, who abstain from all alcohol consumption. The adjacent state of Nevada has a high rate of consumption of all types of alcohol because of the heavy concentration of gambling and other resort activities there.
- Texans may prefer tortilla chips because of the large number of Hispanic Americans there. Westerners may prefer multigrain chips because of greater concern for the nutritional content of snack foods.

Americans may choose particular beverages or snacks in part on the basis of preference for what is produced, grown, or imported locally:

- Wine consumption is relatively high in California, where most of the U.S. production is concentrated, and beer consumption is relatively low there. Beer and spirits consumption are relatively high in the upper Midwest, where much of the grain is grown. Consumption of wine is low in that part of the country, where few grapes are grown.
- Southerners may prefer pork rinds because more hogs are raised there, and northerners may prefer popcorn and potato chips because more corn and potatoes are grown there.

On the other hand, many regional variations are not clearly linked to cultural or environmental factors. For example, the concentration of McDonald's varies among the 50 states from more than 5 to less than 3 per 100,000 population. The Midwest has the highest concentration of McDonald's (Figure 4-26). Dunkin' Donuts is especially popular in the Northeast, Krispy Kreme in the Southeast, White Castle in the Midwest, and In-N-Out Burger in the Southwest, based on the frequency with which they are cited on Twitter, according to Lexicalist (lexicalist.com) (Figure 4-27).

PAUSE & REFLECT 4.2.4

Do your food preferences match the predominant ones in your region?

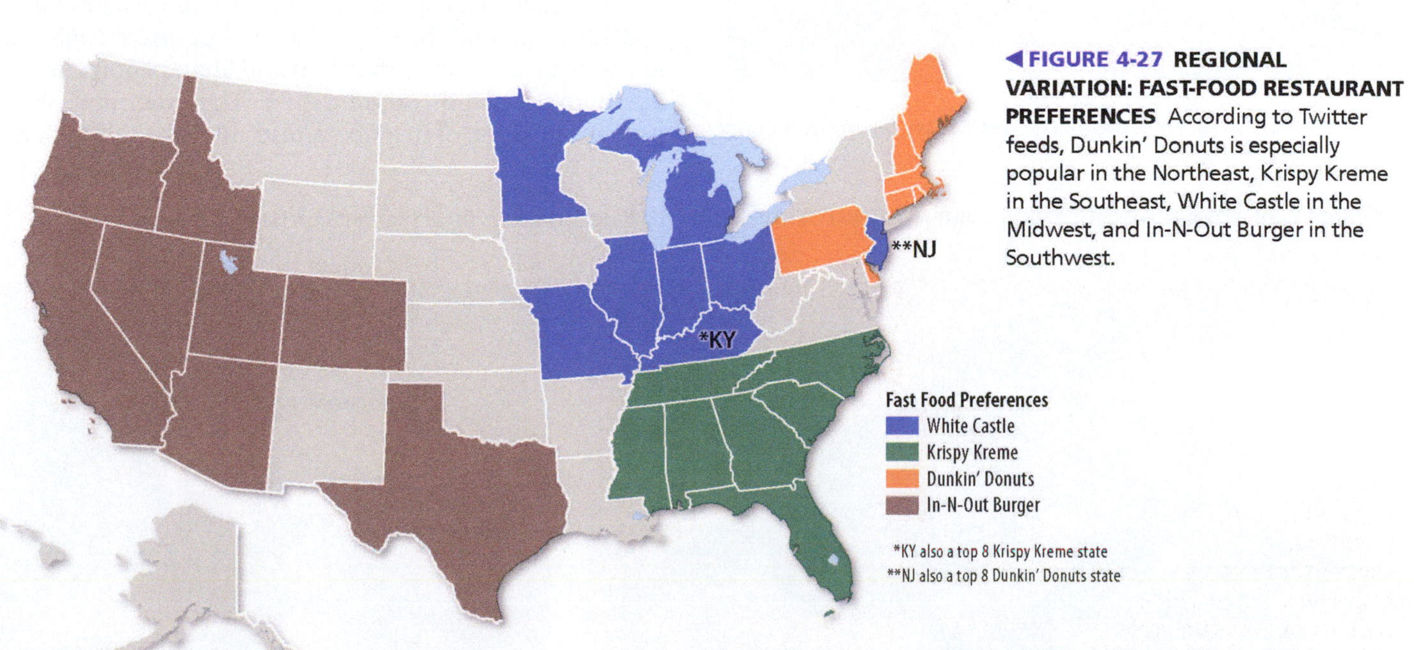

◄ FIGURE 4-27 REGIONAL VARIATION: FAST-FOOD RESTAURANT PREFERENCES According to Twitter feeds, Dunkin' Donuts is especially popular in the Northeast, Krispy Kreme in the Southeast, White Castle in the Midwest, and In-N-Out Burger in the Southwest.

Folk and Popular Housing

LEARNING OUTCOME 4.2.5

Understand factors that influence patterns of folk housing.

French geographer Jean Brunhes, a major contributor to the cultural landscape tradition, viewed the house as being among the essential facts of human geography. It is a product of both cultural traditions and natural conditions. American cultural geographer Fred Kniffen considered the house to be a good reflection of cultural heritage, current fashion, functional needs, and the impact of environment.

FOLK HOUSING

All humans need a place to live. Distinctive environmental and cultural features influence the provision of housing in folk cultures.

ENVIRONMENTAL INFLUENCES. The type of building materials used to construct folk houses is influenced partly by the resources available in the environment. Stone, grass, sod, and skins may be used, but the two most common building materials in the world are wood and brick (Figure 4-28).

(a)

(b)

▲ **FIGURE 4-28 HOUSES UNDER CONSTRUCTION (a)** Wood, **(b)** brick.

The style of construction can also be influenced by the environment. For example, the construction of a pitched roof is important in wet or snowy climates to facilitate runoff and to reduce the weight of accumulated snow. Windows may face south in temperate climates to take advantage of the Sun's heat and light. In hot climates, on the other hand, roofs may be flat, and window openings may be smaller to protect the interior from the full heat of the Sun (Figure 4-29).

CULTURAL INFLUENCES. The distinctive form of folk houses may derive primarily from religious values and other customary beliefs. Some compass directions may be more important than other directions.

Houses may have sacred walls or corners. In the south-central part of the island of Java, for example, the front door always faces south, the direction of the South Sea Goddess, who holds the key to Earth. The eastern wall of a house is considered sacred in Fiji, as is the northwestern wall in parts of China. Sacred walls or corners are also noted in parts of the Middle East, India, and Africa.

In Madagascar, the main door is on the west, which is considered the most important direction, and the northeastern corner is the most sacred. The northern wall is for honoring ancestors; in addition, important guests enter a room from the north and are seated against the northern wall. The bed is placed against the eastern wall of the house, with the head facing north.

The Lao people in northern Laos arrange beds perpendicular to the center ridgepole of the house (Figure 4-30a). Because the head is considered high and noble and the feet low and vulgar, people sleep so that their heads will be opposite their neighbor's heads and their feet opposite their neighbor's feet. There is one principal exception to this arrangement: A child who builds a house next door to his or her parents sleeps with his or her head toward the parents' feet, as a sign of obeying the customary hierarchy.

Although they speak similar Southeast Asian languages and adhere to Buddhism, the Lao do not orient their houses in the same manner as the Yuan and Shan peoples in nearby northern Thailand (Figure 4-30b). The Yuan and Shan ignore the position of neighbors, and all sleep with their heads

▼ **FIGURE 4-29 FOLK HOUSING: ENVIRONMENTAL INFLUENCES (a)** In the desert in Oman, **(b)** in the winter in Haanja, Estonia.

(a)

(b)

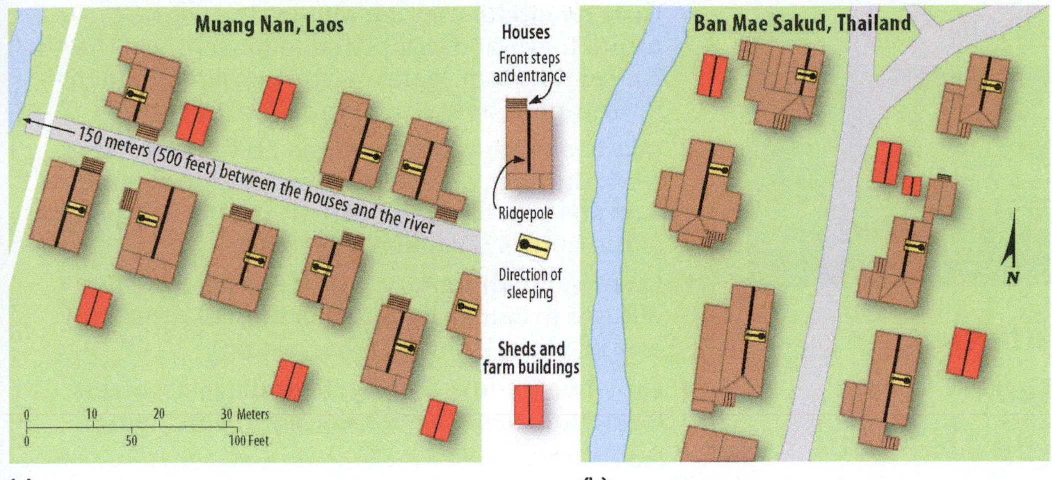

◀ **FIGURE 4-30 FOLK HOUSING: CULTURAL INFLUENCES** **(a)** The fronts of houses of Lao people, such as those in the village of Muang Nan, Laos, face one another across a path, and the backs face each other at the rear. Their ridgepoles (the centerline of the roof) are set perpendicular to the path but parallel to a stream, if one is nearby. Inside adjacent houses, people sleep in the orientation shown, so neighbors are head-to-head or feet-to-feet. **(b)** The houses of Yuan and Shan peoples in the village of Ban Mae Sakud, Thailand, are not set in a straight line because of a belief that evil spirits move in straight lines. Ridgepoles parallel the path, and the heads of all sleeping persons point eastward.

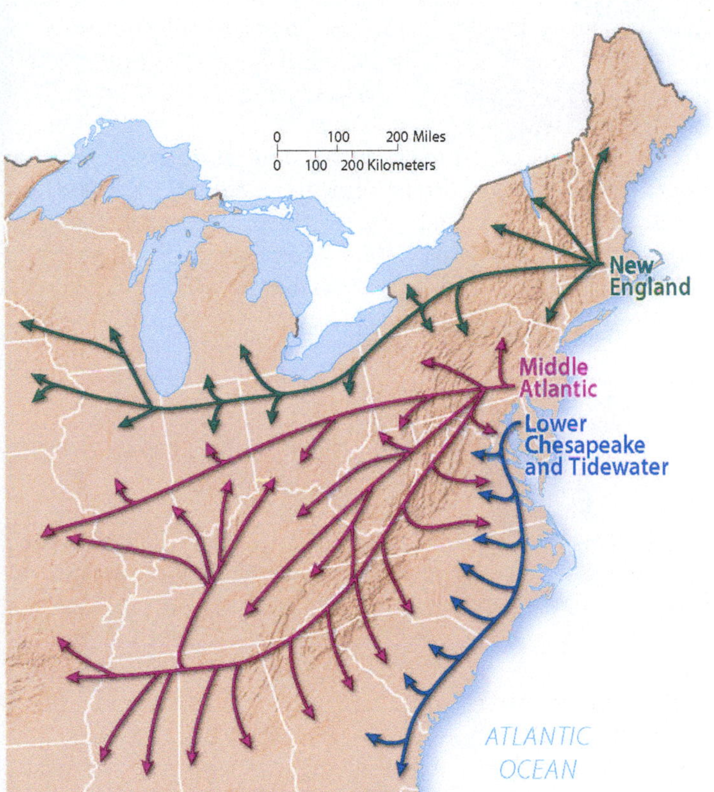

◀ **FIGURE 4-31 FOLK HOUSING: U.S. HEARTHS** U.S. house types in the United States originated in three main source areas and diffused westward along different paths. New England house types can be found throughout the Great Lakes region as far west as Wisconsin because this area was settled primarily by migrants from New England. Its distinctive style was box shaped with a central hall. Middle Atlantic migrants carried their house type westward across the Ohio Valley and southwestward along the Appalachian trails. The principal house type was known as the "I"-house, typically two full stories in height, one room deep and at least two rooms wide. Lower Chesapeake and Tidewater houses were spread by migrants along the Southeast Coast. The style typically comprised one story, with a steep roof and chimneys at either end.

toward the east, which Buddhists consider the most auspicious direction. Staircases must not face west, the least auspicious direction and the direction of death and evil spirits.

PAUSE & REFLECT 4.2.5

What factors were considered in the arrangement of the bed in your bedroom?

U.S. FOLK HOUSES. Older houses in the United States display local folk culture traditions. In contrast, housing built in the United States since the 1940s demonstrates how popular customs vary more in time than in place.

Geographer Fred Kniffen identified three major hearths, or nodes, of folk house forms in the United States (Figure 4-31). When families migrated westward in the 1700s and 1800s, they cut trees to clear fields for planting and used the wood to build houses, barns, and fences. The style of pioneer homes reflected whatever style was prevailing at the place on the East Coast from which they migrated.

CHECK-IN KEY ISSUE 2

Where Are Folk and Popular Material Culture Distributed?

- ✔ **Regional variations in folk food, clothing, and shelter derive from the physical environment, as well as from religion and other cultural values.**
- ✔ **Popular preferences in food, clothing, and shelter vary more in time than in place. However, some regional variations in preferences persist.**

KEY ISSUE 3

Why Is Access to Folk and Popular Culture Unequal?

- ▸ **Diffusion of TV and Internet**
- ▸ **Diffusion of Social Media**
- ▸ **Challenges in Accessing Electronic Media**

LEARNING OUTCOME 4.3.1

Compare the diffusion of TV and the Internet.

Geographic concepts help to understand the distribution and diffusion of popular culture. Popular culture diffuses rapidly around the world in the twenty-first century primarily through electronic media. The latest fashions in material culture and leisure activities can be viewed by anyone in the world who has access to one or more forms of electronic media. Electronic media increase access to popular culture for people who embrace folk culture and at the same time increase access to folk culture for people who are part of the world's popular culture scene.

However, the distribution of popular culture around the world is not uniform. The principal obstacle to popular culture is lack of access to electronic media. Access is limited primarily by lack of income. In some developing countries, access is also limited by lack of electricity.

Diffusion of TV and Internet

The world's most important electronic media format by far is TV. Television supplanted other formats, notably radio and telegraph, during the twentieth century. Into the twenty-first century, other formats have become popular, but they have not yet supplanted TV worldwide.

DISTRIBUTION AND DIFFUSION OF TV

Watching TV is especially important for popular culture for two reasons:

- Watching TV is the most popular leisure activity in the world. The average human watched three hours of TV per day in 2014, and the average American watched five hours.
- TV has been the most important mechanism by which popular culture, such as professional sports, has rapidly diffused across Earth.

Through the second half of the twentieth century, TV diffused from the United States to Europe and other developed countries and then to developing countries (Figure 4-32):

- **Early twentieth century: Multiple hearths.** TV technology was developed simultaneously in the United Kingdom, France, Germany, Japan, and the Soviet Union, as well as in the United States.
- **Mid-twentieth century: United States dominates.** In 1954, the first year that the United Nations published data on the subject, the United States had 86 percent of the world's 37 million TV sets.
- **Late twentieth century: Diffusion to Europe.** TV diffused to Europe by 1970, but most of Africa and Asia had little if any TV broadcasting.
- **Early twenty-first century: Near-universal access.** Ownership rates climbed sharply in developing countries.

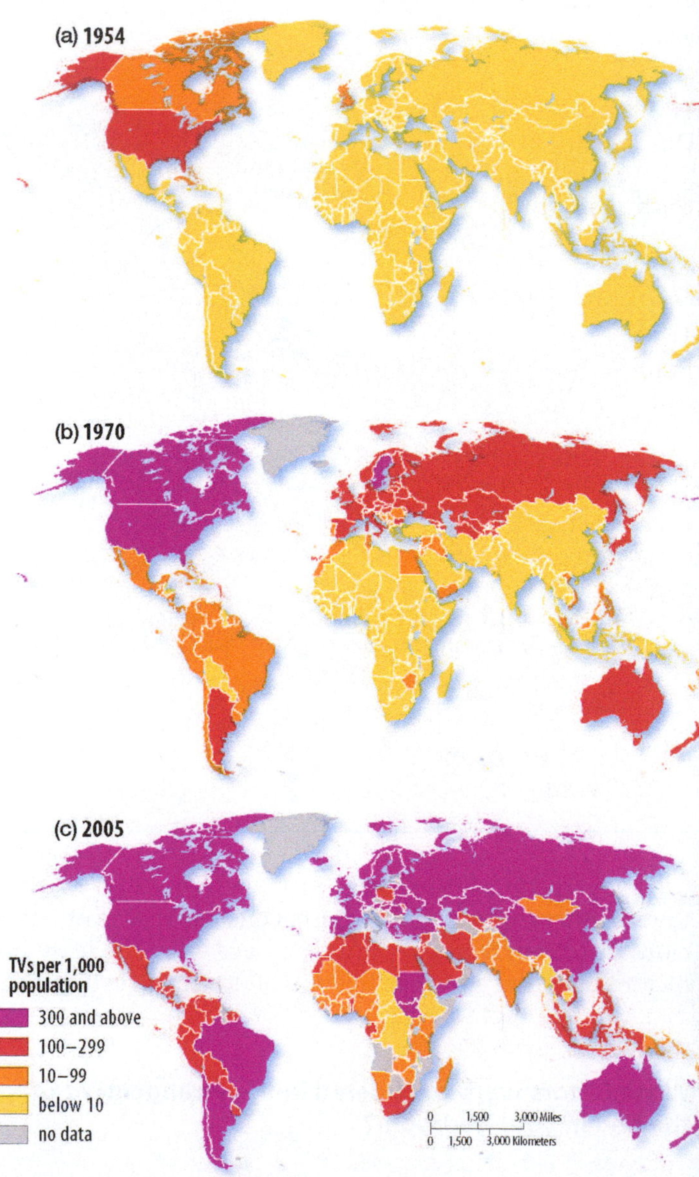

▲ **FIGURE 4-32 DIFFUSION OF TV** Televisions per 1,000 inhabitants in **(a)** 1954, **(b)** 1970, and **(c)** 2005. Television has diffused from North America and Europe to other regions of the world. The United States and Canada had far more TV sets per capita than any other country as recently as the 1970s, but several European countries now have higher rates of ownership.

TABLE 4–1 Changing Distribution And Diffusion Of TV

	1954	1970	2005
Density: Number of U.S. TVs (millions)	32	82	219
U.S. diffusion: TVs per 1,000 population	196	395	882
Global diffusion: U.S. share of world's TVs (%)	86	25	16

Table 4-1 shows changes in the distribution and diffusion of TVs in the United States. The density of TVs in the United States has increased, as TV diffused through the U.S. population. At the same time, TV has diffused to the rest of the world, leaving the United States with an ever-decreasing concentration of the world's TV sets.

PAUSE & REFLECT 4.3.1

The United States has slightly less than 1 TV per person. Does your household have more than 1 TV or less than 1 TV per person? Why might you have more or less than the national average?

DIFFUSION OF THE INTERNET

The changing distribution and diffusion of Internet service follows the pattern established by television a generation earlier, but at a more rapid pace (Figure 4-33 and Table 4-2):

- In 1995, most countries did not have Internet service, and the United States had 63 percent of the world's users.
- Between 1995 and 2000, Internet users increased rapidly in the United States, from 9 percent of the population (25 million people) to 44 percent (124 million people). But the worldwide increase was much greater, so the share of the world's Internet users clustered in the United States declined from 63 percent to 35 percent.
- Between 2000 and 2014, Internet usage continued to increase rapidly in the United States, to 87 percent (280 million people). Again, the U.S. increase was more modest than in the rest of the world, and the share of the world's Internet users in the United States continued to decline, to less than 10 percent in 2014. China now accounts for 22 percent of the world's Internet users.

TABLE 4–2 Changing Distribution And Diffusion Of Internet

	1995	2000	2014
Density: Number of U.S. Internet users (millions)	25	124	280
U.S. diffusion: Internet users per 1,000 population	94	441	868
Global diffusion: U.S. share of world's Internet users (%)	63	35	10

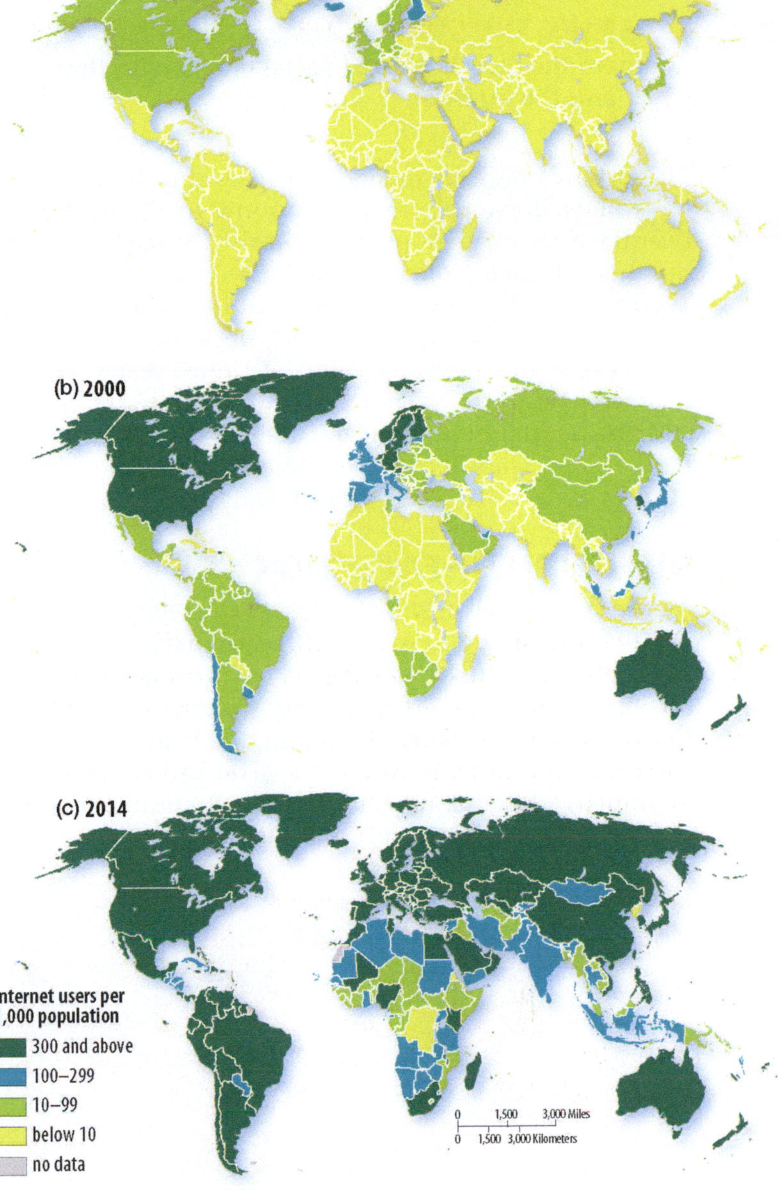

▲ **FIGURE 4-33 DIFFUSION OF THE INTERNET** Internet users per 1,000 inhabitants in **(a)** 1995, **(b)** 2000, and **(c)** 2014. Compare to the diffusion of TV (Figure 4-32). Internet service is following a pattern in the twenty-first century similar to the pattern of diffusion of television in the twentieth century. The United States started out with a much higher rate of usage than elsewhere, until other countries caught up. The difference is that the internet diffused more rapidly than TV.

Note that all six maps in Figures 4-32 and 4-33 use the same intervals. For example, the highest class in all maps is 300 or more per 1,000. And Tables 4-1 and 4-2 use the same units of measurement. What is different between the two figures and two tables is the time interval period. The diffusion of TV from the United States to the rest of the world took a half-century, whereas the diffusion of the Internet took only a decade. Given the history of TV, the Internet is likely to diffuse further in the years ahead at a rapid rate.

Diffusion of Social Media

LEARNING OUTCOME 4.3.2

Compare the distribution of social media with that of TV and the Internet.

The origin of social media in the twenty-first century has followed the pattern of electronic media in the late twentieth century. People based in the United States have dominated the use of social media such as Facebook and Twitter so far. For example, many countries lack information such as Google street view (Figure 4-34).

Social media originating in the United States will undoubtedly diffuse to the rest of the world. However, the rate and extent of diffusion remain to be seen. Will U.S. dominance reduce rapidly, or will people elsewhere in the world embrace other forms of social media instead? Early evidence is mixed.

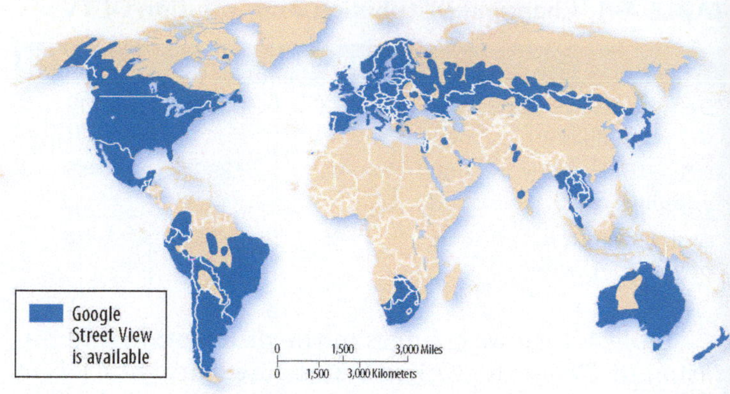

▲ **FIGURE 4-34 AVAILABILITY OF GOOGLE STREET VIEW**

DIFFUSION OF FACEBOOK

Facebook, founded in 2004 by Harvard University students, has diffused rapidly. As with the first few years of TV and the Internet, the United States started out with far more Facebook users than any other country (Figure 4-35). In 2009, five years after Facebook's founding, the United States had 34 percent of all users worldwide. The United States had 55 million Facebook users in 2009, well ahead of second-place United Kingdom, with 18 million.

The number of Facebook users continued to increase in the United States, to 152 million in 2014. But as Facebook has diffused to other countries, the share of users in the United States has declined, to 20 percent of the worldwide total in 2011 and 10 percent in 2014. In 2014, India became the second country to have more than 100 million Facebook users. Behind India were other developing countries, including Brazil, Indonesia, and Mexico. Between 2009 and 2014, the United Kingdom fell from second to sixth in number of users, and Canada fell from third to eleventh.

Notably absent from the list of leading Facebook users is the world's most populous state, China. Also absent is Russia. In the first years of social media, numerous other

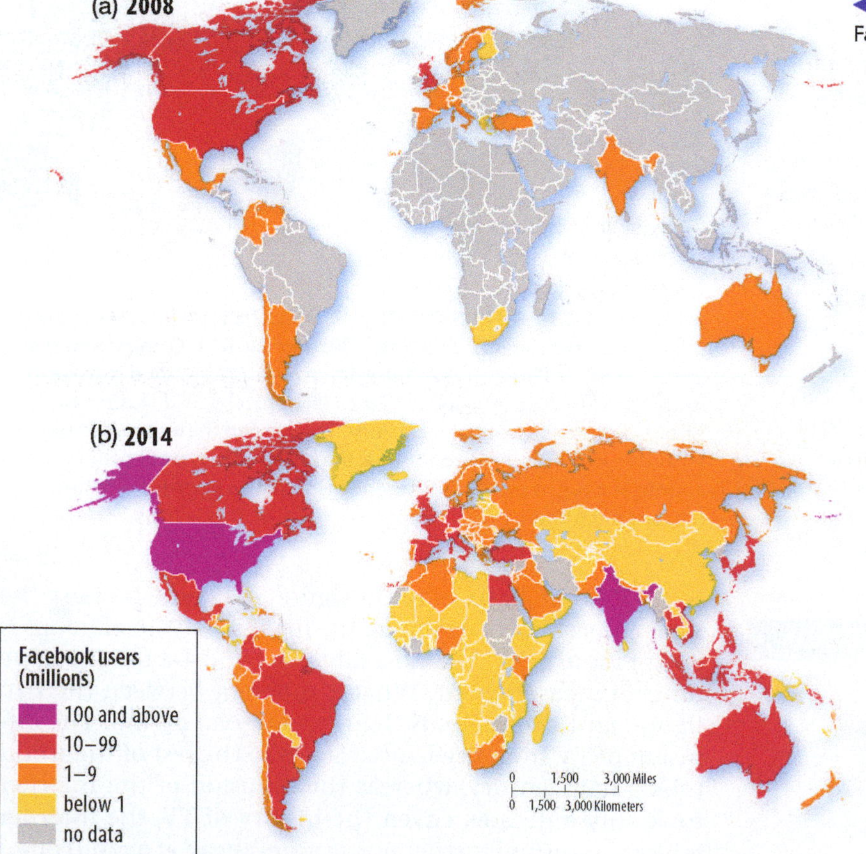

◀ **FIGURE 4-35 DIFFUSION OF FACEBOOK**
Facebook users in **(a)** 2008 and **(b)** 2014.

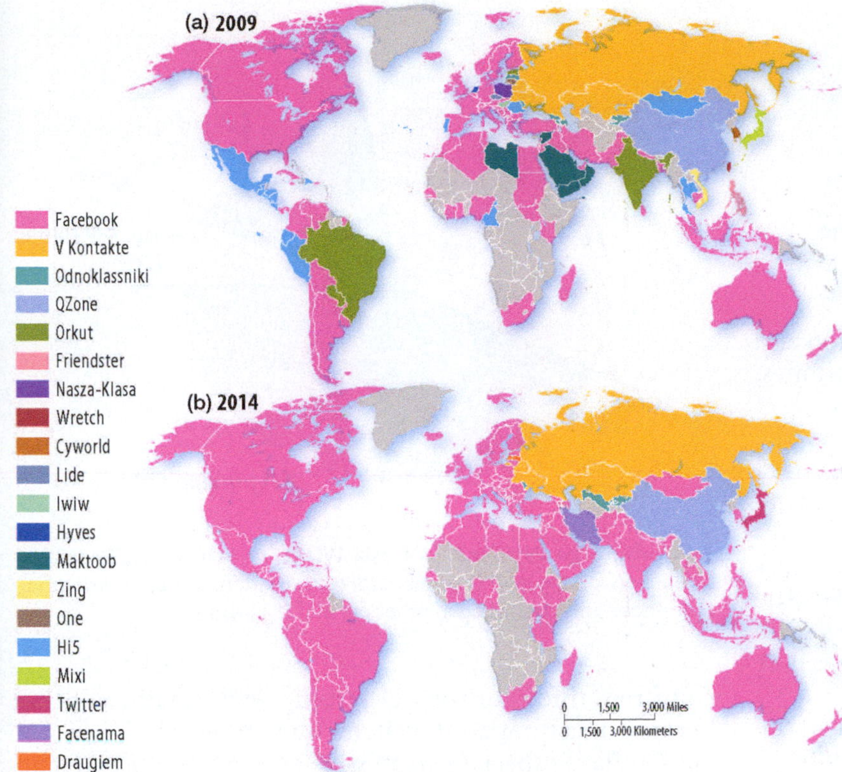

▲ **FIGURE 4-36 MOST POPULAR SOCIAL NETWORK** (a) In 2009, (b) in 2014.

networks were popular in much of the world, especially in developing countries (Figure 4-36). Most of those competing social networks were quickly supplanted by Facebook, especially in Latin America. However, the world's most populous country—China—is a holdout, preferring QZone. China's government has limited the ability of the Chinese people to use Facebook. Restricting the freedom to use the Internet is a major issue in some countries, as discussed on the next page.

DIFFUSION OF TWITTER

The United States was the source of one-third of all Twitter messages in 2014. Another one-third originated in six other countries—India, Japan, Germany, the United Kingdom, Brazil, and Canada (Figure 4-37). The second leading Twitter country is one of the world's poorest, India. This may be a preview of future trends, in which electronic communications advances diffuse rapidly to developing countries, not just to other developed countries (Table 4-3).

Americans dominate the most popular Twitter postings. Eight of the 10 Twitter posters with the largest numbers of followers in 2015 were Americans, including President Obama and seven entertainers (Katy Perry, Justin Bieber, Taylor Swift, Lady Gaga, Britney Spears, Justin Timberlake, and Ellen DeGeneres). The two non-Americans were the entertainer Rihanna (from Barbados) and the football (soccer) player Cristiano Ronaldo (from Portugal).

PAUSE & REFLECT 4.3.2

A recent study of University of Maryland students found that not using any electronics for 24 hours produced anxiety, craving, and other symptoms akin to withdrawal from alcohol or drugs. How do you think you would react to a 24-hour ban on all electronics?

TABLE 4-3 Changing Distribution and Diffusion of Facebook and Twitter

	Twitter 2009	Twitter 2014	Facebook 2009	Facebook 2014
Density: Number of U.S. users (millions)	7	108	55	152
U.S. diffusion: Users per 1,000 population	22	340	179	477
Global diffusion: U.S. share of world's users (%)	51	38	34	10

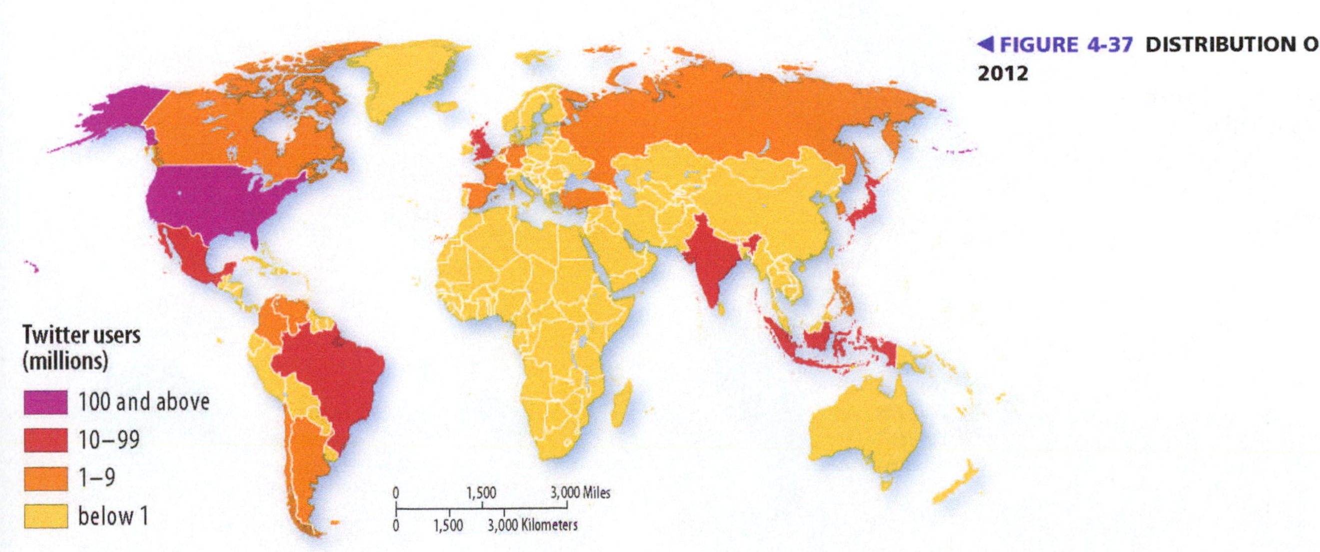

◀ **FIGURE 4-37 DISTRIBUTION OF TWITTER, 2012**

Challenges in Accessing Electronic Media

LEARNING OUTCOME 4.3.3

Understand threats to freedom of use of electronic media.

Most Americans take for granted access to information and communications through the Internet and cell phones. This free access is not found in many other countries.

The organization Freedom on the Net measures the level of Internet and digital media freedom in 65 countries. Excluded are countries with limited Internet connectivity (mostly in Africa) and countries with connectivity but insufficient evidence (mostly in Europe). Each of the 65 countries receives a numerical score from 0 (the most free) to 100 (the least free). Countries are classified "free" if they have a score of 30 or below, "partly free" if they have a score between 31 and 60, and "not free" if they have a score above 60.

Only 19 of the 65 countries surveyed by Freedom on the Net were classified as "free." Thirty-one were "partly free," and 15 were "not free" (Figure 4-38). Freedom on the Net identifies three categories of restrictions on the free use of the Internet: banned technology, blocked content, and violated user rights.

BANNED TECHNOLOGY

Governments can effectively prevent unwanted electronic technology by regulating the underlying technology platforms that are supported by the infrastructure in the country. Some governments prohibit the sale of certain models of phones, tablets, and computers. Devices that are permitted must be configured to exclude certain applications and technologies. Even some travelers between free countries, such as from the United States to Western European countries, find that their electronic devices fail to operate because of incompatible cellular infrastructure.

China is especially aggressive at restricting foreign applications. The small number of Facebook and Twitter users in China, displayed in Figures 4-35 and 4-37, is evidence of those restrictions. Instead, electronic interaction in China is undertaken primarily through homegrown apps, such as QZone and QQ (Figure 4-39).

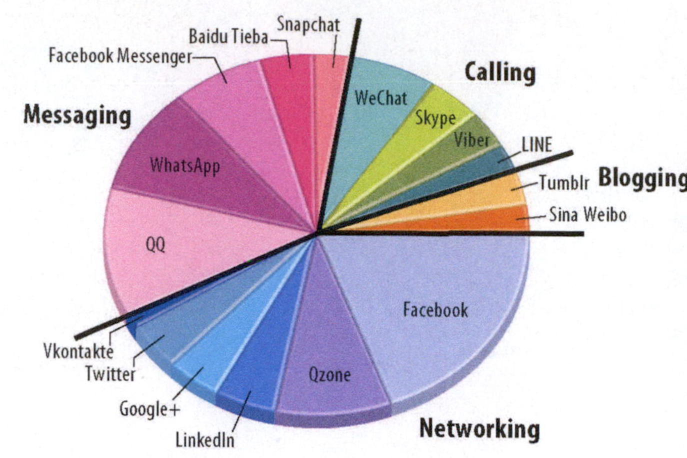

▲ **FIGURE 4-39 DIVERSITY OF THE WORLD'S SOCIAL NETWORKS** Social networking in China is undertaken with different applications than those predominating in other parts of the world.

BLOCKED CONTENT

Some websites are censored or prevented altogether from being seen on devices in a particular country. Blocking Internet content continues a widespread practice with TV.

Three developed countries—Japan, the United Kingdom, and the United States—have dominated TV in developing countries, including supplying much of the programming. Leaders of many developing countries view

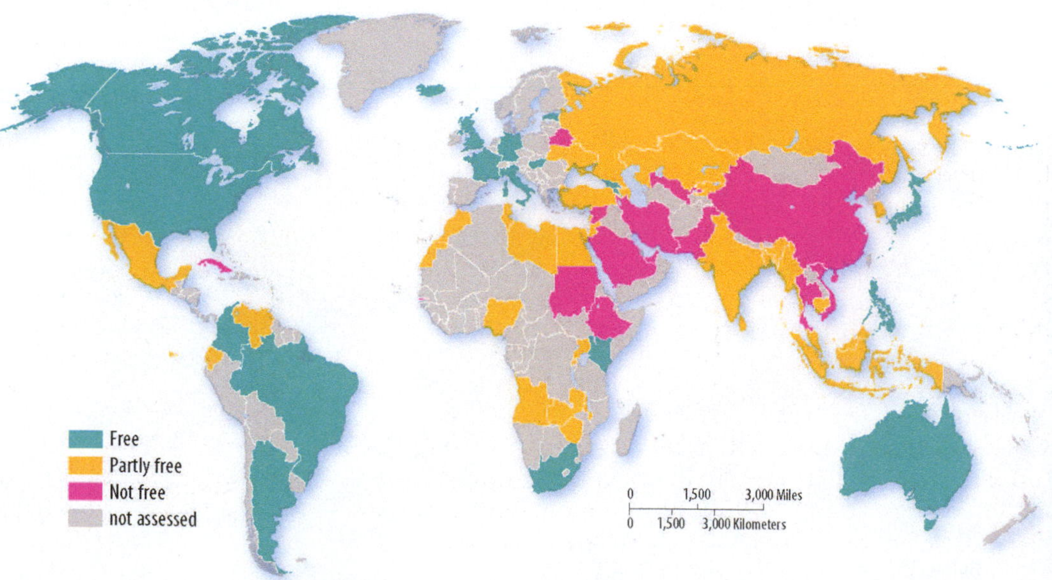

► **FIGURE 4-38 INTERNET FREEDOM 2014** Freedom on the Net determines the degree of Internet freedom based on a combination of inability to use technology, lack of access to electronic sites, and extent of violation of personal rights.

▲ **FIGURE 4-40 SOCIAL MEDIA AND POLITICAL PROTEST** An Egyptian woman documents an antigovernment protest in Cairo in 2012.

American control of much of the world's TV programming as a new method of economic and cultural imperialism. American TV programs present characteristically American beliefs and social forms, such as upward social mobility, relative freedom for women, glorification of youth, and stylized violence. These themes may conflict with and drive out traditional folk culture.

To avoid offending traditional folk culture, many satellite and cable providers in developing countries block offending networks such as MTV and censor unacceptable programs. The entertainment programs that are substituted emphasize family values and avoid controversial or edgy cultural, economic, and political content.

In the twenty-first century, concern with American-produced media content has spilled over into the Internet. OpenNet Initiative has identified three types of Internet content that are routinely censored in other countries.

- Political content that expresses views in opposition to those of the current government or that is related to human rights, freedom of expression, minority rights, and religious movements(Figure 4-40).
- Social content related to sexuality, gambling, and illegal drugs and alcohol, as well as other topics that may be socially sensitive or perceived as offensive.
- Security content related to armed conflicts, border disputes, separatist movements, and militant groups.

Google, the world's most widely used search engine, has come under especially strong criticism for failing to display or provide a link to websites that the government does not wish to be seen by users in that country. Google.cn, Google's search engine for Chinese online users, has been singled out for blocking websites considered unacceptable by the government of China.

VIOLATED USER RIGHTS

Governments are finding it increasingly difficult to stop the diffusion of technology. Their citizens are finding ways to circumvent government restrictions on ownership of hardware, use of software, and viewing of online content.

Instead, according to Freedom on the Net, governments are turning to harassing their citizens through physical attacks and imprisonment because of their Internet activity. In many countries, governments have enacted laws that provide a justification for attacks. For example, a law in The Gambia makes it a criminal offense to use the Internet to criticize public officials. A law in Ethiopia permits the government to search computers, Internet sites, and other social media for anything that it considers damaging to the country.

A number of countries require transnational corporations to maintain a local server in order to do business there. The government has the right to access the data that is stored on the local server.

Freedom on the Net has also determined that women and the LGBTI community have been targeted in a number of countries for their online activities. A woman was stoned to death in Pakistan in 2013 for possessing a cell phone. Gay men in Russia have been lured through social media to in-person meetings, where they have been assaulted by hate groups.

The three worst-offending countries, according to Freedom on the Net, are Iran, Syria, and China. In Iran, especially harsh punishments are imposed on people who promote causes opposed by the government. In Syria, government supporters have launched malware cyberattacks against the computers of thousands of antigovernment protestors. In China, individuals posting antigovernment messages on blogs and websites are arrested.

Russia and Turkey are considered the two countries where Internet freedom has suffered the most severe declines in recent years. In Russia, several laws have been enacted since 2012 to block online content critical of the government, and these laws have been used to arrest antigovernment leaders. In Turkey, government censorship of Internet content has increased, including shutting down YouTube and Twitter, and government protestors and journalists have been arrested for posting critical content online.

PAUSE & REFLECT 4.3.3

If you lived in a "not free" country, what use of electronic media might get you into trouble? Why?

CHECK-IN KEY ISSUE 3

Why Is Access to Folk and Popular Culture Unequal?

- ✔ **Popular culture diffuses primarily through electronic media.**
- ✔ **Many countries limit the ability of their citizens to access electronic media.**

KEY ISSUE 4

Why Do Folk and Popular Culture Face Sustainability Challenges?

- Sustainability Challenges for Folk Culture
- Sustainability Challenges for Popular Culture

LEARNING OUTCOME 4.4.1

Summarize challenges for folk culture from diffusion of popular culture.

Elements of folk and popular culture face challenges in maintaining identities that are sustainable into the future. For folk culture, the challenges are to maintain unique local landscapes in an age of globalization. For popular culture, the challenges derive from the sustainability of practices designed to promote uniform landscapes.

Many fear the loss of folk culture, if rising incomes fuel demand for the possessions typical of popular culture. When people turn from folk to popular culture, they may also turn away from the society's traditional values. And the diffusion of popular culture from developed countries can lead to dominance of Western perspectives.

Sustainability Challenges for Folk Culture

For folk culture, increased connection with popular culture can make it difficult to maintain centuries-old practices. A folk culture group often undergoes a process of **assimilation**, which a process of giving up cultural traditions, such as food and clothing preferences, and adoption of the social customs of the dominant culture of the place.

Instead of assimilation, a folk culture group often undergoes **acculturation**, which a process of adjustment to the dominant culture, while retaining features of a folk culture, or syncretism, which was defined in Chapter 1 as the creation of a new cultural feature through combining elements of two groups. The Amish in the United States is an example of cultural group that is attempting to maintain cultural traditions in the midst of a dominant popular culture. Marriage customs in India are an example of a folk culture practice that is also undergoing acculturation, but many would like to see fuller assimilation into popular culture.

▲ FIGURE 4-41 **AMISH CULTURAL IDENTITY** An Amish buggy shares the road with motor vehicles in Lancaster County, Pennsylvania.

PRESERVING CULTURAL IDENTITY: THE AMISH

The Amish provide an example of a cultural group that has retained distinctive elements of folk culture despite living in a country dominated by popular culture. Shunning mechanical and electrical power, the Amish travel by horse and buggy and continue to use hand tools for farming (Figure 4-41). The Amish have distinctive clothing, farming, religious practices, and other customs. Amish people do not wish to pose for photos because the act of posing is seen as fostering "graven images."

The contemporary distribution of Amish folk culture across the U.S. landscape is explained through relocation diffusion. Several hundred Amish families migrated to North America in two waves. The first group, primarily from Bern and the Palatinate, settled in Pennsylvania in the early 1700s, enticed by William Penn's offer of low-priced land. Because of lower land prices, the second group, from Alsace, settled in Ohio, Illinois, and Iowa in the United States and Ontario, Canada, in the early 1800s. From these core areas, groups of Amish migrated to other locations where inexpensive land was available.

Today Amish communities are visible on the landscape in at least 19 U.S. states (Figure 4-42). Living in rural and frontier settlements relatively isolated from other groups, Amish communities have retained their traditional customs, even as other European immigrants to the United States have adopted new ones. Amish folk culture continues to diffuse slowly through interregional migration within the United States.

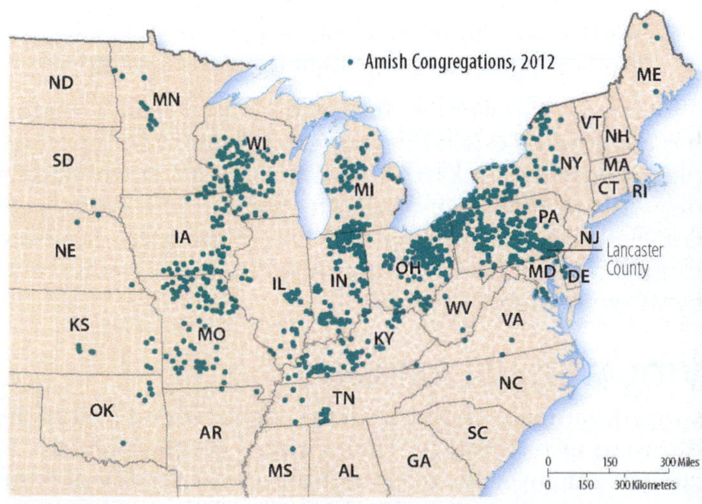

▲ FIGURE 4-42 **DISTRIBUTION OF AMISH** Amish settlements are clustered in Indiana, Ohio, and Pennsylvania.

In recent years, a number of Amish families have sold their farms in Lancaster County, Pennsylvania—the oldest and at one time largest Amish community in the United States—and migrated to southwestern Kentucky.

According to Amish tradition, every son is given a farm when he is an adult, but land suitable for farming is expensive and hard to find in Lancaster County because of its proximity to growing metropolitan areas. With the average price of farmland in southwestern Kentucky less than one-fifth that in Lancaster County, an Amish family can sell its farm in Pennsylvania and acquire enough land in Kentucky to provide adequate farmland for all the sons. Amish families are also migrating from Lancaster County to escape the influx of tourists who come from the nearby metropolitan areas to gawk at the distinctive folk culture.

Amish people have learned ways to retain a distinctive cultural identity while living in a country dominated by popular culture. An Amish person will drive in a car for an important purpose, such as visiting a distant doctor or family member. A telephone is not used for social chats but is available to summon a doctor or convey important information to distant relatives. The Amish do not use social media such as the Internet for business or personal communication but will permit friends and neighbors who are not Amish to use social media on their behalf.

PAUSE & REFLECT 4.4.1

In what ways might Amish people need to interact with popular culture?

CHALLENGING CULTURAL VALUES: DOWRIES IN INDIA

Rapid changes in long-established cultural values can lead to instability and even violence in a society. This threatens not just the institutions of folk culture but the sustainability of the society as a whole.

The global diffusion of popular culture has challenged the subservience of women to men that is embedded in some folk customs. Women may have been traditionally relegated to performing household chores, such as cooking and cleaning, and to bearing and raising large numbers of children. Those women who worked outside the home were likely to be obtaining food for the family, either through agricultural work or by trading handicrafts.

Contact with popular culture has also had negative impacts for women in developing countries. Prostitution has increased in some developing countries to serve men from developed countries traveling on "sex tours." These tours, primarily from Japan and Northern Europe, include airfare, hotels, and the use of a predetermined number of women. Leading destinations include the Philippines, Thailand, and South Korea. International prostitution is encouraged in these countries as a major source of foreign currency. Through this form of global interaction, popular culture may regard women as essentially equal at home but as objects that money can buy in foreign folk societies.

▲ FIGURE 4-43 **ATTACK ON WOMAN IN INDIA** This woman's father-in-law was responsible for burning her and for throwing her and her son out of the family's house.

Global diffusion of popular social customs has had an unintended negative impact for women in India: an increase in demand for dowries. Traditionally, a dowry was a small gift from the groom to the bride's family, as a sign of respect. In the twentieth century, the custom reversed, and the family of a bride was expected to provide a substantial dowry to the husband's family.

The government of India enacted anti-dowry laws in 1961, but the ban is widely ignored. Dowries have become much larger in modern India and an important source of income for the groom's family. A dowry can take the form of cash or expensive consumer goods, such as cars and electronics. The alternative to paying a dowry—having an unwed daughter living with her parents—may be considered shameful.

India's government has tried to ban dowries because of the adverse impact on women. If the bride's family is unable to pay a promised dowry or installments, the groom's family may cast the bride out on the street, and her family may refuse to take her back. Husbands and in-laws angry over the small size of dowry payments have killed an average of 8,000 women per year in India since 2000. Disputes over dowries have led to 100,000 cases per year of torture and cruelty toward women by men, few of whom are prosecuted, let alone convicted, for their attacks (Figure 4-43).

To raise awareness of dowry abuses, shaadi.com, an Indian matrimonial website with several million members, created an online game called Angry Brides. Each groom has a high price tag. Every time the player hits the groom, money is added to the player's Anti-Dowry Fund on her Facebook page (Figure 4-44).

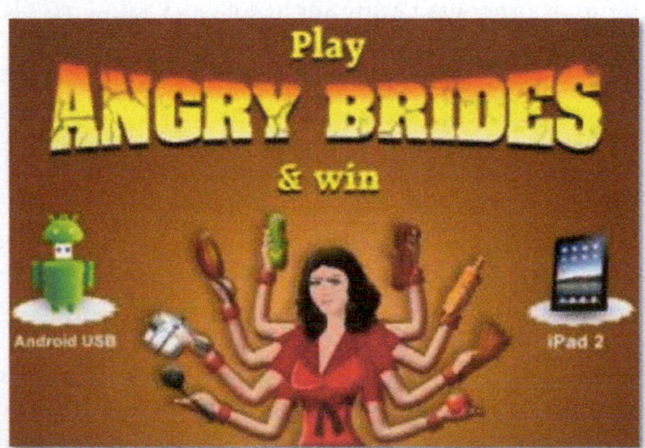

▲ FIGURE 4-44 **PROTESTING DOWRY: ANGRY BRIDES COMPUTER GAME**

Sustainability Challenges for Popular Culture

LEARNING OUTCOME 4.4.2

Summarize two principal ways that popular culture can adversely affect the environment.

Popular culture can significantly modify or control the environment, with little regard for local environmental conditions, such as climate and soil. It may be imposed on the environment rather than spring forth from it, as with many folk customs. The diffusion of some popular customs can adversely impact environmental quality in two ways: pollution of the landscape and depletion of scarce natural resources.

LANDSCAPE POLLUTION

For many popular customs, the environment is modified to enhance participation in a leisure activity or to promote the sale of a product. The desired result is often the creation of a uniform landscape. Even if the resulting built environment looks "natural," it is actually the deliberate creation of people in pursuit of popular social customs.

UNIFORM LANDSCAPES. The spatial expression of a popular custom in one location will be similar to another. To create a uniform landscape, hills may be flattened and valleys filled in. The same building and landscaping materials may be employed regardless of location.

Promoters of popular culture actually want a uniform appearance to generate "product recognition" and greater consumption (Figure 4-45). Uniformity in the appearance of the landscape is promoted by a wide variety of other popular structures in North America, such as gas stations, supermarkets, and motels. These structures are designed so that both local residents and visitors immediately recognize the purpose of a building, even if not the name of the company.

The diffusion of fast-food restaurants is a good example of the uniform landscape resulting from popular culture. Such restaurants are usually organized as franchises. A franchise is a company's agreement with businesspeople in a local area to market that company's product. The franchise agreement lets the local outlet use the company's name, symbols, trademarks, methods, and architectural styles. To both local residents and travelers, the buildings are immediately recognizable as part of a national or multinational company. A uniform sign is prominently displayed.

Much of the attraction of fast-food restaurants comes from the convenience of the product and the use of the building as a low-cost socializing location for teenagers or families with young children. At the same time, the success of fast-food restaurants depends on large-scale mobility: People who travel or move to another city immediately recognize a familiar place. Newcomers to a particular place know what to expect in the restaurant because the establishment does not reflect strange and unfamiliar local customs that could be uncomfortable.

Fast-food restaurants were originally developed to attract people who arrived by car. The buildings generally were brightly colored, even gaudy, to attract motorists. Recently built fast-food restaurants are more subdued, with brick facades, pseudo-antique fixtures, and other stylistic details. To facilitate reuse of the structure in case the restaurant fails, company signs are often freestanding rather than integrated into the building design.

Physical expression of uniformity in popular culture has diffused from North America to other parts of the world. American motels and fast-food chains have opened in other countries. These establishments appeal to North American travelers, yet most customers are local residents who wish to sample American customs they have seen on television.

▼ **FIGURE 4-45 UNIFORM LANDCAPE** U.S. Route 1, Miami, Florida.

PAUSE & REFLECT 4.4.2

What steps might fast-food restaurants take to reduce adverse impacts on the environment?

DEPLETION OF NATURAL RESOURCES

Increased demand for the products of popular culture can strain the capacity of the environment. Creation of golf courses and consumption of animal products are two examples.

Golf courses consume large quantities of land (80 hectares [200 acres]). In Scotland, where golf originated as part of folk culture, courses are designed to modify the environment as little as possible (see Sustainability & Our Environment feature). Elsewhere in the world, courses may be designed partially in response to local physical conditions. Grass species may be selected to thrive in the local climate and still be suitable for the needs of greens, fairways, and roughs. Existing trees and native vegetation may be retained if possible.

However, more often golf courses are designed to remake the environment by creating or flattening hills, carting in or digging up sand for traps, and draining or expanding bodies of water to create hazards. The courses are planted with nonnative grass species, fertilizers and pesticides are applied to the grass to ensure an appearance considered suitable for the game, and a large quantity of water is needed to maintain the grass, even in the desert.

SUSTAINABILITY & OUR ENVIRONMENT Golf Courses

The modern game of golf originated as a folk custom in Scotland in the fifteenth century or earlier and diffused to other countries during the nineteenth century. In this respect, the history of golf is like that of soccer, described earlier in this chapter. Early Scottish golf courses were primarily laid out on sand dunes adjacent to bodies of water.

Largely because of golf's origin as a local folk custom, golf courses in Scotland do not modify the environment to the same extent as those constructed in more recent years in the United States and other countries, where hills, sand, and grass are imported, often with little regard for local environmental conditions (Figure 4-46). The severe drought in the U.S. West in recent years has brought into question the environmental sustainability of using scarce water supplies for golf courses. Around Las Vegas, Nevada, for example, golf courses account for 20 percent of water usage.

Modern golf also departs from its folk culture roots by being an economically unsustainable sport to play in most places because of high cost.

(a)

(b)

▲ **FIGURE 4-46 SCOTLAND AND U.S. GOLF COURSES** **(a)** Scotland's Royal Troon Golf Club was built into a seaside dune with little alteration of the landscape. **(b)** Bear's Best Course, located in the desert near Las Vegas, Nevada, uses much of the region's scarce water supplies to create grassy fairways and greens.

Diffusion of some popular customs increases demand for animal products, ranging from rare wildlife to common domesticated animals. Some animals are killed for their skins, which can be shaped into fashionable clothing and sold to people living thousands of kilometers from the animals' habitat. The skins of the mink, lynx, jaguar, kangaroo, and whale have been heavily consumed for various articles of clothing, to the point that the survival of these species is endangered. This makes unsustainable the ecological systems of which the animals are members. Folk culture may also encourage the use of animal skins, but the demand is usually smaller than for popular culture.

Increased meat consumption in popular culture has not caused extinction of cattle and poultry; we simply raise more. But animal consumption is an inefficient way for people to acquire calories: It is 90 percent less efficient than if people simply ate grain directly. To produce 1 kilogram (2.2 pounds) of beef sold in the supermarket, nearly 10 kilograms (22 pounds) of grain are consumed by the animal. For every kilogram of chicken, nearly 3 kilograms (6.6 pounds) of grain are consumed by the fowl. This grain could be fed to people directly, bypassing the inefficient meat-production step. With a large percentage of the world's population undernourished, some question the inefficient use of grain to feed animals for eventual human consumption.

CHECK-IN KEY ISSUE 4

Why Do Folk and Popular Culture Face Sustainability Challenges?

- ✔ **Folk culture faces loss of traditional values in the face of rapid diffusion of popular culture.**
- ✔ **Popular culture can cause two environmental concerns: pollution of the landscape and depletion of scarce resources.**

Summary & Review

KEY ISSUE 1

Where are folk and popular leisure activities distributed?

Culture can be divided into folk and popular culture. Folk culture is traditionally practiced primarily by small, homogeneous groups living in isolated rural areas. Popular culture is found in large, heterogeneous societies that share certain habits despite differences in other personal characteristics.Folk culture is transmitted relatively slowly and primarily through relocation diffusion.Popular culture typically diffuses rapidly through a process of hierarchical diffusion.

THINKING GEOGRAPHICALLY

1. In what ways does age affect the distribution of leisure activities in folk or popular culture?

▲ FIGURE 4-47 **FOLK LEISURE CULTURE AND AGE** Older man plays boules (also known aspétanque and bocce), Provence, France.

KEY ISSUE 2

Where are folk and popular material culture distributed?

Material elements of folk culture typically have unknown or multiple origins among groups living in relative isolation, and they diffuse slowly to other locations through the process of relocation diffusion. Popular clothing, food, and shelter vary more in time than in place.

THINKING GEOGRAPHICALLY

2. In what ways might gender affect the diffusion of material culture in folk or popular culture?

▲ FIGURE 4-48 **POPULAR MATERIAL CULTURE AND GENDER** What do you typically wear to class?

KEY ISSUE 3

Why is access to folk and popular culture unequal?

Popular culture diffuses rapidly around the world in the twenty-first century primarily through electronic media. TV is by far the world's most important electronic media format. The Internet and social media appear to be following similar patterns of diffusion. Access to electronic media is not equal around the world, and in many places governments are trying to prevent or limit access to what is available.

THINKING GEOGRAPHICALLY

3. Why do many governments consider it important to limit the freedom to use social media?

▲ FIGURE 4-49 **PROTESTING CHINA'S INTERNET CENSORSHIP** A woman wears a monitor to protest internet censorship in China, in front of an international computer expo in Hanover, Germany.

KEY ISSUE 4

Why do folk and popular culture face sustainability challenges?

Elements of folk and popular culture face challenges in maintaining identities that are sustainable into the future. For folk culture, the challenges are to maintain unique local landscapes in an age of globalization. For popular culture, the challenges derive from the sustainability of practices designed to promote uniform landscapes.

THINKING GEOGRAPHICALLY

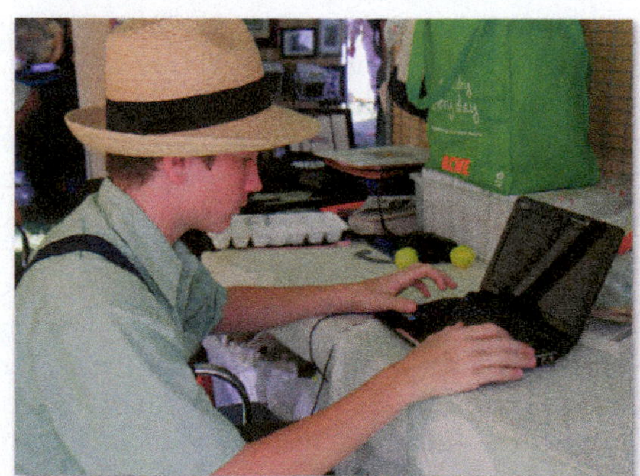

4. What types of folk customs might be able to be communicated through social media?

▲ FIGURE 4-50 **FOLK CULTURE MEETS POPULAR CULTURE** Amish boy uses a computer.

KEY TERMS

Acculturation *(p. 138)* The process of adjustment to the dominant culture.

Assimilation *(p. 138)* The process of giving up cultural traditions and adopting the social customs of the dominant culture of a place.

Custom *(p. 112)* The frequent repetition of an act, to the extent that it becomes characteristic of the group of people performing the act.

Folk culture *(p. 112)* Culture traditionally practiced by a small, homogeneous, rural group living in relative isolation from other groups.

Habit *(p. 112)* A repetitive act performed by a particular individual.

Popular culture *(p. 112)* Culture found in a large, heterogeneous society that shares certain habits despite differences in other personal characteristics.

Taboo *(p. 127)* A restriction on behavior imposed by social custom.

Terroir *(p. 126)* The contribution of a location's distinctive physical features to the way food tastes.

GeoVideo | Log in to the **MasteringGeography** Study Area to view this video.

Bhutan

A small kingdom in the Himalaya Mountains between India and China, Bhutan is known for its distinctive folk culture.

1. How is the fact that mountain climbing is forbidden in Bhutan a reflection of the country's folk culture?
2. Based on the video, how prevalent is global, popular culture in Bhutan? Explain.
3. List and discuss at least three reasons for the survival of folk culture in Bhutan.

EXPLORE

Use **Google Earth** to explore the place in London where the 2012 Summer Olympics were held.

1. Fly to *Olympic Stadium, London*.
2. Click Historical Imagery. Move the time slider to 9/1999. What sort of structures occupied the site of the stadium then?
3. Click *More*, then *Transportation*. Zoom out until you see the nearest subway station. What is the straight-line distance from the station to the stadium? What feature prevents walking in a straight line from the station to the stadium?
4. Click *View in Google Maps*. What is the function of the football-shaped building immediately to the east of the stadium?

MasteringGeography™

Looking for additional review and test prep materials?

Visit the Study Area in **MasteringGeography**™ to enhance your geographic literacy, spatial reasoning skills, and understanding of this chapter's content by accessing a variety of resources, including MapMaster interactive maps, videos, *In the News* RSS feeds, flashcards, web links, self-study quizzes, and an eText version of *The Cultural Landscape*.
www.masteringgeography.com

5

Languages

Language is an important part of culture. It is the means through which other cultural values, such as religion and ethnicity, are communicated. Language is a source of pride to a people, a symbol of cultural unity. As a culture develops, language is both a cause of that development and a consequence.

English has achieved an unprecedented globalization because people around the world are learning it to participate in a global economy and culture. At the same time, people are trying to preserve local diversity in language because language is one of the basic elements of cultural identity.

Learning English in Malaysia.

LOCATIONS IN THIS CHAPTER

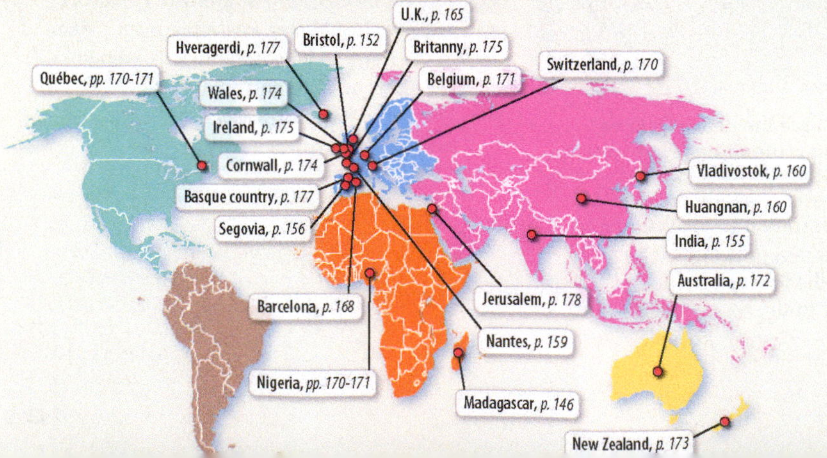

KEY ISSUES

1 Where Are the World's Languages Distributed?

Most languages can be classified as belonging to a family. Individual languages and language families cluster in distinctive *regions*.

2

Where Did English and Related Languages Originate and Diffuse?

People in two locations speak the same language because of migration from one of the locations to another. If the two groups have few ***connections*** with each other after the migration, the languages spoken by the two groups will begin to differ.

3

Why Do Individual Languages Vary Among Places?

Distinctive languages form as people migrate to new ***places*** and incorporate new words into their language while holding on to some words brought from their previous place of residence.

4

Why Do Local Languages Survive?

Languages display contradictory trends of ***scale***. On the one hand, individual languages remain clustered in ***space*** as an expression of cultural identity. On the other hand, languages such as English have achieved unprecedented globalization because people around the world are learning them to participate in a global economy and culture.

KEY ISSUE 1

Where Are the World's Languages Distributed?

- Introducing Languages
- Language Families
- Two Largest Language Families
- Other Large Language Families

LEARNING OUTCOME 5.1.1

Understand how languages are classified.

Language is a system of communication through speech, a collection of sounds that a group of people understands to have the same meaning. Language is an important part of culture, which, as shown in Chapter 1, has two main meanings—people's values and their tangible artifacts. Chapter 4 looked at the material objects of culture. This chapter and the next two discuss three traits that distinguish cultural values: language, religion, and ethnicity.

We start our study of the geographic elements of cultural values with language in part because it is the means through which other cultural values, such as religion and ethnicity, are communicated.

Introducing Languages

How many languages do you speak? If you are Dutch, you were required to learn at least two foreign languages in high school. For those of you who do not happen to be Dutch, the number is probably a bit lower.

Most people in the United States know only English. In the United States, only 8 percent of college students and 18 percent of high school students take a foreign language. In contrast, 69 percent of graduates from Dutch high schools have learned at least two foreign languages. Across Europe as a whole, 75 percent of elementary school students and 94 percent of high school students learn English.

LANGUAGE AND MIGRATION

The study of language follows logically from migration because the contemporary distribution of languages around the world is largely a result of past migrations of peoples. People in two locations speak the same language because of migration from one of the locations to another.

For example, the people of Madagascar (the large island off the east coast of Africa) speak a language belonging to the same family as the languages of most of Indonesia and the Philippines (Figure 5-1). The shared language family between Indonesia and Madagascar is strong evidence of migration a long time ago between these two places. Researchers have concluded that migrants sailed the 3,000 kilometers across the Indian Ocean from Indonesia to Madagascar approximately 2,000 years ago. Imagine sailing across 3,000 kilometers of ocean in tiny boats 1,500 years before Columbus sailed 6,000 kilometers across the Atlantic Ocean.

If the two groups have few connections with each other after the migration, the languages spoken by the two groups will begin to differ. After a long period without contact, the two groups will speak languages that are so different that they are classified as separate languages. The interplay between interaction and isolation helps explain the distribution of individual languages and entire language families.

PAUSE & REFLECT 5.1.1

What forms of power would have moved boats 2,000 years ago?

Language is like luggage: People carry it with them when they move from place to place. They incorporate new words into their own language when they reach new places, and they contribute words brought with them to the existing

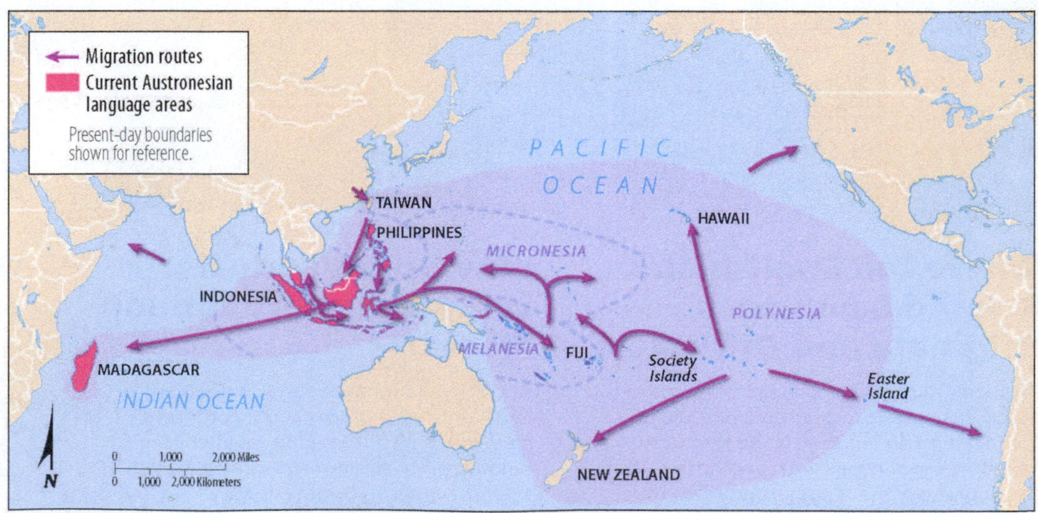

▶ **FIGURE 5-1 AUSTRONESIAN LANGUAGES** The current distribution of Austronesian languages is a function of migration of Austronesian people in the past.

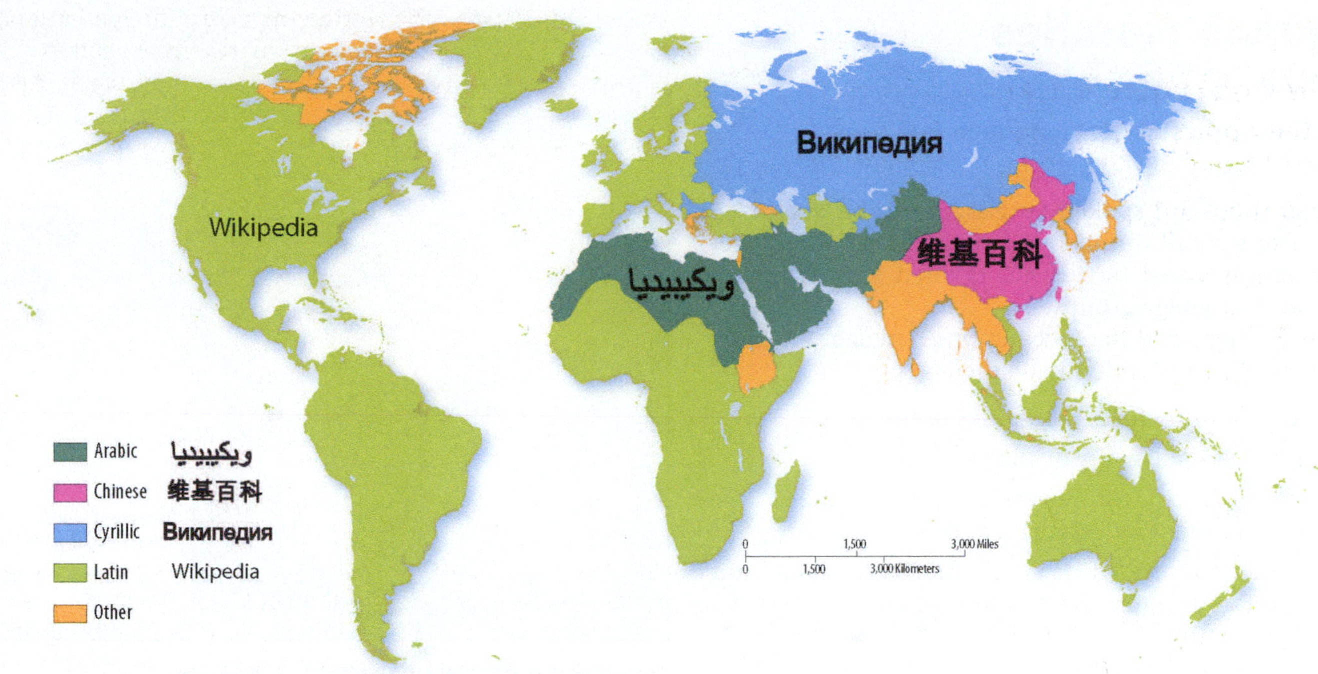

▲ **FIGURE 5-2 WRITING SYSTEMS** The map shows the word "Wikipedia" in several writing systems.

language at the new location. Geographers look at the similarities among languages to understand the diffusion and interaction of people around the world.

CLASSIFYING LANGUAGES

Earth's heterogeneous collection of languages is one of its most obvious examples of cultural diversity. *Ethnologue*, one of the most authoritative sources of languages (online at ethnologue.com), estimates that the world has 7,102 languages, including 90 spoken by at least 10 million people, 304 spoken by between 1 and 10 million people, and 6,708 spoken by fewer than 1 million people. The distribution of some of these languages is easy for geographers to document, whereas with others—especially in Africa and Asia—it is difficult, if not impossible.

Ethnologue classifies languages as institutional, developing, vigorous, in trouble, and dying. Of the world's 7,102 languages, 578 are institutional, 1,598 developing, 2,479 vigorous, 1,531 in trouble, and 916 dying.

An **institutional language** is used in education, work, mass media, and government. Many countries designate at least one institutional language as an **official language**, which is used by the government for laws, reports, and public objects, such as road signs, money, and stamps. Logically, an official language would be understood by most if not all of the country's citizens, but some countries that were once British colonies designate English as an official language, even though few of their citizens can speak it. Some countries have more than one official language and require all public documents to be in all of the official languages.

An institutional language also has a **literary tradition**, which means it is written as well as spoken. The system of written communication includes a method of writing and rules of grammar. The world's languages with literary traditions make use of more than one alphabet (Figure 5-2). Thousands of spoken languages lack a literary tradition. The lack of written records is one reason it is difficult to document the total number of languages or their distribution.

A **developing language** is spoken in daily use by people of all ages, from children to elderly individuals. A developing language also has a literary tradition, though literature in the language may not be widely distributed. A **vigorous language** is spoken in daily use by people of all ages, but it lacks a literary tradition. Languages in trouble and dying, as the names imply, are considered by *Ethnologue* to be in various stages of disappearing from use. Some of these endangered languages are being saved, as discussed later in this chapter.

ORGANIZING LANGUAGE FAMILIES

The world's languages can be organized into families, branches, and groups:

- A **language family** is a collection of languages related through a common ancestral language that existed long before recorded history.
- A **language branch** is a collection of languages within a family related through a common ancestral language that existed several thousand years ago; differences are not as extensive or as old as between language families, and archaeological evidence can confirm that the branches derived from the same family.
- A **language group** is a collection of languages within a branch that share a common origin in the relatively recent past and display many similarities in grammar and vocabulary.

Language Families

LEARNING OUTCOME 5.1.2

Identify the world's largest language families.

The several thousand spoken languages can be organized logically into a small number of language families. Larger language families can be further divided into language branches and language groups.

Figure 5-3 depicts differences among language families, branches, groups, and individual languages:

- Language families form the trunks of the trees.
- Individual languages are displayed as leaves.
- Some trunks divide into several branches, which logically represent language branches.
- The branches representing Germanic, Balto-Slavic, and Indo-Iranian in Figure 5-3 divide a second time into language groups.

The larger the trunks and leaves are, the greater the number of speakers of those families and languages.

PAUSE & REFLECT 5.1.2

Based on Figure 5-3, which four other languages with at least 5 million speakers belong to the same language family, branch, and group as English?

Figure 5-3 displays each language family as a separate tree at ground level because differences among families predate recorded history. Some linguists speculate that language families were joined together as a handful of superfamilies tens of thousands of years ago. Superfamilies are shown as roots below the surface because their existence is highly controversial and speculative.

A biologist in New Zealand, Quentin Atkinson, carries the speculation further, arguing that all languages can be ultimately traced to Africa (Figure 5-4). According to Atkinson, languages are most complex and diverse in Africa.

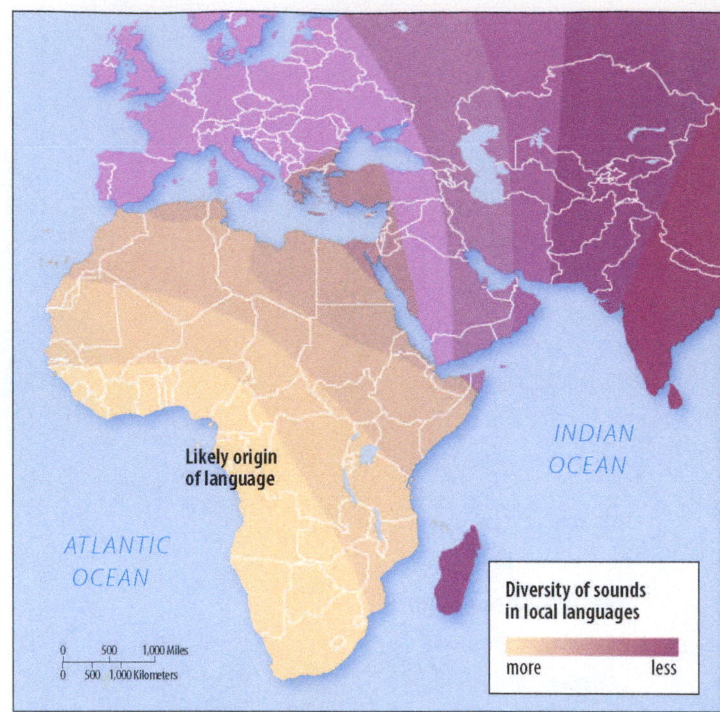

▲ **FIGURE 5-4 ORIGIN OF LANGUAGE** Biologist Quentin Atkinson argues that languages originated in western Africa. Further from western Africa, languages are less diverse with fewer phonetic sounds.

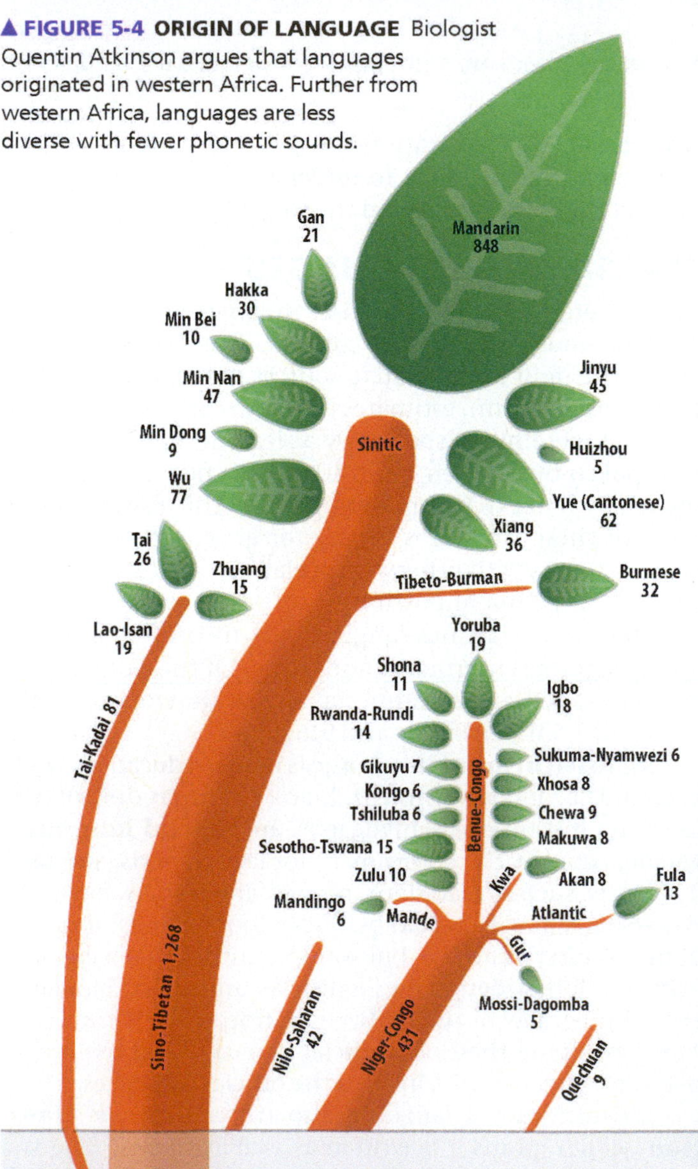

► **FIGURE 5-3 LANGUAGE FAMILY TREES** Language families with at least 9 million speakers, according to *Ethnologue*, are shown as trunks of trees. Individual languages that have more than 5 million speakers are shown as leaves. Some trunks divide into several branches, which logically represent language branches. The branches representing Germanic, Balto-Slavic, and Indo-Iranian divide a second time into language groups. Some linguists speculate that language families were joined together as a handful of superfamilies tens of thousands of years ago. Superfamilies are shown as roots below the surface because their existence is highly controversial and speculative.

Indo- European 46.5%
Sino-Tibetan 20.1%
Niger-Congo 6.9%
Afro-Asiatic 6.0%
Austronesian 5.5%
Dravidian 3.6%
Altaic 2.6%
Japanese 2.0%
Austro-Asiatic 1.6%
Others 5.2%

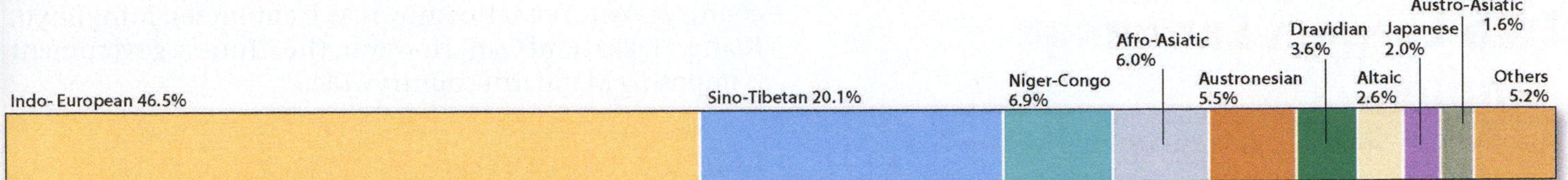

▲ **FIGURE 5-5 WORLD'S MAJOR LANGUAGE FAMILIES** The graph shows the percentage of people who speak a language from each major family.

Atkinson thinks humans outside Africa display less linguistic diversity because their languages have had a shorter time in which to evolve into new languages than have African languages.

Ethnologue identifies 142 language families. The 14 families depicted in Figure 5-3 are used by 99 percent of the world's population. Two language families—Indo-European and Sino-Tibetan—are used by more than 1 billion people, seven language families by between 100 million and 500 million people, and 5 by between 9 million and 100 million (Figure 5-5).

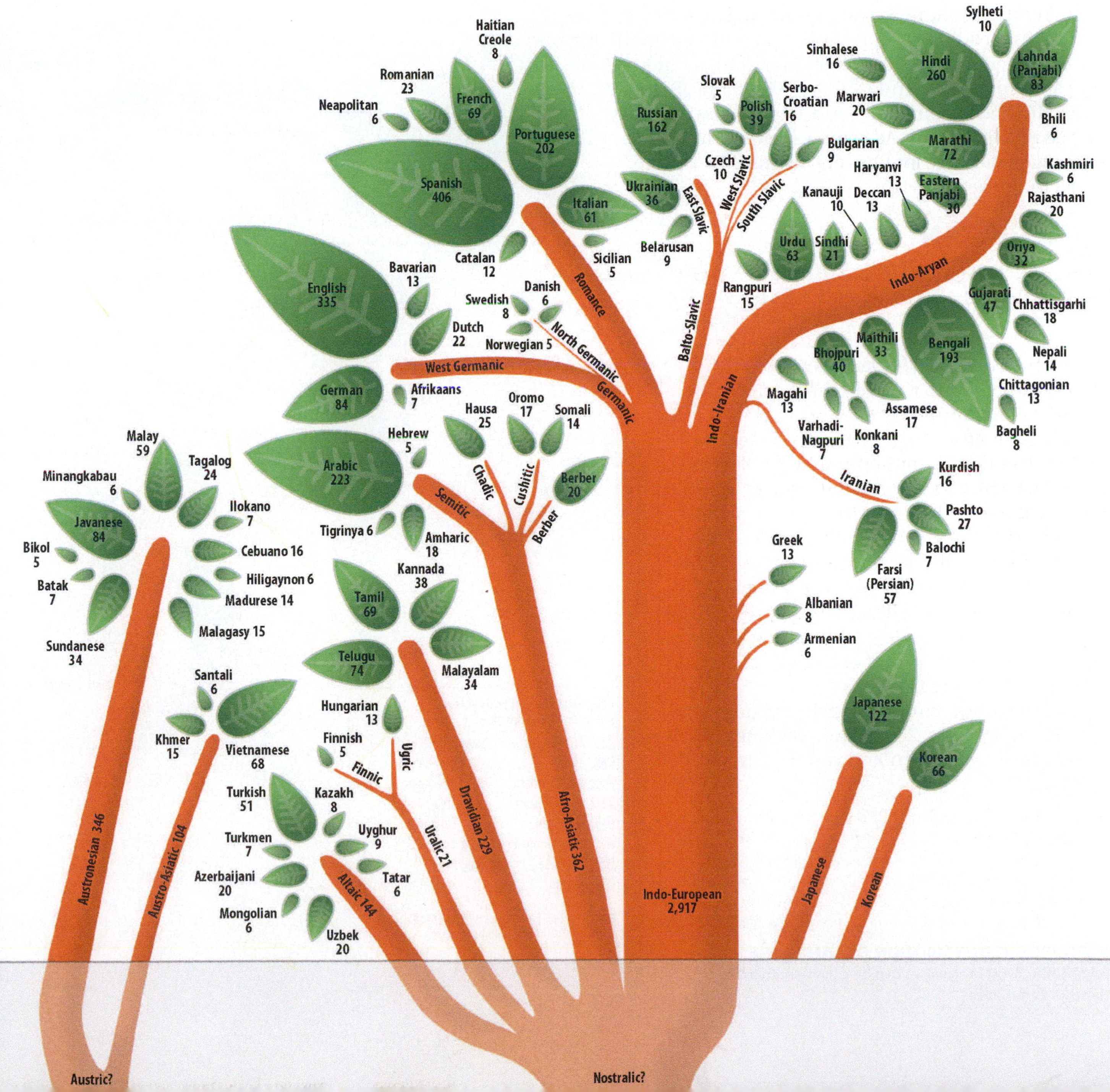
Haitian Creole 8
Romanian 23
Neapolitan 6
French 69
Portuguese 202
Spanish 406
Italian 61
Sicilian 5
Catalan 12
Romance
Russian 162
Ukrainian 36
Belarusan 9
East Slavic
Slovak 5
Polish 39
Czech 10
West Slavic
Serbo-Croatian 16
Bulgarian 9
South Slavic
Balto-Slavic
Rangpuri 15
Sinhalese 16
Marwari 20
Hindi 260
Sylheti 10
Lahnda (Panjabi) 83
Bhili 6
Marathi 72
Kashmiri 6
Haryanvi 13
Eastern Panjabi 30
Rajasthani 20
Kanauji 10
Deccan 13
Urdu 63
Sindhi 21
Oriya 32
Indo-Aryan
Gujarati 47
Chhattisgarhi 18
Maithili 33
Bhojpuri 40
Bengali 193
Nepali 14
Chittagonian 13
Magahi 13
Assamese 17
Bagheli 8
Varhadi-Nagpuri 7
Konkani 8
Indo-Iranian
Iranian
Kurdish 16
Pashto 27
Balochi 7
Farsi (Persian) 57
Greek 13
Albanian 8
Armenian 6
English 335
Bavarian 13
Dutch 22
Swedish 8
Danish 6
Norwegian 5
North Germanic
West Germanic
Germanic
German 84
Afrikaans 7
Hausa 25
Oromo 17
Somali 14
Chadic
Cushitic
Berber 20
Berber
Hebrew 5
Arabic 223
Semitic
Tigrinya 6
Amharic 18
Malay 59
Tagalog 24
Minangkabau 6
Ilokano 7
Javanese 84
Cebuano 16
Bikol 5
Hiligaynon 6
Batak 7
Madurese 14
Malagasy 15
Sundanese 34
Kannada 38
Tamil 69
Telugu 74
Malayalam 34
Santali 6
Khmer 15
Vietnamese 68
Hungarian 13
Finnish 5
Ugric
Finnic
Uralic 21
Turkish 51
Kazakh 8
Turkmen 7
Uyghur 9
Azerbaijani 20
Tatar 6
Mongolian 6
Uzbek 20
Altaic 144
Dravidian 229
Afro-Asiatic 362
Indo-European 2,917
Austronesian 346
Austro-Asiatic 104
Japanese 122
Korean 66
Japanese
Korean
Austric?
Nostralic?

Two Largest Language Families

LEARNING OUTCOME 5.1.3

Identify the distribution of Indo-European and Sino-Tibetan, the two largest families.

Language families with at least 9 million native speakers are shown in Figure 5-6. Individual languages with at least 50 million speakers are named on the map. The two language families with the most speakers are Indo-European and Sino-Tibetan.

INDO-EUROPEAN

Indo-European, the most widely used language family, is the predominant one in Europe, South Asia, and North and Latin America (Figure 5-7). It has eight branches, including four that are widely used (Indo-Iranian, Germanic, Romance, and Balto-Slavic) and four that are spoken by relatively few people (Albanian, Armenian, Celtic, and Greek). Its origin and distribution are discussed in more detail in the next key issue.

SINO-TIBETAN

The Sino-Tibetan family encompasses languages spoken in the People's Republic of China—the world's most populous state, at 1.3 billion—and in several smaller countries in Southeast Asia. The languages of China generally belong to the Sinitic branch of the Sino-Tibetan family.

There is no single Chinese language. Rather, the most commonly used is Mandarin, which the Chinese call *Putonghua* ("common speech"). Spoken by approximately three-fourths of the Chinese people, Mandarin is by a wide margin the most-used language in the world. Once the language of emperors in Beijing, Mandarin is now the official language of both the People's Republic of China and Taiwan, and it is one of the six official languages of the United Nations.

Seven other Sinitic branch languages are spoken by at least 20 million each in China, mostly in the southern and eastern parts of the country—Wu, Yue (also known as Cantonese), Min, Jinyu, Xiang, Hakka, and Gan. However, the Chinese government is imposing Mandarin countrywide.

PAUSE & REFLECT 5.1.3

Sino-Tibetan is the second-most widely used language family, but it appears on Figure 5-6 to encompass a smaller land area than several other language families. Why is this the case?

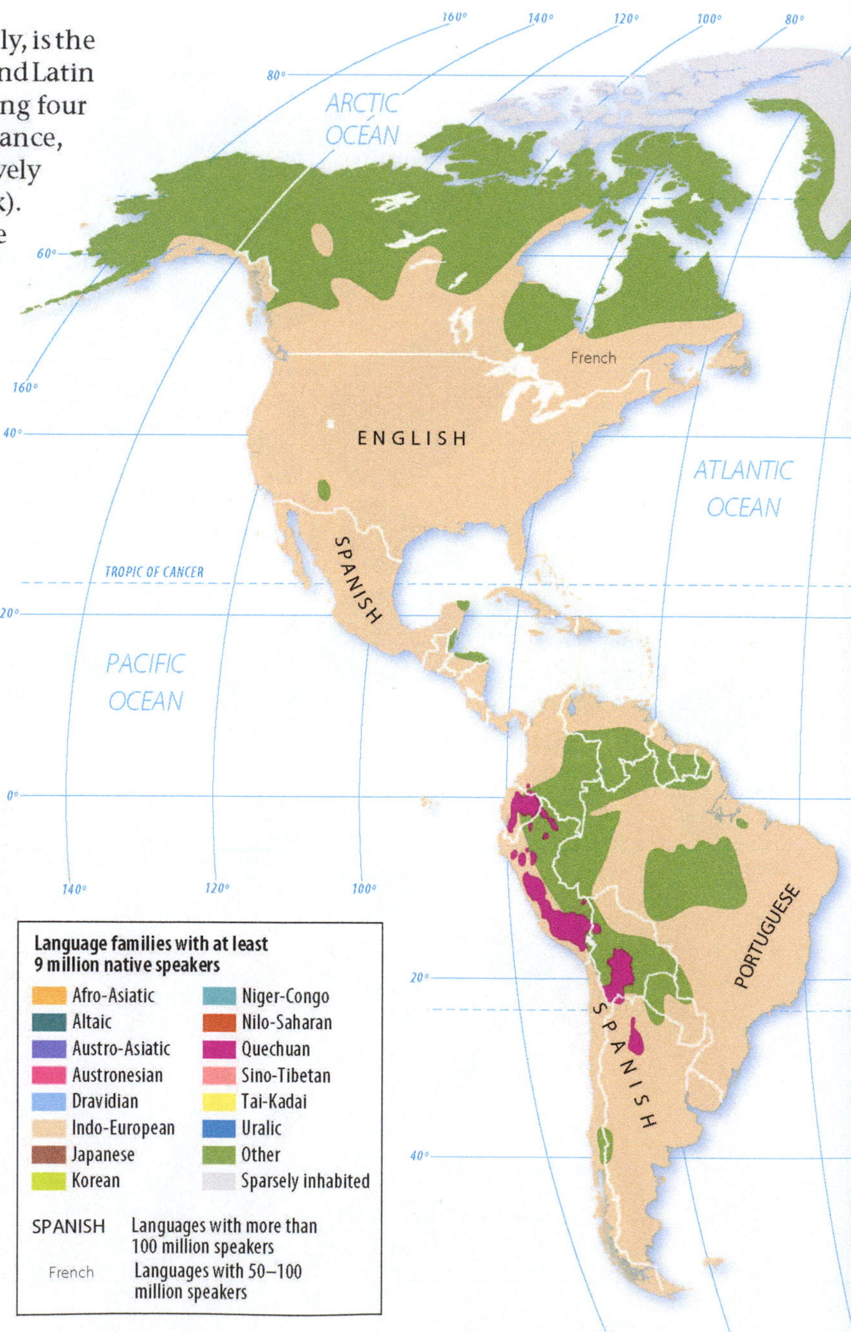

▶ **FIGURE 5-6 DISTRIBUTION OF LANGUAGE FAMILIES** Most languages can be classified into one of a handful of families.

▶ **FIGURE 5-7 BRANCHES OF THE INDO-EUROPEAN LANGUAGE FAMILY** Most Europeans speak languages from the Indo-European language family.

Other Large Language Families

LEARNING OUTCOME 5.1.4

Identify the distribution of the largest language families other than Indo-European and Sino-Tibetan.

Figures 5-3 and 5-6 show 14 language families. The 2 largest—Indo-European and Sino-Tibetan—were introduced on the previous page. This page summarizes the distribution of the other 12 (Figure 5-8).

SOUTHEAST ASIA LANGUAGE FAMILIES

The three largest language families of Southeast Asia are Austronesian, Austro-Asiatic, and Tai-Kadai.

AUSTRONESIAN. These languages are spoken by about 6 percent of the world's people, who are mostly in Indonesia, the world's fourth-most-populous country. With its inhabitants dispersed among thousands of islands, Indonesia has many distinct languages and dialects; *Ethnologue* identifies 706 living languages in Indonesia. Indonesia's most widely used first language is Javanese, spoken by 84 million people, mostly on the island of Java, where two-thirds of the country's population is clustered.

AUSTRO-ASIATIC. This family is spoken by about 2 percent of the world's population, Vietnamese, the most-spoken tongue of the family, is written with our familiar Roman alphabet, with the addition of a large number of diacritical marks above the vowels. The Vietnamese alphabet was devised in the seventeenth century by Roman Catholic missionaries.

◀ FIGURE 5-8 WELCOME IN 33 LANGUAGES The sign is attached to a school in Bristol, United Kingdom.

TAI-KADAI. The Tai-Kadai family was once classified as a branch of Sino-Tibetan. The principal languages of this family are spoken in Thailand and neighboring portions of China. Similarities with the Austronesian family have led some linguistic scholars to speculate that people speaking these languages may have migrated from the Philippines.

EAST ASIA LANGUAGE FAMILIES

The two most widely used language families outside of China are Japanese and Korean.

JAPANESE. Written in part with Chinese characters, Japanese also uses two systems of phonetic symbols, used either in place of Chinese characters or alongside them. Chinese cultural traits have diffused into Japanese society, including the original form of writing Japanese. But the structures of the two languages differ. Foreign terms may be written with one of these sets of phonetic symbols.

KOREAN. Unlike Sino-Tibetan languages and Japanese, Korean is written in a system known as *hankul* (also called *hangul* or *onmun*). In this system, each letter represents a sound, as in Western languages. More than half of the Korean vocabulary derives from Chinese words. In fact, Chinese and Japanese words are the principal sources for creating new words to describe new technology and concepts.

OTHER ASIAN LANGUAGE FAMILIES

Dravidian is the principal language family of South Asia, in addition to Indo-European. The Altaic and Uralic language families were once thought to be linked as one family, but recent studies point to geographically distinct origins.

DRAVIDIAN. Dravidian languages are the principal ones in southern India. The two most widely used are Telugu and Tamil. The origin of Dravidian is unknown, and it has been studied less than other widely used language families. When speakers of Indo-European reached India, speakers of Dravidian languages were already present.

ALTAIC. Altaic languages are thought to have originated in the steppes bordering the Qilian Shan and Altai mountains between Tibet and China. Present distribution covers an 8,000-kilometer band of Asia. The Altaic language that has by far the most speakers is Turkish. When the Soviet Union governed most of the Altaic-speaking region of Central Asia, use of Altaic languages was suppressed. With the dissolution of the Soviet Union in the early 1990s, Altaic languages became official in several new countries, including Azerbaijan, Kazakhstan, Kyrgyzstan, Turkmenistan, and Uzbekistan.

URALIC. Uralic languages are traceable back to a common language, first used 7,000 years ago by people living in the Ural Mountains of present-day Russia. Migrants carried the Uralic languages to Europe, carving out homelands in the midst of Germanic and Slavic-speaking peoples and retaining their language as a major element of cultural identity.

▲ **FIGURE 5-9 AFRICA'S LANGUAGE FAMILIES** The great number of languages results from at least 5,000 years of minimal interaction among the thousands of cultural groups inhabiting the African continent.

Estonians, Finns, and Hungarians speak languages that belong to the Uralic family.

AFRICAN LANGUAGE FAMILIES

No one knows the precise number of languages in Africa, and scholars disagree on classifying them into families. In the 1800s, European missionaries and colonial officers recorded African languages using the Roman or Arabic alphabet. *Ethnologue* lists 2,146 languages in Africa; only 699 have a literary tradition. The world's third- and fourth-largest language families are based in Africa: Afro-Asiatic in North Africa and Niger-Congo in sub-Saharan Africa.

AFRO-ASIATIC. Arabic is the major language of the Afro-Asiatic family, an official language in two dozen countries of Southwest Asia & North Africa, and one of six official languages of the United Nations. According to *Ethnologue*, 206 million people speak and write the official language Arabic. Most also speak a second language that is distinct from official Arabic. For example, 54 million people use Egyptian Spoken Arabic. *Ethnologue* identifies 34 distinct Arabic languages in addition to the official one. A large percentage of the world's 1 billion Muslims have at least some knowledge of Arabic because Islam's holiest book, the Quran (Koran), was written in that language in the seventh century. The Afro-Asiatic family also includes Hebrew, the original language of Judaism's Bible and Christianity's Old Testament.

NIGER–CONGO. More than 95 percent of the people in sub-Saharan Africa speak languages of the Niger-Congo family (Figure 5-9). The three most widely spoken Niger-Congo languages are Yoruba, Igbo, and Swahili. Yoruba and Igbo are among the many languages of Nigeria (look ahead to Figure 5-35). Swahili is an official language only in Tanzania, but it is the first language of 15 million people and is spoken as a second language by another 25 million Africans. Especially in rural areas, the local language is used to communicate with others from the same village, and Swahili is used to communicate with outsiders. Swahili originally developed through interaction among African groups and Arab traders, so its vocabulary has strong Arabic influences. It is one of the few African languages with an extensive literature.

NILO-SAHARAN. Languages of the Nilo-Saharan family are spoken by 43 million people in north-central Africa, immediately north of the Niger-Congo language region. Divisions within the Nilo-Saharan family exemplify the problem of classifying African languages. Despite having relatively few speakers, the Nilo-Saharan family is divided into six branches, plus numerous groups and subgroups. The total number of speakers of each individual Nilo-Saharan language is extremely small.

AMERICA'S OTHER LANGUAGE FAMILY: QUECHUAN

Quechuan is the most widely used language family in the Western Hemisphere other than Indo-European. Its speakers live primarily in the Andes Mountains of western South America. *Ethnologue* estimates that around 9 million people use a Quechuan language. *Ethnologue* identifies 44 distinct Quechuan languages. Quechua Cusco is the only one with more than 1 million speakers. According to *Ethnologue*, most speakers of a Quechuan language use Spanish first. Aymara, another indigenous language family in the Andes, has 3 million speakers, mostly in Bolivia.

PAUSE & REFLECT 5.1.4

Most languages are named for regions or countries. For example, based on their names, how would you expect the distributions of Austronesian and Austro-Asiatic to differ?

CHECK-IN KEY ISSUE 1

Where Are the World's Languages Distributed?

- ✔ **Languages are classified as institutional, developing, vigorous, in trouble, and dying.**
- ✔ **Languages are organized into families and branches.**
- ✔ **Eighteen language families are used by at least 9 million people.**

KEY ISSUE 2

Where Did English and Related Languages Originate and Diffuse?

- Distribution of Indo-European Branches
- Origin and Diffusion of Indo-European
- Origin and Diffusion of English
- Global Importance of English
- Official Languages

LEARNING OUTCOME 5.2.1

Identify the origin, diffusion, and current distribution of Indo-European branches.

Nearly one-half of the world's people speak a language belonging to the Indo-European language family. Indo-European languages have a common ancestor that predates recorded history.

Distribution of Indo-European Branches

Indo-European is divided into eight branches. Four are widely spoken, and four much less so.

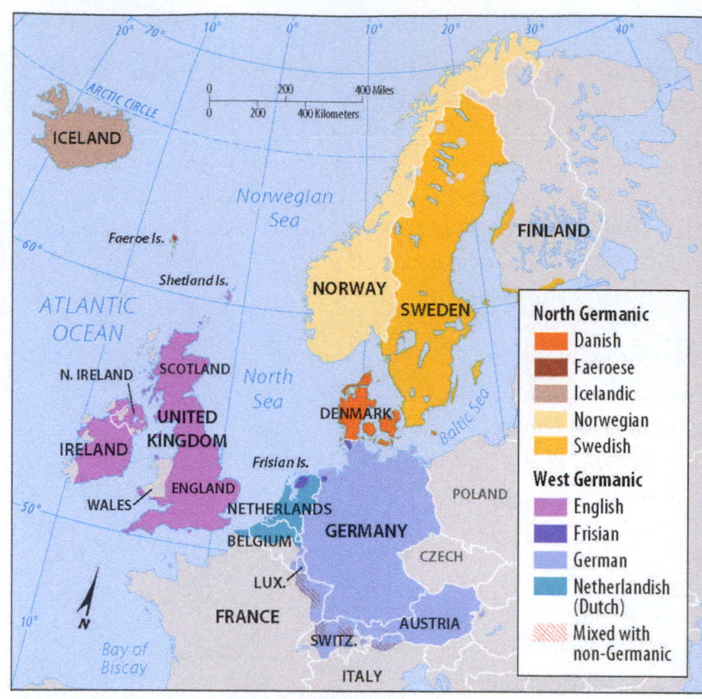

▲ **FIGURE 5-10 GERMANIC BRANCH OF INDO-EUROPEAN**

GERMANIC BRANCH

English belongs to the West Germanic group of the Germanic language branch of the Indo-European family (Figure 5-10). German and Dutch are also West Germanic group languages.

The Germanic language branch also includes languages in the North Germanic group, spoken in Scandinavia. The four Scandinavian languages—Swedish, Danish, Norwegian, and Icelandic—all derive from Old Norse, which was the principal language spoken throughout Scandinavia before A.D. 1000. Four distinct languages emerged after that time because of migration and the political organization of the region into four independent and isolated countries.

ROMANCE BRANCH

The four most widely used contemporary Romance languages are Spanish, Portuguese, French, and Italian (Figure 5-11). The

▼ **FIGURE 5-11 ROMANCE BRANCH OF INDO-EUROPEAN** Romance branch languages predominate in southwestern Europe.

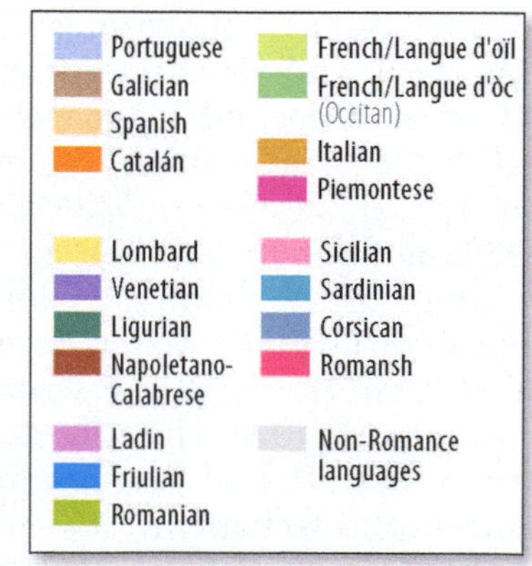

European regions in which these four languages are spoken correspond somewhat to the boundaries of the modern states of Spain, Portugal, France, and Italy. Rugged mountains serve as boundaries among these four countries. The fifth most widely used Romance language, Romanian, is the principal language of Romania and Moldova. It is separated from the other Romance-speaking European countries by Slavic-speaking peoples.

INDO-IRANIAN BRANCH

The branch of the Indo-European language family with the most speakers is Indo-Iranian. The branch is divided into the Iranian, or Western, group and the Indic, or Eastern, group. The major Iranian group languages include Persian (sometimes called Farsi) in Iran, Pashto in eastern Afghanistan and western Pakistan, and Kurdish, used by the Kurds of western Iran, northern Iraq, and eastern Turkey. These languages are written in the Arabic alphabet.

The most widely used languages in South Asia belong to the Indo-European language family and, more specifically, to the Indic group of the Indo-Iranian branch of Indo-European. One of the main elements of cultural diversity among the 1.2 billion residents of India is language (Figure 5-12). *Ethnologue* identifies 461 languages currently spoken in India, including 29 languages spoken by at least 1 million people.

The official language of India is Hindi, which is an Indo-European language. Originally a variety of Hindustani spoken in the area of New Delhi, Hindi grew into a national language in the nineteenth century, when the British encouraged its use in government.

After India became an independent state in 1947, Hindi was proposed as the official language, but speakers of other languages strongly objected. Consequently, English—the language of the British colonial rulers—has been retained as a secondary official language. Speakers of different Indian languages who wish to communicate with each other sometimes use English as a common language.

India also recognizes 22 so-called scheduled languages, including 15 Indo-European 4 Dravidian, 2 Sino-Tibetan, and 1 Austro-Asiatic. The government of India is obligated to encourage the use of these languages.

Hindi is spoken in many different ways, but there is only one official way to write Hindi, using a script called Devanagari. For example, the word for sun is written in Hindi as सूरज़ (pronounced "surya").

Adding to the complexity, Urdu is spoken very much like Hindi, but it is recognized as a distinct language. Urdu is written with the Arabic alphabet, a legacy of the fact that most of its speakers are Muslims, and their holiest book (the Quran) is written in Arabic.

BALTO-SLAVIC BRANCH

Balto-Slavic languages predominate in Eastern Europe. Slavic was once a single language, but differences developed in the seventh century A.D., when several groups of Slavs migrated from Asia to different areas of Eastern Europe

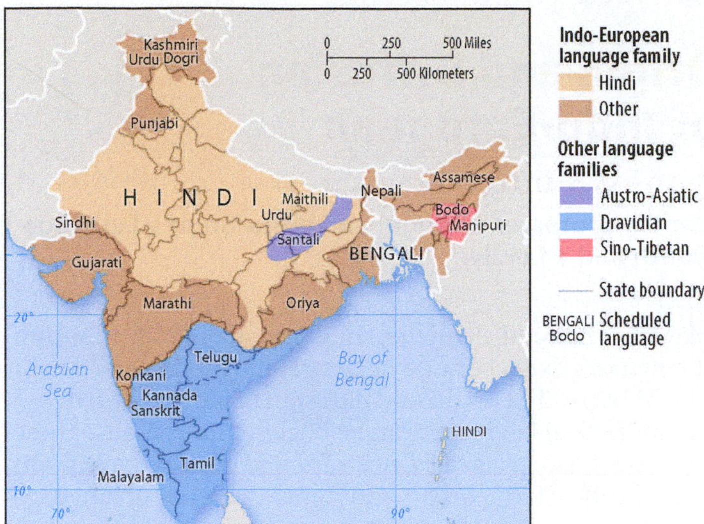

▲ **FIGURE 5-12 LANGUAGE FAMILIES IN INDIA** India's principal official language is Hindi, which has many dialects. The country has 22 scheduled languages that the government is required to protect.

and thereafter lived in isolation from one other. As a result, this branch can be divided into East, West, and South Slavic groups, as well as a Baltic group. Figure 5-7 shows the widespread area populated with Balto-Slavic speakers.

The most widely used Slavic languages are the eastern ones, primarily Russian, which is spoken by more than 80 percent of Russian people. The importance of Russian increased with the Soviet Union's rise to power. Soviet officials forced native speakers of other languages to learn Russian as a way of fostering cultural unity among the country's diverse peoples. In Eastern European countries that were dominated by the Soviet Union, Russian was taught as the second language.

The most spoken West Slavic language is Polish, followed by Czech and Slovak. The latter two are quite similar, and speakers of one can understand the other. The most widely used South Slavic language is the one spoken in Bosnia & Herzegovina, Croatia, Montenegro, and Serbia. When they were all part of Yugoslavia, the language was called Serbo-Croatian. This name now offends Bosnians and Croatians because it recalls when they were once in a country that was dominated by Serbs. Instead, the names Bosnian, Croatian, and Serbian are preferred by people in these countries, to demonstrate that each language is unique, even though linguists consider them one. Bosnians and Croats write the language in the Roman alphabet (what you are reading now), whereas Montenegrins and Serbs use the Cyrillic alphabet (for example, Serbia is written **Србија**).

Differences among all of the Slavic languages are relatively small. However, because language is a major element in a people's cultural identity, relatively small differences among Slavic as well as other languages are being preserved and even accentuated in recent movements.

PAUSE & REFLECT 5.2.1

Which branch predominates to the north in Europe, which to the south, and which to the east?

Origin and Diffusion of Indo-European

LEARNING OUTCOME 5.2.2

Identify processes of origin and diffusion of a language branch and a family.

Like other cultural elements, the contemporary distribution of languages exists because of geographic processes of origin and diffusion. The origin and diffusion of language branches and individual languages can be documented because these processes have occurred since recorded history began. On the other hand, the origin and initial diffusion of language families predate recorded history, so we can only speculate about them. The origin and diffusion of the Romance language branch and Indo-European family are examples.

ORIGIN AND DIFFUSION OF ROMANCE LANGUAGES

The Romance languages belong to a single branch because they originated from Latin, the "Romans' language." The rise in the importance of the city of Rome 2,000 years ago brought a diffusion of its Latin language. At its height in the second century A.D., the Roman Empire extended from the Atlantic Ocean on the west to the Black Sea on the east and encompassed all lands bordering the Mediterranean Sea (Figure 5-13). The empire's boundary is shown in Figure 8-8. As the conquering Roman armies occupied the provinces of this vast empire, they brought the Latin language with them. In the process, the languages spoken by the natives of the provinces were either extinguished or suppressed in favor of the language of the conquerors.

Even during the period of the Roman Empire, Latin varied to some extent from one province to another. The empire grew over a period of several hundred years, and the Latin used in each province was based on that spoken by the Roman army at the time of occupation. The Latin spoken in each province also integrated words from the language formerly spoken in the area. The Latin that people in the provinces learned was not the standard literary form but a spoken form, known as **Vulgar Latin**, from the Latin word referring to "the masses" of the populace.

▼ **FIGURE 5-13 ROMAN AQUEDUCT, SEGOVIA SPAIN** The Romans built this 17-kilometer (11-mile) aqueduct to bring water to one of their westernmost outposts.

Following the collapse of the Roman Empire in the fifth century A.D., communication among the former provinces declined, creating still greater regional variation in spoken Latin. By the eighth century, regions of the former empire had been isolated from each other long enough for distinct languages to evolve.

In the past, when migrants were unable to communicate with speakers of the same language back home, major differences emerged between the languages spoken in the old and new locations, leading to the emergence of distinct, separate languages. This was the case with the migration of Latin speakers 2,000 years ago.

Romance branch languages have achieved worldwide importance because of the colonial activities of their European speakers. Spanish is the official language of 18 Latin American states, and fewer than 10 percent of the speakers of Spanish live in Spain. Portuguese is spoken in Brazil, which has a population of 200 million, compared to only 10 million in Portugal. The division of Central and South America into Portuguese- and Spanish-speaking regions resulted from a 1493 decision by Pope Alexander VI to give the western portion of the New World to Spain and the eastern part to Portugal. The Treaty of Tordesillas, signed a year later, carried out the papal decision.

COMMON ANCESTRY OF INDO-EUROPEAN LANGUAGES

If Germanic, Romance, Balto-Slavic, and Indo-Iranian languages are all part of the same Indo-European language family, then they must be descended from a single common ancestral language. Unfortunately, the existence of a single ancestor—which can be called Proto-Indo-European—cannot be proved with certainty.

Because the origin of language families predates recorded history, the evidence that Indo-European originated with a single language comes primarily from words related to the physical environment. For example:

- Individual Indo-European languages share common words for winter and snow but not for ocean. Therefore, linguists conclude that original Proto-Indo-European speakers probably lived in a cold climate or one that had a winter season but did not come in contact with oceans.
- Individual Indo-European languages share words for some animals and trees (such as beech, oak, bear, deer, pheasant, and bee), but other words are unshared (such as elephant, camel, rice, and bamboo).

Therefore, linguists conclude that original Proto-Indo-European speakers lived in a place where the shared animals

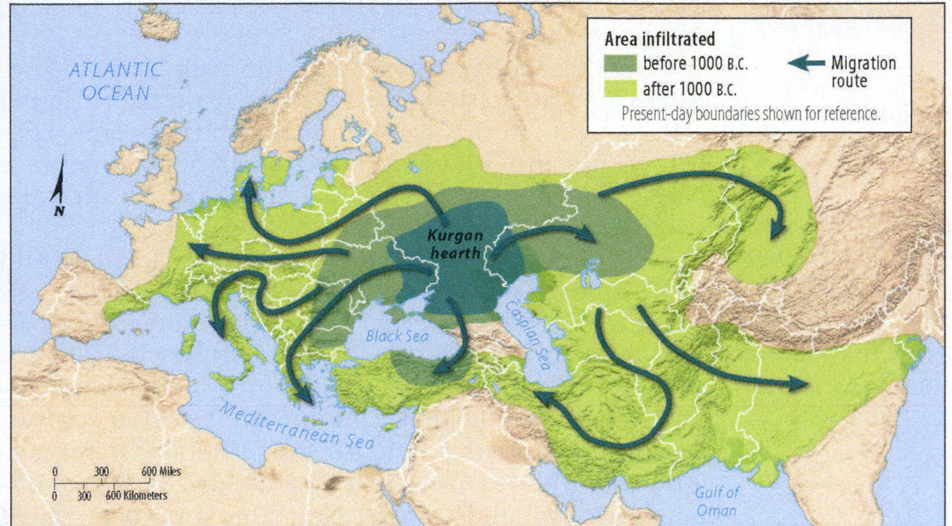

◀ **FIGURE 5-14 ORIGIN AND DIFFUSION OF INDO-EUROPEAN: NOMADIC WARRIOR THEORY** The Kurgan homeland was north of the Caspian Sea, near the present-day border between Russia and Kazakhstan. According to this theory, the Kurgans may have infiltrated into Eastern Europe beginning around 4000 B.C. and into central Europe and Southwest Asia beginning around 2500 B.C.

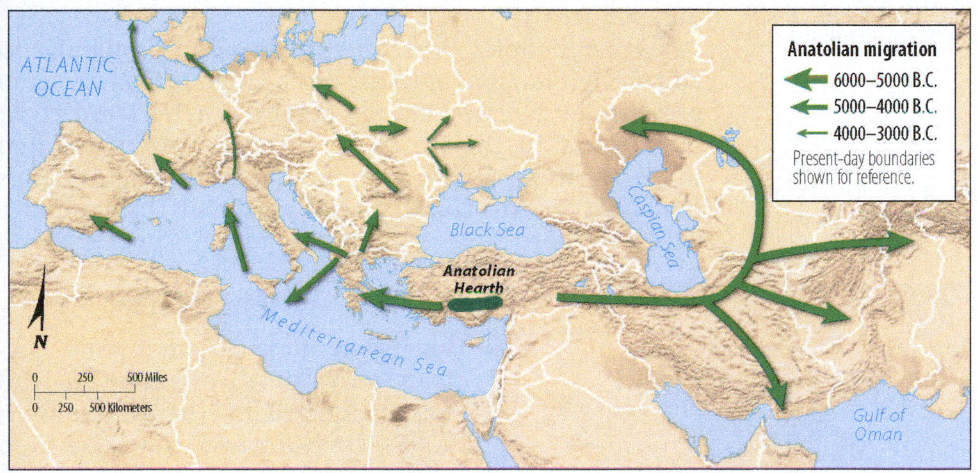

◀ **FIGURE 5-15 ORIGIN AND DIFFUSION OF INDO-EUROPEAN: SEDENTARY FARMER THEORY** Indo-European may have originated in present-day Turkey 2,000 years before the Kurgans. According to this theory, the language diffused along with agricultural innovations west into Europe and east into Asia.

and trees are found, whereas the unshared words were added later, after the original language split into branches.

NOMADIC WARRIOR THEORY. Linguists and anthropologists disagree on when and where Proto-Indo-European originated and the process and routes by which it diffused. The first Proto-Indo-European speakers were the Kurgan people, according to archaeologist Marija Gimbutas. The earliest archaeological evidence of the Kurgans dates to around 4300 B.C., near the border between present-day Russia and Kazakhstan.

Among the first people to domesticate horses and chariots, the Kurgans migrated in search of grasslands for their animals. This took them westward through Europe, eastward to Siberia, and southeastward to Iran and South Asia. Between 3500 and 2500 B.C., Kurgan warriors, using their domesticated horses and chariots to establish military superiority, conquered much of Europe and South Asia (Figure 5-14).

SEDENTARY FARMER THEORY. Archaeologist Colin Renfrew argues that the first speakers of Proto-Indo-European lived 2,000 years before the Kurgans, in the eastern part of present-day Turkey (Figure 5-15). Supporting Renfrew, biologist Russell D. Gray dates the first speakers even earlier, at around 6700 B.C. This hypothesis argues that Indo-European diffused into Europe and South Asia along with agricultural practices rather than by military conquest. The language triumphed because its speakers became more numerous and prosperous by growing their own food instead of relying on hunting.

Thus, the diffusion of Indo-European speaks to a fundamental question for humanity: Do cultural elements such as language diffuse primarily through warfare and conquest or primarily through peaceful sharing of food? Regardless of how Indo-European diffused, communication was poor among different peoples, whether warriors or farmers. After many generations of complete isolation, individual groups evolved increasingly distinct languages.

PAUSE & REFLECT 5.2.2

Which hypothesis appeals more to you: the "war" hypothesis or the "peace" hypothesis? Why?

Origin and Diffusion of English

LEARNING OUTCOME 5.2.3

Understand processes underlying current distribution of English.

The British Isles have been inhabited for thousands of years, but we know nothing of their early languages until tribes called the Celts arrived around 2000 B.C., speaking languages we call Celtic. Around A.D. 450, tribes from mainland Europe invaded, pushing the Celts into the remote northern and western parts of Britain, including Cornwall and the highlands of Scotland and Wales.

GERMAN INVASION

The tribes invading the British Isles were known as the Angles, Jutes, and Saxons. The Jutes came from northern Denmark, the Angles from southern Denmark, and the Saxons from northwestern Germany (Figure 5-16). Modern English has evolved primarily from the language spoken by the Angles, Saxons, and Jutes.

The three tribes who brought the beginnings of English to the British Isles shared a language similar to that of other peoples in the region from which they came. English people and others who trace their cultural heritage back to England are often called Anglo-Saxons, after the two larger tribes.

The name England comes from Angles' Land. In Old English, Angles was spelled Engles, and the Angles' language was known as *englisc*. The Angles came from a corner, or angle, of Germany known as Schleswig-Holstein. At some time in history, all Germanic people spoke a common language, but that time predates written records. The common origin of English with other Germanic languages can be reconstructed by analyzing language differences that emerged after Germanic groups migrated to separate territories and lived in isolation from each other, allowing their languages to continue evolving independently.

Other peoples subsequently invaded England and added their languages to the basic English. Vikings from present-day Norway landed on the northeastern coast of England in the ninth century. Although defeated in their effort to conquer the islands, many Vikings remained in the country, and enriched the language with new words.

▼ **FIGURE 5-16 INVASIONS OF ENGLAND** The first speakers of the language that became known as English were invaders from present-day Germany and Denmark. Later invasions by Vikings and Normans brought new words to the language spoken in the British Isles. The Normans were the last successful invaders of England.

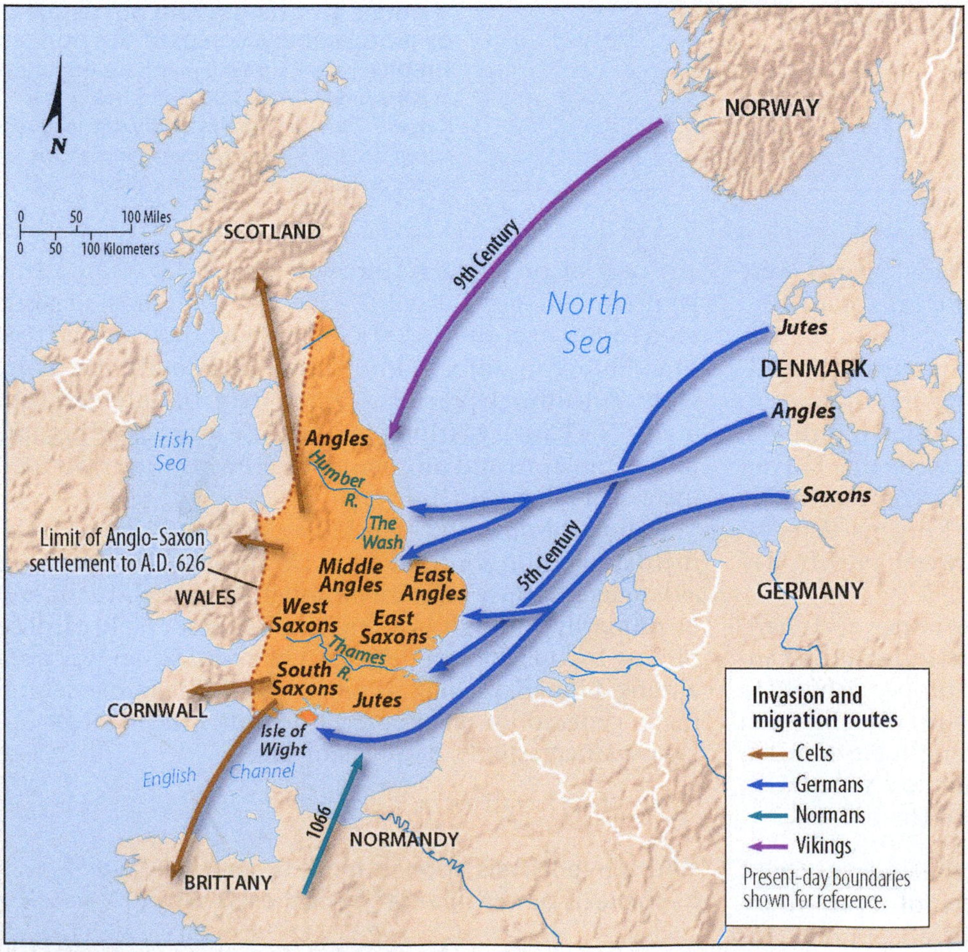

NORMAN INVASION

English is quite different from German today primarily because England was conquered by the Normans in 1066. The Normans, who came from present-day Normandy in France, spoke French, which they established as England's official language for the next 300 years. The leaders of England, including the royal family, nobles, judges, and clergy, spoke French. However, the majority of the people, who had little education, did not know French, so they continued to speak English to each other.

England lost control of Normandy in 1204, and entered a long period of conflict with France. As a result, fewer people in England wished to speak French, and English again became the country's unchallenged dominant language. Recognizing that nearly everyone in England was speaking English, Parliament enacted the Statute of Pleading in 1362 to change the official language of court business from French to English. Parliament continued to conduct business in French until 1489.

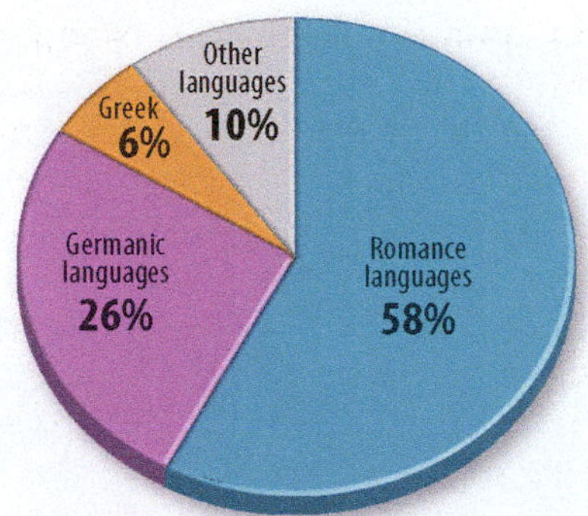

▲ **FIGURE 5-17 ORIGIN OF ENGLISH WORDS** Although classified in the Germanic branch, English actually has a higher percentage of Romance branch words.

During the 300-year period that French was the official language of England, the Germanic language used by the common people and the French used by the leaders mingled to form a new language. Modern English owes its simpler, straightforward words, such as sky, horse, man, and woman, to its Germanic roots, and more elegant words, such as celestial, equestrian, masculine, and feminine, to its French invaders (Figure 5-17).

PAUSE & REFLECT 5.2.3

***School* and *university*: Which word entered the English language through the German invasions and which through the Norman invasion?**

DIFFUSION OF ENGLISH

The contemporary distribution of English speakers around the world exists because the people of England migrated with their language when they established colonies over the course of four centuries.

English first diffused west from England to North America in the seventeenth century. The first successful English colonies were built in North America, beginning with Jamestown, Virginia, in 1607, and Plymouth, Massachusetts, in 1620. After England defeated France in a battle to dominate the North American colonies during the eighteenth century, the position of English as the principal language of North America was assured.

Similarly, the British took control of Ireland in the seventeenth century, South Asia in the mid-eighteenth century, the South Pacific in the late eighteenth and early nineteenth centuries, and southern Africa in the late nineteenth century. In each case, English became an official language, even if only the colonial rulers and a handful of elite local residents could speak it.

Into the twentieth century, the United States was responsible for diffusing English to several places, most notably the Philippines, which Spain ceded to the United States in 1899, a year after losing the Spanish-American War. After gaining full independence in 1946, the Philippines retained English as one of its official languages, along with Filipino.

COMBINING ENGLISH WITH OTHER LANGUAGES

New languages are being created through mixing English with other languages. Here are several examples:

- **Franglais** is the mix of French and English. French is an official language in 29 countries and for hundreds of years served as the lingua franca for international diplomats. Traditionally, language has been an especially important source of national pride and identity in France (Figure 5-18).
- **Spanglish** is the mix of Spanish and English. English is diffusing into the Spanish language spoken by 34 million Hispanics in the United States. New words have been invented in Spanglish that do not exist in English but would be useful if they did. For example, *textear* is a verb derived from the English and is less awkward than the Spanish *mandar un mensajito*.
- **Denglish** is the mix of German and English. The D stands for Deutsch, the German word for German. For example, the German telephone company Deutsche Telekom uses the German word *Deutschlandverbindungen* for "long distance" and the Denglish word *Cityverbindungen* for "local" (rather than the German word *Ortsverbindungen*).

▼ **FIGURE 5-18 FRANGLAIS** The name of the cafe in Nantes, France, is a mix of English ("death") and French ("porc").

Global Importance of English

LEARNING OUTCOME 5.2.4

Understand the concept of lingua franca.

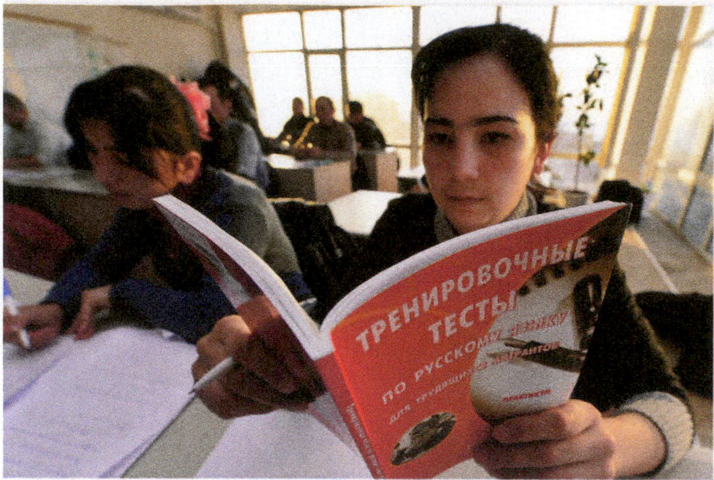

▲ **FIGURE 5-20 LEARNING RUSSIAN** School in Vladivostok, Russia.

One of the most fundamental needs in a global society is a common language for communication. In the modern world, the most important language of international communication is English. A Polish airline pilot who flies over Spain speaks to the traffic controller on the ground in English. Swiss bankers speak German among themselves, but with German bankers they prefer to speak English. English is the official language at an aircraft factory in France and an appliance company in Italy.

The dominance of English as an international language has facilitated the diffusion of popular culture and science and the growth of international trade. However, people who forsake their native language must weigh the benefits of using English against the cost of losing a fundamental element of local cultural identity.

LINGUA FRANCA

A language of international communication, such as English, is known as a **lingua franca**. To facilitate trade, speakers of two different languages create a lingua franca by mixing elements of the two languages into a simple common one. The term, which means "language of the Franks," was originally applied by Arab traders during the Middle Ages to the language they used to communicate with Europeans, whom they called Franks.

People in smaller countries need to learn English to participate more fully in the global economy and culture. All children learn English in the schools of countries such as the Netherlands and Sweden to facilitate international communication. This may seem culturally unfair, but obviously it is more likely that several million Dutch people will learn English than that a half-billion English speakers around the world will learn Dutch.

The rapid growth in importance of English is reflected in the percentage of students learning English as a second language in school (Figure 5-19). More than 90 percent of students in the European Union learn English in middle or high school, not just in smaller countries such as Denmark and the Netherlands but also in populous countries such as France, Germany, and Spain. The Japanese government, having determined that fluency in English is mandatory in a global economy, has even considered adding English as a second official language.

Foreign students increasingly seek admission to universities in countries that teach in English rather than in German, French, or Russian. Students around the world want to learn in English because they believe it is the most effective way to work in the global economy and participate in the global culture.

A group that learns English or another lingua franca may learn a simplified form, called a **pidgin language**. To communicate with speakers of another language, two groups construct a pidgin language by learning a few of the grammar rules and words of a lingua franca and mixing in some elements of their own languages. A pidgin language has no native speakers; it is always spoken in addition to one's native language.

Other than English, modern lingua franca languages include Swahili in East Africa, Hindi in South Asia, Indonesian in Southeast Asia, and Russian in the former Soviet Union (Figure 5-20). A number of African and Asian

▲ **FIGURE 5-19 TEACHING ENGLISH** Huangnan, China.

countries that became independent in the twentieth century adopted English or Swahili as an official language for government business, as well as for commerce, even if the majority of the people couldn't speak it.

In view of the global dominance of English, many U.S. citizens do not recognize the importance of learning other languages. One of the best ways to learn about the beliefs, traits, and values of people living in other regions is to learn their language. The lack of effort by Americans to learn other languages is a source of resentment among people elsewhere in the world, especially when Americans visit or work in other countries.

The inability to speak other languages is also a handicap for Americans who try to conduct international business. Successful entry into new overseas markets requires knowledge of local culture, and officials who can speak the local language are better able to obtain important information. Japanese businesses that wish to expand in the United States send English-speaking officials, but American businesses that wish to sell products to the Japanese are rarely able to send a Japanese-speaking employee.

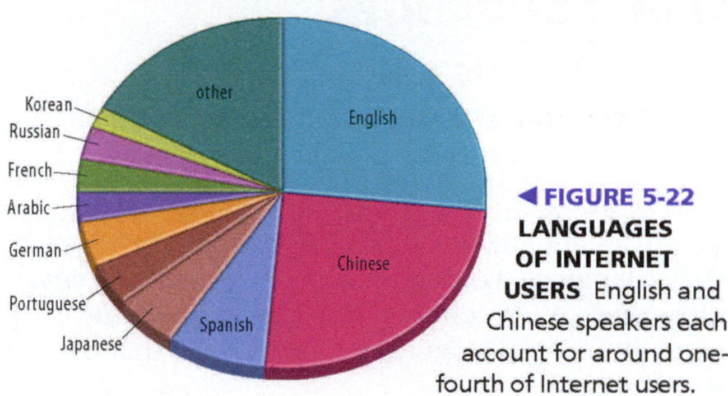

◀ **FIGURE 5-22 LANGUAGES OF INTERNET USERS** English and Chinese speakers each account for around one-fourth of Internet users.

ENGLISH ON THE INTERNET

The emergence of the Internet as an important means of communication has further strengthened the dominance of English. More than one-half of all Internet content is in English (Figure 5-21). Because a majority of the material on the Internet is in English, knowledge of English is essential for Internet users around the world.

The dominance of English-language websites persists despite the fact that a decreasing percentage of Internet users are English speakers. English was the language for only 27 percent of Internet users in 2015, a substantial decline from 71 percent in 1998 (Figure 5-22). The early dominance of English on the Internet was partly a reflection of the fact that the most populous English-speaking country, the United States, had a head start on the rest of the world in making the Internet available to most of its citizens (refer to Figure 4-33). Meanwhile, Chinese (Mandarin) language online users increased from 2 percent of the world total in 1998 to 25 percent in 2015.

English may be less dominant as the language of the Internet later in the twenty-first century. But the United States—and with it the English language—remains the Internet leader in key respects. The United States created the English-language nomenclature for the Internet that the rest of the world has followed. The designation "www," which English speakers recognize as an abbreviation of "World Wide Web," is awkward in other languages, some of which do not have an equivalent sound to the English *w*.

The U.S.-based Internet Corporation for Assigned Names and Numbers (ICANN) has been responsible for assigning domain names and for the suffixes following the dot, such as "com" and "edu." Domain names in the rest of the world include a two-letter suffix for the country, such as "fr" for France and "jp" for Japan, whereas U.S.-based domain names don't need the suffix. Reflecting the globalization of the languages of the Internet, ICANN agreed in 2009 to permit domain names in characters other than Latin. Arabic, Chinese, and other characters may now be used.

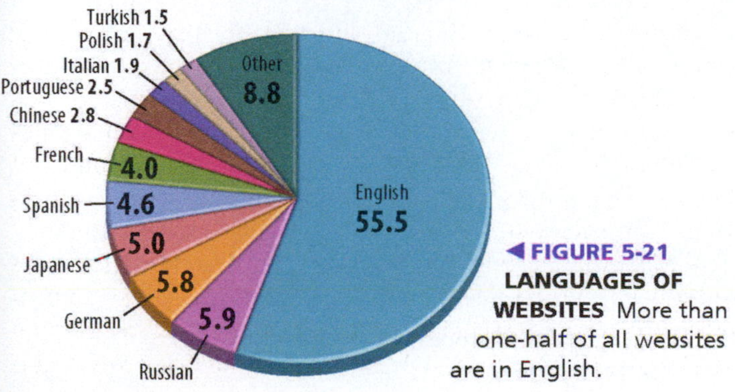

◀ **FIGURE 5-21 LANGUAGES OF WEBSITES** More than one-half of all websites are in English.

CHINESE: THE NEXT LINGUA FRANCA?

The future leadership of Chinese in social media comes in part from the large number of people worldwide who speak Chinese languages. The attraction of Chinese languages also comes from the way they are written. Rather than sounds (as in English), Chinese languages are written primarily with **logograms**, which are symbols that represent words or meaningful parts of words. Ability to read a book requires understanding several thousand logograms. Most logograms are compounds; words related to bodies of water, for example, include a symbol that represents a river, plus additional strokes that alter the river in some way.

Chinese is thus an attractive language to use in Twitter and other social media that restrict the number of characters. An English message that uses the maximum 140 symbols permitted by Twitter could be written in Chinese in only around 70 characters.

PAUSE & REFLECT 5.2.4

What is the title of the book in Figure 5-20? Type the letters into Google Translate.

Official Languages

LEARNING OUTCOME 5.2.5

Understand the official status of English and other languages.

Some countries designate one or more languages as official. The official language is used by the government to enact legislation and other conduct other public business. In some cases, the official language is the only one that is used in public schools.

Europe has a large number of official languages (Figure 5-23). The European Union recognizes 24 official and working languages: Bulgarian, Croatian, Czech, Danish, Dutch, English, Estonian, Finnish, French, German, Greek, Hungarian, Irish, Italian, Latvian, Lithuanian, Maltese, Polish, Portuguese, Romanian, Slovak, Slovenian, Spanish, and Swedish.

English is an official language in 58 countries, more than any other language (Figure 5-24). Two billion people live in a country where English is an official language, even if they cannot speak it. In addition, English is the predominant but not official language in several of the most prominent English-speaking countries, including Australia, the United Kingdom, and the United States (see Debate It feature).

As the primary language of the United States, English is used for all official documents, but it does not have an official status. On the other hand, 28 states have English-only laws (Figure 5-25).

PAUSE & REFLECT 5.2.5

Is English an official language in your state?

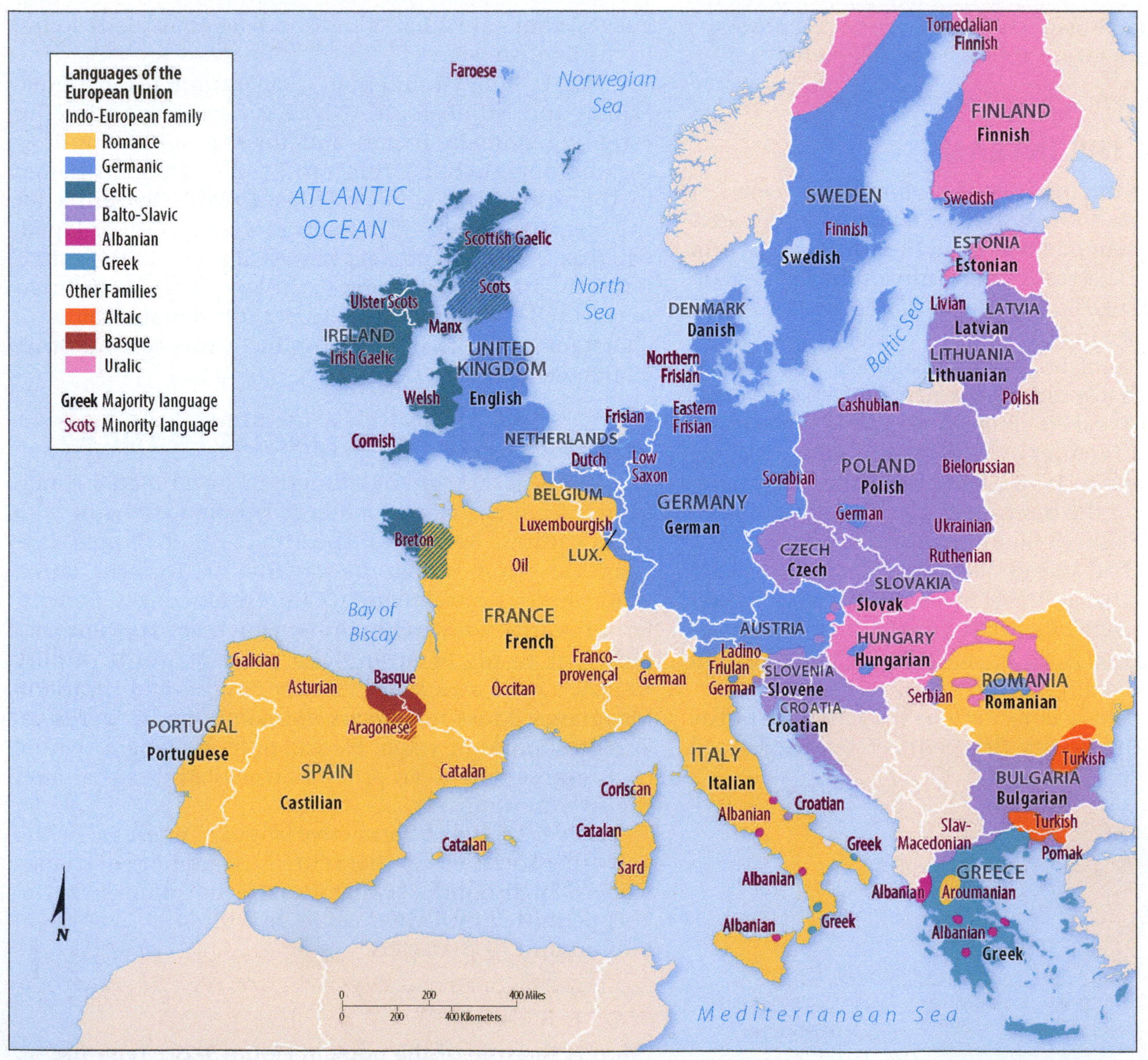

▲ **FIGURE 5-23 EUROPE'S OFFICIAL AND MINORITY LANGUAGES** The official languages of the European Union are written in black. Languages that the European Union wishes to see preserved and protected are in purple.

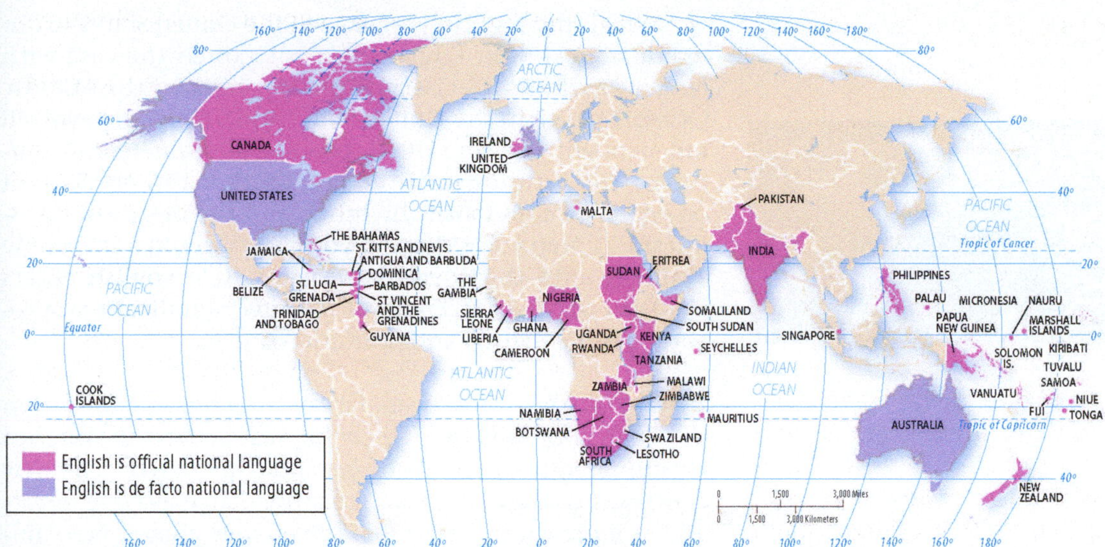

◀ **FIGURE 5-24 ENGLISH-SPEAKING COUNTRIES**
English is an official language in 58 countries and the predominant language in several other countries.

DEBATE IT! Should English be the official language of the United States?

The English-only movement advocates federal legislation that would require use of only English in U.S. documents.

MAKE ENGLISH THE OFFICIAL U.S. LANGUAGE

- 58 countries and 28 U.S. states already make English official.
- Requiring use of English is a symbol of national unity.
- Knowledge of English is essential for survival in the United States.
- Providing services to non-English speakers is expensive.

DO NOT MAKE ENGLISH THE OFFICIAL U.S. LANGUAGE

- A law is unnecessary because English is already the primary language of the United States.
- The government should not interfere with people's language rights.
- Advocating English-only attacks immigrants from non-English-speaking countries.

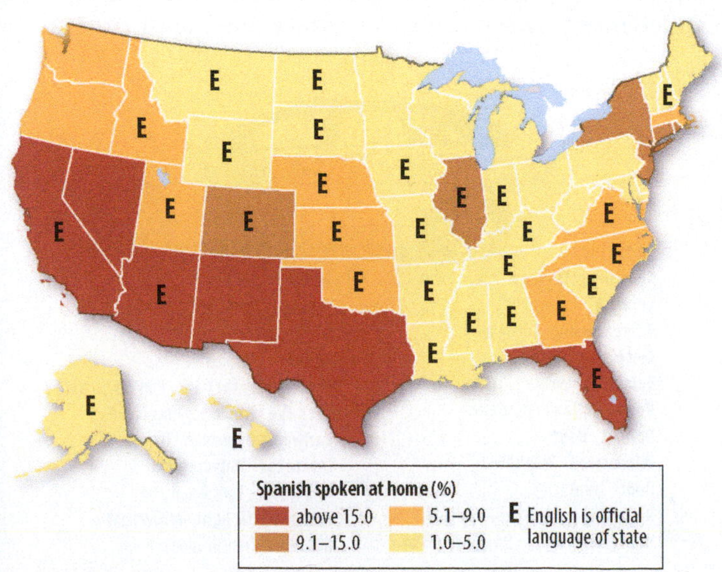

▲ **FIGURE 5-25 STATUS OF ENGLISH IN U.S. STATES**

CHECK-IN KEY ISSUE 2

Where Did English and Related Languages Originate and Diffuse?

- ✔ The Indo-European family has four widely spoken branches.
- ✔ The origin and early diffusion of language families such as Indo-European is speculative because these language families existed before recorded history.
- ✔ Individual languages, such as English and languages of the Romance branch, have documented places of origin and patterns of diffusion.
- ✔ English has become the world's most important lingua franca, especially in the Internet era.

KEY ISSUE 3

Why Do Individual Languages Vary Among Places?

- English Dialects
- U.S. Dialects
- Dialect or Language?
- Multilingual Places

LEARNING OUTCOME 5.3.1

Understand the ways that dialects vary.

A **dialect** is a regional variation of a language distinguished by distinctive vocabulary, spelling, and pronunciation. Generally, speakers of one dialect can understand speakers of another dialect. A **subdialect** is a subdivision of a dialect. Two subdialects of the same dialect share relatively few differences, primarily in pronunciation and a small amount of vocabulary. Geographers are especially interested in differences in dialects and subdialects because they reflect distinctive features of the environments in which groups live.

English Dialects

When speakers of a language migrate to other locations, various dialects of that language may develop. This was the case with the migration of English speakers to North America several hundred years ago. Because of its large number of speakers and widespread distribution, English has an especially large number of dialects and subdialects. North Americans are well aware that they speak English differently from the British, not to mention people living in India, Pakistan, Australia, and other English-speaking countries.

AMERICAN AND BRITISH ENGLISH

The English language was brought to the North American continent by colonists from England who settled along the Atlantic Coast beginning in the seventeenth century. The early colonists naturally spoke the language they had been using in England at the time. Later immigrants from other countries found English already implanted here. Although they made significant contributions to American English, they became acculturated into a society that already spoke English. Therefore, the earliest colonists were most responsible for the dominant language patterns that exist today in the English-speaking part of the Western Hemisphere.

Why is the English language in the United States so different from that in England? As is so often the case with languages, the answer is isolation. Separated by the Atlantic Ocean, English in the United States and in England evolved independently during the eighteenth and nineteenth centuries, with little influence on one another. Few residents of one country could visit the other, and the means to transmit the human voice over long distances would not become available until the twentieth century. U.S. English differs from the English of England in three significant ways—vocabulary, spelling, and pronunciation.

VOCABULARY. The vocabulary of U.S. English differs from the English of England largely because settlers in America encountered many new objects and experiences. The new continent contained physical features, such as large forests and mountains, that had to be given new names. New animals were encountered, including the moose, raccoon, and chipmunk, all of which were given names borrowed from Native Americans. Indigenous American "Indians" also enriched American English with names for objects such as canoe, moccasin, and squash. As new inventions appeared, they acquired different names on either side of the Atlantic. For example, the elevator is called a lift in England, and the flashlight is known as a torch. The British call the hood of a car the bonnet and the trunk the boot (Figure 5-26).

SPELLING. American spelling diverged from the British standard because of a strong national feeling in the United States for an independent identity. Noah Webster, the creator of the first comprehensive American dictionary and grammar books, was not just a documenter of usage; he had an agenda. Webster was determined to develop a uniquely American dialect of English. He either ignored or was unaware of recently created rules of grammar and spelling developed in England. Webster argued that spelling and grammar reforms would help establish a national language, reduce cultural dependence on England, and inspire national pride. The spelling differences between British and American English, such as the

▲ **FIGURE 5-26 U.S. AND U.K. DIALECTS** Many words related to cars and motoring differ. (British words are in bold.)

▲ **FIGURE 5-27 LONDON BUS** A vintage AEC K-type bus takes part in a parade in London. 1. In approximately what year was the bus manufactured? Hint: Use your Internet search engine to learn about the AEC K-type bus. 2. What type of product does Schweppes produce? 3. What is a lemon squash?

elimination of the u from the British spelling of words such as *honour* and *colour* and the substitution of s for c in *defence*, are due primarily to the diffusion of Webster's ideas inside the United States (Figure 5-27).

PRONUNCIATION. From the time of their arrival in North America, colonists began to pronounce words differently from the British. Such divergence is normal, for interaction between the two groups was largely confined to exchange of letters and other printed matter rather than direct speech. Americans pronounce unaccented syllables with more clarity than do British English speakers. The words *secretary* and *necessary* have four syllables in American English but only three in British (*secret'ry* and *necess'ry*). Surprisingly, pronunciation has changed more in England than in the United States. A single dialect of southern English did not emerge as the British national standard until the late eighteenth century, after the American colonies had declared independence and were politically as well as physically isolated from England. Thus people in the United States do not speak "proper" English because when the colonists left England, "proper" English was not what it is today. Furthermore, few colonists were drawn from the English upper classes.

DIALECTS IN THE UNITED KINGDOM

English varies by regions within individual countries. In both the United States and England, northerners sound different from southerners.

As already discussed, English originated with three invading groups from Northern Europe who settled in different parts of Britain—the Angles in the north, the Jutes in the southeast, and the Saxons in the southwest. The language each spoke was the basis of distinct regional dialects of Old English (Figure 5-28a).

In a language with multiple dialects, one dialect may be recognized as the **standard language**, which is a dialect that is well established and widely recognized as the most acceptable for government, business, education, and mass communication. In the case of England, the standard language is known as **Received Pronunciation (RP)**. It is well known around the world as the dialect commonly used by politicians, broadcasters, and actors.

RP was the dialect used by upper-class residents in the capital city of London and the two important university cities of Cambridge and Oxford. The diffusion of the upper-class London and university dialects was encouraged by the introduction of the printing press to England in 1476. Grammar books and dictionaries printed in the eighteenth century established rules for spelling and grammar that were based on the London dialect. These frequently arbitrary rules were then taught in schools throughout the country.

Despite the current dominance of RP, strong regional differences persist in English dialects spoken in the United Kingdom, especially in rural areas. They can be grouped into three main ones—Northern, Midland, and Southern (Figure 5-28b).

The boundaries between English dialects have been moving (Figure 5-28c). The changes reflect patterns of migration. The emergence of a subdialect in London reflects migration of people from other countries into the capital city, and the northern expansion of the southeastern subdialect reflects the outmigration of Londoners.

PAUSE & REFLECT 5.3.1

In what ways do you learn about differences between American and British English?

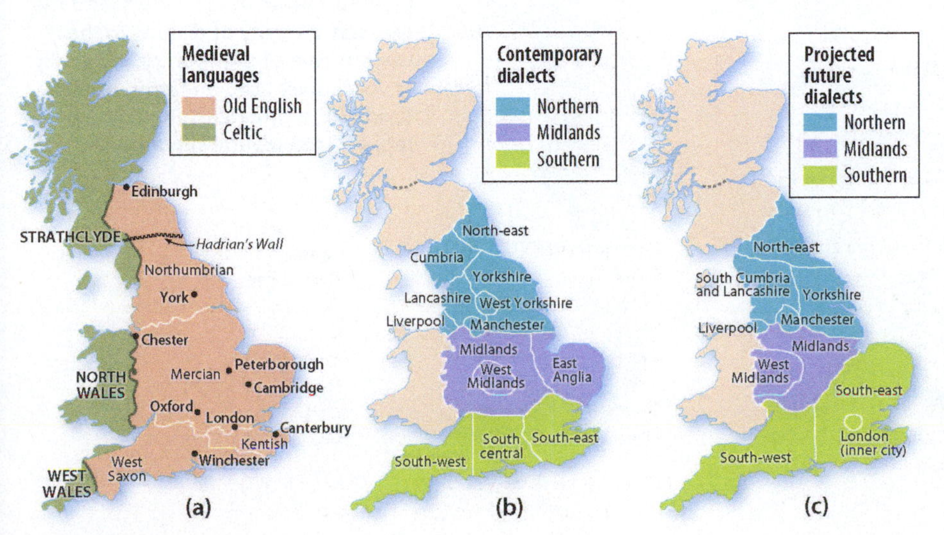

◀ **FIGURE 5-28 DIALECTS AND SUBDIALECTS IN ENGLAND** The Southeast dialect based in London is forecast to expand because of migration of Londoners.

U.S. Dialects

LEARNING OUTCOME 5.3.2

Understand the distribution of principal U.S. dialects.

The distribution of dialects is documented through the study of particular words. Every word that is not used nationally has some geographic extent within the country and therefore has boundaries. Such a word-usage boundary, known as an **isogloss**, can be constructed for each word. Isoglosses are determined by collecting data directly from people, particularly natives of rural areas. People are shown pictures to identify or are given sentences to complete with a particular word. Although every word has a unique isogloss, boundary lines of different words coalesce in some locations to form regions.

DISTRIBUTION OF U.S. DIALECTS

The United States has four major dialect regions: North, Midland, South, and West (Figure 5-29). The three eastern dialect regions can also be divided into several subdialects. The regional dialects display some familiar differences in pronunciation. For example:

- The South dialect includes making such words as *half* and *mine* into two syllables ("ha-af" and "mi-yen").
- The North dialect is well known for dropping the /r/ sound, so that *heart* and *lark* are pronounced "hot" and "lock."

The current distribution of U.S. dialects can be traced to differences in the origin of the English colonists along the East Coast. Three distinct dialect regions developed in the early colonies:

- **North.** Two-thirds of the New England colonists were Puritans from East Anglia in southeastern England, and only a few came from the north of England. The characteristic dropping of the /r/ sound is shared with speakers from the south of England.
- **South.** About half came from southeastern England, although they represented a diversity of social-class backgrounds, including deported prisoners, indentured servants, and political and religious refugees.
- **Midland.** These immigrants were more diverse. The early settlers of Pennsylvania were predominantly Quakers from the north of England. Scots and Irish also went to Pennsylvania, as well as to New Jersey and Delaware. The Middle Atlantic colonies also attracted many German, Dutch, and Swedish immigrants who learned their English from the English-speaking settlers in the area.

The English dialect spoken by the first colonists, who arrived in the seventeenth century, determined the future speech patterns for their communities because later immigrants adopted the language used in their new homes when they arrived. The language may have been modified somewhat by the new arrivals, but the distinctive elements brought over by the original settlers continued to dominate.

The diffusion of particular English dialects across the United States is a result of the westward movement of colonists from the three East Coast dialect regions. The North and South accents sound unusual to the majority of Americans because the standard pronunciation throughout the American West comes from the Midland rather than the North and South regions. This pattern occurred because most western settlers came from the Midland.

Many words that were once regionally distinctive are now national in distribution. Mass media, especially TV,

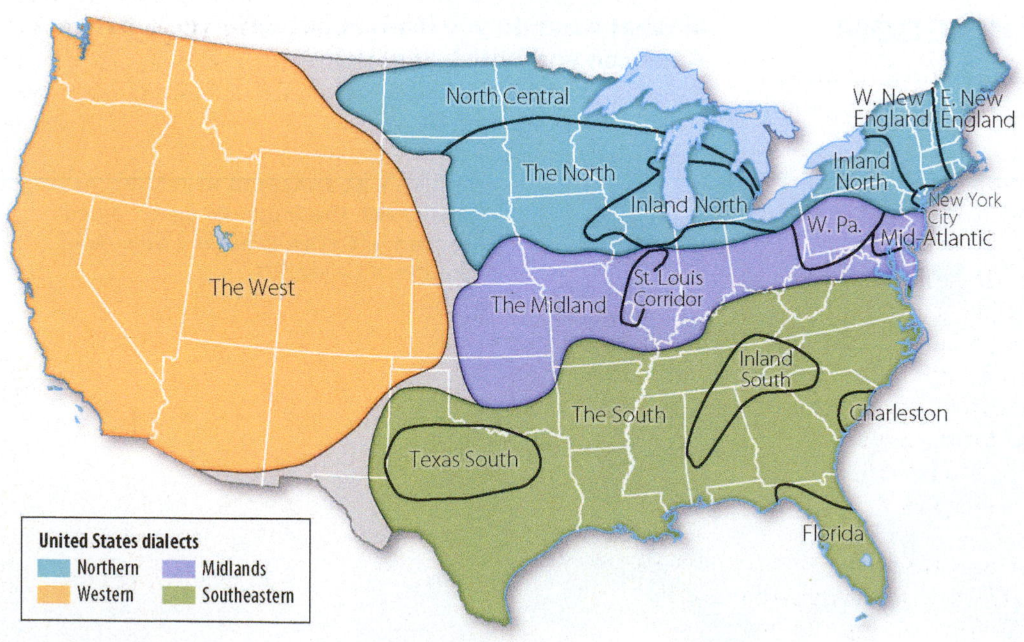

◄ **FIGURE 5-29 U.S. DIALECTS AND SUBDIALECTS** The United States has four major U.S. dialect regions. The most comprehensive classification of dialects in the United States was made by Hans Kurath in 1949. He found the greatest diversity of dialects in the eastern part of the country, especially in vocabulary used on farms. Kurath divided the three eastern U.S. dialect regions into several subdialects.

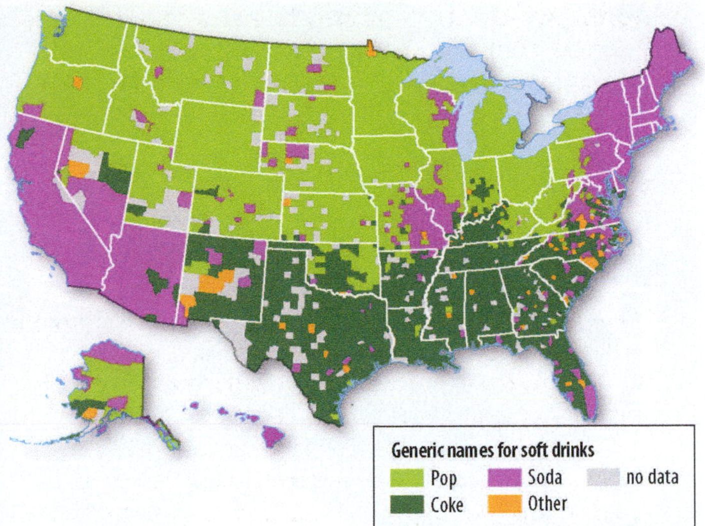

▲ **FIGURE 5-30 SOFT-DRINK DIALECTS** Soft drinks are called *soda* in the Northeast and Southwest, *pop* in the Midwest and Northwest, and *Coke* in the South. The map reflects voting at popvssoda.com.

influence the adoption of the same words throughout the country. Nonetheless, regional dialect differences persist in the United States. For example, the word for soft drink varies. Most people in the Northeast and Southwest, as well as the St. Louis area, use *soda* to describe a soft drink. Most people in the Midwest, Great Plains, and Northwest prefer *pop*. Southerners refer to all soft drinks as *Coke* (Figure 5-30).

PAUSE & REFLECT 5.3.2

Does your English fall into one of these dialects? Why or why not?

AFRICAN AMERICAN ENGLISH

Some African Americans speak a dialect of English heavily influenced by the group's distinctive heritage of forced migration from Africa during the eighteenth century to be slaves in the southern colonies. African American slaves preserved a distinctive dialect in part to communicate in a code not understood by their white masters. Black dialect words such as *gumbo* and *jazz* have long since diffused into the standard English language.

In the twentieth century, many African Americans migrated from the South to the large cities in the Northeast and Midwest (see Chapter 7). Living in racially segregated neighborhoods within northern cities and attending segregated schools, many African Americans preserved their distinctive dialect. That dialect has been termed African American Vernacular English (AAVE). Since 1996, the term **Ebonics**, a combination of ebony and phonics, has sometimes been used as a synonym for AAVE.

The American Speech, Language and Hearing Association classifies AAVE as a distinct dialect, with recognized vocabulary, grammar, and word meaning. Among the distinctive elements of Ebonics are the use of double negatives, such as "I ain't going there no more," and such sentences as "She be at home" instead of "She is usually at home."

Use of AAVE is controversial within the African American community. On one hand, some regard it as substandard, a measure of poor education, and an obstacle to success in the United States. Others see AAVE as a means for preserving a distinctive element of African American culture and an effective way to teach African Americans who otherwise perform poorly in school.

APPALACHIAN ENGLISH

Natives of Appalachian communities, such as in rural West Virginia, also have a distinctive dialect, pronouncing *hollow* as "holler" and *creek* as "crick," for example. Distinctive grammatical practices include the use of the double negative as in Ebonics and adding "a" in front of verbs ending in "ing," such as *a-sitting*.

As with Ebonics, speaking an Appalachian dialect produces both pride and challenges. An Appalachian dialect is a source of regional identity but has long been regarded by other Americans as a sign of poor education and an obstacle to obtaining employment in other regions of the United States. Some Appalachian residents are "bidialectic": They speak "standard" English outside Appalachia and slip back into their regional dialect at home.

CREOLE LANGUAGES

A **creole**, or creolized language, is a language that results from the mixing of a colonizer's language with the indigenous language of the people being dominated. A creolized language forms when the colonized group adopts the language of the dominant group but makes some changes, such as simplifying the grammar and adding words from the former language (Figure 5-31). Examples include French Creole in Haiti, Papiamento (creolized Spanish) in Netherlands Antilles (West Indies), and Portuguese Creole in the Cape Verde Islands off the African coast. These creole languages spoken in former colonies are classified as separate languages because they differ substantially from the original introduced by European colonizers.

▼ **FIGURE 5-31 CREOLE** Local election ballot in Miami is written in English, Spanish, and Spanish Creole.

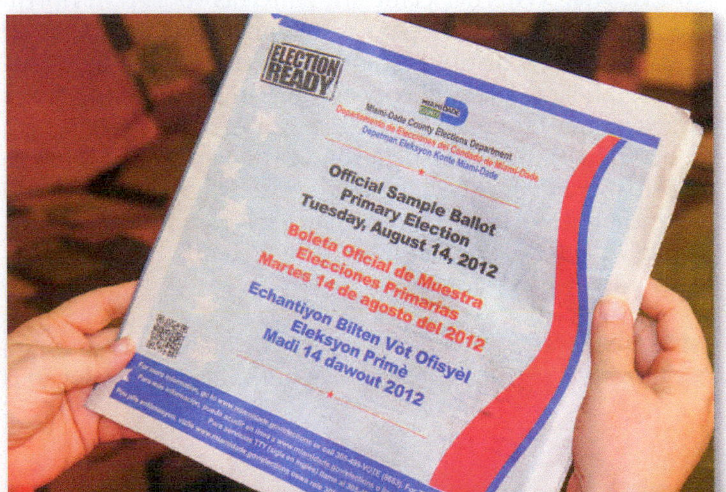

Dialect or Language?

LEARNING OUTCOME 5.3.3

Understand challenges in distinguishing between some languages and dialects.

Distinguishing between dialects and distinct languages is a good example of global–local tensions. Migration, increased interaction, and other globalization processes have resulted in strengthening of standard languages and suppression of dialects. On the other hand, desire for more local cultural identity has resulted in the emergence of distinct languages that were once considered dialects.

DIALECTS BECOME LANGUAGES

It is sometimes difficult to distinguish between a language and a dialect. The Romance branch offers several examples.

CATALÁN-VALENCIAN-BALEAR. Catalán was once regarded as a dialect of Spanish, but linguists now agree that it is a separate Romance language (Figure 5-32). Like other Romance languages, Catalán can be traced to Vulgar Latin, and it developed as a separate language after the collapse of the Roman Empire.

Catalán is the official language of Andorra, a tiny country of 79,000 inhabitants situated in the Pyrenees Mountains between Spain and France. Catalán is also spoken by 5 million people in eastern Spain and is the official language of Spain's highly autonomous Catalonia province, centered on the city of Barcelona.

With the status of Catalán settled as a separate language, linguists are identifying its principal dialects. Linguists agree that Balear is a dialect of Catalán that is spoken in the Balearic Islands, which include Ibiza and Majorca. More controversial is the status of Valencian, which is spoken mostly in and around the city of Valencia. Most linguists consider Valencian a dialect of Catalán. However, many in Valencia, including the Valencian Language Institute, consider Valencian a separate language, because it contains words derived from people who lived in the region before the Roman conquest. *Ethnologue* now calls the language Catalán-Valencian-Balear.

▲ **FIGURE 5-32 CATALÁN** Store in Barcelona advertises "sale" written in Catalán. In Spanish, it would be written "rebajas."

GALICIAN. Whether Galician, which is spoken in northwestern Spain and northeastern Portugal, is a dialect of Portuguese or a distinct language is debated among speakers of Galician. The Academy of Galician Language considers it a separate language and a symbol of cultural independence. The Galician Association of the Language prefers to consider it a dialect because as a separate language, it would be relegated to a minor and obscure status, whereas as a dialect of Portuguese, it can help influence one of the world's most widely used languages.

MOLDOVAN. Generally classified as a dialect of Romanian, Moldovan is the official language of Moldova. Moldovan is written, like Russian, in Cyrillic letters, a legacy of Moldova being a part of the Soviet Union, whereas Romanian is written in Roman letters.

ITALY'S LANGUAGES. Several languages in Italy that have been traditionally considered dialects of Italian are now viewed by *Ethnologue* as sufficiently different to merit classification as languages distinct from Italian. These include (with the number of speakers in parentheses) Lombard (3.9 million), Napoletano-Calebrese (5.7 million), Piemontese (1.6 million), Sicilian (4.7 million), and Venetian (3.9 million). These languages do not have official national status but are recognized by regional governments within Italy. See Figure 5-11 for the distribution of these languages within Italy.

OCCITAN. Occitan is spoken by about 2 million people in southern France and adjacent countries. The name derives from the French region of Aquitaine, which in French has a similar pronunciation to Occitan. The French government has established bilingual elementary and high schools called *calandretas* in the Occitan region. These schools teach both French and Occitan, according to a curriculum established by the national ministry of education. Still, many people living in southern France want to see more efforts by the government of France to encourage the use of Occitan.

STANDARDIZING LANGUAGES

Governments have long promoted the designation of a single dialect as the official or standard language in order to promote cultural unity. For example, the standard form of French derives from Francien, which was once a dialect of the Île-de-France region of the country. Francien became the standard form of French because the region included Paris, which became the capital and largest city of France. Francien French became the country's official language in the sixteenth century, and local dialects tended to disappear as a result of the capital's longtime dominance over French political, economic, and social life.

The Portuguese and Spanish languages spoken in the Western Hemisphere differ from their European versions, as is the case with U.S. and British English. To unify Spanish, the members of the Spanish Royal Academy meet every week in a mansion in Madrid to clarify rules for the vocabulary, spelling, and pronunciation of the Spanish language around the world. The academy's official dictionary, published in 1992, has added hundreds of "Spanish" words that originated either in the regional dialects of Spain or the Indian languages of Latin America.

To unify Portuguese, Brazil, Portugal, and several Portuguese-speaking countries in Africa agreed in 1994 to standardize the way their common language is written. Many people in Portugal are upset that the new standard language more closely resembles the Brazilian version, which eliminates some of the accent marks—such as tildes (as in São Paulo), cedillas (as in Alcobaça), circumflexes (as in Estância), and hyphens—and the agreement recognizes as standard thousands of words that Brazilians have added to the language, such as words for flowers, animals, and other features of the natural environment found in Brazil but not in Portugal.

The standardization of Spanish and Portuguese is a reflection of the level of interaction that is possible in the modern world between groups of people who live tens of thousands of kilometers apart. Books and television programs produced in one country diffuse rapidly to other countries where the same language is used.

PAUSE & REFLECT 5.3.3

Does your Internet search engine show tildes, cedillas, and circumflexes?

SUSTAINABILITY & OUR ENVIRONMENT Gender Differences in Language

The principle of gender equality is a foundation of of the cultural and economic pillars of sustainability. Languages vary in their ability to absorb and to reflect a cultural change, such as gender equality.

Languages vary by gender in two principal ways:

- The grammar of the language may distinguish between masculine and feminine.
- Men and women use different words and converse differently.

Languages are divided about evenly between those that distinguish between masculine and feminine and those that do not. Of the 20 most widely used languages, 12 make gender distinctions, and 8 do not. Nouns and verbs may carry distinctive forms or endings depending on gender, such as adding "ova" to designate feminine names in Russian. Others use distinctive articles, such as the French "la" in front of a feminine noun and "le" in front of a masculine one.

Austronesian, Altaic, and Uralic are considered genderless language families. All of the widely spoken languages in the Indo-European family other than English distinguish between masculine and feminine nouns and verbs. The gender-specific exception in English is singular pronouns (he/she and his/hers). Careful construction of sentences can avoid the handful of gender-specific pronouns in English, but gender neutrality is not possible in many other languages.

Though the English language is largely gender neutral, men and women have been found to speak the language differently. Women are more likely to use hedge words like "probably" or "kind of," intensive adverbs like "very" and "extremely," fillers like "uh" and "I mean," and tag endings like "isn't it." Older women are less likely than men to use strong expletives, though younger people do not show differences.

Underlying these formal differences in language may be differences in respect and status for men and women. A recent study of conversations between one man and one woman found that the man interrupted the woman far more often than the woman interrupted the man. Furthermore, a woman was less likely to interrupt a man than to interrupt another woman, and a man was more likely to interrupt a woman than to interrupt another man.

Multilingual Places

LEARNING OUTCOME 5.3.4

Understand how some countries embrace more than one language.

Multiple languages coexist in some countries, with varying degrees of success. One cultural group living in a region of the country may speak one language, while a group elsewhere in the country uses another language. In other countries, speakers of various languages intermingle. Some countries have devised strategies to promote peaceful coexistence among speakers of different languages, whereas others face challenges among the cultural groups.

SWITZERLAND: INSTITUTIONALIZED DIVERSITY

Figure 5-7 shows that the boundary between the Romance and Germanic branches runs through the middle of two small European countries, Belgium and Switzerland. Belgium has had more difficulty than Switzerland in reconciling the interests of the different language speakers.

Switzerland has four official languages: German (used by 65 percent of the population), French (18 percent), Italian (10 percent), and Romansh (1 percent). These four languages predominate in different parts of the country (Figure 5-33). Swiss voters made Romansh an official language in a 1938 referendum, despite the small percentage of people who use the language.

Switzerland peacefully exists with multiple languages. The Swiss, relatively tolerant of citizens who speak other languages, have institutionalized cultural diversity by creating a form of government that places considerable power in small communities. The key is a long tradition of decentralized government, in which local authorities hold most of the power, and decisions are frequently made by voter referenda.

▲ **FIGURE 5-33 LANGUAGE DIVERSITY IN SWITZERLAND** Switzerland has four official languages that predominate in different regions of the country.

CANADA: BILINGUAL AUTONOMY

French is one of Canada's two official languages, along with English. French speakers comprise one-fourth of the country's population and are clustered in Québec, where they account for more than three-fourths of the province's speakers (Figure 5-34). Colonized by the French in the seventeenth century, Québec was captured by the British in 1763, and in 1867 it became one of the provinces in the Confederation of Canada.

The Québec government has made the use of French mandatory in many daily activities. Québec's Commission de Toponymie renamed in French towns, rivers, and mountains that had been given English-language names. French must be the predominant language on all commercial signs.

Until the late twentieth century, Québec was one of Canada's poorest and least-developed provinces. Its economic and political activities were dominated by an English-speaking minority, and the province suffered from cultural isolation and lack of French-speaking leaders. To promote French-language cultural values, the Parti Québécois—one of the province's leading political parties—advocates sovereignty (effectively independence from Canada), but voters have thus far not supported it.

Confrontation has been replaced in Québec by increased cooperation between French and English speakers. The neighborhoods of Montréal, Québec's largest city, have become more linguistically mixed, and one-third of Québec's native English speakers have married French speakers in recent years.

NIGERIA: SPATIAL COMPROMISE

Africa's most populous country, Nigeria, provides an example of the tensions that can arise from the presence of many speakers of many languages. Nigeria has 529 distinct languages, according to *Ethnologue*, but only 3 (Hausa, Igbo, and Yoruba) are used by more than 10 percent of the country's population and only 4 others (Adamawa Fulfulde, Kanura, Nigerian Fulfulde, and Tiv) by between 1 and 10 percent of the population (Figure 5-35). Further splitting the country, the north is predominantly Muslim, and the south is predominantly Christian.

Groups living in different regions of Nigeria have often battled. The southern Igbo attempted to secede from Nigeria during the 1960s, and northerners have repeatedly claimed that the Yoruba discriminate against them. To reduce these regional tensions, the government has moved the capital from Lagos in the Yoruba-dominated southwest to Abuja in the center of the country, where none of the three major languages or two major religions predominates.

PAUSE & REFLECT 5.3.4

What is the most widely used language in the same family as Hausa? What religion's holiest book is written in that language?

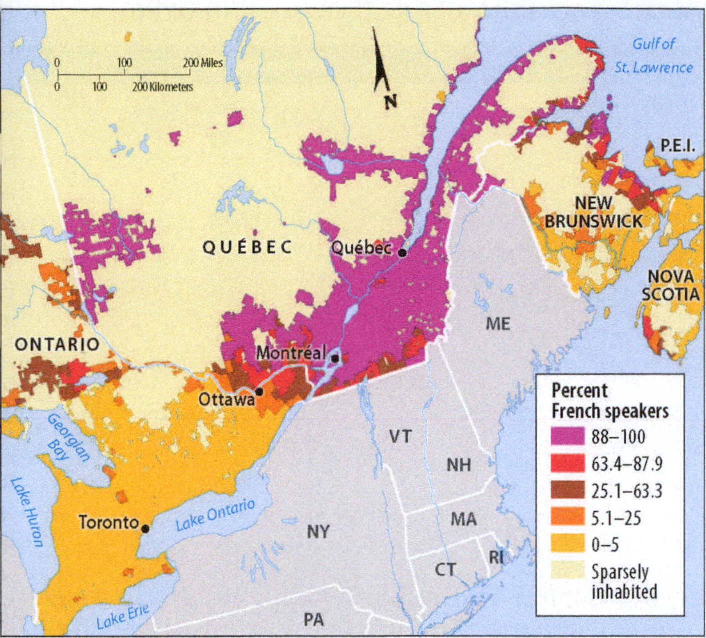

▲ **FIGURE 5-34 LANGUAGE DIVERSITY IN CANADA** Canada has two official languages. French predominates in Québec and English elsewhere.

▲ **FIGURE 5-36 LANGUAGE DIVERSITY IN BELGIUM** Flemings in the north speak Flemish, a Dutch dialect. Walloons in the south speak French.

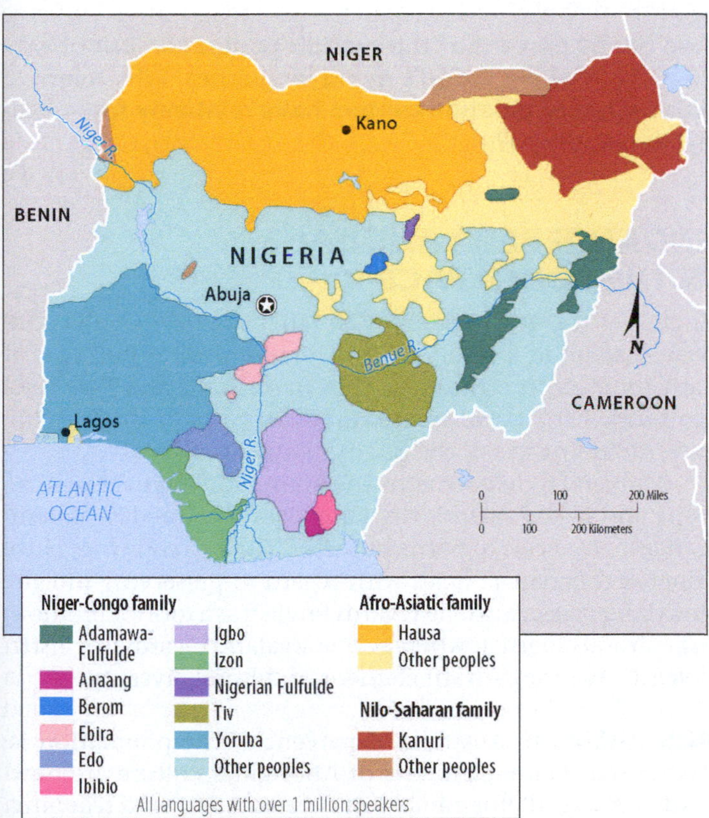

▲ **FIGURE 5-35 LANGUAGE DIVERSITY IN NIGERIA** Hausa, Igbo, and Yoruba are the most widely used languages.

BELGIUM: BARELY SPEAKING

A language boundary sharply divides the small country of Belgium into two regions. Southern Belgians (known as Walloons) speak French, whereas northern Belgians (known as Flemings) speak Flemish, a dialect of the Germanic language Dutch (Figure 5-36). Brussels, the capital city, is officially bilingual, and signs there are in both French and Flemish.

Antagonism between the Flemings and Walloons is aggravated by economic and political differences. Historically, the Walloons dominated Belgium's economy and politics, and French was the official state language. But in recent years Flanders has been much more prosperous than Wallonia, and the Flemish-speaking northerners do not wish to see their taxes spent in the poorer south.

In response to pressure from Flemish speakers, Belgium has been divided into two autonomous regions, Flanders and Wallonia. Each elects an assembly that controls cultural affairs, public health, road construction, and urban development in its region. But for many in Flanders, regional autonomy is not enough. They want to see Belgium divided into two independent countries. Were that to occur, Flanders would be one of Europe's richest countries and Wallonia one of the poorest.

CHECK-IN KEY ISSUE 3

Why Do Individual Languages Vary Among Places?

- ✔ **A dialect is a regional variation of a language.**
- ✔ **The United States has several major dialects.**
- ✔ **The distinction between a dialect and an entirely different language is not always clear-cut.**
- ✔ **Some countries more peacefully embrace multiple languages than do others.**

KEY ISSUE 4

Why Do Local Languages Survive?

- Endangered Languages
- Preserving Languages
- Isolated and Extinct Languages
- New and Growing Languages

LEARNING OUTCOME 5.4.1

Understand the classification of languages by severity of threat to their survival.

The distribution of a language is a measure of the fate of a cultural group. English has diffused around the world from a small island in northwestern Europe because of the dominance of England and the United States over other territory on Earth's surface. Icelandic remains a little-used language because of the isolation of the Icelandic people.

As in other cultural traits, language displays the two competing geographic trends of globalization and local diversity. English has become the principal language of communication and interaction for the entire world. At the same time, local languages endangered by the global dominance of English are being protected and preserved.

Endangered Languages

Many languages are threatened with extinction. *Ethnologue* considers 2,447 of the world's 7,102 living languages to be endangered. The United Nations *UNESCO Atlas of the World's Languages in Danger* has a slightly different total of 2,348 endangered languages.

Ethnologue classifies 916 of the 2,447 endangered languages as dying because the childbearing generation is not capable of teaching the language to their children, and the only remaining fluent speakers are older people. Another 1,531 of the endangered languages are considered to be in trouble because parents who are fluent in the language are no longer teaching it to their children. Using slightly different definitions, UNESCO considers 598 languages to be vulnerable, 646 definitely endangered, 528 severely endangered, and 576 critically endangered.

Regardless of the precise count and method of definition, by any measure, a large percentage of the world's languages are endangered. Because parents are no longer able to teach them to children, these languages will survive only through concerted community efforts.

The world regions with the largest numbers of dying languages are the South Pacific, Latin America, and North America (Table 5-1). The South Pacific and North America are regions dominated by only one widely used institutional language (English), and Latin America has only two widely used institutional languages (Spanish and Portuguese).

TABLE 5–1 LANGUAGES IN TROUBLE AND DYING

	In Trouble	Dying
North America	84	154
Europe	50	51
Latin America	225	185
Sub-Saharan Africa	209	117
Southwest Asia & North Africa	67	27
South Asia	129	29
Central Asia	4	1
East Asia	112	33
Southeast Asia	417	111
South Pacific	234	208
Total	1,531	916

PAUSE & REFLECT 5.4.1

Asia has 60 percent of the world's population but only 20 percent of the world's dying languages. Why might Asia's large population centers have relatively few dying languages?

ENDANGERED LANGUAGES IN THE SOUTH PACIFIC

English is the most widely used language in Australia and New Zealand as a result of British colonization during the early nineteenth century. Settlers in Australia and New Zealand established and maintained outposts of British culture, including use of the English language.

Though English remains the dominant language of Australia and New Zealand, the languages that predate British settlement survive in both countries. The two countries have adopted different policies with regard to preserving indigenous languages. Australia regards English as a tool for promoting cultural diversity, whereas New Zealand regards linguistic diversity as an important element of cultural diversity.

AUSTRALIA. In Australia, 1 percent of the population is Aboriginal. Many elements of Aboriginal culture are now being preserved. But education is oriented toward teaching English rather than maintaining local languages. English is the language of instruction throughout Australia, and others are relegated to the status of second language. As a result, Australia has 211 living indigenous languages in addition to English, but each of them has fewer than 10,000 speakers.

An essential element in maintaining British culture was restriction of immigration from non-English-speaking places during the nineteenth and early twentieth centuries. Fear of

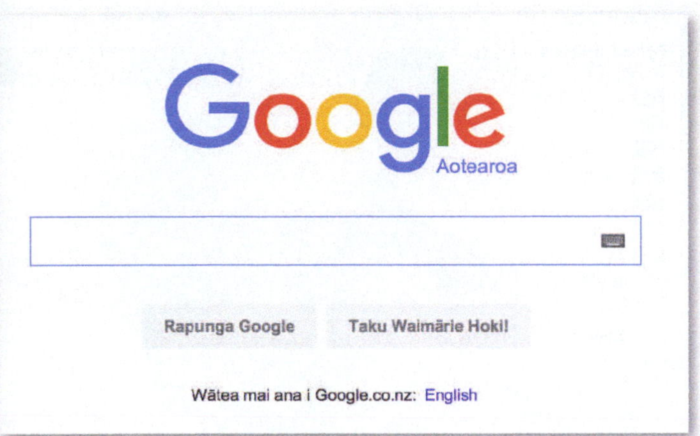

▲ FIGURE 5-37 **GOOGLE HOME PAGE IN MAORI**

immigration was especially strong in Australia because of its proximity to Asian countries. Under a "White Australia" policy, every prospective immigrant was required to write 50 words of a European language dictated by an immigration officer. The dictation test was not eliminated until 1957. The Australian government now merely requires that immigrants learn English.

NEW ZEALAND. In New Zealand, 14 percent of the population is Maori, descendants of Polynesian people who migrated there around 1,000 years ago. In contrast with Australia, New Zealand has adopted policies to preserve the Maori language. Most notably, Maori is one of New Zealand's three official languages, along with English and sign language (Figure 5-37). A Maori Language Commission was established to preserve the language. Despite official policies, only 4 percent of New Zealanders are fluent in Maori, and most of them are over age 50. Preserving the language requires skilled teachers and the willingness to endure inconvenience compared to using the world's lingua franca, English. Consequently, despite its official status, Maori is classified by *Ethnologue* as a threatened language.

Although promoting language diversity through protecting Maori, New Zealand has imposed a language requirement for immigrants that is more stringent than the one in Australia. In most circumstances, immigrants must already be fluent in English, although free English lessons are available to immigrants for the exceptions. More remote from Asian landmasses, New Zealand has attracted fewer Asian immigrants.

ENDANGERED LANGUAGES IN NORTH AMERICA

The United States has 61 languages in trouble and 142 languages dying, according to *Ethnologue*. The United Nations has a somewhat different count of 25 definitely endangered, 35 severely endangered, and 74 critically endangered languages in the United States (Figure 5-38). The endangered languages of the United States were once used by large numbers of Native Americans, but as older speakers pass away, younger speakers are not taught the languages.

Five endangered Native American languages are reawakening, according to *Ethnologue*. One of the reawakening languages is Myaamia (listed in *Ethnologue* as Miami). The Miami Native American Tribe, which is located in northeastern Oklahoma, traditionally spoke the Myaamia language, but no one has spoken it as their first language since the 1960s—until recently. Daryl Baldwin, a Miami Tribe member and director of the Myaamia Center at Miami University in Ohio, has acquired fluency in the Myaamia language as his second language, and his two children are learning it as their first language—its only two first language speakers in the world, according to *Ethnologue* (Figure 5-39). As part of the revival of the Myaamia language, an online dictionary has been created, accessible at myaamiadictionary.org.

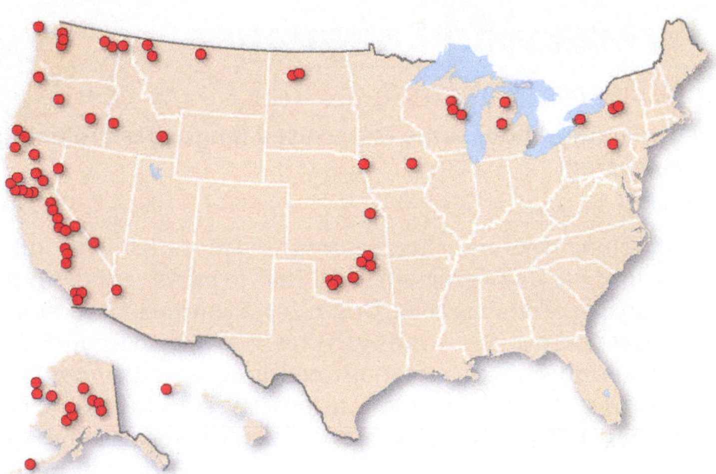

▲ FIGURE 5-38 **CRITICALLY ENDANGERED U.S. LANGUAGES** The dots point to the approximate locations where the recently-extinct languages were last spoken.

▼ FIGURE 5-39 **REVIVING THE MYAAMIA LANGUAGE**

Preserving Languages

LEARNING OUTCOME 5.4.2

Understand how some lesser-used languages are being protected.

Some languages are being preserved and protected. The United Nations has had a program since 2003 to preserve endangered languages. The European Union has identified 60 local languages that people are trying to preserve (refer to Figure 5-23).

Recent efforts have bolstered languages belonging to the Celtic branch of Indo-European. Preserving Celtic languages is of particular interest to English speakers because these languages offer insights into the cultural heritage of places that now speak English. Celtic languages were spoken in much of present-day Germany, France, northern Italy, and the British Isles 2,000 years ago. Today, Celtic languages survive only in remote parts of Scotland, Wales, and Ireland and on the Brittany peninsula of France.

BRYTHONIC CELTIC

The Celtic language branch is divided into Goidelic (Gaelic) and Brythonic groups. Speakers of Brythonic (also called Cymric or Britannic) fled westward during the Germanic invasions to Wales, southwestward to Cornwall, or southward across the English Channel to the Brittany peninsula of France.

WELSH. Wales—the name derived from the Germanic invaders' word for foreign—was conquered by the English in 1283. Welsh remained dominant in Wales until the nineteenth century, when many English speakers migrated there to work in coal mines and factories. A 2014 census found 580,000 Welsh speakers in Wales, 23 percent of the population. Another 150,000 Welsh speakers live across the border in England. In some isolated communities in the northwest, especially in the county of Gwynedd, two-thirds speak Welsh (Figure 5-40).

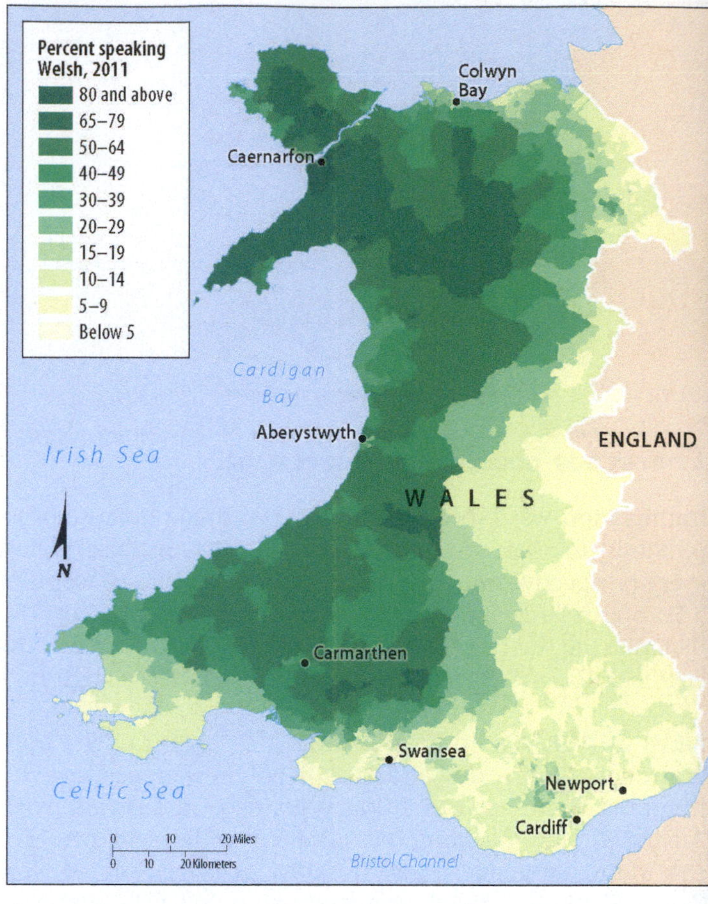

▲ **FIGURE 5-40 WELSH** The percentage speaking Welsh increases with distance from the border with England.

Cymdeithas yr Iaith Gymraeg (Welsh Language Society) has been instrumental in preserving the language. Britain's 1988 Education Act made Welsh language training a compulsory subject in all schools in Wales, and Welsh history and music have been added to the curriculum. In 2011, the government of the United Kingdom made Welsh the official language in Wales. All local governments and utility companies are obliged to provide services in Welsh. Welsh-language road signs have been posted throughout Wales, and the British Broadcasting Corporation (BBC) produces Welsh-language television and radio programs. Knowledge of Welsh is now required for many jobs, especially in public service, media, culture, and sports. Nonetheless, despite these efforts, 73 percent of people living in Wales reported on the 2011 census that they had no Welsh language skills.

CORNISH. Cornish was the Celtic language spoken in Cornwall, the southwesternmost county of the United Kingdom. Its last-known native speaker of Cornish, died in 1777.

Cornish was revived in the twentieth century. A major step forward was an agreement in 2008 among several Cornish language advocacy groups on a standard system of writing (Figure 5-41). A 2011 census found 557 people in the United Kingdom who claimed fluency in the language. It is taught in grade schools and adult evening courses and is used

▼ **FIGURE 5-41 CORNISH** *Kernow* means "Cornwall," *a'gas* means "you," and *dynergh* means "welcomes."

▲ **FIGURE 5-42 BRETON** The stones at Karnag (Breton) or Carnac French) are the largest collection in the world (more than 3,000). They were probably erected around 3300 B.C.

in some church services. Some banks accept checks written in Cornish.

BRETON. Brittany, like Cornwall, is an isolated peninsula that juts out into the Atlantic Ocean. As part of France, Brittany has a language containing more words borrowed from French than do the other Celtic languages (Figure 5-42).

The number of Breton speakers has declined from around 1 million in 1950 to around 200,000 today, and three-fourths of the remaining speakers are over age 65. The decline in speakers has placed Breton on the lists of endangered languages maintained by *Ethnologue* and the United Nations. Around 15,000 students learn Breton in schools, but the government of France requires French to be the principal language of instruction in public schools.

▼ **FIGURE 5-43 IRISH** The percentage of Irish speakers is highest in the most remote portions of the country.

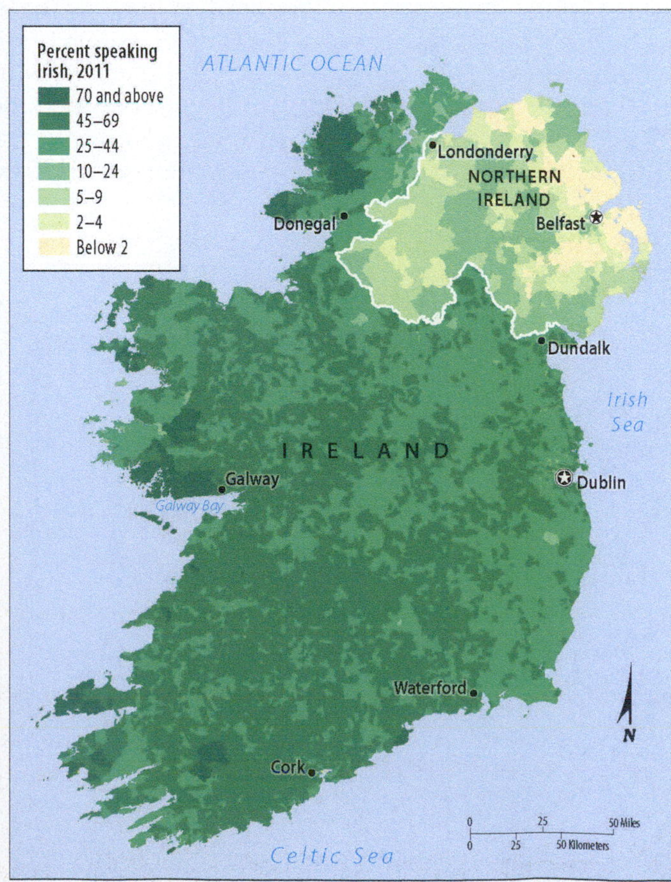

▲ **FIGURE 5-44 SCOTTISH GAELIC** An anti-littering sign says "STOP!" at top; (left) "Don't do this" at left, "Do this" at right.

GOIDELIC CELTIC

Two Goidelic languages survive: Irish Gaelic and Scottish Gaelic.

IRISH. Irish is one of two official languages in the Republic of Ireland, along with English. According to a 2011 census, Irish is spoken by 94,000 people on a daily basis, and 1.3 million say that they can speak it and use it at least occasionally. As with Welsh, the percentage of people speaking Irish is higher in the more remote areas of the country (Figure 5-43).

An Irish-language TV station began broadcasting in 1996. English road signs were banned from portions of western Ireland in 2005. The revival is being led by young Irish living in other countries who wish to distinguish themselves from the English (in much the same way that Canadians traveling abroad often make efforts to distinguish themselves from U.S. citizens). Irish singers, including many rock groups, have begun to record and perform in Gaelic. In the 1300s, the Irish were forbidden to speak their own language in the presence of their English masters.

SCOTTISH GAELIC. In Scotland 59,000, or 1 percent of the people, speak Scottish Gaelic (Figure 5-44). An extensive body of literature exists in Gaelic languages, including the Robert Burns poem *Auld Lang Syne* ("old long since"), the basis for the popular New Year's Eve song. Gaelic was carried from Ireland to Scotland about 1,500 years ago.

Meanwhile, *Ethnologue* considers Scots to be an Indo-European Germanic branch language separate from English, not merely a dialect of English. Scots is said to be the first language for 90,000 people in Scotland and the second language for 1.5 million.

PAUSE & REFLECT 5.4.2

Google Translate includes Irish and Welsh. Type something in English and see whether similar words are used in Irish and in Welsh and whether the words are similar to English.

Isolated and Extinct Languages

LEARNING OUTCOME 5.4.3

Understand geographic factors resulting in isolated and extinct languages.

Similarities and differences between languages—our main form of communication—are a measure of the degree of interaction among groups of people. Some languages bear no similarity to other languages. Isolation from others has helped to preserve some of these languages, but in other cases it may hasten their demise.

ISOLATED LANGUAGES

An **isolated language** is a language that is unrelated to any other and therefore not attached to any language family. An isolated language arises because the speakers of that language have limited interaction with speakers of other languages.

An isolated language is considered vigorous if it is in full use in the community and is being learned by children as their first language. *Ethnologue* to be in vigorous use considers only 6 of 82 isolated languages (Figure 5-45). A vigorous language is used for face-to-face communication by both young and elderly people. As a result of its usage in daily life, the language is judged by *Ethnologue* to be sustainable—that is, likely to survive at least in the near future. The other isolated languages are considered to be endangered.

Basque is the only example of a vigorous isolated language in Europe. It is the first language of 600,000 people in the Pyrenees Mountains of northern Spain and southwestern France (refer to Figure 5-11, the gray area in northern Spain). No attempt to link Basque to the common origin of the other European languages has been successful.

Basque was probably once spoken over a wider area but was abandoned where its speakers came in contact with Indo-Europeans. It is considered the only language currently spoken in Europe that survives from the period before the arrival of Indo-European speakers. Basque's lack of connection to other languages reflects the isolation of the Basque people in their mountainous homeland. This isolation has helped them preserve their language in the face of the wide diffusion of Indo-European languages (Figure 5-46).

The diffusion of Indo-European languages demonstrates that a common ancestor dominated much of Europe before recorded history. No attempt to link Basque to the common origin of Indo-European languages has been successful. Basque's lack of connection to other languages reflects the isolation of the Basque people in their mountainous homeland. This isolation has helped them preserve their language in the face of the wide diffusion of Indo-European languages.

AN UNCHANGING LANGUAGE. Icelandic is not an isolated language because it is in the North Germanic group of the Germanic branch of the Indo-European family. Icelandic's significance is that over the past 1,000 years, it has changed less than any other language in

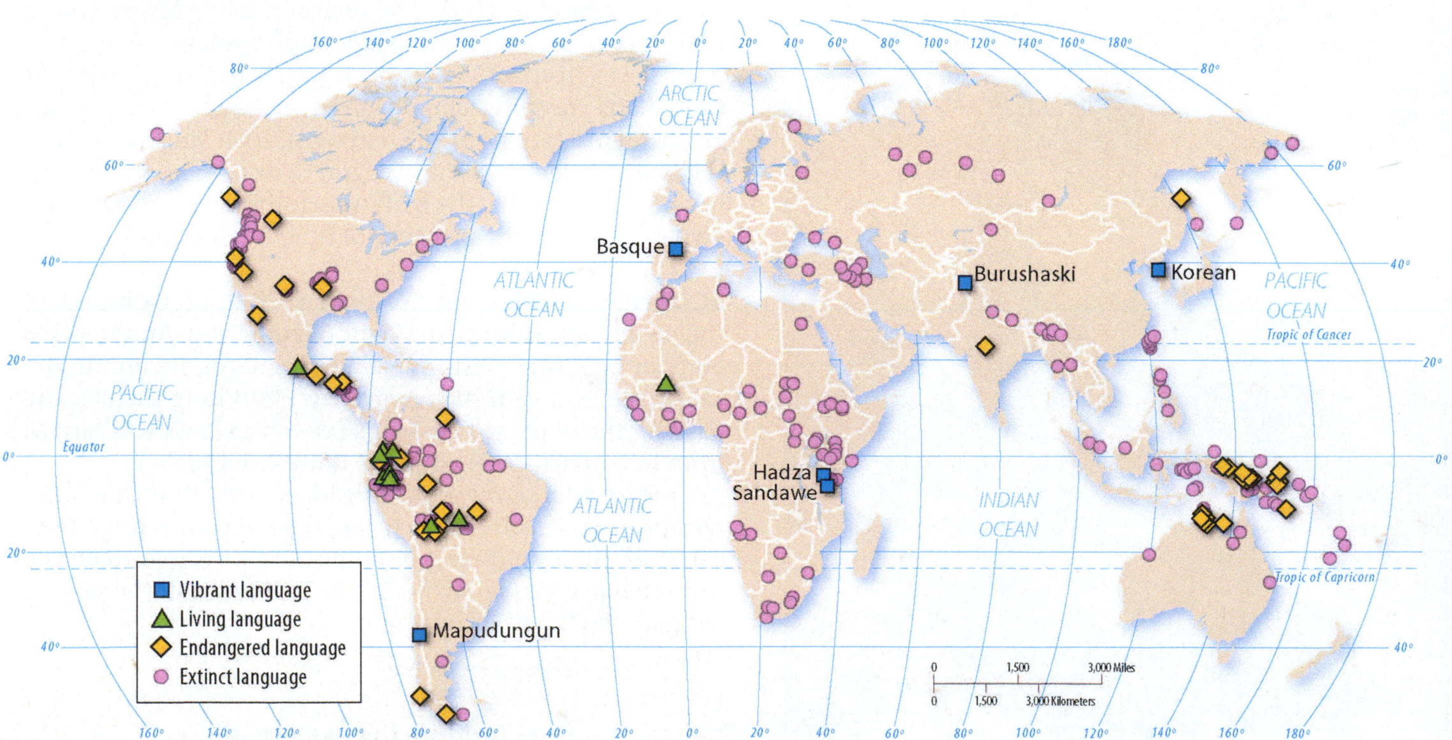

▲ **FIGURE 5-45 ISOLATED LANGUAGES**

◀ **FIGURE 5-46 BASQUE** Marchers in the Basque region capital Bilbao protest budget cuts. The banner says in Spanish at left "for a social Europe" and in Basque at right "no restrictions."

the Germanic branch (Figure 5-48). As was the case with England, people in Iceland speak a Germanic language because their ancestors migrated to the island from the east, in this case from Norway. Norwegian settlers colonized Iceland in A.D. 874.

When an ethnic group migrates to a new location, it takes along the language spoken in the former home. The language spoken by most migrants—such as the Germanic invaders of England—changes in part through interaction with speakers of other languages. But in the case of Iceland, the Norwegian immigrants had little contact with speakers of other languages when they arrived in Iceland, and they did not have contact with speakers of their language back in Norway. After centuries of interaction with other Scandinavians, Norwegian and other North Germanic languages had adopted new words and pronunciation, whereas the isolated people of Iceland had less opportunity to learn new words.

PAUSE & REFLECT 5.4.3

Why do you think the sign in Figure 5-47 is written in both Icelandic and English?

EXTINCT LANGUAGES

An **extinct language** is a language that was once used by people in daily activities but is no longer in use. *Ethnologue* estimates that 367 languages have become extinct since 1950, a rate of 6 per year. The United Nations identifies 231 recently extinct languages (refer to Figure 5-45).

Two examples of recently extinct languages are Liv and Clallam. Liv, a language belonging to the Uralic family, became extinct on June 5, 2013, when its last speaker, Grizelda Kristina, died in Latvia. Around 200 ethnic Livonians live along the northwestern coast of Latvia, but none of them speak Liv.

Clallam, a language once spoken in Washington's Olympic Peninsula and Vancouver Island, Canada, became extinct on February 4, 2014, when its last-known speaker, Hazel Sampson, died at age 103. The language is taught at Port Angelos High School in Washington, and six youths are reported to be able to speak some words of Clallam as their second language. A dictionary of Clallam was published in 2012. If someday a child is taught Clallam as his or her first language, perhaps by one of the six who know some words, then—like Myaamia—the language might be reclassified as reawakening.

When Spanish missionaries reached the eastern Amazon region of Peru in the sixteenth century, they found more than 500 languages. Only 92 survive today, according to *Ethnologue*, and 14 of them face immediate extinction because fewer than 100 speakers remain. Of Peru's 92 surviving indigenous languages, only Cusco, a Quechuan language, is currently used by more than 1 million people. *Ethnologue* lists 74 languages based in the United States that are now extinct, and UNESCO lists 54. These are languages once spoken by groups of Native Americans, especially in the West.

The loss of many languages is a reflection of globalization. To be part of a global economy and culture, people choose to use a widely used language, leaving their traditional or indigenous language to disappear.

▼ **FIGURE 5-47 ICELANDIC** The warning sign in Icelandic and English is located in Hveragerdi, Iceland.

New and Growing Languages

LEARNING OUTCOME 5.4.4

Understand processes of creation of new languages.

While the number of languages in the world is declining, a handful of languages are being invented or revived. In other cases, endangered languages are being preserved before they become extinct. These efforts reflect the importance that groups place on language as an element of local culture.

NEW LANGUAGES

Isolated languages continue to be identified and documented, and entirely new languages are invented (see Doing Geography feature). For example, a research team from Oregon's Living Tongues Institute for Endangered Languages was in India to study rarely spoken languages. They heard people in the area speaking a language that was not listed in *Ethnologue*. The researchers concluded that what they were hearing was a distinct language belonging to the Tibeto-Burman branch of the Sino-Tibetan family. The language, now known as Koro Aka, is now listed in *Ethnologue* as a language of northeastern India with 1,500 speakers.

GROWING LANGUAGES

Hebrew is an example of a language that was once rarely used but is more commonly used now. Most of the Jewish Bible and Christian Old Testament were written in Hebrew. A language of daily activity in biblical times, Hebrew diminished in use in the fourth century B.C. and was thereafter retained primarily for Jewish religious services. At the time of Jesus, most people in present-day Israel spoke Aramaic, which in turn was replaced by Arabic.

When Israel was established as an independent country in 1948, Hebrew became one of the new country's two official languages, along with Arabic (Figure 5-48). Hebrew was chosen because the Jewish population of the State of Israel consisted of refugees and migrants from many countries who spoke many languages. Because Hebrew was still used in Jewish prayers, no other language could so symbolically unify the disparate cultural groups in the new country.

The task of making Hebrew a modern language was formidable. Words had to be created for thousands of objects and inventions unknown in biblical times, such as telephones, cars, and electricity. The revival effort was initiated by Eliezer Ben-Yehuda, who lived in Palestine before the creation of the State of Israel and refused to speak any language other than Hebrew. Ben-Yehuda is credited with the invention of 4,000 new Hebrew words—related when possible to ancient ones—and the creation of the first modern Hebrew dictionary.

CHECK-IN KEY ISSUE 4

Why Do Local Languages Survive?

- ✓ **Many languages have become extinct and others are threatened with extinction.**
- ✓ **Some endangered languages are being preserved and protected.**
- ✓ **Some lesser-used languages are growing in number of speakers. Other languages are being invented, in some cases through combination with English.**

◀ **FIGURE 5-48 HEBREW** The street sign in Jerusalem is in (top) Hebrew, (middle) Arabic, and English.

DOING GEOGRAPHY Finding a New Language

Some linguists spend time behaving like geographers: They head off to remote places to study endangered languages and unusual dialects. Every so often, linguists acting like geographers uncover something more very special: a language previously unknown to the rest of the world.

Even rarer is the discovery of a new language that was recently invented. This happened with a language now called Warlpiri rampaku, or Light Warlpiri. It is the first language of about one-half of the 700 residents of the village of Lajamanu in Australia's Northern Territory. No one outside the village knew about the language until a University of Michigan researcher, Carmel O'Shannessy, heard it in 2013.

Isolation is the perfect breeding ground for an entirely new language, and Lajamanu is about as isolated as a village can be. Lajamanu has a single shop that is restocked once a week by a truck. It is 557 kilometers from the nearest town offering other services. A community-owned airplane gets other items and people into and out of the village.

Young people invented the new language. It started with parents speaking to their babies in a combination of three languages: Strong Warlpiri, English, and a Creole language that combines English and Strong Warlpiri. As they got older, the children combined the three languages into an entirely new one through actions such as inventing new verb endings and tenses. For example, Light Warlpiri has a verb tense that means present or past but not future.

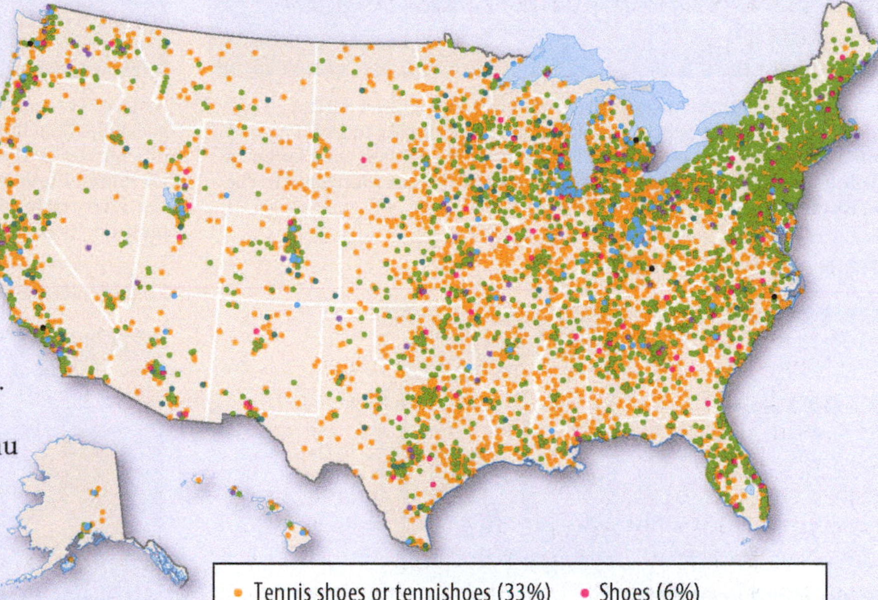

▲ **FIGURE 5-49 TERMS FOR RUBBER-SOLED SHOES WORN FOR ATHLETICS** The two most common terms are tennis shoes (especially in the Midwest) and sneakers (especially in the northeast).

What's Your Language Geography?

Inventing New Words Young people rarely invent entirely new languages, as they have done in Lajamanu. But they constantly invent new words and even distinct dialects.

1. Search YouTube.com for "Monster Story in Light Warlpiri" videos.

   What Light Warlpiri words do you see in the videos that are clearly borrowed from English?
2. What examples of new words can you find from music or from the use of electronic devices?
3. Take a quiz to determine whether you have a strong regional dialect. At nytimes.com, search for either "Harvard Dialect Survey Quiz" or "How Y'all, Youse and You Guys" (Figure 5-49). Does the quiz accurately identify where you are from?
4. Are any indicators of dialect especially important in identifying your distinctive use of language? Why or why not?
5. Does this map from the dialect quiz accurately reflect what you call your shoes?

Summary & Review

KEY ISSUE 1

Where are the world's languages distributed?

Languages are organized into families, branches, and groups. All but 1 percent of the world's people speak a language belonging to 1 of 14 language families. Nearly one-half of the world's people speak a language in the Indo-European family, and another one-fifth speak a language in the Sino-Tibetan family.

THINKING GEOGRAPHICALLY

▲ **FIGURE 5-50 CHINESE ELECTRONICS** This is the registration page for Twitter in Chinese.

1. As Chinese languages become increasingly important as lingua franca, what distinctive elements make it especially attractive to use, and which make it especially difficult to use?

KEY ISSUE 2

Where did English and related languages originate and diffuse?

English is an Indo-European language. Indo-European includes four branches with widely used languages and four other branches. English originated following the invasion of England by Germanic tribes. More recent patterns of migration have diffused English to other places. All Indo-European languages have a common prehistoric ancestor, although its place of origin and process of diffusion are disputed.

THINKING GEOGRAPHICALLY

2. The two theories of the origin of Indo-European depend on different views underlying the reasons that people migrate. Do you think the "war" theory or the "peace" theory better explains the migration of early humans? Why?

▲ **FIGURE 5-51 ORIGIN OF INDO-EUROPEAN?** Karain Cave in Turkey was first occupied 200,000 years ago. The first speakers of Indo-European may have lived near here.

KEY ISSUE 3

Why do individual languages vary among places?

A dialect is a regional variation of a language. Dialects can vary in terms of vocabulary, spelling, and pronunciation. Some dialects become sufficiently different to justify classifying them as separate languages. This is important for some cultural groups as a symbol of cultural identity. Some countries peacefully embrace multiple languages, whereas in other places the multiplicity of languages is a source of tension.

THINKING GEOGRAPHICALLY

3. According to Figure 5-28c, which dialects of British English are forecast to expand, and which are expected to contract by 2030?
4. What geographic factors might account for this changing distribution?

▲ **FIGURE 5-52 ENGLISH IN LONDON**

KEY ISSUE 4

Why do local languages survive?

Many languages are endangered because not enough people continue to use them or teach them to children. Lack of interaction has helped a few isolated languages to survive, while interaction with other languages has caused the extinction of others. Some threatened languages, such as the Celtic branch of Indo-European, are being preserved with the help of governments.

THINKING GEOGRAPHICALLY

5. Myaamia is an example of a reawakening language. What would have to happen for the language to become even more vigorous in the future?

▲ **FIGURE 5-53 REAWAKENING LANGUAGE** *Ethnologue* classifies Ohlone as one of five reawakening languages in the United States. A reconstructed Ohlone village house sites in Coyote Hills Regional Park, California.

KEY TERMS

Creole *(or creolized)* language *(p. 167)* A language that results from the mixing of a colonizer's language with the indigenous language of the people being dominated.

Denglish *(p. 159)* A combination of *Deutsch* (the German word for German) and *English*.

Developing language *(p. 147)* A language spoken in daily use with a literary tradition that is not widely distributed.

Dialect *(p. 164)* A regional variety of a language distinguished by vocabulary, spelling, and pronunciation.

Ebonics *(p. 167)* A dialect spoken by some African Americans.

Extinct language *(p. 177)* A language that was once used by people in daily activities but is no longer used.

Franglais *(p. 159)* A combination of *français* and *anglais* (the French words for French and English, respectively).

Institutional language *(p. 147)* A language used in education, work, mass media, and government.

Isogloss *(p. 166)* A boundary that separates regions in which different language usages predominate.

Isolated language *(p. 176)* A language that is unrelated to any other languages and therefore not attached to any language family.

Language *(p. 146)* A system of communication through the use of speech, a collection of sounds understood by a group of people to have the same meaning.

Language branch *(p. 147)* A collection of languages related through a common ancestor that can be confirmed through archaeological evidence.

Language family *(p. 147)* A collection of languages related to each other through a common ancestor long before recorded history.

Language group *(p. 147)* A collection of languages within a branch that share a common origin in the relatively recent past and display relatively few differences in grammar and vocabulary.

Lingua franca *(p. 160)* A language mutually understood and commonly used in trade by people who have different native languages.

Literary tradition *(p. 147)* A language that is written as well as spoken.

Logogram *(p. 161)* A symbol that represents a word rather than a sound.

Official language *(p. 147)* The language adopted for use by a government for the conduct of business and publication of documents.

Pidgin language *(p. 160)* A form of speech that adopts a simplified grammar and limited vocabulary of a lingua franca; used for communications among speakers of two different languages.

Received Pronunciation (RP) *(p. 165)* The dialect of English associated with upper-class Britons living in London and now considered standard in the United Kingdom.

Spanglish *(p. 159)* A combination of *Spanish* and *English* spoken by Hispanic Americans.

Standard language *(p. 165)* The form of a language used for official government business, education, and mass communications.

Subdialect *(p. 164)* A subdivision of a dialect.

Vigorous language *(p. 147)* A language that is spoken in daily use but that lacks a literary tradition.

Vulgar Latin *(p. 156)* A form of Latin used in daily conversation by ancient Romans, as opposed to the standard dialect, which was used for official documents.

EXPLORE

Wentworth Avenue, Chicago

Use Google Earth to explore *Wentworth Avenue in Chicago*.

Fly to 2230 S *Wentworth, Chicago, IL*.

Use Street View to click on *Wentworth Avenue*.

1. What language other than English do you see?
2. What factors probably determine the amount of English used on signs along Wentworth Avenue?

GeoVideo Log in to the **MasteringGeography** Study Area to view this video.

Israel: Reinventing Hebrew

The ancient language of Hebrew has been reinvented as a modern, living language.

1. What were the advantages to using Hebrew as the national language of Israel?
2. What are the disadvantages to using an ancient language in the modern world?
3. How does the video explain using a modernization of an ancient term (such as for "battery") and for using an English term (such as for "puncture")?

MasteringGeography™

Looking for additional review and test prep materials?

Visit the Study Area in **MasteringGeography™** to enhance your geographic literacy, spatial reasoning skills, and understanding of this chapter's content by accessing a variety of resources, including MapMaster interactive maps, videos, *In the News* RSS feeds, flashcards, web links, self-study quizzes, and an eText version of *The Cultural Landscape*.

www.masteringgeography.com

6

Religions

Religions interest geographers because understanding them is essential for recognizing spatial patterns underlying how humans occupy Earth. Many people care deeply about their religion and draw from religion their core values and beliefs, an essential element of the definition of culture.

Geographers document the places where various religions are located in the world and offer explanations for why some religions have widespread distributions and others are highly clustered in particular places. The predominant religion varies among regions of the world, as well as among regions within North America.

Interfaith rally, Dearborn, Michigan.

LOCATIONS IN THIS CHAPTER

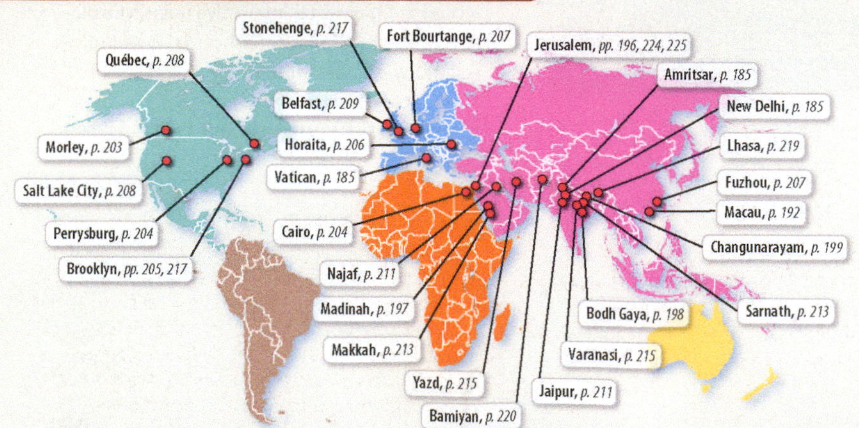

KEY ISSUES

1 Where Are the World's Religions Distributed?

Religions have distinctive distributions across the ***space*** of Earth. Some religions are highly clustered in one or two of Earth's ***regions*** whereas others are distributed throughout the world. Individual religions also have distinctive distributions within regions.

2

Why Do Religions Have Distinctive Distributions?

Geographers study spatial ***connections*** in religions. Some religions have well-documented places of origin, whereas others have unknown origins. Some religions have diffused extensively from their places of origin, whereas others have experienced limited diffusion.

3

Why Do Religions Organize Space in Distinctive Patterns?

Religions display distinctive patterns on the landscape of particular ***places***. Religions derive distinctive meaning from the physical landscape and construct places for worship and other religious practices.

4

Why Do Territorial Conflicts Arise Among Religious Groups?

Adherents of various religions have disputes over Earth's territory. The attempt by adherents of one religion to organize a portion of Earth's surface can create conflicts with governments or with other religious groups. Intense disputes occur at all ***scales***, from extensive world regions to a few square meters.

KEY ISSUE 1

Where Are the World's Religions Distributed?

- Introducing Religions
- Global Distribution of Religions
- Distribution of Christians
- Distribution of Muslims and Buddhists
- Distribution of Ethnic Religions
- Distribution of Other Religions

LEARNING OUTCOME 6.1.1

Identify the world's major religions.

Religion, like language, can be a source of pride and a means of identification with a distinct culture. As with language, migrants take their religion with them to new locations, but language and religion have important geographic differences. Although most migrants learn the language of the new location, they typically retain their religion. Furthermore, people can learn a globally important language such as English and at the same time still speak the language of their local culture, but most (though not all) religions require exclusive adherence, so adopting a new religion could require turning away from the former one.

Introducing Religions

Only a few religions can claim the adherence of large numbers of people. This section identifies the major religions and their distributions.

Reliable statistics on the number of adherents are difficult to obtain. No official count of religious membership is taken in the United States or in many other countries. China conducted a census of religion in 2007 but gave respondents only five choices and excluded several important possibilities. Statistics on the number of followers of religions can be controversial; adherents may feel that their religion has been undercounted and therefore accorded less prominence than deserved in world and regional data.

PAUSE & REFLECT 6.1.1

How should a child's religion be classified if the two parents adhere to different religions?

Most international statistics in this chapter come from *Adherents.com,* the Association of Statisticians of American Religious Bodies, the Pew Research Center, and World Religion Database. None of these organizations is affiliated with a particular religion. The statistics are built on self-identification—that is, whatever people themselves view as the particular religious groups to which they belong. They do not measure how actively an individual practices a religion.

The world's religions can be grouped as follows:

- **Four largest religions.** These four religions together claim the adherence of 77 percent of the world's people (Figure 6-1): Christianity (Figure 6-2), Islam, Hinduism, and Buddhism. Pew estimates that 2.2 billion people in the world view themselves as Christian, 1.6 billion as Muslim, 1 billion as Hindu, and 500 million as Buddhist.
- **Folk religions.** A number of religions grouped by Pew under the title "folk religions" account for an estimated 6 percent of the world's population, although the number of adherents of these religions is especially difficult to measure. *Adherents.com* identifies the three largest groups of folk religions as Chinese traditional, primal-indigenous, and African traditional.
- **Other religions.** Another 1 percent of the world's people adhere to a number of other religions. The four most numerous of these other religions—Juchte, Judaism, Sikhism (Figure 6-3), and Spiritism—have between 14 and 23 million adherents each. Six other religions have between 1 and 10 million adherents: Bahá'í (Figure 6-4), Cao Dai, Jainism, Shinto, Tenrikyo, and Zoroastrianism. Many other religions have fewer than 1 million adherents.
- **Unaffiliated.** The remaining 16 percent of the world's population are unaffiliated with a religion. According to *Adherents.com,* most people in this category affirm neither belief nor lack of belief in God or some other

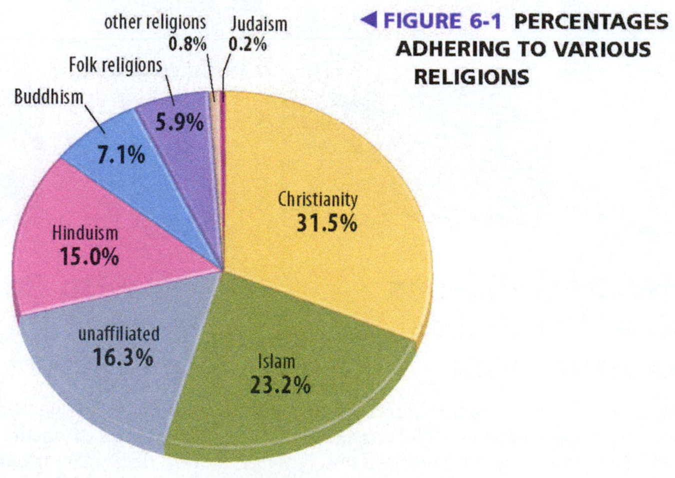

FIGURE 6-1 PERCENTAGES ADHERING TO VARIOUS RELIGIONS

▲ **FIGURE 6-2 CHRISTIANITY** Saint Peter's Basilica in the Vatican, one of the holiest shrines for Roman Catholics.

Higher Power. In the United States, many classified as unaffiliated believe in God and attend a religious service at least on occasion, but they do not have a formal association with a religious institution. In some countries, the unaffiliated are primarily people who express no religious interest or preference and do not participate in any organized religious activity. Some people in this group espouse **atheism,** which is belief that God does not exist, or **agnosticism,** which is belief that the existence of God can't be proven empirically.

CLASSIFYING RELIGIONS

Geographers distinguish between two types of religions:

- **Universalizing religions** attempt to be global—to appeal to all people, wherever they may live in the world, not just to those of one culture or location.
- **Ethnic religions** appeal primarily to one group of people living in one place.

▲ **FIGURE 6-4 BAHÁ'Í HOUSE OF WORSHIP** New Delhi, India.

Geographers consider the distinction between universalizing and ethnic religions to be significant because the two types of religions tend to display different spatial characteristics, including origin, diffusion, and distribution. In reality, the distinction between the two types of religions is not absolute because most religions display both universalizing and ethnic elements.

Among the five largest religious groups, Christianity, Islam, and Buddhism are considered universalizing, whereas Hinduism and folk religions are considered ethnic. Most of the religions with fewer adherents are ethnic, but several are universalizing, and others display features of both and so are especially difficult to classify.

▶ **FIGURE 6-3 GOLDEN TEMPLE, AMRITSAR**

Global Distribution of Religions

LEARNING OUTCOME 6.1.2

Describe the distribution of major religions.

At first glance, the global distribution of religions appears to be straightforward. In all but a handful of countries, the religion with the largest number of adherents is either Christianity or Islam (Figure 6-5). Christianity is the most widely practiced religion in every country of the Western Hemisphere and in most countries of Europe and sub-Saharan Africa. Islam is the most widely practiced religion in nearly every country of Southwest Asia & North Africa, as well as in Central Asia. Asia also has countries where the most widely practiced religion is Buddhism or Hinduism. In several countries, including China, the largest number of people are unaffiliated with any religion. Judaism is the most widely practiced religion in the State of Israel.

Figure 6-5 must be viewed with caution because the map of religions is in reality more complex, especially at regional and local scales. Figure 6-5 does not display the variety of other religions found in most countries in addition to the most numerous one, nor does it account for regional variations within individual countries. Furthermore, many faiths are divided into branches that have distinctive spatial distributions, as will be discussed in more detail beginning in the next section.

According to Pew, 27 percent of the world's people live in countries where their religion is in the minority. This includes 3 percent of Hindus, 13 percent of Christians, 27 percent of Muslims, 29 percent of unaffiliated, 59 percent of Jews, 72 percent of Buddhists, 99 percent of folk religionists, and 100 percent of other religious groups. In addition, a large percentage of Christians and Muslims live in countries where their branch of the religion is the minority.

REGIONAL DIVERSITY OF RELIGIONS

Some world regions have a large majority of adherents of one religion, whereas others have more of a mix. More than 90 percent of Latin Americans and more than 75 percent of Europeans and North Americans identify themselves as adhering to Christianity. Muslims comprise more than 90 percent of the population of Central Asia and of Southwest Asia & North Africa (refer to the pie charts in Figure 6-5).

The four regions with more diverse religious composition are East Asia, South Asia, Southeast Asia, and sub-Saharan Africa:

- In East Asia, more than one-half of the people are unaffiliated with any religion. Most of the other one-half are divided about equally between Buddhism and folk religions.
- In South Asia, around two-thirds of the people are Hindus and one-third Muslims. However, adherents are sharply divided by country; India is 80 percent Hindu, whereas Bangladesh and Pakistan are more than 90 percent Muslim.

▼ FIGURE 6-5 MOST NUMEROUS RELIGIONS BY COUNTRY AND WORLD REGION. A darker shading represents a greater prevalence of the majority religion.

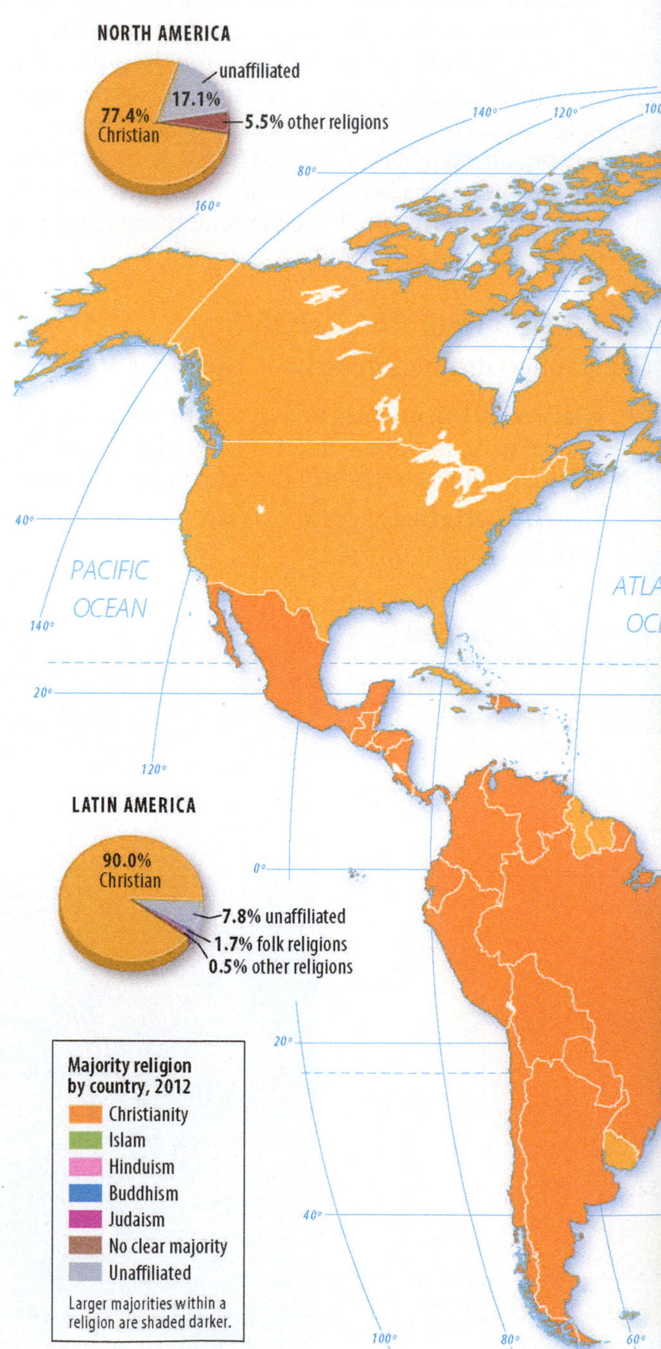

- In Southeast Asia, around 40 percent are Muslims, 24 percent Buddhists, and 21 percent Christians. Indonesia (the region's most populous country) is 87 percent Muslim, Cambodia and Thailand are more than 90 percent Buddhist, and the Philippines is more than 90 percent Christian.
- In sub-Saharan Africa, around two-thirds are Christian and one-third Muslim.

PAUSE & REFLECT 6.1.2

In what ways is the distribution of religion and language families similar?

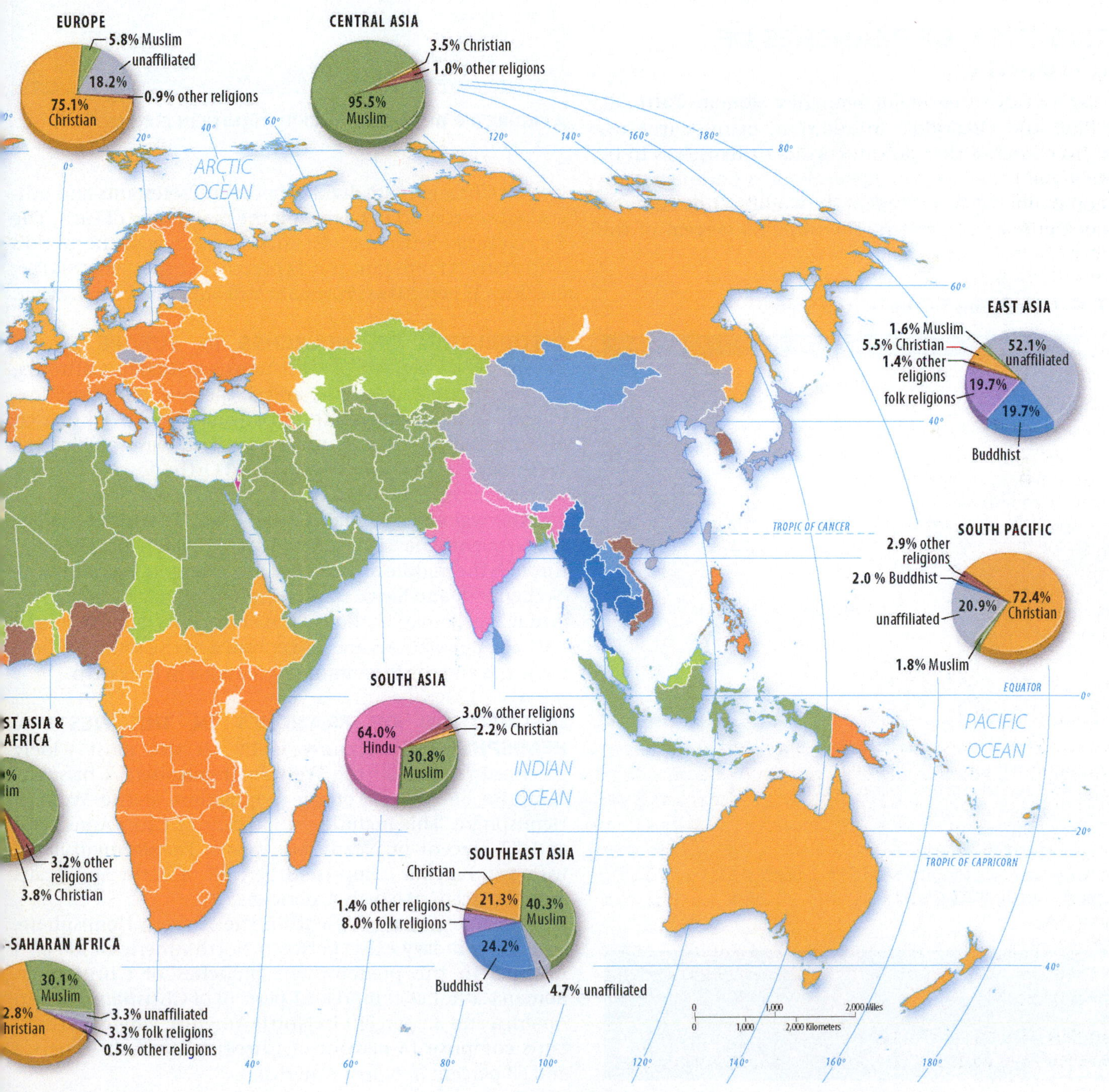

Distribution of Christians

LEARNING OUTCOME 6.1.3

Describe regional variations in the distribution of Christian branches.

Many religions, including the three most widely practiced universalizing religions, are divided into congregations, denominations, and branches. A **congregation** is a local assembly of persons brought together for common religious worship. A **denomination** unites a number of local congregations in a single legal and administrative body. A **branch** is a large and fundamental division within a religion.

DISTRIBUTION OF BRANCHES OF CHRISTIANITY

Christianity has three major branches: Roman Catholic, Protestant, and Orthodox. In addition, many Christians belong to churches that do not consider themselves to be within any of these three branches. Roman Catholics comprise approximately 50 percent of the world's Christians and Orthodox around 12 percent. The other 38 percent of the world's Christians are divided between Protestants and others, but sources do not agree on the magnitude of each. The *Encyclopaedia Britannica* classifies 24 percent of the world's Christians as Protestant and 14 percent as other, whereas Pew classifies 37 percent as Protestant and only 1 percent as other.

TABLE 6–1 Religions of the United States

Christian			78.5
Roman Catholic		29.3	
Evangelical Protestant Churches		26.3	
Southern Baptist Convention	6.7		
Independent Baptist in the Evangelical Tradition	2.5		
Pentecostal in the Evangelical Tradition (other than Assemblies of God)	2.0		
Other Baptist in the Evangelical Tradition	1.6		
Lutheran in the Evangelical Tradition	1.8		
Assemblies of God (Pentecostal in the Evangelical Tradition)	1.4		
Restorationist in the Evangelical Tradition (primarily Church of Christ)	1.7		
Other Evangelical Protestant Traditions	8.6		
Mainline Protestant Churches		18.1	
Baptist in the Mainline Tradition	1.9		
Methodist in the Mainline Tradition	5.4		
Lutheran in the Mainline Tradition	2.8		
Presbyterian in the Mainline Tradition	1.9		
Anglican/Episcopal in the Mainline Tradition	1.4		
Other Mainline Protestant Traditions	4.7		
Historically Black Churches		6.9	
National Baptist Convention	1.8		
Other Baptist in the Historically Black Tradition	2.6		
Other Historically Black Churches	2.5		
Mormon		1.7	
Other Christian		1.6	
Jewish			1.7
Buddhist			0.7
Muslim			0.6
Hindu			0.4
Other religions or don't know			2.0
Unaffiliated			16.1

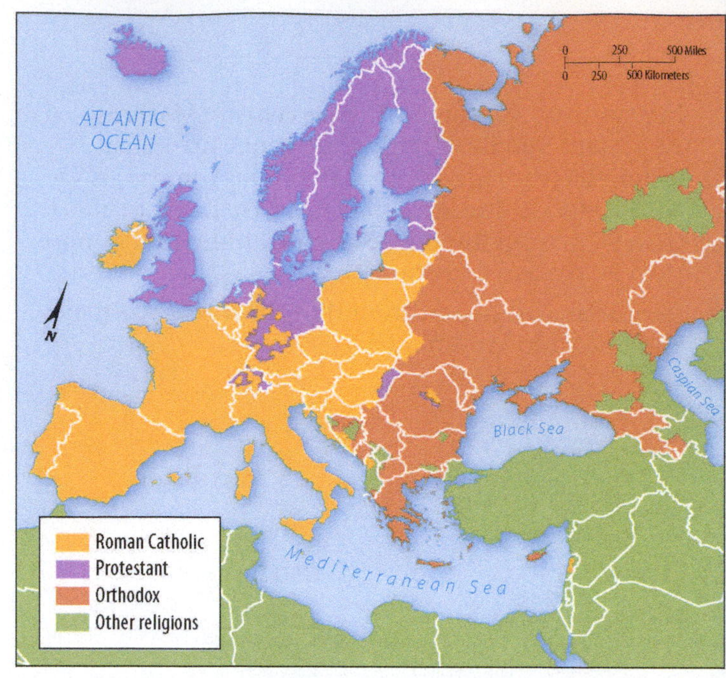

▲ **FIGURE 6-6 BRANCHES OF CHRISTIANITY IN EUROPE**

DISTRIBUTION OF BRANCHES IN EUROPE. Overall in Europe, 47 percent of Christians are Roman Catholics, 18 percent are Protestants, and 35 percent are Orthodox. Roman Catholicism is the most widely practiced branch of Christianity in the southwest and east of Europe, Protestantism in the northwest, and Orthodoxy in the east and southeast (Figure 6-6).

The regions of Roman Catholic and Protestant majorities frequently have sharp boundaries, even when they run through the middle of countries. For example, Germany, the Netherlands and Switzerland have approximately equal percentages of Roman Catholics and Protestants, but the Roman Catholic populations are concentrated in the south of these countries and the Protestant populations in the north.

DISTRIBUTION OF BRANCHES IN THE WESTERN HEMISPHERE. Christianity is by far the most widely practiced religion in the Western Hemisphere. Christians comprise 86 percent of the population of the Western Hemisphere. This includes 90 percent of Latin Americans and 77 percent of North Americans. People unaffiliated with any religion comprise 8 percent of Latin Americans and 17 percent of North Americans.

At the regional scale within the Western Hemisphere, a sharp boundary exists between North America and Latin America in the predominant branches of Christianity. Roman Catholics comprise 81 percent of Christians in Latin America and 32 percent in North America, whereas Protestants comprise 18 percent of Christians in Latin America and 63 percent in North America.

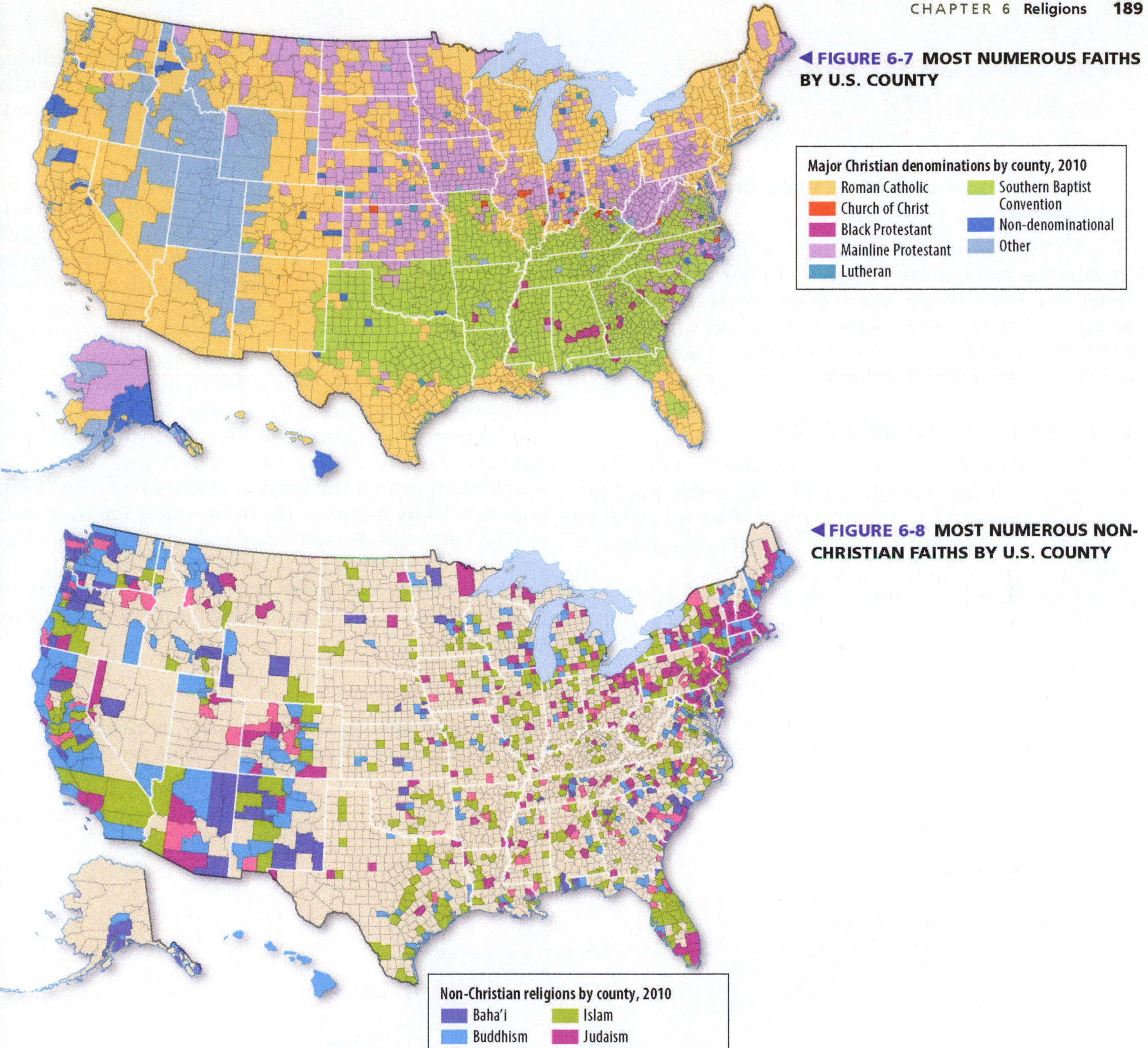

◀ **FIGURE 6-7 MOST NUMEROUS FAITHS BY U.S. COUNTY**

◀ **FIGURE 6-8 MOST NUMEROUS NON-CHRISTIAN FAITHS BY U.S. COUNTY**

The diversity of faiths in the United States is displayed in Table 6-1. Roman Catholics comprise 29 percent of the U.S. population, Evangelical Protestants 26 percent, Mainline Protestants 18 percent, historically black churches 7 percent, other Christians 3 percent, and other faiths 5 percent. Southern Baptist Convention is the most numerous Evangelical Protestant church, and Methodist is the most numerous Mainline Protestant church. Roman Catholics are more numerous in the Northeast and Southwest, whereas Evangelical Protestants are most numerous in the Southeast (Figure 6-7). Sixteen percent of Americans are unaffiliated, and most of them respond that they are "nothing in particular."

Other less numerous faiths have distinctive distributions within the United States. Members of The Church of Jesus Christ of Latter-day Saints (Mormons) regard their church as separate from the three branches of Christianity. About 2 percent of Americans are members of the Latter-day Saints, and a large percentage is clustered in Utah and surrounding states. Other less numerous faiths have distinctive distributions within the United States. For example, Jews are more likely to be in the Northeast and Buddhists on the West Coast (Figure 6-8).

PAUSE & REFLECT 6.1.3

Based on what you see in Table 6-1, what are some of the largest Christian denominations in the United States that do not have highly clustered distributions shown in Figure 6-7?

Distribution of Muslims and Buddhists

LEARNING OUTCOME 6.1.4

Describe the distribution of the major branches of Islam and Buddhism.

Islam is the predominant religion of Central Asia and of Southwest Asia & North Africa. Buddhism is clustered primarily in East Asia and Southeast Asia. Like Christianity, Islam and Buddhism are divided into major branches with distinctive geographic distributions.

DISTRIBUTION OF MUSLIMS

The word *Islam* in Arabic means "submission to the will of God," and it has a similar root to the Arabic word for "peace." An adherent of the religion of Islam is known as a *Muslim*, which in Arabic means "one who surrenders to God."

On a standard world map, such as Figure 6-5, Islam predominates in Central Asia and in Southwest Asia & North Africa. However, on a cartogram, most of the world's Muslims live further east, in South and Southeast Asia. The countries with the most Muslims are Indonesia, Pakistan, India, and Bangladesh. These four countries together are home to more than 40 percent of the world's Muslims (Figure 6-9).

ISLAM'S BRANCHES. Islam is divided into two principal branches: Sunni and Shiite. The word *Sunni* comes from the Arabic for "people following the tradition of Muhammad." The word *Shiite* (sometimes spelled *Shia*) comes from the Arabic word for "party" or "support group."

Sunnis comprise 88 percent of Muslims and are the most numerous branch in most Muslim countries in Southwest Asia & North Africa, as well as in Southeast Asia. Sunnis follow various schools of thought and religious law, which have distinctive regional distributions (Figure 6-10). The Hanafi, Hanbali, Maliki, and Shafi'i schools of thought and religious law are named for their founders.

Shiites are the largest branch in Azerbaijan, Bahrain, Iran, Iraq, Lebanon, and Yemen. Nearly 40 percent of all Shiites live in Iran, 15 percent in Pakistan, 12 percent in India, and 10 percent in Iraq. Shiite Islam is divided into three principal schools of thought, based in part on disputes over leadership after the Prophet Muhammad. The largest, known as Ithna Ashari, is the most widely followed tradition in Azerbaijan, Bahrain, Iran, Iraq, and Lebanon. Other traditions include the Ismaili and Zaidi. Ismailis are clustered in Pakistan and Zaidiyyahs in Yemen. A third branch of Islam, Ibadi, is the predominant form of Islam adhered to in Oman.

PAUSE & REFLECT 6.1.4

What countries in Figure 6-10 appear to have large concentrations of both Sunni and Shiite Muslims?

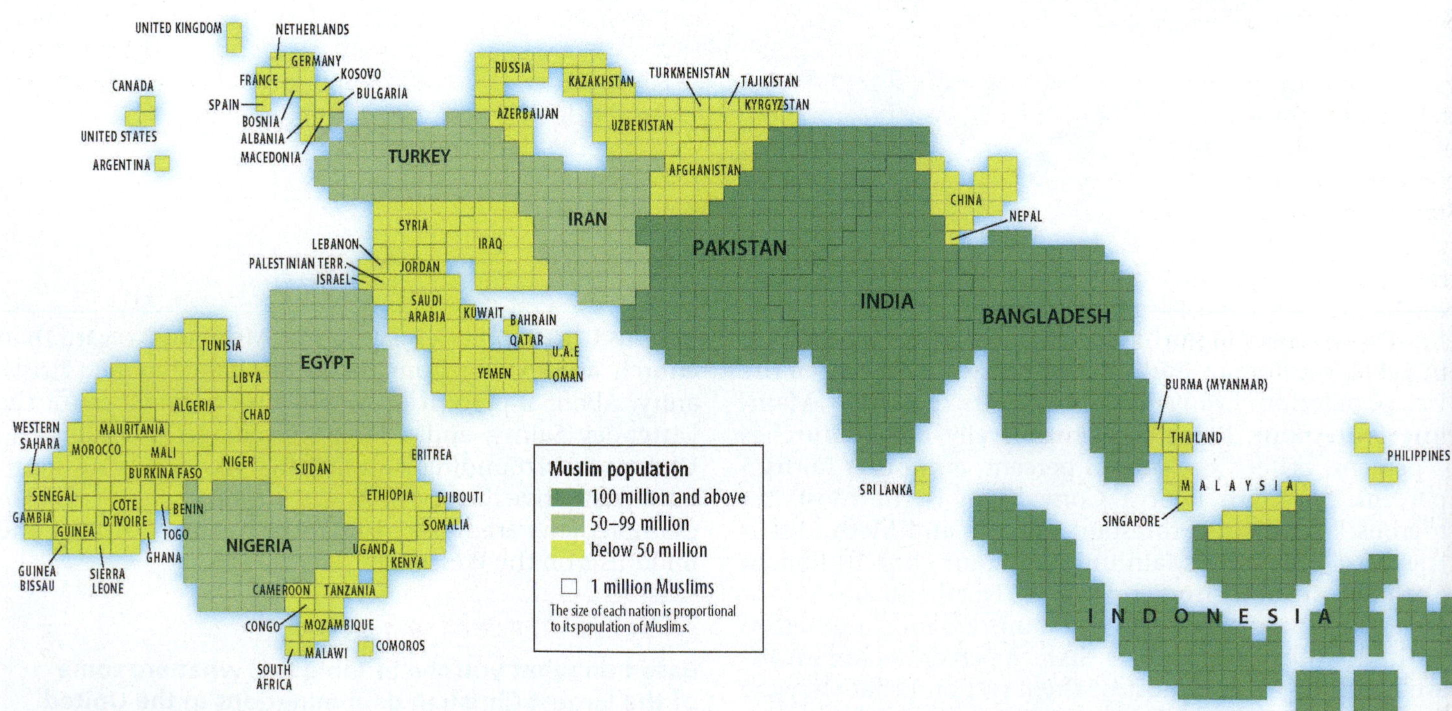

▲ **FIGURE 6-9 CARTOGRAM OF THE DISTRIBUTION OF WORLD'S MUSLIMS** Indonesia ranks only 15th in land area but is home to more Muslims than any other country.

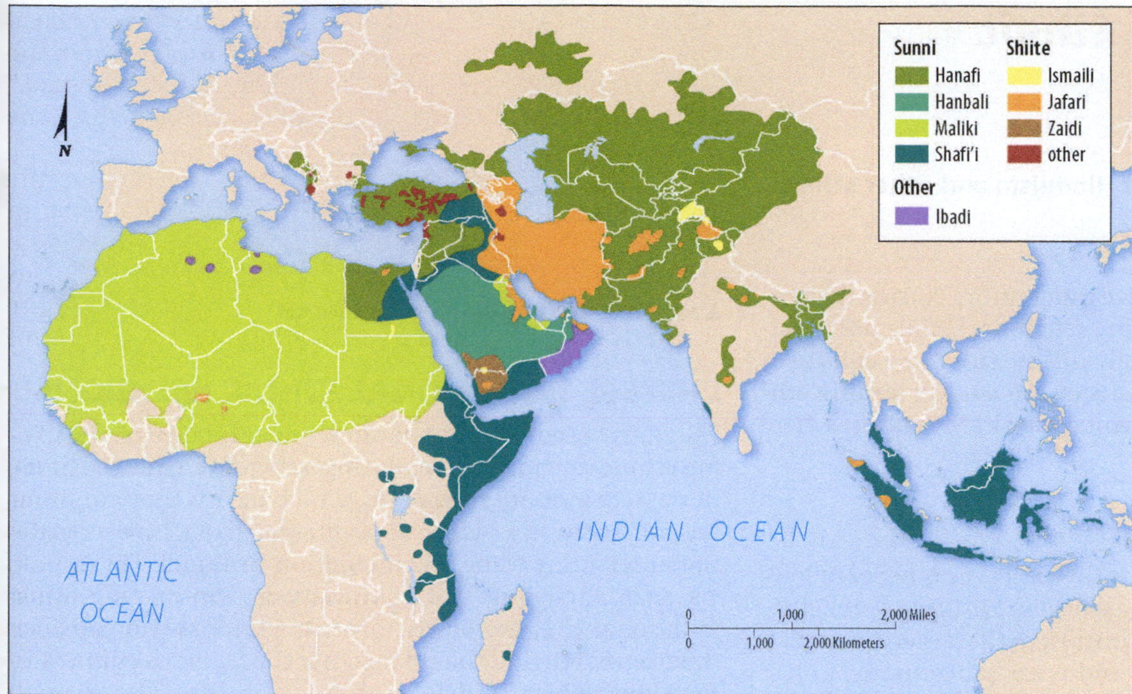

◀ **FIGURE 6-10 DISTRIBUTION OF BRANCHES OF ISLAM**

ISLAM IN EUROPE AND NORTH AMERICA. The Muslim populations of North America and Europe have increased rapidly in recent years. In Europe, Muslims account for 5 percent of the population. France has the largest Muslim population, about 4 million, a legacy of immigration from predominantly Muslim former colonies in North Africa. Germany has about 3 million Muslims, also a legacy of immigration, in Germany's case primarily from Turkey. In Southeast Europe, Albania, Bosnia & Herzegovina, and Serbia each have about 2 million Muslims.

Estimates of the number of Muslims in North America vary widely, from 1 million to 5 million, but in any event, the number has increased dramatically from only a few hundred thousand in 1990. Muslims in the United States come from a variety of backgrounds. According to the U.S. State Department, approximately one-third of U.S. Muslims trace their ancestry to Pakistan and other South Asian countries and one-fourth to Arab countries of Southwest Asia & North Africa. Many of these Muslims immigrated to the United States during the 1990s. Another one-fourth are African Americans.

DISTRIBUTION OF BUDDHISTS

Buddhism, the third of the world's major universalizing religions, is clustered primarily in East Asia and Southeast Asia. Like the other two universalizing religions, Buddhism split into more than one branch, as followers disagreed on interpreting statements by the founder, Siddhartha Gautama. The three main branches are Mahayana, Theravada, and Vajrayana (Figure 6-11). Mahayanists account for about 56 percent of Buddhists, primarily in China, Japan, and Korea. Theravadists comprise about 38 percent of Buddhists, especially in Cambodia, Laos, Myanmar, Sri Lanka, and Thailand. Vajrayanists, also known as Lamaists and Tantrayanists, comprise about 6 percent and are found primarily in Tibet and Mongolia.

An accurate count of Buddhists is especially difficult because only a few people participate in Buddhist institutions. Religious functions are performed primarily by monks rather than by the general public. The number of Buddhists is also difficult to count because Buddhism, although a universalizing religion, differs in significant respects from the Western concept of a formal religious system. Someone can be both a Buddhist and a believer in other Eastern religions, whereas Christianity and Islam both require exclusive adherence. Most Buddhists in China and Japan, in particular, believe at the same time in an ethnic religion.

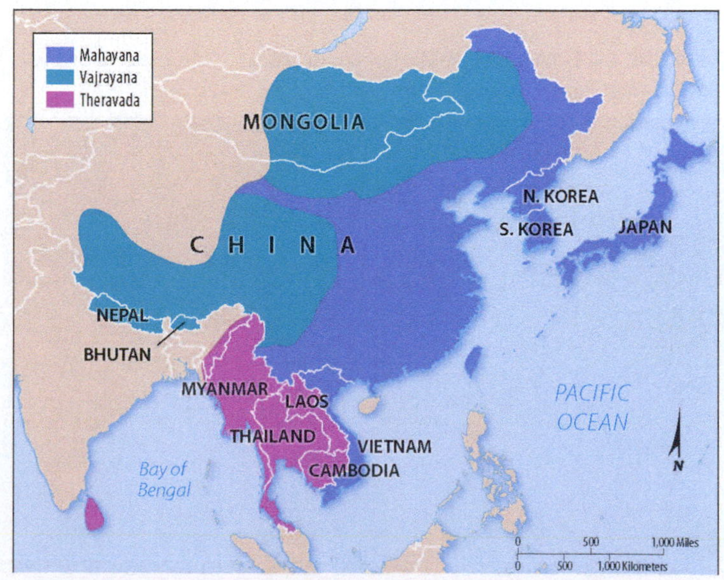

▲ **FIGURE 6-11 DISTRIBUTION OF BRANCHES OF BUDDHISM**

Distribution of Ethnic Religions

LEARNING OUTCOME 6.1.5

Describe the distribution of Hinduism and other ethnic religions.

Ethnic religions typically have much more clustered distributions than do universalizing religions. Unlike universalizing religions, which typically diffuse from one culture to another, most of the adherents of the world's leading ethnic religions have remained embedded in the culture where they originated.

DISTRIBUTION OF HINDUS

The ethnic religion with by far the largest number of followers is Hinduism, which is the world's third-largest religion, with 1 billion adherents. In contrast to the large universalizing religions, 97 percent of Hindus are concentrated in just one country (India), 2 percent are in Nepal, 1 percent are in Bangladesh, and small numbers are elsewhere. Hindus comprise more than 80 percent of the population of India and Nepal, while about 9 percent live in Bangladesh, and a small minority are found in every other country (Figure 6-12).

The average Hindu has allegiance to a particular god or concept within a broad range of possibilities. The manifestation of God with the largest number of adherents—an estimated 80 percent—is Vaishnavism, which worships the god Vishnu, a loving god incarnated as Krishna. The second-largest is Shaivism, dedicated to Shiva, a protective and destructive god.

PAUSE & REFLECT 6.1.5

What is the second-most-numerous religion in India?

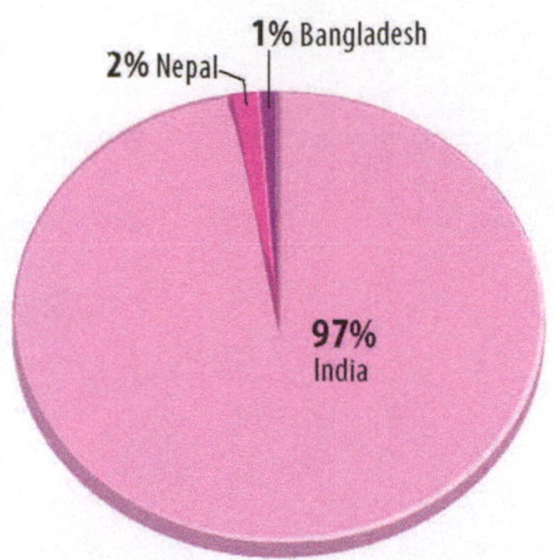

▲ **FIGURE 6-12 DISTRIBUTION OF HINDUS**

CHINESE TRADITIONAL ETHNIC RELIGIONS

Religions based in East Asia show the difficulty of classifying ethnic religions and counting adherents. Chinese traditional religions are **syncretic,** which means they combine several traditions. *Adherents.com* considers Chinese traditional religions to be a combination of Buddhism (a universalizing religion) and Confucianism, Taoism, and other traditional Chinese practices. Most Chinese who consider themselves religious blend together the religious cultures of these multiple traditions.

CONFUCIANISM. Confucius (551–479 B.C.) was a philosopher and teacher in the Chinese province of Lu. His sayings, which were recorded by his students, emphasized the importance of the ancient Chinese tradition of *li*, which can be translated roughly as "propriety" or "correct behavior." Confucianism prescribed a series of ethical principles for the orderly conduct of daily life in China, such as following traditions, fulfilling obligations, and treating others with sympathy and respect. These rules applied to China's rulers as well as to their subjects.

TAOISM. Lao-Zi (604–531? B.C., also spelled Lao Tzu) organized Taoism. Although a government administrator by profession, Lao-Zi's writings emphasized the mystical and magical aspects of life rather than the importance of public service, which Confucius had emphasized. *Tao*, which means "the way" or "the path," cannot be comprehended by reason and knowledge because not everything is knowable (Figure 6-13). It emphasizes the importance of studying nature to find one's place in the world instead of striving to change the world.

▼ **FIGURE 6-13 TAOISM** A-Ma Temple in Macau, China, is one of the oldest Taoist temples, dating from 1488.

PRIMAL-INDIGENOUS ETHNIC RELIGIONS

Several hundred million people practice what *Adherents.com* has grouped into the category primal-indigenous religions. Most of these people reside in Southeast Asia or on South Pacific islands, especially in Vietnam and Laos (Figure 6-14).

Followers of primal-indigenous religions believe that because God dwells within all things, everything in nature is spiritual. Narratives concerning nature are specific to the physical landscape where they are told.

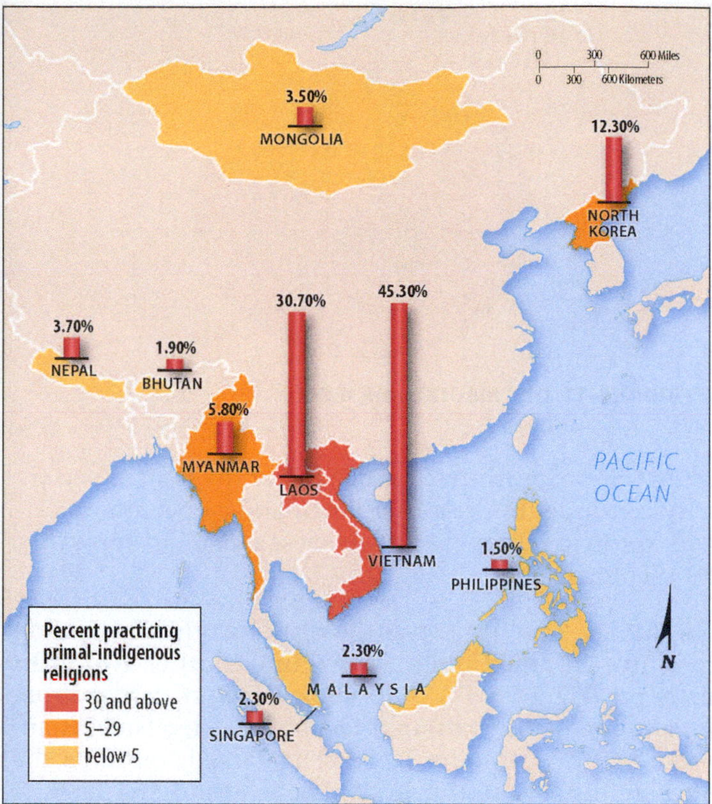

▲ **FIGURE 6-14 DISTRIBUTION OF PRIMAL-INDIGENEOUS RELIGIONS IN SOUTHEAST ASIA**

Included in this group are Shamanism and Paganism. According to Shamans, invisible forces or spirits affect the lives of the living. "Pagan" used to refer to the practices of ancient peoples, such as the Greeks and Romans, who had multiple gods with human forms. The term is currently expanded to also include other beliefs that originated with religions that predate Christianity and Islam.

AFRICAN TRADITIONAL FOLK RELIGIONS

Approximately 27 million Africans, 2 percent of the continent's people, are estimated by Pew to follow folk religions, sometimes called animism. According to **animism,** inanimate objects such as plants and stones, or natural events such as thunderstorms and earthquakes, are "animated," or have discrete spirits and conscious life.

Today Africa is 51 percent Christian—split about evenly among Roman Catholic, Protestant, and other—and 43 percent Muslim. This distribution is in sharp contrast with the past. In 1900, more than 70 percent of Africans followed traditional folk religions (Figure 6-15a). As recently as 1980, one-half of Africans—around 200 million people—were still classified as folk religionists. The growth in the two universalizing religions at the expense of ethnic religions reflects fundamental geographic differences between the two types of religions. Remaining folk religionists in Africa are clustered primarily in a belt that separates predominantly Muslim North Africa from what has become predominantly Christian sub-Saharan Africa (Figure 6-15b).

▼ **FIGURE 6-15 DISTRIBUTION OF AFRICAN TRADITIONAL RELIGIONS** **(a)** 1900, **(b)** 2010.

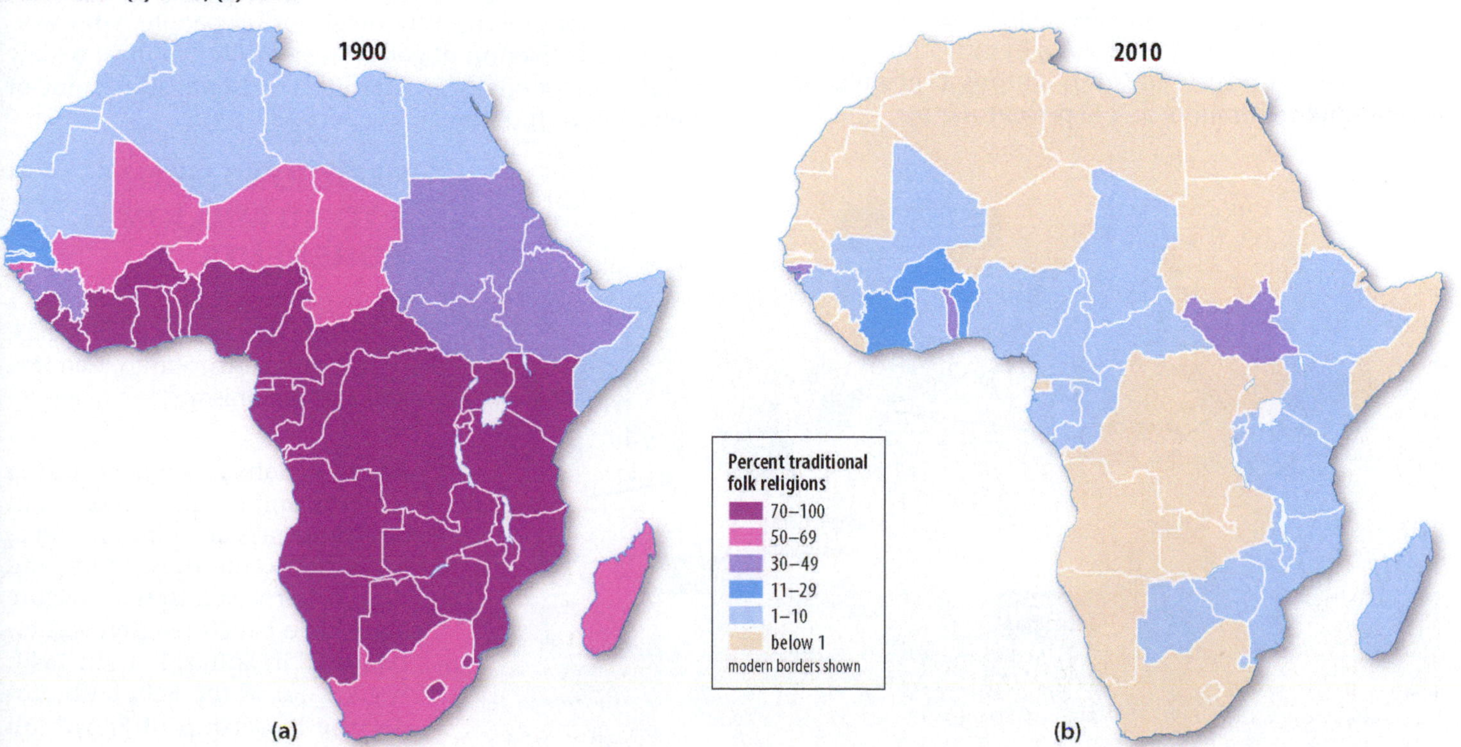

Distribution of Other Religions

LEARNING OUTCOME 6.1.6

Describe the distribution of religions other than the most numerous ones.

Ten religions are briefly described in this section in order of the estimated numbers of adherents. The adherents of 9 of these 10 religions are highly clustered in one or two countries (Figure 6-16). The exception is Bahá'í.

RELIGIONS WITH 14 TO 25 MILLION ADHERENTS

Four religions have an estimated 14 to 25 million adherents: Sikhism, Juche, Spiritism, and Judaism.

SIKHISM. All but 3 million of the world's 23 million Sikhs are clustered in the Punjab region of India. The founder of Sikhism, Guru Nanak (1469–1538), lived in a village near the city of Lahore, in present-day Pakistan. God was revealed to Nanak as The One Supreme Being, or Creator, who rules the universe by divine will. Nanak traveled widely through South Asia around 500 years ago, preaching his new faith, and his many followers became known as *Sikhs* (Hindu for "disciples"). Nine other gurus succeeded Guru Nanak. In 1604, Arjan, the fifth guru, compiled and edited the Guru Granth Sa-hib (the Holy Granth of Enlightenment), which became the book of Sikh holy scriptures.

JUCHE. Most North Koreans are classified by *Adherents.com* as following Juche, which is Korean for "self-reliance." Juche was organized by Kim Il-sung, the leader of North Korea between 1948 and his death in 1994. Other sources regard Juche as a government ideology or philosophy rather than a religion. Pew classifies 71 percent of North Koreans as unaffiliated, 12 percent as folk religionists, and 13 percent as other.

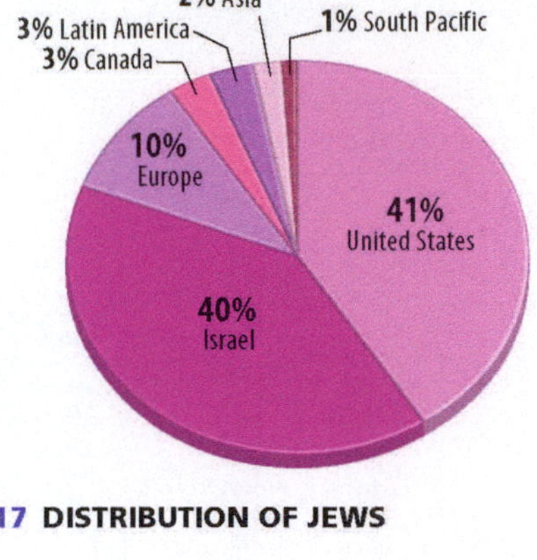

▲ **FIGURE 6-17 DISTRIBUTION OF JEWS**

SPIRITISM. Spiritism is the belief that the human personality continues to exist after death and can communicate with the living through the agency of a medium or psychic. Most Spiritists reside in Brazil.

JUDAISM. Roughly two-fifths of the world's 14 million Jews live in the United States and another two-fifths in Israel (Figure 6-17). The name *Judaism* derives from Judah, one of the patriarch Jacob's 12 sons; Israel is another biblical name for Jacob. The Tanakh recounts the ancient history of the Jewish people and the laws of the Jewish faith. Tanakh is an acronym for Torah (also known as the Five Books of Moses), Nevi'im ("Prophets") and Ketuvim ("Writings"). Judaism plays a more substantial role in Western civilization than its number of adherents would suggest. Judaism is the first recorded religion to espouse **monotheism,** belief that there is only one God. Fundamental to Judaism is belief in one all-powerful God. Judaism offered a sharp contrast to the **polytheism** practiced by neighboring people, who worshipped a collection of gods. The world's two most widely practiced religions—Christianity and Islam—find some of their roots in Judaism.

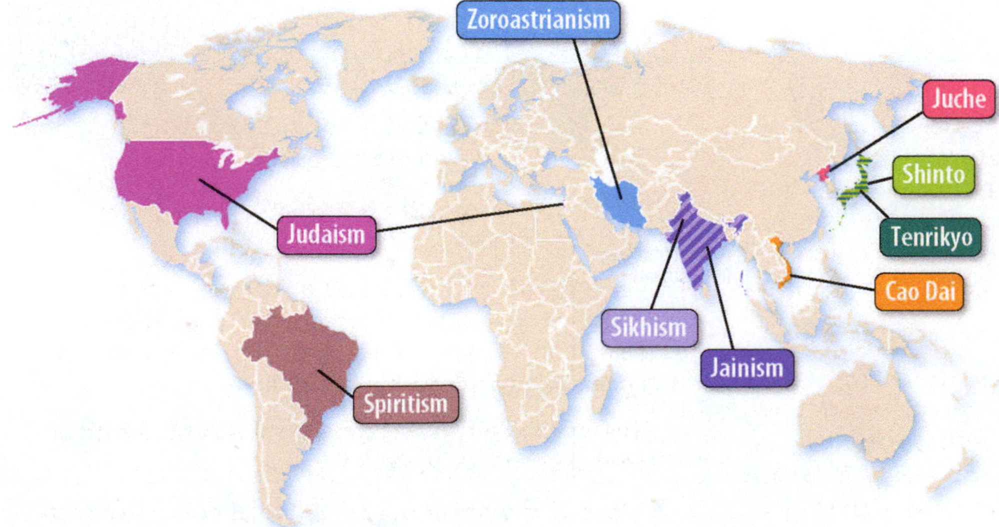

▲ **FIGURE 6-16 CLUSTERED RELIGIONS WITH AT LEAST 1 MILLION ADHERENTS**

RELIGIONS WITH 1 TO 10 MILLION ADHERENTS

Six religions have an estimated 1 to 10 million adherents: Bahá'í, Tenrikyo, Jainism, Shinto, Cao Dai, and Zoroastrianism.

BAHÁ'Í. Bahá'í is a universalizing religion, and roughly equal numbers of Bahá'ís are found in India, other Asian countries, Africa, and the Western Hemisphere (Figure 6-18). The Bahá'í religion was established in Shíráz, Iran, in 1844. It grew out of the Bábi faith, under the leadership of Siyyid 'Ali Muhammad, known as the Báb

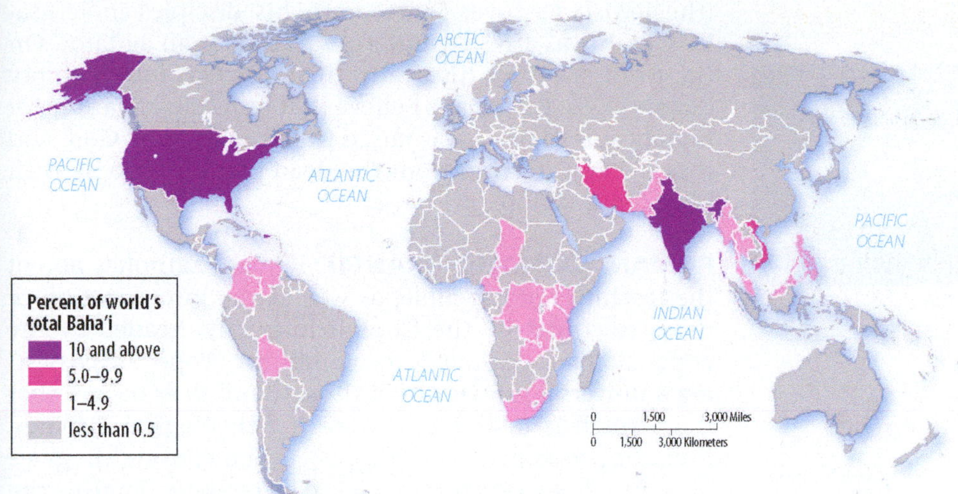

▲ **FIGURE 6-18 DISTRIBUTION OF BAHÁ'ÍS** India is home to 26 percent of the world's Bahá'ís, and the United States 12 percent.

(Persian for "gateway"). Bahá'í provoked strong opposition from Shiite Muslims, and the Báb was executed in 1850, as were 20,000 of his followers. Bahá'ís believe that one of the Báb's disciples, Husayn 'Ali Nuri, known as Bahá'u'lláh (Arabic for "Glory of God"), was the prophet and messenger of God. Bahá'u'lláh's function was to overcome the disunity of religions and establish a universal faith through abolition of racial, class, and religious prejudices. Bahá'u'lláh was arrested and then exiled. In 1863, his claim that he was the messenger of God anticipated by the Báb was accepted by other followers. Before he died in 1892, Bahá'u'lláh appointed his eldest son, 'Abdu'l-Bahá (1844–1921), to be the leader of the Bahá'í community and the authorized interpreter of his teachings.

TENRIKYO. Originally regarded as a branch of Shinto, Tenrikyo was organized as a separate religion in 1854 by a woman named Nakayama Miki (1798–1887). Followers of Tenrikyo believe that God expressed the divine will through Nakayama's role as the Shrine of God. *Adherents.com* reports 2 million adherents around the world, 95 percent of whom are in Japan.

JAINISM. Jainism originated in South Asia around 2,500 years ago. Its importance declined with the rise in importance of Buddhism and Hinduism in the region, especially since the eighth century A.D. Jains believe that nonviolence and self-control are the means to achieve liberation. India is the home to 95 percent of the world's 4 million Jains, although Jain centers are located in 25 of the 50 U.S. states.

SHINTO. Shinto, Japan's ethnic religion, is strongly rooted in the cultural history of the country. Japanese government statistics report around 100 million Shintos, or 78 percent of the country's population. However, in opinion polls only 4 million Japanese, or 3 percent of the population, identify themselves as Shinto. The large discrepancy stems in part from the fact that a seventeenth-century law in Japan assigns Shinto organizations with the task of maintaining records on Japanese citizens. The discrepancy also stems from the perception by some Japanese people that Shinto is a cultural feature rather than a religion.

CAO DAI. Cao Dai was founded in Vietnam during the 1920s. The name refers to belief in God as the Supreme Being, Creator, and Ultimate Reality of the Universe. The religion found itself in opposition to a succession of rulers of Vietnam, including the French colonial administration and the Communists. Since Vietnam's Communist government granted Cao Dai legal status in 1997, the number of adherents has increased to an estimated 4 million, nearly all of whom live in Vietnam.

ZOROASTRIANISM. The Prophet Zoroaster (or Zarathustra) founded the religion that bears his name around 3,500 years ago. The religion was more formally organized around 1,500 years ago in the Persian Empire (present-day Iran) and was the state religion for several ancient empires in Central Asia. The number of adherents declined after Muslims came to dominate Central Asia. Current records show 70,000 Zoroastrians in India, 25,000 in Iran, and 20,000 in the United States. However, *Adherents.com* says that a more realistic worldwide count is between 2 and 3 million because Zoroastrians are said to be reticent about identifying themselves.

PAUSE & REFLECT 6.1.6

Most of the religions with at least 1 million adherents are clustered in which continent?

KEY ISSUE 1

Where Are the World's Religions Distributed?

- ✔ **Religions can be classified into two major categories: universalizing and ethnic.**
- ✔ **The three largest universalizing religions are Christianity, Islam, and Buddhism.**
- ✔ **Christianity predominates in Europe, North America, and Latin America; Islam in Southeast Asia, Central Asia, and Southwest Asia & North Africa; and Buddhism in East Asia.**
- ✔ **The largest ethnic religion is Hinduism, which is found primarily in South Asia.**
- ✔ **Most of the other major religions have clustered distributions.**

KEY ISSUE 2

Why Do Religions Have Distinctive Distributions?

- ▸ **Origin of Christianity and Islam in Southwest Asia**
- ▸ **Origin of Buddhism and Hinduism in South Asia**
- ▸ **Historical Diffusion of Religions**
- ▸ **Recent Migration of Christians**
- ▸ **Migration of Muslims and Jews**

LEARNING OUTCOME 6.2.1

Describe the origins of Christianity and Islam.

The current global and regional distributions of religions and their branches result from geographic processes of origin and diffusion. The most widely followed universalizing religions—Buddhism, Christianity, and Islam—have well-defined places of origin and widespread and well-documented patterns of diffusion. An ethnic religion such as Hinduism has unknown origins and limited diffusion.

Origin of Christianity and Islam in Southwest Asia

The two religions with the most adherents—Christianity and Islam—both originated in Southwest Asia. It is typical of universalizing religions such as Christianity and Islam that their places of origin are known and derived from events in the life of a man.

ORIGIN OF CHRISTIANITY

Christianity was founded upon the teachings of Jesus, who was born in Bethlehem between 8 and 4 B.C. and died on a cross in Jerusalem about A.D. 30. Raised as a Jew, Jesus gathered a small band of disciples and preached the coming of the Kingdom of God. The four Gospels of the Christian Bible—Matthew, Mark, Luke, and John—document miracles and extraordinary deeds that the writers believed Jesus performed. He was referred to as *Christ*, from the Greek word for the Hebrew word *messiah*, which means "anointed."

According to the Gospels, in the third year of his mission, Jesus was betrayed to the authorities by one of his companions, Judas Iscariot. After sharing the Last Supper (the Jewish Passover Seder) with his disciples in Jerusalem, Jesus was arrested and put to death as an agitator. On the third day after his death, his tomb was found empty (Figure 6-19). Christians believe that Jesus died to atone for human sins, that he was raised from the dead by God, and that his Resurrection from the dead provides people with hope for salvation.

ROMAN CATHOLIC BRANCH. Roman Catholics accept the teachings of the Bible, as well as the interpretation of those teachings by the Church hierarchy, headed by the Pope. Roman Catholics recognize the Pope as possessing a universal primacy or authority, and they believe that the Church is infallible in resolving theological disputes. According to Roman Catholic belief, God conveys His grace directly to humanity through seven sacraments: Baptism, the Eucharist (the partaking of bread and wine that repeats the actions of Jesus at the Last Supper), Penance, Confirmation, Matrimony, Holy Orders, and Anointing the Sick.

ORTHODOX BRANCH. Orthodoxy comprises the faith and practices of a collection of churches that arose in the eastern part of the Roman Empire. The split between the Roman and Eastern churches dates to the fifth century, and it was a result of rivalry between the Pope of Rome and the Patriarchy of Constantinople, which was especially intense after the collapse of the Roman Empire. The split between the two churches became final in 1054, when Pope Leo IX condemned the Patriarch of Constantinople. Orthodox Christians accept the seven sacraments but reject doctrines that the Roman Catholic Church has added since the eighth century.

PROTESTANT BRANCH. Protestantism originated with the principles of the Reformation in the sixteenth century. The Reformation movement is regarded as beginning when Martin Luther (1483–1546) posted 95 theses on the door of the church at Wittenberg on October 31, 1517. According to Luther, individuals have primary responsibility for achieving personal salvation through direct

▼ **FIGURE 6-19 ORIGIN OF CHRISTIANITY** This tomb in the center of the Church of the Holy Sepulchre in Jerusalem was erected at the place Christians accept to be where Jesus was buried and resurrected. Armenian Orthodox clergy are in procession around the Tomb.

▲ **FIGURE 6-20 ORIGIN OF ISLAM** Muhammad is buried under the green dome in the Mosque of the Prophet in Madinah, Saudi Arabia. The mosque, built on the site of Muhammad's house, is the second holiest in Islam and the second-largest mosque in the world.

communication with God. Grace is achieved through faith rather than through sacraments performed by the Church.

ORIGIN OF ISLAM

Like other universalizing religions, Islam arose from the teachings of a historical founder. The core of Islamic belief involves performing five acts, known as five pillars of faith:

1. *Shahadah*, which means that Muslims frequently recite their belief that there is no deity worthy of worship except the one God, the source of all creation, as well as their belief that Muhammad is the messenger of God.
2. *Salat*, which means that five times daily, a Muslim prays, facing the city of Makkah (Mecca), as a direct link to God.
3. *Zakat*, which means that a Muslim gives generously to charity as an act of purification and growth.
4. *Sawm of Ramadan*, which means that a Muslim fasts during the month of Ramadan as an act of self-purification.
5. *Hajj*, which means that if physically and financially able, a Muslim makes a pilgrimage to Makkah.

Islam traces its origin to the same narrative as Judaism and Christianity. All three religions consider Adam to have been the first man and Abraham to have been one of his descendants. According to the Jewish Torah and Christian Old Testament narrative, Abraham married Sarah, who did not bear children. As polygamy was a custom of the culture, Abraham then married Hagar, who bore a son, Ishmael. Sarah's fortunes changed, and she bore a son, Isaac.

Jews and Christians trace their story through Abraham's original wife Sarah and her son Isaac. Muslims trace their story through his second wife, Hagar, and her son Ishmael. The Islamic tradition tells that Abraham brought Hagar and Ishmael to Makkah (spelled Mecca on many English-language maps), in present-day Saudi Arabia. Centuries later, according to the Muslim narrative, one of Ishmael's descendants, Muhammad, became the Prophet of Islam.

PROPHET MUHAMMAD. Muhammad was born in Makkah about 570. At age 40, while engaged in a meditative retreat, Muhammad is believed by Muslims to have received his first revelation from God through the Angel Gabriel. The Quran, the holiest book in Islam, is accepted by Muslims to be a record of God's words, as revealed to the Prophet Muhammad through Gabriel. Arabic is the lingua franca, or language of communication, within the Muslim world, because it is the language in which the Quran is written.

Islam teaches that as he began to preach the truth that God had revealed to him, Muhammad and his followers suffered persecution, and in 622 he was commanded by God to emigrate. His migration from Makkah to the city of Yathrib—an event known as the Hijra (from the Arabic word for "migration," sometimes spelled hegira)—marks the beginning of the Muslim calendar. Yathrib was subsequently renamed Madinah, Arabic for "the City" (Figure 6-20). After several years, Muhammad and his followers returned to Makkah and established Islam as the city's religion. By Muhammad's death, in 632 at about age 63, Islam had spread through most of present-day Saudi Arabia.

PAUSE & REFLECT 6.2.1

The government of Saudi Arabia prefers to spell some place names differently than is common in English (such as Makkah instead of Mecca). Given your knowledge of the principal language used in predominantly Muslim countries including Saudi Arabia, what might account for this preference?

SHIITE AND SUNNI BRANCHES. Differences between the two main branches of Islam—Shiite and Sunni—go back to the earliest days of the religion and reflect disagreement over the line of succession in Islamic leadership. Muhammad had no surviving son and no agreed-upon successor. His successor was his father-in-law, Abu Bakr (573–634), an early supporter from Makkah, who became known as caliph ("successor of the prophet"). The next two caliphs, Umar (634–644) and Uthman (644–656), expanded the territory under Muslim influence to Egypt and Persia.

Uthman was a member of a powerful Makkah clan that had initially opposed Muhammad before the clan's conversion to Islam. The more ardent converts criticized Uthman for seeking compromises with other formerly pagan families in Makkah. Uthman's opponents found a leader in Ali (600?–661), a cousin and son-in-law of Muhammad, and thus Muhammad's nearest male heir. When Uthman was murdered, in 656, Ali became caliph, although five years later he, too, was assassinated.

Ali's descendants claim leadership of Islam, and Shiites support this claim. But Shiites disagree among themselves about the precise line of succession from Ali to modern times. They acknowledge that the chain of leadership was broken, but they dispute the date and events surrounding the disruption.

Origin of Buddhism and Hinduism in South Asia

LEARNING OUTCOME 6.2.2

Describe the origin of Buddhism and the reason the origin of Hinduism is unknown.

Buddhism, like other universalizing religions, has a precise place of origin, based on events in the life of a man. Ethnic religions like Hinduism have unknown or unclear origins, not tied to single historical individuals.

ORIGIN OF BUDDHISM

The founder of Buddhism, Siddhartha Gautama, was born about 563 B.C. in Lumbinī in present-day Nepal, near the border with India. The son of a lord, he led a privileged existence, sheltered from life's hardships. Gautama had a beautiful wife, palaces, and servants.

According to Buddhist tradition, Gautama's life changed after a series of four trips. He encountered a decrepit old man on the first trip, a disease-ridden man on the second trip, and a corpse on the third trip. After witnessing these scenes of pain and suffering, Gautama began to feel he could no longer enjoy his life of comfort. On a fourth trip, Gautama saw a monk, who taught him about withdrawal from the world.

At age 29 Gautama left his palace one night and lived in a forest for the next 6 years, thinking and experimenting with forms of meditation (Figure 6-21). Gautama emerged as the Buddha, the "awakened or enlightened one," and spent 45 years preaching his views across India. In the process, he trained monks, established orders, and preached to the public.

The foundation of Buddhism is represented by these concepts, known as the Four Noble Truths:

1. All living beings must endure suffering.
2. Suffering, which is caused by a desire to live, leads to reincarnation (repeated rebirth in new bodies or forms of life).
3. The goal of all existence is to escape suffering and the endless cycle of reincarnation into Nirvana (a state of complete redemption), which is achieved through mental and moral self-purification.
4. Nirvana is attained through an Eightfold Path: rightness of belief, resolve, speech, action, livelihood, effort, thought, and meditation.

THERAVADA BUDDHISM. Theravada is the older of the two largest branches of Buddhism. The word means "the way of the elders," indicating the Theravada Buddhists' belief that they are closer to Buddha's original approach. Theravadists believe that Buddhism is a full-time occupation, so to become a good Buddhist, one must renounce worldly goods and become a monk.

MAHAYANA BUDDHISM. Mahayana split from Theravada Buddhism about 2,000 years ago. Mahayana is translated as "the great vehicle," and Mahayanists call Theravada Buddhism by the name Hinayana, or "the lesser vehicle." Mahayanists claim that their approach to Buddhism can help more people because it is less demanding and all-encompassing. Theravadists emphasize Buddha's life of self-help and years of solitary introspection, and Mahayanists emphasize Buddha's later years of teaching and helping others. Theravadists cite Buddha's wisdom and Mahayanists his compassion.

◀ **FIGURE 6-21 ORIGIN OF BUDDHISM** Mahabodhi (Great Awakening) Temple was built in Bōdh Gayā, the site where Siddhartha Gautama Buddha attained enlightenment.

VAJRAYANA BUDDHISM. Vajrayanas emphasize the practice of rituals, known as Tantras, which have been recorded in texts. Vajrayanas believe that Buddha began to practice Tantras during his lifetime, although other Buddhists regard Vajrayana as an approach to Buddhism that evolved from Mahayana Buddhism several centuries later.

PAUSE & REFLECT 6.2.2

Rank the origins of the three large universalizing religions from earliest to most recent.

UNKNOWN ORIGIN OF HINDUISM

As an ethnic religion based in India, Hinduism has unknown origins. The word *Hinduism* originated in the sixth century B.C. to refer to people living in what is now India, but religious practices existed prior to recorded history.

Elements of Hinduism may have originated in the Indus Valley civilization, which flourished between approximately 2500 B.C. and 1500 B.C. in the valley along the Indus River in present-day eastern Pakistan. Archaeological evidence of possible rituals from that era includes bathing rituals, animal sacrifices, and sculptures that may depict Hindu gods (Figure 6-22).

Aryan tribes from Central Asia invaded South Asia about 1400 B.C. and brought with them Indo-European languages, as discussed in Chapter 5. In addition to their language, the Aryans brought their religion. The Aryans first settled in the area now called the Punjab in northwestern India and later migrated east to the Ganges River valley, as far as Bengal. Centuries of intermingling with the Dravidians already living in the area modified their religious beliefs. The earliest surviving Hindu texts, known as Vedas, were written around 1100 B.C. Some rituals from this era survive in contemporary Hinduism, including worship of various gods representing Earth, atmosphere, and sky.

▲ **FIGURE 6-23 UNKNOWN ORIGIN OF HINDUISM** Changu Narayan is the oldest surviving Hindu temple in Nepal. It was built in 325 A.D. and was dedicated to Lord Vishnu.

By around 2,000 years ago, key texts were composed, and rituals were developed that remain central to contemporary Hinduism. The earliest surviving Hindu temples were constructed between 1,500 and 2,000 years ago (Figure 6-23).

◀ **FIGURE 6-22 BATHING IN THE GANGES** Hindus bathe in the Ganges River to wash away their sins.

Historical Diffusion of Religions

LEARNING OUTCOME 6.2.3

Describe the process of diffusion of universalizing religions.

The universalizing religions have diffused from specific hearths, or places of origin, to other regions of the world. In contrast, most ethnic religions have typically remained clustered.

The hearths where each of the three largest universalizing religions originated are based on the events in the lives of the three key individuals. All three hearths are in Asia (Christianity and Islam in Southwest Asia, Buddhism in South Asia). Followers transmitted the messages preached in the hearths to people elsewhere, diffusing them across Earth's surface along distinctive paths, as shown in Figure 6-24.

DIFFUSION OF CHRISTIANITY

Christianity's diffusion has been rather clearly recorded since Jesus first set forth its tenets in the Roman province known at the time as Judea. Consequently, geographers can examine its diffusion by reconstructing patterns of communications, interaction, and migration.

In Chapter 1 two processes of diffusion were identified: relocation (diffusion through migration) and expansion (diffusion through an additive effect). Within expansion diffusion, we distinguished between hierarchical diffusion (diffusion through key leaders) and contagious diffusion (widespread diffusion). Christianity diffused through a combination of these forms of diffusion.

RELOCATION DIFFUSION OF CHRISTIANITY. Christianity first spread from its hearth in Judea through relocation diffusion. **Missionaries**—individuals who help to transmit a universalizing religion through relocation diffusion—carried the teachings of Jesus along the Roman Empire's protected sea routes and excellent road network to people in other locations (Figure 6-25).

People in commercial towns and military settlements that were directly linked by the communications network received the message first from Paul and other missionaries. But Christianity also spread widely within the Roman Empire through contagious diffusion—daily contact between believers in the towns and nonbelievers in the surrounding countryside.

EXPANSION DIFFUSION OF CHRISTIANITY. The dominance of Christianity throughout the Roman Empire was assured during the fourth century through the two types of expansion diffusion:

- **Hierarchical diffusion.** The spread of Christianity was encouraged when the Roman Empire's key elite figure, Emperor Constantine (274?–337), embraced the religion in 313, and Emperor Theodosius proclaimed it the empire's official religion in 380. In subsequent centuries, Christianity further diffused into Eastern Europe through conversion of kings or other elite figures.
- **Contagious diffusion.** Since the year 1500, contagious diffusion, especially through migration and missionary activity by Europeans, has extended Christianity around the world. Through permanent resettlement of Europeans, Christianity became the dominant religion in North and South America, Australia, and New Zealand. Christianity's dominance was further achieved by conversion of indigenous populations and by intermarriage. In recent decades, Christianity has further diffused to Africa, where it is now the most widely practiced religion.

DIFFUSION OF ISLAM

Muhammad's successors organized followers who extended the region of Muslim control over an extensive area of Africa, Asia, and Europe (Figure 6-26). Within a century of Muhammad's death, Muslims controlled Palestine, the Persian Empire, and much of India, resulting in the conversion of many non-Arabs to Islam, often through intermarriage.

To the west, Muslims diffused across North Africa, crossed the Strait of Gibraltar, and retained much of present-day Spain, until 1492. During the same century in which the Christians regained all of Western Europe,

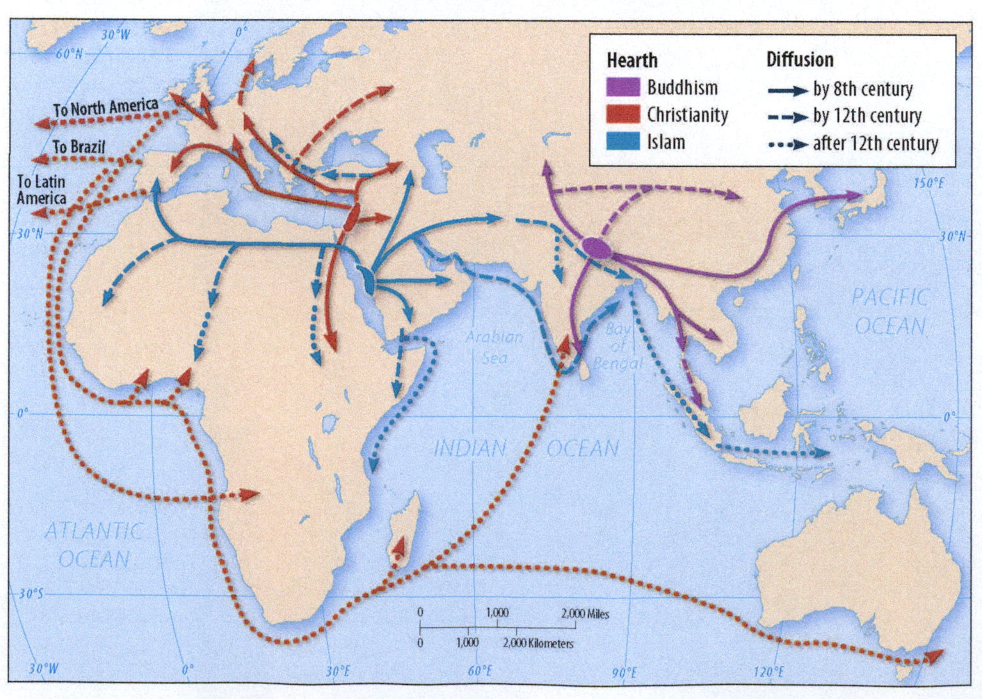

◀ **FIGURE 6-24 DIFFUSION OF UNIVERSALIZING RELIGIONS**

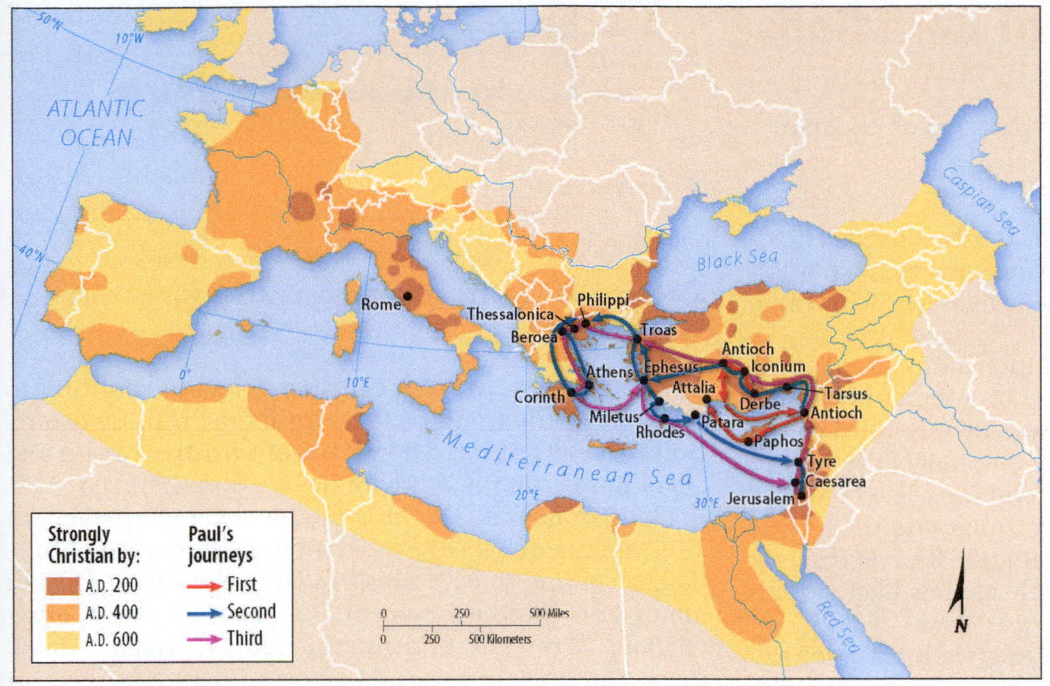

FIGURE 6-25 EARLY DIFFUSION OF CHRISTIANITY Christianity began to diffuse throughout Europe during the time of the Roman Empire and continued after the empire's collapse. Paul of Tarsus, a disciple of Jesus, traveled especially extensively through the Roman Empire as a missionary.

Muslims took control of much of southeastern Europe and Turkey.

As was the case with Christianity, Islam, as a universalizing religion, diffused well beyond its hearth in Southwest Asia through relocation diffusion of missionaries to portions of sub-Saharan Africa and Southeast Asia. Although it is spatially isolated in Southeast Asia from the Islamic core region, Indonesia is predominantly Muslim. The world's fourth-most-populous country is home to more Muslims than any other country because Arab traders took the religion there in the thirteenth century.

PAUSE & REFLECT 6.2.3

Does the diffusion of Islam provide a good example of hierarchical diffusion, relocation diffusion, or both?

DIFFUSION OF BUDDHISM

Buddhism did not diffuse rapidly from its point of origin in northeastern India. Most responsible for the spread of Buddhism was Asoka, emperor of the Magadhan Empire from about 273 to 232 B.C. The Magadhan Empire formed the nucleus of several powerful kingdoms in South Asia between the sixth century B.C. and the eighth century A.D.

About 257 B.C., at the height of the Magadhan Empire's power, Asoka became a Buddhist and thereafter attempted to put into practice Buddha's social principles. A council organized by Asoka at Pataliputra decided to send missionaries to territories neighboring the Magadhan Empire. Emperor Asoka's son, Mahinda, led a mission to the island of Ceylon (now Sri Lanka), where the king and his subjects were converted to Buddhism. As a result, Sri Lanka is the country that claims the longest continuous tradition of practicing Buddhism. Missionaries were also sent in the third century B.C. to Kashmir, the Himalayas, Burma (Myanmar), and elsewhere in India.

In the first century A.D., merchants along the trading routes from northeastern India introduced Buddhism to China. Many Chinese were receptive to the ideas brought by Buddhist missionaries, and Buddhist texts were translated into Chinese languages. Chinese rulers allowed their people to become Buddhist monks during the fourth century A.D., and in the following centuries Buddhism turned into a genuinely Chinese religion. Buddhism further diffused from China to Korea in the fourth century and from Korea to Japan two centuries later. During the same era, Buddhism lost its original base of support in India.

FIGURE 6-26 DIFFUSION OF ISLAM Islam diffused rapidly from its point of origin in present-day Saudi Arabia. Within 200 years, Muslims controlled much of Southwest Asia & North Africa, as well as southwestern Europe.

Recent Migration of Christians

LEARNING OUTCOME 6.2.4

Understand distinctive migration patterns of Christian groups in modern times.

The distribution of religions continues to be altered in modern times by migration. Although most people do not migrate primarily for religious reasons, when they do migrate, they take their religion with them.

The religious composition of international migrants does not match the overall share of adherents of various religions. Christians comprise only one-third of the world's population but account for one-half of the world's international migrants. Muslims and Jews also comprise higher percentages of the world's migrants than their shares of the world's population. Buddhists, Hindus, folk religionists, and unaffiliated people are less likely to migrate (Figure 6-27).

Jews are much more likely to migrate internationally than their share of the world's population would suggest. Around 25 percent of all Jews have migrated from one country to another at some point in their lives. In comparison, only 3 percent of all people alive today have migrated internationally. Among the more numerous religious groups, around 5 percent of all Christians and 4 percent of all Muslims have migrated, compared to only 1 or 2 percent of other religious groups (Figure 6-28).

MIGRATION PATTERNS OF CHRISTIANS

The destinations of international migrants who are Christian do not match the distribution of Christians. North America is home to 12 percent of the world's Christians but is the destination for 34 percent of migrating Christians, including 30 percent entering the United States and 4 percent entering Canada.

Europe is home to 26 percent of the world's Christians but is the destination for 38 percent of migrating Christians. On the other hand, Latin America and sub-Saharan Africa attract relatively few Christian migrants compared to the share of the world's Christians living in those two regions. The largest migration flows of Christians are in and out of Russia and the United States (Figure 6-29).

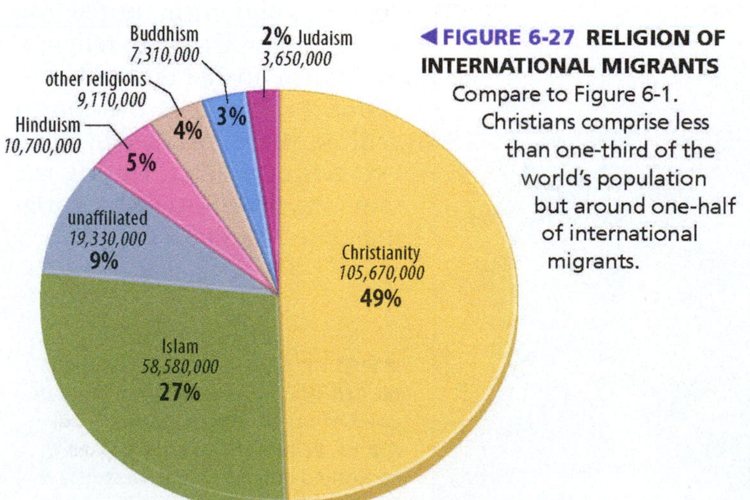

◀ **FIGURE 6-27 RELIGION OF INTERNATIONAL MIGRANTS** Compare to Figure 6-1. Christians comprise less than one-third of the world's population but around one-half of international migrants.

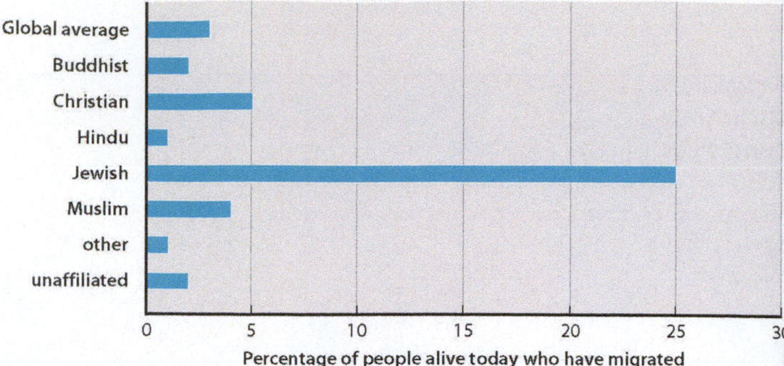

▲ **FIGURE 6-28 PERCENTAGE OF PEOPLE ALIVE TODAY WHO HAVE MIGRATED**

Most immigrants to the United States are Christians. In 2012, approximately 61 percent of immigrants were Christian, according to Pew. Muslims comprised 10 percent of immigrants, Hindus 7 percent, Buddhists 6 percent, other religions 3 percent, and unaffiliated 14 percent. Unauthorized immigrants were 83 percent Christian, 7 percent other religions, and 9 percent unaffiliated. The percentage of immigrants to the United States who are Christians declined from 68 percent in 1992 to 61 percent in 2012. The decline has been offset by increases in the percentage who are Muslims and Hindus.

The percentage of Christian immigrants to Canada declined from 88 percent in 1971 to 66 percent in 2011. Most of the increase has been in people unaffiliated with religions.

Canada (except Québec) and the United States have Protestant majorities because their early colonists came primarily from Protestant England. Some regions and localities within the United States and Canada are predominantly Roman Catholic because of immigration from Roman Catholic countries (refer to Figure 6-7). Immigration from Mexico and other Latin American countries has concentrated Roman Catholics in the southwest, whereas French settlement from the seventeenth century, as well as recent immigration, has produced a predominantly Roman Catholic Québec.

Similarly, geographers trace the distribution of other Christian denominations within the United States to the fact that migrants came from different parts of Europe, especially during the nineteenth century. Followers of The Church of Jesus Christ of Latter-day Saints, popularly known as Mormons, settled at Fayette, New York, near the hometown of their founder Joseph Smith. During Smith's life, the group moved several times in search of religious freedom. Eventually, under the leadership of Brigham Young, they migrated to the sparsely inhabited Salt Lake Valley in the present-day state of Utah.

The distribution of Christian branches in Canada is also a function of migration. Roman Catholics form a majority in Québec, a legacy of its settlement by French migrants in the seventeenth century (Figure 6-30). Protestants form a majority in other provinces, especially as a result of westward migration from Ontario (Figure 6-31).

▶ **FIGURE 6-29 MIGRATION OF CHRISTIANS** **(a)** Largest flows of Christian migrants **(b)** Origin and destination of Christian migrants.

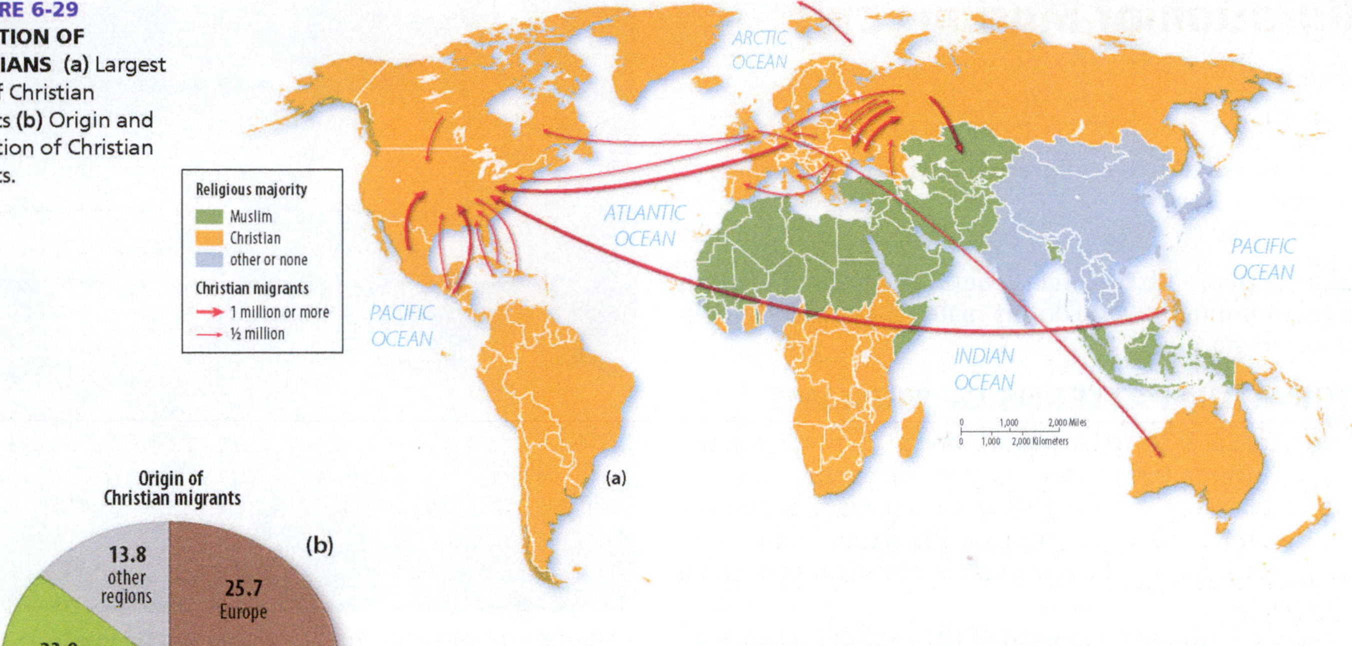

▼ **FIGURE 6-30 CHRISTIANITY IN CANADA** McDougall Memorial Church, built in 1875 in Morley, Alberta, is the oldest surviving Protestant church in the province.

◀ **FIGURE 6-31 MOST NUMEROUS FAITHS IN CANADA**

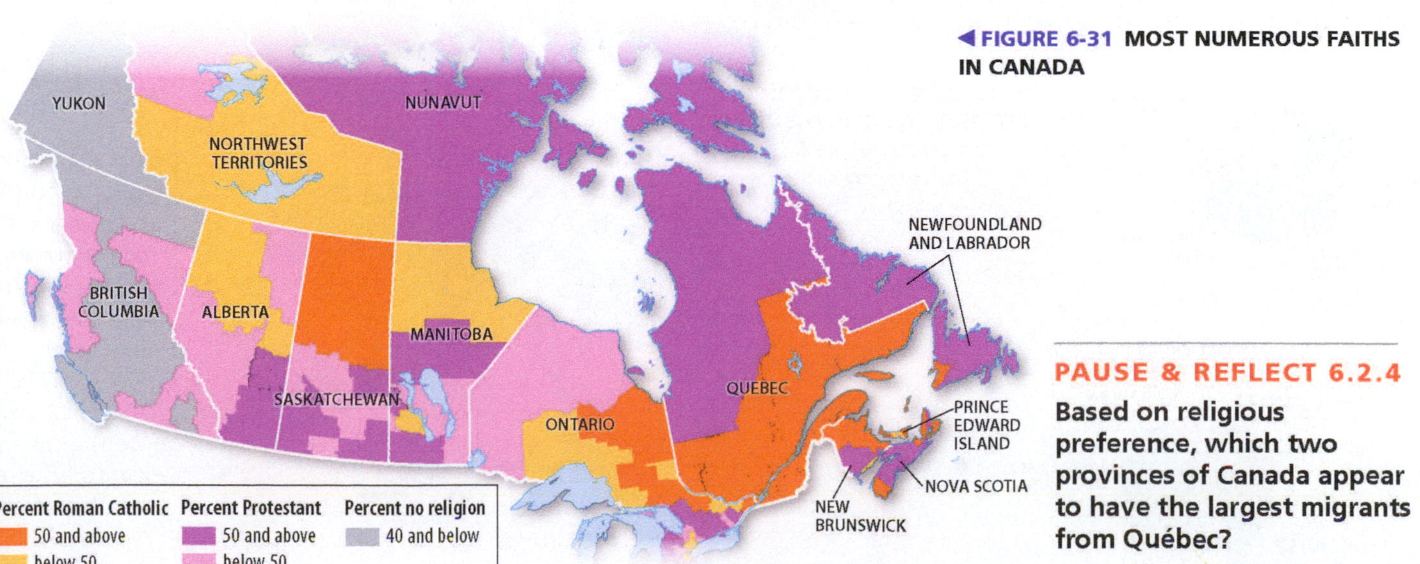

PAUSE & REFLECT 6.2.4

Based on religious preference, which two provinces of Canada appear to have the largest migrants from Québec?

Migration of Muslims and Jews

LEARNING OUTCOME 6.2.5

Understand distinctive migration patterns of Muslims and Jews.

Even more than with Christians, the distributions of Muslims and Jews around the world do not match the pattern of international migration.

MIGRATION PATTERNS OF MUSLIMS

Southwest Asia & North Africa is home to 20 percent of the world's Muslims but attracts 34 percent of the migrants. Saudi Arabia is the country that attracts the largest number of Muslim migrants, accounting for 10 percent of the world total. Egypt sends the largest number of migrants to Saudi Arabia (Figure 6-32).

Europe, with only 3 percent of the world's Muslims, attracts 34 percent of Muslims who migrate internationally. Russia, Germany, and France are all world leaders in attracting Muslims (Figure 6-33). The largest numbers of Muslims who have migrated to Europe have come from Turkey to Germany and from Algeria to France. Morocco has been the origin of a large number of Muslims, divided between France and Spain. North America receives only around 1 percent of Muslim migrants (Figure 6-34).

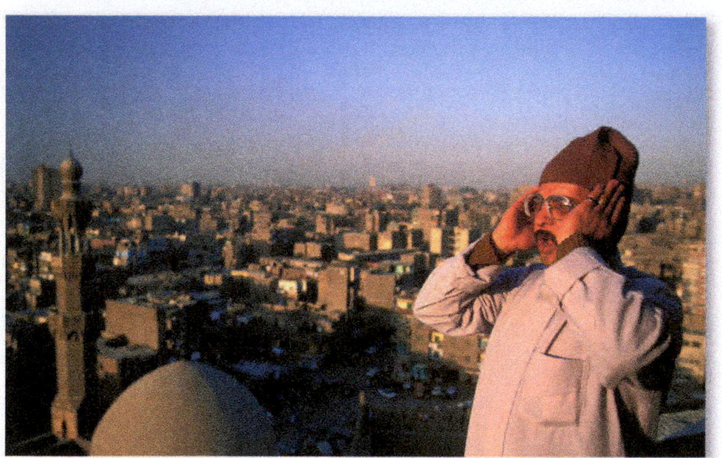

▲ **FIGURE 6-32 CALLING MUSLIMS TO PRAYER** Ibn Tulun Mosque, Cairo, Egypt.

▲ **FIGURE 6-34 MOSQUE** The Islamic Center of Greater Toledo, Perrysburg, Ohio.

On the other hand, South Asia, with 30 percent of the world's Muslims, attracts only 6 percent of the migrants. Central Asia and Southeast Asia also attract fewer migrants than their share of the world's Muslim population. The four countries with the largest Muslim populations—Indonesia, Pakistan, Bangladesh, and India—are home to 43 percent of the world's Muslims but attract only 6 percent of Muslim migrants (4 percent of whom are heading for Pakistan).

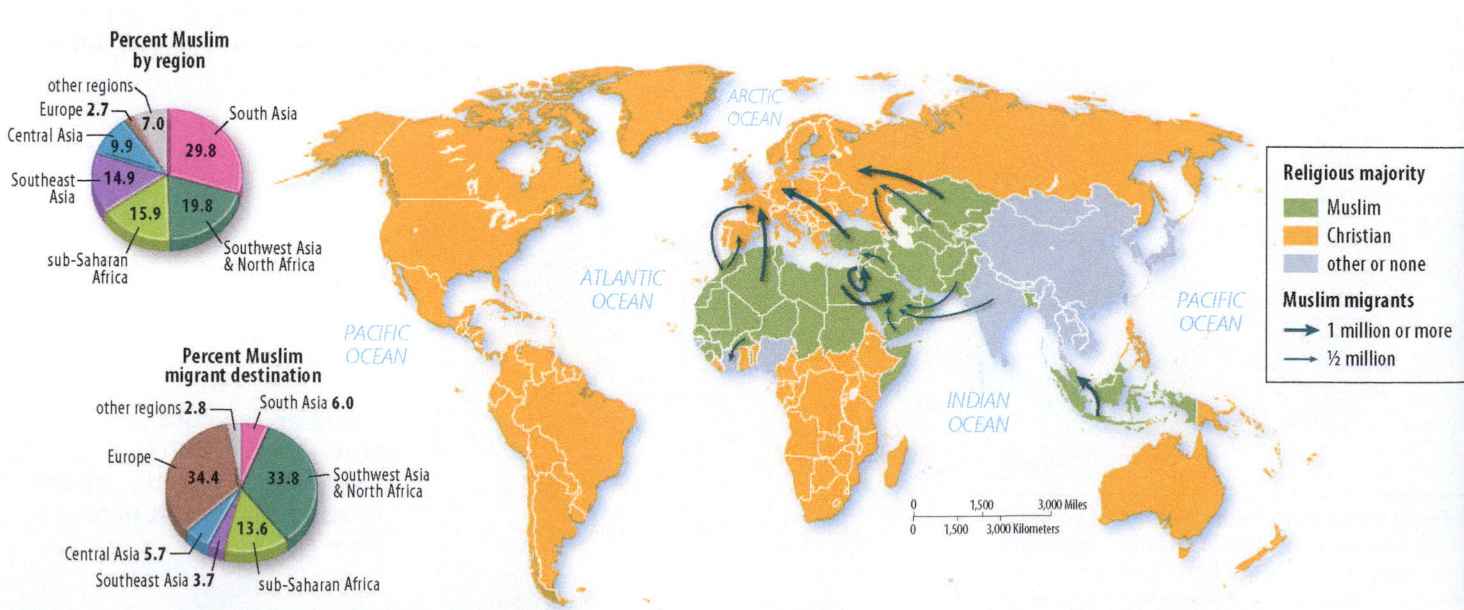

▲ **FIGURE 6-33 LARGEST FLOWS OF MUSLIM MIGRANTS**

MIGRATION PATTERNS OF JEWS

Israel is the destination for 73 percent of Jews who migrate internationally. As the world's only state with a Jewish majority, Israel offers an especially strong pull factor for Jewish migrants (see Key Issue 4). The United States is the destination for 10 percent of Jewish international migrants and Canada for 4 percent (Figure 6-35).

Only since the creation of the State of Israel in 1948 has a significant percentage of the world's Jews lived in that territory. Most Jews have not lived there since A.D. 70, when the Romans forced them to disperse throughout the world, an action known as the *diaspora*, from the Greek word for "dispersion."

Most Jews migrated from the eastern Mediterranean to Europe. Having been exiled from the home of their religion, Jews lived among other nationalities, retaining separate religious practices but adopting other cultural characteristics of the host country, such as language.

Other nationalities often persecuted the Jews living in their midst. Historically, the Jews of many European countries were forced to live permanently in **ghettos,** defined as city neighborhoods set up by law to be inhabited only by Jews. The term *ghetto* originated during the sixteenth century in Venice, Italy, as a reference to the city's former copper foundry or metal-casting district, where Jews were forced to live. Ghettos were frequently surrounded by walls, and the gates were locked at night to prevent escape.

Beginning in the 1930s, but especially during World War II (1939–1945), the Nazis systematically rounded up a large percentage of European Jews, transported them to concentration camps, and exterminated them in the Holocaust. About 4 million Jews died in the camps and 2 million in other ways. Many of the survivors migrated to territory that became the State of Israel. Today, less than 15 percent of the world's 14 million Jews live in Europe, compared to 90 percent a century ago (Figure 6-36).

PAUSE & REFLECT 6.2.5

What country had the largest Jewish population in 1910?

▲ **FIGURE 6-35 SYNAGOGUE** Brooklyn, New York.

CHECK-IN KEY ISSUE 2

Why Do Religions Have Distinctive Distributions?

- ✔ **Universalizing religions have well-documented places of origin, based on events in the lives of the founders.**
- ✔ **Ethnic religions typically have unknown or unclear origins.**
- ✔ **Universalizing religions typically diffuse widely from their place of origin, whereas ethnic religions typically do not.**
- ✔ **The contemporary migration patterns of some religious groups vary from the distribution of those religions.**

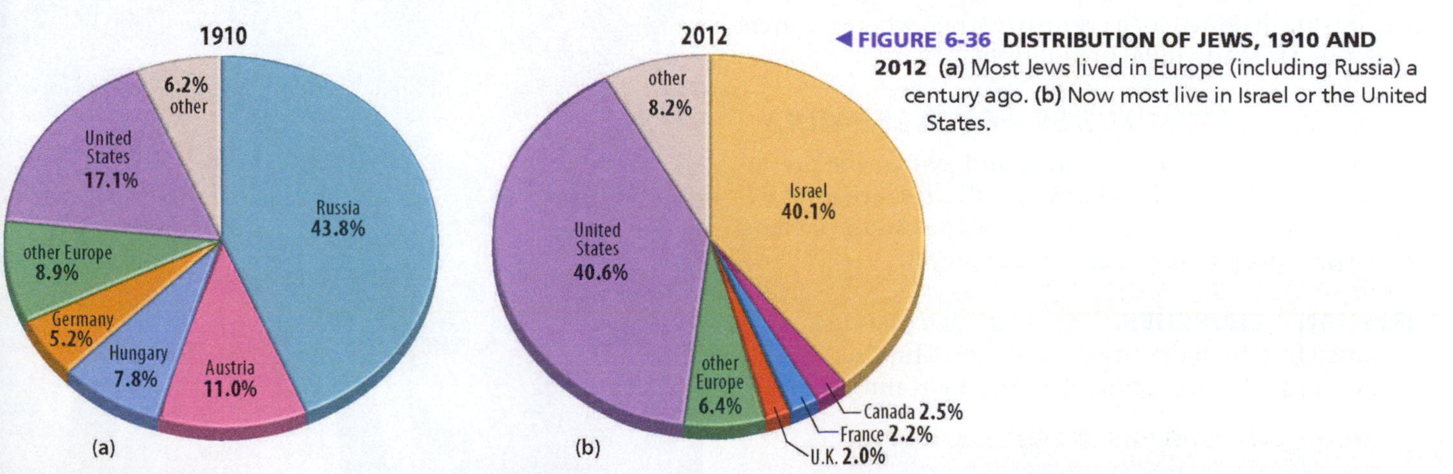

◀ **FIGURE 6-36 DISTRIBUTION OF JEWS, 1910 AND 2012 (a)** Most Jews lived in Europe (including Russia) a century ago. **(b)** Now most live in Israel or the United States.

KEY ISSUE 3

Why Do Religions Organize Space in Distinctive Patterns?

- ▸ Places of Worship
- ▸ Religious Settlements and Toponyms
- ▸ Administration of Space
- ▸ Sacred Space in Universalizing Religions
- ▸ The Landscape in Ethnic Religions
- ▸ Religious Calendars

LEARNING OUTCOME 6.3.1

Describe places of worship in various religions.

Geographers study the major impact on the landscape made by religions. In large cities and small villages around the world, regardless of the region's prevailing religion, the tallest, most elaborate buildings are often religious structures.

The distribution of religious elements on the landscape reflects the importance of religion in people's values. The impact of religion on the landscape is particularly profound, for many religious people believe that their life on Earth ought to be spent in service to God. The impact of religion is clearly seen in the arrangement of human activities on the landscape at several scales, from relatively small parcels of land to entire communities. How each religion distributes its elements on the landscape depends on its beliefs.

Places of Worship

Sacred structures are physical "anchors" of religion. Some structures are designed for a group to gather, whereas others are intended for individual meditation.

RELIGIOUS STRUCTURES FOR ASSEMBLY

Church, basilica, mosque, temple, and synagogue are familiar names that identify places for collective religious expression in various religions. They share the feature of being places where people come together for prayer.

CHRISTIAN CHURCHES. The Christian landscape is dominated by a high density of churches. The church is an expression of religious principles, an environment in the image of God. The word *church* derives from a Greek term meaning "lord," "master," and "power." *Church* also refers to a gathering of believers, as well as the building at which the gathering occurs.

The church is relatively prominent in Christianity in part because attendance at a collective service of worship is considered extremely important by all three major Christian branches. The prominence of churches on the landscape also stems from their style of construction and location. In some communities, the church was traditionally the largest and tallest building and was placed at an important square or other prominent location. Although such characteristics may no longer apply in large cities, they are frequently still true for small towns and neighborhoods within cities.

Since Christianity split into many denominations, no single style of church construction has dominated. Churches reflect both the cultural values of the denomination and the region's architectural heritage. Orthodox churches follow an architectural style that developed in the Byzantine Empire during the fifth century. Byzantine-style Orthodox churches tend to be highly ornate, topped by prominent domes (Figure 6-37). Many Protestant churches in North America, on the other hand, are simple, with little ornamentation. This austerity is a reflection of the Protestant conception of a church as an assembly hall for preaching to the congregation (Figure 6-38).

Availability of building materials also influences church appearance. In the United States, early churches were most frequently built of wood in the Northeast, brick in the Southeast, and adobe in the Southwest. Stucco and stone predominated in Latin America. This diversity reflected differences in the most common building materials found by early settlers.

MUSLIM MOSQUES. The word *mosque* derives from the Arabic for "place of worship." Muslims consider a mosque to be a location for the community to gather together for worship. Mosques are found primarily in larger cities of the Muslim world; simple structures may serve as places of prayer in rural villages.

▼ FIGURE 6-37 **ORTHODOX CHURCH** Horaita Monastery, Romania.

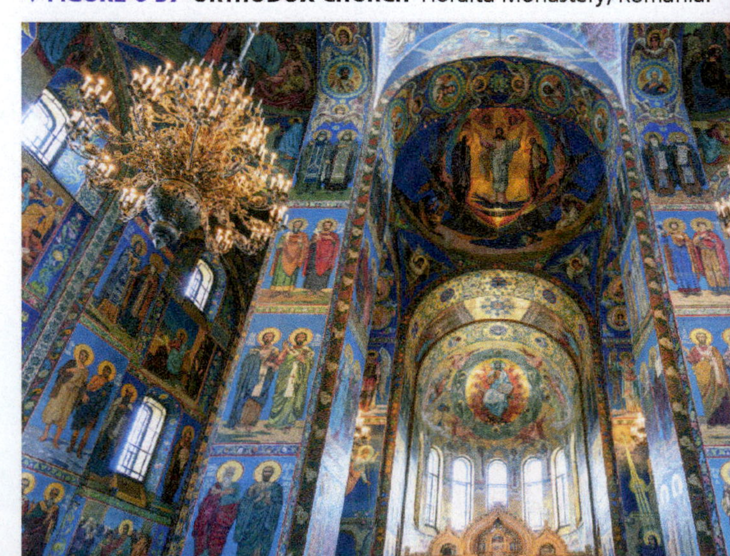

▲ **FIGURE 6-38 PROTESTANT CHURCH** Fort Bourtange, Netherlands.

A mosque is organized around a central courtyard—traditionally open-air, although it may be enclosed in harsher climates. The pulpit is placed at the end of the courtyard facing Makkah, the direction toward which all Muslims pray. Surrounding the courtyard is a cloister used for schools and nonreligious activities. A distinctive feature of the mosque is the minaret, a tower where a man known as a muezzin summons people to worship (refer to Figure 6-32).

SIKH GURDWARAS. Sikhs come together for worship at a gurdwara. The most important gurdwara is the Harmandir Sahib, or Golden Temple, in Amritsar, India (refer to Figure 6-3). The Golden Temple houses Sikhism's holiest book, the Guru Granth Sahib. Most gurdwaras imitate the layout of the Golden Temple. A gurdwara is identified by the Sikh flag flying from a tall flagpole.

JEWISH SYNAGOGUES. The word *synagogue* derives from the Greek word for "assembly." The building is often referred to by the Yiddish word *shul*, similar to the German word for "school." As these words suggest, a synagogue is a place for study and public assembly, as well as for prayer. The origin of the synagogue is unknown; it possibly came to be during the sixth century B.C., when Jews were living in exile in Babylonia, after the destruction of the First Temple in 586 B.C. Synagogues took on more importance as the place for communal prayer after the destruction of the Second Temple in A.D. 70 (refer to Figure 6-35).

▼ **FIGURE 6-39 BUDDHIST PAGODA** White Pagoda, Fuzhou, China.

BAHÁ'Í HOUSES OF WORSHIP. Bahá'ís have built Houses of Worship in every continent to dramatize that Bahá'í is a universalizing religion with adherents all over the world. Houses of Worship have been built in Wilmette, Illinois, in 1953; Sydney, Australia, in 1961; Kampala, Uganda, in 1962; Lagenhain, near Frankfurt, Germany, in 1964; Panama City, Panama, in 1972; Tiapapata, near Apia, Samoa, in 1984; and New Delhi, India, in 1986 (refer to Figure 6-4). The first Bahá'í House of Worship, built in 1908 in Ashgabat, Russia, now the capital of Turkmenistan, was turned into a museum by the Soviet Union and demolished in 1962 after a severe earthquake. New Houses of Worship are planned in Tehran, Iran; Santiago, Chile; and Haifa, Israel. In addition, several holy places related to the Prophet Bahá'u'lláh are located in Israel. All Bahá'í Houses of Worship are required to be built in the shape of a nonagon (nine-sided).

RELIGIOUS STRUCTURES FOR INDIVIDUAL MEDITATION

For some religions, a structure is a place of reflection for an individual or a small group.

BUDDHIST PAGODAS. Pagodas contain relics that Buddhists believe to be a portion of Buddha's body or clothing (Figure 6-39). After Buddha's death, his followers scrambled to obtain these relics. As part of the process of diffusing the religion, Buddhists carried these relics to other countries and built pagodas for them. Pagodas typically include tall, many-sided towers arranged in a series of tiers, balconies, and slanting roofs. Pagodas are not designed for congregational worship. Individual prayer or meditation is more likely to be undertaken at an adjacent temple, at a remote monastery, or in a home.

HINDU TEMPLES. Important Hindu religious functions are more likely to take place at home within the family. A Hindu temple is a structure designed to bring individuals closer to their gods. It serves as a shrine to one or more gods and as a place for individual reflection and meditation, in accordance with one's personal practices within the faith (refer to Figure 6-23).

As with many other elements of Hinduism, the time and place of origin of temples are unknown. Detailed evidence of the existence of temples dates from the first century B.C. Size and number of temples are determined by local preferences and commitment of resources rather than standards imposed by religious doctrine.

PAUSE & REFLECT 6.3.1

What differences appear in the images of a Hindu Temple and Christian churches?

Religious Settlements and Toponyms

LEARNING OUTCOME 6.3.2

Describe examples of religious settlements and of religious toponyms.

Buildings for worship and burial places are smaller-scale manifestations of religion on the landscape, but there are larger-scale examples—entire settlements. Most human settlements serve an economic purpose (see Chapter 12), but some are established primarily for religious reasons.

UTOPIAN SETTLEMENTS

A utopian settlement is an ideal community built around a religious way of life. Buildings are sited and economic activities organized to integrate religious principles into all aspects of daily life. An early utopian settlement in the United States was Bethlehem, Pennsylvania, founded in 1741 by Moravians, Christians who had emigrated from the present-day Czechia. By 1858, some 130 different utopian settlements had been established in the United States, in conformance with a group's distinctive religious beliefs. Examples include Oneida, New York; Ephrata, Pennsylvania; Nauvoo, Illinois; and New Harmony, Indiana.

The culmination of the utopian movement in the United States was the founding of Salt Lake City by the Latter-day Saints (Mormons), beginning in 1848. The layout of Salt Lake City is based on a plan of the city of Zion given to the church elders in 1833 by the Mormon prophet Joseph Smith. The city has a regular grid pattern, unusually broad boulevards, and church-related buildings situated at strategic points (Figure 6-40).

Most utopian communities declined in importance or disappeared altogether. Some disappeared because the inhabitants were celibate and could not attract immigrants; in other cases, residents moved away in search of better economic conditions. The utopian communities that have not been demolished are now inhabited by people who are not members of the original religious sect, although a few have been preserved as museums.

▲ **FIGURE 6-40 SALT LAKE CITY** The Salt Lake Temple of The Church of Jesus Christ of Latter-day Saints is positioned at the center of the city. To the right is the Salt Lake Tabernacle.

▲ **FIGURE 6-41 RELIGIOUS TOPONYMS** Place names near Québec's boundaries with Ontario and the United States show the imprint of religion on the landscape. In Québec, a province with a predominantly Roman Catholic population, a large number of settlements are named for saints, whereas relatively few religious toponyms are found in predominantly Protestant Ontario, New York, and Vermont.

Although most colonial settlements were not planned primarily for religious purposes, religious principles affected many of the designs. Most early New England settlers were Protestants called Puritans. The Puritans generally migrated together from England and preferred to live near each other in clustered settlements rather than on dispersed, isolated farms. Reflecting the importance of religion in their lives, New England settlers placed the church at the most prominent location in the center of the settlement, usually adjacent to a public open space known as a common, because it was for common use by everyone.

RELIGIOUS PLACE NAMES

Roman Catholic immigrants have frequently given religious place names, or toponyms, to their settlements in the New World, particularly in Québec and the U.S. Southwest. Québec's boundaries with Ontario and the United States clearly illustrate the difference between toponyms selected by Roman Catholic and Protestant settlers. Religious place names are common in Québec but rare in its two neighbors (Figure 6-41).

PAUSE & REFLECT 6.3.2

What examples of religious toponyms can you find in your community?

DOING GEOGRAPHY Distribution of Religions

Geographers can create maps of the distribution of adherents of different religions, but only for places where information about religious preferences is collected. Maps can show the extent to which different religious groups intermingle or live apart from each other.

Belfast, Northern Ireland, is an example of a place where it has been important for geographers to map the distribution of religious groups. The city is about evenly divided between Roman Catholics and Protestants. But the two groups are highly segregated. Nearly all Roman Catholics live on the west side of the River Lagan, whereas Protestants are clustered on the east side plus one neighborhood on the west side.

Belfast is located on the island of Eire (Ireland). The entire island was an English colony for many centuries and was made part of the United Kingdom in 1801. Following a succession of bloody confrontations, Ireland achieved independence in 1937. However, a majority in the northern part of Eire, including Belfast, voted to remain in the United Kingdom.

Within Northern Ireland, Roman Catholics were long victimized by discriminatory practices, such as exclusion from higher-paying jobs and better schools. Demonstrations by Roman Catholics in Belfast to protest discrimination began in 1968 and resulted in the deaths of more than 3,000 people. A peace agreement, which provided for the two religious groups to share power, has reduced the level of violence since 1999. However, most Roman Catholics and Protestants continue to live in segregated neighborhoods within Belfast (Figure 6-42).

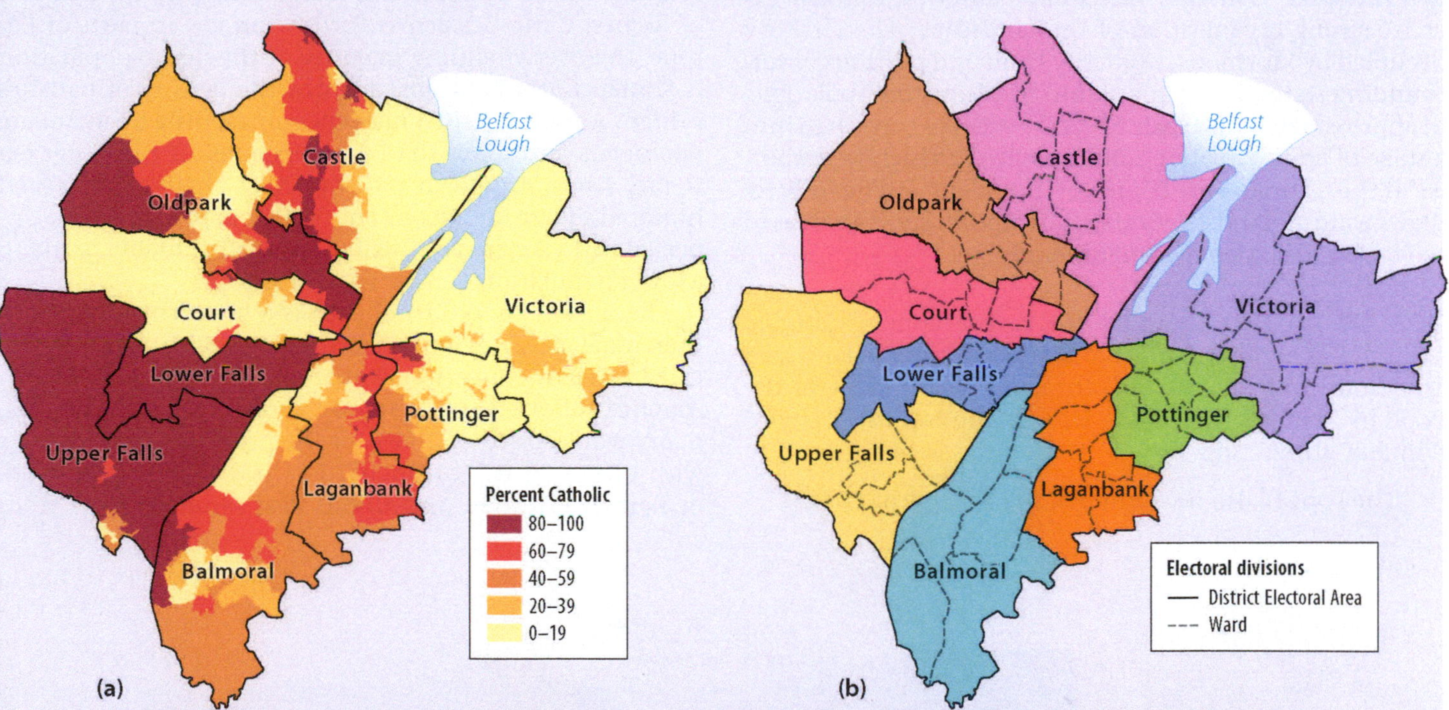

▲ **FIGURE 6-42 RELIGIOUS SEGREGATION IN BELFAST, NORTHERN IRELAND** **(a)** The distribution of Catholics vs. **(b)** the electoral divisions in Belfast.

What's Your Geography of Religions?

The U.S. Census does not collect information on people's religion, so the rendering of an accurate local-scale census-derived map such as the one for Belfast is not possible in the United States.

1. Does your immediate family have multiple faiths?
2. Do your immediate neighbors share the same faith as you?
3. What denomination or congregation are the five nearest places of worship to your house?
4. How many of the five nearest places of worship are part of the most widely practiced faith in your county?
5. Does the most widely practiced faith in your family and immediate neighbors match the county map, or does it differ?
6. Do you share the same faith, denomination, or congregation with your closest friends?

Administration of Space

LEARNING OUTCOME 6.3.3

Compare the administrative organization of hierarchical and locally autonomous religions.

Followers of a universalizing religion must be connected in order to ensure communication and consistency of doctrine. The method of interaction varies among universalizing religions, branches, and denominations. Ethnic religions tend not to have organized, central authorities.

HIERARCHICAL RELIGIONS

A **hierarchical religion** has a well-defined geographic structure and organizes territory into local administrative units. The Church of Jesus Christ of Latter-day Saints and Roman Catholicism are examples of hierarchical religions.

LATTER-DAY SAINTS. Latter-day Saints (Mormons) exercise strong organization of the landscape. The territory occupied by Mormons, primarily Utah and portions of surrounding states, is organized into wards, with populations of approximately 750 each. Several wards are combined into a stake of approximately 5,000 people. The highest authority in the Church—the board and president—frequently redraws ward and stake boundaries in rapidly growing areas to reflect the ideal population standards.

ROMAN CATHOLIC HIERARCHY. The Roman Catholic Church has organized much of Earth's inhabited land into an administrative structure ultimately accountable to the Pope in Rome (Figure 6-43). Here is the top hierarchy of Roman Catholicism:

- The Pope is also the bishop of the Diocese of Rome.
- Archbishops report to the Pope. Each heads a province, which is a group of several dioceses. The archbishop also is bishop of one diocese within the province, and some distinguished archbishops are elevated to the rank of cardinal.
- Bishops report to an archbishop. Each administers a diocese, which is the basic unit of geographic organization in the Roman Catholic Church. The bishop's headquarters, called a "see," is typically the largest city in the diocese.
- Priests report to bishops. A diocese is spatially divided into parishes, each headed by a priest.

PAUSE & REFLECT 6.3.3

What are the different spatial units of administration in the Roman Catholic Church?

The area and population of parishes and dioceses vary according to historical factors and the distribution of Roman Catholics across Earth's surface. In parts of Europe, the overwhelming majority of the dense population is Roman Catholic. Consequently, the density of parishes is high. A typical parish may encompass only a few square kilometers and fewer than 1,000 people. At the other extreme, Latin American parishes may encompass several hundred square kilometers and 5,000 people. The more dispersed Latin American distribution is attributable partly to a lower population density than in Europe.

Because Roman Catholicism is a hierarchical religion, individual parishes must work closely with centrally located officials concerning rituals and procedures. If Latin America followed the European model of small parishes, many would be too remote for the priest to communicate with others in the hierarchy. The less intensive network of Roman Catholic institutions also results in part from

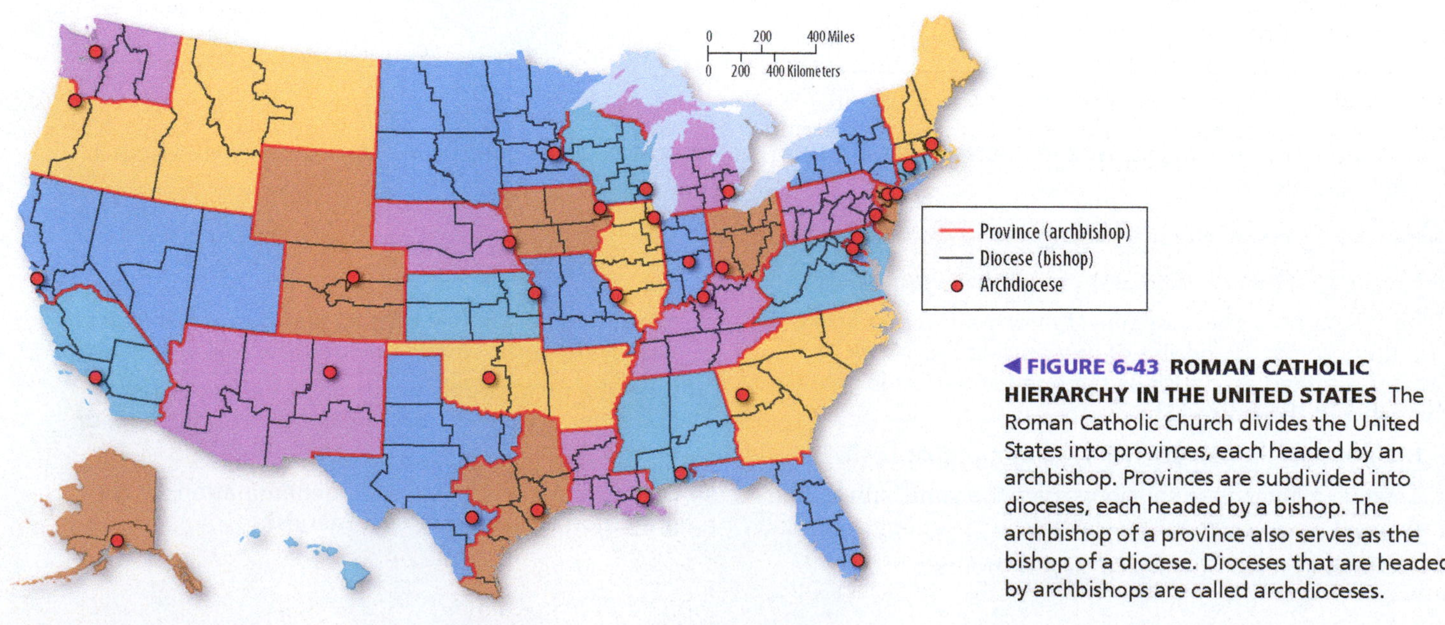

FIGURE 6-43 ROMAN CATHOLIC HIERARCHY IN THE UNITED STATES The Roman Catholic Church divides the United States into provinces, each headed by an archbishop. Provinces are subdivided into dioceses, each headed by a bishop. The archbishop of a province also serves as the bishop of a diocese. Dioceses that are headed by archbishops are called archdioceses.

▲ **FIGURE 6-44 LOCALLY AUTONOMOUS ISLAM** The shrine of Imam Ali, in Najaf, Iraq, contains the tomb of Ali, from whom traces the Shiite branch.

colonial traditions, for both Portuguese and Spanish rulers discouraged parish development in Latin America.

The Roman Catholic population is growing rapidly in the U.S. Southwest and in suburbs of some large North American and European cities. Some of these areas have a low density of parishes and dioceses compared to the population, so the Church must adjust its territorial organization. New local administrative units can be created, although funds to provide the desired number of churches, schools, and other religious structures might be scarce. Conversely, the Roman Catholic population is declining in inner cities and rural areas. Maintaining services in these areas is expensive, but the process of combining parishes and closing schools is very difficult.

LOCALLY AUTONOMOUS RELIGIONS

Some universalizing religions are highly **autonomous religions,** or self-sufficient, and interaction among communities is confined to little more than loose cooperation and shared ideas. Islam and some Protestant denominations are good examples.

LOCAL AUTONOMY IN ISLAM. Among the three large universalizing religions, Islam provides the most local autonomy. Like other locally autonomous religions, Islam has neither a religious hierarchy nor a formal territorial organization. A mosque is a place for public prayer, led by an imam (Figure 6-44), but everyone is expected to participate equally in the rituals and is encouraged to pray privately.

In the absence of a hierarchy, the only formal organization of territory in Islam is through the coincidence of religious territory with secular states. Governments in some predominantly Islamic countries include in their bureaucracy people who administer Islamic institutions. These administrators interpret Islamic law and run welfare programs.

Strong unity in the Islamic world is maintained by a relatively high degree of communication and migration, such as the pilgrimage to Makkah. In addition, uniformity is fostered by Islamic doctrine, which offers more explicit commands than most other religions.

PROTESTANT DENOMINATIONS. Protestant Christian denominations vary in geographic structure from extremely autonomous to somewhat hierarchical. The Episcopalian, Lutheran, and most Methodist churches have hierarchical structures, somewhat comparable to the Roman Catholic Church but led by bishops, not by a single leader such as the pope. Extremely autonomous denominations such as Baptists and United Church of Christ are organized into self-governing congregations. Each congregation establishes the precise form of worship and selects the leadership.

Presbyterian churches represent an intermediate degree of autonomy. Individual churches are united in a presbytery, several of which in turn are governed by a synod, with a general assembly as ultimate authority over all churches. Each Presbyterian church is governed by an elected board of directors with lay members.

ETHNIC RELIGIONS. Hinduism is highly autonomous because worship is usually done alone or with others in the household. Hindus share ideas primarily through undertaking pilgrimages and reading traditional writings. Hinduism has no centralized structure of religious control (Figure 6-45).

▼ **FIGURE 6-45 HINDU TEMPLE** Galta Temple, Jaipur, India.

Sacred Space in Universalizing Religions

LEARNING OUTCOME 6.3.4

Explain why places are sacred in universalizing religions.

Religions may elevate particular places to holy positions. Universalizing and ethnic religions differ on the types of places that are considered holy:

- An ethnic religion typically has a less widespread distribution than a universalizing one in part because its holy places derive from the distinctive physical environment of its hearth, such as mountains, rivers, or rock formations.
- A universalizing religion endows with holiness cities and other places associated with the founder's life. Its holy places do not necessarily have to be near each other, and they do not need to be related to any particular physical environment.

Buddhism and Islam are the universalizing religions that place the most emphasis on identifying shrines. Places are holy because they are the locations of important events in the life of Buddha or Muhammad. Making a **pilgrimage** to these holy places—a journey for religious purposes to a place considered sacred—is incorporated into the rituals of some religions. Hindus and Muslims are especially encouraged to make pilgrimages to visit holy places in accordance with recommended itineraries.

▲ **FIGURE 6-46 HOLY PLACES IN BUDDHISM** Most are clustered in northeastern India and southern Nepal because they were the locations of important events in Buddha's life.

BUDDHIST SHRINES

Eight places are holy to Buddhists because they were the locations of important events in Buddha's life (Figure 6-46). The four most important of the eight places are concentrated in a small area of northeastern India and southern Nepal:

- Lumbinī in southern Nepal, where Buddha was born around 563 B.C., is most important. Many sanctuaries and monuments were built there, but all are in ruins today.
- Bōdh Gayā, 250 kilometers (150 miles) southeast of Buddha's birthplace, is the site of the second great event in his life, where he reached perfect wisdom. A temple has stood near the site since the third century B.C., and part of the surrounding railing built in the first century A.D. still stands. Because Buddha reached perfect Enlightenment while sitting under a bo tree, that tree has become a holy object as well (refer to Figure 6-21). To honor Buddha, the Bodhi (or Bo) tree has been diffused to other Buddhist countries, such as China and Japan.
- Deer Park in Sarnath, where Buddha gave his first sermon, is the third important location. The Dhamek Stupa at Sarnath, built in the third century B.C., is probably the oldest surviving structure in India (Figure 6-47). Nearby is an important library of Buddhist literature, including many works removed from Tibet when Tibet's Buddhist leader, the Dalai Lama, went into exile.
- Kuśinagara, the fourth holy place, is where Buddha died at age 80 and passed into Nirvana, a state of peaceful extinction. Temples built at the site are currently in ruins.

Four other sites in northeastern India are particularly sacred because they were the locations of Buddha's principal miracles:

- Srāvastī is where Buddha performed his greatest miracle. Before an assembled audience of competing religious leaders, Buddha created multiple images of himself and visited heaven. Srāvastī became an active center of Buddhism, and one of the most important monasteries was established there.
- Sāmkāśya, the second miracle site, is where Buddha is said to have ascended to heaven, preached to his mother, and returned to Earth.
- Rajagrha, the third site, is holy because Buddha tamed a wild elephant there, and shortly after Buddha's death, it became the site of the first Buddhist Council.
- Vaisālā, the fourth location, is the site of Buddha's announcement of his impending death and the second Buddhist Council.

All four miracle sites are in ruins today, although excavation activity is under way.

▲ **FIGURE 6-47 DEER PARK** Dhamek stupa at Sarnath, site of Buddha's first sermon.

HOLY PLACES IN ISLAM

The holiest locations in Islam are in cities associated with the life of Muhammad. The holiest city for Muslims is Makkah (Mecca), the birthplace of Muhammad. The word *mecca* now has a general meaning in the English language as a goal sought or a center of activity.

Now a city of 1.3 million inhabitants, Makkah contains the holiest object in the Islamic landscape, namely al-Ka'ba, a cubelike structure encased in silk, which stands at the center of the Great Mosque, Masjid al-Haram, Islam's largest mosque (Figure 6-48). The Ka'ba, thought by Muslims to have been built by Abraham and Ishmael, contains a black stone that Muslims believe was given to Abraham by Gabriel as a sign of a covenant with Ishmael and the Muslim people.

The Ka'ba had been a religious shrine in Makkah for centuries before the origin of Islam. After Muhammad defeated the local people, he captured the Ka'ba, cleared it of idols, and rededicated it to the all-powerful Allah (God). The Masjid al-Haram mosque also contains the well of Zamzam, considered to have the same source as the water given to Hagar by the Angel Gabriel to quench the thirst of her infant, Ishmael.

The second-most-holy geographic location in Islam is Madinah (Medina), a city of 1.3 million inhabitants, 350 kilometers (220 miles) north of Makkah. Muhammad received his first support from the people of Madinah and became the city's chief administrator. Muhammad's tomb is at Madinah, inside Islam's second-largest mosque (refer to Figure 6-20).

Every healthy Muslim who has adequate financial resources is expected to undertake a pilgrimage, called a hajj, to Makkah (Mecca). Regardless of nationality and economic background, all pilgrims dress alike, in plain white robes, to emphasize common loyalty to Islam and the equality of people in the eyes of Allah. A precise set of rituals is practiced, culminating in a visit to the Ka'ba. The hajj attracts millions of Muslims annually to Makkah. Hajj visas are issued by the government of Saudi Arabia according to a formula of 1 per 1,000 Muslims in a country. Roughly 80 percent come from Southwest Asia & North Africa and 20 percent from elsewhere in Asia. In 2015, more than 1,000 died in a stampede outside the Great Mosque. Although Indonesia is the country with the most Muslims, it has not sent the largest number of pilgrims to Makkah because of the relatively long travel distance.

PAUSE & REFLECT 6.3.4

Based on the lives of the Buddha and the prophet Muhammad, what types of sites are likely to be goals of pilgrimage for the followers of a universalizing religion?

▲ **FIGURE 6-48 MASJID AL-HARAM (GREAT MOSQUE), MAKKAH, SAUDI ARABIA** Al-Ka'ba stands at the center.

The Landscape in Ethnic Religions

LEARNING OUTCOME 6.3.5

Analyze the importance of the physical geography in ethnic religions.

One of the principal reasons that ethnic religions are highly clustered is that they are closely tied to the physical geography of a particular place. Pilgrimages are undertaken to view these physical features.

HINDUISM'S SOUTH ASIAN LANDSCAPE

As an ethnic religion of India, Hinduism is closely tied to the physical geography of India. According to a survey conducted by the geographer Surinder Bhardwaj, the natural features most likely to rank among the holiest shrines in India are riverbanks and coastlines. Hindus consider a pilgrimage, known as a tirtha, to be an act of purification. Although not a substitute for meditation, the pilgrimage is an important act in achieving redemption.

Hindu holy places are organized into a hierarchy. Particularly sacred places attract Hindus from all over India, despite the relatively remote locations of some; less important shrines attract primarily local pilgrims. Because Hinduism has no central authority, the relative importance of shrines is established by tradition, not by doctrine. For example, many Hindus make long-distance pilgrimages to Mt. Kailās, located at the source of the Ganges in the Himalayas, which is holy because Shiva lives there. Other mountains may attract only local pilgrims: Local residents may consider a nearby mountain to be holy if Shiva is thought to have visited it at one time.

Hindus believe that they achieve purification by bathing in holy rivers. The Ganges is the holiest river in India because it is supposed to spring forth from the hair of Shiva, one of the main deities. Indians come from all over the country to Hardwār, the most popular location for bathing in the Ganges (refer to Figure 6-22).

The remoteness of holy places from population clusters once meant that making a pilgrimage required major commitments of time and money as well as undergoing considerable physical hardship. Recent improvements in transportation have increased the accessibility of shrines. Hindus can now reach holy places in the Himalaya Mountains by bus or car, and Muslims from all over the world can reach Makkah by airplane.

COSMOGONY AND SPIRITS

Cosmogony is a set of religious beliefs concerning the origin of the universe. The universalizing religions Christianity and Islam consider that God (or Allah, for Islam) created the universe, including Earth's physical environment and human beings. A religious person can serve the Creator by cultivating the land, draining wetlands, clearing forests, building new settlements, and otherwise making productive use of natural features that the Creator made.

The cosmogony underlying Chinese traditional ethnic religions, is that the universe is made up of two forces, yin and yang, which exist in everything. The force of yin (earth, darkness, female, cold, depth, passivity, and death) interacts with the force of yang (heaven, light, male, heat, height, activity, and life) to achieve balance and harmony. An imbalance results in disorder and chaos.

To primal-indigenous animists, the powers of the universe are mystical, and only a few people on Earth can harness these powers for medical or other purposes. Spirits or gods can be placated, however, through prayer and sacrifice. Rather than attempt to transform the environment, animists accept environmental hazards as normal and unavoidable.

DISPOSING OF THE DEAD

A prominent example of religiously inspired arrangement of land at a smaller scale is burial practices. Climate, topography, and religious doctrine combine to create differences in practices to shelter the dead.

BURIAL. Christians, Muslims, and Jews usually bury their dead in a specially designated area called a cemetery. The Christian burial practice can be traced to the early years of the religion. In ancient Rome, underground passages known as catacombs were used to bury early Christians (and to protect the faithful when the religion was still illegal).

After Christianity became legal, Christians buried their dead in the yard around the church. As these burial places became overcrowded, separate burial grounds had to be established outside the city walls. Public health and sanitation considerations in the nineteenth century led to public management of many cemeteries. Some cemeteries are still operated by religious organizations. The remains of the dead are customarily aligned in some traditional direction. Some Christians bury the dead with the feet toward Jerusalem so that they may meet Christ there on the Day of Judgment.

Cemeteries may consume significant space in a community, increasing the competition for scarce space. In congested urban areas, Christians and Muslims have traditionally used cemeteries as public open space. Before the widespread development of public parks in the nineteenth century, cemeteries were frequently the only green space in rapidly growing cities. Cemeteries are still used as parks in Muslim countries, where the idea faces less opposition than in Christian societies.

Traditional burial practices in China have put pressure on agricultural land. By burying dead relatives, rural residents have removed as much as 10 percent of the land from productive agriculture. The government in China has ordered the practice discontinued, even urging farmers to plow over old burial mounds. Cremation is encouraged instead.

◀ **FIGURE 6-49 HINDU CREMATION** Hindus consider the Manikarnika Ghat (the steps leading to the Ganges at Varanasi, India) to be the most auspicious place for a cremation.

OTHER METHODS OF DISPOSING OF BODIES. Not all faiths bury their dead. Hindus generally practice cremation rather than burial (Figure 6-49). The body is washed with water from the Ganges River and then burned with a slow fire on a funeral pyre. Burial is reserved for children, ascetics, and people with certain diseases. Cremation is considered an act of purification, although it tends to strain India's wood supply.

Motivation for cremation may have originated from unwillingness on the part of nomads to leave their dead behind, possibly because of fear that the body could be attacked by wild beasts or evil spirits, or even return to life. Cremation could also free the soul from the body for departure to the afterworld and provide warmth and comfort for the soul as it embarked on the journey to the afterworld. Cremation was the principal form of disposing of bodies in Europe before Christianity. It is still practiced in parts of Southeast Asia, possibly because of Hindu influence.

To strip away unclean portions of the body, Zoroastrians (Parsis) traditionally exposed the dead to scavenging birds and animals. The ancient Zoroastrians did not want the body to contaminate the sacred elements of fire, earth, or water. The dead were exposed in a circular structure called a dakhma, or tower of silence (Figure 6-50). Tibetan Buddhists also practiced exposure for some dead, with cremation reserved for the most exalted priests.

Disposal of bodies at sea is used in some parts of Micronesia, but the practice is much less common than in the past. The bodies of lower-class people would be flung into the sea; elites could be set adrift on a raft or boat. Water burial was regarded as a safeguard against the living being contaminated by the dead.

PAUSE & REFLECT 6.3.5

What are some of the cultural or religious factors that influence methods of disposing of bodies other than burial?

◀ **FIGURE 6-50 ZOROASTRIAN TOWER OF SILENCE** Yazd, Iran.

Religious Calendars

LEARNING OUTCOME 6.3.6

Understand the roles of holidays and calendars in various religions.

Universalizing and ethnic religions have different approaches to the calendar. An ethnic religion typically is more clustered than a universalizing religion, in part because its holidays are based on the distinctive physical geography of the homeland. In universalizing religions, major holidays relate to events in the life of the founder rather than to the changing seasons of one particular place.

A prominent feature of ethnic religions is celebration of the seasons—the calendar's annual cycle of variation in climatic conditions. Knowledge of the calendar is critical to successful agriculture, whether for sedentary crop farmers or nomadic animal herders. The seasonal variations of temperature and precipitation help farmers select the appropriate times for planting and harvesting and make the best choice of crops. Rituals are performed to pray for favorable environmental conditions or to give thanks for past success.

THE CALENDAR IN UNIVERSALIZING RELIGIONS

The principal purpose of the holidays in universalizing religions is to commemorate events in the founder's life. Examples can be found in the various universalizing religions.

ISLAM. Islam uses a lunar calendar. In a 30-year cycle, the Islamic calendar has 19 years with 354 days and 11 years with 355 days. As a result, Muslim holidays arrive in different seasons from generation to generation.

At the moment, the start of Ramadan is occurring in the Northern Hemisphere spring—for example, May 27, 2017, on the western Gregorian calendar. In A.D. 2010, Ramadan started on August 11, and in A.D. 2020 Ramadan will start on April 24. Because Ramadan occurs at different times of the solar year in different generations, the number of hours of the daily fast varies widely because the amount of daylight varies by season and by location on Earth's surface.

Observance of Ramadan can be a hardship because it can interfere with critical agricultural activities, depending on the season. However, as a universalizing religion with 1.6 billion adherents worldwide, Islam is practiced in various climates and latitudes. If Ramadan were fixed at the same time of the Southwest Asia's agricultural year, Muslims in various places of the world would need to make different adjustments to observe Ramadan.

CHRISTIANITY. Christians commemorate the resurrection of Jesus on Easter, observed on the first Sunday after the first full Moon following the spring equinox in late March. But not all Christians observe Easter on the same day because Protestant and Roman Catholic branches calculate the date on the Gregorian calendar, whereas Orthodox churches use the Julian calendar.

Christians associate their holidays with seasonal variations in the calendar, but climate and the agricultural cycle are not central to the liturgy and rituals. In Southern Europe, Easter is a joyous time of harvest. Northern Europe and North America do not have a major Christian holiday at harvest time, which would be placed in the fall.

Most Northern Europeans and North Americans associate Christmas, the birthday of Jesus, with winter conditions, such as low temperatures, snow cover, and the absence of vegetation except for needle-leaf evergreens. But for Christians in the Southern Hemisphere, December 25 is the height of summer, with warm days and abundant sunlight.

BUDDHISM. All Buddhists celebrate as major holidays Buddha's birth, Enlightenment, and death. However, not all Buddhists observe them on the same days. Japanese Buddhists celebrate Buddha's birth on April 8, his Enlightenment on December 8, and his death on February 15; Theravadist Buddhists observe all three events on the same day, usually in April.

SIKHISM. The major holidays in Sikhism are the births and deaths of the religion's 10 gurus. The tenth guru, Gobind Singh, declared that after his death, instead of an eleventh guru, Sikh-ism's highest spiritual authority would be the holy scriptures the Guru Granth Sahib. A major holiday in Sikhism is the day when the Holy Granth was installed as the religion's spiritual guide. Commemorating historical events distinguishes Sikhism as a universalizing religion, in contrast to India's major ethnic religion, Hinduism, which glorifies the physical geography of India.

BAHÁ'Í. The Bahá'ís use a calendar established by the Báb and confirmed by Bahá'u'lláh, in which the year is divided into 19 months of 19 days each, with the addition of 4 extra days (5 in leap years). The year begins on the first day of spring, March 21, one of several holy days in the Bahá'í calendar. Bahá'ís attend the Nineteen Day Feast, held on the first day of each month of the Bahá'í calendar, to pray, read scriptures, and discuss community activities.

THE CALENDAR IN JUDAISM

Judaism's major holidays are based on events in the agricultural calendar of the religion's homeland in present-day Israel. These agricultural holidays later gained importance because they also commemorated events in the Exodus of the Jews from Egypt, as recounted in the Bible. The reinterpretation of natural holidays in light of historical events has been especially important for Jews in North America, Europe, and other regions who are unfamiliar with the agricultural calendar of Southwest Asia.

Major Jewish holidays include:

- Pesach (Passover) derives from traditional agricultural practices in which farmers offered God the first fruits of the new spring harvest. It also recalls the liberation of the Jews from slavery in Egypt and the miracle of their successful flight under the leadership of Moses.

- Sukkot celebrates the final gathering of fruits for the year, and prayers, especially for rain, are offered to bring success in the upcoming agricultural year (Figure 6-51). It derives from the Hebrew word for the "booths," or "temporary shelters," occupied by Jews during their wandering in the wilderness for 40 years after fleeing Egypt.
- Shavuot (Feast of Weeks) comes at the end of the grain harvest. It is also considered the date during the wandering when Moses received the Ten Commandments from God.
- Rosh Hashanah (New Year) and Yom Kippur (Day of Atonement), the two most holy and solemn days in the Jewish calendar, come in the autumn, which is the season when grain crops are planted in the Mediterranean agricultural region and therefore a time of hope and worry over whether the upcoming winter's rainfall will be sufficient. Today they are days of repentance and prayer to be inscribed in the Book of Life.

The solar calendar has 12 months, each containing 30 or 31 days, taking up the astronomical slack with 28 or 29 days in February. But Judaism and Islam use a lunar calendar rather than a solar calendar. The Jewish calendar inserts an extra month every few years to match the agricultural and solar calendars, whereas Islam retains a strict lunar calendar.

THE SOLSTICE

The **solstice** has special significance in some ethnic religions. A major holiday in some pagan religions is the winter solstice, December 21 or 22 in the Northern Hemisphere. The winter solstice is the shortest day and longest night of the year, when the Sun appears lowest in the sky, and the apparent movement of the Sun's path north or south comes to a stop before reversing direction (*solstice* comes from the Latin to "stand still"). Stonehenge, a collection of stones erected in southwestern England some 4,500 years ago (Figure 6-52), is a prominent remnant of a pagan structure apparently aligned so the Sun rises between two stones on the summer and winter solstices.

▼ **FIGURE 6-51 AGRICULTURAL HOLIDAYS IN JUDAISM** Jews bless the *etrog* (citron) and *lulav* (date palm branches) during the holiday of Sukkot at a synagogue in Brooklyn, New York.

▲ **FIGURE 6-52 STONEHENGE** The 30 enormous stones that comprise Stonehenge were erected around 4,500 years ago. The stones replaced earlier circular arrangements made originally of timber, probably around 5,000 years ago. Stonehenge was constructed by people who left no written records. So we can only speculate on how the stones were transported from a quarry, and the reason for the arrangement. **1.** Sunrise on the summer solstice hits precisely between two of the stones. Can you think of a reason why the people who built Stonehenge would use this arrangement? **2.** What is the meterological significance of the summer solstice? **3.** How might the arrangement of the stones, and the position of the stones with respect to the solstice, carry religious meaning for an ancient people?

PAUSE & REFLECT 6.3.6

Why do some religions organize their annual calendars according to the lunar cycle?

CHECK-IN KEY ISSUE 3

Why Do Religions Organize Space in Distinctive Patterns?

- ✔ **Religious structures, such as churches and mosques, are prominent features of the landscape.**
- ✔ **Some religions have hierarchical administrative structures, whereas others emphasize local autonomy.**
- ✔ **Universalizing religions typically celebrate events in the life of the founder or prophet.**
- ✔ **Ethnic religions are more closely tied to their local physical environment than are universalizing religions.**
- ✔ **The calendar typically revolves around the physical environment in ethnic religions and the founder's life in universalizing religions.**

KEY ISSUE 4

Why Do Territorial Conflicts Arise Among Religious Groups?

- Challenges for Religions in South and East Asia
- Challenges for Religions in Central and Southwest Asia
- Geographic Perspectives in the Middle East
- Jerusalem's Challenging Geography

LEARNING OUTCOME 6.4.1

Understand reasons for geographic conflicts between religious and secular cultural groups.

Religion is an element of cultural diversity that has led to disputes in some places. The attempt by intense adherents of one religion to organize Earth's surface can conflict with the spatial expression of other faiths or of nonreligious ideas. Conflicts arise in all continents, but disputes involving religious groups have been especially intense across Asia.

Challenges for Religions in South and East Asia

Changing cultural, political, and economic customs can sometimes conflict with traditional religious values. In South Asia, Hinduism has been forced to react to secular ideas from the West, and in East Asia Buddhism has been challenged by Communist perspectives to diminish the importance of religion in society.

INDIA: HINDUISM AND SOCIAL EQUALITY

Hinduism has been strongly challenged since the 1800s, when British colonial administrators introduced their social and moral concepts to India. The most vulnerable aspect of the Hindu religion was its rigid **caste** system, which indicated the class or distinct hereditary order into which a Hindu was born, according to religious law.

The caste system apparently originated around 1500 B.C., when Aryans invaded India from the west. The Aryans divided themselves into four castes that developed strong differences in social and economic position:

- Brahmans, the priests and top administrators.
- Kshatriyas, or warriors.
- Vaisyas, or merchants.
- Shudras, or agricultural workers and artisans.

The Shudras occupied a distinctly lower status than the other three castes. Below the four castes were the Dalits—outcasts or untouchables—who did work considered too dirty for other castes. In theory, the untouchables were descended from the indigenous people who dwelled in India prior to the Aryan conquest.

Over the centuries, these original castes split into thousands of subcastes. Until recently, social relations among the castes were limited, and the rights of non-Brahmans, especially Dalits, were restricted. British administrators and Christian missionaries pointed out the shortcomings of the caste system, such as neglect of the untouchables' health and economic problems.

The type of Hinduism practiced depends in part on an individual's caste. A high-caste Brahman may practice a form of Hinduism based on knowledge of relatively obscure historical texts. At the other end of the caste system, a low-caste illiterate in a rural village may perform religious rituals without a highly developed set of written explanations for them.

The rigid caste system has been considerably relaxed in recent years. The Indian government classifies untouchables, Shudras, and other historically discriminated castes as "scheduled castes." They comprise 16 percent of India's total population and are now often called Dalit (Figure 6-53). Consciousness of caste persists: A government plan to devise a quota system designed to give untouchables more places in the country's universities generated strong opposition. People looking for a marriage partner advertise their caste and the castes they are willing to consider for a spouse.

▼ **FIGURE 6-53 DALIT** An Indian Dalit child makes the finishing touch to a statue of B.R. Ambedkar, who was born a low-caste Hindu and became a leader in the fight against caste-based discrimination.

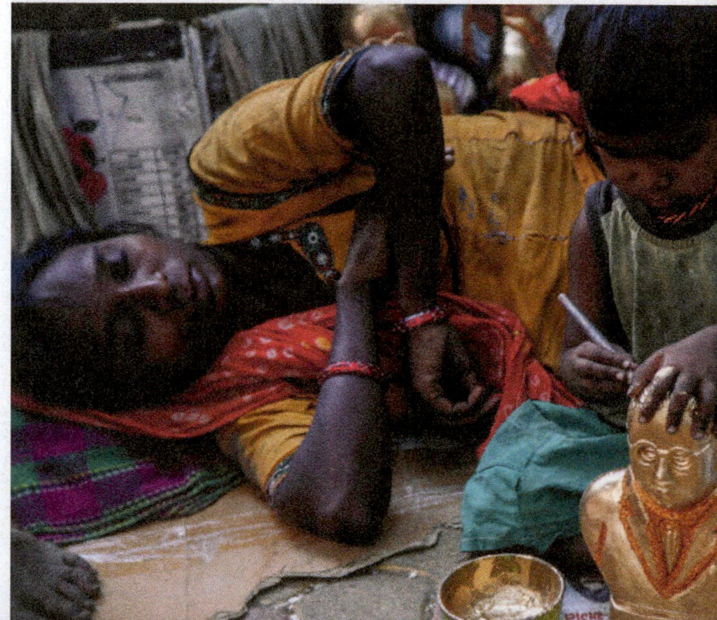

COMMUNISM AND RELIGION

Organized religion was challenged in the twentieth century by the rise of communism in Europe and Asia. Communist regimes generally discouraged religious belief and practice.

Following the 1917 Bolshevik Revolution, the Communist government of the Soviet Union pursued especially strong antireligious programs. Karl Marx had called religion "the opium of the people," a view shared by V. I. Lenin and other early Communist leaders. Marxism became the official doctrine of the Soviet Union, so religious doctrine was a potential threat to the success of the revolution. People's religious beliefs could not be destroyed overnight, but the role of organized religion in Soviet life could be reduced—and it was.

In 1721, Czar Peter the Great had made the Russian Orthodox Church a part of the Russian government. The Soviet government in 1918 eliminated the official church–state connection that Peter the Great had forged. All church buildings and property were nationalized and could be used only with government permission. The Orthodox religion retained adherents in the Soviet Union, especially among the elderly, but younger people generally had little contact with the church beyond attending a service perhaps once a year. With religious organizations prevented from conducting social and cultural work, religion dwindled in daily life.

The end of Communist rule in the late twentieth century brought a religious revival in Eastern Europe, especially where Roman Catholicism is the most prevalent branch of Christianity, including Croatia, Czechia, Hungary, Lithuania, Poland, Slovakia, and Slovenia. Property confiscated by the Communist governments reverted to Church ownership, and attendance at church services increased.

In countries in Central Asia that were once part of the Soviet Union—Kazakhstan, Kyrgyzstan, Tajikistan, Turkmenistan, and Uzbekistan—most people are Muslims. These newly independent countries are struggling to determine the extent to which laws should be rewritten to conform to Islamic custom rather than to the secular tradition inherited from the Soviet Union.

Conflict between communism and religion remains especially acute in China. The dispute centers around the role of Buddhism and its spiritual leader the Dalai Lama in the territory known as Tibet in most of the world but called Xizang by the government of China (see Sustainability & Our Environment feature).

PAUSE & REFLECT 6.4.1

Why would the Chinese Communists feel it important to dismantle the religious institutions of a poor remote country?

SUSTAINABILITY & OUR ENVIRONMENT

Communist China and the Dalai Lama

When the Dalai Lama dies, Tibetan Buddhists believe that his spirit enters the body of a child. In 1937, a group of priests located and recognized a 2-year-old child named Tenzin Gyatso as the fourteenth Dalai Lama. The child was brought to Lhasa in 1939 when he was 4, enthroned a year later, and trained by priests to assume leadership (Figure 6-54).

Before the Communist takeover, daily life in Tibet was traditionally dominated by Buddhist rites. China, which had ruled Tibet from 1720 until its independence in 1911, invaded the rugged, isolated country in 1950, turned it into a province named Xizang in 1951, and installed a Communist government in Tibet in 1953. The Chinese Communists sought to reduce the domination of Buddhist monks in the country's daily life by attacking the economic and cultural pillars of sustainability for Buddhism in TIbet. For example they destroyed monasteries and temples. Farmers were required to join agricultural communes that were economically unsustainable for their nomadic style of raising livestock, especially yaks.

After crushing a rebellion in 1959, China executed or imprisoned tens of thousands and forced another 100,000, including the Dalai Lama, to emigrate. Buddhist temples were closed and demolished, and religious artifacts and scriptures were destroyed. Since then, the Dalai Lama and other Tibetan Buddhist leaders have been based in Dharamsala, India.

The Communist government of China has declared that the next Dalai Lama will be found in Tibet, and they will be responsible for finding him. Buddhist leaders are accused of plotting to turn Tibet once again into an independent country. In response, Buddhist leaders assert that they will identify the next Dalai Lama, and he may well be living somewhere else in the world.

◀ **FIGURE 6-54 POTALA PALACE, LHASA, TIBET** The spiritual leader of Tibetan Buddhists, the Dalai Lama, lived in the palace from 1649 until 1959.

Challenges for Religions in Central and Southwest Asia

LEARNING OUTCOME 6.4.2

Understand the role of a fundamentalist group such as the Taliban.

Intense religious disputes in Central and Southwest Asia have become entwined with ethnic and political conflicts, as discussed in Chapters 7 and 8. Two particular conflicts with religious undertones are discussed here. Ethnic and political dimensions of these regions' conflicts are discussed in Chapters 7 and 8.

CENTRAL ASIA: TALIBAN AND WESTERN VALUES

More than 99 percent of Afghans are Muslims. Much of the country's unrest has stemmed from competing visions of the role of Islam. Several civil wars in recent years in Afghanistan have also involved multiple ethnicities, as discussed in Chapter 7.

Contributing to intense religious conflict in Afghanistan and elsewhere has been a resurgence of religious **fundamentalism,** which is a literal interpretation and a strict and intense adherence to what the fundamentalists define as the basic principles of a religion (or a religious branch, denomination, or sect). In Afghanistan, the Taliban are an example of a fundamentalist Islamic group.

Fundamentalism is one of the most important ways in which a group can maintain a distinctive cultural identity in a world increasingly dominated by a global culture and economy. A fundamentalist group like the Taliban, convinced that its religious view is the only correct one, may intrude upon the territory controlled by other adherents, as well as those unaffiliated with religions, sometimes brutally, as has been the case with the Taliban.

When the Taliban gained power in Afghanistan in 1996, many Afghans welcomed them as preferable to the corrupt and brutal warlords who had been running the country. U.S. and other Western officials also welcomed them as strong defenders against a possible new invasion by Russia. The Taliban (which means "religious students") have run Islamic Knowledge Movement (religious) schools, mosques, shrines, and other religious and social services since the seventh century A.D., shortly after the arrival of Islam in Afghanistan.

Once in control of Afghanistan's government in the 1990s, the Taliban imposed very strict laws inspired by Islamic values as the Taliban interpreted them:

- "Western, non-Islamic" leisure activities were banned, such as playing music, flying kites, watching television, and surfing the Internet.
- Soccer stadiums were converted to settings for executions and floggings.
- Men were beaten for shaving their beards and women stoned for committing adultery.
- Homosexuals were buried alive, and prostitutes were hanged in front of large audiences.
- Thieves had their hands cut off, and women wearing nail polish had their fingers cut off.

Western values were not the only targets: Enormous Buddhist statues as old as the second century A.D. were destroyed in 2001 because they were worshipped as "graven images," in violation of Islam (Figure 6-55). The Ministry for the Promotion of Virtue and the Prevention of Vice enforced the laws. The Taliban believed that they had been called by Allah to purge Afghanistan of sin and violence and make it a pure Islamic state. Islamic scholars criticized the Taliban as poorly educated in Islamic law and history and for misreading the Quran.

(a) 1998

(b) 2001

▶ **FIGURE 6-55**
TALIBAN DESTRUCTION
(a) A 1998 image of a 55-meter (180-foot) statue of Buddha in Bamiyan, Afghanistan; **(b)** the empty niche after the Taliban destroyed the statue in 2001.

A U.S.-led coalition overthrew the Taliban in 2001 and replaced it with a democratically elected government. However, the Taliban was able to regroup and has regained control parts of Afghanistan and Pakistan (see Chapter 8).

PAUSE & REFLECT 6.4.2

Why did the Taliban destroy priceless artistic works from Afghanistan's ancient past?

JEWS, CHRISTIANS, AND MUSLIMS IN SOUTHWEST ASIA

Conflict in the portion of Southwest Asia often referred to as the Middle East or the Eastern Mediterranean is among the world's longest standing and most intractable. Jews, Christians, and Muslims have fought for many centuries to control the same small strip of land (Figure 6-56).

To some extent, tensions among Christians, Muslims, and Jews in the Middle East stem from their similar heritage. All three groups trace their origins to Abraham in the Hebrew Bible narrative, but the religions diverged in ways that have made it difficult for them to share the same territory:

- Judaism makes a special claim to the territory it calls the Promised Land. The major events in the development of Judaism took place there, and the religion's customs and rituals acquired meaning from the agricultural life of the ancient Israelite tribes. Descendants of 10 of Jacob's sons, plus 2 of his grandsons, constituted the 12 tribes of Israelites who emigrated from Egypt in the Exodus narrative. Each received a portion of the land. After the Romans gained control of this land, which they later renamed the province of Palestine, they expelled most of the Jews, and only a handful were permitted to live in the region until the twentieth century.
- Christianity considers Palestine the Holy Land and Jerusalem the Holy City because the major events in Jesus's life, death, and resurrection were concentrated there. Most inhabitants of Palestine accepted Christianity after the religion was officially adopted by the Roman Empire and before the Muslim army conquest in the seventh century.
- Islam became the most widely practiced religion in Palestine after the Muslim army conquered it in the seventh century A.D. Muslims regard Jerusalem as their third-holiest city, after Makkah and Madinah, because it is the place from which Muhammad is thought by Muslims to have ascended to heaven.

In the seventh century, Muslims from the Arabian Peninsula came and captured most of the Middle East, including Palestine and Jerusalem. The Arab Muslim presence diffused the Arabic language across the Middle East and subsequently converted most of the people in that region from Christianity to Islam.

To recapture the Holy Land from its Muslim conquerors, European Christians launched a series of military campaigns, known as Crusades, over a 200-year period. Crusaders captured Jerusalem from the Muslims in 1099 during the First Crusade, lost it in 1187 (which led to the Third Crusade), regained it in 1229 as part of a treaty ending the Sixth Crusade, and lost it again in 1244. The Crusades ended when the Muslims defeated the Christians in 1291 at Acre, the last Christian stronghold.

The Muslim Ottoman Empire controlled the Palestine region for most of the four centuries between 1516 and 1917. Inspired by other nationalist movements and the rise of anti-Semitism in the late nineteenth century, Jews began returning in larger numbers to their historic homeland. Upon the Ottoman Empire's defeat in World War I, the United Kingdom took over Palestine, under a mandate from the League of Nations. For a few years, the British allowed some Jews to return to the Palestine Mandate, but immigration was restricted again during the 1930s, in response to intense pressure by Arabs in the region. As violence initiated by both Jewish and Muslim settlers escalated after World War II, the British announced their intention to withdraw from Palestine, setting the scene for the region's contemporary disputes.

▲ **FIGURE 6-56 PHYSICAL GEOGRAPHY OF THE EASTERN MEDITERRANEAN** The physical geography consists of narrow coastal lowlands and interior highlands interrupted by the Jordan River valley.

Geographic Perspectives in the Middle East

LEARNING OUTCOME 6.4.3

Understand reasons for conflict in the Middle East.

The conflict in the Middle East is now played out primarily among various countries and groups of people aspiring to control territory. But differences in religious traditions described on the previous page and their uses in nationalist ideologies underlie the origins of the conflicts and the challenges in peacefully resolving them.

WARS BETWEEN ISRAEL AND NEIGHBORS

The United Nations voted in 1947 to partition the United Kingdom's Palestine Mandate into two independent states, one Jewish and one Arab. Jerusalem was to be an international city, open to all religions, and run by the U.N. (Figure 6-57a). When the British withdrew in 1948, Jews declared an independent State of Israel within the boundaries prescribed by the U.N. resolution. Over the next quarter-century, Israel fought four wars with its neighbors:

- **1948–1949 Independence War.** The day after Israel declared independence, five Arab states began a war. Israel survived the attack, and the combatants signed an armistice in 1949. Israel's boundaries were extended beyond the U.N. partition, including the western suburbs of Jerusalem. Jordan gained control of the West Bank and East Jerusalem, including the Old City, where holy places are clustered. Egypt took control of the Gaza Strip (Figure 6-57b).
- **1956 Suez War.** Egypt nationalized the Suez Canal, a key shipping route between Europe and Asia, that had been built and controlled up until then by France and the United Kingdom. Egypt also blockaded international waterways near its shores that Israeli ships were using. Israel, France, and the United Kingdom attacked Egypt and got the waterways reopened, although Egypt retained control of the Suez Canal.
- **1967 Six-Day War.** Israel's neighbors massed a quarter-million troops along the borders and again blocked Israeli ships from using international waterways. In retaliation, Israel launched a surprise attack, destroying the coalition's air forces. Israel captured territory (Figure 6-57c):
  - From Jordan, the Old City of Jerusalem and the West Bank (the territory west of the Jordan River taken by Jordan in the 1948–1949 war).
  - From Syria, the Golan Heights.
  - From Egypt, the Gaza Strip and Sinai Peninsula.
- **1973 Yom Kippur War.** A surprise attack on Israel by its neighbors took place on the holiest day of the year for Jews. The war ended without a change in boundaries.

Egypt's President Anwar Sadat and Israel's Prime Minister Menachem Begin signed a peace treaty in 1979, following a series of meetings with U.S. President Jimmy Carter at Camp

(a)

(b)

(c)

▲ **FIGURE 6-57 TERRITORIAL CHANGES IN ISRAEL AND ITS NEIGHBORS** **(a)** The 1947 U.N. partition plan, **(b)** After the 1948–1949 war, and **(c)** After the 1967 Six-Day War.

David, Maryland. Israel returned the Sinai Peninsula to Egypt, and in return Egypt recognized Israel's right to exist. Sadat was assassinated by Egyptian soldiers, who were extremist Muslims opposed to compromising with Israel, but his successor Hosni Mubarak carried out the terms of the treaty. Over a half-century after the Six-Day War, the status of the other territories occupied by Israel has still not been settled.

COMPETING ISRAELI AND PALESTINIAN PERSPECTIVES

After the 1973 war, the Palestinians emerged as Israel's principal opponent. Egypt and Jordan eventually renounced their claims to the Gaza Strip and the West Bank, respectively, and recognized the Palestinians as the legitimate rulers of these territories. The Palestinians in turn saw themselves as the legitimate rulers of Palestine, which they defined as territory including the State of Israel. Palestinian and Israeli perspectives over the future have not been reconciled over the past half century.

ISRAELI PERSPECTIVES. In dealing with its neighbors, Israel considers two elements of the local landscape especially meaningful:

- Israel is a small country (smaller than New Hampshire), with a Jewish majority, surrounded by a region of hostile neighbors. In contrast, Muslim Arab countries in the region encompass more than 25 million square kilometers (10 million square miles). Nearly all Israelis live within 20 kilometers (12 miles) of an international border, making them vulnerable to attack.
- The land between the Mediterranean Sea and the Jordan River is divided into three narrow, roughly parallel physical regions (refer to Figure 6-56):
  - A coastal plain along the Mediterranean Sea.
  - A series of hills reaching elevations above 1,000 meters (3,300 feet).
  - The Jordan River valley, much of which is below sea level.

The U.N. plan for the partition of the Palestine Mandate in 1947 (as modified slightly by the armistice ending the 1948–1949 war) allocated most of the coastal plain to Israel, whereas Jordan took most of the hills between the coastal plain and the Jordan River valley, a region generally called the West Bank (of the Jordan River). Farther north, Israel's territory extended eastward to the Jordan River valley, but Syria controlled the highlands east of the valley (refer to Figure 6-57).

Jordan and Syria used the hills between 1948 and 1967 as staging areas to attack Israeli settlements on the coastal plain and in the Jordan River valley. Israel captured these highlands during the 1967 war to stop attacks on the lowland population. Israel still has military control over the Golan Heights and West Bank, and attacks by Palestinians against Israeli citizens have continued.

After capturing the West Bank from Jordan in 1967, Israel permitted Jewish settlers to construct settlements in the territory. Some Israelis built settlements in the West Bank because they regarded the territory as an integral part of the biblical Jewish homeland, known as Judea and Samaria. Others migrated to the settlements because of a shortage of affordable housing inside Israel's pre-1967 borders. Jewish settlers comprise about 17 percent of the West Bank population, and Palestinians see their immigration as a hostile act. To protect the settlers, Israel has military control over most of the West Bank.

Israeli Jews have been divided between those who wished to retain some of the West Bank and those who wished to make compromises with the Palestinians in return for formal recognition and a stable peace. In recent years, a large majority of Israelis have supported construction of a barrier to deter Palestinian attacks.

PALESTINIAN PERSPECTIVES. Five groups of people consider themselves Palestinians:

- People living in the West Bank, Gaza, and East Jerusalem territories captured by Israel in 1967.
- Some citizens of Israel who are Arabs.
- People who fled from Israel to other countries after the 1948–1949 war.
- People who fled from the West Bank or Gaza to other countries after the 1967 Six-Day War.
- Some citizens of other countries, especially Jordan, Lebanon, Syria, Kuwait, and Saudi Arabia.

The Palestinian fight against Israel was coordinated by the Palestine Liberation Organization (PLO), under the longtime leadership of Yassir Arafat, until his death in 2004. Israel has permitted the organization of a limited form of government in much of the West Bank and Gaza, called the Palestinian Authority, but Palestinians are not satisfied with either the territory or the power they have received thus far.

The Palestinians have been divided by sharp differences, reflected in a struggle for power between the Fatah and Hamas parties. Some Palestinians, especially those aligned with the Fatah Party, are willing to recognize the State of Israel with its Jewish majority in exchange for return of all territory taken by Israel in the 1967 Six-Day War. Other Palestinians, especially those aligned with the Hamas Party, do not recognize the right of Israel to exist and want to continue fighting for control of the entire territory between the Jordan River and the Mediterranean Sea. The United States, European countries, and Israel consider Hamas to be a terrorist organization.

PAUSE & REFLECT 6.4.3

What is the difference in elevation between Hebron (the largest city in the West Bank) and Tel Aviv (the largest city in Israel)? Refer to Figure 6-56.

Jerusalem's Challenging Geography

LEARNING OUTCOME 6.4.4

Explain the importance of Jerusalem to Jews and Muslims.

At the heart of the conflicts among religious groups in Southwest Asia lies the city of Jerusalem, an important place for all three religions (Figure 6-58). Geography makes it difficult to settle long-standing claims to Jerusalem by Jews and Muslims. The challenge is that the most sacred space in Jerusalem for Muslims was literally built on top of the most sacred space for Jews.

JUDAISM'S JERUSALEM

Jerusalem is especially holy to Jews as the location of the Temple, their center of worship in ancient times. The First Temple, built by King Solomon in approximately 960 B.C., was destroyed by the Babylonians in 586 B.C. After the Persian Empire, led by Cyrus the Great, gained control of Jerusalem, Jews were allowed to build a Second Temple in 516 B.C. The Romans destroyed the Jewish Second Temple in A.D. 70. The Western Wall of the Temple survives.

Christians and Muslims call the Western Wall the Wailing Wall because for many centuries Jews were allowed to visit the surviving Western Wall only once a year to lament the Temple's destruction. After Israel captured the entire city of Jerusalem during the 1967 Six-Day War, officials removed the barriers that had prevented Jews from visiting and living in the Old City of Jerusalem, including the Western Wall. The Western Wall soon became a site for daily prayers by observant Jews.

To deter Palestinian suicide bombers from entering Jerusalem and the rest of Israel, the Israeli government has constructed barriers along the West Bank, the Gaza Strip, and suburbs of Jerusalem. The West Bank barrier is especially controversial because it places on Israel's side around 10 percent of the land, home to between 10,000 and 50,000 Palestinians, according to various sources (see Debate It feature). Naming the structure is controversial. Israel calls the barrier a "security fence," and Palestinians call it a "racial segregation wall." Neutral sources call it a "separation barrier."

ISLAM'S JERUSALEM. The most important Muslim structure in Jerusalem is the Dome of the Rock, built in A.D. 691. Muslims believe that the large rock beneath the building's dome is the place from which Muhammad ascended to heaven, as well as the altar on which Abraham prepared to sacrifice his son Isaac (according to Jews and Christians) or his son Ishmael (according to Muslims). Immediately south of the Dome of the Rock is the al-Aqsa Mosque. The challenge facing Jews and Muslims is that al-Aqsa Mosque was built on the site of the ruins of the Jewish Second Temple. Thus, the surviving Western Wall of the Jewish Temple is situated immediately beneath holy Muslim structures.

Israel allows Muslims unlimited access to that religion's holy structures in Jerusalem and some control over them.

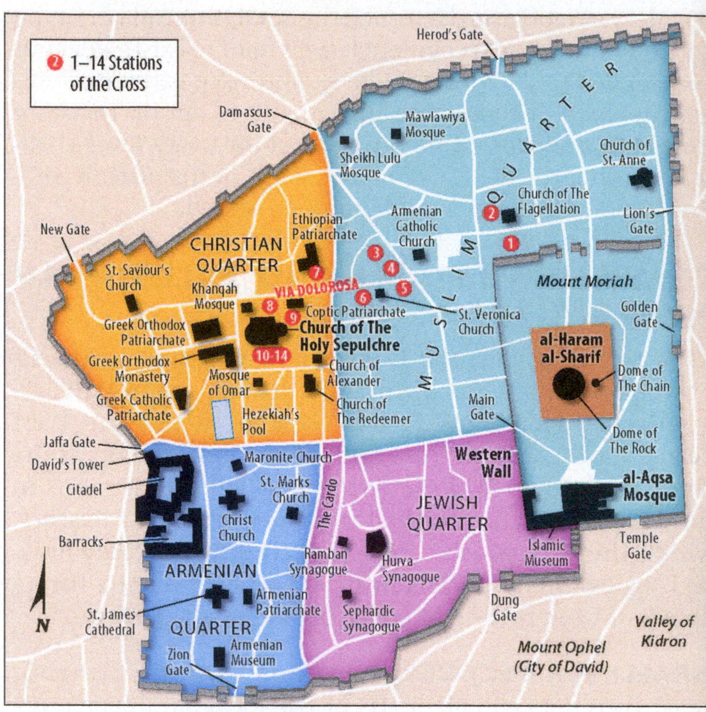

▲ **FIGURE 6-58 OLD CITY OF JERUSALEM** The Old City of Jerusalem is less than 1 square kilometer (0.4 square miles). It is divided into four quarters.

A ramp and passages patrolled by Palestinian guards provide Muslims access to the Dome of the Rock and the al-Aqsa Mosque without requiring them to walk in front of the Western Wall, where Jews are praying (the ramp is visible in the chapter opener image). However, because the holy Muslim structures sit literally on top of the holy Jewish structure, the two sets of holy structures cannot be logically divided by a line on a map (Figure 6-59).

PAUSE & REFLECT 6.4.4

Why is the Western Wall important in Judaism, and why is the Dome of the Rock important in Islam?

CHECK-IN KEY ISSUE 4

Why Do Territorial Conflicts Arise Among Religious Groups?

- ✔ **Religious groups have opposed government policies, especially those of Communist governments.**
- ✔ **Religious principles seen as representing Western social values have been opposed by groups in Asia.**
- ✔ **An especially long-standing and intractable conflict among religious has been centered in Israel/Palestine, an area considered holy by Jews, Christians, and Muslims.**

◀ **FIGURE 6-59 WESTERN WALL AND DOME OF THE ROCK** The Western Wall, which remains from the Jewish Temple, is situated immediately below the mount containing Islam's Dome of the Rock.

DEBATE IT! How do perspectives on the separation barrier differ?

Israel has built a security fence along 70 percent of the boundary with the West Bank.

KEEP THE SECURITY FENCE

- Israel is a very small country with a Jewish majority surrounded by a very large region of hostile Arabs.
- After repeated attacks by its neighbors, Israel has protected its citizens by constructing a fence near its borders to help keep out attackers (Figure 6-60).
- Israel has made numerous adjustments to the location of the fence in response to humanitarian concerns.

▲ **FIGURE 6-60 SEPARATION BARRIER**

REMOVE THE SEGREGATION WALL

- The wall has helped Jewish settlers to increase the territory under their control.
- Palestinians living in the West Bank consider Israel's construction of settlements and the wall to be hostile acts.
- The wall that Israel built prevents some Palestinians from reaching their fields and workplaces (Figure 6-61).

▲ **FIGURE 6-61 SEPARATION BARRIER IN JERUSALEM** Predominantly Palestinian East Jerusalem is in the foreground, and predominantly Jewish West Jerusalem is in the background.

Summary & Review

KEY ISSUE 1

Where are the world's religions distributed?

Religions are classified as universalizing or ethnic. The three universalizing religions with the largest number of adherents are Christianity, Islam, and Buddhism. Each is divided into branches and denominations. Hinduism is the largest ethnic religion. Other ethnic religions with the largest numbers of followers are clustered primarily elsewhere in Asia.

THINKING GEOGRAPHICALLY

1. Only 3 percent of Hindus live in countries where their religion is in the minority, compared with 72 percent of Buddhists. What differences between these two religions account for this sharp difference in the percentage living as a minority?

▲ **FIGURE 6-62 HINDUS OUTSIDE OF INDIA** Hindus living in London, England, celebrate Ratha Yatra (Chariot) Festival.

KEY ISSUE 2

Why do religions have distinctive distributions?

A universalizing religion has a known origin and clear patterns of diffusion, whereas ethnic religions typically have unknown origins and little diffusion. The three most widely practiced universalizing religions originated with single historical individuals and have diffused from their places of origin to other regions of the world. An ethnic religion like Hinduism typically has unknown origins and a more clustered distribution.

THINKING GEOGRAPHICALLY

2. This image shows refugees from Myanmar being rescued by Indonesian fishermen. These refugees are part of a Muslim group known as Rohingya. Based on Figures 6-9, 6-10, and 6-11, what would be the position of Muslims in Myanmar compared with other religions, and why would the Rohingya seek to reach Indonesia?

▲ **FIGURE 6-63 FORCED MIGRATION FOR RELIGIOUS REASONS** Rohingya Muslims from Myanmar are rescued by Indonesian fishermen.

KEY ISSUE 3

Why do religions organize space in distinctive patterns?

The holiest places in universalizing religions typically relate to events in the life of the founder. In an ethnic religion, holy places typically derive from the physical geography where the religion's adherents are clustered. Holidays typically derive from the physical geography where an ethnic religion is clustered, whereas they derive from the founder's life in a universalizing religion.

THINKING GEOGRAPHICALLY

3. The Roman Catholic Church created the Archdiocese of Mobile, Alabama, in 1980, and the Archdiocese of Galveston-Houston, Texas, in 2004. What geographic patterns and processes might underlie those decisions?

▲ **FIGURE 6-64 NEW ARCHDIOCESE** The Co-Cathedral of the Sacred Heart was constructed between 2005 and 2008 for the newly established Archdiocese of Galveston-Houston.

KEY ISSUE 4

Why do territorial conflicts arise among religious groups?

Some religions have competed for control of territory with nonreligious ideas, notably communism and economic modernization. Conflict in the Middle East goes back many centuries. Jews, Muslims, and Christians have fought for control of land in the Middle East that is now part of Israel/Palestine. The most sacred space in Jerusalem for Muslims was built on top of the most sacred space for Jews.

THINKING GEOGRAPHICALLY

4. What are some similarities and differences in the reasons underlying construction of the West Bank separation barrier and the U.S.–Mexico border fence?

▲ **FIGURE 6-65 SEPARATION BARRIER** The minaret tower of a mosque in East Jerusalem can be seen in the image.

KEY TERMS

Agnosticism *(p. 185)* The belief that the existence of God can't be proven empirically.

Animism *(p. 193)* The belief that objects, such as plants and stones, or natural events, like thunderstorms and earthquakes, have a discrete spirit and conscious life.

Atheism *(p. 185)* The belief that God does not exist.

Autonomous religion *(p. 211)* A religion that does not have a central authority but shares ideas and cooperates informally.

Branch *(p. 188)* A large and fundamental division within a religion.

Caste *(p. 218)* The class or distinct hereditary order into which a Hindu is assigned, according to religious law.

Congregation *(p. 188)* A local assembly of persons brought together for common religious worship.

Cosmogony *(p. 214)* A set of religious beliefs concerning the origin of the universe.

Denomination *(p. 188)* A division of a branch that unites a number of local congregations into a single legal and administrative body.

Ethnic religion *(p. 185)* A religion with a relatively concentrated spatial distribution whose principles are likely to be based on the physical characteristics of the particular location in which its adherents are concentrated.

Fundamentalism *(p. 220)* Literal interpretation and strict adherence to basic principles of a religion (or a religious branch, denomination, or congregation).

Ghetto *(p. 205)* During the Middle Ages, a neighborhood in a city set up by law to be inhabited only by Jews; now used to denote a section of a city in which members of any minority group live because of social, legal, or economic pressure.

Hierarchical religion *(p. 210)* A religion in which a central authority exercises a high degree of control.

Missionary *(p. 200)* An individual who helps to diffuse a universalizing religion.

Monotheism *(p. 194)* The doctrine of or belief in the existence of only one God.

Pilgrimage *(p. 212)* A journey to a place considered sacred for religious purposes.

Polytheism *(p. 194)* Belief in or worship of more than one god.

Solstice *(p. 217)* An astronomical event that happens twice each year, when the tilt of Earth's axis is most inclined toward or away from the Sun, causing the Sun's apparent position in the sky to reach it most northernmost or southernmost extreme, and resulting in the shortest and longest days of the year.

Syncretic *(p. 192)* Combining several religious traditions.

Universalizing religion *(p. 185)* A religion that attempts to appeal to all people, not just those living in a particular location.

GeoVideo

Log in to the **MasteringGeography** Study Area to view this video.

Christians of the Holy Land

Christians living in Israel and the West Bank represent a community that dates backs to Christianity's early centuries but today is experiencing stress.

1. How has the number of Christians in Jerusalem's Old City and Bethlehem changed in recent decades? How does the video explain this change?
2. What does the video imply about the size of the Christian population of Nazareth, and what possible explanations are discussed?

EXPLORE

Makkah, Saudi Arabia

Use Google Earth to explore *Masjid al-Haram*, Islam's largest mosque in Makkah, Saudi Arabia. Millions of Muslims make a pilgrimage to Makkah each year and gather at Masjid al-Haram Mosque.

Fly to: *Masjid al-Haram Mosque, Makkah, Saudi Arabia.*

Drag to: *Enter Street View* to the square in the middle of the mosque

Click to look around so that North is at the bottom.

Click *3D buildings*.

Continue to look around to see the tall building with the clock tower immediately south of the Mosque.

1. What is in the tall building?
2. Why would this building be located immediately next to the mosque?

MasteringGeography™

Visit the Study Area in **MasteringtGeography**™ to enhance your geographic literacy, spatial reasoning skills, and understanding of this chapter's content by accessing a variety of resources, including MapMaster interactive maps, videos, *In the News* RSS feeds, flashcards, web links, self-study quizzes, and an eText version of *The Cultural Landscape*.

www.masteringgeography.com

7

Ethnicities

Ethnicity is a source of pride to people, a link to the experiences of ancestors and to cultural traditions, such as food and music preferences. The ethnic group to which one belongs has important measurable differences, such as average income, life expectancy, and infant mortality rate. Ethnicity also matters in places with a history of discrimination by one ethnic group against another.

World Cross Country Championship senior women's race, Amman, Jordan.

LOCATIONS IN THIS CHAPTER

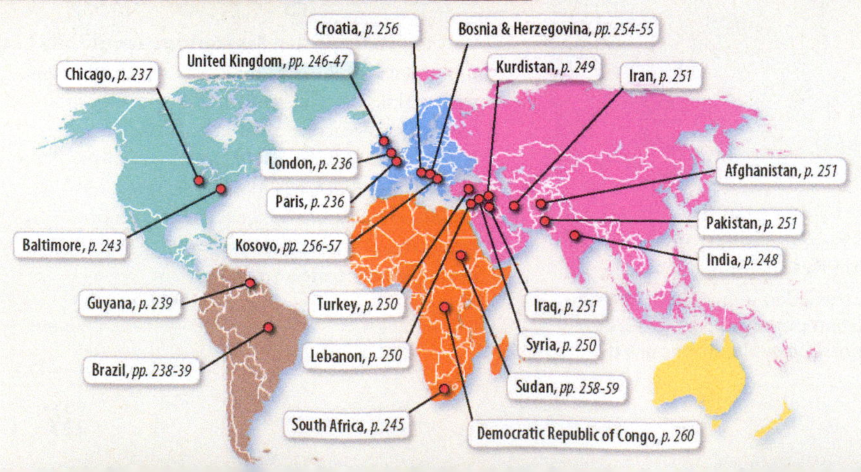

KEY ISSUES

1 Where Are Ethnicities Distributed?

Ethnicities are distributed in distinctive ***places***, including in the United States.

2 Why Do Ethnicities Have Distinctive Distributions?

Ethnicities display distinctive spatial distributions that derive from patterns of ***connections*** and isolation.

3 Why Might Ethnicities Face Conflicts?

Geographic factors underly conflicts among ethnicities. Conflicts result in many places when more than one ethnic group fights to occupy the same ***space***.

4 Why Do Ethnic Cleansing And Genocide Occur?

In some ***regions*** of conflicts among ethnicities, large-***scale*** forced migration and killings have resulted.

KEY ISSUE 1

Where Are Ethnicities Distributed?

- Introducing Ethnicities
- Ethnicity and Race
- Distribution of U. S. Ethnicities
- Ethnic Enclaves
- Ethnically Complex Brazil

LEARNING OUTCOME 7.1.1

Introduce the principal ethnicities in the United States.

The meaning of *ethnicity*, the subject of this chapter, is frequently confused with definitions of *race* and *nationality*:

- **Ethnicity** is identity with a group of people who share the cultural traditions of a particular homeland or hearth.
- **Race** is identity with a group of people who are perceived to share a physiological trait, such as skin color.
- **Nationality** is identity with a group of people who share legal attachment to a particular country.

In principle, the three cultural traits are distinct: Nationality is a person's country of citizenship, race is a person's skin color, and ethnicity is a person's place of cultural heritage. In actuality, differences among the three concepts are not always clear-cut. Some individuals are not in a position to identify one or more of their personal traits. For example, an individual's two parents may identify with different ethnicities, races, and nationalities. Furthermore, many societies muddle the three concepts, either through unclear language or deliberate decisions.

The title of this chapter is "Ethnicities," because it is the cultural feature that is place-based. "Nationality," which is also place-based, is introduced in this chapter and is discussed in more detail in Chapter 8. Though not place-based, "race" is an important concept because it is often misused as a synonym for "ethnicity."

President Barack Obama illustrates the complexity of designating ethnicity, race, and nationality (Figure 7-1). President Obama's father, Barack Obama, Senior, was born in the village of Kanyadhiang, Kenya. He was a member of Kenya's third-largest ethnicity, known as Luo. President Obama's mother, Ann Dunham, was born in Kansas. Most of her ancestors migrated to the United States from England in the nineteenth century. President Obama's stepfather—his mother's second husband, Lolo Soetoro—was born in the village of Yogyakarta, Indonesia. He was a member of Indonesia's most numerous ethnicity, known as Javanese. The son of a white mother and a black father, President Obama has chosen to identify as African American.

(a)

(b)

(c)

▲ **FIGURE 7-1 ETHNIC DIVERSITY: BARACK OBAMA'S FAMILY** Barack Obama with **(a)** his mother Ann Dunham, **(b)** his father Barack Obama, Senior, and **(c)** his stepfather, mother, and stepsister.

PAUSE & REFLECT 7.1.1

President Obama has self-identified his ethnicity as African American. Based on his parents' ethnicities, what other ways might his ethnicity be identified?

Introducing Ethnicities

Ethnicity is an important cultural element of local diversity because our ethnic identity is immutable. We can suppress our ethnicity, but we cannot change it in the same way we can speak a new language or practice a different religion. If our parents come from two ethnic groups or our grandparents from four, our ethnic identity may be complex.

Geographers are interested in where ethnicities, like other elements of culture, are distributed. An ethnic group is tied to a particular place because members of the group—or their ancestors—were born and raised there. The cultural traits displayed by an ethnicity derive from particular conditions and practices in the group's homeland.

The significance of ethnic diversity is controversial in the United States:

- To what extent does discrimination persist against minority ethnicities, especially African Americans and Hispanics?
- Should preferences be given to minority ethnicities to correct past patterns of discrimination?
- To what extent should the distinct cultural identity of ethnicities be encouraged or protected?

Ethnicity is especially important to geographers because in the face of globalization trends in culture and economy, ethnicity stands as the strongest bulwark for the preservation of local diversity. Despite efforts to preserve local languages, it is not far-fetched to envision a world in which virtually all educated people speak English. And universalizing religions continue to gain adherents around the world. But no ethnicity is attempting or even aspiring to achieve global dominance, although ethnic groups are fighting with each other to control specific places. Even if globalization engulfs language, religion, and other cultural elements, regions of distinct ethnic identity will remain.

ETHNICITIES IN THE UNITED STATES

The three most numerous U.S. ethnicities are Hispanic American, African American, and Asian American. Approximately 17 percent of Americans say they are Hispanic, 12 percent African American, and 5 percent Asian American. In addition, 2 percent of Americans identify their ethnicity as American Indian, Native Hawaiian, or Alaska Native.

HISPANIC AMERICANS. A Hispanic or Hispanic American is a person who has migrated (or whose ancestors have migrated) to the United States from a Spanish-speaking country in Latin America. The terms Latino (for males) and Latina (for females) are generally used interchangeably with Hispanic. The U.S. government adopted the term Hispanic in 1973 because it was considered an inoffensive label that could be applied to all people from Spanish-speaking countries while avoiding the gender-specific limitation of the term Latino. The 1980 U.S. Census was the first to classify some Americans as Hispanic.

A 2013 survey by the Pew Research Center found that 33 percent of Americans of Latin American descent preferred the term Hispanic and 15 percent Latino/Latina, leaving 52 percent who didn't care. The Pew survey found that only 20 percent of Americans of Latin American descent actually used either of the terms to identify themselves.

Instead, most Americans of Latin American heritage prefer to identify with a more specific ethnicity or national origin. Nearly two-thirds come from Mexico and one-fourth from Caribbean islands (Figure 7-2). Mexican Americans are sometimes called Chicanos (males) or Chicanas (females). Originally these terms were considered insulting, but in the 1960s Mexican American youths in Los Angeles began to call themselves Chicanos and Chicanas with pride.

ASIAN AMERICANS. The term Asian American encompasses Americans who trace their heritage to a number of countries in Asia. Only 19 percent of Asian Americans identify their ethnicity as Asian American, whereas 62 percent identify with their ethnicity as the country of origin of themselves or their ancestors (Figure 7-3).

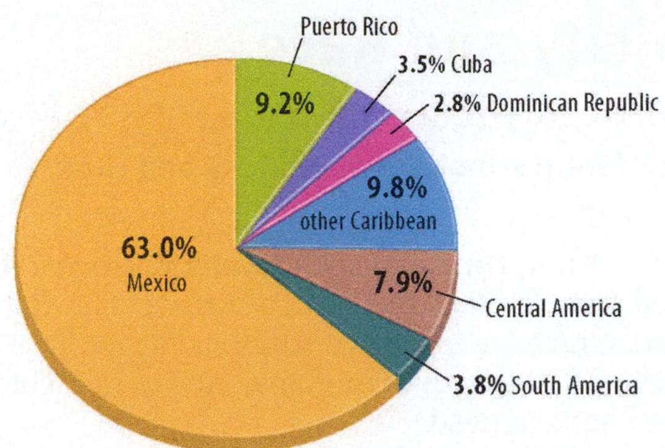

▲ **FIGURE 7-2 HISPANIC AMERICANS BY COUNTRY OF ORIGIN**

AFRICAN AMERICANS. It is harder to pinpoint more precisely the ethnic origins of African Americans. The ancestors of most African Americans arrived around 300 years ago from places in Africa that were not yet organized into independent countries. Records were not kept of the ethnic origin of African Americans who arrived as slaves.

DNA testing is now helping to narrow the ethnic heritage of African Americans. The ancestors of most African Americans came from three areas in West Africa. The modern-day countries of these three areas are:

- Senegal, Mali, Gambia, Guinea, Sierra Leone, and Liberia.
- Southern Ghana, Togo, Benin, and Nigeria and southeastern Côte d'Ivoire.
- Western Democratic Republic of Congo and Angola.

DESCENDANTS OF INDIGENOUS PEOPLES. The descendants of people who lived in North America prior to the arrival of Europeans comprise numerous ethnicities. They are grouped in the United States into three principal ethnic identities: Native American, Alaska Native, and Native Hawaiian. The most numerous are Native Hawaiians, Cherokee, Navajo, Chippewa, Sioux, and Choctaw. The largest numbers of Native Americans, however, do not specify a group.

Indigenous peoples comprise 4 percent of the population of Canada, where they are known as aboriginals. Canada's aboriginals are grouped into three main ethnic identities: First Nations, Inuit, and Métis. The Inuit settled in northern Canada around 3,000 years ago, and the First Nations peoples further south around 2,500 years ago. Métis are descendants of indigenous peoples who married Europeans beginning 400 years ago.

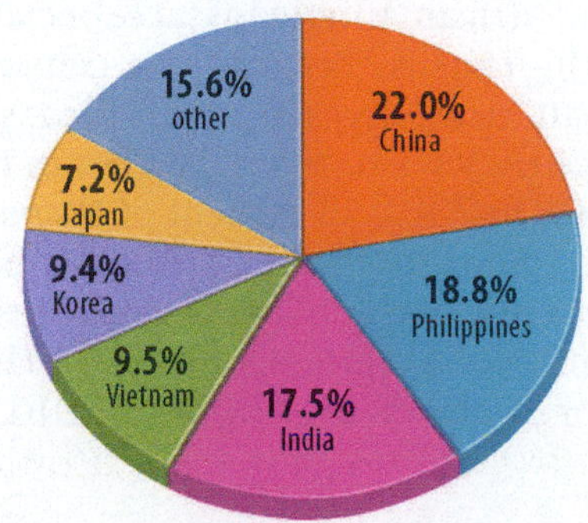

▲ **FIGURE 7-3 ASIAN AMERICANS BY COUNTRY OF ORIGIN**

Ethnicity and Race

LEARNING OUTCOME 7.1.2

Clarify differences between ethnicity and race.

Ethnicity is often confused with race. A person's ethnicity is derived from a place on Earth's surface. The traits that characterize race are not place specific. Rather, race relates to physiological features of humans, such as skin color, hair type, and shape of head.

Features of race were once thought to be scientifically classifiable into a handful of groups. Contemporary geographers and other scientists reject the entire biological basis of classifying humans into a handful of races. These biological features are so highly variable among members of a race that any prejudged classification is meaningless. Perhaps many tens or hundreds of thousands of years ago, early "humans" (however they emerged as a distinct species) lived in such isolation from other early "humans" that they were truly distinct genetically. But the degree of isolation needed to keep biological features distinct genetically vanished when the first human crossed a river or climbed a hill.

However, one feature of customary racial classifications is of interest to geographers: the color of skin. The identification of persons by skin color matters to geographers because it is one factor by which people in many societies sort out where they reside, attend school, spend their leisure time, and perform many other activities of daily life.

At worst, biological classification by race is the basis for **racism,** which is the belief that race is the primary determinant of human traits and capacities and that racial differences produce an inherent superiority of a particular race (Figure 7-4). A **racist** is a person who subscribes to the beliefs of racism.

▲ FIGURE 7-4 ANTI-RACISM MARCH, LONDON

RACE AND ETHNICITY IN THE UNITED STATES

The U.S. census shows the difficulty in distinguishing between ethnicity and race (see Doing Geography feature and Table 7-1a). The three most numerous U.S. ethnicities—Asian American, African American, and Hispanic American—illustrate the difficulty. The census regards Hispanic American as an ethnicity, but it regards Asian American and African American as races even though the two are also logically place-based ethnicities.

Hispanics and African Americans face special challenges in completing the census. Because the census considers Hispanic to be an ethnicity, Hispanics also get to identify with any race they wish. Many Hispanics have a difficult time doing so. In 2010, 53 percent of Hispanics picked white, 37 percent some other race, 6 percent more than one box, and 4 percent one of the 13 other categories.

Nearly 10 million Americans changed their answer to question 8 or 9 between 2000 and 2010. The most frequent changes involved Hispanics who changed their race from white to something else or from something else to white.

African Americans also face a distinctive challenge in responding to the census. Although African American is an ethnicity—for a person whose ancestry can be traced to Africa—the 2010 census grouped "Black, African Am., or Negro" as a race. Most black Americans are descended from African immigrants and therefore also belong to an African American ethnicity. Some American blacks, however, trace their cultural heritage to regions other than Africa, including Latin America, Asia, and Pacific islands. The term African American identifies a group with an extensive cultural tradition, whereas the term black in principle denotes nothing more than dark skin. Because many Americans make judgments about the values and behavior of others simply by observing skin color, black is substituted for African American in daily language.

Today, many Americans are of mixed ancestry and may or may not choose to identify with a single race or ethnicity. Other Americans trace their heritage to places in Europe, such as Ireland and Italy, that are not included in the two race and ethnicity census questions.

PAUSE & REFLECT 7.1.2

What might be benefits and challenges of changing the census questions about ethnicity and race from multiple choice to short answer format?

DOING GEOGRAPHY Ethnicity and Race and the Census

Every 10 years, the U.S. Bureau of the Census asks people to classify themselves according to the ethnicity and race with which they most closely identify. Americans are asked to identify themselves by answering two questions (Table 7-1a).

The U.S. census shows the difficulty in distinguishing between ethnicity and race. Note that while the census uses the term *race* for question 9, the options in that question are a mix of skin color and ethnicity (that is, place of origin). The census permits people to check more than one box, and 3 percent did that in 2010. "Other race" was selected for question 9 by 6 percent of respondents in 2010; according to Census Bureau research, most were Hispanics.

Census officials have concluded that many Americans can't find themselves in questions 8 and 9. One option under consideration for the 2020 census is to combine the two ethnicity and race questions into one (Table 7-1b).

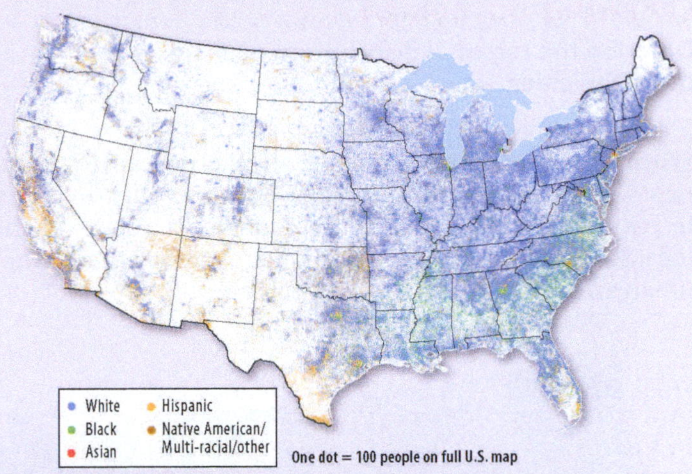

▲ **FIGURE 7-5 ETHNIC MAP OF THE UNITED STATES** 1 dot equals 1 person on this map based on 2010 Census Block Data.

TABLE 7-1 U.S. Census Questions About Race and Ethnicity

(a) These questions were asked on the 2010 census.

8. Is Person 1 of Hispanic, Latino, or Spanish origin?
   - ☐ **No,** not of Hispanic, Latino, or Spanish origin
   - ☐ Yes, Mexican, Mexican Am., Chicano
   - ☐ Yes, Puerto Rican
   - ☐ Yes, Cuban
   - ☐ Yes, another Hispanic, Latino, or Spanish origin
9. What is Person 1's race?
   - ☐ White
   - ☐ Black, African Am., or Negro
   - ☐ American Indian or Alaska Native
   - ☐ Asian Indian ☐ Chinese
   - ☐ Filipino ☐ Japanese
   - ☐ Korean ☐ Vietnamese
   - ☐ Other Asian
   - ☐ Native Hawaiian
   - ☐ Guamanian or Chamorro
   - ☐ Samoan ☐ Other Pacific Islander
   - ☐ Some other race

(b) This question is being considered for the 2020 census.

8. What is Person 1's race or origin? Mark one or more boxes and write in the specific race(s) or origin(s).
   - ☐ White [for example, German, Irish, Lebanese, Egyptian, and so on]
   - ☐ Black, African Am., or Negro [for example, African American, Haitian, Nigerian, and so on.]
   - ☐ Hispanic, Latino, or Spanish origin [for example, Mexican, Mexican Am., Puerto Rican, Cuban, Argentinian, Colombian, Dominican, Nicaraguan, Salvadorian, Spaniard, and so on]
   - ☐ American Indian or Alaska Native [for example, Navajo, Mayan, Tingit, and so on]
   - ☐ Asian [for example, Asian Indian, Chinese, Filipino, Japanese, Korean, Vietnamese, Hmong, Laotian, Thai, Pakistani, Cambodian, and so on]
   - ☐ Native Hawaiian or Other Pacific Islander [for example, Native Hawaiian, Guamanian or Chamorro, Samoan, Fijian, Tongan, and so on]
   - ☐ Some other race or origin

What's Your Ethnic Geography?

1. Are you able to answer questions 8 and 9 on the 2010 census? Are you able to answer the questions for your mother? For your father?
2. Does the proposed 2020 census form change the ease of answering for you? Why or why not? Does the proposed form change the answers for your mother and your father?
3. What changes would need to be made to the 2010 census to clarify the difference between ethnicity and race, as defined by geographers? Does the proposed 2020 census help to clarify or muddle the difference between ethnicity and race? Why?
4. Figure 7-5 depicts the most prevalent ethnicity or race in each location in the United States. Does the color on the map for where you live correspond with your race/ethnicity?
5. You can zoom in on the dot map at demographics.coopercenter.org/DotMap. As you zoom in on where you live, how does the map change? When you zoom in all the way, does the map reflect your race/ethnicity?

Distribution of U. S. Ethnicities

LEARNING OUTCOME 7.1.3

Describe the regional distribution of major U.S. ethnicities.

Within a country, clustering of ethnicities can occur on two scales. At the regional scale, ethnic groups may live in particular regions within the country or a state. At the local scale, they may live in particular communities within urban areas.

Within the United States, ethnicities are clustered at both the regional and urban scales. On a regional scale, ethnicities have distinctive distributions within the United States:

- **Hispanics.** Clustered in the Southwest, Hispanics exceed one-third of the population of Arizona, New Mexico, and Texas and one-quarter of California (Figure 7-6). California is home to one-third of all Hispanics, Texas one-fifth, and Florida and New York one-sixth each.
- **African Americans.** Clustered in the Southeast, African Americans comprise at least one-fourth of the

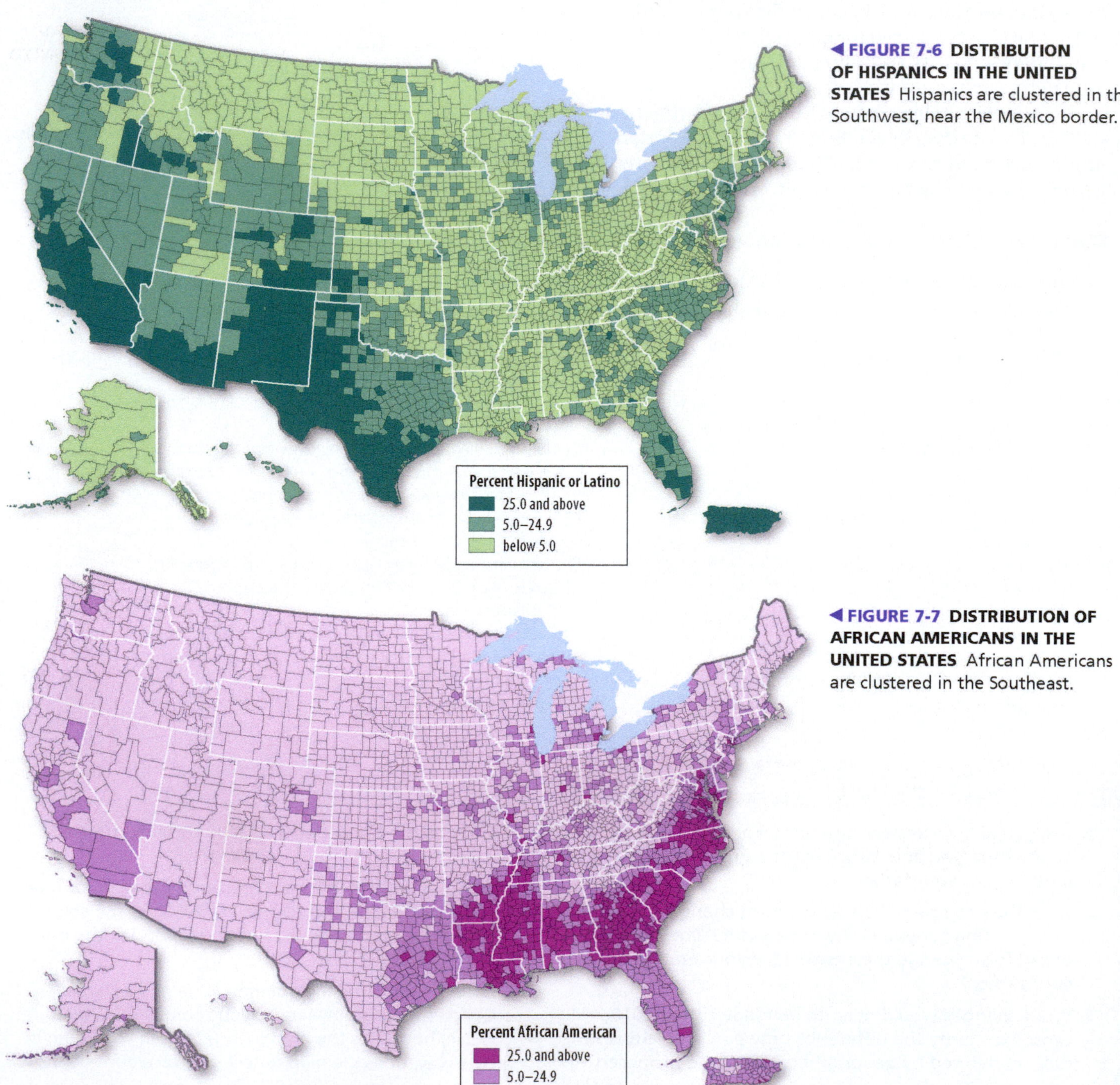

FIGURE 7-6 DISTRIBUTION OF HISPANICS IN THE UNITED STATES Hispanics are clustered in the Southwest, near the Mexico border.

FIGURE 7-7 DISTRIBUTION OF AFRICAN AMERICANS IN THE UNITED STATES African Americans are clustered in the Southeast.

population in Alabama, Georgia, Louisiana, Maryland, and South Carolina and more than one-third in Mississippi (Figure 7-7). Concentrations are even higher in selected counties. At the other extreme, nine states in upper New England and the West have less than 1 percent African Americans.

- **Asian Americans.** Clustered in the West, Asian Americans comprise more than 40 percent of the population of Hawaii (Figure 7-8). One-half of all Asian Americans live in California, where they comprise 12 percent of the population.
- **Native Americans.** Clustered in the southwest and north-central regions of the United States, as well as Alaska (Figure 7-9), Native Americans exceed 10 percent of the population of Alaska, Oklahoma, New Mexico, and South Dakota.

PAUSE & REFLECT 7.1.3

What region of the United States has low concentrations of all four of the ethnicities?

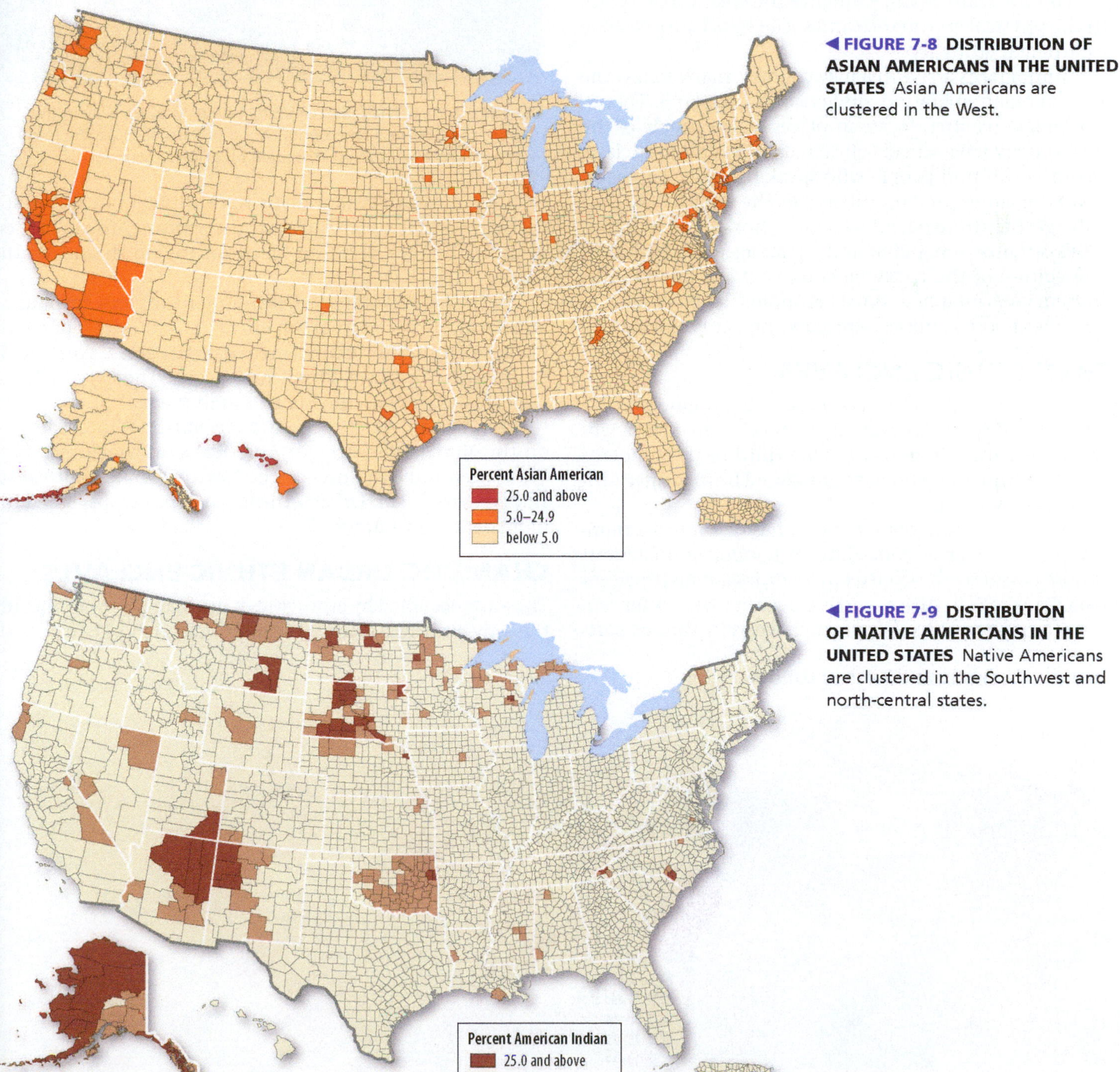

◀ **FIGURE 7-8 DISTRIBUTION OF ASIAN AMERICANS IN THE UNITED STATES** Asian Americans are clustered in the West.

◀ **FIGURE 7-9 DISTRIBUTION OF NATIVE AMERICANS IN THE UNITED STATES** Native Americans are clustered in the Southwest and north-central states.

Ethnic Enclaves

LEARNING OUTCOME 7.1.4

Describe the distribution of ethnicities within urban areas.

An **ethnic enclave** is a place with a high concentration of an ethnic group that is distinct from those in the surrounding area. Most ethnic enclaves are neighborhoods within large cities. Ethnic enclaves with distinctive physical appearances and social structures typically form through migration.

Ethnicities are defined in part by their possession of distinct cultural features, such as languages, religions, and art. These cultural features can influence the creation of a place with the physical appearance and social structure reflective of a particular ethnicity.

As immigrants arrive in a new country, many follow the process of chain migration, discussed in Chapter 3. That is, new immigrants often locate in places where people of the same ethnicity have already clustered. In an ethnic enclave, newcomers can find people who speak the same language, practice the same religion, and prepare the same foods. They can also get help from people who know how to fill out forms, obtain assistance from public and private agencies, and adapt to the culture of the receiving country. Most importantly, ethnic enclaves offer newcomers economic support, such as employment opportunities, affordable housing, and loans.

URBAN ETHNIC ENCLAVES

The clustering of ethnicities is especially pronounced at the neighborhood scale. An example is the Goutte d'Or neighborhood in Paris (Figure 7-10). One-third of the residents belong to ethnicities who have emigrated from former African colonies of France.

In the United Kingdom, London has developed a number of ethnic enclaves. One-third of London's inhabitants were born outside of the United Kingdom, including 18 percent in Asia, 7 percent in Africa, and 6 percent in the Caribbean. These ethnic groups have formed enclaves in various parts of London. South Asia Indians have clustered in the west, Pakistanis and Bangladeshis in the northeast, African blacks in the east, and Caribbean blacks in the north and south (Figure 7-11).

▲ **FIGURE 7-10 ETHNIC ENCLAVE IN PARIS** Muslims pray in the street in the Goutte d'Or neighborhood of Paris because the local mosque is too small to accommodate all of the worshipers.

In the United States, African Americans and Hispanics are highly clustered in urban enclaves. Around 90 percent of these ethnicities live in metropolitan areas, compared to around 75 percent for all Americans. The City of Chicago, for example, consists of roughly equal numbers of whites, African Americans, and Hispanics. Whites cluster on the North Side, African Americans on the South and West sides, and Hispanics on the Northwest and Southwest sides (Figure 7-12). Urban ethnic patterns are discussed in more detail in Chapter 13.

CHANGING URBAN ETHNIC ENCLAVES

The areas occupied by ethnicities have changed over time. In the early twentieth century, Chicago, Cleveland, Detroit, and

▼ **FIGURE 7-11 ETHNIC ENCLAVES IN LONDON**

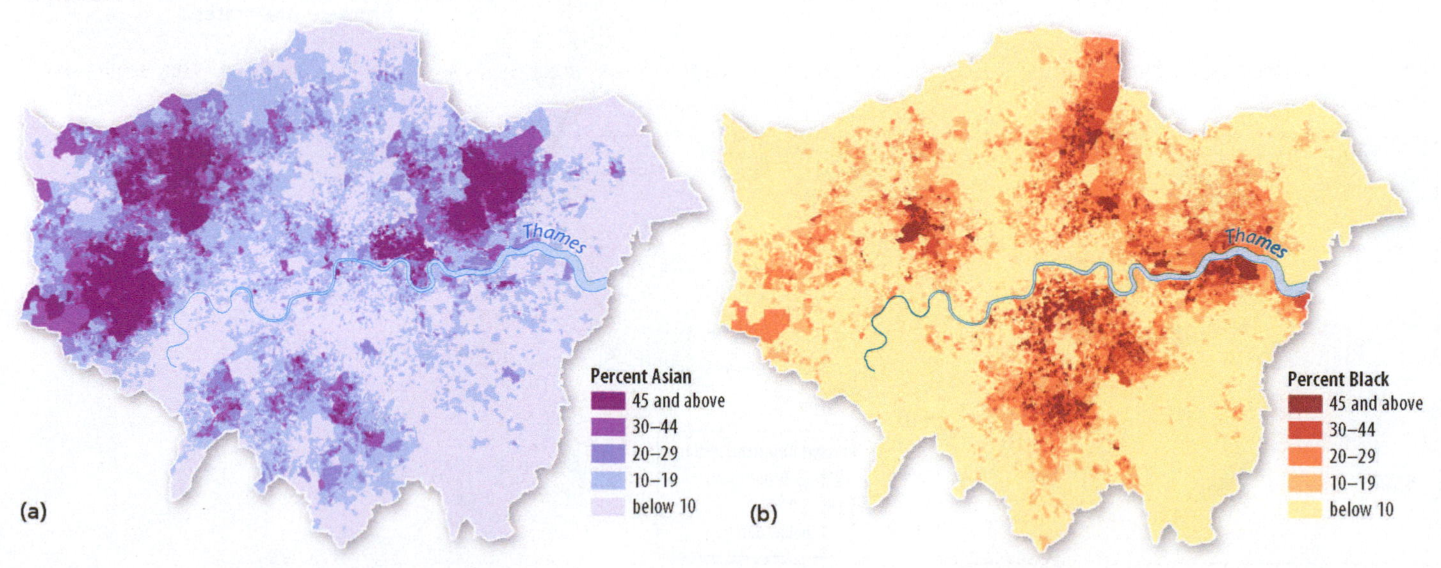

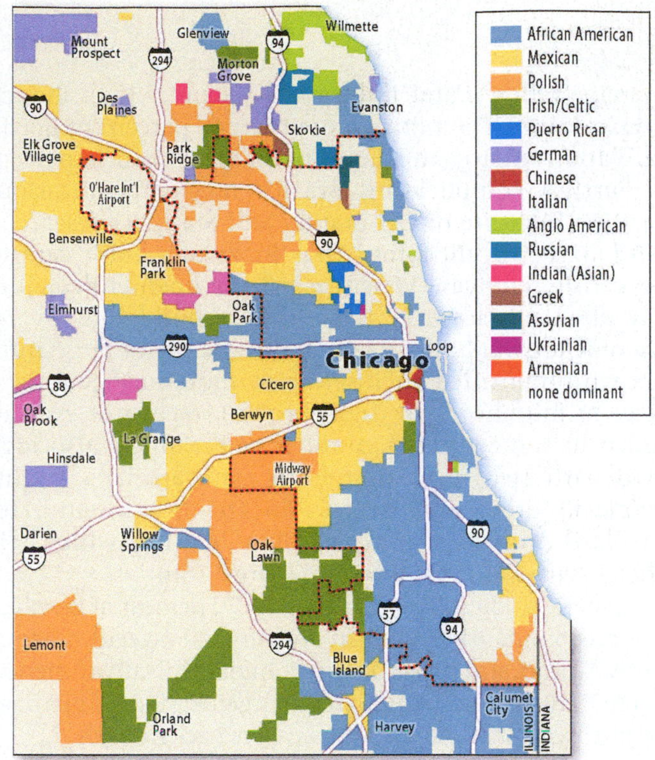

▲ **FIGURE 7-12 ETHNIC ENCLAVES IN CHICAGO, 2000**

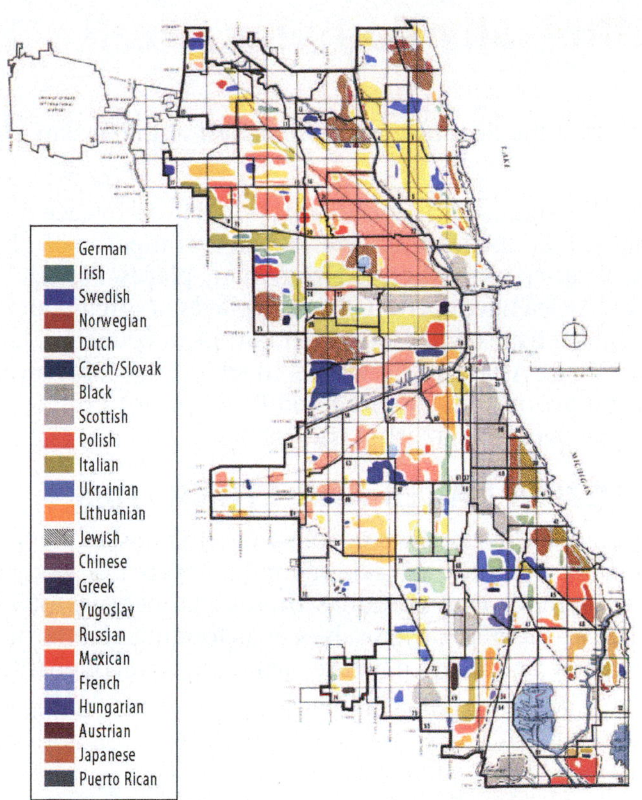

▲ **FIGURE 7-13 ETHNIC ENCLAVES IN CHICAGO, 1950**

other Midwest cities attracted ethnic groups primarily from Southern and Eastern Europe to work in the rapidly growing steel, automotive, and related industries. In 1910, when Detroit's auto production was expanding, three-fourths of the city's residents were immigrants and children of immigrants. Southern and Eastern European ethnic groups clustered in neighborhoods named for their predominant ethnicities, such as Detroit's Greektown and Poletown.

As recently as the middle of the twentieth century, large U.S. cities still had ethnic enclaves established by European immigrants (Figure 7-13). By the late twentieth century, most of the children and grandchildren of European immigrants had moved out of the inner-city enclaves to suburbs, in some cases forming ethnoburbs. An **ethnoburb** is a suburban area with a cluster of a particular ethnic population. For descendants of European immigrants, ethnic identity is more likely to be retained through religion, food, and other cultural traditions than through location of residence. A visible remnant of early-twentieth-century European ethnic neighborhoods is the clustering of restaurants in such areas as Little Italy and Greektown.

Cartograms depict the change in Chicago. In 1910, most residents of the city were descendants of immigrants from Europe, but by 1990, descendants of immigrants from Latin America and Asia were comparable in number (Figure 7-14).

PAUSE & REFLECT 7.1.4

Can you give an example of an ethnoburb in or near your community?

▼ **FIGURE 7-14 CARTOGRAMS OF ETHNIC GROUPS IN CHICAGO** (a) In 1910, most immigrants were from Europe. (b) In 1990, origins were more diverse.

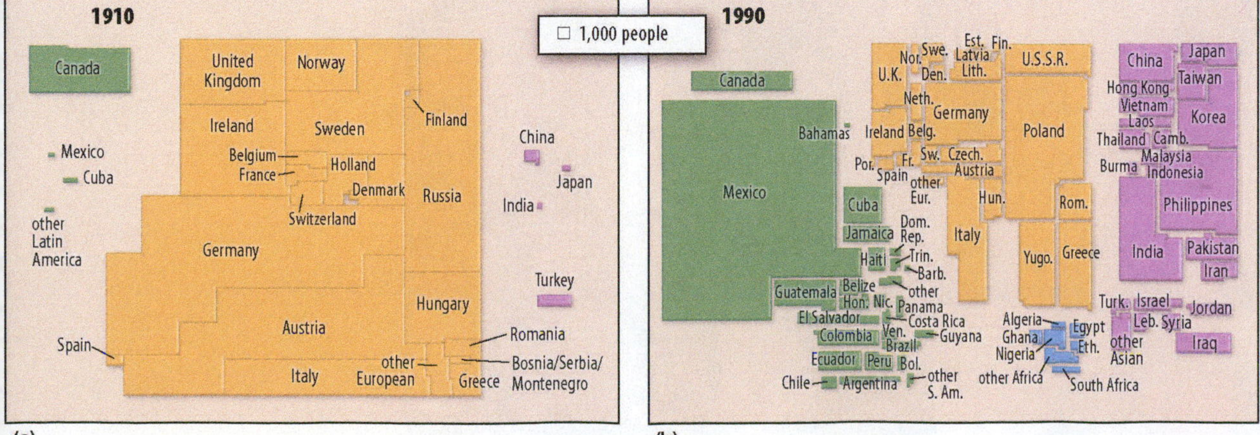

Ethnically Complex Brazil

LEARNING OUTCOME 7.1.5

Describe the distribution of ethnicities within urban areas.

Brazil struggles with defining its population by race or ethnicity. Like the United States, Brazil is composed of people whose ancestors emigrated from many places. Portugal and West Africa have been the leading places of origin, but large numbers have come from other European countries, Japan, Southwest Asia, and elsewhere. In addition, a large number of indigenous people inhabited Brazil prior to the emigration of people from other continents.

BRAZIL'S RACES AND ETHNICITIES

Brazil's census classifies people according to skin color. The Brazilian Institute of Geography and Statistics, a government agency that conducts the official census, asks Brazilians to identify themselves as belonging to one of five so-called races: branco (white), pardo (brown), preto (black), amarelo (yellow), and indigenous (Figure 7.15a). Brancos and pardos each comprise more than 40 percent of Brazil's population and together account for more than 90 percent.

Further complicating Brazil's racial classifications, Brazilian Institute of Geography and Statistics researcher José Luiz Petrucelli found that Brazilians don't care for the census choices. When asked an open-ended question about their race, Brazilians responded with 143 different answers. Most significant were the large numbers who identified themselves as moreno (brunette or olive) or moreno claro (light brown), races not even included in the census. A number of Brazilians also identified with two races that both translate into English as (black) preto and negro. On the other hand, few Brazilians considered themselves pardo, which the official census uses for brown (Figure 7.15b).

Genetic studies show that roughly 70 percent of Brazilians have predominantly European ancestry, 20 percent predominantly African, and 10 percent predominantly Native American (Figure 7-16). However, through many generations of marriages and births, most Brazilians have a mix of backgrounds.

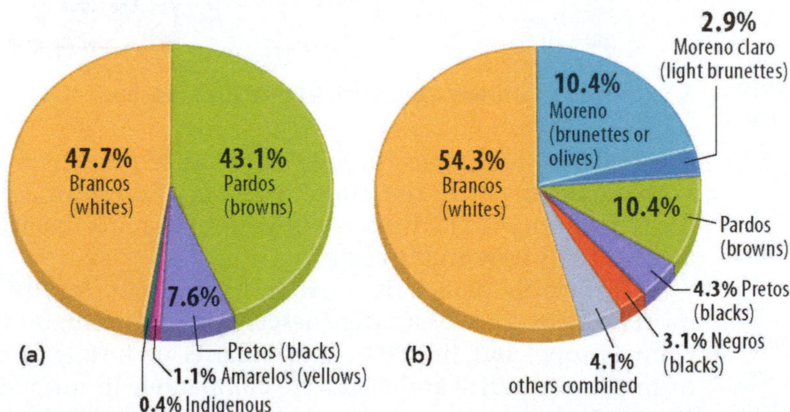

◀ **FIGURE 7-15 RACES IN BRAZIL (a)** Official results from the 2010 census. **(b)** Results of a self-identification survey by Brazilian Institute of Geography and Statistics.

◀ **FIGURE 7-16 MULTIRACIAL BRAZIL**

CLUSTERING OF RACES IN BRAZIL

Brazil also displays sharp regional differences in the distribution of races (Figure 7-17):

- **South.** Whites are clustered in the south. Brazil's four southernmost states, including the largest city, São Paulo, are approximately 70 percent white, compared to 40 percent in the rest of the country. The south was a major destination for immigrants from Portugal during colonial times and from other European countries after independence.
- **North.** Brazil's interior north, which is covered primarily by the Amazon tropical rain forest, has the highest percentage of indigenous people, who are classified by Brazil's census as brown. Relatively few European immigrants reached the interior.
- **Northeast.** Brazil's northern coast is also populated primarily by persons classified as brown. The region also received the largest number of blacks forced to migrate from Africa to be slaves. Nearly one-half of Brazil's blacks are clustered along the east coast between Bahia and Rio de Janeiro.
- **West-central.** This region has a mix of white and brown population. It was sparsely inhabited until Brasília was constructed in the region to be the capital of Brazil beginning in 1960. Brazilians of all races have migrated to the region to work in the capital.

PAUSE & REFLECT 7.1.5

If you had to fill out Brazil's census, what race would you select for yourself? Why?

▲ **FIGURE 7-17 DISTRIBUTION OF RACES IN BRAZIL** Whites are clustered in the south and browns in the north.

DISTRIBUTION OF ETHNICITIES IN GUYANA

Guyana, a country of only 800,000 inhabitants, is divided among several ethnicities. Around 30 percent of Guyanans are descended from Africans brought over as slaves in the seventeenth century, when it was a Dutch colony. Around 43 percent are descended from people brought over from India as indentured servants in the nineteenth century, when it was a British colony. Guyana also contains 9 percent indigenous people and 17 percent with mixed ethnicity.

Although the country is only 215,000 square kilometers (83,000 square miles) (the size of Kansas), the principal ethnicities cluster in different parts of Guyana. The India-descended people are clustered along the coast, the Africa-descended people are clustered in the interior north, and the indigenous people form the majority in the sparsely inhabited far west (Figure 7-18)

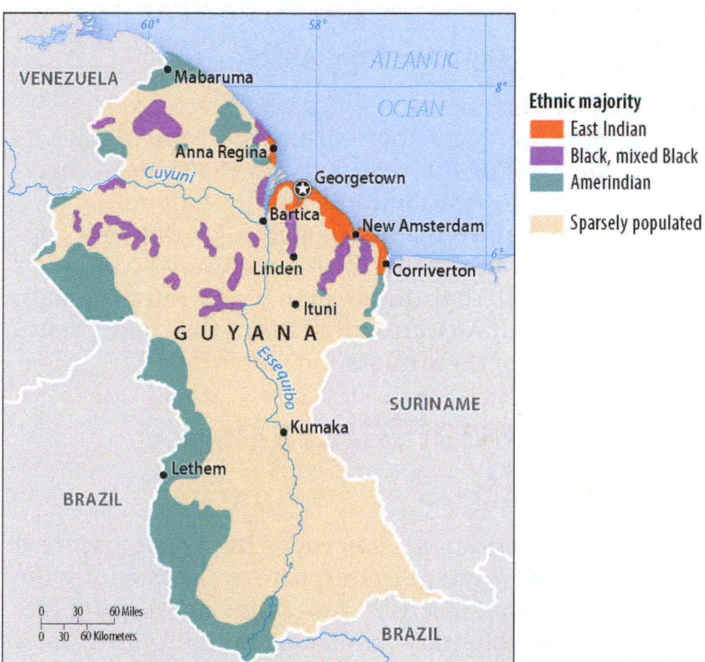

▲ **FIGURE 7-18 DISTRIBUTION OF ETHNICITIES IN GUYANA** Ethnic South Asian Indians are clustered along the coast, Africans in the north, and Amerindians in the west.

CHECK-IN KEY ISSUE 1

Where Are Ethnicities Distributed?

- ✓ **Ethnicity, race, and nationality are frequently confused.**
- ✓ **The most numerous ethnicities in the United States are Hispanic, African American, and Asian American.**
- ✓ **The three most numerous U.S. ethnicities have distinctive distributions at regional, state, and urban scales.**

KEY ISSUE 2

Why Do Ethnicities Have Distinctive Distributions?

- International Migration of Ethnicities
- Internal Migration of African Americans
- Segregation by Race

LEARNING OUTCOME 7.2.1

Describe forced migration from Africa.

The clustering of ethnicities within the United States is partly a function of the same process that helps geographers explain the distribution of other cultural factors, such as language and religion—namely migration. In Chapter 3, migration was divided into international (voluntary or forced) and internal (interregional and intraregional). The distribution of African Americans in the United States demonstrates all of these migration patterns.

International Migration of Ethnicities

Most African Americans are descended from Africans forced to migrate to the Western Hemisphere as slaves during the eighteenth century. Most Asian Americans and Hispanics are descended from voluntary immigrants to the United States during the late twentieth and early twenty-first centuries, although some—notably many Vietnamese and Cubans—felt compelled for political reasons to come to the United States.

FORCED MIGRATION FROM AFRICA

Slavery is a system whereby one person owns another person as a piece of property and can force that slave to work for the owner's benefit. The first Africans brought to the American colonies as slaves arrived at Jamestown, Virginia, on a Dutch ship in 1619. During the eighteenth century, the British shipped about 400,000 Africans to the 13 colonies that later formed the United States (Figure 7-19). In 1808 the United States banned bringing in additional Africans as slaves, but an estimated 250,000 were illegally imported during the next half-century.

Slavery was widespread during the time of the Roman Empire, about 2,000 years ago. During the Middle Ages, slavery was replaced in Europe by a feudal system, in which laborers working the land (known as serfs) were bound to the land and not free to migrate elsewhere. Serfs had to turn over a portion of their crops to the lord and provide other services, as demanded by the lord.

▲ **FIGURE 7-19 SLAVE SHIP** This drawing made around 1845 for a French magazine shows the high density and poor conditions of Africans transported to the Western Hemisphere to become slaves.

Although slavery was rare in Europe, Europeans were responsible for diffusing the practice to the Western Hemisphere. Europeans who owned large plantations in the Americas turned to African slaves as an abundant source of labor that cost less than paying wages to other Europeans.

At the height of the slave trade between 1710 and 1810, at least 10 million Africans were uprooted from their homes and sent on European ships to the Western Hemisphere for sale in the slave markets. During that period, the British and Portuguese each shipped about 2 million slaves to the Western Hemisphere, with most of the British slaves going to Caribbean islands and the Portuguese slaves to Brazil.

The forced migration began when people living along the east and west coasts of Africa, taking advantage of their superior weapons, captured members of other groups living farther inland and sold the captives to Europeans. Europeans in turn shipped the captured Africans to the Americas, selling them as slaves either on consignment or through auctions. The Spanish and Portuguese first participated in the slave trade in the early sixteenth century, and the British, Dutch, and French joined in during the next century.

Different European countries operated in various regions of Africa, each sending slaves to different destinations in the Americas (Figure 7-20). At the height of the eighteenth-century slave demand, a number of European countries adopted the **triangular slave trade**, an efficient triangular trading pattern.

PAUSE & REFLECT 7.2.1

Which area of Africa appears to have been the place of origin for most slaves sent to the North American colonies?

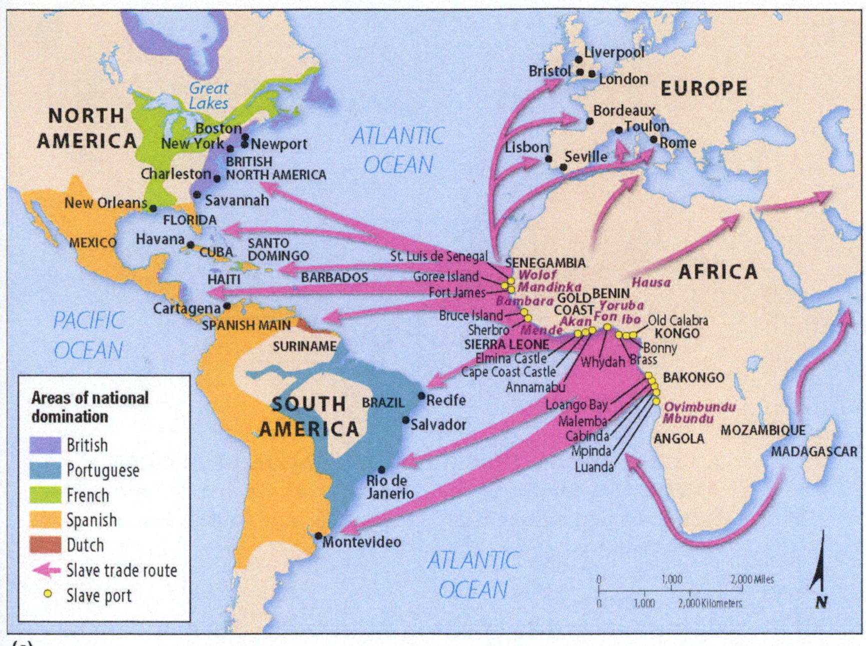

(a)

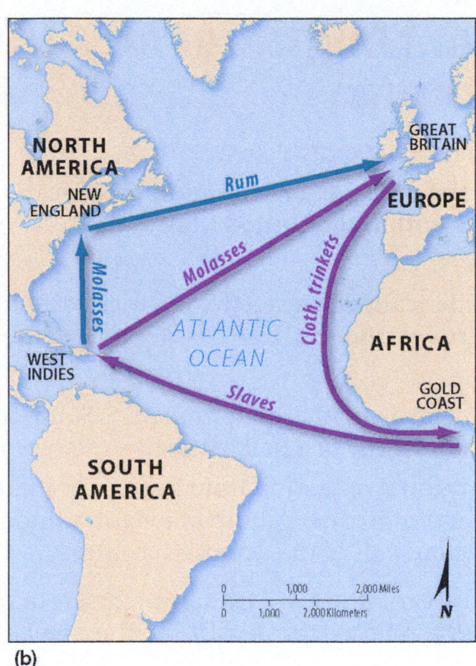

(b)

▲ **FIGURE 7-20 TRIANGULAR SLAVE TRADE** Ships sailed from Africa to the Western Hemisphere, from the Americas to Europe, and from Europe to Africa: **(a)** From Africa, slaves and gold were transported to the Western Hemisphere, primarily to the Caribbean islands. **(b)** From the Americas, the same ships took sugar and molasses to Europe. Some ships took molasses from the Caribbean to North American colonies, and rum was transported from the colonies to Europe. From Europe, ships carried cloth and other trade goods to Africa to buy slaves.

The large-scale forced migration of Africans caused them unimaginable hardship, separating families and destroying villages. Traders generally seized the stronger and younger villagers, who could be sold as slaves for the highest price. The Africans were packed onto ships at extremely high density, kept in chains, and provided with minimal food and sanitary facilities. Approximately one-fourth died crossing the Atlantic.

In the 13 colonies that later formed the United States, most of the large plantations in need of labor were located in the South, primarily those growing cotton and tobacco. Consequently, nearly all Africans shipped to the 13 colonies ended up in the Southeast.

Attitudes toward slavery dominated U.S. politics during the nineteenth century. During the early 1800s, when new states were carved out of western territory, anti-slavery northeastern states and pro-slavery southeastern states bitterly debated whether to permit slavery in the new states. The Civil War (1861–1865) was fought to prevent 11 pro-slavery Southern states from seceding from the Union. In 1863, during the Civil War, Abraham Lincoln issued the Emancipation Proclamation, freeing the slaves in the 11 Confederate states. The Thirteenth Amendment to the Constitution, adopted eight months after the South surrendered, outlawed slavery.

VOLUNTARY MIGRATION FROM LATIN AMERICA AND ASIA

Until the late twentieth century, quotas limited the number of people who could immigrate to the United States from Latin America and Asia, as discussed in Chapter 3. After the immigration laws were changed during the 1960s and 1970s, the population of Hispanics and Asian Americans in the United States increased rapidly. Initially, most Hispanics and Asian Americans were recent immigrants who came to the United States in search of work, but in the twenty-first century, most Americans who identify themselves as Hispanics or Asian Americans are children or grandchildren of immigrants (Figure 7-21).

▼ **FIGURE 7-21 CHINESE ETHNIC ENCLAVE, NEW YORK CITY** Spectators watch the Lunar New Year parade in the Flushing neighborhood of Queens, New York, which has one of the largest populations of ethnic Chinese people in the United States.

Internal Migration of African Americans

LEARNING OUTCOME 7.2.2

Describe the patterns of migration of African Americans within the United States.

African Americans displayed two distinctive internal migration patterns within the United States during the twentieth century:

- Interregional migration from the U.S. South to northern cities during the first half of the twentieth century.
- Intraregional migration from inner-city ghettos to outer-city and inner suburban neighborhoods during the second half of the twentieth century.

The spatial distribution of African Americans in the twenty-first century reflects the legacy of these two migration patterns.

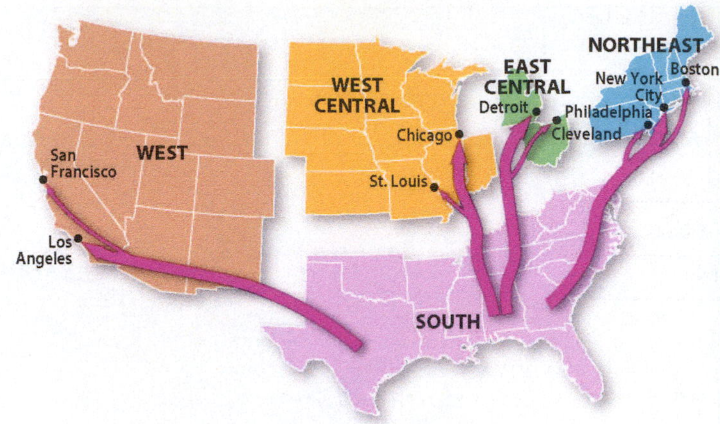

▲ **FIGURE 7-23 INTERREGIONAL MIGRATION OF AFRICAN AMERICANS** Migration followed four distinct channels along the East Coast and east-central, west-central, and southwestern regions of the country.

INTERREGIONAL MIGRATION

At the close of the Civil War, most African Americans were concentrated in the rural South. Today, as a result of interregional migration, many African Americans live in cities throughout the Northeast, Midwest, and West as well. Freed from slavery, most African Americans remained in the rural South during the late nineteenth century, working as sharecroppers (Figure 7-22).

A **sharecropper** works fields rented from a landowner and pays the rent by turning over to the landowner a share of the crops. To obtain seed, tools, food, and living quarters, a sharecropper gets a line of credit from the landowner and repays the debt with yet more crops. The sharecropper system burdened poor African Americans with high interest rates and heavy debts. Instead of growing food that they could eat, sharecroppers were forced by landowners to plant extensive areas of crops such as cotton that could be sold for cash.

Sharecropping became less common into the twentieth century, as the introduction of farm machinery and a decline in land devoted to cotton reduced demand for labor. At the same time that sharecroppers were being pushed off the farms, they were being pulled by the prospect of jobs in the booming industrial cities of the North.

▼ **FIGURE 7-22 SHARECROPPER** Near Lake Dick, Arkansas, around 1935.

African Americans migrated out of the South along several clearly defined channels (Figure 7-23). Most traveled by bus and car along the major two-lane long-distance U.S. roads that were paved and signposted in the early decades of the twentieth century and have since been replaced by interstate highways:

- **East Coast.** From the Carolinas and other South Atlantic states north to Baltimore, Philadelphia, New York, and other northeastern cities, along U.S. Route 1 (parallel to present-day I-95).
- **East central.** From Alabama and eastern Tennessee north to either Detroit, along U.S. Route 25 (present-day I-75), or Cleveland, along U.S. Route 21 (present-day I-77).
- **West central.** From Mississippi and western Tennessee north to St. Louis and Chicago, along U.S. routes 61 and 66 (present-day I-55).
- **Southwest.** From Texas west to California, along U.S. routes 80 and 90 (present-day I-10 and I-20).

Southern African Americans migrated north and west in two main waves, the first in the 1910s and 1920s, before and after World War I, and the second in the 1940s and 1950s, before and after World War II. The world wars stimulated expansion of factories in the 1910s and 1940s to produce war materiel, while the demands of the armed forces created shortages of factory workers. After the wars, during the 1920s and 1950s, factories produced steel, motor vehicles, and other goods demanded in civilian society.

INTRAREGIONAL MIGRATION

Intraregional migration—migration within cities and metropolitan areas—also changed the distribution of African Americans and people of other ethnicities. When they reached the big cities, African American immigrants

(a)

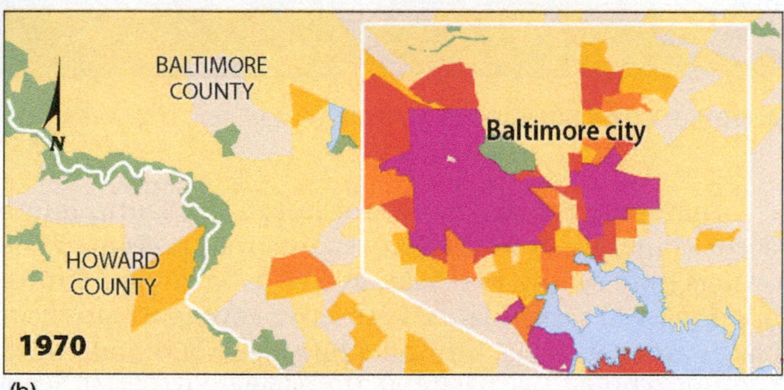

(b)

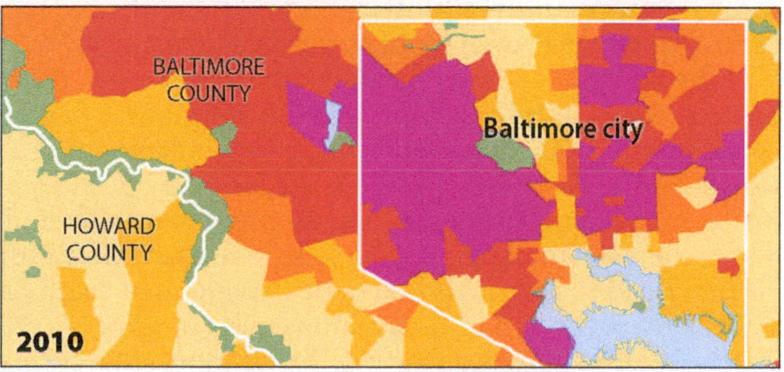

(c)

▲ **FIGURE 7-24 EXPANSION OF THE GHETTO IN BALTIMORE**

clustered in the one or two neighborhoods where the small numbers who had arrived in the nineteenth century were already living. These areas became known as ghettos, after the term for neighborhoods in which Jews were forced to live in the Middle Ages (see Chapter 6).

EXPANSION OF THE GHETTO. African Americans moved from the tight ghettos into immediately adjacent neighborhoods during the 1950s and 1960s. Expansion of the ghetto typically followed major avenues that radiated out from the center of the city.

In Baltimore, for example, most of the city's quarter-million African Americans in 1950 were clustered in a 3-square-kilometer (1-square-mile) neighborhood northwest of downtown (Figure 7-24). The remainder were clustered east of downtown or in a large isolated housing project on the south side built for black wartime workers in port industries.

Densities in the ghettos were high, with 40,000 inhabitants per square kilometer (100,000 per square mile) common. Contrast that density with the current level found in typical American suburbs of 2,000 inhabitants per square kilometer (5,000 per square mile). Because of the shortage of housing in the ghettos, families were forced to live in one room. Many dwellings lacked bathrooms, kitchens, hot water, and heat.

Baltimore's west-side African American ghetto expanded from 3 square kilometers (1 square mile) in 1950

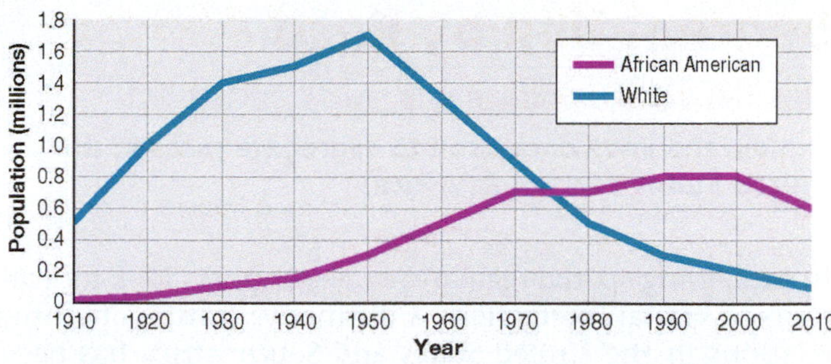

▲ **FIGURE 7-25 ETHNIC POPULATION CHANGE IN DETROIT**
Between 1950 and 2010, the white population of Detroit declined from 1.7 million to 100,000, whereas the African American population increased from 300,000 to 600,000.

to 25 square kilometers (10 square miles) in 1970, and a 5-square-kilometer (2-square-mile) area on the east side became mainly populated by African Americans. Expansion of the ghetto continued to follow major avenues to the northwest and northeast in subsequent decades.

"WHITE FLIGHT." The expansion of the black ghettos in American cities was made possible by "white flight," the emigration of whites from an area in anticipation of blacks immigrating into the area. Rather than integrate, whites fled.

Detroit provides an example. African Americans poured into Detroit in the early twentieth century (Figure 7-25). Many found jobs in the rapidly growing auto industry. Immigration into Detroit from the South subsided during the 1950s, but as legal barriers to integration crumbled, whites left Detroit. Detroit's white population dropped by about 1 million between 1950 and 1975 and by another half million between 1975 and 2000. The overall population of Detroit declined from a historic peak of nearly 2 million in 1950 to around 700,000 in the early twenty-first century.

White flight was encouraged by unscrupulous real estate practices, especially **blockbusting**. Under blockbusting, real estate agents convinced white homeowners to sell their houses at low prices, preying on their fears that black families would soon move into the neighborhood and cause property values to decline. The agents then sold the houses at much higher prices to black families desperate to escape the overcrowded ghettos. Through blockbusting, a neighborhood could change from all-white to all-black in a matter of months, and real estate agents could start the process all over again in the next white area.

The National Advisory Commission on Civil Disorders, known as the Kerner Commission, wrote in 1968 that U.S. cities were divided into two separate and unequal societies, one black and one white. A half-century later, despite serious efforts to integrate and equalize the two, segregation and inequality persist.

PAUSE & REFLECT 7.2.2

Which has changed the most in Detroit, the number of African Americans or the number of whites?

Segregation by Race

LEARNING OUTCOME 7.2.3

Explain the laws once used to segregate races in the United States and South Africa.

In explaining spatial regularities, geographers look for patterns of spatial interaction. A distinctive feature of ethnic relations in the United States and South Africa has been the strong discouragement of spatial interaction in the past through legal means. Although segregation laws are no longer in effect, their legacy remains a feature of the geography of ethnicity in both countries.

▲ **FIGURE 7-27 CIVIL RIGHTS MARCH** Washington, D.C., 1963.

UNITED STATES: "SEPARATE BUT EQUAL"

The concept of "separate but equal" was upheld by the U.S. Supreme Court in 1896. Louisiana had enacted a law that required black and white passengers to ride in separate train cars. In *Plessy v. Ferguson*, the Supreme Court stated that Louisiana's law was constitutional because it provided separate, but equal, treatment of blacks and whites, and equality did not mean that whites had to mix socially with blacks.

Once the Supreme Court permitted "separate but equal" treatment of races, southern states enacted a comprehensive set of laws to segregate blacks from whites as much as possible (Figure 7-26). These were called "Jim Crow" laws, named for a nineteenth-century song-and-dance act that depicted blacks offensively. Blacks had to sit in the backs of buses, and shops, restaurants, and hotels could choose to serve only whites. Separate schools were established for blacks and whites. This was equal, white southerners argued, because the bus got blacks sitting in the rear to the destination at the same time as the whites in the front, some commercial establishments served only blacks, and all the schools had teachers and classrooms.

Throughout the country, not just in the South, house deeds contained restrictive covenants that prevented the owners from selling to blacks, as well as to Roman Catholics or Jews in some places. Restrictive covenants also kept blacks from moving into all-white neighborhoods. And because schools, especially at the elementary level, were located to serve individual neighborhoods, most were segregated in practice, even if not by legal mandate.

The landmark Supreme Court decision *Brown v. Board of Education of Topeka, Kansas*, in 1954, found that having separate schools for blacks and whites was unconstitutional because no matter how equivalent the facilities, racial separation branded minority children as inferior and therefore was inherently unequal. A year later, the Supreme Court further ruled that schools had to be desegregated "with all deliberate speed." A nationwide movement in favor of civil rights forced elimination of U.S. segregation laws during the 1950s and 1960s (Figure 7-27).

Civil Rights Acts during the 1960s outlawed racial discrimination. However, segregation still exists in American cities. Many African Americans, recent immigrants, and those of other ethnicities remain clustered in urban neighborhoods because of economic and cultural factors. In urban schools, black and other "minority" students are now often the majority, as white residents have moved out.

PAUSE & REFLECT 7.2.3

Why might schools in cities like Baltimore and Detroit still be racially segregated (refer to Figures 7-24 and 7-25)?

◄ **FIGURE 7-26 "SEPARATE BUT EQUAL" PRACTICE** Until the 1960s in the U.S. South, whites and blacks had to use separate drinking fountains, restrooms, bus seats, hotel rooms, and other public facilities.

SOUTH AFRICA: APARTHEID

Discrimination by race reached its peak in the late twentieth century in South Africa. The cornerstone of the South African policy was the creation of **apartheid,** which was the legal separation of races into different geographic areas. Under apartheid, a newborn baby was classified as being one of four government-designated races—black, white, colored (mixed white and black), or Asian.

To ensure geographic isolation of these groups, the South African government designated 10 so-called homelands for blacks (Figure 7-28). The white minority government expected every black to become a citizen of one of the homelands and to move there. More than 99 percent of the population in the 10 homelands was black. The apartheid laws determined where different races could live, attend school, work, shop, travel, and own land (Figure 7-29). Blacks were restricted to certain occupations and were paid far lower wages than were whites for similar work. They could not vote or run for political office in national elections.

The apartheid system was created by descendants of whites who arrived in South Africa from the Netherlands in 1652 and settled in Cape Town, at the southern tip of the territory. They were known either as Boers, from the Dutch word for "farmer," or Afrikaners, from the word "Afrikaans," the name of their language, which is a dialect of Dutch. The British seized the Dutch colony in 1795 and controlled South Africa's government until 1948, when the Afrikaner-dominated Nationalist Party won elections. The Afrikaners vowed to resist pressures to turn over South Africa's government to blacks, and the Nationalist Party created the apartheid laws in the next few years to perpetuate white dominance of the country.

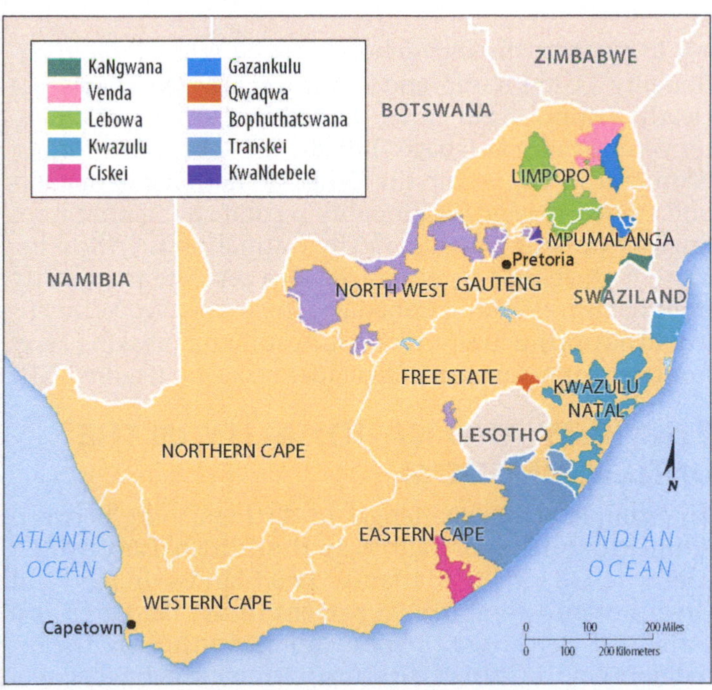

▲ **FIGURE 7-28 SOUTH AFRICA'S APARTHEID HOMELANDS** As part of its apartheid system, the government of South Africa designated 10 homelands. The government declared 4 of these homelands to be independent countries, but no other country recognized the action. With the end of apartheid and the election of a black majority government, the homelands were abolished, and South Africa was reorganized into 9 provinces.

▲ **FIGURE 7-29 APARTHEID** The train station in South Africa had separate areas for whites only and non-whites ("nie-blankes").

The white-minority government of South Africa repealed the apartheid laws in 1991. The principal anti-apartheid organization, the African National Congress, was legalized, and its leader, Nelson Mandela, was released from jail after more than 27 years of imprisonment. When all South Africans were permitted to vote in national elections for the first time, in 1994, Mandela was overwhelmingly elected the country's first black president.

Though South Africa's apartheid laws have been repealed, the legacy of apartheid will linger for many years. South Africa's blacks have achieved political equality, but they are much poorer than white South Africans. For example, South Africa is a major producer of wine, but only a small number of wineries are owned by blacks.

CHECK-IN **KEY ISSUE 2**

Why Do Ethnicities Have Distinctive Distributions?

- ✔ **Ancestors of some African Americans immigrated to the United States as slaves.**
- ✔ **Large numbers of African Americans migrated from the South to the North and West during the early twentieth century.**
- ✔ **In the United States, as well as in South Africa, segregation of races was legal for much of the twentieth century.**

KEY ISSUE 3

Why Might Ethnicities Face Conflicts?

- Ethnicities and Nationalities
- Dividing Ethnicities
- Ethnic Diversity in Asia

LEARNING OUTCOME 7.3.1

Understand differences between ethnicities and nationalities.

Nationality was defined at the beginning of this chapter as identity with a group of people who share legal attachment and personal allegiance to a particular country. In principle, the cultural values shared with others of the same ethnicity derive from religion, language, and material culture, whereas those shared with others of the same nationality derive from voting, obtaining a passport, and performing civic duties. However, this distinction is not always clear.

Ethnicities and Nationalities

Sorting out ethnicity and nationality can be challenging. Consider golfers Rory McIlroy and Tiger Woods (Figure 7-30). Woods's U.S. nationality is clear, but his ethnicity is less clear. His father was a mix of African American, Native American, and possibly Chinese, and his mother is a mix of Thai, Chinese, and Dutch. Woods invented the term "Cablinasian" to describe his complex ethnicity.

McIlroy has the opposite challenge. He identifies his ethnicity as Irish Catholic, but his nationality is complex. He was born in Northern Ireland, which is a part of the United Kingdom, so that is his nationality by birth, and he carries a U.K. passport. But McIlroy states that he shares ethnic identity with the people of the Republic of Ireland. The Republic of Ireland accepts as citizens anyone from Northern Ireland who so chooses, so McIlroy is permitted to identify as nationality only the U.K., only Ireland, or both.

ETHNICITY AND NATIONALITY IN NORTH AMERICA

Nationality is generally kept reasonably distinct from ethnicity and race in common usage in the United States:

- Nationality identifies citizens of the United States of America, including those born in the country and those who immigrated and became citizens.
- Ethnicity identifies groups with distinct ancestry and culture, such as African Americans, Mexican Americans, Chinese Americans, or Polish Americans.

▲ FIGURE 7-30 **COMPLEX ETHNICITY AND NATIONALITY: RORY MCILROY AND TIGER WOODS** Tiger Woods's nationality is clearly the United States, but his ethnicity is complex. Rory McIlroy's ethnicity is Irish Catholic, but his nationality is complex.

In Canada, the distinction between ethnicity and nationality is less clear and controversial. Québécois are clearly distinct from other Canadians in cultural traditions, especially language. But do the Québécois form a distinct ethnicity within the Canadian nationality or a second French-speaking nationality separate altogether from English-speaking Canadian? The distinction is critical because if Québécois is recognized as a separate nationality from English-speaking Canadian, the Québec government would have a much stronger justification for breaking away from Canada to form an independent country (Figure 7-31).

ETHNICITY AND NATIONALITY IN THE UNITED KINGDOM

An example of the challenges in distinguishing between ethnicities and nationalities is the British Isles, which comprise 133 inhabited islands. Two islands—Ireland and Great Britain—comprise 96 percent of the land area and contain 99 percent of the population. Sorting out the inhabitants of the British Isles into nationalities has become controversial.

The British Isles are divided between two countries—the United Kingdom of Great Britain and Northern Ireland and the Republic of Ireland. The Republic of Ireland comprises the southern 84 percent of the island of Ireland (called Eire in Irish), and the United Kingdom comprises Great Britain and the northern 16 percent of Eire. The United Kingdom

▲ **FIGURE 7-31 QUÉBEC INDEPENDENCE SUPPORTERS** The flag of Québec is waved at a pro-independence rally in Montréal.

is divided into four main parts: England, Northern Ireland, Scotland, and Wales (Figure 7-32).

The nationality of the citizens of the Republic of Ireland is clearly Irish. However, the nationality of the citizens of the United Kingdom is disputed. Does the United Kingdom contain one nationality (called British)? Or does it contain four nationalities—English, Scottish, Welsh, and the Irish of Northern Ireland?

- **English.** The English are descendants of Germanic tribes who crossed the North Sea and invaded the country in the fifth century. These invasions were summarized in Chapter 5, in the discussion of the origin of the English language.
- **Welsh.** The Welsh were Celtic people conquered by England in 1282 and formally united with England through the Act of Union of 1536. Welsh laws were abolished, and Wales became a local government unit.
- **Scots.** The Scots were Celtic people who had an independent country for more than 700 years, until 1603, when Scotland's King James VI also became King James I of England, thereby uniting the two countries. The Act of Union in 1707 formally merged the two governments, although Scotland was allowed to retain its own systems of education and local laws.

◀ **FIGURE 7-32 ETHNICITIES AND NATIONALITIES IN THE BRITISH ISLES** The British Isles comprise two countries: The Republic of Ireland and the United Kingdom.

- **Irish.** The Irish were Celtic people who were ruled by England until the twentieth century, when most of the island became the independent country of Ireland. The northern portion remained part of the United Kingdom.

International sports organizations permit each of the four parts of the United Kingdom to field its own separate national soccer team in major tournaments, such as the World Cup. The most important annual international rugby tournament, known as the Six Nations Championship, includes teams from England, Scotland, and Wales, as well as from the Republic of Ireland, Italy, and France. Given the history of English conquest, the other nationalities often root against England when it is playing teams from other countries.

The government of the United Kingdom has established separate governments for Northern Ireland, Scotland, and Wales. The government of Scotland exercises considerable authority, whereas the one in Wales has much less authority. Scots voted in 2014 to remain part of the United Kingdom, but political differences between Scotland and England may eventually result in an independent Scotland (Figure 7-33).

NATIONALISM

Nationalism is loyalty and devotion to a nationality. Nationalism typically promotes a sense of national consciousness that exalts one nation above all others and emphasizes its culture and interests as opposed to those of other nationalities. People display nationalism by supporting a country that preserves and enhances the culture and attitudes of their nationality. States foster nationalism by promoting symbols of the country, such as flags and songs.

Nationalism is an important example of a **centripetal force,** which is an attitude that tends to unify people and enhance support for a state. (The word *centripetal* means "directed toward the center"; it is the opposite of *centrifugal,* which means "to spread out from the center.") Most countries find that the best way to achieve citizen support is to emphasize shared attitudes that unify the people.

▼ **FIGURE 7-33 PRO-INDEPENDENCE SCOTS** Rally in Glasgow in favor of vote for independence.

Dividing Ethnicities

LEARNING OUTCOME 7.3.2

Describe how ethnicities can be divided among more than one nationality.

Few ethnicities inhabit an area that matches the territory of a nationality. Ethnicities are sometimes divided among more than one nationality. Several examples can be found in Asia.

ETHNICITIES IN SOUTH ASIA

South Asia provides vivid examples of what happens when independence comes to colonies that contain two major ethnicities. When the British ended their colonial rule of the Indian subcontinent in 1947, they divided the colony into two irregularly shaped countries—India and Pakistan (Figure 7-34). Pakistan comprised two noncontiguous areas, West Pakistan and East Pakistan, 1,600 kilometers (1,000 miles) apart, separated by India. East Pakistan became the independent state of Bangladesh in 1971. An eastern region of India was also practically cut off from the rest of the country, attached only by a narrow corridor north of Bangladesh that is less than 13 kilometers (8 miles) wide in some places.

The basis for separating West and East Pakistan from India was ethnicity. The people living in the two areas of Pakistan were predominantly Muslim; those in India were predominantly Hindu. Antagonism between the two religious groups was so great that the British decided to place the Hindus and Muslims in separate states. Hinduism has become a great source of national unity in India. In modern India, with its hundreds of languages and ethnic groups, Hinduism has become the cultural trait shared by the largest percentage of the population.

▲ **FIGURE 7-34 ETHNIC DIVISION OF SOUTH ASIA** In 1947, British India was partitioned into two independent states, India and Pakistan, which resulted in the forced migration of an estimated 17 million people. The creation of Pakistan as two territories nearly 1,600 kilometers (1,000 miles) apart proved unstable, and in 1971 East Pakistan became the independent country of Bangladesh.

Muslims have long fought with Hindus for control of territory, especially in South Asia. After the British took over India in the early 1800s, a three-way struggle began, with the Hindus and Muslims fighting each other as well as the British rulers. Mahatma Gandhi, a Hindu advocate of nonviolence and reconciliation with Muslims, was assassinated in 1948, ending the likelihood of creating a single state in which Muslims and Hindus could live together peacefully.

The partition of South Asia into two states resulted in massive migration because the two boundaries did not correspond precisely to the territory inhabited by the two ethnicities. Approximately 17 million people caught on the wrong side of a boundary felt compelled to migrate during the late 1940s. Some 6 million Muslims moved from India to West Pakistan and about 1 million from India to East Pakistan. Hindus who migrated to India included approximately 6 million from West Pakistan and 3.5 million from East Pakistan. As they attempted to reach the other side of the new border, Hindus in Pakistan and Muslims in India were killed by people from the rival religion. Extremists attacked small groups of refugees traveling by road and halted trains to massacre the passengers.

Pakistan and India never agreed on the location of the boundary separating the two countries in the northern region of Kashmir (Figure 7-35). Since 1972, the two countries have maintained a "line of control," with Pakistan administering the northwestern portion and India the southeastern portion. Muslims, who comprise a majority in both portions, have fought to secure reunification

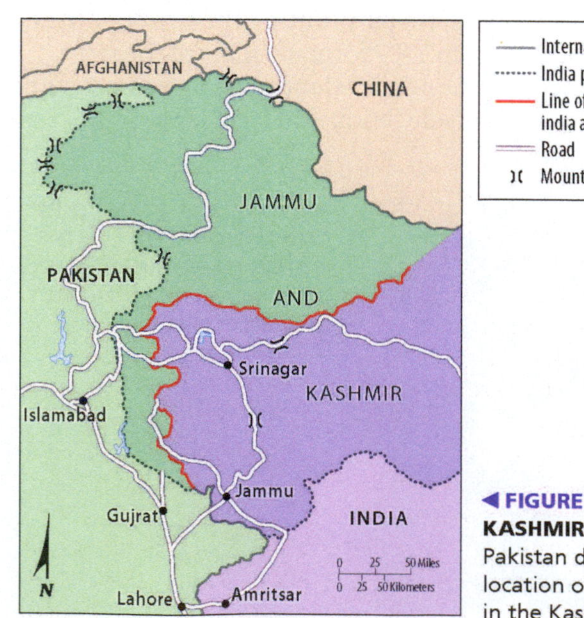

◀ **FIGURE 7-35 KASHMIR** India and Pakistan dispute the location of their border in the Kashmir region.

of Kashmir, either as part of Pakistan or as an independent country. India blames Pakistan for the unrest and vows to retain its portion of Kashmir. Pakistan argues that Kashmiris on both sides of the border should choose their own future in a vote, confident that the majority Muslim population would break away from India.

India's religious unrest is further complicated by the presence of 23 million Sikhs, who have long resented that they were not given their own independent country when India was partitioned (see Chapter 6). Although they constitute only 2 percent of India's total population, Sikhs comprise a majority in the Indian state of Punjab, along the border with Pakistan. Sikh extremists have fought for more control over the Punjab or even complete independence from India.

DIVIDING THE KURDS

A prominent example of an ethnicity divided among several countries in Asia is the Kurds, who live in the Caucasus Mountains. The Kurds are Sunni Muslims who speak a language in the Iranian group of the Indo-Iranian branch of Indo-European and have distinctive literature, dress, and other cultural traditions.

When the victorious European allies carved up the Ottoman Empire after World War I, they created an independent state of Kurdistan to the south and west of Van Gölü (Lake Van) under the 1920 Treaty of Sèvres. Before the treaty was ratified, however, the Turks, under the leadership of Mustafa Kemal (later known as Kemal Ataturk), fought successfully to expand the territory under their control beyond the small area the Allies had allocated to them. The Treaty of Lausanne in 1923 established the modern state of Turkey, and three years later the League of Nations determined that much of Kurdistan would become part of Turkey.

Today the 30 million Kurds are split among several countries; 14.5 million live in eastern Turkey, 6 million in western Iran, 5.5 million in northern Iraq, 2 million in Syria, and 1.5 million in other countries (primarily Germany). Kurds comprise 18 percent of the population in Turkey, 17 percent in Iraq, 9 percent in Syria, and 8 percent in Iran (Figure 7.36a).

To foster the development of Turkish nationalism, the Turks have tried repeatedly to suppress Kurdish culture. Use of the Kurdish language was illegal in Turkey until 1991, and laws banning its use in broadcasts and classrooms remain in force. Kurdish nationalists, for their part, have waged a guerrilla war since 1984 against the Turkish army. In recent years, Turkey has permitted Kurds to practice more of their cultural traditions (Figure 7-36b).

Iraq's Kurds have made several unsuccessful attempts to gain independence, including in the 1930s, 1940s, and 1970s. A few days after Iraq was defeated in the 1991 Gulf War, the country's Kurds launched another unsuccessful rebellion. The United States and its allies decided not to resume their recently concluded fight against Iraq on behalf of the Kurdish rebels, but after the revolt was crushed, they sent troops to protect the Kurds from further attacks by the Iraqi army. After the United States attacked Iraq and deposed Saddam Hussein in 2003, Iraqi Kurds achieved even more autonomy. The Kurdistan Regional Parliament governs northern Iraq, with little control exercised by the Iraqi national government.

(a)

(b)

▲ **FIGURE 7-36 KURDISTAN (a)** The area on the map is the territory that Kurds wish to consolidate into an independent Kurdistan. **(b)** Kurds in Diyarbakir, Turkey, celebrate Nawroz (Kurdish New Year), which occurs in March on the equinox.

Ethnic Diversity in Asia

LEARNING OUTCOME 7.3.3

Identify and describe the principal ethnicities in western areas of Asia.

The lack of correspondence between the territories occupied by ethnicities and by nationalities is especially severe in areas of Southwest Asia and Central Asia. The Kurds, discussed on the previous page, provide just one example. Dozens of ethnicities inhabit the area, allocated among seven nationalities (Figure 7-37). A map of the western portions of Asia would look very different if national boundaries were drawn to separate the major ethnicities.

ETHNICITIES IN TURKEY

Ethnic Turks comprise approximately three-fourths of the population of Turkey. Turks are descended from migrants to present-day Turkey around 1,000 years ago. The most populous minority is the Kurds, 18 percent of Turkey's population.

ETHNICITIES IN LEBANON

Lebanon is divided among around 54 percent Muslims, 41 percent Christians, and 5 percent Druze. Lebanon's most numerous Christians are Maronites, who consider themselves Roman Catholic, followed by various Orthodox denominations (Figure 7-38). Lebanon's Muslims are estimated to be divided about equally between Sunnis and Shiites. The Druze are sometimes classified as Muslims, but they do not follow Islam's five pillars of faith described in Chapter 6 and therefore do not self-identify as Muslims. Lebanon's Constitution officially recognizes 18 religions, including some Christian and Muslim denominations and congregations with very few adherents.

When Lebanon became independent in 1943, the constitution required that each religion be represented in the Chamber of Deputies according to its percentage in the 1932 census. By unwritten convention, the president of Lebanon was a Maronite Christian, the premier a Sunni Muslim, the speaker of the Chamber of Deputies a Shiite Muslim, and the foreign minister a Greek Orthodox Christian. Other cabinet members and civil servants were similarly apportioned among the various faiths.

Lebanon's religious groups have tended to live in different regions of the country. Maronites are concentrated in the west-central, Sunnis in the northwest, and Shiites in the south and east. Beirut, the capital and largest city, has been divided between a Christian eastern zone and a Muslim western zone. During a civil war between 1975 and 1990, each religious group formed a private militia to guard its territory. The territory controlled by each militia changed according to battles with other religious groups.

ETHNICITIES IN SYRIA

On the surface, Syria has a simple ethnic composition, with 90 percent Arabs and 9 percent Kurds. However, the Arab population is divided among 64 percent Sunni Muslim, 11 percent Alawi Muslim, 10 percent Christian, 3 percent Druze, and 2 percent other Muslim. The most numerous Christian denominations are Greek Orthodox

▼ FIGURE 7-37 ETHNICITIES IN WESTERN AREAS OF ASIA The complex mix of ethnicities across western areas of Asia, and the lack of correspondence between their distribution and the boundaries of countries, has been a major source of conflict in the region.

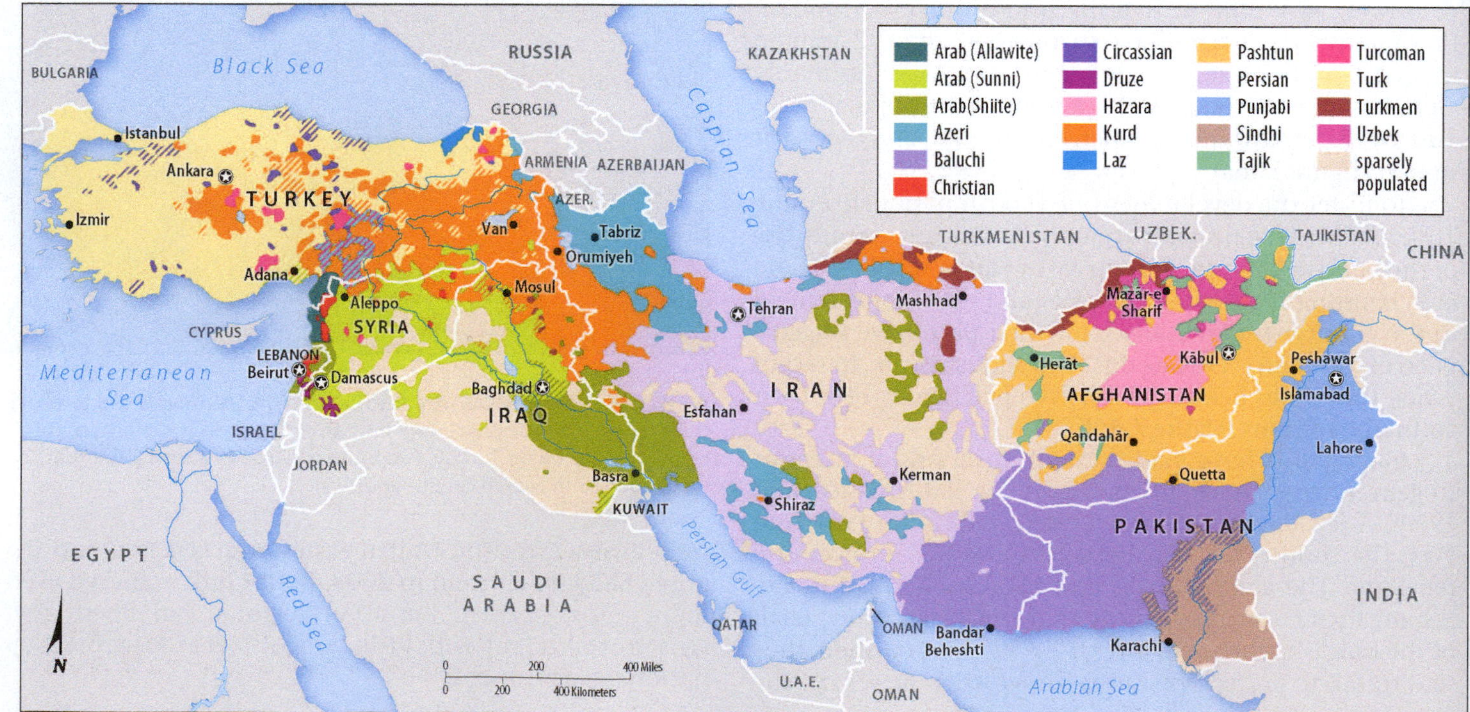

◀ **FIGURE 7-38 LEBANESE CHRISTIANS** Palm Sunday is marked by a walk at St. George Cathedral in Beirut, Lebanon.

and Greek Catholic. The Alawis, who adhere to a denomination of Shiite Islam, have held power in Syria since 1970. A civil war has raged for several years between supporters and opponents of the Alawi-dominated government.

ETHNICITIES IN IRAQ

The most numerous ethnicities in Iraq include 55 percent Shiite Muslim Arabs, 21 percent Kurds, and 19 percent Sunni Muslim Arabs. The United States led an attack against Iraq in 2003 that resulted in the removal and death of the country's longtime president, Saddam Hussein. U.S. officials justified removing Hussein because he ran a brutal dictatorship, created weapons of mass destruction, and allegedly had close links with terrorists. Having invaded Iraq and removed Hussein from power, the United States became embroiled in a complex and violent struggle among ethnicities. The political situation in Iraq is discussed in more detail in Chapter 8.

ETHNICITIES IN IRAN

Most Iranians are Persians, believed to be descendants of Indo-European tribes that migrated from Central Asia into what is now Iran several thousand years ago (see Chapter 5). Persians constitute the world's largest ethnicity adhering to Shiite Islam. Other important ethnicities are Azeri and Baluchi. Relations between Iran and the United States have been poor since 1979, when a revolution brought to power fundamentalist Shiites, and some of their supporters seized the U.S. Embassy, holding 52 Americans hostage for more than a year. More recently, other countries have struggled to keep Iran from creating nuclear weapons (discussed in more detail in Chapter 8).

▼ **FIGURE 7-39 PASHTUN BOYS, AFGHANISTAN**

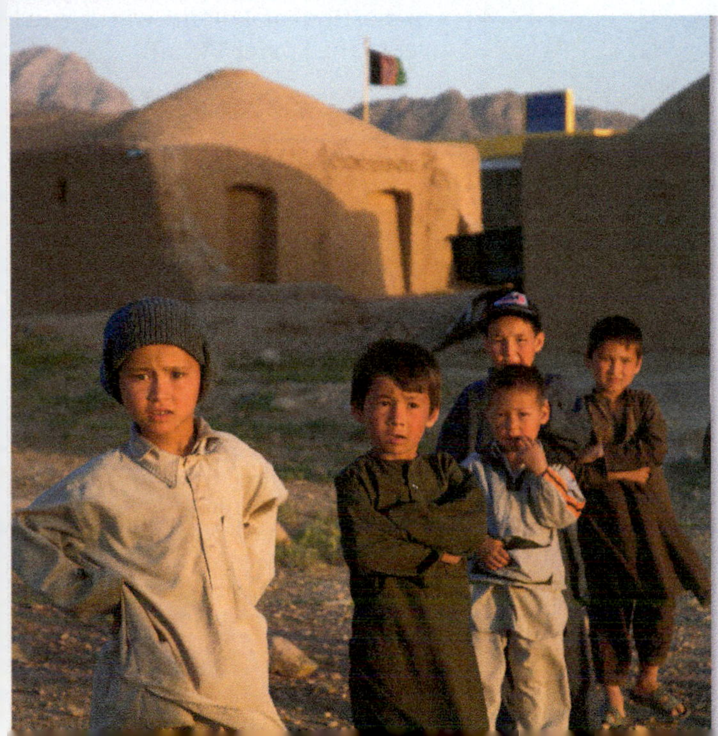

ETHNICITIES IN AFGHANISTAN

The most numerous ethnicities in Afghanistan are Pashtun, Tajik, and Hazara (Figure 7-39). The current unrest among Afghanistan's ethnicities dates from 1979, with the start of a rebellion by several ethnic groups against the government. After years of fighting among ethnicities, a Pashtun faction called the Taliban gained control over most of the country in 1995. The Taliban imposed harsh laws on Afghanistan, according to Islamic values as the Taliban interpreted them (see Chapter 6). The United States invaded Afghanistan in 2001 and overthrew the Taliban-led government because it was harboring terrorists (see Chapter 8). Removal of the Taliban unleashed a new struggle for control of Afghanistan among the country's many ethnicities.

ETHNICITIES IN PAKISTAN

The Punjabi have been the most numerous ethnicity of what is now Pakistan since ancient times, but the mountainous border area with Afghanistan is principally Baluchi and Pashtun. As with the neighboring Pashtun, the Punjabi converted to Islam in the seventh century. The Punjabi remained Sunni Muslims rather than convert to Shiite Islam like the Pashtun.

PAUSE & REFLECT 7.3.3

How do the ethnic complexities in Southwest and Central Asia make it difficult to set up stable democracies?

CHECK-IN KEY ISSUE 3

Why Might Ethnicities Face Conflicts?

- ✔ **Nationality is identity with a group of people who share legal attachment and personal allegiance to a particular country.**
- ✔ **Some ethnicities, such as the Kurds, are divided among more than one nationality.**
- ✔ **Lack of correspondence between ethnicities and nationalities is especially severe in western areas of Asia.**

KEY ISSUE 4

Why Do Ethnic Cleansing and Genocide Occur?

- ▸ **Forced Migration in Europe**
- ▸ **Ethnic Cleansing in Bosnia & Herzegovina**
- ▸ **Ethnic Cleansing Elsewhere in the Balkans**
- ▸ **Ethnic Cleansing and Genocide in Africa**
- ▸ **Ethnic Cleansing and Genocide in Central Africa**

LEARNING OUTCOME 7.4.1

Describe the process of ethnic cleansing.

Throughout history, ethnic groups have been forced to flee from other ethnic groups' more powerful armies. The United Nations defines **ethnic cleansing** as a purposeful policy designed by one ethnic or religious group to remove by violent and terror-inspiring means the civilian population of another ethnic or religious group from certain geographic areas. Ethnic cleansing is especially important for cultural geography because it changes the spatial distribution of ethnicities, and it does so through force and criminal violence.

Ethnic cleansing is undertaken to rid an area of an entire ethnicity so that the surviving ethnic group can be the sole inhabitants. The point of ethnic cleansing is not simply to defeat an enemy or to subjugate them, as was the case in traditional wars. Rather than a clash between armies of male soldiers, ethnic cleansing involves the removal of every member of the less powerful ethnicity—women as well as men, children as well as adults, the frail elderly as well as the strong youth.

Forced Migration in Europe

The largest forced migration occurred during World War II (1939–1945) because of events leading up to the war, the war itself, and postwar adjustments. Especially notorious was the deportation by the German Nazis of millions of Jews, gypsies, and other ethnic groups to the infamous concentration camps, where they exterminated most of them.

After World War II ended, millions of ethnic Germans, Poles, Russians, and other groups were forced to migrate as a result of boundary changes (Figure 7-40). For example, when a portion of eastern Germany became part of Poland, the Germans living in the region were forced to move west to Germany, and Poles were allowed to move into the area. Similarly, Poles were forced to move when the eastern portion of Poland was turned over to the Soviet Union.

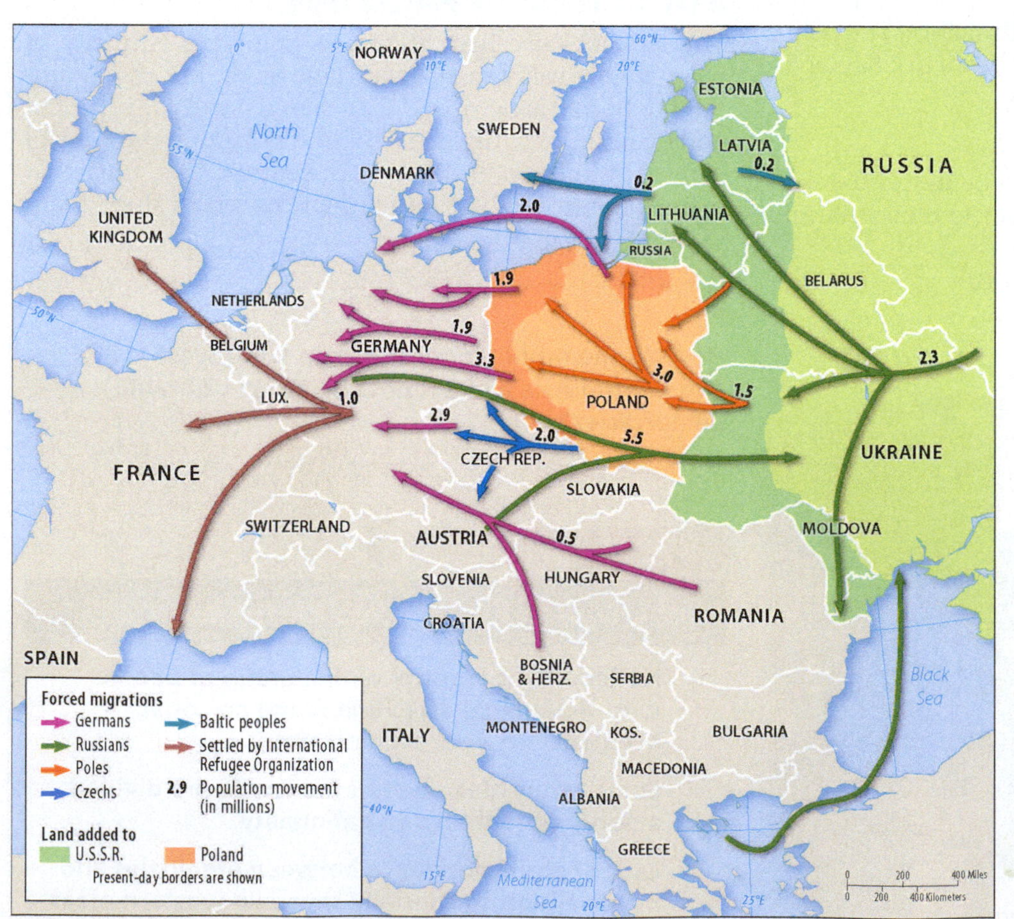

◀ **FIGURE 7-40 FORCED MIGRATION OF ETHNICITIES AFTER WORLD WAR II** The largest number were Poles forced to move from territory occupied by the Soviet Union (now Russia), Germans forced to migrate from territory taken over by Poland and the Soviet Union, and Russians forced to return to the Soviet Union from Western Europe.

▲ **FIGURE 7-41 THE BALKANS IN 1914**

MULTIETHNIC YUGOSLAVIA

The scale of ethnic cleansing during World War II has not been repeated, but ethnic cleansing has occurred more recently in Europe in the Balkans. Ethnic cleansing is also occurring in Africa (discussed later in this chapter).

The Balkans, about the size of Texas, is a region named for the Balkan Mountains (known in Slavic languages as Stara Planina), which extend east–west across the region. The region includes Albania, Bulgaria, Greece, and Romania, as well as several countries that once comprised Yugoslavia. A complex assemblage of ethnicities has long made the Balkans a hotbed of unrest (Figure 7-41). Most profoundly for the rest of the world, the incident that sparked World War I occurred in the Balkans. In June 1914, the heir to the throne of Austria-Hungary was assassinated in Sarajevo by a Serb who sought independence for Bosnia.

PAUSE & REFLECT 7.4.1

What is another example of a country that is inhabited primarily by ethnic Slavs?

After World War I, the Allies created a new country, Yugoslavia, to unite several Balkan ethnicities that spoke similar South Slavic languages. The prefix "Yugo" in the country's name derived from the Slavic word for "south." Longtime leader Josip Broz Tito (prime minister 1943–1963 and president 1953–1980) was instrumental in forging a Yugoslav nationality. Central to Tito's vision of a Yugoslav nationality was acceptance of ethnic diversity in language and religion. Individuals from the five most numerous ethnicities—Croat, Macedonian, Montenegrin, Serb, and Slovene—were allowed to exercise considerable control over the areas they inhabited within Yugoslavia.

Rivalries among ethnicities resurfaced in Yugoslavia during the 1980s after Tito's death, leading ultimately to its breakup into seven small countries (Figure 7-42). Because the boundaries of the new countries did not match the distribution of ethnicities, the breakup of Yugoslavia did not happen peacefully. Several episodes of ethnic cleansing ensued. (See Sustainability & Our Environment feature on page 257, and the next section of this chapter.)

◀ **FIGURE 7-42 BREAKUP OF YUGOSLAVIA** Until its breakup in 1992, the country of Yugoslavia consisted of six republics—Bosnia & Herzegovina, Croatia, Macedonia, Montenegro, Serbia, and Slovenia. These republics exercised considerable control over their local affairs. Kosovo and Vojvodina were considered autonomous regions within Serbia.

Ethnic Cleansing in Bosnia & Herzegovina

LEARNING OUTCOME 7.4.2

Explain the concept of ethnic cleansing in Bosnia & Herzegovina.

Ethnic cleansing after the breakup of Yugoslavia occurred primarily where the territories occupied by the various ethnic groups did not match the boundaries of the new countries. Figure 7-42 shows three places in former Yugoslavia where ethnic and country boundaries were especially poorly matched—Croatia, Bosnia & Herzegovina, and Kosovo.

Ethnic cleansing often follows these steps:

1. Move a large amount of military equipment and personnel into a village that has no strategic value.
2. Round up all the people in the village. Segregate men from women, children, and old people. Place men in detention camps or kill them.
3. Force the rest of the people to leave the village. March them in a convoy to a place outside the territory being ethnically cleansed.
4. Destroy the vacated village, such as by setting it on fire.

ETHNICALLY DIVERSE BOSNIA & HERZEGOVINA

Bosnia & Herzegovina was the most ethnically diverse republic of the former Yugoslavia. At the time of the breakup of Yugoslavia, the population of Bosnia & Herzegovina was 44 percent Bosniaks, 31 percent Serbs, and 17 percent Croats. Bosniaks are frequently called Bosnian Muslims, in recognition of their predominant religion. Rather than live in an independent multiethnic state with a Muslim plurality, Bosnia & Herzegovina's Serbs and Croats fought to unite the portions of the republic that they inhabited with Serbia and Croatia, respectively.

To strengthen their cases for breaking away from Bosnia & Herzegovina, Serbs and Croats engaged in ethnic cleansing of Bosniaks (Figure 7-43). According to the United Nations, ethnic cleansing in Bosnia included "murder, torture, arbitrary arrest and detention, extra-judicial executions, rape and sexual assaults, confinement of civilian population in ghetto areas, forcible removal, displacement and deportation of civilian population, deliberate military attacks or threats of attacks on civilians and civilian areas, and wanton destruction of property."

Ethnic cleansing ensured that areas did not merely have majorities of Bosnian Serbs and Bosnian Croats but were ethnically homogeneous and therefore better candidates for union with Serbia and Croatia. Ethnic cleansing by Serbs against Bosniaks was especially severe because the territory inhabited by Serbs in Bosnia comprised several discontinuous areas, many of which were separated from Serbia by areas with Bosniak majorities. By ethnically cleansing Bosniaks from intervening areas, Bosnian Serbs created one continuous area of Serb domination rather than several discontinuous ones.

(a)

(b)

(c)

▲ **FIGURE 7-43 ETHNIC CLEANSING IN BOSNIA & HERZEGOVINA** **(a)** The Stari Most (old bridge), in the city of Mostar, Bosnia & Herzegovina, was built by the Turks in 1566 across the Neretva River. **(b)** The bridge was blown up by Croats in 1993 as part of their ethnic cleansing against the Bosniaks. **(c)**. With the end of the war in Bosnia & Herzegovina, the bridge was rebuilt in 2004. **1. Why would the Croats consider blowing up a bridge to be an important element in their ethnic cleansing? 2. Why would the Bosniaks consider it important to rebuild the bridge after the war? 3. Compare the rebuilt bridge with the original. Why do you think it was important for the Bosniaks to make the new bridge appear almost identical to the original one?**

(a)

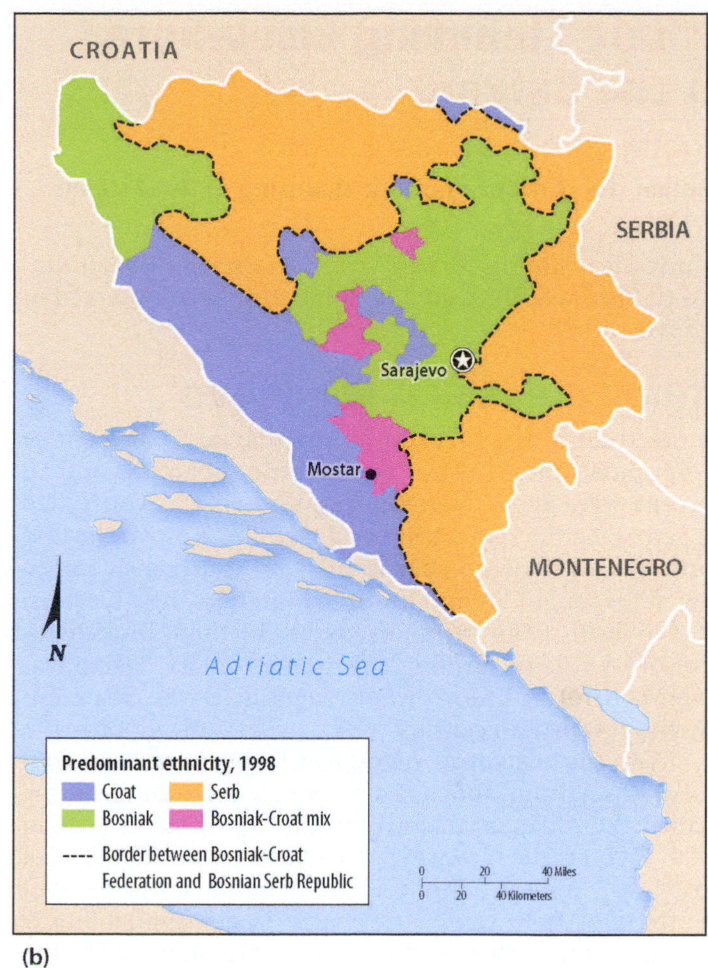

(b)

▲ **FIGURE 7-44 ETHNIC CLEANSING IN BOSNIA & HERZEGOVINA** **(a)** Distribution of ethnicities in Bosnia & Herzegovina before the war during the 1990s. **(b)** Distribution of ethnicities after ethnic cleansing. Ethnic cleansing resulted in a transfer of control of territory from Bosniaks to Serbs and Croats.

Accords reached in Dayton, Ohio, in 1996 by leaders of the various ethnicities divided Bosnia & Herzegovina into three regions, one each dominated, respectively, by Bosniaks, Croats, and Serbs. The Bosniak and Croat regions were combined into a federation, with some cooperation between the two groups, but the Serb region has operated with almost complete independence in all but name from the others.

The International Criminal Tribunal for the former Yugoslavia, established by the United Nations, convicted Croat and Serb leaders of war crimes for the ethnic cleansing of Bosniaks. Yet ethnic cleansing was ultimately successful. The Dayton agreements gave Serbs control of around 50 percent of the land even though they comprised only 40 percent of the population of Bosnia & Herzegovina. Croats got around 20 percent of the land but comprised only 15 percent of the population. Bosniaks, with 45 percent of the population, got only 30 percent of the land (Figure 7-44). Nonetheless, Bosnia & Herzegovina is once again a relatively peaceful place (Figure 7-45).

PAUSE & REFLECT 7.4.2

In which regions within Bosnia & Herzegovina did Serbs gain most of their territory?

▼ **FIGURE 7-45 MOSTAR TODAY** Mostar's Stari Most is now a tourist attraction and hosts the annual International Bridge Jumping Competition.

Ethnic Cleansing Elsewhere in the Balkans

LEARNING OUTCOME 7.4.3

Explain the concept of ethnic cleansing in the Balkans.

Ethnic cleansing in former Yugoslavia has not been limited to Bosnia & Herzegovina. It has also been observed in Kosovo and Croatia.

ETHNIC CLEANSING IN KOSOVO

At the time of the breakup of Yugoslavia, 82 percent of the population of Kosovo were ethnic Albanians and 10 percent were Serbs. Nonetheless, Kosovo was controlled by Serbia. Serbia had a historical claim to Kosovo, having controlled it between the twelfth and fourteenth centuries. Serbs fought an important --though losing --battle in Kosovo against the Ottoman Empire in 1389. In recognition of its role in forming the Serb ethnicity, Serbia was given control of Kosovo when Yugoslavia was created in the early twentieth century.

With the breakup of Yugoslavia, Serbia took direct control of Kosovo and launched a campaign of ethnic cleansing against the Albanian majority (Figure 7-46). At its peak in 1999, Serb ethnic cleansing forced more than 800,000 of Kosovo's 2 million ethnic Albanian residents from their homes, mostly to camps in Albania (see Sustainability & Our Environment feature and Figure 7-47). Outraged by the ethnic cleansing, the United States and Western European countries, operating through the North Atlantic Treaty Organization (NATO), launched an air attack against Serbia. The bombing campaign ended when Serbia agreed to withdraw all of its soldiers and police from Kosovo.

Kosovo declared its independence from Serbia in 2008. Around 108 countries, including the United States, recognize Kosovo as an independent country, but Serbia, Russia, China and their allies oppose it. However, the declaration of independence induced nearly 90 percent of the country's Serbs to leave.

▼FIGURE 7-46 ETHNIC CLEANSING IN KOSOVO Kosovars who were ethnically cleansed by Serbs in 1999 lived in tents in camps across the border in Macedonia.

ETHNIC CLEANSING IN CROATIA

With the breakup of Yugoslavia, Croatia declared its independence in 1991. At the time, ethnic Serbs, who constitute 12 percent of Croatia's population rebelled against the new Croatian government, and tried to form a separate country in the east, which they called the Republic of Serbian Krajina.

To strengthen their case for an independent Krajina, Serbs engaged in ethnic cleansing. The Serbs expelled around 170,000 Croats and other non-Serbs from the eastern part of Croatia. After a four-year war that ended with a Croat victory, around 20,000 Serbs were expelled and 180,000 chose to leave Croatia. The International Criminal Tribunal convicted the Serb leaders of war crimes for their ethnic cleansing of Croats in the eastern part of Croatia.

PAUSE & REFLECT 7.4.3

Which two countries carved out of former Yugoslavia have not been mentioned in these two sections on ethnic cleansing? Why might they not have suffered from ethnic cleansing?

BALKANIZATION

The terms Balkanized and Balkanization were once widely used by world leaders as well as geographers:

- **Balkanized** was defined as a small geographic area that could not successfully be organized into stable countries because it was inhabited by many ethnicities with complex, long-standing antagonisms toward each other.
- **Balkanization** was defined as the process by which a state breaks down through conflicts among its ethnicities.

A century ago, world leaders regarded the process of Balkanization of Balkanized countries as a threat to world peace. They were right: Balkanization led directly to World War I because the various nationalities in the Balkans dragged into the war the larger powers with which they had alliances.

After two world wars and the rise and fall of communism during the twentieth century, the Balkans once again became Balkanized into the twenty-first century. Peace has come to the Balkans in the twenty-first century because ethnic cleansing tragically accomplished its goal. Millions of people were rounded up and killed or forced to migrate because they constituted ethnic minorities. Ethnic homogeneity has become the price of peace in areas that once were multiethnic.

SUSTAINABILITY & OUR ENVIRONMENT

Geographic Evidence of Ethnic Cleansing

Early reports of ethnic cleansing by Serbs in the former Yugoslavia were so shocking that many people dismissed them as journalistic exaggeration or partisan propaganda. It took one of geography's most important analytic tools, aerial-photography interpretation, to provide irrefutable evidence of the process, as well as the magnitude, of ethnic cleansing.

According to the society pillar of sustainability, humans need material resources, such as shelter and food, to survive. Ethnic cleansing denies people these basic needs. And a landscape cannot be sustained environmentally and economically if no people are left living there to farm or provide services.

A series of three photographs taken by NATO air reconnaissance over the village of Glodane, in western Kosovo, illustrated the four steps in ethnic cleansing summarized on p. 254. Figure 7-47 is the first of the three photos:

- Illustrating step 1, the red circles in Figure 7-47 show the location of Serb armored vehicles along the main street of the village. Figure 7-47 shows the village's houses and farm buildings clustered on the left side, with fields on the outskirts of the village, including the center and right portions of the photograph. As discussed in Chapter 12, rural settlements in most of the world have houses and farm buildings clustered together and surrounded by fields rather than in isolated, individual farms typical of North America.
- Illustrating step 2, the farm field immediately to the east of the main north–south road is filled with the villagers. At the scale that the photograph is reproduced in this book, the people appear as a dark mass inside the blue lines labeled "internally displaced persons." The white rectangles to the north of the people are civilian cars and trucks.
- Illustrating step 3, the second photograph of the sequence (not included here) showed the same location a short time later, with one major change: The people and vehicles massed in the field in the first photograph are gone—no people and no vehicles.
- Illustrating step 4, the third photograph (not included here) showed that the buildings in the village had been set on fire.

Aerial photographs such as these not only "proved" that ethnic cleansing was occurring but also provided critical evidence to prosecute Serb leaders for war crimes.

▲ **FIGURE 7-47 EVIDENCE OF ETHNIC CLEANSING IN KOSOVO** Ethnic cleansing by Serbs forced Albanians living in Kosovo to flee in 1999. The village of Glodane is on the west side of the road. The villagers and their vehicles have been rounded up and placed in the field east of the road. The villagers are the dark mass inside the blue area marked internally displaced persons. The red circles show the locations of Serb armored vehicles.

Ethnic Cleansing and Genocide in Africa

LEARNING OUTCOME 7.4.4

Identify the principal recent episodes of genocide in Africa.

Competition among ethnicities can lead in a handful of the most extreme cases to **genocide**, which is the mass killing of a group of people in an attempt to eliminate the entire group from existence. Several areas of Africa have been plagued by conflicts among ethnicities that have resulted in genocide in recent years. Other countries have been either unable or unwilling to stop the genocide.

ETHNICITIES AND NATIONALITIES IN AFRICA

Traditionally, the most important element of cultural identity in Africa was ethnicity rather than nationality. Africa contains several thousand ethnicities with distinct languages, religions, and social customs. The precise number of ethnicities is impossible to determine because boundaries separating them can be hard to define. Further, it is hard to determine whether a particular group forms a distinct ethnicity or is part of a larger collection of similar groups.

During the late nineteenth and early twentieth centuries, European countries carved up the continent into a collection of colonies, with little regard for the distribution of these ethnicities. When the European colonies in Africa became independent states, especially during the 1950s and 1960s, the areas of the new states typically matched the colonial administrative units imposed by the Europeans rather than the historical distribution of ethnicities (Figure 7-48). As a result, most states contained a large collection of often dissimilar ethnicities, and some ethnic groups were divided among more than one state. Conflict among ethnicities is widespread in Africa, largely because the historical distribution of ethnicities bears little relationship to present-day nationalities.

PAUSE & REFLECT 7.4.4

Why might the areas of individual ethnicities be much larger in northern Africa than in the rest of the continent? Refer to Figure 1-47 on page 34 and Figure 2-4 on p. 48.

◀ **FIGURE 7-48 AFRICA'S MANY ETHNICITIES** The territory occupied by ethnic groups rarely matches the boundaries of states.

ETHNIC CLEANSING AND GENOCIDE IN SUDAN

Several civil wars have raged in Sudan since 1983, resulting in genocide and ethnic cleansing. Sudan's two conflicts that have generated the most victims are with Darfur and South Sudan.

Ethnic diversity lies at the heart of Sudan's conflicts. Sudan is around 70 percent Arab and 97 percent Muslim. The remainder belong to a large number of other ethnicities descended from groups living in Sudan prior to the arrival of Arabs in the twelfth century. The non-Arab ethnicities tend to live in the west, south, and east of Sudan.

DARFUR. Resenting discrimination and neglect by the Arab-dominated national government, Darfur's black African ethnicities launched a rebellion in 2003. Marauding Arab nomads, known as Janjaweed, with the support of Sudan's government, crushed Darfur's black population, made up mainly of settled farmers.

An estimated 450,000 people in Darfur have been victims of genocide and another 2.5 million victims of ethnic cleansing. Most of the ethnic cleansing victims live in dire conditions in refugee camps in the harsh desert environment of Darfur (Figure 7-49).

SOUTH SUDAN. A war from 1983 until 2005 between Sudan's northern and southern ethnicities resulted in the death of an estimated 1.9 million Sudanese and the ethnic cleansing of an estimated 700,000. The war ended with the establishment of South Sudan as an independent state in 2011 (Figure 7-50). In contrast to the predominantly Arab Muslim northerners, South Sudan's two largest ethnicities are the predominantly Christian Dinka and the predominantly folk religionist Nuer. The north–south war was sparked by southern ethnicities attempting to resist northerners' attempts to impose a legal system based on Muslim religious practice. Independence from Arab Muslim northerners has not brought peace to southerners, however. South Sudan's diverse ethnicities have not been able to work together to create a stable government.

ABEYI. With the independence of South Sudan in 2011, conflict moved to the areas of Sudan along the new international border with South Sudan. Ethnicities aligned with those in the new country of South Sudan fought with supporters of the government of Sudan. The status of Abeyi, a small border area inhabited by ethnicities aligned with both Sudan and South Sudan, was to be settled by a referendum of the people living there, but the vote was postponed. Until their status is settled, the people of Abeyi are considered citizens of both Sudan and South Sudan. A peacekeeping force from Ethiopia is preventing either Sudan or South Sudan from seizing control of Abeyi.

▲ **FIGURE 7-50 SUDAN AND SOUTH SUDAN**

SOUTH KORDOFAN AND BLUE NILE. Two other border areas, South Kordofan and Blue Nile, also contain large numbers of ethnicities sympathetic to both Sudan and South Sudan. As in Abeyi, a referendum intended to decide whether to place these territories in Sudan or South Sudan was canceled, leaving the future status unsettled.

EASTERN FRONT. Ethnicities in the east have fought Sudanese government forces, with the support of neighboring Eritrea. At issue has been disbursement of profits from oil.

▼ **FIGURE 7-49 DARFUR** Open-air classroom in a refugee camp in Darfur, Sudan. Children lack paper so are learning to write on pieces of wood.

Ethnic Cleansing and Genocide in Central Africa

LEARNING OUTCOME 7.4.5

Identify the principal recent episodes of genocide in Central Africa.

Rwanda and Burundi, tiny countries in central Africa, have suffered from especially severe genocide. Long-standing conflicts between the two countries' two ethnic groups, the Hutus and Tutsis, lie at the heart of ethnic cleansing and genocide in central Africa. The two ethnicities speak the same language, hold similar beliefs, and practice similar social customs, and intermarriage has lessened the physical differences between the two ethnic groups. Yet Hutus and Tutsis have engaged in large-scale ethnic cleansing and genocide:

- Hutus were settled farmers, growing crops in the fertile hills and valleys of present-day Rwanda and Burundi, known as the Great Lakes region of central Africa.
- Tutsis were cattle herders who migrated to present-day Rwanda and Burundi from the Rift Valley of western Kenya beginning 400 years ago.

Relations between settled farmers and herders are often uneasy; this is also an element of the ethnic cleansing in Darfur described above.

Hutus constituted a majority of the population of Rwanda and Burundi historically, but Tutsis controlled kingdoms there for several hundred years and turned the Hutus into their serfs. They became colonies of Germany during the 1880s and then of Belgium between 1924 and 1962. During the colonial period, the Tutsis retained leadership positions.

PAUSE & REFLECT 7.4.5

Why might the European colonial powers have preferred to place in leadership positions members of the minority Tutsis rather than members of the majority Hutus?

When Rwanda became an independent country in 1962, Hutus gained power and undertook ethnic cleansing and genocide against the Tutsis. Descendants of the ethnically cleansed Tutsis invaded Rwanda in 1990, launching a three-year civil war. Meanwhile in Burundi, where the Tutsis has remained in power, a civil war resulted in genocide committed by and against both Hutus and Tutsis.

An agreement to share power in Rwanda was signed in 1993, but genocide resumed after an airplane carrying the presidents of Rwanda and neighboring Burundi—both Hutus—was shot down by a surface-to-air missile in 1994. International intelligence groups and independent researchers have never been able to determine whether the attacker was a Hutu or a Tutsi presumably trying to scuttle the peace agreement. This attack followed an assassination a few months earlier of the previous president of Burundi, the first Hutu to be elected president of that country.

After the assassination of the two presidents in 1994, Hutus launched a genocide campaign, killing an estimated 800,000 Tutsis in Rwanda and 300,000 in Burundi. However, the Tutsis prevailed in both countries, and reprisals by Tutsis added to the total fatalities. Rwanda continues to be governed by Tutsis, but Burundi has been led since 2005 by democratically elected Hutus.

The conflict between Hutus and Tutsis spilled into the Democratic Republic of Congo. The Congo is the region's largest and most populous country, with considerable mineral wealth. It is also one of the most multiethnic countries, estimated to be home to more than 200 distinct ethnicities. Most Congolese are classified as ethnic Bantus, but this encompasses a large number of specific ethnic groups (Figure 7-51).

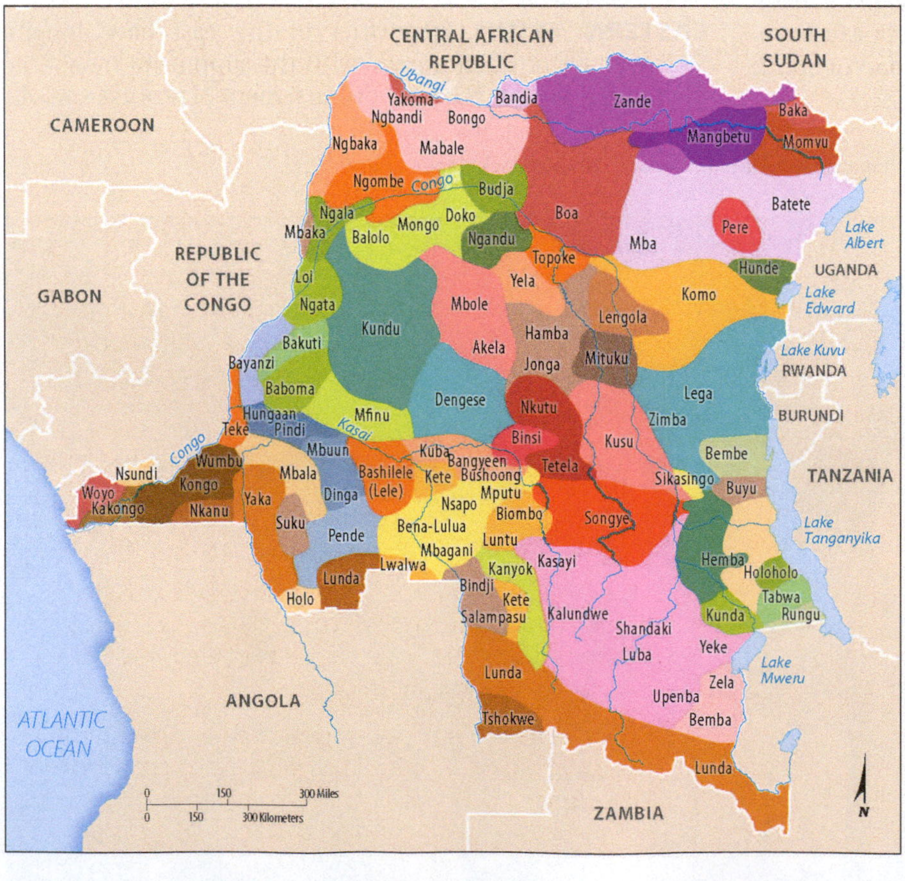

◀ **FIGURE 7-51 ETHNICITIES IN THE DEMOCRATIC REPUBLIC OF CONGO** The distribution of the very large number of ethnicities in the Congo is not easy to place on a map.

DEBATE IT! Should the United States intervene in ethnic conflicts?

U.S. citizens and government officials have debated whether the United States should send its military into places where ethnic cleansing and genocide are occurring.

THE UNITED STATES SHOULD INTERVENE

- The United States should not stand aside and do nothing when innocent women and children are being killed (Figure 7-52).
- If the United States doesn't stop genocide now, others will be emboldened to do it.
- Democratic values compel the United States to help people in trouble.

▲ FIGURE 7-52 **U.S. TROOPS DELIVER HUMANITARIAN AID**

THE UNITED STATES SHOULD NOT INTERVENE

- The United States can't be the world's police force and must leave ethnic groups to solve their problems themselves.
- The United States should intervene only if its national interests are directly threatened.
- Intervening gets the United States entangled in complex disputes that result in hurting the U.S. itself (Figure 7-53).

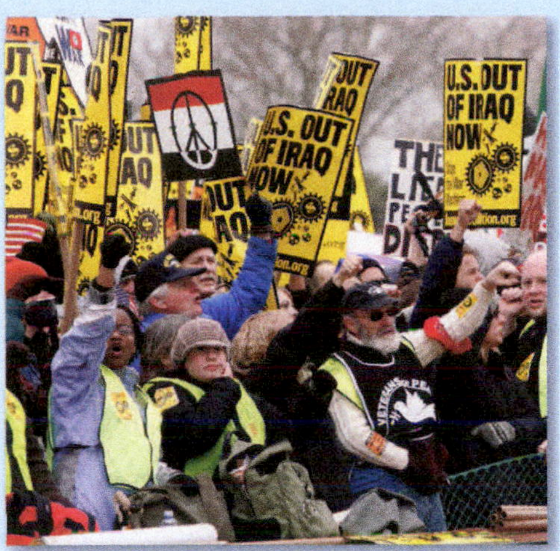

▲ FIGURE 7-53 **OPPONENTS OF U.S. INVOLVEMENT IN IRAQ**

The Congo is considered to have suffered from the world's deadliest wars in the past 70 years. More than 5 million have died in the Congo's ongoing civil wars, mostly from malaria, diarrhea, pneumonia, and malnutrition, aggravated by displacement and unsanitary and overcrowded living conditions. Tutsis were instrumental in the successful overthrow of the Congo's longtime president Joseph Mobutu in 1997. Mobutu had amassed a several-billion-dollar personal fortune from the sale of minerals while impoverishing much of the country. After succeeding Mobutu as president, Laurent Kabila relied heavily on Tutsis and permitted them to kill some of the Hutus who had been responsible for atrocities against Tutsis in the early 1990s. But Kabila soon split with the Tutsis, and the Tutsis once again found themselves offering support to rebels seeking to overthrow Congo's government.

Kabila turned for support to Hutus and other ethnic groups that also hated Tutsis. Armies from Angola, Namibia, Zimbabwe, and other neighboring countries came to Kabila's aid. Kabila was assassinated in 2001 and succeeded by his son, who negotiated an accord with rebels the following year. Despite the accord, conflict among the country's many ethnicities has continued, and casualties have mounted.

CHECK-IN KEY ISSUE 4

Why Do Ethnic Cleansing and Genocide Occur?

- ✓ **Ethnic cleansing is a process in which a more powerful ethnic group forcibly removes a less powerful one in order to create an ethnically homogeneous region.**
- ✓ **Genocide is the mass killing of a group of people in an attempt to eliminate the entire group from existence.**

Summary & Review

KEY ISSUE 1

Where are ethnicities distributed?

Ethnicity is identity with a group of people who share the cultural traditions of a particular homeland or hearth. Ethnicity refers to a person's place of cultural heritage, whereas race refers to biological traits such as skin color. The three most numerous ethnic groups in the United States are Hispanics, African Americans, and Asian Americans.

THINKING GEOGRAPHICALLY

▲ **FIGURE 7-54 MEXICAN FOOD STAND IN LISBON, PORTUGAL**

1. What are examples of ethnic foods that are now regularly consumed by people of other ethnicities?

KEY ISSUE 2

Why do ethnicities have distinctive distributions?

Ethnicities cluster as a result of distinctive patterns of migration and segregation. For example, many African Americans trace their ancestry to forced migration from Africa for slavery. African Americans migrated in large numbers from the South to the North and West in the early twentieth century and clustered in inner-city ghettos that have expanded.

THINKING GEOGRAPHICALLY

▲ **FIGURE 7-55 BLACK-OWNED SOUTH AFRICAN WINERY** M'hudi Wines is one of the few black-owned wineries in South Africa.

2. South Africa is one of the world's leading producers of wine. Despite the end of apartheid, very few wineries and vineyards are owned by blacks. Why might this be the case?

KEY ISSUE 3

Why might ethnicities face conflicts?

Conflicts can arise when a country contains several ethnicities competing with each other for control or dominance. Conflicts also arise when an ethnicity is divided among more than one country. Nationality is identity with a group of people who share legal attachment and personal allegiance to a particular country. Some ethnicities find themselves divided among more than one nationality.

THINKING GEOGRAPHICALLY

▲ **FIGURE 7-56 BALUCHISTAN** The city of Quetta, in northwest Pakistan, would be the largest city in Baluchistan.

3. The Kurds claim to be the world's largest ethnic group not in control of a country. Based on Figure 7-37, what other ethnic groups might have a strong claim to reorganize territorial boundaries so that they could become the majority?

KEY ISSUE 4

Why do ethnic cleansing and genocide occur?

Ethnic cleansing is a process in which a more powerful ethnic group forcibly removes a less powerful one in order to create an ethnically homogeneous region.Genocide is the mass killing of a group of people in an attempt to eliminate the entire group from existence.

THINKING GEOGRAPHICALLY

▲ **FIGURE 7-57 STARI MOST BRIDGE DIVE** An annual diving competition is held on the bridge.

4. The Mostar bridge, pictured in Figures 7-43 and 7-45, was rebuilt to look almost precisely like the original bridge. Why do you think it was important to Bosniaks to replicate the original bridge?

5. Professional divers jump off the bridge as part of competitions. Why might the bridge be an important place to hold professional diving competitions?

KEY TERMS

Apartheid *(p. 245)* Laws (no longer in effect) in South Africa that physically separated different races into different geographic areas.

Balkanization *(p. 256)* A process by which a state breaks down through conflicts among its ethnicities.

Balkanized *(p. 256)* A small geographic area that cannot successfully be organized into stable countries because it is inhabited by many ethnicities with complex, long-standing antagonisms toward each other.

Blockbusting *(p. 243)* A process by which real estate agents convince white property owners to sell their houses at low prices because of fear that persons of color will soon move into the neighborhood.

Centripetal force *(p. 247)* An attitude that tends to unify people and enhance support for a state.

Ethnic cleansing *(p. 252)* A purposeful policy designed by one ethnic or religious group to remove by violent and terror-inspiring means the civilian population of another ethnic or religious group from certain geographic areas.

Ethnic enclave *(p. 236)* A place with a high concentration of an ethnic group that is distinct from those in the surrounding area.

Ethnicity *(p. 230)* Identity with a group of people who share the cultural traditions of a particular homeland or hearth.

Ethnoburb *(p. 237)* A suburban area with a cluster of a particular ethnic population.

Genocide *(p. 258)* The mass killing of a group of people in an attempt to eliminate the entire group from existence.

Nationalism *(p. 247)* Loyalty and devotion to a particular nationality.

Nationality *(p. 230)* Identity with a group of people who share legal attachment to a particular country.

Race *(p. 230)* Identity with a group of people who are perceived to share a physiological trait, such as skin color.

Racism *(p. 232)* The belief that race is the primary determinant of human traits and capacities and that racial differences produce an inherent superiority of a particular race.

Racist *(p. 232)* A person who subscribes to the beliefs of racism.

Sharecropper *(p. 242)* A person who works fields rented from a landowner and pays the rent and repays loans by turning over to the landowner a share of the crops.

Triangular slave trade *(p. 240)* A practice, primarily during the eighteenth century, in which European ships transported slaves from Africa to Caribbean islands, molasses from the Caribbean to Europe, and trade goods from Europe to Africa.

GeoVideo

Log in to the **MasteringGeography** Study Area to view this video.

Battle Over History

This video examines the ongoing controversy over the mass killings of Armenians that occurred 100 years ago, during World War I in the Ottoman Empire, part of which became modern Turkey.

1. What was the relationship between Turks and Armenians in the Ottoman Empire at the time of World War I? Explain.
2. What evidence supports claims that the killing of Armenians during 1915 constituted genocide? What position does the government of Turkey take with regard to these claims?
3. Why have efforts to obtain official U.S. government recognition of the Armenian genocide been controversial?

EXPLORE

Multiethnic Sarajevo

Use Google Earth to see evidence of ethnic diversity in the heart of Sarajevo, capital of Bosnia.

Fly to *Gazi mosque, Sarajevo, Bosnia*.

Click *3D buildings.*

Under *Primary Database*, click *Places*.

1. What other religions are represented by religious symbols in the vicinity of the Gazi mosque?

MasteringGeography™

Looking for additional review and test prep materials?

Visit the Study Area in **MasteringGeography™** to enhance your geographic literacy, spatial reasoning skills, and understanding of this chapter's content by accessing a variety of resources, including MapMaster interactive maps, videos, *In the News* RSS feeds, flashcards, web links, self-study quizzes, and an eText version of *The Cultural Landscape*.

www.masteringgeography.com

8

Political Geography

How many states of the world can you name? Old-style geography sometimes required memorization of countries and their capitals. Human geographers now emphasize a thematic approach. We are concerned with the locations of activities in the world, the reasons for particular spatial distributions, and the significance of the arrangements. Despite this change in emphasis, you still need to know the locations of states. Without such knowledge, you lack a basic frame of reference—knowing where things are.

Waving the flag of Thailand at a political rally in Bangkok.

LOCATIONS IN THIS CHAPTER

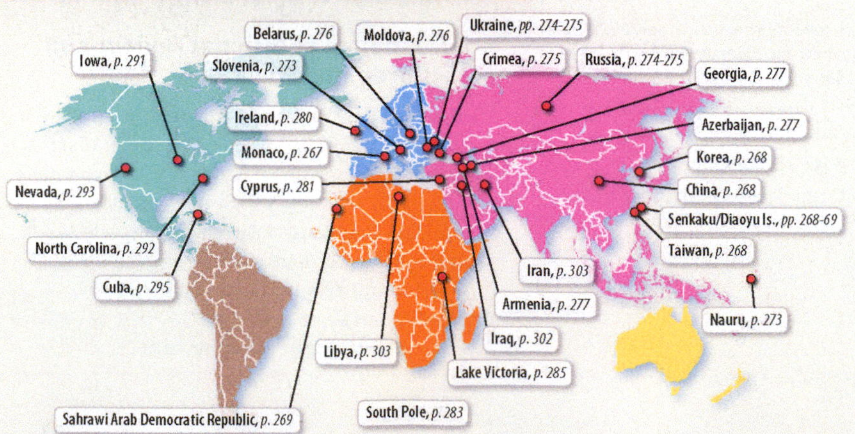

1

Where Are States Distributed?

The *space* of Earth's land area is distributed into states, although what constitutes a state is not always clear-cut.

2 Why Are Nation-States Difficult To Create?

Local diversity has increased in political affairs, as individual cultural groups demand more control over the ***regions*** they inhabit.

3 Why Do Boundaries Cause Problems?

Boundary lines are not painted on the ground, but these specific ***places*** on Earth might as well be. Boundaries are the specific places where states have ***connections*** with each other.

4 Where Do States Face Threats?

Conflicts during the twentieth century were dominated by global ***scale*** war, including two world wars. Into the twenty-first century, many attacks are initiated not by warring states but by terrorist organizations.

KEY ISSUE 1

Where Are States Distributed?

- ▸ **Introducing Political Geography**
- ▸ **Challenges in Defining States**

LEARNING OUTCOME 8.1.1

Understand the difference between a state of the world and a state within the United States.

When looking at satellite images of Earth, we easily distinguish landmasses and water bodies. What we cannot see are where boundaries are located between countries. To many, national boundaries are more meaningful than natural features. One of Earth's most fundamental cultural characteristics—one that we take for granted—is the division of our planet's surface into a collection of spaces occupied by individual countries.

Introducing Political Geography

A **state** is an area organized into a political unit and ruled by an established government that has control over its internal and foreign affairs. It occupies a defined territory on Earth's surface and contains a permanent population. *Country* is a synonym for *state*. The term *state*, as used in political geography, does not refer to the 50 regional governments inside the United States. The 50 states of the United States are subdivisions within a single state—the United States of America.

A map of the world shows that virtually all habitable land is organized into states. But for most of history, until recently, this was not so. As recently as the 1940s, the world contained only about 50 countries, compared to approximately 200 today (Figure 8-1).

The land area occupied by the states of the world varies considerably. The largest state is Russia, which encompasses 17.1 million square kilometers (6.6 million square miles), or 11 percent of the world's entire land area. Other states with more than 5 million square kilometers (2 million square miles) include Canada, the United States, China, Brazil, and Australia.

At the other extreme are about two dozen **microstates**, which are states with very small land areas. If Russia were the size of this page, a microstate would be the size of a single letter on it. The Vatican is the world's smallest microstate, at 0.44 square kilometers (0.17 square miles). The second-smallest microstate and the smallest that is a member of the United Nations is Monaco (Figure 8-2), which is only 1.5 square kilometers (0.6 square miles).

Other U.N. member states that are smaller than 1,000 square kilometers (400 square miles) include Andorra, Antigua and Barbuda, Bahrain, Barbados, Dominica, Grenada,

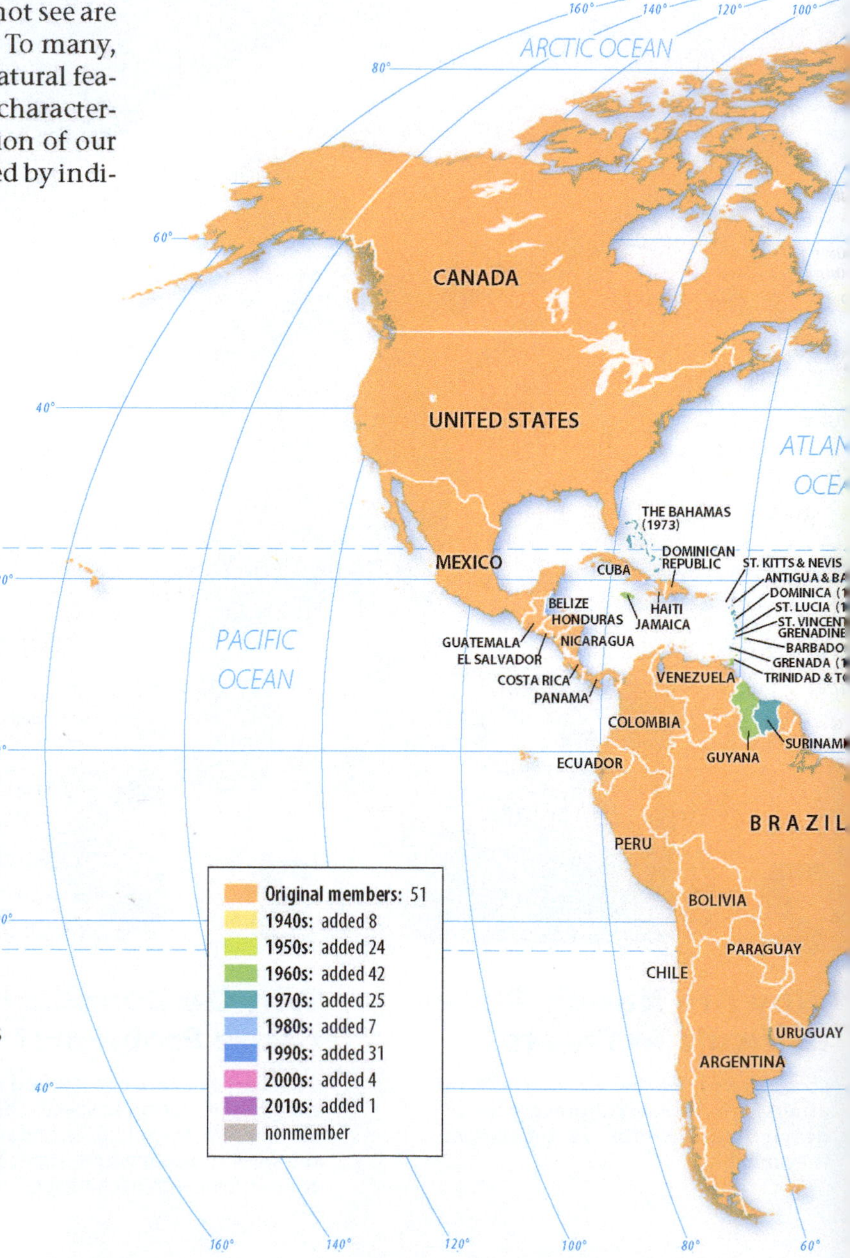

▸ **FIGURE 8-1 STATES OF THE WORLD** All but a handful of states are members of the United Nations.

Kiribati, Liechtenstein, Maldives, Malta, Micronesia, Palau, St. Kitts & Nevis, St. Lucia, St. Vincent & the Grenadines, San Marino, São Tomé e Príncipe, the Seychelles, Singapore, Tonga, and Tuvalu. Many of the microstates are islands, which explains both their small size and sovereignty.

PAUSE & REFLECT 8.1.1

With virtually all of Earth's land now allocated to states, how might the number of states increase in the future?

▲ **FIGURE 8-2 MONACO: WORLD'S SMALLEST U.N. MICROSTATE** The smallest state in the United Nations is the Principality of Monaco.

ARCTIC OCEAN
RUSSIA
NORWAY
FINLAND
SWEDEN
ESTONIA
LATVIA
LITH.
UNITED KINGDOM
DEN.
NETH.
BELG.
LUX.
GER.
POLAND
BELARUS
UKRAINE
LIECH. (1990)
CZH.
SLVK.
AUS.
HUN.
MOL.
SWITZ.
SLOV.
CRO.
ROM.
SERB.
BOS. HER.
KOS.
BUL.
MAC.
FRANCE
ITALY
MON. (1993)
SAN MAR. (1992)
MONT.
ALB.
GREECE
SPAIN
TURKEY
GEORGIA
ARMENIA
AZERBAIJAN
KAZAKHSTAN
UZBEKISTAN
KYRGYZSTAN
TURKMENISTAN
TAJIKISTAN
MONGOLIA
NORTH KOREA
SOUTH KOREA
JAPAN
CHINA
TUNISIA
MALTA (1964)
CYPRUS
LEBANON
ISRAEL
SYRIA
IRAQ
IRAN
AFGHANISTAN
JORDAN
KUWAIT
BAHRAIN (1971)
QATAR
U.A.E.
PAKISTAN
NEPAL
BHUTAN
BANGLADESH
INDIA
ALGERIA
LIBYA
EGYPT
SAUDI ARABIA
OMAN
Taiwan
MYANMAR (BURMA)
LAOS
THAILAND
CAMBODIA
VIETNAM
PHILIPPINES
MALI
NIGER
CHAD
SUDAN
ERITREA
YEMEN
DJIBOUTI
BURKINA FASO
BENIN
NIGERIA
COTE D'IVOIRE
GHANA
TOGO
CAMEROON
CENT. AF. REP.
SOUTH SUDAN
ETHIOPIA
SOMALIA
MALDIVES (1965)
SRI LANKA
BRUNEI
MALAYSIA
PALAU (1994)
PACIFIC OCEAN
MARSHALL IS. (1991)
FED. STATES OF MICRONESIA (1991)
RWANDA
UGANDA
KENYA
GABON
EQUATORIAL GUINEA
REP. OF CONGO
DEM. REP. OF CONGO
BURUNDI
TANZANIA
ANGOLA (CABINDA)
SEYCHELLES (1976)
INDIAN OCEAN
SINGAPORE (1965)
INDONESIA
PAPUA NEW GUINEA
NAURU (1999)
KIRIBATI (1999)
SOLOMON IS. (1978)
TUVALU (2000)
SAMOA (1976)
TIMOR-LESTE
COMOROS (1975)
ANGOLA
ZAMBIA
MALAWI
MADAGASCAR
NAMIBIA
ZIMBABWE
BOTSWANA
MOZAMBIQUE
MAURITIUS (1968)
VANUATU (1981)
FIJI (1970)
TONGA (1999)
AUSTRALIA
SWAZILAND
SOUTH AFRICA
LESOTHO
NEW ZEALAND
Tropic of Cancer
Equator
Tropic of Capricorn
0 1,000 2,000 Miles
0 1,000 2,000 Kilometers

Challenges in Defining States

LEARNING OUTCOME 8.1.2

Explain why it is difficult to determine whether some territories are sovereign states.

A state has **sovereignty**, which means independence from control of its internal affairs by other states. Because the entire area of a state is managed by its national government, laws, army, and leaders, it is a good example of a formal or uniform region.

There is some disagreement about the number of sovereign states. This disagreement is closely tied to the history and geography of the places involved and most often involves neighboring states. Among places that test the definition of *sovereignty* are Korea, China, and Western Sahara (Sahrawi Republic).

KOREA: ONE STATE OR TWO?

The Korean peninsula is divided between the Democratic People's Republic of Korea (North) and the Republic of Korea (South). A colony of Japan for 35 years, Korea was divided into two occupation zones by the United States and the former Soviet Union after they defeated Japan in World War II. The Soviet Union installed a pro-Communist government in the North, while a pro-U.S. government was established in the South. North Korea invaded the South in 1950, sparking a three-year war that ended in a cease-fire. Both governments are committed to reuniting the country into one sovereign state. However, both maintain that they are the one that should exercise sovereignty over the entire Korean peninsula.

North Korea is one of the world's poorest and most isolated countries, and since 1948 it has been governed as a dictatorship by Kim Il-sung, his son Kim Jong-il, and his grandson Kim Jong-un. Further aggravating reconciliation, North Korea has built and tested nuclear weapons, even though the country lacks the ability to provide its citizens with food, electricity, and other basic needs (Figure 8-3).

▲ **FIGURE 8-3 KOREA** A nighttime satellite image recorded by the U.S. Air Force Defense Meteorological Satellite Program shows the illumination of electric lights in South Korea, whereas North Korea has virtually no electric lights. 1. Why might North Korea have virtually no electric lights? 2. Is the absence of electric lights a measure of wealth or poverty? 3. In what ways would the absence of electric lights hinder promotion of democracy in North Korea?

CHINA AND TAIWAN: ONE STATE OR TWO?

Most other countries consider China (officially the People's Republic of China) and Taiwan (officially the Republic of China) as separate and sovereign states. According to China's government, Taiwan is not sovereign but a part of China. The government of Taiwan agrees.

The current status arises from a civil war in China during the late 1940s between the Nationalists and the Communists. After losing in 1949, Nationalist leaders fled to Taiwan, 200 kilometers (125 miles) off the Chinese coast, and proclaimed that they were still the legitimate rulers of the entire country of China. Until some future occasion when they could defeat the Communists and recapture all of China, the Nationalists argued, at least they could continue to govern one island of the country.

The United States had supported the Nationalists during the civil war, so many Americans opposed acknowledging that China was firmly under the control of the Communists. Consequently, the United States continued to regard the Nationalists as the official government of China until the 1970s, when U.S. policy finally changed, and the United Nations voted to transfer China's seat from the Nationalists to the Communists.

SENKAKU / DIAOYU ISLANDS: WHO IS SOVEREIGN?

The People's Republic of China, Taiwan, and Japan all claim sovereignty over several small uninhabited islands in the East China Sea. These islands are known as Diaoyu in China, Diaoyutai in Taiwan, and Senkaku in Japan (Figure 8-4).

▲ **FIGURE 8-4 DISPUTED ISLANDS** The Senkaku / Diaoyu islands are all claimed by China, Taiwan, and Japan.

◀ **FIGURE 8-5 SENKAKU / DIAOYU ISLANDS** The largest of the disputed Senkaku / Diaoyu islands, called Uotsuri-shima in Japanese and Diàoyú Dǎo in Chinese, is only 4.3 square kilometers (1.7 square miles).

The largest of five islands is only 4.32 square kilometers (1.7 square miles) (Figure 8-5). The collection also includes three rock outcroppings, the smallest of which is only 800 square meters (8,600 square feet).

Japan has controlled the islands since 1895, except between 1945 and 1972, when the United States administered them after defeating Japan in World War II. China and Taiwan claim that the islands historically belonged to China until the Japanese government illegally seized them in 1895. Japan's position is that China did not state that it had sovereignty over the uninhabited islands back in 1895, when Japan claimed them. To bolster their claims, China and Japan have both established air zones in the East China Sea with conflicting boundaries.

SAHRAWI REPUBLIC / WESTERN SAHARA: WHO IS SOVEREIGN?

The Sahrawi Arab Democratic Republic, also known as Western Sahara, is considered by most African countries as a sovereign state. Morocco, however, claims the territory and to prove it has built a 2,700-kilometer (1,700-mile) wall around the territory to keep out rebels.

Spain controlled the territory on the continent's west coast between Morocco and Mauritania until withdrawing in 1976. An independent Sahrawi Republic was declared by the Polisario Front and recognized by most African countries, but Morocco and Mauritania annexed the northern and southern portions, respectively. Three years later Mauritania withdrew, and Morocco claimed the entire territory.

Morocco controls most of the populated area, but the Polisario Front operates in the vast, sparsely inhabited deserts, especially the one-fifth of the territory that lies east of Morocco's wall (Figure 8-6). The United Nations has tried but failed to reach a resolution among the parties.

PAUSE & REFLECT 8.1.2

Other than military action, how might the sovereignty of these disputed territories be settled?

CHECK-IN KEY ISSUE 1

Where Are States Distributed?

- ✔ **The world is divided into approximately 200 sovereign states that vary considerably in size.**
- ✔ **The sovereignty of some territories is disputed among states.**

▼ **FIGURE 8-6 SAHRAWI REPUBLIC / WESTERN SAHARA (a)** Morocco built sand walls during the 1980s to isolate Polisario Front rebels fighting for independence. **(b)** Morocco controls the western portion of Western Sahara, and the Polisario Front controls the eastern portion.

(b)

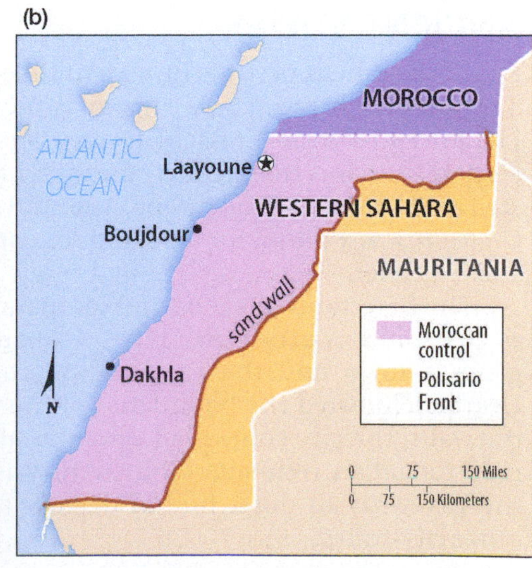

KEY ISSUE 2

Why Are Nation-States Difficult to Create?

- Development of States
- Nation-States and Multinational States
- Russia: The Largest Multiethnic State
- Nation-States in the Former Soviet Union
- Colonies

LEARNING OUTCOME 8.2.1

Understand the development of nation-states.

A **nation-state** is a state whose territory corresponds to that occupied by a particular ethnicity. To preserve and enhance distinctive cultural characteristics, ethnicities seek to govern themselves without interference. The concept of dividing the world into a collection of independent nation-states is recent.

Development of States

The first states emerged in ancient times in Southwest Asia & North Africa. However, most of Earth's surface historically was either unorganized territory or organized in ways other than states, such as empires, kingdoms, and estates controlled by a hereditary class of nobles. In modern times, the concept of a state developed first in Europe.

ANCIENT STATES

The development of states in ancient times can be traced to a region of Southwest Asia known as the Fertile Crescent. The ancient Fertile Crescent formed an arc between the Persian Gulf and the Mediterranean Sea (Figure 8-7). Situated at the crossroads of Europe, Asia, and Africa, the Fertile Crescent was a center for land and sea communications in ancient times.

The first states to evolve in Mesopotamia were known as city-states. A **city-state** is a sovereign state that comprises a town and the surrounding countryside. Walls clearly delineated the boundaries of the city, and outside the walls, the city controlled agricultural land to produce food for urban residents. The countryside also provided the city with an outer line of defense against attack by other city-states.

▲ **FIGURE 8-7 FERTILE CRESCENT** The crescent-shaped area of relatively fertile land was organized into a succession of empires starting several thousand years ago.

Periodically, one city or tribe in Mesopotamia would gain military dominance over the others and form an empire. Mesopotamia was organized into a succession of empires by the Sumerians, Assyrians, Babylonians, and Persians.

The eastern end, Mesopotamia, was centered in the valley formed by the Tigris and Euphrates rivers, in present-day Iraq. The Fertile Crescent then curved westward over the desert and turned southward to encompass the Mediterranean coast through present-day Syria, Lebanon, and Israel. The Nile River valley of Egypt is sometimes regarded as an extension of the Fertile Crescent into North Africa.

PAUSE & REFLECT 8.2.1

What is the importance of the Fertile Crescent in the development of religions, as discussed in Chapter 6? How might the development of ancient states and religions in the region be related?

MEDIEVAL STATES

Political unity in the ancient world reached its height with the establishment of the Roman Empire, which controlled most of Europe and Southwest Asia & North Africa, from modern-day Spain to Iran and from Egypt to England (Figure 8-8). At its maximum extent, the empire comprised 38 provinces, each using the same set of laws that had been created in Rome. Massive walls helped the Roman army defend many of the empire's frontiers.

The Roman Empire collapsed in the fifth century, after a series of attacks by people living on its frontiers and because of internal disputes. The European portion of the Roman Empire was fragmented into a large number of

▲ **FIGURE 8-8 ROMAN EMPIRE A.D. 100** At its height, the Roman Empire controlled much of Europe and Southwest Asia & North Africa.

estates owned by competing monarchs, dukes, barons, and other nobles. A handful of powerful monarchs and emperors emerged as rulers over large numbers of these European estates beginning about A.D. 1100 (Figure 8-9). The consolidation of neighboring estates under the unified control of a monarch or an emperor formed the basis for the development of such modern European states as England, France, and Spain (Figure 8-10).

STATES IN TWENTIETH-CENTURY EUROPE

Into the twentieth century, most of Europe's territory was ruled by a handful of emperors, kings, and queens. After World War I, which engulfed nearly all of Europe, leaders of the victorious countries met at the Versailles Peace Conference to redraw the map of Europe. One of the chief advisers to President Woodrow Wilson, the geographer Isaiah Bowman, played a major role in the decisions.

▲ **FIGURE 8-9 EUROPE, 1300** Much of Europe was fragmented into small estates controlled by nobles.

The goal of the Allied leaders was to divide Europe into a collection of nation-states, using language as the principal criterion for identifying ethnic groups. New states were created and the boundaries of existing states were adjusted to conform as closely as possible to the territory occupied by speakers of different languages (Figure 8-11). This undertaking created some clear-cut examples of nation-states, but many of the states created a century ago in Europe have not survived as nation-states.

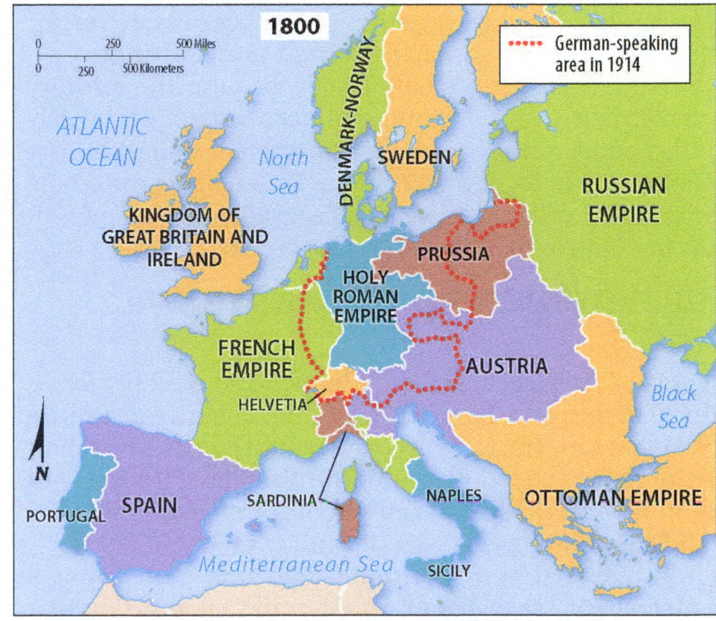

▲ **FIGURE 8-10 EUROPE, 1800** Much of Europe was organized into empires.

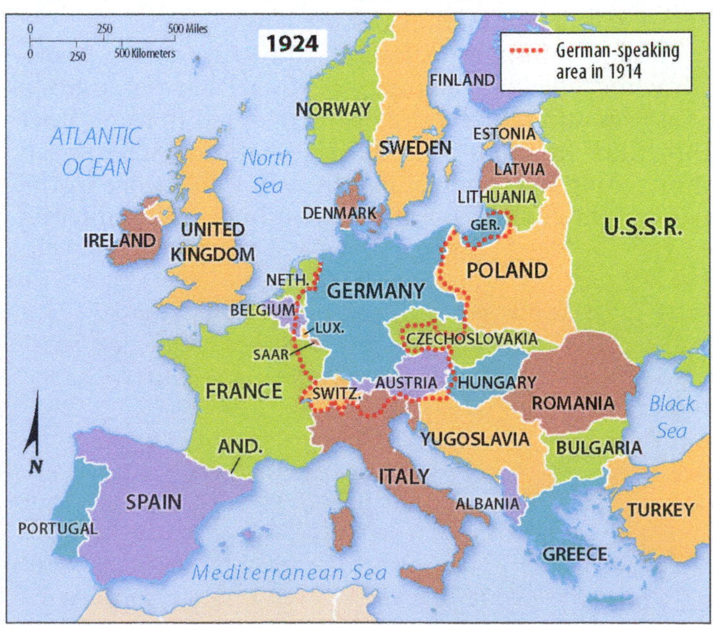

▲ **FIGURE 8-11 EUROPE, 1924** Much of Europe was organized into nation-states.

Nation-States and Multinational States

LEARNING OUTCOME 8.2.2

Understand differences between a nation-state and a multinational state.

To preserve and enhance distinctive cultural characteristics, ethnicities seek to govern themselves without interference. The concept that ethnicities have the right to govern themselves is known as **self-determination**. Ethnic groups have pushed to create nation-states because desire for self-determination is a very important shared attitude.

There is no perfect nation-state because the territory occupied by a particular ethnicity never corresponds precisely to the boundaries of countries. Nonetheless, some states are excellent examples of nation-states. For example, the ethnic composition of Japan is 98.5 percent Japanese, 0.5 percent Korean, 0.4 percent Chinese, and 0.6 percent other.

MULTIETHNIC AND MULTINATIONAL STATES

Figure 8-12 depicts an attempt by political scientist James Fearon to measure the extent of ethnic diversity in a country. States with the least diversity would be the best examples of nation-states. Eleven of the 18 states with the least ethnic diversity are in Europe. On the other hand, 18 of the world's 20 most diverse states are in Africa.

A state that contains more than one ethnicity is a **multiethnic state**. Because no state has a population that is 100 percent a single ethnicity, every state in the world is to a varying degree multiethnic. A **multinational state** is a state that contains more than one ethnicity with traditions of self-determination and self-government.

In some multinational states, ethnicities coexist peacefully while remaining culturally distinct. Each ethnic group recognizes and respects the distinctive traditions of other ethnicities. In some peaceful multinational states, each ethnic group may control governmental functions in the region of the country it inhabits. In other cases, ethnicities all contribute cultural features to the formation of a single nationality. For example, the United States has numerous ethnic groups, all of which consider themselves as belonging to a single U.S. nationality.

MULTIETHNIC REVIVAL IN EUROPE

During the 1930s, German National Socialists (Nazis) claimed that all German-speaking parts of Europe constituted one nationality and should be unified into one state. After many years of appeasing the Nazis' expansion in Central Europe, the United Kingdom and France finally declared war when the Nazis invaded Poland, clearly not a German-speaking state.

Following its defeat in World War II, Germany was divided into two states (Figure 8-13). Two Germanys existed from 1949 until 1990. The present-day state of Germany bears little resemblance to the territory occupied by German-speaking people prior to the upheavals of the twentieth century. However, a massive forced migration of people in Europe after World War II relocated many ethnic groups into the newly demarcated territory of the region's various nation-states. With the end of communism, the German Democratic Republic ceased to exist, and its territory became part of the German Federal Republic. As discussed in Chapter 7, former Yugoslavia has been the principal example of a failed nation-state in Europe.

In other multinational states, one ethnicity tries to dominate another, especially if one is much more numerous than the others. The people of the less numerous ethnicity may be assimilated into the cultural characteristics of the other, sometimes by force.

Europeans thought that ethnicity had been left behind as an insignificant relic, such as wearing costumes to amuse tourists. Karl Marx wrote that nationalism was a means for the dominant social classes to maintain power over workers. He believed that workers would identify with other working-class people instead of with an ethnicity. The attempt after World War I to divide Europe into nation-states was not a recipe for peace, as noted on the previous page.

In the twenty-first century, ethnic identity has once again become important in Europe. The multinational states of Yugoslavia, Czechoslovakia, and the Soviet Union were broken up into multiple states. Some of the new states are good examples of nation-states, but others are not:

- In Czechoslovakia, a multinational state was peacefully transformed in 1993 to two nation-states—Czechia (Czech Republic) and Slovakia. Slovaks comprise only 1 percent of Czechia's population and Czechs less than 1 percent of Slovakia's population (Figure 8-14).

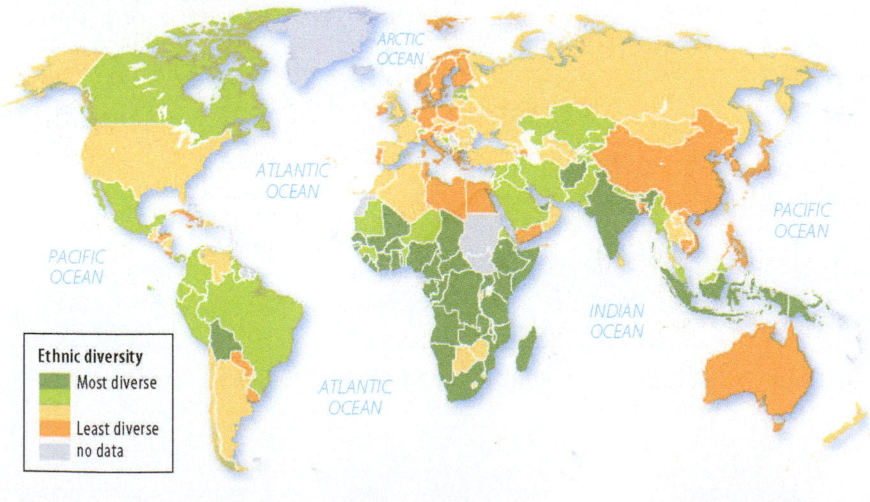

◀ **FIGURE 8-12 ETHNIC DIVERSITY** More than half of the best examples of nation-states are in Europe. The most ethnically diverse countries are primarily in Africa.

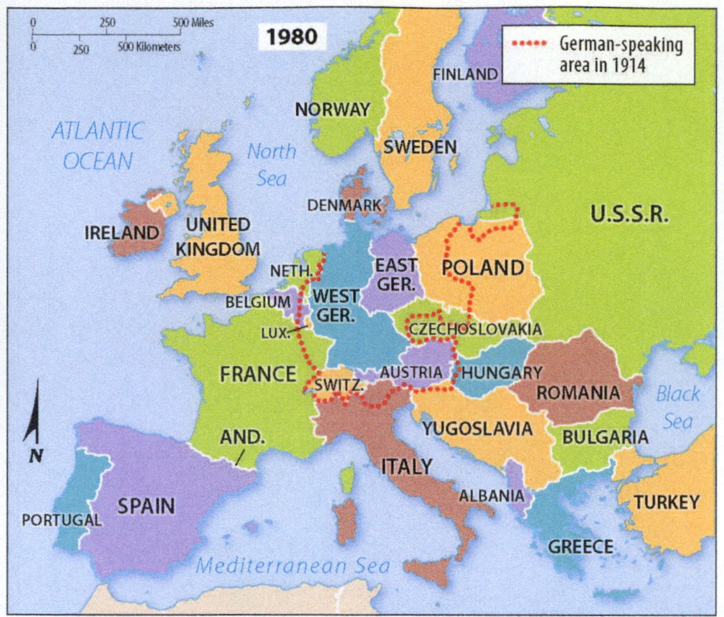

▲ **FIGURE 8-13 EUROPE, 1980** Germany was divided into two states after its defeat in World War II.

- In Yugoslavia, the breakup included a peaceful conversion of Slovenia in 1991 from a republic in multinational Yugoslavia to a nation-state (Figure 8-15). However, other portions of former Yugoslavia became nation-states only after ethnic cleansing and other atrocities, as discussed in the previous chapter.

PAUSE & REFLECT 8.2.2

Are Africa's principal areas of ethnic cleansing and genocide in Africa among the most ethnically diverse? Why might account for similarities or differences?

▲ **FIGURE 8-14 CZECHIA AND SLOVAKIA** Flags of Czechia (left) and Slovakia (right). Poland's flag is in the middle.

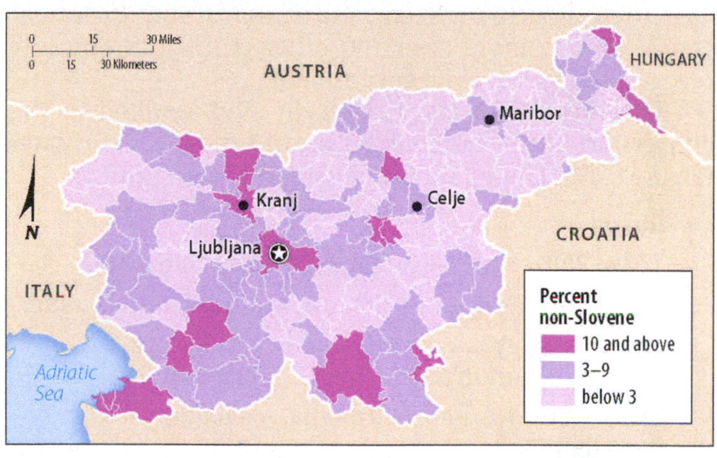

▲ **FIGURE 8-15 SLOVENIA**

SUSTAINABILITY & OUR ENVIRONMENT Rising Oceans and the Future of Nauru

The sustainability of the world's smallest island state, Nauru, as well other island microstates, is in danger due to rising ocean levels. Sea levels rose around 17 centimeters (6.7 inches) during the twentieth century. Scientists working for the United Nations forecast another rise of between 18 and 59 centimeters (between 7 and 23 inches). The rising oceans will submerge a large percentage of the tiny island. Another Pacific Ocean microstate, Kirbati, a collection of approximately 32 small islands, has already witnessed the disappearance of two of its islands under rising oceans.

Nauru, Kiribati, and other Pacific island microstates are atolls—that is, islands made of coral reefs (Figure 8-16). A coral is a small sedentary marine animal that has a horny or calcareous skeleton. Corals form colonies, and the skeletons build up to form coral reefs. Coral is very fragile. Humans are attracted to coral for its beauty and the diversity of species it supports, but handling coral can kill it. The threat of climate change to the sustainability coral is especially severe: Coral stays alive in only a narrow range of ocean temperatures, between 23°C and 25°C (between 73°F and 77°F), so global warming threatens the ecology of the portions of the islands that remain above sea level.

▲ **FIGURE 8-16 NAURU: WORLD'S SMALLEST ISLAND MICROSTATE** Rising sea level because of climate change threatens the future of the island, whose area is only 21 square kilometers (8 square miles).

Russia: The Largest Multiethnic State

LEARNING OUTCOME 8.2.3

Understand Russia's status as a multiethnic state.

RUSSIA: THE LARGEST MULTINATIONAL STATE

During its existence between 1922 and 1991, the Union of Soviet Socialist Republics (U.S.S.R.) was the world's largest state in land area, as well as the world's largest multinational state. The Soviet Union consisted of 15 republics, based on its 15 largest ethnicities. According to estimates a year before the breakup of the Soviet Union, Russians comprised 51 percent of the state's population, Ukrainians 15 percent, Uzbeks 6 percent, and the remaining 28 percent spread among more than 100 other ethnicities officially recognized by the Soviet government.

The breakup of the U.S.S.R. in 1991 resulted in the conversion of the 15 republics into 15 independent states (Figure 8-17). These 15 states consist of five groups:

- Russia.
- Three Baltic states: Estonia, Latvia, and Lithuania.
- Three European states: Belarus, Moldova, and Ukraine.
- Five Central Asian states: Kazakhstan, Kyrgyzstan, Tajikistan, Turkmenistan, and Uzbekistan.
- Three Caucasus states: Armenia, Azerbaijan, and Georgia.

Some of these new states are good examples of nation-states, and some are clearly multiethnic. However, the best examples of nation-states are not necessarily the most stable and peaceful of the new states.

With the breakup of the Soviet Union, Russia is now the world's largest multinational state. Russia comprises 81 percent ethnic Russians, but the government officially recognizes the existence of 39 ethnic groups among the remaining 19 percent.

Russia's ethnicities are clustered in two principal locations (Figure 8-18). Some are located along borders with neighboring states, including Buryats and Tuvinian near Mongolia, and Chechens, Dagestani, Kabardins, and Ossetians near the two former Soviet republics of Azerbaijan and Georgia. Other ethnicities are clustered in the center of Russia, especially between the Volga River basin and the Ural Mountains. Among the most numerous in this region are Bashkirs, Chuvash, and Tatars, who speak Altaic languages similar to Turkish, and Mordvins and Udmurts, who speak Uralic languages similar to Finnish. Most of these groups were conquered by the Russians in the sixteenth century, under the leadership of Ivan IV (Ivan the Terrible).

▲ **FIGURE 8-17 SOVIET UNION** The U.S.S.R. consisted of 15 republics that have become independent states.

Russia's constitution grants autonomy over local government affairs to around two dozen of the most numerous ethnicities. Local government units with a large ethnic population are allowed to designate the ethnic language as an official language in addition to Russian. Nonetheless, independence movements are flourishing among several of Russia's ethnicities.

RUSSIANS IN UKRAINE

After the breakup of the Soviet Union, prospects for a stable nation-state were favorable in independent Ukraine because it possessed economic assets, such as coal deposits, a steel industry, and proximity to the wealthy countries of Western Europe. However, Ukraine's minority Russian population started an uprising in the eastern region of the country, where they were clustered (Figure 8-19). Claiming that the Russian ethnic minority in Ukraine was endangered, Russia invaded eastern Ukraine and seized Crimea.

Crimea, a 27,000-square-kilometer (10,000-square-mile) peninsula, has long been an area of conflict (Figure 8-20). Crimea's population is approximately 60 percent Russian, 24 percent Ukrainian, 10 percent Tatar, and 6 percent other ethnicities.

Russia took control of Crimea in 1783, and in 1921 it became an autonomous republic within the Russian Soviet Federative Socialist Republic, which in turn was a republic within the Soviet Union. In 1954, the Soviet government transferred responsibility for Crimea to the Ukrainian Soviet Socialist Republic, which was then also part of the Soviet Union.

When the Soviet Union broke up in 1991, Crimea became an autonomous republic in the newly independent Ukraine. In 2014, Russia invaded Crimea and annexed it,

▼ **FIGURE 8-18 ETHNICITIES IN RUSSIA** Russians are clustered in western Russia, and the percentage declines to the south and east. The largest numbers of non-Russians are found between the Volga River and the Ural Mountains and near the southern borders.

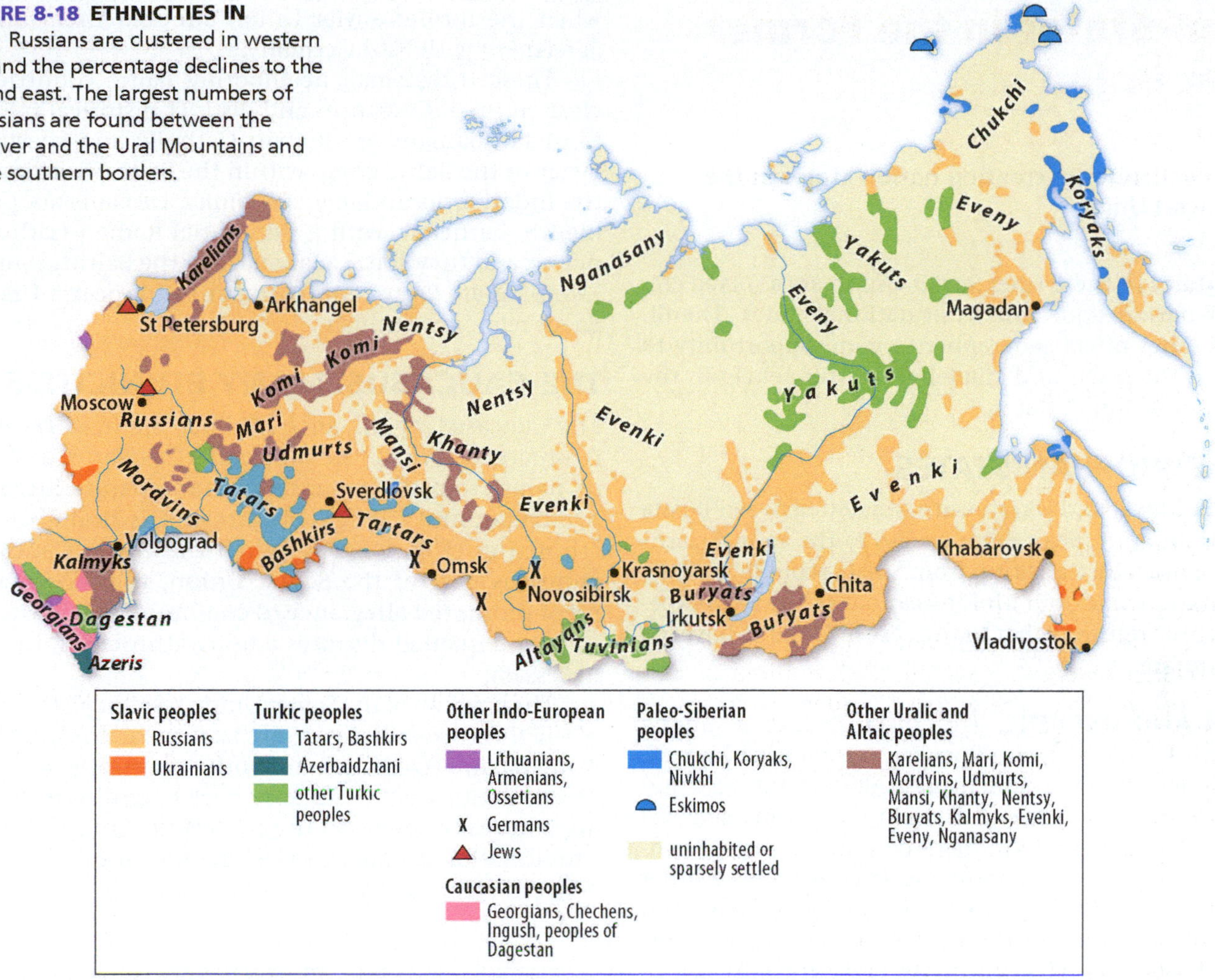

Slavic peoples: Russians; Ukrainians
Turkic peoples: Tatars, Bashkirs; Azerbaidzhani; other Turkic peoples
Other Indo-European peoples: Lithuanians, Armenians, Ossetians; X Germans; ▲ Jews
Caucasian peoples: Georgians, Chechens, Ingush, peoples of Dagestan
Paleo-Siberian peoples: Chukchi, Koryaks, Nivkhi; Eskimos; uninhabited or sparsely settled
Other Uralic and Altaic peoples: Karelians, Mari, Komi, Mordvins, Udmurts, Mansi, Khanty, Nentsy, Buryats, Kalmyks, Evenki, Eveny, Nganasany

claiming that the majority of the Crimean people, who are ethnic Russians, supported the action. Nearly every other country in the world continues to recognize Ukraine's sovereignty over Crimea. However, the international community has not found a way to remove the Russians and restore Crimea to Ukraine.

PAUSE & REFLECT 8.2.3

Why is most of Russia classified as sparsely settled? Compare the map of Russia's ethnicities (Figure 8-18) with the maps of world climate (Figure 1-47) and of population concentrations (Figure 2-4).

▲ **FIGURE 8-19 ETHNICITIES IN UKRAINE** The Russian ethnic minority is clustered in the east of the country, near the border with Russia.

▲ **FIGURE 8-20 CRIMEA** Russia claims sovereignty over Crimea, but most other states consider it legally still part of Ukraine.

Nation-States in the Former Soviet Union

LEARNING OUTCOME 8.2.4

Describe challenges in creating nation-states in the former Soviet Union.

The new states in the former Soviet Union are a mixed collection of nation-states and multinational states. The diversity of states offers geographers a good opportunity to understand the assets and challenges of differences in the ethnic composition of states.

THREE EUROPEAN STATES

Belarus, Moldova, and Ukraine are situated between Russia to the east and European democracies to the west. Belarus has made a peaceful transition from Soviet republic to independent nation-state, but Moldova and Ukraine have experienced ethnic tensions that in the case of Ukraine have led to open warfare.

BELARUS AND UKRAINE. The ethnic distinction among Belarusians, Ukrainians, and Russians is somewhat blurred. The three groups speak similar East Slavic languages and trace their ethnic heritage to the same roots in medieval Europe. Belarusians and Ukrainians became distinct ethnicities from Russians after they were isolated from each other after invasions and conquests by Mongolians, Poles, and Lithuanians beginning in the thirteenth century. Russians conquered Belarus and Ukraine in the late eighteenth century, but after five centuries of exposure to non-Slavic influences, Belarusians and Ukrainians displayed sufficient cultural differences to consider themselves distinct from Russians.

MOLDOVA. Moldovans are ethnically indistinguishable from Romanians, and Moldova (then called Moldavia) was part of Romania until the Soviet Union seized it in 1940. When Moldova changed from a Soviet republic back to an independent country in 1992, many Moldovans pushed for reunification with Romania, both to reunify the ethnic group and to improve the region's prospects for economic development. But it was not to be that simple.

When Moldova became a Soviet republic in 1940, its eastern boundary was the Dniester River. The Soviet government increased the size of Moldova by about 10 percent, transferring from Ukraine a 3,000-square-kilometer (1,200-square-mile) sliver of land on the east bank of the Dniester. The majority of the inhabitants of this area, known as Trans-Dniestria, are Ukrainian and Russian. They, of course, oppose Moldova's reunification with Romania.

THREE BALTIC STATES

Estonia, Latvia, and Lithuania are known as the Baltic states for their location on the Baltic Sea. They were independent countries between the end of World War I in 1918 and 1940, when the former Soviet Union annexed them under an agreement with Nazi Germany.

These three small neighboring Baltic countries have clear cultural differences and distinct historical traditions. Most Lithuanians are Roman Catholic and speak a language of the Baltic group within the Balto-Slavic branch of the Indo-European language family. Latvians are predominantly Lutheran, with a substantial Roman Catholic minority, and they speak a language of the Baltic group. Most Estonians are Protestant (Lutheran) and speak a Uralic language related to Finnish.

THE CAUCASUS: MANY ETHNICITIES

The Caucasus region, an area about the size of Colorado, is situated between the Black and Caspian seas and gets its name from the mountains that separate Russia from Azerbaijan and Georgia. The region is home to several ethnicities (Figure 8-21). When the entire Caucasus region was part of the Soviet Union, the Soviet government promoted allegiance to communism and the Soviet state and quelled disputes among ethnicities, by force if necessary.

The breakup of the Soviet Union resulted in the creation of the three small states Armenia, Azerbaijan, and Georgia. Armenia and Azerbaijan are both statistically good examples of nation-states, but they have fought over demarcating boundaries between the two ethnic groups. Georgia is a multinational state experiencing uprisings and independence movements by several of its ethnic groups.

▲ **FIGURE 8-21 ETHNICITIES IN THE CAUCASUS** The boundaries of the states of Armenia, Azerbaijan, and Georgia do not match the territories occupied by the Armenian, Azeri, and Georgian ethnicities. The Abkhazians, Chechens, Kurds, and Ossetians are examples of ethnicities in this region that have not been able to organize nation-states.

ARMENIA. More than 3,000 years ago Armenians controlled an independent kingdom in the Caucasus. Converted to Christianity in 303, they lived for many centuries as an isolated Christian enclave under the rule of Turkish Muslims. A century ago, an estimated 1 million Armenians were killed by the Turks in actions now classified by most observers as genocide. After World War I the Allies created an independent state of Armenia, but it was soon swallowed by its neighbors. In 1921, Turkey and the Soviet Union agreed to divide Armenia between them. Armenians comprise 98 percent of the population in Armenia, making it the most ethnically homogeneous country in the region.

AZERBAIJAN. Azeris trace their roots to Turkish invaders who migrated from Central Asia in the eighth and ninth centuries and merged with the existing Persian population. An 1828 treaty allocated northern Azeri territory to Russia and southern Azeri territory to Persia (now Iran). The western part of Azerbaijan, Nakhichevan (named for the area's largest city), is separated from the rest of Azerbaijan by a 40-kilometer (25-mile) corridor that belongs to Armenia.

Armenians and Azeris both have achieved long-held aspirations of forming nation-states, but after their independence from the Soviet Union, the two went to war over the boundaries between them. The war concerned possession of Nagorno-Karabakh, a 5,000-square-kilometer (2,000-square-mile) enclave within Azerbaijan that is inhabited primarily by Armenians but placed under Azerbaijan's control by the Soviet Union during the 1920s. A 1994 cease-fire has left Nagorno-Karabakh technically part of Azerbaijan, but in reality it acts as an independent republic called Artsakh. Numerous clashes have occurred since then between Armenia and Azerbaijan.

GEORGIA. The population of Georgia is more diverse than that in Armenia and Azerbaijan. Ethnic Georgians comprise 71 percent of the population. The country also includes about 8 percent Armenian, 6 percent each Azeri and Russian, 3 percent Ossetian, and 2 percent each Abkhazian, Greek, and other ethnicities. Georgia's cultural diversity has been a source of unrest, especially among the Ossetians and Abkhazians.

During the 1990s, the Abkhazians fought for control of the northwestern portion of Georgia and declared Abkhazia to be an independent state. In 2008, the Ossetians fought a war with the Georgians that resulted in the Ossetians declaring the South Ossetia portion of Georgia to be independent. Russia has recognized Abkhazia and South Ossetia as independent countries and has sent troops there. Only a handful of other countries recognize the independence of Abkhazia and South Ossetia, although the two operate as if they were independent of Georgia.

CENTRAL ASIAN STATES

The five states in Central Asia carved out of the former Soviet Union display varying degrees of conformance to the principles of a nation-state (Figure 8-22). Together the five provide an important reminder that multinational states can be more peaceful than nation-states.

Turkmenistan and Uzbekistan are relatively stable nationstates. Altaic language. In contrast, Tajikistan is a nation-state that has suffered from a civil war between Tajiks who were former Communists, and an unusual alliance of Muslim fundamentalists and Western-oriented intellectuals. Fifteen percent of the population was made homeless during a civil war that lasted between 1992 and 1997. A U.N. peacekeeping force has helped to prevent a recurrence.

Kazakhstan is a relatively peaceful multinational state divided between Kazakhs, who comprise 67 percent of the population, and Russians, at 18 percent. Kazakhs are Muslims who speak an Altaic language similar to Turkish. In contrast, Kyrgyzstan is a multinational state that has suffered from ethnic conflict. The country comprises 69 percent Kyrgyz, 15 percent Uzbek, and 9 percent Russian. The Kyrgyz and Uzbeks are both Muslims who speak Altaic languages. Nonetheless, conflict between the two ethnicities led to the overthrow of successive presidents in the first decade of the twenty-first century, as well as violence in 2010 that included charges of ethnic cleansing of hundreds of thousands of Uzbeks by Kyrgyz.

PAUSE & REFLECT 8.2.4

If Abkhazia, Artsakh, and South Ossetia were widely recognized independent states, how would they compare in size to the microstates described earlier in this chapter?

▲ **FIGURE 8-22 ETHNICITIES IN CENTRAL ASIA** Turkmenistan and Uzbekistan are relatively stable nation-states, Kazakhstan is a relatively peaceful multinational state, Tajikistan is a relatively unstable nation-state, and Kyrgyzstan is a multinational state that has had conflict.

Colonies

LEARNING OUTCOME 8.2.5

Explain the concept of colonies and describe their current distribution.

Nearly all of Earth's land area has been allocated to a collection of around 200 sovereign states, but some territories remain that have not achieved self-determination and statehood. A **colony** is a territory that is legally tied to a sovereign state rather than being completely independent. In some cases, a sovereign state runs only the colony's military and foreign policy. In others, it also controls the colony's internal affairs.

DISTRIBUTION OF COLONIES

The United Nations identifies 17 places in the world that it calls "non-self-governing territories" (Figure 8-23). Of the 17, Western Sahara is by far the most extensive (266,000 square kilometers [103,000 square miles]) and most populous (around 500,000). The two next most populous are French Polynesia and New Caledonia, both controlled by France, with around 250,000 inhabitants each. All but Western Sahara are islands.

The least-populated colony is Pitcairn Island, a 36-square-kilometer (14-square-mile) possession of the United Kingdom. The island in the South Pacific was settled in 1790 by British mutineers from the ship *Bounty*, commanded by Captain William Bligh. Its 50 islanders survive by selling fish as well as postage stamps to collectors.

The U.N. list does not include territories that are uninhabited, such as Baker and Midway islands, controlled by the United States. The U.N. also does not list inhabited territories that it considers to have considerable autonomy in self-governing. For example, the U.N. does not classify these territories as colonies:

- **Puerto Rico.** A commonwealth of the United States. Puerto Ricans are citizens of the United States, but they do not participate in U.S. elections or have a voting member of Congress.
- **Greenland.** An autonomous unit within the Kingdom of Denmark. Greenland runs its internal affairs, but Denmark control foreign affairs and defense.
- **Hong Kong and Macao.** Attached to the mainland of China as special administrative regions within the People's Republic of China. Hong Kong was a colony of the United Kingdom until it returned to China in 1997, and a year later Portugal returned its colony of Macao. The two have some autonomy in economic matters, but China controls foreign affairs and defense.

PAUSE & REFLECT 8.2.5

What would need to change for Puerto Rico to no longer be classified as a colony of the United States?

COLONIALISM

At one time, colonies were widespread over Earth's surface. European states came to control much of the world through **colonialism,** which is an effort by one country to establish settlements in a territory and to impose its political, economic, and cultural principles on that territory (Figure 8-24). European states established colonies elsewhere in the world for three basic reasons: Gold, gold, and glory.

- To promote Christianity.
- To extract useful resources and to serve as captive markets for their products.
- To establish relative power through the number of colonies claimed.

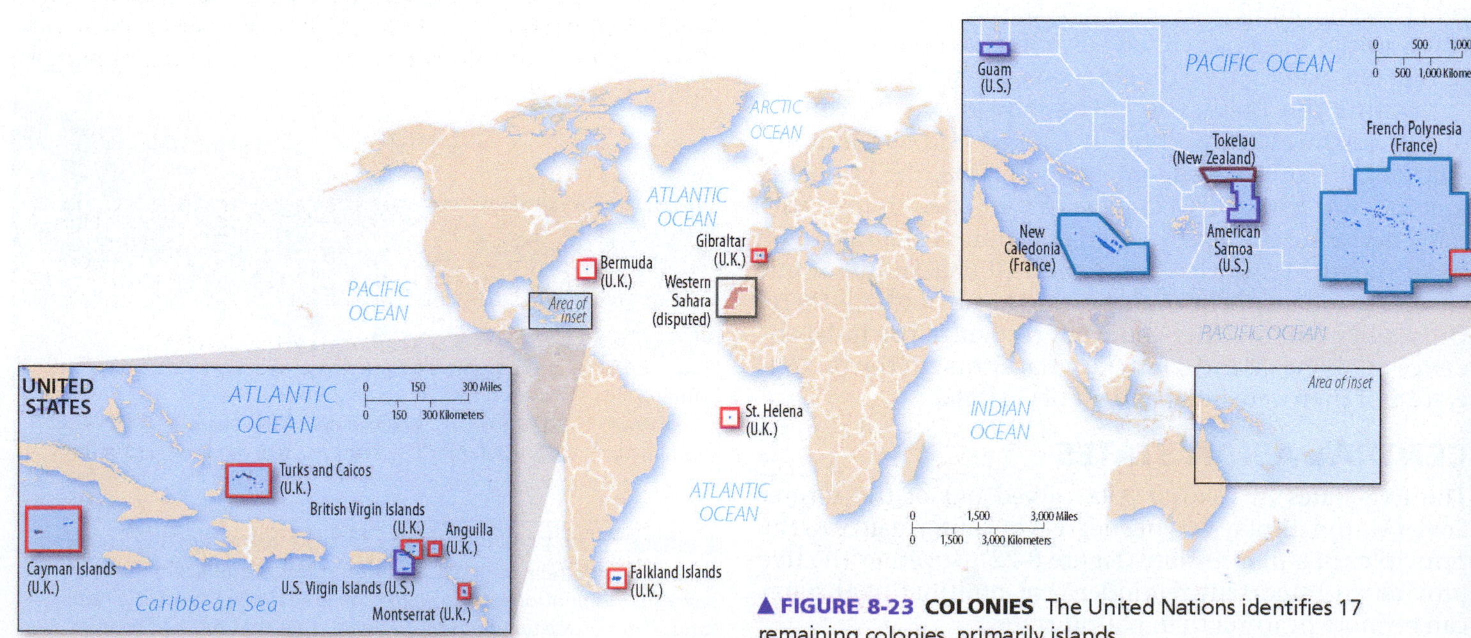

▲ **FIGURE 8-23 COLONIES** The United Nations identifies 17 remaining colonies, primarily islands.

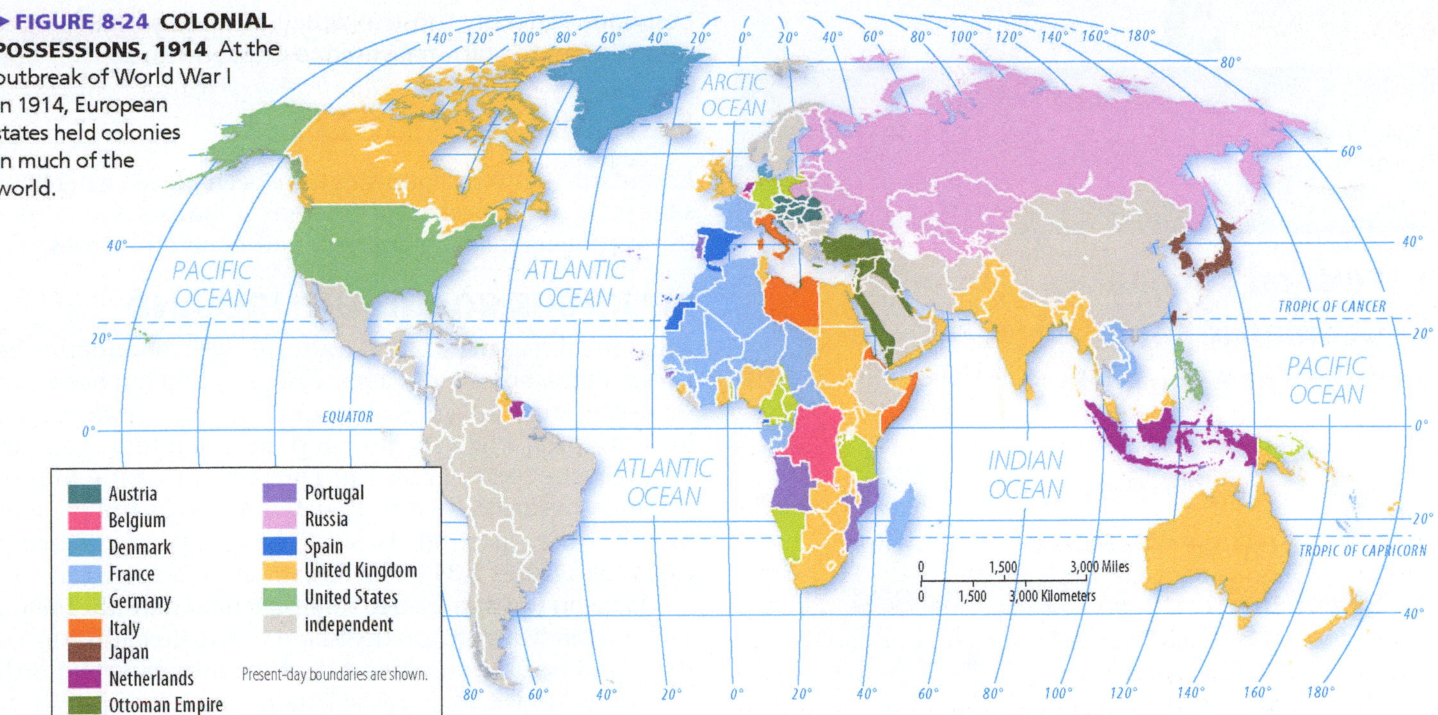

▶ FIGURE 8-24 COLONIAL POSSESSIONS, 1914 At the outbreak of World War I in 1914, European states held colonies in much of the world.

The colonial era began in the 1400s, when European explorers sailed westward for Asia but encountered and settled in the Western Hemisphere instead. Eventually, the European states lost most of their Western Hemisphere colonies: Independence was declared by the United States in 1776 and by most Latin American states between 1800 and 1824. European states then turned their attention to Africa and Asia.

▼ FIGURE 8-25 BRITISH EMPIRE The British Field Marshal Edmund Allenby entered Jerusalem in 1917, after the city was captured by British Empire troops.

The United Kingdom planted colonies on every continent, including much of eastern and southern Africa, South Asia, the Middle East, Australia, and Canada. With by far the largest colonial empire, the British proclaimed that the "Sun never set" on their empire (Figure 8-25).

France had the second-largest overseas territory, primarily in West Africa and Southeast Asia. France attempted to assimilate its colonies into French culture and educate an elite group to provide local administrative leadership. After independence, most of these leaders retained close ties with France.

Most African and Asian colonies became independent after World War II. Only 15 African and Asian states were members of the United Nations when it was established in 1945, compared to 106 in 2012. The boundaries of the new states frequently coincide with former colonial provinces, although not always.

CHECK-IN KEY ISSUE 2

Why Are Nation-States Difficult to Create?

- ✔ **Good examples of nation-states can be identified, though none are perfect.**
- ✔ **The Soviet Union was once the world's largest multinational state; with its breakup, Russia is now the largest.**
- ✔ **Much of Earth's land area once comprised colonies, but only a few colonies remain.**

KEY ISSUE 3

Why Do Boundaries Cause Problems?

- Cultural Boundaries
- Geometric Boundaries
- Physical Boundaries
- Shapes of States
- Governing States
- Electoral Geography
- Geography of Gerrymandering

LEARNING OUTCOME 8.3.1

Describe the types of cultural boundaries between states.

A state is separated from its neighbors by a **boundary,** an invisible line that marks the extent of a state's territory. Boundaries completely surround an individual state to mark the outer limits of its territorial control and to give it a distinctive shape. Boundaries interest geographers because the process of selecting their location is frequently difficult.

Historically, frontiers rather than boundaries separated states. A **frontier** is a zone where no state exercises complete political control. It is a tangible geographic area, whereas a boundary is an infinitely thin line. Frontier areas were either uninhabited or sparsely settled. Frontiers between states have been replaced by boundaries. Modern communications systems permit countries to monitor and guard boundaries effectively, even in previously inaccessible locations.

Boundaries are of three types.

- Cultural boundaries follow the distribution of cultural features.
- Geometric boundaries are based on human constructs, such as straight lines.
- Physical boundaries coincide with significant features of the natural landscape.

No other type of boundary is better or more "natural" than the others, and many boundaries are a combination of types.

Boundary locations can generate conflict, both within a country and with its neighbors. A boundary line, which must be shared by more than one state, is the only location where direct physical contact must take place between two neighboring states. Therefore, the boundary has the potential to become the focal point of conflict between them. The best boundaries are those to which all affected states agree, regardless of the rationale used to draw the line.

Cultural Boundaries

Boundaries between countries have been placed where possible to separate speakers of different languages, followers of different religions, or members of different ethnicities.

RELIGIOUS BOUNDARY: IRELAND

Religious differences often coincide with boundaries between states, but in only a few cases has religion been used to select boundary lines. The most notable example was in South Asia, when the British partitioned India into two states on the basis of religion. The predominantly Muslim portions were allocated to Pakistan, whereas the predominantly Hindu portions became the independent state of India (see Figures 7-34 and 7-35).

Religion was also used to some extent to draw the boundary between two states on the island of Eire (Ireland). Most of the island became an independent country, but the northeast—now known as Northern Ireland—remained part of the United Kingdom. Roman Catholics comprise 84 percent of the population in the 26 counties that joined the Republic of Ireland, compared with only 41 percent in the six counties of Northern Ireland (Figure 8-26).

Language is an important cultural characteristic for drawing boundaries, especially in Europe. France, Portugal, and Spain are examples of European states that coalesced around distinctive languages before the nineteenth century. Germany and Italy emerged in the nineteenth century as states unified by language.

PAUSE & REFLECT 8.3.1

Referring to Chapters 5 and 6, what other cultural boundaries run through the middle of other states in Europe?

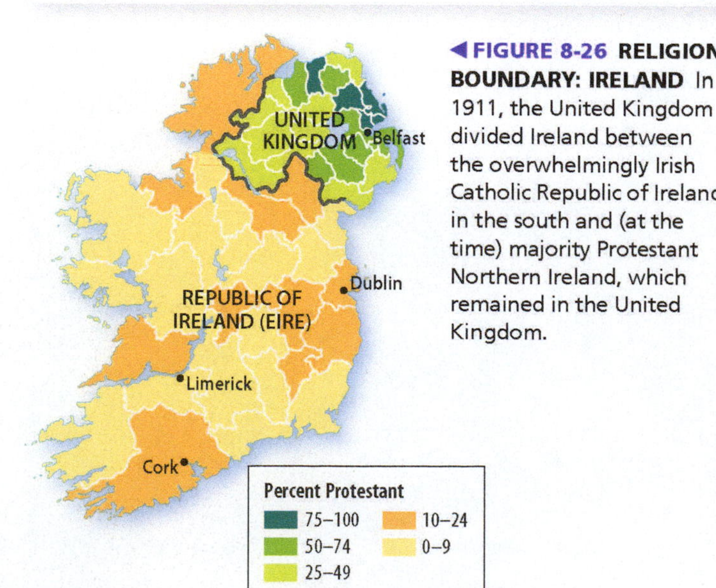

◄ **FIGURE 8-26 RELIGION BOUNDARY: IRELAND** In 1911, the United Kingdom divided Ireland between the overwhelmingly Irish Catholic Republic of Ireland in the south and (at the time) majority Protestant Northern Ireland, which remained in the United Kingdom.

▲ **FIGURE 8-27 ETHNIC BOUNDARY: CYPRUS** Since 1974, Cyprus has been divided into Greek and Turkish areas, separated by a United Nations buffer zone. The United Kingdom, the colonial ruler of Cyprus until 1960, maintains two military bases on the island.

ETHNIC BOUNDARY: CYPRUS

Cyprus, the third-largest island in the Mediterranean Sea, contains two nationalities: Greek and Turkish. Although the island is physically closer to Turkey, Turks comprise only 24 percent of the country's population, whereas Greeks account for 63 percent. When Cyprus gained independence from Britain in 1960, its constitution guaranteed the Turkish minority a substantial share of elected offices and control over its own education, religion, and culture. But Cyprus has never peacefully integrated the Greek and Turkish nationalities.

Several Greek Cypriot military officers who favored unification of Cyprus with Greece seized control of the government in 1974. Shortly after the coup, Turkey invaded Cyprus to protect the Turkish Cypriot minority. The Greek coup leaders were removed within a few months, and an elected government was restored, but the Turkish army remained on Cyprus. The northern 36 percent of the island controlled by Turkey declared itself the independent Turkish Republic of Northern Cyprus in 1983, but only Turkey recognizes it as a separate state (Figure 8-27).

A wall was constructed between the two areas, and a buffer zone patrolled by the United Nations was delineated across the entire island (Figure 8-28). Traditionally, the Greek and Turkish Cypriots had mingled, but after the wall and buffer zone were established, the two nationalities became geographically isolated. The northern part of the island is now overwhelmingly Turkish, and the southern part is overwhelmingly Greek. Approximately one-third of the island's Greeks were forced to move from the region controlled by the Turkish army, whereas nearly one-fourth of the Turks moved from the region now regarded as the Greek side.

The two sides have been brought closer in recent years. A portion of the wall was demolished, and after three decades, each nationality could again cross to the other sides. The European Union accepted the entire island of Cyprus as a member in 2004.

▶ **FIGURE 8-28 ETHNIC BOUNDARY: CYPRUS GREEN LINE** A Greek Cypriot soldier guards the Greek Cypriot side of the Green Line. The U.N. buffer zone is behind the white barrels. The Greek graffiti reads "Our Border Is Not Here."

Geometric Boundaries

LEARNING OUTCOME 8.3.2

Describe types of geometric boundaries between states.

A world map, such as Figure 8-1, shows two regions where geometric boundaries are especially prominent: North America and North Africa. These boundaries are peaceful now but were not always so in the past.

GEOMETRIC BOUNDARIES: NORTH AMERICA

Part of the northern U.S. boundary with Canada is a 2,100-kilometer (1,300-mile) straight line along 49° north latitude, running from Lake of the Woods between Minnesota and Manitoba to the Strait of Georgia between Washington State and British Columbia (Figure 8-29). Québec's boundary with New York and Vermont is also geometric, along 45° north latitude. The two countries share an additional 1,100-kilometer (700-mile) geometric boundary between Alaska and the Yukon Territory along the north - south arc of 141° west longitude.

The U.S.-Canada boundary was established through a series of treaties between 1783 and 1903 between the United States and the United Kingdom, which then still controlled Canada. During the 1840s, many Americans called for the border between the Rockies and the Pacific to be set further north, at 54°40' north latitude (the southern tip of Alaska). Advocates of the northern line used the slogan "54 -40 or fight," but the Oregon Treaty settled the dispute peacefully in 1846.

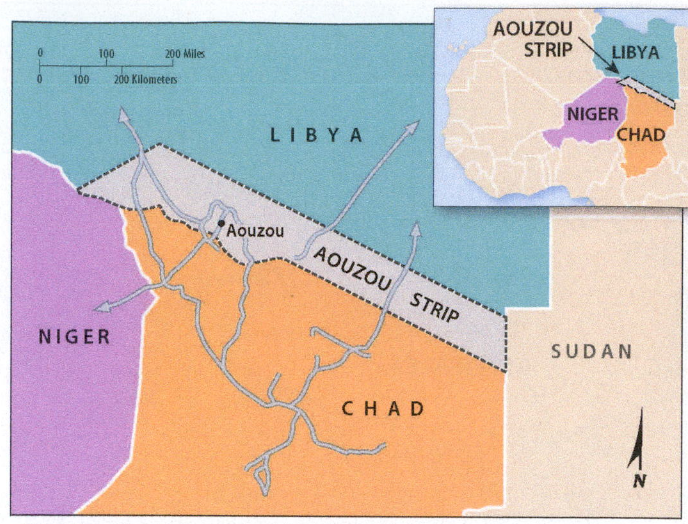

▲ **FIGURE 8-30 GEOMETRIC BOUNDARY: NORTH AFRICA** The boundary between Chad and Libya is a straight line, drawn by European states early in the twentieth century, when the area comprised a series of colonies. Libya, however, claims that the boundary should be located 100 kilometers (60 miles) to the south and that it should have sovereignty over the Aouzou Strip.

GEOMETRIC BOUNDARY: NORTH AFRICA

Boundaries between Algeria, Libya, and Egypt on the north and Mali, Niger, Chad, and Sudan on the south are for the most part geometric. Many of these boundaries are a legacy of treaties among European countries to divide up much of Africa into colonies. For example, the 1,000-kilometer (600-mile) boundary between Chad and Libya is a straight line drawn across the desert in 1899 by the French and British to set the northern limit of French colonies in Africa.

As an independent country, Libya claimed that the straight line should be 100 kilometers (60 miles) to the south, to include territory known as the Aouzou Strip (Figure 8-30). Citing an agreement in 1935 between France and Italy, which then controlled much of Libya, Libya seized the Aouzou Strip in 1973. Chad regained control of the strip in 1987, and Libya withdrew its troops after the International Court of Justice ruled in favor of Chad's claim in 1994.

◀ **FIGURE 8-29 GEOMETRIC BOUNDARY: NORTH AMERICA** The boundary between Canada and the United States runs through Waterton-Glacier International Peace Park. The line of cut trees is the boundary between the United States (left) and Canada (right).

PAUSE & REFLECT 8.3.2

Where is the U.S.–Canada boundary not based on geometry?

GEOMETRIC BOUNDARY: SOUTH POLE

The South Pole region contains the only large landmasses on Earth's surface that are not part of a state. Seven states claim portions of the South Pole region: Argentina, Australia, Chile, France, New Zealand, Norway, and the United Kingdom. These claims are divided geometrically, following meridians that converge on the South Pole (Figure 8-31). Some of the territorial claims are overlapping and conflicting, and on the other hand some of the South Pole region is unclaimed by any state. The United States, Russia, and a number of other states do not recognize the claims of any country to Antarctica. The Antarctic Treaty, signed in 1959 by 47 states, provides a legal framework for managing Antarctica. States may establish research stations there for scientific investigations, but no military activities are permitted.

The South Pole region has not offered a useful precedent for settling claims around the North Pole. Canada, Denmark, Iceland, Norway, Russia, and the United States make conflicting claims. The heart of the dispute is the Lomonosov Ridge, which runs 1,800 kilometers (1,100 miles) across the polar region (Figure 8-32). The ridge rises several thousand meters above the floor of Arctic Sea and in some places is only a few hundred meters below the current sea level. Russia and Denmark (which controls Greenland) both claim the Lomonosov Ridge as an extension of their respective landmasses. The territorial claims around the North Pole are based on differing interpretations of the Law of the Sea, discussed in the next section.

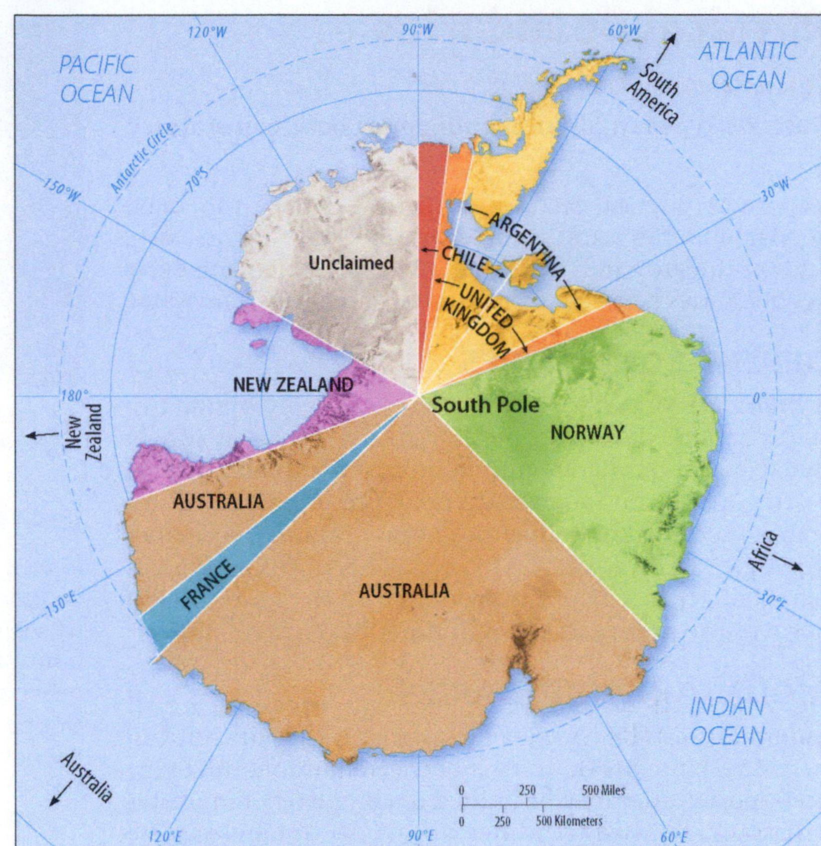

▲ **FIGURE 8-31 NATIONAL CLAIMS TO ANTARCTICA** Portions of Antarctica are claimed by Argentina, Australia, Chile, France, New Zealand, Norway, and the United Kingdom; claims by Argentina, Chile, and the United Kingdom are conflicting.

► **FIGURE 8-32 NATIONAL CLAIMS TO THE ARCTIC** (Under the Law of the Sea Treaty of 1982, countries have submitted claims to territory inside the Arctic Circle. Some of these claims overlap.

Physical Boundaries

LEARNING OUTCOME 8.3.3

Describe types of physical boundaries between states.

Important physical features on Earth's surface can make good boundaries because they are easily seen, both on a map and on the ground. Three types of physical elements serve as boundaries between states: deserts, mountains, and water.

DESERT BOUNDARIES

A boundary drawn in a desert can effectively divide two states because deserts are hard to cross and sparsely inhabited. Desert boundaries are common in Africa and Asia. In North Africa, the Sahara has in most cases proved to be a stable physical (as well as geometric) boundary separating Algeria, Libya, and Egypt on the north from Mauritania, Mali, Niger, Chad, and the Sudan on the south. Desert borders also are found in South America (Figure 8-33).

MOUNTAIN BOUNDARIES

Mountains can be effective boundaries if they are difficult to cross (Figure 8-34). Contact between nationalities living on opposite sides may be limited or completely impossible if passes are closed by winter storms. Mountains are also useful boundaries because they are rather permanent and are usually sparsely inhabited.

Mountains do not always provide for the amicable separation of neighbors. Argentina and Chile agreed to be divided by the crest of the Andes Mountains but could not decide on the precise location of the crest. Was the crest a jagged line, connecting mountain peak to mountain peak? Or was it a curving line following the continental divide (the continuous ridge that divides rainfall and snowmelt between flow toward the Atlantic or Pacific)? The two countries almost fought a war over the boundary line. But with the help of U.S. mediators, they finally decided on the line connecting adjacent mountain peaks.

▲ **FIGURE 8-34 MOUNTAIN BOUNDARY** The border between Argentina and Chile runs through the Andes Mountains.

WATER BOUNDARIES

Rivers, lakes, and oceans are the physical features most commonly used as boundaries. Water boundaries are readily visible on maps and aerial imagery. Historically, water boundaries offered good protection against attack from another state because an invading state had to transport its troops by air or ship and secure a landing spot in the country being attacked. The state being invaded could concentrate its defense at the landing point.

Water boundaries are especially common in East Africa:

- The boundary between the Democratic Republic of Congo and Uganda runs through Lake Albert.
- The boundary separating Kenya, Tanzania, and Uganda runs through Lake Victoria (Figure 8-35).
- The boundary separating Burundi, the Democratic Republic of Congo, Tanzania, and Zambia runs through Lake Tanganyika.
- The boundary between Malawi and Mozambique runs through Lake Nyasa, which is also known as Lake Malawi.

Water boundaries may seem to be set permanently, but the precise position of water may change over time.

◀ **FIGURE 8-33 DESERT BOUNDARY** The border between Bolivia and Chile runs through the Atacama Desert.

▲ **FIGURE 8-35 WATER BOUNDARY** The borders separating Kenya, Tanzania, and Uganda run in part through Lake Victoria.

Rivers, in particular, can slowly change their course. The Rio Grande, the river separating the United States and Mexico, has frequently meandered from its previous course since it became part of the boundary in 1848. Land that had once been on the U.S. side of the boundary came to be on the Mexican side and vice versa. The United States and Mexico have concluded treaties that restore land affected by the shifting course of the river to the country in control at the time of the original nineteenth-century delineation. The International Boundary and Water Commission, jointly staffed by the United States and Mexico, oversees the border treaties and settles differences.

THE LAW OF THE SEA

The Law of the Sea, signed by 165 countries, identifies three types of water boundaries (Figure 8-36):

- **Territorial waters.** Up to 12 nautical miles from shore (about 22 kilometers or 14 land miles), a state may set laws regulating passage by ships registered in other states.
- **Contiguous zone.** Between 12 and 24 nautical miles from shore, a state may enforce laws concerning pollution, taxation, customs, and immigration.
- **Exclusive economic zone.** Between 24 and 200 nautical miles, a state has the sole right to the fish and other marine life.

Disputes can be taken to a tribunal for the Law of the Sea or to the International Court of Justice.

Through enforcement of the exclusive economic zone, states bordering an ocean are able to claim vast areas of the ocean for control of valuable resources. This has become especially important in the delineation of boundaries in the polar regions.

PAUSE & REFLECT 8.3.3

What examples have you seen thus far in this chapter of physical features that have not served as peaceful boundaries?

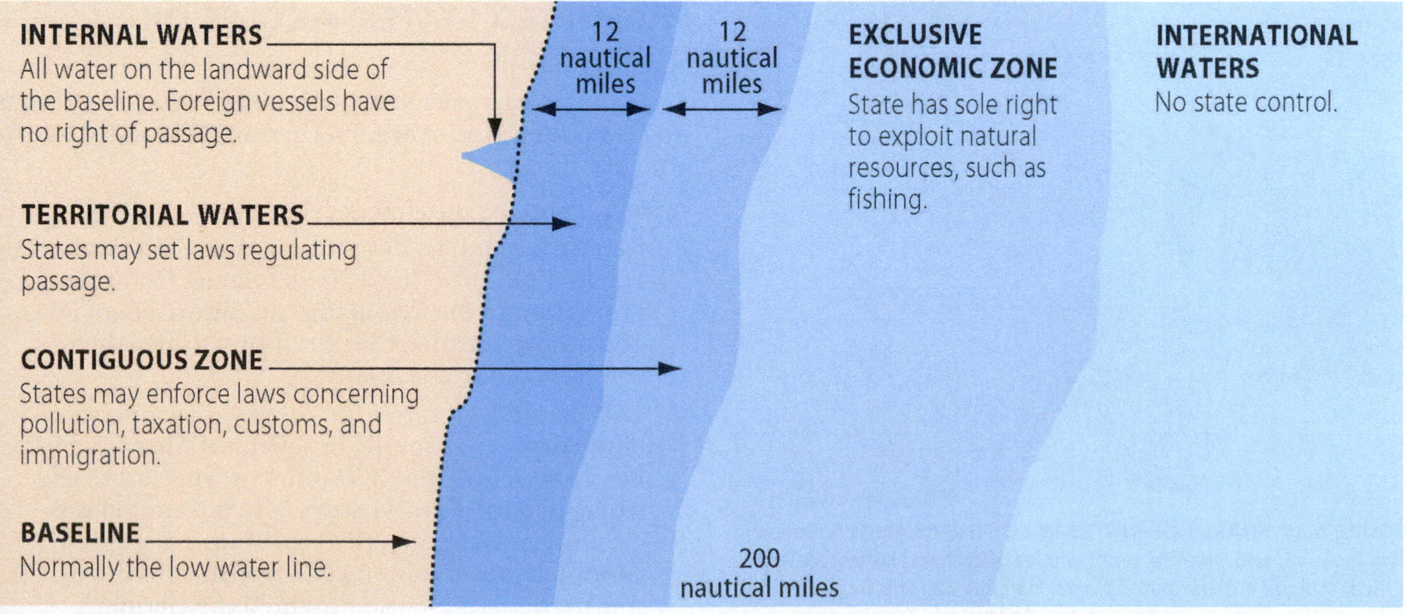

▲ **FIGURE 8-36 THE LAW OF THE SEA**

Shapes of States

LEARNING OUTCOME 8.3.4

Describe the five shapes of states.

The shape of a state controls the length of its boundaries with other states. The shape therefore affects the potential for communication and conflict with neighbors. The shape also, as in the outline of the United States or Canada, is part of its unique identity. Beyond its value as a centripetal force, the shape of a state can influence the ease or difficulty of internal administration and can affect social unity.

Countries have one of five basic shapes—compact, elongated, prorupted, fragmented, or perforated—and examples of each can be seen in sub-Saharan Africa (Figure 8-37). Each shape displays distinctive characteristics and challenges.

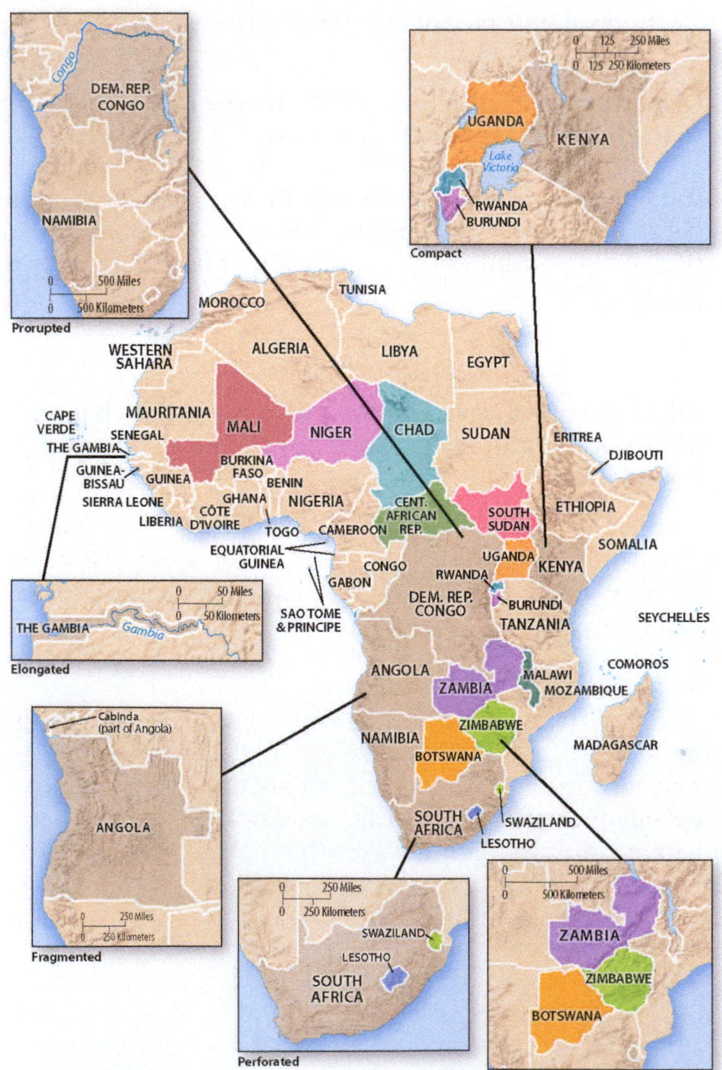

▲ **FIGURE 8-37 SHAPES OF STATES IN SOUTHERN AFRICA** Burundi, Kenya, Rwanda, and Uganda are examples of compact states. Malawi and Mozambique are elongated states. Namibia and the Democratic Republic of Congo are prorupted states. Angola and Tanzania are fragmented states. South Africa is a perforated state. The countries in color are landlocked African states.

COMPACT STATE: EFFICIENT

In a **compact state,** the distance from the center to any boundary does not vary significantly. The ideal theoretical compact state would be shaped like a circle, with the capital at the center and with the shortest possible boundaries to defend.

Compactness can be beneficial for smaller states because good communications can be more easily established with all regions, especially if the capital is located near the center. However, compactness does not necessarily mean peacefulness, as compact states are just as likely as others to experience civil wars and ethnic rivalries.

ELONGATED STATES: POTENTIAL ISOLATION

An **elongated state** has a long and narrow shape. Elongated states in sub-Saharan Africa include:

- Malawi, which measures about 850 kilometers (530 miles) north–south but only 100 kilometers (60 miles) east–west.
- The Gambia, which extends along the banks of the Gambia River about 500 kilometers (300 miles) east–west but is only about 25 kilometers (15 miles) north–south.

Chile, a prominent example in South America, stretches north–south for more than 4,000 kilometers (2,500 miles) but rarely exceeds an east–west distance of 150 kilometers (90 miles). Chile is wedged between the Pacific Coast of South America and the rugged Andes Mountains, which rise more than 6,700 meters (20,000 feet).

Elongated states may suffer from poor internal communications. A region located at an extreme end of the elongation might be isolated from the capital, which is usually placed near the center.

PRORUPTED STATES: ACCESS OR DISRUPTION

An otherwise compact state with a large projecting extension is a **prorupted state.** Proruptions are created for two principal reasons:

- To provide a state with access to a resource, such as water. For example, in southern Africa, the Democratic Republic of Congo has a 500-kilometer (300-mile) proruption to the west along the Zaire (Congo) River. The Belgians created the proruption to give their colony access to the Atlantic.
- To separate two states that otherwise would share a boundary. For example, in southern Africa, Namibia has a 500-kilometer (300-mile) proruption to the east called the Caprivi Strip. When Namibia was a colony of Germany, the proruption disrupted communications among the British colonies of southern Africa. It also provided the Germans with access to the Zambezi, one of Africa's most important rivers.

PERFORATED STATE: SOUTH AFRICA

A state that completely surrounds another one is a **perforated state.** In this situation, the state that is surrounded may face problems of dependence on, or interference from, the surrounding state. In sub-Saharan Africa, For example, South Africa completely surrounds the state of Lesotho. Lesotho must depend almost entirely on South Africa for the import and export of goods. Dependency on South Africa was especially difficult for Lesotho when South Africa had a government controlled by whites who discriminated against the black majority population. Elsewhere in the world, Italy surrounds the Holy See (the Vatican) and San Marino.

FRAGMENTED STATES: PROBLEMATIC

A **fragmented state** includes several discontinuous pieces of territory. Technically, all states that have offshore islands as part of their territory are fragmented. However, fragmentation is particularly significant for some states.

There are two kinds of fragmented states: those separated by water and those separated by an intervening state. Both may face problems and costs associated with communications and maintaining national unity.

A fragmented state separated by water in sub-Saharan Africa is Tanzania, which was created in 1964 as a union of the island of Zanzibar with the mainland territory of Tanganyika. Although home to different ethnic groups, the two entities agreed to join together because they shared common development goals and political priorities. Elsewhere in the world, Indonesia comprises 13,677 islands that extend more than 5,000 kilometers (3,000 miles) between the Indian and Pacific oceans. The fragmentation hinders communications and makes integration of people living on remote islands nearly impossible. To foster national integration, the Indonesian government has encouraged migration to sparsely inhabited islands from Java and Sumatra, where more than 80 percent of the population is clustered.

A fragmented state separated by an intervening state in sub-Saharan Africa is Angola, which is divided into two fragments by the Congo proruption described above. An independence movement is trying to detach Cabinda as a separate state from Angola, with the justification that its population belongs to distinct ethnic groups. Elsewhere in the world, Russia has a fragment called Kaliningrad (Konigsberg), a 16,000-square-kilometer (6,000-square-mile) entity 400 kilometers (250 miles) west of the remainder of Russia, separated by the states of Lithuania and Belarus. The area was part of Germany until the end of World War II, when the Soviet Union seized it after the German defeat. The German population fled westward after the war, and virtually all of the area's 430,000 residents are Russians. Russia wants Kaliningrad because it has the country's largest naval base on the Baltic Sea.

LANDLOCKED STATES

A **landlocked state** lacks a direct outlet to a sea because it is completely surrounded by several other countries

▲ **FIGURE 8-38 LANDLOCKED STATES AND RAIL LINES IN AFRICA** Landlocked states must import and export goods by land-based transportation, primarily rail lines, to reach ocean ports in cooperating neighbor states. South Africa is the only state in Africa with a dense rail network.

(or only one country, in the case of Lesotho). Landlocked states are most common in Africa, where 15 of the continent's 55 states have no direct ocean access (Figure 8-38). The prevalence of landlocked states in Africa is a remnant of the colonial era, when Britain and France controlled extensive regions. The European powers built railroads to connect the interior of Africa with the sea. Railroads moved minerals from interior mines to seaports, and in the opposite direction, rail lines carried mining equipment and supplies from seaports to the interior.

Now that the British and French empires are gone, and former colonies have become independent states, some important colonial railroad lines pass through several independent countries. These new landlocked states must cooperate with neighboring states that have seaports. Direct access to an ocean is critical to states because it facilitates international trade. Bulky goods, such as petroleum, grain, ore, and vehicles, are normally transported long distances by ship. This means that a country needs a seaport where goods can be transferred between land and sea. To send and receive goods by sea, a landlocked state must arrange to use another country's seaport.

PAUSE & REFLECT 8.3.4

Name a state outside of Africa that is landlocked.

Governing States

LEARNING OUTCOME 8.3.5

Describe differences among the three regime types.

A state has two types of government: a national government and local governments. At the national scale, a government can be more or less democratic. At the local scale, the national government can determine how much power to allocate to local governments.

NATIONAL SCALE: REGIME TYPES

Some national governments are better able than others to provide the leadership needed to promote peace and prosperity. A corrupt repressive government embroiled in wars is less able to respond effectively to economic challenges.

National governments can be classified as democratic, autocratic, or anocratic (Figure 8-39):

- A **democracy** is a country in which citizens elect leaders and can run for office.
- An **autocracy** is a country that is run according to the interests of the ruler rather than the people.
- An **anocracy** is a country that is not fully democratic or fully autocratic but displays a mix of the two types.

According to the Center for Systemic Peace, democracies and autocracies differ in three essential elements: selection of leaders, citizen participation, and checks and balances (Table 8-1).

The world has become more democratic (Figure 8-40). The Center for Systemic Peace cites these reasons:

- The replacement of increasingly irrelevant and out-of-touch monarchies with elected governments that broaden individual rights and liberties.
- The widening of participation in policymaking to all citizens through rights to vote and to serve in government.
- The diffusion of democratic government structures created in Europe and North America to other regions.

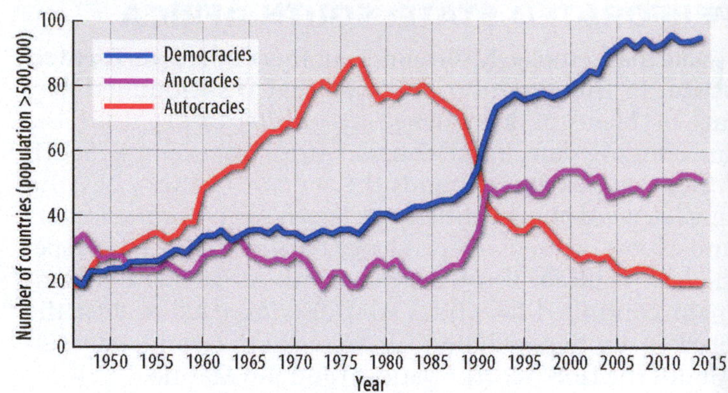

▲ **FIGURE 8-40 TREND TOWARD DEMOCRACY** The number of autocracies has declined sharply since the 1980s. The most rapid increase in democracies came after the breakup of Communist states in Europe.

PAUSE & REFLECT 8.3.5

What region of the world appears to have the greatest concentration of autocratic regimes?

The State Fragility Index, calculated by the Center for Systemic Peace, measures the effectiveness of the government, as well as its perceived legitimacy, to govern a country. The index combines several factors, including extent of regional unrest among disaffected citizens, ability of legal system to enforce contracts and property rights, level of compliance with paying taxes, and freedom to express diverse political views (Figure 8-41).

The most fragile states are clustered in sub-Saharan Africa. This is not surprising, as we have already seen that the region has the world's highest population growth and poorest health (Chapter 2), the greatest extent of ethnic cleansing and genocide (Chapter 7), and the most problematic

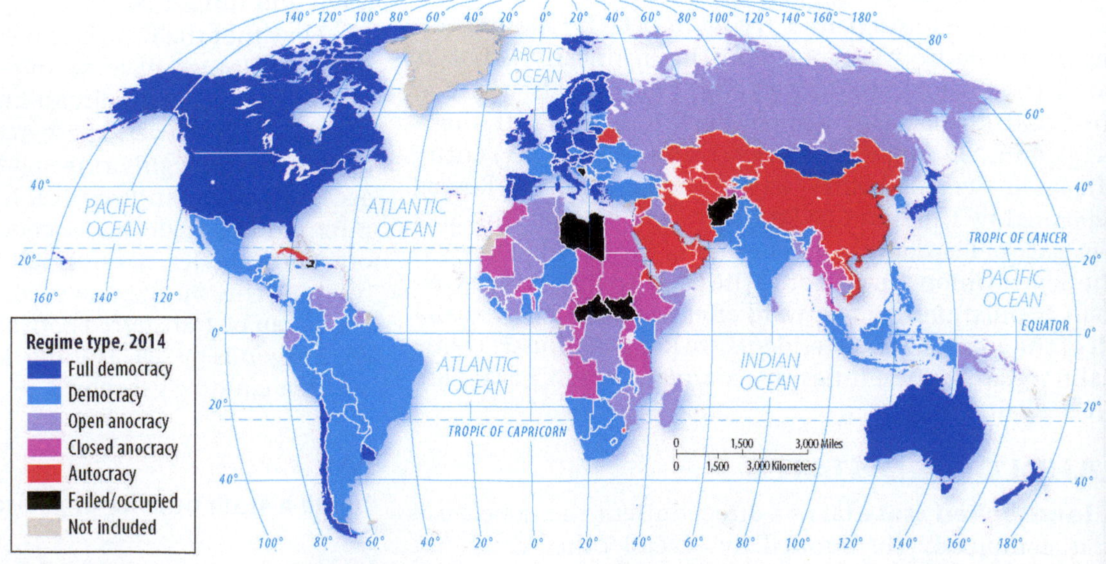

▶ **FIGURE 8-39 REGIME TYPE, 2014** Most states are either democratic, autocratic, or anocratic. In a few "failed" states, such as Libya and South Sudan, government institutions have broken down because of civil war, extreme poverty, or natural disasters—or some combination of the three.

TABLE 8-1 Comparing Democracy And Autocracy

Element	Democracy	Autocracy
Selection of leaders	Institutions and procedures through which citizens can express effective preferences about alternative policies and leaders.	Leaders are selected according to clearly defined (often hereditary) rules of succession from within the established political elite.
Citizen participation	Institutionalized constraints on the exercise of power by the executive.	Citizens' participation is sharply restricted or suppressed.
Checks and balances	Guarantee of civil liberties to all citizens in their daily lives and in acts of political participation.	Leaders exercise power with no meaningful checks from legislative, judicial, or civil society institutions.

shapes of states (the previous page). The region also has the largest number of recent civil wars.

LOCAL SCALE: UNITARY AND FEDERAL STATES

The governments of states are organized according to one of two approaches:

- A **unitary state** places most power in the hands of central government officials.
- A **federal state** allocates strong power to units of local government within the country.

UNITARY STATES. In principle, the unitary government system works best in nation-states characterized by few internal cultural differences and a strong sense of national unity. Because the unitary system requires effective communications with all regions of the country, smaller states are more likely to adopt it.

Some multinational states have adopted unitary systems so that the values of one nationality can be imposed on others. In Kenya and Rwanda, for instance, the mechanisms of a unitary state have enabled one ethnic group to extend dominance over weaker groups.

A good example of a nation-state, France has a long tradition of unitary government in which a very strong national government dominates local government decisions. The country's basic local government unit is 96 départements (departments). A second tier of local government in France is the 36,686 communes.

FEDERAL STATES. In a federal state, such as the United States, local governments possess considerable authority to adopt their own laws. Multinational states may adopt a federal system of government to empower different nationalities, especially if they live in separate regions of the country. Under a federal system, local government boundaries can be drawn to correspond with regions inhabited by different ethnicities.

Most of the world's largest states are federal, including Russia, Canada, the United States, Brazil, and India. However, the size of the state is not always an accurate predictor of the form of government: Tiny Belgium is a federal state (to accommodate the two main cultural groups, the Flemish and the Walloons), whereas China is a unitary state (to promote Communist values).

In recent years there has been a strong global trend toward federal government. Unitary systems have been sharply curtailed in a number of countries and scrapped altogether in others. In the face of increasing demands by ethnicities for more self-determination, states have restructured their governments to transfer some authority from the national government to local government units. An ethnicity that is not sufficiently numerous to gain control of the national government may be content with control of a regional or local unit of government.

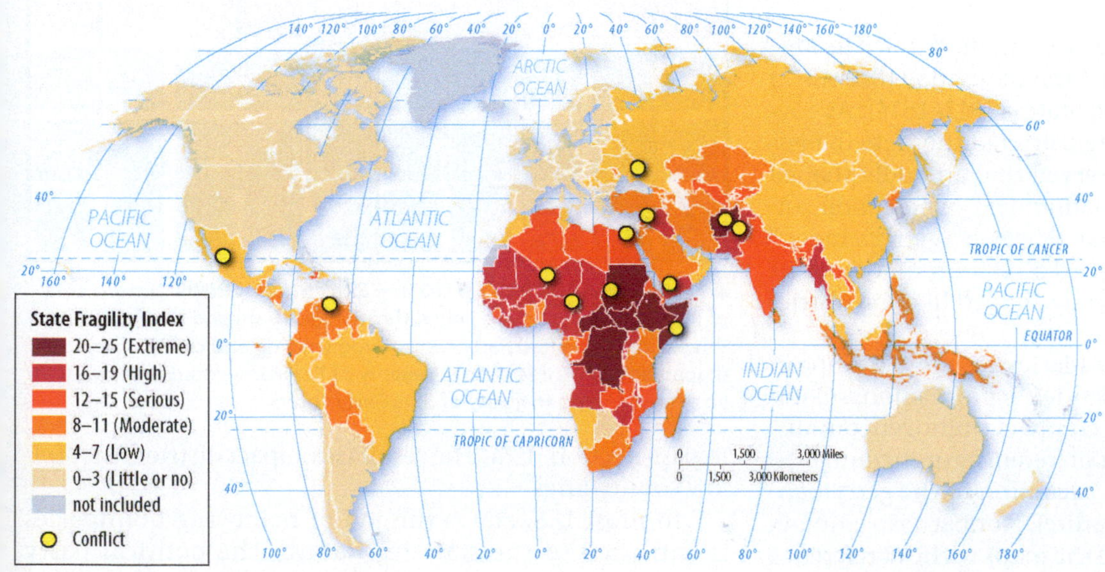

◀ **FIGURE 8-41 STATE FRAGILITY INDEX** The index is based on the extent of regional unrest among disaffected citizens, ability of legal system to enforce contracts and property rights, level of compliance with paying taxes, and freedom to express diverse political views.

Electoral Geography

LEARNING OUTCOME 8.3.6

Explain the concept of gerrymandering and three ways that it is done.

In democracies, politics must follow legally prescribed rules. But all parties to the political process often find ways of bending those rules to their advantage. A case in point is the drawing of legislative district boundaries. The boundaries separating legislative districts within the United States and other countries are redrawn periodically to ensure that each district has approximately the same population. Boundaries must be redrawn because migration inevitably results in some districts gaining population and others losing population. The 435 districts of the U.S. House of Representatives are redrawn every 10 years, following the Census Bureau's release of official population figures.

Redrawing legislative boundaries to benefit the party in power is called **gerrymandering.** Gerrymandering takes three forms:

- **Wasted vote.** Opposition supporters are spread across many districts but in the minority (Figure 8-42).
- **Excess vote.** Opposition supporters are concentrated into a few districts (Figure 8-43).
- **Stacked vote.** Distant areas of like-minded voters are linked through oddly shaped boundaries (Figure 8-44).

▲ **FIGURE 8-42 WASTED VOTE GERRYMANDERING** Opposition supporters are spread across many districts as a minority. Assume there are 50 voters, 20 supporters of the Red Party and 30 supporters of the Blue Party. If the Blue Party controls the redistricting process, it could create a wasted vote gerrymander by creating all five districts with slender majorities of Blue Party voters.

▲ **FIGURE 8-43 EXCESS VOTE GERRYMANDERING** Opposition supporters are concentrated into a few districts. Although only 20 of 50 voters are Red Party supporters, if the Red Party controls the redistricting process, it could create an excess vote gerrymander. Three districts have a slender majority of Red Party voters and two have an overwhelming majority of Blue Party voters.

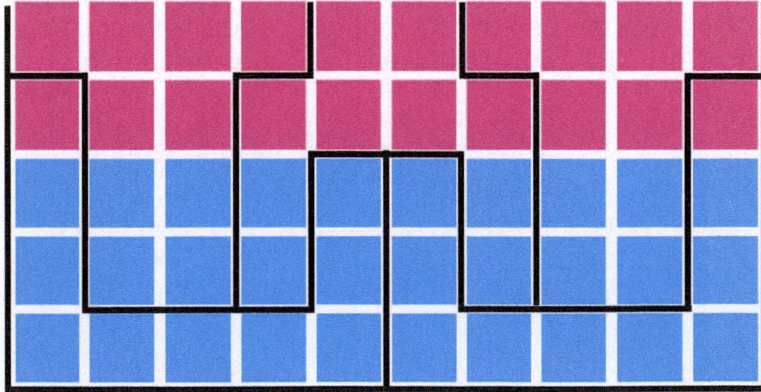

▲ **FIGURE 8-44 STACKED VOTE GERRYMANDERING** Distant areas of like-minded voters are linked through oddly shaped boundaries. In this example, the Red Party controls redistricting and creates five oddly shaped districts, three with a majority of Red Party voters and two with an overwhelming majority of Blue Party voters.

Stacked vote gerrymandering has been especially attractive for creating districts inclined to elect ethnic minorities. Because the two largest ethnic groups in the United States (African Americans and most Hispanics other than Cubans) tend to vote Democratic—in some elections more than 90 percent of African Americans vote Democratic—creating a majority African American district virtually guarantees election of a Democrat. Republicans support a "stacked" Democratic district because they are better able to draw boundaries that are favorable to their candidates in the rest of the state.

The term gerrymandering was named for Elbridge Gerry (1744–1814), governor of Massachusetts (1810–1812) and vice president of the United States (1813–1814). As governor, Gerry signed a bill that redistricted the state to benefit his party. An opponent observed that an oddly shaped new district looked like a "salamander," whereupon another opponent responded that it was a "gerrymander." A newspaper subsequently printed a cartoon of a monster named "gerrymander" with a body shaped like the district (Figure 8-47).

The job of redrawing boundaries in most European countries is entrusted to independent commissions. Commissions typically try to create compact homogeneous districts without regard for voting preferences or incumbents. In the United States, Iowa is an exception to the gerrymandering practice (see Debate It feature). Nonpartisan employees of the state legislature create the maps without reference to past election data. The result is compact districts that follow county lines.

In most U.S. states the job of redrawing boundaries is entrusted to the state legislature. The political party

DEBATE IT! Should independent commissions decide boundaries?

In the United States, most legislative boundaries are delineated by political leaders. In contrast, independent commissions are widely used in other countries to determine legislative boundaries.

DO NOT USE INDEPENDENT COMMISSIONS

- Elected officials, by virtue of having been elected, best represent the will of the people (Figure 8-45).
- Bestowing power in the hands of unelected commissioners makes the process less accountable to the people.
- Politicians help ensure that racial and ethnic minorities constitute a majority in some districts.

▲ **FIGURE 8-45 TOWN MEETING** Fayston, Vermont.

USE INDEPENDENT COMMISSIONS

- Districts will be compact and follow logical boundaries such as cities and counties (Figure 8-46).
- Communities that don't support the majority party will be divided among multiple districts, diluting their power.
- One party can gain a much higher percentage of seats than their share of the total vote would suggest.

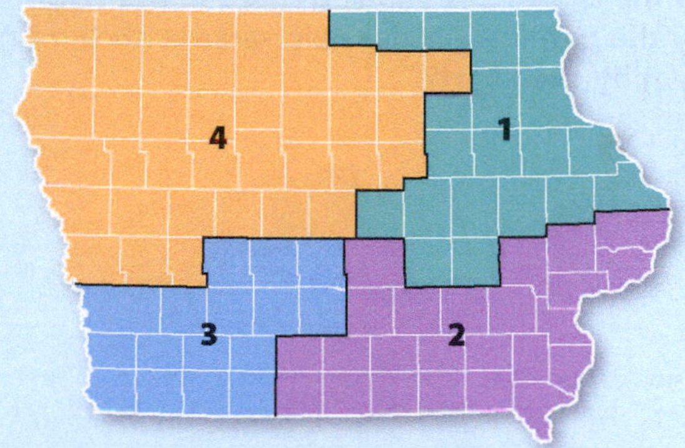

▲ **FIGURE 8-46 INDEPENDENT COMMISSION: IOWA** Iowa does not have gerrymandered congressional districts. Each district is relatively compact, and boundaries coincide with county boundaries. A nonpartisan commission creates Iowa's district each decade, without regard for past boundaries or impact on incumbents.

in control of the state legislature naturally attempts to redraw boundaries to improve the chances of its supporters to win seats (see Doing Geography feature on page 293). Only about one-tenth of congressional seats across the United States are competitive, making a shift of more than a few seats unlikely from one election to another in the United States, except in unusual circumstances. In contrast, all four districts in Iowa typically have competitive races.

The U.S. Supreme Court ruled gerrymandering illegal in 1985 but did not require dismantling of existing oddly shaped districts, and a 2001 ruling allowed North Carolina to add another oddly shaped district that ensured the election of an African American Democrat. A 2015 ruling permitted states to establish independent commissions.

PAUSE & REFLECT 8.3.6

How do the congressional districts in Iowa compare with the three forms of gerrymandering?

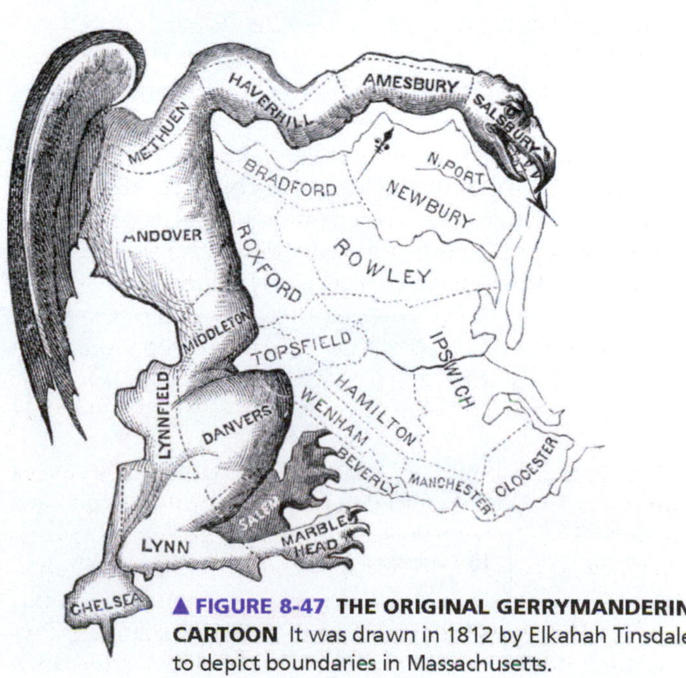

▲ **FIGURE 8-47 THE ORIGINAL GERRYMANDERING CARTOON** It was drawn in 1812 by Elkahah Tinsdale to depict boundaries in Massachusetts.

Geography of Gerrymandering

LEARNING OUTCOME 8.3.7

Describe examples of gerrymandering.

A score was given by the Washington Post to each Congressional district according to the extent of gerrymandering (Figure 8-48). The gerrymandering score was determined by calculating the ratio of the area of the district to the area of a circle with the same perimeter. A district that follows a regular compact shape has a lower score than a district with irregularities. The state judged to have the most gerrymandering is North Carolina (Figure 8-49).

PAUSE & REFLECT 8.3.6

How was the city of Las Vegas treated in the two maps drawn by the political parties compared with the final map drawn by the court?

CHECK-IN KEY ISSUE 3

Why Do Boundaries Cause Problems?

- **Two types of boundaries are physical and cultural.**
- **Deserts, mountains, and water can serve as physical boundaries between states.**
- **Geometry and ethnicity can create cultural boundaries between states.**
- **Five shapes of states are compact, elongated, prorupted, perforated, and fragmented.**
- **The governance of states can be classified as democratic, anocratic, or autocratic; democracies have been increasing.**
- **Boundaries dividing electoral districts within countries can be gerrymandered in several ways to favor one political party.**

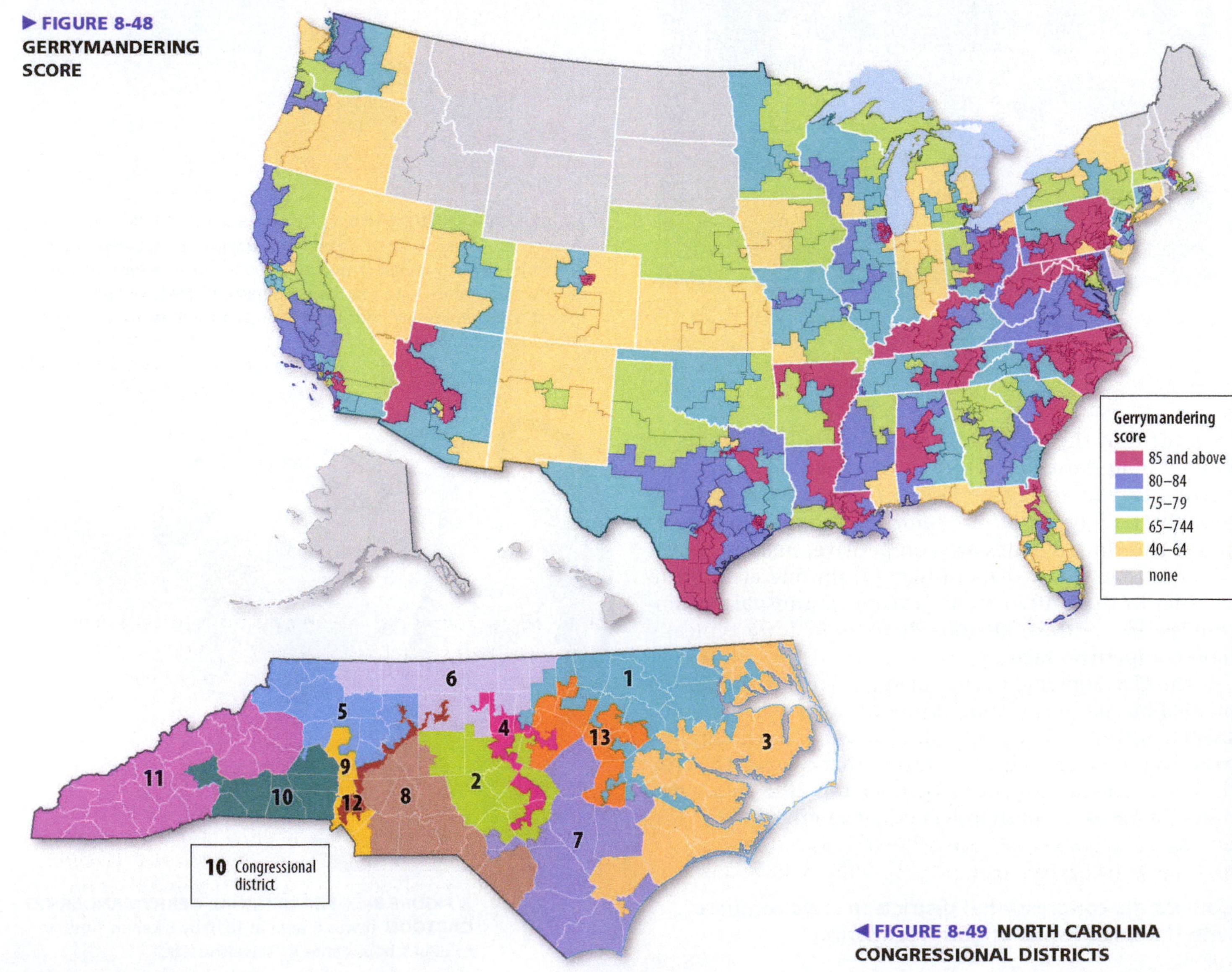

▶ **FIGURE 8-48 GERRYMANDERING SCORE**

◀ **FIGURE 8-49 NORTH CAROLINA CONGRESSIONAL DISTRICTS**

DOING GEOGRAPHY Dueling Gerrymandering in Nevada

Competing plans by Democrats and Republicans to draw boundaries for Nevada's four congressional districts in 2010 illustrate all three forms of gerrymandering (Figure 8-50).

Wasted vote gerrymandering (Figure 8-50a): the Democratic plan. Although Nevada as a whole has slightly more registered Democrats than Republicans (43 percent to 37 percent), the Democratic plan made Democrats more numerous than Republicans in three of the four districts.

Excess vote gerrymandering (Figure 8-50b): the Republican plan. By clustering a large share of the state's registered Democrats in District 4, the Republican plan gave Republicans the majority of registered voters in two of the four districts.

Stacked vote gerrymandering (Figure 8-50a and Figure 8-50b): both plans. In the Republican plan, District 4 had a majority Hispanic population and was surrounded by a C-shaped District 1. The Democratic plan created a long, narrow District 3.

Without gerrymandering (Figure 8-50c): nonpartisan plan. The Nevada Court rejected both parties' maps and created regularly shaped districts that minimized gerrymandering. Districts 1 and 4 have more registered Democrats, and Districts 2 and 3 more Republicans.

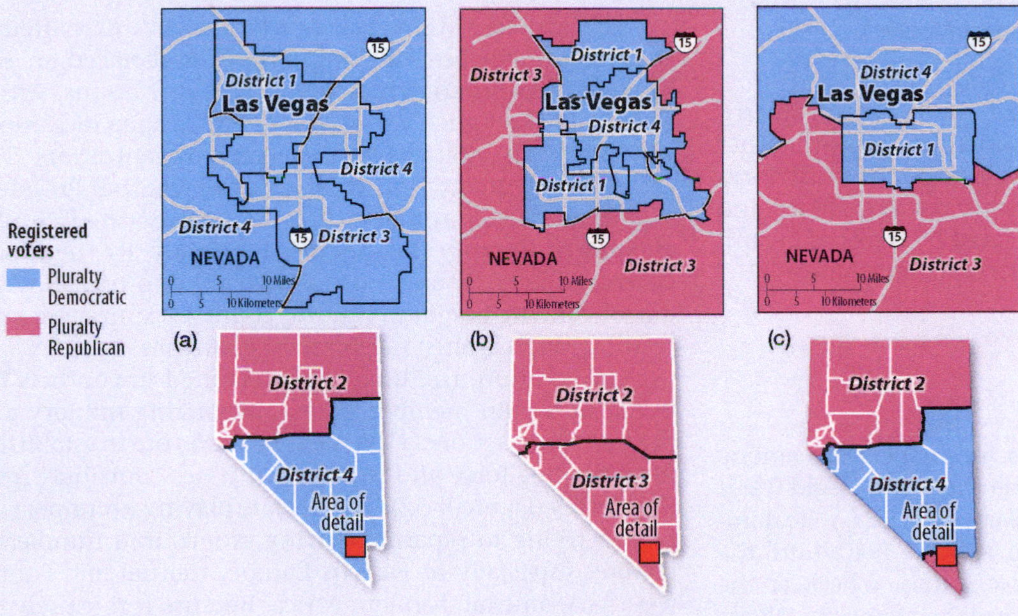

FIGURE 8-50 REDISTRICTING PROPOSALS IN NEVADA The three maps on top show the Las Vegas area in more detail. **(a)** Wasted vote and stacked vote gerrymandering. **(b)** Excess vote and stacked vote gerrymandering. **(c)** Nonpartisan court-imposed map.

Your Political Geography

Political Boundaries in Your Community

Check out the shape of your state's legislative districts.

1. In your search engine, enter "[your state] congressional district map." If you live in a state with only one state-wide at-large Representative (Alaska, Delaware, Montana, North Dakota, South Dakota, Vermont, and Wyoming), enter "[your state] state legislature district map." If you live in a state with more than one Representative, you can enter your state legislature map in addition to or instead of the congressional map.
2. Is your legislative district compact and geometrically shaped, or is it irregularly shaped? If irregularly shaped, can you see a geographical reason for the shape, perhaps a natural feature such as a body of water, or a cultural boundary such as between ethnicities?
3. A gerrymander score has been calculated for each Congressional district. Use your Internet browser to search for "gerrymander score." The higher the score, the more severe the gerrymandering. What is the gerrymander score for your Congressional district? Did you expect your district to have a higher score or a lower score? Why?

KEY ISSUE 4

Where Do States Face Threats?

- ▸ Global Cooperation and Competition
- ▸ Competition and Cooperation in Europe
- ▸ Terrorist Attacks Against the United States
- ▸ Terrorist Organizations
- ▸ State Support for Terrorism

LEARNING OUTCOME 8.4.1

Describe the functions of the United Nations.

States compete for many reasons, including control of territory, access to trade and resources, and influence over other states. To further their competitive goals, states may form alliances with other states. In recent years, violence and wars have increasingly been instigated by terrorist organizations not affiliated with particular states or alliances.

Global Cooperation and Competition

The most important global forum for cooperation among states is the United Nations, created at the end of World War II by the victorious Allies. The early years of the U.N. were dominated by the start of the Cold War era (late 1940s until the early 1990s), a period of competition and tension between the United States and the Soviet Union and their respective allies.

THE UNITED NATIONS

The United Nations was organized in 1945 with 51 original members, including 49 sovereign states plus Byelorussia (now Belarus) and Ukraine, then part of the Soviet Union. The number of U.N. members reached 193 in 2011. The U.N. membership has increased rapidly on three occasions (Figure 8-51):

- **1955.** Sixteen countries joined in 1955, mostly European countries that had been liberated from Nazi Germany during World War II.
- **1960.** Seventeen new members were added in 1960, all but one a former African colony of Britain or France. Only four African states were original members of the United Nations—Egypt, Ethiopia, Liberia, and South Africa—and only six more joined during the 1950s.

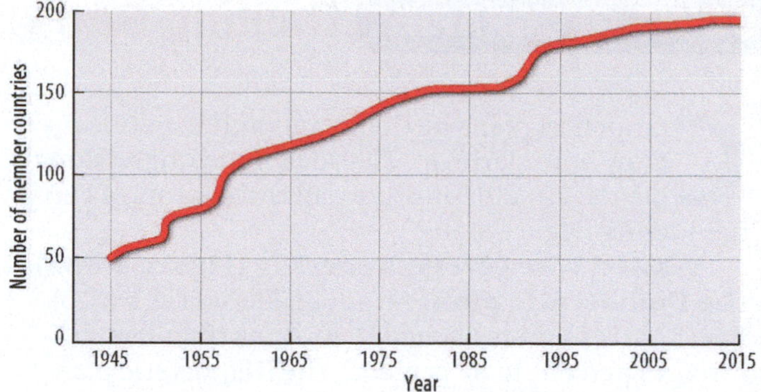

▲ **FIGURE 8-51 GROWTH IN U.N. MEMBERSHIP** U.N. membership has increased from its original 51 to 193 in 2011.

- **1990–1993.** Twenty-six countries were added between 1990 and 1993, primarily due to the breakup of the Soviet Union and Yugoslavia. U.N. membership also increased in the 1990s because of the admission of several microstates.

The United Nations was not the world's first attempt at international peacemaking. The U.N. replaced an earlier organization known as the League of Nations, which was established after World War I. The League of Nations was never an effective peacekeeping organization. The United States did not join it, despite the fact that President Woodrow Wilson initiated the idea, because the U.S. Senate refused to ratify the membership treaty. By the 1930s, Germany, Italy, Japan, and the Soviet Union had all withdrawn, and the League of Nations could not stop aggression by these states against neighboring countries.

On occasion, the U.N. has intervened in conflicts between or within member states, authorizing military and peacekeeping actions. U.N. members can vote to establish a peacekeeping force and request that states contribute military forces (Figure 8-52). The U.N. is playing an important role in trying to separate warring groups in a number of regions, especially in Eastern Europe, Central and Southwest Asia, and sub-Saharan Africa. Because it must rely on individual countries to supply troops, the U.N. often lacks enough of them to keep peace effectively. The U.N. tries to

▼ **FIGURE 8-52 U.N. PEACEKEEPERS** Patrolling the only crossing between Syria and Israel.

maintain strict neutrality in separating warring factions, but this has proved difficult in places such as Bosnia & Herzegovina, where most of the world sees two ethnicities (Bosnia's Serbs and Croats) as aggressors undertaking ethnic cleansing against weaker victims (Bosniaks).

However, any one of the five permanent members of the Security Council—China, France, Russia (formerly the Soviet Union), the United Kingdom, and the United States—can veto a peacekeeping operation. During the Cold War era, the United States and the Soviet Union used the veto to prevent undesired U.N. intervention, and it was only after the Soviet Union's delegate walked out of a Security Council meeting in 1950 that the U.N. voted to send troops to support South Korea. More recently, the opposition of China and Russia has made it difficult for the international community to prevent Iran from developing nuclear weapons.

Despite its shortcomings, though, the U.N. represents a forum where, during this era of rapid changes in states and their relationships, for the first time in history virtually all states of the world can meet and vote on issues without resorting to war. More importantly, the U.N. has played a major role in the promotion of international cooperation to address global economic problems, promote human rights, and provide humanitarian relief.

THE COLD WAR

During the Cold War era, the United States and the Soviet Union were the world's two superpowers. As very large states, both superpowers could quickly deploy armed forces in different regions of the world. To maintain strength in regions that were not contiguous to their own territory, the United States and the Soviet Union established military bases in other countries. From these bases, ground and air support were in proximity to local areas of conflict. Naval fleets patrolled the major bodies of water.

Both superpowers repeatedly demonstrated that they would use military force if necessary to prevent an ally from becoming too independent. The Soviet Union sent its armies into Hungary in 1956 and Czechoslovakia in 1968 to install more sympathetic governments. Because these states were clearly within the orbit of the Soviet Union, the United States chose not to intervene militarily. Similarly, the United States sent troops to the Dominican Republic in 1965, Grenada in 1983, and Panama in 1989 to ensure that they would remain allies.

Before the Cold War, the world typically contained more than two superpowers. For example, before the outbreak of World War I in the early twentieth century, there were eight great powers: Austria, France, Germany, Italy, Japan, Russia, the United Kingdom, and the United States. When a large number of states ranked as great powers of approximately equal strength, no single state could dominate. Instead, major powers joined together to form temporary alliances. A condition of roughly equal strength between opposing alliances is known as a **balance of power.** In contrast, the post–World War II balance of power was bipolar between the United States and the Soviet Union. Because the power of these two states was so much greater than the power of all other states, the world comprised two camps, each under the influence of one of the superpowers. Other states lost the ability to tip the scales significantly in favor of one or the other superpower. They were relegated to a new role of either ally or satellite.

CUBAN MISSILE CRISIS. A major confrontation during the Cold War between the United States and Soviet Union came in 1962, when the Soviet Union secretly began to construct missile-launching sites in Cuba, less than 150 kilometers (90 miles) from U.S. territory. President John F. Kennedy went on TV to demand that the missiles be removed, and he ordered a naval blockade to prevent additional Soviet material from reaching Cuba. At the United Nations, immediately after Soviet Ambassador Valerian Zorin denied that his country had placed missiles in Cuba, U.S. Ambassador Adlai Stevenson dramatically revealed aerial photographs taken by the U.S. Department of Defense, clearly showing preparations for them (see examples in Figure 8-53). Faced with irrefutable evidence that the missiles existed, the Soviet Union ended the crisis by dismantling them.

PAUSE & REFLECT 8.4.1

Why have only a small handful of states joined the U.N. since 2000?

(a)

(b)

▲ **FIGURE 8-53 CUBAN MISSILE CRISIS** Aerial photographs to show the Soviet buildup in Cuba. **(a)** Three Soviet ships with missile equipment being unloaded at Mariel naval port in Cuba. **(b)** Within the outline (enlarged and rotated) are Soviet missile transporters, fuel trailers, and oxidizer trailers for the combustion of missile fuel.

Competition and Cooperation in Europe

LEARNING OUTCOME 8.4.2

Describe the principal military and economic alliances in Europe.

During the Cold War that followed World War II, two military alliances and two economic alliances formed in Europe. In the twenty-first century, one of the military alliances and one of the economic alliances continues, whereas the other two have been disbanded.

COLD WAR–ERA MILITARY ALLIANCES

After World War II, most European states joined one of two military alliances (Figure 8-54):

- **North Atlantic Treaty Organization (NATO).** A military alliance among 16 democratic states, including the United States and Canada plus 14 European states.
- **The Warsaw Pact.** A military agreement among Communist Eastern European countries. The Warsaw Pact disbanded in 1991 following the end of communism in Eastern Europe.

NATO and the Warsaw Pact were designed to maintain a bipolar balance of power in Europe. For NATO allies, the principal objective was to prevent the Soviet Union from overrunning West Germany and other smaller countries. The Warsaw Pact provided the Soviet Union with a buffer of allied states between it and Germany to discourage a third German invasion of the Soviet Union in the twentieth century. Some of Hungary's leaders in 1956 asked for the help of Warsaw Pact troops to crush an uprising that threatened Communist control of the government. Warsaw Pact troops also invaded Czechoslovakia in 1968 to depose a government committed to reforms.

▲ **FIGURE 8-54 NATO AND WARSAW PACT, 1949–1991** During the Cold War era, most European states joined either the pro-U.S. North Atlantic Treaty Organization (NATO) or the pro-Soviet Warsaw Pact.

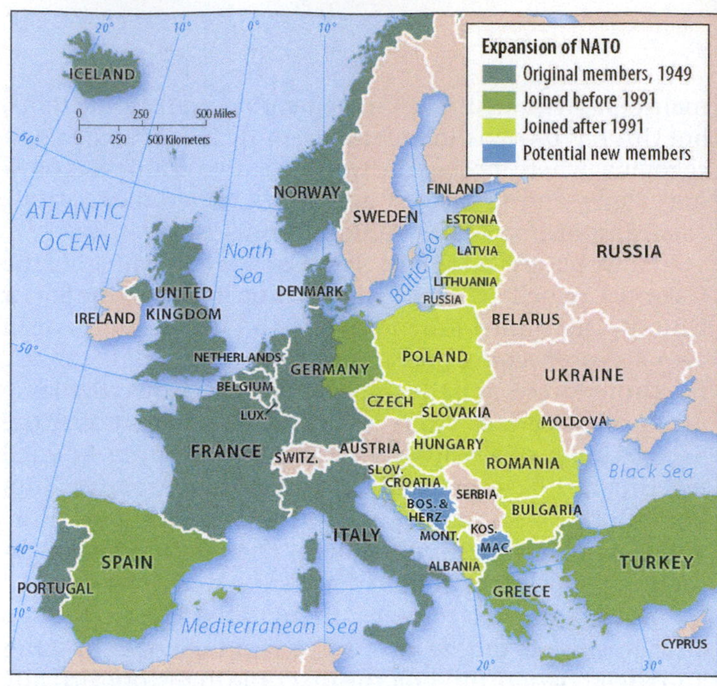

▲ **FIGURE 8-55 NATO** Membership in NATO has extended to 26 European countries, plus Canada and the United States.

In a Europe no longer dominated by military confrontation between two blocs, the Warsaw Pact was disbanded, and the number of troops under NATO command was sharply reduced. NATO expanded its membership to 29 states by adding the former Warsaw Pact countries except for Russia, as well as several states formerly republics within the Soviet Union (Figure 8-55). Membership in NATO has offered Eastern European countries a sense of security against any future Russian threat, as well as participation in a common united Europe. Russia's annexation of Crimea in 2014 and continued support for rebels in eastern Ukraine have heightened fears of renewed confrontation.

COLD WAR–ERA ECONOMIC ALLIANCES

During the Cold War, two economic alliances formed in Europe:

- **European Union (EU).** The EU (formerly known as the European Economic Community, the Common Market, and the European Community) formed in 1958 with six members. The EU was designed to heal scars from World War II (which had ended only 13 years earlier).
- **Council for Mutual Economic Assistance (COMECON).** COMECON formed in 1949 with six members in 1960. Mongolia, Cuba, and Vietnam also joined. COMECON was designed to promote trade and sharing of natural resources in Communist Eastern Europe. Like the Warsaw Pact, COMECON disbanded in 1991.

THE EUROPEAN UNION IN THE TWENTY-FIRST CENTURY

With the end of the Cold War, economic cooperation throughout Europe has become increasingly important.

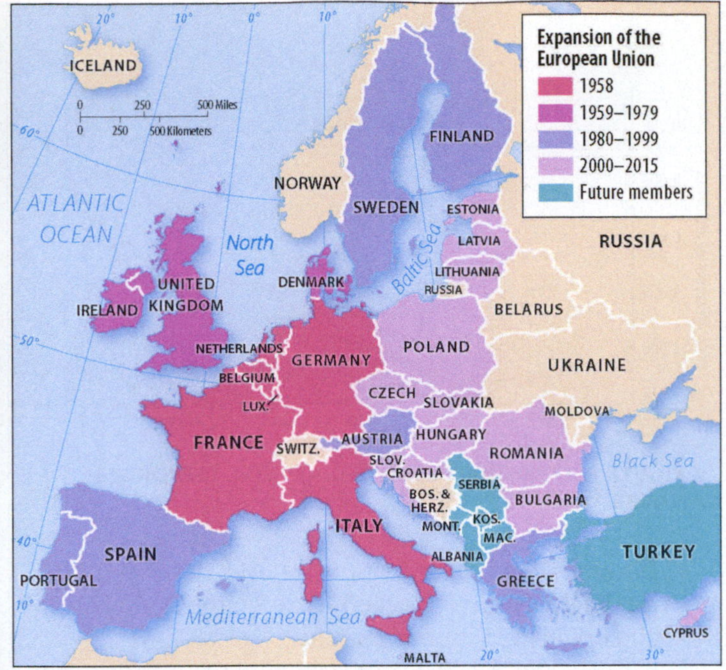

▲ **FIGURE 8-56 EUROPEAN UNION**

The EU expanded from its original 6 countries to 12 during the 1980s and 28 during the first decade of the twenty-first century (Figure 8-56). The most recent additions have been former members of COMECON.

The main task of the European Union is to promote development within the member states through economic and political cooperation:

- A European Parliament is elected by the people in each of the member states simultaneously.
- Subsidies are provided to farmers and to economically depressed regions.
- Most goods move across borders of member states in trucks and trains without stopping.
- With a few exceptions, a citizen of one EU member state is permitted to work in other states.
- A bank or retailer can open branches in any member country with supervision only by the corporation's home country.

The most dramatic step taken toward integrating Europe's nation-states into a regional organization was the creation of the eurozone. A single bank, the European Central Bank, was given responsibility for setting interest rates and minimizing inflation throughout the eurozone.

Most importantly, a common currency, the euro, was created for electronic transactions beginning in 1999 and in notes and coins beginning in 2002. France's franc, Germany's mark, and Italy's lira—powerful symbols of sovereign nation-states—have disappeared. Twenty-five countries use the euro, including 19 of the 28 EU members.

European leaders bet that every country in the region would be stronger economically if it replaced its national currency with the euro. For the first few years that was the case, but the future of the euro has been called into question by the severe global recession that began in 2008. The economically weaker countries within the eurozone, such as Greece, Ireland, Italy, and Spain, have been forced to implement harsh and unpopular policies, such as drastically cutting services and raising taxes, whereas the economically strong countries, especially Germany, have been forced to subsidize the weaker states. Greece has been in the forefront of the confrontation between north and south within the EU. Germany and other northern European states charge that Greece has failed to implement economic reforms.

Despite the north–south tensions within the EU, future enlargements are possible: Albania, Macedonia, Montenegro, Serbia, and Turkey are in various stages of negotiations.

PAUSE & REFLECT 8.4.2

What might be the benefits and disadvantages for additional countries to join the European Union? What might be the benefits and disadvantages for the European Union to let in new members?

ALLIANCES IN OTHER REGIONS

Economic cooperation has been an important factor in the creation of international organizations that now can be found far beyond Western Europe. Other prominent regional organizations include:

- **Organization on Security and Cooperation in Europe (OSCE).** The OSCE's 57 members include the United States, Canada, Russia, all European states, and all states formerly part of the Soviet Union. When founded in 1975, the Organization on Security and Cooperation was composed primarily of Western European countries and played only a limited role. With the end of the Cold War in the 1990s, the renamed OSCE became a more active forum for countries concerned with ending conflicts in Europe, especially in the Balkans and Caucasus.
- **Organization of American States (OAS).** All 35 states in the Western Hemisphere are members of the OAS. The OAS promotes social, cultural, political, and economic links among member states. Cuba is a member but was suspended from most OAS activities in 1962. As Cuba and the United States move towards normalization of relations, Cuba is in the process of being reintegrated into OAS activities.
- **African Union (AU).** Established in 2002, the AU encompasses 54 countries in Africa. The AU replaced an earlier organization called the Organization of African Unity, founded in 1963 primarily to seek an end to colonialism and apartheid in Africa. The new organization has placed more emphasis on promoting economic integration in Africa.
- **Commonwealth.** The Commonwealth includes the United Kingdom and 52 other states that were once British colonies, including Australia, Bangladesh, Canada, India, Nigeria, and Pakistan. Most other members are African states or island countries in the Caribbean or Pacific. Commonwealth members seek economic and cultural cooperation.

Terrorist Attacks Against the United States

LEARNING OUTCOME 8.4.3

Explain the concept of terrorism and cite U.S. examples.

Terrorism is the systematic use of violence by a group calculated to create an atmosphere of fear and alarm among a population or to coerce a government into actions it would not otherwise undertake or refrain from actions it wants to take. Distinctive characteristics of terrorists include:

- Trying to achieve their objectives through organized acts that spread fear and anxiety among the population, such as bombing, kidnapping, hijacking, taking of hostages, and assassination.
- Viewing violence as a means of bringing widespread publicity to goals and grievances that are not being addressed through peaceful means.
- Believing in a cause so strongly that they do not hesitate to attack despite knowing that they will probably die in the act.

The term *terror* (from the Latin "to frighten") was first applied to the period of the French Revolution between March 1793 and July 1794, known as the Reign of Terror. In the name of protecting the principles of the revolution, the Committee of Public Safety, headed by Maximilien Robespierre, guillotined several thousand of its political opponents. In modern times, the term *terrorism* has been applied to actions by groups operating outside government rather than to groups of official government agencies, although some governments provide military and financial support for terrorists.

Terrorism differs from assassination and other acts of political violence in that attacks are aimed at ordinary people rather than at military targets or political leaders. Other types of military action can result in civilian deaths—bombs can go astray, targets can be misidentified, or an enemy's military equipment can be hidden in civilian buildings—but average individuals are unintended victims rather than principal targets in most conflicts. Terrorists consider all citizens responsible for the actions they oppose, so they view civilians as legitimate targets.

The number of terrorist incidents between 2000 and 2013 has been documented by the Institute for Economics and Peace. The number of terrorist incidents increased from around 1,000 in 2000 to around 10,000 in 2013 (Figure 8-57). The number of deaths from terrorist attacks increased from around 3,500 in 2000 to 18,000 in 2013 (Figure 8-58). Around two-thirds of the terrorist incidents and fatalities have occurred in five countries: Iraq, Afghanistan, Pakistan, Nigeria, and Syria.

TERRORISM AGAINST AMERICANS

The United States has suffered several terrorist attacks since 1988. Among the most destructive:

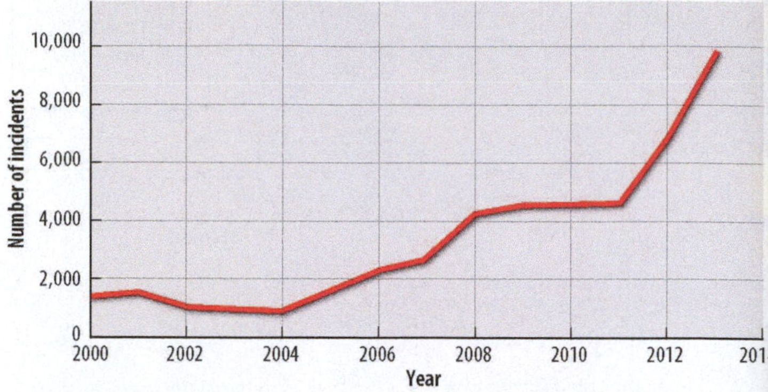

▲ **FIGURE 8-57 NUMBER OF INTERNATIONAL TERRORISM ATTACKS**

- **December 21, 1988.** A terrorist bomb destroyed Pan Am Flight 103 over Lockerbie, Scotland, killing all 259 aboard, plus 11 on the ground.
- **February 26, 1993.** A car bomb parked in the underground garage damaged New York's World Trade Center, killing 6 and injuring about 1,000.
- **April 19, 1995.** A car bomb killed 168 people in the Alfred P. Murrah Federal Building in Oklahoma City.
- **June 25, 1996.** A truck bomb blew up an apartment complex in Dhahran, Saudi Arabia, killing 19 U.S. soldiers who lived there and injuring more than 100 people.
- **August 7, 1998.** U.S. embassies in Kenya and Tanzania were bombed, killing 190 and wounding nearly 5,000.
- **October 12, 2000.** The *USS Cole* was bombed while in the port of Aden, Yemen, killing 17 U.S. service personnel.
- **April 15, 2013.** Two bombs were detonated near the finish line of the Boston Marathon, killing 3 and injuring more than 180 (Figure 8-59).
- **April 2, 2014.** Four people were killed and fourteen injured in an attack on the military base at Fort Hood, Texas.
- **July 16, 2015.** Five U.S. military personnel were killed in an attack on two installations in Chattanooga, Tennessee.
- **December 2, 2015.** A mass shooting at the Inland Regional Center in San Bernardino, California, killed 14 and injured 21.

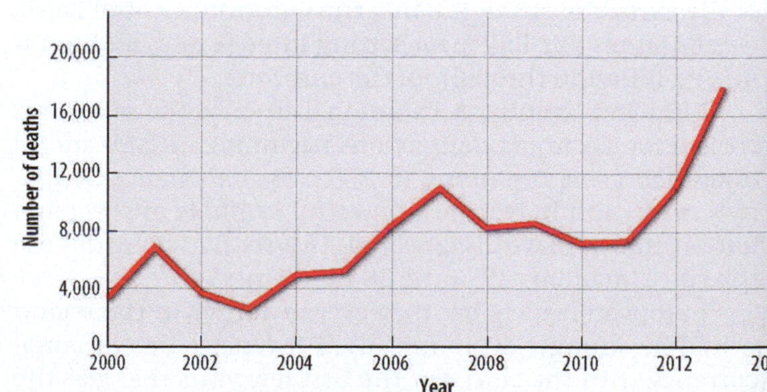

▲ **FIGURE 8-58 NUMBER OF DEATHS FROM INTERNATIONAL TERRORISM**

▲ **FIGURE 8-59 TERRORIST ATTACK ON THE BOSTON MARATHON, APRIL 15, 2013**

Some of the terrorists during the 1990s were American citizens operating alone or with a handful of others:

- Theodore J. Kaczynski, known as the Unabomber, was convicted of killing 3 people and injuring 23 others by sending bombs through the mail during a 17-year period. His targets were mainly academics in technological disciplines and executives in businesses whose actions he considered to be adversely affecting the environment.
- Timothy J. McVeigh was convicted and executed for the Oklahoma City bombing. For assisting McVeigh, Terry I. Nichols was convicted of conspiracy and involuntary manslaughter but not executed. McVeigh claimed that his terrorist act was provoked by rage against the U.S. government for such actions as the Federal Bureau of Investigation's 51-day siege of the Branch Davidian religious compound near Waco, Texas, culminating with an attack on April 19, 1993, that resulted in 80 deaths.
- Dzhokhar Tsarnaev was convicted and sentenced to death for the Boston Marathon bombing. His older brother Tamerlan Tsarnaev was killed in a shootout. The Tsarnaev brothers were of Chechen ethnicity. Before his capture, Dzhokhar left a note criticizing U.S. actions in Iraq and Afghanistan against Muslims.

SEPTEMBER 11, 2001, ATTACKS

The most dramatic terrorist attacks against the United States occurred on September 11, 2001 (Figure 8-60). The tallest buildings in the United States, the 110-story twin towers of the World Trade Center in New York City, were destroyed, and the Pentagon, near Washington, D.C., was damaged. The attacks resulted in 2,977 civilian fatalities:

- 88 (77 other passengers and 11 crew members) on American Airlines Flight 11, which crashed into World Trade Center Tower 1 (North Tower).
- 60 (51 passengers and 9 crew members) on United Airlines Flight 175, which crashed into World Trade Center Tower 2 (South Tower).
- 2,605 on the ground at the World Trade Center.
- 59 (53 passengers and 6 crew members) on American Airlines Flight 77, which crashed into the Pentagon.
- 125 on the ground at the Pentagon.
- 40 (33 passengers and 7 crew members) on United Airlines Flight 93, which crashed near Shanksville, Pennsylvania, after passengers fought with terrorists on board, preventing an attack on another Washington, D.C., target.

In addition, 19 terrorists died on the four hijacked airplanes. Responsible or implicated in most of the anti-U.S. terrorism during the 1990s, as well as the September 11, 2001, attack, was the al-Qaeda network.

PAUSE & REFLECT 8.4.3

How has travel in the United States been affected by the 9/11 attacks?

(a)

(b)

▲ **FIGURE 8-60 TERRORIST ATTACK ON THE WORLD TRADE CENTER, SEPTEMBER 11, 2001** **(a)** At 9:03 A.M., United Flight 175 approaches World Trade Center Tower 2 and **(b)** crashes into it. Tower 1 is already burning from the crash of American Flight 11 at 8:45 A.M.

Terrorist Organizations

LEARNING OUTCOME 8.4.4

Describe the major terrorist organizations.

Some terrorist attacks are the work of one or two individuals who are not formally associated with terrorist organizations. Most recent attacks, however, have been carried out by members belonging to terrorist organizations. Three prominent terrorist organizations in recent years are al-Qaeda, the Islamic State, and Boko Haram.

AL-QAEDA

Al-Qaeda (an Arabic word meaning "the foundation," or "the base") was founded around 1990 by Osama bin Laden to unite several groups of fighters in Afghanistan, as well as his supporters elsewhere in Southwest Asia. Osama's father, Mohammed bin Laden, a native of Yemen, established a construction company in Saudi Arabia and became a billionaire through close connections to the royal family. Osama bin Laden, one of about 50 children fathered by Mohammed with several wives, used his several-hundred-million-dollar inheritance to fund al-Qaeda.

Bin Laden moved to Afghanistan during the mid-1980s to support the fight against the Soviet army and the country's Soviet-installed government. Calling the anti-Soviet fight a holy war, or *jihad*, bin Laden recruited militant Muslims from Arab countries to join the cause. After the Soviet Union withdrew from Afghanistan in 1989, bin Laden returned to Saudi Arabia, but he was expelled in 1991 for opposing the Saudi government's decision permitting the United States to station troops there during the 1991 war against Iraq. Bin Laden moved to Sudan but was expelled in 1994 for instigating attacks against U.S. troops in Yemen and Somalia, so he returned to Afghanistan, where he lived as a "guest" of the Taliban-controlled government.

Bin Laden issued a declaration of war against the United States in 1996 because of U.S. support for Saudi Arabia and Israel. In a 1998 *fatwa* ("religious decree"), bin Laden argued that Muslims have a duty to wage a holy war against U.S. citizens because the United States was responsible for maintaining the Saud royal family as rulers of Saudi Arabia and the State of Israel. He claimed that destruction of the Saudi monarchy and the Jewish state would liberate from their control Islam's three holiest sites of Makkah (Mecca), Madinah, and Jerusalem.

Al-Qaeda's most deadly attacks since 9/11 have been in Iraq, especially between 2007 and 2011. Al-Qaeda has also been implicated in terrorist attacks in Egypt, Iraq, Jordan, Pakistan, Saudi Arabia, Turkey, and the United Kingdom.

Al-Qaeda is not a single unified organization, and the number involved in al-Qaeda is unknown. Bin Laden was advised by a small leadership council, which has several committees that specialize in areas such as finance, military, media, and religious policy. In addition to the original organization founded by Osama bin Laden responsible for the World Trade Center attack, al-Qaeda also encompasses local franchises concerned with country-specific issues, as well as imitators and emulators ideologically aligned with al-Qaeda but not financially tied to it.

In recent years, al-Qaeda's most active affiliate has been in Yemen. al-Qaeda's Yemen affiliate controls a large portion of Yemen's territory (Figure 8-61). Its most notorious terrorist attack was on January 7, 2015, at the Paris offices of the French satirical newspaper *Charlie Hebdo*. Two members of al-Qaeda in Yemen killed 11 people in the office and 6 more elsewhere in Paris, and injured 22. The attackers were protesting cartoons depicting Muhammad in the newspaper. Several cartoonists were among the victims (Figure 8-62).

PAUSE & REFLECT 8.4.4

Translate "je suis Charlie." Why might people in Paris carry signs saying this?

ISLAMIC STATE (ISIS/ISIL)

The Islamic State originated in 1999 and became an affiliate of al-Qaeda in 2004. However, in 2014, the two organizations split apart because of lack of agreement on how to cooperate and consult with each other. The Islamic State is also known as the Islamic State of Iraq and Syria (ISIS) and the Islamic State of Iraq and the Levant (ISIL).

Members of the Islamic State are Sunni Muslims who seek to impose strict religious laws throughout Southwest Asia. They have maintained control of territory through human rights violations, such as beheadings, massacres, and torture. The organization claims that it has authority to rule Muslims around the world. The Islamic State has had

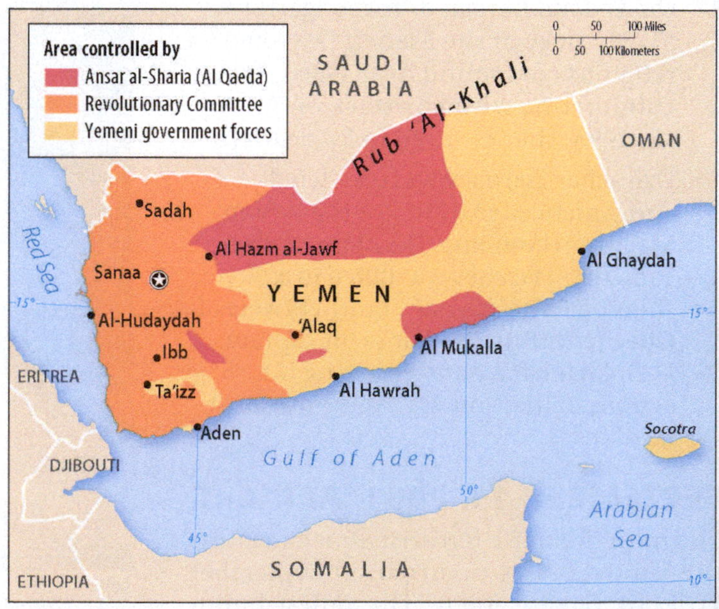

▲ **FIGURE 8-61 AL-QAEDA IN YEMEN** al-Qaeda controls a large portion of Yemen's desert interior, as well as a portion along the Gulf of Aden.

▲ **FIGURE 8-62 AFTERMATH OF AL-QAEDA ATTACK** Marchers in Paris show support for victims of the al-Qaeda terrorist attack on the newspaper.

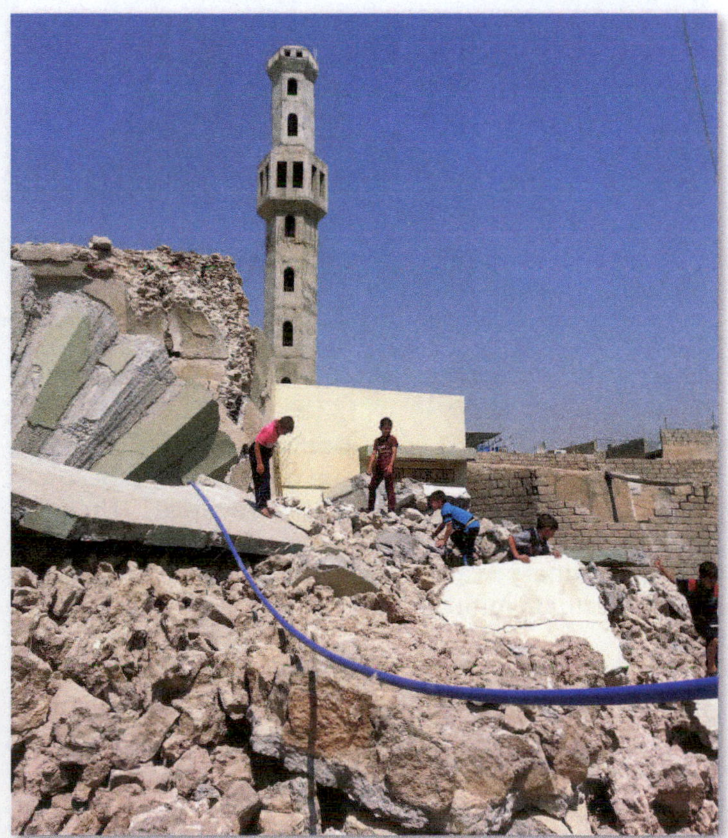

▲ **FIGURE 8-63 ISLAMIC STATE DESTRUCTION** Islamic State terrorists have destroyed historic graves in Mosul, Iraq, because they disagree with the teachings of the clerics buried there.

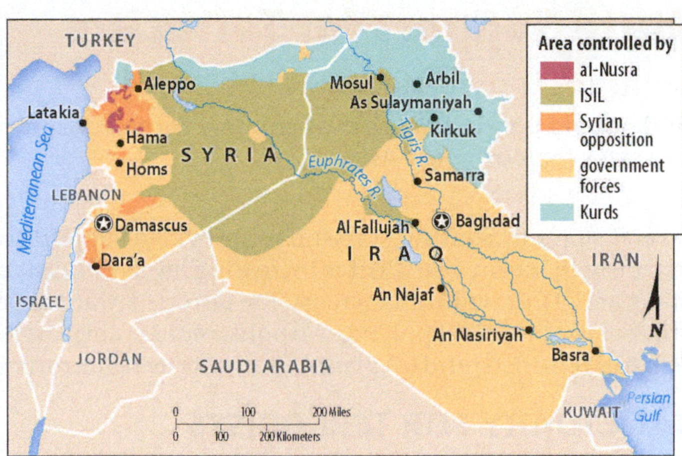

▲ **FIGURE 8-64 ISLAMIC STATE CONTROL OF IRAQ AND SYRIA** The Islamic State claims control over much of eastern Syria and western Iraq.

success recruiting members through Internet and social media communications that show beheadings and destruction of sites of historical importance, such as Shiite Muslim shrines (Figure 8-63).

The Islamic State controls much of northern Iraq and eastern Syria (Figure 8-64). It has also launched attacks in Europe, including in Paris in 2015.

BOKO HARAM

Boko Haram, which is Arabic for "Western education is forbidden," was founded in 2002 in northeastern Nigeria. It seeks to transform Nigeria into an Islamic state, and it opposes adoption of Western cultural practices, especially by Christians in the south of the country.

During its first seven years, the organization peacefully ran a religious complex and school that attracted poor Muslim families. A violent uprising in July 2009 resulted in the arrest of several hundred followers and the death of its founder Mohammed Yusuf. Six hundred incarcerated Boko Haram members broke out of prison the following year.

Since then, Boko Haram has resorted to terrorist tactics. Its leader, Abubakar Shekau, who had been Yusuf's deputy, initially aligned the group with al-Qaeda, but in 2014 it allied with the Islamic State. It now calls itself Islamic State's West Africa Province. Like the Islamic State, Boko Haram is considered to be especially adept at the use of social media to promote its views.

The use of religion by groups such as al-Qaeda, the Islamic State, and Boko Haram to justify attacks has posed challenges to Muslims and non-Muslims alike. For many Muslims, the challenge has been to express disagreement with the policies of governments in the United States and Europe yet disavow the use of terrorism. For many Americans and Europeans, the challenge has been to distinguish between the peaceful but unfamiliar principles and practices of the world's 1.6 billion Muslims and the misuse and abuse of Islam by Islamic terrorist groups.

State Support for Terrorism

LEARNING OUTCOME 8.4.5

Understand varying degrees of support for terrorism by states in Southwest Asia.

Several states in Southwest Asia have provided support for terrorism in recent years. Three increasing levels of involvement are (1) providing sanctuary for terrorists wanted by other countries; (2) supplying weapons, money, and intelligence to terrorists; and (3) planning attacks using terrorists.

SANCTUARY FOR TERRORISTS

Countries known to provide sanctuary for terrorists include Afghanistan and Pakistan. The United States, with the cooperation of several other countries, attacked Afghanistan in 2001 when its leaders, known as the Taliban, sheltered al-Qaeda leaders including bin Laden after 9/11. Removing the Taliban from power was considered a necessary step before going after al-Qaeda leaders, who were living in rugged mountains near Afghanistan's border with Pakistan.

The Taliban (which means "religious students" in the Pashtun language) had gained power in Afghanistan in 1995 and had imposed strict Islamic fundamentalist law on the population. Afghanistan's Taliban leadership treated women especially harshly. Women were prohibited from attending school, working outside the home, seeking health care, or driving cars. They were permitted to leave home only if fully covered by clothing and escorted by a male relative.

Removal of the Taliban unleashed a new struggle for control of Afghanistan among the country's many ethnic groups. When U.S. attention shifted to Iraq and Iran, the Taliban were able to regroup and resume an insurgency against the U.S.-backed Afghanistan government. The United States committed more than 30,000 troops to Afghanistan to keep the Taliban from regaining control of the entire country.

After the U.S.-led attack in eastern Afghanistan, al-Qaeda's leaders, including bin Laden, were able to escape across the border into Pakistan. After searching without success for nearly a decade, U.S. intelligence finally tracked bin Laden to a house in Abbottabad, Pakistan, where he was killed in 2011 (Figure 8-65). The United States believed that Pakistan security had to be aware that bin Laden had been living in the compound for at least five years. The compound was located only 6 kilometers (4 miles) from the Pakistan Military Academy, the country's principal institution for training military officers.

▼ **FIGURE 8-65 OSAMA BIN LADEN'S COMPOUND, PAKISTAN** While in this compound in Abbottabad, Pakistan, Osama bin Laden was killed by U.S. Navy SEALS.

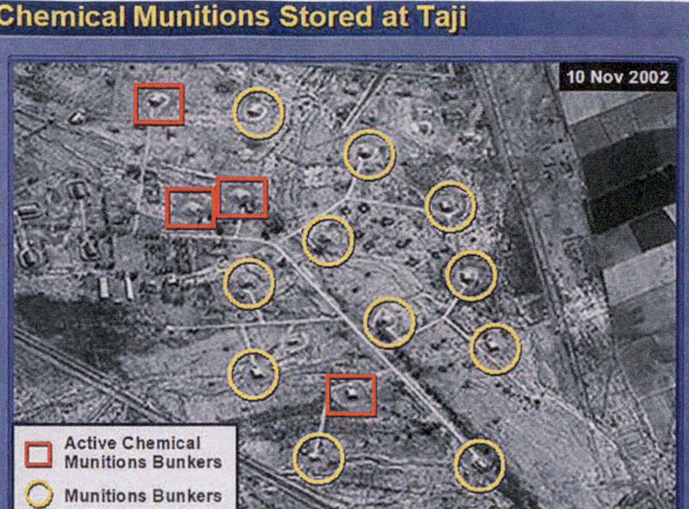

(a)

Sanitization of Ammunition Depot at Taji

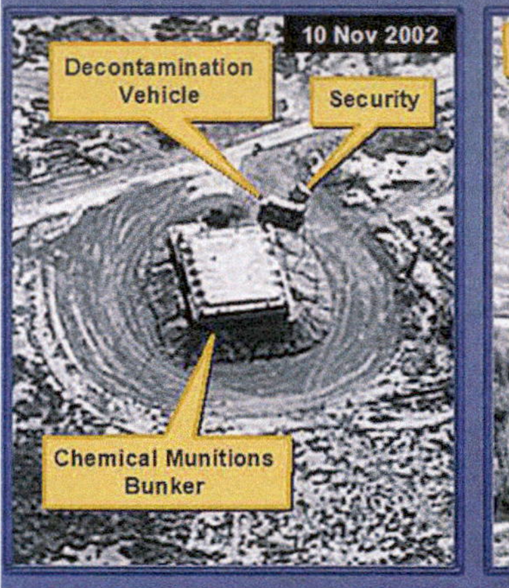

(b)

▲ **FIGURE 8-66 AIR PHOTOS ALLEGING IRAQ'S PREPARATIONS FOR CHEMICAL WARFARE (a)** U.S. satellite image purporting to show 15 munitions bunkers in Taji, Iraq. **(b)** Close-up of alleged munitions bunker outlined in red near the bottom of the left image. The truck labeled "decontamination vehicle" turned out to be a water truck. **(c)** Close-up of the two bunkers, outlined in red part **(a)**, allegedly sanitized.

SUPPLYING TERRORISTS

Iraq and Iran have both been accused of providing material and financial support for terrorists. The extent of their involvement in terrorism is controversial.

IRAQ. The United States led an attack against Iraq in 2003 in order to depose Saddam Hussein, the country's longtime president. The 1991 U.S.-led Gulf War, known as Operation Desert Storm, had driven Iraq out of Kuwait, but it had failed to remove Hussein from power. Desert Storm was supported by nearly every country in the United Nations because the purpose was to end one country's unjustified invasion and attempted annexation of another. In contrast, few countries supported the U.S.-led attack in 2003.

As the United States moved toward war with Iraq in 2003, Secretary of State Colin Powell scheduled a speech at the U.N. to present evidence to the world justifying military action against Iraq. Recalling the Cuban Missile Crisis (refer to Figure 8-53), Powell displayed a series of air photos designed to prove that Iraq possessed weapons of mass destruction. However, the photos did not provide clear evidence (Figure 8-66).

The U.S. assertion that Hussein had weapons of mass destruction, as well as close links with al-Qaeda, was refuted by most other countries, as well as ultimately by U.S. intelligence agencies. The United States argued instead that Hussein's quarter-century record of brutality justified replacing him with a democratically elected government. Since Hussein was deposed, Iraq has been mired in a civil war, especially between Shiites and Sunnis. Much of the country is under the control of the terrorist organization the Islamic State.

IRAN. Hostility between the United States and Iran dates from 1979, when a revolution forced abdication of Iran's pro-U.S. Shah Mohammad Reza Pahlavi. Iran's majority Shiite supporters of exiled fundamentalist Shiite Muslim leader Ayatollah Ruholiah Khomeini then proclaimed Iran an Islamic Republic. Militant supporters of the ayatollah seized the U.S. embassy on November 4, 1979, and held 52 Americans hostage until January 20, 1981.

The United States has accused Iran of harboring al-Qaeda members and of trying to gain influence in Iraq, where, as in Iran, the majority of the people are Shiites. In addition, Iran has long provided funding to Hezbollah, an organization based in Lebanon that the United States classifies as a terrorist organization because it seeks Israel's destruction.

The United States and most other countries have opposed Iran's aggressive development of a nuclear program (Figure 8-67). Iran has claimed that its nuclear program is for civilian purposes, but other countries have evidence that it is intended to develop weapons. Prolonged negotiations produced an agreement to degrade Iran's nuclear capabilities without resorting to yet another war in the region.

▲ **FIGURE 8-67 IRAN'S NUCLEAR FACILITIES**

▲ **FIGURE 8-68 LIBYAN STATE TERRORISM** The wreckage of Pan Am Flight 103 landed in Lockerbie, Scotland.

PAUSE & REFLECT 8.4.5

What events have occurred in Iran since this book was published?

STATE TERRORIST ATTACKS: LIBYA

Libya has been an active sponsor of terrorist attacks. Examples include (1) a 1986 bombing of a nightclub popular with U.S. military personnel in Berlin, killing 3 people (including 1 U.S. soldier); (2) planting of bombs on Pan Am Flight 103, which blew up over Lockerbie, Scotland, in 1988, killing 270 (Figure 8-68); and (3) planting of bombs on UTA Flight 772, which blew up over Niger in 1989, killing 170.

Libya's long-time leader Muammar el-Qaddafi (1942–2011, ruler 1969–2011) renounced terrorism in 2003 and provided compensation for victims of Pan Am 103. But his brutal attacks on Libyan protestors in 2011 again brought most other states of the world into active opposition to Qaddafi's regime, which was ultimately overthrown. Qaddafi was captured and killed. Since then, Libya has been plunged into a civil war, in which the Islamic State is actively involved.

CHECK-IN KEY ISSUE 4

Where Do States Face Threats?

- ✔ **During the Cold War, the world was divided into two alliances led by superpowers.**
- ✔ **With the end of the Cold War, economic alliances have become more important.**
- ✔ **Terrorism by individuals and organizations has included the 9/11 attacks on the United States.**
- ✔ **Some states have provided support for terrorism.**

Summary & Review

KEY ISSUE 1

Where are states distributed?

Nearly all of Earth's land area is divided among around 200 sovereign states. "State" is a synonym for country, an area organized into a political unit. The fifty U.S. states are a form of local government. Some disputes exist concerning whether an area belongs to one state or to two.

THINKING GEOGRAPHICALLY

1. Why might China, Japan, and Taiwan all wish to control the Senkaku/Diaoyu islands, despite their extremely small size?

▲ **FIGURE 8-69 SENKAKU/DIAOYU PROTEST** Protesters in Taiwan burn an effigy representing a Japanese warship during a protest against what the Taiwanese claim is Japan's illegal occupation of the islands.

KEY ISSUE 2

Why are nation-states difficult to create?

A nation-state is a state whose territory corresponds to that occupied by a particular ethnicity. Ancient and medieval states were organized as city-states, kingdoms, and other forms of government. The creation of nation-states began in Europe in modern times. A colony is territory legally tied to a state. The world once contained a large number of colonies, but only a few remain.

THINKING GEOGRAPHICALLY

2. Why might some states in Europe split in the future into multiple states, as occurred in the 1990s in Yugoslavia and the Soviet Union?

▲ **FIGURE 8-70 SUPPORTERS OF INDEPENDENCE FOR CATALONIA** Marchers in the streets of Barcelona show their support for Catalonia becoming independent of Spain.

KEY ISSUE 3

Why do boundaries cause problems?

States are separated from each other by boundaries. Boundaries can be cultural or physical. Countries are considered democracies, autocracies, or anocracies. The trend has been towards more democracies. Legislative districts are sometimes gerrymandered to benefit the party in power.

THINKING GEOGRAPHICALLY

3. What factors might be promoting the decline of autocracies?

▲ **FIGURE 8-71 ASTANA, CAPITAL OF KAZAKHSTAN** The Center for Systemic Peace classifies Kazakhstan as an autocracy.

KEY ISSUE 4

Where do states face threats?

Nearly all states are members of the United Nations. Europe was divided into two sets of military and economic alliances during the Cold War. Now most European states are united in the European Union. Terrorism is the systemic use of violence against the entire population. The world's three most active terrorist organizations are currently al-Qaeda, the Islamic State, and Boko Haram.

THINKING GEOGRAPHICALLY

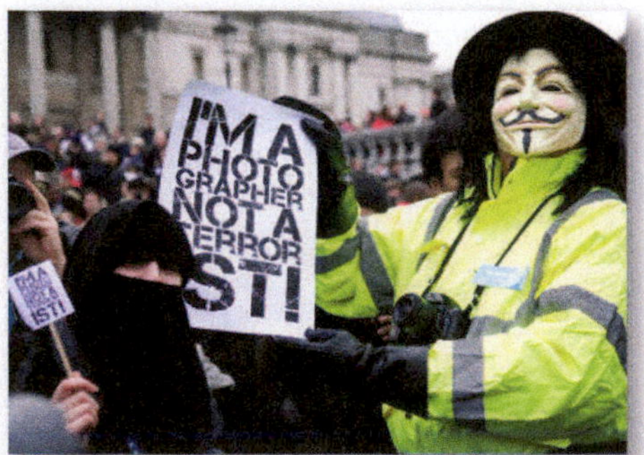

4. In what ways might the diffusion of personal electronic devices and social media help promote pro-democratic forces, and in what ways might they provide assistance to terrorists?

▲ **FIGURE 8-72 ANTI-TERRORIST PROTEST** The protest is against laws in the United Kingdom limiting members of the public from taking photographs at events where the police believe that a threat of a terrorist attack may exist.

KEY TERMS

Anocracy *(p. 288)* A country that is not fully democratic or fully autocratic but rather displays a mix of the two types.
Autocracy *(p. 288)* A country that is run according to the interests of the ruler rather than the people.
Balance of power *(p. 295)* A condition of roughly equal strength between opposing countries or alliances of countries.
Boundary *(p. 280)* An invisible line that marks the extent of a state's territory.
City-state *(p. 270)* A sovereign state comprising a city and its immediately surrounding countryside.
Colonialism *(p. 278)* An attempt by one country to establish settlements and to impose its political, economic, and cultural principles in another territory.
Colony *(p. 278)* A territory that is legally tied to a sovereign state rather than completely independent.
Compact state *(p. 286)* A state in which the distance from the center to any boundary does not vary significantly.
Democracy *(p. 288)* A country in which citizens elect leaders and can run for office.
Elongated state *(p. 286)* A state with a long, narrow shape.
Federal state *(p. 289)* An internal organization of a state that allocates strong power to units of local government.
Fragmented state *(p. 287)* A state that includes several discontinuous pieces of territory.
Frontier *(p. 280)* A zone separating two states in which neither state exercises political control.
Gerrymandering *(p. 290)* The process of redrawing legislative boundaries for the purpose of benefiting the party in power.
Landlocked state *(p. 287)* A state that does not have a direct outlet to the sea.
Microstate *(p. 266)* A state that encompasses a very small land area.
Multiethnic state *(p. 272)* A state that contains more than one ethnicity.
Multinational state *(p. 272)* A state that contains two or more ethnic groups with traditions of self-determination that agree to coexist peacefully by recognizing each other as distinct nationalities.
Nation-state *(p. 270)* A state whose territory corresponds to that occupied by a particular ethnicity.
Perforated state *(p. 287)* A state that completely surrounds another one.
Prorupted state *(p. 286)* An otherwise compact state with a large projecting extension.
Self-determination *(p. 272)* The concept that ethnicities have the right to govern themselves.
Sovereignty *(p. 268)* Ability of a state to govern its territory free from control of its internal affairs by other states.
State *(p. 266)* An area organized into a political unit and ruled by an established government that has control over its internal and foreign affairs.
Terrorism *(p. 298)* The systematic use of violence by a group calculated to create an atmosphere of fear and alarm among a population or to coerce a government into actions it would not otherwise undertake or refrain from actions it wants to take.
Unitary state *(p. 289)* An internal organization of a state that places most power in the hands of central government officials.

EXPLORE

Osama bin Laden's Hideout

Use Google Earth to see Osama bin Laden's hideout in *Abbottabad, Pakistan*.

Fly to *Osama bin Laden's compound*.

Set the imagery date to 5/8/2010.

Zoom in to eye alt 4,500 ft.

Click *3D buildings* to see the extent of bin Laden's compound.

Deselect *3D buildings* and move the imagery date to 3/22/2001.

Then move the imagery date to 6/14/2005.

1. What changed inside the compound boundaries between 3/22/2001 and 6/14/2005? Did bin Laden's World Trade Center attack take place before, between, or after these dates?

   Move the imagery date to 4/7/2013.

2. How has the compound changed since 6/14/2005? What happened to bin Laden between 2005 and 2013?

MG GeoVideo

Log in to the **MasteringGeography** Study Area to view this video.

Iraq & Saudi Arabia: Defining the Border

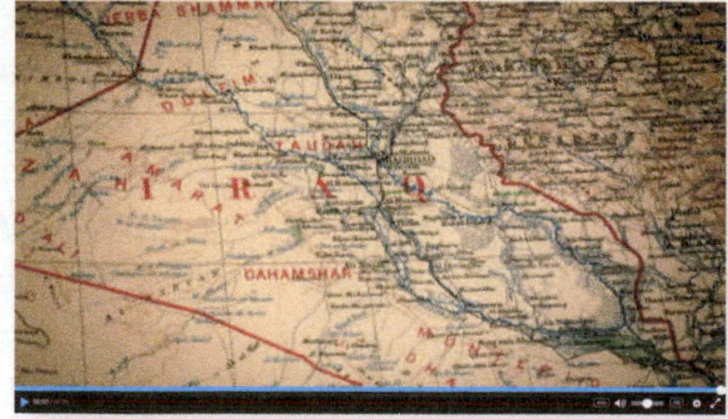

The state of Iraq was created after WWI following the fall of the Ottoman Empire

1. Before WWI, there were no linear boundaries in the middle east. Explain why.
2. After WWI, what were the two proposals for creating borders for the new state of Iraq? Describe each.
3. The video implies that some of the problems facing Iraq today can be traced back to the decisions about its borders made in 1922. Do you agree? Why or why not?

MasteringGeography™

Visit the Study Area in **MasteringtGeography™** to enhance your geographic literacy, spatial reasoning skills, and understanding of this chapter's content by accessing a variety of resources, including MapMaster interactive maps, videos, *In the News* RSS feeds, flashcards, web links, self-study quizzes, and an eText version of *The Cultural Landscape*.
www.masteringgeography.com

9

Food and Agriculture

All humans need food to survive. We have two choices in obtaining our food: Buy it or produce it ourselves. In developed countries people purchase nearly all of their food, which is produced primarily through commercial agriculture. In developing countries people produce much of their food themselves through subsistence agriculture.

Fair trade coffee farmer harvesting her crop in Uganda.

LOCATIONS IN THIS CHAPTER

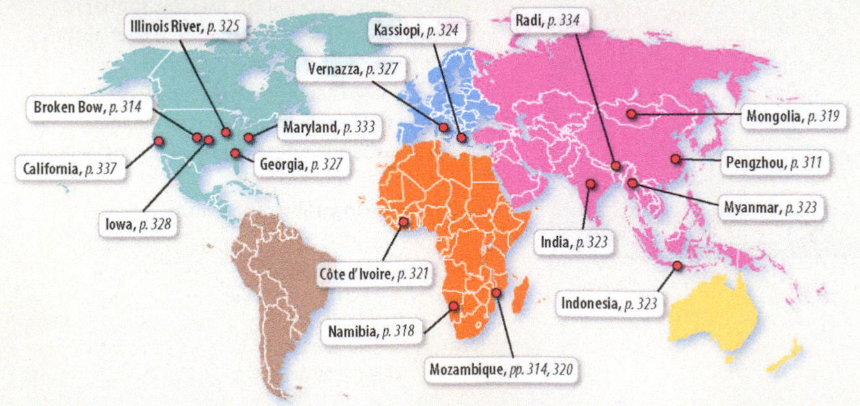

KEY ISSUES

OLIVE
Southwest Asia
Central Asia
South Asia
Sub-Saharan Africa
CHICKEN
FINGER MILLET

1 Where Did Agriculture Originate?

Agriculture was invented around 10,000 years ago in multiple ***places***. Prior to the agricultural revolution, people survived by hunting and gathering food.

2

Why Do People Consume Different Foods?

Humans display a distinctive distribution across ***space*** of food consumption. People vary in the quantity of food they consume and the variety.

3

Where Is Agriculture Distributed?

Farming practices vary around the world depending on cultural and environmental factors. Distinctive ***regions*** can be identified in developing countries and in developed countries.

4

Why Do Farmers Face Sustainability Challenges?

Agriculture is a global business with many ***connections*** between farms and other industries, as well as connections through international trade. At the same time, the local ***scale*** is increasingly important in farming practices and obtaining food.

KEY ISSUE 1

Where Did Agriculture Originate?

- Introducing Food and Agriculture
- Subsistence and Commercial Agriculture

LEARNING OUTCOME 9.1.1

Understand the origin of agriculture.

Agriculture is deliberate modification of Earth's surface through cultivation of plants and rearing of animals to obtain sustenance or economic gain (Figure 9-1). Agriculture originated when humans domesticated plants and animals for their use. The word *cultivate* means "to care for," and a **crop** is any plant cultivated by people.

Introducing Food and Agriculture

The origin of agriculture cannot be documented with certainty because it began before recorded history. Scholars try to reconstruct a logical sequence of events based on fragments of information about ancient agricultural practices and historical environmental conditions. Improvements in cultivating plants and domesticating animals evolved over thousands of years. This section offers an explanation for the origin and diffusion of agriculture.

INVENTION OF AGRICULTURE

The **agricultural revolution** was the time when human beings first domesticated plants and animals and no longer relied entirely on hunting and gathering. Geographers and other scientists believe that the agricultural revolution occurred around the year 8000 B.C. because the world's population began to grow at a more rapid rate than it had in the past. By growing plants and raising animals, human beings created larger and more stable sources of food, so more people could survive.

Scientists do not agree on whether the agricultural revolution originated primarily because of environmental factors or cultural factors. Probably a combination of both factors contributed:

- **Environmental factors.** The first domestication of crops and animals coincided with climate change. This marked the end of the last ice age, when permanent ice cover receded from Earth's mid-latitudes to the polar regions, resulting in a massive redistribution of humans, other animals, and plants at that time.
- **Cultural factors.** A preference for living in a fixed place rather than as nomads may have led hunters and gatherers to build permanent settlements and to store surplus vegetation there. In gathering wild vegetation, people inevitably cut plants and dropped berries, fruits, and seeds. These hunters probably observed that, over time, damaged or discarded food produced new plants. They may have deliberately cut plants or dropped berries on the ground to see if they would produce new plants. Subsequent generations learned to pour water over the site and to introduce manure and other soil improvements. Over thousands of years, plant cultivation apparently evolved from a combination of accident and deliberate experiment.

▼ **FIGURE 9-1 AGRICULTURE** Agriculture involves the deliberate modification of Earth's surface.

AGRICULTURE HEARTHS

Scientists also do not agree on how agriculture diffused or why most nomadic groups converted from hunting, gathering, and fishing to agriculture. Scientists agree that planting of crops originated in multiple hearths around the world (Figure 9-2). Animals were also domesticated in multiple hearths at various dates. Important hearths include:

- **Southwest Asia.** The earliest crops domesticated in Southwest Asia around 10,000 years ago are thought to have been barley, wheat, lentil, and olive. Southwest Asia is also thought to have been the hearth for the domestication of the largest number of animals that would prove to be most important for agriculture, including cattle, goats, pigs, and sheep, between 8,000 and 9,000 years ago. Domestication of the dog is thought to date even earlier, around 12,000 years ago. From this hearth, cultivation diffused west to Europe and east to Central Asia.
- **East Asia.** Rice is thought to have been domesticated in East Asia more than 10,000 years ago, along the Yangtze River in eastern China. Millet was cultivated at an early date along the Yellow River.
- **Central and South Asia.** Chickens are thought to have diffused from South Asia around 4,000 years ago. The horse is considered to have been domesticated in Central Asia. Diffusion of the domesticated horse is thought to be associated with the diffusion of the Indo-European language, as discussed in Chapter 5.
- **Sub-Saharan Africa.** Sorghum was domesticated in central Africa around 8,000 years ago. Yams may have been domesticated even earlier. Millet and rice may have been domesticated in sub-Saharan Africa independently of the hearth in East Asia. From central Africa, domestication of crops probably diffused further south in Africa.
- **Latin America.** Two important hearths of crop domestication are thought to have emerged in Mexico and Peru around 4,000 to 5,000 years ago. Mexico is considered a hearth for beans and cotton, and Peru for potato. The most important contribution of the Americas to crop domestication, maize (corn), may have emerged in the two hearths independently around the same time. From these two hearths, cultivation of maize and other crops diffused northward into North America and southward into tropical South America. Some researchers place the origin of squash in the southeastern portion of present-day United States.

PAUSE & REFLECT 9.1.1

Which crops appear to have reached the present-day United States first, according to Figure 9-2?

That agriculture had multiple origins means that, from earliest times, people have produced food in distinctive ways in different regions. This diversity derives from a unique legacy of wild plants, climatic conditions, and cultural preferences in each region. Improved communications in recent centuries have encouraged the diffusion of some plants to varied locations around the world. Many plants and animals thrive across a wide portion of Earth's surface, not just in their place of original domestication. Only after 1500, for example, were wheat, oats, and barley introduced to the Western Hemisphere and maize to the Eastern Hemisphere.

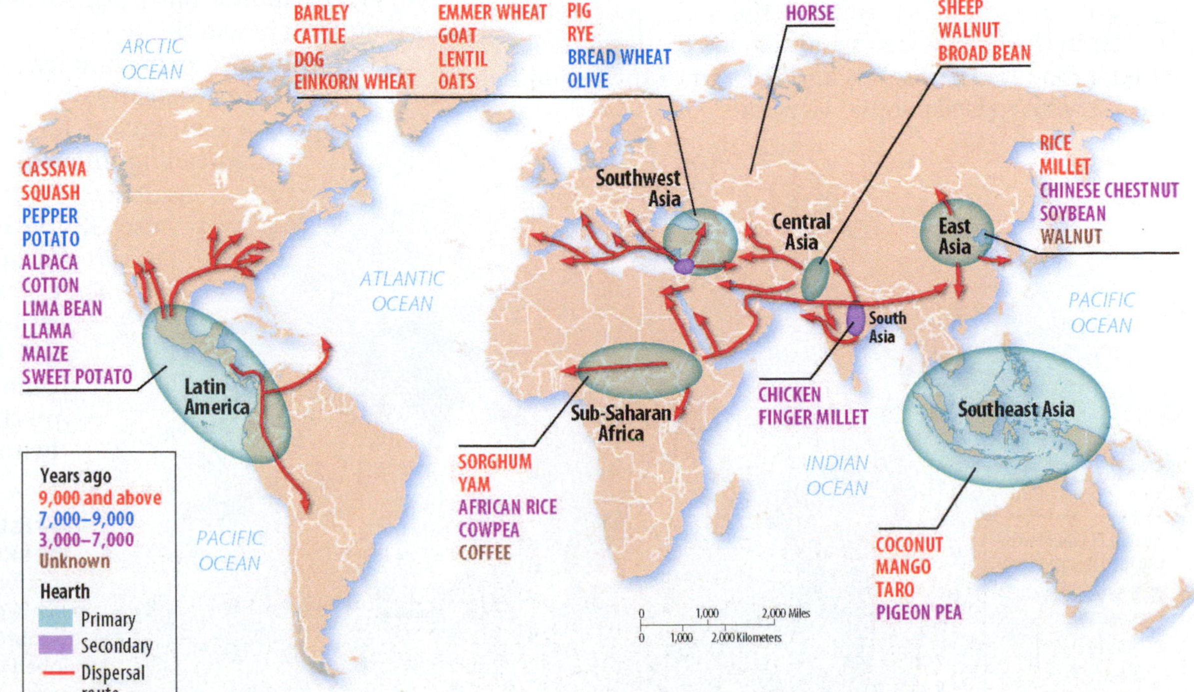

▶ **FIGURE 9-2 AGRICULTURE HEARTHS** Planting of crops and domestication of animals originated in multiple hearths.

Subsistence and Commercial Agriculture

LEARNING OUTCOME 9.1.2

Describe the principal differences between subsistence and commercial agriculture.

The most fundamental differences in agricultural practices are between those in developing countries and those in developed countries. Farmers in developing countries generally practice subsistence agriculture, whereas farmers in developed countries practice commercial agriculture. **Subsistence agriculture**, found in developing countries, is the production of food primarily for consumption by the farmer's family. **Commercial agriculture**, found in developed countries, is the production of food primarily for sale off the farm. The main features that distinguish commercial agriculture from subsistence agriculture include the percentage of farmers in the labor force, the use of machinery, and farm size.

PERCENTAGE OF FARMERS

In developed countries, around 3 percent of workers are engaged directly in farming, compared to around 42 percent in developing countries (Figure 9-3). A priority for all people is to secure the food needed to survive. In developing countries a large percentage of people are subsistence farmers who work in agriculture to produce the food they and their families require. In developed countries the relatively few people engaged in farming are commercial farmers, and most people buy food with money earned by working in factories or offices or by performing other services.

The percentage of farmers is even lower in North America—only around 2 percent. Yet the small percentage of farmers in the United States and Canada produces not only enough food for themselves and the rest of the region but also a surplus to feed people elsewhere.

The number of farmers declined dramatically in developed countries during the twentieth century. The United States had about 60 percent fewer farms and 85 percent fewer farmers in 2000 than in 1900. The number of farms in the United States declined from about 6 million in 1940 to 4 million in 1960 and 2 million in 1980. Both push and pull migration factors have been responsible for the decline: People were pushed away from farms by lack of opportunity to earn a decent income, and at the same time they were pulled to higher-paying jobs in urban areas. The number of U.S. farmers has stabilized since 1980, at around 2 million.

ROLE OF MACHINERY, SCIENCE, AND TECHNOLOGY

In developed countries, a small number of commercial farmers can feed many people because they rely on machinery to perform work rather than on people or animals (Figure 9-4). In developing countries, subsistence farmers do much of the work with hand tools and animal power.

Traditionally, the farmer or local craftspeople made equipment from wood, but beginning in the late eighteenth century, factories produced farm machinery. The first all-iron plow was made in the 1770s and was followed in the nineteenth and twentieth centuries by inventions that made farming less dependent on human or animal power. Today, farmers use tractors, combines, corn pickers, planters, and other factory-made farm machines to increase productivity.

Experiments conducted in university laboratories, industry, and research organizations generate new fertilizers, herbicides, hybrid plants, animal breeds, and farming practices, which lead to higher crop yields and healthier animals. Access to other scientific information has enabled farmers to make more intelligent decisions concerning proper agricultural practices. Some farmers conduct their own on-farm research.

Electronics also help commercial farmers. Farmers use Global Positioning System (GPS) devices to determine the precise coordinates for planting seeds and for spreading different types and amounts of fertilizers. On large ranches, they also use GPS devices to monitor the locations of cattle and tractors. They use satellite imagery to measure crop progress and yield monitors attached to combines to determine the precise number of bushels being harvested.

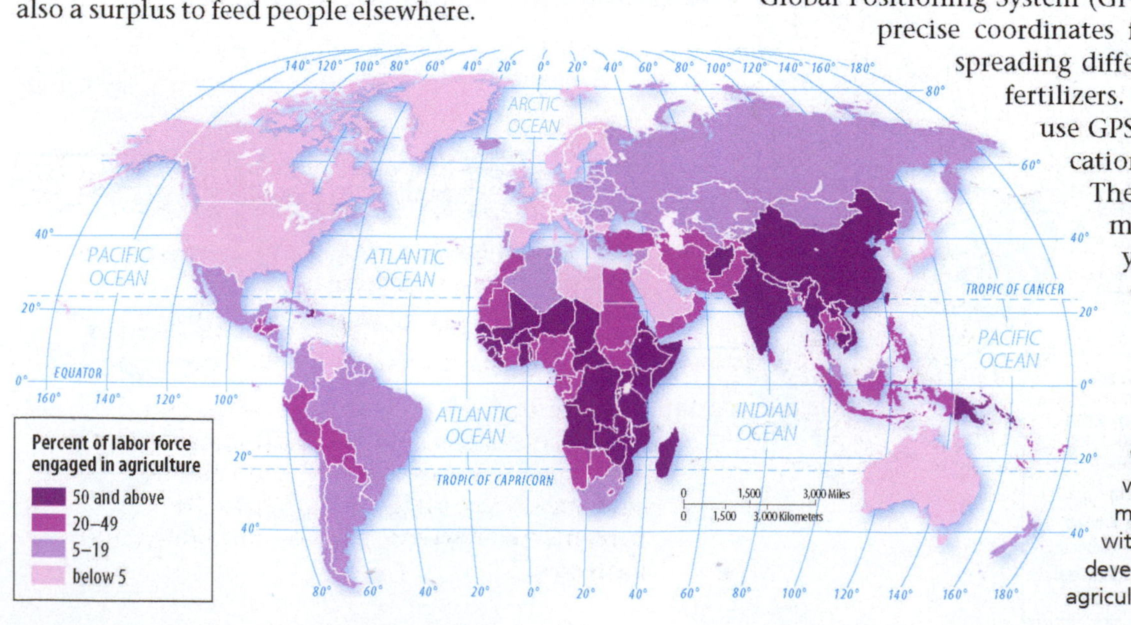

FIGURE 9-3 AGRICULTURAL WORKERS The percentage of the workforce engaged in agriculture is much higher in developing countries with subsistence agriculture than in developed countries with commercial agriculture.

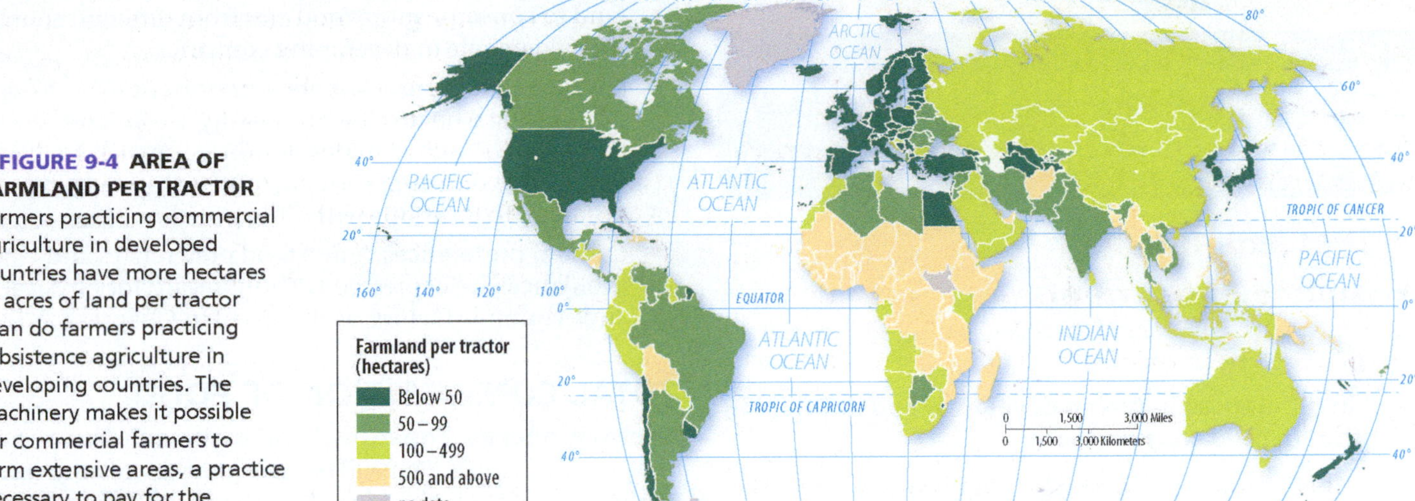

▶ **FIGURE 9-4 AREA OF FARMLAND PER TRACTOR** Farmers practicing commercial agriculture in developed countries have more hectares or acres of land per tractor than do farmers practicing subsistence agriculture in developing countries. The machinery makes it possible for commercial farmers to farm extensive areas, a practice necessary to pay for the expensive machinery.

FARM SIZE

The average farm is relatively large in commercial agriculture. Farms average 178 hectares (441 acres) in the United States, compared to about 1 hectare (2.5 acres) in China (Figure 9-5). Farm size partly depends on mechanization. Combines, pickers, and other machinery perform most efficiently at very large scales, and their considerable expense cannot be justified on a small farm. As a result of the large size and the high level of mechanization, commercial agriculture is an expensive business. Farmers spend hundreds of thousands of dollars to buy or rent land and machinery before beginning operations. This money is frequently borrowed from a bank and repaid after output is sold.

▼ **FIGURE 9-5 SMALL CHINA FARM WITH HAND TOOLS** Digging up sweet potato vines, Pengzhou, China.

Commercial agriculture is increasingly dominated by a handful of large farms. In the United States, the largest 5 percent of farms produce 75 percent of the country's total agriculture. Despite their size, most commercial farms in developed countries—90 percent in the United States—are family owned and operated. Commercial farmers frequently expand their holdings by renting nearby fields.

Although the United States had fewer farms and farmers in 2000 than in 1900, the amount of land devoted to agriculture increased by 13 percent, primarily due to irrigation and reclamation. However, in the twenty-first century, the United States has been losing 1.2 million hectares (3 million acres) per year of its 400 million hectares (1 billion acres) of farmland, primarily because of the expansion of urban areas.

PAUSE & REFLECT 9.1.2

What other electronics, in addition to GPS devices, might help a farmer on a very large farm?

CHECK-IN KEY ISSUE 1

Where Did Agriculture Originate?

- **Before the invention of agriculture, most humans were hunters and gatherers.**
- **Agriculture was invented in multiple hearths beginning approximately 10,000 years ago.**
- **Modern agriculture is divided between subsistence agriculture in developing countries and commercial agriculture in developed countries. They differ according to the percentage of farmers, use of machinery, and farm size.**

KEY ISSUE 2

Why Do People Consume Different Foods?

- ▸ Diet and Nutrition
- ▸ Source of Nutrients

LEARNING OUTCOME 9.2.1

Explain differences between developed and developing countries in food consumption.

When you buy food in a supermarket, are you reminded of a farm? Not likely. The meat is carved into pieces that no longer resemble an animal and is wrapped in paper or plastic film. Often the vegetables are canned or frozen. The milk and eggs are in cartons.

The food industry in the United States and Canada is vast, but only a few people are full-time farmers, and they may be more familiar with the operation of computers and advanced machinery than the typical factory or office worker. The mechanized, highly productive American or Canadian farm contrasts with the subsistence farm found in much of the world. The most "typical" human—if there is such a person—is an Asian farmer who grows enough food to survive, with little surplus. This sharp contrast in agricultural practices constitutes one of the most fundamental differences between the more developed and less developed countries of the world.

Diet and Nutrition

Everyone needs food to survive. Consumption of food varies around the world, both in total amount and source of nutrients. The variation results from a combination of:

- **Level of development.** People in developed countries tend to consume more food and from different sources than do people in developing countries.
- **Physical conditions.** Climate is important in influencing what can be most easily grown and therefore consumed in developing countries. In developed countries, though, food is shipped long distances to locations with different climates.
- **Cultural preferences.** Some food preferences and avoidances are expressed without regard for physical and economic factors, as discussed in Chapter 4.

TOTAL CONSUMPTION OF FOOD

Dietary energy consumption is the amount of food that an individual consumes. The unit of measurement of dietary energy is the kilocalorie (kcal), or Calorie in the United States. One gram (or ounce) of each food source delivers a kilocalorie level that nutritionists can measure.

Most humans derive most of their kilocalories through consumption of **cereal grain,** or simply cereal, which is a grass that yields grain for food. **Grain** is the seed from a cereal grass. The three leading cereal grains—wheat, rice, and maize (corn in North America)—together account for nearly 90 percent of all grain production and more than 40 percent of all dietary energy consumed worldwide (Figure 9-6):

- **Wheat.** The principal cereal grain consumed in the developed regions of Europe and North America is wheat, which is consumed in bread, pasta, cake, and many other forms. It is also the most consumed grain in the developing regions of Central and Southwest Asia, where relatively dry conditions are more suitable for growing wheat than other grains.
- **Rice.** The principal cereal grain consumed in the developing regions of East, South, and Southeast Asia is rice. It is the most suitable crop for production in tropical climates.

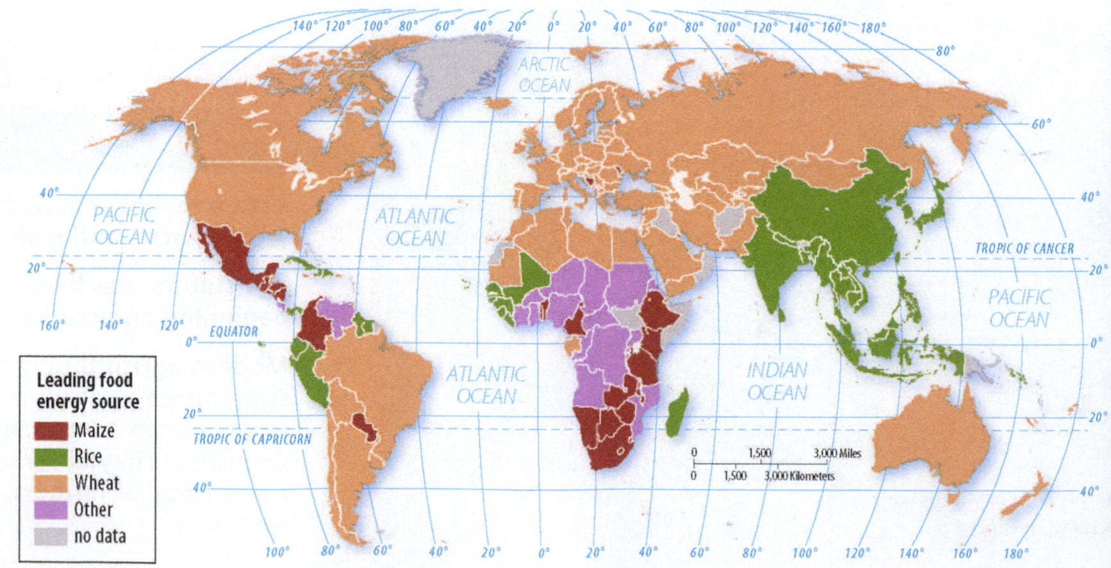

▸ FIGURE 9-6 DIETARY ENERGY BY SOURCE Wheat, rice, and maize are the three main sources of kilocalories.

- **Maize.** The leading crop in the world is maize (called corn in North America), though much of it is grown for purposes other than direct human consumption, especially as animal feed. It is the leading crop in some countries of sub-Saharan Africa.
- **Other crops.** A handful of countries obtain the largest share of dietary energy from other crops, especially in sub-Saharan Africa. These include cassava, sorghum, millet, plantains, sweet potatoes, and yams. Sugar is the leading source of dietary energy in Venezuela.

DIETARY ENERGY NEEDS

To maintain a moderate level of physical activity, according to the U.N. Food and Agricultural Organization, an average individual needs to consume on a daily basis at least 1,844 kcal. Average consumption worldwide is 2,902 kcal per day, well above the recommended minimum. Thus, most people get enough food to survive (Figure 9-7).

People in developed countries are consuming on average nearly twice the recommended minimum, 3,400 kcal per day. The United States has the world's highest consumption, 3,800 kcal per day per person. The consumption of so much food is one reason that obesity is more prevalent than hunger in the United States, as well as in other developed countries.

In developing regions, average daily consumption is 2,800 kcal, still above the recommended minimum. However, the average in sub-Saharan Africa is only 2,400, an indication that a large percentage of Africans are not getting enough to eat. Diets are more likely to be deficient in countries where people have to spend a high percentage of their income to obtain food (Figure 9-8).

PAUSE & REFLECT 9.2.1

Many restaurants now tell you how many Calories (kilocalories) are in their meals. Does this information influence your choice of meal? Why or why not?

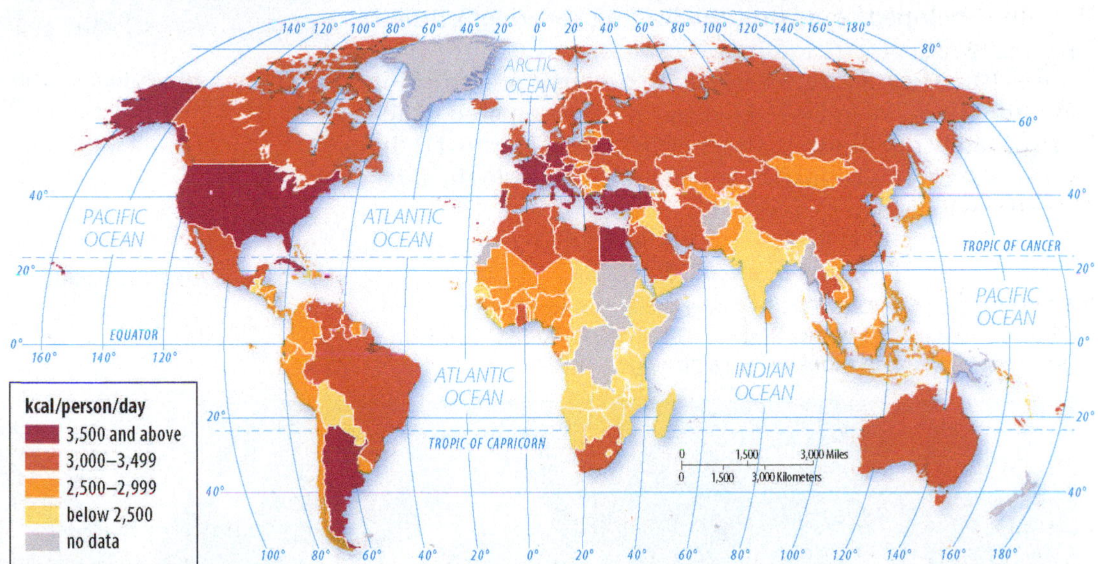

▶ **FIGURE 9-7 DIETARY ENERGY CONSUMPTION**

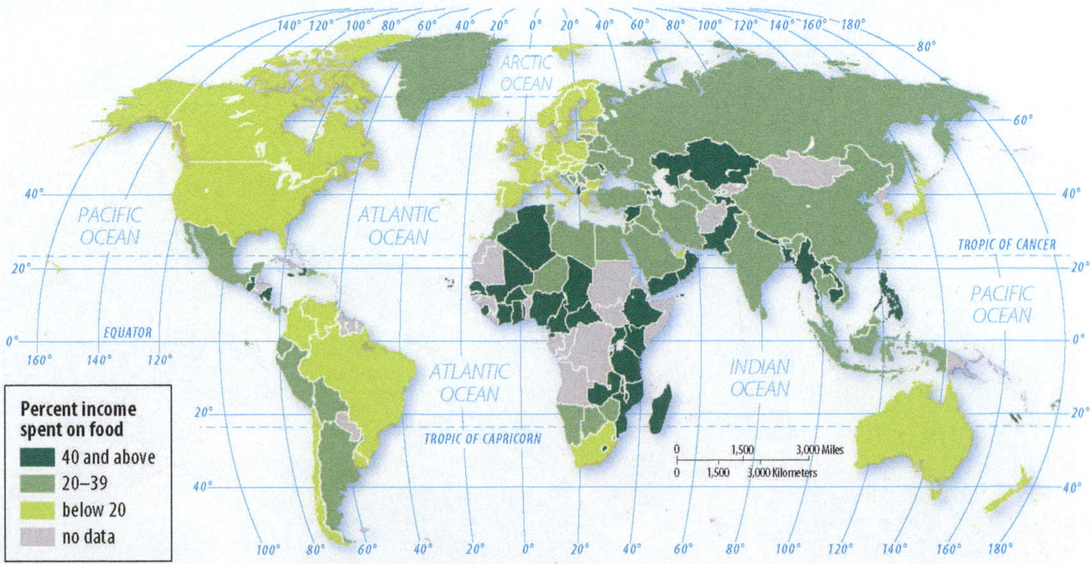

▶ **FIGURE 9-8 INCOME SPENT ON FOOD**

Source of Nutrients

LEARNING OUTCOME 9.2.2

Explain differences between developed and developing countries in source of nutrients.

The United Nations defines **food security** as physical, social, and economic access at all times to safe and nutritious food sufficient to meet dietary needs and food preferences for an active and healthy life. By this definition, around 10 percent of the world's inhabitants do not have food security.

PROTEIN

Protein is a nutrient needed for growth and maintenance of the human body. Many food sources provide protein of varying quantity and quality. One of the most fundamental differences between developed and developing regions is the primary source of protein (Figure 9-9).

In developed countries, the leading source of protein is meat products, including beef, pork, and poultry (Figure 9-10). Meat accounts for approximately one-third of all protein intake in developed countries, compared to approximately one-tenth in developing ones (Figure 9-11). In most developing countries, cereal grains provide the largest share of protein (Figure 9-12).

▼ **FIGURE 9-9 AFRICAN FOOD** This group in Mozambique is preparing cassavas. At home, these roots will be pounded to break up the fibrous texture and cooked into a porridge.

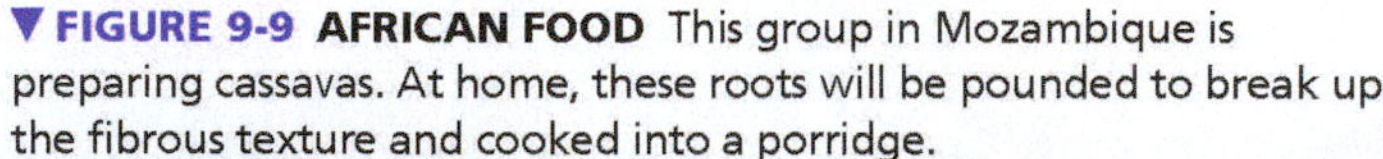

▲ **FIGURE 9-10 WORLD'S LARGEST CATTLE FEEDLOT** Cattle are fattened in Broken Bow, Nebraska, before being processed into cuts of beef.

PAUSE & REFLECT 9.2.2

What is your main source of protein?

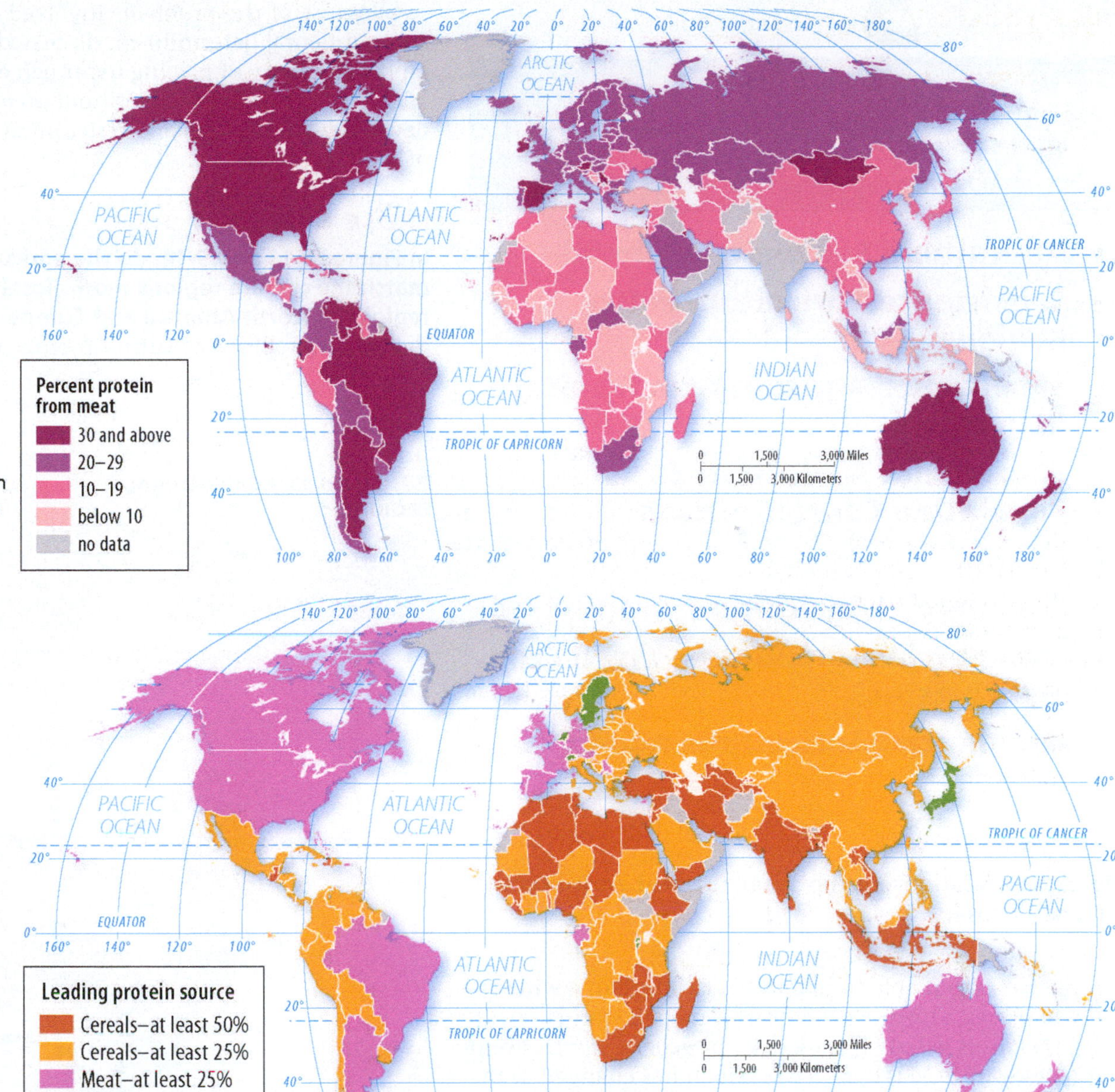

▶ **FIGURE 9-11**
PROTEIN FROM MEAT The percentage of protein from meat is much higher for people in developed countries than for those in developing countries.

▶ **FIGURE 9-12**
PROTEIN BY SOURCE People get most of their protein from meat in developed countries and from cereals in developing countries.

CHECK-IN KEY ISSUE 2

Why Do People Consume Different Foods?

- ✔ **Most food is consumed in the form of cereal grains, especially wheat, rice, and maize.**
- ✔ **People in developed countries consume more total calories and a higher percentage through animal products.**
- ✔ **Most humans consume more than the recommended minimum calories, and the number who are undernourished is declining, but undernourishment is still common in Asia and sub-Saharan Africa.**

KEY ISSUE 3

Where Is Agriculture Distributed?

- ▸ **Agricultural Regions and Climate**
- ▸ **Subsistence Agriculture in Dry Regions**
- ▸ **Subsistence Agriculture in Tropical Regions**
- ▸ **Subsistence Agriculture in Population Concentrations**
- ▸ **Fishing**
- ▸ **Commercial Agriculture: Crop-based**
- ▸ **Commercial Agriculture: Mixed Crop and Livestock**
- ▸ **Commercial Agriculture: Animal-based**

LEARNING OUTCOME 9.3.1

Recognize relationships between maps of agriculture and of climate.

People have been able to practice agriculture in a wide variety of places. The most widely used map of world agricultural regions is based on work done by geographer Derwent Whittlesey in 1936 (Figure 9-13). Much has changed in the world since Whittlesey's work, but his broad division of the world into agriculture regions is still used.

Agricultural Regions and Climate

Similarities between the agriculture and climate maps are striking (Figure 9-14). For example, pastoral nomadism is the predominant type of agriculture in Southwest Asia & North Africa, which has a dry climate, whereas shifting cultivation is the predominant type of agriculture in sub-Saharan Africa, which has a tropical climate. Note the division between southeastern China (warm mid-latitude climate, intensive subsistence agriculture with wet-rice dominant) and northeastern China (cold mid-latitude climate, intensive subsistence agriculture with wet rice not dominant).

In the United States, much of the West is distinguished from the rest of the country according to climate (dry) and agriculture (livestock ranching). Thus, agriculture varies between the drylands and the tropics within developing countries—as well as between the drylands of developing and developed countries.

Because of the problems involved with the concept of environmental determinism, discussed in Chapter 1, geographers are wary of placing too much emphasis on the role of climate. Differences in cultural preferences and levels of development also explain agricultural differences in areas of similar climate.

PAUSE & REFLECT 9.3.1

In Figures 9-13 and 9-14, do the agricultural regions match the climate regions more closely in the developed regions of North America and Europe or in the developing regions of Latin America, Africa, and Asia?

▼ **FIGURE 9-13 AGRICULTURAL REGIONS**

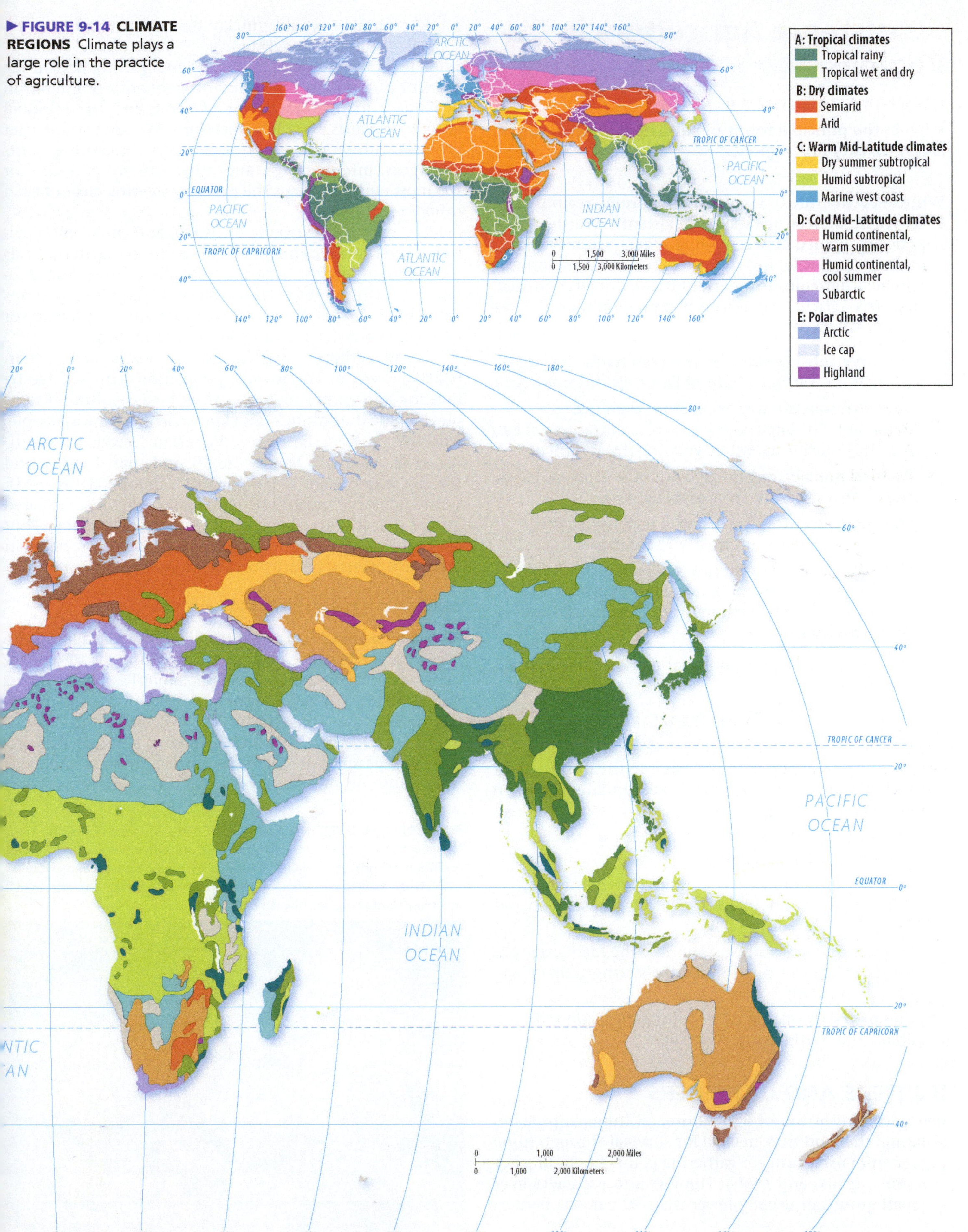

▶ FIGURE 9-14 CLIMATE REGIONS Climate plays a large role in the practice of agriculture.

Subsistence Agriculture in Dry Regions

LEARNING OUTCOME 9.3.2

Explain the principal forms of subsistence agriculture in lower-density dry regions.

Whittlesey identified 11 main agricultural regions, plus areas where agriculture was nonexistent. Whittlesey's 11 regions are divided between 5 that are important forms of agriculture in developing countries and 6 that are forms of commercial agriculture important in developed countries.

The five agricultural regions that predominate in developing countries are:

- **Intensive subsistence, wet-rice dominant.** The large population concentrations of East Asia and South Asia.
- **Intensive subsistence, crops other than rice dominant.** The large population concentrations of East Asia and South Asia, where growing rice is difficult.
- **Pastoral nomadism.** The drylands of Southwest Asia & North Africa, Central Asia, and East Asia.
- **Shifting cultivation.** The tropical regions of Latin America, sub-Saharan Africa, and Southeast Asia.
- **Plantation.** A form of commercial agriculture found in tropical and subtropical developing countries of Latin America, sub-Saharan Africa, South Asia, and Southeast Asia.

The six agricultural regions that predominate in developed countries are:

- **Mixed crop and livestock.** The U.S. Midwest and central Europe.
- **Dairying.** Near population clusters in the northeastern United States, southeastern Canada, and northwestern Europe.
- **Grain.** The north-central United States, south-central Canada, and Eastern Europe.
- **Ranching.** The drylands of western North America, southeastern Latin America, Central Asia, sub-Saharan Africa, and the South Pacific.
- **Mediterranean.** Lands surrounding the Mediterranean Sea, the western United States, the southern tip of Africa, and Chile.
- **Commercial gardening.** The southeastern United States and southeastern Australia.

HUNTERS AND GATHERERS

Before the invention of agriculture, all humans probably obtained the food they needed for survival through hunting for animals, fishing, or gathering plants (including berries, nuts, fruits, and roots). Hunters and gatherers lived in small groups of usually fewer than 50 persons because a larger number would quickly exhaust the available resources within walking distance.

The group traveled frequently, establishing new home bases or camps. The direction and frequency of migration depended on the movement of game and the seasonal growth of plants at various locations. We can assume that groups communicated with each other concerning hunting rights, intermarriage, and other specific subjects. For the most part, they kept the peace by steering clear of each other's territory.

The men hunted game or fished, and the women collected berries, nuts, and roots. This division of labor sounds like a stereotype but is based on evidence from archaeology and anthropology. They collected food often, perhaps daily. The food search might have taken only a short time or much of the day, depending on local conditions.

Today, perhaps a quarter-million people, or less than 0.005 percent of the world's population, still survive by hunting and gathering rather than by agriculture. Examples include the Spinifex (also known as Pila Nguru) people, who live in Australia's Great Victorian Desert; the Sentinelese people, who live in India's Andaman Islands; and the San, who live in Botswana and Namibia (Figure 9-15). Contemporary hunting and gathering societies are isolated groups that live on the periphery of world settlement, but they provide insight into human customs that prevailed in prehistoric times, before the invention of agriculture.

PASTORAL NOMADISM

Pastoral nomadism is a form of subsistence agriculture based on the herding of domesticated animals in dry climates, where planting crops is impossible. Pastoral nomads, unlike other subsistence farmers, depend primarily on animals rather than crops for survival. The Bedouins of Saudi Arabia and North Africa and the Masai of East Africa are examples of pastoral nomads. The animals provide milk, and their skins and hair are used for clothing and tents. Their

▼ **FIGURE 9-15 HUNTING AND GATHERING** Two members of Namibia's San People hunt with bow and arrow 1. Before the invention of agriculture, 8,000 years ago, why would a bow and arrow be used rather than ammunition? 2. Why might a bow and arrow be used by hunters today, rather than a rifle?

animals are usually not slaughtered, although dead ones may be consumed. To nomads, the size of their herd is both an important measure of power and prestige and their main security during adverse environmental conditions.

Like other subsistence farmers, though, pastoral nomads consume mostly grain rather than meat. To obtain grain, many present-day nomads do raise crops or obtain them by trading animal products.

Nomads select the type and number of animals for the herd according to local cultural and physical characteristics. The choice depends on the relative prestige of animals and the ability of species to adapt to a particular climate and vegetation:

- Camels are well suited to arid climates because they can go long periods without water, carry heavy baggage, and move rapidly, but they are particularly bothered by flies and sleeping sickness and have a relatively long gestation period—12 months from conception to birth.
- Goats need more water than do camels but are tough and agile and can survive on virtually any vegetation, no matter how poor (Figure 9-16).
- Sheep are relatively slow moving and affected by climatic changes; they require more water than camels and goats and are more selective about which plants they will eat.

Pastoral nomads do not wander randomly across the landscape but have a strong sense of territoriality. Every group controls a piece of territory and will invade another group's territory only in an emergency or if war is declared. The goal of each group is to control a territory large enough to contain the forage and water needed for survival. The actual amount of land a group controls depends on its wealth and power. The precise migration patterns evolve from intimate knowledge of the area's physical and cultural characteristics.

Some pastoral nomads practice **transhumance,** which is seasonal migration of livestock between mountains and lowland pasture areas. Pasture is grass or other plants grown for feeding grazing animals, as well as land used for grazing. Sheep or other animals may pasture in alpine meadows in the summer and be herded back down into valleys for winter pasture.

Agricultural experts once regarded pastoral nomadism as a stage in the evolution of agriculture—between the hunters and gatherers who migrated across Earth's surface in search of food and sedentary farmers who cultivated grain in one place. Because they had domesticated animals but not plants, pastoral nomads were considered more advanced than hunters and gatherers but less advanced than settled farmers.

Pastoral nomadism is now generally recognized as an offshoot of sedentary agriculture, not as a primitive precursor of it. It is simply a practical way of surviving on land that receives too little rain for cultivation of crops.

Today, pastoral nomadism is a declining form of agriculture, partly a victim of modern technology. Before recent transportation and communications inventions, pastoral nomads played an important role as carriers of goods and information across the sparsely inhabited drylands. They used to be the most powerful inhabitants of the drylands, but now, with modern weapons, national governments can control nomadic population more effectively.

Governments force groups to give up pastoral nomadism because they want the land for other uses. Land that can be irrigated is converted from nomadic to sedentary agriculture. In some instances, the mining and petroleum industries now operate in drylands formerly occupied by pastoral nomads. Some nomads are encouraged to try sedentary agriculture or to work for mining or petroleum companies. Others are still allowed to move about, but only within ranches of fixed boundaries. In the future, pastoral nomadism will be increasingly confined to areas that cannot be irrigated or that lack valuable raw materials.

FIGURE 9-16
PASTORAL NOMADISM
Milking goats, Gobi Desert, Mongolia.

Subsistence Agriculture in Tropical Regions

LEARNING OUTCOME 9.3.3

Explain the principal forms of subsistence agriculture in lower-density tropical regions.

Shifting cultivation is practiced in much of the world's tropical, or A, climate regions, which have relatively high temperatures and abundant rainfall. It is practiced by roughly 250 million people across 36 million square kilometers (14 million square miles), especially in the tropical rain forests of Latin America, sub-Saharan Africa, and Southeast Asia. **Plantation** farming is also found.

SHIFTING CULTIVATION

Two distinctive features of shifting cultivation are:

- Farmers clear land for planting by slashing vegetation and burning the debris; shifting cultivation is sometimes called **slash-and-burn agriculture.**
- Farmers grow crops on a cleared field for only a few years, until soil nutrients are depleted, and then leave it fallow (with nothing planted) for many years so the soil can recover.

People who practice shifting cultivation generally live in small villages and grow food on the surrounding land, which the village controls. Each year villagers designate for planting an area surrounding the settlement. Before planting, they must remove the dense vegetation that typically covers tropical land. Using axes, they cut down most of the trees, sparing only those that are economically useful. The undergrowth is cleared away with a machete or other long knife. On a windless day the debris is burned under carefully controlled conditions. The rains wash the fresh ashes into the soil, providing needed nutrients (Figure 9-17).

▲ **FIGURE 9-17 SHIFTING CULTIVATION: SLASH AND BURN** A field is burned prior to planting in Mozambique.

Before planting, the cleared area, known by a variety of names in different regions, including **swidden**, lading, milpa, chena, and kaingin, is prepared by hand, perhaps with the help of a simple implement such as a hoe; plows and animals are rarely used (Figure 9-18). The only fertilizer generally available is potash (potassium) from burning the debris when the site is cleared. Little weeding is done the first year that a cleared patch of land is farmed; weeds may be cleared with a hoe in subsequent years.

The cleared land can support crops only briefly, usually three years or less. In many regions, the most productive harvest comes in the second year after burning. Thereafter, soil nutrients are rapidly depleted and the land becomes too infertile to nourish crops. Rapid weed growth also contributes to the abandonment of a swidden after a few years. When the swidden is no longer fertile, villagers identify a new site and begin clearing it. They leave the old site uncropped for many years, allowing it to become overrun again by natural vegetation. The field is not actually abandoned; the villagers will return to the site someday, perhaps as few as 6 years or as many as 20 years later, to begin the process of clearing the land again. In the meantime, they may still care for fruit-bearing trees on the site.

If a cleared area outside a village is too small to provide food for the population, then some of the people may establish a new village and practice shifting cultivation there. Some farmers may move temporarily to another settlement if the field they are clearing that year is distant.

CROPS OF SHIFTING CULTIVATION. The crops grown by each village vary by local custom and taste. The predominant crops include upland rice in Southeast Asia, maize (corn) and manioc (cassava) in South America, and millet and sorghum in Africa. Yams, sugarcane, plantain, and vegetables are also grown in some regions. These crops have originated in one region of shifting cultivation and have diffused to other areas in recent years.

The Kayapo people of Brazil's Amazon tropical rain forest do not arrange crops in the rectangular fields and rows that are familiar to us. They plant in concentric rings. At first they plant sweet potatoes and yams in the inner area. In successive rings go corn and rice, manioc, and more yams. In subsequent years the inner area of potatoes and yams expands to replace corn and rice. The outermost ring contains plants that require more nutrients, including papaya, banana, pineapple, mango, cotton, and beans. It is here that the leafy crowns of cut trees fall when the field is cleared, and their rotting releases more nutrients into the soil.

Most families grow only for their own needs, so one swidden may contain a large variety of intermingled crops, which are harvested individually at the best time. In shifting cultivation a "farm field" appears much more chaotic than do fields in developed countries, where a single crop such as corn or wheat may grow over an extensive area. In some cases, families may specialize in a few crops and trade with villagers who have a surplus of others.

OWNERSHIP AND USE OF LAND IN SHIFTING CULTIVATION. Traditionally, land was owned by the village as a whole rather than separately by each resident. The chief or ruling council allocated a patch of land to each family and allowed it to retain the output. Individuals may also have had the right to own or protect specific trees surrounding the village. Today, private individuals now own the land in some communities, especially in Latin America.

Shifting cultivation occupies approximately one-fourth of the world's land area, a higher percentage than any other type of agriculture. However, less than 5 percent of the world's people engage in shifting cultivation. The gap between the percentage of people and land area is not surprising, because the practice of moving from one field to another every couple of years requires more land per person than do other types of agriculture.

FUTURE OF SHIFTING CULTIVATION. Land devoted to shifting cultivation is declining in the tropics at the rate of about 75,000 square kilometers (30,000 square miles), or 0.2 percent, per year according to the United Nations (Figure 10-25). The amount of Earth's surface allocated to tropical rain forests has already been reduced to less than half of its original area, for until recent years the World Bank supported deforestation with loans to finance development schemes that required clearing forests. Shifting cultivation is being replaced by logging, cattle ranching, and the cultivation of cash crops. Selling timber to builders or raising beef cattle for fast-food restaurants are more effective development strategies than maintaining shifting cultivation. Developing countries also see shifting cultivation as an inefficient way to grow food in a hungry world. Indeed, compared to other forms of agriculture, shifting cultivation can support only a small population in an area without causing environmental damage.

To its critics, shifting cultivation is at best a preliminary step in economic development. Pioneers use shifting cultivation to clear forests in the tropics and to open land for development where permanent agriculture never existed. People unable to find agricultural land elsewhere can migrate to the tropical forests and initially practice shifting cultivation. Critics say it then should be replaced by more sophisticated agricultural techniques that yield more per land area. Defenders of shifting cultivation consider it the most environmentally sound approach for the tropics. Practices used in other forms of agriculture, such as fertilizers and pesticides and permanently clearing fields, may damage the soil, cause severe erosion, and upset balanced ecosystems.

Large-scale destruction of the rain forests also may contribute to global warming. When large numbers of trees are cut, their burning and decay release large volumes of carbon dioxide. This gas can build up in the atmosphere, acting like the window glass in a greenhouse to trap solar energy in the atmosphere, resulting in the "greenhouse effect," discussed in Chapter 11. Elimination of shifting cultivation could also upset the traditional local diversity of cultures in the tropics. The activities of shifting cultivation are intertwined with other social, religious, political, and various folk customs. A drastic change in the agricultural economy could disrupt other activities of daily life.

As the importance of tropical rain forests to the global environment has become recognized, developing countries have been pressured to restrict further destruction of them. In one innovative strategy, Bolivia agreed to set aside 1.5 million hectares (3.7 million acres) in a forest reserve in exchange for cancellation of $650,000,000 of its debt to developed countries. Meanwhile, in Brazil's Amazon rain forest, deforestation has increased from 2.7 million hectares (7 million acres) per year during the 1990s to 3.1 million hectares (8 million acres) since 2000.

PAUSE & REFLECT 9.3.3

How would rapid population growth made it difficult to practice shifting cultivation?

PLANTATION FARMING

Most plantations are located in the tropics and subtropics, especially in Latin America, Africa, and Asia. Although generally situated in developing countries, plantations are often owned or operated by Europeans or North Americans, and they grow crops for sale primarily to developed countries. Among the most important crops grown on plantations are cotton, sugarcane, coffee, rubber, and tobacco. Also produced in large quantities are cocoa, jute, bananas, tea, coconuts, and palm oil.

Until the Civil War, plantations were important in the U.S. South, where the principal crop was cotton, followed by tobacco and sugarcane. Slaves brought from Africa performed most of the labor until the abolition of slavery and the defeat of the South in the Civil War. Thereafter, plantations declined in the United States; they were subdivided and either sold to individual farmers or worked by tenant farmers.

▼ **FIGURE 9-18 SHIFTING CULTIVATION: PREPARING THE LAND** This field in Côte d'Ivoire is being prepared for planting.

Subsistence Agriculture in Population Concentrations

LEARNING OUTCOME 9.3.4

Explain the principal forms of agriculture in higher-density developing regions.

Three-fourths of the world's people live in developing countries, and the form of subsistence agriculture that feeds most of them is **intensive subsistence agriculture**. The term intensive implies that farmers must work intensively to subsist on a parcel of land.

CHARACTERISTICS OF INTENSIVE SUBSISTENCE FARMING

In developing countries, most people produce food for their own consumption. This is especially true in intensive subsistence agriculture, the most widely practiced form of agriculture in densely populated East, South, and Southeast Asia.

Intensive subsistence agriculture involves careful agricultural practices refined over thousands of years in response to local environmental and cultural patterns. Because the agricultural density—the ratio of farmers to arable land—is so high in parts of East and South Asia, families must produce enough food for their survival from a very small area of land. Most of the work is done by hand or with animals rather than with machines, in part due to abundant labor but largely from lack of funds to buy equipment.

The typical farm in Asia's intensive subsistence agriculture regions is much smaller than farms elsewhere in the world. Many Asian farmers own several fragmented plots, frequently a result of dividing individual holdings among several children over several centuries. To maximize food production, intensive subsistence farmers waste virtually no land. Corners of fields and irregularly shaped pieces of land are planted rather than left idle. Paths and roads are kept as narrow as possible to minimize the loss of arable land. Livestock are rarely permitted to graze on land that could be used to plant crops, and little grain is grown to feed the animals.

Land is used even more intensively in parts of Asia by obtaining two harvests per year from one field, a process known as **double cropping.** Double cropping is common in places that have warm winters, such as southern China and Taiwan, but is relatively rare in India, where most areas have dry winters. Normally, double cropping involves alternating between wet rice, grown in the summer when precipitation is higher, and wheat, barley, or another dry crop, grown in the drier winter season. Crops other than rice may be grown in the wet-rice region in the summer on nonirrigated land.

WET-RICE DOMINANT

The intensive agriculture region of Asia can be divided between areas where wet rice dominates and areas where it does not (refer to Figure 9-13). The term **wet rice** refers to rice planted on dry land in a nursery and then moved as seedlings to a flooded field to promote growth. Wet rice occupies a relatively small percentage of Asia's agricultural land but is the region's most important source of food.

Intensive wet-rice farming is the dominant type of agriculture in southeastern China, East India, and much of Southeast Asia. China and India account for nearly 50 percent of the world's rice production, and more than 90 percent is produced in East, South, and Southeast Asia (Figure 9-19).

Successful production of large yields of rice is an elaborate process that is time-consuming and done mostly by hand. The consumers of the rice also perform the work, and all family members, including children, contribute to the effort. Growing rice involves four principal steps. As the name implies, all four steps are intensive:

1. The field is prepared, typically using animal power (Figure 9-20). Flat land is needed to grow rice, so hillsides are terraced.
2. The field is flooded with water (Figure 9-21). The flooded field is called a **sawah** in Indonesia and is

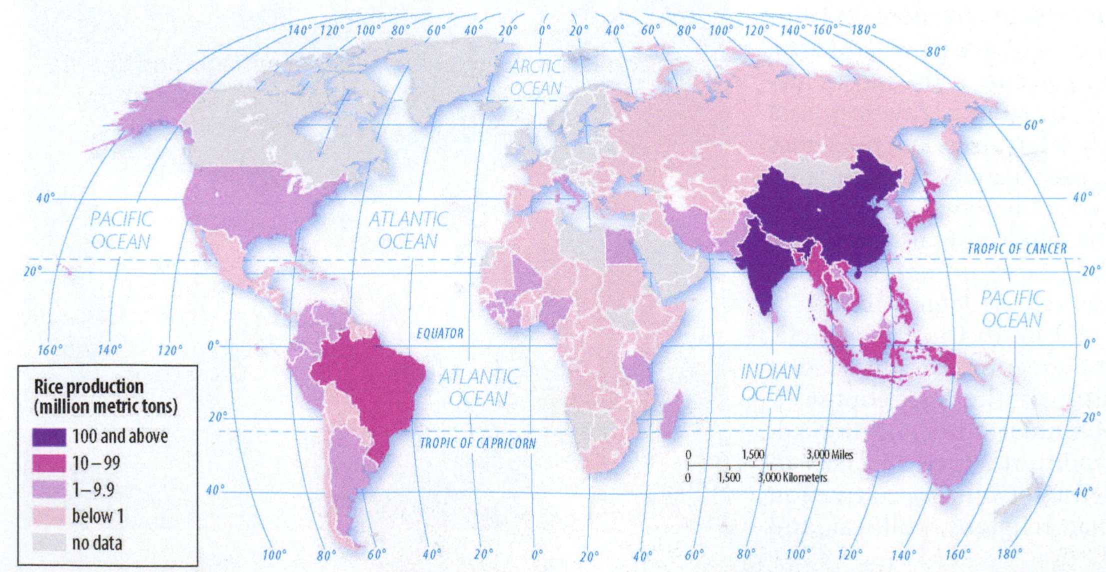

FIGURE 9-19 RICE PRODUCTION China and India produce one-half of the world's rice.

▲ **FIGURE 9-20 GROWING RICE: PREPARING THE FIELD** Cows pull a level in India.

▲ **FIGURE 9-21 GROWING RICE: FLOODING THE FIELD** Flooded fields in Indonesia.

increasingly referred to a **paddy,** which is actually the Malay word for wet rice.

3. Rice seedlings grown for the first month in a nursery are transplanted into the flooded field (Figure 9-22).
4. Rice plants are harvested with knives (Figure 9-23). The chaff (husks) is separated from the seeds by threshing (beating) the husks on the ground. The threshed rice is placed in a tray for winnowing, in which the lighter chaff is allowed to be blown away by the wind.

▲ **FIGURE 9-22 GROWING RICE: TRANSPLANTING PLANTS** Rice plants are transplanted by hand in India.

WET RICE NOT DOMINANT

Climate discourages farmers from growing wet rice in portions of Asia, especially where summer precipitation levels are too low and winters are too harsh (refer to Figure 9-13). Agriculture in much of the interior of India and northeastern China is devoted to crops other than wet rice. Wheat is the most important crop, followed by barley. Other grains and legumes are grown for household consumption, including millet, oats, corn, sorghum, and soybeans. In addition, some crops are grown in order to be sold for cash, such as cotton, flax, hemp, and tobacco.

Aside from what is grown, this region shares most of the features of the wet-rice region. Land is used intensively and worked primarily by human power, with the assistance of some hand implements and animals. In milder parts of the region where wet rice does not dominate, more than one harvest can be obtained some years through skilled use of **crop rotation,** which is the practice of rotating use of different fields from crop to crop each year to avoid exhausting the soil. In colder climates, wheat or another crop is planted in the spring and harvested in the fall, but no crops can be sown through the winter.

PAUSE & REFLECT 9.3.4

What challenges might farmers in East and South Asia face if they wish to adopt farming practices that are less labor intensive than growing wet rice?

◀ **FIGURE 9-23 GROWING RICE: HARVESTING** Harvesting and thrashing rice by hand in Myanmar.

Fishing

LEARNING OUTCOME 9.3.5

Describe the contribution of fishing to the world food supply.

Food acquired from Earth's waters includes fish, crustaceans (such as shrimp and crabs), mollusks (such as clams and oysters), and aquatic plants (such as watercress). Water-based food is acquired in two ways:

- **Fishing,** which is the capture of wild fish and other seafood living in the waters.
- **Aquaculture,** or **aquafarming,** which is the cultivation of seafood under controlled conditions.

Fishing and aquaculture are practiced in both subsistence and commercial agriculture.

Oceans are vast, covering nearly three-fourths of Earth's surface and lying near most population concentrations. Historically the sea has provided only a small percentage of the world food supply. So at first glance, increased use of food from the sea is attractive. However, overfishing has reduced fish supplies in many regions.

FISH PRODUCTION

The world's oceans are divided into 18 major fishing regions, including seven each in the Atlantic and Pacific oceans, three in the Indian Ocean, and one the Mediterranean (Figure 9-24). Fishing is also conducted in inland waterways, such as lakes and rivers. The areas with the largest yields are the Pacific Northwest and Asia's inland waterways.

During the past half-century, global fish production has increased from approximately 36 to 158 million metric tons. The growth has resulted entirely from expansion of aquaculture (Figure 9-25). The capture of wild fish in the oceans and lakes has stagnated since the 1990s, despite population growth and increased demand to consume fish. The reason that production is higher than human consumption is that a large portion of the fish that is caught is converted to fish meal and fed to poultry and hogs. Only two-thirds of the fish caught from the ocean is consumed directly by humans (see Sustainability & Our Environment feature).

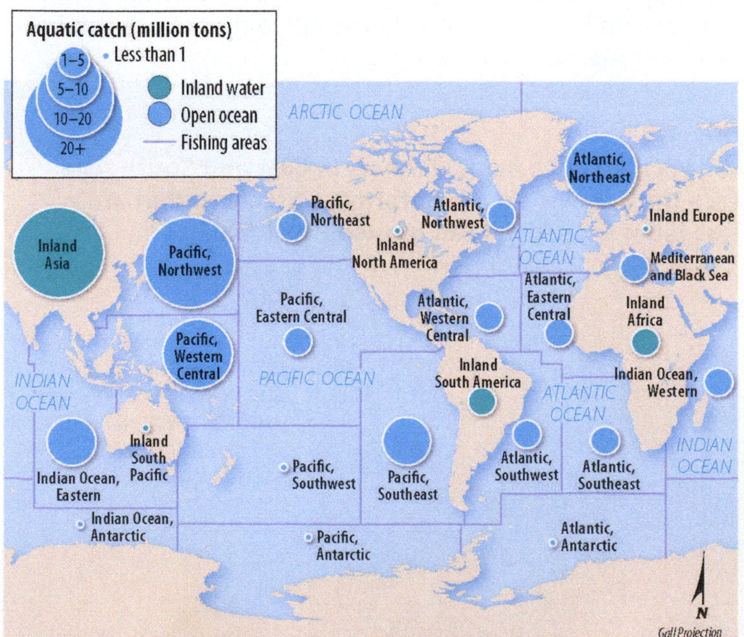

▲ **FIGURE 9-24 MAJOR FISHING REGIONS**

▲ **FIGURE 9-25 AQUACULTURE** Fish are raised inside the containers at this fish farm in Kassiopi, Corfu, Greece.

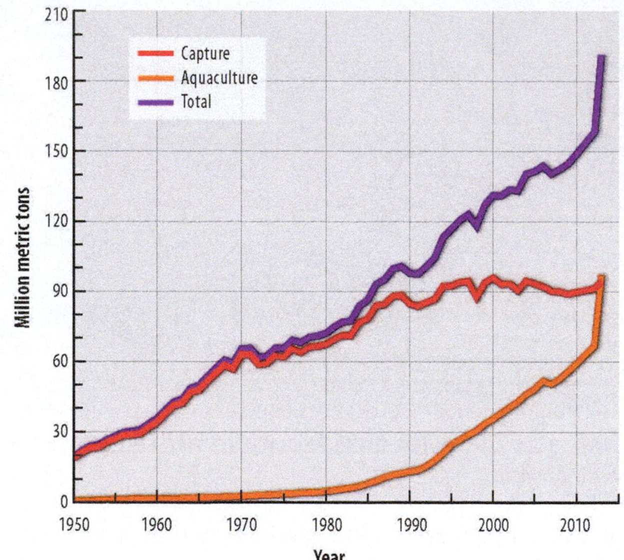

▲ **FIGURE 9-26 GROWTH IN FISH PRODUCTION**

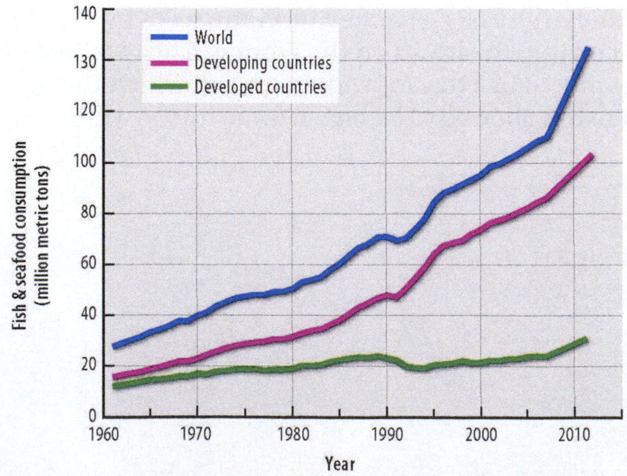

▲ **FIGURE 9-27 GROWTH IN HUMAN CONSUMPTION OF FISH**

SUSTAINABILITY & OUR ENVIRONMENT Asian Carp and Chicago's Economy

The growth of aquaculture has led to the farming of nonnative species. One example is the Asian carp, which were imported to the United States in the 1970s to stock a fish farm in Arkansas. Flooding allowed the carp to escape the farm and enter U.S. waterways. Fast-growing and voracious eaters, Asian carp can grow to over 45 kilograms (100 pounds) (Figure 9-28). Once in the waterways, the extremely aggressive Asian carp compete successfully with native fish for food and habitat threatening the sustainability of native species. They have even attacked people fishing in small boats.

Asian carp have traveled up the Mississippi and Illinois rivers, and they now constitute 97 percent of the fish in these rivers. Now the Asian carp threaten to reach the Great Lakes. The most likely point of entry into the Great Lakes for the Asian carp is through Chicago-area waterways. The U.S. Army Corps of Engineers has installed electric barriers to try to keep the Asian carp from traveling through the canals to Lake Michigan.

In the long run, the only effective way to keep the carp out of the Great Lakes is to shut the canals. But the canals play a major role, in sustaining Chicago's economy, so shutting them could devastate the city's economy.

▲ **FIGURE 9-28 ASIAN CARP** In the Illinois River.

FISH CONSUMPTION

Human consumption of fish and seafood increased from 27 million metric tons in 1960 to 132 million metric tons in 2012. Developing countries are responsible for five-sixths of the increase.

Fish consumption has increased more rapidly than population growth. During the past half-century, per capita consumption of fish has nearly doubled in both developed and developing countries. Still, fish and seafood account for only 1 percent of all calories consumed by humans.

A comparison of Figures 9-26 and 9-27 shows that production of fish is considerably higher than human consumption of it. Around 85 percent of the fish is consumed directly by humans. The remainder is converted to fish meal and fed to poultry and hogs.

OVERFISHING

China is responsible for one-third of the world's yield of fish (Figure 9-29). The other leading countries are naturally those with extensive ocean boundaries, such as Chile, Indonesia, and Peru.

The populations of some fish species in the oceans and lakes have declined because of **overfishing**, which is capturing fish faster than they can reproduce. Hope grew during the mid-twentieth century that increased fish consumption could meet the needs of a rapidly growing global population. However, the populations of some fish species declined because of overfishing. The U.N. estimates that one-quarter of fish stocks have been overfished and one-half fully exploited, leaving only one-fourth underfished.

PAUSE & REFLECT 9.3.5

The average resident in a developed country consumes around 6 ounces (170 grams) of fish per week. Why might your consumption be higher or lower than the average?

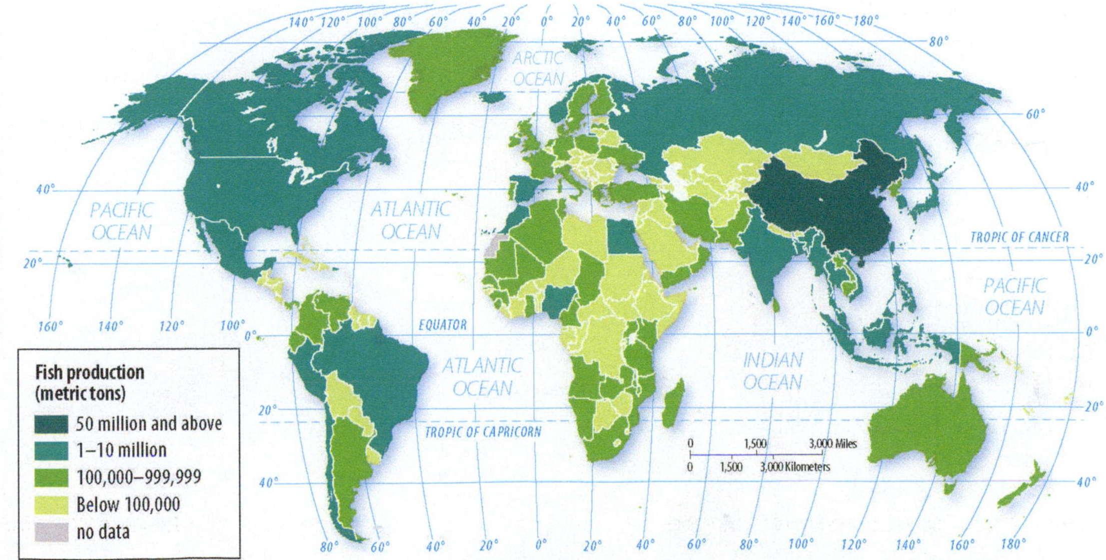

▶ **FIGURE 9-29 FISH PRODUCTION**

Commercial Agriculture: Crop Based

LEARNING OUTCOME 9.3.6

Describe the basic principles of several forms of crop-based commercial agriculture.

The system of commercial farming found in developed countries is called **agribusiness** because farming is integrated into a large food-production industry. Farmers account for less than 2 percent of the U.S. labor force, but around 20 percent of U.S. laborers work in food production and services related to agribusiness—food processing, packaging, storage, distribution, and retailing. Agribusiness also encompasses such diverse enterprises as tractor manufacturing, fertilizer production, and seed distribution.

Most farms are owned by individual families, but most other aspects of agribusiness are controlled by large corporations. Agricultural products are not sold directly to consumers but to food-processing companies. Large processors, such as General Mills and Kraft, typically sign contracts with commercial farmers to buy their crops and animals.

Commercial agriculture in developed countries can be divided into six main types: grain, Mediterranean, commercial gardening and fruit, mixed crop and livestock, dairy, and ranching. The first three derive most of their income from the sale of crops and the second three from the sale of animal products.

GRAIN FARMING

The major crop on most farms is grain, such as wheat, corn, oats, barley, rice, and millet. Commercial grain agriculture is distinguished from mixed crop and livestock farming because crops on a grain farm are grown primarily for consumption by humans rather than by livestock. Large-scale grain production, like other commercial farming ventures in developed countries, is heavily mechanized, conducted on large farms, and oriented to consumer preferences.

Commercial grain farms sell their output to manufacturers of food products, such as breakfast cereal and bread. The most important crop grown is wheat, used to make bread flour. Wheat generally can be sold for a higher price than other grains, such as rye, oats, and barley, and it has more uses as human food. It can be stored relatively easily without spoiling and can be transported a long distance. Because wheat has a relatively high value per unit weight, it can be shipped profitably from remote farms to markets.

As has been the case with milk production, the share of world production of wheat in developing countries has increased rapidly. Much of this increased production results from growth in large-scale commercial agriculture. Developing countries accounted for more than one-half of world wheat production in 2013, compared to only one-fourth in 1960. The United States is by far the largest producer of wheat among developed countries, but it now ranks third among all countries, behind China and India (Figure 9-30). China has been the world leader since 1983, and India has been second since 1999.

Commercial grain farms are generally located in regions that are too dry for mixed crop and livestock agriculture. Within North America, large-scale grain production is concentrated in three areas:

- The winter wheat belt through Kansas, Colorado, and Oklahoma, where planting is in the autumn and harvesting in the early summer.
- The spring wheat belt through the Dakotas, Montana, and southern Saskatchewan in Canada, where planting is in the spring and harvesting in the late summer.
- The Palouse region of Washington State.

Wheat's significance extends beyond the amount of land or number of people involved in growing it. Unlike other agricultural products, wheat is grown to a considerable extent for international trade, and it is the world's

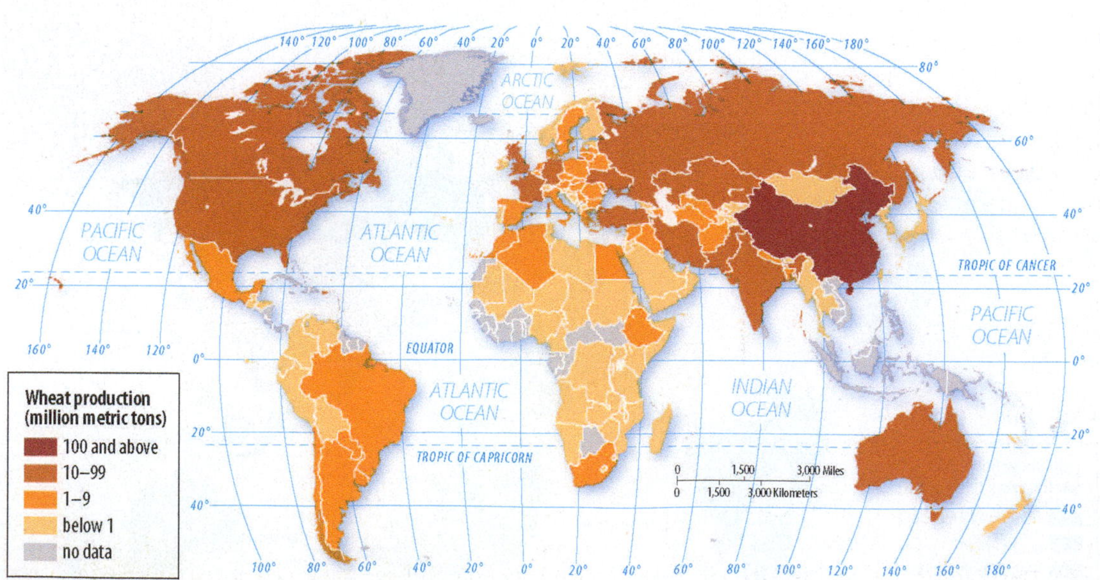

◀ **FIGURE 9-30 WHEAT PRODUCTION** China and India are the leading wheat producers, followed by the United States.

▲ **FIGURE 9-31 MEDITERRANEAN AGRICULTURE** Vineyard on a hillside overlooking the Mediterranean Sea, Vernazza, Italy.

leading export crop. The United States and Canada account for about one-fourth of the world's wheat exports. The ability to provide food for many people elsewhere in the world is a major source of economic and political strength for these two countries.

MEDITERRANEAN AGRICULTURE

Mediterranean agriculture exists primarily on the lands that border the Mediterranean Sea in Southern Europe, North Africa, and western Asia. Farmers in California, central Chile, the southwestern part of South Africa, and southwestern Australia practice Mediterranean agriculture as well.

These Mediterranean areas share a similar physical environment (refer to Figure 9-13). Every Mediterranean area borders a sea (Figure 9-31). Prevailing sea winds provide moisture and moderate the winter temperatures. Summers are hot and dry, but sea breezes provide some relief. The land is very hilly, and mountains frequently plunge directly to the sea, leaving very narrow strips of flat land along the coast.

Most crops in Mediterranean lands are grown for human consumption rather than for animal feed. **Horticulture**—which is the growing of fruits, vegetables, and flowers—and tree crops form the commercial base of Mediterranean farming. In the lands bordering the Mediterranean Sea, the two most important cash crops are olives and grapes (refer to Figure 4-16). A large portion of California farmland is devoted to fruit and vegetable horticulture, which supplies a large portion of the citrus fruits, tree nuts, and deciduous fruits consumed in the United States.

COMMERCIAL GARDENING AND FRUIT FARMING

Commercial gardening and fruit farming is the predominant type of agriculture in the southeastern United States. The region has a long growing season and humid climate, and it is accessible to the large number of consumers in the northeast United States. This type of agriculture is frequently called **truck farming**, from the Middle English word truck, meaning "barter" or "exchange of commodities."

Truck farms grow many of the fruits and vegetables that consumers in developed countries demand, such as apples, asparagus, cherries, lettuce, mushrooms, and tomatoes (Figure 9-32). Some of these fruits and vegetables are sold fresh to consumers, but most are sold to large processors for canning or freezing. Farms tend to specialize in a few crops, and a handful of farms may dominate national output of some fruits and vegetables.

A form of truck farming called specialty farming has spread to New England, among other places. Farmers are profitably growing crops that have limited but increasing demand among affluent consumers, such as asparagus, peppers, mushrooms, strawberries, and nursery plants. Specialty farming represents a profitable alternative for New England farmers at a time when dairy farming is declining because of relatively high operating costs and low milk prices.

PAUSE & REFLECT 9.3.6

Why might wheat be easier to export than produce from truck farming?

▼ **FIGURE 9-32 TRUCK FARMING** Harvesting peanuts, Georgia.

Commercial Agriculture: Mixed Crop and Livestock

LEARNING OUTCOME 9.3.7

Describe the basic principles of several forms of mixed crop and livestock agriculture.

Mixed crop and livestock farming is the most common form of commercial agriculture in the United States west of the Appalachians and east of 98° west longitude and in much of Europe from France to Russia.

MIXED CROP AND LIVESTOCK

The most distinctive characteristic of **mixed crop and livestock farming** is the integration of crops and livestock. Most of the crops are fed to animals rather than consumed directly by humans. In turn, the livestock supply manure to improve soil fertility to grow more crops.

A typical mixed crop and livestock farm devotes nearly all land area to growing crops but derives more than three-fourths of its income from the sale of animal products, such as beef, milk, and eggs. In the United States pigs are often bred directly on the farms, whereas cattle may be brought in to be fattened on corn.

Mixing crops and livestock permits farmers to distribute the workload more evenly through the year. Fields require less attention in the winter than in the spring, when crops are planted, and in the fall, when they are harvested. Meanwhile, livestock require year-round attention. A mix of crops and livestock also reduces seasonal variations in income; most income from crops comes during the harvest season, but livestock products can be sold throughout the year.

In the United States, corn (maize) is the crop most frequently planted in the mixed crop and livestock region because it generates higher yields per area than do other crops (Figure 9-33). Some of the corn is consumed by people as oil, margarine, and other food products, but most is fed to pigs and cattle. The most important mixed crop and livestock farming region in the United States—extending from Ohio to the Dakotas, with its center in Iowa—is frequently called the Corn Belt because around half of the cropland is planted in corn (Figure 9-34).

▲ **FIGURE 9-34 MIXED CROP AND LIVESTOCK** Harvesting corn in Iowa.

Soybeans have become the second-most-important crop in the U.S. mixed commercial farming region. Like corn, soybeans are mostly used to make animal feed. Tofu (made from soybean milk) is a major food source, especially for people in China and Japan. Soybean oil is widely used in U.S. foods, as a hidden ingredient.

IMPORTANCE OF ACCESS TO MARKETS

The use of land in developed countries is determined primarily by market forces of supply and demand. The value of the land affects the form of commercial agriculture practiced

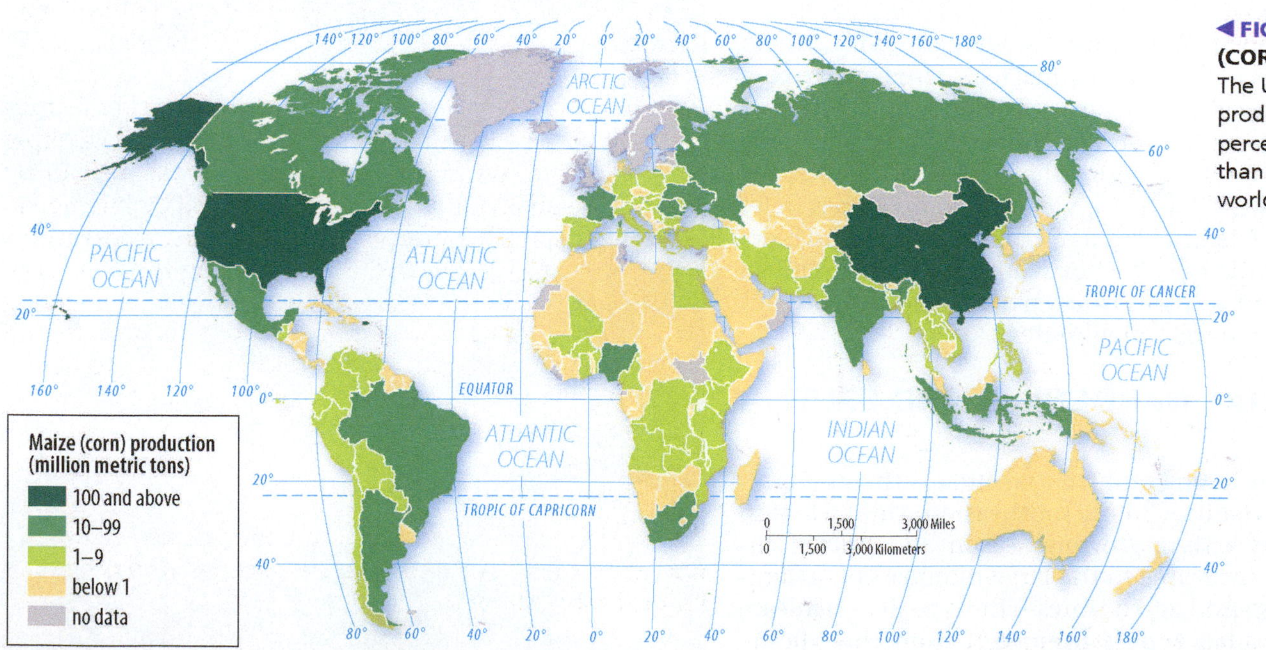

◄ **FIGURE 9-33 MAIZE (CORN) PRODUCTION** The United States produces nearly 40 percent and China more than 20 percent of the world's maize.

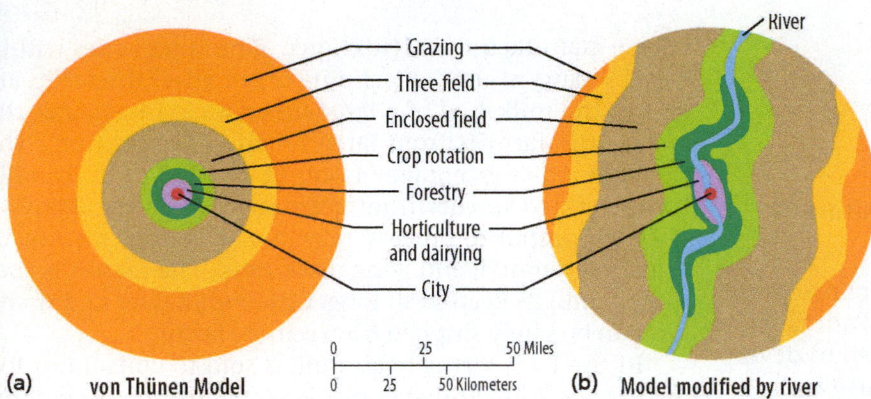

◄ **FIGURE 9-35 VON THÜNEN MODEL (a)** According to the von Thünen model, in the absence of topographic factors, different types of farming are conducted at different distances from a city, depending on the cost of transportation and the value of the product. **(b)** von Thünen recognized that his model would be modified by site factors, such as a river in this sketch, which changes the accessibility of different land parcels to the market center. Agricultural uses that seek highly accessible locations need to locate nearer the river.

on it. Agricultural land is lost altogether to some other use if someone is willing to pay to use it for another purpose.

Because the purpose of commercial farming is to sell produce off the farm, the distance from the farm to the market influences the farmer's choice of crop to plant. Geographers use the von Thünen model to help explain the importance of proximity to market in the choice of crops on commercial farms. Johann Heinrich von Thünen, an estate owner in northern Germany, first proposed the model in 1826, in a book titled *The Isolated State*. According to this model, which geographers later modified, a commercial farmer initially considers which crops to cultivate and which animals to raise based on market location. In choosing an enterprise, the farmer compares two costs: the cost of the land and the cost of transporting products to market.

Von Thünen based his general model of the spatial arrangement of different crops on his experience as the owner of a large estate in northern Germany during the early nineteenth century. He found that specific crops were grown in different rings around the cities in the area (Figure 9-35):

- **First ring.** Market-oriented gardens and milk producers were located in the first ring out from the cities. These products are expensive to deliver and must reach the market quickly because they are perishable.
- **Second ring.** The next ring out from the cities contained wood lots, where timber was cut for construction and fuel; closeness to market is important for this commodity because of its weight.
- **Third ring.** The next ring was used for various crops and for pasture; the specific commodity was rotated from one year to the next.
- **Fourth ring.** The outermost ring was devoted exclusively to animal grazing, which requires lots of space.

The model assumed that all land in a study area had similar site characteristics and was of uniform quality, although von Thünen recognized that the model could vary according to topography and other distinctive physical conditions. For example, a river might modify the shape of the rings because transportation costs change when products are shipped by water routes rather than over roads. The model also failed to consider that social customs and government policies influence the attractiveness of plants and animals for a commercial farmer.

Table 9-1 illustrates the influence of transportation cost on the profitability of growing wheat. The example in Table 9-1 shows that a farmer would make a profit by growing wheat on land located less than 10,000 kilometers from the market. Beyond 10,000 kilometers, wheat is not profitable because the cost of transporting it exceeds the gross profit. These calculations demonstrate that farms located closer to market tend to select crops with higher transportation costs per hectare of output, whereas more distant farms are more likely to select crops that can be transported less expensively.

TABLE 9–1 Transportation in the von Thünen Model

The influence of transportation cost on the profitability of growing wheat, according to the von Thünen model:
• Gross profit from sale of wheat grown on 1 hectare of land *not including* transportation costs:
a. Wheat can be sold for $250 per metric ton. b. Yield per hectare of wheat is 4 tons. c. Gross profit is $1,000 per hectare ($250 per ton × 4 tons).
• Net profit from sale of wheat grown on 1 hectare of land *including* transportation costs:
a. Cost of transporting 4 tons of wheat to market is $0.10 per kilometer. b. Net profit from the sale of 4 tons of wheat grown on a farm located 1,000 kilometers from the market is $900 ($1,000 gross profit – $100 for 1,000 kilometers of transport costs). c. Net profit from sale of 1,000 kilograms of wheat grown on a farm located 10,000 kilometers from the market is $0 ($1,000 gross profit – $1,000 for 10,000 kilometers of transport costs).

PAUSE & REFLECT 9.3.7

If the price of wheat dropped to $200 per ton, what would be the maximum distance that the wheat could be profitably shipped?

Commercial Agriculture: Animal-based

LEARNING OUTCOME 9.3.8

Describe dairy and ranching commercial agriculture.

Dairy farming is the most important agriculture practiced near large urban areas in developed countries. Ranching is adapted to semiarid or arid land and is practiced in developed countries where the vegetation is too sparse and soil too poor to support crops.

DAIRY

A **dairy farm** specializes in the production of milk and other dairy products. Because milk is highly perishable, dairy farms must be closer to their markets than other products. The ring surrounding a city from which milk can be supplied without spoiling is known as the **milkshed.**

Dairy farms located farther from consumers are more likely to sell their output to processors that make butter, cheese, or dried, evaporated, and condensed milk. The reason is that these products keep fresh longer than milk does and therefore can be safely shipped from remote farms.

Traditionally most milk was produced and consumed in developed countries (Figure 9-36). However, the share of the world's dairy farming conducted in developing countries has risen dramatically in recent years, and it now surpasses the total in developed countries (Figure 9-37). Rising incomes permit urban residents to buy more milk products.

Dairy farmers, like other commercial farmers, usually do not sell their products directly to consumers. Instead, they generally sell milk to wholesalers, who distribute it in turn to retailers. Retailers then sell milk to consumers in shops or at home. Farmers also sell milk to butter and cheese manufacturers. The choice of product varies within the U.S. dairy region, depending on whether the farms are within the milkshed of a large urban area. In general, the farther the farm is from large urban concentrations, the smaller is the percentage of output devoted to fresh milk. Farms located farther from consumers are more likely to sell their output to processors who make butter, cheese, or dried, evaporated, and condensed milk. The reason is that these products keep fresh longer than milk does and therefore can be safely shipped from remote farms.

In the East, virtually all milk is sold to consumers living in New York, Philadelphia, Boston, and the other large urban areas. Farther west, most milk is processed into cheese and butter. Most of the milk in Wisconsin is processed, for example, compared to only 5 percent in Pennsylvania. The proximity of northeastern farmers to several large markets accounts for these regional differences.

Countries likewise tend to specialize in certain products. New Zealand, the world's largest per capita producer of dairy products, devotes about 5 percent to liquid milk, compared to more than 50 percent in the United Kingdom. New Zealand farmers do not sell much liquid milk, because the country is too far from North America and Northwest Europe, the two largest relatively wealthy population concentrations.

Like other commercial farmers, dairy farmers face economic difficulties because of declining revenues and rising costs. Dairy farmers who have quit most often cite lack of profitability and excessive workload as reasons for getting out of the business. Distinctive features of dairy farming have exacerbated the economic difficulties:

- **Labor-intensive.** Cows must be milked twice a day, every day; although the actual milking can be done by machines, dairy farming nonetheless requires constant attention throughout the year.

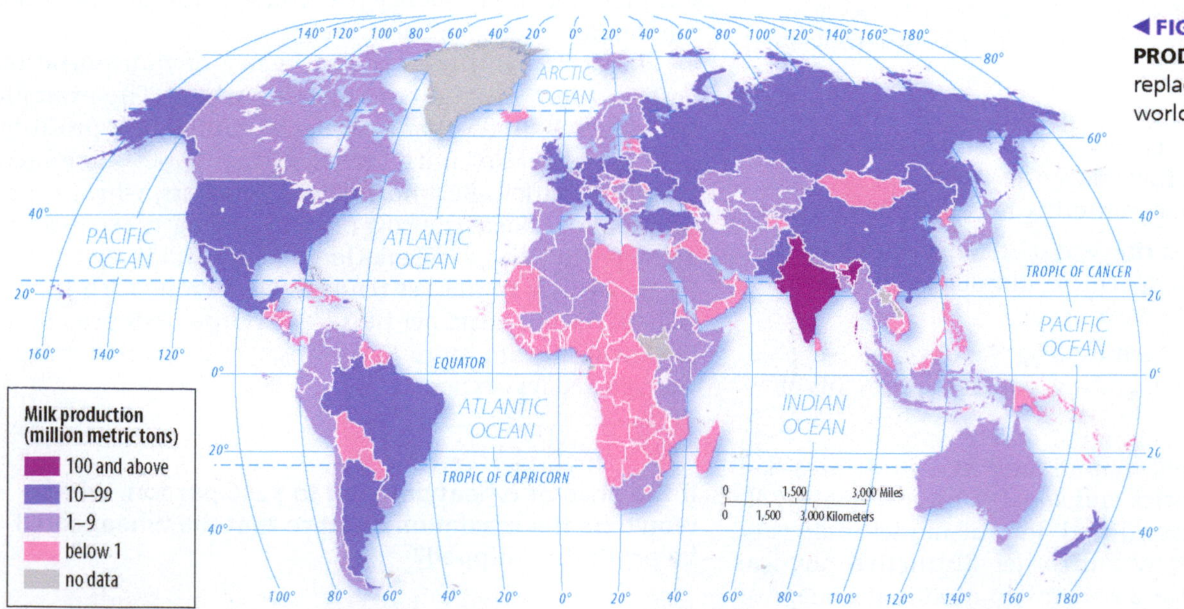

◀ **FIGURE 9-36 MILK PRODUCTION** India has replaced the United States as the world's leading milk producer.

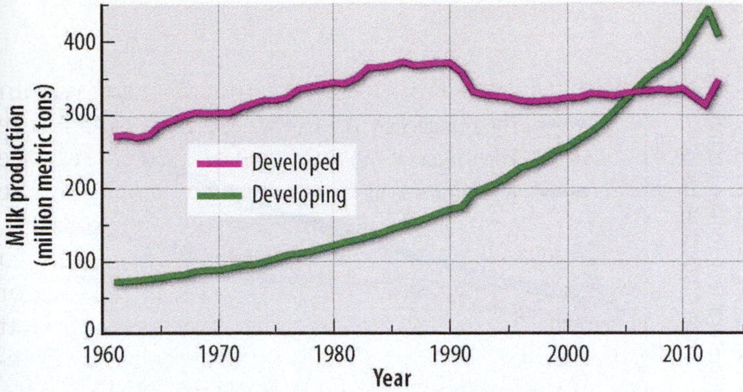

▲ **FIGURE 9-37 CHANGING MILK PRODUCTION** Developing countries now produce more milk than developed countries.

- **Winter Feed.** Dairy farmers face the expense of feeding the cows in the winter, when they may be unable to graze on grass. In Northwest Europe and in the Northeastern United States, farmers generally purchase hay or grain for winter feed. In the western part of the U.S. dairy region, crops are more likely to be grown in the summer and stored for winter feed on the same farm.

LIVESTOCK RANCHING

Ranching is the commercial grazing of livestock over an extensive area. Commercial ranching is conducted in several developed countries besides the United States and, increasingly, in developing countries. For example, the interior of Australia was opened for grazing in the nineteenth century, although sheep are more common there than cattle.

The United States is the leading producer of chicken and beef, but as with other forms of commercial agriculture, the growth in ranching has been in developing countries. Developed countries were responsible for only one-third of world meat production in 2013, compared to two-thirds in 1980. China is the leading producer of meat, ahead of the United States, and Brazil is third (Figure 9-38). China passed the United States as the world's leading meat producer in 1990 and now produces twice as much. In South America, a large portion of the pampas of Argentina, southern Brazil, and Uruguay is devoted to grazing cattle and sheep. The cattle industry grew rapidly in Argentina in part because the land devoted to ranching was relatively accessible to the ocean, making it possible for meat to be transported to overseas markets.

Meanwhile, due to the spread of irrigation techniques and hardier crops, land in the United States has been converted from ranching to crop growing. Ranching generates lower income per area of land, although it has lower operating costs. Cattle are still raised on ranches but are frequently sent for fattening to farms or to local feed lots along major railroad and highway routes rather than directly to meat processors.

PAUSE & REFLECT 9.3.8

What are the two most important ranched animals, according to Figure 9-38?

CHECK-IN KEY ISSUE 3

Where Is Agriculture Distributed?

- ✔ **Agriculture can be divided into 11 major regions, including 5 subsistence and 6 commercial regions.**
- ✔ **In subsistence regions, pastoral nomadism is prevalent in drylands, shifting cultivation in tropical forests, and intensive subsistence in regions with high population concentrations.**
- ✔ **In commercial regions, mixed crop and livestock is the most common form of agriculture. Dairy, commercial gardening, grain, Mediterranean, and livestock ranching are also important.**

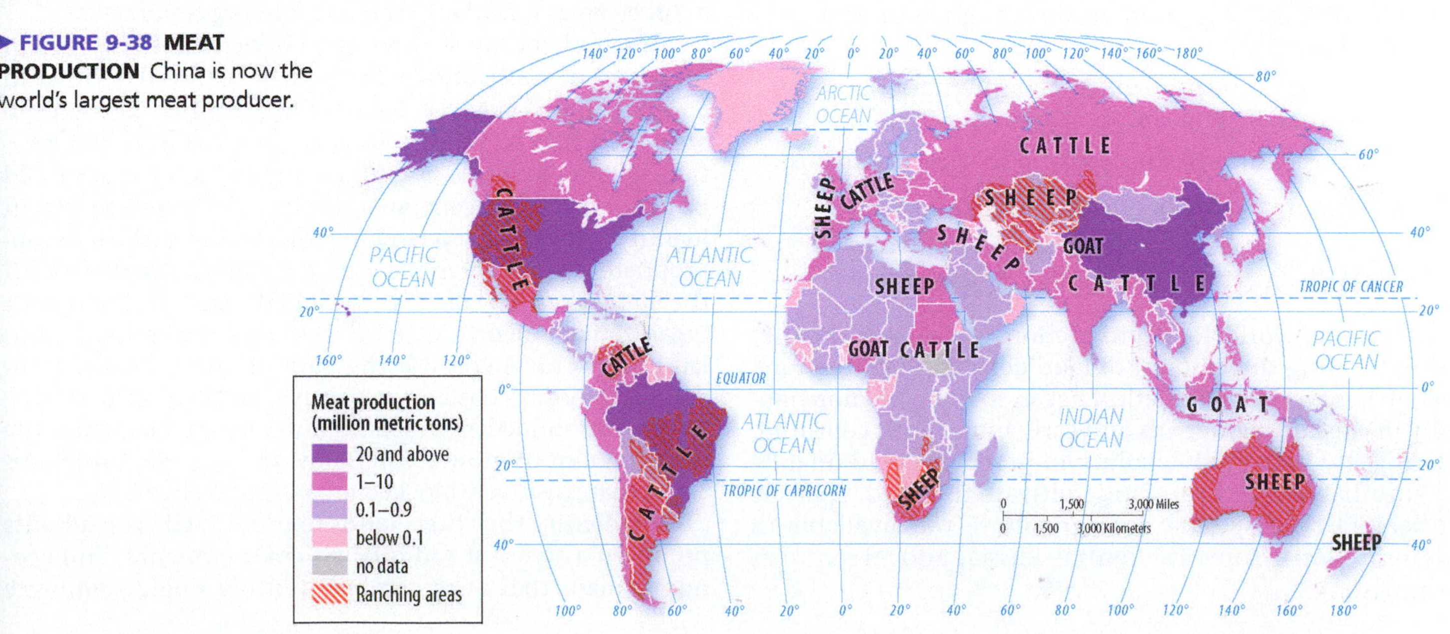

▶ **FIGURE 9-38 MEAT PRODUCTION** China is now the world's largest meat producer.

KEY ISSUE 4

Why Do Farmers Face Sustainability Challenges?

- Losing Agricultural Land
- Improving Agricultural Productivity
- Conserving Agricultural Resources
- Applying Biotechnology to Agriculture
- Global Food Trade
- Global Agriculture and Undernourishment
- Sustainable Agriculture

LEARNING OUTCOME 9.4.1

Explain reasons for loss of farmland.

Both subsistence and commercial agriculture faces several challenges to produce more food for a growing and hungry world in ways that preserve and protect Earth's agricultural resources for the future. This key issue considers seven challenges for agriculture:

- Losing agricultural land to competing uses.
- Improving the productivity of existing farmland.
- Conserving scarce resources, such as water and top soil.
- Identifying the appropriate role in agriculture for biotechnology.
- Balancing production of food for international trade rather than for consumption at home.
- Meeting the needs of people who are undernourished.
- Making greater use of organic farming.

Losing Agricultural Land

Historically, world food production increased primarily by expanding the amount of land devoted to agriculture. When the world's population began to increase more rapidly in the late eighteenth and early nineteenth centuries, during the Industrial Revolution, pioneers could migrate to uninhabited territory and cultivate the land. Sparsely inhabited land suitable for agriculture was available in western North America, central Russia, and Argentina's pampas.

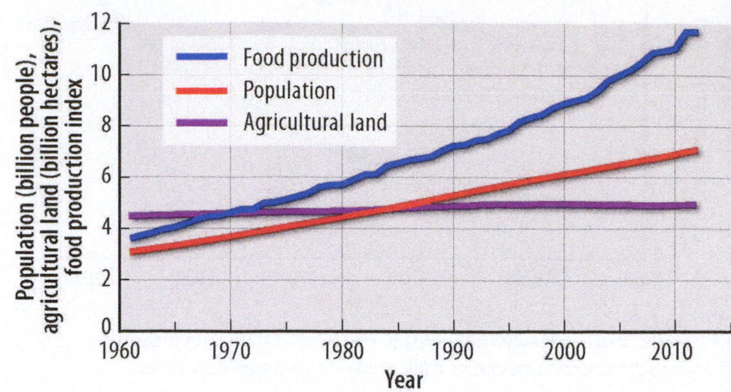

▲ **FIGURE 9-39 WORLD POPULATION GROWTH, AGRICULTURAL LAND, AND FOOD PRODUCTION** The food production index is set at 10 in the year 2005.

Two centuries ago, people believed that good agricultural land would always be available for willing pioneers. Today few scientists believe that further expansion of agricultural land can feed the growing world population. At first glance, new agricultural land appears to be available because only 11 percent of the world's land area is currently cultivated. However, in recent decades, population has increased much more rapidly than agricultural land (Figure 9-39).

LOSS OF FARMLAND TO URBANIZATION

The von Thünen model shows the influence of proximity to urban markets in the form of agriculture practiced on a piece of land. The expansion of urban areas has contributed to reducing agricultural land.

Loss of farmland to urban growth is especially severe at the edge of the string of large metropolitan areas along the East Coast of the United States. As urban areas grow in population and land area, farms on the periphery are replaced by urban land uses. A serious problem in the United States has been the loss of 200,000 hectares (500,000 acres) of the most productive farmland, known as **prime agricultural land,** as urban areas sprawl into the surrounding countryside.

Some of the most threatened agricultural land lies in Maryland, a small state where two major cities—Washington and Baltimore—have coalesced into a continuous built-up area (see Chapter 13). In Maryland, a geographic information system (GIS) was used to identify which farms should be preserved. Maps generated through GIS were essential in identifying agricultural land to protect because the most appropriate farms to preserve were not necessarily those with the highest-quality soil. Why should the state and nonprofit organizations spend scarce funds to preserve "prime" farmland that is nowhere near the path of urban sprawl? Conversely, why purchase an expensive, isolated farm already totally surrounded by residential developments, when the same amount of money could buy several large contiguous farms that effectively blocked urban sprawl elsewhere?

To identify the "best" lands to protect, GIS consultants produced a series of soil quality, environmental, and economic maps that were combined into a single composite

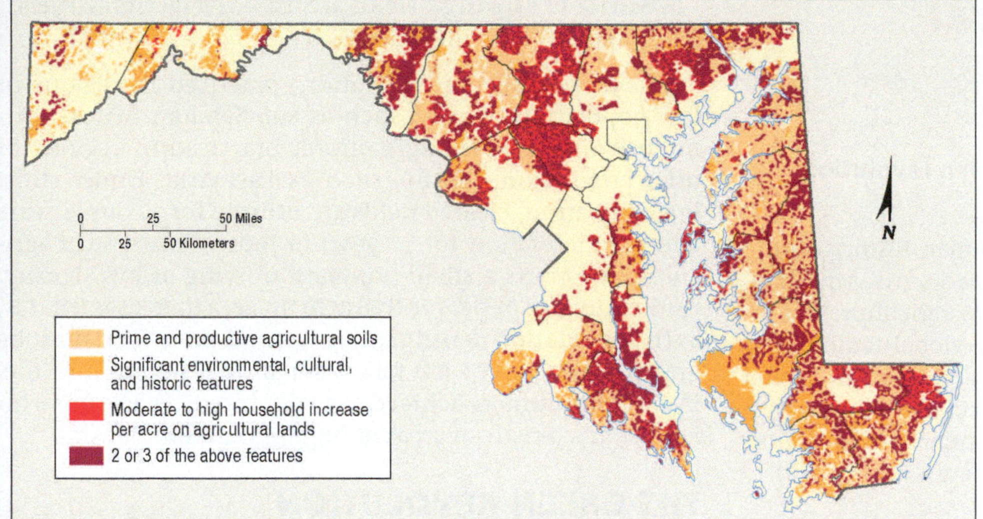

◄ **FIGURE 9-40 PROTECTING FARMLAND IN MARYLAND** Prime farmland is typically flat and well drained. Significant environmental features include water quality, flood control, species habitats, historic sites, and especially attractive scenery.

map (Figure 9-40). The map shows that 4 percent of the state's farmland had prime soils, significant environmental features, and high projected population growth, and 25 percent had two of the three factors.

DESERTIFICATION

In some regions, farmland is abandoned for lack of water. Especially in semiarid regions, human actions are causing land to deteriorate to a desertlike condition, a process called **desertification** (or, more precisely, semiarid land degradation). Semiarid lands that can support only a handful of pastoral nomads are overused because of rapid population growth. Excessive crop planting, animal grazing, and tree cutting exhaust the soil's nutrients and preclude agriculture.

The Earth Policy Institute estimates that 2 billion hectares (5 million acres) of land have been degraded around the world (Figure 9-41). Overgrazing is thought to be responsible for 34 percent of the total, deforestation for 30 percent, and agricultural use for 28 percent. The U.N. estimates that desertification removes 27 million hectares (70 million acres) of land from agricultural production each year, an area roughly equivalent to Colorado.

Excessive water threatens other agricultural areas, especially drier lands that receive water from human-built irrigation systems. If the irrigated land has inadequate drainage, the underground water level rises to the point where roots become waterlogged. The U.N. estimates that 10 percent of all irrigated land is waterlogged, mostly in Asia and South America. If the water is salty, it can damage plants. The ancient civilization of Mesopotamia may have collapsed in part because of waterlogging and excessive salinity in its agricultural lands near the Tigris and Euphrates rivers.

PAUSE & REFLECT 9.4.1

Have you seen loss of farmland near where you live? For what new purpose is the land used?

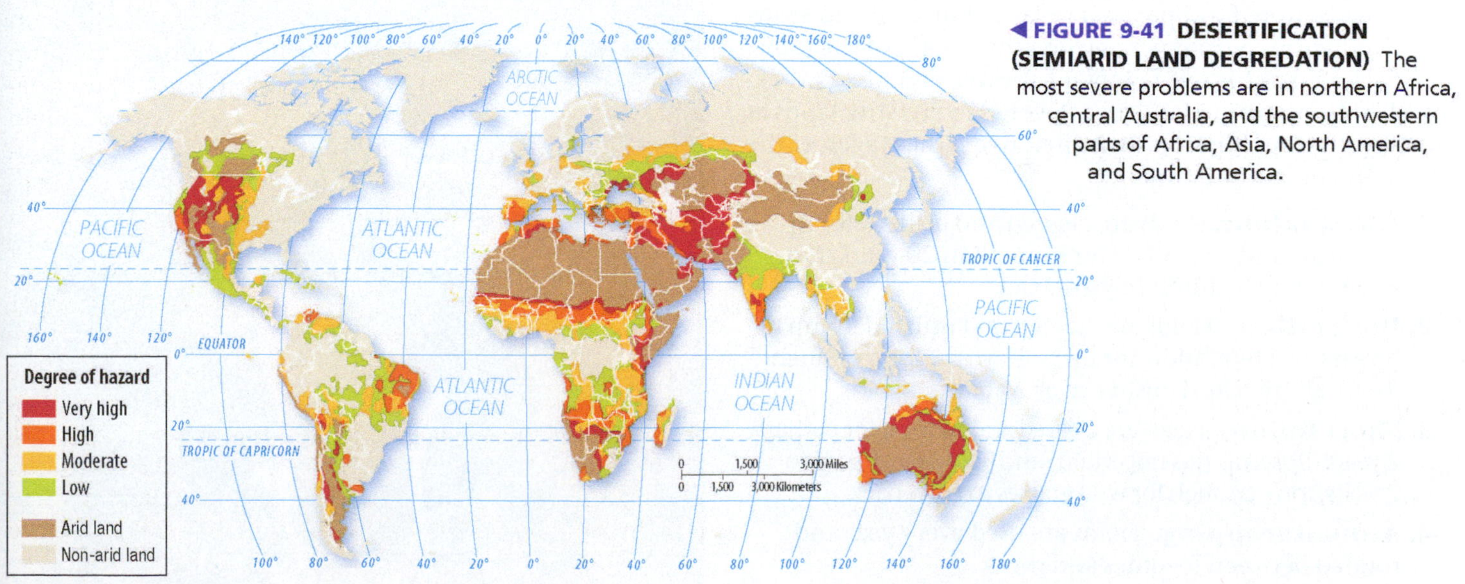

◄ **FIGURE 9-41 DESERTIFICATION (SEMIARID LAND DEGREDATION)** The most severe problems are in northern Africa, central Australia, and the southwestern parts of Africa, Asia, North America, and South America.

Improving Agricultural Productivity

LEARNING OUTCOME 9.4.2

Understand the importance of the green revolution.

Population grew at the fastest rate in human history during the second half of the twentieth century, as discussed in Chapter 2. With the amount of land devoted to agriculture not increasing, many experts forecast massive global famine. But these dire predictions did not come true. Instead, increased productivity has resulted in an expansion of food supply. New agricultural practices have permitted farmers worldwide to achieve much greater yields from the same amount of land.

INTENSIFICATION BY SUBSISTENCE FARMERS

For hundreds if not thousands of years, subsistence farming in developing countries yielded enough food for people living in rural villages to survive, assuming that no drought, flood, or other natural disaster occurred. Suddenly in the late twentieth century, subsistence farming practices needed to provide enough food for a rapidly increasing population as well as for the growing number of urban residents who cannot grow their own food.

Population growth influences the distribution of types of subsistence farming, according to economist Ester Boserup. It compels subsistence farmers to consider new farming approaches that produce enough food to take care of the additional people. According to Boserup, subsistence farmers increase the supply of food through intensification of production, achieved in two ways. First, new farming methods are adopted. Plows replace axes and sticks. More weeding is done, more manure is applied, more terraces are carved out of hillsides, and more irrigation ditches are dug (Figure 9-42). The additional labor needed to perform these operations comes from the population growth. The farmland yields more food per area of land, but with the growing population, output per person remains about the same.

Second, land is left fallow for shorter periods. This expands the amount of land area devoted to growing crops at any given time. Boserup identified five basic stages in the reduction of fallow farmland:

1. **Forest fallow.** Fields are cleared and utilized for up to 2 years and left fallow for more than 20 years, long enough for the forest to grow back.
2. **Bush fallow.** Fields are cleared and utilized for up to 8 years and left fallow for up to 10 years, long enough for small trees and bushes to grow back.
3. **Short fallow.** Fields are cleared and utilized for perhaps 2 years (Boserup was uncertain) and left fallow for up to 2 years, long enough for wild grasses to grow back.
4. **Annual cropping.** Fields are used every year and rotated between legumes and roots.
5. **Multi-cropping.** Fields are used several times a year and never left fallow.

Contrast shifting cultivation, practiced in regions of low population density, such as sub-Saharan Africa, with intensive subsistence agriculture, practiced in regions of high population density, such as East Asia. Under shifting cultivation, cleared fields are utilized for a couple years and then left fallow for 20 years or more. This type of agriculture supports a small population living at low density. As the number of people living in an area increases (that is, as the population density increases) and more food must be grown, fields will be left fallow for shorter periods of time. Eventually, farmers achieve the very intensive use of farmland characteristic of areas of high population density.

THE GREEN REVOLUTION

The invention and rapid diffusion of more productive agricultural techniques during the 1970s and 1980s is called the **green revolution.** The green revolution involves two main practices: the introduction of new higher-yield seeds and the expanded use of fertilizers. Because of the green revolution, agricultural productivity at a global scale has increased faster than population growth.

Scientists began an intensive series of experiments during the 1950s to develop a higher-yield form of wheat. A decade later, the "miracle wheat seed" was ready. Shorter and stiffer than traditional breeds, the new wheat was less sensitive to variation in day length, responded better to fertilizers, and matured faster. The Rockefeller and Ford foundations sponsored many of the studies, and the program's director, Dr. Norman Borlaug, won the Nobel Peace Prize in 1970.

▼ **FIGURE 9-42 INTENSIVE FARMING** Hillsides in Radi, Bhutan, are terraced into fields for intensive planting of rice.

▲ **FIGURE 9-43 INTERNATIONAL RICE RESEARCH INSTITUTE, HOME OF THE GREEN REVOLUTION** "Miracle" high-yield seeds have been produced through laboratory experiments at the International Rice Research Institute (IRRI). The IRRI is testing rice varieties in the Philippines.

The International Rice Research Institute, established in the Philippines by the Rockefeller and Ford foundations, worked to create a miracle rice seed (Figure 9-43). During the 1960s, their scientists introduced a hybrid of Indonesian rice and Taiwan dwarf rice that was hardier and that increased yields. More recently, scientists have developed new high-yield maize (corn).

The new miracle seeds were diffused rapidly around the world. India's wheat production, for example, more than doubled in five years. After importing 10 million tons of wheat annually in the mid-1960s, India had a surplus of several million tons by 1971. Other Asian and Latin American countries recorded similar productivity increases. The green revolution was largely responsible for preventing a food crisis in these regions during the 1970s and 1980s. But will these scientific breakthroughs continue in the twenty-first century?

To take full advantage of the new miracle seeds, farmers must use more fertilizer and machinery. Farmers have known for thousands of years that application of manure, bones, and ashes somehow increases, or at least maintains, the fertility of the land. Not until the nineteenth century did scientists identify nitrogen, phosphorus, and potassium (potash) as the critical elements in these substances that improve fertility. Today these three elements form the basis for fertilizers—products that farmers apply to their fields to enrich the soil by restoring lost nutrients.

Nitrogen, the most important fertilizer, is a ubiquitous substance. China is the leading producer of nitrogen fertilizer. Europeans most commonly produce a fertilizer known as urea, which contains 46 percent nitrogen. In North America, nitrogen is available as ammonia gas, which is 82 percent nitrogen but more awkward than urea to transport and store. Both urea and ammonia gas combine nitrogen and hydrogen. The problem is that the cheapest way to produce both types of nitrogen-based fertilizers is to obtain hydrogen from natural gas or petroleum. When fossil fuel prices increase, so do the prices for nitrogen-based fertilizers, which then become too expensive for many farmers in developing countries.

In contrast to nitrogen, phosphorus and potash reserves are not distributed uniformly across Earth's surface. Phosphate rock reserves are clustered in China, Morocco, and the United States. Proven potash reserves are concentrated in Canada, Russia, and Ukraine.

Farmers need tractors, irrigation pumps, and other machinery to make the most effective use of the new miracle seeds. In developing countries, farmers cannot afford such equipment and cannot, in view of high energy costs, buy fuel to operate the equipment. To maintain the green revolution, governments in developing countries must allocate scarce funds to subsidize the cost of seeds, fertilizers, and machinery.

PAUSE & REFLECT 8.4.2

What would be the impact on the green revolution of a decline in energy prices?

INCREASED PRODUCTIVITY: COMMERCIAL FARMERS

Productivity has also increased among commercial farmers in recent years. New seeds, fertilizers, pesticides, mechanical equipment, and management practices have enabled commercial farmers to obtain greatly increased yields per area of land.

The experience of dairy farming in the United States demonstrates the growth in productivity. The number of dairy cows in the United States decreased from 10.8 million to 9.3 million between 1980 and 2014. But milk production increased from 58 to 93 million metric tons. Thus, yield per cow increased 78 percent during this 34-year period, from 5.4 to 10.1 metric tons per cow (Figure 9-44).

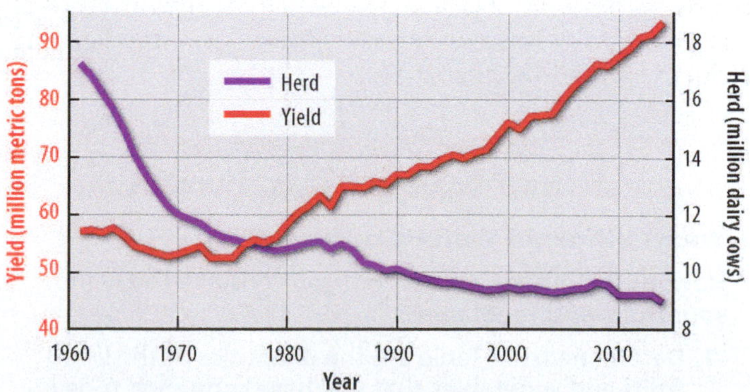

▲ **FIGURE 9-44 U.S. DAIRY PRODUCTIVITY** The amount of milk produced per cow has increased rapidly in the United States, especially since the 1980s.

Conserving Agricultural Resources

LEARNING OUTCOME 9.4.3

Understand the importance of water in agriculture.

Plants and animals need water to survive and thrive. Lack of water is causing stress on agriculture in many regions. Too much water can cause soil erosion.

AGRICULTURE AND WATER IN CALIFORNIA

California's limited water supply comes from two main sources:

- Surface water, which is water that travels or gathers on the ground, like rivers, streams, and lakes.
- Groundwater, which is water that is pumped out from the ground.

After several years of drought, the volume of water in California's rivers, streams, and lakes has been severely reduced. In normal years 70 percent of California's water comes from surface water, but after several years of drought the share dropped to 40 percent.

The distribution of water in California does not match the distribution of demand. Most of California's water is in the north, whereas most of the demand is in the central and southern parts of the state. Urban areas use around 20 percent of the water that California distributes, and most of the demand is in Los Angeles and adjacent smaller metropolitan areas in the south. Agriculture uses around 80 percent of California's distributed water, and the bulk of that demand is in the center of the state (Figure 9-45).

DOING GEOGRAPHY California Agriculture and Water

California's extended extreme drought is stressing agriculture, which uses 80 percent of the state's distributed water. Homeowners and businesses in California have been required to make substantial cuts in their water usage. California farmers produce one-third of U.S. vegetables and two-thirds of fruits and nuts. It takes a lot of water to grow these fruits and vegetables. So if you are living in any of the 50 U.S. states, you are consuming California water indirectly through consuming produce. In fact, the average American consumes around 40 gallons of California water per day. Table 9-2 has examples of the amounts of California water that go into growing some fruits and vegetables.

TABLE 9–2 Amount of Water Needed to Grow Selected Fruits and Vegetables in California

Fruits and nuts	Gallons	Your produce consumption	Your water consumption
1 apple, peach, pear, or plum, ¼ melon	7.0		
5 strawberries	3.0		
1 almond	1.0		
1 walnut	5.0		
3 grapes	1.0		
1 lemon, orange, grapefruit, or clementine	20		
1 avocado	40		
Vegetables			
1 broccoli or cauliflower floret	0.5		
Lettuce, cabbage, spinach [salad portion]	1.0		
1 carrot or celery stalk	0.5		
1 slice tomato, onion, or potato	0.5		

What's Your Food and Agriculture Geography?

Your California Water Consumption

How much California water did you consume today in your fruits and vegetables?

1. Determine from Table 9-2 the quantities of the listed fruits and vegetables that you have consumed today (or another day specified for your class).
2. What was your total consumption of California water from eating produce?
3. How does your total consumption compare to the national average of 40 gallons?
4. What factors might account for having consumption that is higher or lower than the national average?

◀ **FIGURE 9-45 AGRICULTURE AND DROUGHT IN CALIFORNIA** The contrast between agricultural land being irrigated and the dry land not irrigated.

SUSTAINABLE LAND MANAGEMENT

In the U.S. Midwest, farm fields are more likely to be damaged by too much precipitation than by drought. Heavy rains can wash away the protective layer of high-quality top soil and deposit it in bodies of water. After harvesting, conventional farming clears away crop residue, such as corn stalks. The soil is churned up or tilled before the next year's seeds are planted. This practice loosens the soil particles, making them susceptible to being washed away by rain or blown away by wind.

Conservation tillage is a method of soil cultivation that reduces soil erosion and runoff. Under conservation tillage, some or all of the previous harvest is left on the fields through the winter. **No tillage,** as the name implies, leaves all of the soil undisturbed, and the entire residue of the previous year's harvest is left untouched on the fields (Figure 9-46). **Ridge tillage** is a system of planting crops on ridge tops. Crops are planted on 10- to 20-centimeter (4- to 8-inch) ridges that are formed during cultivation or after harvest. A crop is planted on the same ridges, in the same rows, year after year.

PAUSE & REFLECT 9.4.3

According to Figure 9-13, what is the predominant form of agriculture in California?

▶ **FIGURE 9-46 NO-TILL AGRICULTURE** Soybeans are growing in a field on top of stubble from last year's corn harvest.

Applying Biotechnology to Agriculture

LEARNING OUTCOME 9.4.4

Understand the debate over the planting of GMO seeds.

One of the most challenging issues in contemporary agriculture is the extent to which genetically modified seeds should be planted. Farmers have been manipulating crops and livestock for thousands of years. The very nature of agriculture is to deliberately manipulate nature. Humans control selective reproduction of plants and animals in order to produce a larger number of stronger, hardier survivors.

Beginning in the nineteenth century, the science of genetics expanded understanding of how to manipulate plants and animals to secure dominance of the most favorable traits. However, genetic modification, which became widespread in the late twentieth century, marks a sharp break with the agricultural practices of the past several thousand years. The genetic composition of an organism is not merely studied, it is actually altered.

GENETICALLY MODIFIED ORGANISMS

A **genetically modified organism (GMO)** is a living organism that possesses a novel combination of genetic material obtained through the use of modern biotechnology. GMO seeds are genetically modified to survive when herbicides and insecticides are sprayed on fields to kill weeds and insects. These are known as "Roundup-ready" seeds because their creator, Monsanto, sells its weed killers under the brand name Roundup. A GMO mixes genetic material of two or more species that would not otherwise mix in nature.

Worldwide, 160 million hectares—10 percent of all farmland—was devoted to genetically modified crops in 2010; 77 percent of the world's soybeans, 49 percent of cotton, and 26 percent of maize were genetically modified in 2010. Genetic modification is especially widespread in the United States: 94 percent of soybeans, 90 percent of cotton, and 88 percent of maize; usage increased rapidly during the first decade of the twenty-first century (Figure 9-47). Three-fourths of the processed food that Americans consume has at least one genetically modified ingredient. North America was responsible for one-half of the world's genetically modified foods, and developing countries—especially in Latin America—were responsible for the other one-half.

The United States has urged developing countries, especially in sub-Saharan Africa, to increase their food supply in part through increased use of GMOs. Africans are divided on whether to accept GMOs. The positives of genetic modification are higher yields, increased nutrition, and more resistance to pests. Genetically modified foods are also better tasting to some palates. Despite these benefits, opposition to GMOs is strong in Africa for several reasons:

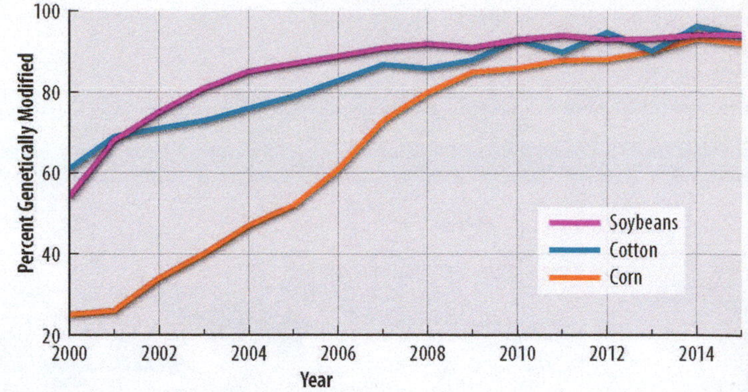

▲ **FIGURE 9-47 GENETICALLY MODIFIED CROPS IN THE UNITED STATES** Approximately 90 percent of major crops in the United States are grown from GMO seeds.

- **Health problems.** Consuming large quantities of genetically modified foods may reduce the effectiveness of antibiotics and could destroy long-standing ecological balances in local agriculture.
- **Export problems.** European countries, the main markets for Africa's agricultural exports, require genetically modified foods to be labeled. Europeans are especially strongly opposed to GMOs because they believe genetically modified food is not as nutritious as food from traditionally bred crops and livestock. Because European consumers shun genetically modified food, African farmers fear that if they are no longer able to certify their exports as being not genetically modified, European customers will stop buying them.
- **Increased dependence on the United States.** U.S.-based transnational corporations, such as Monsanto, manufacture most of the GMO seeds. Africans fear that the biotech companies could—and would—introduce a so-called "terminator" gene in the GMO seeds to prevent farmers from replanting them after harvest and require them to continue to purchase seeds year after year from the transnational corporations.

"We don't want to create a habit of using genetically modified maize that the country cannot maintain," explained Mozambique's prime minister. If agriculture is regarded as a way of life, not just a food production business, GMO represents for many Africans an unhealthy level of dependency on developed countries.

Many countries, including most European countries, China, and India, require GMO labeling (Figure 9-48). The debate in the United States has been especially fierce concerning whether GMO food should be labeled (see Debate It feature and Figures 9-49 and 9-50).

PAUSE & REFLECT 9.4.4

Does your family avoid foods made with GMO seeds? Why or why not?

▶ **FIGURE 9-48 COUNTRIES THAT REQUIRE GMO LABELS** North America is a holdout in GMO labeling laws.

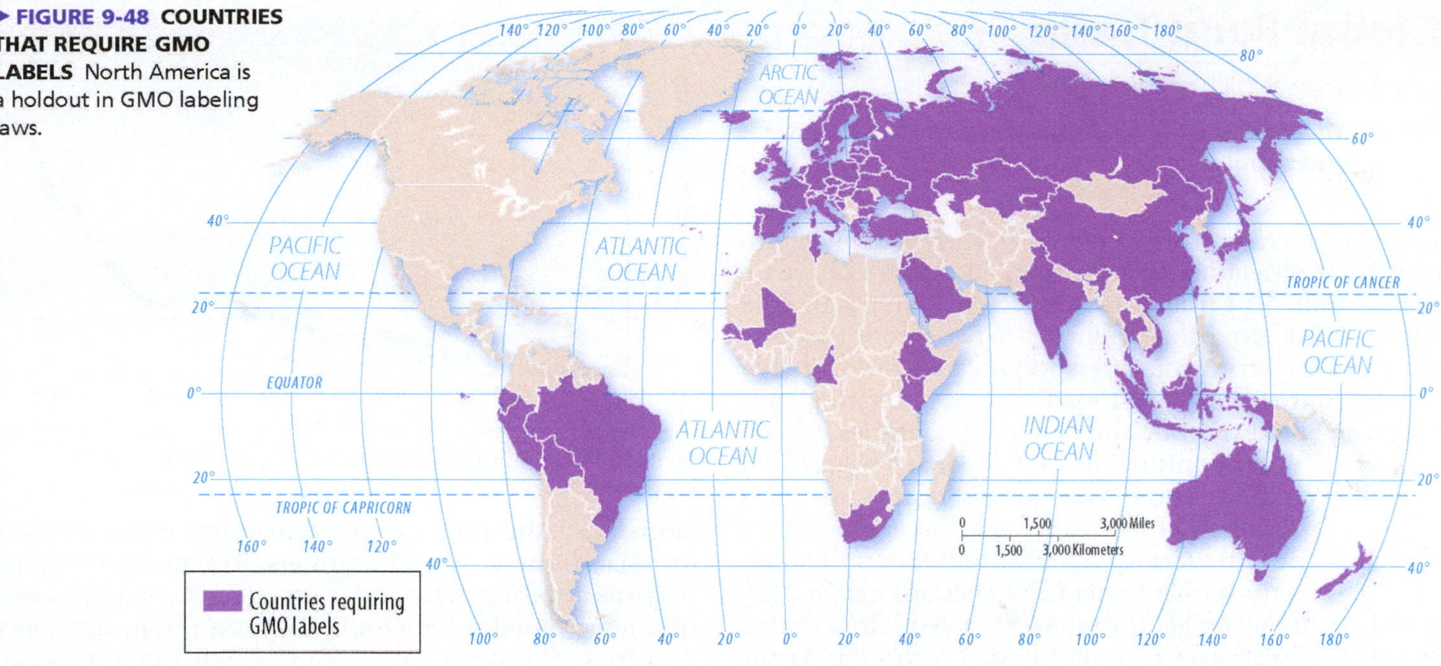

DEBATE IT! Should GMOs be labeled?

Labeling is required for most GMOs in Europe (Figures 9-49 and 9-50). In the United States, the food industry opposes GMO labeling, and voters are divided.

GMOs SHOULD BE LABELED

- Mandatory labeling of GMO products would give consumers the information necessary to choose whether or not to consume GMOs.
- Most countries other than the United States have signed agreements to regulate GMOs, including labels.
- U.S. consumers may wish to cut back on their consumption of GMOs until more is learned about their long-term effects on ecosystems and health.

▲ **FIGURE 9-49 GMO LABEL IN EUROPE**

GMOs SHOULD NOT BE LABELED

- Labeling would unnecessarily spook consumers because labeling is for health and safety, not type of seed. GMOs have comparable nutrition content to GMO-free food.
- Mandatory labeling would severely disrupt U.S. agriculture because GMO products are already widespread in the food system.
- The private sector is increasingly labeling GMO-free products, so requiring GMO labeling is unnecessary.

▲ **FIGURE 9-50 GMO-FREE LABEL**

Global Food Trade

LEARNING OUTCOME 9.4.5

Explain the contribution of expanding exports to world food supply.

Trade in food has increased rapidly in the twenty-first century. Total agricultural exports from all countries have increased from $0.4 trillion in 2000 to $1.3 trillion in 2012 (Figure 9-51). Exporting countries benefit from the revenues, and importing countries meet the food needs of their people. However, increased food trade is generating challenges for both exporting and importing countries.

Europe was the only major food importing region prior to World War II. Historically, European countries used their colonies as suppliers of food; after they became independent countries, the former colonies sold food to Europe. East Asia and the former Soviet Union became net food importers in the 1950s, Southwest Asia & North Africa during the 1970s, South Asia and sub-Saharan Africa during the 1980s, and Central Asia in 2008. Food production was unable to keep up with rapid population growth in these regions, and as they embraced the international trade path of development, agriculture was increasingly devoted to growing export crops for sale in developed countries. Japan is by far the leading importer of food, followed by the United Kingdom, China, and Russia.

In response to the increasing global demand for food imports, the United States passed Public Law 480, the Agricultural, Trade, and Assistance Act of 1954 (referred to as P.L.-480). Title I of the act provided for the sale of grain at low interest rates, and Title II gave grants to needy groups of people. The United States remains the world's leading exporter of grain, including nearly one-half of the world's maize exports.

GLOBAL TRADE PATTERNS

On a global scale, agricultural products are moving primarily from the Western Hemisphere to the Eastern Hemisphere. Latin America, led by Brazil and Argentina, is the by far the leading region for export of agricultural products; North America, Southeast Asia, and the South Pacific are the other major exporting regions (Figure 9-52). The overall share of exports accounted for by the United States has declined rapidly, from 18 to 19 percent of the world total in the 1970s to 10 to 11 percent in the twenty-first century. Agricultural exports from the United States have continued to increase rapidly, but developing regions—especially Latin America and Southeast Asia—have had more rapid increases.

To expand production, subsistence farmers need higher-yield seeds, fertilizer, pesticides, and machinery. Some needed supplies can be secured by trading food with urban dwellers. For many African and Asian countries, though, the main way to obtain agricultural supplies is to import them from other countries. However, subsistence

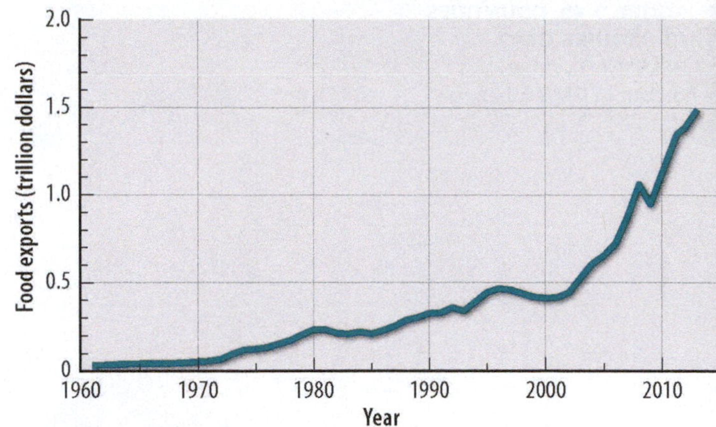

▲ **FIGURE 9-51 AGRICULTURAL EXPORTS** Agricultural trade increased from $400 billion in 2000 to $1.3 trillion in 2012.

farmers lack the money to buy agricultural equipment and materials from developed countries. To generate the funds they need to buy agricultural supplies, developing countries must produce something they can sell in developed countries. The developing countries sell some manufactured goods (see Chapter 11), but most raise funds through the sale of crops in developed countries. Consumers in developed countries are willing to pay high prices for fruits and vegetables that would otherwise be out of season or for crops such as coffee and tea that cannot be grown in developed countries because of the climate.

In a developing country such as Kenya, families may divide by gender between traditional subsistence agriculture and contributing to international trade. Women practice most of the subsistence agriculture—that is, growing food for their families to consume—in addition to the tasks of cooking, cleaning, and carrying water from wells. Men may work for wages, either growing crops for export or at jobs in distant cities. Because men in Kenya frequently do not share the wages with their families, many women try to generate income for the household by making clothes, jewelry, baked goods, and other objects for sale in local markets.

The sale of export crops brings a developing country foreign currency, a portion of which can be used to buy agricultural supplies. But governments in developing countries face a dilemma: The more land that is devoted to growing export crops, the less that is available to grow crops for domestic consumption. Rather than help to increase productivity, the funds generated through the sale of export crops may be needed to feed the people who switched from subsistence farming to growing export crops.

DRUG TRADE

The export crops grown in some developing countries, especially in Latin America and Asia, are those that can be converted to drugs. Cocaine and heroin, the two leading, especially dangerous drugs, are abused by 16 to 17 million people each, and marijuana, the most popular drug, is estimated to be used by 140 million worldwide:

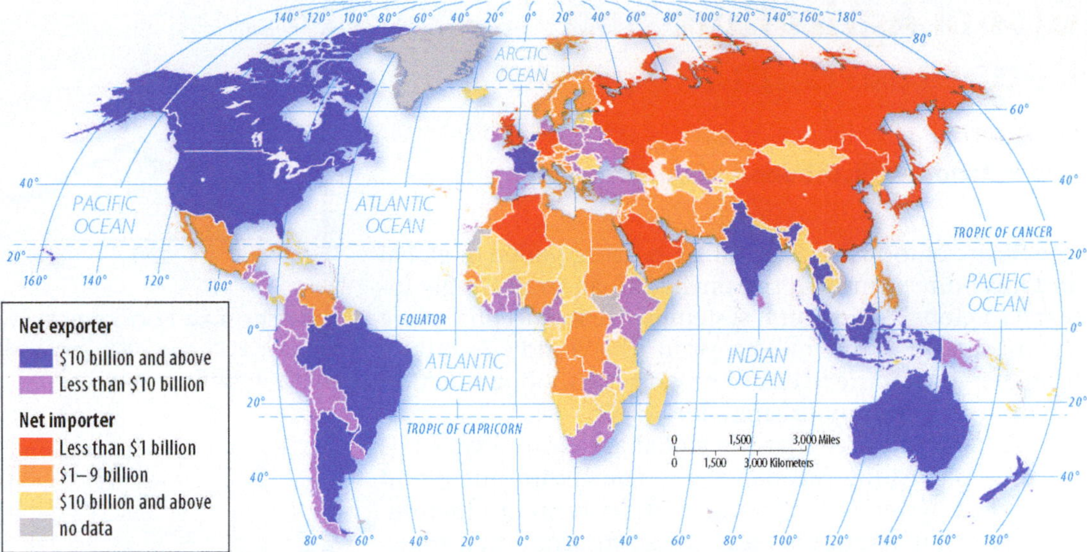

▶ **FIGURE 9-52 TRADE IN AGRICULTURE** The principal flow of agriculture in the world is from the Western Hemisphere to Europe and Asia.

- Cocaine is derived from coca leaf, most of which is grown in Colombia or the neighboring countries Peru and Bolivia. Most consumers are located in developed countries, especially in North America. The principal shipping route is from Colombia by sea to Mexico or other Central American countries and then by land through Mexico to the United States (Figure 9-53).
- Heroin is derived from raw opium gum, which is produced by the opium poppy plant. Afghanistan is the source of nearly 90 percent of the world's opium; most of the remainder is grown in Myanmar (Burma) and Laos. Most traffic flows from Afghanistan through Iran, Turkey, and the Balkans to Western Europe, where the largest numbers of the world's users live. A second route goes through Central Asia to Russia.
- Marijuana, produced from the *Cannabis sativa* plant, is cultivated widely around the world. The overwhelming majority of the marijuana that reaches the United States is grown in Mexico. Cultivation of *C. sativa* is not thought to be expanding worldwide, whereas cultivation of opium poppies and coca leaf are.

PAUSE & REFLECT 9.4.5

Why does most consumption of cocaine and heroin occur in developed countries?

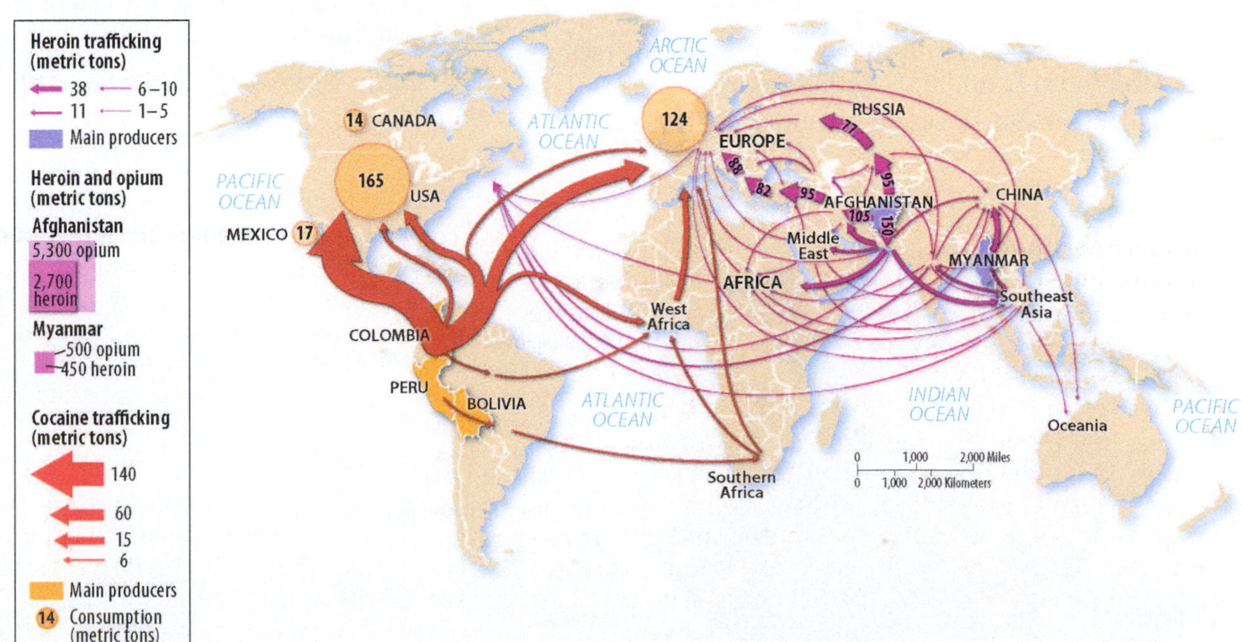

▶ **FIGURE 9-53 INTERNATIONAL DRUG TRAFFICKING** The main routes for heroin are from Afghanistan through Southwest Asia to Europe and through Central Asia to Russia. The main routes for cocaine are from Colombia to North America through Mexico and to Europe by sea.

Global Agriculture and Undernourishment

LEARNING OUTCOME 9.4.6

Understand the distribution of undernourishment.

The future of food and agriculture is being pulled in global and local directions. On the one hand, an increasingly integrated global agricultural system is devoted to producing the most food at the lowest cost for the world's 7+ billion humans. And in the twenty-first century, food production is higher and undernourishment is lower.

But critics charge that the global agriculture system is causing major long-run damage to the environment and to local ecosystems for the sake of short-term production. They say that Roundup-ready seeds, international trade, deforestation, and other practices documented in this chapter are not sustainable ways to meet humanity's need for food.

Meanwhile, the biggest increase in demand in developed countries is for locally grown food produced through sustainable farming methods. But criticism of the local and organic food movements is also fierce. Critics charge that the local and organic movements are not capable of providing affordable food for the world's 7+ billion humans

GLOBAL SCALE: SUPPLY AND DEMAND

The greatest challenge to world food supply in the twenty-first century has been food prices rather than food supply. Food prices more than doubled between 2006 and 2008, remained at record high levels through 2014, and declined sharply in 2015 (Figure 9-54). The U.N. attributes the record high food prices through 2014 to four factors:

- Poor weather, especially in major crop-growing regions of the South Pacific and North America.
- Higher demand, especially in China and India.
- Smaller growth in productivity, especially without major new "miracle" breakthroughs.
- Use of crops as biofuels instead of food, especially in Latin America.

The sharp decline in 2015 was attributed to increased supplies.

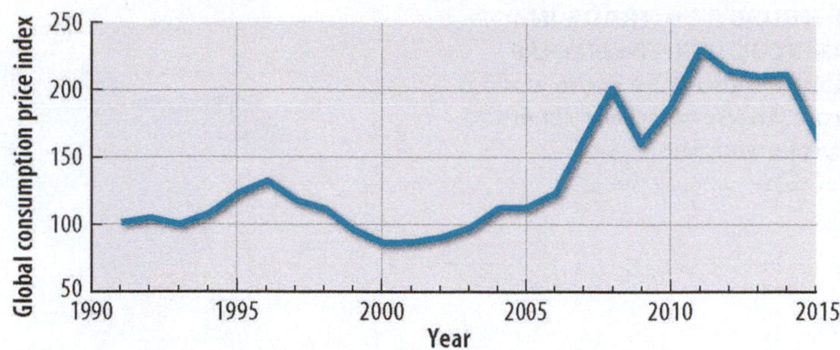

▲ **FIGURE 9-54 FOOD PRICES** Worldwide food prices rose rapidly between 2006 and 2008, remained high through 2014, and declined rapidly in 2015.

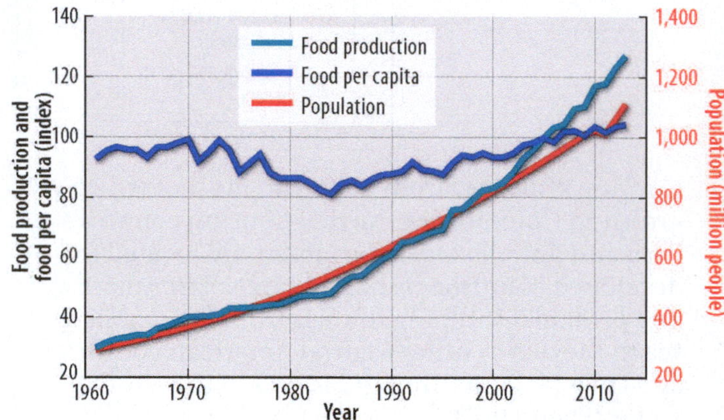

▲ **FIGURE 9-55 POPULATION AND FOOD IN AFRICA** Food production and food per capita indexes are set at 100 in the year 2005. Food production is increasing at about the same rate as population in Africa. As a result, food production per capita is staying about the same.

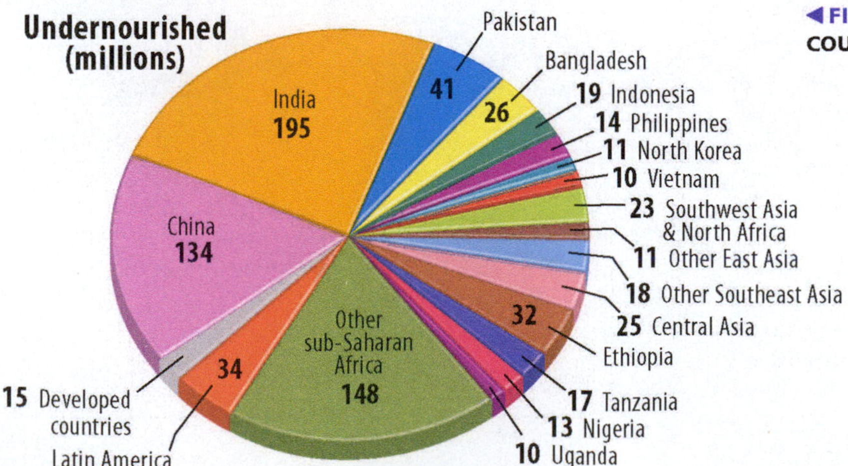

◀ **FIGURE 9-56 DISTRIBUTION OF UNDERNOURISHMENT BY COUNTRY**

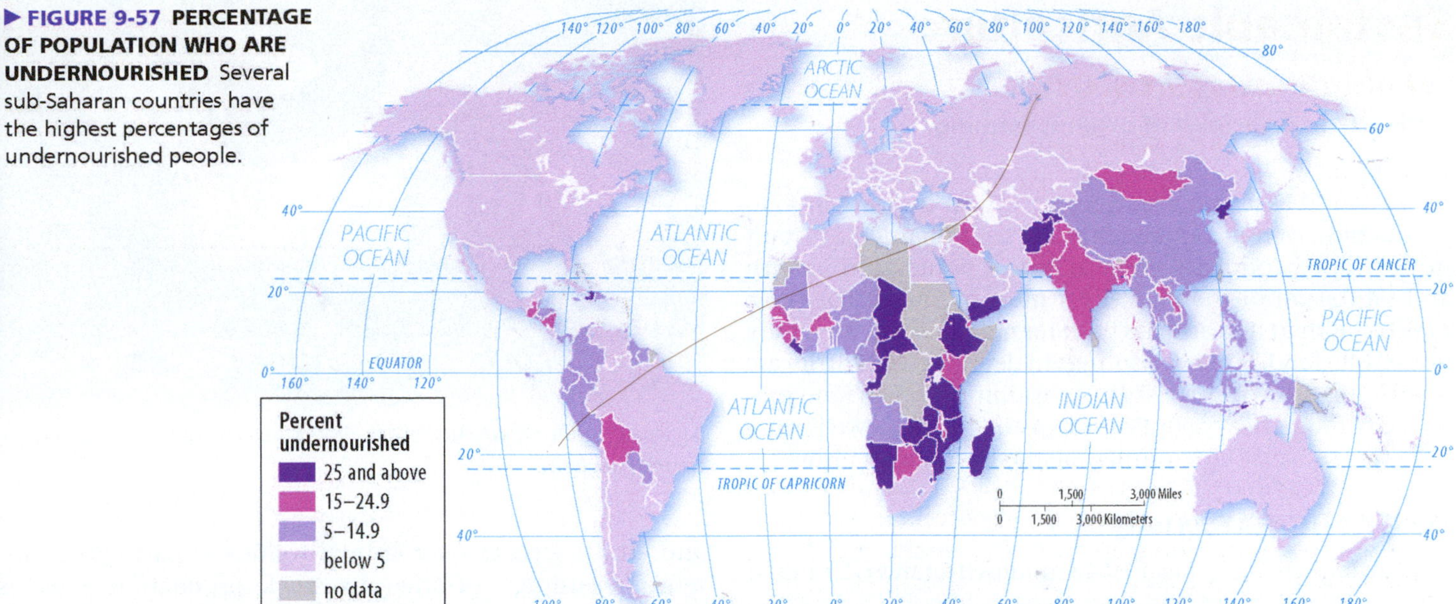

▶ **FIGURE 9-57 PERCENTAGE OF POPULATION WHO ARE UNDERNOURISHED** Several sub-Saharan countries have the highest percentages of undernourished people.

On the other side of the coin, record-high food prices have stimulated record-high prices for prime agricultural land. Adjusting for inflation, the price of farmland in Iowa doubled from around $2,500 per acre in 2000 to $5,000 in 2010.

Sub-Saharan Africa is struggling to keep food production ahead of population growth. Since 1961, food production has increased substantially in sub-Saharan Africa, but so has population (Figure 9-55). As a result, food production per capita has changed little in a half-century.

The threat of famine is particularly severe in the Horn of Africa and the Sahel. Traditionally, this region supported limited agriculture. With rapid population growth, farmers overplanted, and herd size increased beyond the capacity of the land to support the animals. Animals overgrazed the limited vegetation and clustered at scarce water sources.

UNDERNOURISHMENT

Undernourishment is dietary energy consumption that is continuously below that needed for healthy life and carrying out light physical activity. The U.N. estimates that 795 million people in the world are undernourished, one-half of them in South Asia and East Asia. The largest numbers are in India and China, the world's two most populous countries. India has one-fourth of the world's undernourished people (Figure 9-56). Overall, all but 2 percent of the world's undernourished people are in developing countries.

As a percentage of a country's population, undernourishment is most prevalent in sub-Saharan Africa and South Asia. One-fourth of the population in sub-Saharan Africa and one-fifth in South Asia are undernourished (Figure 9-57).

The world as a whole has made progress in reducing hunger during the twenty-first century. Between 2000 and 2015, the number of undernourished people declined from 924 million to 795 million, and the percentage of people who are undernourished declined from 15 percent in 2000 to 11 percent (Figure 9-58). East Asia, led by China, has had by far the largest decrease in the number and percentage undernourished, followed by Southeast Asia. Sub-Saharan Africa has had an increase in the number undernourished and a modest decrease in the percentage.

PAUSE & REFLECT 9.4.6

Why does China appear as a leader in undernourishment in Figure 9-56 but not in Figures 9-57 and 9-58?

▼ **FIGURE 9-58 CHANGE IN UNDERNOURISHMENT**

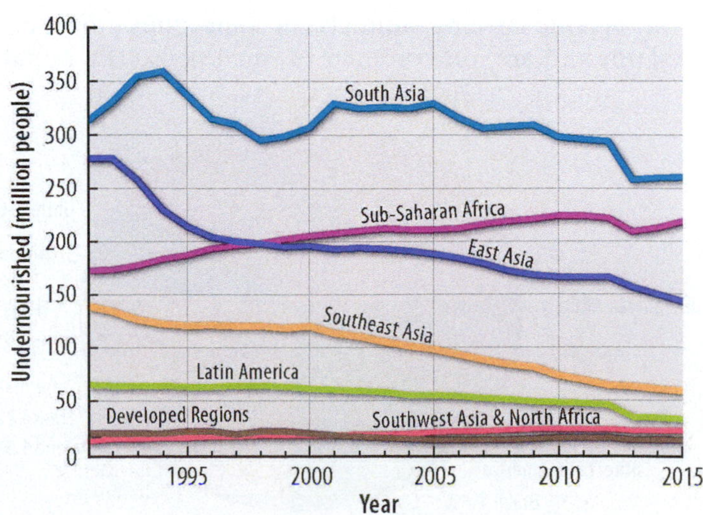

Sustainable Agriculture

LEARNING OUTCOME 9.4.7

Understand principles of organic farming.

As the world as a whole is experiencing an increase in food production, overall demand for food has remained constant in developed countries because of low population growth and saturation of the market for most products. The most rapid growth in demand has been for organic food, including non-GMO food. Some consumers in developed countries are questioning the health risks in consuming biotech food produced through heavy application of chemicals, as well as the long-term adverse environmental impacts for agriculture.

ORGANIC FARMING

Worldwide, the U.N. classified 43 million hectares (75 million acres), or 1 percent of farmland, as organic in 2013. Australia was the leader, with 17 million hectares, or 40 percent of the worldwide total (Figure 9-59). Argentina accounted for 8 percent of the worldwide total, and the United States and China for 5 percent each. USDA economists reported that organic food sales spiked from $3.4 billion in 1997 to an estimated $35.9 billion in 2014. Organic food accounted for 5 percent of food purchases in the United States in 2014, compared with less than 1 percent in 1997.

Organic farming is sensitive to the complexities of biological and economic interdependencies between crops and livestock. Growing crops and raising livestock is integrated as much as possible at the level of the individual farm. This integration reflects a return to the historical practice of mixed crop and livestock farming, in which growing crops and raising animals were regarded as complementary activities on the farm. This was the common practice for centuries, until the mid-1900s, when technology, government policy, and economics encouraged farmers to become more specialized.

In organic farming, crops are grown without application of herbicides and pesticides to control weeds. GMO seeds are not used.

In organic farming, animals consume crops grown on the farm and are not confined to small pens. The moral and ethical debate over animal welfare is particularly intense regarding confined livestock production systems (Figure 9-60). Confining livestock leads to surface water and groundwater pollution, particularly where the density of animals is high. If animals are not confined, manure can contribute to soil fertility.

▲ **FIGURE 9-60 ORGANIC FARM** Free-range chickens, Hampshire, United Kingdom.

In organic farming, antibiotics are administered to animals only for therapeutic purposes. Many conventional livestock farms have fed animals antibiotics to foster weight gain. The European Union has banned the use of antibiotics in livestock for reasons other than medical. The United States has permitted the practice, but the U.S. Food & Drug Administration has ordered the practice to be phase out.

HOW CLEAN IS OUR PRODUCE?

The U.S. Department of Agriculture tests samples of fruits and vegetables for pesticides. Nearly two-thirds of the 3,015 produce samples tested by the USDA in 2013 contained residues of at least one of 165 different pesticides. These pesticides linger on produce even if washed or peeled.

The five most pesticide-ridden fruits are apples, peaches, nectarines, strawberries, and grapes. The USDA found pesticide residue in 99 percent of apples, 98 percent of peaches, and 97 percent of nectarines. Leafy greens (kale and collards) and hot peppers have especially high concentrations of highly toxic insecticides. The five cleanest of the 50 types of produce

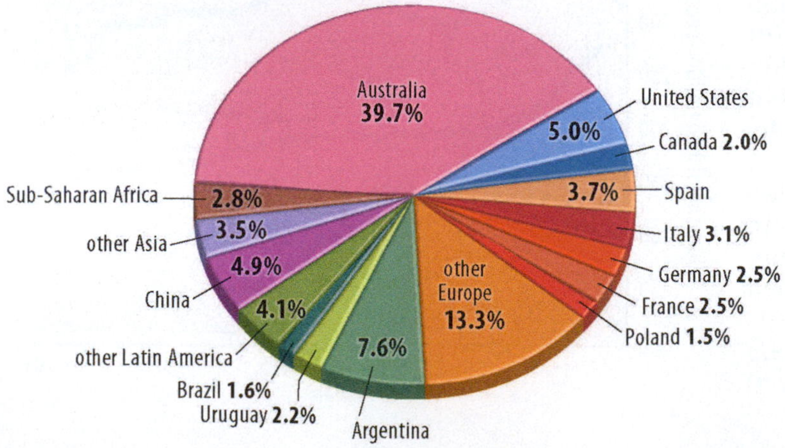

◀ **FIGURE 9-59 DISTRIBUTION OF ORGANIC FARMING** Australia has the largest share of the world's organic farms.

tested are avocados, corn, pineapples, cabbage, and frozen sweet peas. Relatively few pesticides were detected on these foods; pesticides were detected on only 1 percent of avocados.

What features differentiate the five cleanest from the five dirtiest? Think about the outermost portion. How protected from the environment is the edible portion of the fruit or vegetable?

The Environmental Working Group publishes the findings annually at ewg.org. The 12 most infected are called "the dirty dozen," and the 15 least infected are called "the clean fifteen." The Environmental Working Group urges consumers to consider organic versions of the dirty dozen (Table 9-3).

GOVERNMENT POLICIES

Government policies have aggravated the food-shortage crisis. To make food affordable for urban residents, governments keep agricultural prices low. Constrained by price controls, farmers are unable to sell their commodities at a profit and therefore have little incentive to increase production.

The U.S. government has three agriculture policies designed to improve the financial position of farmers:

- Farmers are encouraged to avoid producing crops that are in excess supply. Because soil erosion is a constant threat, the government encourages planting fallow crops, such as clover, to restore nutrients to the soil and to help hold the soil in place. These crops can be used for hay or forage for pigs, or to produce seeds for sale.
- The government pays farmers when certain commodity prices are low. The government sets a target price for a commodity and pays farmers the difference between the price they receive in the market and the target price set by the government as a fair level for the commodity. The target prices are calculated to give farmers the same price for the commodity today as in the past, when compared to other consumer goods and services.
- The government buys surplus production and sells or donates it to foreign governments. In addition, low-income Americans receive food stamps in part to stimulate their purchase of additional food.

The United States has averaged about $20 billion a year on farm subsidies in recent years. Annual spending varies considerably from one year to the next. Subsidy payments are lower in years when market prices rise and production is down, typically as a result of poor weather conditions in the United States or political problems in other countries.

Farming in Europe is subsidized even more than in the United States. More farmers receive subsidies in Europe, and they receive more than American farmers. The high subsidies are a legacy of a long-standing commitment by the European Union to maintain agriculture in its member states, especially in France. Supporters point to the preservation of rural village life in parts of Europe, while critics charge that Europeans pay needlessly high prices for food as a result of the subsidies.

TABLE 9-3 The Dirty Dozen and Clean Fifteen, 2015

Dirty Dozen	Clean Fifteen
Apples	Avocados
Peaches	Sweet corn
Nectarines	Pineapples
Strawberries	Cabbage
Grapes	Frozen sweet peas
Celery	Onions
Spinach	Asparagus
Sweet bell peppers	Mangoes
Cucumbers	Papayas
Cherry tomatoes	Kiwi
Imported snap peas	Eggplant
Potatoes	Grapefruit
	Cantaloupe
* Kale	Cauliflower
* Collard greens	Sweet potatoes

*Not ranked but buy organic because nonorganic versions have trace levels of highly toxic chemicals.

Government policies in developed countries point out a fundamental irony in worldwide agricultural patterns: In developed regions such as North America and Europe, farmers are encouraged to grow less food, whereas developing countries struggle to increase food production to match the rate of growth in the population.

PAUSE & REFLECT 9.4.7

Does your family go out of its way to get local or organic food? Why or why not?

CHECK-IN KEY ISSUE 4

Why Do Farmers Face Sustainability Challenges?

- ✔ **Agricultural land is being lost to competing uses, such as urbanization.**
- ✔ **The green revolution has improved the productivity of farming in some countries.**
- ✔ **Some agricultural regions face a severe shortage of water.**
- ✔ **GMO crops are increasingly planted in some countries, especially in the United States.**
- ✔ **International trade in food is increasing, but in some places at the expense of producing food for domestic consumption.**

Summary & Review

KEY ISSUE 1

Where did agriculture originate?

Before the invention of agriculture, humans were hunters and gatherers. Agriculture originated in multiple hearths and diffused to numerous places independently. Subsistence agriculture is the production of food primarily for consumption by the farmer's family. Commercial agriculture is the production of food primarily for sale off the farm. Commercial agriculture involves larger farms, fewer farmers, and more mechanization than does subsistence agriculture.

THINKING GEOGRAPHICALLY

▲ **FIGURE 9-61 HORSE SCULPTURE, ASTANA KAZAKHSTAN**

1. Several large sculptures honoring the horse are erected in the center of Kazakhstan's capital Astana. Why might the horse be specially honored there?

KEY ISSUE 2

Why do people consume different foods?

What people eat is influenced by a combination of level of development, cultural preferences, and environmental constraints. Most humans derive most of their dietary energy from grains, but the amount consumed varies widely. People in developed countries consume more protein through animal products.One in 10 humans is undernourished. The percentage of undernourished people has been declining.

THINKING GEOGRAPHICALLY

▲ **FIGURE 9-62 FARMERS MARKET** Union Square Greenmarket, New York City.

2. Review what you have eaten today. How conscious were you of where the food was grown or raised?
3. Why or why are you not paying attention to the origin of your food?
4. Do you have access to a farmers' market or an organic grocery store? If so, do you go to it? Why or why not?

KEY ISSUE 3

Where is agriculture distributed?

The distribution of major agricultural regions is closely related to the distribution of climate regions.The principal forms of subsistence agriculture are intensive subsistence, pastoral nomadism, and shifting cultivation. Commercial agriculture in developed countries includes three principal crop-based forms (grain, Mediterranean, and commercial gardening and fruit farming) and three principal animal-based forms (mixed crop and livestock, dairy, and ranching).

THINKING GEOGRAPHICALLY

▲ **FIGURE 9-63** Elderly Japanese man works in rice field.

5. Although one of the world's wealthiest countries, Japan is shown in Figure 9-13 as specializing intensive subsistence farming with wet rice dominant. The rice farming is done primarily by elderly people. Why might Japan be retaining traditional rice production despite its wealthy status?

KEY ISSUE 4

Why do farmers face sustainability challenges?

Subsistence and commercial farmers are challenged to produce food in sustainable ways. The global food production system is expanding and feeding more people. The green revolution has increased production in subsistence agriculture through higher-yield seeds and expanded use of fertilizers.At the same time, demand is increasing for local and organic foods.

THINKING GEOGRAPHICALLY

▲ **FIGURE 9-64 SCHOOL CAFETERIA** Napa High School serves California-grown food.

6. The food services at a number of schools and universities now offer some healthy, local, and organic food choices. If yours does, why do you or don't you select these healthy and local choices? If yours does not, have you asked your school to provide healthier choices? Why have you or have you not asked?

KEY TERMS

Agribusiness *(p. 326)* Commercial agriculture characterized by the integration of different steps in the food-processing industry, usually through ownership by large corporations.

Agricultural revolution *(p. 308)* The time when human beings first domesticated plants and animals and no longer relied entirely on hunting and gathering.

Agriculture *(p. 308)* The deliberate effort to modify a portion of Earth's surface through the cultivation of crops and the raising of livestock for sustenance or economic gain.

Aquaculture (or aquafarming) *(p. 324)* The cultivation of seafood under controlled conditions.

Cereal grain *(p. 312)* A grass that yields grain for food.

Commercial agriculture *(p. 310)* Agriculture undertaken primarily to generate products for sale off the farm.

Crop *(p. 308)* Any plant gathered from a field as a harvest during a particular season.

Crop rotation *(p. 323)* The practice of rotating use of different fields from crop to crop each year to avoid exhausting the soil.

Dairy farm *(p. 330)* A form of commercial agriculture that specializes in the production of milk and other dairy products.

Desertification *(p. 333)* Degradation of land, especially in semiarid areas, primarily because of human actions such as excessive crop planting, animal grazing, and tree cutting. Also known as semiarid land degradation.

Dietary energy consumption *(p. 312)* The amount of food that an individual consumes, measured in kilocalories (Calories in the United States).

Double cropping *(p. 322)* Harvesting twice a year from the same field.

Fishing *(p. 324)* The capture of wild fish and other seafood living in the waters.

Food security *(p. 314)* Physical, social, and economic access at all times to safe and nutritious food sufficient to meet dietary needs and food preferences for an active and healthy life.

Genetically modified organism (GMO) *(p. 338)* A living organism that possesses a novel combination of genetic material obtained through the use of modern biotechnology.

Grain *(p. 312)* Seed of a cereal grass.

Green revolution *(p. 324)* Rapid diffusion of new agricultural technology, especially new high-yield seeds and fertilizers.

Horticulture *(p. 327)* The growing of fruits, vegetables, and flowers.

Intensive subsistence agriculture *(p. 322)* A form of subsistence agriculture characteristics of Asia's major population concentrations in which farmers must expend a relatively large amount of effort to produce the maximum feasible yield from a parcel of land.

Milkshed *(p. 330)* The area surrounding a city from which milk is supplied.

Mixed crop and livestock farming *(p. 328)* Commercial farming characte rized by integration of crops and livestock; most of the crops are fed to animals rather than consumed directly by humans.

No tillage *(p. 337)* A farming practice that leaves all of the soil undisturbed and the entire residue of the previous year's harvest left untouched on the fields.

Overfishing *(p. 325)* Capturing fish faster than they can reproduce.

Paddy *(p. 323)* The Malay word for wet rice, increasingly used to describe a flooded field.

Pastoral nomadism *(p. 318)* A form of subsistence agriculture based on herding domesticated animals.

Plantation *(p. 320)* A large farm in tropical and subtropical climates that specializes in the production of one or two crops for sale, usually to a more developed country.

Prime agricultural land *(p. 332)* The most productive farmland.

Ranching *(p. 331)* A form of commercial agriculture in which livestock graze over an extensive area.

Ridge tillage *(p. 337)* A system of planting crops on ridge tops in order to reduce farm production costs and promote greater soil conservation.

Sawah *(p. 322)* A flooded field for growing rice.

Shifting cultivation *(p. 320)* A form of subsistence agriculture in which people shift activity from one field to another; each field is used for crops for a relatively few years and left fallow for a relatively long period.

Slash-and-burn agriculture *(p. 320)* Another name for shifting cultivation, so named because fields are cleared by slashing the vegetation and burning the debris.

Subsistence agriculture *(p. 310)* Agriculture designed primarily to provide food for direct consumption by the farmer and the farmer's family.

Swidden *(p. 320)* A patch of land cleared for planting through slashing and burning.

Transhumance *(p. 319)* The seasonal migration of livestock between mountains and lowland pastures.

Truck farming *(p. 327)* Commercial gardening and fruit farming, so named for the Middle English word *truck*, meaning "barter" or "exchange of commodities."

Undernourishment *(p. 343)* Dietary energy consumption that is continuously below the minimum requirement for maintaining a healthy life and carrying out light physical activity.

Wet rice *(p. 322)* Rice planted on dry land in a nursery and then moved to a deliberately flooded field to promote growth.

MG GeoVideo

Log in to the **MasteringGeography** Study Area to view this video.

Organic Farming in the U.K.

Organic farming is increasing in many countries. Farmers feel better about organic practices and they are responding to growing consumer demand.

1. What were some of the things the farmer needed to do to convert the farm to organic?
2. What are some of the environmental and economic benefits of organic farming?
3. What are some of the challenges in converting to organic farming?

EXPLORE

Growing Rice

Use Google Earth to see rice growing outside a village in Asia.

Fly to *Banaue, Philippines*.

Drag to Street *View* on the balloon marking the location of the village of Banaue.

1. Describe the landscape around the village.
2. According to Figure 9-17, what is the principal form of agriculture in the Philippines?
   Exit ground level view. Zoom in immediately to the north of balloon marking the village.
   A series of parallel strips can be seen immediately north of the village.
3. Why would the villagers have created these terraces?

MasteringGeography™

Visit the Study Area in **MasteringtGeography**™ to enhance your geographic literacy, spatial reasoning skills, and understanding of this chapter's content by accessing a variety of resources, including MapMaster interactive maps, videos, *In the News* RSS feeds, flashcards, web links, self-study quizzes, and an eText version of *The Cultural Landscape*.

www.masteringgeography.com

10

Development

The average person in the world is better off today than three decades ago. The average person possesses more wealth, has had more education, and will live longer than someone 30 years ago. But not everyone is better off. Development is a process of improving a person's prospects of leading a long and healthy life, acquiring knowledge, and obtaining adequate resources.

An important element of development: Road construction, Kenya.

LOCATIONS IN THIS CHAPTER

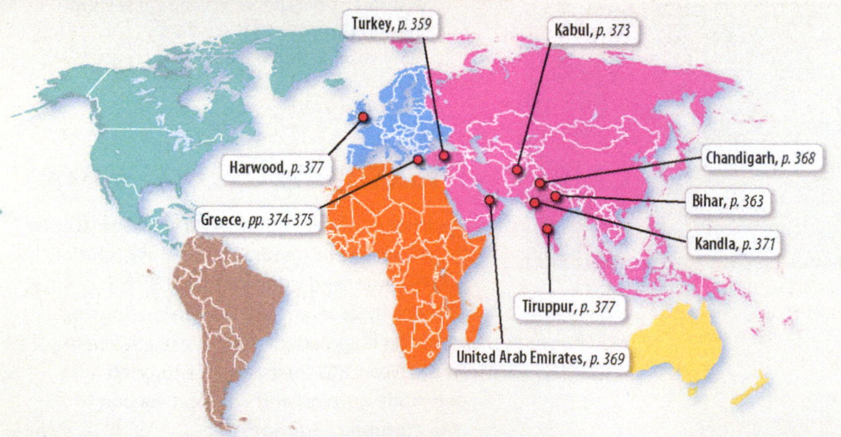

KEY ISSUES

1 Why Does Development Vary Among Countries?

Geographers divide the world into nine ***regions***, according to their level of development. Developed countries cluster in some ***spaces*** and developing countries in other spaces, based on their level of development.

2 Where Are Inequalities in Development Found?

Development varies by gender and economic groups. The ***scale*** of inequality between men and women and between rich and poor are important elements of a country's level of development.

3 Why Do Countries Face Challenges to Development?

A country may pursue development through isolation policies, but most countries in recent years have adopted policies that require them to have increased ***connections*** with other countries.

4 Why Are Countries Making Progress in Development?

Most countries have made considerable progress in overall indicators of development. Increased consideration is now being given to making connections between ***places*** more fair and equitable, especially for producers and workers in developing countries, as well as for consumers in developed countries.

KEY ISSUE 1

Why Does Development Vary Among Countries?

- Introducing Development
- A Decent Standard of Living
- Access to Knowledge
- Health and Wealth

LEARNING OUTCOME 10.1.1

Understand the Human Development Index.

Earth's nearly 200 countries can be classified according to their level of **development,** which is the process of improving the conditions of people through diffusion of knowledge and technology. Every place lies at some point along a continuum of development. The development process is continuous, involving never-ending actions to constantly improve the health and prosperity of the people.

Introducing Development

Because many countries cluster at the high and low ends of the continuum of development, they can be divided into two groups:

- A **developed country,** also known as a more developed country (MDC) and referred to by the U.N. as a very high developed country, has progressed further along the development continuum.
- A **developing country,** also frequently called a less developed country (LDC), has made some progress toward development, though less than the developed countries.

Recognizing that progress has varied widely among developing countries, the U.N. divides them into high, medium, and low developing.

HUMAN DEVELOPMENT INDEX

To measure the level of development of every country, the U.N. created the **Human Development Index (HDI).** The U.N. has computed HDIs for countries every year since 1980, although it has occasionally modified the method of computation. The highest HDI possible is 1.0, or 100 percent.

The U.N.'s HDI considers development to be a function of three factors. Each country gets an overall HDI score based on these three factors (Figure 10-1):

- A decent standard of living.
- A long and healthy life.
- Access to knowledge.

Development is a process of enlarging people's ability to lead a long and healthy life, to acquire knowledge, and to have access to resources needed for a decent standard of living. With access to these three dimensions, people have greater opportunities to be creative and productive and to enjoy personal self-respect and guaranteed human rights.

DEVELOPMENT REGIONS

Geographers divide the world into two developed regions and seven developing regions (Figure 10-2). Each region has

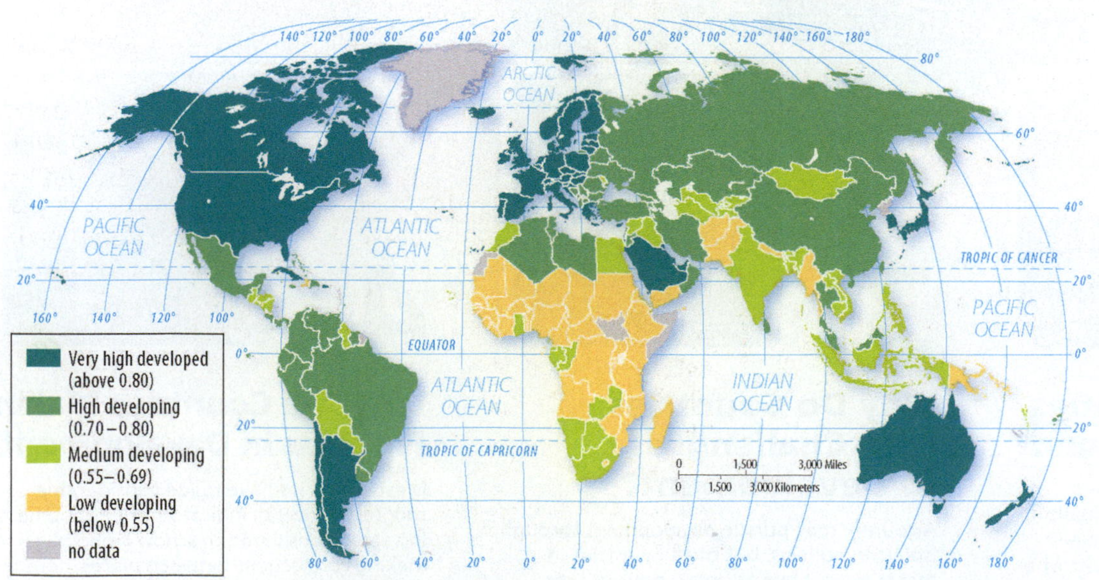

▲ **FIGURE 10-1 HUMAN DEVELOPMENT INDEX (HDI)** Developed countries are those with very high HDI scores. The other three classes are for developing countries.

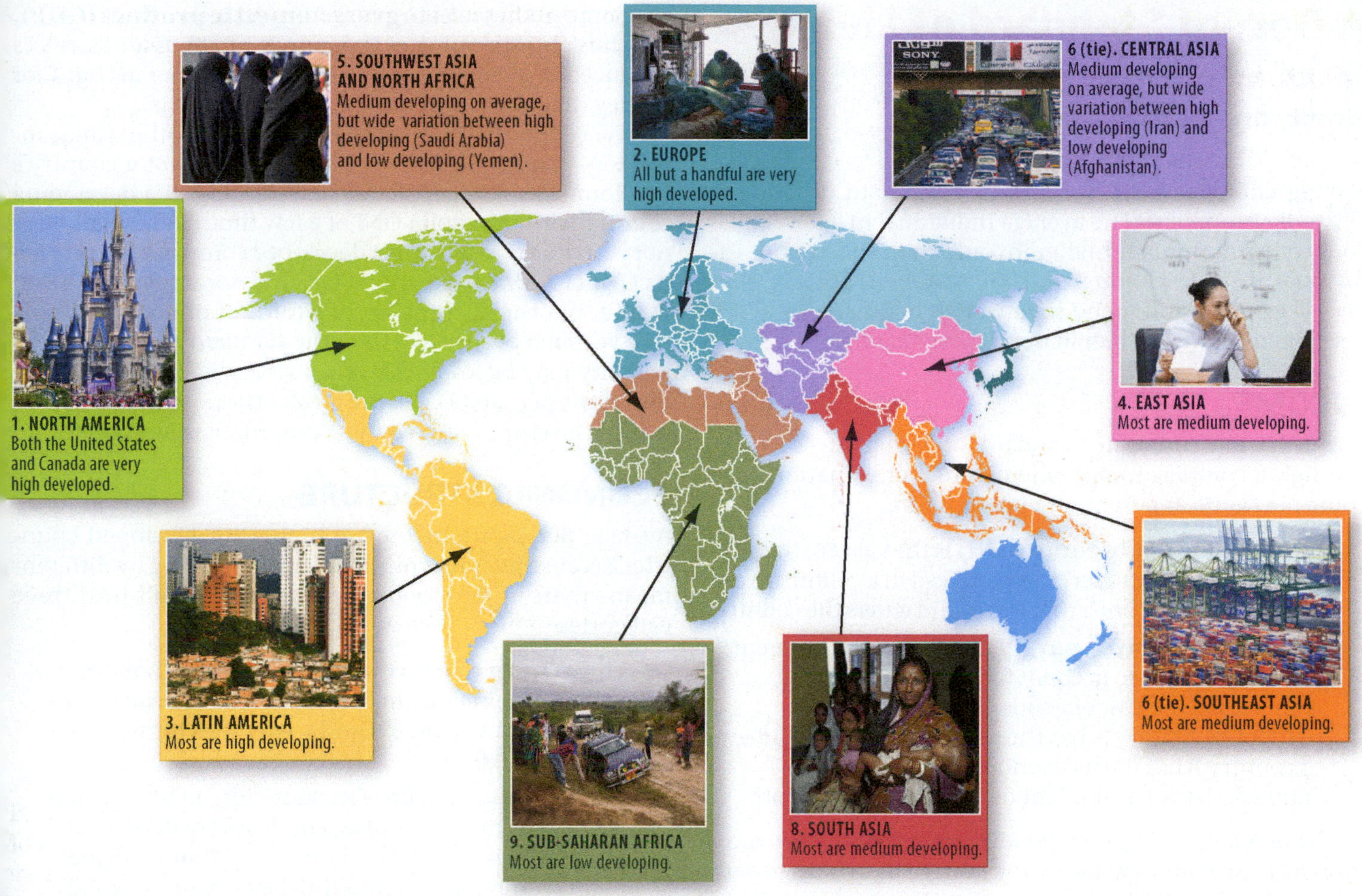

5. SOUTHWEST ASIA AND NORTH AFRICA
Medium developing on average, but wide variation between high developing (Saudi Arabia) and low developing (Yemen).
2. EUROPE
All but a handful are very high developed.
6 (tie). CENTRAL ASIA
Medium developing on average, but wide variation between high developing (Iran) and low developing (Afghanistan).
1. NORTH AMERICA
Both the United States and Canada are very high developed.
4. EAST ASIA
Most are medium developing.
3. LATIN AMERICA
Most are high developing.
9. SUB-SAHARAN AFRICA
Most are low developing.
8. SOUTH ASIA
Most are medium developing.
6 (tie). SOUTHEAST ASIA
Most are medium developing.

▲ **FIGURE 10-2 DEVELOPMENT REGIONS** The nine regions are ranked from highest (1) to lowest (9) level of development, according to the UN's HDI.

HDI
1.0
0.9
0.8
0.7
0.6
0.5
0.4
North America
AVERAGE FOR DEVELOPED COUNTRIES, Japan & South Korea
Europe
South Pacific
Russia
Latin America
East Asia
Southwest Asia & North Africa
Southeast Asia, Central Asia
AVERAGE FOR DEVELOPING COUNTRIES
South Asia
Sub-Saharan Africa
Developed
Developing

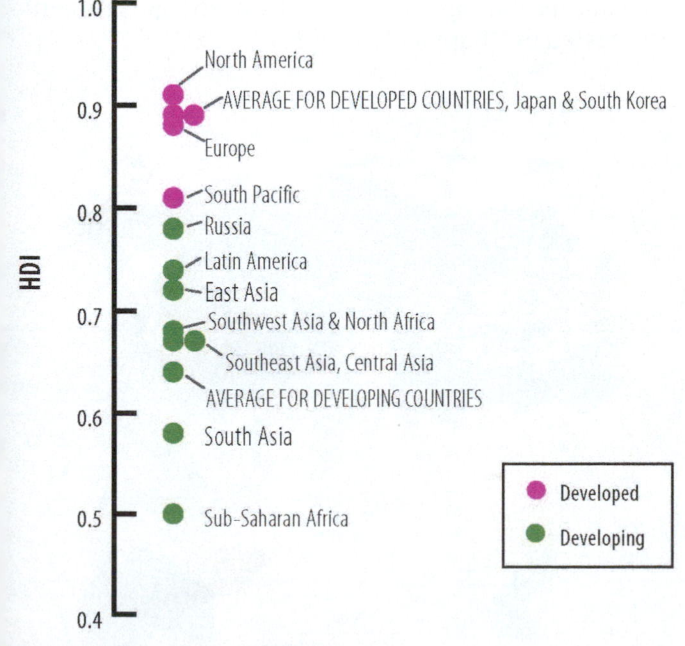

▲ **FIGURE 10-3 HDI BY REGION**

an overall HDI score (Figure 10-3). Sub-Saharan Africa and South Asia are the regions with the two lowest HDI scores.

In addition to the nine regions, three other distinctive areas can be identified. Japan and South Korea are classified separately rather than included with the rest of East Asia because their level of development is much higher than that of their neighbors. The South Pacific is a much less populous area than the nine development regions; Australia (its most populous country) and New Zealand are developed, but the area's other countries are developing. The U.N. previously classified Russia as a developed country, but because of its limited progress in development both under and since communism, the U.N. now classifies Russia as a high developing country.

PAUSE & REFLECT 10.1.1

Which developing regions appear to have relatively high diversity in the HDIs of individual countries?

A Decent Standard of Living

LEARNING OUTCOME 10.1.2

Identify the HDI standard of living factor.

Having enough wealth for a decent standard of living is key to development. The average individual in a developed country earns a much higher income than the average individual in a developing one. Geographers observe that people generate and spend their wealth in different ways in developed countries than in developing countries.

INCOME

The U.N. measures the standard of living in countries through a complex index called annual gross national income per capita at purchasing power parity:

- **Gross national income (GNI)** is the value of the output of goods and services produced in a country in a year, including money that leaves and enters the country.
- **Purchasing power parity (PPP)** is an adjustment made to the GNI to account for differences among countries in the cost of goods. For example, if a resident of country A has the same income as a resident of country B but must pay more for a Big Mac or a Starbucks latte, the resident of country B is better off.

By dividing GNI by total population, it is possible to measure the contribution made by the average individual toward generating a country's wealth in a year. For example, GNI in the United States was approximately $17.8 trillion in 2014, and the population was approximately 319 million, so GNI per capita was approximately $55,860. In 2014, per capita GNI was approximately $40,000 in developed countries compared to approximately $10,000 in developing countries (Figure 10-4).

Some studies refer to **gross domestic product (GDP)**, which is also the value of the output of goods and services produced in a country in a year. GDP does not account for money that leaves and enters the country.

Per capita GNI—or, for that matter, any other single indicator—cannot measure perfectly the level of a country's development. Few people may be starving in a developing country with per capita GNI of a few thousand dollars. And not everyone is wealthy in a developed country with per capita GNI of $40,000. Per capita GNI measures average (mean) wealth, not the distribution of wealth. If only a few people receive much of the GNI, then the standard of living for the majority may be lower than the average figure implies. The higher the per capita GNI, the greater the potential for ensuring that all citizens can enjoy a comfortable life.

ECONOMIC STRUCTURE

Average per capita income is higher in developed countries because people typically earn their living by different means than in developing countries. Jobs fall into three categories:

- The **primary sector** includes activities that directly extract materials from Earth through agriculture (as discussed in Chapter 9) and sometimes by mining, fishing, and forestry.
- The **secondary sector** includes manufacturers that process, transform, and assemble raw materials into useful products, as well as industries that fabricate manufactured goods into finished consumer goods (discussed in Chapter 11).
- The **tertiary sector** involves the provision of goods and services to people in exchange for payment, such as retailing, banking, law, education, and government (discussed in Chapters 12 and 13).

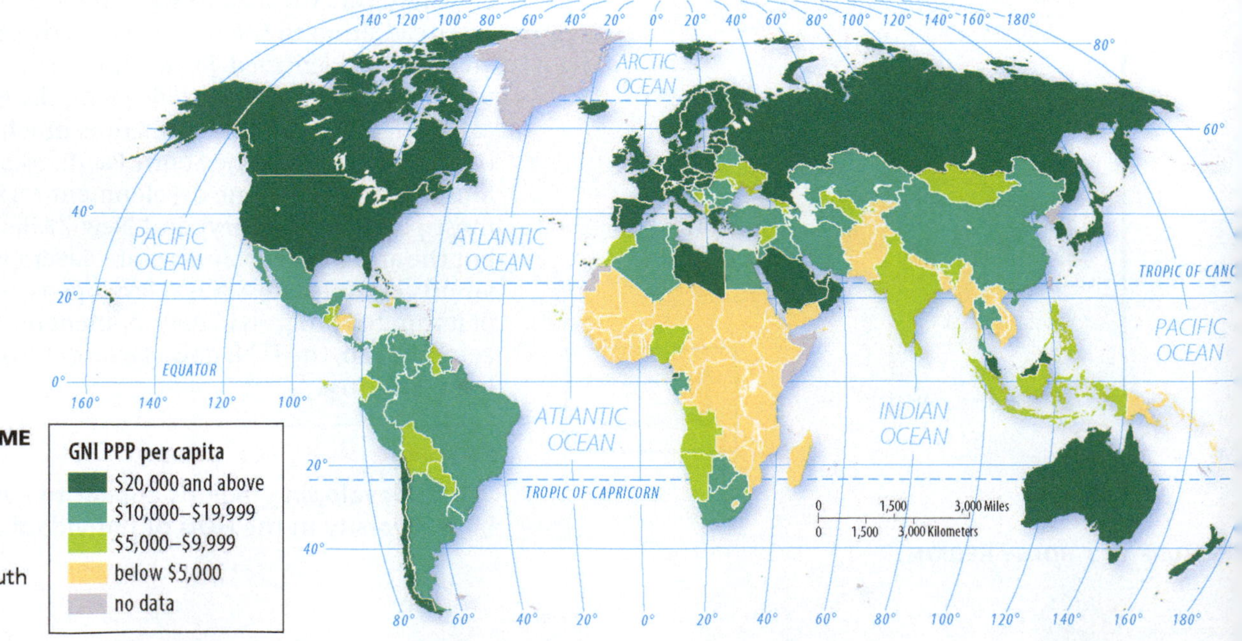

▶ FIGURE 10-4 INCOME GNI per capita PPP is highest in developed countries. The lowest figures are in sub-Saharan Africa and South Asia.

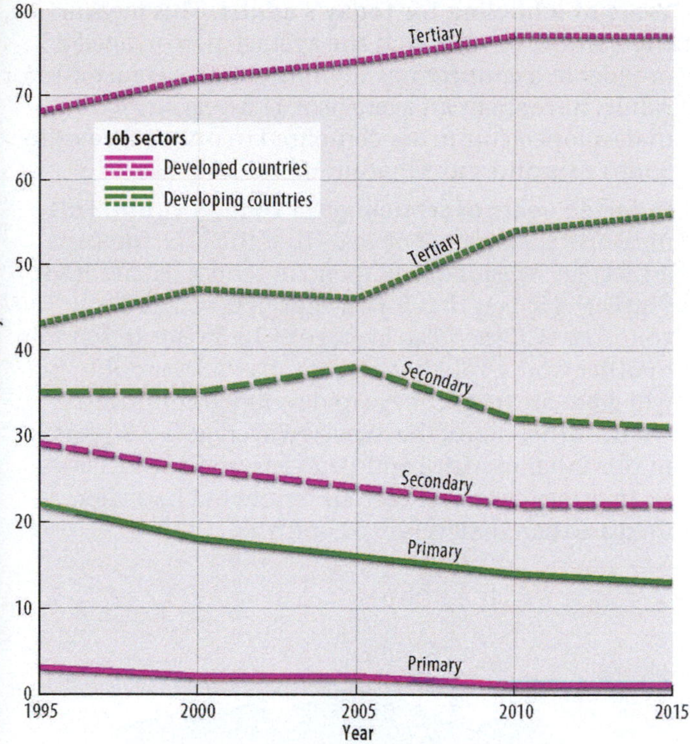

▲ **FIGURE 10-5 ECONOMIC STRUCTURE** The graph shows the changing contribution to GNI by the three sectors of jobs.

The contribution to GNI among primary, secondary, and tertiary sectors varies between developed and developing countries (Figure 10-5):

- The share of GNI accounted for by the primary sector has decreased in developing countries, but it remains higher than in developed countries. The low share in developed countries indicates that a handful of farmers produce enough food for the rest of society.
- The share of GNI accounted for by the secondary sector has decreased sharply in developed countries and is now less than in developing countries.
- The share of GNI accounted for by the tertiary sector is relatively high in developed countries, and is now growing in developing countries.

PAUSE & REFLECT 10.1.2

Figure 9-3 shows the percentage of workers engaged in agriculture. Does a country with a high percentage of agricultural workers, as shown in Figure 9-3, typically have a high HDI or low HDI?

PRODUCTIVITY

Workers in developed countries are more productive than those in developing countries. **Productivity** is the value of a particular product compared to the amount of labor needed to make it. Productivity can be measured by the value added per capita. The **value added** in manufacturing is the gross value of a product minus the costs of raw materials and energy. The value added per capita is around $67 per hour in the United States compared to around $16 in Mexico (Figure 10-6).

Workers in developed countries produce more with less effort because they have access to more machines, tools, and equipment to perform much of the work. On the other hand, production in developing countries relies more on human and animal power. The larger per capita GNI in developed countries in part pays for the manufacture and purchase of machinery, which in turn makes workers more productive and generates more wealth.

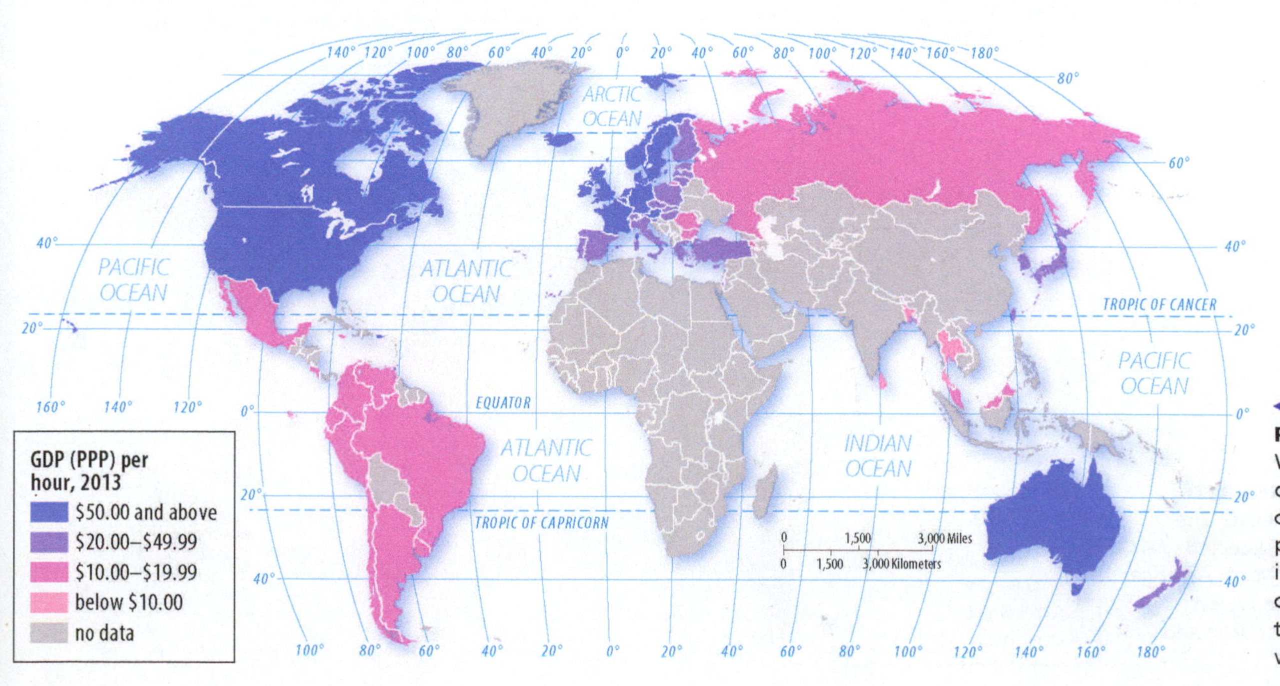

◀ **FIGURE 10-6 PRODUCTIVITY** Workers in developed countries are more productive than in developing countries because they have a higher value added.

Access to Knowledge

LEARNING OUTCOME 10.1.3

Identify the HDI education factors.

Development is about more than wealth. The U.N. believes that development is about people becoming healthier and wiser, not just wealthier.

The U.N. considers years of schooling to be the most critical measure of the ability of an individual to gain access to knowledge needed for development. The assumption is that no matter how poor the school, the longer the pupils attend, the more likely they are to learn something.

To form the access to knowledge component of HDI, the U.N. combines two measures of years of schooling:

- **Years of schooling for today's adults.** This measures the number of years that the average person aged 25 or older in a country has spent in school (Figure 10-7). Adults have spent an average of 11.5 years in school in developed countries, compared to only 4.7 years in South Asia and sub-Saharan Africa.
- **Expected years of schooling for today's youth.** This measures the number of years that the U.N. forecasts an average 5-year-old will spend in school (Figure 10-8). The U.N. expects that 5-year-olds in developed countries will spend an average of 16.3 years in school; in other words, roughly half of today's 5-year-olds will graduate from college in developed countries. On the other hand, the expected average is 9.3 years in sub-Saharan Africa and 10.2 years in South Asia—an improvement over current figures but a smaller increase than in developed countries.

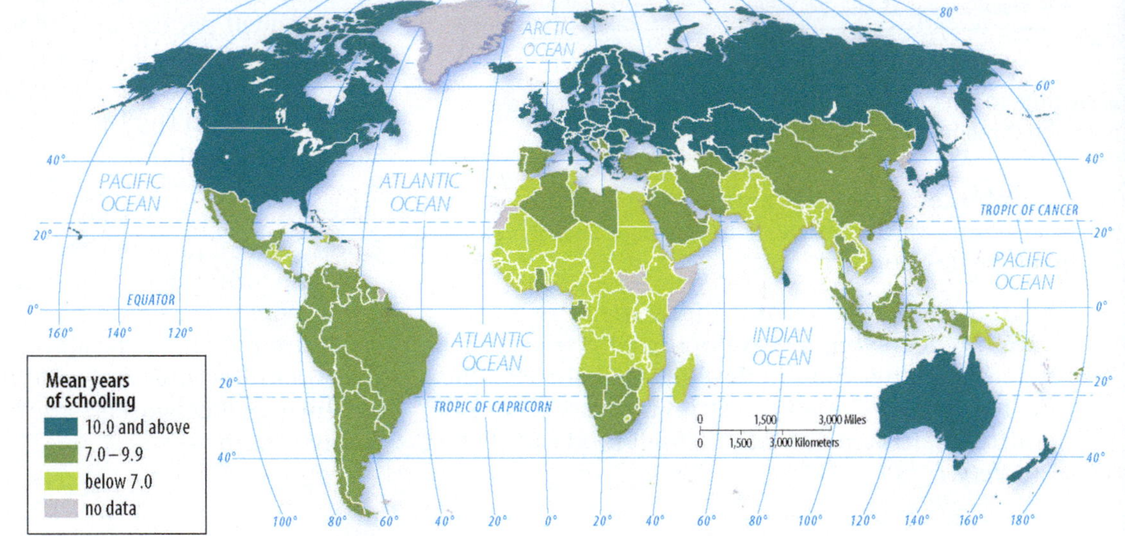

▶ **FIGURE 10-7 MEAN YEARS OF SCHOOLING** The highest number of years of schooling is in North America, and the lowest numbers are in South Asia and sub-Saharan Africa.

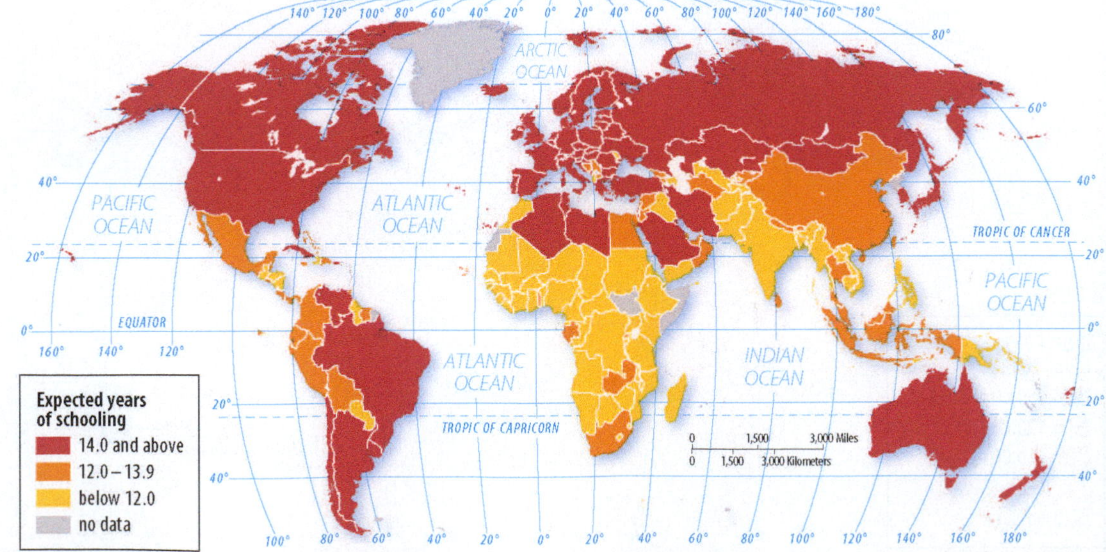

▶ **FIGURE 10-8 EXPECTED YEARS OF SCHOOLING** The highest number of expected years of schooling are in North America and Europe, and the lowest numbers are in South Asia and sub-Saharan Africa.

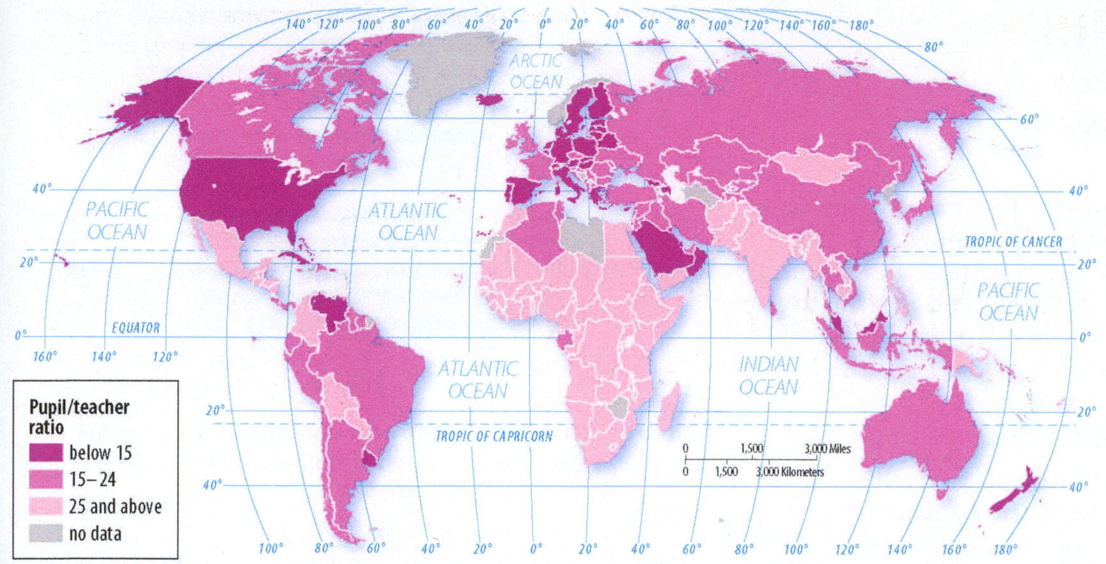

◀ **FIGURE 10-9 PUPIL/TEACHER RATIO** The lowest pupil/teacher ratio is in North America, and the highest is in sub-Saharan Africa.

Other indicators can measure regional variations in access to knowledge:

- **Pupil/teacher ratio.** The **pupil/teacher ratio** is the number of enrolled students divided by the number of teachers. The fewer pupils a teacher has, the more likely that each student will receive effective instruction (Figure 10-9).
- **Literacy rate.** The **literacy rate** is the percentage of a country's people who can read and write (Figure 10-10).

Improved education is a major goal of many developing countries, but funds are scarce. Education may receive a higher percentage of GNI in developing countries, but those countries' GNI is far lower to begin with, so they spend far less per pupil than do developed countries.

Adding to the challenge of teaching and learning in developing countries, most books, newspapers, and magazines are published in developed countries, in part because more of their citizens can read and write and can afford to buy them. Developed countries dominate scientific and nonfiction publishing worldwide. (This textbook is an example.) Students in developing countries must learn technical information from books that usually are not in their native language.

PAUSE & REFLECT 10.1.3

The United States has a lower pupil/teacher ratio than Canada. Does that mean that the pupil/teacher ratio in the United States is more favorable or less favorable than in Canada?

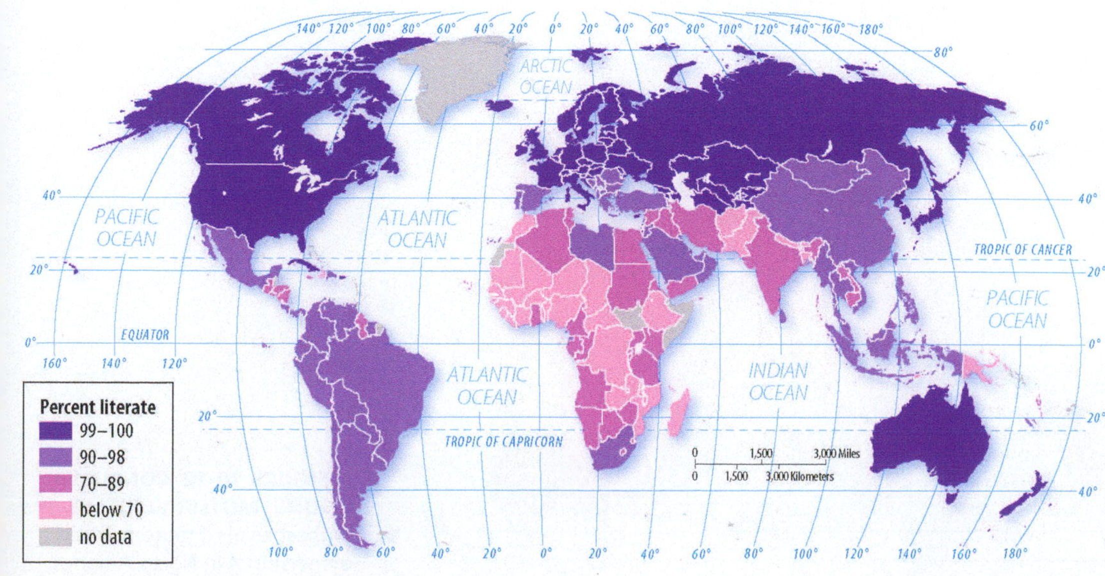

◀ **FIGURE 10-10 LITERACY RATE** Literacy is nearly 100 percent in developed countries. The lowest rates are in sub-Saharan Africa and South Asia.

Health and Wealth

LEARNING OUTCOME 10.1.4

Identify the HDI health factors.

The U.N. considers good health to be as an important a measure of development as wealth and knowledge. A goal of development is to provide the nutrition and medical services needed for people to lead long and healthy lives. Chapter 2 discussed in detail the many differences worldwide in health and medical services.

A LONG AND HEALTHY LIFE

From the many health and medical indicators, the U.N. has selected life expectancy at birth as the contributor to the HDI (Figure 10-11). Life expectancy at birth was defined in Chapter 2 as the average number of years a newborn infant can expect to live at current mortality levels. A baby born this year is expected to live on average to age 71 worldwide, to 80 in developed countries, and to only 57 in sub-Saharan Africa (refer to Figure 2-11).

People are healthier in developed countries than in developing ones. When people in developed countries get sick, these countries possess the resources to care for them. Developed countries use part of their wealth to protect people who, for various reasons, are unable to work. In these countries, some public assistance is offered to those who are sick, elderly, poor, disabled, orphaned, veterans of wars, widows, unemployed, or single parents. Better health and welfare in developed countries permit people to live longer.

With longer life expectancies, developed countries have a higher percentage of older people who have retired and receive public support and a lower percentage of children under age 15 who are too young to work and must also be supported by employed adults. The number of young people is six times higher than the number of older people in developing countries, whereas the two are nearly the same in developed countries.

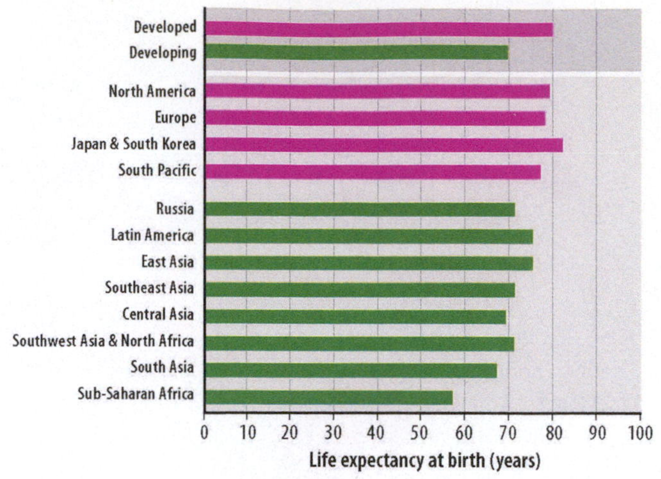

▲ **FIGURE 10-11 LIFE EXPECTANCY AT BIRTH** The highest life expectancy among the nine development regions is in North America, and the lowest is in sub-Saharan Africa.

Better health and welfare also permit babies to survive infancy in developed countries. About 94 percent of infants survive and 6 percent die in developing countries, whereas in developed countries more than 99.5 percent survive. The infant mortality rate is greater in developing countries for several reasons. Babies may die from malnutrition or lack of medicine needed to survive illness, such as dehydration and diarrhea. They may also die from poor medical practices that arise from lack of education.

CONSUMER GOODS

Part of the wealth generated in developed countries is used to purchase goods and services. Especially important are goods and services related to transportation and communications, including motor vehicles, telephones, and computers.

- Motor vehicles provide individuals with access to jobs and services and permit businesses to distribute their products (Figure 10-12). The number of motor vehicles per 1,000 persons is approximately 170 in the world as a whole, 630 in developed countries and 80 in developing countries.

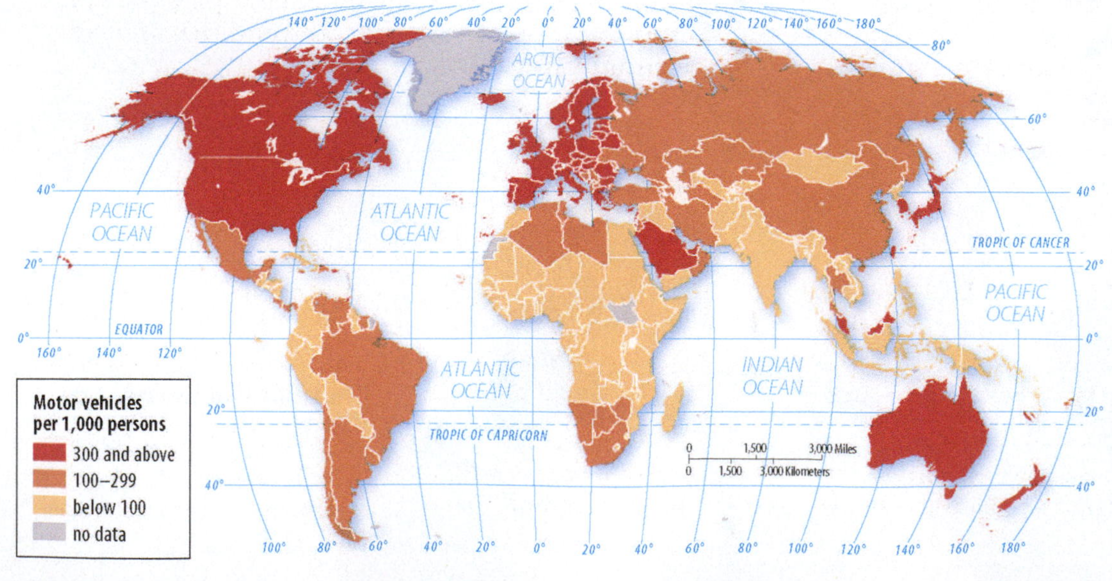

◄ **FIGURE 10-12 CONSUMER GOODS: MOTOR VEHICLES** The highest level of motor vehicle ownership is in North America, and the lowest is in South Asia.

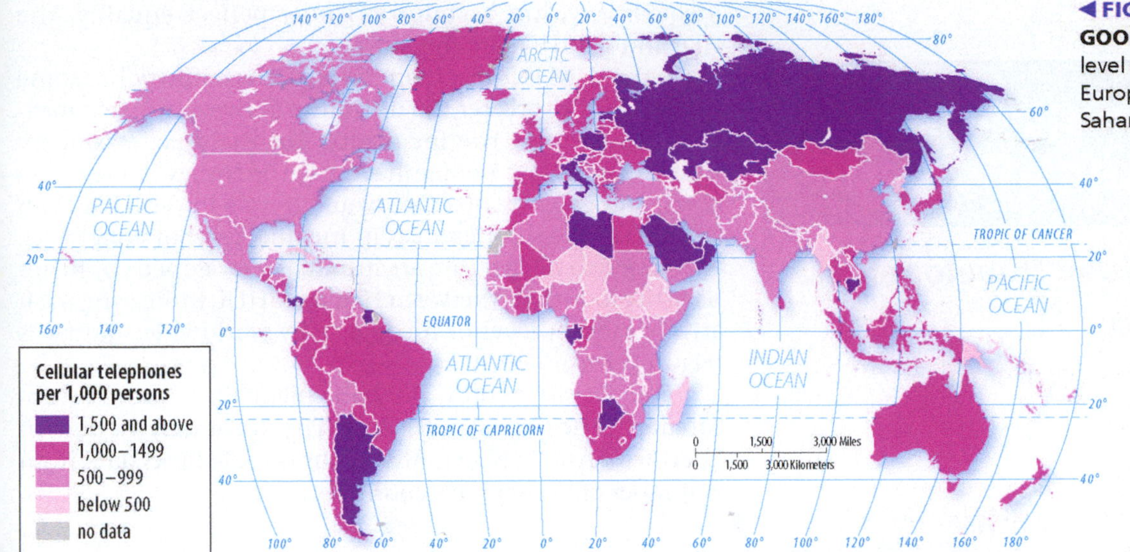

◀ **FIGURE 10-13 CONSUMER GOODS: CELL PHONES** The highest level of cell phone ownership is in Europe, and the lowest is in sub-Saharan Africa.

- Telephones enhance interaction with providers of raw materials and customers for goods and services (Figure 10-13). The number of cell phones per 1,000 persons is approximately 800 in the world as a whole, 1,100 in developed countries and 700 in developing countries.
- Computers facilitate the sharing of information with other buyers and suppliers (Figure 10-14). The number of Internet users per 1,000 persons is approximately 300 in the world as a whole, 700 in developed countries and 200 in developing countries.

Products that promote better transportation and communications are accessible to virtually all residents in developed countries and are vital to the economy's functioning and growth. In contrast, in developing countries these products do not play a central role in daily life for many people. Motor vehicles, computers, and telephones are not essential to people who live in the same village as their friends and relatives and work all day growing food in nearby fields.

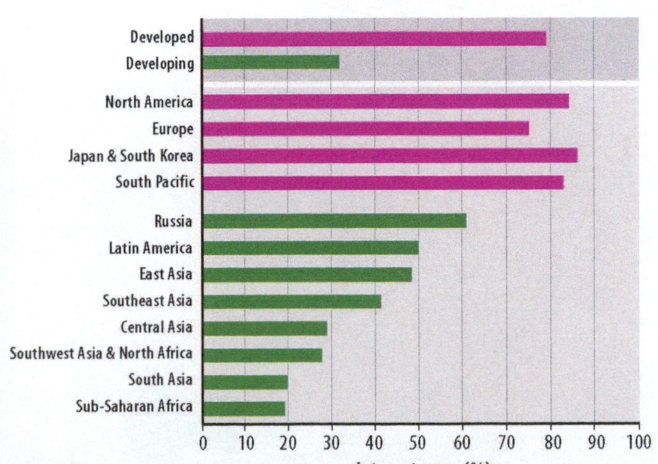

▲ **FIGURE 10-14 CONSUMER GOODS: INTERNET USERS** The highest level of Internet users is in North America, and the lowest is in South Asia.

But most people in developing countries are familiar with these goods, even if they cannot afford them, and may desire them as symbols of development.

Because possession of consumer goods is not universal in developing countries, a gap can emerge between the "haves" and the "have-nots." The minority of people who have these goods may include government officials, business owners, and other elites, whereas their lack among the majority who are denied access may provoke political unrest. In many developing countries, those who have these products are concentrated in urban areas; those who do not live in the countryside. Technological innovations tend to diffuse from urban to rural areas. Access to these goods is more important in urban areas because of the dispersion of homes, factories, offices, and shops.

Technological change is helping to reduce the gap between developed and developing countries in access to communications. Cell phone ownership, for example, is expanding rapidly in developing countries because these phones do not require the costly investment of connecting wires to each individual building and more individuals can obtain service from a single tower or satellite.

PAUSE & REFLECT 10.1.4

In addition to cell phones, what other electronic devices might diffuse rapidly to developing countries because of low cost of equipment and lack of need for costly infrastructure?

CHECK-IN KEY ISSUE 1

Why Does Development Vary Among Countries?

- ✔ **The Human Development Index (HDI) measures the level of development of each country.**
- ✔ **HDI is based on three factors: a decent standard of living, a long and healthy life, and access to knowledge.**

KEY ISSUE 2

Where Are Inequalities in Development Found?

- Unequal and Uneven Development
- Gender Inequality
- Gender Empowerment and Employment
- Reproductive Health
- HDI and Gender Inequality

LEARNING OUTCOME 10.2.1

Describe the U.N.'s measures of inequality.

The U.N. believes that every person should have access to decent standards of living, knowledge, and health. At least some degree of inequality is found in every country.

Unequal and Uneven Development

Inequality within countries can exist in terms of income, gender, and region. Inequality also exists between countries in the core and periphery of development.

INEQUALITY-ADJUSTED HDI

To measure the extent of inequality, the U.N. has created the **Inequality-adjusted Human Development Index (IHDI).** The IHDI modifies the HDI to account for inequality within a country. Under perfect equality, the HDI and the IHDI are the same.

If the IHDI is lower than the HDI, the country has some inequality; the greater the difference between the two measures, the greater the inequality. For example, a country where only a few people have high incomes, college degrees, and good health care would have a lower IHDI than a country where differences in income, level of education, and access to health care are minimal. Developed countries have the lowest gap between HDI and IHDI, indicating a relatively modest level of inequality by worldwide standards (Figure 10-15).

The lowest scores (highest inequality) are in sub-Saharan Africa and South Asia. The score may be low in Southwest Asia & North Africa, but the UN lacks data from a number of the region's countries.

PAUSE & REFLECT 10.2.1

In the United States, the HDI is .914 and the IHDI is .755. In Canada, the HDI is .902 and the IHDI is .833. Which country has greater inequality?

INEQUALITY WITHIN DEVELOPING COUNTRIES

Brazil and Turkey are among the world's largest and most populous countries. At the national scale, the two countries fall somewhere in the middle of the pack in terms of HDI. Among the 186 countries with HDI scores, Turkey ranks 69th and Brazil ranks 79th.

The extent of inequality within these countries can be seen in two ways. First is the difference between the HDI and IHDI. The two countries have similar HDI scores, but Brazil has a lower IHDI, indicating more inequality in Turkey.

Inequality can also be seen through differences in GNI per capita among states or provinces within the countries.

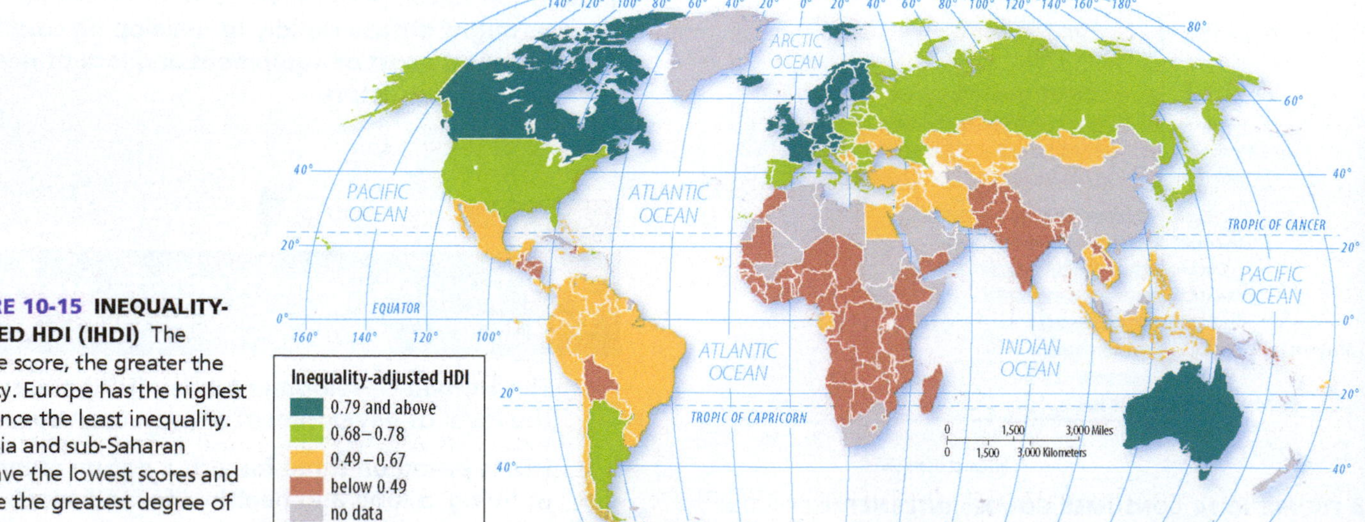

▶ **FIGURE 10-15 INEQUALITY-ADJUSTED HDI (IHDI)** The lower the score, the greater the inequality. Europe has the highest score, hence the least inequality. South Asia and sub-Saharan Africa have the lowest scores and therefore the greatest degree of inequality.

◀ **FIGURE 10-16 INEQUALITY AMONG REGIONS WITHIN TURKEY**

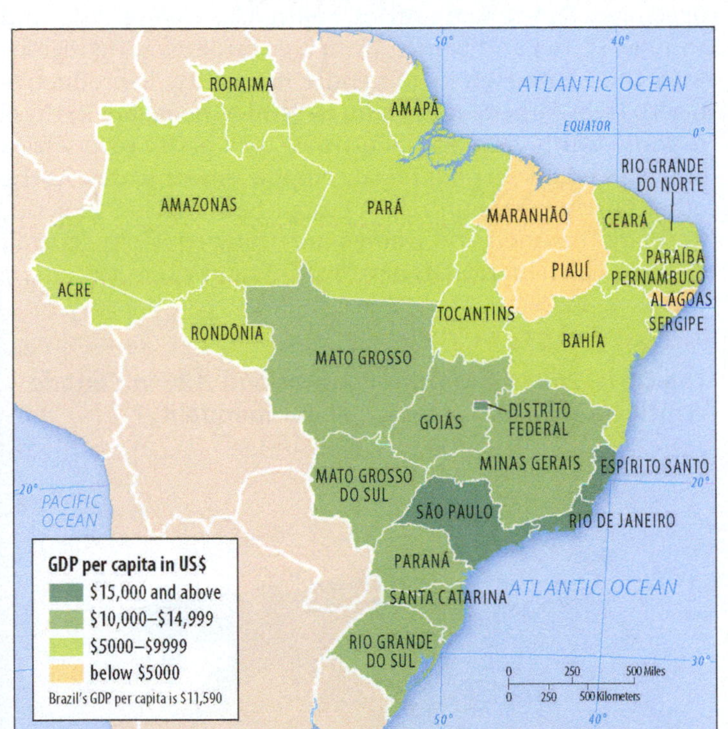

◀ **FIGURE 10-17 INEQUALITY AMONG REGIONS WITHIN BRAZIL**

In Turkey, wealth is much higher in the western part of the country, closest to Europe (Figure 10-16). Wealth is much lower in the east, home to Kurds, many of whom seek independence (see Chapter 7). In Brazil, wealth is highest along the east coast, especially in the country's largest city, São Paulo (Figure 10-17).

INEQUALITY WITHIN DEVELOPED COUNTRIES

Developed countries have regional internal variations in GNI per capita that are less extreme than in developing countries. In the United States, for example, the GNI per capita is 122 percent of the national average in the wealthiest region (New England) and 90 percent of the national average in the poorest region (Southeast).

Through most of the twentieth century, the gap between rich and poor narrowed in developed countries. Inequality was reduced because developed countries used some of their wealth to extend health care and education to more people and to provide some financial assistance to poorer people. Since 1980, however, inequality has increased in most developed countries, including the United States and the United Kingdom (Figure 10-18).

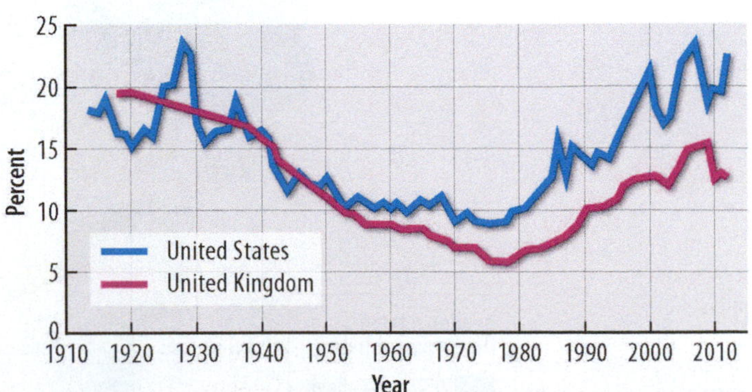

▲ **FIGURE 10-18 WIDENING INEQUALITY** The share of the national wealth held by the richest 1 percent declined in the United States and the United Kingdom for most of the twentieth century but has increased in recent decades.

Gender Inequality

LEARNING OUTCOME 10.2.2

Describe the U.N.'s measures of gender inequality.

A country's overall level of development can mask inequalities in the status of men and women. The quest for an improved standard of living, access to knowledge, health, and a sustainable future are aspirations of people in all countries. Yet long-standing cultural and legal obstacles can limit women's participation in development and access to its benefits.

The U.N. uses two indexes to measure gender inequality: the Gender Inequality Index (GII) and the Gender-related Development Index (GDI). The U.N. has not found a single country in the world where the women are treated as well as the men according to the GII and only a handful according to the GDI. At best, women have achieved near-equality with men in some countries, but in other countries, the level of development for women lags far behind the level for men. The U.N. argues that inequality between men and women is a major factor that keeps a country from achieving a higher level of development.

GENDER-RELATED DEVELOPMENT INDEX

The **Gender-related Development Index (GDI)** measures the gender gap in the level of achievement for the three dimensions of the Human Development Index: income, education, and life expectancy. The GDI uses the same methodology as the HDI, described in the previous section. Countries are ranked based on their deviation from gender parity in the three dimensions of the HDI.

If females and males had precisely the same HDI scores, the GDI would be 1.000. In fact, the overall GDI in the world is .920, which means that the average HDI for all females in the world (.655) is .920, or 92 percent of the average HDI for all males (.712). The average in developed regions is .975, indicating close to parity, whereas the average in developing countries is .904. The lowest scores are in South Asia, sub-Saharan Africa, and Southwest Asia & North Africa (Figure 10-19).

GENDER INEQUALITY INDEX

The **Gender Inequality Index (GII)** measures the gender gap in the level of achievement in three dimensions: reproductive health, empowerment, and the labor market. The GII uses similar methodology to the IHDI discussed on the previous page. The higher the GII, the greater the inequality between men and women (Figure 10-20). A score of 0 would mean that men and women fare equally, and a score of 1.0 would mean that women fare as poorly as possible in all measures.

The GII is higher in developing countries than in developed ones. Sub-Saharan Africa, South Asia, Central Asia, and Southwest Asia and North Africa are the developing regions with the highest levels of gender inequality. Reproductive health is the largest contributor to gender inequality in these regions. South Asia and Southwest Asia and North Africa also have relatively poor female empowerment scores. At the other extreme, 10 countries in Europe have GIIs less than 0.1, meaning that men and women are nearly equal. In general, countries with high HDIs have low GIIs and vice versa.

PAUSE & REFLECT 10.2.2

The GII is .262 in the United States and .136 in Canada. Which country has greater gender inequality?

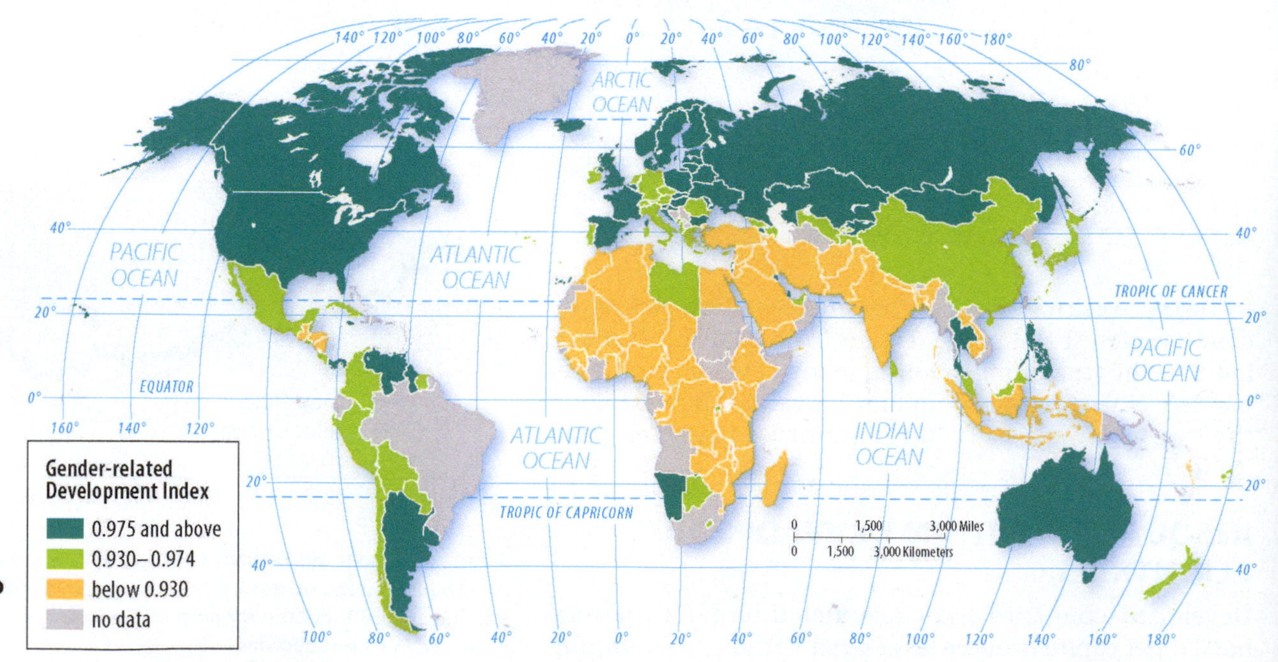

▶ FIGURE 10-19 GENDER-RELATED DEVELOPMENT INDEX

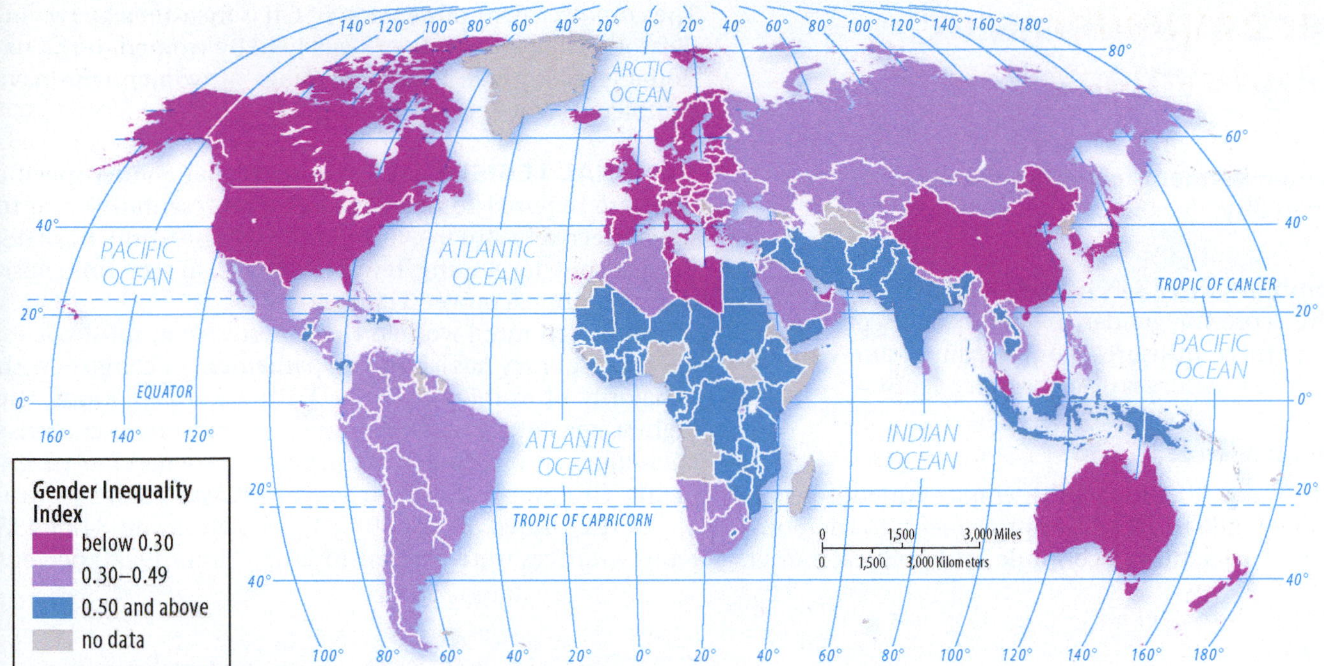

▲ **FIGURE 10-20 GENDER INEQUALITY INDEX** The lowest GII numbers and therefore the least inequality are in Europe, and the highest numbers are in sub-Saharan Africa and South Asia.

GII OVER TIME

The U.N. has found that gender inequality has declined since the 1990s in all but 4 of 138 countries for which time-series data are available (Figure 10-21). The greatest improvement has been in Southwest Asia & North Africa.

The improvement in gender inequality has been relatively modest in the United States. Furthermore, the United States has a GII rank of only 47, although it ranks fifth in the world on HDI. The U.N. points to two factors accounting for the relatively low U.S. GII ranking:

- Compared with other very high HDI countries, the United States has a much higher birth rate among teenage women and a higher mortality rate among women during childbirth.
- The percentage of women in the national legislature is lower in the United States than in other high HDI countries.

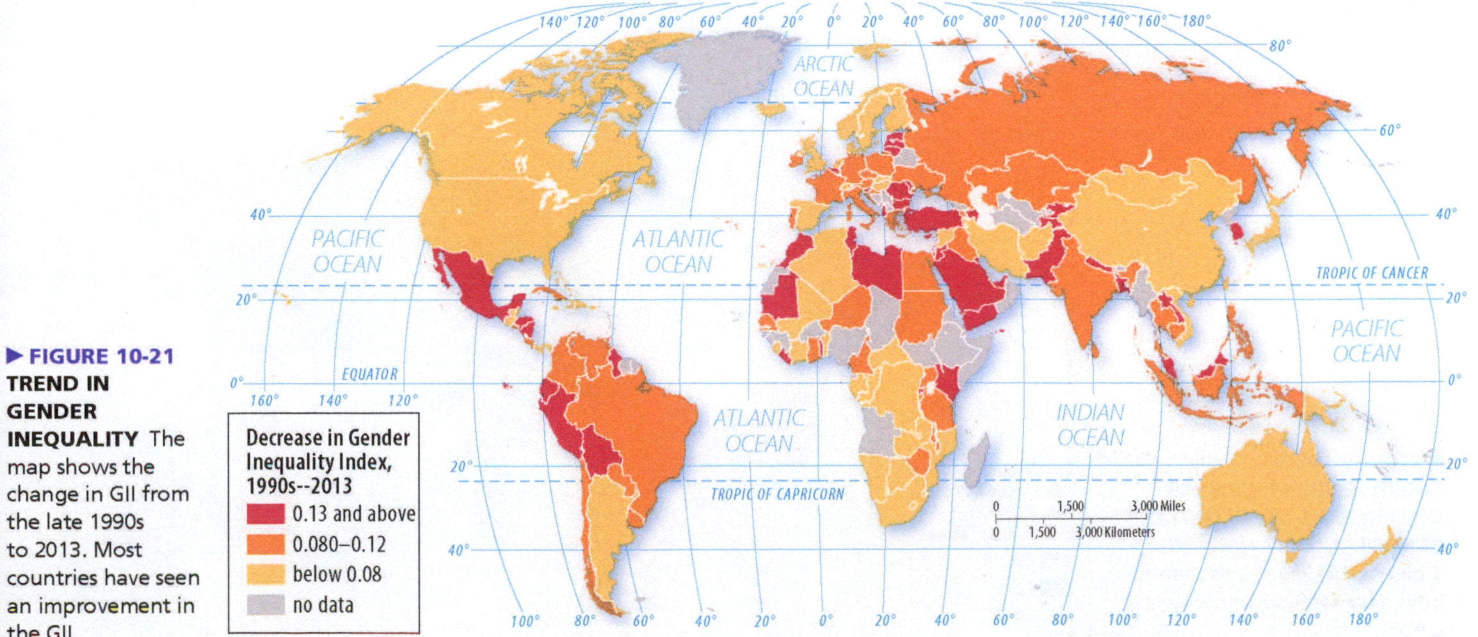

► **FIGURE 10-21 TREND IN GENDER INEQUALITY** The map shows the change in GII from the late 1990s to 2013. Most countries have seen an improvement in the GII.

Gender Empowerment and Employment

LEARNING OUTCOME 10.2.3

Describe empowerment-related components of gender inequality.

The GII combines three sets of measures to come up with a composite score for gender inequality. This page looks at two of the three measures: empowerment and employment. The third measure is discussed on page 364.

EMPOWERMENT

In the context of gender inequality, empowerment refers to the ability of women to achieve improvements in their own status—that is, to achieve economic and political power. The empowerment dimension of GII is measured by two indicators: The percentage of seats held by women in the national legislature and the percentage of women who have completed some secondary school.

NATIONAL LEGISLATURE. No particular gender-specific skills are required to be elected as a representative and to serve effectively. But in every country of the world, both developed and developing, fewer women than men hold positions of political power (Figure 10-22).

Although more women than men vote in most places, only one country has a national parliament or congress with a majority of women: Rwanda. With a few exceptions, the highest percentages are in Europe, where women comprise approximately one-fourth of the members of national parliaments. The lowest rates are in Southwest Asia & North Africa.

In the United States, 20 of 100 senators and 84 of 435 representatives were women in 2015. These 19–20 percent

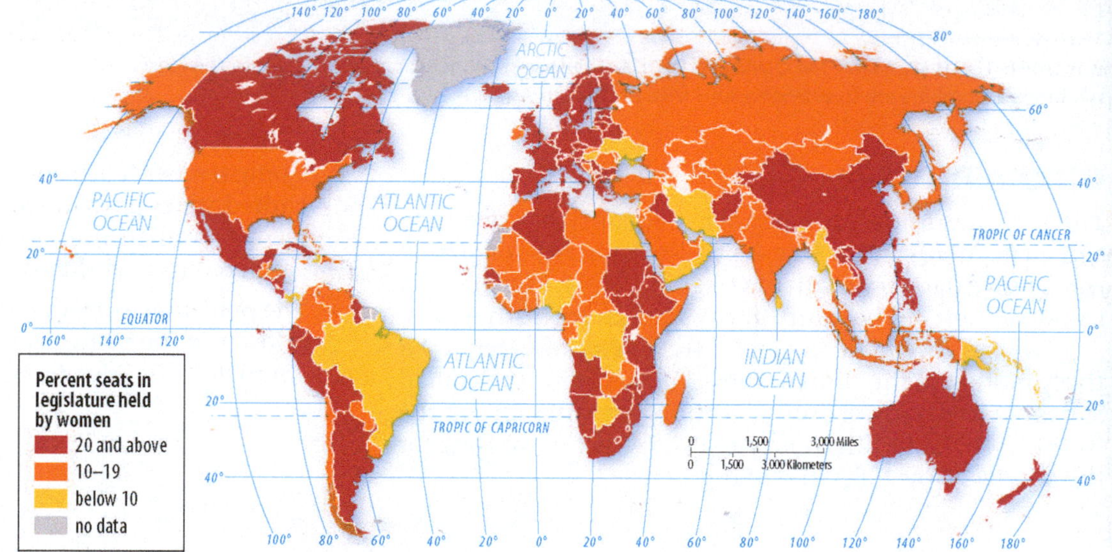

▶ FIGURE 10-22
EMPOWERMENT: WOMEN IN THE NATIONAL LEGISLATURE The highest percentages are in Europe, and the lowest are in Southwest Asia & North Africa.

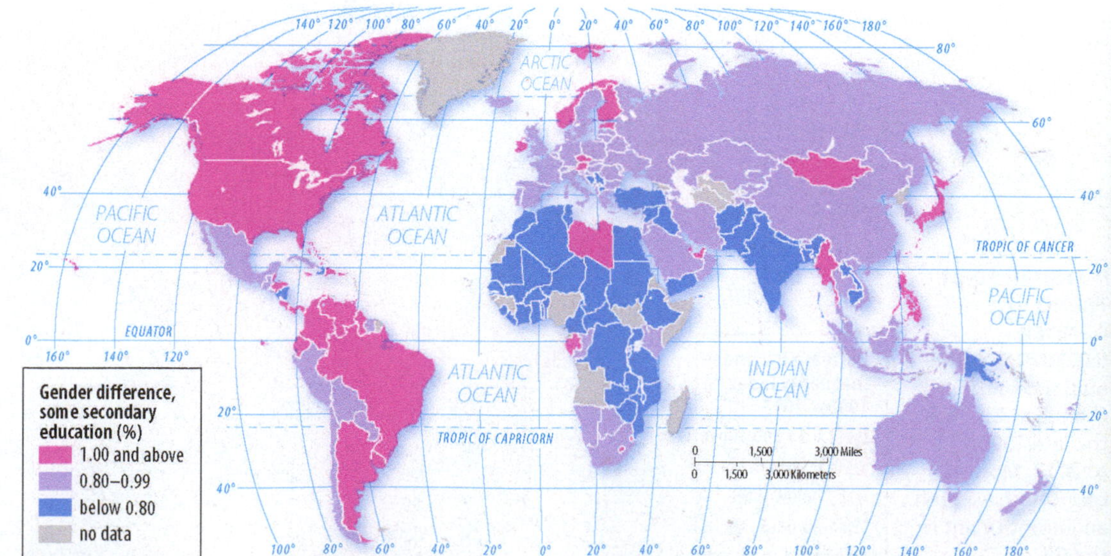

▶ FIGURE 10-23
EMPOWERMENT: GENDER DIFFERENCES IN SECONDARY SCHOOL A figure about 1 means that more girls than boys have some secondary school.

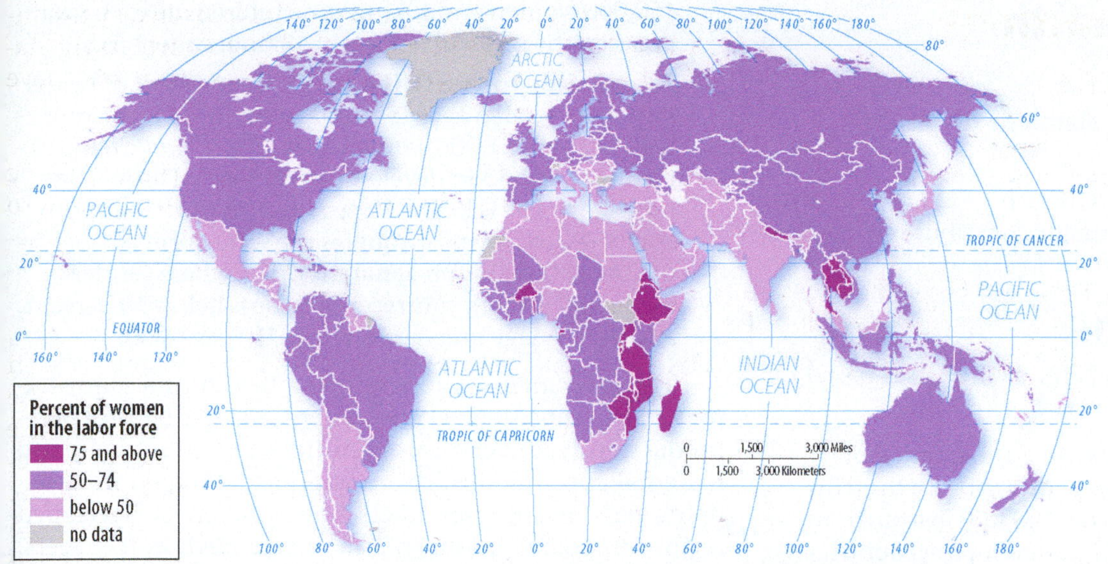

◀ **FIGURE 10-24 LABOR FORCE PARTICIPATION** A lower number means that relatively few women participate in the labor force compared with men.

figures are much lower than the average for developed countries and slightly below average for all countries in the world. In Canada, for example, 30 of 83 senators (excluding vacancies) and 88 of 338 members of parliament in the House of Commons were women in 2015.

PAUSE & REFLECT 10.2.3

What is the gender ratio on the governing body of your local community?

SECONDARY SCHOOL. Worldwide, 54 percent of women have completed some secondary (high) school, compared to 64 percent of men. In North America, girls are more likely than boys to complete some high school, and boys are slightly ahead in Europe. In developing countries, boys are much more likely than girls to be high school graduates. For every 10 boys who attend high school in developing countries, only 6 girls attend (Figure 10-23). The gap in education between girls and boys is especially high in South Asia.

EMPLOYMENT

The **female labor force participation rate** is the percentage of women holding full-time jobs outside the home. Worldwide, 51 percent of women work outside the home, compared to 77 percent of men. In general, women in developed countries are more likely than women in developing countries to hold full-time jobs outside the home. Figures vary widely among developing regions (Figure 10-24). South Asia and Southwest Asia & North Africa have substantial gaps between male and female labor participation, whereas East Asia and sub-Saharan Africa have smaller gaps (Figure 10-25). Women hold jobs in agriculture or services in sub-Saharan Africa, even while they have the world's highest fertility rates.

◀ **FIGURE 10-25 EMPLOYMENT FOR WOMEN** A representative of a bank meets with women who are receiving loans to start businesses in Bihar State, India.

Reproductive Health

LEARNING OUTCOME 10.2.4

Describe reproductive health elements of the GII.

The third component of the GII is reproductive health. Poor reproductive health is a major contributor to gender inequality around the world.

REPRODUCTIVE HEALTH

The reproductive health dimension of the GII is based on two indicators:

- The **maternal mortality rate** is the number of women who die giving birth per 100,000 births. The rate is 16 deaths of mothers per 100,000 live births in developed countries and 171 in developing countries (Figure 10-26). The highest rates (most deaths per births) are in sub-Saharan Africa. The U.N. estimates that 150,000 women and 1.6 million children die each year between the onset of labor and 48 hours after birth.
- The **adolescent fertility rate** is the number of births per 1,000 women ages 15 to 19 (Figure 10-27). The rate is 19 births per 1,000 women ages 15 to 19 in developed countries and 53 in developing countries. The teenage pregnancy rate is below 10 per 1,000 in most European countries, where most couples use some form of contraception. In sub-Saharan Africa, where gender inequality is high, contraceptive use is below 10 percent, and the teenage pregnancy rate is 110 per 1,000.

The U.N. includes reproductive health as a contributor to GII because in countries where effective control of reproduction is universal, women have fewer children, and maternal and child health are improved. Women in developing regions are more likely than women in developed regions to die in childbirth and to give birth as teenagers. Every country that offers women a full range of reproductive health options has a very low total fertility rate.

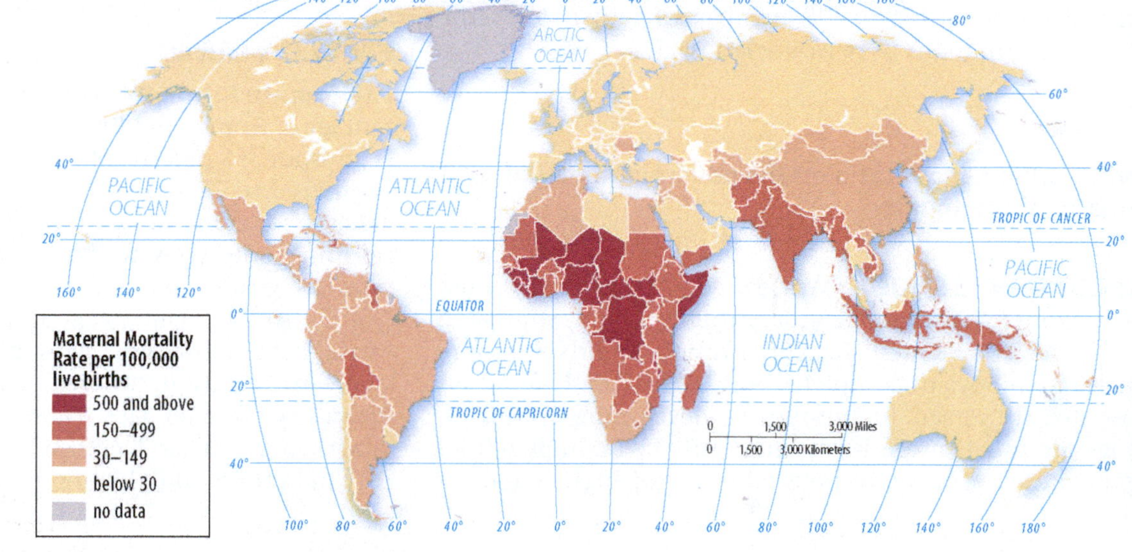

▶ **FIGURE 10-26 MATERNAL MORTALITY RATE** The maternal mortality ratio is the number of deaths of mothers in childbirth compared to the number of live births.

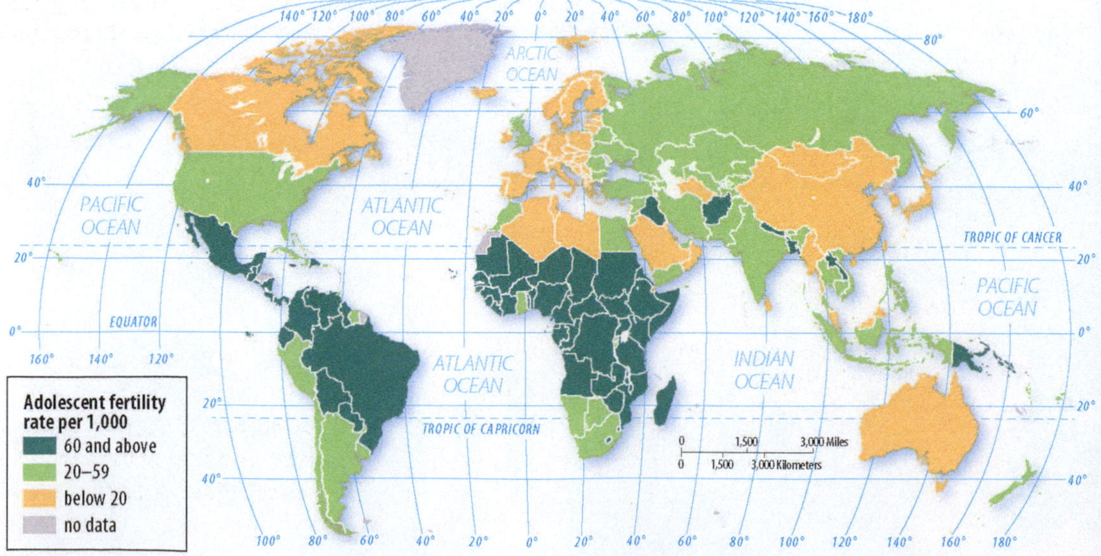

▶ **FIGURE 10-27 ADOLESCENT FERTILITY RATE** The adolescent fertility rate is the number of births per 1,000 women ages 15 to 19.

CORE AND PERIPHERY

The relationship between developed countries and developing countries is often described as a north–south split because most of the developed countries are north of the equator, whereas many developing countries are south (Figure 10-28). Immanuel Wallerstein, a U.S. social scientist, describes the relationship between developed and developing countries as one of "core" and "periphery." According to Wallerstein's world-systems analysis, in an increasingly unified world economy, developed countries form an inner core area, whereas developing countries occupy peripheral locations.

Uneven development was defined in Chapter 1 as the increasing gap in economic conditions between core and peripheral regions as a result of the globalization of the economy. North America, Europe, Japan, and South Korea account for a high percentage of the world's economic activity and wealth. Developing countries in the periphery have less access to the world centers of consumption, communications, wealth, and power, which are clustered in the core.

Since the original formulation of world systems theory, an increasingly important element has been the emergence of countries classified as semi-periphery. These are countries that are either intermediate in level of economic development or situated close to both core and periphery regions.

Particular core and periphery regions have special connections. The development prospects of Latin America are tied to governments and businesses primarily in North America, those of Africa and Eastern Europe to Western Europe, and those of Asia to Japan and to a lesser extent Europe and North America. As countries like China, India, and Brazil develop, relationships between core and periphery are changing, and the line between core and periphery may need to be redrawn.

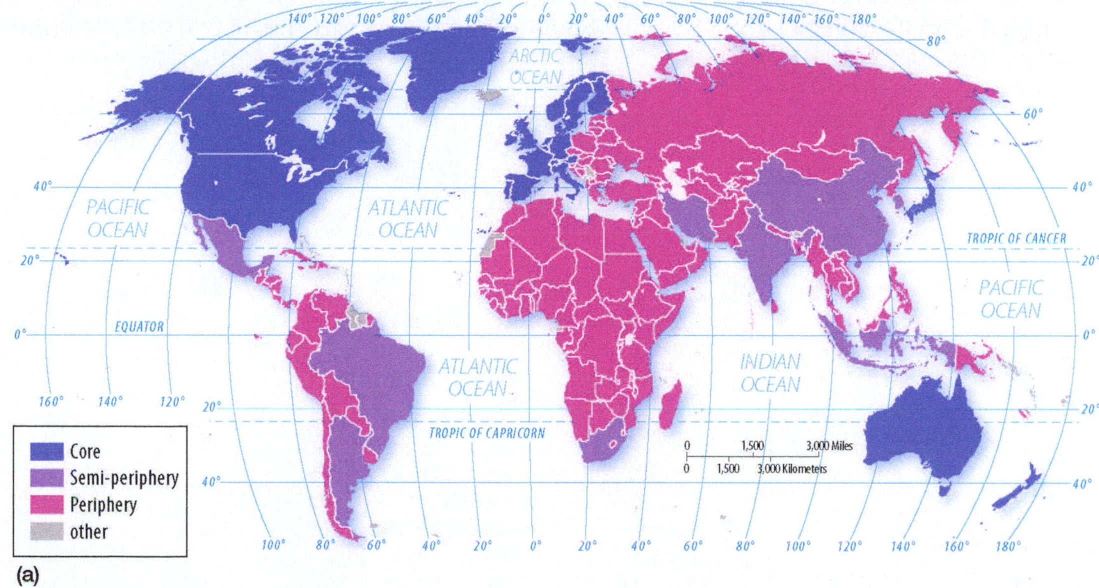

(a)

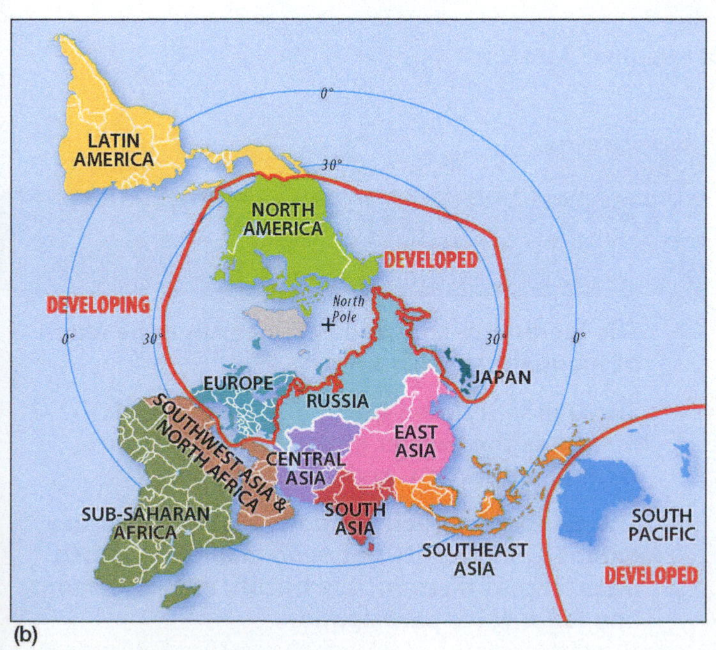

(b)

◀ **FIGURE 10-28 CORE AND PERIPHERY (a)** In world systems theory, developed regions are classified as core and most developing regions as periphery. A handful of developing countries are now classified as semi-periphery. **(b)** This unorthodox world map projection emphasizes the central role that developed countries play at the core of the world economy.

PAUSE & REFLECT 10.2.4

What combination of development features might be making several countries in Latin America and Asia be considered semi-periphery instead of periphery?

HDI and Gender Inequality

LEARNING OUTCOME 10.2.5

Compare HDI and GII for selected countries.

The HDI is a measure of a country's level of development, with a higher score indicating a higher level of development. The GII measures the extent of a country's gender inequality, with a higher score indicating a higher level of inequality. In general, development analysts expect that a more developed country will have less gender inequality than a developing one. The Doing Geography feature discusses the extent to which this is the case.

DOING GEOGRAPHY Comparing HDI and GII

The extent to which the expected relationship between development and equality actually holds true can be depicted in several ways. One approach is a graph. Figure 10-29 compares the HDI and GII for 9 of the 10 most populous countries (excluding Nigeria, which does not have a GII score). The dot for each country is placed at the HDI score along the x-axis and the GII score along the y-axis.

A second way to depict the relationship between HDI and GII is in a map. Figure 10-30 shows that 7 of the 9 countries follow an expected pattern. Three countries have low HDI and high GII, 2 have medium HDI and GII, and 2 have high HDI and low GII. The outliers are Russia (which has a higher-than-expected GII) and China (which has a lower-than-expected GII).

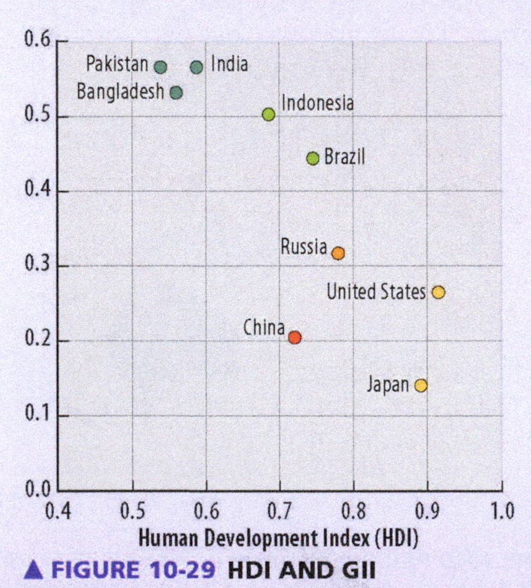

▲ **FIGURE 10-29 HDI AND GII COMPARED** IIn graph format.

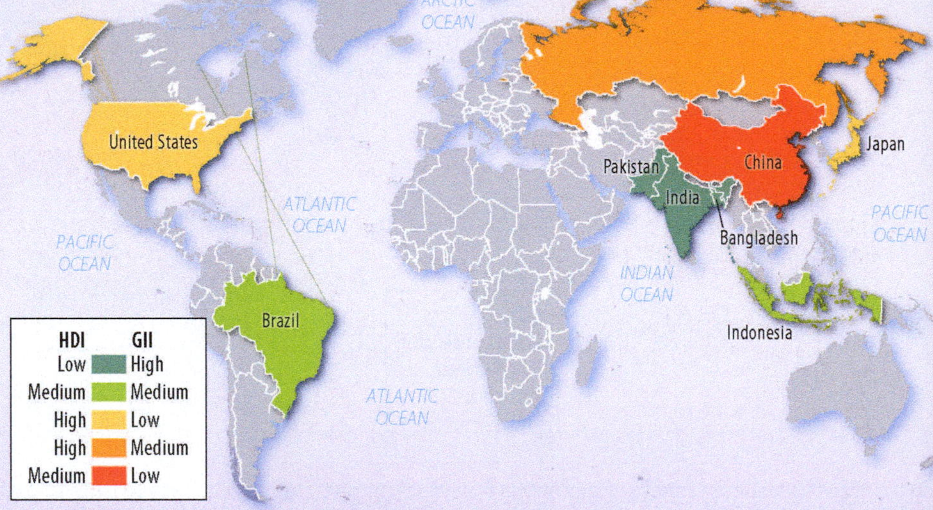

▲ **FIGURE 10-30 HDI AND GII COMPARED** In map format.

PAUSE & REFLECT 10.2.5

In summary, how close is the relationship between development and gender inequality?

CHECK-IN KEY ISSUE 2

Where Are Inequalities in Development Found?

- ✔ **The inequality-adjusted HDI measures the extent of inequality in a country.**
- ✔ **The Gender-related Development Index (GDI) measures the gender gap in achievement of income, education, and life expectancy.**
- ✔ **The Gender Inequality Index (GII) measures the extent of inequality between men and women according to reproductive health, empowerment, and labor force participation.**

What's Your Development Geography?

The United Nations computes development measures such as GII for entire countries rather than for individual communities within countries. How would the GII of your community rank compared with that of the United States as a whole or other countries? The GII combines measures of reproductive rights, empowerment, and labor participation. You can find indicators of all three factors for your community.

1. **Empowerment.** Women accounted for 18.2 percent of the U.S. Congress, according to the United Nations.
   a. What percentage of your state's legislature are women?
   b. What percentage of your local community's elected body are women?
   c. Are these percentages higher or lower than the national average?
   d. What might account for the variation?
2. **Labor force.** The percentage of women in the U.S. labor force is 56.8 percent. To learn the percentage of women in your community's labor force, go to American Factfinder (factfinder.census.gov). Enter the name of your community or zip code. Select Income, then Occupation by Sex and Median Earnings in the Past 12 Months. The percentage of women in your local labor force appears in the row Civilian Employed Population 16 Years and Over and the column Female Estimate.
   a. Is the figure higher or lower than the national average?
   b. What might account for the variation?
   c. Other columns in the same row show the earnings for male and female workers. Which gender has higher earnings?
3. **Reproductive rights.** The rate of teen births in the United States was 29 per 1,000 in 2013. The rate varies widely among U.S. states, from a low of 13 to a high of 46. The 10 highest and 10 lowest states are shown in Figure 10-31. Go to cdc.gov/nchs/fastats/teen-births or search online for CDC FastStats Teen Births. Select National and State Patterns of Teen Births in the United States 1940–2013. Go to Table 5 on page 20.
   a. Is your state higher or lower than the national average?
   b. What might account for the variation?
4. **Your community's GII.**
   a. Does a higher percentage of female legislators contribute to a higher GII or a lower GII? Why? Does your community have a higher or lower percentage of female legislators than the national average?
   b. Does a higher female labor force participation contribute to a higher GII or a lower GII? Why? Is the labor force participation in your community higher or lower than the national average?
   c. Does a higher teen birth rate contribute to a higher GII or a lower GII? Why? Is the teen birth rate in your state higher or lower than the national average?
   d. Putting these three measures together, do you consider your community to have a higher GII than the U.S. average or a lower GII?

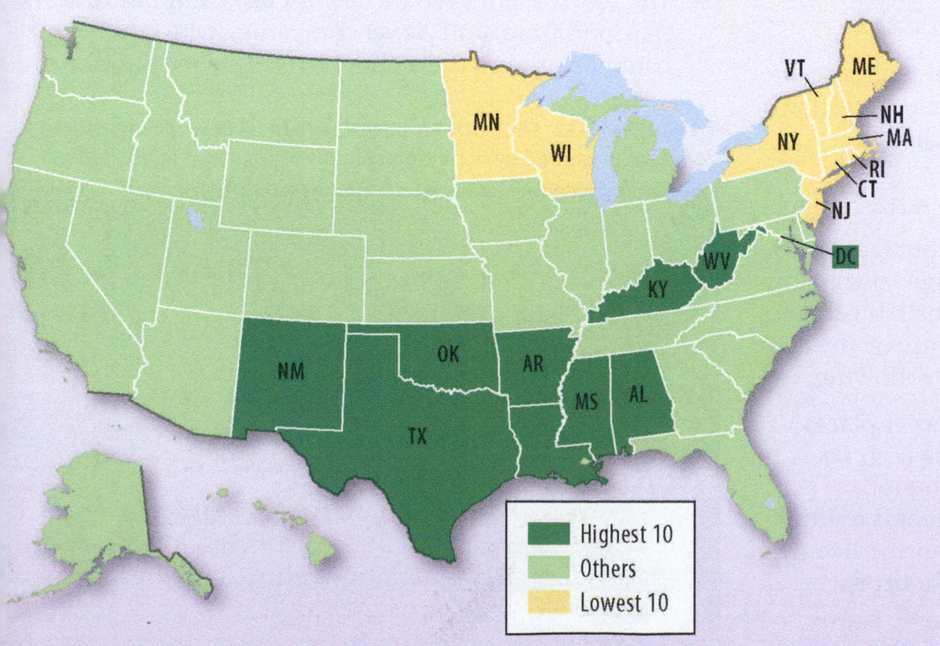

◀ **FIGURE 10-31 TEEN BIRTH RATES BY U.S. STATE**

KEY ISSUE 3

Why Do Countries Face Challenges to Development?

- Two Paths to Development
- World Trade
- Financing Development
- Development Challenges During Hard Times

LEARNING OUTCOME 10.3.1

Summarize the two paths to development.

The gap between rich and poor countries is substantial. Poorer countries lack much of what people in richer countries take for granted, such as access to electricity, safe drinking water, and paved roads. To reduce disparities between rich and poor countries, developing countries must develop more rapidly. This means increasing per capita GNI more rapidly and using the additional funds to make more rapid improvements in social and economic conditions.

Developing countries face two fundamental obstacles in trying to encourage more rapid development:

- Adopting policies that successfully promote development.
- Finding funds to pay for development.

Two Paths to Development

To promote development, developing countries choose one of two models: self-sufficiency or international trade. Each has important advantages and faces serious challenges.

SELF-SUFFICIENCY PATH

In the self-sufficiency model, countries encourage domestic production of goods, discourage foreign ownership of businesses and resources, and protect their businesses from international competition. Key elements of the self-sufficiency path to development include the following:

- Barriers limit the import of goods from other places. Three widely used barriers include setting high taxes (tariffs) on imported goods to make them more expensive than domestic goods, fixing quotas to limit the quantity of imported goods, and requiring licenses in order to restrict the number of legal importers.
- Fledgling businesses are nursed to success by being isolated from competition with large international corporations. Such insulation from the potentially adverse impacts of decisions made by businesses and governments in developed countries encourages a country's fragile businesses to achieve independence.
- Investment is spread as equally as possible across all sectors of a country's economy and in all regions.
- Incomes in the countryside keep pace with those in the city, and reducing poverty takes precedence over encouraging a few people to become wealthy consumers.

CASE STUDY: INDIA. For several decades after it gained independence from Britain in 1947, India was a leading example of the self-sufficiency strategy. Policies included limiting foreign companies from importing into India and exercising strong control over companies operating in India.

Here are examples of limitations on imports:

- To import goods into India, most foreign companies had to secure a license, which was a long and cumbersome process because several dozen government agencies had to approve each request.
- A company holding an import license was severely restricted in how much it could actually import into India.
- Heavy taxes on imported goods raised the prices that consumers had to pay.
- Indian money could not be converted to other currencies.

Here are examples of controls over Indian companies:

- A business needed government permission to sell a new product, modernize a factory, expand production, set prices, hire or fire workers, and change the job classification of existing workers (Figure 10-32).
- An unprofitable business received government subsidies, such as cheap electricity, or elimination of debts.
- The government itself owned not just communications, transportation, and power companies, which is common around the world, but also owned businesses

▼ FIGURE 10-32 SELF-SUFFICIENCY: INDIA A government official signs files presented by his secretary in his office, Chandigarh, India.

▲ **FIGURE 10-33 INTERNATIONAL TRADE: UNITED ARAB EMIRATES** Shopper in a supermarket featuring foods imported from Europe.

such as insurance companies and automakers, which are left to the private sector in most countries.

INTERNATIONAL TRADE PATH

In the international trade model, countries open themselves to foreign investment and international markets. For most of the twentieth century, self-sufficiency, or balanced growth, was the more popular of the development alternatives. International trade became more popular beginning in the late twentieth century.

The international trade model of development calls for a country to identify its distinctive or unique economic assets. What animal, vegetable, or mineral resources does the country have in abundance that other countries are willing to buy? What product can the country manufacture and distribute at a higher quality and a lower cost than other countries? According to the international trade approach, a country can develop economically by concentrating scarce resources on expansion of its distinctive local industries. The sale of these products in the world market brings into the country funds that can be used to finance other development.

ROSTOW MODEL. A pioneering advocate of the international trade approach was W. W. Rostow, who in the 1950s proposed a five-stage model of development. According to the international trade model, each country is in one of these five stages of development:

1. **Traditional society.** A traditional society has not yet started a process of development. It contains a very high percentage of people engaged in agriculture and a high percentage of national wealth allocated to what Rostow called "nonproductive" activities, such as the military and religion.
2. **Preconditions for takeoff.** An elite group initiates innovative economic activities. Under the influence of these well-educated leaders, the country starts to invest in new technology and infrastructure, such as water supplies and transportation systems. Support from international funding sources often emphasizes the importance of constructing new infrastructure. These projects will ultimately stimulate an increase in productivity.
3. **Takeoff.** Rapid growth is generated in a limited number of economic activities, such as textiles or food products. These few takeoff industries achieve technical advances and become productive, whereas other sectors of the economy remain dominated by traditional practices.
4. **Drive to maturity.** Modern technology, previously confined to a few takeoff industries, diffuses to a wide variety of industries, which then experience rapid growth comparable to the growth of the takeoff industries. Workers become more skilled and specialized.
5. **Age of mass consumption.** The economy shifts from production of heavy industry, such as steel and energy, to consumer goods, such as motor vehicles and refrigerators.

INTERNATIONAL TRADE EXAMPLES. When most developing countries were following the self-sufficiency approach during the twentieth century, two groups of countries chose the international trade approach:

- **The Four Asian Dragons.** Among the first places to adopt the international trade path were South Korea, Singapore, Taiwan, and Hong Kong, known as the "four dragons." Singapore and Hong Kong, British colonies until 1965 and 1997, respectively, were large cities surrounded by very small amounts of rural land and had virtually no natural resources. Lacking many natural resources, the four dragons promoted development by concentrating on producing a handful of manufactured goods, especially clothing and electronics. Low labor costs enabled these countries to sell products inexpensively in developed countries.
- **Petroleum-rich Arabian Peninsula states.** The Arabian Peninsula includes Saudi Arabia, the region's largest and most populous country, as well as Kuwait, Bahrain, Oman, and the United Arab Emirates. Once among the world's least developed countries, they were transformed overnight into some of the wealthiest countries, thanks to escalating petroleum prices beginning in the 1970s. Arabian Peninsula countries used petroleum revenues to finance large-scale projects, such as housing, highways, hospitals, airports, universities, and telecommunications networks. Their steel, aluminum, and petrochemical factories competed on world markets with the help of government subsidies. The landscape of these countries has been further changed by the diffusion of consumer goods, such as motor vehicles and electronics. Supermarkets in Arabian Peninsula countries are stocked with food imported from Europe and North America (Figure 10-33).

PAUSE & REFLECT 10.3.1

Many countries that have adopted the international trade model are relatively small states (see Chapter 8). Why might a nation's size be a factor in the early adoption of the international trade path?

World Trade

LEARNING OUTCOME 10.3.2

Analyze reasons for the triumph of the international trade approach to development.

Most countries embraced the international trade approach as the preferred alternative for stimulating development. Long-time advocates of self-sufficiency converted to international trade, especially during the 1990s.

INTERNATIONAL TRADE TRIUMPHS

Trade increased more rapidly than wealth (as measured by GDP) during the late twentieth and early twenty-first centuries. This is a measure of the growing importance of the international trade approach, especially in developing countries (Figure 10-34).

Optimism about the benefits of the international trade development model was based on three observations:

- Developed countries in Europe and North America were joined by others in Southern and Eastern Europe and Japan during the second half of the twentieth century. If they could become more developed by following this model, why couldn't other countries?
- Developing countries contained an abundant supply of many raw materials sought by manufacturers and producers in developed countries. In the past, European colonial powers extracted many of these resources without paying compensation to the colonies. In a global economy, the sale of these raw materials could generate funds for developing countries with which they could promote development.
- A country that concentrates on international trade benefits from exposure to the demands, needs, and preferences of consumers in other countries. To remain competitive, the takeoff industries constantly evaluate changes in international consumer preferences, marketing strategies, production engineering, and design technologies. Concern for international competitiveness in the exporting takeoff industries can filter through other sectors of the economy.

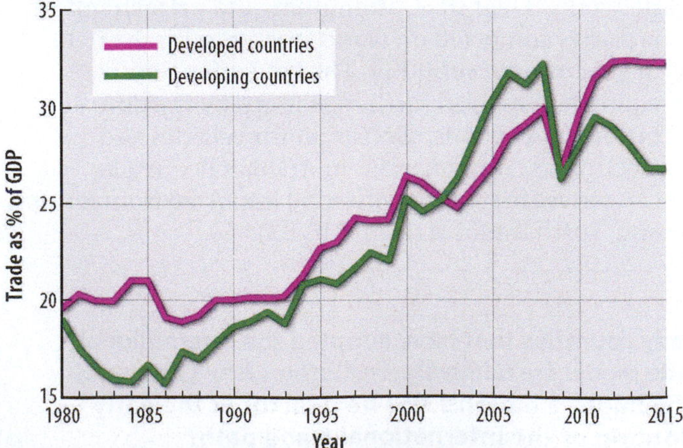

▲ **FIGURE 10-34 WORLD TRADE AS A PERCENTAGE OF GDP**

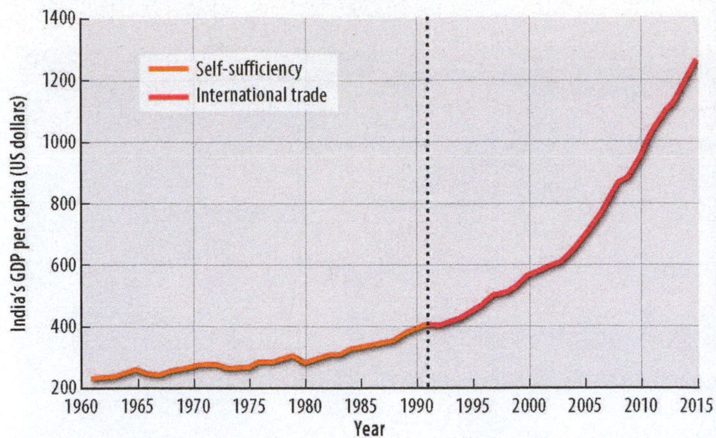

▲ **FIGURE 10-35 GDP PER CAPITA CHANGE IN INDIA** India's per capita GDP has grown more rapidly since the country converted from the self-sufficiency path to the international trade path.

Longtime advocates of the self-sufficiency approach converted to international trade during the 1990s. India, for example, dismantled its formidable collection of barriers to international trade:

- Foreign companies were allowed to set up factories and sell in India.
- Tariffs and restrictions on the import and export of goods were reduced or eliminated.
- Monopolies in communications, insurance, and other industries were eliminated.

With increased competition, Indian companies have improved the quality of their products. For example, during the self-sufficiency era, India's auto industry was dominated by Maruti-Udyog Ltd., which was controlled by the Indian government. Nursed by import duties that rose from 15 percent in 1984 to 66 percent in 1991, Maruti captured more than 80 percent of the Indian market by selling cars that would be considered out-of-date in other countries. In the international trade era, the government sold control of Maruti to the Japanese company Suzuki, which now holds only 45 percent of India's market.

Countries like India converted from self-sufficiency to international trade during the 1990s because of overwhelming evidence at the time that international trade better promoted development (Figure 10-35). After converting to international trade, India's GNI per capita increased on average 6.5 percent per year, compared to 1.8 percent per year under self-sufficiency. Worldwide, GNI increased more than 4 percent annually in countries strongly oriented toward international trade compared with less than 1 percent in countries strongly oriented toward self-sufficiency. However, conversion to international trade has not been without controversy (see Debate It feature).

WORLD TRADE ORGANIZATION

To promote the international trade development model, countries representing 97 percent of world trade established the World Trade Organization (WTO) in 1995. The WTO works to reduce barriers to international trade in two principal ways. First, through the WTO, countries negotiate reduction or elimination of international trade restrictions on manufactured goods, such as government subsidies for

▲ **FIGURE 10-36 WORLD TRADE ORGANIZATION PROTEST** Protestors outside the Department of Agriculture in the Philippines demonstrate during a speech delivered by the director of the World Trade Organization in 2015.

exports, quotas for imports, and tariffs on both imports and exports. Also reduced or eliminated are restrictions on the international movement of money by banks, corporations, and wealthy individuals.

The WTO also promotes international trade by enforcing agreements. One country can bring to the WTO an accusation that another country has violated a WTO agreement. The WTO is authorized to rule on the validity of the charge and order remedies. The WTO also protects intellectual property in the age of the Internet. An individual or a corporation can also bring charges to the WTO that someone in another country has violated a copyright or patent, and the WTO can order illegal actions to stop.

Critics have sharply attacked the WTO. Protesters routinely gather in the streets outside high-level meetings of the WTO (Figure 10-36). Progressive critics charge that the WTO is antidemocratic because decisions made behind closed doors promote the interests of large corporations rather than poor people. Conservatives charge that the WTO compromises the power and sovereignty of individual countries because it can order changes in taxes and laws that it considers unfair trading practices.

PAUSE & REFLECT 10.3.2

Top WTO officials meet every two years in a so-called ministerial conference. Where was the most recent conference held? Use the Internet to search for "WTO ministerial conference" to see if there were protests at the conference.

DEBATE IT! Should countries adopt the international trade approach?

The international trade approach has been embraced by most countries in recent years.

DO NOT EMBRACE INTERNATIONAL TRADE

- Some countries that are dependent on the sale of one commodity have suffered because the price of that commodity has not risen as rapidly as the cost of the products they need to buy.
- Building up a handful of takeoff industries that sell to people in developed countries has forced some developing countries to cut back on production of food, clothing, and other necessities for their own people.
- Countries that depend on selling low-cost manufactured goods have found that the world market for many products has declined in recent years.

▲ **FIGURE 10-37 SELF-SUFFICIENCY IN INDIA** Old Maruti car still in use as a taxi.

EMBRACE INTERNATIONAL TRADE

- Under self-sufficiency, businesses have little incentive to improve quality, lower production costs, reduce prices, or increase production.
- Companies protected from international competition are not pressured to keep up with rapid technological changes.
- The complex system needed to administer the controls requires many government employees and encourages inefficiency, abuse, and corruption.
- Potential entrepreneurs find that struggling to produce goods or offer services is less rewarding financially than advising others how to get around the complex government regulations.

▲ **FIGURE 10-38 INTERNATIONAL TRADE IN INDIA** Containers are loaded on a ship in Kandla, India's largest port.

Financing Development

LEARNING OUTCOME 10.3.3

Identify the main sources of financing development.

Developing countries lack money to fund development, so they obtain financial support from developed countries. Finance comes from two primary sources: direct investment by transnational corporations and loans from banks and international organizations.

FOREIGN DIRECT INVESTMENT

International trade requires corporations based in a particular country to invest in other countries. Investment made by a foreign company in the economy of another country is known as **foreign direct investment (FDI).** FDI has grown rapidly since the 1990s, from $130 billion in 1990 and $1.5 trillion in 2000 to $16.4 trillion in 2013 (Figure 10-39).

Investment does not flow equally around the world (Figure 10-40). Only one-third of FDI went to developing countries in 2013, while the other two-thirds went to developed countries. And FDI is not evenly distributed among developing countries. In 2013, one-third of all FDI destined for developing countries went to China, and another one-third to Singapore, Brazil, Russia, and Mexico.

The major sources of FDI are transnational corporations that invest and operate in countries other than the one in which the company headquarters are located. The United States was home to the 8 largest companies in 2014 (Apple, Exxon Mobil, Microsoft, Alphabet (Google), Berkshire Hathaway, Johnson & Johnson, Wells Fargo, and General Electric), as well as 47 of the 100 largest and 203 of the 500 largest. Another 211 of the 500 largest companies had headquarters in other developed countries. Developing countries were the location of only 86 of the 500 largest companies; 32 of them were located in China (including Hong Kong).

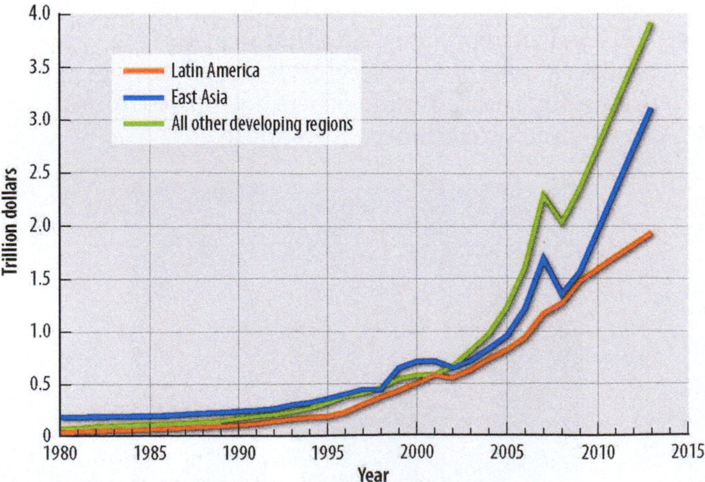

▲ **FIGURE 10-39 GROWTH IN FDI** East Asia and Latin America have the highest levels of FDI among developing regions.

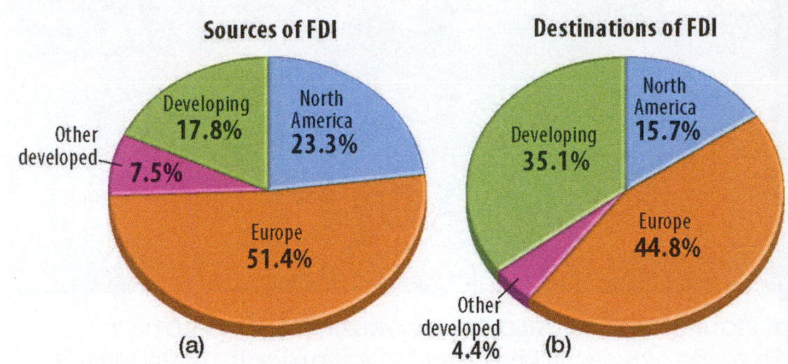

▲ **FIGURE 10-40 SOURCES AND DESTINATIONS OF FDI** Most FDI **(a)** originates in and **(b)** is destined for developed countries.

PAUSE & REFLECT 10.3.3

Why might Apple, Microsoft, and Alphabet rank as three of the world's largest transnational corporations?

LOANS

The two major lenders to developing countries are the World Bank and the International Monetary Fund (IMF):

- **World Bank.** The World Bank includes the International Bank for Reconstruction and Development (IBRD) and the International Development Association (IDA). The IBRD provides loans to countries to reform public administration and legal institutions, develop and strengthen financial institutions, and implement transportation and social service projects. The IDA provides support to poor countries considered too risky to qualify for IBRD loans. The IBRD lends money raised from sales of bonds to private investors; the IDA lends money from government contributions. Around $270 billion worth of loans are outstanding, of which around 14 percent are to India and around 5 percent each to China, Mexico, Turkey, Brazil, Indonesia, Pakistan, Bangladesh, and Vietnam (Figure 10-41).
- **International Monetary Fund (IMF).** The IMF provides loans to countries experiencing balance-of-payments problems that threaten expansion of international trade. IMF assistance is designed to help a country rebuild international reserves, stabilize currency exchange rates, and pay for imports without the imposition of harsh trade restrictions or capital controls that could hamper the growth of world trade. Unlike development banks, the IMF does not lend for specific projects. Funding of the IMF is based on each member country's relative size in the world economy.

The World Bank and IMF were conceived at a 1944 United Nations Monetary and Financial Conference in Bretton Woods, New Hampshire, to promote economic development and stability after the devastation of World War II and to avoid a repetition of the disastrous economic policies that contributed to the Great Depression of the 1930s. The IMF and World Bank became specialized agencies of the U.N. when it was established in 1945.

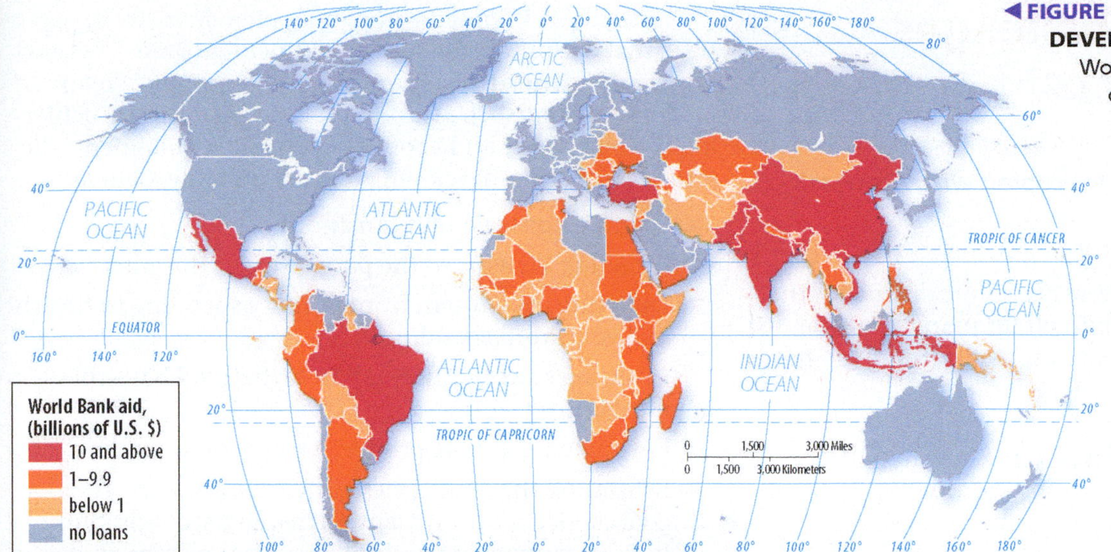

◀ **FIGURE 10-41 WORLD BANK DEVELOPMENT ASSISTANCE** Most World Bank assistance goes to countries in Asia.

Developing countries borrow money to build new infrastructure, such as hydroelectric dams, electric transmission lines, flood-protection systems, water supplies, roads, and hotels (Figure 10-42). The theory is that new infrastructure will make conditions more favorable for domestic and foreign businesses to open or expand. After all, no business wants to be located in a place that lacks paved roads, running water, and electricity.

In principle, new or expanded businesses are attracted to an area because improved infrastructure will contribute additional taxes that the developing country will use in part to repay the loans and in part to improve its citizens' living conditions. In reality, the World Bank itself has judged half of the projects it has funded in Africa to be failures. Common reasons include the following:

- Projects don't function as intended because of faulty engineering.
- Recipient nations squander or spend aid on armaments, or the aid is stolen.
- New infrastructure does not attract other investment.

▲ **FIGURE 10-42 WORLD BANK PROJECT** A road funded with support from the World Bank is constructed in Kabul, Afghanistan. **1.** Why is it important to build roads in countries with low HDIs, such as Afghanistan? **2.** Why do you think a lot of people are watching the road being built? **3.** Once the road is completed, few of the people watching are likely to have their own cars. What types of transport would you expect to see on the road after it is open?

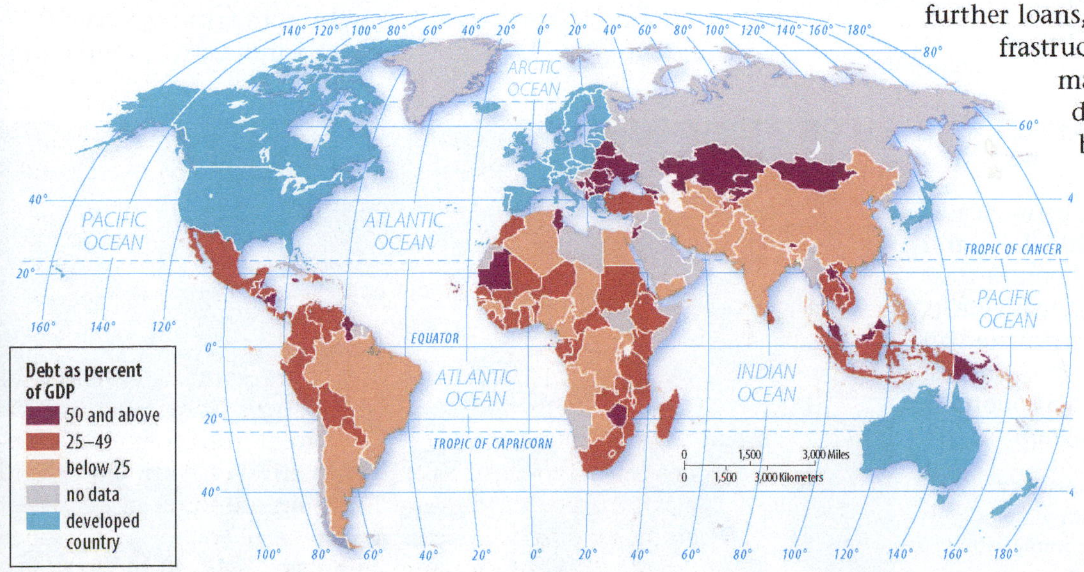

Some countries have been unable to repay the interest on their loans, let alone the principal. Debt actually exceeds annual income in a number of countries (Figure 10-43). When countries cannot repay their debts, financial institutions in developed countries refuse to make further loans, so construction of needed infrastructure stops. The inability of many countries to repay loans also damages the financial stability of banks in developed countries.

◀ **FIGURE 10-43 DEBT AS A PERCENTAGE OF GNI** Developed countries have joined developing countries in accumulating substantial debts.

Development Challenges During Hard Times

LEARNING OUTCOME 10.3.4

Explain alternate strategies for coping with economic downturns.

The lingering effects of the severe economic downturn that began in 2008 have made it difficult for many countries to make progress in development. Especially hard hit have been countries in Southern Europe.

STIMULUS OR AUSTERITY?

Political leaders and independent analysts have been sharply divided on the optimal strategy for fighting economic downturns:

- **Stimulus strategy.** Proponents of stimulus argue that during a downturn, governments should spend more money than they collect in taxes. Governments should stimulate the economy by putting people to work building bridges and other needed infrastructure projects. Once the economy recovers, they say, people and businesses will be in a position to pay more taxes and pay off the debt.
- **Austerity strategy.** Proponents of austerity argue that government should sharply reduce taxes so that people and businesses can revive the economy by spending their tax savings. Spending on government programs should be sharply cut as well in order to keep the debt from swelling and hampering the economy in the future.

In the United States, the stimulus strategy was initially employed by Presidents Bush and Obama to help the country recovery from the severe economic downtown in 2008. After the success of Tea Party candidates in 2010, more attention was paid to the austerity strategy. European countries divided between supporting stimulus and austerity. The lack of agreement has led to serious difficulties in Europe and may possibly result in the demise of the euro currency (Figure 10-44).

STRUCTURAL ADJUSTMENT PROGRAMS

Not every country gets to choose between austerity and stimulus. The IMF, World Bank, and economically healthy developed countries fear that granting, canceling, or refinancing debts without strings attached will perpetuate bad habits in economically struggling countries. Therefore, to apply for debt relief, a country is required to adopt an austerity program.

Austerity is imposed through a policy framework paper (PFP) that outlines a structural adjustment program. A **structural adjustment program** contains economic "reforms" or "adjustments," such as economic goals, strategies for achieving the objectives, and external financing requirements. Reforms required of a developing country typically include:

- Spending only what it can afford.
- Directing benefits to the poor, not just the elite.
- Diverting investment from military spending to health and education spending.
- Investing scarce resources where they will have the most impact.
- Encouraging a more productive private sector.
- Reforming the government, including making the civil service more efficient, increasing accountability in fiscal management, implementing more predictable rules and regulations, and disseminating more information to the public.

Critics charge that poverty worsens under structural adjustment programs. By placing priority on reducing government spending and inflation, structural adjustment programs may result in the following:

- Cuts in health, education, and social services that benefit the poor.
- Higher unemployment.
- Loss of jobs in state enterprises and civil service.
- Less support for those most in need, such as poor pregnant women, nursing mothers, young children, and elderly people.

In short, structural reforms allegedly punish Earth's poorest people for actions they did not commit, such as waste, corruption, misappropriation, and military buildup.

International organizations respond that the poor suffer more when a country does not undertake reforms. Economic growth is what benefits the poor the most in the long run. Nevertheless, in response to criticisms, the IMF and the World Bank now encourage innovative programs to

▶ **FIGURE 10-44 ANTI-AUSTERITY PROTEST**
Workers in Greece protest government measures to cut wages and benefits in order to comply with European Union austerity requirements.

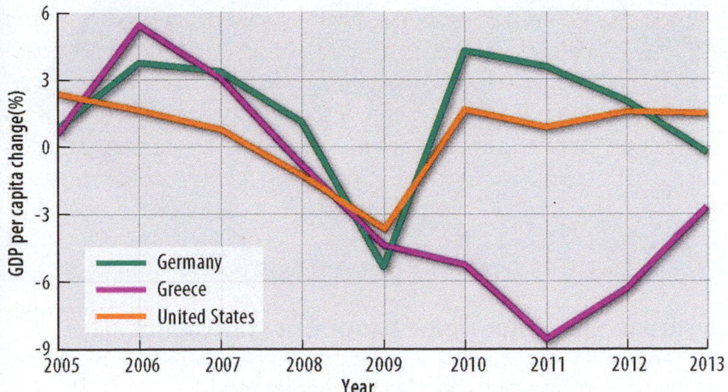

▲ **FIGURE 10-45 GDP PER CAPITA CHANGE** Prior to the severe recession in 2008, Germany and Greece showed similar economic patterns. Under austerity, Greece's economy has contracted every year since 2008. Germany's increased at first but has fallen back to virtually no growth.

reduce poverty and corruption and consult more with average citizens. A safety net must be included to ease short-term pain experienced by poor people.

PAUSE & REFLECT 10.3.4

Do the terms of a structural adjustment program seem harsh for a country or fair? Why?

EUROPE'S SOVEREIGN DEBT CRISIS

Europe has faced an especially difficult challenge in responding to the sharp economic slowdown of the early twenty-first century. Economic difficulties call into question the region's ability to continue supporting the international trade development path.

Most European countries adopted the euro as their common currency in 1999. Europeans believed that if every country in the region operated with the same currency, trade within the region would be enhanced. In reality, once the severe economic downturn hit, having each country saddled with the same currency proved to be a burden for the countries in Europe that had weaker economies.

Consider Germany and Greece. Germany has a strong economy, with businesses producing cars, electronics, and other goods at higher quality and lower cost than can be done in Greece. The export of many of Germany's products brings additional revenue flowing into the country. If Germany and Greece had two different currencies, as in the past, Greece could lower the value of its drachma currency compared to the German mark. The result would have made German goods more expensive and Greek goods less expensive, thereby stimulating economic activity in Greece. But with both Greece and Germany using the same currency, the euro, Greece no longer has the option of devaluing its currency against Germany's currency.

The Northern European countries argue that the Southern European countries with weaker economies need to adopt austerity programs. The Southern European countries argue that Northern European countries with stronger economies should fund stimulus programs that would in the long run lead to more prosperity through Europe as a whole.

Figure 10-45 shows the impact of the north–south split in Europe. Prior to the severe recession in 2008, Germany and Greece had similar patterns of economic growth. Both suffered sharp declines in GDP per capita during the severe recession. Under austerity programs, Greece's economic condition worsened; GDP per capita has declined every year since 2008. At first Germany recovered, but then it fell back to virtually no growth. Meanwhile, the United States has displayed steady though modest economic growth since 2009.

MICROFINANCE FOR DEVELOPMENT

Many would-be business owners in developing countries are too poor to qualify for regular bank loans. An alternative source of loans is **microfinance,** which is provision of small loans and other financial services to individuals and small businesses in developing countries that are unable to obtain loans from commercial banks (refer to Figure 10-25).

A prominent example of microfinance is the Grameen Bank, which was established in 1977. Based in Bangladesh, Grameen specializes in making loans to women, who make up three-fourths of the borrowers. Approximately two-thirds of the artisans providing fair trade handcrafted products are women. Often these women are mothers and the sole wage earners in the home. Women have borrowed money to buy cows, make perfume, bind books, and sell matches, mirrors, and bananas. For founding the bank, Muhammad Yunus was awarded the Nobel Peace Prize in 2006.

The Grameen Bank has made several hundred thousand loans to women in Bangladesh and neighboring South Asian countries, and only 1 percent of the borrowers have failed to make their weekly loan repayments, an extraordinarily low percentage for a bank. Several million loans have also been provided to women by the Bangladesh Rural Advancement Committee. The average loan is about $60. The smallest loan the bank has made was $1, to a woman who wanted to sell plastic bangles door to door.

CHECK-IN KEY ISSUE 3

Why Do Countries Face Challenges to Development?

- ✔ **Two paths to development are self-sufficiency and international trade.**
- ✔ **International trade has been embraced by most countries in recent decades.**
- ✔ **Development is financed through foreign direct investment, as well as loans from international organizations.**
- ✔ **Austerity and stimulus are two strategies for fighting economic downturns. International lending agencies frequently require countries to choose austerity.**

KEY ISSUE 4

Why Are Countries Making Progress in Development?

- Fair Trade Standards
- Measuring Progress

LEARNING OUTCOME 10.4.1
Explain principles of fair trade.

Fair trade has been proposed as a variation of the international trade model of development that promotes sustainability. **Fair trade** is international trade that provides greater equity to workers, small businesses, and consumers. Fair trade products are made and traded according to standards that protect workers and small businesses in developing countries.

Fair Trade Standards

The fair trade movement focuses primarily on products exported from developing countries to developed countries. Sustainability is promoted by offering better trading and working conditions for producers and workers in developing countries. Fair trade organizations, backed by consumers, raise awareness of deficiencies in conventional international production and trade and the role of fair trade in improving economic, social, and environmental conditions for producers and workers.

Three sets of standards distinguish fair trade: One set applies to producers, a second set to workers on farms and in factories, and a third set to consumers, especially those living in developed countries.

FAIR TRADE FOR PRODUCERS

Critics of international trade charge that only a tiny percentage of the price a consumer pays for a product reaches the individuals in the developing country who are responsible for making or growing it. A Haitian sewing clothing for the U.S. market, for example, earns less than 1 percent of the retail price of the garment, according to the National Labor Committee. The rest goes to wholesalers, importers, distributors, advertisers, retailers, and others who did not actually make the item. In contrast, fair trade returns on average one-third of the price to the producer in the developing country.

▲ **FIGURE 10-46 FAIR TRADE PRODUCER STANDARDS** Equal Exchange is an example of a company that is responsible for marketing of fair trade coffee, tea, chocolate, and other food.

Fair trade is a set of business practices designed to advance a number of economic, social, and environmental goals. These include:

- Raising the incomes of small-scale farmers and artisans by eliminating some of the intermediaries.
- Distributing the profits and risks associated with production and sale of goods more fairly among producers, distributors, retailers, and financiers.
- Increasing the entrepreneurial and management skills of the producers.
- Promoting safe and sustainable farming methods as well as working conditions, such as by prohibiting the use of dangerous pesticides and herbicides and by promoting the production of certified organic crops (Figure 10-46).

International fair trade organizations set standards for implementing these principles, and they monitor, audit, and certify that practices comply with the standards.

FAIR TRADE FOR WORKERS

Protection of workers' rights is not a high priority in the international trade development approach, according to its critics. Critics charge that:

- Oversight of workers' conditions by governments and international lending agencies is minimal.
- Workers allegedly work long hours in poor conditions for low pay.
- Children or forced labor may be in the workforce.
- Health problems may result from poor sanitation and injuries from inadequate safety precautions.
- Injured, ill, or laid-off workers are not compensated.

Fair trade requires the following:

- Workers must be paid fair wages—at least enough to cover food, shelter, education, health care, and other basic needs (Figure 10-47).
- Workers must be permitted to organize a union and to have the right to collective bargaining.

▲ **FIGURE 10-47 FAIR TRADE WORKER STANDARDS** Workers in a factory in Tiruppur, India, produce fair trade clothing using organically grown cotton.

- Workers must be protected by high environmental and safety standards.

Cooperatives also enable small-scale farmers and artisans to undertake fair trade production. Producer and worker cooperatives offer several advantages:

- A cooperative can qualify for credit so that funds can be borrowed to buy equipment and invest in improving farms.
- Materials can be purchased at a lower cost.
- The people who grow or make the products democratically manage allocation of resources and assure safe and healthy working conditions.
- Profits are reinvested in the community instead of going to absentee corporate owners.

FAIR TRADE FOR CONSUMERS

Most fair trade products are food, including coffee, tea, bananas, chocolate, cocoa, juice, wine, sugar, and honey. Fair trade products reach consumers primarily through cooperatively owned groceries.

A **cooperative store** is a member-owned, member-governed business that operates for the benefit of its members according to common principles agreed upon by the international cooperative community (Figure 10-48). According to international co-op principles, a co-op business:

- Is owned by the people who shop in the store rather than by a corporation.
- Is democratically governed by a member-elected board rather than appointed by a corporation owner.

▼ **FIGURE 10-48 FAIR TRADE AND CONSUMERS** A grocery store in Harwood, United Kingdom, is cooperatively owned by local shoppers.

- Is financed primarily by investments made by the member-owners rather than by commercial lenders.
- Is controlled by people living in the community rather than by distant shareholders.
- Educates the public on the importance of sustainable development rather than hiding information about poor working conditions.
- Helps other cooperatives through sharing expertise and sharing of sources of fair trade products rather than competing with other cooperatives.
- Helps to improve the community in which the cooperative is locating rather than seeking to extract the maximum profit out of the community.

The consumer-owned cooperative movement originated in the nineteenth century, in reaction to poor working conditions and inequalities during the Industrial Revolution. During the past century, consumer-owned cooperatives have flourished in three eras. These eras coincided with periods when many people felt that traditional profit-making retailers and other service providers were not meeting their needs:

- The first era was during the Great Depression of the 1930s, when people living in desperate poverty banded together to improve their lives through self-help. Cooperatives were especially important in bringing electricity to rural areas where population density was too low for private companies to serve profitably.
- The second era came in the 1960s and 1970s, as part of the counterculture movement. Young people who felt alienated from profit-making corporations banded together to open grocery stores and craft shops.
- The third—and current—era began around 2000. Hundreds of cooperatively owned grocery stores have been opened to provide places to buy healthier food than is available in large supermarkets. Local farmers, who are typically unable to sell through the large supermarket chains, have a place to sell their locally raised and grown meat and produce, as well as national brands produced without chemicals and GMOs.

Buying fair trade products from cooperatives helps consumers connect more directly with the companies and workers responsible for the items. Fair trade products do not necessarily cost the consumer more than conventionally grown or produced alternatives. Because fair trade organizations bypass exploitative intermediaries and work directly with producers, they are able to cut costs and return a greater percentage of the retail price to the producers. The cost remains the same as for traditionally traded goods, but the distribution of the cost of the product is different because the large percentage taken by intermediaries is removed from the equation.

PAUSE & REFLECT 10.4.1

Are you aware of the availability of fair trade products in your community? Is there a co-op in your community?

Measuring Progress

LEARNING OUTCOME 10.4.2

Describe ways in which differences in development have narrowed or increased.

Since the U.N. began measuring HDI in 1980, both developed and developing regions have made progress in the HDI, as well as the variables contributing to the HDI. However, progress varies among the three variables contributing to the HDI.

INDICATORS OF PROGRESS

Here is a summary of the progress made in the HDI and its components:

- **HDI.** The gap in HDI between developed countries and developing countries has narrowed since 1980 (Figure 10-49). The HDI has increased more rapidly in developing regions than in developed ones (Figure 10-50).
- **GNI per capita.** Since 1980, GNI per capita has increased much more rapidly in developed countries than in developing countries (refer to Figure 1-35).

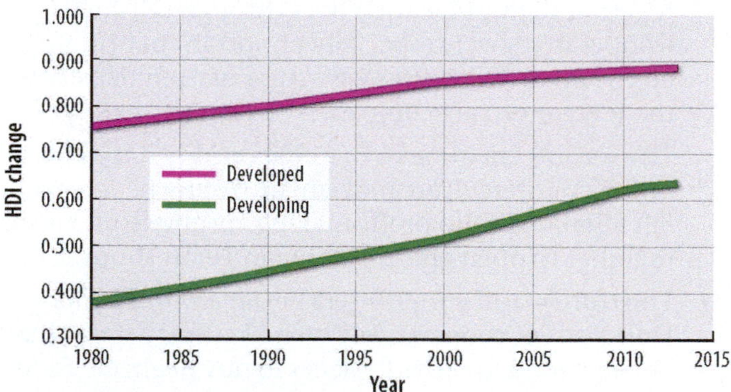

▲ **FIGURE 10-49 CHANGE IN HDI** HDI has increased in both developed and developing countries.

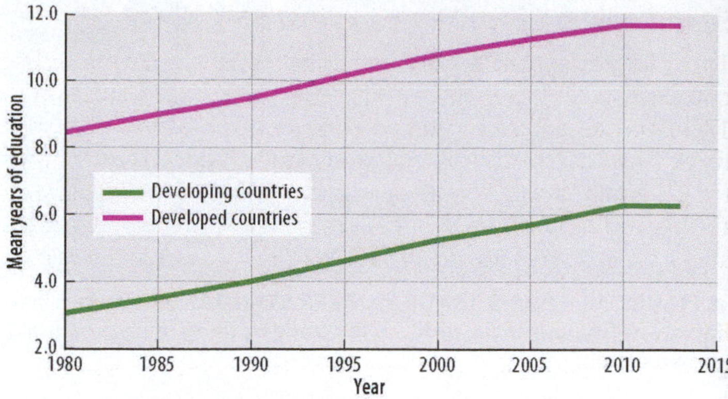

▲ **FIGURE 10-51 CHANGE IN EDUCATION** Mean years of education have increased by around three years in both developed and developing countries.

- **Education.** Since 1980, mean years of education has increased by around the same number in developed and developing countries (Figure 10-51).
- **Life expectancy.** Since 1980, life expectancy has increased by around the same number of years in developed and developing countries (Figure 10-52).

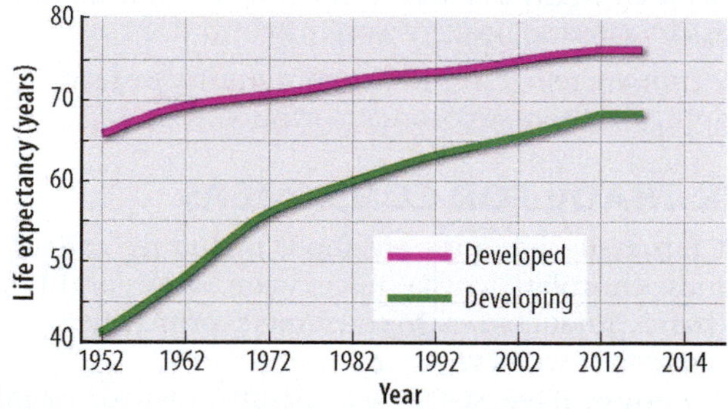

▲ **FIGURE 10-52 CHANGE IN LIFE EXPECTANCY** Life expectancy has increased in both developed and developing countries.

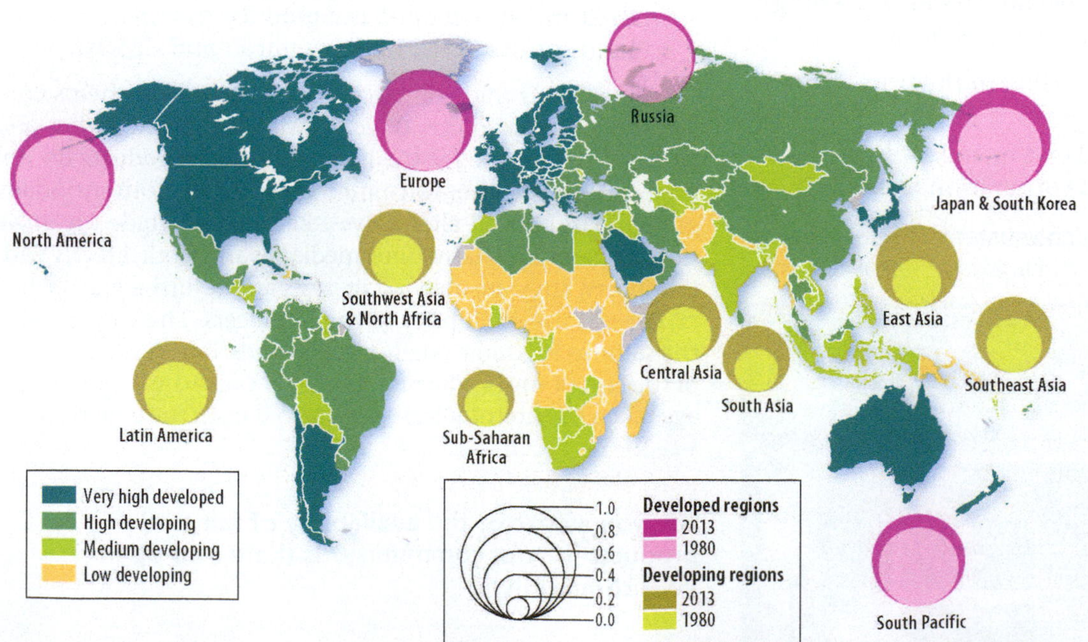

◀ **FIGURE 10-50 HDI CHANGE BY REGION** HDI has increased in all regions. Data are not available for Russia and Central Asia.

SUSTAINABLE DEVELOPMENT GOALS

To reduce disparities between developed and developing countries, the 193 members of the United Nations adopted 17 Sustainable Development Goals in 2015 (see Sustainability & Our Environment). All UN members agreed to achieve these goals by 2030. The **Sustainable Development Goals** replaced eight **Millennium Development Goals** adopted in 2002 with the goal of achieving them by 2015.

PAUSE & REFLECT 10.4.2

Which Sustainable Development Goals appear to be making the most limited progress? Why might that be the case?

CHECK-IN KEY ISSUE 4

Why Are Countries Making Progress in Development?

- ✔ **Fair trade is an alternative to the international trade model that provides greater equity to workers, small businesses, and consumers.**
- ✔ **Progress has been made in achieving development in most regions, but substantial gaps in performance between developed and developing countries persist.**

SUSTAINABILITY & OUR ENVIRONMENT Sustainable Development Goals

Following are the 17 Sustainable Development Goals that were adopted by the UN in 2015.

1. End poverty in all its forms everywhere. Poverty was cut in half between 1990 and 2010, but more than 1 billion people still live in extreme poverty.
2. End hunger, achieve food security and improved nutrition and promote sustainable agriculture. By 2030, the UN hopes to end all forms of malnutrition, double agricultural output, and ensure sustainable food production.
3. Ensure healthy lives and promote well-being for all at all ages. Infant mortality rates have declined in most regions, except sub-Saharan Africa. The maternal mortality rate declined by one-half between 1990 and 2013, but the UN hopes to see a further reduction. The number of HIV infections declined by 44 percent between 2001 and 2012, as discussed in Chapter 2, and the UN hopes to see an end to the epidemic by 2030.
4. Ensure inclusive and equitable quality education and promote lifelong learning opportunities for all. The percentage of children enrolled in primary school increased from 82 percent to 90 percent in the first decade of the twenty-first century. The goal is to have all children in school.
5. Achieve gender equality and empower all women and girls. Gender inequalities have been reduced but linger in all regions, as discussed in Key Issue 2. The UN goal is to eliminate all forms of discrimination and violence against women and girls.
6. Ensure availability and sustainable management of water and sanitation for all.
7. Ensure access to affordable, reliable, sustainable and modern energy for all. The UN goal is to increase use of alternative energy sources, as discussed in Chapter 11.
8. Promote inclusive and sustainable economic growth and full and productive employment. The UN wants an increase in GDP per capita in developing countries of at least 7 percent per year.
9. Build resilient infrastructure, promote inclusive and sustainable industrialization, and foster innovation.
10. Reduce inequality within and among countries. The UN goal is for the income of the poorest 40 percent of the population to increase faster than average.
11. Make cities and human settlements inclusive, safe, resilient and sustainable. Urban settlements are discussed in detail in Chapters 12 and 13.
12. Ensure sustainable consumption and production, such as reducing food waste and increasing recycling.
13. Take urgent action to combat climate change and its impacts. Global emissions of carbon dioxide increased by more than 50 percent between 1990 and 2015, as discussed in Chapter 11.
14. Conserve the oceans, seas and marine resources for sustainable development. Progress has been made in water quality, but overfishing is a problem.
15. Protect, restore and promote sustainable use of terrestrial ecosystems, sustainably manage forests, combat desertification, and halt and reverse land degradation and halt biodiversity loss.
16. Promote peaceful and inclusive societies for sustainable development, provide access to justice for all and build effective, accountable and inclusive institutions at all levels.
17. Strengthen implementation and revitalize the global partnership for sustainable development. Overall aid has increased from developed countries to developing countries, but little of it is targeted for the least developed countries.

Summary & Review

KEY ISSUE 1

Why does development vary among countries?

The United Nations uses the Human Development Index (HDI) to measure the level of development of every country. The HDI is based on three factors: a decent standard of living, a long and healthy life, and access to knowledge. In a more developed country, people are more likely to have adequate incomes, more schooling, and longer life expectancy.

THINKING GEOGRAPHICALLY

▲ **FIGURE 10-53 HUMAN DEVELOPMENT TREE** This winning visualization of development sheds the leaves of all countries that do not meet the development standards selected by the user.

1. The United Nations held the Human Development Data Visualization Competition in 2015. Participants were encouraged to create innovative ways to depict HDI data. The winner was the "Human Development Tree" created by Jurjen Verhagen, from the Netherlands (Figure 10-53). To view the "Human Development Tree" and the other eight finalists, enter Human Development Data Visualization Competition in your search engine, or go to http://hdr.undp.org/en/dataviz-competition. Which visualization do you think is the most effective? Is the winning visualization effective? Why?

KEY ISSUE 2

Where are inequalities in development found?

At least some degree of gender and economic inequality is found in every country. The inequality-adjusted HDI (IHDI) modifies the HDI to account for inequality within a country. The GDI measures the gender gap in the level of achievement for the three dimensions of the HDI. The GII measures the gender gap in the level of achievement in terms of reproductive health, empowerment, and the labor market.

THINKING GEOGRAPHICALLY

▲ **FIGURE 10-54 DEVELOPMENT INEQUALITY** Rickshaw and BMW share the road in Delhi, India.

2. Based on the world maps included in Key Issue 2, which two of the nine world regions appear to have the highest levels of inequality? Do these two regions have high or low HDIs?

3. What factors might account for the high levels of inequality and the distinctive HDIs in these two regions?

4. In Figure 10-54, what might be the reaction of the rickshaw driver to sharing the road with the BMW? What might be the reaction of the BMW driver to sharing the road with the rickshaw?

KEY ISSUE 3

Why do countries face challenges to development?

To promote development, developing countries choose between self-sufficiency and international trade paths. Most countries have adopted the international trade path, influenced by evidence that it gets better results than self-sufficiency. Development is financed through FDI and through international development agencies. To obtain development aid, countries are often required to adopt austerity measures that may cause short-term hardship for the population.

THINKING GEOGRAPHICALLY

▲ **FIGURE 10-55 EVIDENCE OF DEVELOPMENT** The Trans-Kalahari Corridor includes several new highways across Namibia, Botswana, and South Africa.

5. The Trans-Kalahari Corridor consists of several new roads and rail lines connecting Namibia with Botswana and South Africa. In what condition does the road appear?

6. What important elements of development can be seen in Figure 10-55, in addition to the road? Why would it be important for Namibia to have connections with South Africa?

7. As a very poor country, Namibia lacks the resources to build this infrastructure, so how do you think it got built?

KEY ISSUE 4

Why are countries making progress in development?

Countries have made considerable progress in achieving United Nations Millennium Development Goals for the first 15 years of the twenty-first century. The HDI has increased more rapidly in developing countries, whereas the GNI per capita has increased more rapidly in developed countries. Developed and developing countries have made comparable progress in education and life expectancy.

THINKING GEOGRAPHICALLY

▲ **FIGURE 10-56 DEVELOPMENT PROGRESS** A woman in Monrovia, Liberia, has opened a shop to sell fish with the help of a microfinance loan.

8. The U.N. lets you change the numbers used to calculate the HDI to see the impact on a country's level of development. Go to hdr.undp.org/and search for the *Calculating the Indices using Excel* tool. Select the HDI worksheet. Country A is an example of a high developing country. The U.N.'s cutoff for a developed country is 0.80. Change one or more of the four data columns until you get an HDI above 0.80.

9. Which of the four columns needs to change the most in order for Country A to be reclassified from developing to developed? Which of the four columns needs to change the most to reclassify Country D as developed?

KEY TERMS

Adolescent fertility rate *(p. 364)* The number of births per 1,000 women ages 15 to 19.
Cooperative store *(p. 377)* A member-owned, member-governed business that operates for the benefit of its members according to common principles agreed upon by the international cooperative community.
Developed country *(p. 350)* A country that has progressed relatively far along a continuum of development.
Developing country *(p. 350)* A country that is at a relatively early stage in the process of development.
Development *(p. 350)* A process of improvement in the conditions of people through diffusion of knowledge and technology.
Fair trade *(p. 376)* An alternative to international trade that provides greater equity to workers, small businesses, and consumers, focusing primarily on products exported from developing countries to developed countries.
Female labor force participation rate *(p. 363)* The percentage of women holding full-time jobs outside the home.
Foreign direct investment (FDI) *(p. 372)* Investment made by a foreign company in the economy of another country.
Gender-related Development Index (GDI) *(p. 360)* An indicator constructed by the U.N. to measure the gender gap in the level of achievement in terms of income, education, and life expectancy.
Gender Inequality Index (GII) *(p. 360)* An indicator constructed by the U.N. to measure the extent of each country's gender inequality in terms of reproductive health, empowerment, and the labor market.
Gross domestic product (GDP) *(p. 352)* The value of the total output of goods and services produced in a country in a year, not accounting for money that leaves and enters the country.
Gross national income (GNI) *(p. 352)* The value of the output of goods and services produced in a country in a year, including money that leaves and enters the country.
Human Development Index (HDI) *(p. 350)* An indicator constructed by the U.N. to measure the level of development for a country through a combination of income, education, and life expectancy.
Inequality-adjusted Human Development Index (IHDI) *(p. 358)* A modification of the HDI to account for inequality.
Literacy rate *(p. 355)* The percentage of a country's people who can read and write.
Maternal mortality rate *(p. 364)* The number of women who die giving birth per 100,000 births.
Microfinance *(p. 375)* Provision of small loans and financial services to individuals and small businesses in developing countries.
Millennium Development Goals *(p. 379)* Eight goals adopted by the U.N. in 2002 to reduce disparities between developed and developing countries by 2015.
Primary sector *(p. 352)* The portion of the economy concerned with the direct extraction of materials from Earth, generally through agriculture.
Productivity *(p. 353)* The value of a particular product compared to the amount of labor needed to make it.
Pupil/teacher ratio *(p. 355)* The number of enrolled students divided by the number of teachers.
Purchasing power parity (PPP) *(p. 352)* The amount of money needed in one country to purchase the same goods and services in another country.
Secondary sector *(p. 352)* The portion of the economy concerned with manufacturing useful products through processing, transforming, and assembling raw materials.
Structural adjustment program *(p. 374)* Economic policies imposed on less developed countries by international agencies to create conditions that encourage international trade.
Sustainable Development Goals *(p. 379)* Seventeen goals adopted by the U.N. in 2015 to reduce disparities between developed and developing countries by 2030.
Tertiary sector *(p. 352)* The portion of the economy concerned with transportation, communications, and utilities, sometimes extended to the provision of all goods and services to people in exchange for payment.
Value added *(p. 353)* The gross value of a product minus the costs of raw materials and energy.

GeoVideo

Log in to the **MasteringGeography** Study Area to view this video.

China: Economic Growth and Communism

In just a few decades, China has become an industrial powerhouse by following a model of growth very different from that of the United States and most other industrial countries.

1. What has been the relationship between market forces and government policy throughout China's recent rapid economic growth?
2. Describe the role of the city of Wenzhou in China's "new Industrial Revolution."
3. According to the video, what two problems must China solve to ensure continued economic growth and social stability?

EXPLORE

Niger: The World's Lowest HDI

In some years, Niger has the lowest HDI of any country.

Fly to *Arlit, Niger*.

1. Nearly the entire landscape is a desert. For a city to exist in the desert, it needs water. Where is Arlit's water being stored?

   Zoom in to around 3,000 feet eye alt.

2. What is the material of the roads and open space? Do you see any grass?
3. Set imagery date to 2004. What differences do you see in 2004 and 2014? In which direction has the town grown most rapidly?

   Zoom out to around 45,000 feet eye alt.

Niger is one of the world's leading sources of uranium. The reason for Arlit's existence is the uranium mine at the top of the image.

Click *Roads*, center the image on the uranium mine, and zoom in to 32,000 feet eye alt.

4. How might the uranium mine affect Arlit's level of development in transportation, income, employment, and environmental quality?

MasteringGeography™

Visit the Study Area in **MasteringtGeography™** to enhance your geographic literacy, spatial reasoning skills, and understanding of this chapter's content by accessing a variety of resources, including MapMaster interactive maps, videos, *In the News* RSS feeds, flashcards, web links, self-study quizzes, and an eText version of *The Cultural Landscape*.
www.masteringgeography.com

11

Industry and Energy

Industry was once highly clustered in a handful of communities within a handful of developed countries, but industry has diffused to many communities in many developing countries. Communities around the world view manufacturing jobs as a special asset, and they mourn when factories close and rejoice when they open.

Indian schoolchildren from Suraj Memorial High Senior Secondary School in Delhi, India wear oxygen masks to raise awareness regarding the dangers of air pollution.

LOCATIONS IN THIS CHAPTER

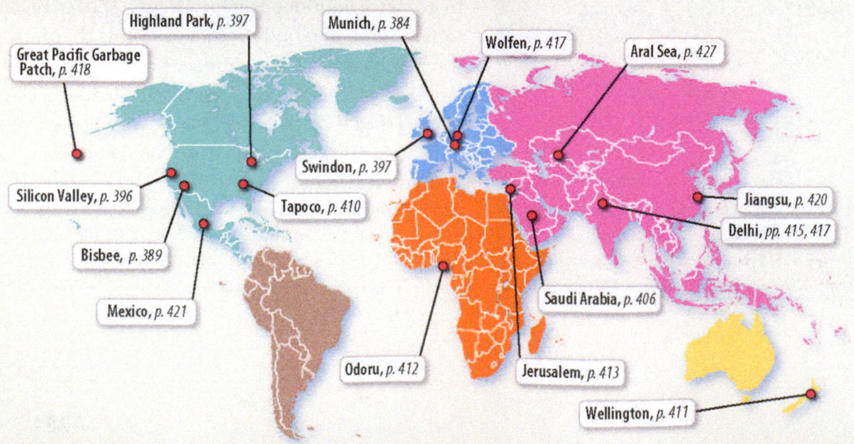

KEY ISSUES

1 Where Is Industry Distributed?

Industry is highly clustered in several ***regions***, especially North America, Europe, and East Asia.

2

Why Are Situation and Site Factors Important?

Industry is distributed in a distinctive pattern. Factors relate to a combination of the unique characteristics of a *place* and the *connections* between places.

3

Why Do Industries Face Resource Challenges?

Industries are major users of energy and major generators of pollution. Resource issues arise because the distribution of demand for them across *space* varies from the distribution of their supply.

4

Why Are Industries Changing Locations?

Industries are changing locations at two *scales*: Industries are growing in some developing regions, and they are growing within developed regions in nontraditional locations.

KEY ISSUE 1

Where Is Industry Distributed?

- Introducing Industry and Energy
- Industrial Regions

LEARNING OUTCOME 11.1.1

Understand the causes of the Industrial Revolution.

The title of this chapter refers to the manufacturing of goods in a factory. The word *industry* is appropriate because it also means *persistence* or *diligence* in creating value. A factory utilizes a large number of people as well as considerable machinery and money to turn out valuable products.

Introducing Industry and Energy

The hearth of modern industry—meaning the manufacturing of goods in a factory—was in northern England and southern Scotland during the second half of the eighteenth century. From that hearth, industry diffused to Europe and to North America in the nineteenth century and to other regions in the twentieth century.

THE INDUSTRIAL REVOLUTION

The Industrial Revolution was defined in Chapter 2 as a series of improvements in industrial technology that transformed the process of manufacturing goods. Prior to the Industrial Revolution, industry was geographically dispersed across the landscape. People made household tools and agricultural equipment in their own homes or obtained them in the local village. Home-based manufacturing was known as the **cottage industry** system.

The catalyst of the Industrial Revolution was technology, and several inventions transformed the way in which goods were manufactured. The invention most important to the development of factories was the steam engine, patented in 1769 by James Watt, a maker of mathematical instruments in Glasgow, Scotland (Figure 11-1). The large supply of steam power available from Watt's steam engines induced firms to concentrate all their process steps in one building attached to a single power source.

The revolution in industrial technology created an unprecedented expansion in productivity and resulted in substantially higher standards of living. For example, the Industrial Revolution was cited in Chapter 2 as a principal cause of population growth in stage 2 of the demographic transition. Watt's engine and other inventions enabled the United Kingdom to become the world's dominant industrial power during the nineteenth century.

The term *Industrial Revolution* is somewhat misleading:

- The transformation was far more than industrial; it resulted in new social, economic, and political inventions, not just industrial ones.
- The changes involved a gradual diffusion of new ideas and techniques over decades rather than an instantaneous revolution.

Nonetheless, the term is commonly used to define the process that began in the United Kingdom in the late 1700s.

Among the first industries impacted by the Industrial Revolution were:

- **Iron.** The first industry to benefit from Watt's steam engine was the iron tool industry. The usefulness of iron had been known for centuries, but it was difficult to produce because ovens had to be constantly heated, which was difficult before the steam engine.
- **Transportation.** First canals and then railroads enabled factories to attract large numbers of workers, bring in bulky raw materials such as iron ore and coal, and ship finished goods to consumers (Figure 11-2).
- **Textiles.** Textile production was transformed from a dispersed cottage industry to a concentrated factory system during the late eighteenth century. In 1768, Richard Arkwright, a barber and wigmaker in Preston, England, invented machines to untangle cotton prior to spinning. Too large to fit inside a cottage, spinning frames were placed inside factories near sources of rapidly flowing water, which supplied the power.

▼FIGURE 11-1 **JAMES WATT'S STEAM ENGINE** Watt built the first useful steam engine, which could pump water far more efficiently than the watermills then in common use, let alone human or animal power. Steam injected in the cylinder pushes a piston attached to a crankshaft (top of the engine) that turns the large wheel (right side of the engine), which in turn drives machinery. This Watt steam engine is on display in the Deutsches Museum in Munich, Germany.

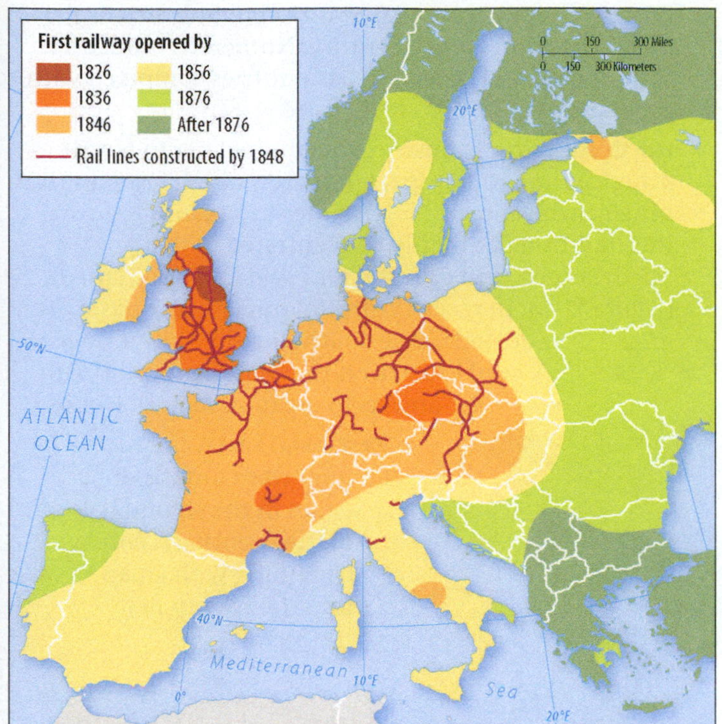

▲ **FIGURE 11-2 DIFFUSION OF THE INDUSTRIAL REVOLUTION** The construction of railroads in the United Kingdom and on the European continent reflects the diffusion of the Industrial Revolution. Europe's political problems impeded the diffusion of the railroad. Cooperation among small neighboring states was essential to build an efficient rail network and to raise money for constructing and operating the system. Because such cooperation could not be attained, railroads in some parts of Europe were delayed 50 years after their debut in the U.K.

- **Chemicals.** The chemical industry was created to bleach and dye cloth. In 1746, John Roebuck and Samuel Garbett established a factory to bleach cotton with sulfuric acid obtained from burning coal. When combined with various metals, sulfuric acid produced another acid called vitriol, which was useful for dying clothing.
- **Food processing.** In 1810, French confectioner Nicolas Appert started canning food in glass bottles sterilized in boiling water. Canned food was essential to feed the factory workers who no longer lived on farms.

FOSSIL FUELS

Manufacturing depends on the availability of abundant low-cost energy. Developed countries use large quantities of energy to run factories, as well as to produce food, keep homes comfortable, and transport people and goods. Developing countries expect to use more energy to improve the lives of their citizens.

Historically, people relied primarily on **animate power,** which is power supplied by animals or by people themselves. Animate power was supplemented by **biomass fuel** (such as wood, plant material, and animal waste), which is burned directly or converted to charcoal, alcohol, or methane gas. Biomass remains an important source of fuel in some developing countries, but during the past 200 years, developed countries have converted primarily to energy from fossil fuels.

A **fossil fuel** is an energy source formed from the residue of plants and animals buried millions of years ago. As sediment accumulated over these remains, intense pressure and chemical reactions slowly converted them into the fossil fuels that are currently used. When these substances are burned, energy that was stored in plants and animals millions of years ago is released. Five-sixths of the world's energy needs are supplied by three fossil fuels (Figure 11-3):

- **Coal.** As North America and Europe developed rapidly in the late 1800s, coal supplanted wood as the leading energy source in these regions.
- **Petroleum.** First pumped in 1859, petroleum did not become an important source of energy until the diffusion of motor vehicles in the twentieth century.
- **Natural gas.** Originally burned off as a waste product of petroleum drilling, natural gas is now used to heat homes and to produce electricity.

In Chapter 1, we distinguished between renewable resources (those produced in nature more rapidly than they are consumed by humans) and nonrenewable resources (those produced in nature more slowly than they are consumed by humans). Because the fossil fuels are nonrenewable, sustainable development will necessitate increased reliance in the future on renewable energy instead of fossil fuels.

PAUSE & REFLECT 11.1.1

The use of one of the three fossil fuels is increasing in the United States in the twenty-first century. Which one is it? Why might that be the case?

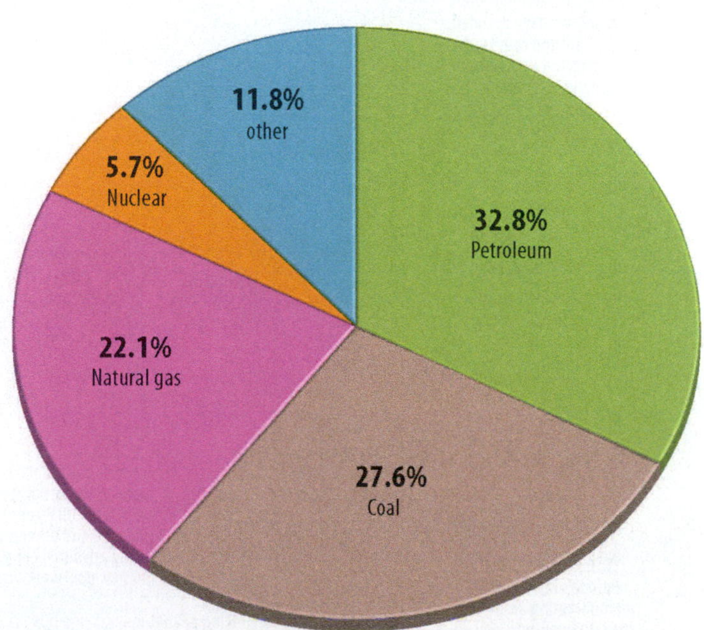

▲ **FIGURE 11-3 FOSSIL FUEL DEMAND** Petroleum, coal, and natural gas account for most of the world's energy consumption.

Industrial Regions

LEARNING OUTCOME 11.1.2

Describe the locations of the three principal industrial regions.

Industry is not distributed uniformly around the world (Figure 11-4). Rather, it is concentrated in three of the nine world regions discussed in Chapter 10:

- **Europe.** Europe was the first region to industrialize, during the nineteenth century. Numerous industrial centers emerged in Europe as countries competed with each other for supremacy (Figure 11-5).
- **North America.** Industry arrived a bit later in North America than in Europe, but it grew much faster in the nineteenth century. North America's manufacturing was traditionally highly concentrated in the northeastern United States and southeastern Canada (Figure 11-6). In recent years, manufacturing has relocated to the South, lured by lower wages and legislation that has made it difficult for unions to organize factory workers.
- **East Asia.** East Asia became an important industrial region in the second half of the twentieth century, beginning with Japan. Into the twenty-first

ARCTIC OCEAN
PACIFIC OCEAN
ATLANTIC OCEAN
ATLANTIC OCEAN
INDIAN OCEAN
PACIFIC OCEAN
TROPIC OF CANCER
EQUATOR
TROPIC OF CAPRICORN
0 1,500 3,000 Miles
0 1,500 3,000 Kilometers
Industrial output
$1 trillion and above
$1 billion–$999 billion
below $1 billion

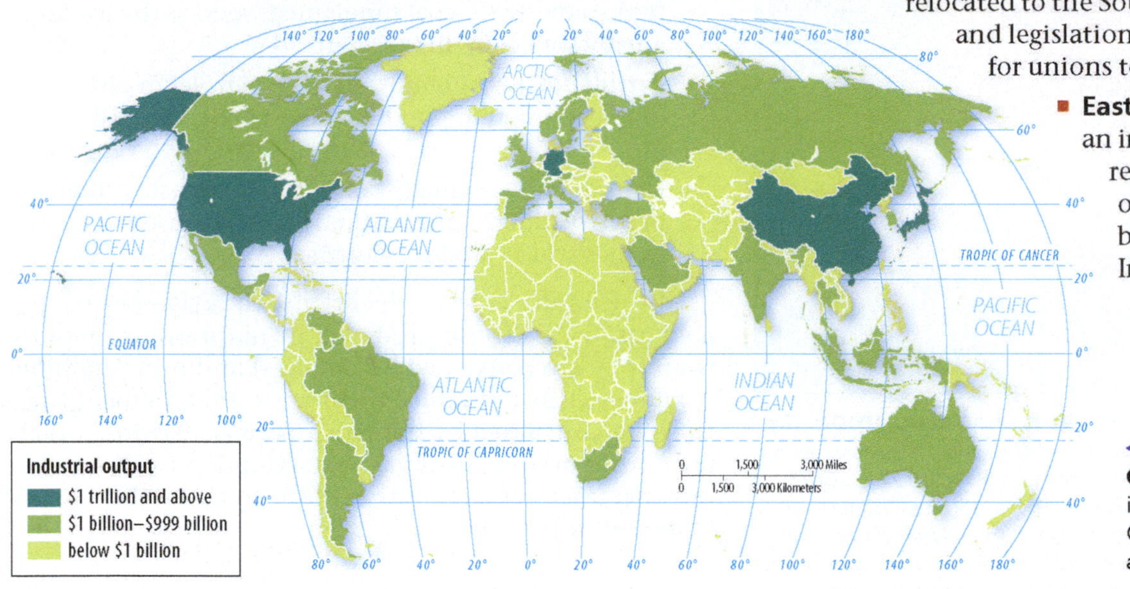

◀ **FIGURE 11-4 INDUSTRIAL OUTPUT** One-half of the world's industrial output is clustered in China, the United States, Japan, and Germany.

▼ **FIGURE 11-5 EUROPE'S INDUSTRIAL AREAS**

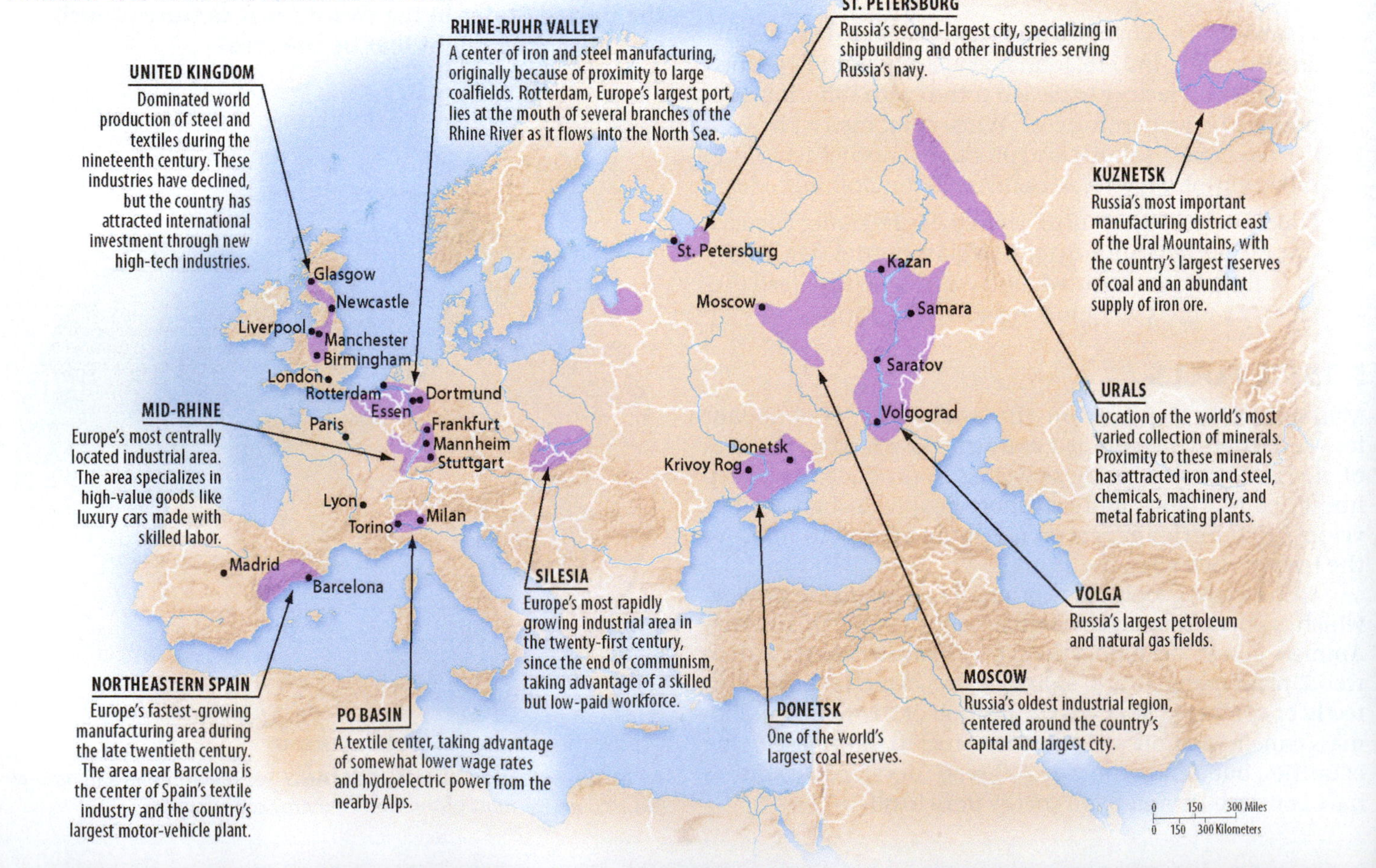

UNITED KINGDOM
Dominated world production of steel and textiles during the nineteenth century. These industries have declined, but the country has attracted international investment through new high-tech industries.
RHINE-RUHR VALLEY
A center of iron and steel manufacturing, originally because of proximity to large coalfields. Rotterdam, Europe's largest port, lies at the mouth of several branches of the Rhine River as it flows into the North Sea.
ST. PETERSBURG
Russia's second-largest city, specializing in shipbuilding and other industries serving Russia's navy.
KUZNETSK
Russia's most important manufacturing district east of the Ural Mountains, with the country's largest reserves of coal and an abundant supply of iron ore.
MID-RHINE
Europe's most centrally located industrial area. The area specializes in high-value goods like luxury cars made with skilled labor.
URALS
Location of the world's most varied collection of minerals. Proximity to these minerals has attracted iron and steel, chemicals, machinery, and metal fabricating plants.
VOLGA
Russia's largest petroleum and natural gas fields.
SILESIA
Europe's most rapidly growing industrial area in the twenty-first century, since the end of communism, taking advantage of a skilled but low-paid workforce.
NORTHEASTERN SPAIN
Europe's fastest-growing manufacturing area during the late twentieth century. The area near Barcelona is the center of Spain's textile industry and the country's largest motor-vehicle plant.
PO BASIN
A textile center, taking advantage of somewhat lower wage rates and hydroelectric power from the nearby Alps.
DONETSK
One of the world's largest coal reserves.
MOSCOW
Russia's oldest industrial region, centered around the country's capital and largest city.
Glasgow
Newcastle
Liverpool
Manchester
Birmingham
London
Rotterdam
Essen
Dortmund
Paris
Frankfurt
Mannheim
Stuttgart
Lyon
Torino
Milan
Madrid
Barcelona
St. Petersburg
Moscow
Kazan
Samara
Saratov
Volgograd
Donetsk
Krivoy Rog
0 150 300 Miles
0 150 300 Kilometers

century, China has emerged as the world's leading manufacturing country by most measures (Figure 11-7).

Four countries account for one-half of the world's industrial output: China, the United States, Japan, and Germany.

PAUSE & REFLECT 11.1.2

Compare the world's principal industrial areas with population distribution (Figure 2-4). Are industrial areas generally in regions of high population density or low? Why?

CHECK-IN KEY ISSUE 1

Where Is Industry Distributed?

- ✔ **The Industrial Revolution involved a series of improvements that transformed manufacturing. Most of the improvements occurred first in the United Kingdom.**
- ✔ **Industry depends on access to abundant and affordable suppliers of fossil fuels.**
- ✔ **The world's three principal industrial regions are Europe, North America, and East Asia.**

▶ FIGURE 11-6 NORTH AMERICA'S INDUSTRIAL REGIONS

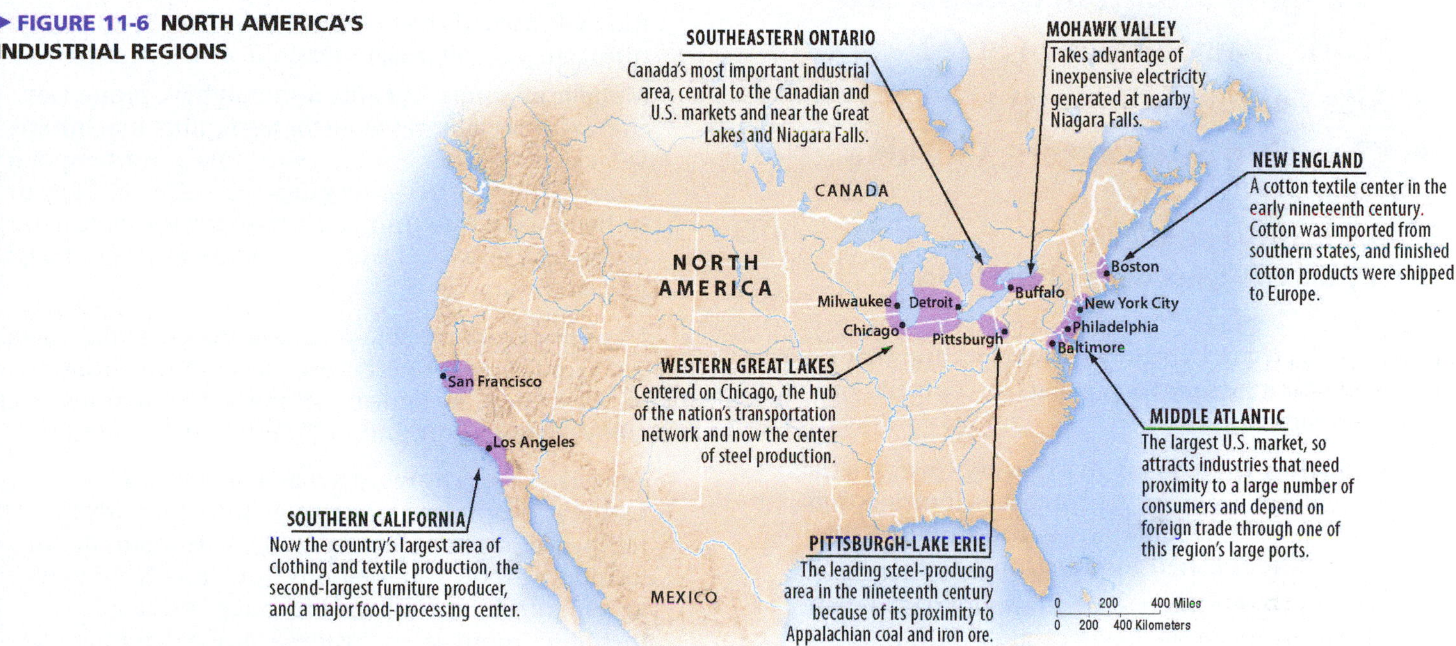

◀ FIGURE 11-7 EAST ASIA'S INDUSTRIAL REGIONS

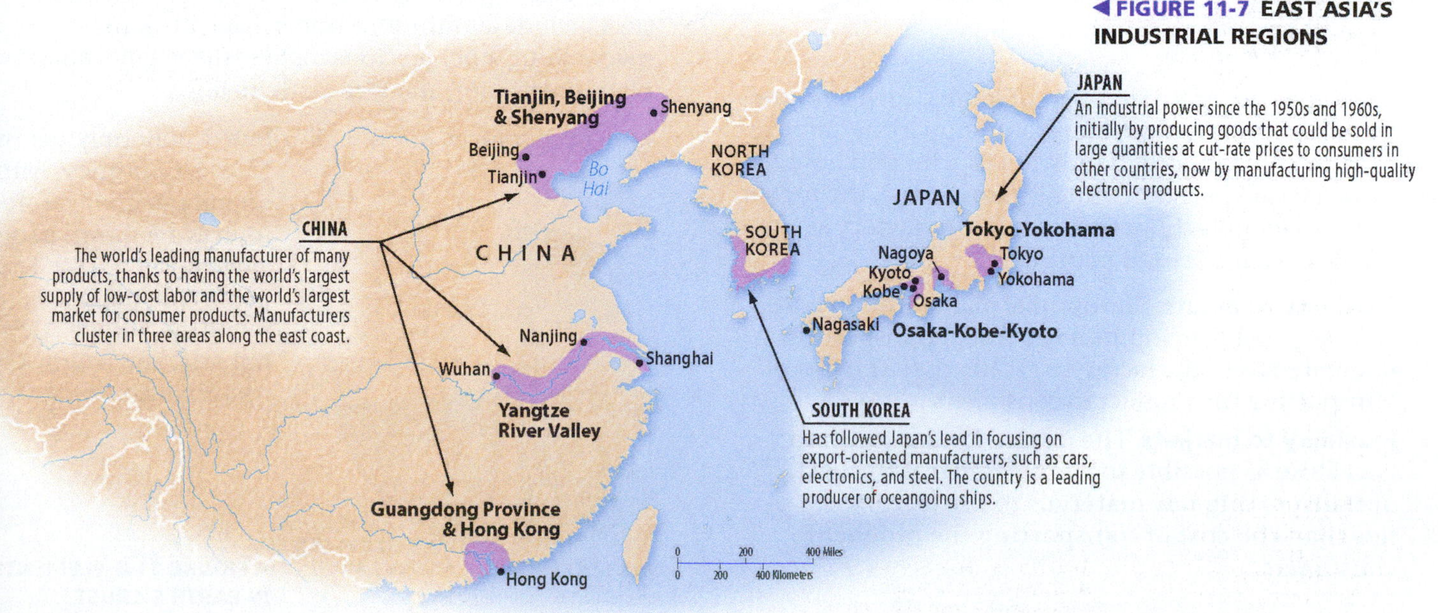

KEY ISSUE 2

Why Are Situation and Site Factors Important?

- Situation Factors: Proximity to Inputs
- Situation Factors: Proximity to Markets
- Changing Situation Factors: Steel
- Truck, Train, Ship, or Plane?
- Site Factors in Industry
- Changing Site Factors: Clothing

LEARNING OUTCOME 11.2.1

Identify the two types of situation factors.

Geographers try to explain why one location may be more profitable for a factory than others. A company ordinarily faces two geographic costs:

- **Situation factors** involve transporting materials to and from a factory. A firm seeks a location that minimizes the cost of transporting inputs to the factory and finished goods to consumers.
- **Site factors** result from the unique characteristics of a location. These are labor, capital, and land.

Situation Factors: Proximity to Inputs

Manufacturers buy from suppliers of inputs, such as minerals, materials, energy, machinery, and supporting services. They sell to companies and individuals who purchase the products. The farther something is transported, the higher the cost, so a manufacturer tries to locate its factory as close as possible to its inputs and markets:

- **Proximity to inputs.** The optimal plant location is as close as possible to inputs if the cost of transporting raw materials to the factory is greater than the cost of transporting the product to consumers.
- **Proximity to markets.** The optimal plant location is as close as possible to the customer if the cost of transporting raw materials to the factory is less than the cost of transporting the product to consumers.

MINERAL RESOURCES

Minerals are especially important inputs for many industries. Earth has 92 natural elements, but about 99 percent of the crust is composed of 8 of them (Figure 11-8). The 8 most common elements combine with thousands of rare ones to form approximately 3,000 different minerals, all with their own properties of hardness, color, and density, as well as spatial distribution. Many of these minerals have important industrial uses.

Minerals are either nonmetallic or metallic:

- **Nonmetallic minerals.** By weight, more than 90 percent of the minerals that humans use are nonmetallic. Important nonmetallic minerals include building stones, gemstones such as diamonds, and minerals used in the manufacture of fertilizers such as nitrogen, phosphorus, potassium, calcium, and sulfur.
- **Metallic minerals.** Metallic minerals have properties that are especially valuable for fashioning machinery, vehicles, and other essential elements of contemporary society. They are to varying degrees malleable (able to be hammered into thin plates) and ductile (able to be drawn into fine wire) and are good conductors of heat and electricity.

Many metals are capable of combining with other metals to form alloys with distinctive properties that are important for industry. Alloys are known as ferrous or nonferrous. A ferrous alloy contains iron, and a nonferrous one does not:

- **Ferrous alloys.** The word *ferrous* comes from the Latin for "iron." Iron is extracted from iron ore, by far the world's most widely used ore. Humans began fashioning tools and weapons from iron 4,000 years ago. Important metals used to make ferrous alloys include chromium, manganese, molybdenum, nickel, tin, titanium, and tungsten.
- **Nonferrous alloys.** Important metals utilized to manufacture products that don't contain iron and steel include aluminum, copper, lead, lithium, magnesium, zinc, precious metals (silver, gold, and the platinum group), and rare earth metals.

Mineral resources are not distributed uniformly across Earth. Countries with important mineral resources are shown in orange in Figure 11-9. Few important minerals are found in Europe, Central Asia, and Southwest Asia & North Africa.

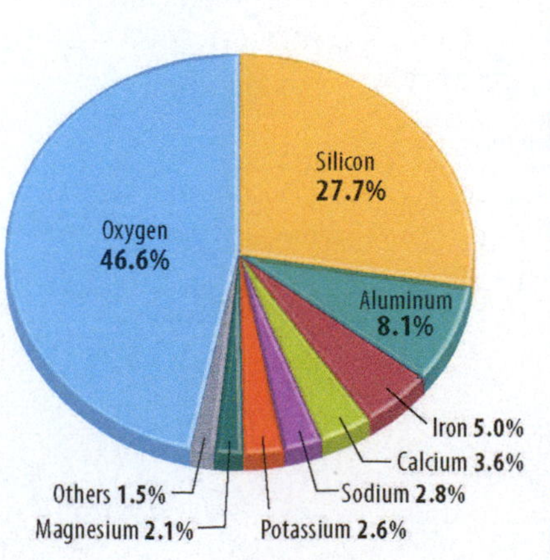

◀ **FIGURE 11-8 ELEMENTS IN EARTH'S CRUST**

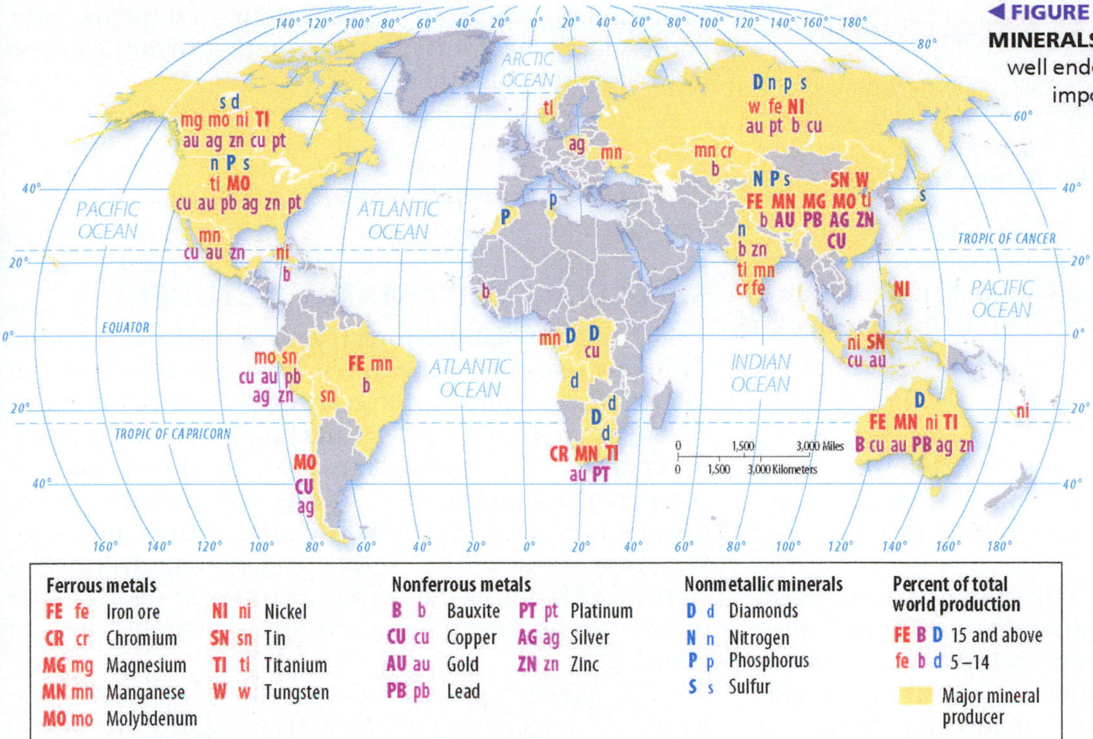

FIGURE 11-9 DISTRIBUTION OF MINERALS Australia and China are especially well endowed with minerals that are important for industry.

PROXIMITY TO INPUTS: COPPER INDUSTRY

An industry in which the inputs weigh more than the final products is a **bulk-reducing industry.** To minimize transport costs, a bulk-reducing industry locates near its sources of inputs. An example of a bulk-reducing industry is most of the steps of copper production.

Copper production involves several steps:

1. **Mining.** Mining in general is bulk reducing because the heavy, bulky ore extracted from mines is mostly waste, known as gangue (Figure 11-10).
2. **Concentration.** The ore is crushed and ground into fine particles, mixed with water and chemicals, filtered, and dried. Concentration mills are near copper mines because concentration transforms the heavy, bulky copper ore into a product of much higher value per weight.
3. **Smelting.** The concentrated copper becomes the input for smelters, which remove more impurities. Because smelting is a bulk-reducing industry, smelters are built near their main inputs—the concentration mills—again to minimize transportation costs.

FIGURE 11-10 COPPER MINE Bisbee, Arizona.

4. **Refining.** The purified copper produced by smelters is treated at refineries to produce copper cathodes, about 99.99 percent pure copper. Most refineries are located near smelters.
5. **Manufacturing.** Copper that is ready for use in other products is produced in foundries.

Figure 11-11 shows the distribution of the U.S. copper industry. Mining, concentration, smelting, and refining are bulk-reducing industries; because two-thirds of U.S. copper is mined in Arizona, the state also has most of the concentration mills, smelters, and refineries. The first four steps are good examples of bulk-reducing activities that need to be located near their sources of inputs. The fifth step—manufacturing—is not bulk reducing, so foundries are located near markets on the East and West coasts rather than near inputs.

PAUSE & REFLECT 11.2.1

What is an example of a product purchased by consumers that is made of copper?

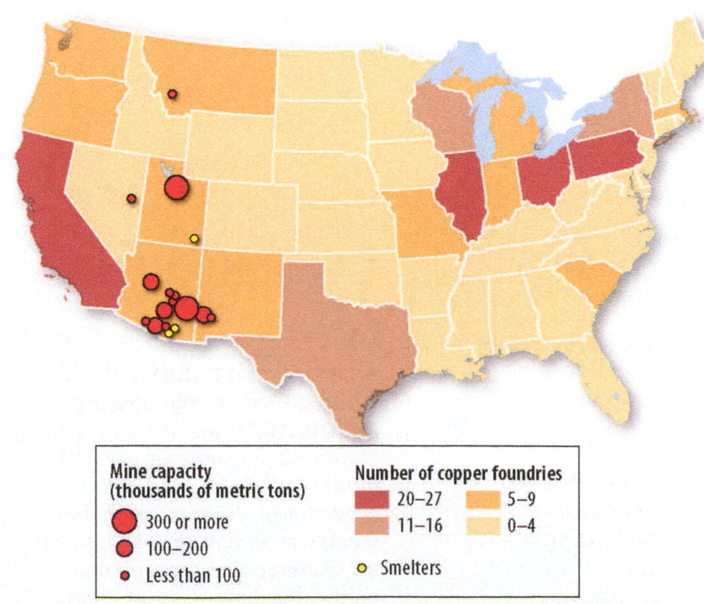

FIGURE 11-11 DISTRIBUTION OF COPPER PRODUCTION

Situation Factors: Proximity to Markets

LEARNING OUTCOME 11.2.2

Explain why some industries locate near markets.

For many firms, the optimal location is close to customers. Proximity to markets is a critical locational factor for three types of industries: bulk-gaining industries, single-market manufacturers, and perishable-products companies.

BULK-GAINING INDUSTRIES

A **bulk-gaining industry** makes something that gains volume or weight during production. To minimize transport costs, a bulk-gaining industry needs to locate near where the product is sold.

A prominent example of a bulk-gaining industry is the fabrication of parts and machinery from steel and other metals. A fabricated-metal factory brings together metals such as steel and previously manufactured parts as the main inputs and transforms them into a more complex product. Fabricators shape individual pieces of metal using such processes as bending, forging (hammering or rolling metal between two dies), stamping (pressing metal between two dies), and forming (pressing metal against one die). Separate parts are joined together through welding, bonding, and fastening with bolts and rivets.

Beverage bottling is another good example of a bulk-gaining industry. In this case, the product gains weight (Figure 11-12). Because water is the principal ingredient in beer or cola, a filled container is much heavier than an empty one. Shipping filled containers is more expensive than shipping empty ones, so to minimize shipping costs, bottlers locate near their customers rather than the manufacturers of the containers.

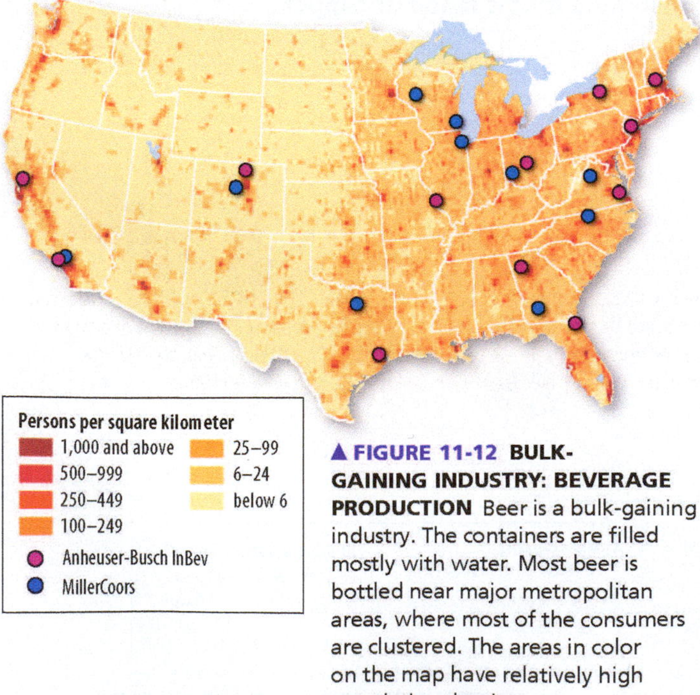

▲ **FIGURE 11-12 BULK-GAINING INDUSTRY: BEVERAGE PRODUCTION** Beer is a bulk-gaining industry. The containers are filled mostly with water. Most beer is bottled near major metropolitan areas, where most of the consumers are clustered. The areas in color on the map have relatively high population density.

PAUSE & REFLECT 11.2.2

Why isn't wine bottled near the market, as beer and cola are?

SINGLE-MARKET MANUFACTURERS

A single-market manufacturer is a specialized manufacturer with only one or two customers. The optimal location for these factories is often in close proximity to the customers.

An example of a single-market manufacturer is a producer of buttons, zippers, clips, pins, or other specialized components attached to clothing. The clothing manufacturer may need additional supplies of these pieces on very short notice. The world's largest manufacturer of zippers, YKK, for example, has factories in 68 countries in order to be near its customers, the manufacturers of clothing.

The makers of parts for motor vehicles are another example of specialized manufacturers with only one or two customers—the major motor vehicle producers, such as GM and Toyota. Carmakers' assembly plants account for only around 30 percent of the value of the vehicles that bear their names. Independent parts makers supply the other 70 percent of the value. In the past, most motor vehicle parts were made in Michigan and shipped to nearby warehouses and distribution centers maintained in that state by the major producers. From the warehouses, the producers sent the parts to plants around the country where the vehicles were assembled. Parts makers now ship most of their products directly to assembly plants and are therefore more likely than in the past to cluster near the final assembly plants.

PERISHABLE-PRODUCTS COMPANIES

To deliver their products to consumers as rapidly as possible, perishable-products industries must be located near their markets. Because few people want stale bread or sour milk, food producers such as bakers and milk bottlers must locate near their customers to assure rapid delivery. Processors of fresh food into frozen, canned, and preserved products can, however, locate far from their customers. Cheese and butter, for example, are manufactured in Wisconsin because rapid delivery to the urban markets is not critical for products with a long shelf life, and the area is well suited agriculturally for raising dairy cows.

MOTOR VEHICLE PRODUCTION AND SALES

The motor vehicle is a prominent example of a fabricated metal product that is likely to be built near its market. Around 90 million new vehicles are sold annually worldwide. China accounts for 23.5 million or 27 percent of those sales, other

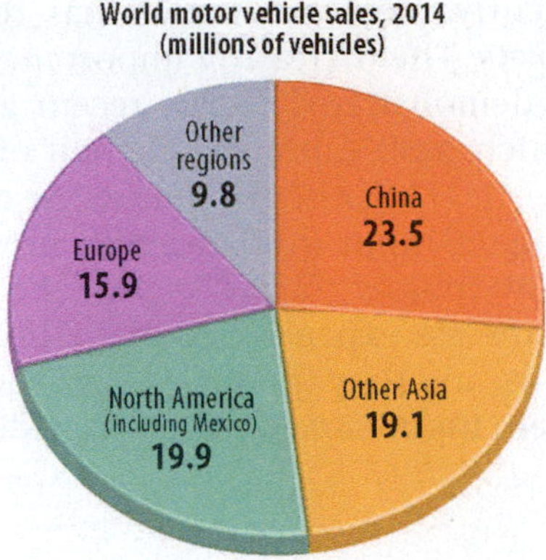

▲ **FIGURE 11-13 MOTOR VEHICLE SALES** China, the rest of Asia, North America, and Europe account for nearly 90 percent of world vehicles sales.

Asian countries 22 percent, North America (including Mexico) 23 percent, and Europe 17 percent (Figure 11-13).

Not surprisingly, in view of the importance of producing vehicles near their markets, the regional distribution of production is extremely close to that for sales (Figure 11-14). China and the rest of Asia each have 26 percent of world production, and Europe and North America (including Mexico) each have 19 percent. For example, around 80 percent of vehicles sold in North America are produced in North America. Around 10 percent come from Japan and 10 percent from elsewhere (Figure 11-15). Similarly, most vehicles sold in Europe are assembled in Europe, most vehicles sold in Japan are assembled in Japan, and most vehicles sold in China are assembled in China.

While most vehicles are produced near where they are sold, the nationality of the manufacturers is less likely to be local. Eight carmakers sold at least 4 million vehicles in 2014 and together accounted for 70 percent of the world's sales. These were:

- Two based in North America: Ford and GM.
- Three based in Europe: Germany's Volkswagen, France's Renault (which controls Nissan), and Italy's Fiat Chrysler.
- Three based in East Asia: Japan's Toyota and Honda, and South Korea's Hyundai.

These carmakers operate assembly plants in all three major industrial regions.

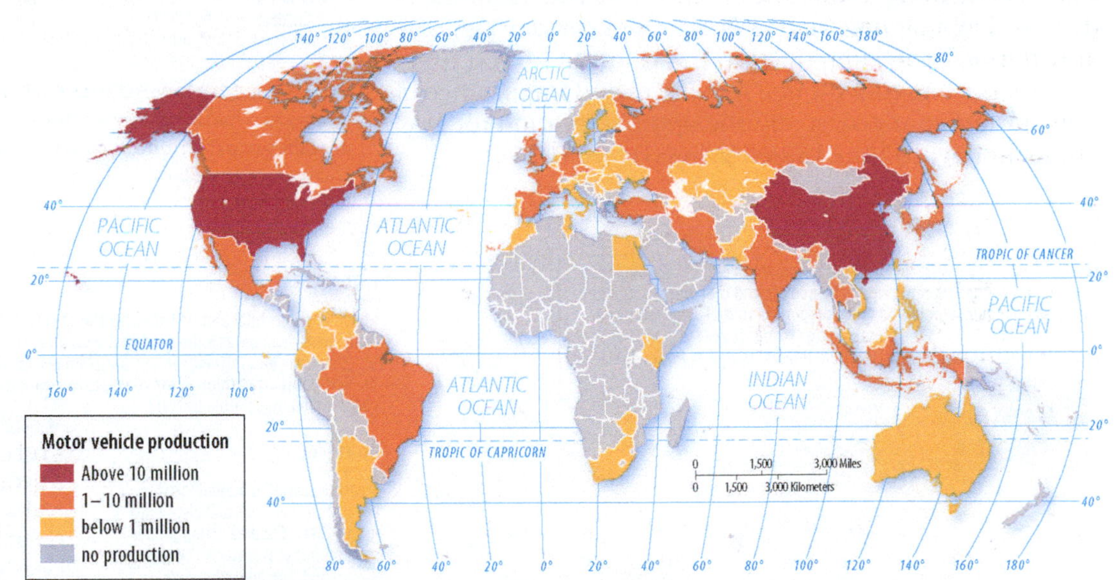

► **FIGURE 11-14 MOTOR VEHICLE PRODUCTION** The regional distribution of vehicle production closely matches the regional distribution of sales.

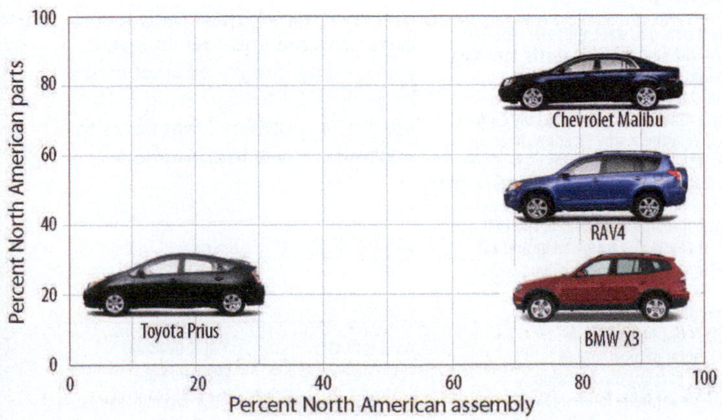

◄ **FIGURE 11-15 "AMERICAN" AND "FOREIGN" CARS** The x-axis shows the percentage of these vehicles sold in North America that were assembled in North America in 2015. The y-axis shows the percentage of North American parts in these vehicles.

Changing Situation Factors: Steel

LEARNING OUTCOME 11.2.3

Describe how the optimal location for steel production has changed.

Steel is an alloy of iron that is manufactured by removing impurities in iron, such as silicon, phosphorus, sulfur, and oxygen, and adding desirable elements, such as manganese and chromium. The two principal inputs in steel production are iron ore and coal. Most steel was produced at large integrated mill complexes. They processed iron ore, converted coal into coke, converted the iron into steel, and formed the steel into sheets, beams, rods, or other shapes.

CHANGING DISTRIBUTION OF U.S. STEEL

In the past, steel production was a good example of a bulk-reducing industry that located near its inputs. Because of the need for large quantities of bulky, heavy iron ore and coal, steelmaking traditionally clustered near sources of the two key raw materials. Within the United States, the distribution of steel production has changed several times because of changing inputs. As sources and importance of these inputs changed, so did the optimal location for steel production within the United States (Figure 11-16).

More recently, steel production has relocated to be closer to markets. The increasing importance of proximity to markets is demonstrated by the recent growth of steel minimills, which have captured one-half of the U.S. steel market (Figure 11-17). Rather than iron ore and coal, scrap metal is the main input into minimill production. Minimills, generally limited to one step in the process—steel production—are less expensive to build and operate than integrated mills, and they can locate near their markets because their main input—scrap metal—is widely available.

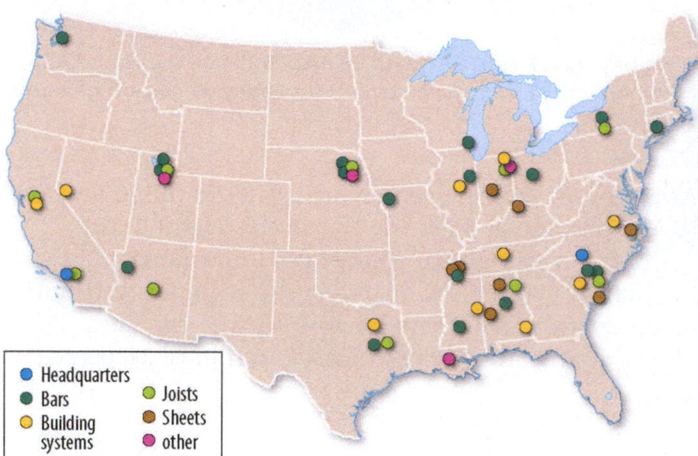

▲ **FIGURE 11-17 MINIMILLS** Minimills, which produce steel from scrap metal, are more numerous than integrated steel mills, and they are distributed around the country near local markets. Shown are the plants of Nucor, the largest minimill operator in the United States.

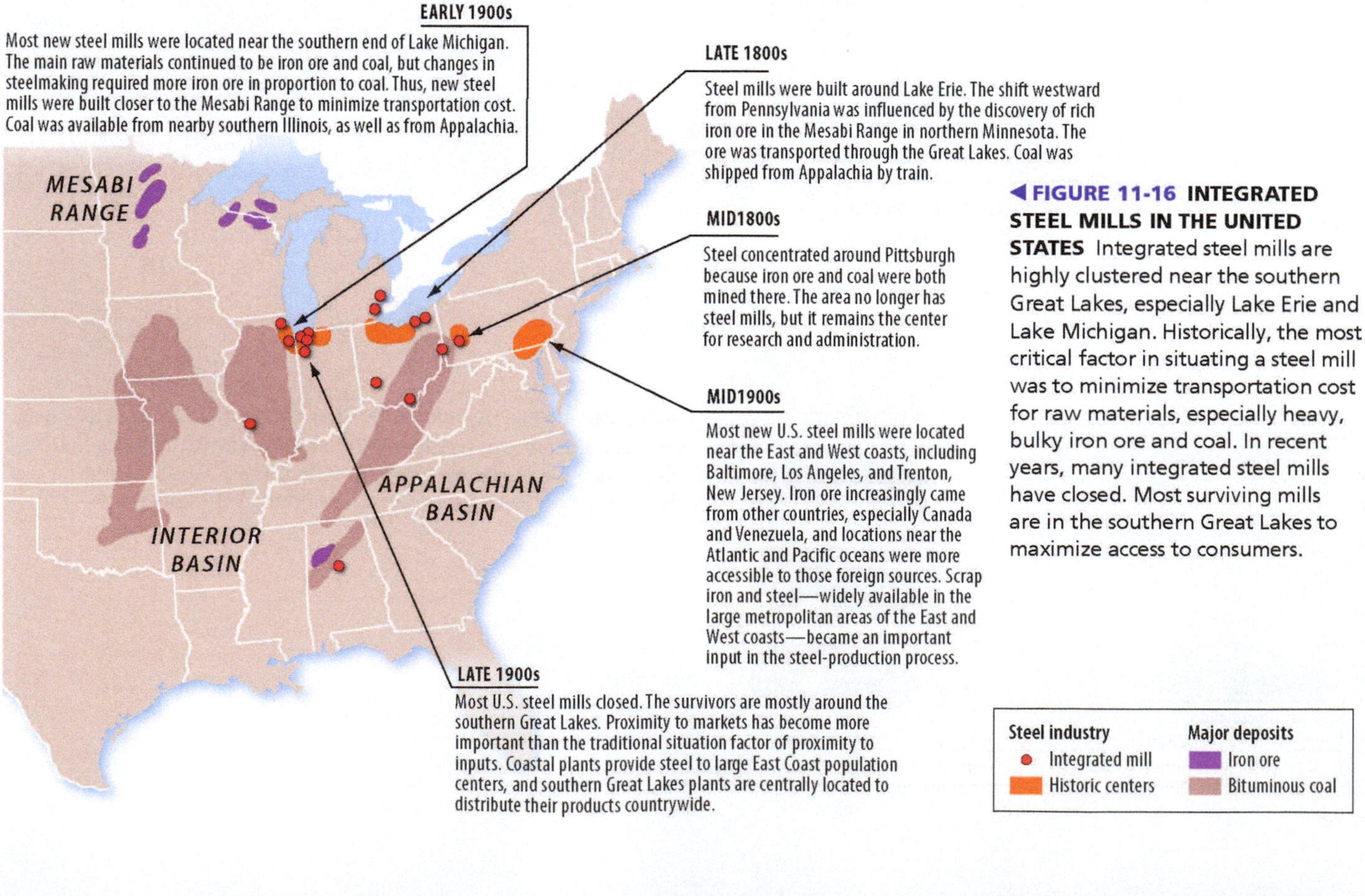

EARLY 1900s

Most new steel mills were located near the southern end of Lake Michigan. The main raw materials continued to be iron ore and coal, but changes in steelmaking required more iron ore in proportion to coal. Thus, new steel mills were built closer to the Mesabi Range to minimize transportation cost. Coal was available from nearby southern Illinois, as well as from Appalachia.

LATE 1800s

Steel mills were built around Lake Erie. The shift westward from Pennsylvania was influenced by the discovery of rich iron ore in the Mesabi Range in northern Minnesota. The ore was transported through the Great Lakes. Coal was shipped from Appalachia by train.

MID1800s

Steel concentrated around Pittsburgh because iron ore and coal were both mined there. The area no longer has steel mills, but it remains the center for research and administration.

MID1900s

Most new U.S. steel mills were located near the East and West coasts, including Baltimore, Los Angeles, and Trenton, New Jersey. Iron ore increasingly came from other countries, especially Canada and Venezuela, and locations near the Atlantic and Pacific oceans were more accessible to those foreign sources. Scrap iron and steel—widely available in the large metropolitan areas of the East and West coasts—became an important input in the steel-production process.

LATE 1900s

Most U.S. steel mills closed. The survivors are mostly around the southern Great Lakes. Proximity to markets has become more important than the traditional situation factor of proximity to inputs. Coastal plants provide steel to large East Coast population centers, and southern Great Lakes plants are centrally located to distribute their products countrywide.

MESABI RANGE

INTERIOR BASIN

APPALACHIAN BASIN

Steel industry: Integrated mill; Historic centers

Major deposits: Iron ore; Bituminous coal

◀ **FIGURE 11-16 INTEGRATED STEEL MILLS IN THE UNITED STATES** Integrated steel mills are highly clustered near the southern Great Lakes, especially Lake Erie and Lake Michigan. Historically, the most critical factor in situating a steel mill was to minimize transportation cost for raw materials, especially heavy, bulky iron ore and coal. In recent years, many integrated steel mills have closed. Most surviving mills are in the southern Great Lakes to maximize access to consumers.

PAUSE & REFLECT 11.2.3

Pittsburgh's football team is named "Steelers," but based on Figure 11-16, what city's team might be a more appropriate choice for the nickname?

CHANGING DISTRIBUTION OF WORLD STEEL PRODUCTION

The shift of world manufacturing to new industrial regions can be seen clearly in steel production. In 1980, 81 percent of world steel was produced in developed countries and 19 percent in developing countries (Figure 11-18). Between 1980 and 2013, the share of world steel production declined to 27 percent in developed countries and increased to 73 percent in developing countries (Figure 11-19).

World steel production more than doubled between 1980 and 2013, from 0.7 billion to 1.5 billion metric tons. China was responsible for 0.7 billion of the 0.8 billion metric ton increase, and other developing countries (primarily India and South Korea) were responsible for another 0.2 billion. Production in developed countries declined by 0.1 billion metric tons. China's steel industry has grown in part because of access to the primary inputs iron ore and coal. However, the principal factor in recent years has been increased demand by growing industries in China that use a lot of steel, such as motor vehicles.

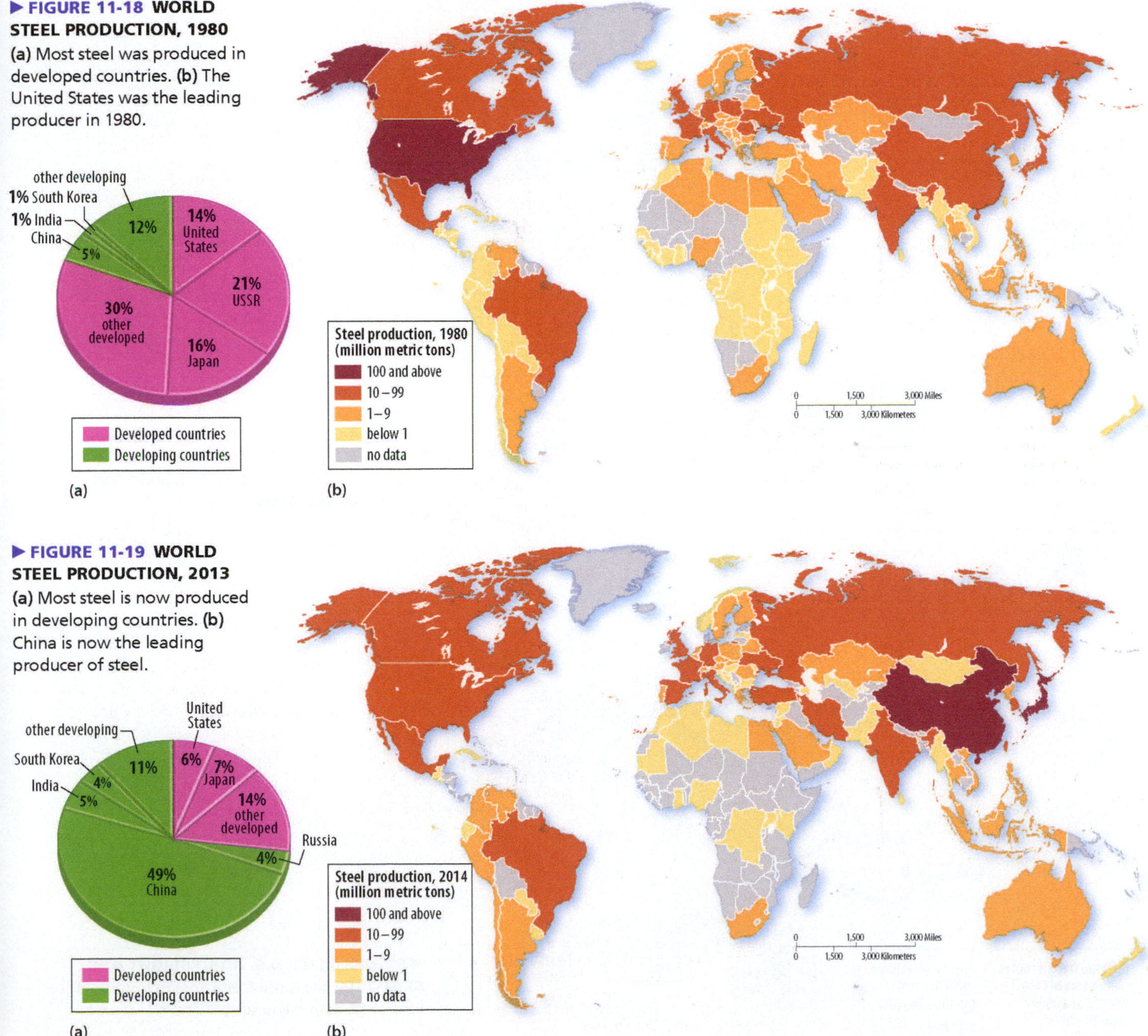

▶ **FIGURE 11-18 WORLD STEEL PRODUCTION, 1980**
(a) Most steel was produced in developed countries. **(b)** The United States was the leading producer in 1980.

▶ **FIGURE 11-19 WORLD STEEL PRODUCTION, 2013**
(a) Most steel is now produced in developing countries. **(b)** China is now the leading producer of steel.

Truck, Train, Ship, or Plane?

LEARNING OUTCOME 11.2.4

Explain why industries use different modes of transportation.

Inputs and products are transported in one of four ways: via ship, rail, truck, or air. Firms seek the lowest-cost mode of transport, but which of the four alternatives is cheapest changes with the distance that goods are being sent.

The farther something is transported, the lower the cost per kilometer (or mile). Longer-distance transportation is cheaper per kilometer in part because firms must pay workers to load goods on and off vehicles, whether the material travels 10 kilometers or 10,000. The cost per kilometer decreases at different rates for each of the four modes because the loading and unloading expenses differ for each mode:

- Trucks are most often used for short-distance delivery because they can be loaded and unloaded quickly and cheaply. Truck delivery is especially advantageous if the driver can reach the destination within one day, before having to stop for an extended rest (Figure 11-20).
- Trains are often used to ship to destinations that take longer than one day to reach, such as between the East and West coasts of the United States. Loading trains takes longer than loading trucks, but once under way, trains aren't required to make daily rest stops like trucks (Figure 11-21).

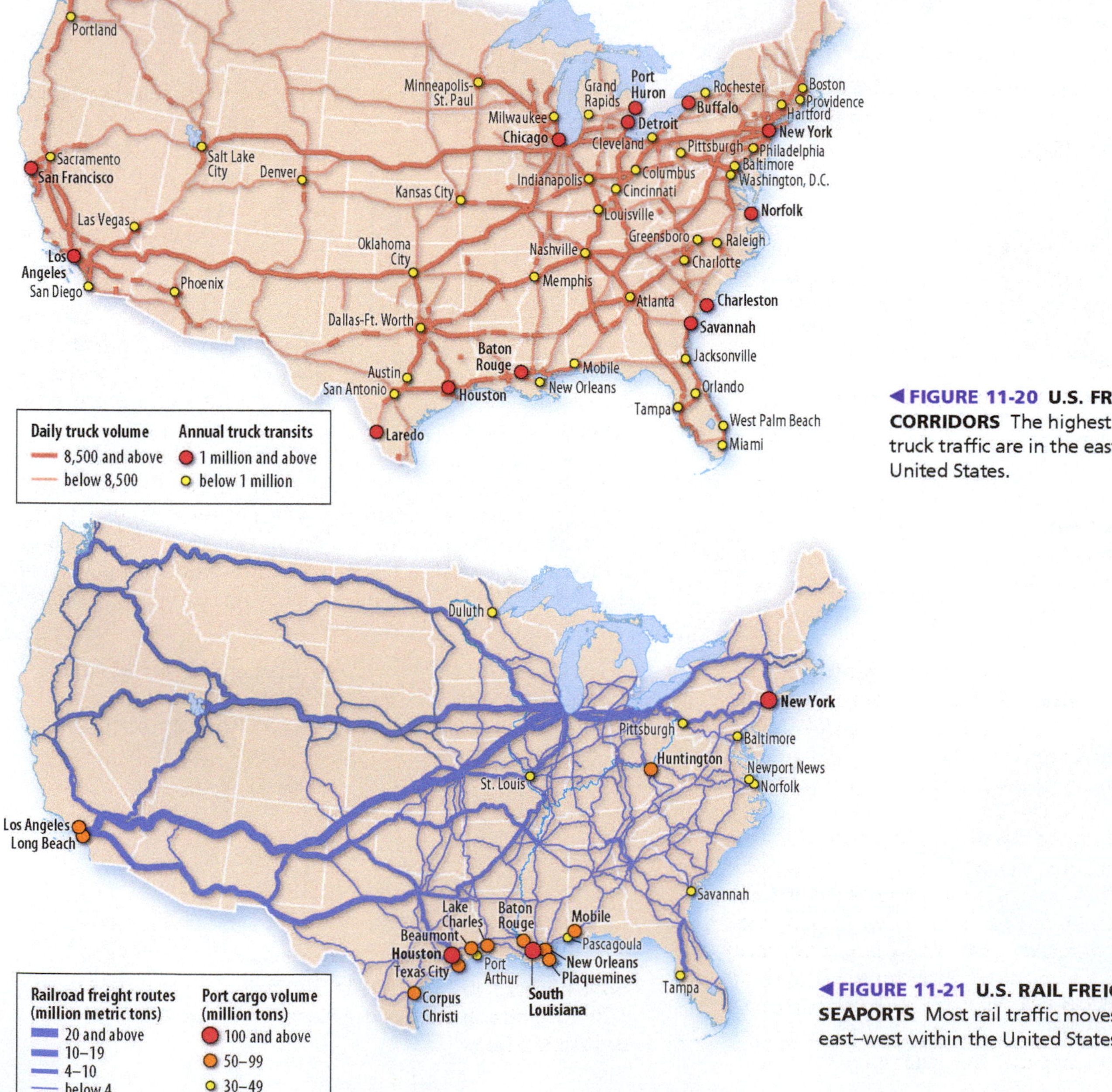

◀ FIGURE 11-20 U.S. FREIGHT CORRIDORS The highest volumes of freight truck traffic are in the eastern portion of the United States.

◀ FIGURE 11-21 U.S. RAIL FREIGHT AND SEAPORTS Most rail traffic moves long distances east–west within the United States.

◄ **FIGURE 11-22 WORLD SHIPPING ROUTES** Most freight is moved between continents by ship.

- Ships are attractive for transport over very long distances because the cost per kilometer is very low. Ships are slower than land-based transportation, but unlike trains or trucks, they can cross oceans, such as to North America from Europe or Asia (Figure 11-22).
- Airplanes are most expensive for all distances so are usually reserved for speedy delivery of small-bulk, high-value packages.

BREAK-OF-BULK POINTS

Mixed modes of delivery are often used. For example, air-freight companies pick up packages in the afternoon and transport them by truck to the nearest airport. Late at night, planes filled with packages are flown to a central hub airport in the interior of the country, such as Memphis, Tennessee, or Louisville, Kentucky. The packages are transferred to other planes, flown to airports nearest their destination, transferred to trucks, and delivered the next morning.

Many companies that use multiple transport modes locate at a **break-of-bulk point,** which is a location where transfer among transportation modes is possible. Important break-of-bulk points include seaports and airports. For example, a steel mill in Gary, Indiana, may receive iron ore by ship via Lake Michigan and coal by train from Appalachia.

Containerization has facilitated transfer of packages between modes. Containers may be packed into a rail car, transferred quickly to a container ship to cross the ocean, and unloaded onto trucks at the other end. Large ships have been specially built to accommodate large numbers of rectangular box-like containers.

Regardless of transportation mode, cost rises each time inputs or products are transferred from one mode to another. For example, workers must unload goods from a truck and then reload them onto a plane. The company may need to build or rent a warehouse to store goods temporarily after unloading from one mode and before loading to another mode. Some companies may calculate that the cost of one mode is lower for some inputs and products, whereas another mode may be cheaper for other goods.

JUST-IN-TIME DELIVERY

Proximity to market has become more important in recent years because of the rise of just-in-time delivery. As the name implies, **just-in-time delivery** is shipment of parts and materials to arrive at a factory moments before they are needed. Just-in-time delivery is especially important for delivery of inputs, such as parts and raw materials, to manufacturers of fabricated products, such as cars and computers.

Under a just-in-time system, parts and materials arrive at a factory frequently, in many cases daily or even hourly. Suppliers of the parts and materials are told a few days in advance how much will be needed over the next week or two, and first thing each morning, they are told exactly what will be needed at precisely what time that day. To meet a tight timetable, a supplier of parts and materials must locate factories near its customers. If given only an hour or two of notice, a supplier has no choice but to locate a factory within around one hour of the customer.

Just-in-time delivery reduces the money that a manufacturer must tie up in wasteful inventory. Manufacturers also save money through just-in-time delivery by reducing the size of the factory because space does not have to be wasted on piling up a mountain of inventory.

Leading computer manufacturers have eliminated inventory altogether. They build computers only in response to customer orders placed primarily over the Internet or by telephone.

Just-in-time delivery means that producers have lower inventory to cushion against disruptions in the arrival of needed parts. Three kinds of disruptions can result from reliance on just-in-time delivery:

- **Natural hazards.** Poor weather conditions can affect deliveries anywhere in the world. For example, blizzards can close highways, rail lines, and airports.
- **Traffic.** Deliveries may be delayed when traffic is slowed by accident, construction, or unusually heavy volume.
- **Labor unrest.** A strike at one supplier plant can shut down the entire production within a couple days.

PAUSE & REFLECT 11.2.4

How might weather conditions influence the choice of a factory site between the North and the South in the United States?

Site Factors in Industry

LEARNING OUTCOME 11.2.5

Understand the three types of site factors.

Site factors are industrial location factors related to the costs of factors of production inside a plant. For some firms, site factors are more important than situation factors in locating a factory. The three production factors that may vary among locations are labor, capital, and land.

LABOR

The most important site factor on a global scale is labor. Minimizing labor costs is important for some industries, and the variation of labor costs around the world is large. Worldwide, around one-half billion workers are engaged in industry, according to the UN International Labor Organization (ILO). China has around one-fourth of the world's manufacturing workers, India around one-fifth, and all developed countries combined around one-fifth.

A **labor-intensive industry** is an industry in which wages and other compensation paid to employees constitute a high percentage of expenses. Labor constitutes an average of 11 percent of overall manufacturing costs in the United States, so a labor-intensive industry in the United States would have a much higher percentage than that. The reverse case, an industry with a much lower-than-average percentage of expenditures on labor, is considered capital intensive.

The average wage paid to manufacturing workers is approximately $35 per hour in developed countries and exceeds $40 per hour in parts of Europe (Figure 11-23). Health-care, retirement pensions, and other benefits add substantially to the compensation. In China and India, average wages are less than $2 per hour and include limited additional benefits. For some manufacturers—but not all—the difference between paying workers $2 and $35 per hour is critical.

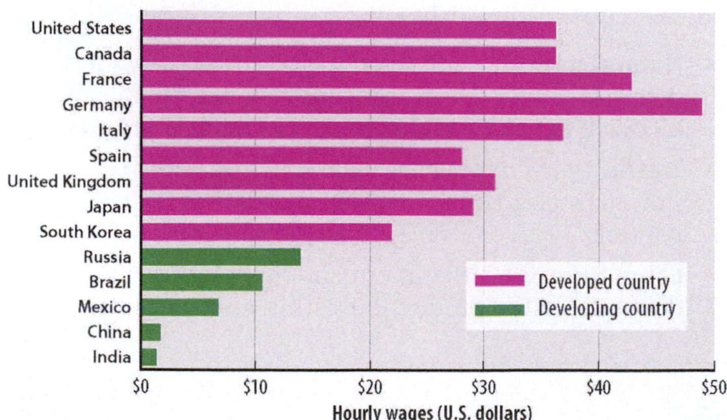

▲ **FIGURE 11-23 SITE FACTOR: LABOR** The chart shows average hourly wages for workers in manufacturing for 2012 in 14 of the 15 countries with the largest industrial production in 2014.

A labor-intensive industry is not the same as a high-wage industry. "Labor-intensive" is measured as a percentage, whereas "high-wage" is measured in dollars or other currencies. For example, motor-vehicle workers are paid much higher hourly wages than textile workers, yet the textile industry is labor intensive, and the auto industry is not. Although auto workers earn relatively high wages, most of the value of a car is accounted for by the parts and the machinery needed to put together the parts. On the other hand, labor accounts for a large percentage of the cost of producing a towel or shirt compared with materials and machinery.

PAUSE & REFLECT 11.2.5

Labor accounts for around 5 percent of the cost of manufacturing a car. Does this mean that motor vehicle manufacturing is a labor-intensive industry?

CAPITAL

Manufacturers typically borrow capital—the funds to establish new factories or expand existing ones. The U.S. motor-vehicle industry concentrated in Michigan early in the twentieth century largely because that region's financial institutions were more willing than eastern banks to lend money to the industry's pioneers.

The most important factor in the clustering of high-tech industries in California's Silicon Valley—even more important than proximity to skilled labor—was the availability of capital. Banks in Silicon Valley have long been willing to provide money for new software and communications firms, even though lenders elsewhere have hesitated. High-tech industries have been risky propositions—roughly two-thirds of them fail—but Silicon Valley financial institutions have continued to lend money to engineers who have good ideas so that they can buy the software, communications, and networks they need to get started (Figure 11-24). One-fourth of all capital in the United States is spent on new industries in Silicon Valley.

▼ **FIGURE 11-24 SITE FACTOR: CAPITAL** High-tech industries, such as Google, have clustered in Silicon Valley, California, primarily because of availability of capital.

The ability to borrow money has become a critical factor in the distribution of industry in developing countries. Financial institutions in many developing countries are short of funds, so new industries must seek loans from banks in developed countries. But enterprises may not get loans if they are located in a country that is perceived to have an unstable political system, a high debt level, or ill-advised economic policies.

LAND

Land suitable for constructing a factory can be found in many places. If considered to encompass energy and other natural resources in addition to terra firma, "land" is an especially critical site factor. Energy inputs into production are discussed in the next section.

Early factories located inside cities due to a combination of situation and site factors. A city offered an attractive situation—proximity to a large local market and convenience in shipping to a national market by rail. A city also offered an attractive site—proximity to a large supply of labor as well as to sources of capital. The site factor that cities have always lacked is abundant land. To get the necessary space in cities, early factories were typically multistory buildings. Raw materials were hoisted to the upper floors to make smaller parts, which were then sent downstairs on chutes and pulleys for final assembly and shipment (Figure 11-25). Water was stored in tanks on the roof.

Contemporary factories operate most efficiently when laid out in one-story buildings (see, for example, Figure 11-26). Raw materials are typically delivered at one end and moved through the factory on conveyors or forklift trucks. Products are assembled in logical order and shipped out at the other end. The land needed to build one-story factories is now more likely to be available in suburban and rural locations. Also, land is much cheaper in suburban and rural locations than near the center of a city.

In addition to providing enough space for one-story buildings, locations outside cities are also attractive because they facilitate delivery of inputs and shipment of products. In the past, when most material moved in and out of a factory by rail, a central location was attractive because rail lines converged there. With trucks now responsible for transporting most inputs and products, proximity to major highways is more important for a factory. Especially attractive is the proximity to the junction of a long-distance route and the beltway, or ring road, that encircles most cities. Thus, factories cluster in industrial parks located near suburban highway junctions.

▲ **FIGURE 11-25 HISTORIC MULTISTORY FACTORY** In the early twentieth century, Ford assembled Model T cars in a multistory factory in Highland Park, Michigan. The bodies were slid down the ramp from the second floor and attached to the chassis, which rolled out of the first floor 1. When Ford built this factory in the early twentieth century, why were the exterior walls made almost entirely of glass? 2. What were the advantages and disadvantages of assembling large objects like cars in a multistory factory a century ago? 3. What technological advances during the past century have influenced replacement of multistory factories with one-story factories?

▼ **FIGURE 11-26 MODERN ONE-STORY FACTORY** The Honda assembly plant in Swindon, United Kingdom, is spread out over one story.

Changing Site Factors: Clothing

LEARNING OUTCOME 11.2.6

Explain the distribution of clothing production.

Production of textiles (woven fabrics) and apparel (clothing) is a prominent example of an industry that generally requires less-skilled, low-cost workers. The textile and apparel industry accounts for 6 percent of the dollar value of world manufacturing but a much higher 14 percent of world manufacturing employment, an indicator that it is a labor-intensive industry. The percentage of the world's women employed in this type of manufacturing is even higher.

Textile and apparel production involves three principal steps:

- Spinning of fibers to make yarn.
- Weaving or knitting of yarn into fabric.
- Assembly of fabric into products.

Spinning, weaving, and assembly are all labor intensive compared to other industries, but the importance of labor varies somewhat among them. Their global distributions are not identical because the three steps are not equally labor intensive.

SPINNING

Fibers can be spun from natural or synthetic elements. The principal natural fiber is cotton, and synthetics now account for three-fourths of world thread production. Because it is a labor-intensive industry, spinning is done primarily in low-wage countries (Figure 11-27). China produces one-fourth and India one-fifth of the world's cotton thread.

WEAVING

For thousands of years, fabric has been woven or laced together by hand on a loom, which is a frame on which two sets of threads are placed at right angles to each other. One set of threads, called the warp, is strung lengthwise. A second set of threads, called the weft, is carried in a shuttle that is inserted over and under the warp. Because the process of

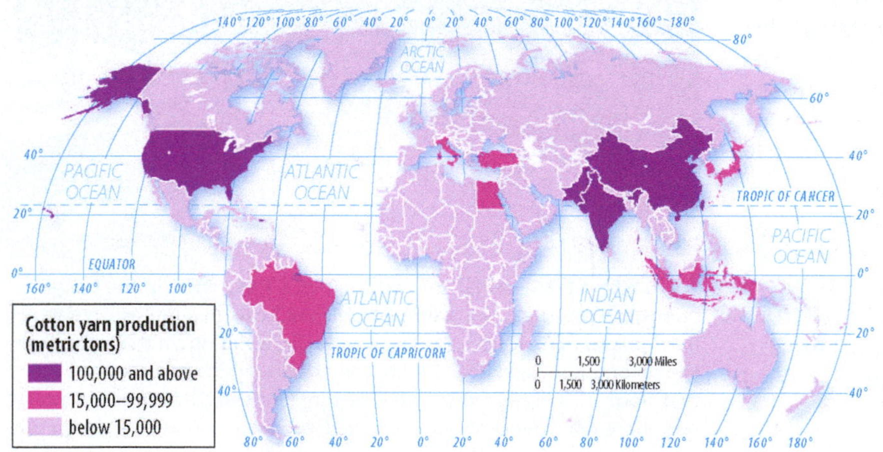

(b)

▲ **FIGURE 11-27 COTTON SPINNING** **(a)** Nearly one-half of world cotton yarn is spun in China and India. **(b)** Factory in Indore, India, spins yarn from organic and fair trade cotton.

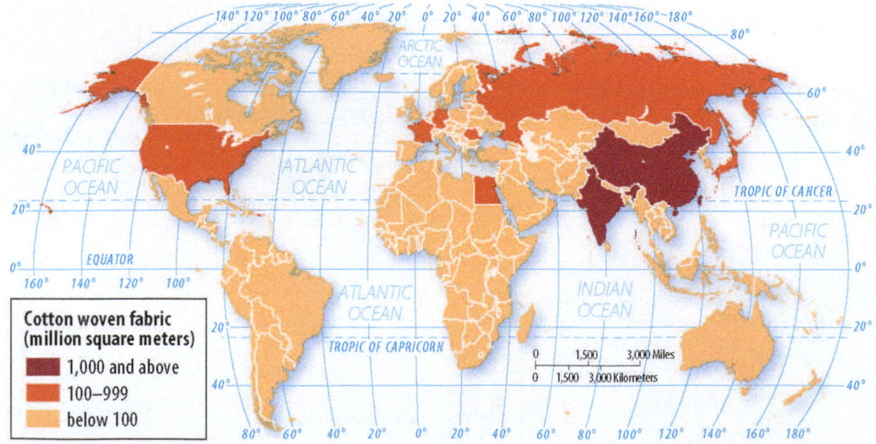

(b)

▲ **FIGURE 11-28 COTTON WEAVING** **(a)** China and India account for 90 percent of world cotton weaving. **(b)** Factory in India weaves textiles.

weaving by hand is physically hard work, weavers were traditionally men.

For mechanized weaving, labor constitutes a high percentage of the total production cost. Consequently, weaving is highly clustered in low-wage countries (Figure 11-28). Despite their remoteness from European and North American markets, China and India have become the dominant fabric producers because their lower labor costs offset the expense of shipping inputs and products long distances. China accounts for nearly 60 percent of the world's woven cotton fabric production and India 30 percent.

ASSEMBLY

Sewing by hand is a very old human activity. Needles made from animal horns or bones date back tens of thousands of years, and iron needles date from the fourteenth century. The first functional sewing machine was invented by French tailor Barthelemy Thimonnier in 1830. Isaac Singer manufactured the first commercially successful sewing machine in the United States during the 1850s.

Textiles are cut and sewn to be assembled into four main types of products: garments, carpets, home products such as bed linens and curtains, and industrial items such as headliners for motor vehicles. Developed countries play a larger role in assembly than in spinning and weaving because most of the consumers of assembled products are located in developed countries (Figure 11-29). Nonetheless, most clothing is assembled in developing countries. Overall production costs are generally lower in developing countries because substantially lower labor costs compared to developed countries offset higher shipping and taxation costs (Figure 11-30).

PAUSE & REFLECT 11.2.6

Check the label on the shirt you are wearing at this moment. Where was it made?

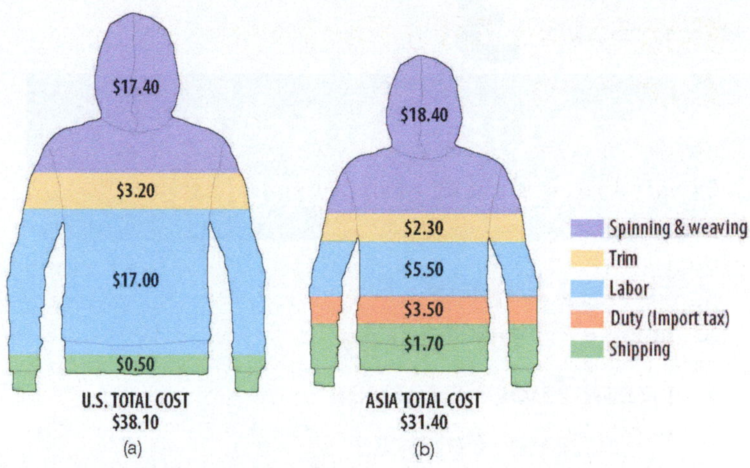

▲ **FIGURE 11-30 HOW MUCH IT COSTS TO MAKE A HOODIE SOLD IN THE UNITED STATES** The cost of production is lower **(b)** in Asia than **(a)** in the United States because of lower wages paid to clothing workers.

CHECK-IN KEY ISSUE 2

Why Are Situation and Site Factors Important?

- ✔ **Situation factors involve transporting materials to and from a factory.**
- ✔ **Bulk-reducing industries locate near their sources of inputs.**
- ✔ **Bulk-gaining, single-market, and perishable-products industries locate near their markets.**
- ✔ **Site factors derive from distinctive features of a particular place, including labor, capital, and land.**

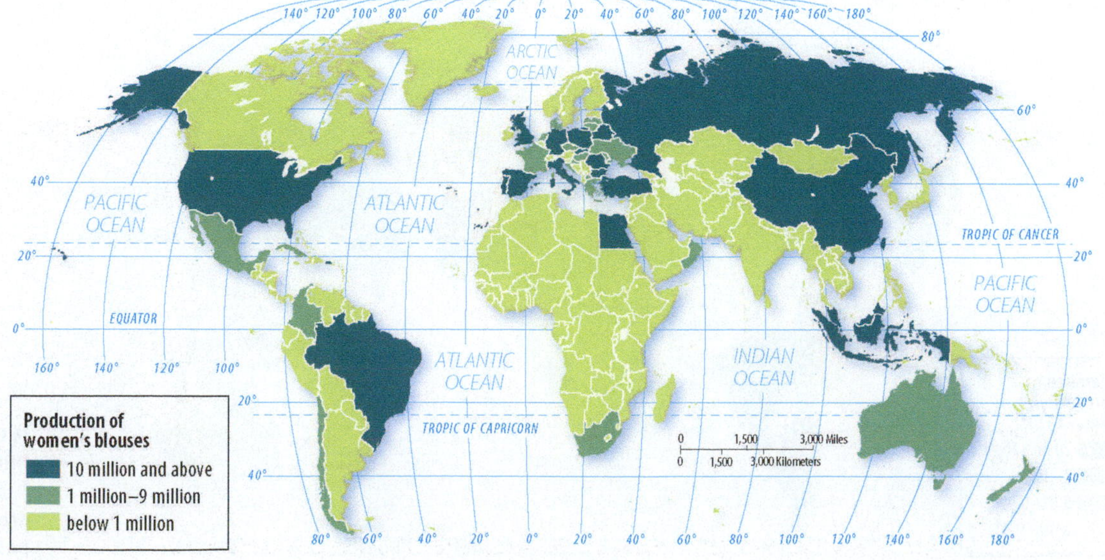

► **FIGURE 11-29 WOMEN'S BLOUSE PRODUCTION** The United States is the leading producer of women's blouses.

KEY ISSUE 3

Why Do Industries Face Resource Challenges?

- Energy Supply
- Demand for Energy
- Fossil Fuel Reserves
- Petroleum Futures
- Nuclear Energy
- Energy Alternatives
- Solar Energy
- Air Pollution
- Water Pollution
- Solid Waste Pollution

LEARNING OUTCOME 11.3.1

Describe the distribution of production of the three fossil fuels.

Earth offers a large menu of resources available for people to use in factories, homes, and vehicles. A resource was defined in Chapter 1 as a substance in the environment that is useful to people, is economically and technologically feasible to access, and is socially acceptable to use. Geographers observe two challenges regarding resources:

- We deplete scarce resources, especially petroleum, natural gas, and coal, for energy production.
- We destroy resources through pollution of air, water, and soil.

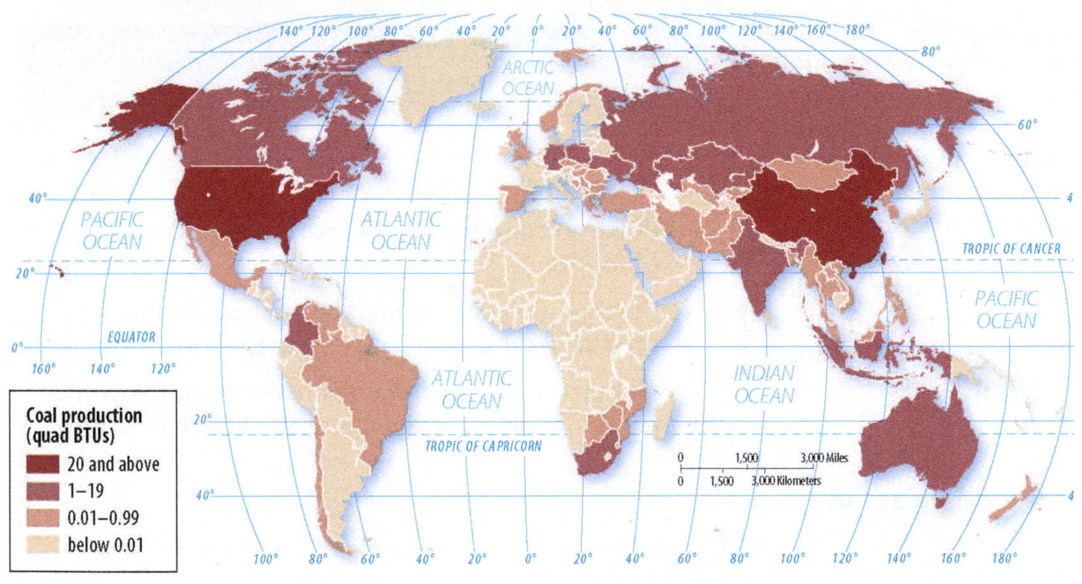

◀ **FIGURE 11-31 COAL PRODUCTION** China is the world's leading producer of coal, followed by the United States.

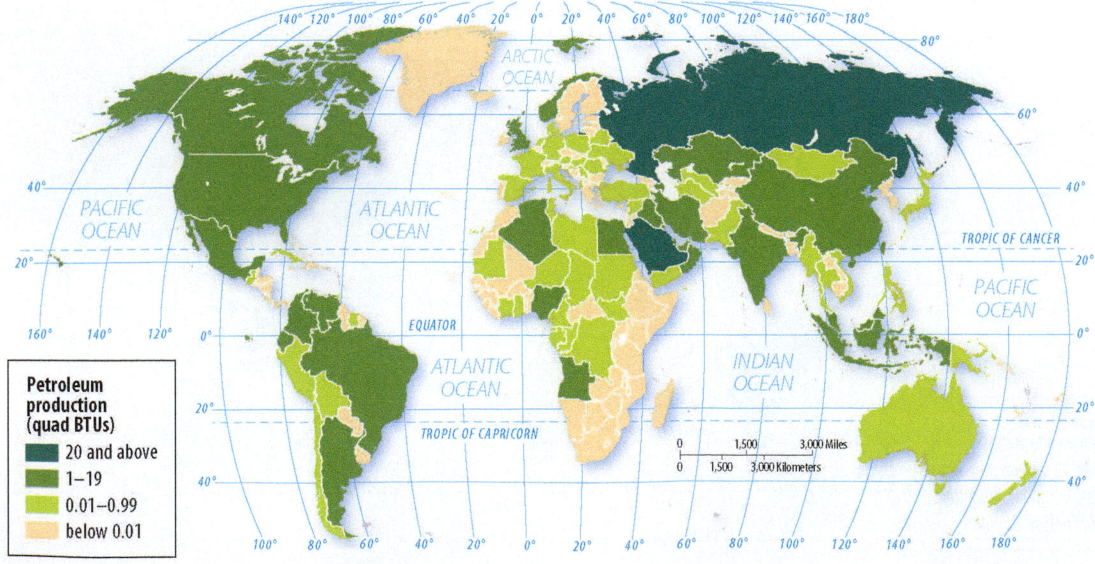

◀ **FIGURE 11-32 PETROLEUM PRODUCTION** Russia, Saudi Arabia, and the United States are the leading producers of petroleum.

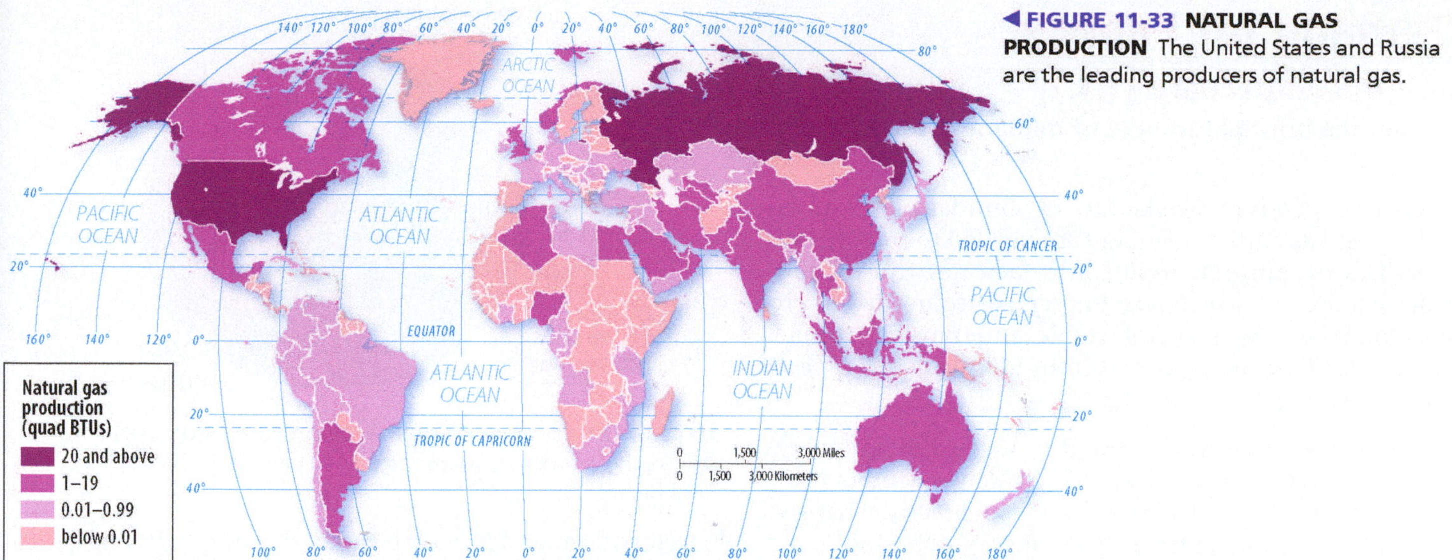

◀ **FIGURE 11-33 NATURAL GAS PRODUCTION** The United States and Russia are the leading producers of natural gas.

Energy Supply

Fossil fuels are not distributed evenly. Some regions are well endowed with one or more fossil fuels, whereas other regions have little. The uneven distribution of fossil fuels partly reflects how fossil fuels form:

- **Coal.** Coal formed in tropical locations, in lush, swampy areas rich in plants. Thanks to the slow movement of Earth's drifting continents, the tropical swamps of 250 million years ago relocated to the mid-latitudes. As a result, today's main reserves of coal are in mid-latitude countries rather than in the tropics. China produces nearly one-half of the world's coal, other developing countries one-fourth, and developed countries (primarily the United States) the remaining one-fourth (Figure 11-31).
- **Petroleum.** Petroleum formed millions of years ago from residue deposited on the seafloor. Some still lies beneath such seas as the Persian Gulf and the North Sea, but other reserves are located beneath land that was under water millions of years ago. Russia and Saudi Arabia together supply one-fourth of the world's petroleum, other developing countries (primarily in Southwest and Central Asia) one-half, and developed countries (primarily the United States) the remaining one-fourth (Figure 11-32).
- **Natural gas.** Like petroleum, natural gas formed millions of years ago from sediment deposited on the seafloor. One-third of natural gas production is supplied by Russia and Southwest Asia, one-third by other developing regions, and one-third by developed countries (primarily the United States) (Figure 11-33).

Figures 11-31 11-32, and 11-33 use the same units (quad BTUs), as well as the same classes. Quad is short for quadrillion (1 quadrillion = 1,000,000,000,000,000), and BTU is short for British thermal unit. One quad BTU equals approximately 8 billion U.S. gallons of gasoline, which would fill the tanks of one-half million cars.

The United States is highly dependent on the three fossil fuels (Figure 11-34). Wood was the principal energy source in the nineteenth century. Coal became most important during the late nineteenth century, and petroleum and natural gas gained prominence during the second half of the twentieth century.

PAUSE & REFLECT 11.3.1

Which country produces at least 20 quad BTUs of all three fossil fuels?

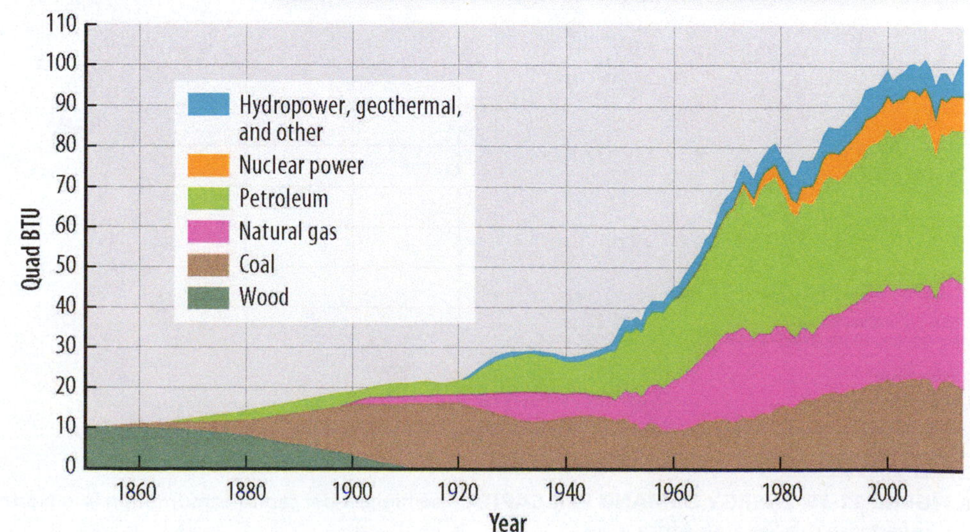

▶ **FIGURE 11-34 CHANGING U.S. ENERGY SUPPLY** Petroleum, natural gas, and coal account for nearly 90 percent of all energy supplied in the United States.

Demand for Energy

LEARNING OUTCOME 11.3.2

Explain the principal sources of demand for fossil fuels.

Industry depends on availability of abundant low-cost energy. Large quantities of energy are needed to run factories as well as to transport inputs into factories and products from factories to consumers. Energy is also needed to produce food, keep homes comfortable, and transport people.

Demand for energy comes from four principal types of consumption in the United States:

- **Industries.** Factories use roughly 40 percent natural gas and 30 percent each coal and petroleum. Most of the natural gas and petroleum is burned directly, whereas coal is consumed primarily through purchasing electricity.
- **Transportation.** Almost all transportation systems run on petroleum products.
- **Homes.** Natural gas and coal provide roughly equal shares of home needs. Natural gas is the principal source of home heating and air conditioning, whereas electricity generated primarily from coal is the principal source of electricity.
- **Commercial.** Stores and offices have uses and sources similar to those for homes.

Developing countries accounted for more energy usage than developed countries for the first time in 2006 (Figure 11-35). The United States has long led in demand for energy, but China is now the leader. The highest per capita consumption of energy remains in developed countries (Figure 11-36). Consumption of fossil fuels has been increasing at a much faster rate in developing countries—around 3 percent per year, compared to 1 percent per year in developed countries. So the gap in demand between developing and developed countries will widen considerably in the years ahead (Figure 11-37).

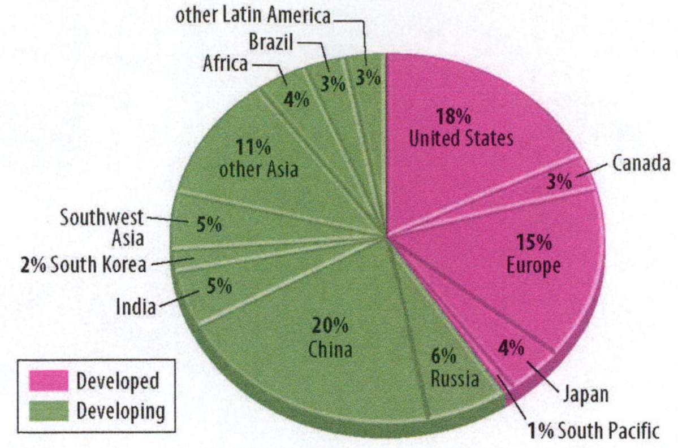

▲ **FIGURE 11-35 WORLD ENERGY DEMAND BY COUNTRY** Developing countries consume around 60 percent of the world's fossil fuels.

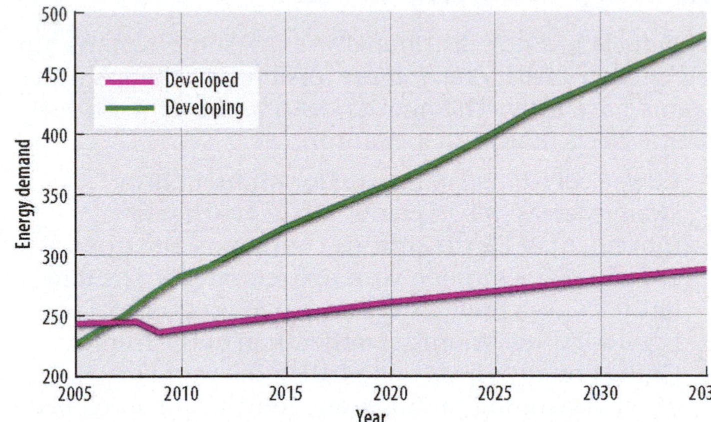

▲ **FIGURE 11-37 FUTURE ENERGY DEMAND** Developing countries will account for an increasing share of world energy usage in the years ahead.

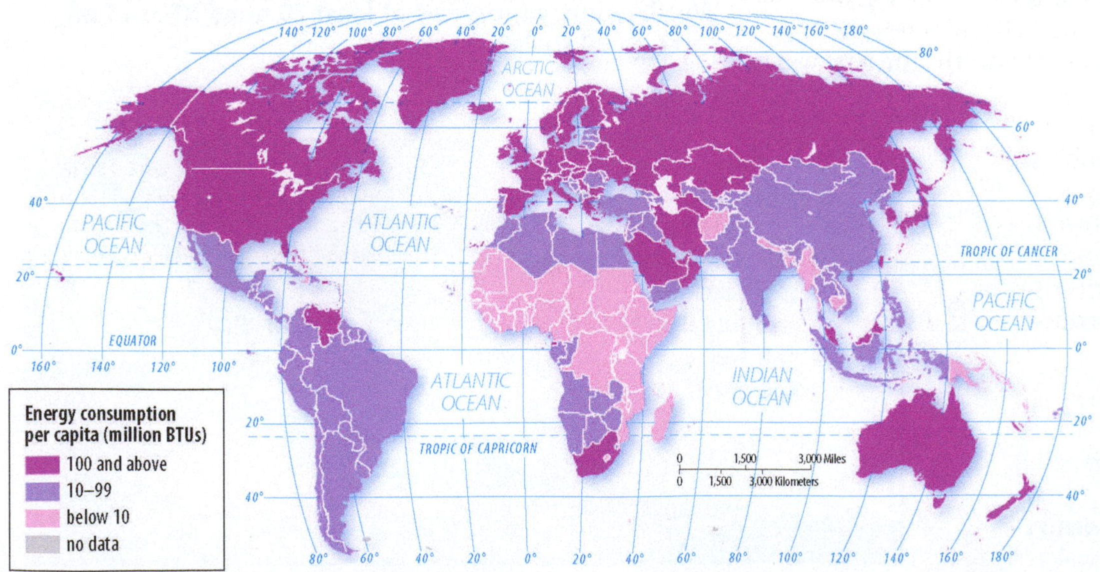

▲ **FIGURE 11-36 ENERGY DEMAND PER CAPITA** The highest per capita consumption is in North America, and the lowest is in sub-Saharan Africa.

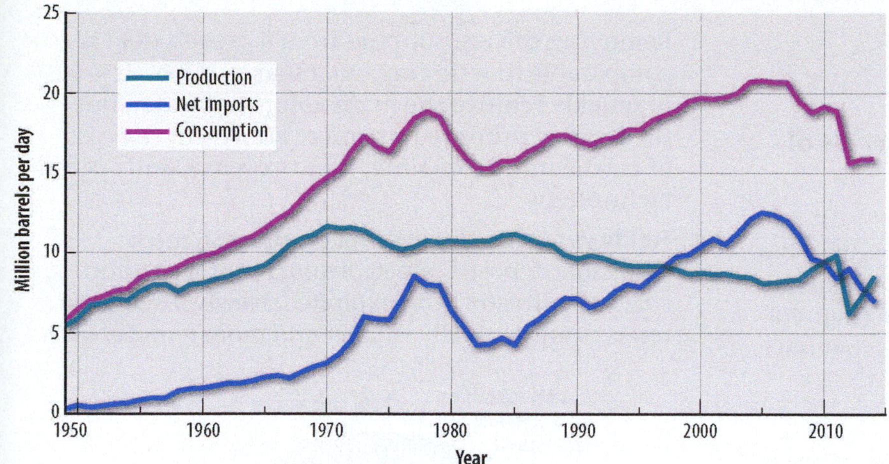

◀ **FIGURE 11-38 U.S. PETROLEUM SUPPLY AND DEMAND** The United States imports more than it produces in order to meet demand for consumption.

PETROLEUM CHALLENGES

Meeting world demand for petroleum has been especially challenging in recent years. Most of the world's petroleum is produced in Southwest Asia & North Africa and Central Asia, the two regions at the center of religious, ethnic, and political conflicts discussed in Chapters 6 through 8.

Other developed countries have always depended on foreign petroleum because of limited domestic supplies, but the United States produced more petroleum than it consumed during the first half of the twentieth century. Beginning in the 1950s, the handful of large transnational companies then in control of international petroleum distribution determined that extracting petroleum in the United States was more expensive than importing it from Southwest and Central Asia. U.S. petroleum imports increased from 14 percent of total consumption in 1955 to 60 percent in 2005, before declining to 44 percent in 2014 (Figure 11-38).

Several developing countries possessing substantial petroleum reserves, primarily in Southwest Asia & North Africa, created the Organization of the Petroleum Exporting Countries (OPEC) in 1960. OPEC was originally formed to enable oil-rich countries to gain more control over their resource. U.S. and European transnational companies, which had originally explored and exploited the oil fields, were selling the petroleum at low prices to consumers in developed countries and keeping most of the profits. Countries possessing the oil reserves nationalized or more tightly controlled the fields, and prices were set by governments rather than by petroleum companies.

Under OPEC control, world oil prices have increased sharply on several occasions, especially during the 1970s and 1980s and in the early twenty-first century. Developed countries entered the twenty-first century optimistic that oil prices would remain low for some time. But in 2008, prices hit a record high, in both real terms and accounting for inflation. The 2008 oil shock contributed to the severe global recession that began that year. Gas prices are currently low in the United States, especially when adjusted for inflation (Figure 11-39). However, the average price paid for a gallon of petroleum exceeds $8 in most other developed countries.

PAUSE & REFLECT 11.3.2

Referring to Figure 11-36, what world region of developing countries appears to have the highest per capita demand for energy?

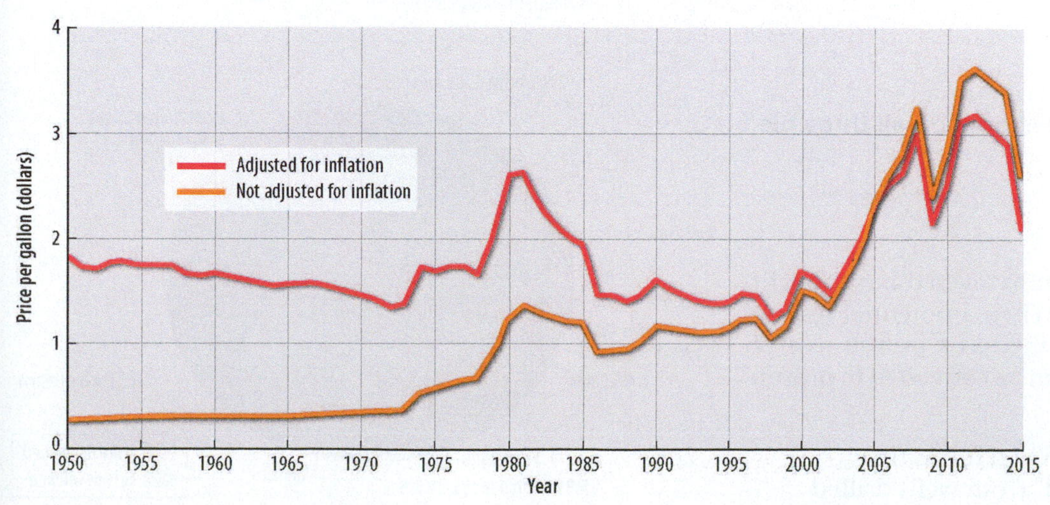

◀ **FIGURE 11-39 U.S. GASOLINE PRICES** Fuel prices increased sharply in the 1970s and again in the 2000s but declined a few years after both periods.

Fossil Fuel Reserves

LEARNING OUTCOME 11.3.3

Understand the distinctive distributions of reserves of the three fossil fuels.

Because petroleum, natural gas, and coal are deposited beneath Earth's surface, considerable technology and skill are required to locate these substances and estimate their volume. The supply of energy remaining in deposits that have been discovered is called a **proven reserve.**

PROVEN RESERVES

Developed countries have historically possessed a disproportionately high supply of the world's proven fossil fuel reserves (Figure 11-40):

- **Coal.** World reserves are approximately 1 trillion metric tons (23,000 quad BTUs). At current demand, proven coal reserves would last 130 years. Developed and developing regions each have about one-half of the supply of proven reserves. The United States has approximately one-fourth of the proven reserves, and other developed countries have one-fourth. Most of the developing regions' coal reserves are in Russia and China.
- **Natural gas.** World reserves are approximately 200 trillion cubic meters (7,000 quad BTUs). At current demand, proven natural gas reserves would last 56 years. Less than 10 percent of natural gas reserves are in developed countries, primarily the United States. Russia has 25 percent and Iran and Qatar together 30 percent of the world's proven natural gas reserves.
- **Petroleum.** World reserves are approximately 1.6 trillion barrels (10,000 quad BTUs). At current demand, proven petroleum reserves would last 55 years. Developing countries possess 87 percent of the proven petroleum reserves, most of which is in Southwest Asia & North Africa and Central Asia. Venezuela, Saudi Arabia, Canada, Iran, and Iraq together have nearly two-thirds of the world's proven petroleum reserves.

PAUSE & REFLECT 11.3.3

What are the only two countries named on all three pie charts of proven reserves?

POTENTIAL RESERVES

The supply in deposits that are undiscovered but thought to exist is a **potential reserve.** When a potential reserve is actually discovered, it is reclassified as a proven reserve (Figure 11-41). Potential reserves can be converted to proven reserves in several ways:

- **Fields yet to be developed.** When it was first exploited, petroleum "gushed" from wells drilled into rock layers saturated with it. Coal was quarried in open pits. But now extraction is more difficult. Removing the last supplies from a proven field is comparable to wringing out a soaked towel. It is easy to quickly remove the main volume of water, but the last few drops require more effort—in the case of petroleum, more time, more expense, and special technology.
- **Fields yet to be discovered.** The largest, most accessible deposits of petroleum, natural gas, and coal have already been exploited. Newly discovered reserves are generally smaller and more remote, such

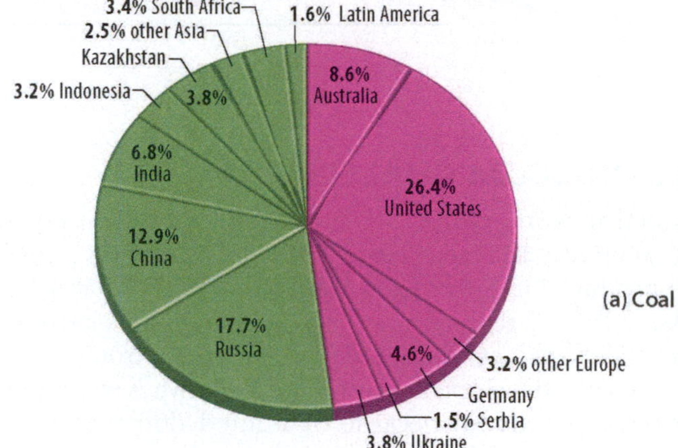

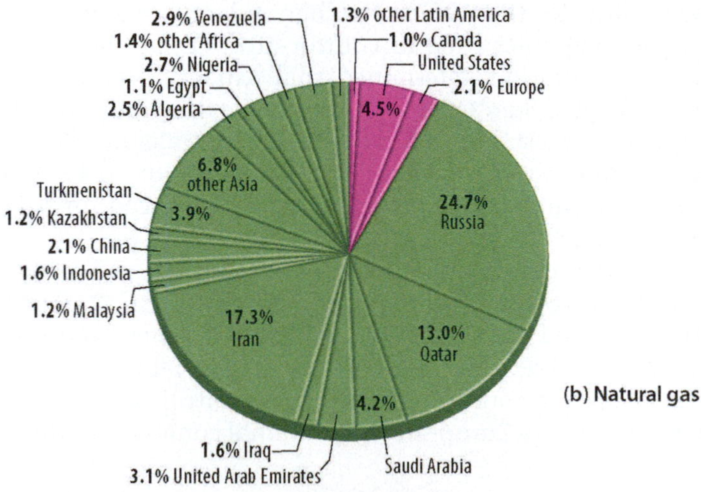

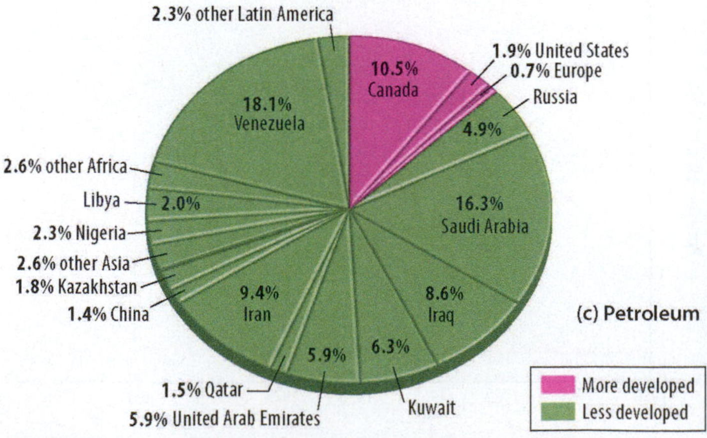

▲ **FIGURE 11-40 PROVEN RESERVES OF FOSSIL FUELS** (a) Coal. (b) Natural gas. (c) Petroleum.

as beneath the seafloor, and extraction is costly. Exploration costs have increased because methods are more elaborate, and the probability of finding new reserves is lower. But as energy prices climb, exploration costs may be justified.

UNCONVENTIONAL RESOURCES

Resources are considered unconventional if we lack economically feasible or environmentally sound technology with which to extract them. As demand increases for a resource and prices rise, exploiting an unconventional source can become profitable. Here are two current examples:

- **Oil sands.** Abundant oil sands are found in Alberta, Canada, as well as in Venezuela and Russia. Oil sands are saturated with thick petroleum commonly called tar because of its dark color and strong odor. The mining of Alberta oil sands has become profitable, and extensive deposits of oil in Alberta oil sands have been reclassified from potential to proven reserves in recent years. As a result, Canada is now thought to have 11 percent of world's petroleum proven reserves, second behind Saudi Arabia.
- **Hydraulic fracturing.** Rocks break apart naturally, and gas can fill the space between the rocks. Hydraulic fracturing, commonly called *fracking*, involves pumping water at high pressure to further break apart rocks and thereby release more gas that can be extracted. The United States has extensive natural gas fields, some of which are now being exploited through fracking. Opponents of fracking fear environmental damage from pumping high-pressure water beneath Earth's surface. Safety precautions can minimize the environmental threat, but fracking does require the use of a large supply of water, and water is in high demand for other important uses, such as human consumption and agriculture. Within the United States, the principal natural gas fields are in Texas, Oklahoma, and the Appalachian Mountains (Figure 11-42).

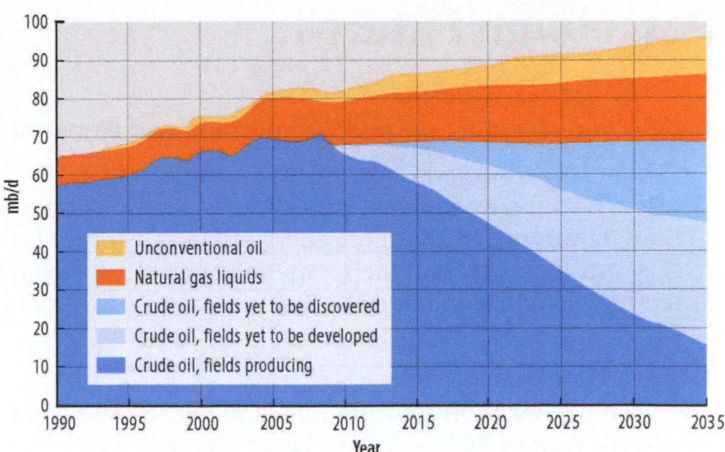

▲ **FIGURE 11-41 PETROLEUM PRODUCTION OUTLOOK** The International Energy Agency forecasts that potential reserves will be converted to proven reserves through discovery and development of new fields at about the same rate as already proven reserves are depleted.

▼ **FIGURE 11-42 NATURAL GAS FIELDS IN THE UNITED STATES** The principal natural gas fields are in Oklahoma, Texas, and the Appalachians.

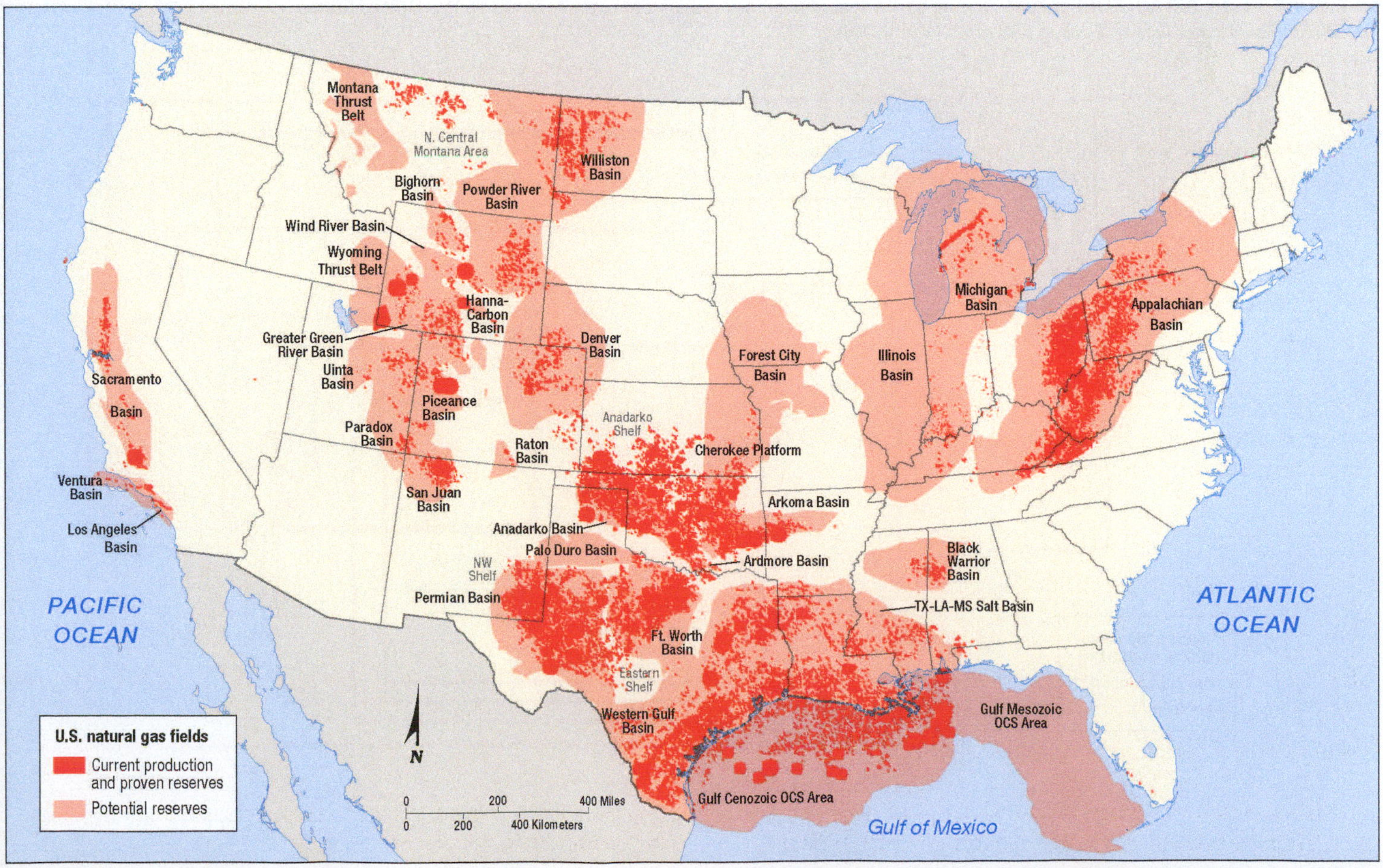

Petroleum Futures

LEARNING OUTCOME 11.3.4

Understand changing patterns of oil trade and demand.

Developed countries supply a large share of the world's fossil fuels, but they demand more energy than they produce, so they must import fossil fuels, especially petroleum, from developing countries (Figure 11-43).

▲ **FIGURE 11-43 OIL REFINERY, SAUDI ARABIA**

WORLD OIL TRADE

The largest flows of oil are from Russia to Europe and from Southwest Asia to Europe and to Japan (Figure 11-44). The United States and Europe import more than half their petroleum, and Japan imports more than 90 percent.

The countries from which the United States imports petroleum have changed since 1973. Canada and Saudi Arabia now supply much higher shares than in the past (Figure 11-45). With demand increasing rapidly in developing countries, developed countries face greater competition in obtaining the world's remaining supplies of fossil fuels. Many of the developing countries with low HDIs also lack energy resources, and they lack the funds to pay for importing them.

DECLINING DEMAND

Demand for petroleum has been dampened in developed countries through conservation. Factories have reduced their demand for petroleum, primarily by consuming more natural gas. The average vehicle driven in the United States got 14 miles per gallon in 1975 and 22 miles per gallon in 1985, and it will get an anticipated 54 miles per gallon in 2025. A government mandate known as Corporate Average Fuel Economy (CAFE) has been responsible for the higher standard.

Other countries also mandate more fuel-efficient vehicles (see Chapter 13). The U.S. Department of Energy forecasts that energy consumption in the United States will

▼ **FIGURE 11-44 PETROLEUM TRADE** The largest flows of petroleum are from Latin America to the United States and from Southwest Asia to Europe and elsewhere in Asia.

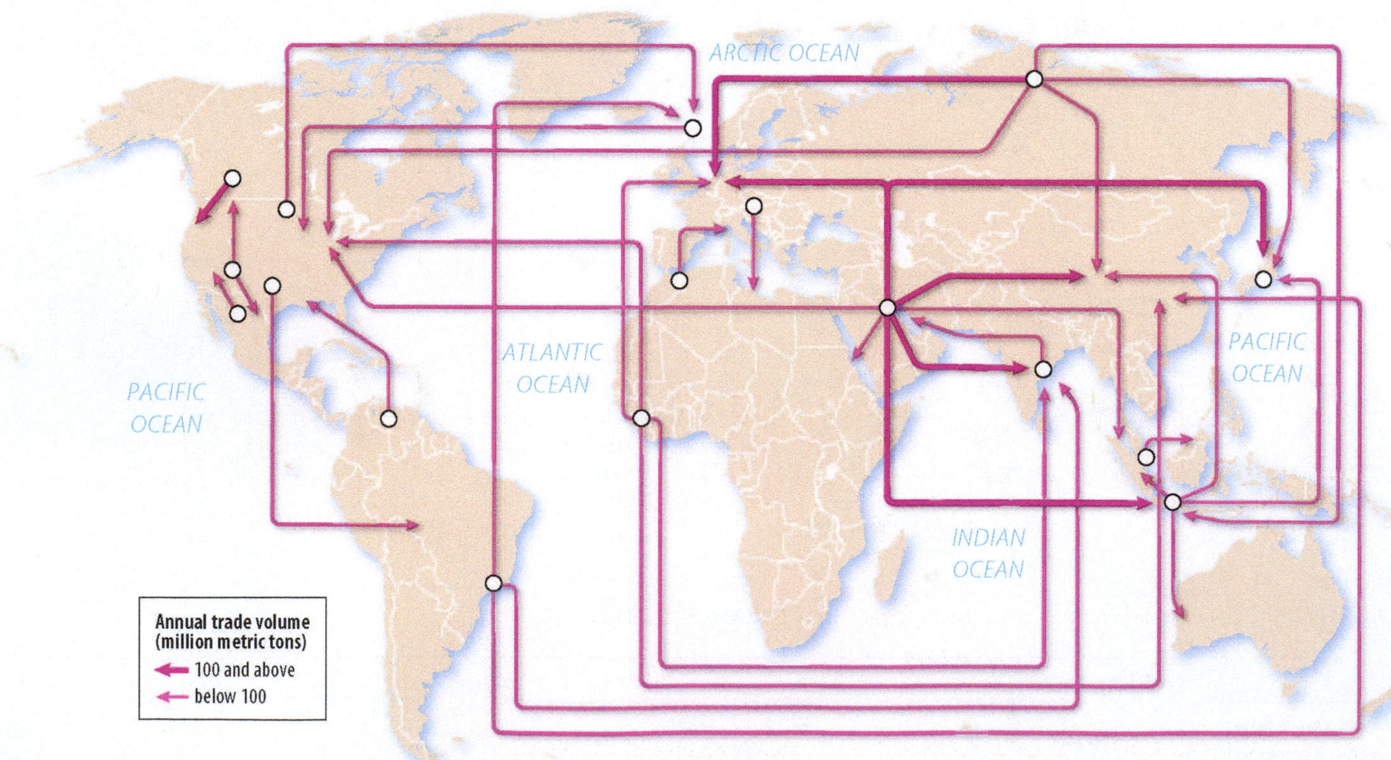

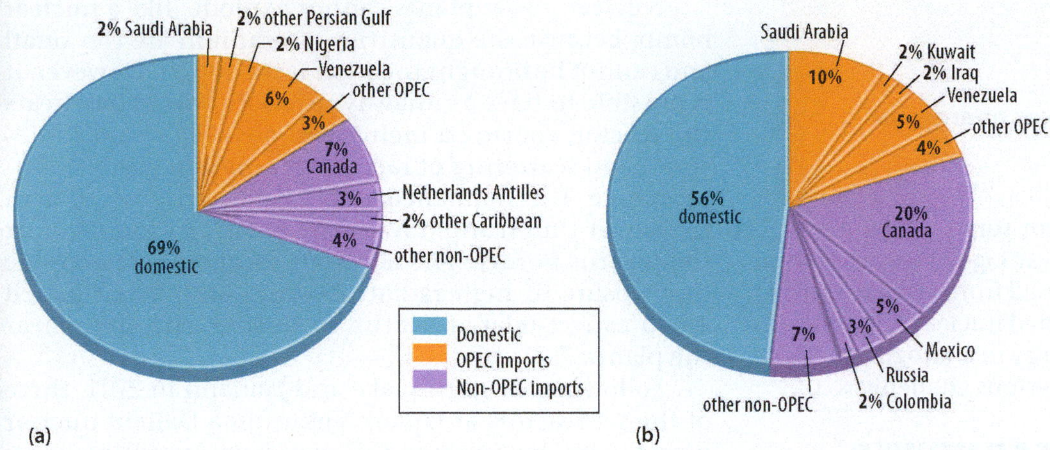

▲ **FIGURE 11-45 U.S. PETROLEUM SOURCES** The United States imports a higher percentage of petroleum **(a)** now than **(b)** in the 1970s. The largest increases have come from Canada and Saudi Arabia.

increase from 97.1 quad BTU in 2013 to 105.7 quad BTU in 2040. Because of population increase, consumption per person is forecast to decline by 10 percent. Consumption of natural gas and energy from sources other than fossil fuels are expected to increase, whereas consumption of petroleum and coal are expected to remain virtually unchanged. However, demand for petroleum can fall if the price rises or increase if the price declines (Figure 11-46).

The world will not literally "run out" of petroleum during the twenty-first century. However, at some point, extracting the remaining petroleum reserves will be so expensive and environmentally damaging that use of alternative energy sources will accelerate, and dependency on petroleum will diminish. The issues for the world are whether dwindling petroleum reserves are handled wisely and other energy sources are substituted peacefully. Given the massive growth in petroleum consumption expected in developing countries such as China and India, the United States and other developed countries may have little influence over when prices rise and when supplies decline. In this challenging environment, all countries will need to pursue sustainable development strategies based on increased reliance on renewable energy sources.

PAUSE & REFLECT 11.3.4

Do you consider fuel efficiency when you or your family buy a new vehicle? Why or why not?

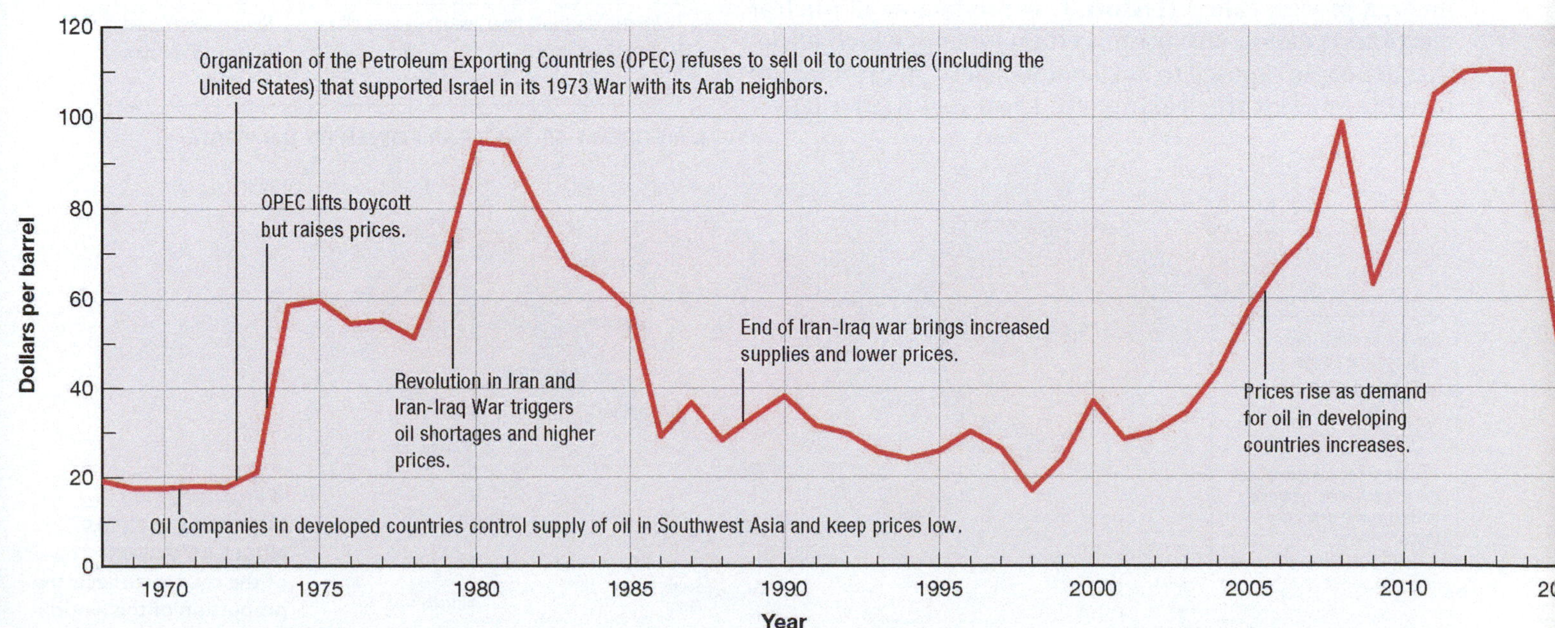

▲ **FIGURE 11-46 OIL PRICE HISTORY** Oil prices have changed sharply on several occasions.

Nuclear Energy

LEARNING OUTCOME 11.3.5

Describe the distribution of nuclear energy and challenges in using it.

Nuclear power is not renewable, but some view it as an alternative to fossil fuels. The big advantage of nuclear power is the large amount of energy released from a small amount of material. One kilogram of enriched nuclear fuel contains more than 2 million times the energy in 1 kilogram of coal. However, nuclear power presents serious challenges.

DISTRIBUTION OF NUCLEAR POWER

Nuclear power supplies 14 percent of the world's electricity. Two-thirds of the world's nuclear power is generated in developed countries, with Europe and North America responsible for generating one-third each. Only 30 of the world's nearly 200 countries make some use of nuclear power, including 19 developed countries and only 11 developing countries. The countries most highly dependent on nuclear power are clustered in Europe (Figure 11-47), where it supplies 80 percent of all electricity in France and more than 50 percent in Belgium, Slovakia, and Ukraine.

Dependency on nuclear power varies widely among U.S. states (Figure 11-48). Nuclear power accounts for more than 70 percent of electricity in Vermont and more than one-half in Connecticut, New Jersey, and South Carolina. At the other extreme, 20 states and the District of Columbia have no nuclear power plants.

POTENTIAL ACCIDENTS

A nuclear power plant produces electricity from energy released by splitting uranium atoms in a controlled environment, a process called **fission**. One product of all nuclear reactions is radioactive waste, certain types of which are lethal to people exposed to it. Elaborate safety precautions are taken to prevent the leaking of nuclear fuel from a power plant.

Nuclear power plants cannot explode, like a nuclear bomb, because the quantities of uranium are too small and cannot be brought together fast enough. However, it is possible to have a runaway reaction, which overheats the reactor, causing a meltdown, possible steam explosions, and scattering of radioactive material into the atmosphere. This happened in 1986 at Chernobyl, then in the Soviet Union and now in the north of Ukraine, near the Belarus border. The accident caused 56 deaths due to exposure to high radiation doses and an estimated 4,000 cancer-related deaths to people who lived near the plant.

Following an earthquake and tsunami in 2011, three of the six reactors at Japan's Fukushima Daiichi nuclear power plant experienced full meltdown, resulting in release of radioactive materials. Three workers died; the death toll among nearby residents exposed to high levels of radioactivity won't be known for years.

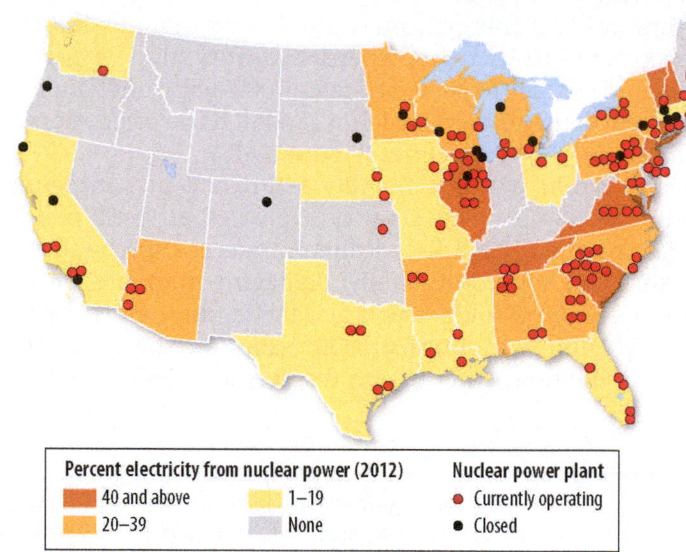

▲ **FIGURE 11-48 NUCLEAR POWER BY U.S. STATE**

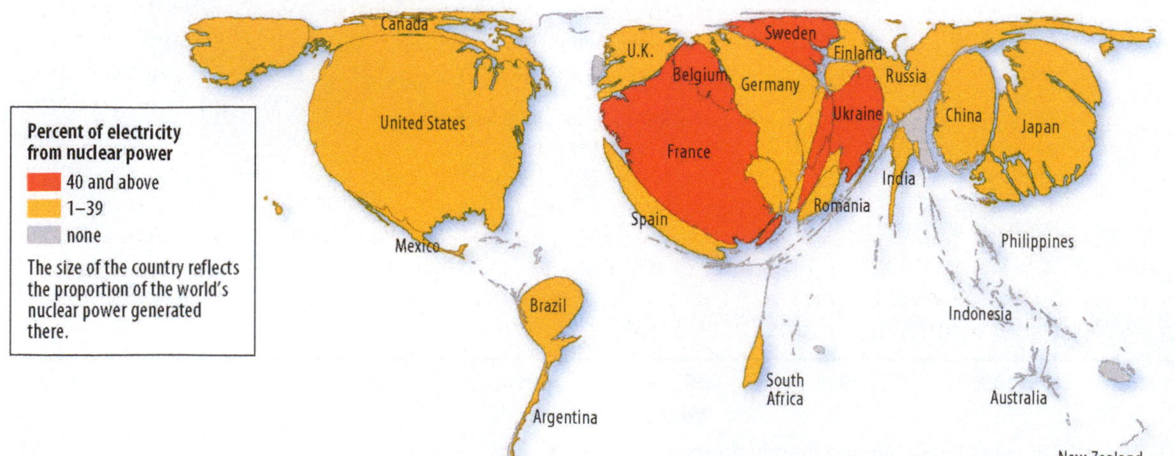

◀ **FIGURE 11-47 ELECTRICITY FROM NUCLEAR POWER** The size of the country reflects the proportion of the world's nuclear power generated there.

RADIOACTIVE WASTE

The waste from nuclear fission is highly radioactive and lethal, and it remains so for many years. Plutonium for making nuclear weapons can be harvested from this waste. Pipes, concrete, and water near the fissioning fuel also become "hot" with radioactivity. No one has yet devised permanent storage for radioactive waste. The waste cannot be burned or chemically treated, and it must be isolated for several thousand years until it loses its radioactivity. Spent fuel in the United States is stored "temporarily" in cooling tanks at nuclear power plants, but these tanks are nearly full.

The United States is Earth's third-largest country in land area, yet it has failed to find a suitable underground storage site because of worry about groundwater contamination. In 2002, the U.S. Department of Energy approved a plan to store the waste in Nevada's Yucca Mountains. But soon after taking office in 2009, the Obama administration reversed the decision and halted construction on the nearly complete repository.

BOMB MATERIAL

Nuclear power has been used in warfare twice, in August 1945, when the United States dropped atomic bombs on Hiroshima and Nagasaki, Japan, ending World War II. No government has dared to use these bombs in a war since then because leaders recognize that a full-scale nuclear conflict could terminate human civilization.

The United States and Russia (previously the Soviet Union) each have several thousand nuclear weapons. China, France, and the United Kingdom have several hundred nuclear weapons each, India and Pakistan several dozen each, and North Korea a handful. Israel is suspected of possessing nuclear weapons but has not admitted to it, and Iran has been developing the capability. Other countries have initiated nuclear programs over the years but have not advanced to the weapons stage. The diffusion of nuclear programs to countries sympathetic to terrorists has been particularly worrying to the rest of the world and has been a major factor in long-time tensions between Iran and other countries that do not want Iran to gain the capability of building a nuclear weapon.

PAUSE & REFLECT 11.3.5

Should dependency on nuclear energy be increased or decreased? Why?

LIMITED RESERVES

Like fossil fuels, uranium is a nonrenewable resource. Proven uranium reserves will last about 124 years at current rates of use. And they are not distributed uniformly around the world: Australia has 23 percent of the world's proven uranium reserves, Kazakhstan 15 percent, and Russia 10 percent (Figure 11-49).

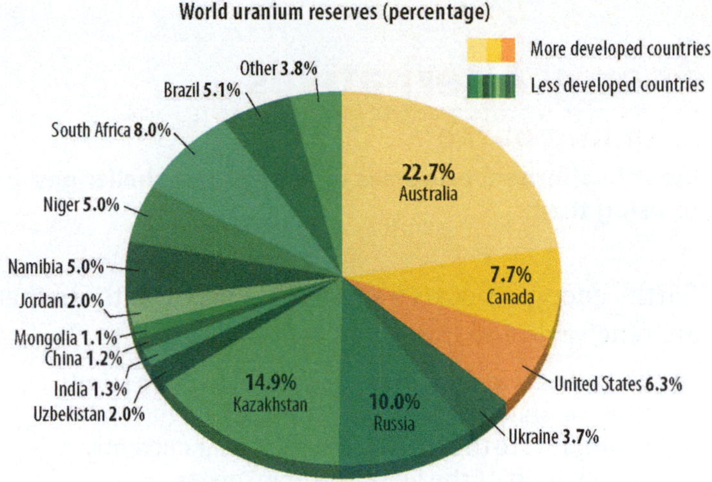

▲ **FIGURE 11-49 WORLD URANIUM RESERVES**

The chemical composition of natural uranium further aggravates the scarcity problem. Uranium ore naturally contains only 0.7 percent U-235; a greater concentration is needed for power generation.

A breeder reactor turns uranium into a renewable resource by generating plutonium, also a nuclear fuel. However, plutonium is more lethal than uranium and could cause more deaths and injuries in an accident. It is also easier to fashion into a bomb. Because of these risks, few breeder reactors have been built, and none are in the United States.

HIGH COST

Nuclear power plants cost several billion dollars to build, primarily because of the elaborate safety measures required. Without double and triple backup systems at nuclear power plants, nuclear energy would be too dangerous to use. Uranium is mined in one place, refined in another, and used in still another. As with coal, mining uranium can pollute land and water and damage miners' health. The complexities of safe transportation add to the cost. As a result, generating electricity from nuclear plants is much more expensive than from coal-burning plants. The future of nuclear power has been seriously hurt by the high costs associated with reducing risks.

Some nuclear power issues might be addressed through nuclear **fusion**, which is the fusing of hydrogen atoms to form helium. Fusion can occur only at very high temperatures (millions of degrees) that cannot been generated on a sustained basis in a power-plant reactor with current technology.

Energy Alternatives

LEARNING OUTCOME 11.3.6

Identify alternative sources of energy and challenges to using them.

Earth's energy resources are divided between those that are renewable and those that are not:

- **Nonrenewable energy** resources form so slowly that for practical purposes, they cannot be renewed. Examples are the three fossil fuels that currently supply most of the world's energy needs.
- **Renewable energy** resources have an essentially unlimited supply and are not depleted when used by people. Water, wind, and the Sun provide sources of renewable energy.

HYDROELECTRIC POWER

Generating electricity from the movement of water is called hydroelectric power (Figure 11-50). Hydroelectric is now the world's second-most-popular source of electricity, after coal. Two-thirds of the world's hydroelectric power is generated in developing countries and one-third in developed countries. A number of developing countries depend on hydroelectric power for most of their electricity (Figure 11-51).

The most populous country to depend primarily on hydroelectric power is Brazil. Brazil has made considerable progress toward sustainable development by generating 80 percent of its electricity from hydroelectric power and 7 percent from other renewable sources.

▲ **FIGURE 11-50 HYDROELECTRIC POWER** Cheoah Dam in Tapoco, Tennessee, provides electricity for Alcoa's nearby aluminum factory.

Among developed countries, Canada gets 60 percent of its electricity from hydroelectric power. The United States is a leading producer of hydroelectric power, but obtains only 8 percent of its electricity from that source. This percentage may decline because few acceptable sites to build new dams remain.

BIOMASS

Biomass fuel is fuel derived from plant material and animal waste. Biomass energy sources include wood and crops. When carefully harvested in forests, wood is a renewable resource that can be used to generate electricity and heat. The waste from processing wood, such as for building construction and demolition, is also available. And crops such as sugarcane, corn, and soybeans can be processed into motor-vehicle fuels. Worldwide production of biomass fuel is approximately 3 quad BTUs, including one-third each

▲ **FIGURE 11-51 ELECTRICITY FROM HYDROELECTRIC POWER**

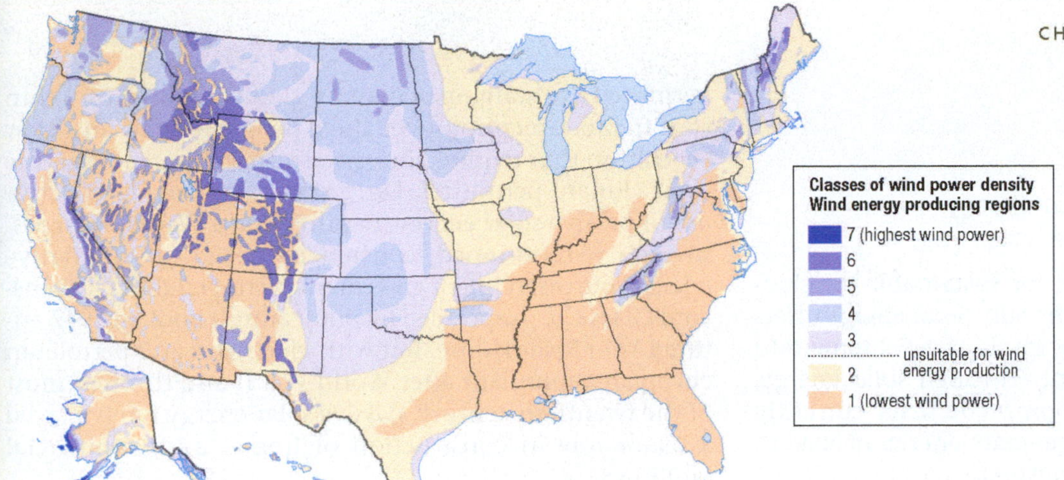

◀ **FIGURE 11-52 WIND POWER** Winds are especially strong enough to support generation of power in the U.S. Plains states.

in North America, Europe, and developing regions. Brazil makes extensive use of biomass to fuel its cars and trucks.

The potential for increasing the use of biomass for fuel is limited, for several reasons:

- Burning biomass may be inefficient because the energy used to produce the crops may be as much as the energy supplied by the crops.
- Biomass already serves essential purposes other than energy, such as providing much of Earth's food, clothing, and shelter.
- When wood is burned for fuel instead of being left in the forest, the fertility of the forest may be reduced.

WIND POWER

The benefits of wind-generated power seem irresistible. Construction of a wind turbine modifies the environment less severely than does construction of a dam. And wind power has greater potential for increased use because only a small portion of the resource has been harnessed.

Despite its attractions, wind power has been harnessed in only a few places. China, North America, and Europe together account for 90 percent of total world production. A significant obstacle for developing countries is the high cost of constructing wind turbines.

Hundreds of wind "farms" consisting of large numbers of wind turbines have been constructed across the United States. One-third of the country is considered windy enough to make wind power economically feasible (Figure 11-52).

Wind power has divided the environmental community. Some oppose construction of wind turbines because they can be noisy and lethal for birds and bats. Some also find them a visual blight when they're constructed in places of outstanding beauty (Figure 11-53).

▼ **FIGURE 11-53 WIND TURBINE FARM** Makara Wind Farm, near Wellington, New Zealand.

GEOTHERMAL ENERGY

Energy from this hot water or steam is called **geothermal energy.** Natural nuclear reactions make Earth's interior hot. Toward the surface, in volcanic areas, this heat is especially pronounced. The hot rocks can encounter groundwater and produce heated water or steam that can be tapped by wells.

Harnessing geothermal energy is most feasible at sites along Earth's surface where crustal plates meet, which are also the sites of many earthquakes and volcanoes. The United States, the Philippines, and Indonesia are the leading producers of geothermal power (Figure 11-54). Ironically, in Iceland, named for its glaciers, nearly all homes and businesses in the capital Reykjavik are heated with geothermal steam.

PAUSE & REFLECT 11.3.6

Does your local power company offer options for using alternative energy? Check the company's website.

▼ **FIGURE 11-54 DISTRIBUTION OF GEOTHERMAL ENERGY**

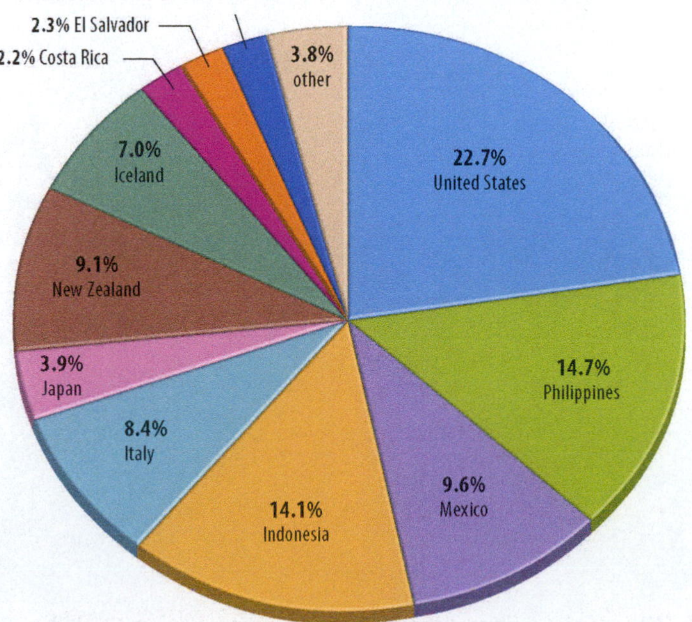

Solar Energy

LEARNING OUTCOME 11.3.7

Compare passive and active solar energy.

The ultimate renewable resource for sustainable development is solar energy supplied by the Sun. Solar energy offers the possibility for countries at low levels of development to promote sustainable development. Through solar energy, people and businesses in developing countries currently unable to obtain electricity can generate energy needed to operate businesses, schools, and hospitals.

Solar sources currently supply the United States with only 1 percent of electricity, but the potential for growth is limitless. The Sun's remaining life is estimated at 5 billion years, and humans appear to be incapable of destroying or depleting that resource. The Sun's energy is free and ubiquitous and cannot be exclusively owned, bought, or sold by any particular individual or enterprise. Utilizing the Sun as a resource does not damage the environment or cause pollution, as does the extraction and burning of nonrenewable fossil fuels.

PASSIVE SOLAR ENERGY

Solar energy is harnessed through either passive or active means. **Passive solar energy systems** capture energy without using special devices. These systems use south-facing windows and dark surfaces to heat and light buildings on sunny days. The Sun's rays penetrate the windows and are converted to heat. Humans act as passive solar energy collectors when they are warmed by sunlight. And since dark objects absorb more energy than light ones, wearing dark clothing warms a person exposed to sunlight even more.

Reliance on passive solar energy increased during the nineteenth century when construction innovations first permitted the hanging of massive glass "curtains" on a thin steel frame. Greenhouses enabled people to grow and view vegetation that required more warmth to flourish than the local climate permitted. Early skyscrapers made effective use of passive solar energy. During World War II when fossil fuels were rationed, consumers looked for alternative energy sources. A major glass manufacturer, Libbey-Owens-Ford Glass Co., responded by publishing a book in 1947 entitled Your Solar House. But with electricity and petroleum cheap and abundant after World War II and through most of the twentieth century, passive solar energy rarely played a major role in construction of homes and commercial buildings.

In recent years, building construction and remodeling have made more use of passive solar energy through advances in glass technology. Double- and triple-pane windows have higher insulating values, and low-E (low emissivity) glass can be coated to let heat in but not out. Window panes made with this glass are filled with argon or other gases that increase their insulating values beyond that of windows that have just air between the panes. Phase-change technologies can also switch the glass from opaque to translucent when a voltage is applied.

PAUSE & REFLECT 11.3.7

Why are people warned not to leave a dog or child unattended in a parked car during the summer?

ACTIVE SOLAR ENERGY

Active solar energy systems collect solar energy and convert it either to heat energy or to electricity (Figure 11-55). The conversion can be accomplished either directly or indirectly.

In direct electric conversion, solar radiation is captured with photovoltaic cells, which convert light energy

◀ **FIGURE 11-55**
SOLAR ENERGY
Solar panels generate electricity in Odoru, Nigeria.

to electrical energy. Bell Laboratories invented the photovoltaic cell in 1954. Each cell generates only a small electric current, but large numbers of these cells wired together produce significant electricity. These cells are made primarily of silicon (also used in computers), the second most abundant element in Earth's crust. When the silicon is combined with one or more other materials, it exhibits distinctive electrical properties in the presence of sunlight, known as the photovoltaic effect. Electrons excited by the light move through the silicon, producing direct current (DC) electricity.

In indirect electric conversion, solar radiation is first converted to heat and then to electricity. The Sun's rays are concentrated by reflectors onto a pipe filled with synthetic oil. The heat from the oil-filled pipe generates steam to run turbines. In heat conversion, solar radiation is concentrated with large reflectors and lenses to heat water or rocks. These store the energy for use at night and on cloudy days. A place that receives relatively little sunlight can use solar energy by using more reflectors and lenses and larger storage containers.

SOLAR POWERED ELECTRICITY

Solar power can be produced at a central station and distributed by an electric company, as coal- and nuclear-generated electricity are now supplied. However, with coal still relatively cheap and investment in nuclear facilities already substantial, public and private utility companies have had little interest in solar technology.

In developed countries, solar-generated electricity is used in spacecraft, light-powered calculators, and at remote sites where conventional power is unavailable, such as California's Mojave Desert. Solar energy is used primarily as a substitute for electricity in heating water. Rooftop devices collect, heat, and store water for apartment buildings in Israel and Japan and individual homes in the United States (Figure 11-56). The initial cost of installing a solar water heater is higher than hooking into the central system but may be justified if an individual plans to stay in the same house for a long time.

▲ **FIGURE 11-56 SOLAR PANELS** Solar panels installed on apartment rooftops in the Old City of Jerusalem are used to heat water, which is stored in the adjacent tanks. The domes are the Church of the Holy Sepulchre, built at the site where Jesus is believed to have been crucified, buried, and resurrected (see Chapter 6).

Electricity was popular in early motor vehicles. Of the 4,000 cars sold in the United States in 1900, 38 percent were powered by electricity, 40 percent by steam, and only 22 percent by gasoline. The electric car was especially popular in 1900 in large cities of the Northeast, such as New York and Philadelphia, where their relative quietness and cleanliness made them popular as taxicabs. Women also preferred electric cars because they were easier to start than gasoline- or steam-powered ones.

The main shortcomings of the electric car in the early 1900s remain unchanged a century later. Compared to gasoline power, the electric-powered vehicle has a more limited range and costs more to operate. Recharging the battery can take several hours. To address these issues, carmakers offer a variety of vehicles that combine electric and gasoline power. Hybrid vehicles conserve gasoline by running on electricity at low speeds. Other vehicles operate exclusively on battery-powered electricity and use the gasoline engine to recharge the battery (see Chapter 13).

In developing countries, the largest and fastest-growing market for photovoltaic cells includes the 2 billion people who lack electricity, especially residents of remote villages. For example, in sub-Saharan Africa, more homes have been electrified in recent years using photovoltaic cells than by hooking up to the central power grid. In Morocco, solar panels are sold in bazaars and open markets, next to carpets and tinware.

Solar energy currently accounts for only 0.3 quad BTU worldwide. The cost of cells must drop and their efficiency must improve for solar power to expand rapidly, with or without government support. Solar energy will become more attractive as other energy sources become more expensive. A bright future for solar energy is indicated by the fact that petroleum companies now own the major U.S. manufacturers of photovoltaic cells.

Air Pollution

LEARNING OUTCOME 11.3.8

Describe causes and effects of air pollution at global, regional, and local scales.

Industry is a major polluter of air, water, and land. People rely on air, water, and land to remove and disperse waste from factories as well as from other human activities. **Pollution** occurs when more waste is added than air, water, and land resources can handle.

At ground level, Earth's average atmosphere is made up of about 78 percent nitrogen, 21 percent oxygen, and less than 1 percent argon. The remaining 0.04 percent includes several trace gases, some of which are critical. **Air pollution** is a concentration of trace substances at a greater level than occurs in average air. Concentrations of these trace gases in the air can damage property and adversely affect the health of people, other animals, and plants.

Most air pollution is generated from factories and power plants, as well as from motor vehicles. Factories and power plants produce sulfur dioxides and solid particulates, primarily from burning coal. Burning petroleum in motor vehicles produces carbon monoxide, hydrocarbons, and nitrogen oxides.

GLOBAL-SCALE AIR POLLUTION

Air pollution concerns geographers at three scales: global, regional, and local. Two global-scale issues are climate change and ozone damage.

CLIMATE CHANGE. The average temperature of Earth's surface increased by 0.89°C (1.6°F) between 1880 and 2014 (Figure 11-57). An international team of UN scientists has concluded that this temperature increase is directly linked to human actions, especially the burning of fossil fuels in factories and vehicles.

When fossil fuels are burned, carbon dioxide is discharged into the atmosphere. Plants and oceans absorb much of the discharge, but increased fossil fuel burning during the past 200 years has caused the level of carbon dioxide in the atmosphere to rise by more than one-fourth, according to the UN scientists (refer to Figure 11-57). A concentration of trace gases in the atmosphere can delay the return of some of the heat leaving Earth's surface heading for space, thereby raising Earth's temperatures. The increase in Earth's temperature, caused by carbon dioxide trapping some of the radiation emitted by the surface, is called the **greenhouse effect.**

As a country's per capita income increases, its per capita carbon dioxide emissions generally increase. Some of the wealthiest countries, located primarily n Europe, with GNI per capita between $30,000 and 50,000, show declines in pollution. However, the rld's richest countries, including the United States l several countries in Southwest Asia, display the est pollution levels (Figure 11-58).

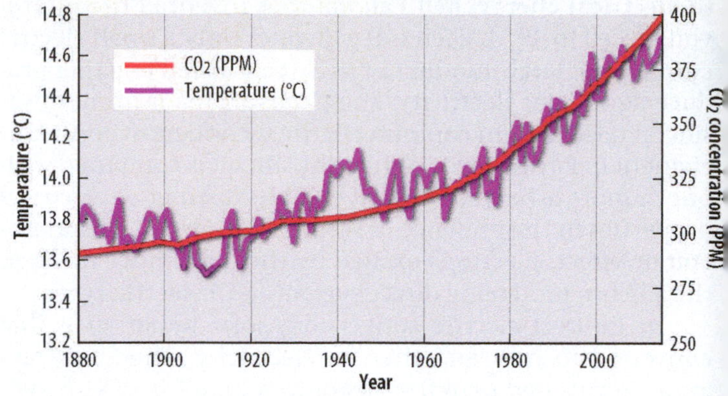

▲ **FIGURE 11-57 CLIMATE CHANGE AND CO_2 CONCENTRATIONS** Since 1880, carbon dioxide concentration has increased by more than one-third, and Earth has warmed by about 1°C (2°F).

OZONE DAMAGE. **Ozone** is a gas that absorbs ultraviolet radiation in the stratosphere, a zone between 15 and 50 kilometers (9 to 30 miles) above Earth's surface. Were it not for the ozone in the stratosphere, UV rays would damage plants, cause skin cancer, and disrupt food chains.

Earth's protective ozone layer is threatened by pollutants called **chlorofluorocarbons (CFCs).** CFCs such as Freon were once widely used as coolants in refrigerators and air conditioners. When they leak from these appliances, the CFCs are carried into the stratosphere, where they break down Earth's protective layer of ozone gas. In 2007, virtually all countries of the world agreed to cease using CFCs by 2020 in developed countries and by 2030 in developing countries.

PAUSE & REFLECT 11.3.8

What gas is now most commonly used as a coolant instead of CFC? Use a search engine to find the answer.

REGIONAL-SCALE AIR POLLUTION

At the regional scale, air pollution may damage a region's vegetation and water supply through acid deposition. The world's three principal industrial regions are especially affected by acid deposition.

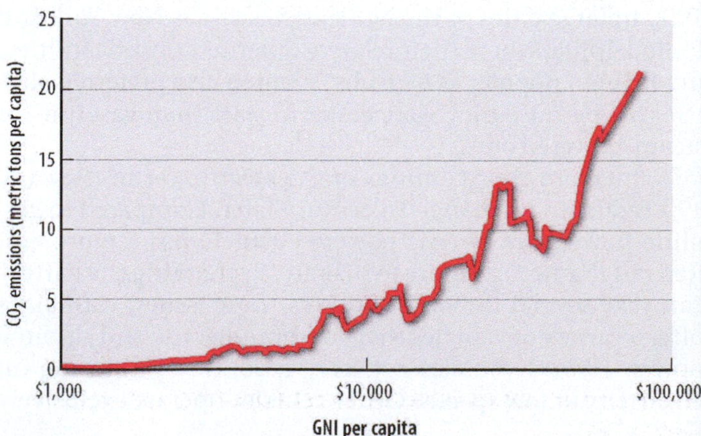

▲ **FIGURE 11-58 GNI AND POLLUTION**

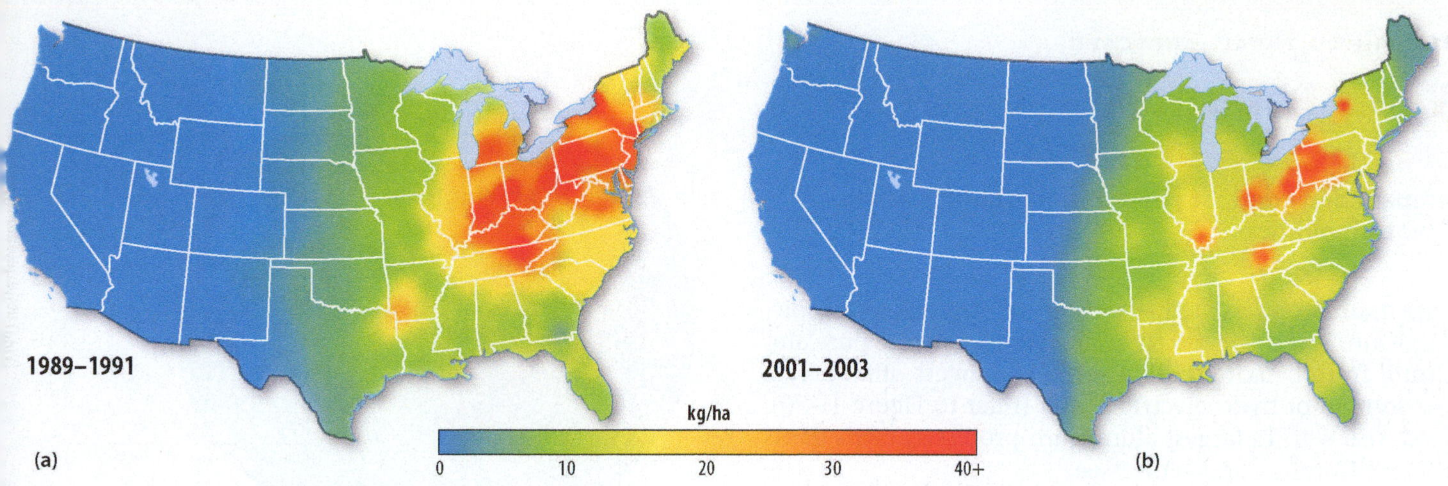

▲ **FIGURE 11-59 ACID DEPOSITION** The worst effects of acid deposition are experienced further east than the principal sources of emissions.

Acid deposition is the accumulation of acids on Earth's surface. Sulfur oxides and nitrogen oxides, emitted by burning fossil fuels, enter Earth's atmosphere, combine with oxygen and water to form sulfuric acid and nitric acid, and are deposited on Earth's surface. **Acid precipitation** is the conversion of sulfur oxides and nitrogen oxides to acids that return to Earth as rain, snow, or fog. The acids can also be deposited in dust.

Acid precipitation damages lakes, killing fish and plants. On land, concentrations of acid in the soil can injure plants by depriving them of nutrients and can harm worms and insects. Buildings and monuments made of marble and limestone have suffered corrosion from acid rain. The United States has reduced sulfur dioxide emissions significantly in recent years.

Geographers are particularly interested in the effects of acid precipitation because the worst damage may not be experienced at the same location as the emission of the pollutants. Before they reach the surface, these acidic droplets might be carried hundreds of kilometers. Within the United States, the major generators of acid deposition are in Ohio and other industrial states along the southern Great Lakes, but the severest effects of acid rain are felt in several areas farther east (Figure 11-59).

LOCAL-SCALE AIR POLLUTION

At the local scale, air pollution is especially severe in places where emission sources are concentrated, such as in urban areas. The air above urban areas may be polluted because a large number of factories, motor vehicles, and other polluters emit residuals in a concentrated area.

Urban air pollution has three basic components:

- **Carbon monoxide.** Breathing carbon monoxide reduces the oxygen level in blood, impairs vision and alertness, and threatens those with breathing problems.
- **Hydrocarbons.** Hydrocarbons and nitrogen oxides in the presence of sunlight form **photochemical smog,** which causes respiratory problems, stinging in the eyes, and an ugly haze over cities.
- **Particulates.** These pollutants include dust and smoke particles. The dark plume of smoke from a factory stack and the exhaust of a diesel truck are examples of particulate emission.

According to the World Health Organization, most of the 20 most polluted cities are in South Asia (Figure 11-60). The city with the world's most polluted air (measured as the highest level of airborne particulates) is Delhi, India (Figure 11-61). The pollution level in Delhi is six times what the WHO considers the safe maximum.

Black Sea
Caspian Sea
Igdir
TURKEY
TURKMENISTAN
SYRIA
IRAQ
Khoramabad
IRAN
Kabul
AFGHANISTAN
Peshawar
Rawalpindi
Amritsar
CHINA
PAKISTAN
Delhi
NEPAL
Agra
Firozabad
Jodhpur
Gwalior
Patna
QATAR
Doha
Karachi
Allahabad
Gazipur
Dhaka
SAUDI ARABIA
Ahmedabad
INDIA
Narayonganj
Red Sea
OMAN
Arabian Sea
Raipur
BANGLADESH
0 300 600 Miles
0 300 600 Kilometers
N

▲ **FIGURE 11-60 CITIES WITH THE MOST POLLUTED AIR** Most are in South Asia.

▼ **FIGURE 11-61 AIR POLLUTION** Delhi, India.

Water Pollution

LEARNING OUTCOME 11.3.9

Compare and contrast point and nonpoint sources of water pollution.

Some manufacturers are heavy users of water. For example, a large amount of electricity is needed to separate pure aluminum from bauxite so aluminum producers often locate near sources of hydroelectric power (refer to Figure 11-50). Alcoa, the world's largest aluminum producer, owns dams in North Carolina and Tennessee.

Water serves many other human purposes. People must drink water to survive. It is used for cooking and bathing. Water provides a location for boating, swimming, fishing, and other recreational activities. It is home to fish and other aquatic life. Water is used for most economic activities, including agriculture and services, as well as manufacturing.

These uses depend on fresh, clean, unpolluted water. But that is not always available because people also use water for purposes that pollute it. Pollution is widespread because it is easy to dump waste into a river and let the water carry it downstream, where it becomes someone else's problem. Water can decompose some waste without adversely impacting other activities, but the volume of waste often exceeds the capacity that many rivers and lakes can accommodate.

DEMAND FOR WATER

Humans use around 3.6 trillion cubic meters (950 billion gallons) of water per year, or around 500 cubic meters (132,000 gallons) per capita. The heaviest demand is for electricity, followed by agriculture and municipal sewage systems (Figure 11-62).

Water usage is either nonconsumptive or consumptive:

- **Nonconsumptive water usage** is use of water that is returned to nature as a liquid. Most industrial and municipal uses of water are nonconsumptive because the wastewater is primarily discharged into lakes and streams.
- **Consumptive water usage** is use of water that evaporates rather than being returned to nature as a liquid. Most agricultural uses are consumptive because the water is used primarily to supply plants that transpire it and therefore cannot be treated and reused.

North America has the world's highest per capita consumption of water, more than three times the worldwide average (Figure 11-63). Water usage is extremely high in the United States primarily because of agriculture rather than industry. U.S. farmers raise a large number of animals to meet the high demand for meat that the average American consumes, as discussed in Chapter 10. These animals drink a lot of water in their lifetimes. A large amount of water is also needed in U.S. agriculture to irrigate fields of crops.

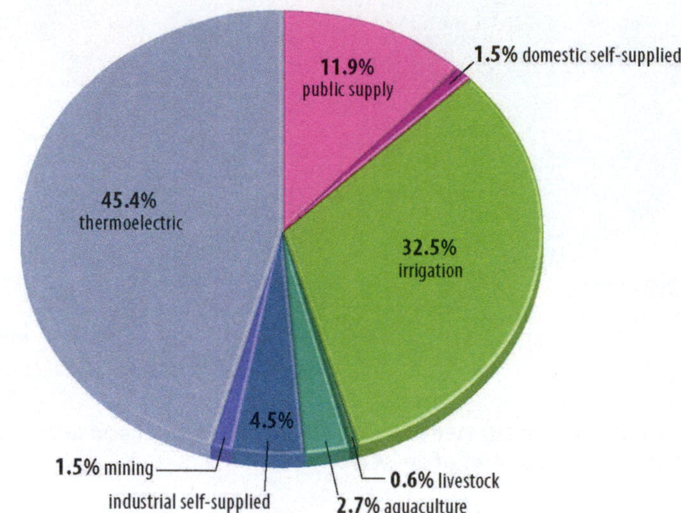

▲ **FIGURE 11-62 WATER WITHDRAWAL BY USE**

IMPACT ON AQUATIC LIFE

Polluted water can harm aquatic life. A commonly used measure of water pollution is **biochemical oxygen demand (BOD)**, which is the amount of oxygen required by aquatic bacteria to decompose a given amount of organic waste. Aquatic plants and animals consume oxygen, and so does the decomposing organic waste that humans dump in the water.

If too much waste is discharged into water, the water becomes oxygen starved, and fish die. This condition is typical when water becomes loaded with municipal sewage or industrial waste. The sewage and industrial pollutants consume so much oxygen that the water can become unlivable for normal plants and animals, creating a "dead" stream or lake (Figure 11-64).

Similarly, when runoff carries fertilizer from farm fields into streams or lakes, the fertilizer nourishes excessive aquatic plant production—a "pond scum" of algae—that consumes too much oxygen. Either type of pollution reduces the normal oxygen level, threatening aquatic plants and animals. Some of the residuals may become concentrated in the fish, making them unsafe for human consumption.

▼ **FIGURE 11-63 PER CAPITA WATER WITHDRAWL** North America has by far the highest per capita use of water.

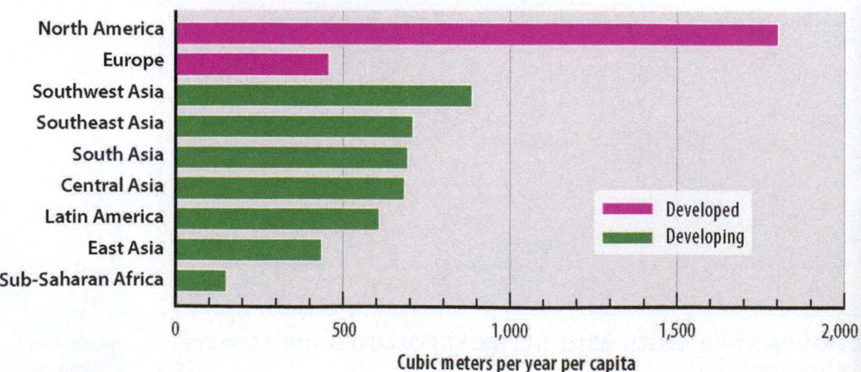

▲ **FIGURE 11-64 POLLUTED RIVER, INDIA**

PAUSE & REFLECT 11.3.9

How might a change in the sources of energy used in factories help improve water quality?

POINT SOURCE POLLUTION

The sources of pollution can be divided into point sources and nonpoint sources:

- **Point source pollution** enters a body of water at a specific location.
- **Nonpoint source pollution** comes from a large, diffuse area.

Point source pollutants are usually smaller in quantity and much easier to control than nonpoint source pollutants.

▼ **FIGURE 11-65 POINT SOURCE POLLUTION** From a factory in Wolfen, Germany.

▲ **FIGURE 11-66 NONPOINT SOURCE POLLUTION** Delhi, India.

Point source water pollution originates from a specific point, such as a pipe from a wastewater treatment plant. These are the two main point sources of water pollution:

- **Water-using manufacturers.** Many factories and power plants use water for cooling and then discharge the warm water back into the river or lake. The warm water may not be polluted with chemicals, but it raises the temperature of the body of water it enters. Fish adapted to cold water, such as salmon and trout, might not be able to survive in the warmer water. Steel, chemicals, paper products, and food processing are major industrial polluters of water. Each requires a large amount of water in the manufacturing process and generates a lot of wastewater.
- **Municipal sewage.** In developed countries, sewers carry wastewater from sinks, bathtubs, and toilets to a municipal treatment plant, where most—but not all—of the pollutants are removed. The treated wastewater is then typically dumped back into a river or lake. In developing countries, sewer systems are rare, and wastewater often drains, untreated, into rivers and lakes. The drinking water, usually removed from the same rivers, may be inadequately treated as well. The combination of untreated water and poor sanitation makes drinking water deadly in some developing countries (Figure 11-65). Waterborne diseases such as cholera, typhoid, and dysentery are major causes of death.

NONPOINT SOURCE POLLUTION

Nonpoint sources usually pollute in greater quantities and are much harder to control than point sources of pollution (Figure 11-66). The principal nonpoint source is agriculture. Fertilizers and pesticides spread on fields to increase agricultural productivity are carried into rivers and lakes by irrigation systems or natural runoff. Expanded use of these products may help avoid a global food crisis, but it may destroy aquatic life by polluting rivers and lakes.

Solid Waste Pollution

LEARNING OUTCOME 11.3.10

Describe principal strategies for disposal of solid waste.

The average American generates about 2 kilograms (4 pounds) of solid waste per day. Overall, residences generate around 60 percent of the solid waste and businesses 40 percent. Paper products, such as corrugated cardboard and newspapers, account for the largest share of solid waste in the United States, especially among residences and retailers. Manufacturers discard large quantities of metals as well as paper.

SANITARY LANDFILL

A **sanitary landfill** is the most common place for disposal of solid waste in the United States. More than one-half of the country's waste is trucked to landfills and buried under soil (Figure 11-67). This strategy is the opposite of our disposal of gaseous and liquid wastes: We disperse air and water pollutants into the atmosphere, rivers, and eventually the ocean, but we concentrate solid waste in thousands of landfills. Concentration would seem to eliminate solid-waste pollution, but it may only hide it—temporarily. Chemicals released by the decomposing solid waste can leak from the landfill into groundwater. This can contaminate water wells, soil, and nearby streams.

The use of landfills has declined sharply in the United States. The percentage of solid waste disposed in a landfill decreased from 89 percent in 1980 to 54 percent in 2010, and the number of landfills in the United States declined from 7,683 in 1986 to 1,908 in 2009. Thousands of small-town "dumps" have been closed and replaced by a small number of large regional ones. Better compaction methods, along with expansion in the land area of some of the large regional dumps, have resulted in expanded landfill capacity.

▲ **FIGURE 11-68 GREAT PACIFIC GARBAGE PATCH** Albatross birds stand beside plastic waste, part of the Great Pacific Garbage Patch that has washed ashore on Midway Atoll.

Some communities now pay to use landfills elsewhere. New York, New Jersey, and Illinois are the states that export the most solid waste. Pennsylvania, Virginia, and Michigan are the leading importers. Low transportation costs mean it is often cheaper to ship waste to states with landfills that charge low fees for accepting solid waste than to dispose of it locally.

The world's largest solid waste disposal site is thought to be the Great Pacific Garbage Patch, which floats in the Pacific Ocean. It is estimated to be twice the size of the state of Texas. The patch is created by slow-moving currents that converge in the area. Trash is collected from all over the world. Plastic constitutes 90 percent of the trash floating in the oceans (Figure 11-68).

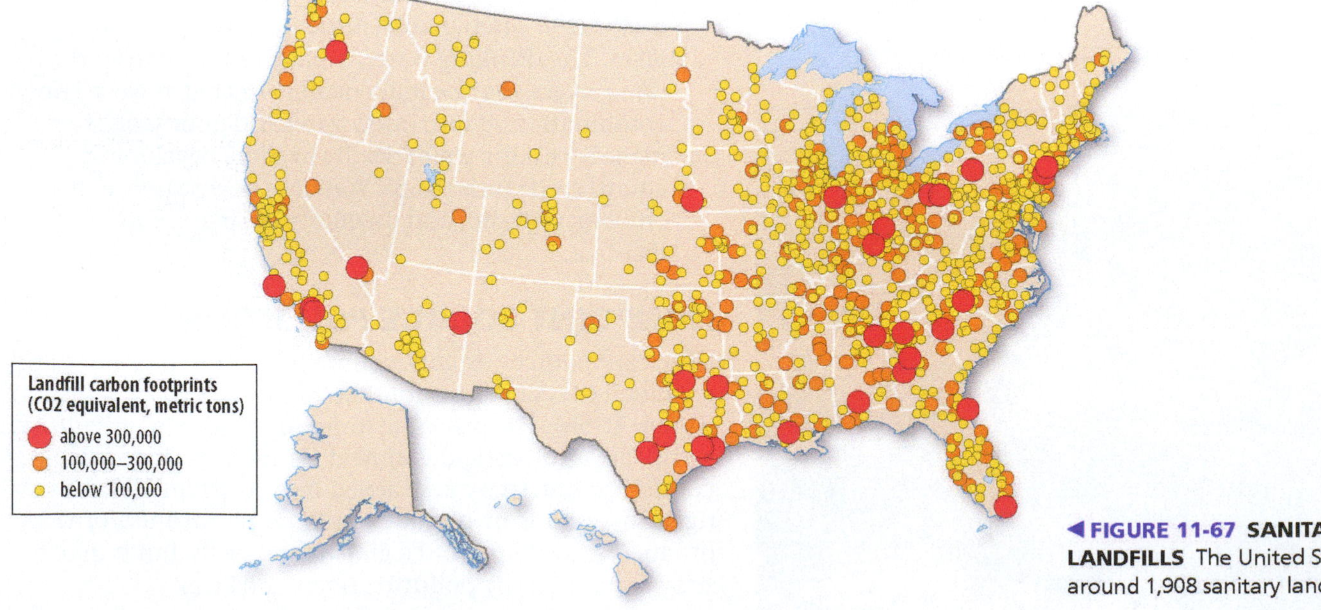

◄ **FIGURE 11-67 SANITARY LANDFILLS** The United States has around 1,908 sanitary landfills.

HAZARDOUS WASTE

Disposing of hazardous waste is especially difficult. Hazardous wastes include heavy metals (including mercury, cadmium, and zinc), PCB oils from electrical equipment, cyanides, strong solvents, acids, and caustics. These may be unwanted byproducts generated in manufacturing or waste to be discarded after usage.

According to the toxic waste inventory published by the U.S. Environmental Protection Agency (EPA), 1.25 billion kilograms (2.75 billion pounds) of toxic chemicals were released into the land in 2013 by the 100 most polluting sites (Figure 11-69). The 4 largest polluters were Red Dog Operations zinc mine in Kotzebue, Alaska; Newmont Mining Corporation's Twin Creeks gold and copper mine in Golconda, Nevada; BHP Billiton's San Manuel copper mine in San Manuel, Arizona; and Kennecott Utah Copper's concentrator in Copperton, Utah.

If poisonous industrial residuals are not carefully placed in protective containers, the chemicals may leach into the soil and contaminate groundwater or escape into the atmosphere. Breathing air or consuming water contaminated with toxic wastes can cause cancer, mutations, chronic ailments, and even immediate death.

PAUSE & REFLECT 11.3.10

Where does your solid waste go?

CHECK-IN KEY ISSUE 3

Why Do Industries Face Resource Challenges?

- ✔ **Supplies of the three fossil fuels—coal, natural gas, and petroleum—are not distributed uniformly.**
- ✔ **Industries are one of four major users of energy, along with transportation, residences, and commercial activities.**
- ✔ **Alternative energy sources include hydroelectricity, wind, nuclear, solar, and geothermal.**
- ✔ **Industry is a major polluter of air, land, and water.**
- ✔ **Air pollution can occur at global, regional, and local scales.**
- ✔ **Water pollution can have point or nonpoint sources.**
- ✔ **Solid waste that is "thrown away" is either transported to landfills or incinerated.**

▼ FIGURE 11-69 TOXIC CHEMICAL RELEASE SITES Ohio has the most sites, although the largest sites are mines in the West.

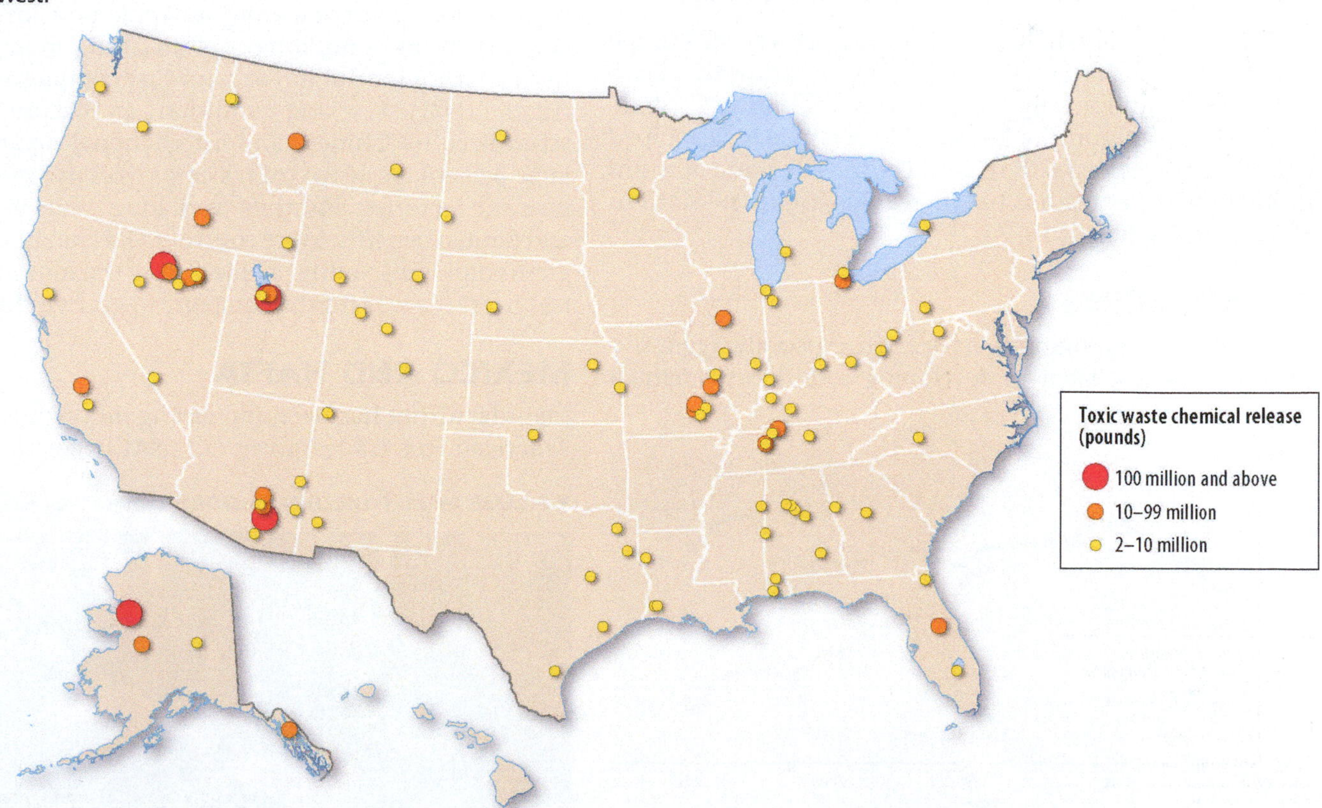

KEY ISSUE 4

Why Are Industries Changing Locations?

- Emerging Industrial Regions
- Industrial Change in Developed Countries
- Skilled or Unskilled Labor?
- Recycling and Remanufacturing

LEARNING OUTCOME 11.4.1

Explain reasons for the emergence of new concentrations of industries.

Industry is on the move around the world. Site factors, especially labor costs, have stimulated industrial growth in new regions, both internationally and in developed regions. Situation factors, especially proximity to growing markets, have also played a role in the emergence of new industrial regions.

Emerging Industrial Regions

In 1970, nearly one-half of world industry was in Europe and nearly one-third was in North America; now these two regions account for only one-fourth each. Industry's share of total economic output has steadily declined in developed countries since the 1970s (Figure 11-70). The share of world industry in other regions has increased—from one-sixth in 1970 to one-half in 2010.

OUTSOURCING

Transnational corporations have been especially aggressive in using low-cost labor in developing countries. To remain competitive in the global economy, they carefully review their production processes to identify steps that can be performed by low-paid, low-skilled workers in developing countries.

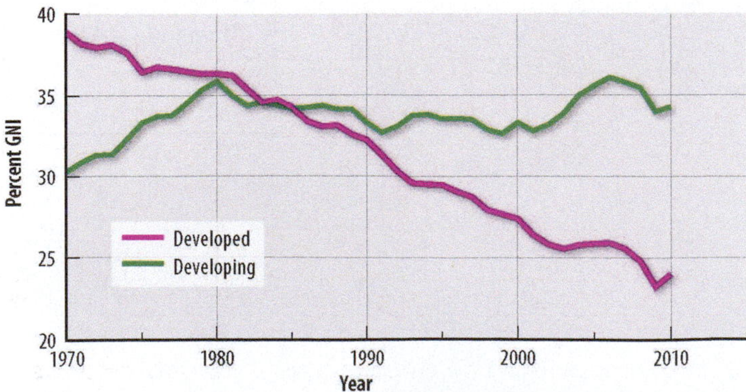

▲ **FIGURE 11-70 MANUFACTURING VALUE AS A PERCENTAGE OF GNI** Manufacturing has accounted for a higher share of GNI in developing countries than in developed countries since the 1990s.

Despite the greater transportation cost, transnational corporations can profitably transfer some work to developing countries, given their substantially lower wages compared to those in developed countries. At the same time, operations that require highly skilled workers remain in factories in developed countries. This selective transfer of some jobs to developing countries is known as the **new international division of labor.**

Transnational corporations allocate production to low-wage countries through **outsourcing,** which is turning over much of the responsibility for production to independent suppliers. Outsourcing has had a major impact on the distribution of manufacturing because each step in the production process is now scrutinized closely in order to determine the optimal location.

Outsourcing contrasts with the approach typical of traditional mass production, called **vertical integration,** in which a company controls all phases of a highly complex production process. Vertical integration was traditionally regarded as a source of strength for manufacturers because it gave them the ability to do and control everything. Carmakers once made nearly all their own parts, for example, but now most of this operation is outsourced to other companies that are able to make the parts cheaper and better.

Outsourcing is especially important in the electronics industry. The world's largest electronics contractor is Foxconn, a major supplier of chips and other electronics components for such companies as Apple and Intel. Foxconn employs around 1 million people in China, including several hundred thousand at its Foxconn City complex in Shenzhen (Figure 11-71). Working conditions at Foxconn have been scrutinized by Chinese and international organizations. A large percentage of Foxconn's employees live in dormitories near the factories, and they work long hours for low wages and limited benefits. More controversial is an internship program employing young people during the summers that critics charge is a way for the company to get free child labor.

MEXICO AND NAFTA

Manufacturing has been increasing in Mexico. The North American Free Trade Agreement (NAFTA), which went into

▼ **FIGURE 11-71 FOXCONN FACTORY** Jiangsu Province, China.

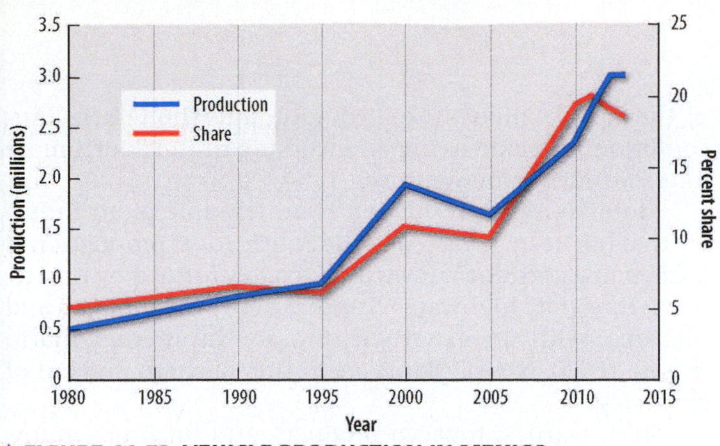

▲ **FIGURE 11-72 VEHICLE PRODUCTION IN MEXICO**

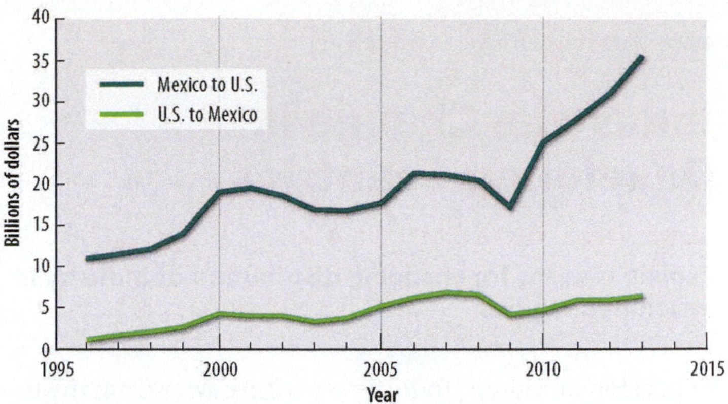

▲ **FIGURE 11-73 U.S.–MEXICO TRADE IN VEHICLES**

effect in 1994, eliminated most barriers to moving goods among Mexico, the United States, and Canada. Because it is the nearest low-wage country to the United States, Mexico attracts industries concerned with both a site factor (low-cost labor) and a situation factor (proximity to U.S. markets).

Nearly all of the growth of motor vehicle production in North America, for example, has been located in Mexico rather than the United States or Canada (Figures 11-72 and 11-73). Plants in Mexico near the U.S. border are known as **maquiladoras.** The term originally applied to a tax when Mexico was a Spanish colony. Under U.S. and Mexican laws, companies receive tax breaks if they ship materials from the United States, assemble components at a maquiladora plant in Mexico, and export the finished product back to the United States. More than 1 million Mexicans are employed at more than 3,000 maquiladoras.

Integration of North American industry has generated fear in the United States and Canada. Labor leaders fear that more manufacturers will relocate production to Mexico to take advantage of lower wage rates. Labor-intensive industries such as food processing, electronics, and textile manufacturing are especially attracted to regions where prevailing wage rates are lower.

Environmentalists fear that NAFTA encourages firms to move production to Mexico because laws there governing air- and water-quality standards are less stringent than in the United States and Canada. Mexico has adopted regulations to reduce air pollution in Mexico City, but environmentalists charge that environmental protection laws are still not strictly enforced in Mexico.

Mexico faces its own challenges. Although much lower than in the United States, Mexican wages are higher than in China. Despite the higher site costs, however, Mexico still competes effectively with China because it has much lower shipping costs to the United States and Europe than does China. Mexico has also signed free trade agreements with many other countries in addition to the United States.

PAUSE & REFLECT 11.4.1

Take a close look at an iPhone (yours or a friend's or relation's). Can you find any evidence that it was made at Foxconn in China?

BRIC COUNTRIES

Much of the world's future growth in manufacturing is expected to cluster in a handful of countries known as BRIC, which is an acronym coined by the investment banking firm Goldman Sachs for Brazil, Russia, India, and China. The foreign ministers of these four countries started meeting in 2006.

The four BRIC countries together encompass one-fourth of the world's land area and contain 3 billion of the world's 7 billion inhabitants, but the four countries combined account for only one-sixth of world GDP (Figure 11-74). Their economies rank second (China), seventh (Brazil), ninth (Russia), and eleventh (India). China is expected to pass the United States as the world's largest economy around 2020, and India is expected to become second around 2050.

China and India have the two largest labor forces, whereas Russia and Brazil are especially rich in inputs critical for industry. China, India, and Russia could form a contiguous industrial region, but long-standing animosity among them has limited their economic interaction so far. Brazil, is of course, not contiguous to the other three. Still, the BRIC concept is that if the four giants work together, they can be the world's dominant industrial bloc in the twenty-first century.

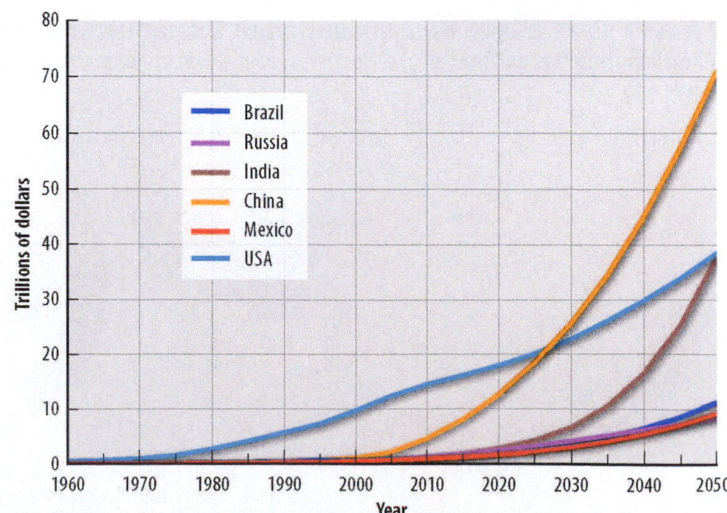

▲ **FIGURE 11-74 GDP FOR BRIC COUNTRIES, MEXICO, AND THE UNITED STATES** China is expected to have the world's largest GDP during the 2020s, and India is expected to be second in the 2050s.

Industrial Change in Developed Countries

LEARNING OUTCOME 11.4.2

Explain reasons for changing distribution of industry in developed regions.

In developed regions, industry is shifting away from the traditional industrial areas of northwestern Europe and the northeastern United States. In the United States, industry has shifted from the Northeast toward the South and West. In Europe, government policies have encouraged relocation toward economically distressed peripheral areas.

INTRAREGIONAL SHIFTS IN NORTH AMERICA

The United States lost 2.3 million manufacturing jobs between 1950 and 2015. The Northeast lost 3.7 million manufacturing jobs during the period, and the Great Lakes states lost 1.6 million. On the other hand, the South added 1.3 million manufacturing jobs, and the far western states gained 1.1 million (Figure 11-75).

Industrialization during the late nineteenth and early twentieth centuries largely bypassed the South, which had not recovered from losing the Civil War. The South lacked the infrastructure needed for industrial development: Road and rail networks were less intensively developed in the South, and electricity was less common than in the North. As a result, the South was the poorest region of the United States. Industrial growth in the South since the 1930s has been stimulated in part by government policies to reduce historical disparities. The Tennessee Valley Authority brought electricity to much of the rural South, and roads were constructed in previously inaccessible sections of the Appalachians, the Piedmont, and the Ozarks. Air-conditioning made living and working in the South more tolerable during the summer.

Motor vehicle production is an example of an industry that has been attracted to the South. Most production is located in a corridor known as auto alley formed by north–south interstate highways 65 and 75 between Michigan and Alabama, with an extension into southwestern Ontario (Figure 11-76). Newer plants are in the southern portion of auto alley.

Steel, textiles, tobacco products, and furniture industries have become dispersed through smaller communities in the South, many in search of a labor force willing to work for less pay than in the North and forgo joining a union. The Gulf Coast has become an important industrial area because of its access to oil and natural gas. Along the Gulf Coast are oil refining, petrochemical manufacturing, food processing, and aerospace product manufacturing.

The principal lure for many manufacturers has been right-to-work laws. A **right-to-work law** requires a factory to maintain a "open shop" and prohibits a "closed shop." In a closed shop, a company and a union agree that everyone must join the union to work in the factory. In an open shop, a union and a company may not negotiate a contract that requires workers to join a union as a condition of employment.

Twenty-five U.S. states (refer to Figure 11-75) have right-to-work laws that make it much more difficult for unions to organize factory workers, collect dues, and bargain with employers from a position of strength. Right-to-work laws send a powerful signal that antiunion attitudes will be tolerated and perhaps even actively supported. As a result, the percentage of workers who are members of a union is much lower in the South than elsewhere in the United States. More importantly, the region has been especially attractive for companies working hard to keep out unions altogether.

▼FIGURE 11-75 CHANGING DISTRIBUTION OF U.S. INDUSTRY (a) 1950. (b) 2015. Manufacturing has declined in the Northeast and Great Lakes states while increasing in the South and West.

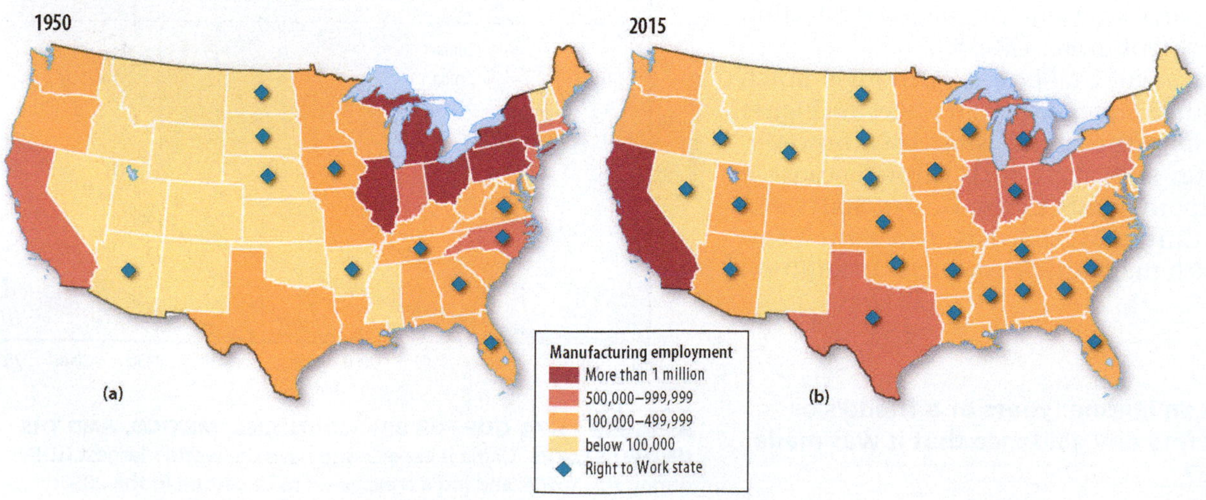

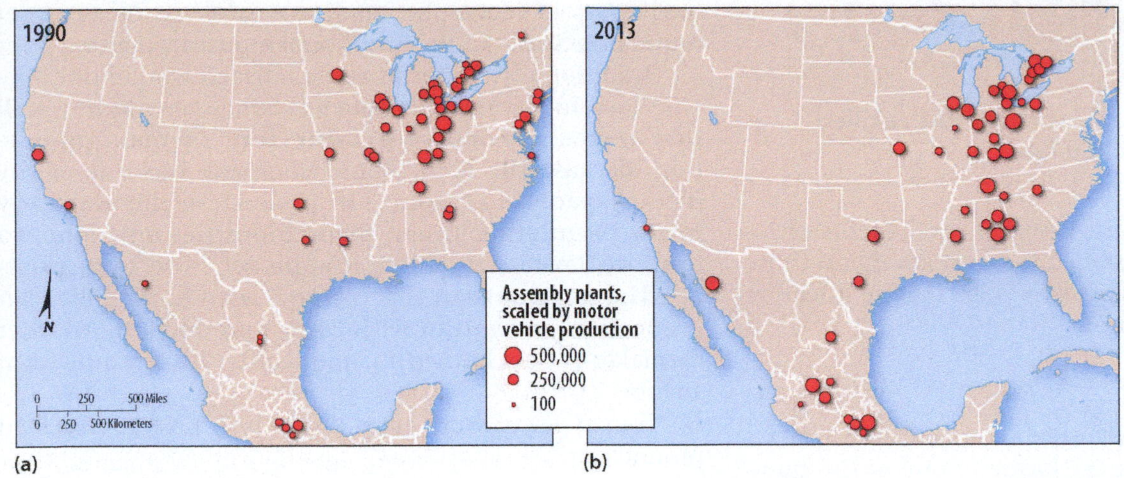

◀ **FIGURE 11-76 MOTOR VEHICLE PRODUCTION IN NORTH AMERICA** **(a)** 1990. **(b)** 2013. The size of the circles represents the volume of production at the assembly plant.

INTRAREGIONAL SHIFTS IN EUROPE

Manufacturing has diffused from traditional industrial centers in northwestern Europe toward Southern and Eastern Europe. To some extent, European government policies have encouraged this industrial relocation. The European Union Structural Funds provide assistance to what it calls convergence regions, competitive, and cooperative regions:

- Convergence regions are primarily in Eastern and Southern Europe, where incomes lag behind Europe's average.
- Competitive and employment regions are primarily Western Europe's traditional core industrial areas, which have experienced substantial manufacturing job losses in recent years.
- Territorial cooperation regions are near borders between European Union countries and neighboring countries.

The Western European country with the most rapid manufacturing growth during the late twentieth century was Spain, especially after its admission to the European Union in 1986. Until then, Spain's manufacturing growth had been impeded by physical and political isolation. Spain's motor-vehicle industry has grown into the second largest in Europe, behind only Germany's, although it is entirely foreign owned. Spain's leading industrial area is Catalonia, in the northeast, centered on the city of Barcelona. The region has the country's largest motor-vehicle plant and is the center of Spain's textile industry as well.

Several European countries situated east of Germany and west of Russia have become major centers of industrial investment since the fall of communism in the early 1990s. Poland, Czechia, Hungary, and Slovakia have had the most industrial development. Central Europe offers manufacturers an attractive combination of two important site and situation factors: labor and market proximity. Central Europe's workers offer manufacturers good value for money; they are less skilled but much cheaper than in Western Europe, and they are more expensive but much more skilled than in Asia and Latin America. At the same time, the region offers closer proximity to the wealthy markets of Western Europe than other emerging industrial centers. Central Europe has attracted an increasing share of the region's motor vehicle production (Figure 11-77).

PAUSE & REFLECT 11.4.2

Motor vehicle production has increased in Turkey and Morocco. What might be advantages and challenges of producing vehicles in these countries instead?

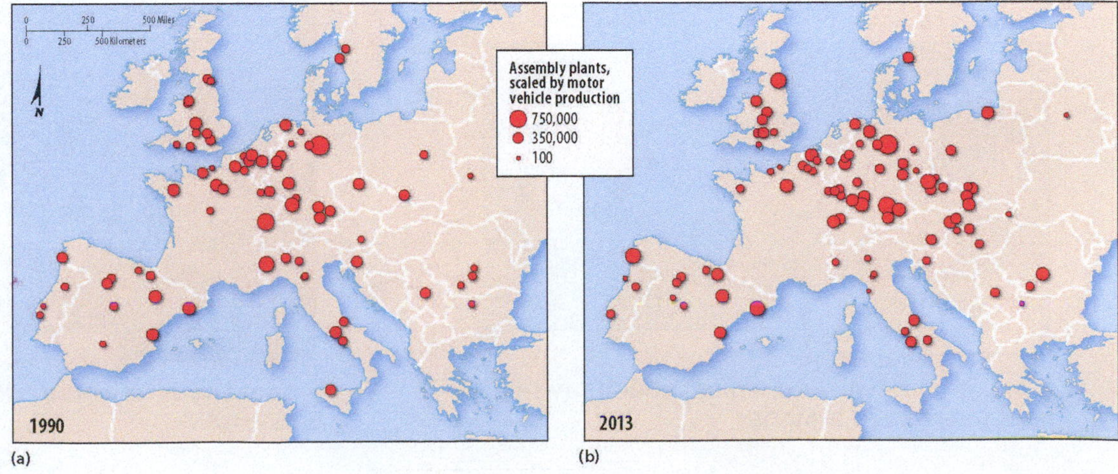

▶ **FIGURE 11-77 MOTOR VEHICLE PRODUCTION IN EUROPE** **(a)** 1990. **(b)** 2013. The size of the circles represents the volume of production at the assembly plant.

Skilled or Unskilled Labor?

LEARNING OUTCOME 11.4.3

Understand the attraction of locations with skilled labor and those with unskilled labor.

Many manufacturers are torn between the benefits of locating in regions with low-skilled low-cost labor and those with highly skilled higher-cost labor (See Debate It feature and Doing Geography feature).

SKILLED LABOR

Given the strong lure of low-cost labor in new industrial regions, why would any industry locate in one of the traditional regions, especially in the northeastern United States or northwestern Europe? Two location factors influence industries to remain in these traditional regions: availability of skilled labor and rapid delivery to market.

Henry Ford boasted that he could take people off the street and put them to work with only a few minutes of training. That has changed for some industries, which now want skilled workers instead. The search for skilled labor has important geographic implications because it is an asset found principally in the traditional industrial regions.

Traditionally, factories assigned each worker one specific task to perform repeatedly. Some geographers call this approach **Fordist production**, or mass production, because the Ford Motor Company was one of the first companies to organize its production this way early in the twentieth century. Many industries now follow a lean, or flexible, production approach. The term **post-Fordist production** is sometimes used to describe lean production, in contrast with Fordist production. Another carmaker is best known for pioneering lean production: Toyota.

Four types of work rules distinguish post-Fordist lean production:

- **Teams.** Workers are placed in teams and told to figure out for themselves how to perform a variety of tasks. Companies are locating production in communities where workers are willing to adopt more flexible work rules.
- **Problem solving.** A problem is addressed through consensus after consulting with all affected

DEBATE IT! Should the U.S. government buy domestic- or foreign-made clothes?

The U.S. government shops for clothing bargains, wherever they can be found in the world. Most uniforms as well as clothing sold in military bases are made in either Latin America or Asia rather than in the United States.

BUY CLOTHING ANYWHERE

- Buying overseas saves taxpayers money.
- It's hard to find U.S. manufacturers for all of the government's needs.
- The government is monitoring conditions in the factories where it gets the clothes.

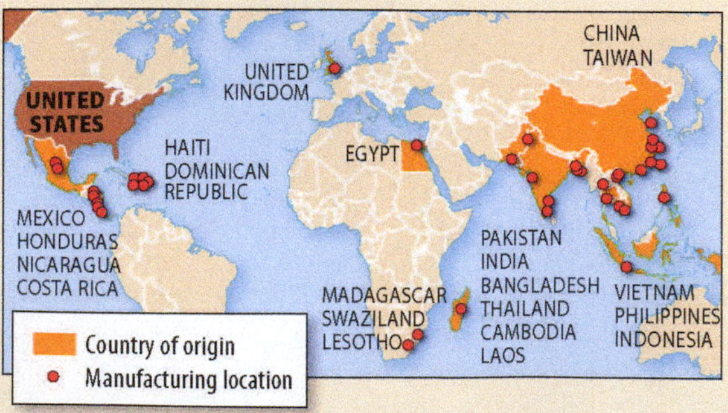

▲ **FIGURE 11-78 WHERE CLOTHING BOUGHT BY THE U.S. GOVERNMENT IS MADE**

BUY U.S.–MADE CLOTHING

- Supporting U.S. businesses will create jobs in the United States.
- The government should set an example by buying domestic products.
- Workplace conditions are poor in some overseas factories.

▲ **FIGURE 11-79 CLOTHING WORKERS IN BANGLADESH**

DOING GEOGRAPHY National Origin of Clothing

Most clothing sold in the United States has been assembled in other countries. The top five countries of origin for clothing sold in the United States are China, Vietnam, Bangladesh, Indonesia, and Honduras.

As recently as the 1990s, most clothing sold in the United States was assembled in the United States, but now very little is U.S. made (Figure 11-80). As apparel from other countries became less expensive and less complicated to import into the United States, mills in the Southeast paying wages of $10 to $15 per hour were unable to compete with manufacturers in countries paying less than $1 per hour. The number of clothing workers in the United States declined from 900,000 in 1990 to 500,000 in 2000 and to 150,000 in 2010.

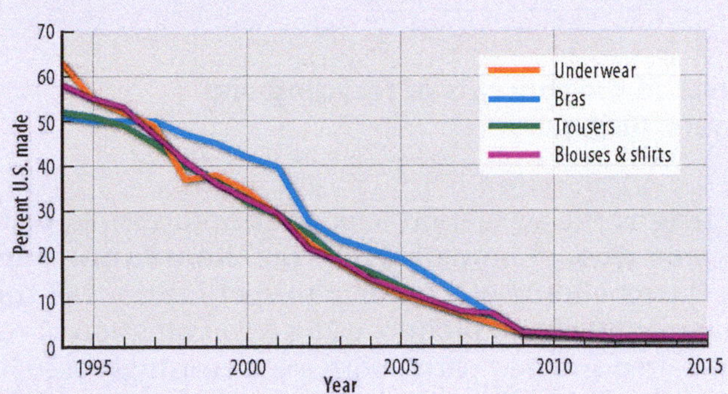

▲ **FIGURE 11-80 DOMESTIC AND IMPORTED U.S. CLOTHING** The percentage of clothing sold in the United States that was made in the United States has declined from around 50 percent in the 1990s to 3 percent today.

What's Your Industrial Geography?

National Origin of Your Clothing

Clothing sold in the United States has a label showing national origin (Figure 11-81a).

1. Take a look at the labels in 10 of your t-shirts. For each t-shirt, record the world region in which it was made (Africa, Asia, Europe, Latin America, or North America). Record the total number of t-shirts made in each of the five regions.
2. Sum the total t-shirts made in each of the five regions. Which region has the highest number? Which has the lowest?
3. Combine the results of everyone in your class. Record on a world map, such as Figure 11-81b. How do your results compare with those of the overall class?
4. Which of the three developing regions (Africa, Asia, or Latin America) has the highest total? Which has the lowest total? What might account for the sharp difference between these developing regions?

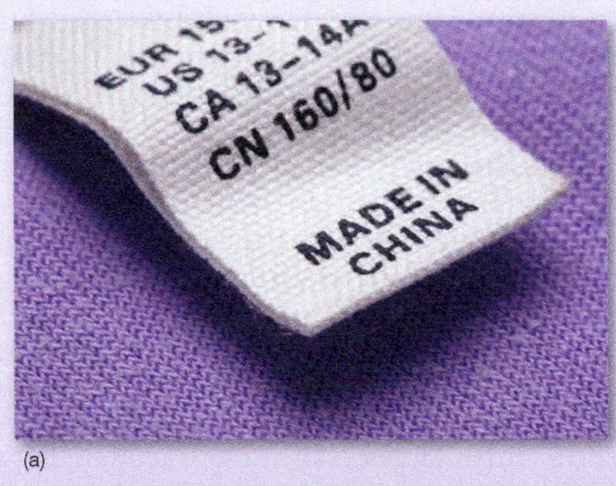

(a)

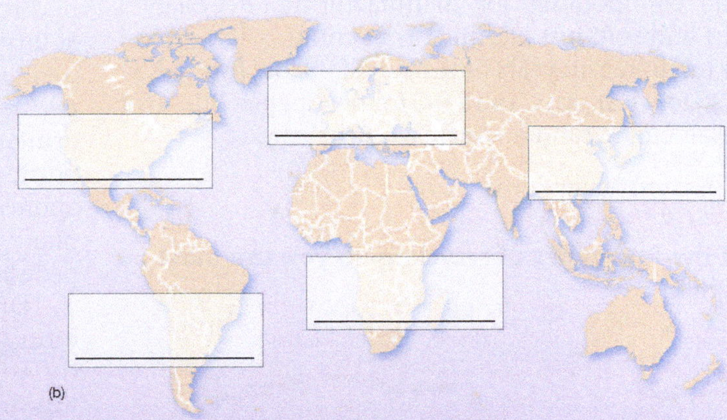

(b)

▲ **FIGURE 11-81 PLACE OF ORIGIN OF YOUR CLOTHING** (a) Example of a clothing label. (b) Regional origin of your t-shirts.

parties rather than through filing a complaint or grievance.

- **Leveling.** Factory workers are treated alike, and managers and veterans do not get special treatment; they wear the same uniform, eat in the same cafeteria, park in the same lot, and participate in the same athletic and social activities.
- **Productivity.** Factories have become more productive through introduction of new machinery and processes. Rather than requiring physical strength, these new machines and processes require skilled operators, typically with college degrees.

Recycling and Remanufacturing

LEARNING OUTCOME 11.4.4

Understand the concepts of recycling and remanufacturing.

Recycling is the separation, collection, processing, marketing, and reuse of unwanted material. **Remanufacturing** is the rebuilding of a product to specifications of the original manufactured product using a combination of reused, repaired and new parts. Both are increasingly used in industry as ways to promote a more sustainble production process.

RECYCLING

Recycling increased in the United States from 7 percent of all solid waste in 1970 to 10 percent in 1980, 17 percent in 1990, and 34 percent in 2013 (Figure 11-82). As a result of recycling, about 87 million of the 254 million tons of solid waste generated in the United States in 2013 did not have to go to landfills and incinerators, compared to 34 million of the 200 million tons generated in 1990.

The percentage of materials recovered by recycling varies widely by product: 50 percent of paper products and 24 percent of yard trimmings are recycled, compared to less than 10 percent for other sources of solid waste (Figure 11-83). Materials that would otherwise be "thrown away" are collected and sorted, in four principal ways: Curbside programs, drop-off centers, buy-back centers, and deposit programs. Regardless of the collection method, recyclables are sent to a materials recovery facility to be sorted and prepared as marketable commodities for manufacturing. Recyclables are bought and sold just like any other commodity; typical prices per ton in recent years have been \$300 for clear plastic bottles, \$30 for clear glass, and \$100 for newspaper. Prices for the materials change and fluctuate with the market.

PAUSE & REFLECT 11.4.4

Which of the four types of recycling do you use most often? Why?

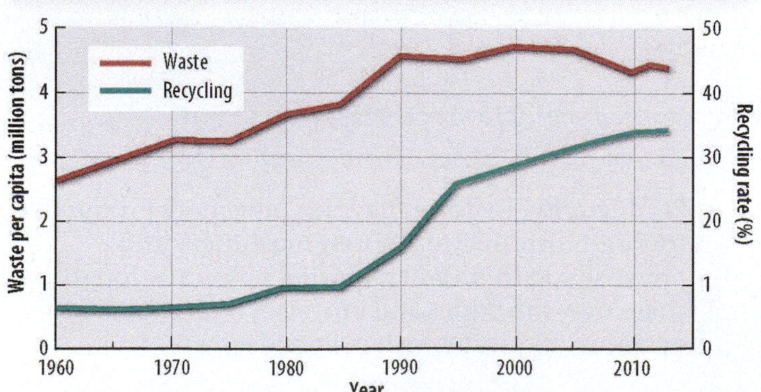

▲ **FIGURE 11-82 RECYCLING IN THE UNITED STATES** Recycling has increased from around 10 percent in 1980 to around 35 percent today.

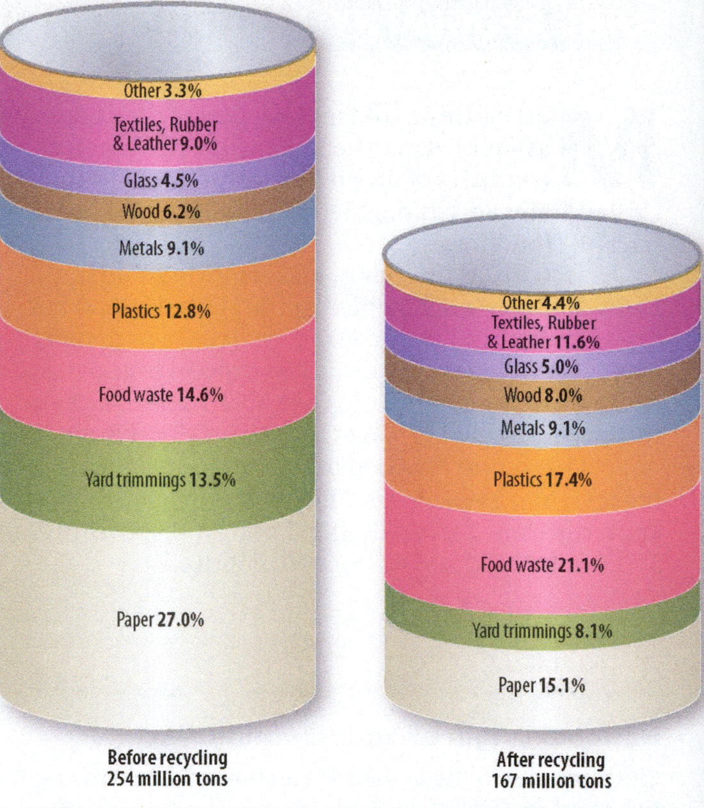

▲ **FIGURE 11-83 SOURCES OF SOLID WASTE BEFORE AND AFTER RECYCLING** Around one-half of paper and one-fourth of yard waste are recycled, whereas 10 percent of other sources are recycled.

REMANUFACTURING

Recycled materials can be remanufactured into new products. Four major manufacturing sectors accounted for more than half of the recycling activity: paper mills, steel mills, plastic converters, and iron and steel foundries (see Sustainability & Our Environment feature). Common household items that contain remanufactured materials include newspapers and paper towels; aluminum, plastic, and glass soft-drink containers; steel cans; and plastic laundry detergent bottles. Recycled materials are also used in such industrial applications as recovered glass in roadway asphalt ("glassphalt") and recovered plastic in carpet, park benches, and pedestrian bridges.

One of the world's most extreme instances requiring a form of remanufacturing is the Aral Sea, divided between the countries of Kazakhstan and Uzbekistan. The Aral Sea was the world's fourth-largest lake in 1960, at 68,000 square kilometers (26,000 square miles). It had shrunk to approximately 5,000 square kilometers (2,000 square miles) in 2010, and it could disappear altogether by 2020 (Figure 11-84). The shrinking has been captured in aerial photos and satellite imagery.

The Aral Sea died because beginning in 1954, the Soviet Union diverted its tributary rivers, the Amu Dar'ya and the Syr Dar'ya, to irrigate cotton fields. Ironically, the cotton now is withering because winds pick up salt from the exposed lakebed and deposit it on the cotton fields. Carp, sturgeon, and other fish species have disappeared; the last

(a) 1975

(b) 1989

(c) 2003

(d) 2014

◀ **FIGURE 11-84 THE DISAPPEARING ARAL SEA** (a) 1975. (b) 1989. (c) 2003. (d) 2014.

fish died in 1983. Large ships lie aground in salt flats that were once the lakebed, outside abandoned fishing villages that now lay tens of kilometers from the rapidly receding shore.

CHECK-IN KEY ISSUE 4

Why Are Industries Changing Locations?

- ✓ **Industry is growing in places that are not part of traditional industrial regions, primarily because of low-cost labor.**
- ✓ **Within developed regions, industry is also growing in nontraditional locations.**

SUSTAINABILITY & OUR ENVIRONMENT Remanufacturing

Remanufacturing contributes to a more sustainable environment. The principal challenge is to increase its economic sustainability.

- **Paper.** Most types of paper can be recycled. Newspapers have been recycled profitably for decades, and recycling of other paper, especially computer paper, is growing. Rapid increases in virgin paper pulp prices have stimulated construction of more plants capable of using waste paper. The key to recycling is collecting large quantities of clean, well-sorted, uncontaminated, dry paper.
- **Plastic.** The plastic industry has developed a system of numbers marked inside triangles. Symbols 2 (milk jugs), 4 (shopping bags), and 5 (such as yogurt containers) are considered to be safest for recycling. The plastics in symbols 3 (such as food wrap), 6 (Styrofoam), and 7 (such as iPad cases) may contain carcinogens. Symbol 1 (soda and water bottles) can allow bacteria to accumulate.
- **Aluminum.** The principal source of recycled aluminum is beverage containers. Aluminum cans began to replace glass bottles for beer during the 1950s and for soft drinks during the 1960s. Aluminum scrap is readily accepted for recycling, although other metals are rarely accepted.

▲ **FIGURE 11-85 REMANUFACTURING** Junked cars await shredding so that the steel can be reused.

- **Glass.** Glass can be used repeatedly with no loss in quality and is 100 percent recyclable. The process of creating new glass from old is extremely efficient, producing virtually no waste or unwanted by-products. Though unbroken clear glass is valuable, mixed-color glass is nearly worthless, and broken glass is hard to sort.

Summary & Review

KEY ISSUE 1

Where is industry distributed?

Industry is not distributed uniformly around the world. It is highly dependent on availability of energy, especially fossil fuels. The concept of manufacturing goods in a factory originated with the Industrial Revolution in the United Kingdom. Most of the world's industry is clustered in the three regions: Europe, North America, and East Asia.

THINKING GEOGRAPHICALLY

▲ **FIGURE 11-86 CLOSED FACTORY, UNITED KINGDOM**

1. What situation and site factors hurt the ability of the United Kingdom to maintain the leadership in industry that it gained in the nineteenth century?

KEY ISSUE 2

Why are situation and site factors important?

A company ordinarily considers a combination of situation and site factors. Situation factors involve transporting inputs into a factory and transporting manufactured goods to the markets. The three site factors are labor, capital, and land. Inputs and products are transported by ship, rail, truck, or air. The optimal choice depends on the distance something is being transported.

THINKING GEOGRAPHICALLY

▲ **FIGURE 11-87 VOLKSWAGEN ASSEMBLY PLANT, CHATTANOOGA, TENNESSEE**

2. Volkswagen received $577 million in government incentives to locate a factory in Chattanooga, Tennessee, in 2012. Do you think it's worth it for governments to provide large incentives like this? Why or why not?

KEY ISSUE 3

Why do industries face resource challenges?

Energy and other resources are critical for industry. Energy is currently derived primarily through the three fossil fuels: coal, petroleum, and natural gas. Reserves are highly concentrated in a handful of countries. Demand for energy comes from four main sources: industries, transportation, homes, and commercial activities. Alternatives to the fossil fuels include hydroelectric, nuclear, wind, geothermal, and solar. Industry is a major polluter of air, water, and land.

THINKING GEOGRAPHICALLY

▲ **FIGURE 11-88 RECYCLING, OAK PARK, ILLINOIS**

3. What strategies does your community employ to encourage recycling? What additional strategies might be effective in your community? Why?

KEY ISSUE 4

Why are industries changing locations?

The location of manufacturing has changed. Industry is increasing in some developing countries and in some areas within developed regions. Site factors, especially labor costs, have stimulated industrial growth in new regions, especially the BRIC countries. Within North America and Europe, manufacturing is growing in areas where it had not traditionally clustered.

THINKING GEOGRAPHICALLY

▲ **FIGURE 11-89 TRUCKS FROM MEXICO BACKED UP AT THE U.S. BORDER**

4. What has been the impact on Canada of the changes in the distribution of vehicle assembly plants in North America (refer to Figure 11-76)?
5. Why might situation and site factors be placing Canadian plants at a disadvantage compared to those in Mexico?

KEY TERMS

Acid deposition *(p. 415)* The accumulation of acids on Earth's surface.

Acid precipitation *(p. 415)* Conversion of sulfur oxides and nitrogen oxides to acids that return to Earth as rain, snow, or fog.

Active solar energy *(p. 412)* Solar radiation captured with photovoltaic cells that convert light energy to electrical energy.

Air pollution *(p. 414)* Concentration of trace substances, such as carbon monoxide, sulfur dioxide, nitrogen oxides, hydrocarbons, and solid particulates, at a greater level than occurs in average air.

Animate power *(p. 385)* Power supplied by animals or by people.

Biomass fuel *(p. 385)* Fuel derived from wood, plant material, or animal waste.

Biochemical oxygen demand (BOD) *(p. 416)* The amount of oxygen required by aquatic bacteria to decompose a given load of organic waste; a measure of water pollution.

Break-of-bulk point *(p. 395)* A location where transfer is possible from one mode of transportation to another.

Bulk-gaining industry *(p. 390)* An industry in which the final product weighs more or comprises a greater volume than the inputs.

Bulk-reducing industry *(p. 389)* An industry in which the final product weighs less or comprises a lower volume than the inputs.

Chlorofluorocarbon (CFC) *(p. 414)* A gas used as a solvent, a propellant in aerosols, a refrigerant, and in plastic foams and fire extinguishers.

Consumptive water usage *(p. 416)* The use of water that evaporates rather than being returned to nature as a liquid.

Cottage industry *(p. 384)* Manufacturing based in homes rather than in factories, most common prior to the Industrial Revolution.

Fission *(p. 408)* The splitting of an atomic nucleus to release energy.

Fordist production *(p. 424)* A form of mass production in which each worker is assigned one specific task to perform repeatedly.

Fossil fuel *(p. 385)* An energy source formed from the residue of plants and animals buried millions of years ago.

Fusion *(p. 409)* Creation of energy by joining the nuclei of two hydrogen atoms to form helium.

Geothermal energy *(p. 411)* Energy from steam or hot water produced from hot or molten underground rocks.

Greenhouse effect *(p. 414)* The anticipated increase in Earth's temperature caused by carbon dioxide (emitted by burning fossil fuels) trapping some of the radiation emitted by the surface.

Just-in-time delivery *(p. 395)* Shipment of parts and materials to arrive at a factory moments before they are needed.

Labor-intensive industry *(p. 396)* An industry for which labor costs comprise a high percentage of total expenses.

Maquiladora *(p. 421)* A factory built by a U.S. company in Mexico near the U.S. border, to take advantage of the much lower labor costs in Mexico.

New international division of labor *(p. 420)* Transfer of some types of jobs, especially those requiring low-paid, less-skilled workers, from more developed to less developed countries.

Nonconsumptive water usage *(p. 416)* The use of water that is returned to nature as a liquid.

Nonpoint source pollution *(p. 417)* Pollution that originates from a large, diffuse area.

Nonrenewable energy *(p. 410)* A source of energy that has a finite supply capable of being exhausted.

Outsourcing *(p. 420)* A decision by a corporation to turn over much of the responsibility for production to independent suppliers.

Ozone *(p. 414)* A gas that absorbs ultraviolet solar radiation and is found in the stratosphere, a zone 15 to 50 kilometers (9 to 30 miles) above Earth's surface.

Passive solar energy systems *(p. 412)* Solar energy systems that collect energy without the use of mechanical devices.

Photochemical smog *(p. 415)* An atmospheric condition formed through a combination of weather conditions and pollution, especially from motor vehicle emissions.

Point source pollution *(p. 417)* Pollution that enters a body of water from a specific source.

Pollution *(p. 414)* Concentration of waste added to air, water, or land at a greater level than occurs in average air, water, or land.

Post-Fordist production *(p. 424)* Adoption by companies of flexible work rules, such as the allocation of workers to teams that perform a variety of tasks.

Potential reserve *(p. 404)* The amount of a resource in deposits not yet identified but thought to exist.

Proven reserve *(p. 404)* The amount of a resource remaining in discovered deposits.

Recycling *(p. 426)* The separation, collection, processing, marketing, and reuse of unwanted material.

Remanufacturing *(p. 426)* The rebuilding of a product to specifications of the original manufactured product using a combination of reused, repaired and new parts.

Renewable energy *(p. 410)* A resource that has a theoretically unlimited supply and is not depleted when used by people.

Right-to-work law *(p. 422)* A U.S. law that prevents a union and a company from negotiating a contract that requires workers to join the union as a condition of employment.

Sanitary landfill *(p. 418)* A place to deposit solid waste, where a layer of earth is bulldozed over garbage each day to reduce emissions of gases and odors from the decaying trash, to minimize fires, and to discourage vermin.

Site factors *(p. 388)* Location factors related to the costs of factors of production inside a plant, such as land, labor, and capital.

Situation factors *(p. 388)* Location factors related to the transportation of materials into and from a factory.

Vertical integration *(p. 420)* An approach typical of traditional mass production in which a company controls all phases of a highly complex production process.

GeoVideo | Log in to the **MasteringGeography** Study Area to view this video.

Human Impacts on Water Resources

Humans use water for many purposes, including manufacturing, agriculture, and recreation, as well as direct consumption. Access to fresh clean water is not possible for many people in the world. The poor condition of infrastructure restricts access to fresh clean water for some people. Other people live in arid locations.

1. What are the principal uses of water resources other than direct consumption by people and animals?
2. Given that the world's total supply of water is constant, how might we increase the world's supply of water suitable as a resource for use by people?
3. What steps, if any, are being taken in your school or community to conserve water?

EXPLORE

An Early Industrial Town

Saltaire is a village in the United Kingdom built in the early years of the Industrial Revolution by Sir Titus Salt. Surrounding his factory, Salt built houses for the workers.

Fly to *Saltaire*.

Zoom in to eye alt around 3,500 ft. Salt's original factory is the large building immediately north of the balloon.

Click *More*, then turn on *Transportation* layer. The factory was built before the invention of cars and trucks.

1. What two modes of transportation from the nineteenth century (still visible immediately to the north and south of the factory) would have been used to move materials and products?

   Turn on *Photos*. Move the cursor over one of the photos on top of the factory.

2. What was Salt's factory used for? (Hint: Turn on *Gallery* in the *Primary Database* and click on the "i" (for information) icon near the north side of the factory.)

   Drag *street view* to the southwest side of the factory complex.

3. Why might the structure no longer be suitable for manufacturing?

MasteringGeography™

Visit the Study Area in **MasteringtGeography**™ to enhance your geographic literacy, spatial reasoning skills, and understanding of this chapter's content by accessing a variety of resources, including MapMaster interactive maps, videos, *In the News* RSS feeds, flashcards, web links, self-study quizzes, and an eText version of *The Cultural Landscape*.

www.masteringgeography.com

12

Services and Settlements

In developed countries, most people work in services, such as shops, offices, restaurants, universities, and hospitals. And most people obtain what they need from service providers. Services are closely linked with settlements because services are located in settlements. Geography plays an especially important role in the provision of services because geographic principles determine the optimal location for a service.

A market in Kashgar, Xinjiang Province, China.

LOCATIONS IN THIS CHAPTER

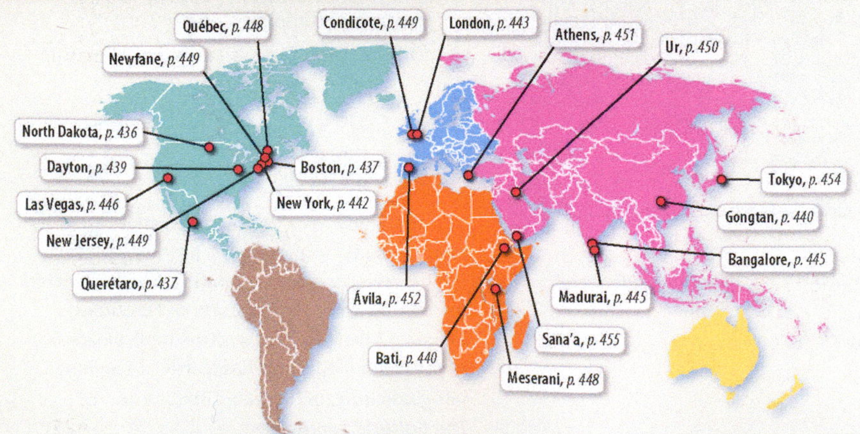

KEY ISSUES

ARCTIC OCEAN

1 Where Are Services Distributed?

Services are divided into three types: consumer, business, and public. These have different distributions in *space*.

2

Where Are Consumer Services Distributed?

In developed countries, consumer services follow a regular distribution, based on the number of people in a market area with a particular radius. The optimal location for a consumer service often has a very local *scale*, such as one corner of an intersection.

3

Where Are Business Services Distributed?

Business services are disproportionately clustered in a relatively small number of global cities that have *connections* to each other in a global economy.

4

Why Do Services Cluster In Settlements?

Settlements are either rural or urban *places*. A higher percentage of people live in urban settlements in developed *regions*, but the world's most populous urban settlements are now in developing regions.

KEY ISSUE 1

Where Are Services Distributed?

▶ Introducing Services and Settlements

LEARNING OUTCOME 12.1.1

Describe the three types of services.

Most people in developed countries work in such places as shops, offices, restaurants, universities, and hospitals. These are examples of the tertiary, or service, sector of the economy. A **service** is any activity that fulfills a human want or need and returns money to those who provide it. A smaller number of people work on farms or in factories, the primary and secondary sectors.

Introducing Services and Settlements

Services generate more than two-thirds of GDP in most developed countries, compared to less than one-half in most developing countries (Figure 12-1). Logically, the distribution of service workers is opposite that of the percentage of primary workers (refer to Figure 9-3). If services were located merely where people lived, then China and India would have the most, rather than the United States and other developed countries. Services cluster in developed countries because more people who are able to buy services live there. Within developed countries, larger cities offer a larger scale of services than do small towns because more customers reside there.

Geographers see a close link between services and settlements because services are located in settlements. A **settlement** is a permanent collection of buildings where people reside, work, and obtain services. They occupy a very small percentage of Earth's surface, well under 1 percent, but settlements are home to nearly all humans because few people live in isolation.

Explaining why services are clustered in settlements is at one level straightforward for geographers. In geographic terms, only one locational factor is critical for a service: proximity to the market. The optimal location of industry, described in Chapter 11, requires balancing a number of site and situation factors, but the optimal location for a service is simply near its customers.

On the other hand, locating a service calls for far more precise geographic skills than locating a factory. The optimal location for a factory may be an area of several hundred square kilometers, whereas the optimal location for a service may be a very specific place, such as a street corner.

THREE TYPES OF SERVICES

The service sector of the economy is subdivided into three types: consumer services, business services, and public services. Each of these sectors is divided into several major subsectors (Figure 12-2):

- **Consumer services.** The principal purpose of **consumer services** is to provide services to individual consumers who desire them and can afford to pay for them. Around one-half of all jobs in the United States are in consumer services. Four main types of consumer services are retail, health, education, and leisure.
- **Business services.** The principal purpose of **business services** is to facilitate the activities of other businesses. One-fourth of all jobs in the United States are in business services. Professional services, transportation services, and financial services are the three main types of business services.
- **Public services.** The purpose of **public services** is to provide security and protection for citizens and businesses. About 10 percent of all U.S. jobs are in the public sector. Excluding educators, one-sixth of public-sector employees work for the federal government, one-fourth for one of the 50 state governments, and three-fifths for one of the tens of thousands of local governments.

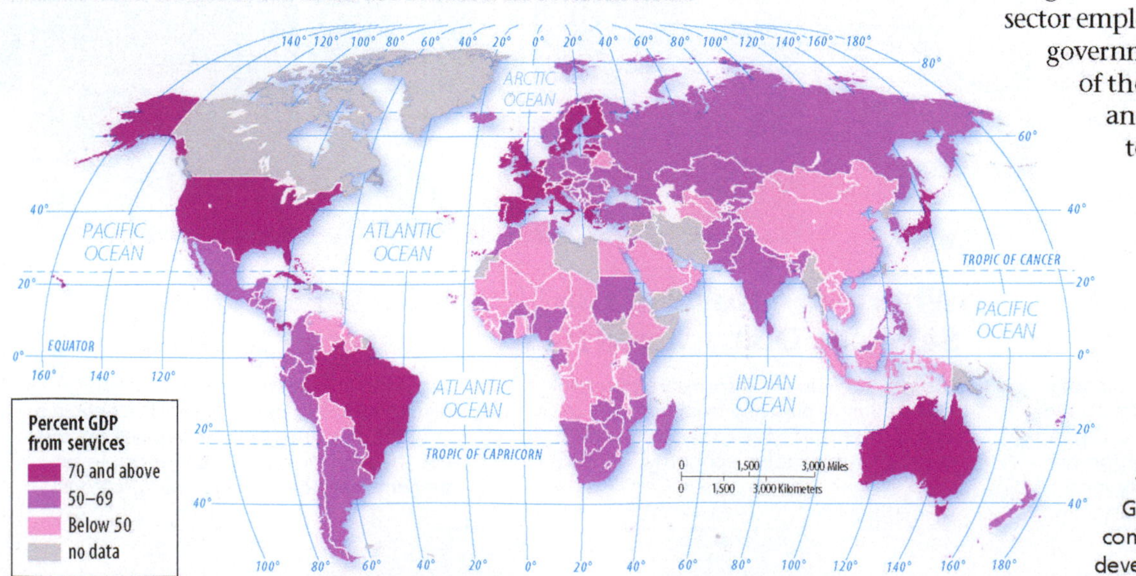

◀ **FIGURE 12-1 PERCENTAGE OF GDP FROM SERVICES** Services account for more than two-thirds of GDP in developed countries, compared to less than one-half in developing countries.

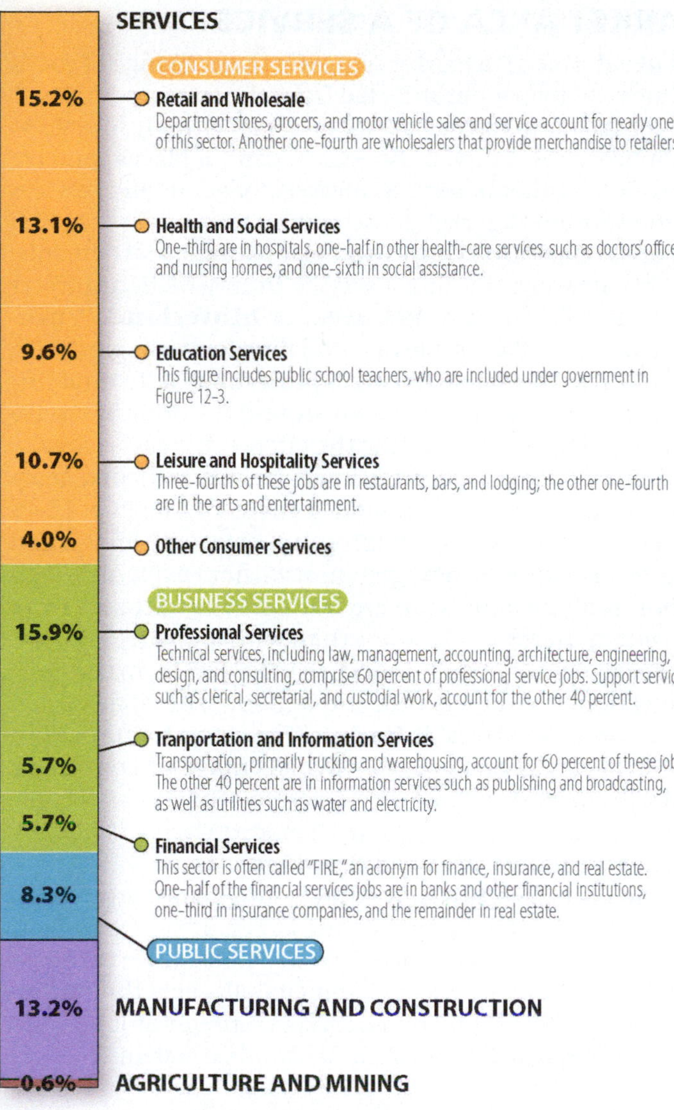

▲ **FIGURE 12-2 EMPLOYMENT IN THE UNITED STATES** Approximately one-half of jobs in the United States are in consumer services and one-fourth in business services.

CHANGING SERVICE EMPLOYMENT

Figure 12-3 shows changes in U.S. employment. All of the growth in U.S. employment has been in services, whereas employment in primary- and secondary-sector activities has declined. Within business services, jobs expanded most rapidly in professional services and more slowly in finance and transportation services because of improved efficiency; fewer workers are needed to run trains and answer phones, for example. On the consumer services side, the most rapid increase was in the provision of health care, education, entertainment, and recreation.

PAUSE & REFLECT 12.1.1

In which sectors of the economy do you or members of your family work? If in the service sector, in which types of services are these jobs?

CHECK-IN KEY ISSUE 1

Where Are Services Distributed?

- ✔ **Three types of services are consumer, business, and public.**
- ✔ **The fastest-growing consumer service is health care, and the fastest-growing business service is professional.**

▶ **FIGURE 12-3 CHANGE IN U.S. EMPLOYMENT** Jobs have increased in the tertiary sector, especially consumer services.

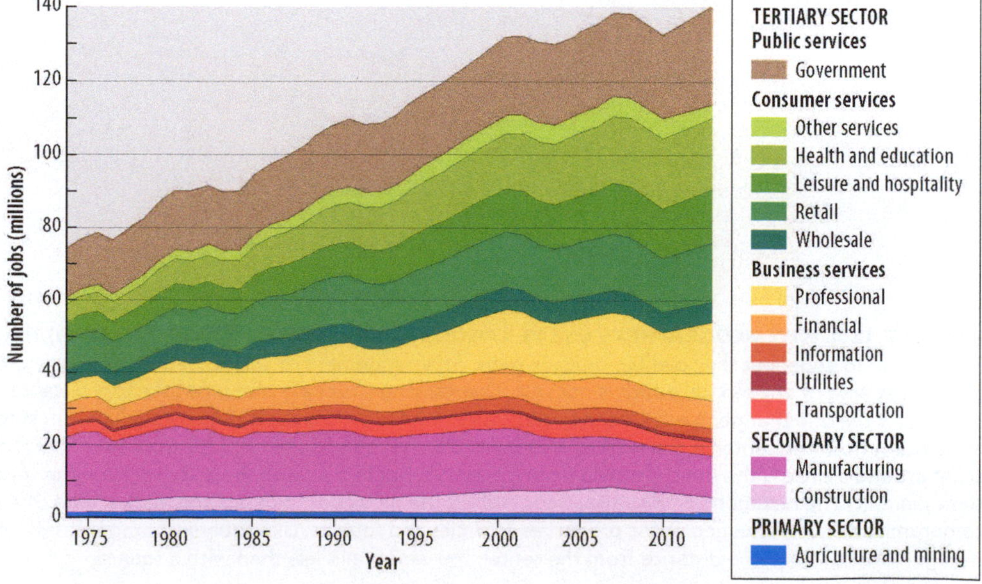

KEY ISSUE 2

Where Are Consumer Services Distributed?

- Central Place Theory
- Hierarchy of Consumer Services
- Market Area Analysis
- Periodic Markets

LEARNING OUTCOME 12.2.1

Explain the concepts of market area, range, and threshold.

Consumer services and business services do not have the same distributions. Consumer services generally follow a regular pattern based on size of settlements, with larger settlements offering more consumer services than smaller ones. The third key issue will describe how business services cluster in specific settlements, creating a specialized pattern.

Central Place Theory

Selecting the right location for a new shop is probably the single most important factor in the profitability of a consumer service. **Central place theory** helps to explain how the most profitable location can be identified.

Central place theory was first proposed in the 1930s by German geographer Walter Christaller, based on his studies of southern Germany. August Lösch in Germany and Brian Berry and others in the United States further developed the concept during the 1950s.

MARKET AREA OF A SERVICE

A **central place** is a market center for the exchange of goods and services by people attracted from the surrounding area. The central place is so called because it is centrally located to maximize accessibility. Businesses in central places compete against each other to serve as markets for goods and services for the surrounding region. According to central place theory, this competition creates a regular pattern of settlements.

The area surrounding a service from which customers are attracted is the **market area,** or **hinterland.** A market area is a good example of a nodal region—a region with a core where the characteristic is most intense. To establish the market area, a circle is drawn around the node of service on a map. The territory inside the circle is its market area.

To represent market areas in central place theory, geographers draw hexagons around settlements (Figure 12-4). Hexagons represent a compromise between circles and squares. Like squares, hexagons nest without gaps. Although all points along a hexagon are not the same distance from the center, the variation is less than with a square.

Because most people prefer to get services from the nearest location, consumers near the center of the circle obtain services from local establishments. The closer to the periphery of the circle, the greater the percentage of consumers who will choose to obtain services from other nodes. People on the circumference of the market-area circle are equally likely to use the service or go elsewhere.

The United States can be divided into market areas based on the hinterlands surrounding the largest urban settlements (Figure 12-5). Studies conducted by C. A. Doxiadis, Brian Berry, and the U.S. Department of Commerce allocated the 48 contiguous states to 171 functional regions centered around commuting hubs, which they called "daily urban systems."

PAUSE & REFLECT 12.2.1

What occurs in nature in the shape of hexagons? You can do an Internet search for "naturally occurring hexagons."

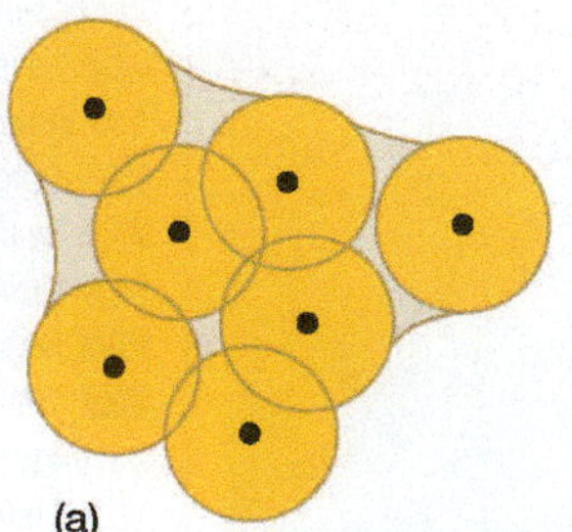

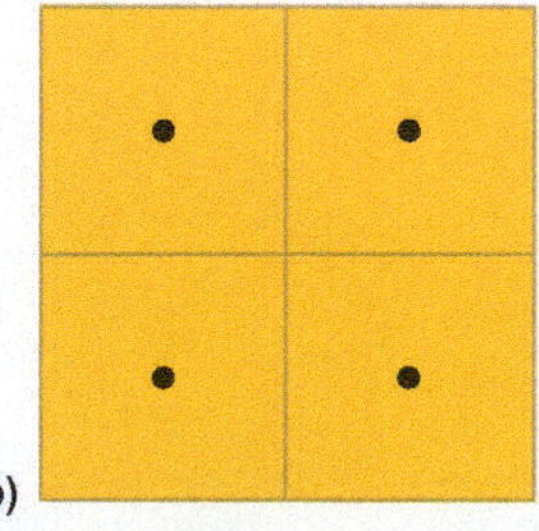

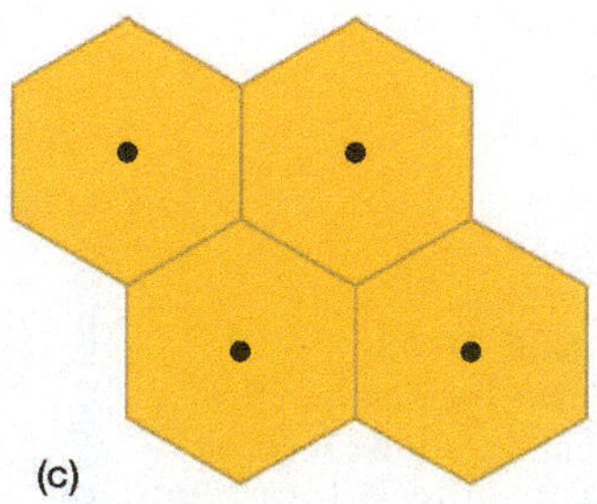

▲ **FIGURE 12-4 WHY GEOGRAPHERS USE HEXAGONS TO DELINEATE MARKET AREAS** **(a) The problem with circles.** Circles are equidistant from center to edge, but they overlap or leave gaps. An arrangement of circles that leaves gaps indicates that people living in the gaps are outside the market area of any service, which is not true. Overlapping circles are also unsatisfactory because only one of the service centers can be the closest and the one that people will tend to patronize. **(b) The problem with squares.** Squares nest together without gaps, but their sides are not equidistant from the center. If the market area is a circle, the radius—the distance from the center to the edge—can be measured because every point around a circle is the same distance from the center. But in a square, the distance from the center varies among points along the side of the square. **(c) The hexagon compromise.** Geographers use hexagons to depict the market area of a good or service because hexagons offer a compromise between the geometric properties of circles and squares. Like squares, hexagons nest without gaps. Although all points along the hexagon are not the same distance from the center, the variation is less than with a square.

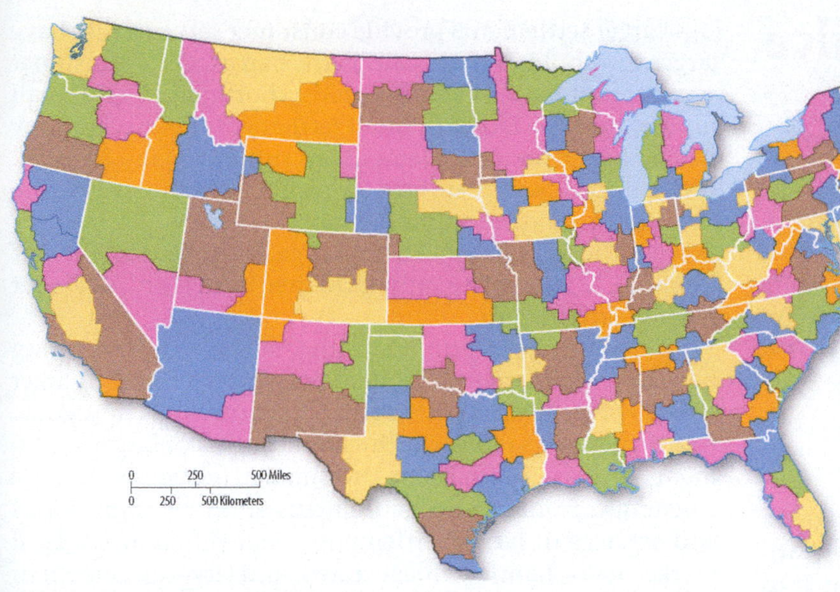

◀ **FIGURE 12-5 DAILY URBAN SYSTEMS** The U.S. Department of Commerce divided the 48 contiguous states into "daily urban systems," delineated by functional ties, especially commuting to the nearest metropolitan area. This division of the country into daily urban systems demonstrates that everyone in the United States has access to services in at least one large settlement.

RANGE OF A SERVICE

Each service has a distinctive market area. To determine the extent of a market area, geographers need two pieces of information about a service: its range and its threshold (Figure 12-6). How far are you willing to travel for a pizza? To see a doctor for a serious illness? To watch a ball game? The **range** is the maximum distance people are willing to travel to use a service. The range is the radius of the circle (or hexagon) drawn to delineate a service's market area.

People are willing to go only a short distance for everyday consumer services, such as groceries and pharmacies. But they will travel longer distances for other services, such as a concert or professional ball game. Thus a convenience store has a small range, whereas a stadium has a large range. In a large urban settlement, for example, the range of a fast-food franchise such as McDonald's is roughly 5 kilometers (3 miles), the range of a casual dining chain such as Steak' n Shake is roughly 8 kilometers (5 miles), and the range of a stadium is more than 100 kilometers (60 miles).

As a rule, people tend to go to the nearest available service. For example, someone in the mood for a McDonald's hamburger is likely to go to the nearest McDonald's. Therefore, the range of a service must be determined from the radius of a circle that is irregularly shaped rather than perfectly round. The irregularly shaped circle takes in the territory for which the proposed site is closer than competitors' sites.

The range must be modified further because most people think of distance in terms of time rather than in terms of a linear measure such as kilometers or miles. If you ask people how far they are willing to travel to a restaurant or a baseball game, they are more likely to answer in minutes or hours than in distance. If the range of a good or service is expressed in travel time, then the irregularly shaped circle must be drawn to acknowledge that travel time varies with road conditions. "One hour" may translate into traveling 100 kilometers (60 miles) while driving on an expressway but only 50 kilometers (30 miles) while driving city streets.

THRESHOLD OF A SERVICE

The second piece of geographic information needed to compute a market area is the **threshold,** which is the minimum number of people needed to support the service. Every enterprise has a minimum number of customers required to generate enough sales to make a profit. So once the range has been determined, a service provider must determine whether a location is suitable by counting the potential customers inside the irregularly shaped circle. Census data help to estimate the potential population within the circle.

How expected consumers inside the range are counted depends on the product. Convenience stores and fast-food restaurants appeal to nearly everyone, whereas other goods and services appeal primarily to certain consumer groups. For example:

- Movie theaters attract younger people; chiropractors attract older folks.
- Poorer people are drawn to thrift stores; wealthier ones might frequent upscale department stores.
- Amusement parks attract families with children; nightclubs appeal to singles.

Developers of shopping malls, department stores, and large supermarkets may count only higher-income people, perhaps those whose annual incomes exceed $50,000. Even though the stores may attract individuals of all incomes, higher-income people are likely to spend more and purchase items that carry higher profit margins for the retailer. A large retailer has many locations to choose from when deciding to build new stores. A suitable site is one with the potential for generating enough sales to justify using the company's scarce capital to build it.

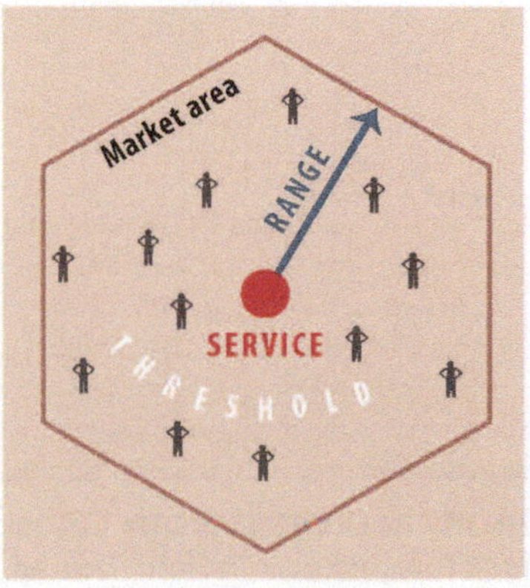

◀ **FIGURE 12-6 MARKET AREA, RANGE, AND THRESHOLD** The market area is the area of the hexagon, the range is the radius, and the threshold is a sufficient number of people inside the area to support the service.

Hierarchy of Consumer Service

LEARNING OUTCOME 12.2.2

Explain the distribution of different-sized settlements.

We spend as little time and effort as possible obtaining consumer services and thus go to the nearest place that fulfills our needs. There is no point in traveling to a distant store if the same merchandise is available at a nearby one. We travel greater distances only if the price is much lower or if the item is unavailable locally.

NESTING OF SERVICES AND SETTLEMENTS

According to central place theory, market areas across a developed country would be a series of hexagons of various sizes, unless interrupted by physical features such as mountains and bodies of water. Developed countries have numerous small settlements with small thresholds and ranges and far fewer large settlements with large thresholds and ranges. In his original study, Walter Christaller showed that the distances between settlements in southern Germany followed a regular pattern.

North-central North Dakota is an example (Figure 12-7). Minot—the largest city in the area, with 46,000 inhabitants—is surrounded by:

- 11 small towns of between 1,000 and 3,000 inhabitants, with average ranges of 30 kilometers (20 miles) and market areas of around 2,800 square kilometers (1,200 square miles).
- 20 villages of between 100 and 999 inhabitants, with ranges of 20 kilometers (12 miles) and market areas of around 1,200 square kilometers (500 square miles).
- 22 hamlets of fewer than 100 inhabitants, with ranges of 15 kilometers (10 miles) and market areas of around 800 square kilometers (300 square miles), including Maxbass, illustrated in Figure 12-8.

Larger settlements provide consumer services that have larger thresholds, ranges, and market areas. Only consumer services that have small thresholds, short ranges, and small market areas are found in small settlements because too few people live in small settlements to support many services. A large store cannot survive in a small settlement because the threshold (the minimum number of people needed) exceeds the population within range of the settlement. For example, Minot is the only settlement in Figure 12-7 that has a Walmart.

The nesting pattern can be illustrated with overlapping hexagons of different sizes. Hamlets with very small market areas are represented by the smallest contiguous hexagons. Larger hexagons represent the market areas of larger settlements and are overlaid on the smaller hexagons because consumers from smaller settlements shop for some goods and services in larger settlements. Four different levels of market area—hamlet, village, town, and city—are shown in Figure 12-9.

Businesses in central places compete against each other to serve as markets for goods and services for the surrounding region. According to central place theory, this competition creates a regular pattern of settlements. Across much of the interior of the United States, a regular pattern of settlements can be observed.

RANK-SIZE DISTRIBUTION

In many developed countries, ranking settlements from largest to smallest (population) produces a regular pattern. This is the **rank-size rule,** in which the country's nth-largest settlement is $1/n$ the population of the largest settlement.

According to the rank-size rule, the second-largest city is one-half the size of the largest, the fourth-largest city is one-fourth the size of the largest, and so on. When plotted on logarithmic paper, the rank-size distribution forms a fairly straight line. In the United States the distribution of settlements closely follows the rank-size rule (Figure 12-10).

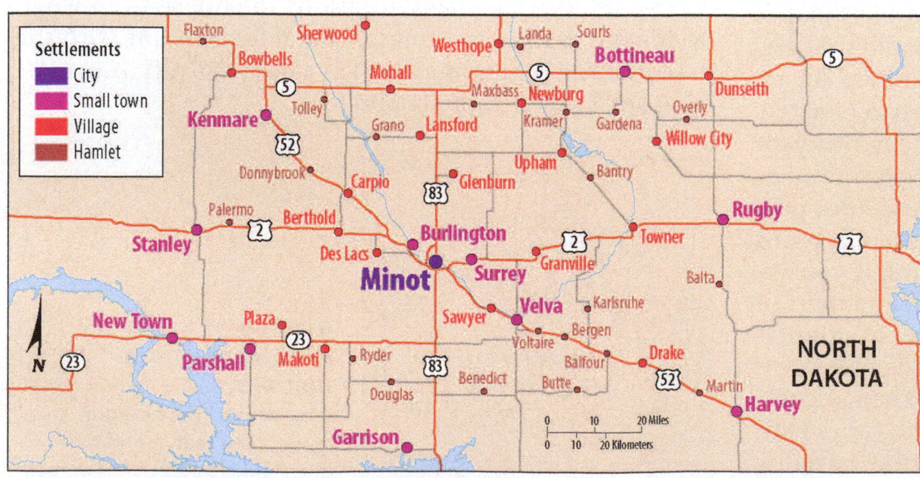

▲ **FIGURE 12-7 CENTRAL PLACE THEORY IN NORTH DAKOTA** Central place theory helps explain the distribution of settlements of varying sizes in North Dakota.

▲ **FIGURE 12-8 HAMLET: MAXBASS, NORTH DAKOTA** North of Minot, near the junction of routes 83 and 5, in Figure 12-9.

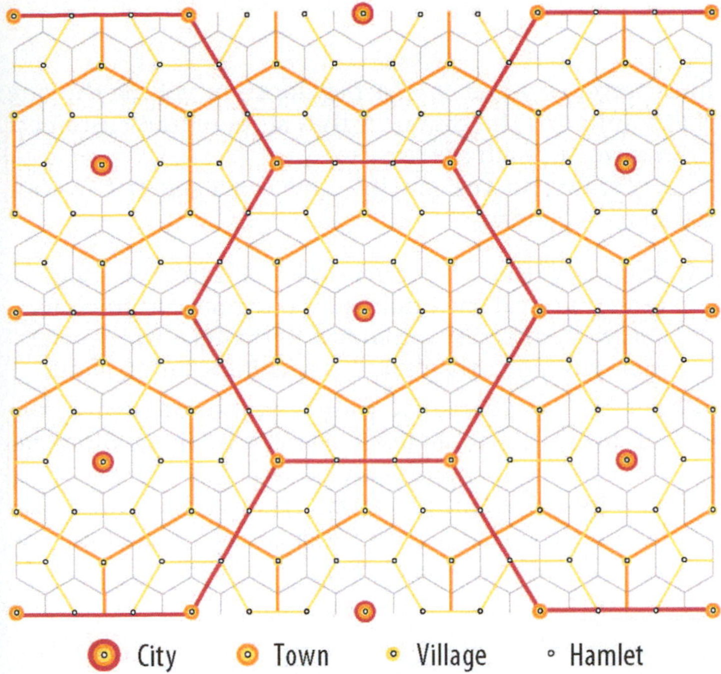

▲ **FIGURE 12-9 CENTRAL PLACE THEORY**

If the settlement hierarchy does not graph as a straight line, then the country does not follow the rank-size rule. Instead, it may follow the **primate city rule,** in which the largest settlement has more than twice as many people as the second-ranking settlement. In this distribution, the country's largest city is called the **primate city.** Mexico is an example of a country that follows the primate city distribution. Its largest settlement, Mexico City, is five times larger than its second-largest settlement, Guadalajara, rather than two times larger (Figure 12-11).

The existence of a rank-size distribution of settlements is not merely a mathematical curiosity. It has a real impact on the quality of life for a country's inhabitants. A regular hierarchy—as in the United States—indicates that the society is sufficiently wealthy to justify the provision of goods and services to consumers throughout the country (Figure 12-12). Conversely, the absence of the rank-size distribution in a developing country indicates that there is not enough wealth in the society to pay for a full variety of services. The absence of a rank-size distribution constitutes a hardship for people who must travel long distances to reach an urban settlement that offers services such as shops and hospitals.

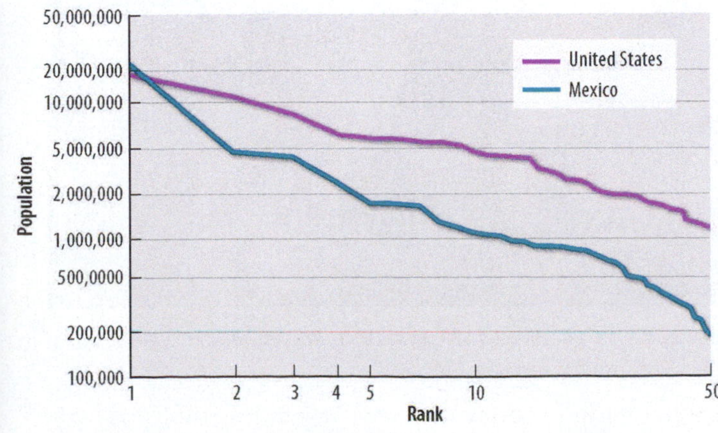

▲ **FIGURE 12-10 RANK-SIZE DISTRIBUTION OF SETTLEMENTS**

▲ **FIGURE 12-11 PRIMATE CITY DISTRIBUTION: MEXICO** Querétaro, Mexico's tenth largest settlement, has a population that is 1/20 the size of the largest settlement Mexico City, instead of 1/10, as the rank-size distribution would suggest.

PAUSE & REFLECT 12.2.2

Does Peru follow the rank-size rule or the primate city rule? Use your search engine to find "most populous cities in Peru."

▲ **FIGURE 12-12 RANK-SIZE DISTRIBUTION: UNITED STATES** Boston, the tenth-largest settlement in the United States, has a population that is one-fourth the size of the largest settlement, New York City, so it is even larger than the rank-size distribution would suggest.

Market Area Analysis

LEARNING OUTCOME 12.2.3

Explain how to use threshold and range to find the optimal location for a service.

Geographers apply central place theory to create market area studies that assist service providers with opening and expanding their facilities (see Doing Geography feature). And in a severe economic downturn, market area analysis helps determine where to close facilities.

Geographers have adapted the gravity model from physics. The **gravity model** predicts that the optimal location of a service is directly related to the number of people in the area and inversely related to the distance people must travel to access it. The best location will be the one that minimizes the distances that all potential customers must travel to reach the service.

According to the gravity model, consumer behavior reflects two patterns:

1. The greater the number of people living in a particular place, the greater the number of potential customers for a service. An area that contains 100 families will generate more customers than a house containing only 1 family.
2. The farther people are from a particular service, the less likely they are to use it. People who live 1 kilometer from a store are more likely to patronize it than people who live 10 kilometers away. The threshold must also be adjusted because the further customers are from the service, the less likely they are to patronize it.

DEBATE IT! Is Walmart good or bad for a settlement?

Walmart is the world's largest provider of consumer services (Figure 12-13). Some of Walmart's business practices have sparked controversy (Figure 12-14). When Walmart comes to town, what are the benefits and drawbacks for the settlement? What do you think?

WALMART BENEFITS THE SETTLEMENT

- Consumers get lower prices on groceries.
- Employment is offered to minimally skilled workers who otherwise have trouble finding jobs.
- Stores are often sited in rural areas and lower-income urban neighborhoods that lack other shopping options.

▲ **FIGURE 12-13 WALMART EMPLOYEE** Miami, Florida.

WALMART HARMS THE SETTLEMENT

- Wage and benefit levels are too low to lift Walmart workers out of poverty.
- Walmart costs Americans jobs because most of the products it sells are made overseas.
- Locally owned shops are forced out of business after Walmart opens.

▲ **FIGURE 12-14 WALMART PROTEST** Protest asking Walmart to pay living wages, Towson, Maryland.

DOING GEOGRAPHY Locating Supermarkets and Food Deserts

Major retailers employ geographers to determine the best locations to build new stores. Here is a typical process for siting a supermarket:

1. **Estimate the range.** Geographers estimate the range as the distance that most people travel to reach an existing store judged to be comparable to the proposed one. The range is calculated as the distance from the existing store where between two-thirds and three-fourths of the customers live. The home addresses are compiled from credit-card records.
2. **Estimate the threshold.** Geographers next determine whether enough people live within the range of the proposed store to justify its construction. The threshold for a large supermarket is about 25,000 people living within a 15-minute range.
3. **Modify the threshold.** The threshold must be modified in two ways:
   - The proposed new supermarket will have to share customers with competitors. The market share of existing stores judged to be comparable is applied to the proposed new store.
   - People are counted only if they have sufficient income to shop regularly at the store. Retailers avoid neighborhoods with below-average incomes. In the Dayton, Ohio, area, for example, the leading supermarket chain, Kroger, has most of its stores in the relatively affluent south and east (Figure 12-15a).

Supermarkets, as well as most other retailers, avoid low-income neighborhoods. This can lead to the existence of food deserts. The U.S. government defines a **food desert** as an area that has a substantial amount of low-income residents and has poor access to a grocery store. Poor access is defined by the government in most cases as further than 1 mile. A distance of 1 mile is not far for people to travel in a car, but it is far for low-income people who do not own cars.

Not surprisingly, food deserts in Dayton are located in areas that lack supermarkets operated by Kroger and its competitors (Figure 12-15b).

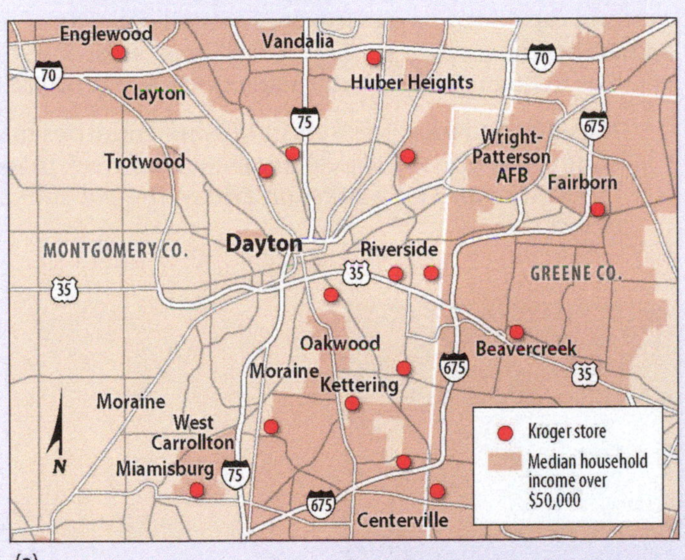

(a)

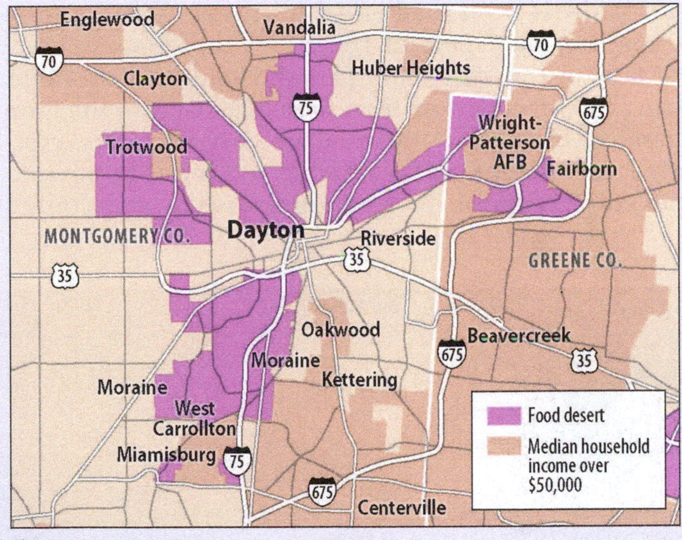

(b)

▲ **FIGURE 12-15 SUPERMARKETS AND FOOD DESERTS IN DAYTON, OHIO** **(a)** Dayton's leading supermarket chain, Kroger, has clustered its stores in and near high-income neighborhoods to the south and east. **(b)** Dayton's food deserts are on the west and north sides, where relatively few supermarkets are located.

WHAT'S YOUR RETAIL GEOGRAPHY?

1. What is the distance from your home to the grocery store most frequently used by your family?
2. By what means of transportation does your family usually shop at the grocery store?
3. Does your family shop at the nearest grocery store? If not, why not?
4. Map the food deserts in your community. Search the internet using key words "Food Access Research Atlas" or go to the U.S. Department of Agriculture's Food Access web page. Zoom in on your community.
5. Is your home located in a food desert? Is your home neighborhood accurately classified as being within or outside a food desert? Why or why not?

Periodic Markets

LEARNING OUTCOME 12.2.4

Describe the concept of periodic markets in developing countries.

Geographers apply central place theory to create market area studies that assist service providers with opening and expanding their facilities. And in a severe economic downturn, market area analysis helps determine where to close facilities.

Services at the lower end of the central place hierarchy may be provided at a periodic market, which is a collection of individual vendors that come together to offer goods and services in a location on specified days. A periodic market typically is set up in a street or other public space early in the morning, taken down at the end of the day, and set up in another location the next day (Figure 12-16).

A periodic market provides goods to residents of developing countries, as well as rural areas in developed countries, where sparse populations and low incomes produce purchasing power too low to support full-time retailing. A periodic market makes services available in more villages than would otherwise be possible, at least on a part-time basis. In urban areas, periodic markets offer residents fresh food brought in that morning from the countryside (Figure 12-17).

Many of the vendors in periodic markets are mobile, driving their trucks from farm to market, back to the farm to restock, then to another market. Other vendors, especially local residents who cannot or prefer not to travel to other villages, operate on a part-time basis, perhaps only a few times a year. Other part-time vendors are individuals who are capable of producing only a small quantity of food or handicrafts.

▲ **FIGURE 12-16 PERIODIC MARKET** The weekly market at Bati is considered the largest in Ethiopia.

The frequency of periodic markets varies by culture:

- **Muslim countries.** Markets in Muslim countries may conform to the weekly calendar—once a week in each of six cities and no market on Friday, the Muslim day of rest.

◀ **FIGURE 12-17 BRINGING FOOD TO THE PERIODIC MARKET** Food is carried to be sold at the periodic market in Gongtan, China.

SUSTAINABILITY & OUR ENVIRONMENT What Types of Services Are Airbnb and Uber?

New companies are challenging the traditional classification of services. Airbnb and Uber are two prominent examples that might be considered a form of periodic services in developed countries. Time will tell whether these sharing services are economically and socially sustainable, or whether they come to be regarded as unsustainable business models.

The Airbnb website and smart phone app match people looking for lodging with people who have rooms and houses for rent (Figure 12-18). Since the company was founded in 2008, it has grown to more than 1 million listings in nearly every country of the world. Airbnb competes with hotels, and the employees of a hotel are clearly classified as a leisure and hospitality consumer service.

How should the employees of Airbnb be classified? Though the purpose of the company is to provide a leisure and hospitality consumer service, the people who actually rent out their rooms and homes are not Airbnb employees. Airbnb employees are primarily computer operators, a type of employment classified under transportation and information business services.

The Uber website and smart phone app match people looking for a ride with people who are willing to transport them. Uber competes with taxis. Most taxis are summoned through a phone call or a hit-or-miss process of flagging one down on the street. In contrast, an Uber car is summoned by using a smart phone app. A driver shows up promptly because the GPS tracking device in your smart phone tells the driver your location.

How should the nearly 200,000 drivers of Uber be classified? Uber claims that each driver is an independent contractor, essentially a one-person business providing a transportation business service. In 2015, the California Labor Commission ruled that Uber drivers should be classified as employees of Uber rather than as independent contractors. The classification matters because critics charge that Uber drivers are not properly screened, trained, and insured. If they are Uber employees, then it is Uber's responsibility to screen, train, and insure them. In some cities, especially in Europe, Uber is banned because the company's alleged insufficient screening, training, and insurance results in unfair competition with licensed taxi services.

(a)

(b)

▲ **FIGURE 12-18 AIRBNB APP (a)** Is it a consumer leisure service (accessed through an app)? **(b)** Or is it a business information service (at Airbnb headquarters, San Francisco)?

- **Rural China.** According to G. William Skinner, rural China has a three-city, 10-day cycle of periodic markets. The market operates in a central market on days 1, 4, and 7; in a second location on days 2, 5, and 8; in a third location on days 3, 6, and 9; and no market on the tenth day. Three 10-day cycles fit in a lunar month.
- **Korea.** Korea has two 15-day market cycles in a lunar month.
- **Africa.** In Africa, the markets occur every 3 to 7 days. Variations in the cycle stem from ethnic differences.

PAUSE & REFLECT 12.2.4

Identify an example of a periodic market in developed countries.

CHECK-IN KEY ISSUE 2

Where Are Consumer Services Distributed?

- ✔ **Central place theory helps determine the most profitable location for a consumer service.**
- ✔ **A central place is surrounded by a market area that has a range and a threshold.**
- ✔ **Market areas of varying sizes nest and overlap.**
- ✔ **Regular patterns of settlements that provide consumer services can be observed, especially in developed countries.**

KEY ISSUE 3

Where Are Business Services Distributed?

- **Hierarchy of Business Services**
- **Business Services in Developing Countries**
- **Economic Specialization of Settlements**

LEARNING OUTCOME 12.3.1

Describe the factors that are used to identify global cities.

Every urban settlement provides consumer services to people in a surrounding area, but not every settlement of a given size has the same number and types of business services. Business services disproportionately cluster in a handful of urban settlements, and individual settlements specialize in particular business services.

Hierarchy of Business Services

Geographers identify a handful of urban settlements known as global cities (also called world cities) that play an especially important role in global business services. Global cities can be subdivided according to a number of criteria.

BUSINESS SERVICES IN GLOBAL CITIES

Global cities are most closely integrated into the global economic system because they are at the center of the flow of information and capital. Business services that concentrate in disproportionately large numbers in global cities include:

▲ **FIGURE 12-19 WALL STREET FINANCIAL INSTITUTIONS** A statue of George Washington stands near the New York Stock Exchange.

- **Financial institutions.** As centers for finance, global cities attract the headquarters of the major banks, insurance companies, and specialized financial institutions where corporations obtain and store funds for expansion of production (Figure 12-19).
- **Headquarters of large corporations.** Shares of these corporations are bought and sold on stock exchanges located in global cities. Obtaining information in a timely manner is essential in order to buy and sell shares at attractive prices. Executives of manufacturing firms meeting far from the factories make key decisions concerning what to make, how much to produce, and what prices to charge. Support staff far from the factory accounts for the flow of money and materials to and from the factories. This work is done in offices in global cities.
- **Lawyers, accountants, and other professional services.** Professional services cluster in global cities to provide advice to major corporations and financial institutions. Advertising agencies, marketing firms, and other services concerned with style and fashion locate in global cities to help corporations anticipate changes in taste and to help shape those changes.

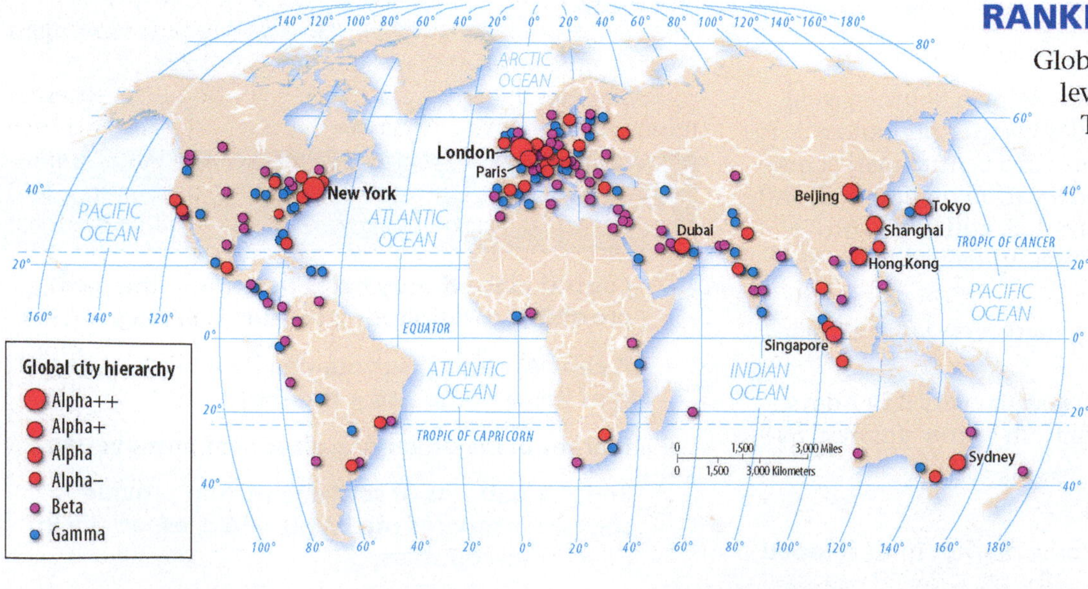

RANKING GLOBAL CITIES

Global cities are divided into three levels: alpha, beta, and gamma. These three levels are further subdivided (Figure 12-20). The

◀ **FIGURE 12-20 GLOBAL CITIES** Global cities are centers for the provision of services in the global economy. London and New York, the two dominant global cities, are ranked as alpha++. Other alpha, beta, and gamma global cities play somewhat less central roles in the provision of services than the two dominant global cities. Cities ranked alpha++ and alpha+ are labeled on the map.

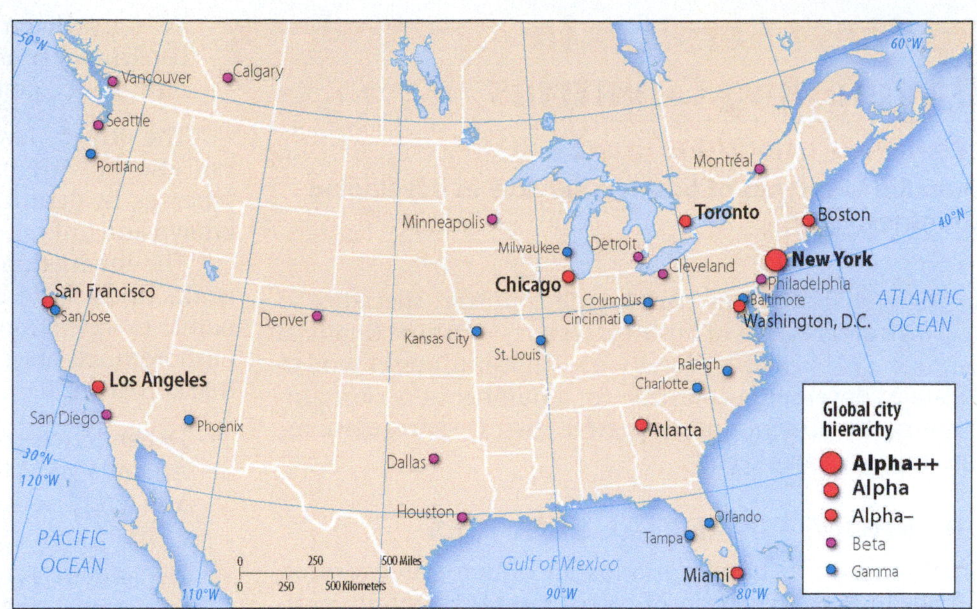

▶ **FIGURE 12-21 GLOBAL CITIES IN NORTH AMERICA** Atop the hierarchy of business services is New York, followed by Chicago, Los Angeles, and Toronto.

same hierarchy of business services can be used within countries or continents (Figure 12-21). A combination of factors is used to identify and rank global cities:

- **Economic factors.** Number of headquarters for multinational corporations, financial institutions, and law firms that influence the global economy.
- **Political factors.** Hosting headquarters for international organizations and capitals of countries that play a leading role in international events.
- **Cultural factors.** Presence of renowned cultural institutions, influential media outlets, sports facilities, and educational institutions.
- **Infrastructure factors.** A major international airport, health-care facilities, and advanced communications systems. Technology was expected to reduce the need for clustering of services in large cities, but it hasn't.
- **Communications.** The telegraph and telephone in the nineteenth century and the computer in the twentieth century made it possible to communicate immediately with coworkers, clients, and customers around the world.
- **Transportation.** The railroad in the nineteenth century and the motor vehicle and airplane in the twentieth century made it possible to deliver people, inputs, and products quickly. Modern transportation and communications enable industry to decentralize, as discussed in Chapter 11, but they reinforce rather than diminish the primacy of global cities in the world economy.

▼ **FIGURE 12-22 LONDON THEATRE DISTRICT** Leisure services, such as theaters, cluster in global cities.

PAUSE & REFLECT 12.3.1

How would you expect an alpha city such as Chicago to differ from a beta city such as Houston or a gamma city such as Phoenix?

CONSUMER AND PUBLIC SERVICES IN GLOBAL CITIES

Because of their large size, global cities have consumer services with extensive market areas, but they may have even more consumer services than large size alone would predict. A disproportionately large number of wealthy people live in global cities, so luxury and highly specialized products are especially likely to be sold there.

Leisure services of national significance are especially likely to cluster in global cities, in part because they require large thresholds and large ranges and in part because of the presence of wealthy patrons. Global cities typically offer the most plays, concerts, operas, night clubs, restaurants, bars, and professional sporting events. They contain the largest libraries, museums, and theaters (Figure 12-22).

Global cities may be centers of national or international political power. Most are national capitals, and they contain mansions or palaces for the head of state, imposing structures for the national legislature and courts, and offices for the government agencies. Also clustered in global cities are offices for groups having business with the government, such as representatives of foreign countries, trade associations, labor unions, and professional organizations.

Unlike other global cities, New York is not a national capital. But as the home of the world's major international organization, the United Nations, it attracts thousands of diplomats and bureaucrats, as well as employees of organizations with business at the United Nations. Brussels is a global city because it is the most important center for European Union activities.

Business Services in Developing Countries

LEARNING OUTCOME 12.3.2

Describe two types of business services in developing countries.

In the global economy, developing countries specialize in two distinctive types of business services: offshore financial services and back-office functions. These businesses tend to locate in developing countries for a number of reasons, including the presence of supportive laws, weak regulations, and low-wage workers.

OFFSHORE FINANCIAL SERVICES

Small countries, usually islands and microstates, exploit niches in the circulation of global capital by offering offshore financial services. Offshore centers provide two important functions in the global circulation of capital:

- **Taxes.** Taxes on income, profits, and capital gains are typically low or nonexistent. Companies incorporated in an offshore center also have tax-free status, regardless of the nationality of the owners. The United States loses an estimated $150 billion in tax revenue each year because companies operating in the country conceal their assets in offshore tax havens.
- **Privacy.** Bank secrecy laws can help individuals and businesses evade disclosure in their home countries. Corporations and people who may be accused of malpractice, such as a doctor or lawyer, or the developer of a collapsed building, can protect some of their assets from lawsuits by storing them in offshore centers. So can a wealthy individual who wants to protect assets in a divorce. Creditors cannot reach such assets in bankruptcy hearings. Short statutes of limitation protect offshore accounts from long-term investigation.

The privacy laws and low tax rates in offshore centers can also provide havens to tax dodges and other illegal schemes. By definition, the extent of illegal activities is unknown and unknowable.

The International Monetary Fund, the Tax Justice Network's Financial Secrecy Index, and the Organisation for Economic Co-operation and Development all maintain lists of offshore financial services centers. Figure 12-23 shows locations that appear on all three organizations' lists. These include:

- Dependencies of the United Kingdom, such as Anguilla and Montserrat in the Caribbean, Isle of Man and Jersey in the English Channel, and Gibraltar off Spain.
- Dependencies of other countries, such as Cook Island (controlled by New Zealand), Aruba and Curaçao (controlled by the Netherlands), and Hong Kong and Macau (controlled by China).
- Independent island countries, such as The Bahamas and Grenada in the Caribbean, Nauru and Vanuatu in the Pacific Ocean, and the Seychelles in the Indian Ocean.
- Other independent countries, such as Liechtenstein and Switzerland in Europe, Belize and Uruguay in Latin America, and Bahrain and Brunei in Asia.

BUSINESS-PROCESS OUTSOURCING

A second distinctive type of business service found in peripheral regions is back-office functions, also known as business-process outsourcing (BPO). Typical back-office functions include insurance claims processing, payroll

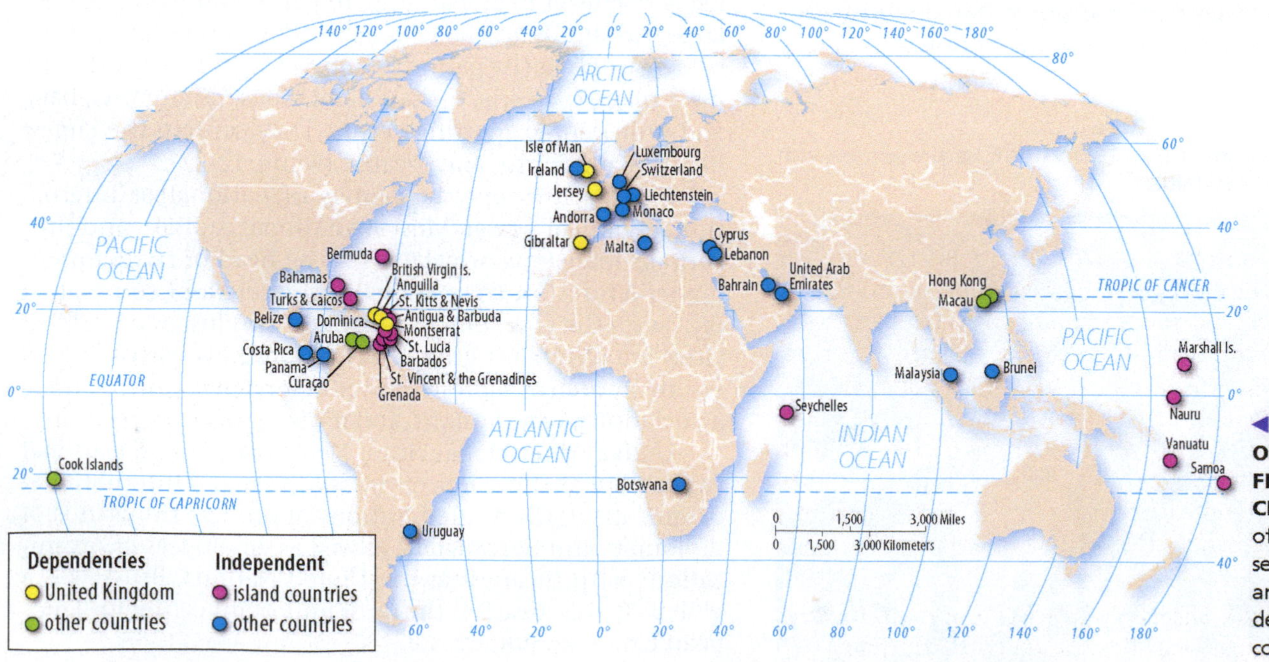

◀ FIGURE 12-23 OFFSHORE FINANCIAL SERVICE CENTERS Most offshore financial service centers are microstates or dependence of other countries.

▲ **FIGURE 12-24 CALL CENTER** Bangalore, India.

management, transcription work, and other routine clerical activities. Back-office work also includes centers for responding to billing inquiries related to credit cards, shipments, and claims, or technical inquiries related to installation, operation, and repair.

Traditionally, companies housed their back-office staff in the same office building downtown as their management staff, or at least in nearby buildings. A large percentage of the employees in a downtown bank building, for example, would be responsible for sorting paper checks and deposit slips. Proximity was considered important to assure close supervision of routine office workers and rapid turnaround of information.

Rising rents downtown have induced many business services to move routine work to lower-rent buildings elsewhere. In most cases, sufficiently low rents can be obtained in buildings in suburbs or nearby small towns. For many business services, improved telecommunications is the most important factor in eliminating the need for spatial proximity.

Selected developing countries have attracted back offices for two reasons related to labor:

- **Low wages.** Most back-office workers earn a few thousand dollars per year—higher than wages paid in most other sectors of the economy but only one-tenth the wages paid for workers performing similar jobs in developed countries. As a result, what is regarded as menial and dead-end work in developed countries may be considered relatively high-status work in developing countries and therefore may be able to attract better-educated, more-motivated employees in developing countries than would be possible in developed countries.
- **Ability to speak English.** Many developing countries offer lower wages than developed countries, but only a handful of developing countries possess a large labor force fluent in English. In Asia, countries such as India, Malaysia, and the Philippines have substantial numbers of workers with English-language skills, a legacy of British and American colonial rule (Figures 12-24 and 12-25). Major multinational companies such as American Express and General Electric have extensive back-office facilities in those countries.

The ability to communicate in English over the telephone is a strategic advantage in competing for back offices with neighboring countries, such as Indonesia and Thailand, where English is less commonly used. Familiarity with English is an advantage not only for literally answering the telephone but also for gaining a better understanding of the preferences of American consumers through exposure to English-language music, movies, and television.

Workers in back offices often must work late at night, when it's daytime in the United States, peak demand for inquiries. Many employees must arrive at work early and stay late because they lack their own transportation and depend on public transportation, which typically does not operate late at night. Sleeping and entertainment rooms may be provided at work to fill the extra hours.

PAUSE & REFLECT 12.3.2

If it is 3 p.m. on a Tuesday where you live, what time and day is it at a call center in India? Refer to Figure 1-15.

▲ **FIGURE 12-25 CALL CENTER TRAINING** Madurai, India.

Economic Specialization of Settlements

LEARNING OUTCOME 12.3.3

Explain the concept of economic base.

Settlements can be classified by the distinctive types of economic activities that take place there. All sectors of the economy—be they the various types of agriculture, the various types of manufacturers, or the various types of services—have distinctive geographic distributions.

ECONOMIC BASE

The economic activities in a settlement can be divided into two types:

- A **basic business** exports primarily to customers outside the settlement.
- A **nonbasic business** serves primarily customers living in the same settlement.

The unique cluster of basic businesses in a settlement is its **economic base.**

A settlement's economic base is important because exporting by the basic businesses brings more money into the local economy, thus stimulating the provision of more nonbasic services for the settlement. It works like this:

- New basic businesses attract new workers to a settlement.
- The new basic business workers bring their families with them.
- New nonbasic services are opened to meet the needs of the new workers and their families.

For example, when a new car assembly plant opens, new supermarkets, restaurants, and other consumer services soon follow. But the opposite doesn't occur: A new supermarket does not induce construction of a new car plant.

Settlements in the United States can be classified by their distinctive collection of basic businesses (Figure 12-26). The concept of basic businesses originally referred to manufacturing, but with the growth of the service sector of the economy, the basic businesses of many communities are in consumer, business, and public services (Figure 12-27).

▲ FIGURE 12-27 ECONOMIC BASE OF LAS VEGAS: GAMING

If a settlement's basic businesses are growing, they attract other basic and nonbasic businesses that can benefit from proximity. The result can be a cluster of businesses that reinforce each other's growth. For example, Boston's basic sector in biotechnology consists of a cluster of business sectors that complement each other (Figure 12-28). Conversely, if a settlement's basic businesses are shedding jobs—such as Detroit's auto industry—then other businesses in the cluster may also decline.

DISTRIBUTION OF TALENT

Individuals possessing special talents are not distributed uniformly among cities. Some cities have a higher percentage of talented individuals such as scientists and professionals

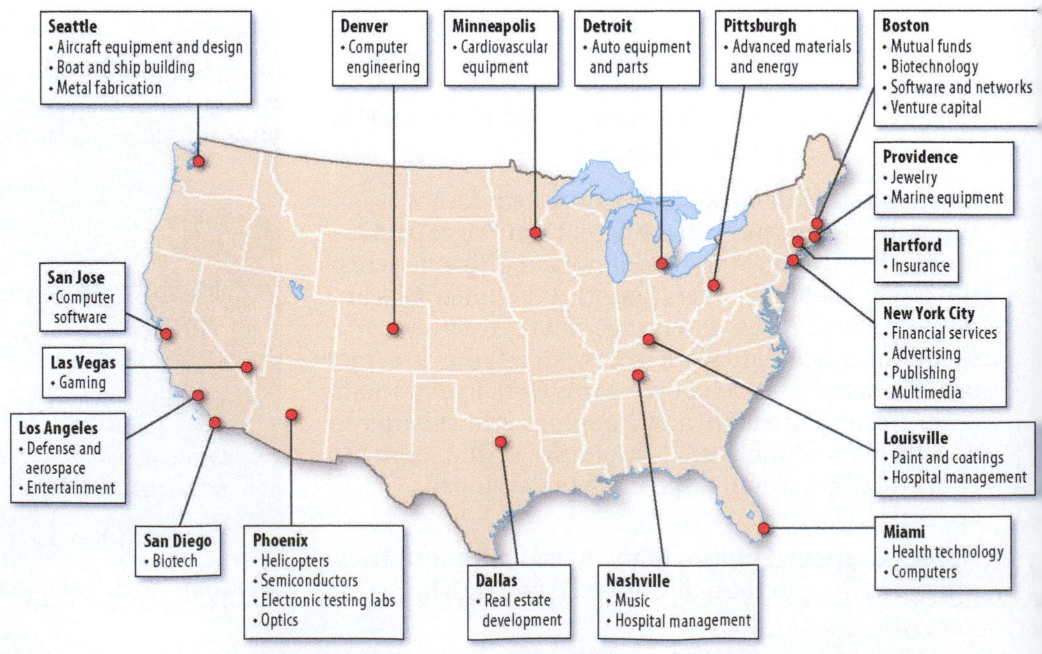

▶ FIGURE 12-26 ECONOMIC BASE OF SELECTED U.S. SETTLEMENTS Settlements specialize in different economic activities.

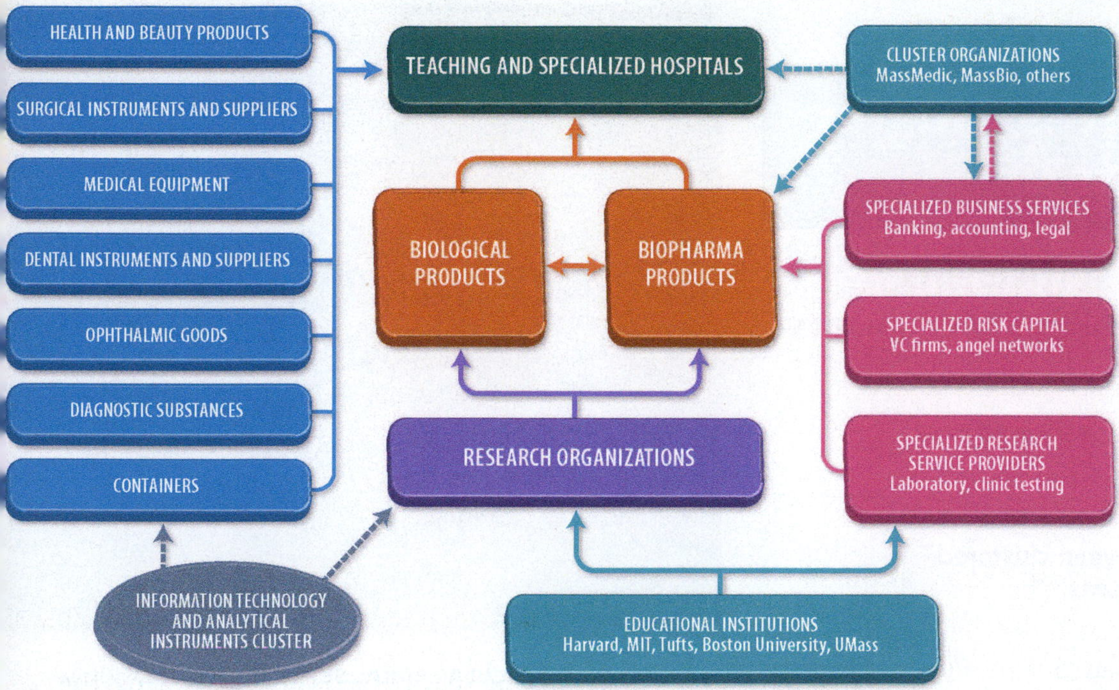

▲ **FIGURE 12-28 ECONOMIC BASE OF BOSTON** Clustering of businesses related to biotechnology.

(Figure 12-29). Attracting talented individuals is important because these individuals are responsible for promoting economic innovation. They are likely to start new businesses and infuse the local economy with fresh ideas.

To some extent, talented individuals are attracted to the cities with the most job opportunities and financial incentives. But the principal enticement for talented individuals is cultural rather than economic, according to research conducted by Richard Florida. Florida found that individuals with special talents gravitate toward cities that offer more cultural diversity. He used a "coolness" index developed by *POV Magazine* that combined the percentage of population in their 20s, the number of bars and other nightlife places per capita, and the number of art galleries per capita (Figure 12-30).

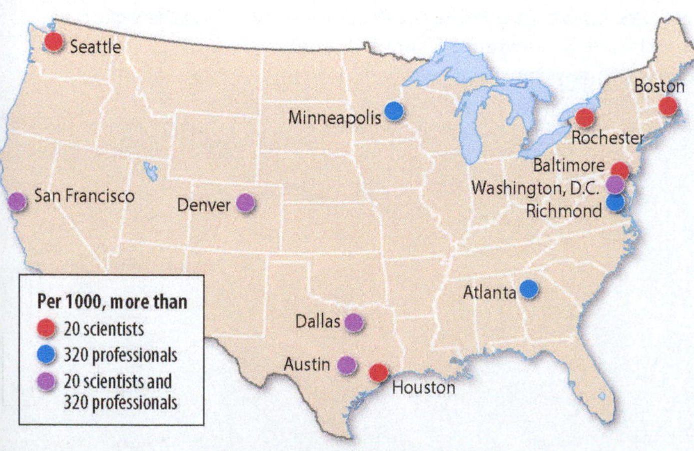

▲ **FIGURE 12-29 DISTRIBUTION OF SCIENTISTS AND PROFESSIONALS** Some cities have higher concentrations than others of scientists and professionals.

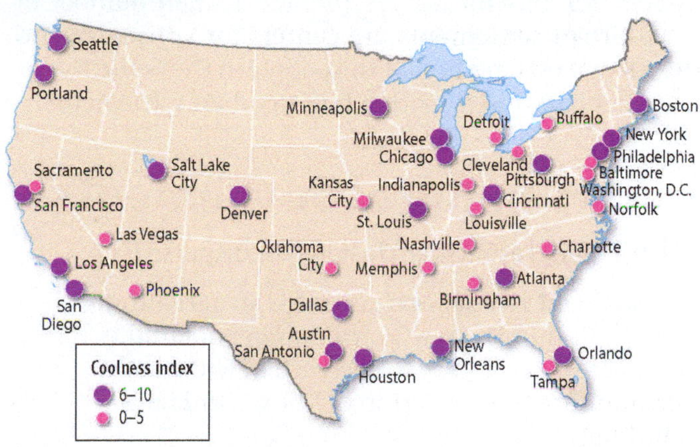

▲ **FIGURE 12-30 COOLNESS INDEX** Based on a "coolness" index developed by *POV Magazine.*

PAUSE & REFLECT 12.3.3

Do settlements with high concentrations of scientists and professionals rank high on "coolness" or low? What might account for that pattern?

CHECK-IN KEY ISSUE 3

Where Are Business Services Distributed?

- ✔ **Business services cluster in global cities.**
- ✔ **Developing countries provide offshore financial services and business-process outsourcing.**
- ✔ **Communities specialize in the provision of particular services; the specialized services constitute a community's economic base.**

KEY ISSUE 4

Why Do Services Cluster in Settlements?

- Services in Rural Settlements
- Services in Early Urban Settlements
- Percent Urban
- Size of Urban Settlements

LEARNING OUTCOME 12.4.1

Describe the difference between clustered and dispersed rural settlements.

Services locate primarily in settlements. Rural settlements are centers for agriculture and provide a small number of services. Urban settlements are centers for consumer and business services. One-half of the people in the world live in rural settlements and the other half in urban settlements.

Services in Rural Settlements

Rural settlements are either clustered or dispersed:

- A **clustered rural settlement** is an agricultural-based community in which a number of families live in close proximity to each other, with fields surrounding the collection of houses and farm buildings.
- A **dispersed rural settlement** is characterized by farmers living on individual farms isolated from neighbors rather than alongside other farmers in settlements.

CLUSTERED RURAL SETTLEMENTS

A clustered rural settlement typically includes homes, barns, tool sheds, and other farm structures, plus consumer services, such as religious structures, schools, and shops. A handful of public and business services may also be present in a clustered rural settlement.

Each person living in a clustered rural settlement is allocated strips of land in the surrounding fields. Homes, public buildings, and fields in a clustered rural settlement are arranged according to local cultural and physical characteristics.

Clustered rural settlements are often arranged in one of two types of patterns, circular or linear:

- **Circular clustered rural settlements.** A circular clustered rural settlement consists of a central open space surrounded by structures. In sub-Saharan Africa, the Maasai people, who are pastoral nomads, build circular settlements known as kraal (Figure 12-31). Women have the principal responsibility for constructing them. The kraal villages have enclosures for livestock in the center, surrounded by a ring of houses. von Thünen observed this circular pattern in Germany in his landmark agricultural studies in the early nineteenth century (refer to Figure 9-35).
- **Linear clustered rural settlements.** Linear rural settlements comprise buildings clustered along a road, river, or dike to facilitate communications. The fields extend behind the buildings in long, narrow strips. Long-lot farms can be seen today along the St. Lawrence River in Québec (Figure 12-32). Québec got the system from the French.

▲ **FIGURE 12-31 CIRCULAR RURAL SETTLEMENT** A kraal village (also known as a boma), Meserani, Tanzania.

▼ **FIGURE 12-32 CLUSTERED LINEAR RURAL SETTLEMENT** Long lots along the St. Lawrence River in Québec.

▲ **FIGURE 12-33 CLUSTERED NEW ENGLAND SETTLEMENT** Newfane, Vermont, includes courthouse and church buildings clustered around a central commons.

▲ **FIGURE 12-35 U.K. RURAL SETTLEMENT** Condicote was originally a clustered rural settlement, but during the enclosure movement, the surrounding fields were consolidated into large farms.

Clustered rural settlements were characteristic of colonial New England. New England colonists typically traveled from England in a group, and they wanted to live close together to reinforce common cultural and religious values. The contemporary New England landscape contains remnants of the old clustered rural settlement pattern. Many New England towns still have a central common surrounded by the church, school, and various houses (Figure 12-33).

▲ **FIGURE 12-34 DISPERSED RURAL SETTLEMENT** Isolated farm, New Jersey.

DISPERSED RURAL SETTLEMENTS

Isolated farms are typical of most of the rural United States. A dispersed pattern developed from the time of initial settlement of the Middle Atlantic colonies because most immigrants to these colonies arrived individually rather than as members of a cohesive group, as in New England. As people moved westward from the Middle Atlantic region, they took with them their preference for isolated individual farms. Land was plentiful and cheap, so people bought as much as they could manage (Figure 12-34).

In Europe, some clustered settlements were converted to dispersed settlements in order to make agriculture more efficient. Clustered rural settlements worked when the population was low, but they had no spare land to meet the needs of a growing population. With the introduction of machinery, farms operated more efficiently at a larger scale. For example, in the United Kingdom, the **enclosure movement** between 1750 and 1850 resulted in the consolidation of individually owned strips of land surrounding villages into large farms owned by single individuals (Figure 12-35). When necessary, the government forced people to give up their holdings. As displaced farmers moved to urban settlements, the population of clustered rural settlements declined drastically. Because the enclosure movement coincided with the Industrial Revolution, villagers displaced from farming became workers in urban factories.

PAUSE & REFLECT 12.4.1

In which sector of the economy would you expect most of the residents of Newfane, Vermont, or Condicote, U.K., to be employed? Why?

Services in Early Urban Settlements

LEARNING OUTCOME 12.4.2

Identify important prehistoric, ancient, and medieval urban settlements.

Before the establishment of permanent settlements as service centers, people lived as nomads, migrating in small groups across the landscape in search of food and water. They gathered wild berries and roots or killed wild animals for food (see Chapter 9). At some point, groups decided to build permanent settlements. Several families clustered together in a rural location and obtained food in the surrounding area. What services would these nomads require? Why would they establish permanent settlements to provide these services?

No one knows the precise sequence of events through which settlements were established to provide services. Based on archaeological research, settlements probably originated to provide consumer and public services. Business services came later.

PREHISTORIC URBAN SETTLEMENTS

Settlements may have originated in Mesopotamia, part of the Fertile Crescent of Southwest Asia (see Figure 8-7), and diffused at an early date west to Egypt and east to China and to South Asia's Indus Valley (Figure 12-36a). Or settlements may have originated independently in each of the four hearths. In any case, from these four hearths, settlements diffused to the rest of the world.

Among the oldest well-documented urban settlements in Mesopotamia is Ur in present-day Iraq. Ur, which means "fire," was where Abraham lived prior to his journey to Canaan in approximately 1900 B.C., according to the Bible. Archaeologists have unearthed ruins in Ur that date from approximately 3000 B.C. (Figure 12-37).

PAUSE & REFLECT 12.4.2

In prehistoric times, before the invention of settlements, why might caves have played an important role?

▲ **FIGURE 12-37 PREHISTORIC URBAN SETTLEMENT: UR** The remains of Ur, in present-day Iraq, provide evidence of early urban civilization. Ancient Ur was compact, perhaps covering 100 hectares (250 acres), and was surrounded by a wall. The most prominent building, the stepped temple, called a ziggurat, was originally constructed around 4,000 years ago. The ziggurat was originally a three-story structure with a base that was 64 by 46 meters (210 by 150 feet), and the upper stories were stepped back. Four more stories were added in the sixth century B.C. Surrounding the ziggurat was a dense network of small residences built around courtyards and opening onto narrow passageways. The excavation site was damaged during the two wars in Iraq. A U.S. Army Black Hawk helicopter hovers above the ancient Ziggurat of Ur in 2009.

EARLY CONSUMER SERVICES. The earliest permanent settlements may have been established to offer consumer services, specifically places to bury the dead. Having established a permanent resting place for the dead, the group might then install priests at the site to perform the service of saying prayers for the deceased. This would have encouraged the building of structures—places for ceremonies and dwellings. By the time recorded history began about 5,000 years ago, many settlements existed, and some of them featured places of worship.

Settlements were places to house families, permitting unburdened males to travel farther and faster in their search for food. Women kept "home and hearth," making household objects, such as pots, tools, and clothing, and educating the children. People also needed tools, clothing, shelter, containers, fuel, and other material goods. Settlements therefore became manufacturing centers. Men gathered the materials needed to make a variety of objects: stones for tools and weapons, grass for containers and matting, animal hair for clothing, and wood for shelter and heat. Women used these materials to manufacture household objects and maintain their dwellings.

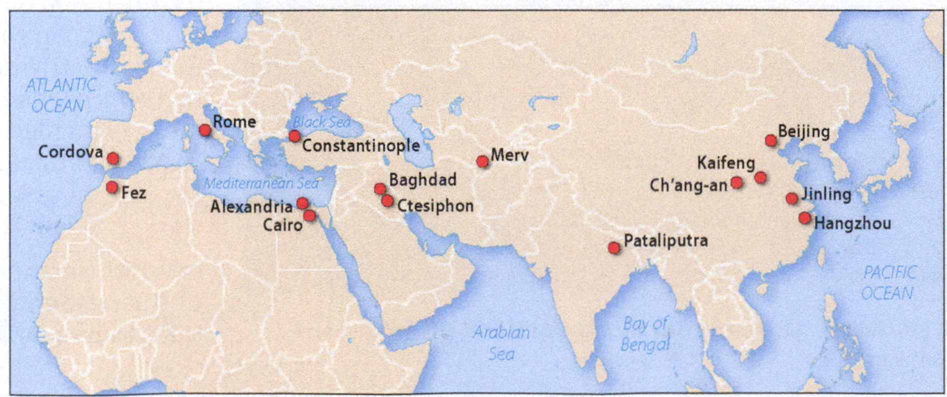

▲ **FIGURE 12-36 LARGEST URBAN SETTLEMENTS IN HISTORY** (a) 3000–500 B.C. (b) 500 B.C.–A.D. 1800.

EARLY BUSINESS SERVICES. Early urban settlements were places where groups could store surplus food and trade with other groups. People brought plants, animals, and minerals, as well as tools, clothing, and containers, to the urban settlements, and exchanged them for items brought by others. To facilitate this trade, officials in the settlement set fair prices, kept records, and created currency.

EARLY PUBLIC SERVICES. Early settlements housed political leaders as well as defense forces to guard the residents of the settlement and defend the surrounding hinterland from seizure by other groups.

ANCIENT URBAN SETTLEMENTS

Settlements were first established in the eastern Mediterranean about 2500 B.C. These settlements were trading centers for the thousands of islands dotting the Aegean Sea and the eastern Mediterranean and provided the government, military protection, and other public services for their surrounding hinterlands. They were organized into city-states, which were defined in Chapter 8 as independent self-governing communities that included the settlement and nearby countryside.

Athens, the largest city-state in ancient Greece (Figure 12-38), made substantial contributions to the development of culture, philosophy, and other elements of Western civilization. The urban settlements provided not only public services but also a concentration of consumer services, notably cultural activities, not found in smaller settlements.

The rise of the Roman Empire encouraged urban settlement. With much of Europe and Southwest Asia & North Africa under Roman rule, settlements were established as centers of administrative, military, and other public services, as well as retail and other consumer services. Trade was encouraged through transportation and utility services, notably construction of many roads and aqueducts, and the security the Roman legions provided. Rome—the empire's center for administration, commerce, culture, and all other services—was the world's most populous city 2,000 years ago and may have been the first city to reach a half million inhabitants.

With the fall of the Roman Empire in the fifth century, urban settlements declined. The empire's prosperity had rested on trading in the secure environment of imperial Rome. But with the empire fragmented under hundreds of rulers, trade diminished. Large urban settlements shrank or were abandoned. For several hundred years, Europe's cultural heritage was preserved largely in monasteries and isolated rural areas.

MEDIEVAL URBAN SETTLEMENTS

After the collapse of the Roman Empire, most of the world's largest urban settlements were clustered in China. Several cities in China are estimated to have been the world's most populous at various times between 600 and 1500 A.D., including Beijing, Ch'ang-an, Hangzhou, Jinling, and Kaifeng (refer to Figure 12-36b).

Urban life began to revive in Europe in the eleventh century, as feudal lords established new urban settlements. The lords gave residents charters of rights with which to establish independent cities in exchange for their military service. Both the lord and the urban residents benefited from this arrangement. The lord obtained people to defend his territory at less cost than maintaining a standing army. For their part, urban residents preferred periodic military service to the burden faced by rural serfs, who farmed the lord's land and could keep only a small portion of their own agricultural output.

With their newly won freedom from the relentless burden of rural serfdom, the urban dwellers set about expanding trade. Surplus from the countryside was brought into the city for sale or exchange, and markets were expanded through trade with other free cities. The trade among different urban settlements was enhanced by new roads and greater use of rivers. By the fourteenth century, Europe was covered by a dense network of small market towns serving the needs of particular lords.

The largest medieval European urban settlements served as power centers for the lords and church leaders, as well as major market centers. The most important public services occupied palaces, churches, and other prominent buildings arranged around a central market square. The tallest and most elaborate structures were usually churches, many of which still dominate the landscape of smaller European towns.

▶ **FIGURE 12-38 ANCIENT URBAN SETTLEMENT: ATHENS** Dominating the skyline of modern Athens is the ancient hilltop site of the city, the Acropolis. Ancient Greeks selected this high place because it was defensible, and they chose it as a place to erect shrines to their gods. The most prominent structure on the Acropolis is the Parthenon, built in the fifth century B.C. to honor the goddess Athena. To the right of the Acropolis is the Propylaea, which was the entrance gate to the Acropolis.

Percent Urban

LEARNING OUTCOME 12.4.3

Explain the two dimensions of urbanization.

The process by which the population of urban settlements grows, known as **urbanization,** has two dimensions:

- An increase in the *percentage* of people living in urban settlements.
- An increase in the *number* of people living in urban settlements.

The distinction between these two factors is important because they occur for different reasons and have different global distributions.

DIFFERENCES BETWEEN URBAN AND RURAL SETTLEMENTS

A century ago, social scientists observed striking differences between urban and rural residents. Louis Wirth argued during the 1930s that an urban dweller follows a different way of life than does a rural dweller. Thus Wirth defined a city as a permanent settlement that has three characteristics: large size, high population density, and socially heterogeneous people. These characteristics produced differences in the social behavior of urban and rural residents.

LARGE SIZE. If you live in a rural settlement, you know most of the other inhabitants and may even be related to many of them. The people with whom you relax are probably the same ones you see in local shops and at church. In contrast, if you live in an urban settlement, you can know only a small percentage of the other residents. You meet most of them in specific roles—your supervisor, your lawyer, your supermarket cashier, your electrician. Most of these relationships are contractual: You are paid wages according to a contract, and you pay others for goods and services. Consequently, the large size of an urban settlement produces different social relationships than those formed in rural settlements.

HIGH DENSITY. High density also produces social consequences for urban residents, according to Wirth. The only way that a large number of people can be supported in a small area is through specialization. Each person in an urban settlement plays a special role or performs a specific task to allow the complex urban system to function smoothly. At the same time, high density also encourages social groups to compete to occupy the same territory.

In medieval times, European urban settlements were usually surrounded by walls even though by then cannonballs could destroy them (Figure 12-39). Dense and compact within the walls, medieval urban settlements lacked space for construction, so ordinary shops and houses nestled into the side of the walls and the large buildings. Most of these modest medieval shops and homes, as well as the walls, have been demolished in modern times, with only the massive churches and palaces surviving. Modern tourists can appreciate the architectural beauty of these medieval churches and palaces, but they do not receive an accurate image of a densely built medieval town.

SOCIAL HETEROGENEITY. The larger the settlement, the greater the variety of people. A person has greater freedom in an urban settlement than in a rural settlement to pursue an unusual profession, sexual orientation, or cultural interest. In a rural settlement, unusual actions might be noticed and scorned, but urban residents are more tolerant of diverse social behavior. Regardless of values and preferences, in a large urban settlement, individuals can find people with similar interests. But despite the freedom and independence of an urban settlement, people may also feel lonely and isolated. Residents of a crowded urban settlement often feel that they are surrounded by people who are indifferent and reserved.

Wirth's three-part distinction between urban and rural settlements may still apply in developing countries. But in developed countries, social distinctions between urban and rural life have blurred. According to Wirth's definition, nearly everyone in a developed country now is urban. All

▲ **FIGURE 12-39 MEDIEVAL WALLED CITY: ÁVILA** The 2.5-kilometer (1.6-mile) wall encircling Ávila de los Caballeros, Spain, was started in A.D. 1090. 1. Why were walls built around cities in ancient and medieval times? 2. What changes in urbanization influenced the decline in walled cities? 3. What technology changes contributed to the decline in walled cities?

▶ **FIGURE 12-40 PERCENTAGE LIVING IN URBAN SETTLEMENTS** Developed countries have higher percentages of urban residents than do developing countries, though Latin America has a comparable percentage to that of developed countries.

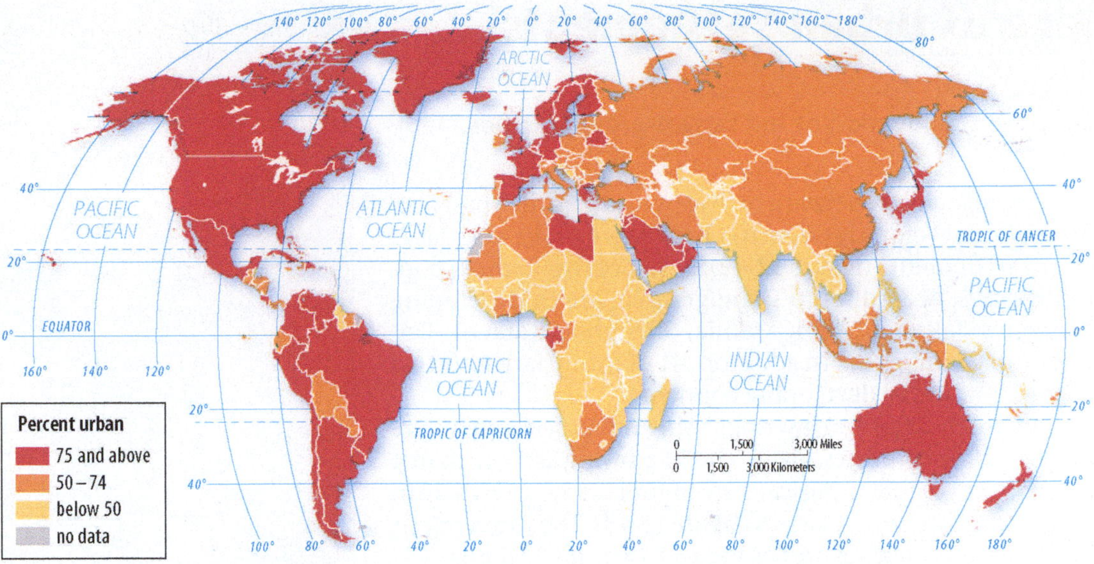

but 1 percent of workers in developed societies hold "urban" types of jobs. Nearly universal ownership of automobiles, telephones, televisions, and other modern communications and transportation has also reduced the differences between urban and rural lifestyles in developed countries. Almost regardless of where you live in a developed country, you have access to urban jobs, services, culture, and recreation.

PERCENTAGE IN URBAN SETTLEMENTS

The percentage of the world's population living in urban settlements has increased rapidly, from 3 percent in 1800 to 6 percent in 1850, 14 percent in 1900, 30 percent in 1950, and 45 percent in 2000. The population of Earth's urban settlements exceeded that of rural settlements for the first time in human history around 2008.

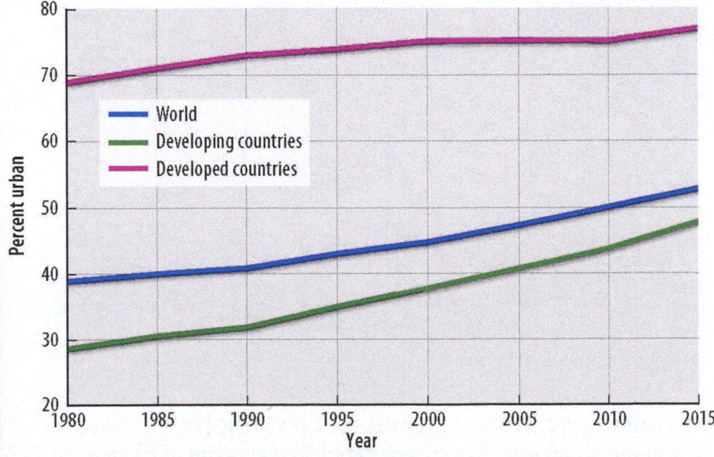

▲ **FIGURE 12-41 CHANGING PERCENTAGE LIVING IN URBAN AREAS** The percentage living in urban areas is increasing more rapidly in developing countries, though the gap between developed and developing regions remains high.

The percentage of people living in urban settlements reflects a country's level of development. In developed countries, 77 percent live in urban areas, compared to 48 percent in developing countries (Figure 12-40). Figure 12-41 shows that the gap in urbanization between developed and developing countries is closing rapidly. In 1980, 69 percent lived in urban areas in developed countries compared to only 29 percent in developing countries. The percentage living in urban areas is comparable in Latin America to the level in developed countries. On the other hand, only 33 percent live in urban areas in South Asia and 38 percent in sub-Saharan Africa, not by coincidence the two developing regions with the lowest HDIs (see Chapter 10).

The higher percentage of urban residents in developed countries is a result of changes in economic structure during the past two centuries—first the Industrial Revolution in the nineteenth century and then the growth of services in the twentieth. The percentage of urban dwellers is high in developed countries because over the past 200 years, rural residents have migrated from the countryside to work in the factories and services that are concentrated in cities.

The need for fewer farm workers has pushed people out of rural areas, and rising employment opportunities in manufacturing and services have lured them into urban areas. Because everyone resides either in an urban settlement or a rural settlement, an increase in the percentage living in urban areas has produced a corresponding decrease in the percentage living in rural areas.

PAUSE & REFLECT 12.4.3

What migration factor discussed in Chapter 3 might help to explain why the percentage of people in developed countries living in urban areas may not continue to increase?

Size of Urban Settlements

LEARNING OUTCOME 12.4.4

Describe the location of the fastest-growing cities.

Developed countries have a higher percentage of urban residents, but developing countries have more of the very large urban settlements (Figure 12-42). Seven of the 10 most populous cities, according to Demographia, are in developing countries: Jakarta, Delhi, Manila, Shanghai, Karachi, Beijing, and Guangzhou. In addition, 41 of the 50 largest urban settlements are in developing countries.

All but 3 of the 100 fastest-growing urban settlements are in developing countries. Five of the 13 growing at more than 4 percent per year are in Africa, 3 are in India, 4 are elsewhere in Asia, and 1 is in Latin America (Figure 12-43). The 3 exceptions in developed countries are Las Vegas, Austin, and Atlanta.

London grabbed the title of world's largest urban settlement during the nineteenth century, as part of the Industrial Revolution. New York held the title briefly during the mid-twentieth century, and Tokyo is now considered to be the world's largest urban settlement (Figure 12-44). However, Tokyo, Seoul, and New York are the only 3 cities in developed

▶ **FIGURE 12-42 URBAN SETTLEMENTS WITH AT LEAST 2 MILLION INHABITANTS** Most of the world's largest settlements are in developing countries, especially in East and South Asia and Latin America.

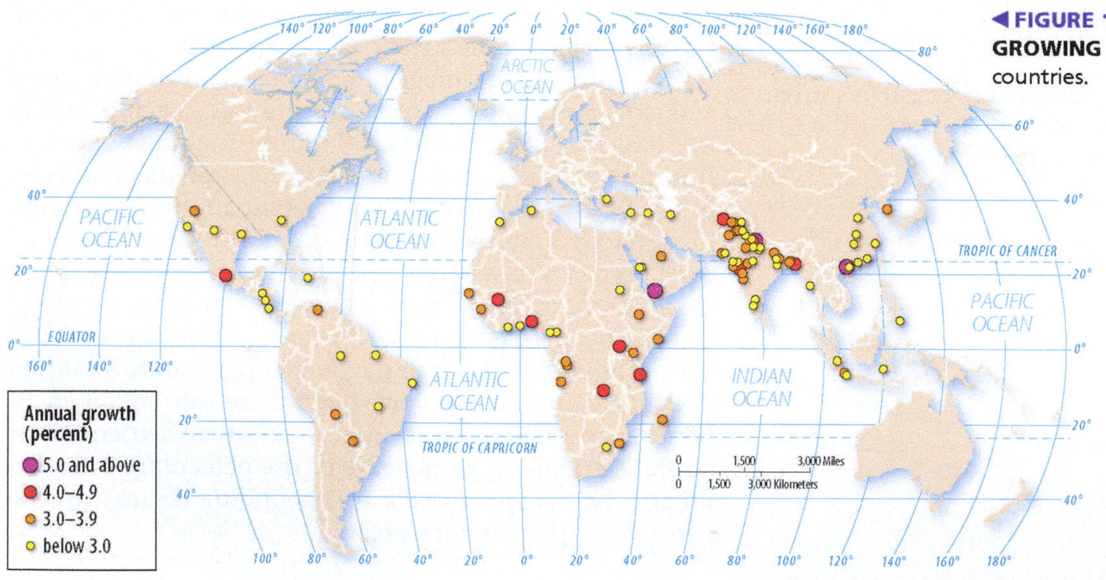

◀ **FIGURE 12-43 WORLD'S 100 FASTEST-GROWING CITIES** Nearly all are in developing countries.

▼ **FIGURE 12-44 WORLD'S LARGEST CITY: TOKYO** The Tokyo-Yokohama urban area has an estimated 38 million inhabitants.

countries ranked among the 10 largest. The three fastest-growing urban settlements are relatively unfamiliar places: Beihai, China, Ghaziabad, India, and Sana'a, Yemen (Figure 12-45).

That developing countries dominate the lists of largest and fastest-growing urban settlements is remarkable because urbanization was once associated with economic development. In 1800, 7 of the world's 10 largest cities were in Asia. In 1900, after diffusion of the Industrial Revolution from the United Kingdom to today's developed countries, all 10 of the world's largest cities were in Europe and North America.

In developing countries, migration from the countryside is fueling half of the increase in population in urban settlements, even though job opportunities may not be available. The other half results from high natural increase rates; in Africa, the natural increase rate accounts for three-fourths of urban growth.

▼ **FIGURE 12-45 THIRD FASTEST-GROWING CITY: SANA'A, YEMEN** Sana'a is adding more than 100,000 inhabitants per year, a rate of 5 percent per year.

CHECK-IN **KEY ISSUE 4**

Why Do Services Cluster in Settlements?

- ✔ **Settlements are either rural or urban; rural settlements, which specialize in agricultural services, may be clustered or dispersed.**
- ✔ **Few humans lived in urban settlements until the nineteenth century.**
- ✔ **Developed countries have higher percentages of urban residents, but developing countries have most of the very large cities.**

Summary & Review

KEY ISSUE 1

Where are services distributed?

Most jobs are in the service sector, especially in developed countries. Three types of services are consumer, business, and public.

THINKING GEOGRAPHICALLY

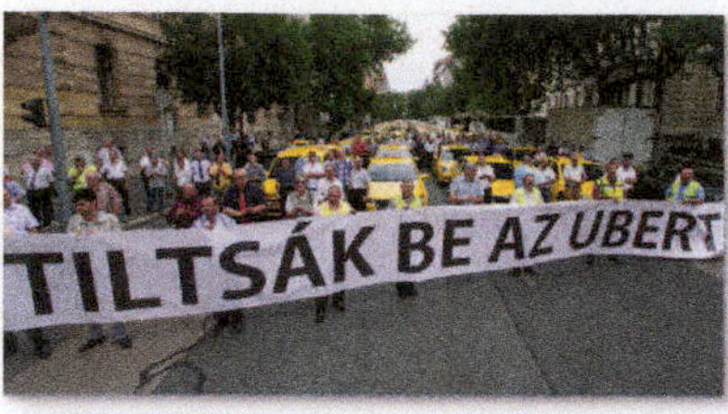

▲ **FIGURE 12-46 UBER PROTEST** Taxi drivers protest against Uber providing services in Budapest, Hungary. The sign in Hungarian says "Ban Uber."

1. Sharing services for housing, such as Airbnb, are illegal in many cities in the United States and Europe. Sharing services for transportation, such as Uber, are strongly opposed in many places. Figure 12-46 shows a protest against Uber by taxi drivers in Budapest, Hungary. Why would cities make these services illegal or difficult to use?

KEY ISSUE 2

Where are consumer services distributed?

Consumer services generally follow a regular pattern, based on size of settlements. The range is the maximum distance people are willing to travel to use a service. The threshold is the minimum number of people needed to support a service. The market area is the area surrounding a service from which customers are attracted. Larger settlements provide consumer services that have larger thresholds, ranges, and market areas.

THINKING GEOGRAPHICALLY

▲ **FIGURE 12-47 CONSUMER SERVICES IN A COLLEGE TOWN** Massachusetts Street, Lawrence, Kansas.

2. Education services frequently attract other types of consumer services to nearby locations.
What are examples of distinctive consumer services that are located near a university in your community? Why might these consumer services pick locations near a university?

KEY ISSUE 3

Where are business services distributed?

Business services disproportionately cluster in global cities, which are centers for global flows of information and capital. Some developing countries attract offshore financial services and back-office operations. A settlement's economic activities that export primarily to businesses and consumers elsewhere are its basic industries. Basic businesses are the principal source of growth and wealth for a settlement.

THINKING GEOGRAPHICALLY

3. What evidence can you find in your community of economic ties to alpha++ or alpha+ global cities? Figure 12-48 offers one example.

▲ **FIGURE 12-48 CONNECTION TO GLOBAL CITY** JPMorganChase, the largest bank based in the United States, has branches across the United States, including this one in Winter Haven, Florida.

KEY ISSUE 4

Why do services cluster in settlements?

Services cluster in rural and urban settlements. Rural settlements may be either clustered or dispersed. Few people lived in urban settlements until modern times. Urbanization involves an increase in the percentage of people living in urban settlements. Developed countries have higher percentages of urban residents than do developing countries. Urbanization also involves an increase in size of settlements. Most very large settlements are in developing countries.

THINKING GEOGRAPHICALLY

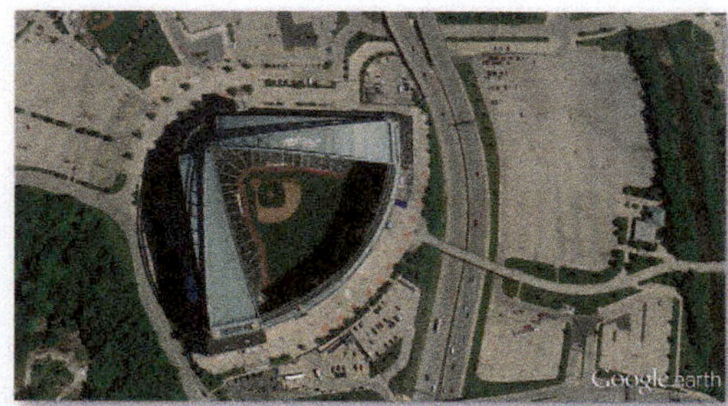

◀ **FIGURE 12-49 SMALLEST MAJOR LEAGUE BASEBALL SETTLEMENT**

4. Professional sports teams are examples of services with large ranges, thresholds, and areas, and consequently they cluster in large urban settlements. Google "Largest U.S. Metropolitan Statistical Areas." Compare the list with the distribution of Major League Baseball teams shown in Figure 1-29.
What are the largest metropolitan areas without a team?
What is the smallest settlement with a team?

KEY TERMS

Basic business *(p. 446)* A business that sells its products or services primarily to consumers outside the settlement.

Business service *(p. 432)* A service that primarily meets the needs of other businesses, including professional, financial, and transportation services.

Central place *(p. 434)* A market center for the exchange of services by people attracted from the surrounding area.

Central place theory *(p. 434)* A theory that explains the distribution of services based on the fact that settlements serve as centers of market areas for services; larger settlements are fewer and farther apart than smaller settlements and provide services for a larger number of people who are willing to travel farther.

Clustered rural settlement *(p. 448)* A rural settlement in which the houses and farm buildings of each family are situated close to each other, with fields surrounding the settlement.

Consumer service *(p. 432)* A service that primarily meets the needs of individual consumers, including retail, education, health, and leisure services.

Dispersed rural settlement *(p. 448)* A rural settlement pattern characterized by isolated farms rather than clustered villages.

Economic base *(p. 446)* A community's collection of basic businesses.

Enclosure movement *(p. 449)* The process of consolidating small landholdings into a smaller number of larger farms in England during the eighteenth century.

Food desert *(p. 439)* An area that has a substantial amount of low-income residents and has poor access to a grocery store, defined in most cases as further than 1 mile.

Gravity model *(p. 438)* A model which holds that the potential use of a service at a particular location is directly related to the number of people in a location and inversely related to the distance people must travel to reach the service.

Hinterland *(p. 434)* The area surrounding a central place from which people are attracted to use the place's goods and services (also known as market area).

Market area *(p. 434)* The area surrounding a central place from which people are attracted to use the place's goods and services (also known as hinterland).

Nonbasic business *(p. 446)* A business that sells its products primarily to consumers in the same settlement.

Primate city *(p. 437)* A city that is the largest settlement in a country and has more than twice as many people as the second-ranking settlement.

Primate city rule *(p. 437)* A pattern of settlements in a country such that the largest settlement has more than twice as many people as the second-ranking settlement.

Public service *(p. 432)* A service offered by the government to provide security and protection for citizens and businesses.

Range (of a service) *(p. 435)* The maximum distance people are willing to travel to use a service.

Rank-size rule *(p. 436)* A pattern of settlements in a country such that the nth largest settlement is $1/n$ the population of the largest settlement.

Service *(p. 432)* Any activity that fulfills a human want or need and returns money to those who provide it.

Settlement *(p. 432)* A permanent collection of buildings and inhabitants.

Threshold *(p. 435)* The minimum number of people needed to support a service.

Urbanization *(p. 452)* An increase in the percentage of and the number of people living in urban settlements.

EXPLORE

West Edmonton Mall

Use Google Earth to explore North America's largest shopping mall.

Fly to *West Edmonton Mall, Alberta.*

Show *Ruler* and measure the area occupied by the mall, including parking lots, inside the rectangle formed by the four perimeter streets. Explore the four streets that form the perimeter of the mall by zooming in to eye alt around 2,500 ft and dragging to Enter Street View.

Large consumer service centers such as a mega-mall often attract other consumer services nearby.

1. Can you see any evidence of consumer services adjacent to the mall? Why might the continent's largest mall have relatively few services immediately adjacent?
2. What is the principal use of land on the other sides of the four perimeter streets?
3. What evidence do you see of transportation services for people to arrive other than in individual passenger cars?

MG GeoVideo

Log in to the **MasteringGeography** Study Area to view this video

Ukraine: Serhiy's Leap

In a Ukrainian village, people raise much of their own food, and jobs other than farming are scarce. Like most of the village's other young people, Serhiy plans to leave when he finishes school.

1. How does Serhiy's family earn its living? What is Serhiy's contribution?
2. How does Serhiy rate the advantages and disadvantages of village life versus city life?
3. Describe Serhiy's plan for his life. Does it seem reasonable? What is the alternative?

MasteringGeography™

Visit the Study Area in **MasteringtGeography**™ to enhance your geographic literacy, spatial reasoning skills, and understanding of this chapter's content by accessing a variety of resources, including MapMaster interactive maps, videos, *In the News* RSS feeds, flashcards, web links, self-study quizzes, and an eText version of *The Cultural Landscape.*

www.masteringgeography.com

13

Urban Patterns

A large city is stimulating and agitating, entertaining and frightening, welcoming and cold. A city has something for everyone, but a lot of those things are for people who are different from you. Urban geography helps to sort out the complexities of familiar and unfamiliar patterns in urban areas.

Dancers perform in downtown Havana, Cuba.

LOCATIONS IN THIS CHAPTER

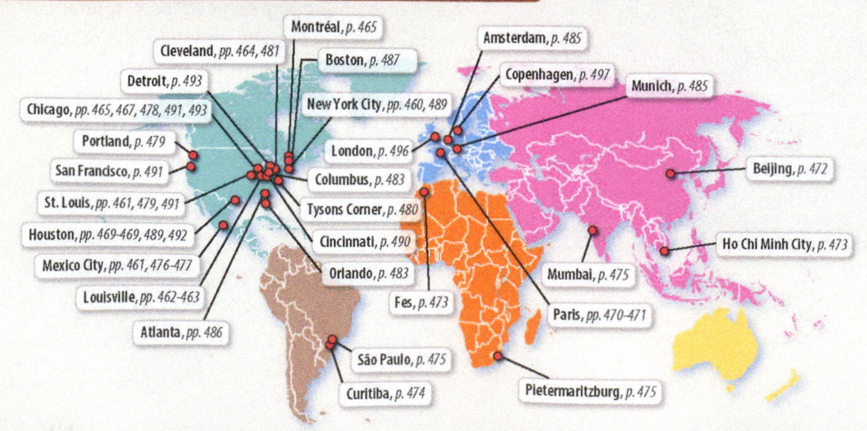

KEY ISSUES

1 Why Are Downtowns Distinctive?

Downtown is the ***place*** where much of the city's business and public services cluster.

2

Where Are People Distributed in Urban Areas?

People occupy distinctive *space* within an urban area.

3

Why Do Urban Areas Expand?

Surrounding cities are suburbs that have *connections* to the rest of the urban *region* through transport systems.

4

Why Do Cities Face Sustainability Challenges?

Because a majority of Earth's inhabitants now live in cities, global *scale* issues of sustainability, such as pollution and resource depletion, are increasingly addressed at the urban scale.

KEY ISSUE 1

Why Are Downtowns Distinctive?

- ▸ Introducing Urban Patterns
- ▸ The Central Business District
- ▸ Competition for Space in CBDs

LEARNING OUTCOME 13.1.1
Understand various definitions of urban settlements.

When you are staring at the Empire State Building, you know you are in a city (Figure 13-1). When you are standing in an Iowa cornfield, you have no doubt that you are in the country. Geographers help explain why urban and rural settlements are different.

Introducing Urban Patterns

Chapter 12 and this chapter are both concerned with urban geography, but at different scales. The previous chapter examined the distribution of urban settlements at national and global scales. This chapter looks at where people and activities are distributed within urban areas. Models have been developed to explain why differences occur within urban areas. In developing countries people are migrating into cities in large numbers, whereas in developed countries people are increasingly likely to be moving out to suburbs.

▼ **FIGURE 13-1 NEW YORK CITY**

CENTRAL CITY

Historically, urban settlements were very small and compact. As these settlements have rapidly grown, however, definitions have been created to characterize their different parts: the central city, the urban area, and the metropolitan area.

A **central city** (or simply **city**) is an urban settlement that has been legally incorporated into an independent, self-governing unit known as a municipality (Figure 13-2). Virtually all countries have a local government system that recognizes cities as legal entities with fixed boundaries. A city has locally elected officials, the ability to raise taxes, and responsibility for providing essential services. The boundaries of the city define the geographic area within which the local government has legal authority.

Population has declined since 1950 by more than one-half in the central cities of Buffalo, Cleveland, Detroit, Pittsburgh, and St. Louis and by at least one-third in more than a dozen other U.S. central cities. In contrast, other definitions of urban settlements reflect increasing population.

URBAN AREA

An **urban area** consists of a central city and its surrounding built-up suburbs (Figure 13-3). The U.S. census recognizes two types of urban areas:

- The **urbanized area** is an urban area with at least 50,000 inhabitants.
- An **urban cluster** is an urban area with between 2,500 and 50,000 inhabitants.

The census identified 486 urbanized areas and 3,087 urban clusters in the United States as of 2013. Approximately 70 percent of the U.S. population lived in one of the 486 urbanized areas, including about 30 percent in central cities and 40 percent in surrounding jurisdictions. Approximately 10 percent of the U.S. population lived in one of the 3,087 urban clusters.

METROPOLITAN AREA

The economic and cultural area of influence of a settlement extends beyond the urban area in the United States as well as other countries (Figure 13-4). The U.S. Bureau of the Census has created a method of measuring the larger functional area of a settlement, known as the **metropolitan statistical area (MSA).** An MSA includes the following:

- An urbanized area with a population of at least 50,000.
- The county within which the city is located. In New England, towns are sometimes used instead of counties.
- Adjacent counties with a high population density and a large percentage of residents working in the central city's county (specifically, a county with a density of 25 persons per square mile and at least 50 percent working in the central city's county).

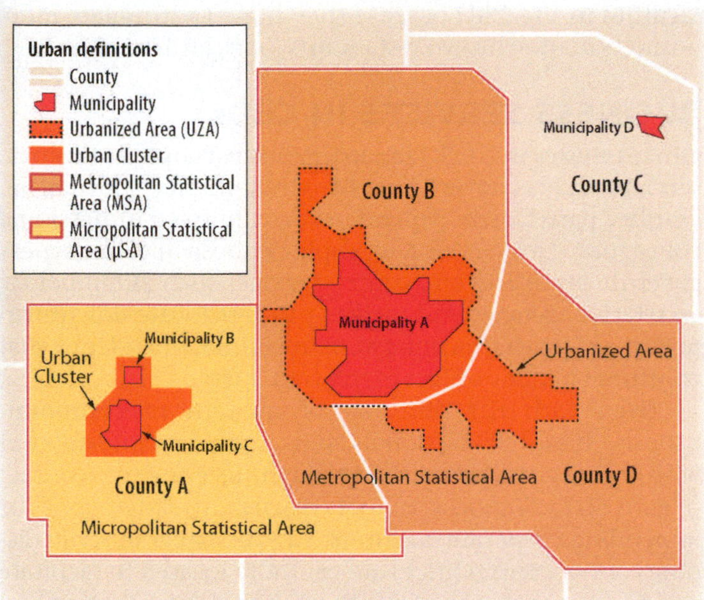

▲ **FIGURE 13-2 DEFINITIONS OF URBAN SETTLEMENTS** The census definitions can be applied to this illustration:

- MSA: Counties B and D.
- μSA: County A.
- CBSAs: The illustration has two (the MSA consisting of Counties B and D and the μSA consisting of County A).
- CSA: A CSA comprising Counties A, B, and D would be designated if the Census Bureau determined that the adjacent MSA and μSA are closely linked.
- PSA: If a CSA existed, that would constitute the PSA; if not, then the MSA and the μSA would be two separate PSAs.

Studies of metropolitan areas in the United States are usually based on information about MSAs. MSAs are widely used because many statistics are published for counties, the basic MSA building block in most states. The Census Bureau designated 388 MSAs as of 2013, encompassing 84 percent of the U.S. population.

The census has also designated smaller urban areas as **micropolitan statistical areas (μSAs).** A μSA includes an urbanized area of between 10,000 and 50,000 inhabitants, the county in which it is located, and adjacent counties tied to the city. The United States had 541 micropolitan statistical areas as of 2013, for the most part found around southern and western communities previously considered rural in character. About 9 percent of Americans live in micropolitan statistical areas.

The census combines MSAs and μSAs in several ways:

- A **core-based statistical area (CBSA)** is any one MSA or μSA (929 as of 2013, including the 388 MSAs and the 541 μSAs).
- A **combined statistical area (CSA)** is two or more contiguous CBSAs tied together by commuting patterns (169 as of 2013).
- A **primary statistical area (PSA)** is a CSA, an MSA not included in a CSA, or a μSA not included in a CSA (574 as of 2013, including the 169 CSAs, plus the 122 MSAs and 283 μSAs not included in a CSA).

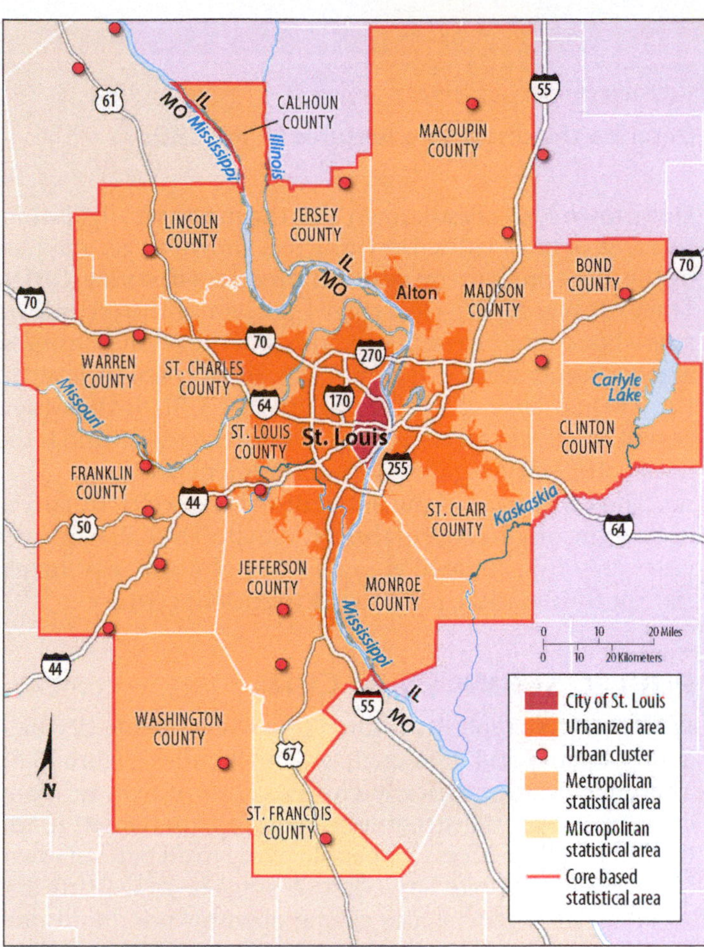

▲ **FIGURE 13-3 DEFINITIONS OF ST. LOUIS** The City of St. Louis comprises only 6 percent of the land area and 11 percent of the population of the MSA.

PAUSE & REFLECT 13.1.1

Do you live inside or outside a central city? An urban area? A metropolitan area?

▼ **FIGURE 13-4 MEXICO CITY METROPOLITAN AREA** Mexico City's metropolitan area extends over nearly 10,000 square kilometers (4,000 square miles).

The Central Business District

LEARNING OUTCOME 13.1.2

Describe the distinctive features of the CBD.

Downtown is the best-known and the most visually distinctive area of most cities. Downtown is known to geographers by the more precise term **central business district (CBD).** The CBD is compact—less than 1 percent of the urban land area—but contains a large percentage of the public, business, and consumer services (Figure 13-5). Services are attracted to the CBD because of its accessibility. The CBD is the easiest part of the city to reach from the rest of the region and is the focal point of the region's transportation network.

The CBD is one of the oldest districts in a city, usually at or near the original site of the settlement. The CBDs of older cities are often situated along a body of water, a principal transportation route prior to the twentieth century.

PUBLIC SERVICES IN CBDs

Public services typically located in a CBD include city hall, courts, county and state agencies, and libraries (Figure 13-6). These facilities historically clustered downtown, in many cases in substantial structures. Today, many remain in the CBD to facilitate access for people living in all parts of town. Similarly, semipublic services such as places of worship and social service agencies also cluster downtown in handsome historic structures.

Sports facilities and convention centers have been constructed or expanded downtown in many cities. These structures attract a large number of people, including many suburbanites and out-of-towners. Cities place these facilities in the CBD because they hope to stimulate more business for downtown restaurants, bars, and hotels.

BUSINESS SERVICES IN CBDs

Offices cluster in a CBD for accessibility. People in business services such as advertising, banking, finance, journalism, and law particularly depend on proximity to professional colleagues. Lawyers, for example, choose locations near government offices and courts. Services such as temporary secretarial agencies and instant printers locate downtown to be near lawyers, forming a chain of interdependency that continues to draw offices to the center city.

Even with the diffusion of modern telecommunications, many professionals still exchange information with colleagues primarily through face-to-face contact (Figure 13-7). Financial analysts discuss attractive stocks or impending corporate takeovers. Lawyers meet to settle disputes out of court. Offices are centrally located to facilitate rapid communication of fast-breaking news through spatial proximity. Face-to-face contact also helps establish a relationship of trust based on shared professional values.

A central location also helps businesses that employ workers from a variety of neighborhoods. Top executives may live in one neighborhood, junior executives in another, secretaries in another, and custodians in still another. Only a central location is readily accessible to all groups. Firms that need highly specialized employees are more likely to find them in the central area, perhaps currently working for another company downtown.

CONSUMER SERVICES IN CBDS

In the past, three types of retail services clustered in a CBD because they required accessibility to everyone in

▼ **FIGURE 13-5 CBD OF LOUISVILLE, KENTUCKY.**

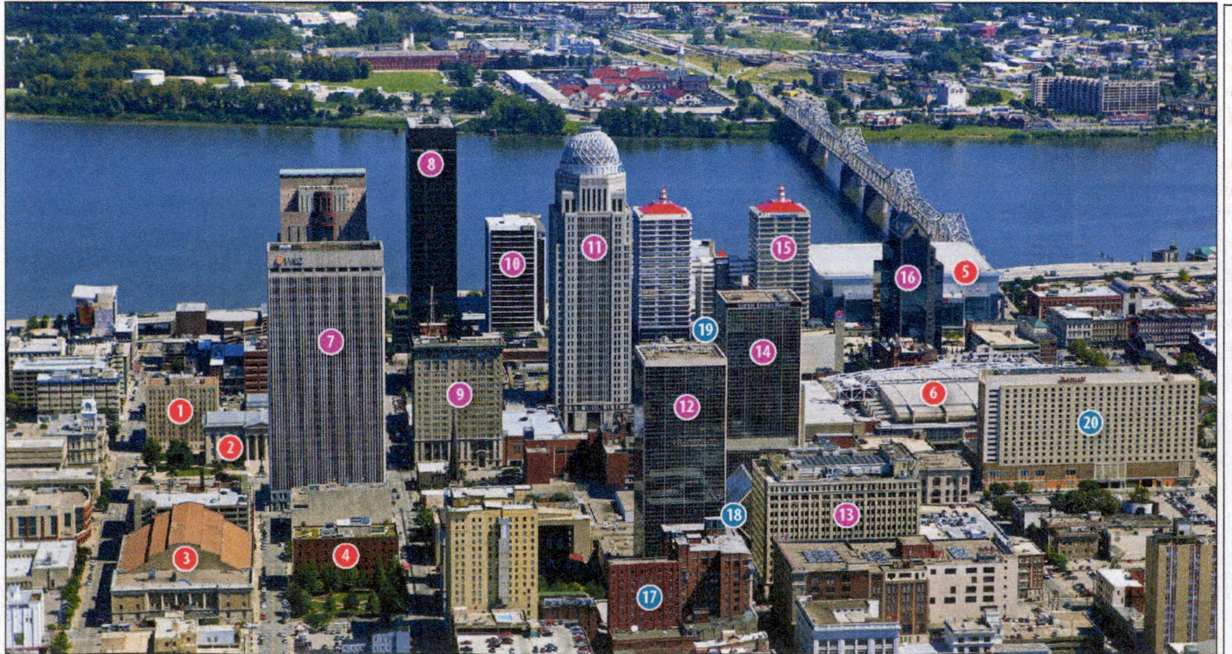

Public services
1. Fiscal Court Building
2. Metro Hall
3. Louisville Gardens
4. Louisville Metro Development Center
5. KFC Yum! Center
6. Kentucky International Convention Center

Business services
7. PNC Plaza
8. National City Tower
9. Kentucky Home Life Building
10. One Riverfront Plaza
11. 400 West Market
12. Meidinger Tower
13. Starks Tower
14. B&W Tower
15. Waterfront Plaza
16. E.ON U.S. Center

Consumer services
17. Seelbach Hilton Hotel
18. 4th Street Live
19. Actors Theater
20. Marriott Hotel

▲ **FIGURE 13-6 PUBLIC SERVICES IN LOUISVILLE'S CBD** Muhammad Ali Center.

the region: retailers with high thresholds, those with high range, and those that served people working in the CBD. Changing shopping habits and residential patterns have reduced the importance of retail services in the CBD.

RETAILERS WITH HIGH THRESHOLDS. Retailers with high thresholds, such as department stores, traditionally preferred a CBD location in order to be accessible to many people. Large department stores in the CBD would cluster near one intersection, which was known as the "100 percent corner." Rents were highest there because that location had the highest accessibility for the most customers.

Most high-threshold shops such as large department stores have closed their downtown branches. CBDs that once boasted three or four stores now have none, or perhaps one struggling survivor. The customers for downtown department stores now consist of downtown office workers, inner-city residents, and tourists. Department stores with high thresholds are now more likely to be in suburban malls.

RETAILERS WITH HIGH RANGES. High-range retailers are often specialists, with customers who patronize them infrequently. These retailers once preferred CBD locations because their customers were scattered over a wide area. For example, a jewelry or clothing store attracted shoppers from all over the urban area, but each customer visited infrequently. Like those with high thresholds, high-range retailers have moved with department stores to suburban locations.

Some retailers with high ranges have located in CBDs because they are visited by tourists. Local residents also patronize shops in the CBD as a leisure activity on evenings and weekends (Figure 13-8).

RETAILERS SERVING CBD WORKERS. A third type of retail activity in the CBD serves the many people who work in the CBD and shop during lunch or working hours. These retailers sell office supplies, computers, and clothing or offer shoe repair, rapid photocopying, dry cleaning, and so on. In contrast to the other two types of retailers, shops that appeal to nearby office workers are expanding in the CBD, in part because the number of downtown office workers has increased and in part because downtown offices require more services.

Patrons of downtown shops tend increasingly to be downtown employees who shop during the lunch hour. Thus, although the total volume of sales in downtown areas has been stable, the pattern of demand has changed. Large department stores have difficulty attracting their old customers, whereas smaller shops that cater to the special needs of the downtown labor force are expanding.

PAUSE & REFLECT 13.1.2

Do you ever spend time in a CBD? If so, for what reasons?

◀ **FIGURE 13-7 BUSINESS SERVICES IN LOUISVILLE'S CBD** 400 West Market (formerly Aegon Center) is Louisville's tallest building.

▶ **FIGURE 13-8 CONSUMER SERVICES IN LOUISVILLE'S CBD** Fourth Street Live restaurant and entertainment district.

Competition for Space in CBDs

LEARNING OUTCOME 13.1.3

Understand the use of vertical space in the CBD and the exclusion of some land uses.

A CBD's accessibility produces extreme competition for the limited available land. As a result, land values are very high in the CBD. In a rural area a hectare of land might cost several thousand dollars. In a suburb it might run tens of thousands of dollars. In the CBD of a global city like London, if a hectare of land were even available, it would cost more than $200 million. If this page were a parcel of land in the CBD of London, it would sell for $1,000.

As a result of intense competition for land, the CBD has distinctive features:

- The CBD has a three-dimensional character, with more space used below and above ground level than elsewhere in the urban area.
- Land uses commonly found elsewhere in the urban area are rare in the CBD.

(a)

(b)

▲ **FIGURE 13-9 MANUFACTURING IN DOWNTOWN CLEVELAND** **(a)** A steel mill and some other factories still line the Cuyahoga River in downtown Cleveland. **(b)** Most of Cleveland's downtown factories have been converted to residences and commercial activities.

ACTIVITIES EXCLUDED FROM THE CBD

High rents and land shortage discourage two principal activities in the CBD—industrial and residential.

LACK OF MANUFACTURING IN THE CBD. Modern factories require large parcels of land to spread operations among one-story buildings. Suitable land is generally available in suburbs. In the past, inner-city factories and retail establishments relied on waterfront CBDs that were once lined with piers for cargo ships to load and unload and warehouses to store the goods (Figure 13-9). Today's large oceangoing vessels are unable to maneuver in the tight, shallow waters of the old CBD harbors. Consequently, port activities have moved to more modern facilities downstream.

Port cities have transformed their waterfronts from industry to commercial and recreational activities. Derelict warehouses and rotting piers have been replaced with new apartments, offices, shops, parks, and museums. As a result, CBD waterfronts have become major tourist attractions in a number of North American cities, including Boston, Toronto, Baltimore, and San Francisco, as well as in European cities such as Barcelona and London. The cities took the lead in clearing the sites and constructing new parks, docks, walkways, museums, and parking lots. They have also built large convention centers to house professional meetings and trade shows. Private developers have added hotels, restaurants, boutiques, and entertainment centers to accommodate tourists and conventioneers.

LACK OF RESIDENTS IN THE CBD. Many people used to live in or near the CBD. Poorer people jammed into tiny, overcrowded apartments, and richer people built mansions downtown. In the twentieth century, most residents abandoned downtown living because of a combination of pull and push factors. They were pulled to suburbs that offered larger homes with private yards and modern schools. And they were pushed from CBDs by high rents that business and retail services were willing to pay and by the dirt, crime, congestion, and poverty that they experienced by living downtown.

In the twenty-first century, however, the population of many U.S. CBDs has increased. New apartment buildings and townhouses have been constructed, and abandoned warehouses and outdated office buildings have been converted into residential lofts. Downtown living is especially attractive to people without school-age children, either "empty nesters" whose children have left home or young professionals who have not yet had children. These two groups are attracted by the entertainment, restaurants, museums, and nightlife that are clustered downtown, and they are not worried about the quality of neighborhood schools. Despite the growth in population in the center of some U.S. cities, some consumer services, such as grocery stores, may still be lacking.

PAUSE & REFLECT 13.1.3

What might be the attractions of living in a former factory near the CBD?

(a) (b)

▲ **FIGURE 13-10 UNDERGROUND MONTRÉAL (a)** Montréal's Underground City is the world's largest underground complex. It is known as RÉSO, which is an abbreviation for Réseau Souterrain (underground network) and a homophone of the French word réseau (network). **(b)** Winter in Montréal discourages outdoor shopping.

VERTICAL FEATURES OF THE CBD

The CBD makes more intensive use of space below and above ground.

THE UNDERGROUND CBD. A vast underground network exists beneath most CBDs. The typical "underground city" includes garages, loading docks for deliveries to offices and shops, and pipes for water and sewer service. Telephone, electric, TV, and broadband cables run beneath the surface as well because not enough space is available in the CBD for the large number of overhead poles that would be needed for such a dense network, and the wires would be unsightly and hazardous. Subway trains run beneath the streets of large CBDs. And cities in cold-weather climates, such as Minneapolis, Montréal, and Toronto, have built extensive underground pedestrian passages and shops. These underground areas segregate pedestrians from motor vehicles and shield them from harsh winter weather (Figure 13-10).

SKYSCRAPERS. Demand for space in CBDs has also made high-rise structures economically feasible. Downtown skyscrapers give a city one of its most distinctive images and unifying symbols. Suburban houses, shopping malls, and factories look much the same from one city to another, but each city has a unique downtown skyline, resulting from the particular arrangement and architectural styles of its high-rise buildings.

The first skyscrapers were built in Chicago in the 1880s, made possible by several inventions, including the elevator, steel girders, and glass structures because they blocked light and air movement. Artificial lighting, ventilation, central heating, and air-conditioning have helped solve these problems. Most North American and European cities enacted zoning ordinances early in the twentieth century in part to control the location and height of skyscrapers.

Skyscrapers are an interesting example of "vertical geography" (Figure 13-11). The nature of an activity influences which floor it occupies in a typical high-rise.

The one large U.S. CBD without skyscrapers is Washington, D.C., where no building is allowed to be higher than the U.S. Capitol dome. Consequently, offices in downtown Washington rise no more than 13 stories. As a result, the typical Washington office building uses more horizontal space—land area—than in other cities. Thus the city's CBD spreads over a much wider area than those in comparable cities.

▲ **FIGURE 13-11 JOHN HANCOCK CENTER AND WATER TOWER PLACE, CHICAGO** The lower floors of the Hancock Center and neighboring buildings, such as Water Tower Place, are devoted to commercial activities. The middle floors are offices, the upper floors are apartments, and the top two floors are commercial activities (observation deck, restaurant, and bar) 1. Why might retail services wish to pay high rents for street-level space? 2. Why might residents prefer the upper floors? 3. In the image of the Hancock Center, note how the windows change between the lower business services floors and the upper residential floors. Why might the two types of land uses prefer different window styles?

CHECK-IN KEY ISSUE 1

Why Are Downtowns Distinctive?

- ✔ **Business, public, and some consumer services cluster in the CBD.**
- ✔ **The CBD has relatively few manufacturers and residents.**
- ✔ **North American CBDs are characterized by vertical features above and below ground level.**

KEY ISSUE 2

Where Are People Distributed in Urban Areas?

- Models of Urban Structure
- Applying the Models in North America
- Applying the Models in Europe
- Pre-modern Cities in Developing Countries
- Applying the Models in Developing Countries
- Changing Urban Structure of Mexico City

LEARNING OUTCOME 13.2.1

Describe the models of internal structure of urban areas.

People are not distributed randomly within an urban area. They concentrate in particular neighborhoods, depending on their social characteristics. Geographers describe where people with particular characteristics are likely to live within an urban area, and they offer explanations for why these patterns occur.

Models of Urban Structure

Sociologists, economists, and geographers have developed three models to help explain where different types of people tend to live in an urban area—the concentric zone, sector, and multiple nuclei models. The peripheral model is a modification of the multiple nuclei model. The three models have been applied to cities in the United States and in other countries with varying degrees of success.

The three models describing the internal social structure of cities were developed in Chicago, a city on a prairie. Chicago includes a CBD known as the Loop because transportation lines (originally cable cars, now El trains) loop around it. Surrounding the Loop are residential suburbs to the south, west, and north. Except for Lake Michigan to the east, few physical features have interrupted Chicago's growth.

CONCENTRIC ZONE MODEL

According to the **concentric zone model,** created in 1923 by sociologist E. W. Burgess, a city grows outward from a central area in a series of concentric rings, like the growth rings of a tree. The precise size and width of the rings vary from one city to another, but the same basic types of rings appear in all cities in the same order. Back in the 1920s, Burgess identified five rings (Figure 13-12).

SECTOR MODEL

According to the **sector model,** developed in 1939 by land economist Homer Hoyt, a city develops in a series of sectors (Figure 13-13). Certain areas of the city are more attractive for various activities, originally because of an environmental factor or even by mere chance. As a city grows, activities expand outward in a wedge, or sector, from the center.

Once a district with high-income housing is established, the most expensive new housing is built on the outer edge of that district, farther out from the center. The

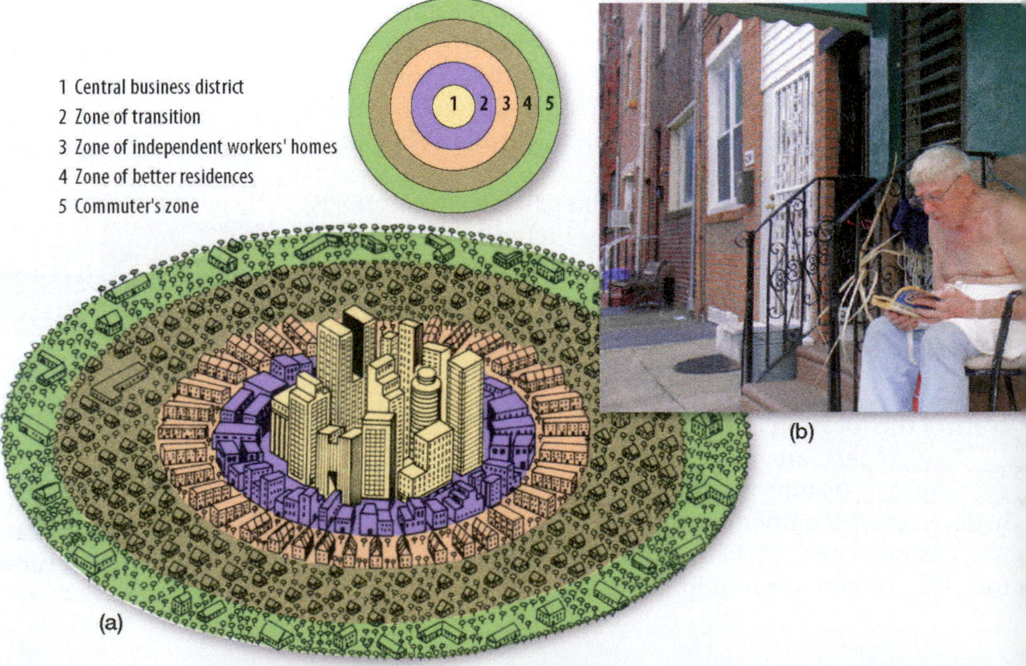

▶ **FIGURE 13-12 CONCENTRIC ZONE MODEL (a)** According to this model, a city grows in a series of rings that surround the CBD: 1. CBD: The innermost ring, where nonresidential activities are concentrated 2. Zone in transition: Industry and poorer-quality housing; immigrants to the city first live in this zone in small dwelling units, frequently created by subdividing larger houses into apartments 3. Zone of working-class homes: Modest older houses occupied by stable, working-class families 4. Zone of better residences: Newer and more spacious houses for middle-class families 5. Commuters' zone: Beyond the continuous built-up area of the city, where people live in small communities and commute to work in the CBD. **(b)** Zone of working-class homes, Philadelphia.

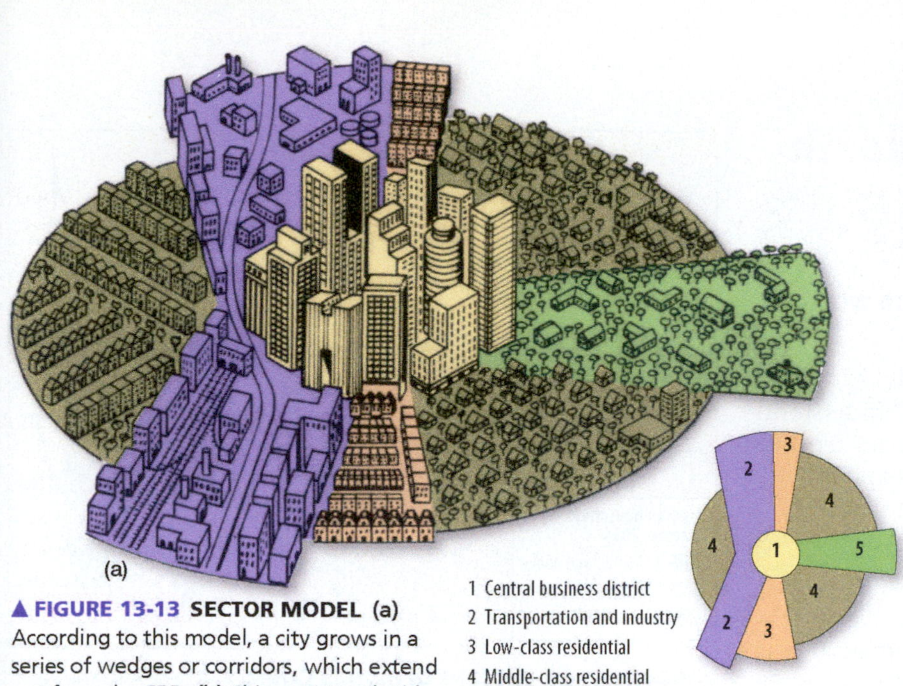

▲ **FIGURE 13-13 SECTOR MODEL (a)** According to this model, a city grows in a series of wedges or corridors, which extend out from the CBD. **(b)** Chicago's North Side, a high-income sector.

best housing is therefore found in a corridor extending from downtown to the outer edge of the city. Industrial and retailing activities develop in other sectors, usually along good transportation lines.

MULTIPLE NUCLEI MODEL

According to the **multiple nuclei model,** developed by geographers C.D. Harris and E.L. Ullman in 1945, a city is a complex structure that includes more than one center around which activities revolve (Figure 13-14). Examples of these nodes include a port, a neighborhood business center, a university, an airport, and a park.

The multiple nuclei theory states that some activities are attracted to particular nodes, whereas others try to avoid them. For example, a university node may attract well-educated residents, pizzerias, and bookstores, whereas an airport may attract hotels and warehouses. On the other hand, incompatible land-use activities avoid clustering in the same locations. Heavy industry and high-income housing, for example, rarely exist in the same neighborhood.

The nodes of consumer and business services around the beltway are called **edge cities.** Edge cities originated as suburban residences for people who worked in the central city, and then shopping malls were built to be near the residents. Now edge cities also contain business services.

PAUSE & REFLECT 13.2.1

If you cut down a large tree, which of the three models will the cross-section resemble? Why is the cross-section of a tree a good analogy for one of the models of urban structure?

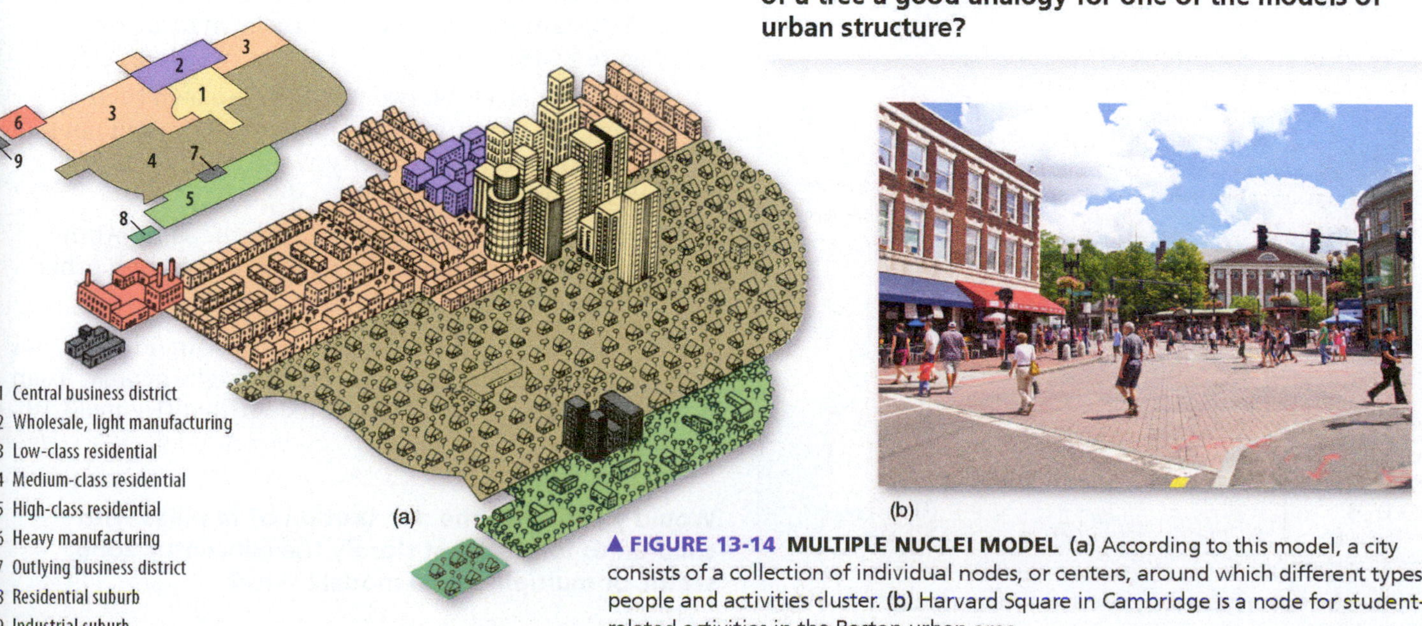

▲ **FIGURE 13-14 MULTIPLE NUCLEI MODEL (a)** According to this model, a city consists of a collection of individual nodes, or centers, around which different types of people and activities cluster. **(b)** Harvard Square in Cambridge is a node for student-related activities in the Boston urban area.

Applying the Models in North America

LEARNING OUTCOME 13.2.2

Analyze how the three models help to explain where people live.

The three models of urban structure help us understand where people with different social characteristics tend to live within an urban area. They can also help explain why certain types of people tend to live in particular places.

SOCIAL AREA ANALYSIS

The study of where people of varying living standards, ethnic background, and lifestyle live within an urban area is **social area analysis.** Social area analysis helps to create an overall picture of where various types of people tend to live, depending on their particular personal characteristics.

Social area analysis depends on the availability of data at the scale of individual neighborhoods. In the United States and many other countries, that information comes from the census. Urban areas in the United States are divided into **census tracts** that each contain approximately 5,000 residents and correspond, where possible, to neighborhood boundaries. The census also divides the entire United States into blocks, which are typically a collection of several dozen houses; inside urban areas, blocks are typically bounded by four streets. A block group, as the name implies, is a collection of several neighboring blocks.

Every decade the U.S. Bureau of the Census publishes data summarizing the characteristics of the residents and the housing in each tract. Annual estimates are issued through the American Fact Finder service of the census's American Community Survey program. Examples of information the census provides at the tract level include the number of nonwhites, the median income of all families, and the percentage of adults who finished high school.

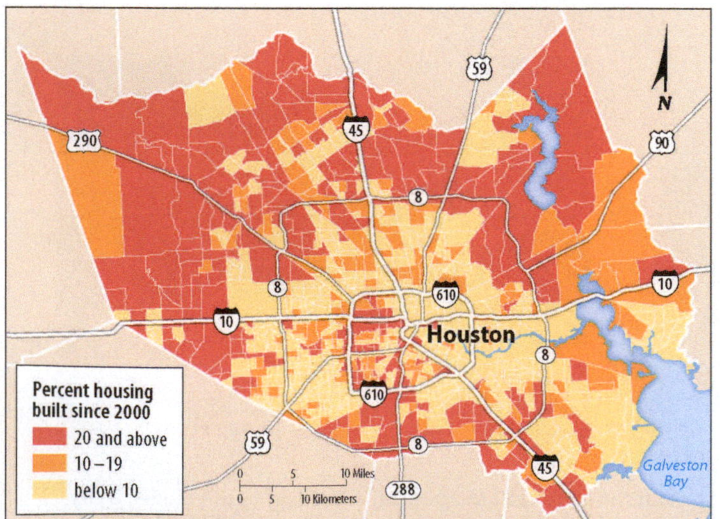

▲ **FIGURE 13-15 CONCENTRIC ZONES IN HARRIS COUNTY** The outer ring has a higher percentage of newer housing.

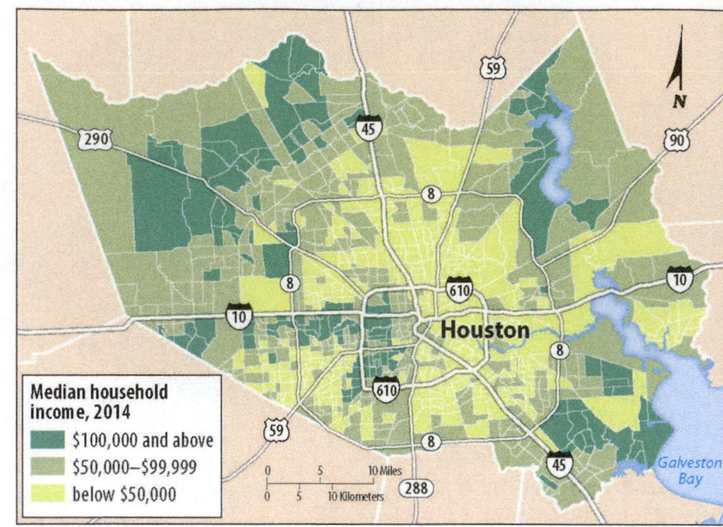

▲ **FIGURE 13-16 SECTORS IN HARRIS COUNTY** A sector to the west and northwest has the highest-income households.

The spatial distribution of any of these social characteristics can be plotted on a map of the community's census tracts. Computers have become invaluable in this task because they permit rapid creation of maps and storage of voluminous data about each census tract. Relatively little is available at the block level because the number of people is so small that publishing the information could violate the privacy of individuals.

Social area analysis suggests the following:

- **Concentric zone model.** Consider two families with the same income and ethnic background. One family lives in a newly constructed home, whereas the other lives in an older one. The family in the newer house is much more likely to live in an outer ring and the family in the older house in an inner ring (Figure 13-15).
- **Sector model.** Given two families who own their homes, the family with the higher income will not live in the same sector of the city as the family with the lower income (Figure 13-16).
- **Multiple nuclei model.** People with the same ethnic or racial background are likely to live near each other (Figure 13-17).

Putting the three models together, we can identify, for example, the neighborhood in which a high-income, Asian American owner-occupant is most likely to live (Figure 13-18).

PAUSE & REFLECT 13.2.2

Would you expect the distribution of families with children to follow most closely the concentric zone, sector, or multiple nuclei model? Why?

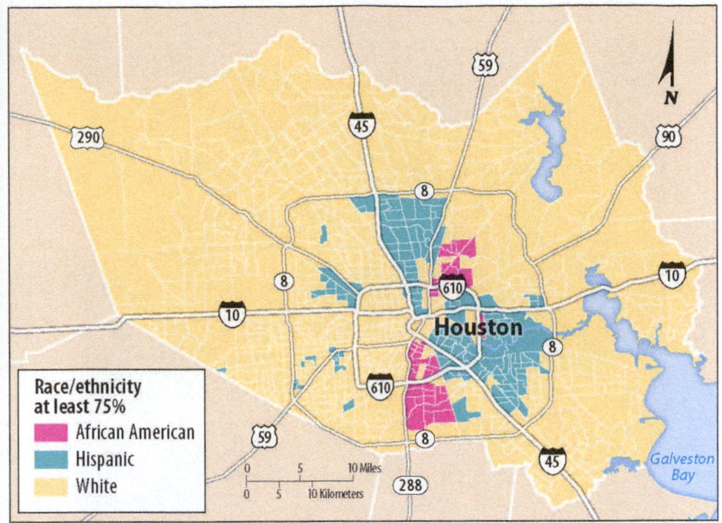

▲ **FIGURE 13-17 MULTIPLE NUCLEI IN HARRIS COUNTY** The largest African American nodes are in the south and northeast. The largest Hispanic nodes are in the north and southeast.

LIMITATIONS OF THE MODELS

None of the three models taken individually completely explains why different types of people live in distinctive parts of a city. Critics point out that the models are too simple and fail to consider the variety of reasons that lead people to select particular residential locations. Because the three models are all based on conditions that existed in U.S. cities between the two world wars, critics also question their relevance to contemporary urban patterns in the United States or in other countries.

If the models are combined rather than considered independently, however, they help geographers describe where different types of people live in a city. People tend to reside in certain locations, depending on their particular personal characteristics. This does not mean that everyone with the same characteristics must live in the same neighborhood, but the models say that most people live near others who have similar characteristics.

(a)

(b)

(c)

◀ **FIGURE 13-18 COMPARING RINGS AND SECTORS IN HARRIS COUNTY** **(a)** House in an outer ring and a high-income sector. **(b)** House in the same ring as (a) but in a different sector. **(c)** Housing in the same sector as the house shown in (a) but in an inner ring.

Applying the Models in Europe

LEARNING OUTCOME 13.2.3

Describe how the three models explain patterns in European cities.

American urban areas differ from those elsewhere in the world. These differences do not invalidate the three models of internal urban structure, but they do point out that social groups in other countries may not have the same reasons for selecting particular neighborhoods within their cities.

CBDs IN EUROPE

Europe's CBDs have a different mix of land uses than those in North America. Differences stem from the medieval origins of many of Europe's CBDs. European cities display a legacy of low-rise structures and narrow streets, built as long ago as medieval times.

- **Residences.** More people live downtown in cities outside North America. The CBD of Paris, which covers around 20 square kilometers (8 square miles), has about 450,000 inhabitants. A comparable area around the CBD of Detroit has around 25,000 inhabitants.
- **Consumer services.** More people live in Europe's CBDs in part because they are attracted to the concentration of consumer services, such as cultural activities and animated nightlife. And with more people living there, Europe's CBDs in turn contain more day-to-day consumer services, such as groceries, bakeries, and butchers (Figure 13-19).
- **Public services.** The most prominent structures in Europe's CBDs are often public and semipublic services, such as churches and former royal palaces, situated on the most important public squares. Parks in Europe's CBDs were often first laid out as private gardens for aristocratic families and later were opened to the public.
- **Business services.** Europe's CBDs contain professional and financial services. However, business services in Europe's CBDs are less likely to be housed in skyscrapers than those in North America. Some European cities try to preserve their historic CBDs by limiting high-rise buildings (Figure 13-20). Although constructing large new buildings is difficult, many shops and offices still wish to be in the center of European cities. The alternative to new construction is renovation of older buildings. However, renovation is more expensive and does not always produce enough space to meet the demand. As a result, rents are much higher in the center of European cities than in U.S. cities of comparable size.

▼ **FIGURE 13-19 CONSUMER SERVICES IN PARIS** Aligre market is one of many street markets in Paris.

▲ **FIGURE 13-20 PUBLIC SERVICES AND BUSINESS SERVICES IN PARIS** École Militaire (Military Academy) in the foreground and Tour Montparnasse office tower in the background. Public outcry over the tower's disfigurement of the city's historic skyline was so great that officials have since set lower height limits for new buildings.

THE THREE MODELS IN EUROPE

The urban structure in Paris can be used to illustrate similarities and differences in the distribution of people in U.S. and European cities:

- **Concentric zones.** As in U.S. urban areas, the newer housing in the Paris region is in outer rings, and the older housing is closer to the center (Figure 13-21). Unlike in U.S. urban areas, though, much of the newer suburban housing is in high-rise apartments rather than single-family homes.
- **Sectors.** Again, as in U.S. urban areas, higher-income people cluster in a sector in the Paris region (Figure 13-22). The wealthy lived near the royal palace (the Louvre) beginning in the twelfth century and the Palace of Versailles from the sixteenth century until the French Revolution in 1789. The preference of Paris's wealthy to cluster in a southwestern sector was reinforced during the Industrial Revolution in the nineteenth century,

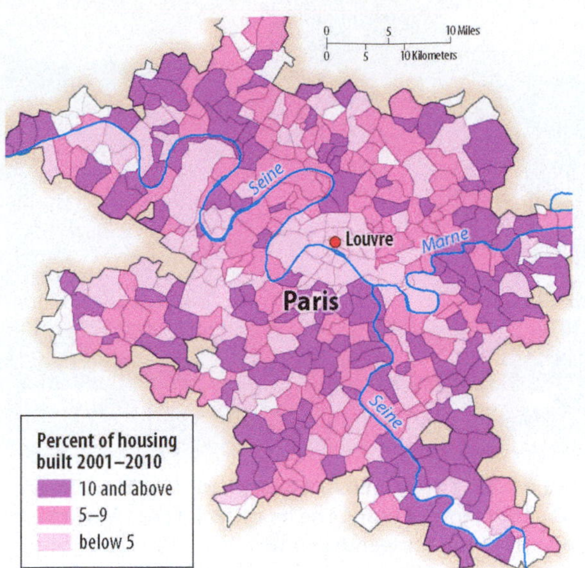

▲ **FIGURE 13-21 CONCENTRIC ZONES IN PARIS** The oldest housing is in the inner ring.

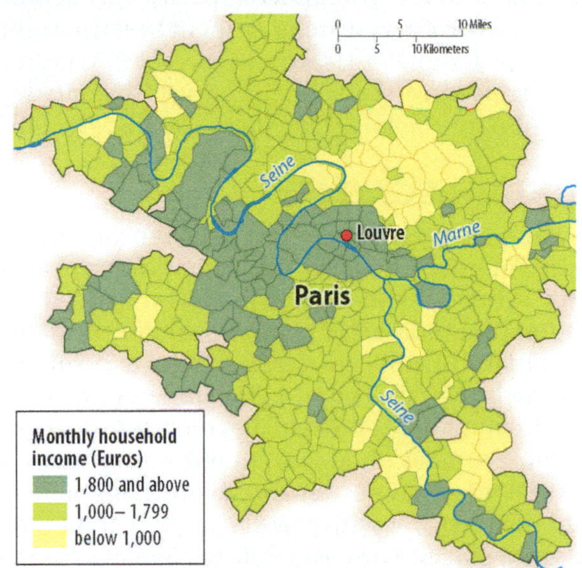

▲ **FIGURE 13-22 SECTORS IN PARIS** The southwest is the highest-income sector, and the northeast is the lowest-income sector.

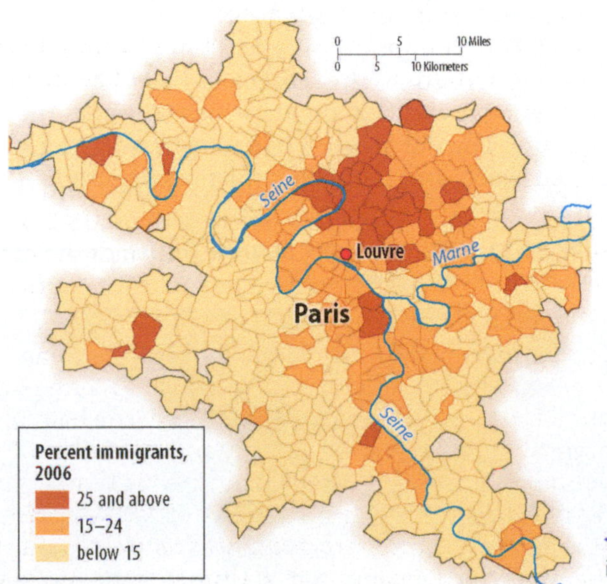

(a)

(b)

(c)

▲ **FIGURE 13-23 COMPARING RINGS AND SECTORS IN PARIS** **(a)** Housing in an outer ring and a high-income sector. **(b)** Housing in the same ring as (a) but in a different sector. **(c)** Housing in the same sector as (a) but in an inner ring.

when factories were built to the south, east, and north, along the Seine and Marne River valleys (Figure 13-23).

- **Multiple nuclei.** European urban areas, including Paris, have experienced a large increase in immigration from other regions of the world (see Chapter 3). In contrast to U.S. urban areas, most ethnic and racial minorities reside in the suburbs of Paris (Figure 13-24).

PAUSE & REFLECT 13.2.3

Are Europe's famous tourist sites located predominantly in inner or outer rings? Why might this be the case?

◀ **FIGURE 13-24 MULTIPLE NUCLEI IN PARIS** The highest percentage of immigrants is in a node in the northern suburbs.

Pre-modern Cities in Developing Countries

LEARNING OUTCOME 13.2.4

Describe patterns in precolonial and colonial cities in developing countries.

Cities in developing countries may date from ancient times. For most of recorded history, the world's largest cities have been in Asia. However, until modern times, most Asians lived in rural settlements. The ancient and medieval structure of these cities was influenced by the cultural values of the indigenous peoples living there. In most cases, these cities passed through a period of restructuring at the hands of European colonial rulers.

ANCIENT AND MEDIEVAL CITY: BEIJING

Archaeological evidence of Beijing dates from 1045 B.C., although the city may have been founded thousands of years earlier. A succession of invaders and dynasties shaped what is now the central area of Beijing. The Yuan and Ming dynasties had especially strong impacts on the early structure of Beijing.

BEIJING DURING THE YUAN DYNASTY. Kubla Khan, founder of the Yuan dynasty, constructed a new city called Dadu beginning in 1267 (Figure 13-25). The Drum Tower was constructed at the center of the city (Figure 13-26). The heart of Dadu was three palaces built on Qionghua Island in the middle of Taiye Lake. The two palaces to the west of the lake housed the imperial family, and the eastern one contained offices. Residential areas were laid out in a checkerboard pattern divided by wider roads and narrower alleys. Three markets were placed in the residential areas. An outer wall surrounded the residential areas, and an inner wall surrounded the palaces.

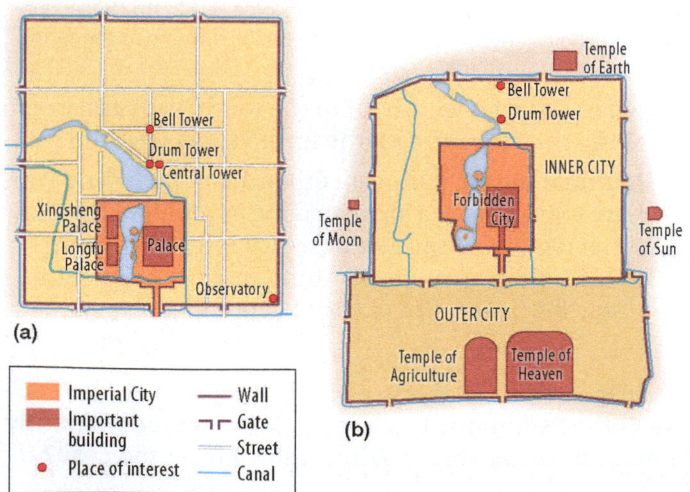

▲ **FIGURE 13-25 HISTORIC BEIJING (a)** Beijing (Dadu) during the Yuan Dynasty. **(b)** Beijing during the Ming Dynasty.

▲ **FIGURE 13-26 YUAN DYNASTY BEIJING** The Drum Tower was built in the thirteenth century. The image was taken from the Bell Tower, also from the thirteenth century.

BEIJING DURING THE MING DYNASTY. After capturing Dadu in 1368, the Ming dynasty reconstructed it over the next several decades. The imperial palace was demolished and replaced with new structures, including the Forbidden City and the Temple of Heaven (Figure 13-27). Other temples were added in the sixteenth century. The city took on the current name Beijing ("Northern Capital") in 1403.

COLONIAL LEGACY

When Europeans gained control of much of Africa, Asia, and Latin America, their colonial policies left a heavy mark on many cities. One feature of European control was the imposition of standardized plans for cities. For example, all Spanish cities in Latin America were built according to the Laws of the Indies, drafted in 1573. The laws explicitly outlined how colonial cities were to be constructed—a gridiron street plan centered on a church and central plaza, walls around individual houses, and neighborhoods built around central, smaller plazas with parish churches or monasteries.

In some places, European colonial powers built a new city next to the existing one. Fès (Fez), Morocco, is an example of a city that consists of two separate and distinct nodes—a precolonial city that existed before the French gained control and one built by the French colonialists (Figure 13-28). The precolonial Muslim city was laid out surrounding a mosque. The center also had a bazaar or marketplace, known as the Medinah, which served as the commercial core. The old quarters had narrow, winding streets, little open space, and cramped residences (Figure 13-29).

▼ **FIGURE 13-27 MING DYNASTY BEIJING** The Temple of Heaven was built in the fifteenth century.

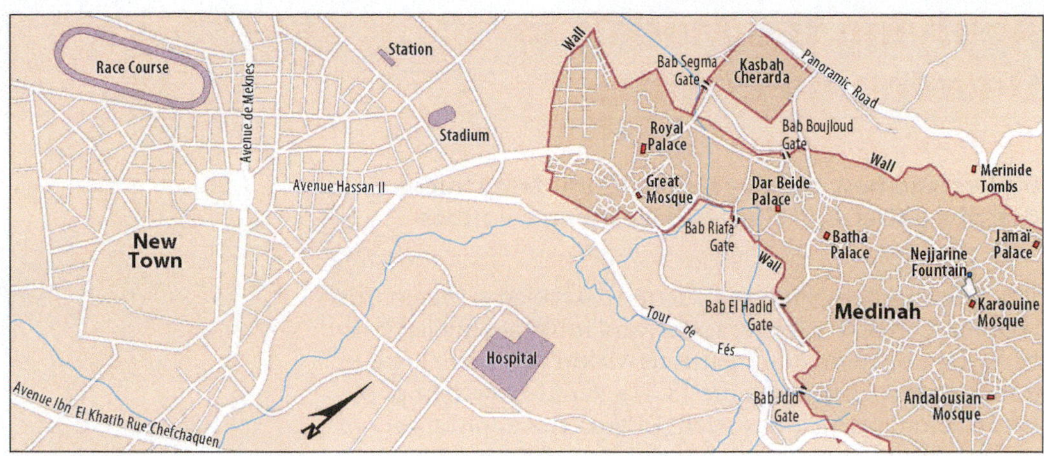

▶ **FIGURE 13-28 FÈS (FEZ) MOROCCO PLAN** The French colonial administration laid out an entirely new district in the west (New Town on the map), with geometrically arranged streets and squares. The precolonial town (Medinah on the map) to the east had narrow, irregularly arranged streets and numerous mosques.

▲ **FIGURE 13-29 FÈS (FEZ) MOROCCO OLD AND NEW TOWNS** The Medinah (Old Town) is in the foreground, and the French-built New Town is in the background. A portion of the wall that encircles the Medinah is visible between the Old Town and New Town.

The new city was the location for colonial services, such as administration, military command, and international trade, as well as housing for European colonists. Compared to the precolonial node, the European district contained wider streets and public squares, larger houses surrounded by gardens, and much lower density. Similarly, the British built New Delhi near the existing city of Delhi, India. Old Delhi was characterized by narrow winding streets and densely packed buildings, whereas New Delhi had broad boulevards and large government structures.

In other cases, European colonial powers simply demolished the precolonial city. For example, the French colonial city of Saigon, Vietnam (now Ho Chi Minh City), was built by completely demolishing the existing city without leaving a trace (Figure 13-30). Mexico City, described later in this Key Issue, is another example.

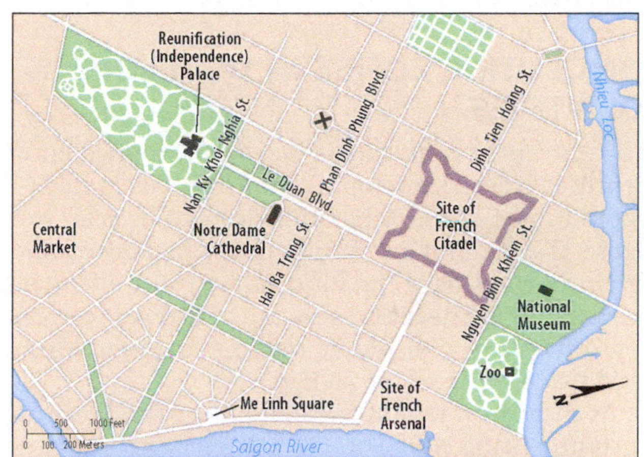

(a)

(b)

▲ **FIGURE 13-30 HO CHI MINH CITY (a)** The French colonial administration demolished the precolonial city and replaced it with one built according to colonial principles, with wide boulevards, public squares, and public and semipublic buildings at key intersections. **(b)** Cathedral of Notre Dame, built by the French in 1863.

PAUSE & REFLECT 13.2.4

Which node in Fès do you think would be more interesting to visit, the Medinah or the French colonial center? Why?

Applying the Models in Developing Countries

LEARNING OUTCOME 13.2.5

Understand how the three models of urban structure describe patterns in cities in developing countries.

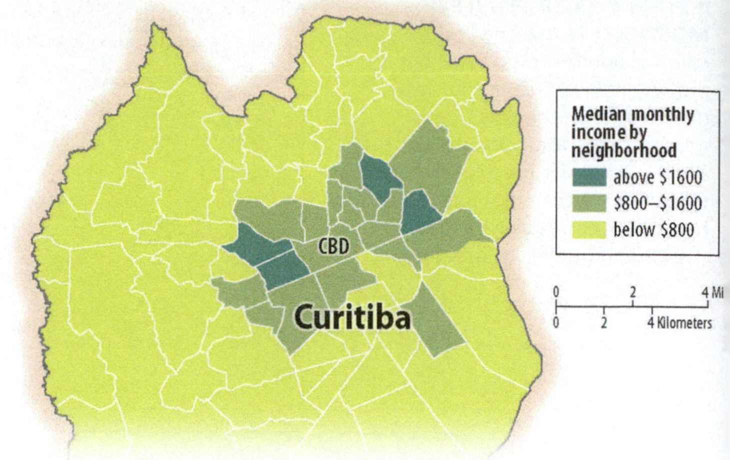

▲ **FIGURE 13-32 CONCENTRIC ZONES IN CURITIBA, BRAZIL** High-income people are more likely to live in an inner ring, whereas low-income people are in outer rings.

The three models of urban structure described earlier in this chapter (concentric zone, sector, and multiple nuclei) help to explain contemporary patterns within the urban areas in developing countries (Figure 13-31). Rapid growth of population and land area has strengthened the applicability of the models in some cities but reduced their usefulness in other cases.

CONCENTRIC ZONES IN DEVELOPING COUNTRIES

The concentric zone model has been applied most frequently to cities in developing countries. Geographer Harm deBlij's model of sub-Saharan African cities is an example (Figure 13-31a). The inner rings house higher-income people. Inner rings have the most attractive residential areas because they are near business and consumer services, and they offer such vital public services as water, electricity, paved roads, and garbage pickup.

As cities grow rapidly in developing countries, rings are constantly being added on the periphery to accommodate immigrants from rural areas attracted by job opportunities (Figure 13-32). Much of the housing in the outer rings is in informal settlements, also known as squatter settlements (Figure 13-33). The United Nations defines an **informal settlement** as a residential area where housing has been built on land to which the occupants have no legal claim or has not been built to the city's standards for legal buildings. Squatter settlements are known by a variety of names, including *barriadas* and *favelas* in Latin America, *bidonvilles* in North Africa, *bastees* in India, *gecekondu* in Turkey, *kampongs* in Malaysia, and *barong-barong* in the Philippines. Estimates of the number of people living in informal settlements worldwide vary widely, between 175 million and 1 billion.

Informal settlements have few services because neither the city nor the residents can afford them. Homes are in primitive shelters made with scavenged cardboard, wood boxes, sackcloth, and crushed beverage cans. Latrines may be designated by the settlement's leaders, and water is carried from a central well or dispensed from a truck. Electricity service may be stolen by running a wire from the nearest power line. In the absence of bus service or available private cars, a resident may have to walk two hours to reach a place of employment.

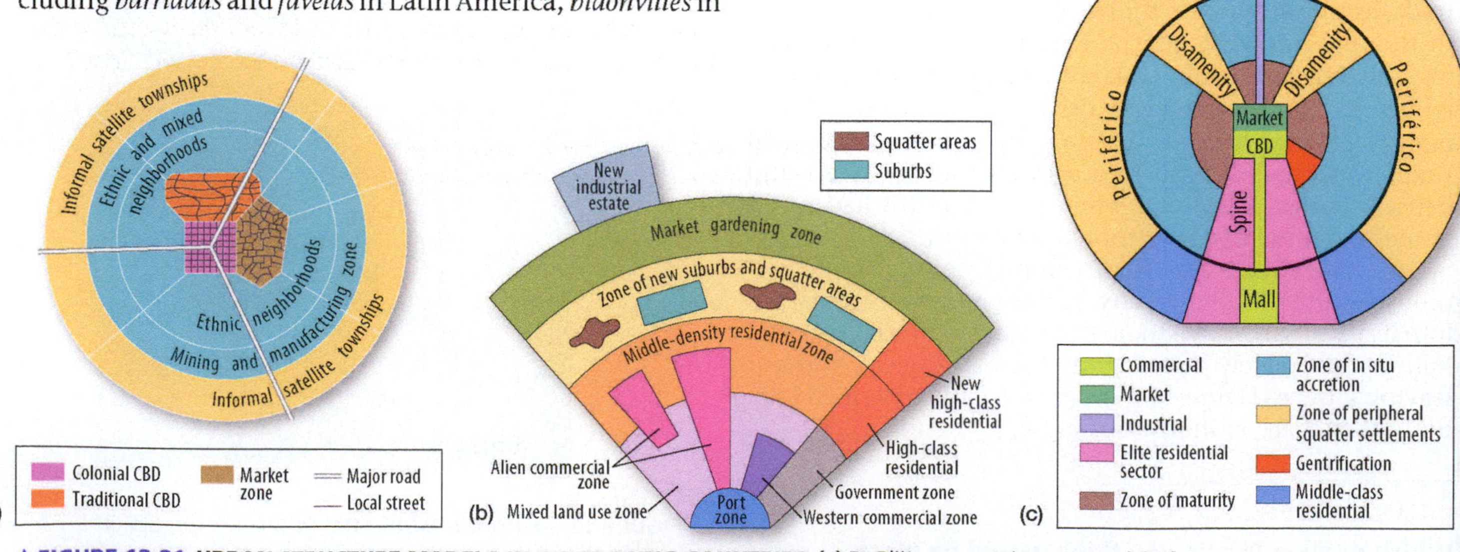

▲ **FIGURE 13-31 URBAN STRUCTURE MODELS IN DEVELOPING COUNTRIES (a)** DeBlij's concentric zone model of a sub-Saharan African city, **(b)** McGee's multiple nuclei model of a Southeast Asian city, **(c)** Griffin-Ford's sector model of a Latin American city.

▲ **FIGURE 13-33 INFORMAL SETTLEMENT IN MUMBAI, INDIA** The informal settlement of Dharavi is located in an outer ring to the north of the CBD.

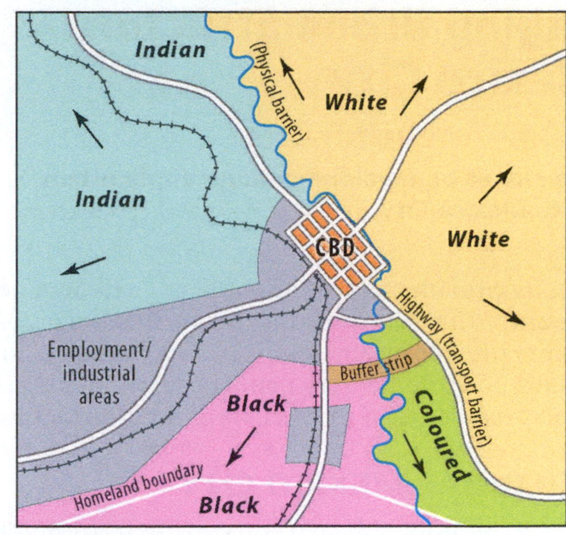

▲ **FIGURE 13-34 MULTIPLE NUCLEI MODEL IN PIETERMARITZBURG, SOUTH AFRICA, DURING APARTHEID.**

MULTIPLE NUCLEI IN DEVELOPING COUNTRIES

T. G. McGee's model of a Southeast Asian city superimposes on concentric zones several nodes of squatter settlements and what he called "alien" zones, where foreigners, usually Chinese, live and work (Figure 13-31b). McGee found that Southeast Asian cities do not typically have a strong CBD. Instead, the various functions of the CBD are dispersed to several nodes.

Cities in some developing countries show evidence of the multiple nuclei model by containing a complex mix of ethnic groups. During the apartheid era (see Chapter 7), South Africa's cities showed especially clear evidence of the multiple nuclei model because each race was segregated into distinct neighborhoods (Figure 13-34).

SECTORS IN DEVELOPING COUNTRIES

Geographers Ernest Griffin and Larry Ford show that in Latin American cities, wealthy people push out from the center in a well-defined elite residential sector. The elite sector forms on either side of a narrow spine that contains offices, shops, and amenities attractive to wealthy people, such as restaurants, theaters, parks, and zoos (Figure 13-31c). The wealthy are also attracted to the center and spine because services such as water and electricity are more readily available and reliable there than elsewhere. Wealthy and middle-class residents avoid living near sectors of "disamenity," which are land uses that may be noisy or polluting or that cater to low-income residents.

For example, São Paulo, Brazil, has an elite sector extending south from the CBD (Figure 13-35a). The U.N.'s Human Development Index (HDI), discussed in Chapter 10, can be used to display multiple factors characteristic of the elite sector, including relatively high incomes, education levels, and life expectancies. At the same time, São Paulo demonstrates elements of the concentric zone model, with most of the informal settlements located in an outer ring (Figure 13-35b). Informal settlements occupy 23 percent of São Paulo's land area and are home to around 1.8 million people, or 17 percent of the city's population.

PAUSE & REFLECT 13.2.5

Would you expect the distribution of families with children in developing countries to follow most closely the concentric zone, sector, or multiple nuclei model? Why?

▼ **FIGURE 13-35 SECTORS AND CONCENTRIC ZONES IN SÃO PAULO, BRAZIL** **(a)** An HDI score has been constructed for each neighborhood in São Paulo. The highest level of development is in a southern sector. **(b)** Informal settlements are in an outer ring.

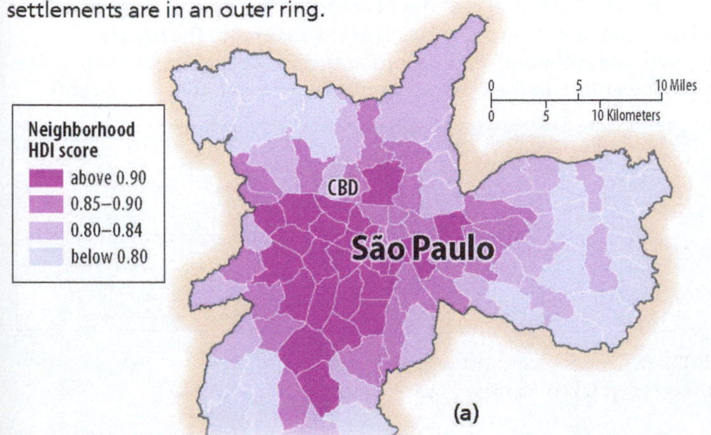

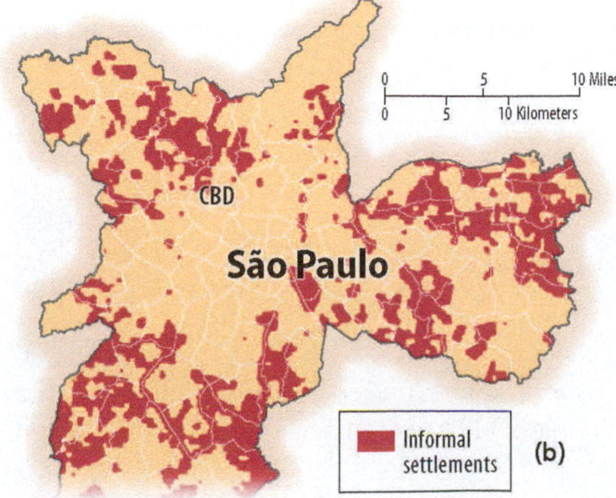

Changing Urban Structure of Mexico City

LEARNING OUTCOME 13.2.6

Describe stages of development and apply urban models to Mexico City.

Mexico City provides a good example of a city in a developing country that has passed through three stages of development: pre-European origin, the European colonial period, and postcolonial independence. The modern city also displays evidence of the models of urban structure.

PRECOLONIAL MEXICO CITY

The Aztecs founded Mexico City—which they called Tenochtitlán—on a hill known as Chapultepec ("the hill of the grasshopper"). When forced by others to leave the hill, they migrated a few kilometers south, near the present-day site of the University of Mexico, and then in 1325 to a marshy 10-square-kilometer island in Lake Texcoco (Figure 13-36).

The node of religious life was the Great Temple (Figure 13-37). Three causeways with drawbridges linked Tenochtitlán to the mainland and also helped control flooding. An aqueduct brought fresh water from Chapultepec. Most food, merchandise, and building materials crossed from the mainland to the island by boat, and the island was laced with canals to facilitate movement. Over the next two centuries, the Aztecs conquered the neighboring peoples and extended their control through much of present-day Mexico. As their wealth and power grew, Tenochtitlán grew to a population of a half-million.

COLONIAL MEXICO CITY

The Spanish conquered Tenochtitlán in 1521, after a two-year siege. They destroyed Tenochtitlán, dispersed or killed most of the inhabitants, and constructed a new city on the site. As in other colonial cities, the Spanish built Mexico City around a main square, called the Zócalo, on the site of the Aztecs' sacred precinct in the center of the island. Streets were laid out in a grid pattern extending from the Zócalo (Figure 13-38).

▲ **FIGURE 13-37 PRECOLONIAL MEXICO CITY CBD** The center of Tenochtitlán was dominated by the Templo Mayor. The twin shrines on the top of the temple were dedicated to the Aztec god of rain and agriculture (in blue) and to the Aztec god of war.

PAUSE & REFLECT 13.2.6

Why might the Spanish wish to destroy the Great Temple? Why might contemporary Mexicans wish to excavate and explore the ruins?

MEXICO CITY SINCE INDEPENDENCE

At independence, Mexico City was a relatively small city. The population grew modestly during the nineteenth century, reaching around 500,000 in 1900. The population grew rapidly during the twentieth century, to 3 million in 1950 and 21 million in the urban area in 2015. Millions of people have migrated to the cities in search of work.

Rapid population growth has resulted in an expansion of the land area of Mexico City (Figure 13-39). In 1903, most of Lake Texcoco was drained by a gigantic canal and tunnel

▼ **FIGURE 13-36 PRECOLONIAL MEXICO CITY SITE AND SITUATION** The Aztec city of Tenochtitlán was built on an island in Lake Texcoco.

▼ **FIGURE 13-38 COLONIAL MEXICO CITY** The main square in downtown Mexico City, the Zócalo, was laid out by the Spanish. The Metropolitan Cathedral is at the near end of the square. The National Palace is to the left, and City Hall is at the far end of the square, opposite the cathedral. Excavations at the site of the Templo Mayor are in the lower left, in the open air and also under the two green roofs.

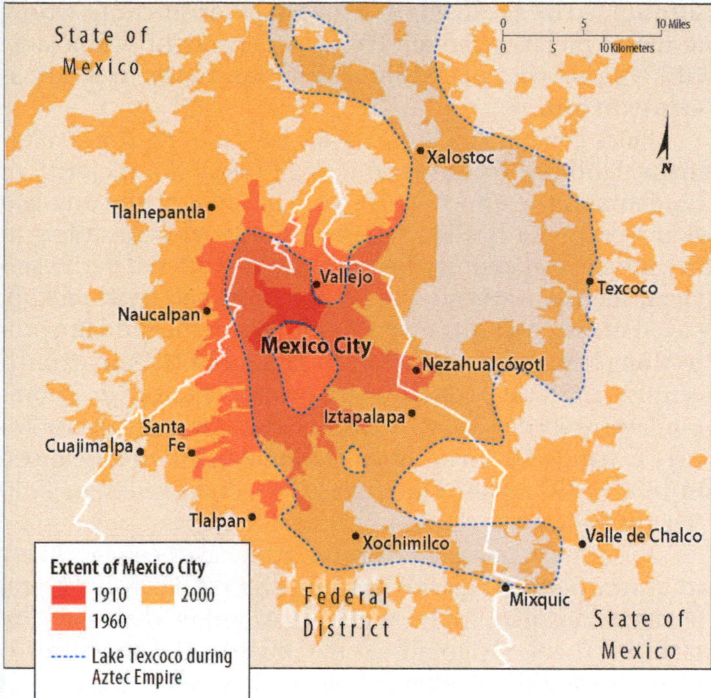

▲ **FIGURE 13-39 EXPANDING LAND AREA OF MEXICO CITY** The land area of Mexico City expanded dramatically in the twentieth century. Most of Lake Texcoco was covered by development.

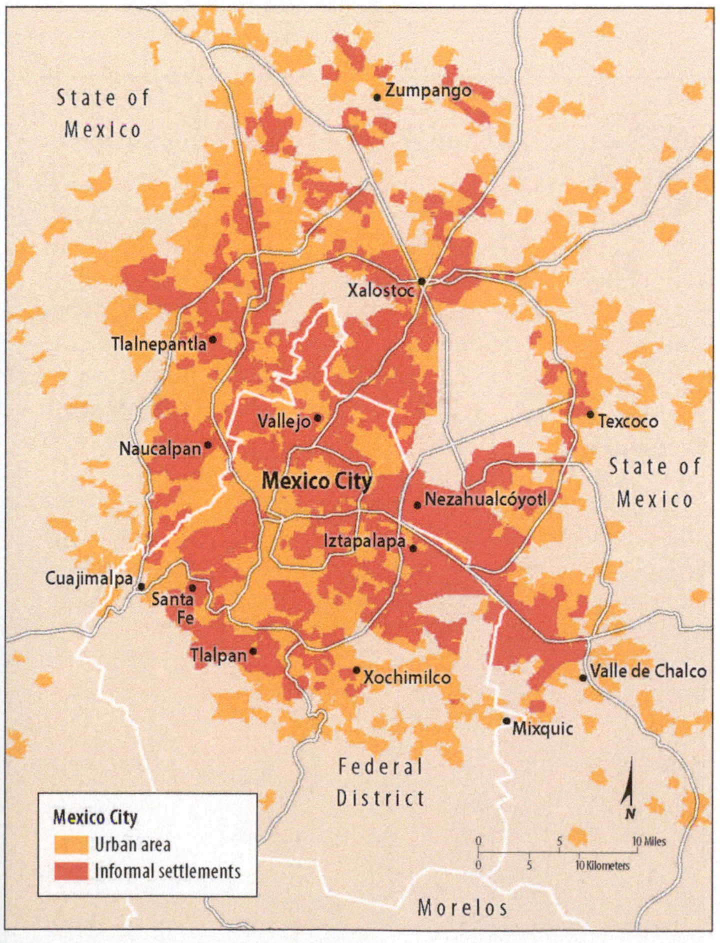

▲ **FIGURE 13-40 CONCENTRIC ZONES IN MEXICO CITY** Most of the informal settlements are in an outer ring.

▲ **FIGURE 13-41 ELITE SECTOR IN MEXICO CITY** The Paseo de la Reforma is the main thoroughfare through the heart of the elite sector.

project, allowing the city to expand to the north and east. The dried-up lakebed is a less desirable residential location than the west side because prevailing winds from the northeast stirred up dust storms. Informal settlements proliferate in outer rings (Figure 13-40).

Emperor Maximilian (1864–1867) designed a 14-lane, tree-lined boulevard patterned after the Champs-Elysées in Paris. The boulevard (now known as the Paseo de la Reforma) extended 3 kilometers southwest from the center to Chapultepec (Figure 13-41). The Reforma between downtown and Chapultepec became the spine of an elite sector. During the late nineteenth century, the wealthy built pretentious palacios (palaces) along it. Physical factors also influenced the movement of wealthy people toward the west, along the Reforma. Because elevation was higher than elsewhere in the city, sewage flowed eastward and northward, away from Chapultepec. As Mexico City's population grew rapidly during the twentieth century, the social patterns inherited from the nineteenth century were reinforced.

CHECK-IN KEY ISSUE 2

Where Are People Distributed in Urban Areas?

- ✔ **Urban areas show evidence of growth in concentric zones, sectors, and multiple nuclei.**
- ✔ **The three models show some similarities and some differences in the patterns within cities of North America and other regions.**
- ✔ **Cities in developing countries are further influenced by colonial history.**

Why Do Urban Areas Expand?

- ▸ **Origin and Growth of Suburbs**
- ▸ **Suburban Sprawl**
- ▸ **Suburban Segregation**
- ▸ **Legacy of Public Transport**
- ▸ **Reliance on Motor Vehicles**

LEARNING OUTCOME 13.3.1

Understand the impact of suburban growth on local government.

A **suburb** is a residential or commercial area situated within an urban area but outside the central city. Suburbs have existed on a small scale since ancient times; residential areas were often located outside the walls surrounding a city. As cities grew rapidly during the nineteenth century, as part of the Industrial Revolution, more extensive suburbs appeared.

Origin and Growth of Suburbs

In 1950, only 20 percent of Americans lived in suburbs compared to 40 percent in cities and 40 percent in small towns and rural areas. The percentage living in suburbs climbed rapidly thereafter. Ten years later, one-third of Americans lived in cities, one-third in suburbs, and one-third in small towns and rural areas. In 2000, 50 percent of Americans lived in suburbs compared to only 30 percent in cities and 20 percent in small towns and rural areas.

Suburbs offer varied attractions—a detached single-family dwelling rather than a row house or an apartment, private land surrounding the house, space to park cars, and a greater opportunity for home ownership. A suburban house provides space and privacy, a daily retreat from the stress of urban living. Families with children are especially attracted to suburbs, which offer more space for play and protection from the high crime rates and heavy traffic that characterize inner-city life. As incomes rose in the twentieth century, first in the United States and more recently in other developed countries, more families were able to afford to buy suburban homes.

ANNEXATION

Until recently in the United States, as cities grew, they expanded by adding peripheral land. Now cities are surrounded by a collection of suburban jurisdictions whose residents prefer to remain legally independent of the large city. The process of legally adding land area to a city is **annexation.**

Rules concerning annexation vary among states. Normally, land can be annexed to a city only if a majority of residents in the affected area vote in favor of the annexation. Peripheral residents generally desired annexation in the nineteenth century because the city offered better services, such as water supply, sewage disposal, trash pickup, paved streets, public transportation, and police and fire protection. Thus, as U.S. cities grew rapidly in the nineteenth century, the legal boundaries frequently changed to accommodate newly developed areas. For example, the city of Chicago expanded from 26 square kilometers (10 square miles) in 1837 to 492 square kilometers (190 square miles) in 1900 (Figure 13-42).

In contrast, in recent decades cities have been less likely to annex peripheral land because the residents have preferred to organize their own services rather than pay city taxes for them. Originally, some of these peripheral jurisdictions were small, isolated towns that had a tradition of independent local government before being swallowed up by urban growth. Others are newly created communities whose residents wish to live close to the large city but not be legally part of it.

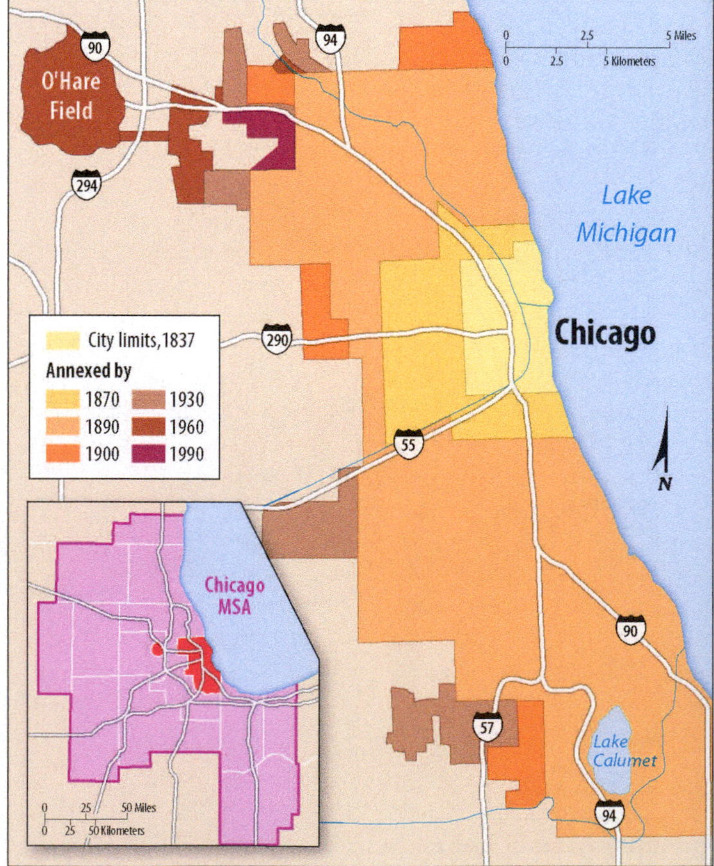

▲ **FIGURE 13-42 ANNEXATION IN CHICAGO**

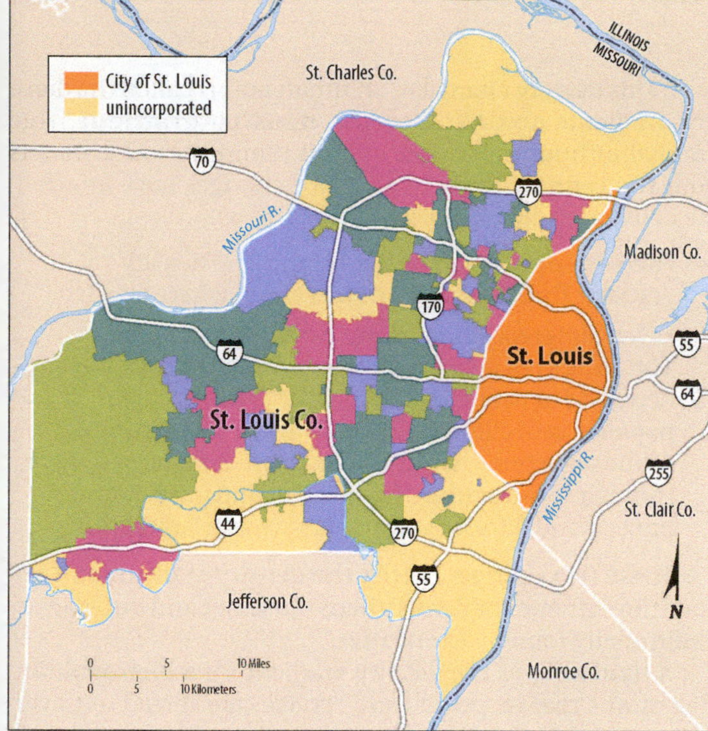

▲ FIGURE 13-43 MUNICIPALITIES IN ST. LOUIS COUNTY

LOCAL GOVERNMENT FRAGMENTATION

Given the difficulty in annexing suburban jurisdictions, local government in the United States is extremely fragmented. According to the 2012 Census of Governments, the United States had 89,004 local governments, including 3,031 counties, 19,522 municipalities, 16,364 townships, 12,884 school districts, and 37,203 special districts. Special districts are organized to provide such services as fire protection, water supply, libraries, and public transportation. Illinois has by far the largest number of local governments (6,968), and Hawaii has the fewest (21).

The larger metropolitan areas have thousands of local governments, with widely varying levels of resources (Figure 13-43). The large number of local government units has led to calls for a metropolitan government that could coordinate—if not replace—the numerous local governments in an urban area. The fragmentation of local government in the United States makes it difficult to solve regional problems of traffic management, solid-waste disposal, and the building of affordable housing.

Most U.S. metropolitan areas have a council of government, which is a cooperative agency consisting of representatives of the various local governments in the region. The council of government may be empowered to do some overall planning for the area that local governments cannot logically do. Strong metropolitan-wide governments have been established in a few places in North America. Two kinds exist:

- **Consolidations of city and county governments.** Examples of consolidations of city and county governments include Indianapolis and Miami. The boundaries of Indianapolis were changed to match those of Marion County, Indiana. Government functions that were handled separately by the city and the county now are combined into a joint operation in the same office building. In Florida, the city of Miami and surrounding Dade County have combined some services, but the city boundaries have not been changed to match those of the county.
- **Federations.** Examples of federations include Toronto and other large Canadian cities. Toronto's metropolitan government was created in 1954, through a federation of 13 municipalities. A two-tier system of government existed until 1998, when the municipalities were amalgamated into a single municipality.

SMART GROWTH

Several U.S. states have taken steps to curb suburban growth. The goal is to produce a pattern of compact and contiguous development and protect rural land for agriculture, recreation, and wildlife. Legislation and regulations to limit suburban growth and preserve farmland has been called **smart growth**. Oregon and Tennessee have defined growth boundaries within which new development must occur (Figure 13-44). Cities can annex only lands that have been included in the urban growth areas. New Jersey, Rhode Island, and Washington were also early leaders in enacting strong state-level smart-growth initiatives. Maryland's smart-growth law discourages the state from funding new highways and other projects that would extend suburban sprawl and destroy farmland. State development money must be allocated to "fill in" already urbanized areas.

PAUSE & REFLECT 13.3.1

How might urban growth boundaries help to slow suburban growth?

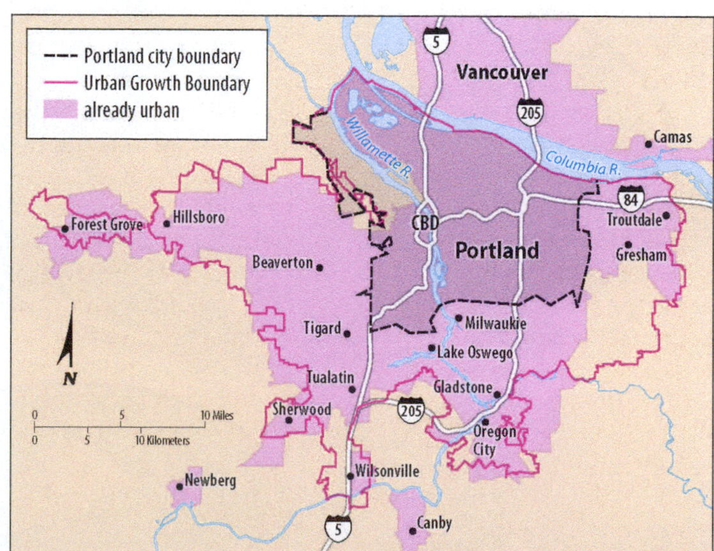

▲ FIGURE 13-44 PORTLAND, OREGON, URBAN GROWTH BOUNDARY New developments must take place inside the boundary.

Suburban Sprawl

LEARNING OUTCOME 13.3.2

Describe suburban sprawl.

Sprawl is the development of suburbs at relatively low density and at locations that are not contiguous to the existing built-up area. When private developers select new housing sites, they seek cheap land that can easily be prepared for construction--land often not contiguous to the existing built-up area. Sprawl is also fostered by the desire of many families to own large tracts of land.

PERIPHERAL MODEL

Chauncey Harris created the peripheral model as a modification of the multiple nuclei model (which he co-authored). According to the **peripheral model**, an urban area consists of an inner city surrounded by large suburban residential and service nodes or nuclei tied together by a beltway or ring road (Figure 13-45).

DENSITY GRADIENT

As you travel outward from the center of a city, you can watch the decline in the density at which people live (Figure 13-46). Inner-city apartments or row houses may pack as many as 250 dwellings on a hectare of land (100 dwellings per acre). Older suburbs have larger row houses, semidetached houses, and individual houses on small lots, at a density of about 10 houses per hectare (4 houses per acre). A detached house typically sits on a lot of 0.25 to 0.5 hectares (0.6 to 1.2 acres) in new suburbs and a lot of 1 hectare or greater (2.5 acres) on the fringe of the built-up area.

This density change in an urban area is called the density gradient. According to the **density gradient**, the number of houses per unit of land diminishes as distance from the center city increases. Two changes have affected the density gradient in recent years:

- **Fewer people living in the center.** The density gradient thus has a gap in the center, where few live.
- **Fewer differences in density within urban areas.** The number of people living on a hectare of land has decreased in the central residential areas through population decline and abandonment of old housing. At the same time, density has increased on the periphery through construction of apartment and town-house projects and diffusion of suburbs across a larger area.

These two changes flatten the density gradient and reduce the extremes of density between inner and outer areas traditionally found within cities.

A flattening of the density gradient for a metropolitan area means that its people and services are spread out over a larger area. U.S. suburbs are characterized by sprawl and segregation.

As long as demand for single-family detached houses remains high, land on the fringe of urban areas will be converted from open space to residential land use. Land is not transformed immediately from farms to housing developments. Instead, developers buy farms for future construction of houses by individual builders. The peripheries of U.S. cities look like Swiss cheese, with pockets of development and gaps of open space.

Sprawl incurs costs:

- Local authorities must spend more money extending roads and utilities to connect developments not contiguous to existing built-up areas.

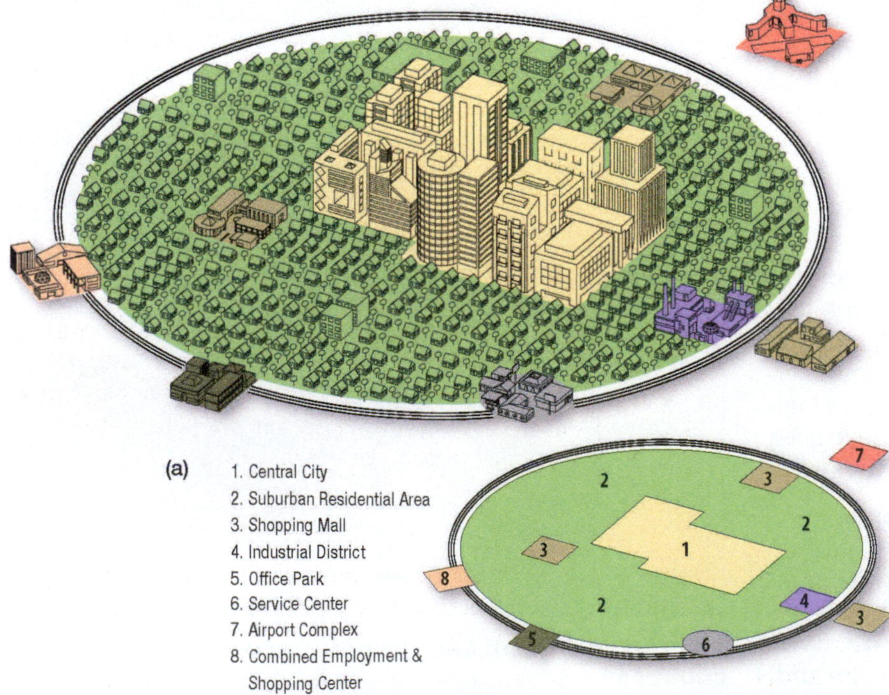

▲ **FIGURE 13-45 PERIPHERAL MODEL (a)** According to this model, the central city is surrounded by a beltway or ring road. Around the beltway are suburban residential areas and nodes, or edge cities, where consumer and business services cluster. **(b)** Tysons Corner, Virginia, is an edge city outside Washington, D.C.

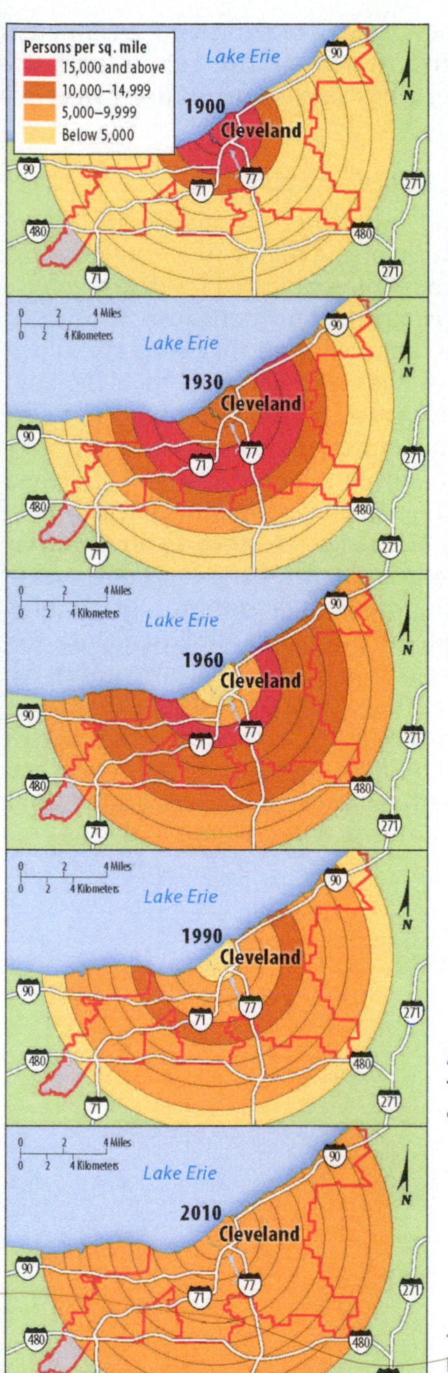

◀ **FIGURE 13-46 DENSITY GRADIENT IN CLEVELAND,** 1900-2010.

▲ **FIGURE 13-47 MEGALOPOLIS** Also known as the Boswash corridor, Megalopolis extends more than 700 kilometers (440 miles) between north of Boston to south of Washington. Megalopolis contains one-fourth of the U.S. population on only 2 percent of the country's total land area.

- More agricultural land is lost through construction of isolated housing developments.
- More energy is expended because trips to work and services must cover longer distances.

MEGALOPOLIS

MSAs in the northeastern United States form one continuous urban complex, extending from north of Boston to south of Washington, D.C. Geographer Jean Gottmann named this region **Megalopolis**, a Greek word meaning "great city" (Figure 13-47). Other U.S. urban complexes include the southern Great Lakes between Milwaukee and Pittsburgh and southern California between Los Angeles and Tijuana. Among examples in other developed regions are the German Ruhr (including the cities of Dortmund, Düsseldorf, and Essen), Randstad in the Netherlands (including the cities of Amsterdam, The Hague, and Rotterdam), and Japan's Tokaido (including the cities of Tokyo and Yokohama).

Within Megalopolis, central cities such as Baltimore, New York, and Philadelphia retain distinctive identities, and the urban areas are visibly separated from each other by parks, military bases, and farms. But at the periphery of the urban areas, the boundaries overlap.

PAUSE & REFLECT 13.3.2

Name a city in Megalopolis that you consider a strong candidate to become part of an MSA in the near future? Why that city?

Suburban Segregation

LEARNING OUTCOME 13.3.3

Explain ways in which suburbs are segregated.

Many suburbs display two forms of segregation:

- **Segregation of social classes.** Housing in a given suburban community is usually built for people of a single social class, with others excluded by virtue of the cost, size, or location of the housing. Segregation by race and ethnicity also persists in some suburbs (see Chapter 7).
- **Segregation of land uses.** Residents are separated from commercial and manufacturing activities that are confined to compact, distinct areas.

U.S. AND U.K. SUBURBS

The supply of land for the construction of new housing is more severely restricted in European urban areas than in the United States (Figure 13-48). Officials try to limit sprawl by designating areas of mandatory open space. Several British cities are surrounded by greenbelts, or rings of open space. New housing is built either in older suburbs inside the greenbelts or in planned extensions to small towns and new towns beyond the greenbelts. On the other hand, restriction of the supply of land on the urban periphery has driven up house prices in Europe.

RESIDENTIAL SEGREGATION

Low-income people and minorities are unable to live in many U.S. suburbs because of the high cost of the housing and unwelcoming attitudes of established residents. Suburban communities discourage the entry of those with lower incomes and minorities because of fear that property values will decline if the high-status composition of the neighborhood is altered. Extensive areas of suburbs have been developed with houses of similar interior dimension, lot size, and cost, appealing to people with similar incomes and lifestyles.

The homogeneity in suburban communities is legally protected through zoning ordinances. A **zoning ordinance** is a law that limits the permitted uses of land and maximum density of development in a community. Zoning ordinances typically identify districts designed only for single-family houses, apartments, industry, or commerce. Low-income families may have difficulty finding affordable housing through provisions such as requiring each house to sit on a large lot and prohibiting apartments. Fences are built around some suburban housing districts, and visitors must check in at a gate house to enter (Figure 13-49).

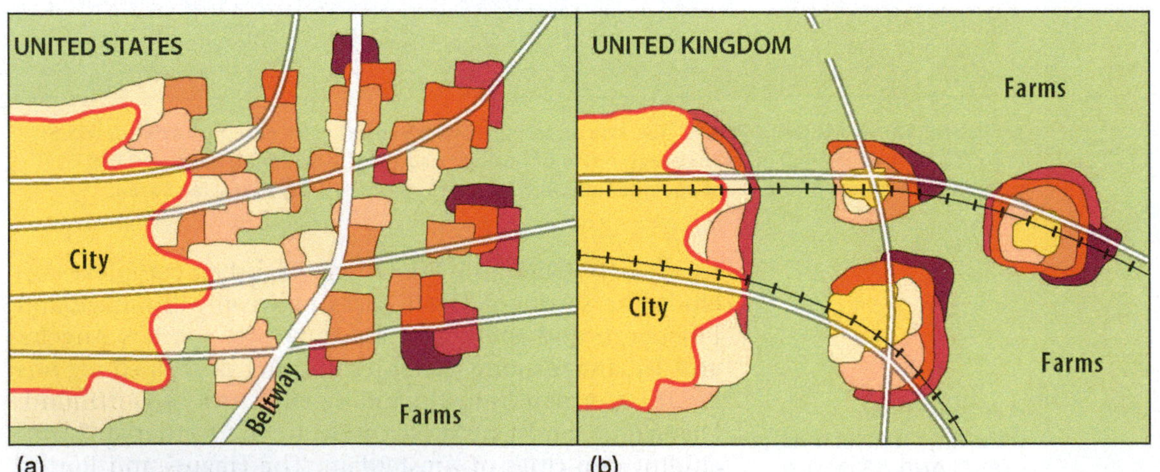

(c)

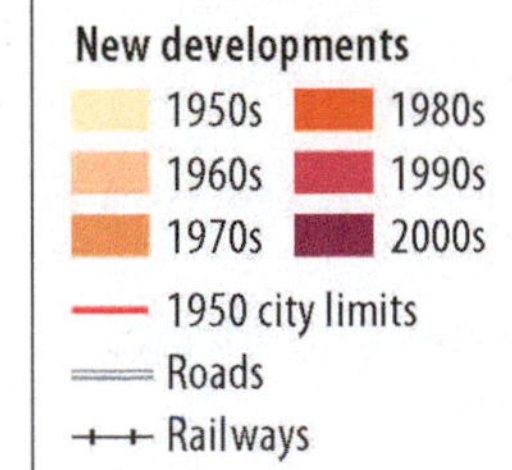

◀ **FIGURE 13-48 SUBURBAN DEVELOPMENT PATTERNS IN THE UNITED KINGDOM AND THE UNITED STATES** **(a)** The United States has much more sprawl than the United Kingdom. **(b)** In the United Kingdom, new housing is more likely to be concentrated in new towns or **(c)** planned extensions of existing small towns.

▲ **FIGURE 13-49 GATED COMMUNITY** Orlando, Florida.

SUBURBAN SERVICES

Consumer and business services are also expanding in suburbs. A number of factors account for this trend.

SUBURBANIZATION OF CONSUMER SERVICES. Consumer services have expanded in the suburbs because most of their customers live there (Figure 13-50). Historically, urban residents bought food and other daily necessities at small neighborhood shops in the midst of housing areas and shopped in the CBD for other products. But since the end of World War II, downtown sales have stagnated, whereas suburban sales have risen at an annual rate of 5 percent.

Suburban retailing is concentrated in shopping malls of varying sizes. Larger malls contain department stores and specialty shops once located only in the CBD. Generous parking lots surround the stores. A shopping mall is built by a developer, who buys the land, builds the structures, and leases space to individual merchants.

Suburban residents no longer wish to make the long journey to shop in the CBD, and corner shops have been replaced by supermarkets in shopping malls. The low density of residential construction discourages people from walking to stores, and restrictive zoning practices often exclude shops from residential areas.

PAUSE & REFLECT 13.3.3

Are you able to walk from your home to consumer services? Would you want to live near shops? Why or why not?

SUBURBANIZATION OF BUSINESS SERVICES. Offices that do not require face-to-face contact are increasingly moving to suburbs, where rents are lower than in the CBD. Executives can drive more easily to their offices and park their cars without charge. Factories and warehouses also increasingly locate in suburbs for more space, cheaper land, and better truck access.

Suburban locations have posed hardships for some employees, especially lower-status workers, such as secretaries and custodians. These employees may live in neighborhoods that are not convenient to where they work. They may not own reliable cars, but public transportation may not serve their place of employment. Some workers miss the stimulation and animation of a central location, particularly at lunchtime.

Polaris Fashion Place
Worthington Square
The Shops at Worthington Place
The Mall at Tuttle Crossing
Easton Town Center
Columbus
Westland Mall
Eastland Mall
0 2 4 Miles
0 2 4 Kilometers
N

(a)

◄ **FIGURE 13-50 SUBURBAN SHOPPING MALLS, COLUMBUS, OHIO (a)** The malls surround the city near the beltway. **(b)** Easton Town Center is a suburban shopping mall designed to look like a CBD.

(b)

Legacy of Public Transport

LEARNING OUTCOME 13.3.4

Describe the benefits and drawbacks of public transport.

Because few people live within walking distance of their place of employment, urban areas are characterized by extensive commuting. The heaviest flow of commuters is into the CBD in the morning and out of it in the evening. **Rush hour,** or peak hour, is the four consecutive 15-minute periods that have the heaviest traffic.

The intense concentration of people in the CBD during working hours strains transportation systems because a large number of people must reach a small area of land at the same time in the morning and disperse at the same time in the afternoon. As much as 40 percent of all trips made into or out of a CBD occur during four hours of the day—two in the morning and two in the afternoon.

PUBLIC TRANSPORT IN THE UNITED STATES

Public transport in the United States is used primarily for commuting by workers into and out of the CBD. Despite its advantages in getting in and out of the CBD, public transport is used for only 5 percent of commuting trips in the United States. One-half of trips to work are by public transport in New York; one-third in Boston, San Francisco, and Washington; and one-fourth in Chicago and Philadelphia (Figure 13-51). But in most other cities, public transport service is minimal or nonexistent. Motorists driving alone account for 76 percent of commuters and carpooling for another 10 percent.

Historically, the growth of suburbs was constrained by poor transportation. People lived in crowded cities because they had to be within walking distance of shops and places of employment. The invention of the railroad in the nineteenth century enabled people to live in suburbs and work in the central city. Many so-called streetcar suburbs built in the nineteenth century still exist and retain unique visual identities. They consist of houses and shops clustered near a station or former streetcar stop at a much higher density than is found in newer suburbs. To accommodate commuters, cities built railroads at street level (called trolleys, streetcars, or trams), above ground (Els), and underground (subways).

The suburban explosion in the twentieth century relied on motor vehicles rather than railroads, especially in the United States. Rail lines restricted nineteenth-century suburban development to narrow ribbons within walking distance of the stations. Cars and trucks permitted large-scale development of suburbs at greater distances from the center, in the gaps between the rail lines. Motor vehicle drivers have much greater flexibility in their choice of residence than was ever before possible.

Public transport reached its peak in the United States in the 1940s, when it carried 23 billion passengers per year. Ridership in the United States declined to a low of around 6 billion in 1970. However, public transport ridership increased from less than 8 billion in 1994 to nearly 11 billion in 2014.

The trolley—now known by the more elegant term fixed light-rail transit—was once relegated almost exclusively to tourist attractions in New Orleans and San Francisco but is making a comeback in North America. Since 1975, 21 new light-rail systems have been opened in the United States. Dallas and Los Angeles—long dependent entirely on motor vehicles—opened light-rail lines during the 1990s and now have the two largest U.S. systems, more than 100 kilometers (60 miles) each. Ten other U.S. cities and two Canadian cities have new systems between 50 and 100 kilometers (30 and 60 miles).

Entirely new subway systems were opened between 1972 and 1993 in six U.S. cities: Atlanta, Baltimore, Los Angeles, Miami, San Francisco, and Washington. Cities with century-old systems such as Boston and Chicago have attracted new passengers through construction of new lines and modernization of existing ones. Chicago has pioneered construction of heavy-rail rapid transit lines in the median strips of expressways. The federal government has permitted Boston, New York, and other cities to use funds

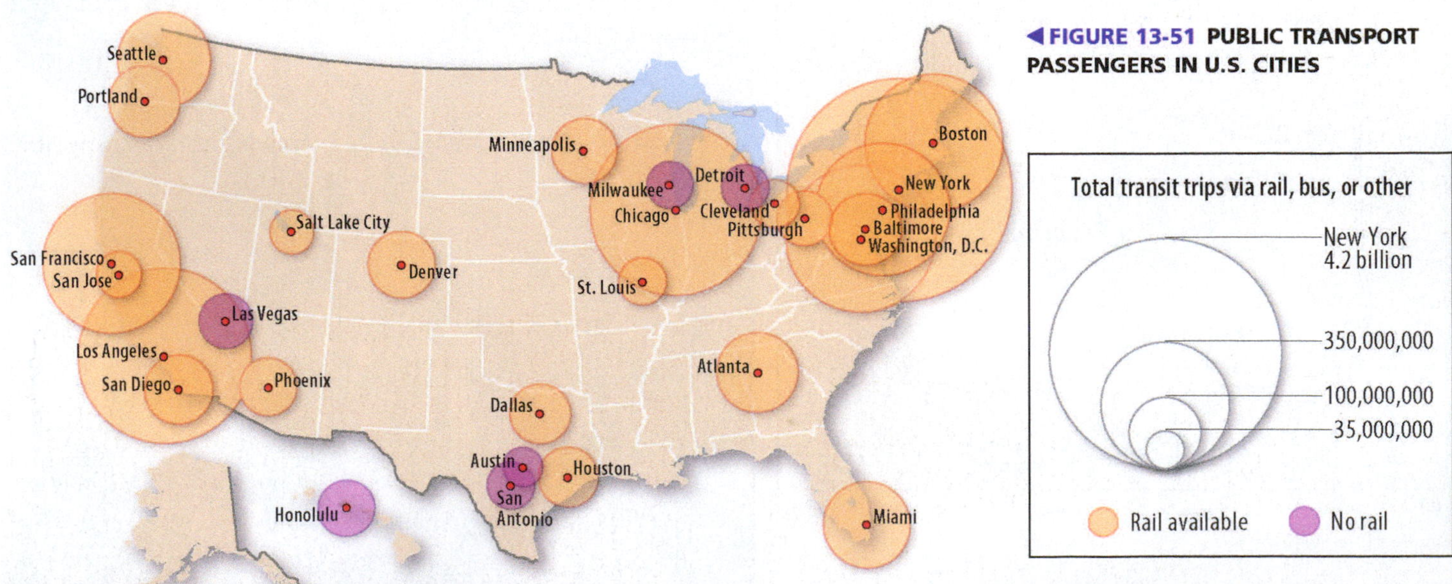

FIGURE 13-51 PUBLIC TRANSPORT PASSENGERS IN U.S. CITIES

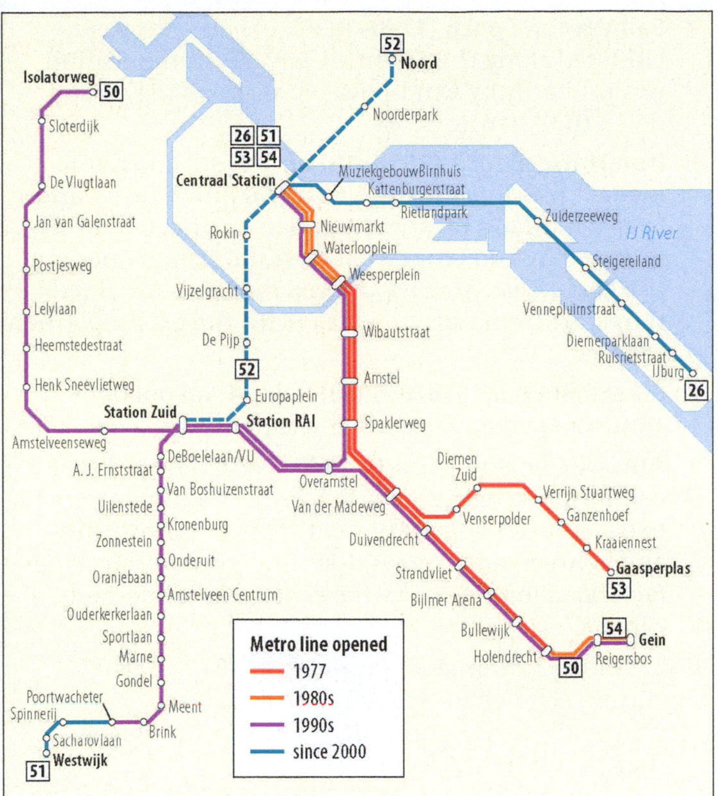

▲ **FIGURE 13-52 AMSTERDAM METRO**

originally allocated for interstate highways to modernize rapid transit service instead. New York's subway cars, once covered with graffiti spray-painted by gang members, have been cleaned so that passengers can ride in a more hospitable environment.

▼ **FIGURE 13-53 PUBLIC TRANSPORT IN MUNICH, GERMANY**
(a) U-Bahn (underground heavy rail), (b) S-Bahn (elevated heavy rail), (c) tram.

(a)

(b)

(c)

Despite modest recent successes, public transit in the United States is caught in a vicious circle because fares do not cover operating costs. As patronage declines and expenses rise, the fares are increased, which drives away passengers and leads to service reduction and still higher fares. Public expenditures to subsidize construction and operating costs have increased, but the United States does not fully recognize that public transportation is a vital utility deserving of subsidy to the degree long assumed by governments in other developed countries, as well as developing countries.

The minimal level of public transit service in most U.S. cities means that low-income people may not be able to reach places of employment. Low-income people tend to live in inner-city neighborhoods, but the job opportunities, especially those requiring minimal training and skill in personal services, are in suburban areas not well served by public transportation. Inner-city neighborhoods have high unemployment rates at the same time that suburban firms have difficulty attracting workers. In some cities, governments and employers subsidize vans to carry low-income inner-city residents to suburban jobs.

PUBLIC TRANSIT IN OTHER COUNTRIES

In hundreds of cities around the world, extensive networks of bus, tram, and subway lines have been maintained, and funds for new construction have been provided in recent years. Wikipedia lists 148 cities with underground heavy-rail systems and 371 cities with light-rail systems in operation as of 2014. Another 36 heavy-rail systems were scheduled to open between 2015 and 2020, including 16 in China. And cities with existing service have been expanding them (Figure 13-52).

The greater importance placed on public transport outside the United States can be seen by comparing Indianapolis, Indiana, with Munich, Germany. Both urban areas have around 1.4 million inhabitants, and both have around 500 kilometers of bus lines. But Indianapolis has no rail service, whereas Munich has 95 kilometers of U-Bahn (underground heavy rail), 442 kilometers of S-Bahn (elevated heavy rail), and 71 kilometers of trams (Figure 13-53).

PAUSE & REFLECT 13.3.4

Do you regularly utilize public transport? Why or why not?

Reliance on Motor Vehicles

LEARNING OUTCOME 13.3.5

Describe the strategies to reduce the impact of motor vehicles in urban areas.

The average American travels 58 kilometers (36 miles) per day. People do not travel aimlessly; their trips have a precise point of origin, destination, and purpose. In the United States, 19 percent of trips are for work, 10 percent for school or church, 28 percent for social and recreational activities, and 43 percent for other personal activities (such as shopping and medical care).

In urban areas, public transport is better suited than motor vehicles to moving large numbers of people because each transit traveler takes up far less space. Public transport is cheaper, less polluting, and more energy efficient than privately operated motor vehicles. It also is particularly suited to rapidly bringing a large number of people into a small area. A bus can accommodate 30 people in the amount of space occupied by one car, whereas a double-track rapid transit line can transport the same number of people as 16 lanes of urban freeway.

Nonetheless, 83 percent of trips in the United States are by car or truck, 12 percent are by walking or biking, 2 percent each are by public transport or school bus, and 1 percent are by other means. Despite the small number of trips by public transport, the mode of travel is an important component of transportation systems in large cities.

TRANSPORTATION EPOCHS

Transportation improvements have played a key role in the changing structure of urban areas. Geographer John Borchert identified five eras of U.S. urban areas resulting from changing transportation systems:

- **Sail-Wagon Epoch (1790–1830).** Urban areas were clustered along the Atlantic Coast. Communication was primarily by wind-powered ships plying up and down the Atlantic Coast.
- **Iron Horse Epoch (1830–1870).** The steam engine made it possible for ships to travel much faster and to reach inland locations. Canals connected newly founded inland cities with the established ones on the East Coast. Steam-powered railroads provided transport from outlying areas into the existing urban centers.
- **Steel Rail Epoch (1870–1920).** Long-haul rail lines connected urban areas around the country.
- **Auto-Air-Amenity Epoch (1920–1970).** The internal combustion engine made it possible for motor vehicles to become the dominant mode of transport within and between urban areas. Gasoline-powered airplanes facilitated long-distance travel between distant urban centers.
- **Satellite-Electronic-Jet Propulsion (1970–?).** The current era is characterized by the ability to communicate electronically, as well as to control transport systems electronically.

Cities have prospered or suffered during the various epochs, depending on their proximity to economically important resources and migration patterns. At the same time, cities retain physical features from the earlier eras that may be assets or liabilities in subsequent eras.

BENEFITS AND COSTS OF MOTOR VEHICLES

There are around 1.2 billion motor vehicles in the world, including 255 million in the United States. The United States actually has more registered motor vehicles than licensed drivers. Motor vehicle ownership is nearly universal among American households, with the exception of some poor families, older individuals, and people living in the centers of large cities such as New York. Motor vehicles offer two principal benefits:

- **Comfort, choice, and flexibility.** Motorists can live wherever they wish and travel whenever they wish. They are not constrained by the timetable of public transport service. A motor vehicle offers comfortable seats, choice of music, and isolation from unpleasant people on a bus or subway.
- **Perceived cost.** Motorists perceive that the cost of using vehicles is less than the cost of using public transport.

◄ **FIGURE 13-54 URBAN FREEWAY** I-75 carves a wide path through downtown Atlanta.

Each time public transport is used a fare must be paid, and the fare is higher than the cost of fuel, at least in the United States. Most of the costs associated with motor vehicles, such as insurance and license, are paid on an annual basis, regardless of the amount of driving that is actually done.

The U.S. government encourages the use of cars and trucks by paying 90 percent of the cost of limited-access, high-speed interstate highways, which stretch for 77,000 kilometers (48,000 miles) across the country. The use of motor vehicles is also supported by policies that keep the price of fuel below the level found in Europe.

Motor vehicles have costs beyond their purchase and operation. These costs are not noticed by most motorists:

- **Consumption of land.** The motor vehicle is an important user of land in the city (Figure 13-54). An average city allocates about one-fourth of its land to roads and parking lots. Multilane freeways cut a 23-meter (75-foot) path through the heart of a city, and elaborate interchanges consume even more space. Valuable land in the central city is devoted to parking cars and trucks, although expensive underground and multistory parking structures can reduce the amount of ground-level space needed. European and Japanese cities have been especially disrupted by attempts to insert new roads and parking areas in or near the medieval central areas.
- **Congestion.** The average American wastes 18 gallons of gasoline and loses 42 hours per year sitting in traffic jams, according to the Urban Mobility Report prepared by the Texas Transportation Institute. In the United States, the total cost of congestion is valued at $160 billion per year. But most Americans still prefer to commute by vehicle. Most people overlook these costs because they place higher value on the privacy and flexibility of schedule offered by a car.

Freeways that once sliced through CBDs have been demolished in a number of cities, including Boston, San Francisco, and Seoul. For example, Boston's Central Artery has been replaced by a park (Figure 13-55).

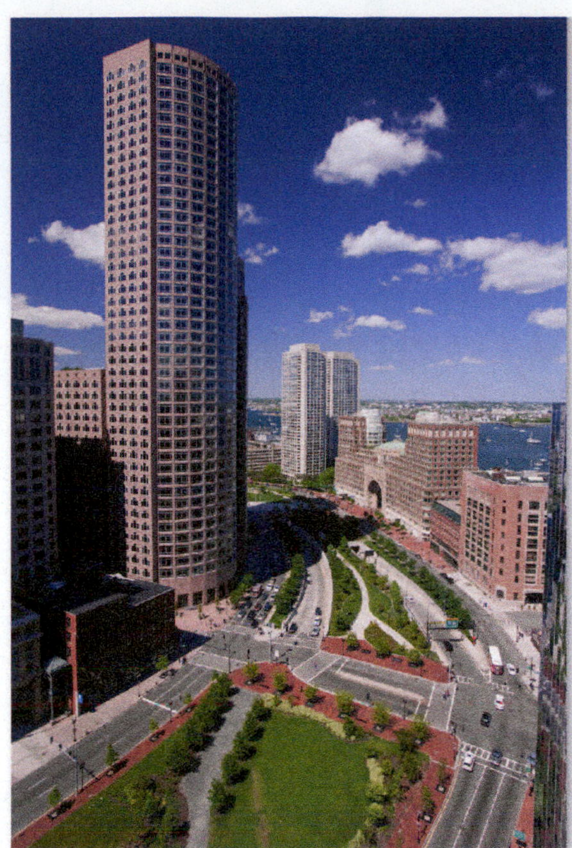

▲ **FIGURE 13-55 FREEWAY DEMOLITION** Boston's Central Artery freeway was demolished and replaced with the Rose Kennedy Greenway.

AUTONOMOUS DRIVING VEHICLES

Future transportation systems are likely to include various forms of autonomous vehicles. Vehicles currently possess technological capabilities supportive of hands-free driving such as sensors and GPS, and they can perform hands-free functions such as automatic braking, parallel parking, and prevention of unsafe lane changes.

Autonomous vehicles are likely to result in fewer accidents caused by human error, provide mobility for people who are too young to drive or have a disability, and decrease the safe distance between vehicles and therefore increase the number of vehicles that can fit on the road.

Still unsettled are many practical problems created by autonomous vehicles, such as liability and insurance. The most significant obstacle to autonomous vehicles may be consumer acceptance. Do you want to give up some or all control of the vehicle? What do your parents and grandparents think? Do autonomous vehicles excite them or frighten them?

PAUSE & REFLECT 13.3.5

What are some benefits and costs of removing urban freeways, such as in Boston?

CHECK-IN KEY ISSUE 3

Why Do Urban Areas Expand?

- ✔ **U.S. cities once expanded by annexing surrounding land, but that practice is now less common.**
- ✔ **Cities are typically surrounded by sprawling independent suburban jurisdictions.**
- ✔ **Suburban sprawl consumes a lot of land and requires investment in a lot of new roads and utilities.**
- ✔ **Suburbs are often segregated by social class and by land use activities.**
- ✔ **Suburban residents are dependent on motor vehicles to get to other places, whereas most cities offer forms of public transit.**

KEY ISSUE 4

Why Do Cities Face Sustainability Challenges?

- The City Challenged
- The City Renewed
- The City Contrasted
- The City Cleaned
- The City Controlled

LEARNING OUTCOME 13.4.1

Understand challenges faced by cities.

The final key issue of the book returns to the themes introduced in the first key issue. The first key issue introduced five basic concepts used by geographers to explain why every place on Earth is in some ways unique and in other ways related to other locations—place, region, scale, space, and connections. These five themes are especially useful in understanding cities.

Place and region help to explain why every city is unique:

- A place is a specific point on Earth, distinguished by a particular characteristic. Every city occupies a unique location on Earth's surface. Furthermore, a city itself contains a collection of unique places, such as the CBD and residential areas occupied by people with distinctive cultural and economic characteristics.
- A region is an area of Earth defined by one or more distinctive characteristics. Urban areas have grown so large that they now constitute regions with widely varying features.

Scale, space, and connections help to explain why different cities are interrelated:

- Scale is the relationship between the portion of Earth being studied and Earth as a whole. Cities reflect the importance of the variety of scales, from local to global. At the local scale, cities are centers of diversity. Living in a city puts you in close proximity to people with different cultural characteristics and economic conditions. At the same time, the economic well-being and cultural vibrancy of a city depends on global economic and cultural patterns and processes.
- Space refers to the physical gap or interval between two objects. People and activities are arranged within a city according to properties of distribution. Density declines with increasing distance from the city (though less than in the past), distinctive groups of people and activities are concentrated in various areas of the city, and the physical structure of the city such as the layout of streets follows a regular pattern.
- Connection refers to relationships among people and objects across the barrier of space. Cities are nodes of connections. They are the centers for the transportation networks that tie together cities, as well as areas within cities.

The City Challenged

One hundred years ago, low-income inner-city neighborhoods in the United States teemed with throngs of recent immigrants from Europe. Such neighborhoods that housed perhaps 100,000 a century ago may contain fewer than 5,000 inhabitants today. Those remaining in these neighborhoods face a variety of distinctive social and physical challenges that are very different from those faced by suburban residents.

SOCIAL CHALLENGES

The **underclass** is a group in society prevented from participating in the material benefits of a more developed society because of a variety of social and economic hardships. A disproportionately large share of the underclass live in inner-city neighborhoods, where they are trapped in an unending cycle of hardships:

- **Inadequate job skills.** Inner-city residents are increasingly unable to compete for jobs. They lack technical skills needed for most jobs because fewer than half complete high school. Despite the importance of education in obtaining employment, many in the underclass live in an atmosphere that ignores good learning habits, such as regular school attendance and completion of homework. In the past, people with limited education could become factory workers or filing clerks, but today these jobs require skills in computing and handling electronics. Inner-city residents do not even have access to the remaining low-skilled jobs, such as custodial and fast-food service jobs because these jobs are increasingly in the distant suburbs and poorly served by public transport.
- **Culture of poverty.** Unwed mothers give birth to two-thirds of the babies in U.S. inner-city neighborhoods, and 80 percent of children in the inner city live with only one parent. Because of inadequate child-care services, single mothers may be forced to choose between working to generate income and staying at home to take care of the children. In principle, government officials would like to see more fathers living with their wives and children, but they provide little incentive for them to do so. Only a small percentage of "deadbeat dads" are tracked down for failing to provide child-care support. If the husband moves back home,

▲ **FIGURE 13-56 HOMELESS PERSON** New York City.

his wife may lose welfare benefits, leaving the couple financially worse off together than apart.

- **Homelessness.** Several million people are homeless in the United States. Most people are homeless because they cannot afford housing and have no regular income. Affordable housing is increasingly difficult to find in cities. Homelessness may have been sparked by family problems or job loss (Figure 13-56). One-fourth of homeless people are children, according to government surveys.
- **Drugs.** Trapped in a hopeless environment, some inner-city residents turn to drugs. Although drug use is a problem in suburbs as well, rates of use have increased most rapidly in inner cities. Some drug users obtain money through criminal activities. Gangs may form in inner-city neighborhoods to control lucrative drug distribution.
- **Crime.** Inner-city neighborhoods have a relatively high share of a metropolitan area's serious crimes, such as murder (Figure 13-57). A relatively high percentage of victims, as well as those arrested for murder, in cities are minorities. Violence may erupt when two gangs fight over the boundaries between their drug distribution areas.
- **Inadequate services.** Inner-city neighborhoods lack adequate police and fire protection, shops, hospitals, clinics, and other health-care facilities. Food deserts are especially common in low-income inner-city areas.
- **Municipal finances.** Low-income residents in inner-city neighborhoods require public services, but they can pay very little of the taxes to support the services. Central cities face a growing gap between the cost of needed services in inner-city neighborhoods and the availability of funds to pay for them.

PAUSE & REFLECT 13.4.1

How might additional investment in education address some of these features of the underclass?

PHYSICAL CHALLENGES

Thousands of vacant houses stand in the inner areas of U.S. cities because the landlords have abandoned them (see Debate It feature). Schools and shops close because they are no longer needed in inner-city neighborhoods with rapidly declining populations.

Filtering is the process of change in the use of a house, from single-family owner-occupancy to rented apartments and ultimately to abandonment. Many inner-city houses built by wealthy families in the nineteenth century have been subdivided by absentee landlords into smaller dwellings for low-income families. Landlords stop maintaining houses when the rent they collect becomes less than the costs of maintenance and taxes. In such a case, the building soon deteriorates and grows unfit for occupancy.

Some financial institutions hastened the abandonment of inner-city housing through redlining. **Redlining** is a process by which financial institutions draw red-colored lines on a map and refuse to lend money for people to purchase or improve property within the lines. As a result of redlining, families who try to fix up houses in the area have difficulty borrowing money. Although redlining is illegal, enforcement of laws against it is frequently difficult. The Community Reinvestment Act requires U.S. banks to document by census tract where they make loans. A bank must demonstrate that inner-city neighborhoods within its service area receive a fair share of its loans.

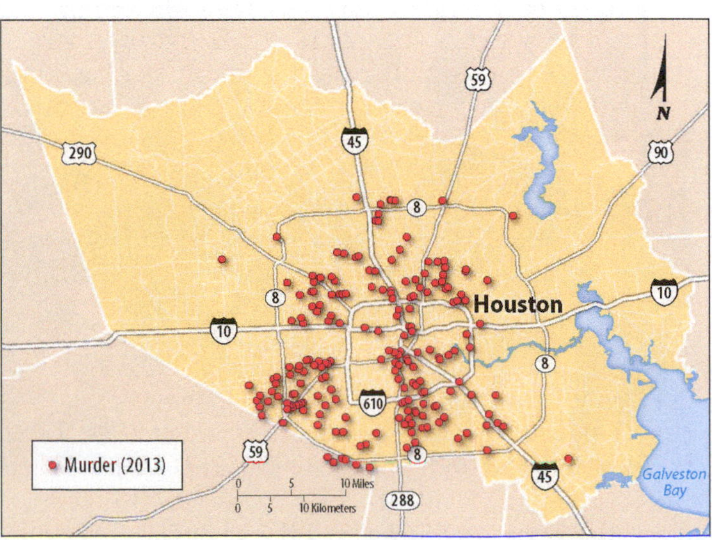

◀ **FIGURE 13-57 MURDERS IN HOUSTON** Compare the distribution of murders with Houston's social areas in Figures 13-15, 13-16, and 13-17.

The City Renewed

LEARNING OUTCOME 13.4.2

Describe the process of gentrification.

Gentrification is the process of converting an urban neighborhood from a predominantly low-income, renter-occupied area to a predominantly middle-class, owner-occupied area. Most cities have at least one substantially renovated inner-city neighborhood that has attracted higher-income residents, especially single people and couples without children who are not concerned with the quality of inner-city schools.

A deteriorated inner-city neighborhood is attractive for several reasons:

- The houses may be larger and more substantially constructed yet less expensive than houses in the suburbs.
- Houses may possess attractive architectural details, such as ornate fireplaces, cornices, high ceilings, and wood trim.
- For people who work downtown, inner-city living eliminates the strain of commuting on crowded freeways or public transport.
- The neighborhoods are near theaters, bars, restaurants, stadiums, and other cultural and recreational facilities.

Because renovating an old inner-city house can be even more expensive than buying a new one in the suburbs, cities encourage the process by providing low-cost loans and tax breaks. Public expenditures for renovation have been criticized as subsidies for the middle class at the expense of people with lower incomes, who are sometimes forced to move out of the gentrified neighborhoods because the rents in the area suddenly become too high for them (Figure 13-58).

REMOVING PUBLIC HOUSING

Given the high cost of housing in cities, governments sometimes step in to actually own or support the management of housing for low-income households. In the United States, **public housing** is government-owned housing rented to low-income individuals, with rents set at 30 percent of the tenant's income. In other countries, local governments or nonprofit organization such as charitable groups build and own much of the housing, aided by subsidies from the national government.

During the mid-twentieth century, many substandard inner-city houses were demolished and replaced with public housing. Several decades later, many of these public housing projects were themselves considered unsatisfactory living environments and in turn have also been demolished (Figure 13-59). Especially unsatisfactory were high-rise public housing projects. The elevators were frequently broken, juveniles terrorized other people in the hallways, and drug use and crime rates were high. Some observers claimed that the high-rise buildings were responsible for the problem because too many low-income families were concentrated into a high-density environment.

With the overall level of funding much lower, the supply of public housing and other government-subsidized housing in the United States diminished by approximately 1 million units between 1980 and 2010. But during the same period, the number of households that needed low-rent dwellings increased by more than 2 million. In Britain, the supply of public housing, known as social housing (formerly council estates), also declined because the government forced local authorities to sell some of the dwellings to the residents. The British also expanded subsidies to nonprofit housing associations that build housing for groups with special needs, including single mothers, immigrants, disabled people, and elderly people, as well as the poor.

(a)

(b)

▲ **FIGURE 13-58 GENTRIFICATION IN CINCINNATI (a)** Into the twenty-first century, the inner-city neighborhood of Over-the-Rhine had many abandoned residences and shops. **(b)** Many of these buildings have been renovated with attractive shops and higher-cost housing.

▲ **FIGURE 13-59 DEMOLISHED PUBLIC HOUSING** One of the most notorious high-rise public housing projects, Pruitt-Igoe in St. Louis, was constructed during the 1950s and demolished during the 1970s.

Cities try to reduce the hardship on poor families forced to move. U.S. law requires that they be reimbursed both for moving expenses and for rent increases over a four-year period. Western European countries have similar laws. Cities are also renovating old houses specifically for lower-income families through public housing or other programs. By renting renovated houses, a city also helps to disperse low-income families throughout the city instead of concentrating them in large inner-city public housing projects. However, some public housing projects were located in neighborhoods that are now gentrifying, so the new housing that is replacing the demolished high rises may be too expensive for the former public housing residents (Figure 13-60).

▲ **FIGURE 13-60 PUBLIC HOUSING AND GENTRIFICATION** A notorious high-rise public housing project in Chicago called Cabrini Green was located in a gentrifying neighborhood. The project has been demolished and replaced with new housing for middle-class families. This image from 2003 shows two Cabrini Green towers awaiting demolition next to new housing.

REVIVING CONSUMER SERVICES

Most consumer services have located in suburbs to be near suburban residents. However, some consumer services are returning to the inner city, in part to meet day-to-day needs of residents of gentrified neighborhoods. Inner-city consumer services are also attracting people looking for leisure activities, such as unusual shops in a dramatic downtown setting or view of a harbor. Several North American CBDs have combined new retail services with leisure services. For example:

- Boston's Faneuil Hall Marketplace is located in renovated eighteenth-century buildings.
- Baltimore's Harbor Place is built in the Inner Harbor, adjacent to waterfront museums, tourist attractions, hotels, and cultural facilities.
- Chicago's Navy Pier, a former cargo dock, has been converted to shops and attractions.
- New York's South Street Seaport integrates the old fish market with retailing and recreational activities.
- Philadelphia's nineteenth-century Reading Terminal Market, which barely survived during the twentieth-century suburbanization movement, has been renovated into a thriving marketplace with individual stalls operated by different merchants.
- San Francisco's Ferry Building, where San Francisco Bay ferries dock, is a gourmet food center (Figure 13-61).

PAUSE & REFLECT 13.4.2

What might be the attractions and the challenges of buying groceries in places like the Reading Terminal or Ferry Building?

▲ **FIGURE 13-61 SAN FRANCISCO'S FERRY BUILDING** The ferry terminal in downtown San Francisco has been renovated into a retail center.

The City Contrasted

LEARNING OUTCOME 13.4.3

Consider contrasting trends in American cities.

Contemporary American cities display ever-sharper contrasts. In some places, young, white, and wealthy people are moving back into the city (Figure 13-63). In other places, cities are in trouble (see Figure 13-64 and Debate It feature).

PAUSE & REFLECT 13.4.3

What similarities and differences do you observe between the maps of Chicago (Figure 13-63) and Paris (Figures 13-21, 13-22, and 13-23)?

DOING GEOGRAPHY Market Segmentation: You Are Where You Live

Marketing geographers identify sectors, rings, and nodes that come closest to matching customers preferred by a retailer. Companies use this information to understand, locate, and reach their customers better and to determine where to put new stores and where advertising should appear.

Segmentation is the process of partitioning markets into groups of potential customers with similar needs and characteristics who are likely to exhibit similar purchasing behavior. A prominent example of geographic segmentation is the Potential Rating Index by Zip Market (PRIZM) clusters created by Nielsen Claritas. As Nielsen Claritas states, "birds of a feather flock together"—in other words, a person is likely to live near people who are similar.

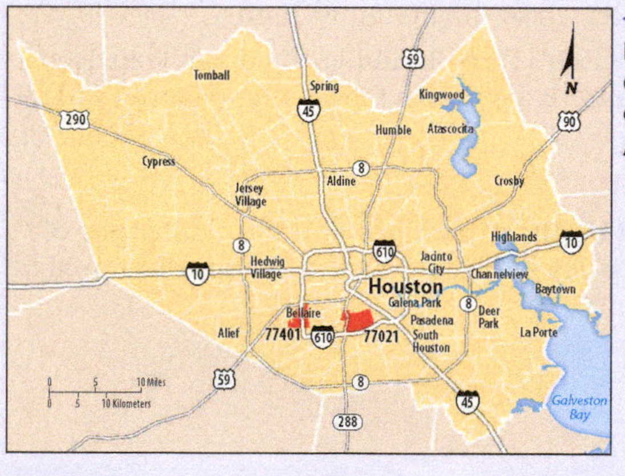

◀ **FIGURE 13-62 HARRIS COUNTY ZIP CODES 77401 AND 77021.**

Nielsen Claritas combines two types of geographic information: distribution of the social and economic characteristics of people obtained from the census and the addresses of purchasers of various products obtained from service providers. The variables are organized into 66 clusters that are given picturesque names. For each zip code in the United States, Nielsen Claritas determines the five clusters that are most prevalent (in alphabetical order). Nielsen Claritas calls this analysis "you are where you live."

For example, the five most common clusters in the suburb of Bellaire (zip code 77401) west of downtown Houston are (Figure 13-62):

- **American Dreams:** Upper-middle-class ethnically diverse families
- **Bohemian Mix:** Upper-middle-class young singles and couples
- **Money & Brains:** Wealthy older highly educated families
- **The Cosmopolitans:** Upper-middle-class older families with children
- **Young Digerati:** Wealthy younger highly educated individuals

What's Your Urban Geography?

You Are Where You Live

Nielsen Claritas will let you check out the PRIZM cluster for any zip code in the United States.

1. Compare the PRIZM cluster for zip code 77401 with that of 77021. Referring to Figures 13-15, 13-16, 13-17, and 13-57, describe similarities and differences in the rings, sectors, and nodes that they occupy within the Houston urban area.
2. The five most common clusters in zip code 77021 are (in alphabetical order):
   - **American Classics:** Older low-income long-time residents
   - **Low-Rise Living:** The lowest income of any PRIZM cluster, including many ethnically diverse single parents
   - **New Beginnings:** Ethnically diverse young adults
   - **Old Glories:** Widows and widowers living on fixed pensions
   - **Young Influentials:** Yuppie apartment-dwelling singles and couples

   Do these clusters appear to be consistent with the patterns shown in Figures 13-15, 13-16, and 13-17? Why or why not?
3. Find the PRIZM cluster for the zip code where you live. Use your search engine to find Nielsen Claritas Prizm website and select ZIP Code look-up. Enter your 5-digit zip code and the security code and click Submit.
   a. What are the five clusters for your zip code?
   b. Click on the names of the clusters to learn more about each of them. Do these five clusters accurately represent your expectations for your zip code? Why or why not?

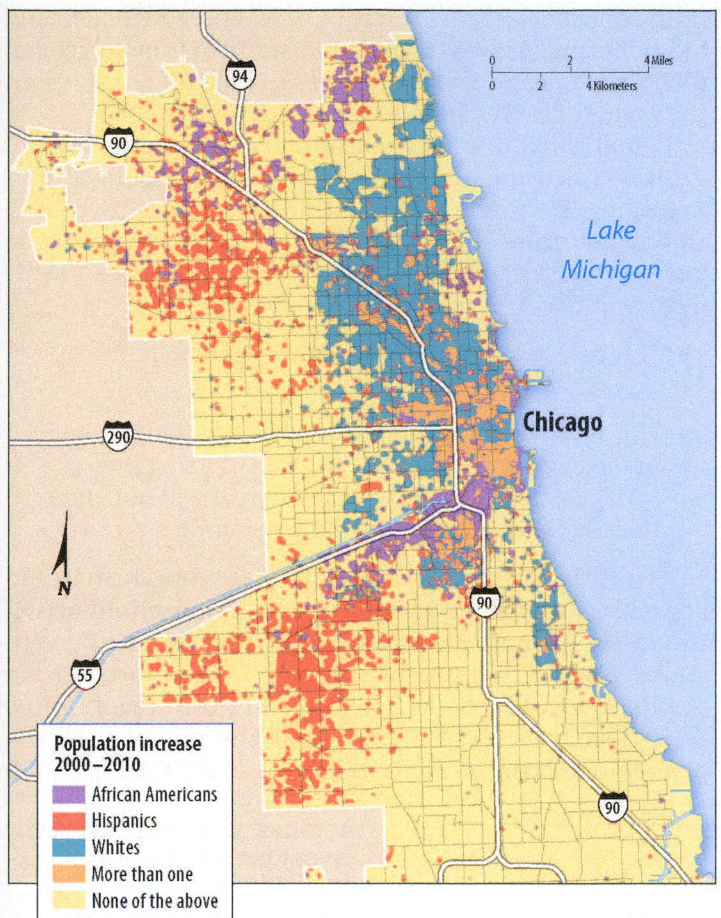

▲ **FIGURE 13-63 RACIAL CHANGE IN CHICAGO** African Americans and Hispanics are moving toward the suburbs, whereas whites are moving back into the city, specifically the North Side, where attractive neighborhoods are clustered.

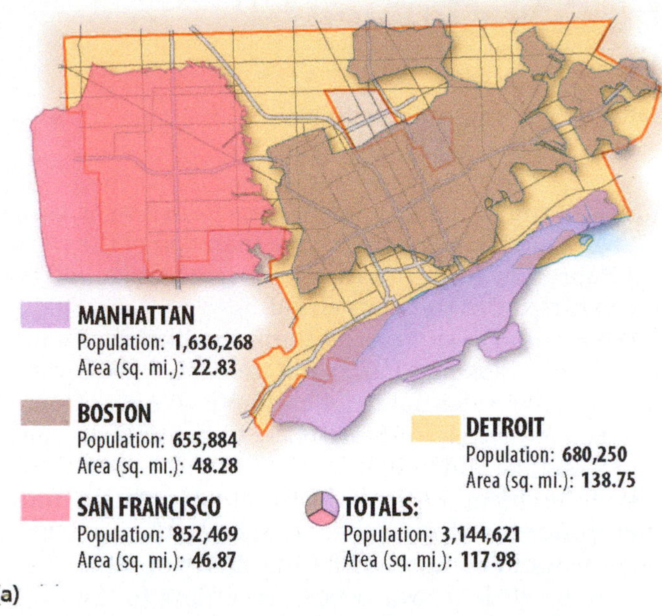

(a)

(b)

▶ **FIGURE 13-64 DETROIT COMPARED TO OTHER CITIES** **(a)** The land area of the city of Detroit is greater than Manhattan, Boston, and San Francisco combined. Its population is less than one-fourth the size of the other three combined. **(b)** Vacant houses or empty lots cover 43 percent of Detroit's land area.

DEBATE IT! Can a declining city be stronger by shrinking?

Detroit has shrunk from 2 million people in 1950 to less than 700,000 now. According to Motor City Mapping, 30 percent of the city's parcels of land are vacant lots, 13 percent have vacant structures, and only 54 percent have occupied buildings. City officials are debating whether to try to help all neighborhoods or concentrate on a handful of them.

HELP ONLY SOME NEIGHBORHOODS

- Detroit doesn't have enough money to improve every neighborhood, so it should pick those with the brightest prospects for improvement.
- The brightest prospects are in neighborhoods with smaller number of vacancies.
- Better police, fire, and garbage collection services can be provided for the remaining neighborhoods.

HELP ALL NEIGHBORHOODS

- The people left behind in the most blighted neighborhoods are too poor to move.
- People would be forced to move against their wishes.
- Vacant houses could be turned over to homeless people who need a place to live.
- Even the most blighted areas need services.

The City Cleaned

LEARNING OUTCOME 13.4.4

Strategies for improving city air.

Where will you be living 10 or 20 years from now? Chances are it will be in an urban area. More than half of the world's inhabitants now live in cities, and the percentage will continue to increase in the years ahead.

What will the city of the future be like? In some respects, it may look familiar. It will probably have commercial areas and residential neighborhoods. People of similar interests and backgrounds are likely to continue to cluster in rings, sectors, and nodes.

Will the city of the future be sustainable? **Sustainable development** is "development that meets the needs of the present without compromising the ability of future generations to meet their own needs," according to the United Nations. The future sustainability of cities—and therefore of Earth—depends largely on how we structure our future transport.

The average American spends more than 600 hours per year driving. That's nearly one month of the year. Our future transport will definitely be more energy-efficient and less polluting. As we conclude our study of human geography, we focus on two key changes in our future transport: the source of power and the control of the vehicle.

What will it take for the world to reduce pollution and fossil fuel dependency in the years ahead (Figure 13-65)? According to the United Nations, strategies will vary among countries (Figure 13-66). The U.S. scientists working with the U.N. offered a strategy with three key elements (Figure 13-67):

- Sharp decrease in the use of the three fossil fuels.
- Increase in the use of renewable energy.
- Use of **carbon capture and storage (CCS),** which involves capturing waste CO_2, transporting it to a storage site, and depositing it where it will not enter the atmosphere, normally underground.

The principal impact on the average American would be reliance on electricity for nearly all household activities and transportation. This electricity would be generated almost exclusively through sources other than the three fossil fuels. As hard as it will be for the United States to reduce its carbon footprint, the challenge is even greater for developing countries, especially China, which

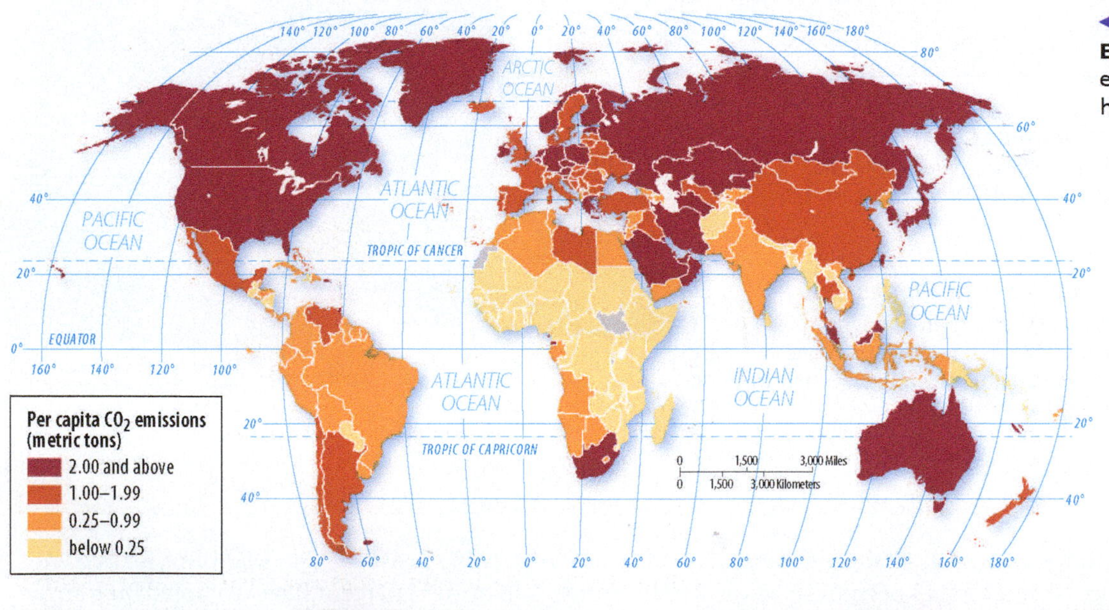

FIGURE 13-65 PER CAPITA CO_2 EMISSIONS Developed countries, especially in North America, have high per capita CO_2 emissions.

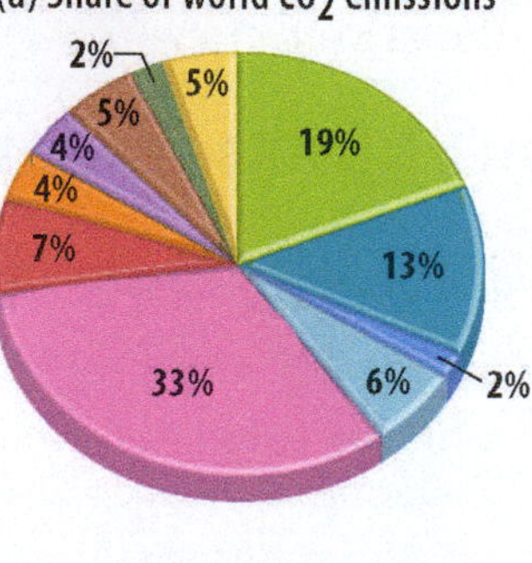

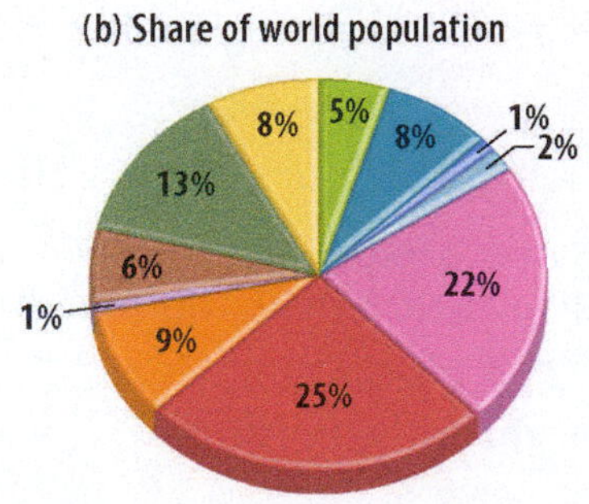

FIGURE 13-66 SHARE OF EMISSIONS AND POPULATION (a) East Asia, especially China, is responsible for the largest share of the world's CO_2 emissions. **(b)** East Asia and North America generate higher percentages of CO_2 emissions than their share of the world's population.

SUSTAINABILITY & OUR ENVIRONMENT Regional Variations in Electricity

Electric-powered vehicles require recharging by being plugged into a source of electricity such as an outlet in the garage. The source of that electricity may or may not be sustainable. Though fossil fuel is not being pumped directly into the tank of the electric-powered vehicle, fossil fuel may be consumed to generate the electricity at the power plant. In fact, the United States as a whole generates around 39 percent of its electricity from coal-burning power plants and around 27 percent from natural gas.

An electric vehicle does reduce consumption of an increasingly scarce and expensive resource—petroleum. But if the electricity is generated by natural gas, then plugging a vehicle into the electric grid may conserve petroleum at the expense of more rapid depletion of natural gas. If electricity is generated by coal, a plug-in may cause more air pollution.

Electricity is generated differently across the 50 U.S. states. In the Pacific Northwest, where hydroelectric is the leading source of electricity, recharging electric vehicles will have much less impact on air quality than will be the case in the Midwest (Figure 13-68). States that depend on farm production may benefit from increased use of ethanol. Thus, the "greenest" alternative varies by location.

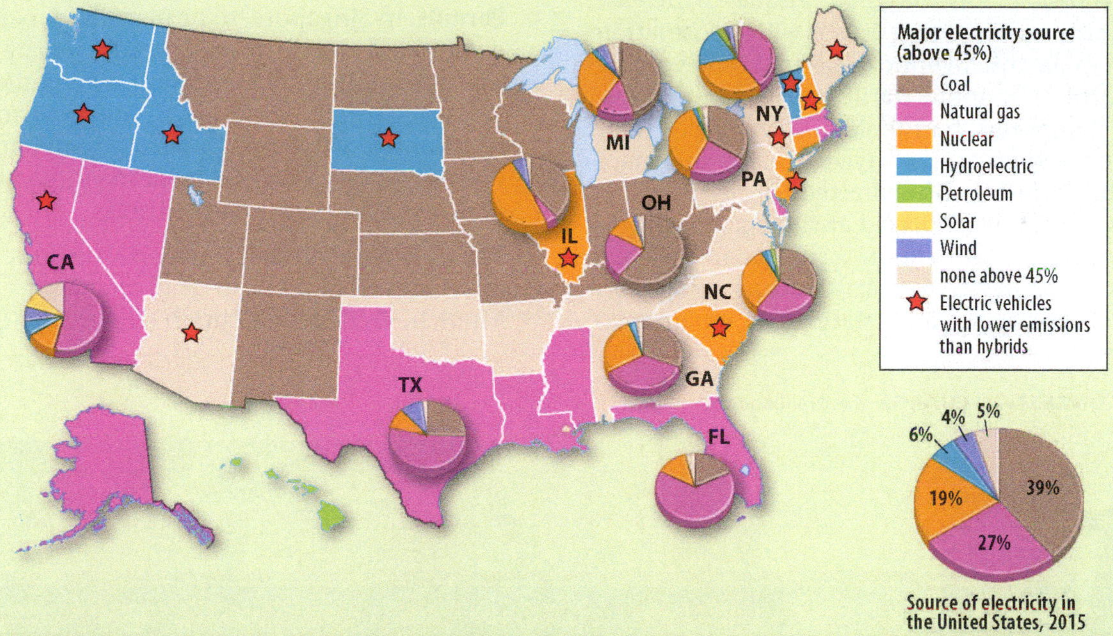

▲ **FIGURE 13-68 ELECTRICITY BY U.S. STATE** The mix of fuels running power plants varies widely among the 50 states. Electric cars generate lower CO_2 emissions than hybrids in some states and higher emissions in others, depending on the source of electricity.

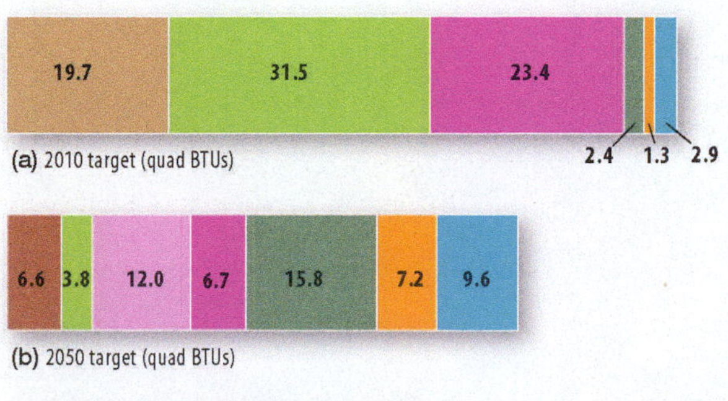

is now the world's leading manufacturing country. International cooperation and coordination will be required to reduce global pollution.

PAUSE & REFLECT 13.4.4

What is happening in or near your community to reduce generation of electricity through fossil fuels?

◀ **FIGURE 13-67 U.N. TARGET FOR CO_2 EMISSIONS REDUCTION IN THE UNITED STATES** The principal source of reduction would be through lower use of fossil fuels.

The City Controlled

LEARNING OUTCOME 13.4.5

Consider alternative future vehicles for the city.

The future health of urban areas depends on relieving traffic congestion. The average American spends around 42 hours a year stuck in traffic jams (but 80 hours in Los Angeles). Geographic tools, including the Global Positioning System (GPS) and electronic mapping, are playing central roles in the design of intelligent transportation systems, either through increasing road capacity or through reducing demand.

CONTROLLING VEHICLES

The current generation of innovative techniques to increase road capacity is aimed at providing drivers with information so that they can make intelligent decisions about avoiding congestion. Information about traffic congestion is transmitted through computers, handheld devices, and vehicle monitors. Traffic hot spots are displayed on electronic maps and images, using information collected through sensors in the roadbeds and cameras placed at strategic locations. An individual wishing to know about a particular route can program an electronic device to receive a congestion alert and to suggest alternatives. Radio stations in urban areas broadcast reports to advise motorists of accidents or especially congested highways.

Demand to use congested roads is being reduced in a number of ways:

- **Congestion charges.** In London, motorists must pay a congestion charge of up to £12 ($18) to drive into the central area between 7 a.m. and 6 p.m. Monday through Friday (Figure 13-69). A similar system exists in Stockholm, where the charge varies depending on the time of day.
- **Tolls.** In Toronto and several California cities, motorists are charged higher tolls to drive on freeways during congested times than at other times. A transponder attached to a vehicle records the time of day it is on the highway. A monthly bill sent to the vehicle's owner reflects the differential tolls.
- **Permits.** In Singapore, to be permitted to drive downtown during the busiest times of the day, a motorist must buy a license and demonstrate ownership of a parking space. The government limits the number of licenses and charges high tolls to drive downtown. Several cities in China intend to require permits to drive in congested areas.
- **Bans.** Cars are banned from portions of the central areas of a number of European cities, including Copenhagen, Munich, Vienna, and Zurich (Figure 13-70).

▼**FIGURE 13-69 CONGESTION CHARGE** Central London.

▲ **FIGURE 13-70 NO-CAR ZONE** Strøget, a main shopping street in Copenhagen, prohibits vehicles and is limited to pedestrians.

ALTERNATIVE FUEL VEHICLES

Consumers in developed countries are reluctant to give up their motor vehicles, and demand for vehicles is soaring in developing countries. One of the greatest challenges to reducing pollution and conserving nonrenewable resources is reliance on petroleum as automotive fuel, so carmakers are scrambling to bring alternative-fuel vehicles to the market. Here are some alternative technologies:

- **Diesel.** Diesel engines burn fuel more efficiently, with greater compression, and at a higher temperature than conventional gas engines. Most new vehicles in Europe are diesel powered, where they are valued for zippy acceleration on crowded roads, as well as for high fuel efficiency. However, diesels have been found to generate high levels of nitrogen oxides. Diesels have made limited inroads in the United States, where they were identified with ponderous heavy trucks, poorly performing versions in the 1980s, and generation of more pollutants. Biodiesel fuel mixes petroleum diesel with biodiesel (typically 5 percent), which is produced from vegetable oils or recycled restaurant grease.
- **Hybrid.** Sales of hybrids increased rapidly during the first decade of the twenty-first century, led by Toyota's success with the hybrid Prius. A gasoline engine powers the vehicle at high speeds, and at low speeds, when the gas engine is at its least efficient, an electric motor takes over. Energy that would otherwise be wasted in coasting and braking is also captured as electricity and stored until needed.
- **Ethanol.** Ethanol is fuel made by distilling crops such as sugarcane, corn, and soybeans. Sugarcane is distilled for fuel in Brazil, where most vehicles run on ethanol. In the United States, corn has been the principal crop for ethanol, but this has proved controversial because the amount of fossil fuels needed to grow and distill the corn is comparable to—and possibly greater than—the amount saved in vehicle fuels. Furthermore, growing corn for ethanol diverts corn from the food chain, thereby allegedly causing higher food prices in the United States and globally. More promising is ethanol distilled from cellulosic biomass, such as trees, grasses, and algae.
- **Full electric.** A full electric vehicle has no gas engine. When the battery is discharged, the vehicle will not run until the battery is recharged by plugging it into an outlet. Motorists can make trips in a local area and recharge the battery at night. Out-of-town trips are difficult because recharging opportunities are scarce. In large cities, a number of downtown garages and shopping malls have recharging stations, but few exist in rural areas.
- **Plug-in hybrid.** In a plug-in hybrid, the battery supplies the power at all speeds. It can be recharged in one of two ways: While the car is moving, the battery can be recharged by a gas engine or, when it is parked, the car can be recharged by plugging into an electrical outlet. The principal limitation of a full electric vehicle has been the short range of the battery before it needs recharging. Using a gas engine to recharge the battery extends the range of the plug-in hybrid to that of a conventional gas engine.
- **Hydrogen fuel cell.** Hydrogen forced through a PEM (polymer electrolyte membrane or proton exchange membrane) can be combined with oxygen from the air to produce an electric charge. The electricity can then be used to power an electric motor. Fuel cells are now widely used in small vehicles such as forklifts. Fuel cell vehicles are being used in a handful of large East Coast and West Coast cities, where hydrogen fueling stations have been constructed.

PAUSE & REFLECT 13.4.5

What strategies are being used at your university or in your school district to promote alternatives to cars and trucks?

CHECK-IN KEY ISSUE 4

Why Do Cities Face Sustainability Challenges?

- ✔ **Cities have large numbers of underclass people who live in a culture of poverty.**
- ✔ **Many cities have areas of gentrification and regeneration.**
- ✔ **Future cities are likely to be dependent on alternative forms of transport.**

Summary & Review

KEY ISSUE 1

Why are downtowns distinctive?

An urban area consists of a central city and its surrounding built-up suburbs. The CBD contains a large percentage of an urban area's public and business services. Some consumer services, especially leisure, are also in the CBD.

THINKING GEOGRAPHICALLY

1. Some professional sports arenas and stadiums are located in the CBD, and some are located in suburbs. What are the advantages and drawbacks for the fans of each location?

▲ **FIGURE 13-71 DOWNTOWN STADIUM** The San Francisco Giants play in AT&T Park, adjacent to downtown San Francisco.

KEY ISSUE 2

Where are people distributed in urban areas?

Three models help to explain where different groups of people live in urban areas. According to the concentric zone model, a city grows outward in rings. According to the sector model, a city grows along transportation corridors. According to the multiple nuclei model, a city grows around several nodes. The models can be used to describe where people of varying characteristics tend to cluster in an urban area.

THINKING GEOGRAPHICALLY

2. Identify the ring, sector, and node in which you (or a friend or relation) live within an urban area. Do conditions in your place fit the overall patterns expected of the three models? Why or why not?

▲ **FIGURE 13-72 SUN CITY, ARIZONA** The city's street pattern consists of a series of concentric circles.

KEY ISSUE 3

Why do urban areas expand?

Urban growth has been primarily focused on suburbs that surround older cities. In the past, cities expanded their land area to encompass outlying areas, but now they are surrounded by independent suburban jurisdictions. Public transport, such as subways and buses, are more suited than private cars to moving large numbers of people into and out of the CBD, but private motor vehicles dominate urban transportation, especially in the United States.

THINKING GEOGRAPHICALLY

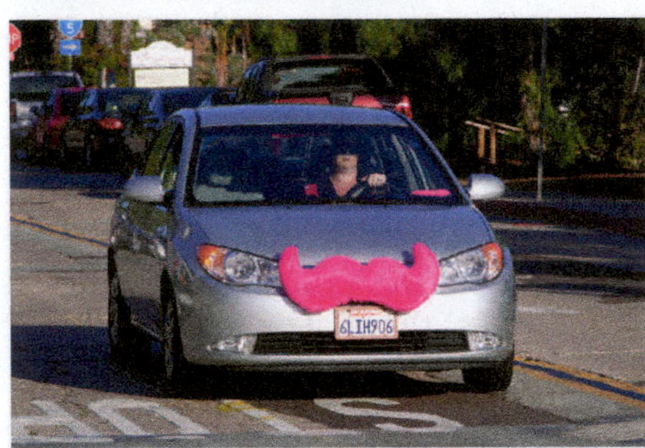

3. What impact might a car-sharing service such as Uber or Lyft have on patterns of urban transport?

▲ **FIGURE 13-73 LYFT CAR-SHARING SERVICE** Lyft drivers identify their cars by attaching a jumbo pink mustache to the front.

KEY ISSUE 4

Why do cities face sustainability challenges?

Cities face physical, social, and economic difficulties, but some improvements have also occurred. The older housing in the inner city can deteriorate through processes of filtering and redlining. Some cities have experienced gentrification, in which higher-income people move in and renovate previously deteriorated neighborhoods.

THINKING GEOGRAPHICALLY

4. What are the impacts of gentrification on low-income inner-city residents? What are some of the benefits and challenges of providing housing for low-income residents in a gentrifying neighborhood?

▲ **FIGURE 13-74 GENTRIFICATION AND AFFORDABLE HOUSING** This building, located in a gentrifying neighborhood of Cincinnati, houses low-income people who were once homeless.

KEY TERMS

Annexation *(p. 478)* Legally adding land area to a city in the United States.

Carbon capture and storage (CCS) *(p. 494)* The process of capturing waste CO_2, transporting it to a storage site, and depositing it where it will not enter the atmosphere, normally underground.

Census tract *(p. 468)* An area delineated by the U.S. Bureau of the Census for which statistics are published; in urban areas, census tracts correspond roughly to neighborhoods.

Central business district (CBD) *(p. 462)* The area of a city where retail and office activities are clustered.

Central city (city) *(p. 460)* An urban settlement that has been legally incorporated into an independent, self-governing unit known as a municipality.

Combined statistical area (CSA) *(p. 461)* In the United States, two or more contiguous CBSAs tied together by commuting patterns.

Concentric zone model *(p. 466)* A model of the internal structure of cities in which social groups are spatially arranged in a series of rings.

Core-based statistical area (CBSA) *(p. 461)* In the United States, any MSA or μSA.

Density gradient *(p. 480)* The change in density in an urban area from the center to the periphery.

Edge city (p. 467) A large node of office and retail activities on the edge of an urban area.

Filtering *(p. 489)* A process of change in the use of a house, from single-family owner occupancy to abandonment.

Gentrification *(p. 490)* A process of converting an urban neighborhood from a predominantly low-income, renter-occupied area to a predominantly middle-class, owner-occupied area.

Informal settlement *(p. 474)* An area within a city in a less developed country in which people illegally establish residences on land they do not own or rent and erect homemade structures.

Megalopolis *(p. 481)* A continuous urban complex in the northeastern United States.

Metropolitan statistical area (MSA) *(p. 460)* In the United States, an urbanized area of at least 50,000 population, the county within which the city is located, and adjacent counties meeting one of several tests indicating a functional connection to the central city.

Micropolitan statistical area (μSA) *(p. 461)* An urbanized area of between 10,000 and 50,000 inhabitants, the county in which it is located, and adjacent counties tied to the city.

Multiple nuclei model *(p. 467)* A model of the internal structure of cities in which social groups are arranged around a collection of nodes of activities.

Peripheral model *(p. 480)* A model of North American urban areas consisting of an inner city surrounded by large suburban residential and business areas tied together by a beltway or ring road.

Primary census area (PSA) *(p. 461)* In the United States, any CSA, any MSA not included in a CSA, or any μSA not included in a CSA.

Public housing *(p. 490)* Government-owned housing rented to low-income individual, with rents set at 30 percent of the tenant's income.

Redlining *(p. 489)* A process by which financial institutions draw red-colored lines on a map and refuse to lend money for people to purchase or improve property within the lines.

Rush hour *(p. 484)* The four consecutive 15-minute periods in the morning and evening with the heaviest volumes of traffic.

Sector model *(p. 466)* A model of the internal structure of cities in which social groups are arranged around a series of sectors, or wedges, radiating out from the central business district.

Smart growth *(p. 479)* Legislation and regulations to limit suburban sprawl and preserve farmland.

Social area analysis *(p. 468)* Statistical analysis used to identify where people of similar living standards, ethnic background, and lifestyle live within an urban area.

Sprawl *(p. 480)* Development of new housing sites at relatively low density and at locations that are not contiguous to the existing built-up area.

Suburb *(p. 478)* A residential or commercial area situated within an urban area but outside the central city.

Sustainable development *(p. 494)* Development that meets the needs of the present without compromising the ability of future generations to meet their own needs.

Underclass *(p. 488)* A group in society prevented from participating in the material benefits of a more developed society because of a variety of social and economic characteristics.

Urban area *(p. 460)* A central city and its surrounding built-up suburbs.

Urban cluster *(p. 460)* In the United States, an urban area with between 2,500 and 50,000 inhabitants.

Urbanized area *(p. 460)* In the United States, an urban area with at least 50,000 inhabitants.

Zoning ordinance *(p. 482)* A law that limits the permitted uses of land and maximum density of development in a community.

GeoVideo Log in to the **MasteringGeography** Study Area to view this video.

Brasília

A planned city completed in 1960, Brazil's capital, Brasília, provides an opportunity to compare a utopian dream with present-day social reality.

1. What values and aspirations motivated the creation of Brasília and shaped its design?
2. How does the form of Brasília reflect the different functions of a city and national capital?
3. Does Brasília today realize its founders' vision of society? Give examples from the video to support your answer.

EXPLORE

Curitiba's transport and housing

Use Google Earth to explore transport and housing in Curitiba, Brazil.

Fly to *Praca GK Gilbran, Curitiba, Brazil*.

Select *Bus* in the *Transportation* menu.

Select *More* then click on the triangle and the box next to *Transportation* and select *Bus*.

Drag to enter *Street View* to the bus stop on the southeast side of the triangle formed by the praca (park).

1. What is unusual about the bus stops, compared to those in other cities?
2. What type of housing structures surround the praca?

Fly to *240 Rua Brasilio Bontorim, Curitiba, Brazil*.

3. Describe differences in the appearance of this suburb compared to a typical one in the United States.

MasteringGeography™

Looking for additional review and test prep materials?

Visit the Study Area in **MasteringGeography™** to enhance your geographic literacy, spatial reasoning skills, and understanding of this chapter's content by accessing a variety of resources, including MapMaster interactive maps, videos, *In the News* RSS feeds, flashcards, web links, self-study quizzes, and an eText version of *The Cultural Landscape*.

www.masteringgeography.com

AFTERWORD

CAREERS IN GEOGRAPHY

Where do you see yourself working in 5 or 10 years after graduation? You could be analyzing customer behavior for a major retailer. You could be evaluating the fairness of redistricting plans for a citizens' oversight committee. You could be documenting the migration of refugees for an international relief organization. Or you could be assessing sustainable business practices for an environmental action group.

What do all these careers have in common? Each offers creative ways to apply core concepts and skills of geography. An increasing number of students recognize that geographic education is practical as well as stimulating. Employment opportunities are expanding for students trained in geography, especially in geospatial technologies, teaching, government service, business, and nonprofit organizations.

The Association of American Geographers cites three distinctive features of geography that especially favor increased employment opportunities for geographers:

- Increased dependency in our daily lives on geographic technologies, such as GPS in cell phones and cars and instantaneous online mapping in handheld devices
- Increased importance of global-scale commonalities and local-scale diversity in cultural features and natural systems across Earth's surface
- Increased emphasis on interdisciplinary studies in the curriculum of schools and universities

GEOSPATIAL TECHNOLOGIES

In the past 20 years the field of geospatial technologies has been making rapid advances, thanks to developments in GIScience, such as geographic information systems, geotagging, volunteered geographic information, and remote sensing. "Readily available, consistent, accurate, complete and current geographic information and the widespread availability and use of advanced technologies offer great job opportunities for people," states the U.S. Department of Labor's Employment and Training Administration website. The number of geospatial technology jobs is increasing 35 percent per year, according to the Labor Department.

Jobs making use of geospatial technologies and GIS can be found in such private- and public-sector areas as environmental consulting, software development, air navigation services (Figure AF-1), spatial database management for mapping companies, location analysis for retailing and real estate, logistical arrangements for the transport of goods, site selection for utilities, and resource management, to name just a few.

▲ **FIGURE AF-1 AIR NAVIGATION ROUTES** Satellite image of North America at night with major air routes visualized.

Recognizing the increasing importance of GIScience jobs, the U.S. Department of Labor has developed a competency model for what it takes to get a job in the field of geospatial technology. Core competencies include understanding Earth's geometry, using tools such as projections, the Global Navigation Satellite System, remote sensing, cartography, geographic information systems, and geospatial information technology.

TEACHING

Doctoral or master's degrees in geography are offered in more than 100 universities in the United States and Canada. A career as a geography teacher is promising because schools are expanding their geography curriculum (Figure AF-2). Educators increasingly recognize geography's role in teaching students about global diversity. A.P. Human Geography is the fastest-growing Advanced Placement discipline in U.S. high schools.

Some university geography departments have emphasized good teaching over research; others are increasingly concerned with research. The Association of American Geographers includes several dozen specialty groups organized around research themes, including agricultural, industrial, medical, and transportation geography.

GOVERNMENT

Geographers contribute their knowledge of the location of activities, the patterns underlying the distribution of various activities, and the interpretation of data from maps and satellite imagery to local, state, and national governments. Employment opportunities with cities, states, provinces, and other units of local government are typically found in departments of planning, transportation, parks and recreation, economic development, housing, zoning, or other similarly titled government agencies. Geographers may be hired to conduct studies of local economic, social, and physical patterns; to prepare information through maps and reports; and to help plan the community's future.

Many federal government agencies employ geographers:

- The Department of Agriculture hires geographers for the Forest Service and Natural Resources Conservation Service to enhance environmental quality.
- The Department of Commerce hires geographers for the Bureau of the Census to study changing population trends and for the Economic Development Administration to promote rural development.
- The Department of Defense hires geographers for the Defense Intelligence Agency and the National Geospatial-Intelligence Agency to analyze satellite imagery.
- The Department of Energy hires geographers for the Office of Environmental Management to administer environmental protection programs.
- The Department of Homeland Security hires geographers in several of its agencies concerned with transportation and infrastructure security, border protection, and disaster preparation and response.
- The Department of Housing and Urban Development hires geographers to help revitalize American cities.
- The Department of the Interior hires geographers for the U.S. Geological Survey to study land use and create topographic maps, for the Office of Environmental Policy and Compliance to administer environmental protection programs, and for the National Park Service to provide technical services.
- The Department of State hires geographers for foreign service.
- The Department of Transportation hires geographers to plan new transportation projects.

BUSINESS

An increasing number of American geographers are finding jobs with private companies. The list of possibilities is long, but here are some common examples:

- Real estate developers hire geographers to find the best locations for new shopping centers.

▲ **FIGURE AF-2 TEACHING GEOGRAPHY**

- Real estate firms hire geographers to assess the value of properties.
- Supermarket chains, department stores, and other retailers hire geographers to determine the potential markets for new stores.
- Banks hire geographers to assess the probability that a loan applicant has planned a successful development.
- Distributors and wholesalers hire geographers to find ways to minimize transportation costs.
- Transnational corporations hire geographers to predict the behavior of consumers and officials in other countries.
- Manufacturers hire geographers to identify new sources of raw materials and markets.
- Utility companies hire geographers to determine future demand at different locations for gas, electricity, and other services.

For more information on careers in geography, consult the Association of American Geographers website, at aag.org, or the National Council for Geographic Education, at ncge.org.

NONPROFITS

More than 1 million nonprofit organizations are registered in the United States to advocate for a wide variety of causes and values. Working for a nonprofit organization is a way to make a difference in the world and to change the world in a positive way. Geographers are especially attracted to nonprofits that advocate for environmental protection, sustainable resource management, and regeneration of communities (Figure AF-3).

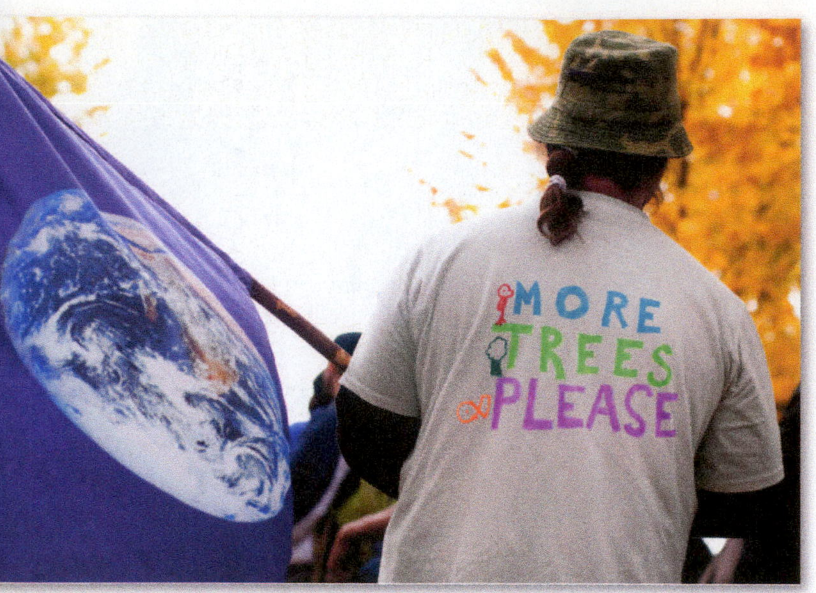

▲ **FIGURE AF-3 ENVIRONMENTAL ACTIVISTS**

CITIZEN GEOGRAPHERS

Regardless of your career, you can practice geography in your daily life. At the very least, you can read—and more importantly—interpret a map on paper or on your electronic device. The features in each chapter of this book—*Doing Geography, What's My Geography?, Sustainability & Our Environment,* and *Debate It*—are all designed to relate geography to choices we make in our daily lives.

Each of us is a citizen geographer living in a community and making personal choices. We can choose where we obtain our food and our clothing, and the amount of resources we consume and throw away.

As citizen geographers living in a country, we can make informed choices about policies and priorities. We see sharp differences in HDI and GII, fences between countries, and gerrymandered districts.

Informed citizen geographers can debate policies for declining cities, for regulating migration, for using languages, and for wearing clothing. We citizen geographers can debate whether Malthus was right or wrong, whether GPS tracking devices should be turned on or off, and whether countries should intervene in ethnic conflicts.

We citizen geographers can all work towards conserving a sustainable Planet Earth.

APPENDIX

MAP SCALE AND PROJECTIONS | PHILLIP C. MUERCKE

Unaided, our human senses provide a limited view of our surroundings. To overcome those limitations, humankind has developed powerful vehicles of thought and communication, such as language, mathematics, and graphics. Each of those tools is based on elaborate rules; each has an information bias, and each may distort its message, often in subtle ways. Consequently, to use those aids effectively, we must understand their rules, biases, and distortions. The same is true for the special form of graphics we call maps: we must master the logic behind the mapping process before we can use maps effectively.

A fundamental issue in cartography, the science and art of making maps, is the vast difference between the size and geometry of what is being mapped—the real world, we will call it—and that of the map itself. Scale and projection are the basic cartographic concepts that help us understand that difference and its effects.

Map Scale

Our senses are dwarfed by the immensity of our planet; we can sense directly only our local surroundings. Thus, we cannot possibly look at our whole state or country at one time, even though we may be able to see the entire street where we live. Cartography helps us expand what we can see at one time by letting us view the scene from some distant vantage point. The greater the imaginary distance between that position and the object of our observation, the larger the area the map can cover, but the smaller the features will appear on the map. That reduction is defined by the *map scale*, the ratio of the distance on the map to the distance on Earth. Map users need to know about map scale for two reasons: so that they can convert measurements on a map into meaningful real-world measures and so that they can know how abstract the cartographic representation is.

REAL-WORLD MEASURES. A map can provide a useful substitute for the real world for many analytical purposes. With the scale of a map, for instance, we can compute the actual size of its features (length, area, and volume). Such calculations are helped by three expressions of a map scale: a word statement, a graphic scale, and a representative fraction.

A *word statement* of a map scale compares X units on the map to Y units on Earth, often abbreviated "X unit to Y units." For example, the expression "1 inch to 10 miles" means that 1 inch on the map represents 10 miles on Earth (Figure A-1). Because the map is always smaller than the area that has been mapped, the ground unit is always the larger number. Both units are expressed in meaningful terms, such as inches or centimeters and miles or kilometers. Word statements are not intended for precise calculations but give the map user a rough idea of size and distance.

A *graphic scale,* such as a bar graph, is concrete and therefore overcomes the need to visualize inches and miles that is associated with a word statement of scale (see Figure A-1). A graphic scale permits direct visual comparison of feature sizes and the distances between features. No ruler is required; any measuring aid will do. It needs only be compared with the scaled bar; if the length of 1 toothpick is equal to 2 miles on the ground and the map distance equals the length of 4 toothpicks, then the ground distance is 4 times 2, or 8 miles. Graphic scales are especially convenient in this age of copying machines, when we are more likely to be working with a copy than with the original map. If a map is reduced or enlarged as it is copied, the graphic scale will change in proportion to the change in the size of the map and thus will remain accurate.

The third form of a map scale is the *representative fraction* (RF). An RF defines the ratio between the distance on the map and the distance on Earth in fractional terms, such as 1/633,600 (also written 1:633,600). The numerator of the fraction always refers to the distance on the map, and the denominator always refers to the distance on Earth. No units of measurement are given, but both numbers must be expressed in the same units. Because map distances are extremely small relative to the size of Earth, it makes sense to use small units, such as inches or centimeters. Thus the RF 1:633,600 might be read as "1 inch on the map to 633,600 inches on Earth."

Herein lies a problem with the RF. Meaningful map-distance units imply a denominator so large that it is impossible to visualize. Thus, in practice, reading the map scale involves an additional step of converting the denominator to a meaningful ground measure, such as miles or kilometers. The unwieldy 633,600 becomes the more manageable 10 miles when divided by the number of inches in a mile (63,360).

On the plus side, the RF is good for calculations. In particular, the ground distance between points can be easily determined from a map with an RF. One simply multiplies the distance between the points on the map by the denominator of the RF. Thus a distance of 5 inches on a map with an RF of 1/126,720 would signify a ground distance of

5 X 126,720, which equals 633,600. Because all units are inches and there are 63,360 inches in a mile, the ground distance is 633,600/ 63,360, or 10 miles. Computation of area is equally straightforward with an RF. Computer manipulation and analysis of maps is based on the RF form of map scale.

GUIDES TO GENERALIZATION. Scales also help map users visualize the nature of the symbolic relation between the map and the real world. It is convenient here to think of maps as falling into three broad scale categories (Figure A-2). (Do not be confused by the use of the words large AND small in this context; just remember that the larger the denominator, the smaller the scale ratio and the larger the area that is shown on the map.) Scale ratios greater than 1:100,000, such as the 1:24,000 scale of U.S. Geological Survey topographic quadrangles, are large-scale maps. Although those maps can cover only a local area, they can be drawn to rather rigid standards of accuracy. Thus they are useful for a wide range of applications that require detailed and accurate maps, including zoning, navigation, and construction.

At the other extreme are maps with scale ratios of less than 1:1,000,000, such as maps of the world that are found in atlases. Those are small-scale maps. Because they cover large areas, the symbols on them must be highly abstract. They are therefore best suited to general reference or planning, when detail is not important. Medium- or intermediate-scale maps have scales between 1:100,000 and 1:1,000,000. They are good for regional reference and planning purposes.

Another important aspect of map scale is to give us some notion of geometric accuracy; the greater the expanse of the real world shown on a map, the less accurate the geometry of that map is. Figure A-3 shows why. If a curve is represented by straight line segments, short segments (*X*) are more similar to the curve than are long segments (*Y*). Similarly, if a plane is placed in contact with a sphere, the difference between the two surfaces is slight where they touch (*A*) but grows rapidly with increasing distance from the point of contact (*B*). In view of the large diameter and slight local curvature of Earth, distances will be well represented on large-scale maps (those with small denominators) but will be increasingly poorly represented at smaller scales. This close relationship between map scale and map geometry brings us to the topic of map projections.

▲ **FIGURE A-1** Common expressions of map scale.

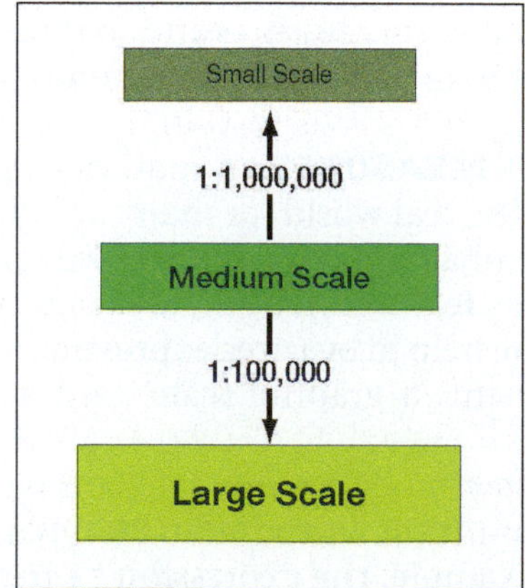

▲ **FIGURE A-2** The scale gradient can be divided into three broad categories.

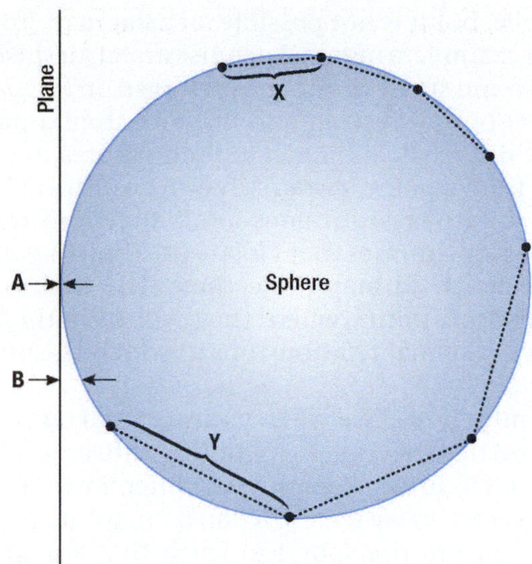

▲ FIGURE A-3 Relationships between surfaces on the round Earth and a flat map.

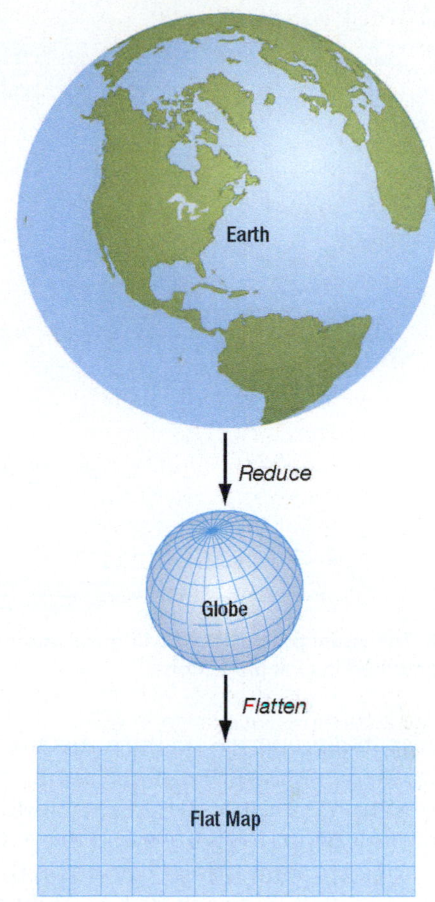

▲ FIGURE A-4 The two-step process of creating a projection.

Map Projections

The spherical surface of Earth is shown on flat maps by means of map projections. The process of "flattening" Earth is essentially a problem in geometry that has captured the attention of the best mathematical minds for centuries. Yet no one has ever found a perfect solution; there is no known way to avoid spatial distortion of one kind or another. Many map projections have been devised, but only a few have become standard. Because a single flat map cannot preserve all aspects of Earth's surface geometry, a mapmaker must be careful to match the projection with the task at hand. To map something that involves distance, for example, a projection should be used in which distance is not distorted. In addition, a map user should be able to recognize which aspects of a map's geometry are accurate and which are distortions caused by a particular projection process. Fortunately, that objective is not too difficult to achieve.

It is helpful to think of the creation of a projection as a twostep process (Figure A-4). First, the immense Earth is reduced to a small globe with a scale equal to that of the desired flat map. All spatial properties on the globe are true to those on Earth. Second, the globe is flattened. Since that cannot be done without distortion, it is accomplished in such a way that the resulting map exhibits certain desirable spatial properties.

PERSPECTIVE MODELS. Early map projections were sometimes created with the aid of perspective methods, but that has changed. In the modern electronic age, projections are normally developed by strictly mathematical means and are plotted out or displayed on computer-driven graphics devices. The concept of perspective is still useful in visualizing what map projections do, however. Thus projection methods are often illustrated by using strategically located light sources to cast shadows on a projection surface from a latitude/longitude net inscribed on a transparent globe.

The success of the perspective approach depends on finding a projection surface that is flat or that can be flattened without distortion. The cone, cylinder, and plane possess those attributes and serve as models for three general classes of map projections: *conic, cylindrical,* and *planar* (or azimuthal). Figure A-5 shows those three classes, as well as a fourth, a false cylindrical class with an oval shape. Although the oval class is not of perspective origin, it appears to combine properties of the cylindrical and planar classes (Figure A-6).

The relationship between the projection surface and the model at the point or line of contact is critical because

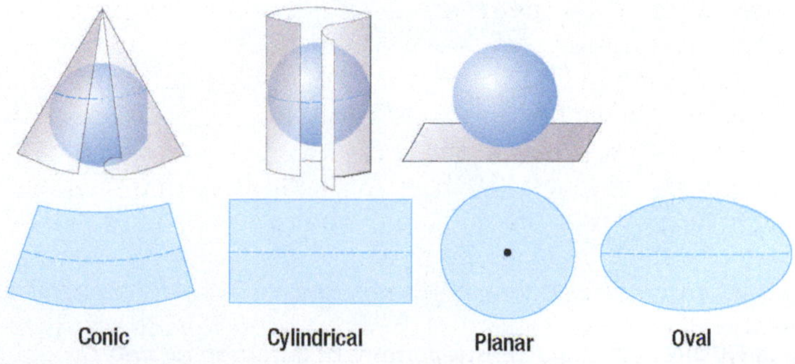

▲ FIGURE A-5 General classes of map projections.

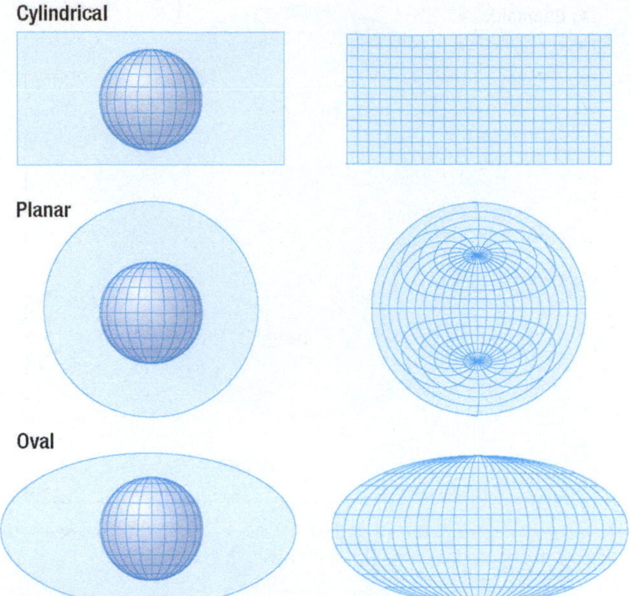

▲ **FIGURE A-6** The visual properties of cylindrical and planar projections combined in oval projections.

distortion of spatial properties on the projection is symmetrical about, and increases with distance from, that point or line. That condition is illustrated for the cylindrical and planar classes of projections in Figure A-7. If the point or line of contact is changed to some other position on the globe, the distortion pattern will be recentered on the new position but will retain the same symmetrical form. Thus centering a projection on the area of interest on Earth's surface can minimize the effects of projection distortion. And recognizing the general projection shape, associating it with a perspective model, and recalling the characteristic distortion pattern will provide the information necessary to compensate for projection distortion.

PRESERVED PROPERTIES. For a map projection to truthfully depict the geometry of Earth's surface, it would have to preserve the spatial attributes of *distance, direction, area, shape,* and *proximity*. That task can be readily accomplished on a globe, but it is not possible on a flat map. To preserve area, for example, a mapmaker must stretch or shear shapes; thus area and shape cannot be preserved on the same map. To depict both direction and distance from a point, area must be distorted. Similarly, to preserve area as well as direction from a point, distance has to be distorted. Because Earth's surface is continuous in all directions from every point, discontinuities that violate proximity relationships must occur on all map projections. The trick is to place those discontinuities where they will have the least impact on the spatial relationships in which the map user is interested.

We must be careful when we use spatial terms, because the properties they refer to can be confusing. The geometry of the familiar plane is very different from that of a sphere; yet when we refer to a flap map, we are in fact making reference to the spherical Earth that was mapped. A shape-preserving projection, for example, is truthful to local shapes—such as the rightangle crossing of latitude and longitude lines—but does not preserve shapes at continental or global levels. A distance-preserving projection can preserve that property from one point on the map in all directions or from a number of points in several directions, but distance cannot be preserved in the general sense that area can be preserved. Direction can also be generally preserved from a single point or in several directions from a number of points but not from all points simultaneously. Thus a shape-, distance-, or direction-preserving projection is truthful to those properties only in part.

Partial truths are not the only consequence of transforming a sphere into a flat surface. Some projections exploit that transformation by expressing traits that are of considerable value for specific applications. One of those is the famous shape-preserving *Mercator projection* (Figure A-8). That cylindrical projection was derived

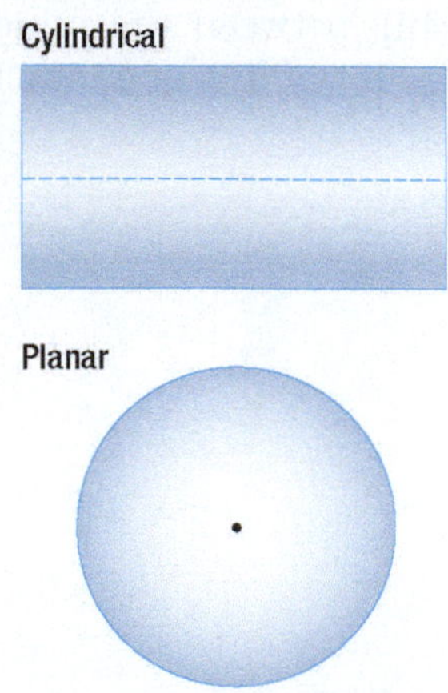

▲ **FIGURE A-7** Characteristic patterns of distortion for two projection classes. Here, darker shading implies greater distortion.

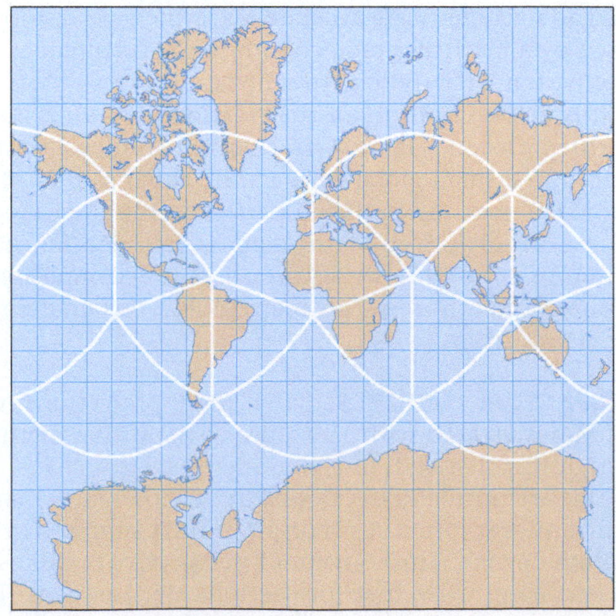

▲ **FIGURE A-8** The useful Mercator projection, showing extreme area distortion in the higher latitudes.

mathematically in the 1500s so that a compass bearing (called rhumb lines) between any two points on Earth would plot as straight lines on the map. That trait let navigators plan, plot, and follow courses between origin and destination, but it was achieved at the expense of extreme areal distortion toward the margins of the projection (see Antarctica in Figure A-8). Although the Mercator projection is admirably suited for its intended purpose, its widespread but inappropriate use for nonnavigational purposes has drawn a great deal of criticism.

The *gnomonic projection* is also useful for navigation. It is a planar projection with the valuable characteristic of showing the shortest (or great circle) route between any two points on Earth as straight lines. Long-distance navigators first plot the great circle course between origin and destination on a gnomonic projection (Figure A-9A). Next they transfer the straight line to a Mercator projection, where it normally appears as a curve (Figure A-9B). Finally, using straight-line segments, they construct an approximation of that course on the Mercator projection. Navigating the shortest course between origin and destination then involves following the straight segments of the course and making directional corrections between segments. Like the Mercator projection, the specialized gnomonic projection distorts other spatial properties so severely that it should not be used for any purpose other than navigation or communications.

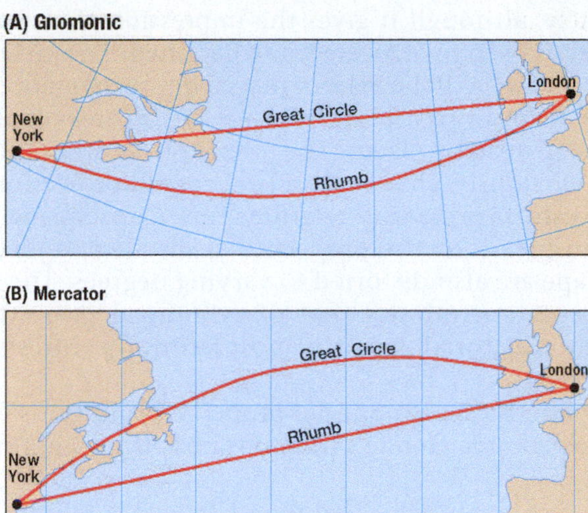

▲ **FIGURE A-9** **(a)** A gnomonic projection and **(b)** a Mercator projection, both of value to long-distance navigators.

PROJECTIONS USED IN TEXTBOOKS. Although a map projection cannot be free of distortion, it can represent one or several spatial properties of Earth's surface accurately if other properties are sacrificed. The two projections used for world maps throughout this textbook illustrate that point well. *Goode's homolosine projection*, shown in Figure A-10, belongs to the oval category and shows area

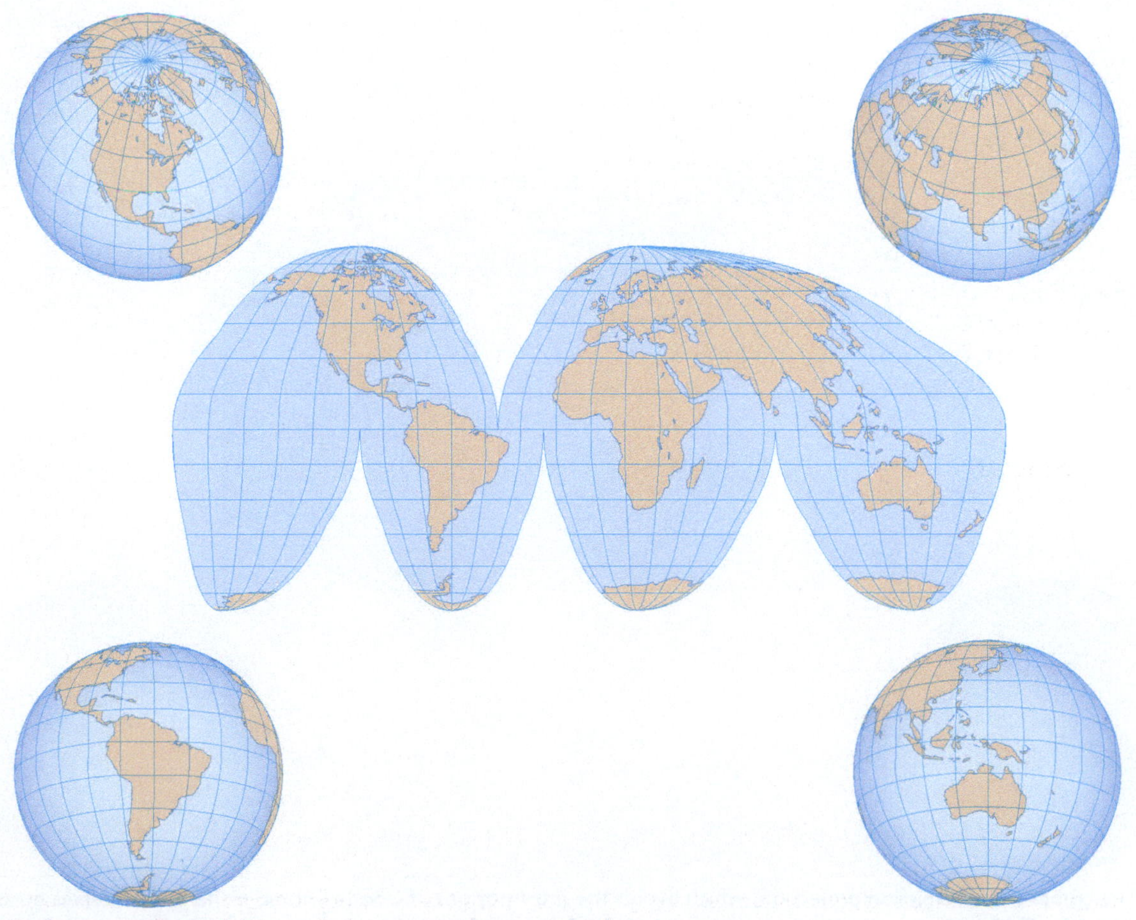

▲ **FIGURE A-10** An interrupted Goode's homolosine, an equal-area projection.

accurately, although it gives the impression that Earth's surface has been torn, peeled, and flattened. The interruptions in Figure A-10 have been placed in the major oceans, giving continuity to the land masses. Ocean areas could be featured instead by placing the interruptions in the continents. Obviously, that type of interrupted projection severely distorts proximity relationships. Consequently, in different locations the properties of distance, direction, and shape are also distorted to varying degrees. The -distortion pattern mimics that of cylindrical projections, with the equatorial zone the most faithfully represented (Figure A-11).

An alternative to special-property projections such as the equal-area Goode's homolosine is the compromise projection. In that case no special property is achieved at the expense of others, and distortion is rather evenly distributed among the various properties, instead of being focused on one or several properties. The *Robinson projection*, which is also used in this textbook, falls into that category (Figure A-12). Its oval projection has a global feel, somewhat like that of Goode's homolosine. But the Robinson projection shows the North Pole and the South Pole as lines that are slightly more than half the length of the equator, thus exaggerating distances and areas near the poles. Areas look larger than they really are in the high latitudes (near the poles) and smaller than they really are in the low latitudes (near the equator). In addition, not all latitude and longitude lines intersect at right angles, as they do on Earth so we know that the Robinson projection does not preserve direction or shape either. However, it has fewer interruptions than the Goode's homolosine does, so it preserves proximity better. Overall, the Robinson projection does a good job of representing spatial relationships, especially in the low to middle latitudes and along the central meridian.

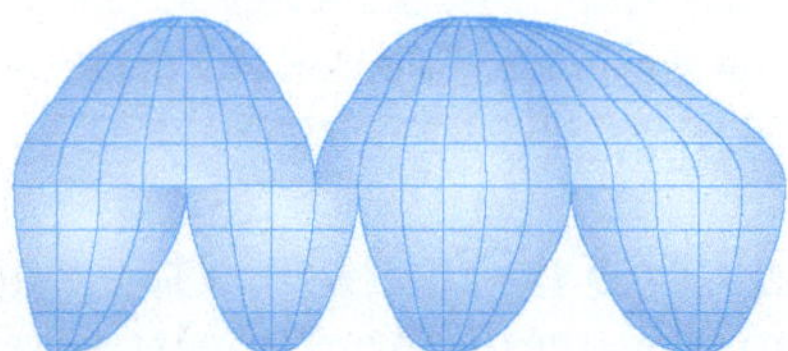

▲ **FIGURE A-11** The distortion pattern of the interrupted Goode's homolosine projection, which mimics that of cylindrical projections.

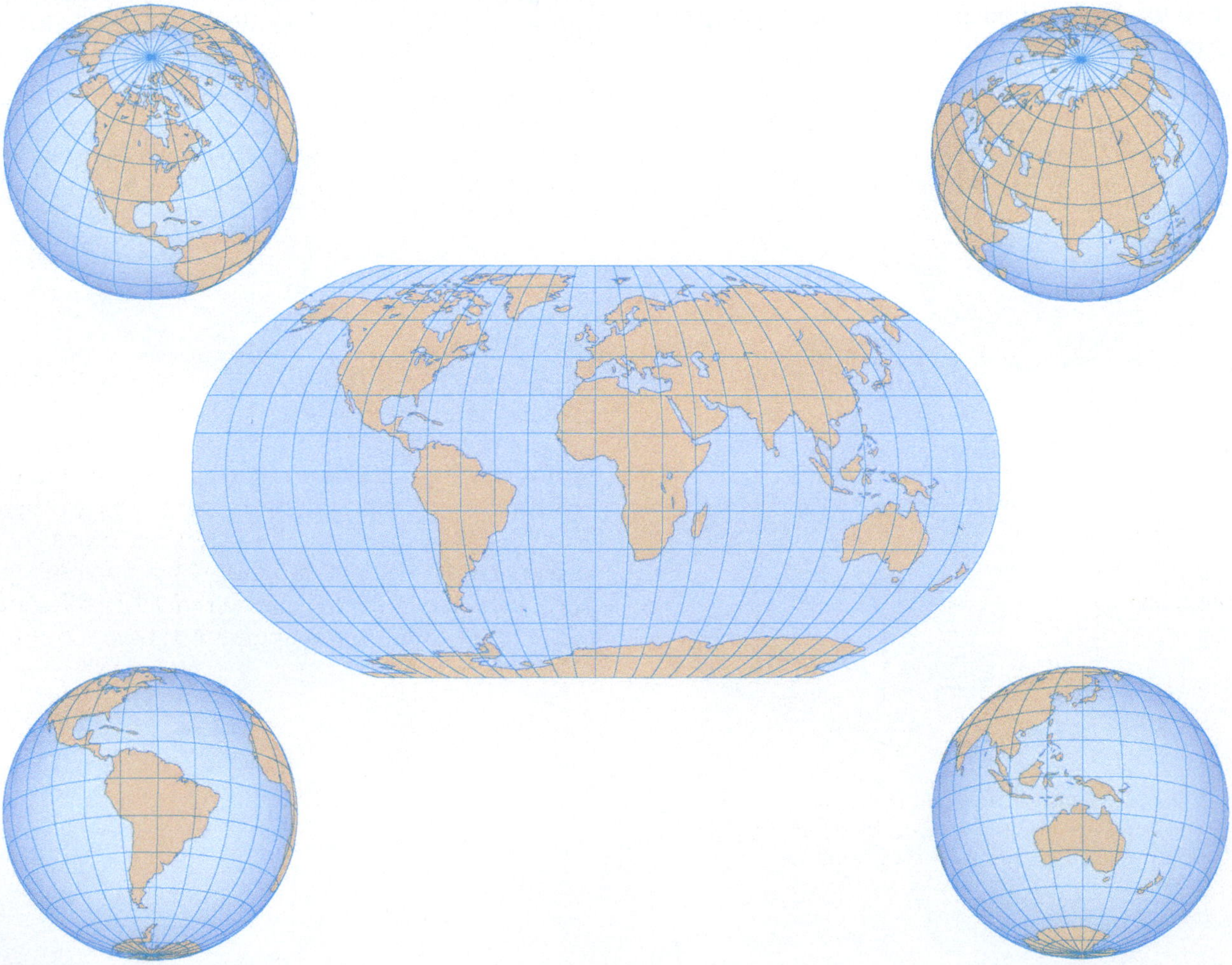

▲ **FIGURE A-12** The compromise Robinson projection, which avoids the interruptions of Goode's homolosine but preserves no special properties. (*Courtesy of ACSM*)

Geospatial Technologies and GIS

Today, user-friendly mapping software enables anyone with a computer to produce maps at a range of scales using a variety of projections. But a challenge remains to geographers and other users of spatial information: how to organize and present in map form the vast amounts of spatial data that are now available.

From data gathered by orbiting satellites or GPS devices to statistical data linked to spatial coordinates, these data provide a more detailed view of Earth's physical and human systems than has ever before been possible. To manage these data, geographers have developed a powerful tool—geographic information systems (GIS), which enable users to manipulate and display spatial data in map form. GIS maps contribute to problem solving in diverse fields such as science and engineering, industry, health care, retail sales, urban planning, environmental protection, law enforcement, and many others.

The power of GIS lies in its ability to map different data sets—called data layers—against each other, revealing relationships that might otherwise be difficult to detect. Figure A-13 shows examples of environmental data organized as GIS data layers.

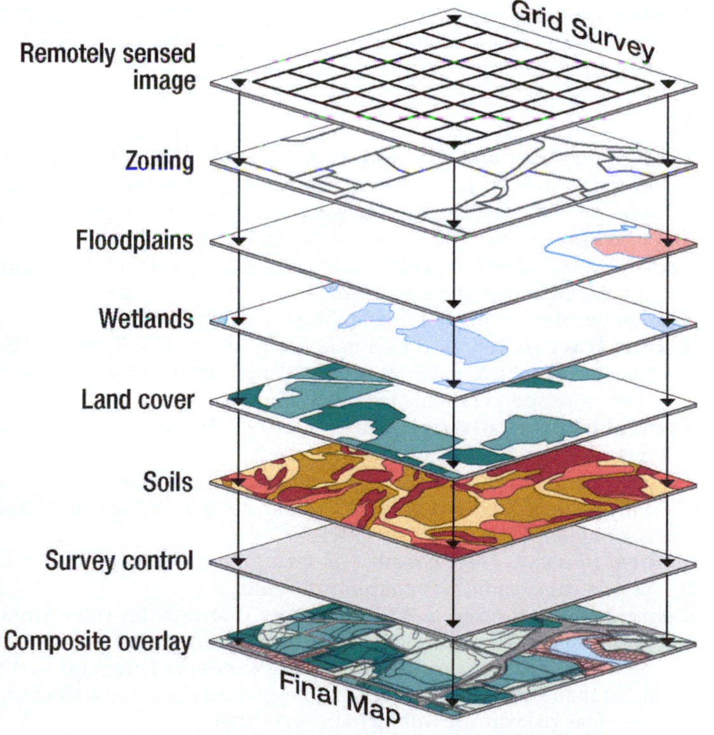

▲ **FIGURE A-13** Within a GIS, environmental data attached to a common terrestrial reference system, such as latitude/longitude, can be stacked in layers for spatial comparison and analysis.

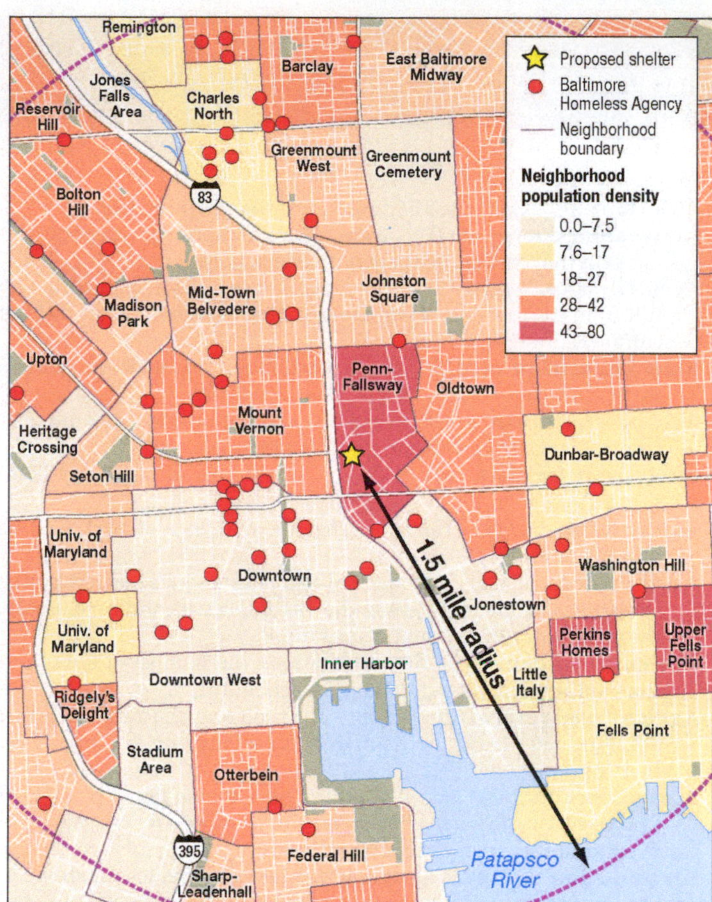

▲ **FIGURE A-14** This GIS map shows that Baltimore's proposed homeless shelter would be near a cluster of service providers. The map's data layers include: neighborhood borders, population density (by neighborhood), and locations of homeless agencies.

GIS can help answer almost any question involving spatial or locational analysis. In one application of GIS, The city of Baltimore, Maryland, wanted to determine the best location for an emergency shelter for homeless people. Among many other factors considered, one main criterion for the shelter was that it be accessible from other facilities providing services to the homeless. As shown in the map in Figure A-14, the site selected was in a densely populated part of the city and at the center of a 1.5-mile-radius circle containing more than 60 percent of the city's providers of homeless services.

GLOSSARY

A

Abiotic *(p. 34)* Composed of nonliving or inorganic matter.

Active solar energy *(p. 412)* Solar radiation captured with photovoltaic cells that convert light energy to electrical energy.

Acculturation *(pp. 28, 138)* The process of changes in culture that result from the meeting of two groups, each of which retains distinct culture features.

Acid deposition *(p. 415)* The accumulation of acids on Earth's surface.

Acid precipitation *(p. 415)* Conversion of sulfur oxides and nitrogen oxides to acids that return to Earth as rain, snow, or fog.

Adolescent fertility rate *(p. 364)* The number of births per 1,000 women ages 15 to 19.

Agnosticism *(p. 185)* The belief that the existence of God can't be proven empirically.

Agribusiness *(p. 326)* Commercial agriculture characterized by the integration of different steps in the food-processing industry, usually through ownership by large corporations.

Agricultural density *(p. 50)* The ratio of the number of farmers to the total amount of land suitable for agriculture.

Agricultural revolution *(p. 308)* The time when human beings first domesticated plants and animals and no longer relied entirely on hunting and gathering.

Agriculture *(p. 308)* The deliberate effort to modify a portion of Earth's surface through the cultivation of crops and the raising of livestock for sustenance or economic gain.

Air pollution *(p. 414)* Concentration of trace substances, such as carbon monoxide, sulfur dioxide, nitrogen oxides, hydrocarbons, and solid particulates, at a greater level than occurs in average air.

Animate power *(p. 385)* Power supplied by animals or by people.

Animism *(p. 193)* The belief that objects, such as plants and stones, or natural events, like thunderstorms and earthquakes, have a discrete spirit and conscious life.

Annexation *(p. 478)* Legally adding land area to a city in the United States.

Anocracy *(p. 228)* A country that is not fully democratic or fully autocratic but rather displays a mix of the two types.

Apartheid *(p. 245)* Laws (no longer in effect) in South Africa that physically separated different races into different geographic areas.

Aquaculture (or aquafarming) *(p. 324)* The cultivation of seafood under controlled conditions.

Arithmetic density *(p. 50)* The total number of people divided by the total land area.

Assimilation *(pp. 28, 138)* The process by which a group's cultural features are altered to resemble those of another more dominant group.

Asylum seeker *(p. 29)* Someone who has migrated to another country in the hope of being recognized as a refugee.

Atheism *(p. 185)* The belief that God does not exist.

Atmosphere *(p. 34)* The thin layer of gases surrounding Earth.

Autocracy *(p. 288)* A country that is run according to the interests of the ruler rather than the people.

Autonomous religion *(p. 211)* A religion that does not have a central authority but shares ideas and cooperates informally.

B

Balance of power *(p. 295)* A condition of roughly equal strength between opposing countries or alliances of countries.

Balkanization *(p. 256)* A process by which a state breaks down through conflicts among its ethnicities.

Balkanized *(p. 256)* A small geographic area that cannot successfully be organized into stable countries because it is inhabited by many ethnicities with complex, long-standing antagonisms toward each other.

Basic business *(p. 446)* A business that sells its products or services primarily to consumers outside the settlement.

Behavioral geography *(p. 26)* The study of the psychological basis for individual human actions in space.

Biochemical oxygen demand (BOD) *(p. 416)* The amount of oxygen required by aquatic bacteria to decompose a given load of organic waste; a measure of water pollution.

Biomass fuel *(p. 385)* Fuel derived from wood, plant material, or animal waste.

Biosphere *(p. 34)* All living organisms on Earth, including plants and animals, as well as microorganisms.

Biotic *(p. 34)* Composed of living organisms.

Blockbusting *(p. 243)* A process by which real estate agents convince white property owners to sell their houses at low prices because of fear that persons of color will soon move into the neighborhood.

Boundary *(p. 280)* An invisible line that marks the extent of a state's territory.

Brain drain *(p. 103)* Large-scale emigration by talented people.

Branch (of a religion) *(p. 188)* A large and fundamental division within a religion.

Break-of-bulk point *(p. 395)* A location where transfer is possible from one mode of transportation to another.

Bulk-gaining industry *(p. 390)* An industry in which the final product weighs more or comprises a greater volume than the inputs.

Bulk-reducing industry *(p. 389)* An industry in which the final product weighs less or comprises a lower volume than the inputs.

Business service *(p. 432)* A service that primarily meets the needs of other businesses, including professional, financial, and transportation services.

C

Carbon capture and storage (CCS) *(p. 494)* The process of capturing waste CO_2, transporting it to a storage site, and depositing it where it will not enter the atmosphere, normally underground.

Cartography *(p. 6)* The science of making maps.

Caste *(p. 218)* The class or distinct hereditary order into which a Hindu is assigned, according to religious law.

Census *(p. 46)* A complete enumeration of a population.

Census tract *(p. 468)* An area delineated by the U.S. Bureau of the Census for which statistics are published; in urban areas, census tracts correspond roughly to neighborhoods.

Central business district (CBD) *(p. 462)* The area of a city where retail and office activities are clustered.

Central city (city) *(p. 460)* An urban settlement that has been legally incorporated into an independent, self-governing unit known as a municipality.

Central place *(p. 434)* A market center for the exchange of services by people attracted from the surrounding area.

Central place theory *(p. 434)* A theory that explains the distribution of services based on the fact that settlements serve as centers of market areas for services; larger settlements are fewer and farther apart than smaller settlements and provide services for a larger number of people who are willing to travel farther.

Centripetal force *(p. 247)* An attitude that tends to unify people and enhance support for a state.

Cereal grain *(p. 312)* A grass that yields grain for food.

Chain migration *(p. 103)* Migration of people to a specific location because relatives or members of the same nationality previously migrated there.

Chlorofluorocarbon (CFC) *(p. 414)* A gas used as a solvent, a propellant in aerosols, a refrigerant, and in plastic foams and fire extinguishers.

Circular migration *(p. 106)* The temporary movement of a migrant worker between home and host countries to seek employment.

Circulation *(p. 78)* Short-term, repetitive, or cyclical movements that recur on a regular basis.
Citizen science *(p. 9)* Scientific research by amateur scientists.
City-state *(p. 451)* A sovereign state comprising a city and its immediate hinterland.
Climate *(p. 34)* The long-term average weather condition at aparticular location.
Clustered rural settlement *(p. 448)* A rural settlement in which the houses and farm buildings of each family are situated close to each other, with fields surrounding the settlement.
Colonialism *(p. 278)* An attempt by one country to establish settlements and to impose its political, economic, and cultural principles in another territory.
Colony *(p. 278)* A territory that is legally tied to a sovereign state rather than completely independent.
Combined statistical area (CSA) *(p. 461)* In the United States, two or more contiguous CBSAs tied together by commuting patterns.
Commercial agriculture *(p. 310)* Agriculture undertaken primarily to generate products for sale off the farm.
Compact state *(p. 286)* A state in which the distance from the center to any boundary does not vary significantly.
Concentration *(p. 22)* The spread of something over a given area.
Concentric zone model *(p. 466)* A model of the internal structure of cities in which social groups are spatially arranged in a series of rings.
Congregation *(p. 188)* A local assembly of persons brought together for common religious worship.
Connection *(p. 32)* Refers to the relationships among people and objects across the barrier of space.
Conservation *(p. 32)* The sustainable management of a natural resource.
Consumer service *(p. 432)* A service that primarily meets the needs of individual consumers, including retail, education, health, and leisure services.
Consumptive water usage *(p. 416)* The use of water that evaporates rather than being returned to nature as a liquid.
Contagious diffusion *(p. 29)* The rapid, widespread diffusion of a feature or trend throughout a population.
Cooperative store *(p. 350)* A member-owned, member-governed business that operates for the benefit of its members according to common principles agreed upon by the international cooperative community.
Core-based statistical area (CBSA) *(p. 461)* In the United States, any MSA or µSA.
Cosmogony *(p. 214)* A set of religious beliefs concerning the origin of the universe.
Cottage industry *(p. 384)* Manufacturing based in homes rather than in factories, most common prior to the Industrial Revolution.
Counterurbanization *(p. 90)* Net migration from urban to rural areas in more developed countries.
Creole *(or creolized)* language *(p. 167)* A language that results from the mixing of a colonizer's language with the indigenous language of the people being dominated.
Crop *(p. 308)* Any plant gathered from a field as a harvest during a particular season.
Crop rotation *(p. 323)* The practice of rotating use of different fields from crop to crop each year to avoid exhausting the soil.
Crude birth rate (CBR) *(p. 54)* The total number of live births in a year for every 1,000 people alive in the society.
Crude death rate (CDR) *(p. 54)* The total number of deaths in a year for every 1,000 people alive in the society.
Cultural ecology *(p. 36)* A geographic approach that emphasizes human—environment relationships.
Cultural landscape *(p. 36)* An approach to geography that emphasizes the relationships among social and physical phenomena in a particular study area.
Culture *(p. 18)* The body of customary beliefs, social forms, and material traits that together constitute a group's distinct tradition.
Custom *(p. 112)* The frequent repetition of an act, to the extent that it becomes characteristic of the group of people performing the act.

D

Dairy farm *(p. 330)* A form of commercial agriculture that specializes in the production of milk and other dairy products.
Democracy *(p. 288)* A country in which citizens elect leaders and can run for office.
Demographic transition *(p. 56)* The process of change in a society's population from a condition of high crude birth and death rates and low rate of natural increase to a condition of low crude birth and death rates, low rate of natural increase, and higher total population.
Demography *(p. 55)* The scientific study of population characteristics.
Denglish *(p. 159)* A combination of *Deutsch* (the German word for German) and *English.*
Denomination *(p. 188)* A division of a branch that unites a number of local congregations into a single legal and administrative body.
Density *(p. 22)* The frequency with which something exists within a given unit of area.
Density gradient *(p. 480)* The change in density in an urban area from the center to the periphery.
Dependency ratio *(p. 60)* The number of people under age 15 and over age 64 compared to the number of people active in the labor force.
Desertification *(pp. 94, 333)* Degradation of land, especially in semi-arid areas, primarily because of human actions such as excessive crop planting, animal grazing, and tree cutting. Also known as semiarid land degradation.
Developed country *(p. 350)* A country that has progressed relatively far along a continuum of development.
Developing country *(p. 350)* A country that is at a relatively early stage in the process of development.
Developing language *(p. 147)* A language spoken in daily use with a literary tradition that is not widely distributed.
Development *(p. 350)* A process of improvement in the conditions of people through diffusion of knowledge and technology.
Dialect *(p. 164)* A regional variety of a language distinguished by vocabulary, spelling, and pronunciation.
Dietary energy consumption *(p. 312)* The amount of food that an individual consumes, measured in kilocalories (Calories in the United States).
Diffusion *(p. 28)* The process of spread of a feature or trend from one place to another over time.
Dispersed rural settlement *(p. 448)* A rural settlement pattern characterized by isolated farms rather than clustered villages.
Distance decay *(p. 30)* The diminished importance and eventual disappearance of a phenomenon with increasing distance from its origin.
Distribution *(p. 22)* The arrangement of something across Earth's surface.
Double cropping *(p. 322)* Harvesting twice a year from the same field.
Doubling time *(p. 52)* The number of years needed to double a population, assuming a constant rate of natural increase.

E

Ebonics *(p. 167)* A dialect spoken by some African Americans.
Ecology *(p. 36)* The scientific study of ecosystems.
Economic base *(p. 446)* A community's collection of basic businesses.
Ecosystem *(p. 36)* A group of living organisms and the abiotic spheres with which they interact.
Ecumene *(p. 49)* The portion of Earth's surface occupied by permanent human settlement.
Edge city *(p. 467)* A large node of office and retail activities on the edge of an urban area.
Elderly support ratio *(p. 61)* The number of working-age people (ages 15 to 64) divided by the number of persons 65 and older.
Elongated state *(p. 286)* A state with a long, narrow shape.
Emigration *(p. 78)* Migration from a location.

Enclosure movement *(p. 449)* The process of consolidating small landholdings into a smaller number of larger farms in England during the eighteenth century.

Environmental determinism *(p. 37)* A nineteenth- and early twentiethcentury approach to the study of geography which argued that the general laws sought by human geographers could be found in the physical sciences. Geography was therefore the study of how the physical environment caused human activities.

Epidemiologic transition *(p. 64)* The process of change in the distinctive causes of death in each stage of the demographic transition.

Epidemiology *(p. 64)* The branch of medical science concerned with the incidence, distribution, and control of diseases that are prevalent among a population at a special time and are produced by some special causes not generally present in the affected locality.

Ethnic cleansing *(p. 252)* A purposeful policy designed by one ethnic or religious group to remove by violent and terror-inspiring means the civilian population of another ethnic or religious group from certain geographic areas.

Ethnic enclave *(p. 236)* A place with a high concentration of an ethnic group that is distinct from those in the surrounding area.

Ethnic religion *(p. 185)* A religion with a relatively concentrated spatial distribution whose principles are likely to be based on the physical characteristics of the particular location in which its adherents are concentrated.

Ethnicity *(p. 230)* Identity with a group of people who share the cultural traditions of a particular homeland or hearth.

Ethnoburb *(p. 327)* A suburban area with a cluster of a particular ethnic population.

Expansion diffusion *(p. 29)* The spread of a feature or trend among people from one area to another in an additive process.

Extinct language *(p. 177)* A language that was once used by people in daily activities but is no longer used.

F

Fair trade *(p. 376)* An alternative to international trade that provides greater equity to workers, small businesses, and consumers, focusing primarily on products exported from developing countries to developed countries.

Federal state *(p. 289)* An internal organization of a state that allocates most powers to units of local government.

Female labor force participation rate *(p. 393)* The percentage of women holding full-time jobs outside the home.

Filtering *(p. 489)* A process of change in the use of a house, from single-family owner occupancy to abandonment.

Fishing *(p. 304)* The capture of wild fish and other seafood living in the waters.

Fission *(p. 408)* The splitting of an atomic nucleus to release energy.

Floodplain *(p. 94)* An area subject to flooding during a given number of years, according to historical trends.

Folk culture *(p. 112)* Culture traditionally practiced by a small, homogeneous, rural group living in relative isolation from other groups.

Food desert *(p. 439)* An area that has a substantial amount of low-income residents and has poor access to a grocery store, defined in most cases as further than 1 mile.

Food security *(p. 314)* Physical, social, and economic access at all times to safe and nutritious food sufficient to meet dietary needs and food preferences for an active and healthy life.

Forced migration *(p. 82)* Permanent movement, usually compelled by cultural factors.

Fordist production *(p. 424)* A form of mass production in which each worker is assigned one specific task to perform repeatedly.

Foreign direct investment (FDI) *(p. 372)* Investment made by a foreign company in the economy of another country.

Formal region (or uniform region) *(p. 16)* An area in which everyone shares in common one or more distinctive characteristics.

Fossil fuel *(p. 385)* An energy source formed from the residue of plants and animals buried millions of years ago.

Fragmented state *(p. 287)* A state that includes several discontinuous pieces of territory.

Franglais *(p. 159)* A combination of *français* and *anglais* (the French words for French and English, respectively).

Frontier *(p. 280)* A zone separating two states in which neither state exercises political control.

Functional region (or nodal region) *(p. 16)* An area organized around a node or focal point.

Fundamentalism *(p. 220)* Literal interpretation and strict adherence to basic principles of a religion (or a religious branch, denomination, or congregation).

Fusion *(p. 409)* Creation of energy by joining the nuclei of two hydrogen atoms to form helium.

G

Gender Inequality Index (GII) *(p. 360)* An indicator constructed by the U.N. to measure the extent of each country's gender inequality in terms of reproductive health, empowerment, and the labor market.

Gender-related Development Index (GDI) *(p. 360)* An indicator constructed by the U.N. to measure the gender gap in the level of achievement in terms of income, education, and life expectancy.

Genetically modified organism (GMO) *(p. 338)* A living organism that possesses a novel combination of genetic material obtained through the use of modern biotechnology.

Genocide *(p. 258)* The mass killing of a group of people in an attempt to eliminate the entire group from existence.

Gentrification *(p. 490)* A process of converting an urban neighborhood from a predominantly low-income, renter-occupied area to a predominantly middle-class, owner-occupied area.

Geographic information science (GIScience) *(p. 9)* The development and analysis of data about Earth acquired through satellite and other electronic information technologies.

Geographic information system (GIS) *(p. 9)* A computer system that stores, organizes, analyzes, and displays geographic data.

Geotagging *(p. 9)* Identification and storage of a piece of information by its precise latitude and longitude coordinates.

Geothermal energy *(p. 411)* Energy from steam or hot water produced from hot or molten underground rocks.

Gerrymandering *(p. 290)* The process of redrawing legislative boundaries for the purpose of benefiting the party in power.

Ghetto *(p. 205)* During the Middle Ages, a neighborhood in a city set up by law to be inhabited only by Jews; now used to denote a section of a city in which members of any minority group live because of social, legal, or economic pressure.

Global Positioning System (GPS) *(p. 8)* A system that determines the precise position of something on Earth through a series of satellites, tracking stations, and receivers.

Globalization *(p. 20)* Actions or processes that involve the entire world and result in making something worldwide in scope.

Grain *(p. 312)* Seed of a cereal grass.

Gravity model *(p. 438)* A model which holds that the potential use of a service at a particular location is directly related to the number of people in a location and inversely related to the distance people must travel to reach the service.

Green revolution *(p. 324)* Rapid diffusion of new agricultural technology, especially new high-yield seeds and fertilizers.

Greenhouse effect *(p. 414)* The anticipated increase in Earth's temperature caused by carbon dioxide (emitted by burning fossil fuels) trapping some of the radiation emitted by the surface.

Greenwich Mean Time (GMT) *(p. 12)* The time in the zone encompassing the prime meridian, or 0° longitude.

Gross domestic product (GDP) *(p. 352)* The value of the total output of goods and services produced in a country in a year, not accounting for money that leaves and enters the country.

Gross national income (GNI) *(p. 352)* The value of the output of goods and services produced in a country in a year, including money that leaves and enters the country.

Guest worker *(p. 106)* A term once used for a worker who migrated to the developed countries of Northern and Western Europe, usually from Southern and Eastern Europe or from North Africa, in search of a higher-paying job.

H

Habit *(p. 112)* A repetitive act performed by a particular individual.

Hearth *(p. 28)* The region from which innovative ideas originate.

Hierarchical diffusion *(p. 29)* The spread of a feature or trend from one key person or node of authority or power to other persons or places.

Hierarchical religion *(p. 210)* A religion in which a central authority exercises a high degree of control.

Hinterland *(p. 434)* The area surrounding a central place from which people are attracted to use the place's goods and services (also known as market area).

Horticulture *(p. 322)* The growing of fruits, vegetables, and flowers.

Human Development Index (HDI) *(p. 350)* An indicator constructed by the U.N. to measure the level of development for a country through a combination of income, education, and life expectancy.

Humanistic geography *(p. 26)* The study of different ways that individuals form ideas about place and give those places symbolic meanings.

Hydrosphere *(p. 34)* All of the water on and near Earth's surface.

I

Immigration *(p. 78)* Migration to a new location.

Industrial Revolution *(p. 56)* A series of improvements in industrial technology that transformed the process of manufacturing goods.

Inequality-adjusted Human Development Index (IHDI) *(p. 358)* A modification of the HDI to account for inequality.

Infant mortality rate (IMR) *(p. 61)* The total number of deaths in a year among infants under 1 year of age for every 1,000 live births in a society.

Informal settlement *(p. 474)* An area within a city in a less developed country in which people illegally establish residences on land they do not own or rent and erect homemade structures.

Institutional language *(p. 147)* A language used in education, work, mass media, and government.

Intensive subsistence agriculture *(p. 322)* A form of subsistence agriculture characteristics of Asia's major population concentrations in which farmers must expend a relatively large amount of effort to produce the maximum feasible yield from a parcel of land.

Internal migration *(p. 82)* Permanent movement within a particular country.

Internally displaced person (IDP) *(p. 92)* Someone who has been forced to migrate for similar political reasons as a refugee but has not migrated across an international border.

International Date Line *(p. 13)* An arc that for the most part follows 180° longitude, although it deviates in several places to avoid dividing land areas. When the International Date Line is crossed heading east (toward America), the clock moves back 24 hours, or one entire day. When it is crossed heading west (toward Asia), the calendar moves ahead one day.

International migration *(p. 80)* Permanent movement from one country to another.

Interregional migration *(p. 82)* Permanent movement from one region of a country to another.

Intervening obstacle *(p. 94)* An environmental or cultural feature of the landscape that hinders migration.

Intraregional migration *(p. 82)* Permanent movement within one region of a country.

Isogloss *(p. 166)* A boundary that separates regions in which different language usages predominate.

Isolated language *(p. 176)* A language that is unrelated to any other languages and therefore not attached to any language family.

J

Just-in-time delivery *(p. 395)* Shipment of parts and materials to arrive at a factory moments before they are needed.

L

Labor-intensive industry *(p. 396)* An industry for which labor costs comprise a high percentage of total expenses.

Landlocked state *(p. 289)* A state that does not have a direct outlet to the sea.

Language *(p. 146)* A system of communication through the use of speech, a collection of sounds understood by a group of people to have the same meaning.

Language branch *(p. 147)* A collection of languages related through a common ancestor that can be confirmed through archaeological evidence.

Language family *(p. 147)* A collection of languages related to each other through a common ancestor long before recorded history.

Language group *(p. 147)* A collection of languages within a branch that share a common origin in the relatively recent past and display relatively few differences in grammar and vocabulary.

Latitude *(p. 12)* The numbering system used to indicate the location of parallels drawn on a globe and measuring distance north and south of the equator (0°).

Life expectancy *(p. 52)* The average number of years an individual can be expected to live, given current social, economic, and medical conditions. Life expectancy at birth is the average number of years a newborn infant can expect to live.

Lingua franca *(p. 160)* A language mutually understood and commonly used in trade by people who have different native languages.

Literacy rate *(p. 355)* The percentage of a country's people who can read and write.

Literary tradition *(p. 147)* A language that is written as well as spoken.

Lithosphere *(p. 34)* Earth's crust and a portion of upper mantle directly below the crust.

Location *(p. 14)* The position of anything on Earth's surface.

Logogram *(p. 161)* A symbol that represents a word rather than a sound.

Longitude *(p. 12)* The numbering system used to indicate the location of meridians drawn on a globe and measuring distance east and west of the prime meridian (0°).

M

Map *(p. 6)* A two-dimensional, or flat, representation of Earth's surface or a portion of it.

Map scale *(p. 10)* The relationship between the size of an object on a map and the size of the actual feature on Earth's surface.

Maquiladora *(p. 421)* A factory built by a U.S. company in Mexico near the U.S. border, to take advantage of the much lower labor costs in Mexico.

Market area *(p. 434)* The area surrounding a central place from which people are attracted to use the place's goods and services (also known as hinterland).

Mashup *(p. 9)* A map that overlays data from one source on top of a map provided by a mapping service.

Maternal mortality rate *(pp. 59, 364)* The annual number of female deaths per 100,000 live births from any cause related to or aggravated by pregnancy or its management (excluding accidental or incidental causes).

Medical revolution *(p. 56)* Medical technology invented in Europe and North America that has diffused to the poorer countries in Latin America, Asia, and Africa. Improved medical practices have eliminated many of the traditional causes of death in poorer countries and enabled more people to live longer and healthier lives.

Megalopolis *(p. 481)* A continuous urban complex in the northeastern United States.

Mental map *(p. 8)* A representation of a portion of Earth's surface based on what an individual knows about a place that contains personal impressions of what is in the place and where the place is located.

Meridian *(p. 12)* An arc drawn on a map between the North and South poles.

Metropolitan statistical area (MSA) *(p. 460)* In the United States, an urbanized area of at least 50,000 population, the county within which the city is located, and adjacent counties meeting one of several tests indicating a functional connection to the central city.

Microfinance *(p. 375)* Provision of small loans and financial services to individuals and small businesses in developing countries.

Micropolitan statistical area (μSA) *(p. 461)* An urbanized area of between 10,000 and 50,000 inhabitants, the county in which it is located, and adjacent counties tied to the city.

Microstate *(p. 266)* A state that encompasses a very small land area.

Migration *(p. 78)* A permanent move to a new location.

Migration transition *(p. 81)* A change in the migration pattern in a society that results from industrialization, population growth, and other social and economic changes that also produces the demographic transition.

Milkshed *(p. 330)* The area surrounding a city from which milk is supplied.

Millennium Development Goals *(p. 379)* Eight goals adopted by the U.N. in 2002 to reduce disparities between developed and developing countries by 2015.

Missionary *(p. 200)* An individual who helps to diffuse a universalizing religion.

Mixed crop and livestock farming *(p. 328)* Commercial farming characterized by integration of crops and livestock; most of the crops are fed to animals rather than consumed directly by humans.

Mobility *(p. 78)* All types of movements between locations.

Monotheism *(p. 194)* The doctrine of or belief in the existence of only one God.

Multiethnic state *(p. 272)* A state that contains more than one ethnicity.

Multinational state *(p. 272)* A state that contains two or more ethnic groups with traditions of self-determination that agree to coexist peacefully by recognizing each other as distinct nationalities.

Multiple nuclei model *(p. 467)* A model of the internal structure of cities in which social groups are arranged around a collection of nodes of activities.

N

Nation-state *(p. 220)* A state whose territory corresponds to that occupied by a particular ethnicity.

Nationalism *(p. 247)* Loyalty and devotion to a particular nationality.

Nationality *(p. 230)* Identity with a group of people who share legal attachment to a particular country.

Natural increase rate (NIR) *(p. 52)* The percentage growth of a population in a year, computed as the crude birth rate minus the crude death rate.

Net migration *(p. 78)* The difference between the level of immigration and the level of emigration.

Network *(p. 30)* A chain of communication that connects places.

New international division of labor *(p. 420)* Transfer of some types of jobs, especially those requiring low-paid, less-skilled workers, from more developed to less developed countries.

No tillage *(p. 377)* A farming practice that leaves all of the soil undisturbed and the entire residue of the previous year's harvest left untouched on the fields.

Nonbasic business *(p. 446)* A business that sells its products primarily to consumers in the same settlement.

Nonconsumptive water usage *(p. 416)* The use of water that is returned to nature as a liquid.

Nonpoint source pollution *(p. 417)* Pollution that originates from a large, diffuse area.

Nonrenewable energy *(p. 10)* A source of energy that has a finite supply capable of being exhausted.

Nonrenewable resource *(p. 32)* Something produced in nature more slowly than it is consumed by humans.

O

Official language *(p. 147)* The language adopted for use by a government for the conduct of business and publication of documents.

Outsourcing *(p. 420)* A decision by a corporation to turn over much of the responsibility for production to independent suppliers.

Overfishing *(p. 325)* Capturing fish faster than they can reproduce.

Overpopulation *(p. 46)* A situation in which the number of people in an area exceeds the capacity of the environment to support life at a decent standard of living.

Ozone *(p. 414)* A gas that absorbs ultraviolet solar radiation and is found in the stratosphere, a zone 15 to 50 kilometers (9 to 30 miles) above Earth's surface.

P

Paddy *(p. 323)* The Malay word for wet rice, increasingly used to describe a flooded field.

Pandemic *(p. 64)* Disease that occurs over a wide geographic area and affects a very high proportion of the population.

Parallel *(p. 12)* A circle drawn around the globe parallel to the equator and at right angles to the meridians.

Participatory GIS (PGIS) *(p. 9)* Community-based mapping, representing local knowledge and information.

Passive solar energy systems *(p. 412)* Solar energy systems that collect energy without the use of mechanical devices.

Pastoral nomadism *(p. 318)* A form of subsistence agriculture based on herding domesticated animals.

Pattern *(p. 23)* The geometric or regular arrangement of something in a particular area.

Perforated state *(p. 286)* A state that completely surrounds another one.

Peripheral model *(p. 480)* A model of North American urban areas consisting of an inner city surrounded by large suburban residential and business areas tied together by a beltway or ring road.

Photochemical smog *(p. 451)* An atmospheric condition formed through a combination of weather conditions and pollution, especially from motor vehicle emissions.

Physiological density *(p. 50)* The number of people per unit area of arable land, which is land suitable for agriculture.

Pidgin language *(p. 160)* A form of speech that adopts a simplified grammar and limited vocabulary of a lingua franca; used for communications among speakers of two different languages.

Pilgrimage *(p. 212)* A journey to a place considered sacred for religious purposes.

Place *(p. 4)* A specific point on Earth, distinguished by a particular characteristic.

Plantation *(p. 318)* A large farm in tropical and subtropical climates that specializes in the production of one or two crops for sale, usually to a more developed country.

Point source pollution *(p. 417)* Pollution that enters a body of water from a specific source.

Polder *(p. 38)* Land that the Dutch have created by draining water from an area.

Pollution *(p. 414)* Concentration of waste added to air, water, or land at a greater level than occurs in average air, water, or land.

Polytheism *(p. 194)* Belief in or worship of more than one god.

Popular culture *(p. 112)* Culture found in a large, heterogeneous society that shares certain habits despite differences in other personal characteristics.

Population pyramid *(p. 60)* A bar graph that represents the distribution of population by age and sex.

Possibilism *(p. 37)* The theory that the physical environment may set limits on human actions, but people have the ability to adjust to the physical environment and choose a course of action from many alternatives.

Post-Fordist production *(p. 424)* Adoption by companies of flexible work rules, such as the allocation of workers to teams that perform a variety of tasks.

Poststructuralist geography *(p. 26)* The study of space as the product of ideologies or value systems of ruling elites.

Potential reserve *(p. 404)* The amount of a resource in deposits not yet identified but thought to exist.

Preservation *(p. 32)* The maintenance of resources in their present condition, with as little human impact as possible.

Primary census area (PSA) *(p. 461)* In the United States, any CSA, any MSA not included in a CSA, or any μSA not included in a CSA.

Primary sector *(p. 352)* The portion of the economy concerned with the direct extraction of materials from Earth, generally through agriculture.

Primate city *(p. 437)* A city that is the largest settlement in a country and has more than twice as many people as the second-ranking settlement.

Primate city rule *(p. 437)* A pattern of settlements in a country such that the largest settlement has more than twice as many people as the second-ranking settlement.

Prime agricultural land *(p. 332)* The most productive farmland.

Prime meridian *(p. 12)* The meridian, designated as 0° longitude, that passes through the Royal Observatory at Greenwich, England.

Productivity *(p. 353)* The value of a particular product compared to the amount of labor needed to make it.

Projection *(p. 11)* A system used to transfer locations from Earth's surface to a flat map.

Prorupted state *(p. 286)* An otherwise compact state with a large projecting extension.

Proven reserve *(p. 404)* The amount of a resource remaining in discovered deposits.

Public housing *(p. 490)* Government-owned housing rented to low-income individual, with rents set at 30 percent of the tenant's income.

Public service *(p. 432)* A service offered by the government to provide security and protection for citizens and businesses.

Pull factor *(p. 92)* A factor that induces people to move to a new location.

Pupil/teacher ratio *(p. 355)* The number of enrolled students divided by the number of teachers.

Purchasing power parity (PPP) *(p. 352)* The amount of money needed in one country to purchase the same goods and services in another country.

Push factor *(p. 92)* A factor that induces people to move out of their present location.

Q

Quota *(p. 102)* In reference to migration, a law that places a maximum limit on the number of people who can immigrate to a country each year.

R

Race *(p. 230)* Identity with a group of people who are perceived to share a physiological trait, such as skin color.

Racism *(p. 232)* The belief that race is the primary determinant of human traits and capacities and that racial differences produce an inherent superiority of a particular race.

Racist *(p. 232)* A person who subscribes to the beliefs of racism.

Ranching *(p. 331)* A form of commercial agriculture in which livestock graze over an extensive area.

Range (of a service) *(p. 435)* The maximum distance people are willing to travel to use a service.

Rank-size rule *(p. 436)* A pattern of settlements in a country such that the *n*th largest settlement is $1/n$ the population of the largest settlement.

Received Pronunciation (RP) *(p. 165)* The dialect of English associated with upper-class Britons living in London and now considered standard in the United Kingdom.

Recycling *(p. 426)* The separation, collection, processing, marketing, and reuse of unwanted material.

Redlining *(p. 489)* A process by which financial institutions draw red-colored lines on a map and refuse to lend money for people to purchase or improve property within the lines.

Refugee *(p. 92)* Someone who is forced to migrate from his or her home country and cannot return for fear of persecution because of his or her race, religion, nationality, membership in a social group, or political opinion.

Region *(p. 4)* An area distinguished by a unique combination of trends or features.

Relocation diffusion *(p. 28)* The spread of a feature or trend through bodily movement of people from one place to another.

Remittance *(p. 97)* Transfer of money by workers to people in the country from which they emigrated.

Remote sensing *(p. 9)* The acquisition of data about Earth's surface from a satellite orbiting the planet or from other long-distance methods.

Renewable energy *(p. 410)* A resource that has a theoretically unlimited supply and is not depleted when used by people.

Renewable resource *(p. 32)* Something produced in nature more rapidly than it is consumed by humans.

Resource *(p. 32)* A substance in the environment that is useful to people, is economically and technologically feasible to access, and is socially acceptable to use.

Ridge tillage *(p. 337)* A system of planting crops on ridge tops in order to reduce farm production costs and promote greater soil conservation.

Right-to-work law *(p. 422)* A U.S. law that prevents a union and a company from negotiating a contract that requires workers to join the union as a condition of employment.

Rush hour *(p. 484)* The four consecutive 15-minute periods in the morning and evening with the heaviest volumes of traffic.

S

Sanitary landfill *(p. 418)* A place to deposit solid waste, where a layer of earth is bulldozed over garbage each day to reduce emissions of gases and odors from the decaying trash, to minimize fires, and to discourage vermin.

Sawah *(p. 322)* A flooded field for growing rice.

Scale *(p. 4)* Generally, the relationship between the portion of Earth being studied and Earth as a whole.

Secondary sector *(p. 352)* The portion of the economy concerned with manufacturing useful products through processing, transforming, and assembling raw materials.

Sector model *(p. 466)* A model of the internal structure of cities in which social groups are arranged around a series of sectors, or wedges, radiating out from the central business district.

Self-determination *(p. 272)* The concept that ethnicities have the right to govern themselves.

Service *(p. 432)* Any activity that fulfills a human want or need and returns money to those who provide it.

Settlement *(p. 432)* A permanent collection of buildings and inhabitants.

Sex ratio *(p. 58)* The number of males per 100 females in the population.

Sharecropper *(p. 242)* A person who works fields rented from a landowner and pays the rent and repays loans by turning over to the landowner a share of the crops.

Shifting cultivation *(p. 318)* A form of subsistence agriculture in which people shift activity from one field to another; each field is used for crops for a relatively few years and left fallow for a relatively long period.

Site *(p. 15)* The physical character of a place.

Site factors *(p. 388)* Location factors related to the costs of factors of production inside a plant, such as land, labor, and capital.

Situation *(p. 15)* The location of a place relative to another place.

Situation factors *(p. 388)* Location factors related to the transportation of materials into and from a factory.

Slash-and-burn agriculture *(p. 320)* Another name for shifting cultivation, so named because fields are cleared by slashing the vegetation and burning the debris.

Smart growth *(p. 479)* Legislation and regulations to limit suburban sprawl and preserve farmland.

Social area analysis *(p. 468)* Statistical analysis used to identify where people of similar living standards, ethnic background, and lifestyle live within an urban area.

Solstice *(p. 217)* An astronomical event that happens twice each year, when the tilt of Earth's axis is most inclined toward or away from the Sun, causing the Sun's apparent position in the sky to reach it most northernmost or southernmost extreme, and resulting in the shortest and longest days of the year.

Sovereignty *(p. 268)* Ability of a state to govern its territory free from control of its internal affairs by other states.

Space *(p. 5)* The physical gap or interval between two objects.

Space–time compression *(p. 30)* The reduction in the time it takes to diffuse something to a distant place as a result of improved communications and transportation systems.

Spanglish *(p. 159)* A combination of Spanish and English spoken by Hispanic Americans.

Spatial association *(p. 18)* The relationship between the distribution of one feature and the distribution of another feature.

Sprawl *(p. 480)* Development of new housing sites at relatively low density and at locations that are not contiguous to the existing built-up area.

Standard language *(p. 165)* The form of a language used for official government business, education, and mass communications.

State *(p. 266)* An area organized into a political unit and ruled by an established government that has control over its internal and foreign affairs.

Stimulus diffusion *(p. 29)* The spread of an underlying principle even though a specific characteristic is rejected.

Structural adjustment program *(p. 374)* Economic policies imposed on less developed countries by international agencies to create conditions that encourage international trade.

Subdialect *(p. 164)* A subdivision of a dialect.

Subsistence agriculture *(p. 310)* Agriculture designed primarily to provide food for direct consumption by the farmer and the farmer's family.

Suburb *(p. 478)* A residential or commercial area situated within an urban area but outside the central city.

Sustainability *(p. 32)* The use of Earth's renewable and nonrenewable natural resources in ways that do not constrain resource use in the future.

Sustainable development *(p. 494)* Development that meets the needs of the present without compromising the ability of future generations to meet their own needs.

Sustainable Development Goals *(p. 379)* Seventeen goals adopted by the U.N. in 2015 to reduce disparities between developed and developing countries by 2030.

Swidden *(p. 320)* A patch of land cleared for planting through slashing and burning.

Syncretism *(p. 28)* The combining of elements of two groups into a new cultural feature.

T

Taboo *(p. 127)* A restriction on behavior imposed by social custom.

Terroir *(p. 126)* The contribution of a location's distinctive physical features to the way food tastes.

Terrorism *(p. 298)* The systematic use of violence by a group calculated to create an atmosphere of fear and alarm among a population or to coerce a government into actions it would not otherwise undertake or refrain from actions it wants to take.

Tertiary sector *(p. 352)* The portion of the economy concerned with transportation, communications, and utilities, sometimes extended to the provision of all goods and services to people in exchange for payment.

Threshold *(p. 435)* The minimum number of people needed to support a service.

Toponym *(p. 14)* The name given to a portion of Earth's surface.

Total fertility rate (TFR) *(p. 55)* The average number of children a woman will have throughout her childbearing years.

Transhumance *(p. 319)* The seasonal migration of livestock between mountains and lowland pastures.

Transnational corporation *(p. 20)* A company that conducts research, operates factories, and sells products in many countries, not just where its headquarters or shareholders are located.

Triangular slave trade *(p. 240)* A practice, primarily during the eighteenth century, in which European ships transported slaves from Africa to Caribbean islands, molasses from the Caribbean to Europe, and trade goods from Europe to Africa.

Truck farming *(p. 327)* Commercial gardening and fruit farming, so named for the Middle English word *truck*, meaning "barter" or "exchange of commodities."

U

Unauthorized immigrant *(p. 100)* A person who enters a country without proper documents to do so.

Underclass *(p. 488)* A group in society prevented from participating in the material benefits of a more developed society because of a variety of social and economic characteristics.

Undernourishment *(p. 343)* Dietary energy consumption that is continuously below the minimum requirement for maintaining a healthy life and carrying out light physical activity.

Uneven development *(p. 27)* The increasing gap in economic conditions between core and peripheral regions as a result of the globalization of the economy.

Unitary state *(p. 289)* An internal organization of a state that places most power in the hands of central government officials.

Universalizing religion *(p. 185)* A religion that attempts to appeal to all people, not just those living in a particular location.

Urban area *(p. 460)* A central city and its surrounding built-up suburbs.

Urban cluster *(p. 460)* In the United States, an urban area with between 2,500 and 50,000 inhabitants.

Urbanization *(p. 452)* An increase in the percentage of and the number of people living in urban settlements.

Urbanized area *(p. 460)* In the United States, an urban area with at least 50,000 inhabitants.

V

Value added *(p. 353)* The gross value of a product minus the costs of raw materials and energy.

Vernacular region (or perceptual region) *(p. 17)* An area that people believe exists as part of their cultural identity.

Vertical integration *(p. 420)* An approach typical of traditional mass production in which a company controls all phases of a highly complex production process.

Vigorous language *(p. 147)* A language that is spoken in daily use but that lacks a literary tradition.

Voluntary migration *(p. 82)* Permanent movement undertaken by choice.

Volunteered geographic information (VGI) *(p. 9)* Creation and dissemination of geographic data contributed voluntarily and for free by individuals.

Vulgar Latin *(p. 156)* A form of Latin used in daily conversation by ancient Romans, as opposed to the standard dialect, which was used for official documents.

W

Wet rice *(p. 322)* Rice planted on dry land in a nursery and then moved to a deliberately flooded field to promote growth.

Z

Zero population growth (ZPG) *(p. 56)* A decline of the total fertility rate to the point where the natural increase rate equals zero.

Zoning ordinance *(p. 482)* A law that limits the permitted uses of land and maximum density of development in a community.

TEXT, PHOTO, AND ILLUSTRATION CREDITS

Chapter 1 Fig. 01-CO SerrNovik/Getty Images; Fig. 01-01 SergiyN/Shutterstock; Fig. Vid. 01-01 Video Still provided by BBC Worldwide Learning; Fig. 01-02 International Mapping/Pearson Education, Inc.; Fig. 01-03a Scirocco340/Shutterstock; Fig. 01-03b Stuart Forster Europe/Alamy; Fig. 01-04 International Mapping/Pearson Education, Inc.; Fig. 01-05 Tatjana Balzer/Fotolia; Fig. 01-06 Mauricio Abreu/JAI/Corbis; Fig. 01-07 North Wind Picture Archives/Alamy; Fig. 01-08 Everett Collection, Inc./Alamy; Fig. 01-09 Prisma Archivo/Alamy; Fig. 01-10 Michael Snell/Alamy; Fig. 01-11 International Mapping/Pearson Education, Inc.; Fig. 01-12 International Mapping/Pearson Education, Inc.; Fig. 01-13 International Mapping/Pearson Education, Inc.; Fig. 01-14 International Mapping/Pearson Education, Inc.; Fig. 01-15 International Mapping/Pearson Education, Inc.; Fig. 01-16 David South/Alamy; Fig. 01-17 International Mapping/Pearson Education, Inc.; Fig. 01-18a International Mapping/Pearson Education, Inc.; Fig. 01-18b Martin Siepmann/Westend61/Corbis; Fig. 01-19 International Mapping/Pearson Education, Inc.; Fig. 01-20 International Mapping/Pearson Education, Inc.; Fig. 01-21 International Mapping/Pearson Education, Inc.; Fig. 01-22 Ton Koene/Alamy; Fig. 01-23a Konstantin L/Fotolia; Fig. 01-23b Delphotostock/Fotolia; Fig. 01-24a International Mapping/Pearson Education, Inc.; Fig. 01-24b International Mapping/Pearson Education, Inc.; Fig. 01-24c International Mapping/Pearson Education, Inc.; Fig. 01-24d International Mapping/Pearson Education, Inc.; Fig. 01-25 International Mapping/Pearson Education, Inc.; Fig. 01-26 International Mapping/Pearson Education, Inc.; Fig. 01-27a Cylonphoto/Alamy; Fig. 01-27b V. Dorosz/Alamy; Fig. 01-28 International Mapping/Pearson Education, Inc.; Fig. 01-29 International Mapping/Pearson Education, Inc.; Fig. 01-30 Jim Wark/AgStock Images/Corbis; Fig. 01-31 International Mapping/Pearson Education, Inc.; Fig. 01-32 International Mapping/Pearson Education, Inc.; Fig. 01-33 International Mapping/Pearson Education, Inc.; Fig. 01-34 International Mapping/Pearson Education, Inc.; Fig. 01-35 International Mapping/Pearson Education, Inc.; Fig. 01-36 International Mapping/Pearson Education, Inc.; Fig. 01-37 International Mapping/Pearson Education, Inc.; Fig. 01-38 International Mapping/Pearson Education, Inc.; Fig. 01-39 International Mapping/Pearson Education, Inc.; Fig. 01-40 International Mapping/Pearson Education, Inc.; Fig. 01-41 International Mapping/Pearson Education, Inc.; Fig. 01-42 International Mapping/Pearson Education, Inc.; Fig. 01-43 Aleksey Boldin/Alamy; Fig. 01-44 Pixellover RM 10/Alamy; Fig. 01-45 International Mapping/Pearson Education, Inc.; Fig. 01-46a Henryk Kotowski/Alamy; Fig. 01-46b LOOK Die Bildagentur der Fotografen GmbH/Alamy; Fig. 01-46c Jeff Morgan 08/Alamy; Fig. 01-47 International Mapping/Pearson Education, Inc.; Fig. 01-48 Xinhua/Alamy; Fig. 01-49 International Mapping/Pearson Education, Inc.; Fig. 01-50 International Mapping/Pearson Education, Inc.; Fig. 01-51a JNP/Shutterstock; Fig. 01-51b Frederic Cirou/PhotoAlto/Alamy; Fig. 01-52 Frans Lemmens/Alamy; Fig. 01-53 International Mapping/Pearson Education, Inc.; Fig. 01-54 International Mapping/Pearson Education, Inc.; Fig. 01-55 Inga Spence/Alamy; Fig. 01-56 International Mapping/Pearson Education, Inc.; Fig. 01-57 Kevin Fleming/Corbis/Flirt/Alamy; Fig. 01-58 Kirk Anderson/Alamy; Fig. 01-59 NASA; Fig. 01-60 PjrStudio/Alamy; Fig. 01-61 Jim Wileman/Alamy; Fig. 01-Explore Map data © 2015 Google;

Chapter 2 Fig. 02-CO Beawiharta/Reuters; Fig. 02-01 Remi Benali/Corbis; Fig. Vid. 02-01 Video still provided by BBC Worldwide Learning; Tbl. 02-01 Pearson Education, Inc.; UNTbl. 02-01 Pearson Education, Inc.; Fig. 02-02 International Mapping/Pearson Education, Inc.; UNTbl. 02-02 Pearson Education, Inc.; Fig. 02-03 International Mapping/Pearson Education, Inc.; UNTbl. 02-03 Malthus, Thomas, An Essay on the Principle of Population; Fig. 02-04 International Mapping/Pearson Education, Inc.; Fig. 02-05 International Mapping/Pearson Education, Inc.; Fig. 02-06 International Mapping/Pearson Education, Inc.; Fig. 02-07 International Mapping/Pearson Education, Inc.; Fig, 02-08 International Mapping/Pearson Education, Inc.; Fig. 02-09 Pearson Education, Inc.; Fig. 02-10 Pearson Education, Inc.; Fig. 02-11 International Mapping/Pearson Education, Inc.; Fig. 02-12 International Mapping/Pearson Education, Inc.; Fig. 02-13 Pearson Education, Inc.; Fig. 02-14 International Mapping/Pearson Education, Inc.; Fig. 02-15 International Mapping/Pearson Education, Inc.; Fig. 02-16 International Mapping/Pearson Education, Inc.; Fig. 02-17 Pearson Education, Inc.; Fig. 02-18 a-c Pearson Education, Inc.; Fig. 02-19 Pearson Education, Inc.; Fig. 02-20 Pearson Education, Inc.; Fig. 02-20a Pearson Education, Inc.; Fig. 02-20b Pearson Education, Inc.; Fig. 02-21 Pearson Education, Inc.; Fig. 02-22 Pearson Education, Inc.; Fig. 02-23 International Mapping/Pearson Education, Inc.; Fig. 02-24 International Mapping/Pearson Education, Inc.; Fig. 02-25 Pearson Education, Inc.; Fig. 02-26 International Mapping/Pearson Education, Inc.; Fig. 02-27 International Mapping/Pearson Education, Inc.; Fig. 02-28 International Mapping/Pearson Education, Inc.; Fig. 02-29 International Mapping/Pearson Education, Inc.; Fig. 02-30 International Mapping/Pearson Education, Inc.; Fig. 02-31 International Mapping/Pearson Education, Inc.; Fig. 02-32 International Mapping/Pearson Education, Inc.; Fig. 02-33 International Mapping/Pearson Education, Inc.; Fig. 02-34 Thomas Malthus, Pearson Education, Inc.; Fig. 02-35 Arturo Rodriguez/VWPics/Alamy; Fig. 02-36 Image Broker/Alamy; Fig. 02-37 Thomas Malthus, Pearson Education, Inc.; Fig. 02-38 Pearson Education, Inc.; Fig. 02-39 Pearson Education, Inc.; Fig. 02-40 Pearson Education, Inc.; Fig. 02-41 Barry Lewis/Alamy; Fig. 02-42 International Mapping/Pearson Education, Inc.; Fig. 02-43 International Mapping/Pearson Education, Inc.; Fig. 02-44 International Mapping/Pearson Education, Inc.; Fig. 02-45 International Mapping/Pearson Education, Inc.; Fig. 02-46 Tim Graham/Getty Images; Fig. 02-47 International Mapping/Pearson Education, Inc.; Fig. 02-48b Office of Travel and Tourism Industries, http://travel.trade.gov/view/m-2013-I-001/index.html; Fig. 02-48 Pearson Education, Inc.; Fig. 02-49 David Bagnall/Alamy; Fig. 02-49a Center for Disease Control and Prevention, http://www.cdc.gov/vhf/ebola/outbreaks/2014-west-africa/distribution-map.html; Fig. 02-49b Pearson Education, Inc.; Fig. 02-49c Pearson Education, Inc.; Fig. 02-50 David Grossman/Alamy; Fig. 02-51 RosaIrene Betancourt 3/Alamy; Fig. 02-52 Morandi Bruno/Hemis/Alamy; Fig. 02-Explore Map data © 2015 Google;

Chapter 3 Fig. 03-CO Mauricio Lima/The New York Times/Redux Pictures; Fig. Vid. 03-01 Video Still provided by BBC Worldwide Learning; Fig. 03-01 Pearson Education, Inc.; Fig. 03-02 International Mapping/Pearson Education, Inc.; Fig. 03-03 International Mapping/Pearson Education, Inc.; Fig. 03-04 International Mapping/Pearson Education, Inc.; Fig. 03-05 Paul Swinney/Alamy; Fig. 03-06 International Mapping/Pearson Education, Inc.; Fig. 03-07 International Mapping/Pearson Education, Inc.; Fig. 03-08 Chico Sanchez/Alamy; Fig. 03-09 ZUMA Press/Alamy; Fig. 03-10 Pearson Education, Inc.; Fig. 03-11 International Mapping/Pearson Education, Inc.; Fig. 03-12 International Mapping/Pearson Education, Inc.; Fig. 03-13 International Mapping/Pearson Education, Inc.; Fig. 03-14 International Mapping/Pearson Education, Inc.; Fig. 03-15 International Mapping/Pearson Education, Inc.; Fig. 03-16 International Mapping/Pearson Education, Inc.; Fig. 03-17b Florian Kopp/Westend61 GmbH/Alamy; Fig. 03-17 International Mapping/Pearson Education, Inc.; Fig. 03-18 Yavuz Sariyildiz/Alamy; Fig. 03-19 International Mapping/Pearson Education, Inc.; Fig. 03-20 Pearson Education, Inc.; Fig. 03-21 International Mapping/Pearson Education, Inc.; Fig. 03-22 International Mapping/Pearson Education, Inc.; Fig. 03-23b Terry Smith Images/Alamy; Fig. 03-23 International Mapping/Pearson Education, Inc.; Fig. 03-24 EPA/Alamy; Fig. 03-25 Anthony Bannister/AGE Fotostock; Fig. 03-25 NASA Archive/Alamy; Fig. 03-26 Joerg Boethling/Alamy; Fig. 03-27 International Mapping/Pearson Education, Inc.; Fig. 03-28 Pearson Education, Inc.; Fig. 03-29 International Mapping/Pearson Education, Inc.; Fig. 03-30 Dubai-UAE/Alamy; Fig. 03-31 International Mapping/Pearson Education, Inc.; Fig. 03-32 International Mapping/Pearson Education, Inc.; Fig. 03-33 David Bagnall/Alamy; Fig. 03-34 International Mapping/Pearson Education, Inc.; Fig. 03-35 ZUMA Press/Alamy; Fig. 03-36 International Mapping/Pearson Education, Inc.; Fig. 03-37 Pearson Education, Inc.; Fig. 03-38 International Mapping/Pearson Education, Inc.; Fig. 03-39 Frans lemmens/Alamy; Fig. 03-40 Xinhua/Alamy; Fig. 03-41b Jim West/Alamy; Fig. 03-41c Jim West/Alamy; Fig. 03-41a International Mapping/Pearson Education, Inc.; Fig. 03-42 Sunpix Travel/Alamy; Fig. 03-43 614 Collection/Alamy; Fig. 03-44 International Mapping/Pearson Education, Inc.; Fig. 03-45 International Mapping/Pearson Education, Inc.; Fig. 03-46 EPA/Alamy; Fig 03-47 Jim West/Alamy; Fig 03-48 Andre Jenny/Alamy; Fig 03-49 Ira Berger/Alamy; Fig 03-50 Jim West/Alamy; Tbl. 03-01 Pearson Education, Inc.; Fig. 03-Explore Map data © 2015 Google;

Chapter 4 Fig. 04-CO Alex Grabchilev/Evgeniya Bakanova/Getty Images; Fig. Vid. 04-01 Video Still provided by BBC Worldwide Learning; Fig. 04-01 Aristidis Vafeiadakis/ZUMA Press/Alamy; Tbl. 04-01 Pearson Education, Inc.; Fig. 04-02 Maurice Joseph/Alamy; Tbl. 04-02 Pearson Education, Inc.; Fig. 04-03 Sergi Reboredo/Alamy; Tbl. 04-03 Internet Live Stats (www.InternetLiveStats.com), Elaboration of data by International Telecommunication Union (ITU), United Nations Population Division, Internet & Mobile Association of India (IAMAI), World Bank.; Fig. 04-04 Van Der Meer Marica/Arterra Picture Library/Alamy; Fig. 04-05a MediaPunch/Rex/AP Images; Fig. 04-06 International Mapping/Pearson Education, Inc.; Fig. 04-07 International Mapping/Pearson Education, Inc.; Fig. 04-08 Pearson Education, inc.; Fig. 04-09 International Mapping/Pearson Education, Inc.; Fig. 04-10 Dai Kurokawa/EPA/Newscom;

Fig. 04-11 International Mapping/Pearson Education, Inc.; Fig. 04-12 International Mapping/Pearson Education, Inc.; Fig. 04-13 International Mapping/Pearson Education, Inc.; Fig. 04-14 Piero Cruciatti/Alamy; Fig. 04-15 Tom Carter/Alamy; Fig. 04-16 International Mapping/Pearson Education, Inc.; Fig. 04-17 International Mapping/Pearson Education, Inc.; Fig. 04-18 Fred Ernst/Reuters; Fig. 04-19a RosalreneBetancourt 10/Alamy; Fig. 04-19b David Noyes/DanitaDelimont/Alamy; Fig. 04-20 Marco Secchi/Alamy; Fig. 04-21 Philip Wolmuth/Alamy; Fig. 04-22 Incamerastock/Alamy; Fig. 04-23b Graham Salter/Imagestate Media Partners Limited/Impact Photos/Alamy; Fig. 04-24 B. Fischer/Arco Images GmbH/Alamy; Fig. 04-25 International Mapping/Pearson Education, Inc.; Fig. 04-26 International Mapping/Pearson Education, Inc.; Fig. 04-27 International Mapping/Pearson Education, Inc.; Fig. 04-28a Huntstock/Alamy; Fig. 04-28b Chris Ryan/ OJO Images, Ltd./Alamy; Fig. 04-29a Moirenc Camille/Hemis/Alamy; Fig. 04-29b Arturs Stalidzans/Alamy; Fig. 04-31 Pearson Education, Inc.; Fig. 04-31 International Mapping/Pearson Education, Inc.; Fig. 04-32 International Mapping/Pearson Education, Inc.; Fig. 04-33 International Mapping/Pearson Education, Inc.; Fig. 04-34 Map data ©2015 Google; Fig. 04-35 International Mapping/Pearson Education, Inc.; Fig. 04-36 International Mapping/Pearson Education, Inc.; Fig. 04-37 International Mapping/Pearson Education, Inc.; Fig. 04-38 International Mapping/Pearson Education, Inc.; Fig. 04-39 Pearson Education, Inc.; Fig. 04-40 Megapress/Alamy; Fig. 04-41 George Sheldon/Alamy; Fig. 04-42 Religious Congregations & Membership Study, U.S. Religion Census, 1952 to 2010, http://www.rcms2010.org/maps2010.php?sel_denom=31&sel_map%5B%5D=2&confirm=confirm; Fig. 04-43 Olaf Krüger/Image Broker/Alamy; Fig. 04-44 Angry Brides is an Anti-dowry initiative by Shaadi.com, the World's No.1 Matchmaking Service; Fig. 04-45 Rosa Betancourt/Alamy; Fig. 04-46a David Ashdown/The Independent/Daily Mail/Rex/Alamy; Fig. 04-46b ZUMA Press/Alamy; Fig. 04-47 Christian Hutter/Image Broker/Alamy; Fig. 04-48 Fredrick Kippe/Alamy; Fig. 04-49 EPA/Alamy; Fig. 04-50 Rosalrene Betancourt/Alamy; Fig. 04-Explore Map data © 2015 Google;

Chapter 5 Fig. 05-CO Bazuki Muhammad/Reuters; Fig. 05-01 International Mapping/Pearson Education, Inc.; Tbl. 05-01 Lewis, M. Paul, Gary F. Simons, and Charles D. Fennig (eds.). 2015. Ethnologue: Languages of the World, Eighteenth edition. Dallas, Texas: SIL International. Online version: http://www.ethnologue.com.; Fig. Vid. 05-01 Video Still provided by BBC Worldwide Learning; Fig. 05-02 International Mapping/Pearson Education, Inc.; Fig. 05-03 Pearson Education, Inc.; Fig. 05-04 International Mapping/Pearson Education, Inc.; Fig. 05-05 International Mapping/Pearson Education, Inc.; Fig. 05-06 International Mapping/Pearson Education, Inc.; Fig. 05-07 International Mapping/Pearson Education, Inc.; Fig. 05-08 Adrian Sherratt/Alamy; Fig. 05-09 International Mapping/Pearson Education, Inc.; Fig. 05-10 International Mapping/Pearson Education, Inc.; Fig. 05-11 International Mapping/Pearson Education, Inc.; Fig. 05-12 International Mapping/Pearson Education, Inc.; Fig. 05-13 Sean Pavone/Alamy; Fig. 05-14 Origina and Diffusion of Indo-European: Nomadic Warrior Theory, Wikipedia, http://upload.wikimedia.org/wikipedia/commons/5/5a/IE_expansion.png; Fig. 05-15 International Mapping/Pearson Education, Inc.; Fig. 05-16 International Mapping/Pearson Education, Inc.; Fig. 05-17 Pearson Education, Inc.; Fig. 05-18 Christopher Canty Photography/Alamy; Fig. 05-19 Xinhua/Alamy; Fig. 05-20 Yury Smityuk/ITAR-TASS Photo Agency/Alamy; Fig. 05-21 International Mapping/Pearson Education, Inc.; Fig. 05-22 Pearson Education, Inc.; Fig. 05-23 International Mapping/Pearson Education, Inc.; Fig. 05-24 International Mapping/Pearson Education, Inc.; Fig. 05-25 International Mapping/Pearson Education, Inc.; Fig. 05-26b Rawpixel/Fotolia; Fig. 05-26 Pearson Education, Inc.; Fig. 05-27 Alan Moore/Alamy; Fig. 05-28 International Mapping/Pearson Education, Inc.; Fig. 05-29 International Mapping/Pearson Education, Inc.; Fig. 05-30 International Mapping/Pearson Education, Inc.; Fig. 05-31 Rosalrene Betancourt 3/ Alamy; Fig. 05-32 Kevin Foy/Alamy; Fig. 05-33 International Mapping/Pearson Education, Inc.; Fig. 05-34 International Mapping/Pearson Education, Inc.; Fig. 05-35 International Mapping/Pearson Education, Inc.; Fig. 05-36 International Mapping/Pearson Education, Inc.; Fig. 05-37 © 2015 Google Inc. All rights reserved. Google and the Google Logo are registered trademarks of Google Inc.; Fig. 05-38 International Mapping/Pearson Education, Inc.; Fig. 05-39 Andrew J. Strack, Courtesy of Myaamia Center Archive; Fig. 05-40 International Mapping/Pearson Education, Inc.; Fig. 05-41 Kevin Britland/Alamy; Fig. 05-41 Nic Hamilton Photographic/Alamy; Fig. 05-42 Adam Eastland/Alamy; Fig. 05-43 International Mapping/Pearson Education, Inc.; Fig. 05-44 UrbanImages/Alamy; Fig. 05-45 International Mapping/Pearson Education, Inc.; Fig. 05-46 Miguel Tona/EPA/Alamy; Fig. 05-47 Ragnar Th Sigurdsson/Arctic Images/Alamy; Fig. 05-47 International Mapping/Pearson Education, Inc.; Fig. 05-48 Zoonar/gkuna/Alamy; Fig. 05-49 International Mapping/Pearson Education, Inc.; Fig. 05-50 Etabeta/Alamy; Fig. 05-51 Images & Stories/Alamy; Fig. 05-52 Peter Phipp/Travelshots.com/Alamy; Fig. 05-53 Robert Clay/Alamy ; Fig. 05-Explore Map data © 2015 Google;

Chapter 6 Fig. 06-CO Jim West/Alamy; Fig. Vid. 06-01 Video Still provided by BBC Worldwide Learning; Fig. 06-01 International Mapping/Pearson Education, Inc.; Fig. 06-02 Eye35/Alamy; Fig. 06-03 F9Photos/Shutterstock; Fig. 06-04 Saiko3P/Shutterstock; Fig. 06-04 International Mapping/Pearson Education, Inc.; Fig. 06-05 International Mapping/Pearson Education, Inc.; Fig. 06-06 International Mapping/Pearson Education, Inc.; Fig. 06-07 International Mapping/Pearson Education, Inc.; Fig. 06-08 "International Mapping/Pearson Education, Inc.; Fig. 06-09 World Distribution of Muslim Population, Pew Research Center's Forum on Religion & Public Life, Mapping the Global Muslim Population. October 2009; Fig. 06-10 International Mapping/Pearson Education, Inc.; Fig. 06-11 International Mapping/Pearson Education, Inc.; Fig. 06-12 Pearson Education, Inc.; Fig. 06-13 Rob Crandall; Fig. 06-13 Pearson Education, Inc.; Fig. 06-14 International Mapping/Pearson Education, Inc.; Fig. 06-15 a-b International Mapping/Pearson Education, Inc.; Fig. 06-16 International Mapping/Pearson Education, Inc.; Fig. 06-17 Pearson Education, Inc.; Fig. 06-18 International Mapping/Pearson Education, Inc.; Fig. 06-19 Jim Hollander/Epa/Alamy; Fig. 06-20 Tmyusof/Alamy; Fig. 06-21 Anthony Asael/Hemis/Alamy; Fig. 06-22 Kat Kallou/Alamy; Fig. 06-23 Roussel Bernard/Alamy; Fig. 06-24 International Mapping/Pearson Education, Inc.; Fig. 06-25 International Mapping/Pearson Education, Inc.; Fig. 06-26 International Mapping/Pearson Education, Inc.; Fig. 06-27 International Mapping/Pearson Education, Inc.; Fig. 06-27 International Mapping/Pearson Education, Inc.; Fig. 06-29 a-b International Mapping/Pearson Education, Inc.; Fig. 06-30 Teithiwr/Alamy; Fig. 06-31 International Mapping/Pearson Education, Inc.; Fig. 06-32 International Mapping/Pearson Education, Inc.; Fig. 06-33 Christine Osborne/World Religions Photo Library/Alamy; Fig. 06-34 Jim West/Alamy; Fig. 06-35 Ira Berger/Alamy; Fig. 06-36 International Mapping/Pearson Education, Inc.; Fig. 06-37 Viacheslav Lopatin/Shutterstock; Fig. 06-38 Johan Pabon/Alamy; Fig. 06-39 Seanpavonephoto/Fotolia; Fig. 06-40 SuperStock/Glow Images; Fig. 06-40 International Mapping/Pearson Education, Inc.; Fig. 06-42 a-b International Mapping/Pearson Education, Inc.; Fig. 06-43 International Mapping/Pearson Education, Inc.; Fig. 06-44 Karim Sahib/AFP/Getty Images; Fig. 06-45 Kat Kallou/Alamy; Fig. 06-46 International Mapping/Pearson Education, Inc.; Fig. 06-47 Gianni Muratore/Alamy; Fig. 06-48 Art Directors & Trip/Alamy; Fig. 06-49 Lee Thomas/Alamy; Fig. 06-50 Horizon Images/Motion/Alamy; Fig. 06-51 Ira Berger/Alamy; Fig. 06-52 Swns/Alamy; Fig. 06-53 Epa European Pressphoto Agency B.V./Alamy; Fig. 06-54 Blaine Harrington Iii/Alamy; Fig. 06-55a Mediacolor's/Alamy; Fig. 06-55b ZUMA Press, Inc./Alamy; Fig. 06-56 International Mapping/Pearson Education, Inc.; Fig. 06-57 a-c International Mapping/Pearson Education, Inc.; Fig. 06-58 International Mapping/Pearson Education, Inc.; Fig. 06-59 Duby Tal/Albatross/Alamy; Fig. 06-60 International Mapping/Pearson Education, Inc.; Fig. 06-61 Galit Seligmann/Alamy; Fig. 06-62 Roger Cracknell 01/classic/Alamy; Fig. 06-63 EPA/Alamy; Fig. 06-64 David R. Frazier/Photolibrary, Inc./Alamy; Fig. 06-65 Rafael Ben-Ari/Alamy; Fig. 06-Explore Map data © 2015 Google;

Chapter 7 Fig. 07-CO Jamal Nasrallah/EPA/Newscom; Fig. 07-08 International Mapping/Pearson Education, Inc.; Fig. 07-01a Reuters; Fig. 07-01b Reuters; Fig. 07-01c Obama Press Office/UPI/Newscom; Fig. Vid. 07-01 Video Still provided by BBC Worldwide Learning; Fig. 07-02 International Mapping/Pearson Education, Inc.; Fig. 07-03 International Mapping/Pearson Education, Inc.; Fig. 07-04 Janine Wiedel Photolibrary/Alamy; Fig. 07-05 International Mapping/Pearson Education, Inc.; Fig. 07-06 International Mapping/Pearson Education, Inc.; Fig. 07-07 International Mapping/Pearson Education, Inc.; Fig. 07-09 2010 Census Redistricting Data (Public Law 94-171) Summary File, Table P1, U.S. Census Bureau, http://www.census.gov/prod/cen2010/briefs/c2010br-10.pdf; Fig. 07-10 Thibault Camus/AP Images; Fig. 07-11 International Mapping/Pearson Education, Inc.; Fig. 07-12 International Mapping/Pearson Education, Inc.; Fig. 07-13 International Mapping/Pearson Education, Inc.; Fig. 07-14 International Mapping/Pearson Education, Inc.; Fig. 07-15 International Mapping/Pearson Education, Inc.; Fig. 07-16 AHowden/Brazil Stock Photography/Alamy; Fig. 07-17 International Mapping/Pearson Education, Inc.; Fig. 07-18 International Mapping/Pearson Education, Inc.; Fig. 07-19 North Wind Picture Archives/Alamy; Fig. 07-21 Europa Newswire/Alamy; Fig. 07-22 Pictorial Press Ltd/Alamy; Fig. 07-23 International Mapping/Pearson Education, Inc.; Fig. 07-24 Pearson Education, Inc.; Fig. 07-25 Pearson Education, Inc.; Fig. 07-26 Elliott Erwitt/Magnum; Fig. 07-27 The Protected Art Archive/Alamy; Fig. 07-28 International Mapping/Pearson Education, Inc.; Fig. 07-29b Evans/Hulton Deutsch Collection/Corbis; Fig. 07-30 Erik S. Lesser/Epa/Alamy; Fig. 07-31 Icpix Can/Alamy; Fig. 07-32 International Mapping/Pearson Education, Inc.; Fig. 07-33 Colin McPherson/Alamy; Fig. 07-34 International Mapping/Pearson Education, Inc.; Fig. 07-35 International Mapping/Pearson Education, Inc.; Fig. 07-36b Roberto Fumagalli/Alamy; Fig. 07-36a International Mapping/Pearson Education, Inc.; Fig. 07-37 International Mapping/Pearson Education, Inc.; Fig. 07-38 Nabil Mounzer/Epa/Alamy; Fig. 07-39 Beth Wald/Aurora Photos/Alamy; Fig. 07-40 International Mapping/Pearson Education, Inc.; Fig. 07-40 International Mapping/Pearson Education, Inc.; N/A Map data © 2015 Google;

Fig. 07-41 International Mapping/Pearson Education, Inc.; Fig. 07-42 International Mapping/Pearson Education, Inc.; Fig. 07-43a Otto Lang/Encyclopedia/Corbis; Fig. 07-43b Nigel Chandler/Sygma/Corbis; Fig. 07-43c Aleksandar Todorovic/Shutterstock; Fig. 07-44 Map data © 2015 Google; Fig. 07-44 International Mapping/Pearson Education, Inc.; Fig. 07-45 Roberto Cornacchia/Alamy; Fig. 07-46 Brianeuro/Alamy; Fig. 07-47 Pearson Education, Inc.; Fig. 07-48 International Mapping/Pearson Education, Inc.; Fig. 07-49 Images & Stories/Alamy; Fig. 07-50 International Mapping/Pearson Education, Inc.; Fig. 07-51 International Mapping/Pearson Education, Inc.; Fig. 07-52 EPA/Alamy; Fig. 07-53 Rafael Suanes/MCT/Newscom; Fig. 07-54 Interface Images/ Food and Drink Photos/Alamy; Fig. 07-55 Frilet Patrick/Hemis/Alamy; Fig. 07-56 Economic images/Alamy; Fig. 07-57 Eric Nathan/Alamy;

Chapter 8 Fig. 08-CO Barbara Lindberg/Rex/AP Images; Fig. Vid. 08-01 Video Still provided by BBC Worldwide Learning; Fig. 08-01 International Mapping/Pearson Education, Inc.; Tbl. 08-01 International Mapping/Pearson Education, Inc.; Fig. 08-02 Andrea Izzotti/Fotolia; Fig. 08-03 Planet Observer/Universal Images Group Limited/Alamy; Fig. 08-03 Planet Observer/Science Source; Fig. 08-04 International Mapping/Pearson Education, Inc.; Fig. 08-06 Ruairidh Villar/Newscom; Fig. 08-06 Michael Fay/National Geographic Image Collection/Alamy; Fig. 08-06b International Mapping/Pearson Education, Inc.; Fig. 08-07 International Mapping/Pearson Education, Inc.; Fig. 08-08 International Mapping/Pearson Education, Inc.; Fig. 08-09 International Mapping/Pearson Education, Inc.; Fig. 08-10 International Mapping/Pearson Education, Inc.; Fig. 08-11 International Mapping/Pearson Education, Inc.; Fig. 08-12 International Mapping/Pearson Education, Inc.; Fig. 08-13 International Mapping/Pearson Education, Inc.; Fig. 08-14 Photocay/Alamy; Fig. 08-15 International Mapping/Pearson Education, Inc.; Fig. 08-17 International Mapping/Pearson Education, Inc.; Fig. 08-18 International Mapping/Pearson Education, Inc.; Fig. 08-19 International Mapping/Pearson Education, Inc.; Fig. 08-20 International Mapping/Pearson Education, Inc.; Fig. 08-21 International Mapping/Pearson Education, Inc.; Fig. 08-22 International Mapping/Pearson Education, Inc.; Fig. 08-23 International Mapping/Pearson Education, Inc.; Fig. 08-24 International Mapping/Pearson Education, Inc.; Fig. 08-25 World History Archive/Alamy; Fig. 08-26 International Mapping/Pearson Education, Inc.; Fig. 08-27 International Mapping/Pearson Education, Inc.; Fig. 08-28 Epa/Alamy; Fig. 08-29 David Muenker/Alamy; Fig. 08-30 International Mapping/Pearson Education, Inc.; Fig. 08-31 International Mapping/Pearson Education, Inc.; Fig. 08-32 International Mapping/Pearson Education, Inc.; Fig. 08-34 Tim Moore/Alamy; Fig. 08-34 AA World Travel Library/Alamy; Fig. 08-34 International Mapping/Pearson Education, Inc.; Fig. 08-35 International Mapping/Pearson Education, Inc.; Fig. 08-36 International Mapping/Pearson Education, Inc.; Fig. 08-37 International Mapping/Pearson Education, Inc.; Fig. 08-38 International Mapping/Pearson Education, Inc.; Fig. 08-39 International Mapping/Pearson Education, Inc.; Fig. 08-40 International Mapping/Pearson Education, Inc.; Fig. 08-41 International Mapping/Pearson Education, Inc.; Fig. 08-42 International Mapping/Pearson Education, Inc.; Fig. 08-43 International Mapping/Pearson Education, Inc.; Fig. 08-44 International Mapping/Pearson Education, Inc.; Fig. 08-45 Sandy Macys/Alamy; Fig. 08-46 International Mapping/Pearson Education, Inc.; Fig. 08-47 Pearson Education, Inc.; Fig. 08-49 International Mapping/Pearson Education, Inc.; Fig. 08-50 International Mapping/Pearson Education, Inc.; Fig. 08-51 International Mapping/Pearson Education, Inc.; Fig. 08-52 Atef Safadi/EPA/Newscom; Fig. 08-53a-b Bettmann/Corbis; Fig. 08-54 International Mapping/Pearson Education, Inc.; Fig. 08-55 International Mapping/Pearson Education, Inc.; Fig. 08-56 International Mapping/Pearson Education, Inc.; Fig. 08-57 International Mapping/Pearson Education, Inc.; Fig. 08-58 International Mapping/Pearson Education, Inc.; Fig. 08-59 Dan Lampariello/Reuters; Fig. 08-60a-b Sean Adair/Reuters; Fig. 08-61 International Mapping/Pearson Education, Inc.; Fig. 08-62 Stephane Gautier/Alamy; Fig. 08-63 EPA/Alamy; Fig. 08-64 Pearson Education, Inc.; Fig. 08-65 Ton Koene/Horizons WWP/Alamy; Fig. 08-66 a-c US State Department; Fig. 08-67 International Mapping/Pearson Education, Inc.; 08-67 Matthias Oesterle/Alamy; N/A Map data © 2015 Google; Fig. 08-68 Martin Cleaver/AP Images; Fig. 08-69 EPA/Alamy; Fig. Fig. 08-71 Raga Jose Fuste/Prisma Bildagenture AG/Alamy; Fig. 08-72 Jeffrey Blackler/Alamy;

Chapter 9 Fig. 09-CO Simon Rawles/Getty Images; Fig. 09-01 Chas53/Fotolia; Fig. Vid. 09-01 Video Still provided by BBC Worldwide Learning; Tbl. 09-01 Pearson Education, Inc.; Fig. 09-02 International Mapping/Pearson Education, Inc.; Tbl. 09-02 Pearson Education, Inc.; Fig. 09-03 International Mapping/Pearson Education, Inc.; Tbl. 09-03 Pearson Education, Inc.; Fig. 09-04 International Mapping/Pearson Education, Inc.; Fig. 09-05 Lee Sninder/Alamy; Fig. 09-06 International Mapping/Pearson Education, Inc.; Fig. 09-07 International Mapping/Pearson Education, Inc.; Fig. 09-08 International Mapping/Pearson Education, Inc.; Fig. 09-09 AfriPics.com/Alamy; Fig. 09-10 Aerial World/GlowImages; Fig. 09-11 International Mapping/Pearson Education, Inc.; Fig. 09-12 International Mapping/Pearson Education, Inc.; Fig. 09-13 International Mapping/Pearson Education, Inc.; Fig. 09-14 International Mapping/Pearson Education, Inc.; Fig. 09-15 Martin Harvey/Alamy; Fig. 09-16 Tuul/Glow Images; Fig. 09-17 AfriPics.com/Alamy; Fig. 09-18 Charles O. Cecil/Alamy; Fig. 09-19 Pearson Education, Inc.; Fig. 09-20 Tim Gainey/Alamy; Fig. 09-21 Edmund Lowe Photography/Shutterstock; Fig. 09-22 Keantian/Fotolia; Fig. 09-23 Keren Su/China Span/Alamy; Fig. 09-24 International Mapping/Pearson Education, Inc.; Fig. 09-25 Terry Foster/Alamy; Fig. 09-26 Pearson Education, Inc.; Fig. 09-27 International Mapping/Pearson Education, Inc.; Fig. 09-28 Jim Weber/ZUMA Press/Newscom; Fig. 09-29 International Mapping/Pearson Education, Inc.; Fig. 09-30 Pearson Education, Inc.; Fig. 09-31 mRGB/Shutterstock; Fig. 09-32 David R. Frazier Photolibrary/Alamy; Fig. 09-33 Pearson Education, Inc.; Fig. 09-34 Belinda Images/SuperStock; Fig. 09-35 International Mapping/Pearson Education, Inc.; Fig. 09-36 International Mapping/Pearson Education, Inc.; Fig. 09-37 International Mapping/Pearson Education, Inc.; Fig. 09-38 International Mapping/Pearson Education, Inc.; Fig. 09-39 International Mapping/Pearson Education, Inc.; Fig. 09-40 International Mapping/Pearson Education, Inc.; Fig. 09-41 International Mapping/Pearson Education, Inc.; Fig. 09-42 Eitan Simanor/Alamy; Fig. 09-44 Joerg Boethling/Alamy; Fig. 09-44 Dairy Data, United States Department of Agriculture Economic Research Service, http://www.ers.usda.gov/data-products/dairy-data.aspx; Fig. 09-45 Ashley Cooper Pics/Alamy; Fig. 09-46 R.Hamilton Smith/Design Pics Inc/Alamy; Fig. 09-47 International Mapping/Pearson Education, Inc.; Fig. 09-48 Pearson Education, Inc.; Fig. 09-49 Norsob/Alamy; Fig. 09-50 Norsob/Alamy; Fig. 09-51 International Mapping/Pearson Education, Inc.; Fig. 09-52 International Mapping/Pearson Education, Inc.; Fig. 09-53 International Mapping/Pearson Education, Inc.; Fig. 09 54 International Mapping/Pearson Education, Inc.; Fig. 09-55 International Mapping/Pearson Education, Inc.; Fig. 09-56 Pearson Education, Inc.; Fig. 09-57 International Mapping/Pearson Education, Inc.; Fig. 09-58 International Mapping/Pearson Education, Inc.; Fig. 09-59 International Mapping/Pearson Education, Inc.; Fig. 09-60 Edd Westmacott/Alamy; Fig. 09-61 Photo20ast/Alamy; Fig. 09-62 Stacy Walsh Rosenstock/Alamy; Fig. 09-63 Urikiri-Shashin-Kan/Alamy; Fig. 09-64 ZUMA Press Inc/Alamy; Fig. 09-Explore Map data © 2015 Google;

Chapter 10 Fig. 10-CO Peter Treanor/Alamy; Fig. Vid. 10-01 Video Still provided by BBC Worldwide Learning; Fig. 10-01 International Mapping/Pearson Education, Inc.; Fig. 10-02a Stephen Searle/Alamy; Fig. 10-02b BSIP SA/Alamy; Fig. 10-02c Celso Diniz/Shutterstock; Fig. 10-02d James Hardy/PhotoAlto/Alamy; Fig. 10-02e Meunierd/Shutterstock; Fig. 10-02f Banana Republic Images/Shutterstock; Fig. 10-02g Marcia Chambers/DBImages/Alamy; Fig. 10-02h Adam Buchanan/Danita Delimont/Alamy; Fig. 10-02I Jake Lyell/Alamy; Fig. 10-02 International Mapping/Pearson Education, Inc.; Fig. 10-03 International Mapping/Pearson Education, Inc.; Fig. 10-04 International Mapping/Pearson Education, Inc.; Fig. 10-05 International Mapping/Pearson Education, Inc.; Fig. 10-06 International Mapping/Pearson Education, Inc.; Fig. 10-07 International Mapping/Pearson Education, Inc.; Fig. 10-08 International Mapping/Pearson Education, Inc.; Fig. 10-09 International Mapping/Pearson Education, Inc.; Fig. 10-10 International Mapping/Pearson Education, Inc.; Fig. 10-11 International Mapping/Pearson Education, Inc.; Fig. 10-12 International Mapping/Pearson Education, Inc.; Fig. 10-13 International Mapping/Pearson Education, Inc.; Fig. 10-14 Pearson Education, Inc.; Fig. 10-15 International Mapping/Pearson Education, Inc.; Fig. 10-16 International Mapping/Pearson Education, Inc.; Fig. 10-17 International Mapping/Pearson Education, Inc.; Fig. 10-18 Pearson Education, Inc.; Fig. 10-19 International Mapping/Pearson Education, Inc.; Fig. 10-20 International Mapping/Pearson Education, Inc.; Fig. 10-21 International Mapping/Pearson Education, Inc.; Fig. 10-22 International Mapping/Pearson Education, Inc.; Fig. 10-23 International Mapping/Pearson Education, Inc.; Fig. 10-25 Jake Lyell/Alamy; Fig. 10-26 Pearson Education, Inc.; Fig. 10-27 International Mapping/Pearson Education, Inc.; Fig. 10-28 Pearson Education, Inc.; Fig. 10-30 International Mapping/Pearson Education, Inc.; Fig. 10-31 Ventura, Stephanie J. M.A.; Hamilton, Brady E. Ph.D.; and Mathews, T.J. M.S, "National and State Patterns of Teen Births in the United States, 1940–2013", Division of Vital Statistics, National Vital Statistics Report, Volume 63, Number 4, August 20, 2014, http://www.cdc.gov/nchs/data/nvsr/nvsr63/nvsr63_04.pdf; Fig. 10-32 Devinder Sangha/Alamy; Fig. 10-33 Rosalrene Betancourt/Alamy; Fig. 10-34 International Mapping/Pearson Education, Inc.; Fig. 10-35 India GOP per capita, Trading Economics, http://www.tradingeconomics.com/india/gdp-per-capita; Fig. 10-36 EPA/Alamy; Fig. 10-37 Radiokafka/Shutterstock; Fig. 10-38 Dinodia Photos/Alamy; Fig. 10-39 International Mapping/Pearson Education, Inc.; Fig. 10-40 International Mapping/Pearson Education, Inc.; Fig. 10-41 IBRD Loans and IDA Credits (DOD, Current US$, The World Bank, http://data.worldbank.org/indicator/DT.DOD.MWBG.CD; Fig. 10-42 EPA/Alamy; Fig. 10-43 Nikolas Georgiou/Alamy; Fig. 10-43 External debt stocks (% of GNI), The World Bank, http://data.worldbank.org/indicator/DT.DOD.DECT.GN.ZS; Fig. 10-45 GDP Per Capita Growth (Annual %), The World Bank, http://data.worldbank.org/indicator/NY.GDP.PCAP.KD.ZG?page=1; Fig. 10-46 Antonio Perez/MCT/

Newscom; Fig. 10-47 Joerg Boethling/Alamy; Fig. 10-48 Yanice Idir/Alamy; Fig. 10-49 International Mapping/Pearson Education, Inc.; Fig. 10-50 International Mapping/Pearson Education, Inc.; Fig. 10-51 International Mapping/Pearson Education, Inc.; Fig. 10-52 International Mapping/Pearson Education, Inc.; Fig. 10-53 Jurjen Verhagen/Zolabo/http://www.zolabo.com/projects/hdi/; Fig. 10-54 DPA Picture Alliance/Alamy; Fig. 10-55 Peter Titmuss/Alamy; Fig. 10-56 Jake Lyell/Alamy; Fig. 10-Explore Map data © 2015 Google;

Chapter 11 Fig. 11-CO Rajat Gupta/EPA/Alamy; Fig. 11-01 Interfoto/Alamy; Fig. Vid. 11-01 Video Still provided by BBC Worldwide Learning; Fig. 11-02 International Mapping/Pearson Education, Inc.; Fig. 11-03 Pearson Education, Inc.; Fig. 11-04 Industrial Production By Country, Quandl, https://www.quandl.com/collections/economics/industrial-production-by-country; Fig. 11-05 International Mapping/Pearson Education, Inc.; Fig. 11-06 International Mapping/Pearson Education, Inc.; Fig. 11-08 International Mapping/Pearson Education, Inc.; Fig. 11-09 International Mapping/Pearson Education, Inc.; Fig. 11-10 Tim Roberts Photography/Shutterstock; Fig. 11-11 International Mapping/Pearson Education, Inc.; Fig. 11-12 International Mapping/Pearson Education, Inc.; Fig. 11-13 International Mapping/Pearson Education, Inc.; Fig. 11-14 Pearson Education, Inc.; Fig. 11-15a Drive Images/Alamy; Fig. 11-15b Drive Images/Alamy; Fig. 11-15c Izmocars/Izmo/Corbis; Fig. 11-15d Drive Images/Alamy; Fig. 11-15 Pearson Education, Inc.; Fig. 11-16 International Mapping/Pearson Education, Inc.; Fig. 11-17 International Mapping/Pearson Education, Inc.; Fig. 11-18a International Mapping/Pearson Education, Inc.; Fig. 11-18b International Mapping/Pearson Education, Inc.; Fig. 11-19a International Mapping/Pearson Education, Inc.; Fig. 11-19b International Mapping/Pearson Education, Inc.; Fig. 11-20 Major Freight Corridors, U.S. Department of Transportation Federal Highway Administration, http://www.ops.fhwa.dot.gov/freight/freight_analysis/nat_freight_stats/mjrfreightcorridors.htm; Fig. 11-21 Tonnage of Trailer-on-Flatcar and Container-on-Flatcar Rail Intermodal Moves: 2011, U.S. Department of Transportation Federal Highway Administration, http://www.ops.fhwa.dot.gov/freight/freight_analysis/nat_freight_stats/intermodalrail2011.htm; Fig. 11-22 World Transportation Patterns, Princeton University, http://qed.princeton.edu/getfile.php?f=World_Transportation_Patterns.jpg; Fig. 11-23 Hourly compensation costs in manufacturing, U.S. dollars, 2012, United States Department of Labor, http://www.bls.gov/fls/ichcc.pdf; Fig. 11-24 Khim Hoe Ng/Alamy; Fig. 11-25 Pictorial Press Ltd/Alamy; Fig. 11-26 Alan Spencer/Powered by Light/Alan Spencer/Alamy; Fig. 11-27a Joerg Boethling/Alamy; Fig. 11-27b Paul Prescott/Dreamstime; Fig. 11-27 International Mapping/Pearson Education, Inc.; Fig. 11-28b Paul Prescott/Shuttterstock; Fig. 11-28 International Mapping/Pearson Education, Inc.; Fig. 11-29 International Mapping/Pearson Education, Inc.; Fig. 11-30 International Mapping/Pearson Education, Inc.; Fig. 11-31 International Mapping/Pearson Education, Inc.; Fig. 11-32 International Mapping/Pearson Education, Inc.; Fig. 11-33 International Mapping/Pearson Education, Inc.; Fig. 11-34 International Mapping/Pearson Education, Inc.; Fig. 11-35 International Mapping/Pearson Education, Inc.; Fig. 11-36 International Mapping/Pearson Education, Inc.; Fig. 11-37 International Mapping/Pearson Education, Inc.; Fig. 11-38 International Mapping/Pearson Education, Inc.; Fig. 11-39 International Mapping/Pearson Education, Inc.; Fig. 11-40 International Mapping/Pearson Education, Inc.; Fig. 11-41 Pearson Education, Inc.; Fig. 11-42 International Mapping/Pearson Education, Inc.; Fig. 11-43 Art Directors & TRIP/Alamy; Fig. 11-44 International Mapping/Pearson Education, Inc.; Fig. 11-45 International Mapping/Pearson Education, Inc.; Fig. 11-46 Pearson Education, Inc.; Fig. 11-47 International Mapping/Pearson Education, Inc.; Fig. 11-48 International Mapping/Pearson Education, Inc.; Fig. 11-49 Pearson Education, Inc.; Fig. 11-50 Ron Buskirk/Alamy; Fig. 11-51 International Mapping/Pearson Education, Inc.; Fig. 11-52 International Mapping/Pearson Education, Inc.; Fig. 11-53 David Wall/Alamy; Fig. 11-54 International Mapping/Pearson Education, Inc.; Fig. 11-55 Jbdodane/Alamy; Fig. 11-56 Rob Lacey/vividstock.net/Alamy ; Fig. 11-57 International Mapping/Pearson Education, Inc.; Fig. 11-58 International Mapping/Pearson Education, Inc.; Fig. 11-59 International Mapping/Pearson Education, Inc.; Fig. 11-60 International Mapping/Pearson Education, Inc.; Fig. 11-61 Marcia Chambers/DB Images/Alamy; Fig. 11-62 International Mapping/Pearson Education, Inc.; Fig. 11-63 International Mapping/Pearson Education, Inc.; Fig. 11-64 Alis Photo/Shutterstock; Fig. 11-65 Photoshot Holdings Ltd./Alamy; Fig. 11-66 Jeremy Graham/dbimages/Alamy; Fig. 11-67 Greenhouse Gas Reporting Program, United States Environmental Protection Agency, http://www.epa.gov/ghgreporting/images/ghgdata/landfills-map.gif; Fig. 11-68 Rosanne Tackaberry/Alamy; Fig. 11-69 International Mapping/Pearson Education, Inc.; Fig. 11-70 International Mapping/Pearson Education, Inc.; Fig. 11-71 Ji Hua/Featurechina/Newscom; Fig. 11-72 International Mapping/Pearson Education, Inc.; Fig. 11-73 International Mapping/Pearson Education, Inc.; Fig. 11-74 International Mapping/Pearson Education, Inc.; Fig. 11-75 International Mapping/Pearson Education, Inc.; Fig. 11-76a Pearson Education, Inc.; Fig. 11-76b Pearson Education, Inc.; Fig. 11-77a Pearson Education, Inc.; Fig. 11-77b Pearson Education, Inc.; Fig. 11-78 International Mapping/Pearson Education, Inc.; Fig. 11-79 Sheikh Rajibul Islam/Alamy; Fig. 11-80 International Mapping/Pearson Education, Inc.; Fig. 11-81a Picsfive/Fotolia; Fig. 11-81b International Mapping/Pearson Education, Inc.; Fig. 11-82 MSW Recycling Rates, 1960 to 2013, United States Environmental Protection Agency, http://www.epa.gov/epawaste/nonhaz/municipal/pubs/2013_advncng_smm_fs.pdf; Fig. 11-83 Total MSW Generation (by material), 2013 254 Millions Tons (before recycling), Total MSW Discards (by material), 2013 167 Million Tons (after recycling and composting), United States Environmental Protection Agency, http://www.epa.gov/epawaste/nonhaz/municipal/pubs/2013_advncng_smm_fs.pdf; Fig. 11-84a NASA; Fig. 11-84d Jesse Allen/NASA Earth Observatory; Fig. 11-84 b,c The Aral Sea Loses Its Eastern Lobe, NASA, http://earthobservatory.nasa.gov/IOTD/view.php?id=84437; Fig. 11-85 Aquariagirl1970/Fotolia; Fig. 11-86 Heiko Kueverling/Shutterstock; Fig. 11-87 DPA Picture Alliance Archive/Alamy; Fig. 11-88 Imagine4sell/Fotolia; Fig. 11-89 Paul H. Nilson/Imperial Valley Press/AP Images; Fig. 11-Explore Map data © 2015 Google;

Chapter 12 Fig. 12-CO Guillem Lopez/Alamy; Fig. Vid. 12-01 Video Still provided by BBC Worldwide Learning; Fig. 12-01 International Mapping/Pearson Education, Inc.; Fig. 12-02 International Mapping/Pearson Education, Inc.; Fig. 12-03 International Mapping/Pearson Education, Inc.; Fig. 12-04 a-c International Mapping/Pearson Education, Inc.; Fig. 12-05 International Mapping/Pearson Education, Inc.; Fig. 12-06 International Mapping/Pearson Education, Inc.; Fig. 12-07 International Mapping/Pearson Education, Inc.; Fig. 12-08 Thomas Roetting/LOOK Die Bildagentur der Fotografen GmbH/Alamy; Fig. 12-09 International Mapping/Pearson Education, Inc.; Fig. 12-10 International Mapping/Pearson Education, Inc.; Fig. 12-11 Ivan Vdovin/Alamy; Fig. 12-12 Jon Bilous/Alamy; Fig. 12-13 RosalreneBetancourt4/Alamy; Fig. 12-14 Philip Scalia/Alamy; Fig. 12-15 a,b Food Access Research Atlas, United States Department of Agriculture Economic Research Service, http://www.ers.usda.gov/data-products/food-access-research-atlas/go-to-the-atlas.aspx; Fig. 12-16 Jim Richardson/National Geographic Image Collection/Alamy; Fig. 12-17 John Behr/Alamy; Fig. 12-18a Urbanmyth/Alamy; Fig. 12-18b DPA Picture Alliance Archive/Alamy; Fig. 12-19 Songquan Den/Shutterstock; Fig. 12-20 International Mapping/Pearson Education, Inc.; Fig. 12-21 International Mapping/Pearson Education, Inc.; Fig. 12-22 Mikecphoto/Shutterstock; Fig. 12-23 International Mapping/Pearson Education, Inc.; Fig. 12-24 S. Forster/Alamy; Fig. 12-25 DB Images/Alamy; Fig. 12-26 International Mapping/Pearson Education, Inc.; Fig. 12-27 Kobby Dagan/Shutterstock; Fig. 12-28 International Mapping/Pearson Education, Inc.; Fig. 12-29 International Mapping/Pearson Education, Inc.; Fig. 12-30 International Mapping/Pearson Education, Inc.; Fig. 12-31 Matthew Oldfield Travel Photography/Alamy; Fig. 12-32 Russ Heinl/All Canada Photos/Alamy; Fig. 12-33 Pegaz/Alamy; Fig. 12-34 Robert Quinlan/Alamy; Fig. 12-35 Cotswolds Photo Library/Alamy; Fig. 12-36 a,b International Mapping/Pearson Education, Inc.; Fig. 12-37 Everett Collection Historical/Alamy; Fig. 12-38 Baloncici/Shutterstock; Fig. 12-39 JLImages/Alamy; Fig. 12-40 International Mapping/Pearson Education, Inc.; Fig. 12-41 International Mapping/Pearson Education, Inc.; Fig. 12-42 International Mapping/Pearson Education, Inc.; Fig. 12-43 Fastest Growing Cities And Urban Areas (1 to 100), City Mayors, http://www.citymayors.com/statistics/urban_growth1.html; Fig. 12-44 Javarman/Fotolia; Fig. 12-45 Michele Falzone/Alamy; Fig. 12-46 Xinhua/Alamy; Fig. 12-47 Philip Scalia/Alamy; Fig. 12-48 Ian Dagnall/Alamy; 12-49 Map data © 2015 Google; Fig 12-Explore Map data © 2015 Google; Fig. 12-Explore Map data © 2015 Google;

Chapter 13 Fig. 13-CO Alexandre Meneghin/Reuters; Fig. 13-01 Gemma Ferrando/Westend61 GmbH/Alamy; Fig. Vid. 13-01 Video Still provided by BBC Worldwide Learning; Fig. 13-02 International Mapping/Pearson Education, Inc.; Fig. 13-03 International Mapping/Pearson Education, Inc.; Fig. 13-04 Aerial Archives/Alamy; Fig. 13-05 Aerial Archives/Alamy ; Fig. 13-06 Dennis MacDonald/Alamy; Fig. 13-07 Aerial Archives/Alamy; Fig. 13-08 Ilene MacDonald/Alamy; Fig. 13-09a Aerial Archives/Alamy; Fig. 13-09b Kathleen Nelson/Alamy; Fig. 13-10a Renault Philippe/Hemis/Alamy; Fig. 13-10b Mario Beauregard/Fotolia; Fig. 13-11 Gerard Lawton/Alamy; Fig. 13-12a International Mapping/Pearson Education, Inc.; Fig. 13-12b Rosalrene Betancourt/Alamy; Fig. 13-13a International Mapping/Pearson Education, Inc.; Fig. 13-13b IDP Chicago Collection/Alamy; Fig. 13-14a International Mapping/Pearson Education, Inc.; Fig. 13-14b Norman Eggert/Alamy; Fig. 13-15 International Mapping/Pearson Education, Inc.; Fig. 13-16 International Mapping/Pearson Education, Inc.; Fig. 13-17 International Mapping/Pearson Education, Inc.; Fig. 13-18a Ian Leonard/Alamy; Fig. 13-18b Miscellaneoustock/Alamy; Fig. 13-18c Witold Skrypczak/Alamy; Fig. 13-19 Sorbis/Shutterstock; Fig. 13-20 Steeve Roche/Fotolia; Fig. 13-21 International Mapping/Pearson Education, Inc.; Fig. 13-22 International Mapping/Pearson Education, Inc.; Fig. 13-23a Colin Matthieu/Hemis/Alamy; Fig. 13-23b Alex Segre/Alamy; Fig. 13-23c Zvonimir Atletic/Shutterstock; Fig. 13-24 International Mapping/Pearson Education, Inc.; Fig. 13-25a International

Mapping/Pearson Education, Inc.; Fig. 13-25b International Mapping/Pearson Education, Inc.; Fig. 13-26 John Woods/Alamy; Fig. 13-27 Lonely Crane/Fotolia; Fig. 13-28 International Mapping/Pearson Education, Inc.; Fig. 13-30a International Mapping/Pearson Education, Inc.; Fig. 13-30b Alexandr/Fotolia; Fig. 13-31a International Mapping/Pearson Education, Inc.; Fig. 13-31b International Mapping/Pearson Education, Inc.; Fig. 13-31c International Mapping/Pearson Education, Inc; Fig. 13-32 International Mapping/Pearson Education, Inc.; Fig. 13-34 International Mapping/Pearson Education, Inc.; Fig. 13-35a International Mapping/Pearson Education, Inc.; Fig. 13-35b International Mapping/Pearson Education, Inc.; Fig. 13-39 International Mapping/Pearson Education, Inc.; Fig. 13-40 International Mapping/Pearson Education, Inc.; Fig. 13-43 International Mapping/Pearson Education, Inc.; Fig. 13-42 International Mapping/Pearson Education, Inc.; Fig. 13-48 a,b International Mapping/Pearson Education, Inc.; Fig. 13-44 International Mapping/Pearson Education, Inc.; Fig. 13-29 Roger Eritja/Alamy; Fig. 13-33 WorldTravel/Alamy; Fig. 13-38 Kenneth Garrett/Alamy; Fig. 13-41 Aerial Archives/Alamy; Fig. 13-47 International Mapping/Pearson Education, Inc.; Fig. 13-48c Stocker1970/Shutterstock; Fig. 13-49 Romakoma/Shutterstock; Fig. 13-53a LH Images/Alamy; Fig. 13-53b Robert Haas/SZ Photo/Alamy; Fig. 13-53c Philip Bird LRPS CPAGB/Shutterstock; Fig. 13-54 Mark Waugh/Alamy; Fig. 13-55 Huntstock, Inc./Alamy; Fig. 13-56 David Grossman/Alamy; Fig. 13-58a Gijsbert Hanekroot/Alamy; Fig. 13-58b Gijsbert Hanekroot/Alamy; Fig. 13-59 ZUMA Press, Inc./Alamy; Fig. 13-60 Chicagoview/Alamy; Fig. 13-61 Photo.ua/Shutterstock; Fig. 13-64b Marvin Dembinsky Photo Associates/Alamy; Fig. 13-69 Matthew Chattle/Alamy; Fig. 13-70 Nico Stengert/Novarc Images/Alamy; Fig. 13-71 Aerial Archives/Alamy; Fig. 13-72 David South/Alamy; Fig. 13-73 DPA Picture Alliance/Alamy; Fig. 13-74 Gijsbert Hanekroot/Alamy; Fig. 13-45a International Mapping/Pearson Education, Inc.; Fig. 13-45b Rob Crandall/Alamy; Fig. 13-62 International Mapping/Pearson Education, Inc.; Fig. 13-50a International Mapping/Pearson Education, Inc.; Fig. 13-50b SG Cityscapes/Alamy; Fig. 13-51 Transit Trips By Mode: 2012, United States Department of Transportation, http://www.rita.dot.gov/bts/sites/rita.dot.gov.bts/files/PassengerTravelFactsFigures.pdf; Fig. 13-52 International Mapping/Pearson Education, Inc.; Fig. 13-57 International Mapping/Pearson Education, Inc.; Fig. 13-64a International Mapping/Pearson Education, Inc.; Fig. 13-63 International Mapping/Pearson Education, Inc.; Fig. 13-65 International Mapping/Pearson Education, Inc.; Fig. 13-66a International Mapping/Pearson Education, Inc.; Fig. 13-66b International Mapping/Pearson Education, Inc.; Fig. 13-67 International Mapping/Pearson Education, Inc.; Fig. 13-68 Pearson Education, Inc.;

Appendix AF-01 Zoonar GmbH/Alamy; AF-02 Franckreporter/Getty; AF-03 Ryan Rodrick Beiler/Alamy

MAP INDEX

Note: "AP-" before a page number indicates the material is found in Appendix A.

Y

SUBJECT INDEX

Note: Page numbers preceded by "AP-" indicate materials found in Appendix A. Page numbers preceded by "AF-" indicate the Afterword.

A

D

E

F

N

Q

R

T

Z

Earth at Night, City Lights

The Americas

These images of Earth at night from NASA's Suomi-NPP "Marble" series use a collection of satellite-based observations, stitched together in a seamless mosaic of our planet. This view is based on instrumentation that observes light emanating from the ground. Notice how strongly major cities show up in the image.